CHILTON'S
TOTAL CAR CARE REPAIR MANUAL

MITSUBISHI CARS AND TRUCKS 1983-89 REPAIR MANUAL

President, Chilton Enterprises	David S. Loewith
Senior Vice President	Ronald A. Hoxter
Publisher and Editor-In-Chief	Kerry A. Freeman, S.A.E.
Managing Editors	Peter M. Conti, Jr. □ W. Calvin Settle, Jr., S.A.E.
Assistant Managing Editor	Nick D'Andrea
Senior Editors	Debra Gaffney □ Ken Grabowski, A.S.E., S.A.E. Michael L. Grady □ Richard J. Rivele, S.A.E. Richard T. Smith □ Jim Taylor Ron Webb
Director of Manufacturing	Mike D'Imperio
Editor	Jim Taylor

Cover vehicle supplied by **Potamkin Mitsubishi, Springfield, PA**

CHILTON BOOK COMPANY
ONE OF THE **DIVERSIFIED PUBLISHING COMPANIES**,
A PART OF **CAPITAL CITIES/ABC, INC.**

Manufactured in USA
© 1990 Chilton Book Company
Chilton Way, Radnor, PA 19089
ISBN 0-8019-7947-1
4567890123 3210987654

Contents

1 — General Information and Maintenance
- 1–2 How to Use this Book
- 1–3 Tools and Equipment
- 1–10 Routine Maintenance and Lubrication
- 1–47 Capacities Chart

2 — Engine Performance and Tune-Up
- 2–2 Tune-Up Specifications
- 2–2 Tune-Up Procedures
- 2–4 Troubleshooting
- 2–11 Firing Orders

3 — Engine and Engine Overhaul
- 3–2 Engine Electrical Systems
- 3–2 Engine Troubleshooting
- 3–26 Engine Service
- 3–33 Engine Specifications
- 3–272 Exhaust Systems

4 — Emission Controls
- 4–2 Gasoline Engine Emission Control System and Service
- 4–13 Diesel Engine Emission Control Systems
- 4–19 Vacuum Diagrams

5 — Fuel System
- 5–2 Carbureted Fuel System
- 5–21 TBI Injection System
- 5–29 MPI Injection System
- 5–48 Diesel Fuel System

Contents

6 Chassis Electrical
6-3	Heating and Air Conditioning	6-75	Lighting
6-50	Instruments and Switches	6-79	Circuit Protection

7 Drive Train
7-2	Manual Transaxle	7-70	Automatic Transmission
7-39	Manual Transmission	7-73	Transfer Case
7-59	Clutch	7-75	Driveshaft and U-Joints
7-64	Automatic Transaxle	7-78	Rear Axle

8 Suspension and Steering
8-2	FWD Front Suspension	8-34	Wheel Alignment Specs.
8-16	Starion Front Suspension	8-36	Rear Suspension
8-22	Pickup and Montero Front Suspension	8-45	Steering

9 Brakes
9-13	Anti-Lock Brake System	9-32	Rear Disc Brakes
9-19	Front Disc Brakes	9-32	Parking Brake
9-26	Drum Brakes	9-45	Brake Specifications

10 Body
10-2	Exterior
10-31	Interior

SAFETY NOTICE

Proper service and repair procedures are vital to the safe, reliable operation of all motor vehicles, as well as the personal safety of those performing repairs. This manual outlines procedures for servicing and repairing vehicles using safe, effective methods. The procedures contain many NOTES, CAUTIONS and WARNINGS which should be followed along with standard safety procedures to eliminate the possibility of personal injury or improper service which could damage the vehicle or compromise its safety.

It is important to note that the repair procedures and techniques, tools and parts for servicing motor vehicles, as well as the skill and experience of the individual performing the work vary widely. It is not possible to anticipate all of the conceivable ways or conditions under which vehicles may be serviced, or to provide cautions as to all of the possible hazards that may result. Standard and accepted safety precautions and equipment should be used when handling toxic or flammable fluids, and safety goggles or other protection should be used during cutting, grinding, chiseling, prying, or any other process that can cause material removal or projectiles.

Some procedures require the use of tools specially designed for a specific purpose. Before substituting another tool or procedure, you must be completely satisfied that neither your personal safety, nor the performance of the vehicle will be endangered

Although information in this manual is based on industry sources and is complete as possible at the time of publication, the possibility exists that some car manufacturers made later changes which could not be included here. While striving for total accuracy, Chilton Book Company cannot assume responsibility for any errors, changes or omissions that may occur in the compilation of this data.

PART NUMBERS

Part numbers listed in this reference are not recommendations by Chilton for any product by brand name. They are references that can be used with interchange manuals and aftermarket supplier catalogs to locate each brand supplier's discrete part number.

SPECIAL TOOLS

Special tools are recommended by the vehicle manufacturer to perform their specific job. Use has been kept to a minimum, but where absolutely necessary, they are referred to in the text by the part number of the tool manufacturer. These tools can be purchased under the appropriate part number, from your Mitsubishi dealer or regional distributor or an equivalent tool can be purchased locally from a tool supplier or parts outlet. Before substituting any tool for the recommended one, read the SAFETY NOTICE at the top of this page.

ACKNOWLEDGMENTS

The Chilton Book Company expresses its appreciation to Mitsubishi Motor Sales of America, Fountain Valley, California and Courtesy Mitsubishi, Turnersville, New Jersey for their generous assistance.

No part of this publication may be reproduced, transmitted or stored in any form or by any means, electronic or mechanical, including photocopy, recording, or by information storage or retrieval system without prior written permission from the publisher.

General Information and Maintenance

QUICK REFERENCE INDEX

Air cleaner	1-10	Fuel filter	1-10
Air Conditioning	1-20	Jump starting	1-41
Automatic Transaxle Application Chart	1-8	Manual Transaxle Application Chart	1-9
Automatic Transmission Application Chart	1-9	Manual Transmission Application Chart	1-9
Capacities Chart	1-47	Oil and filter change (engine)	1-33
Cooling system	1-35	Preventive Maintenance Schedules	1-46
Engine Application Chart	1-7	Windshield wipers	1-25

GENERAL INDEX

Air cleaner	1-10	Chassis greasing	1-36	Differential	1-35
Air conditioning		Coolant	1-35	Engine	1-32
Charging	1-24	Drive axle	1-35	Transfer case	1-34
Discharging	1-24	Engine oil	1-30, 32	Transmission	1-33
Evacuating	1-24	Fuel recommendations	1-30	Outside vehicle maintenance	1-38
Gauge sets	1-21	Manual transmission/transaxle	1-33	PCV valve	1-13
General service	1-20	Master cylinder	1-36	Power steering pump	1-36
Inspection	1-22	Power steering pump	1-36	Preventive Maintenance Charts	1-46
Operation	1-21	Steering gear	1-36	Radiator	1-35
Refrigerant level checks	1-22	Transfer case	1-34	Rear axle	
Safety precautions	1-21	Front drive axle		Lubricant level	1-35
System tests	1-22	Lubricant level	1-35	Routine maintenance	1-10
Tools	1-21	Fuel filter	1-10	Safety measures	1-4
Antifreeze	1-35	Hoses	1-17	Serial number location	1-5
Automatic transmission/transaxle		How to Use This Book	1-2	Special tools	1-3
Application chart	1-8	Identification		Specifications Charts	
Filter change	1-34	Engine	1-5, 7	Capacities	1-47
Fluid change	1-33	Model	1-5	Preventive Maintenance	1-46
Battery		Serial number	1-5, 6	Tires	
Cables	1-16	Transmission		Design	1-27
Charging	1-16	Automatic	1-5,	Inflation	1-27
General maintenance	1-14	Manual	1-5,	Rotation	1-26
Fluid level and maintenance	1-16	Vehicle	1-6	Size chart	1-28
Jump starting	1-41	Jacking points	1-42	Tread depth	1-27
Belts	1-16	Jump starting	1-41	Troubleshooting	1-29
Capacities Chart	1-47	Lubrication		Usage	1-27
Chassis lubrication	1-36	Automatic transmission	1-33	Tools and equipment	1-3
Cooling system	1-35	Body	1-36	Towing	1-42
Crankcase ventilation valve	1-13	Chassis	1-35	Trailer towing	1-39
Drive axle		Differential	1-35	Transfer Case	
Lubricant level	1-35	Engine	1-33	Fluid level	1-34
Evaporative canister	1-14	Manual transmission	1-33	Transmission	
Filters		Transfer case	1-34	Application charts	1-8, 9
Air	1-10	Maintenance Intervals Chart	1-46	Routine maintenance	1-33
Crankcase	1-13	Master cylinder	1-36	Troubleshooting Charts	
Fuel	1-10	Model identification	1-5	Tires	1-29
Oil	1-32	Oil and fuel recommendations	1-32	Wheels	1-29
Fluids and lubricants		Oil and filter change (engine)	1-33	Vehicle identification	1-6
Automatic transmission/transaxle	1-33	Oil level check		Wheels	1-26
Battery	1-16			Windshield wipers	1-25

1-1

1 GENERAL INFORMATION AND MAINTENANCE

HOW TO USE THIS BOOK

Chilton's Repair Manual for Mitsubishi cars, pickups and Montero is designed to help you understand the inner workings of your vehicle and save you money on its upkeep.

The first two Sections will be the most used, since they contain maintenance and tune-up information and procedures. Studies have shown that a properly tuned and maintained vehicle can get at least 10% better gas mileage (which translates into lower operating costs) and periodic maintenance will catch minor problems before they turn into major repair bills. The other Sections deal with the more complex systems of your vehicle. Operating systems from engine through brakes are covered to the extent that the average do-it-yourselfer becomes mechanically involved. This book will not explain such things as rebuilding the differential for the simple reason that the expertise required and the investment in special tools make this task impractical and uneconomical. It will give you the detailed instructions to help you change your own brake pads and shoes, tune-up the engine, replace spark plugs and filters, and do many more jobs that will save you money, give you personal satisfaction and help you avoid expensive problems.

A secondary purpose of this book is a reference guide for owners who want to understand their vehicle and/or their mechanics better. In this case, no tools at all are required. Knowing just what a particular repair job requires in parts and labor time will allow you to evaluate whether or not you're getting a fair price quote and help decipher itemized bills from a repair shop.

Before attempting any repairs or service on your vehicle, read through the entire procedure outlined in the appropriate Section. This will give you the overall view of what tools and supplies will be required. There is nothing more frustrating than having to walk to the bus stop on Monday morning because you were short one gasket on Sunday afternoon. So read ahead and plan ahead. Each operation should be approached logically and all procedures thoroughly understood before attempting any work. Some special tools that may be required can often be rented from local automotive jobbers or places specializing in renting tools and equipment. Check the yellow pages of your phone book.

All Sections contain adjustments, maintenance, removal and installation procedures, and overhaul procedures. When overhaul is not considered practical, we tell you how to remove the failed part and then how to install the new or rebuilt replacement. In this way, you at least save the labor costs. Backyard overhaul of some components (such as the alternator or water pump) is just not practical, but the removal and installation procedure is often simple and well within the capabilities of the average vehicle owner.

Two basic mechanic's rules should be mentioned here. First, whenever the LEFT side of the vehicle or engine is referred to, it is meant to specify the DRIVER'S side of the vehicle. Conversely, the RIGHT side of the vehicle means the PASSENGER'S side. Second, all screws and bolts are removed by turning counterclockwise, and tightened by turning clockwise.

Safety is always the most important rule. Constantly be aware of the dangers involved in working on or around an automobile and take proper precautions to avoid the risk of personal injury or damage to the vehicle. See the entry in this Section, Servicing Your Vehicle Safely, and the SAFETY NOTICE on the acknowledgment page before attempting any service procedures and pay attention to the instructions provided. There are 3 common mistakes in mechanical work:

1. Incorrect order of assembly, disassembly or adjustment. When taking something apart or putting it together, doing things in the wrong order usually just costs you extra time; however it CAN break something. Read the entire procedure before beginning disassembly. Do everything in the order in which the instructions say you should do it, even if you can't immediately see a reason for it. When you're taking apart something that is very intricate (for example a carburetor), you might want to draw a picture of how it looks when assembled at one point in order to make sure you get everything back in its proper position. We will supply exploded views whenever possible, but sometimes the job requires more attention to detail than an illustration provides. When making adjustments (especially tune-up adjustments), do them in order. One adjustment often affects another and you cannot expect satisfactory results unless each adjustment is made only when it cannot be changed by any other.

2. Overtorquing (or undertorquing) nuts and bolts. While it is more common for overtorquing to cause damage, undertorquing can cause a fastener to vibrate loose and cause serious damage, especially when dealing with aluminum parts. Pay attention to torque specifications and utilize a torque wrench in assembly. If a torque figure is not available remember that, if you are using the right tool to do the job, you will probably not have to strain yourself to get a fastener tight enough. The pitch of most threads is so slight that the tension you put on the wrench will be multiplied many times in actual force on what you are tightening. A good example of how critical torque is can be seen in the case of spark plug installation, especially where you are putting the plug into an aluminum cylinder head. Too little torque can fail to crush the gasket, causing leakage of combustion gases and consequent overheating of the plug and engine parts. Too much torque can damage the threads or distort the plug, which changes the spark gap at the electrode. Since more and more manufacturers are using aluminum in their engine and chassis parts to save weight, a torque wrench should be in any serious do-it-yourselfer's tool box.

There are many commercial chemical products available for ensuring that fasteners won't come loose, even if they are not torqued just right (a very common brand is Loctite®). If you're worried about getting something together tight enough to hold, but loose enough to avoid mechanical damage during assembly, one of these products might offer substantial insurance. Read the label on the package and make sure the product is compatible with the materials, fluids, etc. involved before choosing one.

3. Crossthreading. This occurs when a part such as a bolt is screwed into a nut or casting at the wrong angle and forced, causing the threads to become damaged. Crossthreading is more likely to occur if access is difficult. It helps to clean and lubricate fasteners, and to start threading with the part to be installed going straight in, using your fingers. If you encounter resistance, unscrew the part and start over again at a different angle until it can be inserted and turned several times without much effort. Keep in mind that many parts, especially spark plugs, use tapered threads so that gentle turning will automatically bring the part you're threading to the proper angle if you don't force it or resist a change in angle. Don't put a wrench on the part until it's been turned in a couple of times by hand. If you suddenly encounter resistance and the part has not seated fully, don't force it. Pull it back out and make sure it's clean and threading properly.

Always take your time and be patient; once you have some experience, working on your vehicle will become an enjoyable hobby.

GENERAL INFORMATION AND MAINTENANCE

TOOLS AND EQUIPMENT

Naturally, without the proper tools and equipment it is impossible to properly service your vehicle. It would be impossible to catalog each tool that you would need to perform each and every operation in this book. It would also be unwise for the amateur to rush out and buy an expensive set of tools on the theory that he may need one or more of them at sometime.

The best approach is to proceed slowly, gathering together a good quality set of those tools that are used most frequently. Don't be misled by the low cost of bargain tools. It is far better to spend a little more for better quality. Forged wrenches, 6- or 12-point sockets and fine tooth ratchets are by far preferable to their less expensive counterparts. As any good mechanic can tell you, there are few worse experiences than trying to work on a car or truck with bad tools. Your monetary savings will be far outweighed by frustration and mangled knuckles.

Begin accumulating those tools that are used most frequently; those associated with routine maintenance and tune-up.

In addition to the normal assortment of screwdrivers and pliers, you should have the following tools for routine maintenance jobs (your Mitsubishi uses metric fasteners):

• Metric wrenches and sockets, and combination open end/box end wrenches in sizes from 3mm to 19mm; and a spark plug socket ($^{13}/_{16}$ in.)

If possible, buy various length socket drive extensions. One break in this department is that the metric sockets available in the U.S. will all fit the ratchet handles and extensions you may already have ($^{1}/_{4}$ in., $^{3}/_{8}$ in., and $^{1}/_{2}$ in. drive).

• Jackstands for support
• Oil filter wrench
• Oil filter spout for pouring oil
• Grease gun for chassis lubrication
• Hydrometer for checking the battery
• A container for draining oil
• Many rags for wiping up the inevitable mess

In addition to the above items there are several others that are not absolutely necessary, but handy to have around. These include absorbent gravel, a transmission fluid funnel and the usual supply of lubricants, antifreeze and fluids, although these can be purchased as needed. This is a basic list for routine maintenance, but only if your personal needs and desires can accurately determine your list of tools.

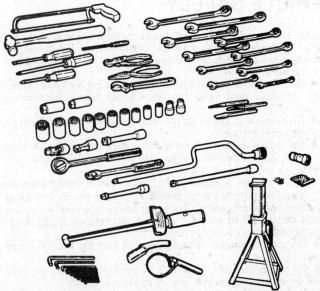

A basic collection of hand tools is necessary for automotive service

The second list of tools is for tune-ups. While the tools involved here are slightly more sophisticated, they need not be outrageously expensive. There are several inexpensive tachometer/dwell meters on the market that are every bit as good for the average mechanic as a $400.00 professional model. Just be sure that the meter scale goes to at least 1,200–1,500 rpm on the tach scale and that it works on 4-cylinder engines. A basic list of tune-up equipment could include:

1. Tach/dwell meter
2. Spark plug wrench
3. Timing light (a DC light that works from the vehicle's battery is best, although an AC light that plugs into 110V house current will suffice at some sacrifice in brightness).
4. Wire spark plug gauge/adjusting tools
5. Set of feeler blades

In addition to these basic tools, there are several other tools and gauges you may find useful. These include:

1. A compression gauge. The screw-in type is slower to use, but eliminates the possibility of a faulty reading due to escaping pressure.
2. A manifold vacuum gauge
3. A test light
4. An induction meter. This is used for determining whether or not there is current in a wire. This is handy for use if a wire is broken somewhere in a wiring harness.

As a final note, you will probably find a torque wrench necessary for all but the most basic work. There are three types of torque wrenches available: deflecting beam type, dial indicator (dial indicator) and click type. The beam and dial indicator models are perfectly adequate, although the click type models are more precise, and allow the user to reach the required torque without having to assume a sometimes awkward position in reading a scale. No matter what type of torque wrench you purchase, have it calibrated periodically to ensure accuracy.

Torque specification for each fastener will be given in the procedure in any case that a specific torque value is required. If no torque specifications are given, use the following values as a guide, based upon fastener size:

Bolts marked 6T
  6mm bolt/nut — 5–7 ft. lbs.
  8mm bolt/nut — 12–17 ft. lbs.
  10mm bolt/nut — 23–34 ft. lbs.
  12mm bolt/nut — 41–59 ft. lbs.
  14mm bolt/nut — 56–76 ft. lbs.

Bolts marked 8T
  6mm bolt/nut — 6–9 ft. lbs.
  8mm bolt/nut — 13–20 ft. lbs.
  10mm bolt/nut — 27–40 ft. lbs.
  12mm bolt/nut — 46–69 ft. lbs.
  14mm bolt/nut — 75–101 ft. lbs.

Special Tools

Normally, the use of special factory tools is avoided for repair procedures, since these are not readily available for the do-it-yourselfer mechanic. When it is possible to perform the job with more commonly available tools, it will be pointed out, but occasionally, a special tool was designed to perform a specific function and should be used. Before substituting another tool, you should be convinced that neither your safety nor the performance of the vehicle will be compromised. Special tools are avaiable at your local Mitsubishi dealer or at your local parts store or jobber market.

1-3

1 GENERAL INFORMATION AND MAINTENANCE

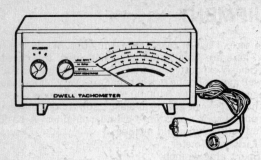

A dwell/tachometer is useful for tune-up work; you won't need the dwell function if the car has electronic ignition

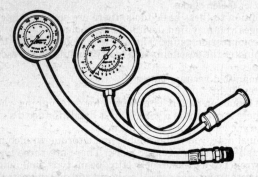

A compression gauge and a combination vacuum/fuel pressure gauge are handy for troubleshooting and tune-up work

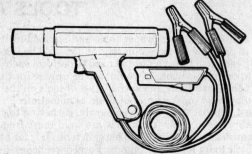

An inductive pickup simplifies timing light connections to the spark plug wire

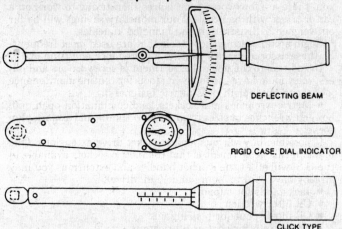

DEFLECTING BEAM

RIGID CASE, DIAL INDICATOR

CLICK TYPE

Views of the three types of torque wrenches

SERVICING YOUR VEHICLE SAFELY

It is virtually impossible to anticipate all of the hazards involved with automotive maintenance and service but care and common sense will prevent most accidents.

The rules of safety for mechanics range from "don't smoke around gasoline," to "use the proper tool for the job." The trick to avoiding injuries is to develop safe work habits and take every possible precaution.

Do's

• Do keep a fire extinguisher and first aid kit within easy reach.

• Do wear safety glasses or goggles when cutting, drilling, grinding or prying. If you wear glasses for the sake of vision, then they should be made of hardened glass that can serve also as safety glasses, or wear safety goggles over your regular glasses.

• Do shield your eyes whenever you work around the battery. Batteries contain sulfuric acid. In case of contact with the eyes or skin, flush the area with water or a mixture of water and baking soda and get medical attention immediately.

• Do use safety stands for any under-vehicle service. Jacks are for raising vehicles; safety stands are for making sure the vehicle stays raised until you want it to come down. Whenever the vehicle is raised, block the wheels remaining on the ground and set the parking brake.

• Do use adequate ventilation when working with any chemicals. Asbestos dust resulting from brake lining wear cause cancer.

• Do disconnect the negative battery cable when working on the electrical system.

• Do follow manufacturer's directions whenever working with potentially hazardous materials. Both brake fluid and antifreeze are poisonous if taken internally.

• Do properly maintain your tools. Loose hammerheads, mushroomed punches and chisels, frayed or poorly grounded electrical cords, excessively worn screwdrivers, spread wrenches (open end), cracked sockets, slipping ratchets, or faulty droplight sockets can cause accidents.

• Do use the proper size and type of tool for the job being done.

• Do when possible, pull on a wrench handle rather than push on it, and adjust you stance to prevent a fall.

• Do be sure that adjustable wrenches are tightly adjusted on the nut or bolt and pulled so that the face is on the side of the fixed jaw.

GENERAL INFORMATION AND MAINTENANCE 1

- Do select a wrench or socket that fits the nut or bolt. The wrench or socket should sit straight, not cocked.
- Do strike squarely with a hammer. avoid glancing blows.
- Do set the parking brake and block the wheels if the work requires that the engine be running.

Don'ts

- Don't run an engine in a garage or anywhere else without proper ventilation — EVER! Carbon monoxide is poisonous. It is absorbed by the body 400 times faster than oxygen. Carbon monoxide is odorless and colorless. Your senses cannot detect its presence. Early symptoms of monoxide poisoning include headache, irritability, improper vision (blurred or hard to focus) and/or drowsiness. When you notice any of these symptoms in yourself or your helpers, stop working immediately and get to fresh, outside air. Ventilate the work area thoroughly before returning to the vehicle. It takes a long time to leave the human body and you can build up a deadly supply of it in your system by simply breathing in a little every day. You may not realize you are slowly poisoning yourself. Always use power vents, windows, fans or open the garage doors.
- Don't work around moving parts while wearing a necktie or other loose clothing. Short sleeves are much safer than long, loose sleeves. Hard-toed shoes with neoprene soles protect your toes and give a better grip on slippery surfaces. Jewelry such as watches, fancy belt buckles, beads, or body adornment of any kind is not safe while working around a vehicle. Long hair should be hidden under a hat or cap.
- Don't use pockets for toolboxes. A fall or bump can drive a screwdriver deep into you body. Even a wiping cloth hanging from the back pocket can wrap around a spinning shaft, pulley or fan.
- Don't smoke when working around gasoline, cleaning solvent or other flammable material.
- Don't smoke when working around the battery. When the battery is being charged, it gives off explosive hydrogen gas.
- Don't use gasoline to wash your hands. There are excellent soaps available. Gasoline may contain lead, and lead can enter the body through a cut, accumulating in the body until you are very ill. Gasoline also removes all the natural oils from the skin so that bone dry hands will suck up oil and grease.
- Don't service the air conditioning system unless you are equipped with the necessary tools and training. The refrigerant, R-12, is extremely cold and when exposed to the air, will instantly freeze any surface it comes in contact with, including your eyes. Although the refrigerant is normally non-toxic, R-12 becomes a deadly poisonous gas in the presence of an open flame. One good whiff of the vapors from burning refrigerant can be fatal.

SERIAL NUMBER IDENTIFICATION

Vehicle

The vehicle identification number (VIN) is located on a plate attached to the left front of the dash panel so it can be seen through the windshield when you stand beside the vehicle, in front of the driver's door. The letters and numbers in the VIN digits can be interpreted according to their positions in the sequence as follows:
1. Manufacturing country.
2. Make.
3. Vehicle type.
4. Type of seat belt system or weight class of truck or multi-purpose vehicle.
5. Vehicle line.
6. Trim Code/Price Class.
7. Body type.
8. Engine displacement.
9. Check digit — a special letter or number code used to verify the serial number. This contains no useful information for the vehicle owner.
10. Model year
    - D = 1983
    - E = 1984
    - F = 1985
    - G = 1986
    - H = 1987
    - J = 1988
    - K = 1989
11. Plant where the vehicle was built.
12. On 1985 and earlier vehicles, the 12th digit represents the transmission code, followed by a 5-digit sequential (serial) number. For 1986 and all later vehicles, all 6 digits are the sequential number.

Engine

The engine model and serial numbers in all cases are stamped on the top edge of the block near the front of the engine. In most cases, they are located on the right side of the engine.

Transmission/Transaxle

On all models, the basic transmission model number is stamped on the Vehicle Information Code Plate, located (Cordia and Tredia) on the front end of the wheel well inner panel on the right side, or (other models) on the engine compartment side of the firewall. The model number is identified, Transmission, or, Transaxle, on the plate.

On Starion models only, the Information Code Plate contains not only the model code, but, following a small box, a four digit representation of the axle ratio.

On the JM600 transmission used on the 1985-89 Starion, there is an identification number tag attached to the right side of the transmission case. The number located there is a seven digit sequence. The first digit represents the year the transmission was built, the second digit represents the month of production (from 1-12), and the final five number sequence represents the serial production number for that month.

The newest families of transmissions have a detailed designation number on the lower left side of the case. The last four digits represent the monthly sequence number.

1-5

1 GENERAL INFORMATION AND MAINTENANCE

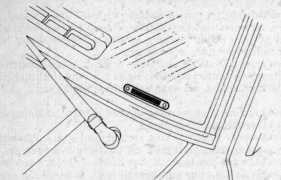

The VIN is located on a plate on the front of the dash panel

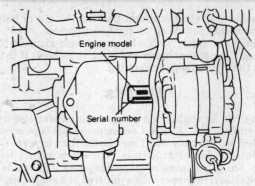

Typical engine serial number

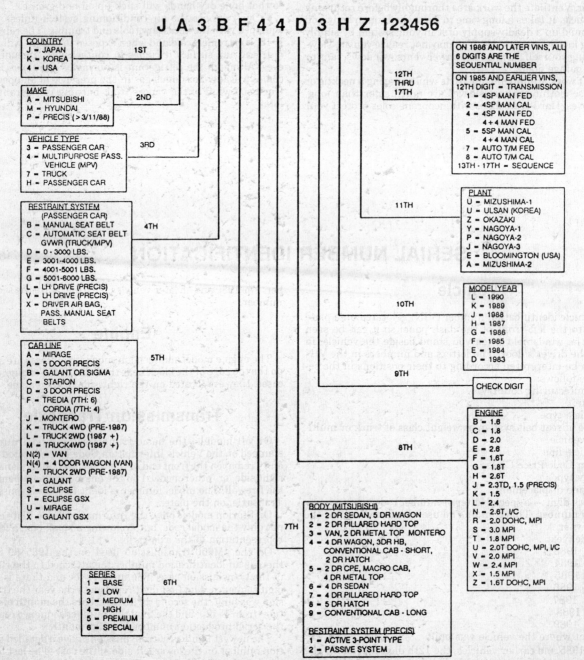

GENERAL INFORMATION AND MAINTENANCE 1

ENGINE IDENTIFICATION

Year	Model	Engine Displacement cu. in. (cc)	Common Name	Engine Series Identification	No. of Cylinders	Engine Type
1983	Cordia	109.5 (1795)	1.8	G62B	4	SOHC
	Tredia	109.5 (1795)	1.8	G62B	4	SOHC
	Starion	155.9 (2555)	2.6	G54B	4	SOHC
	Montero	155.9 (2555)	2.6	G54B	4	SOHC
	RWD Truck	121.9 (1997)	2.0	G63B	4	SOHC
	RWD Truck	155.9 (2555)	2.6	G54B	4	SOHC
	4WD Truck	143.2 (2346)	2.3	4D55	4	SOHC, T, D
1984	Cordia	109.5 (1795)	1.8	G62B	4	SOHC, T
	Cordia	121.9 (1997)	2.0	G63B	4	SOHC
	Tredia	109.5 (1795)	1.8	G62B	4	SOHC, T
	Tredia	121.9 (1997)	2.0	G63B	4	SOHC
	Starion	155.9 (2555)	2.6	G54B	4	SOHC, T
	Montero	155.9 (2555)	2.6	G54B	4	SOHC
	RWD Truck	121.9 (1997)	2.0	G63B	4	SOHC
	RWD Truck	155.9 (2555)	2.6	G54B	4	SOHC
	4WD Truck	143.2 (2346)	2.3	4D55	4	SOHC, T, D
1985	Cordia	109.5 (1795)	1.8	G62B	4	SOHC, T
	Cordia	121.9 (1997)	2.0	G63B	4	SOHC
	Tredia	109.5 (1795)	1.8	G62B	4	SOHC, T
	Tredia	121.9 (1997)	2.0	G63B	4	SOHC
	Starion	155.9 (2555)	2.6	G54B	4	SOHC, T, IC*
	Mirage	89.6 (1468)	1.5	G15B	4	SOHC
	Mirage	97.4 (1597)	1.6	G32B	4	SOHC
	Galant	143.4 (2350)	2.4	G64B	4	SOHC
	Montero	155.9 (2555)	2.6	G54B	4	SOHC
	RWD Truck	121.9 (1997)	2.0	G63B	4	SOHC
	RWD Truck	155.9 (2555)	2.6	G54B	4	SOHC
	4WD Truck	143.2 (2346)	2.3	4D55	4	SOHC, T, D
1986	Cordia	109.5 (1795)	1.8	G62B	4	SOHC, T
	Cordia	121.9 (1997)	2.0	G63B	4	SOHC
	Tredia	109.5 (1795)	1.8	G62B	4	SOHC, T
	Tredia	121.9 (1997)	2.0	G63B	4	SOHC
	Starion	155.9 (2555)	2.6	G54B	4	SOHC, T, IC
	Mirage	89.6 (1468)	1.5	G15B	4	SOHC
	Mirage	97.4 (1597)	1.6	G32B	4	SOHC, T
	Galant	143.4 (2350)	2.4	G64B	4	SOHC
	Montero	155.9 (2555)	2.6	G54B	4	SOHC
	RWD Truck	121.9 (1997)	2.0	G63B	4	SOHC
	4WD Truck	155.9 (2555)	2.6	G54B	4	SOHC
1987	Cordia	109.5 (1795)	1.8	G62B	4	SOHC, T
	Cordia	121.9 (1997)	2.0	G63B	4	SOHC
	Tredia	109.5 (1795)	1.8	G62B	4	SOHC, T
	Tredia	121.9 (1997)	2.0	G63B	4	SOHC
	Starion	155.9 (2555)	2.6	G54B	4	SOHC, T, IC

1 GENERAL INFORMATION AND MAINTENANCE

ENGINE IDENTIFICATION

Year	Model	Engine Displacement cu. in. (cc)	Common Name	Engine Series Identification	No. of Cylinders	Engine Type
1987	Mirage	89.6 (1468)	1.5	G15B	4	SOHC
	Mirage	97.4 (1597)	1.6	G32B	4	SOHC, T
	Galant	143.4 (2350)	2.4	G64B	4	SOHC
	Precis	89.6 (1468)	1.5	G15B	4	SOHC
	Montero	155.9 (2555)	2.6	G54B	4	SOHC
	RWD Truck	121.9 (1997)	2.0	G63B	4	SOHC
	4WD Truck	155.9 (2555)	2.6	G54B	4	SOHC
1988	Cordia	109.5 (1795)	1.8	G62B	4	SOHC, T
	Cordia	121.9 (1997)	2.0	G63B	4	SOHC
	Tredia	109.5 (1795)	1.8	G62B	4	SOHC, T
	Tredia	121.9 (1997)	2.0	G63B	4	SOHC
	Starion	155.9 (2555)	2.6	G54B	4	SOHC, T, IC
	Mirage	89.6 (1468)	1.5	G15B	4	SOHC
	Mirage	97.4 (1597)	1.6	G32B	4	SOHC, T
	Galant	143.4 (2350)	2.4	G64B	4	SOHC
	Galant	181.4 (2972)	3.0	6G72	6	SOHC
	Precis	89.6 (1468)	1.5	G15B	4	SOHC
	Montero	155.9 (2555)	2.6	G54B	4	SOHC
	RWD Truck	121.9 (1997)	2.0	G63B	4	SOHC
	4WD Truck	155.9 (2555)	2.6	G54B	4	SOHC
1989	Starion	155.9 (2555)	2.6	G54B	4	SOHC
	Mirage	89.6 (1468)	1.5	4G15	4	SOHC
	Mirage	97.4 (1597)	1.6	4G61	4	DOHC, T
	Galant	121.9 (1997)	2.0	4G63	4	SOHC & DOHC
	Sigma	181.4 (2972)	3.0	6G72	6	SOHC
	Precis	89.6 (1468)	1.5	G15B	4	SOHC
	Montero	155.9 (2555)	2.6	G54B	4	SOHC
	Montero	181.4 (2972)	3.0	6G72	6	SOHC
	RWD Truck	121.9 (1997)	2.0	G63B	4	SOHC
	4WD Truck	155.9 (2555)	2.6	G54B	4	SOHC

Notes: SOHC—Single Overhead Cam
DOHC—Double Overhead Cam
T—Turbo Charged
IC—Intercooled
*From 1985½
D—Diesel
RWD—Rear Wheel Drive
4WD—Four Wheel Drive

AUTOMATIC TRANSAXLE IDENTIFICATION CHART

Transaxle Types	Years	Models
KM 171 3-sp	1983–85	Cordia/Tredia
KM 172 3-sp	1986–88	Cordia/Tredia
KM 175 4-sp (1)	1985–87	Galant
KM 177 4-sp (1)	1988–89	Galant
KM 171 3-sp	1985–89	Mirage
KM 171 3-sp	1987–89	Precis
KM 177 4-sp (1)	1989	Sigma

GENERAL INFORMATION AND MAINTENANCE 1

MANUAL TRANSAXLE IDENTIFICATION CHART

Transaxle Types	Years	Models
KM 162 5-sp	1983	Cordia/Tredia
KM 163 5-sp	1984–88	Cordia/Tredia
KM 166 4 x 2-sp	1984–85	Cordia/Tredia
KM 210 5-sp	1987–88	Galant
KM 206 5-sp	1989	Galant
KM 161 4-sp	1985–86	Mirage
KM 162 5-sp	1985–86	Mirage
KM 163 5-sp	1985–86	Mirage
KM 200 5-sp	1987	Mirage
KM 201 5-sp	1987–89	Mirage
KM 206 5-sp	1987–88	Mirage
KM 210 5-sp	1989	Mirage
KM 161 4-sp	1987–89	Precis
KM 162 5-sp	1987–89	Precis
KM-210 5-sp	1989	Sigma

MANUAL/AUTOMATIC TRANSMISSION IDENTIFICATION CHART

Transmission Types	Years	Models
KM 132 5-sp (manual)	1983–89	Starion
	1983–89	Truck
JM 600 4-sp (automatic)	1985–89	Starion
KM 130 4-sp (manual)	1983–86	Truck
KM 144 4-sp (manual)	1983–84	Truck
KM 145 5-spd manual (w/transfer case)	1983–89	Truck & Montero
MA 904A Automatic (3-sp)	1983–86	Truck
KM 146 Auto (MA 904 Trans w/KM 145 transfer case)	1984–85	Truck
	1984–86	Montero
AW 372 Auto (4-sp)	1987–89	Truck
KM 148 (AW 372 w/transfer case)	1987	Truck
	1987–89	Montero
V5MT1 5-sp (4-WD)	1989	Montero w/3.0 V-6

*1987–89: 2 versions—KM 132-K for 2.0L and KM 132-I for 2.6L engine

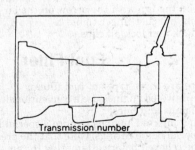

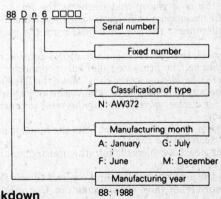

Typical serial number breakdown

1–9

1 GENERAL INFORMATION AND MAINTENANCE

ROUTINE MAINTENANCE

Air Cleaner

An air cleaner is used to keep airborne dirt and dust out of the air flowing through the engine. This material, if allowed to enter the engine, would form an abrasive compound in conjunction with the engine oil and drastically shorten engine life. For this reason, you should never run the engine without the air cleaner in place except for a very brief period in diagnosing a problem. You should also be sure to use the proper replacement part to avoid poor fit and consequent air leakage.

Proper maintenance is important since a clogged air filter will allow somewhat more dirt to enter the engine and, past a certain point, will enrich the fuel/air mixture, causing poor fuel economy, a drastic increase in emissions and even serious damage to the catalytic converter system. The maximum maintenance interval on all models is 30,000 miles. Maximum efficiency is maintained by changing the filter more often.

Diesel engines are particularly sensitive to partly clogged air filters and will lose performance if not well maintained.

In all cases the air cleaner element must be replaced; it cannot be cleaned.

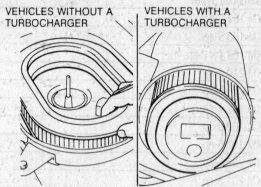

Air cleaner replacement, normally aspirated and turbocharged Cordia and Tredia. Other models similar

REPLACEMENT

1983–86 Starion
1985–86 Galant
All Turbocharged Cordia, Tredia, and Mirage
Diesel Pickup

1. Unsnap the three finger clips and unhook the lower ends. Pull the top of the air cleaner upward or toward you and away from the air cleaner body. The flex hose will give you enough freedom of motion to permit you to do this without disconnecting it from the air cleaner top.
2. Pull the filter cartridge out of the air cleaner body.
3. Install the new filter so that it seats properly in the air cleaner body. Then, install the top squarely on the lower body, hook the clips and lock them. If any of the clips are hard to lock, recheck the position of the top on the lower body. If it is not positioned properly, it will not seal effectively and the clips may be damaged.

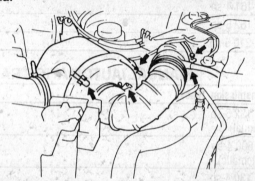

Air cleaner replacement for 1987–89 Starion, 1987–89 Galant and 1989 Sigma

1987–89 Starion
1987–89 Galant
1989 Sigma
Montero with the V6 Engine

1. Loosen the hose clamp that connects the intake flexible hose to the air cleaner cover. Separate the hose from the cover. On Galant, disconnect the breather hose from the rocker arm cover.
2. Disconnect the air flow sensor multi-connectors. On Starion, disconnect the pressure relief valve connector also.
3. Release the air cleaner cover clips. Move the flexible air intake out of the way to allow you to remove the air cleaner cover. Remove the air cleaner cover carefully because it contains the air flow sensor.
4. Remove the air cleaner element from the housing and wipe the housing clean with a rag.
5. Place the new filter element into the housing making sure that it seats properly.
6. Replace the air cleaner cover and connect the air flow sensor and pressure relief valve multi-connetors. Connnect the air intake hose to the air cleaner cover and tighten the hose clamp.

1983–88 Cordia and Tredia without Turbocharger
1985–89 Mirage without Turbocharger
1987–89 Precis
Montero and Pickup with 4-cylinder Gasoline Engine

1. Remove the wing nut with your fingers. Pliers should be used only if it has been tightened excessively.
2. Unsnap and then disconnect the three clips from the upper body.
3. Remove the top of the air cleaner and remove the cartridge.
4. Wipe out the filter housing with a clean rag.
5. Install the new cartridge squarely so that it seats properly. Install the air cleaner lid over the wing nut stud, lining up the arrow on the top with the arrow on the air intake tube. Install the wing nut just finger tight.
6. Hook and lock all clips.

Fuel Filter

There are three types of fuel filters:
- Carbureted engines use a conventional fuel filter with ordinary clamps and lines.
- Fuel injected engines require a special high pressure filter with special banjo fittings. Note that the pressure in the system must be relieved before attempting to remove the filter in this type of system.
- The diesel engine uses a filter assembly that not only filters sediment and water from the fuel but also, under certain conditions, warms the fuel to prevent fuel line blockage. The filter

GENERAL INFORMATION AND MAINTENANCE 1

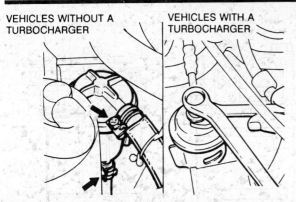

Fuel filters for normal and turbocharged engines

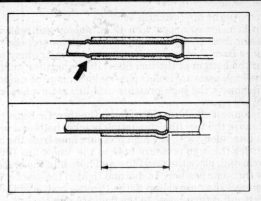

When connecting the fuel lines to the filter for normal engines, make sure the lines are installed far enough onto the connections. The line should overlap the connection for a distance of 0.8–1.0 inch on the engine (outlet) side and 1.0–1.2 inches on the inlet side

housing is equipped with a water level sensor which will illuminate a dash light, warning the operator that the level of trapped water within the filter has risen to capacity. The filter then needs to be drained as soon as possible.

The fuel filter *must* be replaced at least every five years or 50,000 miles (80,000km). The is a maximum replacement period. Good preventive maintenance dictates that the fuel filter be replaced once a year, or every 10,000 miles (16,000km). However, since the amount of dirt and water in the fuel varies greatly, the filter should be replaced whenever you suspect it to be clogged. Typically, the symptoms of a clogged filter are a lack of engine performance under full throttle conditions, especially at high rpm, even though the engine operates normally under moderate driving conditions. With severe clogging, the filter may cause poor running under virtually every operating condition but idle. One way to check out the filter is to follow the steps below for removal and then drain the contents out of the filter through the inlet. Drain the filter into a tin can rather than a glass or styrofoam container. While it is normal for a light concentration of particles or a few drops of water to be trapped in the fuel on the inlet side of the filter, large amounts of water and heavy concentrations of dirt indicate dirty fuel and, probably, a clogged filter. If you find a great deal of dirt trapped in the filter, it is a good idea not only to replace the filter, but to have the fuel tank drained or pumped out as well.

REMOVAL AND INSTALLATION

Carbureted Engines

1. Turn off the engine and allow it to cool. Using a pair of pliers, force open the clamps on the fuel lines and back them well away from the connections.
2. Work the fuel lines back and off the filter connections. If they are difficult to remove, it may help to pull them off with a twisting motion. Remove the filter from its mounting clip.
3. Inspect the fuel lines for cracks or breaks and replace them if necessary.
4. Install the new filter in the same position the old one was in the clamp. Connect the inlet fuel line to the inlet fitting on the bottom of the filter. Connect the outlet to the outlet fitting on top. Make sure the hoses are fully installed over the bulged-out portions of the fittings. Then, with pliers, move the clamps over the filter fittings so they are beyond the bulged-out sections of the fittings but a small distance away from the ends of the hoses.
5. Start the engine and inspection the hose conections for fuel leaks. Correct any fuel leak immediately.

Fuel Injected Engines Except Sigma

1. First, you MUST reduce the pressure in the system as follows:

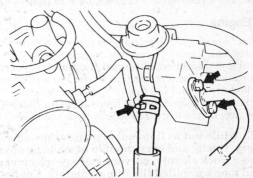

Fuel filter high pressure and delivery hoses on Sigma

  a. Open the trunk and pry up the lid located in the floor which covers the electric fuel pump. On Starion, remove the high floor side panel in the baggage compartment.
  b. Start the engine and allow it to idle.
  c. Disconnect the fuel pump connector. Allow the engine to continue idling until it stalls. Then, turn off the ignition key.
  d. Disconnect the negative battery connector.
2. Remove the air cleaner assembly as required.
3. Using an open-end wrench to hold the fuel filter stationary, loosen the bolt for the banjo type connector on top of the fuel filter with a box wrench. Do exactly the same with the inlet connector on the bottom of the filter.
4. Remove the bolt or nuts attaching the filter to the bracket and remove it.
5. To install the new filter, reverse the above procedure. It's best to torque the bolts for the fuel line banjo fittings. If the banjo fitting washers are damaged, replace them. The outlet fitting is torqued to 18–25 ft. lbs. in all cases. The inlet fitting is torqued, to 25 ft. lbs.
6. Reconnect the fuel pump, start the engine and check for leaks.

Sigma

1. First, you MUST reduce the pressure in the system as follows:
  a. Reach under the rear side of the fuel tank and disconnect the harness connector from the fuel pump.
  b. Start the engine and run it until it stops by itself.
  c. Turn the ignition switch to the OFF position and disconnect the negative battery cable.

GENERAL INFORMATION AND MAINTENANCE

2. Remove the air cleaner assembly.
3. On vehicles equipped with ECS, unbolt and remove the compressor from tis mounting. DO NOT disconnect the refrigerant lines from the compressor as personal injury may result.
4. On the bottom of the fuel filter, disconnect the main pipe from the high pressure (inlet) pipe by loosening the union nut.
5. Disconnect the high pressure and fuel return hoses from the delivery pipe.
6. Disconnect the hose clamp from the throttle body.
7. Remove the fuel filter mounting bracket bolts and remove the fuel filter along with the high pressure hose from the bracket. Transfer the high pressure hose to the nmew fuel filter.
8. To install, place the new filter and hose into the bracket in roughly the same position. Install and tighten the bracket bolts.
9. Connect the hose clamp to the throttle body and the high presssure and return hoses to the delivery pipe.
10. Connect the high pressure hose to the main fuel pipe and tighten the union nut.
11. If equippped with ECS, place the compresor onto its mounting and tighten the mounting bolts. Install the air cleaner assembly.
12. Connect the negative battery cable and fuel pump harness connector. Start the engine and check for fuel leaks.

Diesel Engine

The fuel filter element (cartridge) is contained within the filter canister which must be removed to gain access.
1. Label and disconnect all electrical connectors running to the filter canister. These connectors are for the water level sensor, the fuel heater and the fuel temperature sensor.
2. Carefully disconnect the fuel hoses at the fuel filter. Have a supply of rags handy to catch overflow from the hoses.

NOTE: While not as flammable as gasoline, diesel fuel is slippery, smelly and very capable of staining anything it touches. Work cleanly; prevent spillage whenever possible and mop up spilled fluid immediately. Cat litter or similar products are ideal for dealing with puddles on the floor.

3. Remove the two bolts holding the filter to the body and remove the filter canister.
4. Remove the protector and bracket from the canister.
5. Screw the filter out of the canister body by hand. Carefully remove the water level sensor and the drain plug from the cartridge. It may be handy to lightly clamp the unit in a vise to aid removal; don't damage the sensor.
6. It is possible to clean the filter element with kerosene. Replacement with a new cartridge is highly recommended instead of cleaning.
7. Install the drain plug and and the water level sensor on the new cartridge. Tighten the drain plug to 4 Nm (3 ft. lbs.) and the water sensor to 12 Nm (9 ft. lbs.)
8. Screw the cartridge onto the body. Install the protector and bracket.
9. Install filter assembly onto the vehicle. Install the main fuel hoses. When tightening the clamps, make sure the heads of the clamp bolts face away from the body of the filter.
10. Whenever the fuel supply has run out or the lines have been opened, the system must be bled to eliminate air. The air bleed plug projects at an angle from the top of the filter housing. Loosen it. Wrap the area in rags; fuel will come out of the bleed port.
11. The knob for the hand pump is located on the side of the filter body. Unscrew it and pull the pump lever out of the housing.
12. Pump the hand pump and watch the fuel coming out of the air bleed plug; when there are no air bubbles mixed with the fuel, tighten the air bleed plug.
13. Continue pumping until the operation of the pump lever feels stiff or heavy during each stroke.

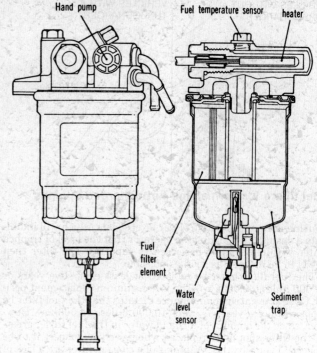

Cross section of the diesel fuel filter

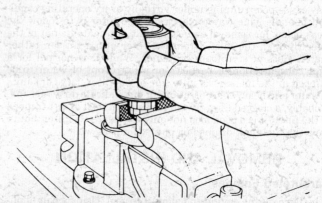

A helpful trick for removing the water level sensor from the diesel fuel filter

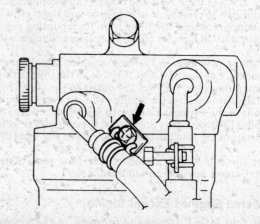

Correct clamp position is important

GENERAL INFORMATION AND MAINTENANCE

14. Push the pump lever all the way in and turn it to the right to lock it in place.
15. Reconnect the wiring connectors to the pump housing terminals. Make certain each is firmly seated.

DRAINING WATER FROM THE DIESEL FUEL FILTER

When water accumulates in the fuel filter, the fuel–water separator light will come on. This indicates that the filter has reached its safe capacity and must be drained. Even if the light has not come on, the wise owner will drain the filter with every oil change. This simple procedure can be done with the filter on the vehicle and can prevent severe engine damage or failure.

1. Use the proper sized wrench to loosen the drain plug on the bottom of the fuel filter.
2. The knob for the hand pump is located on the side of the filter body. Unscrew it and pull the pump lever out of the housing.
3. Pump the hand pump unitl the water is expelled and fuel is being pumped out.
4. Push the pump lever all the way in and turn it to the right to lock it in place.

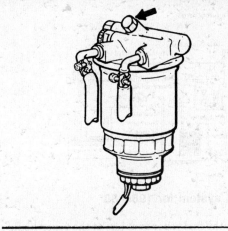
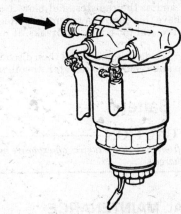

Location of air bleed port (upper) and hand pump for bleeding diesel fuel system

PCV Valve

To prevent combustion blow-by gasses from entering the atmosphere, all Mitsubishi gasoline engines use a closed type crankcase ventilation system. The system uses a PCV valve that is threaded into the valve cover. The PCV system supplies fresh air to the crankcase through the air cleaner through a metered orifice in the valve. Inside the crankcase, fresh air is mixed with the blow-by gasses. This mixture passes through the PCV vavle and into the air induction system.

The diesel engine does not use a PCV system; the crankcase vapors are routed through a breather hose and pipe and introduced into the intake airstream. After passing through the turbocharger, they enter the engine to be reburned.

This system must be checked and cleaned every five years or 50,000 miles, whichever comes first. You should also check the PCV system if your vehicle begins to idle roughly and other causes such as dirty fuel have been investigated. To do this, follow the procedure below.

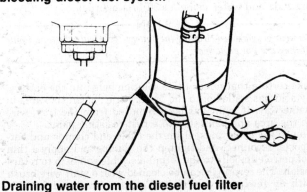

Draining water from the diesel fuel filter

REMOVAL AND INSTALLATION

1. Use a wrench to unscrew the PCV valve at the valve cover. It may be necessary to disconnect the hose at the outer end of the PCV to avoid twisting it too much (see the first part of the

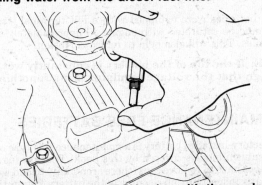

Test for vacuum at the PCV valve with the engine idling

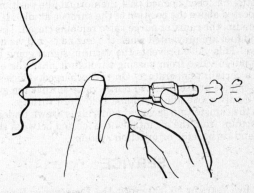

Testing for clogging of the PCV

1–13

1 GENERAL INFORMATION AND MAINTENANCE

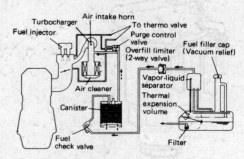

The Cordia and Tredia evaporative emissions control system for 1984–88

next step). If necessary, reconnect the PCV hose. Then, start the engine and run it at idle. You should hear a hissing sound as air is drawn into the valve. Place your finger over the inlet. You should be able to feel a strong vacuum. If either test is failed, the valve will have to be replaced. Stop the engine.

2. Using a pair of pliers, pinch open the clamp that fastens the hose to the outer end of the PCV and then slide the clamp back a few inches. Pull the valve out of the hose. Then, blow through the threaded end of the valve. If air will pass fairly freely, the valve is okay. If not, try to flush it out with spray solvent and if that doesn't work, replace it. PCV valves are not serviceable.

3. Clean both PCV hoses by spraying a safe solvent through them. Inspect the hoses for cracks or excessive stiffness, and replace if necessary.

4. Install the old PCV valve or replace it with a new one in reverse order. Be careful not to crossthread the valve into the aluminum valve cover, and do not overtighten it.

Evaporative Canister
Gasoline Engines Only

The heart of this system is a charcoal canister located in the engine compartment. Fuel vapor that collects in the carburetor float bowl or gas tank and which would ordinarily be released to the atmosphere is stored in the canister because of the affinity of the charcoal for it.

In order to restore the ability of the charcoal to hold fuel, fresh air is drawn through the charcoal under certain operating conditions, thus drawing the fuel back out and burning it in the combustion chambers.

At idle speed, or when the engine is cold, the addition of any fuel vapor to the correct mixture would cause excessive emissions. A port in the carburetor or fuel injection system throttle body allows the fuel to be drawn out of the canister only after the throttle has been opened past the normal idle position (the port is located above the position of the throttle at idle). If there is no vacuum, the canister purge valve remains closed. The flow of air and fuel are prevented when the engine is cold via a thermal valve. This valve prevents the vacuum signal going to the canister purge valve from passing through it until the engine reaches a certain temperature. On turbocharged vehicles, two vacuum signals are sent to the purge valve so that the system will work only above idle speed and below full throttle.

When the canister purge valve opens, air is drawn under very slight vacuum from the air intake hose located between the air cleaner and carburetor or injection system.

SERVICE

Every five years or 50,000 miles, the Evaporative Emissions System should be inspected as follows:

1. Disconnect the fuel vapor vent line at the vapor-liquid separator on the fuel tank and at the canister, and blow it clean with compressed air, or have this done at a repair shop. Remove the filler cap and check the seal for cracks or breaks. If necessary, replace the cap.

2. Replace the canister. To do this, mark and then disconnect all hoses. Then, unclamp and remove the canister. Install in reverse order.

Battery

—————————— CAUTION ——————————
Keep flame or sparks away from the battery! It gives off explosive hydrogen gas, while it is being charged.

GENERAL MAINTENANCE

Periodically, clean the top of the battery with a solution of baking soda and water using a stiff bristle brush.

—————————— CAUTION ——————————
Always wear goggle when cleaning the battery. Acid will splash into your eyes if they are not protected!

Make certain that none of this solution gets into the battery. If any acid has spilled onto the battery tray, clean this area in the same way. If paint has been removed from the tray, wire brush the area and paint it with a rust-resisting paint.

Remove the cable ends, clean the cable end clamps and battery posts, reconnect and tighten the clamps and apply a thin coat of petroleum jelly to the terminals. This will help to retard corrosion. The terminals can be cleaned with a staff wire brush or with an inexpensive terminal cleaner designed for this purpose.

Some batteries were equipped with a felt terminal washer. This should be saturated with engine oil approximately every 6,000 miles. This will also help to retard corrosion.

NOTE: If the top of the battery is constantly wet, it a good sign that the voltage regulator is malfunctioning.

MAINTENANCE FREE BATTERIES

The factory-installed battery is a maintenance free type on all the Mitsubishi models covered by this book. That means that you'll never have to remove caps (there aren't any) to add water. But a yearly inspection and cleaning of the battery, connections, and battery mountings is recommended to guarantee maximum reliability.

GENERAL INFORMATION AND MAINTENANCE

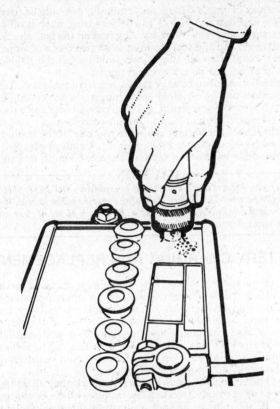

Clean the posts with a wire brush or a terminal cleaner mad for the purpose as shown

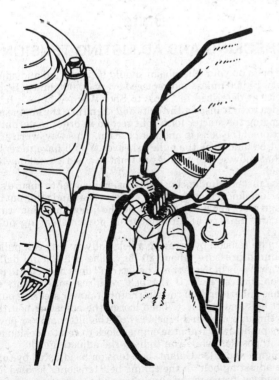

Clean the inside of the clamps with a wire brush or the special tool

Charging necessary

The battery charge indicator is located on the top surface and changes from blue (when adequately charged) to white when the battery is low on charge

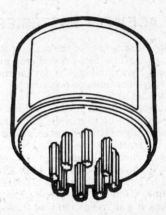

Special tools are also available for cleaning the posts and clamps on side terminal batteries

Testing the Maintenance Free Battery

Maintenance-free batteries do not require normal attention as far as fluid level checks are concerned. However, the terminals require periodic cleaning, which should be performed at least once a year.

The sealed top battery cannot be checked for charge in the normal manner, since there is no provision for access to the electrolyte. To check the condition of the battery:

1. If the battery is equipped with an indicator eye on top of the battery check the color of the eye. If the eye is bright, the battery has enough fluid. If the eye is dark, the electrolyte fluid is too low and the battery must be replaced.
2. If a green dot appears in the middle of the eye, the battery is sufficiently charged. Proceed to Step 4. If no green dot is visible, charge the battery as in Step 3.
3. Charge the battery at this rate:

WARNING: Do not charge the battery for more than 50 amp/hours! If the green dot appears, or if electrolyte squirts out of the vent hole, stop the charge and proceed to Step 4.

1–15

1 GENERAL INFORMATION AND MAINTENANCE

It may be necessary to tip the battery from side to side to get the green dot to appear after charging.

4. Connect a battery load tester and a voltmeter across the battery terminals (the battery cables should be disconnected from the battery). Apply a 300 amp load to the battery for 15 seconds to remove the surface charge. Remove the load.

5. Wait 15 seconds to allow the battery to recover. Apply the appropriate test load, as specified in the accompanying chart. Apply the load for 15 seconds while reading the voltage. Disconnect the load.

6. Check the results against the accompanying chart. If the battery voltage is at or above the specified voltage for the temperature listed, the battery is good. If the voltage falls below what's listed, the battery should be replaced.

REPLACEMENT BATTERIES

Many replacement batteries are of the maintenance free type. For these batteries, follow the procedures above. If the replacemt battery you have purchased is not a maintenance free type, follow these easy maintenance procedures:

Check the battery fluid level at least once a month, more often in hot weather or during extended periods of travel. The electrolyte level should be up to the bottom of the split ring in each cell. If the level is low, add water. Distilled water is good for this purpose, but ordinary tap water can be used.

At least once a year, check the specific gravity of the battery with a hydrometer. It should be between 1.20–1.26 on the hydrometer's scale. Most importantly, all the cells should read approximately the same. If one or more cells read significantly lower than the others, it's an indication that these low cells are shorting out. Replace the battery.

If water is added during freezing weather, the vehicle should be driven several miles to allow the electrolyte and water to mix. Otherwise the battery could freeze.

Filling the Battery

Batteries should be checked for proper electrolyte level at least once a month or more frequently. Keep a close eye on any cell or cells that are unusually low or seem to constantly need water—this may indicate a battery on its last legs, a leak, or a problem with the charging system.

Top up each cell to the bottom of the split ring, or, if the battery has no split ring, about 3/8 in. (9.5mm) above the tops of the plates. Use distilled water where available, or ordinary tap water, if the water in your area isn't too hard. Hard water contains minerals that may slowly damage the plates of your battery.

CABLES AND CLAMPS

Twice a year, the battery terminal posts and the cable clamps should be cleaned. Loosen the clamp bolts (you may have to brush off any corrosion with a baking soda and water solution if they are really messy) and remove the cables, negative cable first. On batteries with posts on top, the use of a battery clamp puller is recommended. It is easy to break off a battery terminal if a clamp gets stuck without the puller. These pullers are inexpensive and available in most auto parts stores or auto departments. Side terminal battery cables are secured with a bolt.

The best tool for battery clamp and terminal maintenance is a battery terminal brush. This inexpensive tool has a female ended wire brush for cleaning terminals, and a male ended wire brush inside for cleaning the insides of battery clamps. When using this tool, make sure you get both the terminal posts and the insides of the clamps nice and shiny. Any oxidation, corrosion or foreign material will prevent a sound electrical connection and inhibit either starting or charging. If your battery has side terminals, there is also a cleaning tool available for these.

Before installing the cables, remove the battery holddown clamp or strap and remove the battery. Inspect the battery casing for leaks or cracks (which unfortunately can only be fixed by buying a new battery). Check the battery tray, wash it off with warm soapy water, rinse and dry. Any rust on the tray should be sanded away, and the tray given at least two coats of a quality anti-rust paint. Replace the battery, and install the holddown clamp or strap, but do not overtighten.

Reinstall your clean battery cables, negative cable last. Tighten the cables on the terminal posts snugly; do not overtighten. Wipe a thin coat of petroleum jelly or grease all over the outsides of the clamps. This will help to inhibit corrosion.

Finally, check the battery cables themselves. If the insulation of the cables is cracked or broken, or if the ends are frayed, replace the cable with a new cable of the same length or gauge.

--- **CAUTION** ---
Batteries give off hydrogen gas, which is explosive. DO NOT SMOKE around the battery! The battery electrolyte contains sulfuric acid. If you should splash any into your eyes or skin, flush with plenty of clear water and get immediate medical help.

BATTERY CHARGING AND REPLACEMENT

Charging a battery is best done by the slow charging method (often called trickle charging), with a low amperage charger. Quick charging a battery can actually "cook" the battery, damaging the plates inside and decreasing the life of the battery drastically. Any charging should be done in a well ventilated area away from the possibility of sparks or flame. The cell caps (not found on maintenance-free batteries) should be unscrewed from their cells, but not removed.

If the battery must be quick-charged, check the cell voltages and the color of the electrolyte a few minutes after the charge is started. If cell voltages are not uniform or if the electrolyte is discolored with brown sediment, stop the quick charging in favor of a trickle charge. A common indicator of an overcharged battery is the frequent need to add water to the battery.

Belts

CHECKING AND ADJUSTING TENSION

The belts on your Mitsubishi should be inspected and adjusted every 15,000 miles, and replaced every 30,000 miles. Inspect the belts as described in "How to Spot Worn V-Belts". It is a good idea to first inspect the belts and then turn the engine over by engaging the starter for a split second in order to permit re-inspection of the belts in another position. This way, worn areas will not be hidden by the pulley sheaves. You'll have to replace belts that show severe wear if you want to avoid a breakdown on the road.

Belts must run under constant tension in order to ensure slip-free performance. Belts that slip will overheat and wear out at a very high rate. Too much tension, on the other hand, may cause premature failure of alternator or power steering pump bearings.

Estimate tension by pressing a belt halfway between pulleys with thumb pressure (about 20 lbs.). The belt should deflect downward 1/4–3/8 in. Note that "deflection" does not refer to play or droop, but stretch. If the belt requires adjustment, use wrenches at both ends of the alternator or power steering pump support bolt and loosen it. Then loosen the adjusting bolt that passes through a slotted bracket. Pull the alternator or power steering pump away from the engine block to create tension and hold it in position while you tighten the adjusting bolt.

On Sigma and 1988 Galant, belt tension is adjusted by turning the adjusting bolt on the timing belt tensioner located just about midway between the alternator and power steering pump pulleys. Once you've reached the proper belt tension, tighten the adjusting bolt locknut to retain the adjustment.

GENERAL INFORMATION AND MAINTENANCE 1

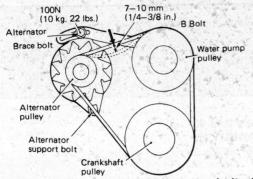

Tensioning the water pump and alternator belt of the Cordia and Tredia

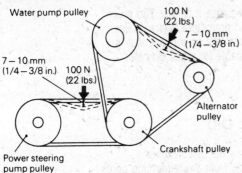

Adjusting the Starion drive belts

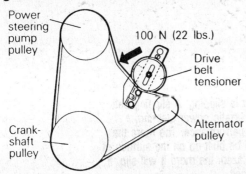

Drive belt tensioner used on Sigma and 1988 Galant

If possible, do not pry on the accessory. If you cannot get enough tension on the belt without prying, make sure to pry on the front of the housing right where the adjusting bolt passes through it. After tightening the adjusting bolt, tighten the support bolt and nut. If you're unsure about how much tension to put on these bolts, you can use a torque wrench and torque the support bolt to 14–18 ft. lbs. and the adjusting bolt to 8.5–11 ft. lbs.

REMOVAL AND INSTALLATION

If you're replacing a belt, use the procedure above, but move the accessory toward the engine block after you've loosened both bolts. Then work the belt off the pulleys. Make sure to move the accessory far enough so that the new belt can be installed without forcing it on. If moving the accessory until it is at the inner end of the adjusting slot still does not permit easy installation, you've probably got the wrong belt. Prying a new belt on with a screwdriver will damage it and substantially shorten its life.

Where there are two belts and you have to replace the one that is located farther back on the crankshaft, you'll have to remove the front belt first. This occurs, for example, in replacing the power steering pump drive belt on the Starion.

Once the belt is over the pulleys, pull the accessory outward and tension it as described above. New belts should be adjusted just a little tighter than ones that have run in. It's a good idea to recheck tension after running the engine for five minutes and then again after a few hundred miles of driving.

Hoses

Radiator hoses are generally of two constructions, the preformed (molded) type, which is custom made for a particular application, and the spring-loaded type, which is made to fit several different applications. Heater hoses are all of the same general construction.

INSPECTION

Cooling system and heater hoses should be inspected carefully once a year for brittleness, cracks, softening and bulging, and heat damage from nearby exhaust system parts. See the accompanying section on "How to Spot Worn Hoses". Mitsubishi recommends that you replace all cooling system and heater hoses, fuel system hoses, and vacuum hoses every five years or 50,000 miles.

In addition, on turbocharged vehicles, the turbocharger intake hoses and the turbocharger oil hose should be inspected and replaced at these intervals. The oil hose to the turbocharger is especially important as oil leaks here could cause immediate failure of the turbo or loss of sufficient oil from the crankcase to cause severe engine damage. Turbocharger cooling hoses should be replaced along with other cooling system and heater hoses.

REMOVAL AND INSTALLATION

There is usually no problem in replacing hoses. Just make sure you allow the engine to cool and then drain the cooling system before starting work. Also make sure to fully loosen clamps and pull them back from the connections before attempting to disconnect hoses. If hoses are tough to break loose from their connections, a twisting motion often works best. If they have been on a hot-running engine for a long time (and you know you are going to replace them), you might want to make a cut with a sharp knife going in the direction of the length of the hose. Cut until you pass the end of the connector on the block or engine accessory. This will loosen the grip the hose has on the fitting and make it easier to remove. Be careful not to cut deeply into relatively soft aluminum fittings. When the hoses are in place on the connections, position the hose clamps about ¼ in. (6mm) from the end of the hose and tighten them down. Hose clamps do not require that much force to form a good seal.

As a final note, don't throw those old hoses away. If they have not been cut or damaged, shake them out to remove the coolant, seal them in a plastic bag and throw them in the trunk. Old hoses will serve as good short term repair if by chance one of your hoses breaks. If you keep up on your maintenance, you may never need them.

Position the hose clamp so that it is about ¼ inch from the end of the hose

1-17

1 GENERAL INFORMATION AND MAINTENANCE

HOW TO SPOT WORN V-BELTS

V–Belts are vital to efficient engine operation—they drive the fan, water pump and other accessories. They require little maintenance (occasional tightening) but they will not last forever. Slipping or failure of the V–belt will lead to overheating. If your V–belt looks like any of these, it should be replaced.

Cracking or Weathering

This belt has deep cracks, which cause it to flex. Too much flexing leads to heat build–up and premature failure. These cracks can be caused by using the belt on a pulley that is too small. Notched belts are available for small diameter pulleys.

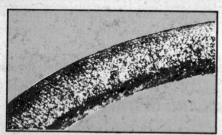

Softening (Grease and Oil)

Oil and grease on a belt can cause the belt's rubber compounds to soften and separate from the reinforcing cords that hold the belt together. The belt will first slip, then finally fail altogether.

Glazing

Glazing is caused by a belt that is slipping. A slipping belt can cause a run-down battery, erratic power steering, overheating or poor accessory performance. The more the belt slips, the more glazing will be built up on the surface of the belt. The more the belt is glazed, the more it will slip. If the glazing is light, tighten the belt.

Worn Cover

The cover of this belt is worn off and is peeling away. The reinforcing cords will begin to wear and the belt will shortly break. When the belt cover wears in spots or has a rough jagged appearance, check the pulley grooves for roughness.

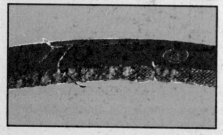

Separation

This belt is on the verge of breaking and leaving you stranded. The layers of the belt are separating and the reinforcing cords are exposed. It's just a matter of time before it breaks completely.

GENERAL INFORMATION AND MAINTENANCE 1

HOW TO SPOT BAD HOSES

Both the upper and lower radiator hoses are called upon to perform difficult jobs in an inhospitable environment. They are subject to nearly 18 psi at under hood temperatures often over 280°F, and must circulate nearly 7500 gallons of coolant an hour—3 good reasons to have good hoses.

Swollen Hose

A good test for any hose is to feel it for soft or spongy spots. Frequently these will appear as swollen areas of the hose. The most likely cause is oil soaking. This hose could burst at any time, when hot or under pressure.

Cracked Hose

Cracked hoses can usually be seen but feel the hoses to be sure they have not hardened; a prime cause of cracking. This hose has cracked down to the reinforcing cords and could split at any of the cracks.

Frayed Hose End (Due to Weak Clamp)

Weakened clamps frequently are the cause of hose and cooling system failure. The connection between the pipe and hose has deteriorated enough to allow coolant to escape when the engine is hot.

Debris in Cooling System

Debris, rust and scale in the cooling system can cause the inside of a hose to weaken. This can usually be felt on the outside of the hose as soft or thinner areas.

GENERAL INFORMATION AND MAINTENANCE

AIR CONDITIONING

The air conditioner is designed so that when the air conditioner switch, installed on the heater control panel is depressed to the ON position and the blower switch is at any other position than OFF the compressor is operated to cool air.

The temperature inside the vehicle is controlled by means of the temperature control lever, whose position determines the opening of the blend-air damper and the resulting mixing ratio of cool and hot air is used to control the outlet temperature.

\With the full-automatic air conditioner, the signals from the various sensors are processed and controlled at the full-automatic air conditioner control unit when the desired temperature is set; thereafter, the system functions for automatic control of the temperature of the air outflow, the amount of air flow, the direction of air outflow and the selection and direction of flow of either outside air or interior (recirculated) air.

With the automatic air conditioner, the signals from the various sensors are processed and controlled at the automatic air conditioner control unit for the selected temperature, thus controlling the temperature of the air outflow as well as the amount of air flow.

General Servicing Procedures

The most important aspect of air conditioning service is the maintenance of a pure and adequate charge of refrigerant in the system. A refrigeration system cannot function properly if a significant percentage of the charge is lost. Leaks are common because the severe vibration encountered in an automobile can easily cause a sufficient cracking or loosening of the air conditioning fittings; as a result, the extreme operating pressures of the system force refrigerant out.

The problem can be understood by considering what happens to the system as it is operated with a continuous leak. Because the expansion valve regulates the flow of refrigerant to the evaporator, the level of refrigerant there is fairly constant. The receiver/drier stores any excess of refrigerant, and so a loss will first appear there as a reduction in the level of liquid. As this level nears the bottom of the vessel, some refrigerant vapor bubbles will begin to appear in the stream of liquid supplied to the expansion valve. This vapor decreases the capacity of the expansion valve very little as the valve opens to compensate for its presence. As the quantity of liquid in the condenser decreases, the operating pressure will drop there and throughout the high side of the system. As the R-12 continues to be expelled, the pressure available to force the liquid through the expansion valve will continue to decrease, and, eventually, the valve's orifice will prove to be too much of a restriction for adequate flow even with the needle fully withdrawn.

At this point, low side pressure will start to drop, and severe reduction in cooling capacity, marked by freeze-up of the evaporator coil, will result. Eventually, the operating pressure of the evaporator will be lower than the pressure of the atmosphere surrounding it, and air will be drawn into the system wherever there are leaks in the low side.

Because all atmospheric air contains at least some moisture, water will enter the system and mix with the R-12 and the oil. Trace amounts of moisture will cause sludging of the oil, and corrosion of the system. Saturation and clogging of the filter/drier, and freezing of the expansion valve orifice will eventually result. As air fills the system to a greater and greater extent, it will interfere more and more with the normal flows of refrigerant and heat.

From this description, it should be obvious that much of the repairman's time will be spent detecting leaks, repairing them, and then restoring the purity and quantity of the refrigerant charge. A list of general precautions that should be observed while doing this follows:

1. Keep all tools as clean and dry as possible.
2. Thoroughly purge the service gauges and hoses of air and moisture before connecting them to the system. Keep them capped when not in use.
3. Thoroughly clean any refrigerant fitting before disconnecting it, in order to minimize the entrance of dirt into the system.
4. Plan any operation that requires opening the system beforehand, in order to minimize the length of time it will be exposed to open air. Cap or seal the open ends to minimize the entrance of foreign material.
5. When adding oil, pour it through an extremely clean and dry tube or funnel. Keep the oil capped whenever possible. Do not use oil that has not been kept tightly sealed.
6. Use only refrigerant 12. Purchase refrigerant intended for use in only automatic air conditioning systems. Avoid the use of refrigerant 12 that may be packaged for another use, such as cleaning, or powering a horn, as it is impure.
7. Completely evacuate any system that has been opened to replace a component, or that has leaked sufficiently to draw in moisture and air. This requires evacuating air and moisture with a good vacuum pump for at least one hour.

If a system has been open for a considerable length of time it may be advisable to evacuate the system for up to 12 hours (overnight).

8. Use a wrench on both halves of a fitting that is to be disconnected, so as to avoid placing torque on any of the refrigerant lines.
9. When overhauling a compressor, pour some of the oil into a clean glass and inspect it. If there is evidence of dirt or metal particles, or both, flush all refrigerant components with clean refrigerant before evacuating and recharging the system. In addition, if metal particles are present, the compressor should be replaced.
10. Schrader valves may leak only when under full operating pressure. Therefore, if leakage is suspected but cannot be located, operate the system with a full charge of refrigerant and look for leaks from all Schrader valves. Replace any faulty valves.

Additional Preventive Maintenance Checks

ANTIFREEZE

In order to prevent heater core freeze-up during air conditioning operation, it is necessary to maintain permanent type antifreeze protection of $+15°F$ ($-9°C$), or lower. A reading of $-15°F$ ($-26°C$) is ideal since this protection also supplies sufficient corrosion inhibitors for the protection of the engine cooling system.

NOTE: The same antifreeze should not be used longer than the manufacturer specifies.

RADIATOR CAP

For efficient operation of an air conditioned vehicle's cooling system, the radiator cap should have a holding pressure which meets manufacturer's specifications. A cap which fails to hold these pressures should be replaced.

CONDENSER

Any obstruction of or damage to the condenser configuration will restrict the air flow which is essential to its efficient operation. It is therefore a good rule to keep this unit clean and in proper physical shape.

NOTE: Bug screens are regarded as obstructions.

CONDENSATION DRAIN TUBE

This single molded drain tube expels the condensation, which accumulates on the bottom of the evaporator housing, into the engine compartment. If this tube is obstructed, the air condi-

GENERAL INFORMATION AND MAINTENANCE 1

tioning performance can be restricted and condensation buildup can spill over onto the vehicle's floor.

Safety Precautions

Because of the importance of the necessary safety precautions that must be exercised when working with air conditioning systems and R-12 refrigerant, a recap of the safety precautions are outlined.

1. Avoid contact with a charged refrigeration system, even when working on another part of the air conditioning system or vehicle. If a heavy tool comes into contact with a section of copper tubing or a heat exchanger, it can easily cause the relatively soft material to rupture.
2. When it is necessary to apply force to a fitting which contains refrigerant, as when checking that all system couplings are securely tightened, use a wrench on both parts of the fitting involved, if possible. This will avoid putting torque on refrigerant tubing. (It is advisable, when possible, to use tube or line wrenches when tightening these flare nut fittings.)
3. Do not attempt to discharge the system by merely loosening a fitting, or removing the service valve caps and cracking these valves. Precise control is possibly only when using the service gauges. Place a rag under the open end of the center charging hose while discharging the system to catch any drops of liquid that might escape. Wear protective gloves when connecting or disconnecting service gauge hoses.
4. Discharge the system only in a well ventilated area, as high concentrations of the gas can exclude oxygen and act as an anesthesia. When leak testing or soldering, this is particularly important, as toxic gas is formed when R-12 contacts any flame.
5. Never start a system without first verifying that both service valves are backseated, if equipped, and that all fittings throughout the system are snugly connected.
6. Avoid applying heat to any refrigerant line or storage vessel. Charging may be aided by using water heated to less than +125°F (+51°C) to warm the refrigerant container. Never allow a refrigerant storage container to sit out in the sun, or near any other source of heat, such as a radiator.
7. Always wear goggles when working on a system to protect the eyes. If refrigerant contacts the eye, it is advisable in all cases to see a physician as soon as possible.
8. Frostbite from liquid refrigerant should be treated by first gradually warming the area with cool water, and then gently applying petroleum jelly. A physician should be consulted.
9. Always keep refrigerant can fittings capped when not in use. Avoid sudden shock to the can which might occur from dropping it, or from banging a heavy tool against it. Never carry a can in the passenger compartment of a vehicle.
10. Always completely discharge the system before painting the vehicle (if the paint is to be baked on), or before welding anywhere near the refrigerant lines.

Air Conditioning Tools and Gauges

Test Gauges

Most of the service work performed in air conditioning requires the use of a set of two gauges, one for the high (head) pressure side of the system, the other for the low (suction) side.

The low side gauge records both pressure and vacuum. Vacuum readings are calibrated from 0 to 30 inches and the pressure graduations read from 0 to no less than 60 psi.

The high side gauge measures pressure from 0 to at least 600 psi.

Both gauges are threaded into a manifold that contains two hand shut-off valves. Proper manipulation of these valves and the use of the attached test hoses allow the user to perform the following services:

1. Test high and low side pressures.
2. Remove air, moisture, and contaminated refrigerant.
3. Purge the system (of refrigerant).
4. Charge the system (with refrigerant).

The manifold valves are designed so they have no direct effect on gauge readings, but serve only to provide for, or cut off, flow of refrigerant through the manifold. During all testing and hook-up operations, the valves are kept in a closed position to avoid disturbing the refrigeration system. The valves are opened only to purge the system of refrigerant or to charge it.

When purging the system, the center hose is uncapped at the lower end, and both valves are cracked open slightly. This allows refrigerant pressure to force the entire contents of the system out through the center hose. During charging, the valve on the high side of the manifold is closed, and the valve on the low side is cracked open. Under these conditions, the low pressure in the evaporator will draw refrigerant from the relatively warm refrigerant storage container into the system.

Service Valves

For the user to diagnose an air conditioning system he or she must gain "entrance" to the system in order to observe the pressures. There are two types of terminals for this purpose, the hand shut off type and the familiar Schrader valve.

The Schrader valve is similar to a tire valve stem and the process of connecting the test hoses is the same as threading a hand pump outlet hose to a bicycle tire. As the test hose is threaded to the service port the valve core is depressed, allowing the refrigerant to enter the test hose outlet. Removal of the test hose automatically closes the system.

Extreme caution must be observed when removing test hoses from the Schrader valves as some refrigerant will normally escape, usually under high pressure. (Observe safety precautions.)

Some systems have hand shut-off valves (the stem can be rotated with a special ratcheting box wrench) that can be positioned in the following three ways:

1. FRONT SEATED—Rotated to full clockwise position.
   a. Refrigerant will not flow to compressor, but will reach test gauge port. COMPRESSOR WILL BE DAMAGED IF SYSTEM IS TURNED ON IN THIS POSITION.
   b. The compressor is now isolated and ready for service. However, care must be exercised when removing service valves from the compressor as a residue of refrigerant may still be present within the compressor. Therefore, remove service valves slowly observing all safety precautions.
2. BACK SEATED—Rotated to full counter clockwise position. Normal position for system while in operation. Refrigerant flows to compressor but not to test gauge.
3. MID-POSITION (CRACKED)—Refrigerant flows to entire system. Gauge port (with hose connected) open for testing.

USING THE MANIFOLD GAUGES

The following are step-by-step procedures to guide the user to correct gauge usage.

1. WEAR GOGGLES OR FACE SHIELD DURING ALL TESTING OPERATIONS. BACKSEAT HAND SHUT-OFF TYPE SERVICE VALVES.
2. Remove caps from high and low side service ports. Make sure both gauge valves are closed.
3. Connect low side test hose to service valve that leads to the evaporator (located between the evaporator outlet and the compressor).
4. Attach high side test hose to service valve that leads to the condenser.
5. Mid-position hand shutoff type service valves.
6. Start engine and allow for warm-up. All testing and charging of the system should be done after engine and system have reached normal operation temperatures (except when using certain charging stations).

1-21

1 GENERAL INFORMATION AND MAINTENANCE

7. Adjust air conditioner controls to maximum cold.
8. Observe gauge readings.

When the gauges are not being used it is a good idea to:
  a. Keep both hand valves in the closed position.
  b. Attach both ends of the high and low service hoses to the manifold, if extra outlets are present on the manifold, or plug them if not. Also, keep the center charging hose attached to an empty refrigerant can. This extra precaution will reduce the possibility of moisture entering the gauges. If air and moisture have gotten into the gauges, purge the hoses by supplying refrigerant under pressure to the center hose with both gauge valves open and all openings unplugged.

SYSTEM INSPECTION

Checking for Refrigerant Leaks

Check the integrity of the air conditioning system whenever the system is suspect of leakage and also after component replacement or after any refrigerant lines have been disconnected.

There are two generally accepted methods of checking the air conditioning system for leaks. One is with the use of a Halide torch and the other with an electronic leak detector. Both are designed to detect small amounts of halogen when placed near a fitting or connection suspected of leaking. The electronic leak detector provides a greater degree of sensitivity and is the most preferred (and expensive) method. When using this equipment, make sure that you follow the manufacturer's instructions carefully.

NOTE: Some leak tests can be performed with a soapy water solution, but there must be at least a ½ lb. charge in the system for a leak to be detected.

If a leak is found, perform the following:

1. Check the tightness of the suspect fitting or connection, and if necessary re-tighten it. Recheck for leaks with a leak detector.
2. If leakage persists after re-tightening the fitting, discharge the refrigerant from the system and disconnect the fitting. Visually inspect the fitting seating surfaces for damage and replace as necessary. Even minor damage will require replacement of the fitting.
3. Check the compressor oil and add more oil as required.
4. Charge the system and perform another leak test. If no leaks are found, discharge, evacuate and recharge the system.

Checking for Oil Leaks

Refrigerant leaks show up only as oily areas on the various components because the compressor oil is transported around the entire system along with the refrigerant. Look for oily spots on all the hoses and lines, and especially on the hose and tube connections. If there are oily deposits, the system may have a leak, and you should have it checked by a qualified repairman.

NOTE: A small area of oil on the front of the compressor is normal and no cause for alarm.

Keep the Condenser Clear

Periodically inspect the front of the condenser for bent fins or foreign material (dirt, bugs, leaves, etc.) If any cooling fins are bent, straighten them carefully with needlenosed pliers. You can remove any debris with a stiff bristle brush or hose.

Operate the System Periodically

A lot of air conditioning problems can be avoided by simply running the air conditioner at least once a week, regardless of the season. Simply let the system run for at least 5 minutes a week (even in the winter), and you'll keep the internal parts lubricated as well as preventing the hoses from hardening.

Checking the Air Conditioner Drive Belt

This belt receives the same kind of attention and inspection as the other drive belts. The only basic difference is that this belt is a heavy duty type and runs with more tension. Deflection should be only about ¾ in. as measured at the center of the very long run between the compressor pulley and tensioner pulley.

To maintain this greater tension, the belt tensioner is positioned by a bolt that runs at right angles to the shaft on which the tensioner pulley runs. To change the belt tension, simply loosen the locknut and then turn the bolt clockwise to tighten or counterclockwise to reduce the tension or remove the bolt. Just be sure to remember to retighten the locknut after completing the adjustment.

REFRIGERANT LEVEL CHECKS

The first order of business when checking the sight glass is to find the sight glass. It is located in the head of the receiver/drier. Once you've found it, wipe it clean and proceed as follows:

1. With the engine and the air conditioning system running with the temperature control lever in the maximum cold position and the blower at maximum speed, look for the flow of refrigerant through the sight glass. If the air conditioner is working properly, you'll be able to see a continuous flow of clear refrigerant through the sight glass, with perhaps an occasional bubble at very high temperatures.
2. Cycle the air conditioner on and off to make sure what you are seeing is clear refrigerant. Since the refrigerant is clear, it is possible to mistake a completely discharged system for one that is fully charged. Turn the system off and watch the sight glass. If there is refrigerant in the system, you'll see bubbles during the off cycle. If you observe no bubbles when the system is running, and the air flow from the unit in the vehicle is delivering cold air, everything is OK.
3. If you observe bubbles in the sight glass while the system is operating, the system is low on refrigerant. Have it checked by a professional.
4. Oil streaks in the sight glass are an indication of trouble. Most of the time, if you see oil in the sight glass, it will appear as series of streaks, although occasionally it may be a solid stream of oil. In either case, it means that part of the charge has been lost.

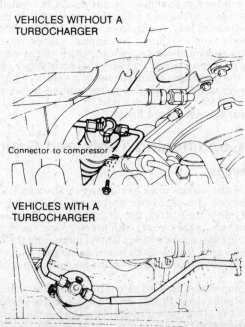

Location of the Cordia and Tredia sight glass

1-22

GENERAL INFORMATION AND MAINTENANCE

Check item \ Amount of refrigerant	Almost no refrigerant	Insufficient	Suitable	Too much refrigerant
Temperature of high pressure and low pressure lines	Almost no difference between high pressure and low pressure side temperature	High pressure side is warm and low pressure side is fairly cold	High pressure side is hot and low pressure side is cold	High pressure side is abnormally hot
State in sight glass	Bubbles flow continuously. **Bubbles will disappear and something like mist will flow when refrigerant is nearly gone**	The bubbles are seen at intervals of 1-2 seconds	Almost transparent. Bubbles may appear when engine speed is raised and lowered. **No clear difference exists between these two conditions**	No bubbles can be seen
Pressure of system	High pressure side is abnormally low	Both pressure on high and low pressure sides are slightly low	Both pressure on high and low pressure sides are normal	Both pressure on high and low pressure sides are abnormally high
Repair	**Stop compressor** and conduct an overall check	Check for gas leakage, repair as required, replenish and charge system		Discharge refrigerant from service valve of low pressure side

Refrigerant level diagnosis chart

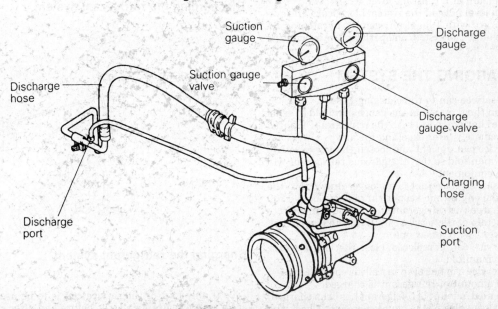

Typical manifold gauge set installation

1-23

1 GENERAL INFORMATION AND MAINTENANCE

DISCHARGING THE SYSTEM

1. Attach the low gauge hose of either a charging station or a manifold gauge set to the low side service fitting on the compressor. Be sure the gauge valves are closed before attaching the hoses to the refrigerant system.
2. Install a long hose to the manifold gauge set connector and position the other end of the hose in a suitable container.
3. Open the compressor discharge and suction line pressure valves and blow the refrigerant into the oil container.
4. When the system has been completely discharged measure the amount of oil collected in the can. The amount of oil measured should be added to the refrigerant system before it is recharged. Dispose of the old oil.

EVACUATING THE SYSTEM

NOTE: Whenever the system has been opened up to the atmosphere it is absolutely necessary that the system be evacuated to remove the air and moisture from the system. Air in the refrigerant system will cause a loss in the systems performance, a high compressor discharge pressure and oxidation of the compressor oil. Moisture in the system will also cause the expansion valve to malfunction.

1. Connect the manifold gauge set to the compressor and a long test hose from the gauge set manifold center connection to a vacuum pump.
2. Open both manifold gauge valves.
3. Start the vacuum pump and operate it until the evaporation suction gauge registers at least 26 in. Hg. of vacuum.

NOTE: If 26 in. Hg. of vacuum cannot be obtained, either there is a refrigerant leak in the system or the vacuum pump is not operating properly. Check the vacuum pump and if the vacuum pump is working properly then there is a refrigerant leak. Charge the system with 1 lb. of refrigerant. Locate and repair all leaks. After repairs have been made, discharge the refrigerant and evacuate the system.

4. Be sure to run the vacuum pump for at least 5 minutes.
5. Close the manifold valves. Turn off the vacuum pump and observe the evaporative suction gauge for 2 minutes. The vacuum level should remain at a constant level.
6. If the vacuum level falls off the system has a leak. The system should be charged with 1 lb. of refrigerant. Locate and repair all leaks. Discharge the system and perform the evacuating procedure again.

CHARGING THE SYSTEM

1. Connect the service can to the charging hose.
2. Slightly loosen the flare nut at the gauge manifold to expel air (from inside the charging hose) with the refrigerant. Tighten the flare nut immediately after expelling the air.
3. Hold the service can upright and loosen the low pressure valve of the gauge manifold so that the gaseous refrigerant will be drawn into the equipment.
4. When the gaseous refrigerant is no longer drawn into the equipment, start the engine and keep it running at about 1100 rpm in order to charge the refrigerant into the equipment.
5. Touch the bottom of the service can. If it is no longer cool, it is empty. Replace it with a new one.
6. When the service can is replaced, close the low pressure valve of the gauge manifold.
7. After a new service can has been installed repeat Steps 1–5 until the specified amount of refrigerant is charged.
8. After the specified amount of refrigerant has been charged, close the low pressure valve of the gauge manifold. Check the condition while observing the pressure gauge.
9. If only part of the service can is to be used, place it on a scale and measure its weight before and after charging.
10. Close the service can valve and remove the gauge manifold.

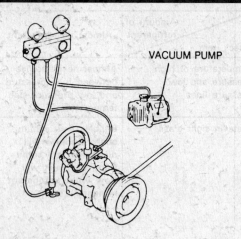

Evacuating the refrigerant system

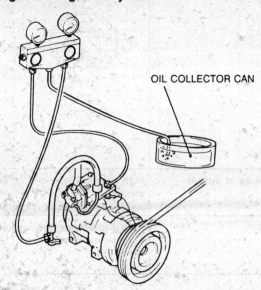

Discharging the refrigerant system

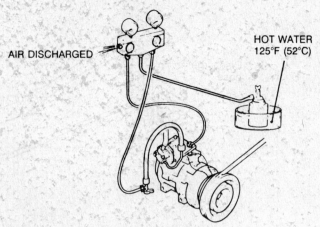

Charging the refrigerant system

GENERAL INFORMATION AND MAINTENANCE

Windshield Wipers

Intense heat from the sun, snow and ice, road oils and the chemicals used in windshield washer solvents combine to deteriorate the rubber wiper refills. The refills should be replaced about twice a year or whenever the blades begin to streak or chatter.

For maximum effectiveness and longest element life, the windshield and wiper blades should be kept clean. Dirt, tree sap, road tar and so on will cause streaking, smearing and blade deterioration if left on the windshield. It is advisable to wash the windshield carefully with a commercial glass cleaner at least once a month. Wipe off the rubber blades with a wet rag afterwards. Do not attempt to move the wipers back and forth by hand! Damage to the motor and drive mechanism will result.

If the blades are found to be cracked, broken or torn they should be replaced immediately. Replacement intervals will vary with usage, although ozone deterioration usually limits blade lift to about one year. If the wiper pattern is smeared or streaked, or if the blade chatters across the glass, the blades should be replaced. It is easiest and most sensible to replace them in pairs.

WIPER REFILL REPLACEMENT

NOTE: For wiper blade and arm service procedures, refer to Section 6.

Normally, if the wipers are not cleaning the windshield properly, only the refill has to be replaced. The blade and arm usually require replacement only in the event of damage. It is necessary only (except on Tridon refills) to remove the arm or the blade to replace the refill (rubber part), though you may have to position the arm higher on the glass. You can do this by turning the ignition switch on and operating the wipers. When they are positioned where they are accessible, turn the ignition switch off.

There are several types of refills and your vehicle could have any kind, since aftermarket blades and arms may not use exactly the same type refill as the original equipment.

The original equipment wiper elements can be replaced as follows:
1. Lift the wiper arm off the glass.
2. Depress the release lever on the center bridge and remove the blade from the arm.

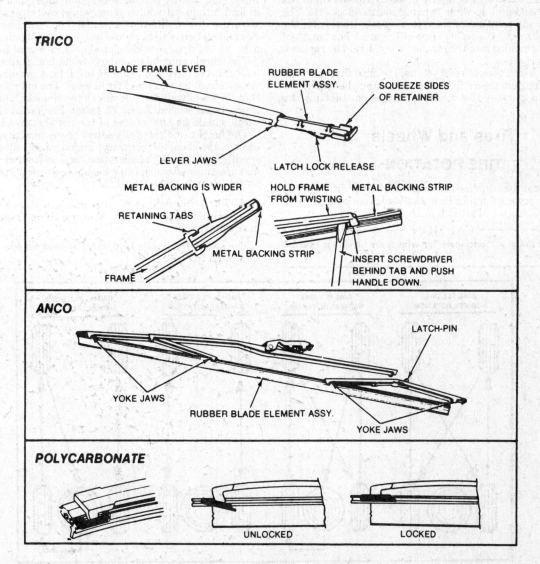

Three types of wiper blade retention

1-25

1 GENERAL INFORMATION AND MAINTENANCE

3. Lift the tab and pinch the end bridge to release it from the center bridge.

4. Slide the end bridge from the wiper blade and the wiper blade from the opposite end bridge.

5. Install a new element and be sure the tab on the end bridge is down to lock the element in place. Check each release point for positive engagement.

Most Trico styles use a release button that is pushed down to allow the refill to slide out of the release jaws. The new refill slide in and locks in place. Some Trico refills are removed by locating where the metal backing strip or the refill is wider. Insert a small screwdriver blade between the frame and the metal backing strip. Press down to release the refill from the retaining tab.

The Anco style is unlocked at one end by squeezing 2 metal tabs, and the refill is slid out of the frame jaws. When the new refill is installed, the tabs will click into place, locking the refill.

The polycarbonate type is held in place by a locking lever that is pushed downward out of the groove in the arm to free the refill. When the new refill is installed, it will lock in place automatically.

The Tridon refill has a plastic backing strip with a notch about an inch from the end. Hold the blade (frame) on a hard surface so that the frame is tightly bowed. Grip the tip of the backing strip and pull up while twisting counterclockwise. The backing strip will snap out of the retaining tab. Do this for the remaining tabs until the refill is free of the arm. The length of these refills is molded into the end and they should be replaced with identical types.

No matter which type of refill you use, be sure that all of the frame claws engage the refill. Before operating the wipers, be sure that no part of the metal frame is contacting the windshield.

Tires and Wheels

TIRE ROTATION

Tire wear can be equalized by switching the position of the tires about every 6000 miles. Including a conventional spare in the rotation pattern can give up to 20% more tire life.

--- CAUTION ---
Do not include the new "Spacesaver" or mini-spare tire in the rotation pattern.

There are certain exceptions to tire rotation, however. Studded snow tires should not be rotated, and radials should be kept on the same side of the vehicle (maintain the same direction of rotation). The belts on radial tires get set in a pattern. If the direction of rotation is reversed, it can cause rough ride and vibration.

NOTE: When radials or studded snows are taken off the vehicle, mark them, so you can maintain the same direction of rotation.

TIRE USAGE

The tires on your car or truck were selected to provide the best all around performance for normal operation when inflated as specified. On trucks, oversize tires (Load Range D) will not increase the maximum carrying capacity of the vehicle, although they will provide an extra margin of tread life. Be sure to check overall height before using larger size tires which may cause interference with suspension components or wheel wells. When replacing conventional tire sizes with other tire size designations, be sure to check the manufacturer's recommendations. Interchangeability is not always possible because of differences in load ratings, tire dimensions, wheel well clearances, and rim size. Also due to differences in handling characteristics, 70 Series and 60 Series tires should be used only in pairs on the same axle. Radial tires should be used only in sets of four.

The wheels must be the correct width for the tire. Tire dealers have charts of tire and rim compatibility. A mismatch can cause sloppy handling and rapid tread wear. The old rule of thumb is that the tread width should match the rim width (inside bead to inside bead) within 1 in. (25.4mm). For radial tires, the rim width should be 80% or less of the tire (not tread) width.

The height (mounted diameter) of the new tires can greatly change speedometer accuracy, engine speed at a given road speed, fuel mileage, acceleration, and ground clearance. Tire manufacturers furnish full measurement specifications. Speedometer drive gears are available for correction.

NOTE: Dimensions of tires marked the same size may vary significantly, even among tires from the same manufacturer.

The spare tire should be usable, at least for low speed operation, with the new tires.

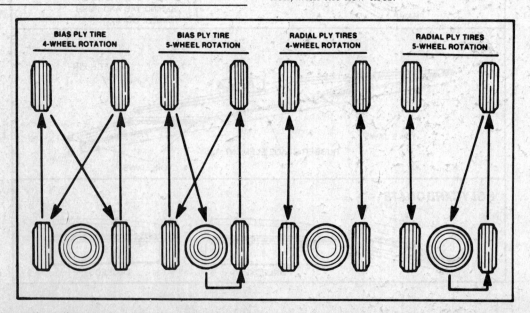

Tire rotation diagrams; note that radials should not be crossed to opposite sides of the car

GENERAL INFORMATION AND MAINTENANCE

TIRE DESIGN

For maximum satisfaction, tires should be used in sets of five. Mixing or different types (radial, bias/belted, fiberglass belted) should be avoided. Conventional bias tires are constructed so that the cords run bead-to-bead at an angle. Alternate plies run at an opposite angle. This type of construction gives rigidity to both tread and sidewall. Bias/belted tires are similar in construction to conventional bias ply tires. Belts run at an angle and also at a 90° angle to the bead, as in the radial tire. Tread life is improved considerably over the conventional bias tire. The radial tire differs in construction, but instead of the carcass plies running at an angle of 90° to each other, they run at an angle of 90° to the bead. This gives the tread a great deal of rigidity and the sidewall a great deal of flexibility and accounts for the characteristic bulge associated with radial tires.

All Mitsubishi vehicles are capable of using radial tires and they are the recommended type for all years. If they are used, tire sizes and wheel diameters should be selected to maintain ground clearance and tire load capacity equivalent to the minimum specified tire. Radial tires should always be used in sets of five, but in an emergency radial tires can be used with caution on the rear axle only. If this is done, both tires on the rear should be of radial design.

NOTE: Radial tires should never be used on only the front axle.

INFLATION PRESSURE

Tire inflation is the most ignored item of auto maintenance. Gasoline mileage can drop as much as 0.8% for every 1 pound per square inch (psi) of under inflation.

Two items should be a permanent fixture in every glove compartment; a tire pressure gauge and a tread depth gauge. Check the tire pressure (including the spare) regularly with a pocket type gauge. Kicking the tires won't tell you a thing, and the gauge on the service station air hose is notoriously inaccurate.

The tire pressures recommended for your vehicle are found on the door post. Ideally, inflation pressure should be checked when the tires are cool. When the air becomes heated it expands and the pressure increases. Every 10° rise (or drop) in temperature means a difference of 1 psi, which also explains why the tire appears to lose air on a very cold night. When it is impossible to check the tires "cold," allow for pressure build-up due to heat. If the "hot" pressure exceeds the "cold" pressure by more than 15 psi, reduce your speed, load or both. Otherwise internal heat is crated in the tire. When the heat approaches the temperature at which the tire was cured, during manufacture, the tread can separate from the body.

---— CAUTION ———
Never counteract excessive pressure build-up by bleeding off air pressure (letting some air out). This will only further raise the tire operating temperature.

Before starting a long trip with lots of luggage, you can add about 2–4 psi to the tires to make them run cooler, but never exceed the maximum inflation pressure on the side of the tire.

TREAD DEPTH

All tires made since 1968, have 8 built-in tread wear indicator bars that show up as ½ in. wide smooth bands across the tire when $1/16$ in. of tread remains. The appearance of tread wear indicators means that the tires should be replaced. In fact, many states have laws prohibiting the use of tires with less than $1/16$ in. tread.

You can check your own tread depth with an inexpensive gauge or by using a Lincoln head penny. Slip the Lincoln penny into several tread grooves. If you can see the top of Lincoln's head in 2 adjacent grooves, the tires have less than $1/16$ in. tread left and should be replaced. You can measure snow tires in the same manner by using the "tails" side of the Lincoln penny. If you can see the top of the Lincoln memorial, it's time to replace the snow tires.

TIRE STORAGE

Store the tires at proper inflation pressure if they are mounted on wheels. All tires should be kept in a cool, dry place. If they are stored in the garage or basement, do not let them stand on a concrete floor; set them on strips of wood.

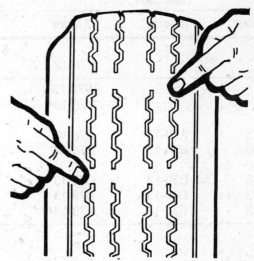
Tread wear indicators will appear when the tire is worn out

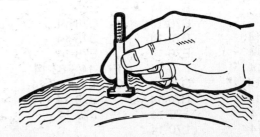
Tread depth can also be checked with an inexpensive gauge

A penny works as well as anything for checking tread depth. If you can see the top of Lincoln's head, the tire is worn out

1 GENERAL INFORMATION AND MAINTENANCE

CARE OF ALUMINUM WHEELS

If your vehicle is equipped with aluminum wheels special attention should be paid to their care and maintenance. Aluminum is very susceptible to the action of akalies often found in various detergents, road and sea salts. If the wheels have been exposed to these type of compounds, wash the wheels with water as soon as possible. After washing the vehicle, coat the wheels with body wax to prevent corrosion.

If steam is used to the clean the vehicle, do not direct the steam at the wheels.

When changing an aluminum rimmed tire, observe the following precations:
- Clean the surface of the hub first
- After the tire is in place, finger tighten the lug nuts and then torque them to the proper value
- DO NOT use an impact wrench to tighten the wheel nuts
- DO NOT use oil on either the nut or stud threads

Tire Size Comparison Chart

"Letter" sizes			Inch Sizes	Metric-inch Sizes		
"60 Series"	"70 Series"	"78 Series"	1965–77	"60 Series"	"70 Series"	"80 Series"
			5.50-12, 5.60-12	165/60-12	165/70-12	155-12
		Y78-12	6.00-12			
		W78-13	5.20-13	165/60-13	145/70-13	135-13
		Y78-13	5.60-13	175/60-13	155/70-13	145-13
			6.15-13	185/60-13	165/70-13	155-13, P155/80-13
A60-13	A70-13	A78-13	6.40-13	195/60-13	175/70-13	165-13
B60-13	B70-13	B78-13	6.70-13	205/60-13	185/70-13	175-13
			6.90-13			
C60-13	C70-13	C78-13	7.00-13	215/60-13	195/70-13	185-13
D60-13	D70-13	D78-13	7.25-13			
E60-13	E70-13	E78-13	7.75-13			195-13
			5.20-14	165/60-14	145/70-14	135-14
			5.60-14	175/60-14	155/70-14	145-14
			5.90-14			
A60-14	A70-14	A78-14	6.15-14	185/60-14	165/70-14	155-14
	B70-14	B78-14	6.45-14	195/60-14	175/70-14	165-14
	C70-14	C78-14	6.95-14	205/60-14	185/70-14	175-14
D60-14	D70-14	D78-14				
E60-14	E70-14	E78-14	7.35-14	215/60-14	195/70-14	185-14
F60-14	F70-14	F78-14, F83-14	7.75-14	225/60-14	200/70-14	195-14
G60-14	G70-14	G77-14, G78-14	8.25-14	235/60-14	205/70-14	205-14
H60-14	H70-14	H78-14	8.55-14	245/60-14	215/70-14	215-14
J60-14	J70-14	J78-14	8.85-14	255/60-14	225/70-14	225-14
L60-14	L70-14		9.15-14	265/60-14	235/70-14	
	A70-15	A78-15	5.60-15	185/60-15	165/70-15	155-15
B60-15	B70-15	B78-15	6.35-15	195/60-15	175/70-15	165-15
C60-15	C70-15	C78-15	6.85-15	205/60-15	185/70-15	175-15
	D70-15	D78-15				
E60-15	E70-15	E78-15	7.35-15	215/60-15	195/70-15	185-15
F60-15	F70-15	F78-15	7.75-15	225/60-15	205/70-15	195-15
G60-15	G70-15	G78-15	8.15-15/8.25-15	235/60-15	215/70-15	205-15
H60-15	H70-15	H78-15	8.45-15/8.55-15	245/60-15	225/70-15	215-15
J60-15	J70-15	J78-15	8.85-15/8.90-15	255/60-15	235/70-15	225-15
	K70-15		9.00-15	265/60-15	245/70-15	230-15
L60-15	L70-15	L78-15, L84-15	9.15-15			235-15
	M70-15	M78-15				255-15
		N78-15				

NOTE: Every size tire is not listed and many size comaprisons are approximate, based on load ratings. Wider tires than those supplied new with the vehicle should always be checked for clearance

GENERAL INFORMATION AND MAINTENANCE 1

Troubleshooting Basic Wheel Problems

Problem	Cause	Solution
The car's front end vibrates at high speed	• The wheels are out of balance • Wheels are out of alignment	• Have wheels balanced • Have wheel alignment checked/adjusted
Car pulls to either side	• Wheels are out of alignment • Unequal tire pressure • Different size tires or wheels	• Have wheel alignment checked/adjusted • Check/adjust tire pressure • Change tires or wheels to same size
The car's wheel(s) wobbles	• Loose wheel lug nuts • Wheels out of balance • Damaged wheel • Wheels are out of alignment • Worn or damaged ball joint • Excessive play in the steering linkage (usually due to worn parts) • Defective shock absorber	• Tighten wheel lug nuts • Have tires balanced • Raise car and spin the wheel. If the wheel is bent, it should be replaced • Have wheel alignment checked/adjusted • Check ball joints • Check steering linkage • Check shock absorbers
Tires wear unevenly or prematurely	• Incorrect wheel size • Wheels are out of balance • Wheels are out of alignment	• Check if wheel and tire size are compatible • Have wheels balanced • Have wheel alignment checked/adjusted

Troubleshooting Basic Tire Problems

Problem	Cause	Solution
The car's front end vibrates at high speeds and the steering wheel shakes	• Wheels out of balance • Front end needs aligning	• Have wheels balanced • Have front end alignment checked
The car pulls to one side while cruising	• Unequal tire pressure (car will usually pull to the low side) • Mismatched tires • Front end needs aligning	• Check/adjust tire pressure • Be sure tires are of the same type and size • Have front end alignment checked
Abnormal, excessive or uneven tire wear See "How to Read Tire Wear"	• Infrequent tire rotation • Improper tire pressure • Sudden stops/starts or high speed on curves	• Rotate tires more frequently to equalize wear • Check/adjust pressure • Correct driving habits
Tire squeals	• Improper tire pressure • Front end needs aligning	• Check/adjust tire pressure • Have front end alignment checked

1–29

1 GENERAL INFORMATION AND MAINTENANCE

FLUIDS AND LUBRICANTS

Fuel Recommendations

GASOLINE ENGINES

Unleaded fuel only must be used in all gasoline models. Leaded fuel will damage the catalytic converter almost immediately. This will increase emissions critically, and may cause the catalyst material to break up and clog the exhaust system.

Use a fuel with an octane rating of 87 (R+M/2). This is an average of the two methods of rating octane—Research and Motor. If a fuel is rated by the Research method only, the required rating is 91.

It always pays to buy a reputable brand of fuel. It is best to purchase fuel from a busy station where a large volume is pumped every day, as this helps protect you from dirt and water in the fuel. If you encounter engine ping or run-on, you might want to try a different brand or a slightly higher octane rated grade of fuel. If the engine exhibits a chronic knock problem which cannot be readily cured by changing the fuel you're using, it is wise to check the ignition timing and reset it, if necessary. If the timing is correct and the engine still exhibits severe knock, there may be internal mechanical problems. Persistent knock is severely damaging to the engine and should be corrected.

If the engine runs on, and changing to a different fuel does not cure the problem, routine checks of ignition timing and idle speed should be made. Persistent run-on can be damaging to such engine parts as the timing chain or belt.

Mitsubishi recommends against the "indiscriminate use of fuel system cleaning agents." Occasional use of solvents added to the fuel tank to remove gum and varnish from the fuel system is permissible; however, continuous use or extremely frequent use can damage gasket and diaphragm parts used in the system.

Gasohol, a mixture of 10% ethanol (or grain alcohol) and 90% unleaded gasoline, may be used in your Mitsubishi. You should switch back to unleaded gasoline or try another brand of gasohol if you experience driveability problems. Some brands may contain special fuel additives designed to overcome certain types of problems that may occur with gasohol use.

DO NOT use gasolines containing methanol (wood alcohol), as they may damage the fuel system.

DIESEL ENGINES

Use quality diesel fuel from reputable suppliers which has been specifically formulated for vehicular use. The diesel engine in the truck line is designed to operate on either No. 1-D, No. 2-D or winterized No 2-D grade fuel. For best economy and performance, a No. 2-D fuel should be used at all temperatures above 20°F. At lower temperatures, either 1-D or winterized 2-D should be used to avoid wax (paraffin) plugging the fuel filter. This can result in the engine not starting on a cold morning.

To avoid severe damage to the engine, do not mix gasoline with the diesel fuel. Do not use household heating oil or any diesel oil intended for use in marine or industrial engines.

The fuel filler cap and lid have labels stating **DIESEL**. In the event of gasoline being pumped into the tank by mistake, DO NOT attempt to start the engine, even to leave the pump area. If the vehicle can be towed and the tank drained before the engine is started, great damage and expense will be avoided.

No solvents or additives should ever be mixed with the diesel fuel. Damage to seals, gaskets and the injection pump can result.

Engine Oil Recommendations

Use only quality oils designated as shown in the chart above. Never use straight mineral or non-detergent oils—that is, oils not equipped with special cleaning agents. You must not only choose the grade of oil, but the viscosity number. Viscosity refers to the thickness of the oil. It's actually measured by how rapidly it flows though a hole of calibrated size. Thicker oil flows more slowly and has higher viscosity numbers — SAE 40 or 50. Thinner oil flows more easily and has lower numbers — SAE 10 or 20.

Mitsubishi recommends the use of what are called "multigrade" oils. These are specially formulated to change their viscosity less with a change in temperature than straight grade oils. The oils are designated by the use of two numbers—the first referring to the thickness of the oil, relative to straight mineral oils, at a low temperature such as 0°F (−18°C); the second to the thickness relative to straight mineral oils at high

*SAE 5W-20 Not recommended for sustained high speed vehicle operation.

**SAE 5W-30 oil should be used only in areas where extremely cold temperatures below −23°C (−10°F) are experienced.

Oil viscosity standards for both normally aspirated and turbocharged engines. Adherence to these standards helps insure long engine life, easier starting and economical operation in cold weather

GENERAL INFORMATION AND MAINTENANCE 1

FLUIDS AND LUBRICANTS

Recommended Lubricants

Engine Oil	SE or SF ①
Manual Transmission/Transaxle	GL-4
Automatic Transmission/Transaxle	DEXRON® II
Rear Axle	GL-5 ②
Wheel Bearings	NLGI Grade #2 EP Multipurpose Grease
Chassis Grease	NLGI Grade #2 Multipurpose Grease
Brake Fluid	DOT3
Clutch Fluid	DOT3
Manual Steering	GL-4
Power Steering	DEXRON® II
Antifreeze	ethylene glycol
Hinges	engine oil

① Applies to 1983–84 models and 1985 Cordia/Tredia. All 1985 models except Cordia/Tredia and all 1986–89 models use SF or SF/CC.
② Applies to standard rear axles only. For limited slip differentials, use Mitsubishi Gear Oil Part #8149630EX, STA-LUBE® API GL-5 High Performance Gear Oil, or equivalent.

temperatures typical of highway driving – in the neighborhood of 200°F (93°C). These numbers are preceded by the designation "SAE" for the society which sets the viscosity standards. For example, use of an SAE 10W-40 oil would give nearly ideal engine operation under almost all operating conditions. The oil would be as thin as a straight 10 weight oil at cold cranking temperatures, and as thick as a straight 40 weight oil at hot running conditions.

Note that turbocharged engines and diesel engines require use of different oils. This is because the oil gets much hotter, partly due to running through the turbocharger bearing, and partly due to the higher frictional and thermal loads of a turbocharged engine. Be careful to adhere to all recommendations strictly for the longest possible engine life and best service. Recommendations are especially critical for turbocharged engines. Oil recommendations are as follows:

Normally Aspirated Engines

- Temperatures ranging from 32°F (0°C) to 120°F (49°C): SAE 20W-20, 20W-40, 20W-50
- Temperatures ranging from −10°F (−23°C) to 120°F (49°C): SAE 10W-30, 10W-40, 10W-50
- Temperatures ranging from −20°F (−29°C) up to 60°F (16°C): SAE 5W-20, 5W-30, 5W-40

NOTE: 5W-20 is not recommended for sustained high speed operation regardless of the weather.

Turbocharged Engines

- Temperatures ranging from 32°F (0°C) to 120°F (49°C): SAE 20W-20, 20W-40
- Temperatures ranging from −10°F (−23°C) to 120°F (49°C): SAE 10W-40
- Temperatures ranging from −20°F (−29°C) to 60°F (16°C): SAE 5W-30

NOTE: 5W-30 oils should be used only in extremely cold areas, where the temperature is consistently below freezing. 5W oils are not recommended for sustained high speed or high rpm driving.

Diesel Engines

- Temperatures ranging from 32°F (0°C) to 104°F (40°C): SAE 30, 20W-40, 15W-40
- Temperatures ranging from 14°F (−10°C) to 120°F (49°C): SAE 20W-40
- Temperatures ranging from 5°F (−15°C) to 120°F (49°C): SAE 15W-40
- Temperatures ranging from −4°F (−20°C) to 104°F (40°C): SAE 10W-30
- Temperatures ranging from 50°F (10°C) to −20°F (−29°C): SAE 5W-30

NOTE: 5W-30 oils should be used only in extremely cold areas, where the temperature is consistently below freezing when starting. 5W oils are not recommended for sustained high speed or high rpm driving.

SYNTHETIC OIL

There are excellent synthetic and fuel-efficient oils available that, under the right circumstances, can help provide better fuel mileage and better engine protection. However, these advantages come at a price, which can be three or four times the cost per quart of conventional motor oils.

Before pouring any synthetic oils into your vehicle's engine, you should consider the condition of the engine and the type of driving you do. Also, check the manufacturer's warranty conditions regarding the use of synthetics.

Generally, it is best to avoid the use of synthetic oil in both brand new and older, high mileage engines. New engines require a proper break-in, and the synthetics are so slippery that they can prevent this. Most manufacturers recommend that you wait at least 5,000 miles before switching to a synthetic oil. Conversely, older engines are looser and tend to use more oil. Synthetics will slip past worn parts more readily than regular oil, and will be used up faster. If your truck already leaks and/or uses oil (due to worn parts and bad seals or gaskets), it will leak and use more with a slippery synthetic inside.

Consider your type of driving. If most of your accumulated mileage is on the highway at higher, steadier speeds, a synthetic oil will reduce friction and probably help deliver fuel mileage. Under such ideal highway conditions, the oil change interval can be extended, as long as the oil filter will operate effectively for the extended life of the oil. If the filter can't do its job for this extended period, dirt and sludge will build up in your engine's crankcase, sump, oil pump and lines, no matter what type of oil is used. If using synthetic oil in this manner, you should continue to change the oil filter at the recommended intervals.

Trucks used under harder, stop-and-go, short hop circumstances should always be serviced more frequently, and for these trucks, synthetic oil may not be a wise investment. Because of the necessary shorter change interval needed for this type of driving, you cannot take advantage of the long recommended change interval of most synthetic oils.

Finally, most synthetic oil are not compatible with conventional oils and cannot be added to them. This means you should always carry a couple of quarts of synthetic oil with you while on a long trip, as not all service stations carry this oil.

1-31

1 GENERAL INFORMATION AND MAINTENANCE

Engine

OIL LEVEL CHECK

As often as you stop for fuel check the engine oil as follows:
1. Park the vehicle on a level surface (if the vehicle is not level, the reading will not be completely accurate).
2. If the vehicle has been running, stop the engine and allow it to sit for a full five minutes. If the engine is cold, check the oil before starting it. It does not matter whether the oil is hot or cold, as long as it has had time to drain out of the engine itself and into the oil pan.
3. Open the hood and locate the dipstick. It is usually in front of transverse engines; it may be on either side of engines that run fore and aft. It consists of a ring-like handle running into a tube which is connected to the engine. Pull the dipstick out and wipe all the oil off the bottom with a clean rag. If this is not done, you will not get an accurate reading of the oil level.
4. Re-insert the dipstick and make sure it goes all the way into the tube. Then, pull it out and read it on the side which bears two lines. If the oil level is above the lower line, the oil level is high enough and you should not add any oil. If it is right near or at the lower line, add one quart. If the oil is below the lower line, add oil, one quart at a time until the level is above the lower line. A level below the lower line indicates that either you are not checking the oil level frequently enough or that the engine is using too much oil. Running the engine with the oil below the lower line may contribute to excessive heat and dirt in the oil, and will leave you with insufficient reserve to allow for normal oil consumption—you could run out on the road. However, you should not add oil to the point where the level is significantly above the upper line. Under these conditions, the rotating crankshaft will cause the oil to foam, which can be damaging to the engine and will sometimes cause valve train noise.
5. To add oil, unscrew the cap on the valve cover on top of the engine and pour the oil into the engine using a pouring spout if you bought the oil in a can. Avoid letting any dirt get into the engine, and make sure to reinstall the cap before starting the engine.

OIL AND FILTER CHANGE

The mileage figures given in the "Maintenance Intervals" chart are the factory recommended intervals for oil and filter changes assuming you are driving your vehicle under average conditions. Make sure to read the footnote concerning decreasing the change interval for certain types of driving and adhere to the recommendations. While Mitsubishi recommends changing the oil filter only every other oil change after the initial change, we recommend changing the filter every time because of the extra insurance this provides. Not only will this help guarantee that the oil filter will always work effectively (the filter bypasses dirty oil directly to engine parts when it gets saturated with dirt); but changing the filter removes at least an additional quart of dirty oil from the old filter and the engine oil passages. Purchase oil which conforms to all the specifications above. Make sure that the filter you purchase is specified for your particular vehicle model, year, and engine type. The filter must be able to withstand 256 psi pressure to conform to factory specifications.

Always drain the oil after the engine has been running long enough to bring it to operating temperature. It's best to actually drive the vehicle past the point where the temperature gauge reaches normal operating temperature to help ensure the oil will be as warm as possible. Hot oil will flow out of the oil pan more easily and will keep contaminants in the engine suspended so that they will be removed with the oil instead of staying in the pan. You will need a large capacity oil pan—usually about 6 qt. capacity is best. Just make sure the capacity of the pan is greater than the capacity of the oil pan and filter as shown in the "Ca-pacities" chart above. You will also need a spout if your oil is in cans, a strap wrench to loosen the filter, and an ordinary set of open-end wrenches. You can purchase tools and supplies at any store which sells automotive parts. It is also necessary for you to have some clean rags available to clean up inevatable spills.

You should also make plans to dispose properly of the used oil. Sometimes a local service station or garage will sell its used oil to a reprocessor. You may be able to add your used oil to his oil drain tank.

Change the oil as follows:
1. Warm the engine up as described above. Turn the engine off and remove the oil filler cap. Then, support the vehicle securely on axle stands or ramps, if they are available. If you can work under the vehicle at its normal height, this is okay provided the wheels are chocked. Just don't try to crawl under the vehicle when it's supported by the jack which is designed for tire changing.
2. Place the drain pan under the oil pan. It should be located where the stream of oil running out of the drain hole will run into the pan—not just directly below the drain hole. Now, loosen the drain plug with an open-end or box wrench. Turn the plug out slowly by hand. By keeping inward pressure on the plug with your fingers as you unscrew it, oil won't escape past the threads and you can remove it without being burned by hot oil. Make sure you keep hold of the plug so that once it is turned past the threads, it does not drop into the pan. Remove the plug and wipe it with a clean rag. Put it in a safe place—one where it won't get kicked or bumped out of sight. As the oil drains, the stream may shift as the level in the pan changes. Keep your eye on the stream and shift the pan as necessary.
3. Once the oil has drained, locate the pan under the oil filter (on some engines, you may be able to place the pan so it will catch oil drained from the filter at the same time it is catching oil drained from the pan). The strap wrench is designed so it will tighten its grip on the filter as you turn it. You must first position (turn) the wrench so it will tighten up as the filter is turned counterclockwise to remove it. Then, position the wrench over the filter at the center with the wrench handle in a position where you can grab it and turn it counterclockwise. Turn the fil-

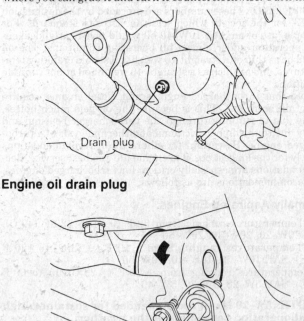

Engine oil drain plug

Removing the oil filter

GENERAL INFORMATION AND MAINTENANCE

ter with the wrench until it loosens up. Now, remove the wrench and turn the filter off by hand. Drain the filter into the drain pan.

4. With a clean rag, wipe off the filter adapter on the engine block. Make sure that no lint from the rag remains on the adapter as it could clog an oil passage. Also make sure the rubber gasket from the old filter did not remain on the adapter; remove it if it did. Thoroughly coat the rubber gasket on the top of the filter with clean oil. Read directions on the side of the filter or on the box it came in as to how far to tighten it. Then, carefully position the filter straight over the screw-type fitting at the center of the filter adapter and turn the filter gently until the threads grab. If you have to force the filter to turn at this point, the threads are probably cross-threaded, and you should turn the filter backwards until it is free and then try to start the threads again, this time at a better angle. Once the threads do start, turn the filter gently until it just touches the engine block; it will suddenly get harder to turn at this point.

5. You may want to mark the filter at this point so you'll know just how far you turn it. By hand, turn it the ½ to ¾ turn usually specified.

6. Wipe the drain plug area on the oil pan and carefully reinstall the drain plug. Just as with the filter, be careful not to cross-thread the plug, but turn it very gently and try very hard to get it to go straight into the hole. It will turn very easily well past the point where the threads have started if it's not cross-threaded.

7. Make sure to tighten the drain plug snugly with a wrench and use a new gasket if so required! More than one engine has been rebuilt because a loose drain plug fell out on the road, draining the oil.

8. Now, pour in oil to the full capacity of the oil pan and filter, as specified in the Capacities chart above. Then, make sure to reinstall the filler cap. Just as a precaution, remove the dipstick and check the oil level. It should be above the full mark.

9. If the engine has enough oil, start the engine, preferably without touching the throttle, as there will be no oil pressure for 10 seconds or more while the oil pump fills the filter and engine oil passages. Allow the engine to idle at the lowest possible speed until the oil light goes out or the gauge shows that oil pressure has been established. Of course, if you don't get oil pressure soon, less than 30 seconds, stop the engine and investigate.

10. Once oil pressure is established, leave the engine running and inspect the filter and oil plug for leaks. If there is slight leakage around the filter, you might want to try to tighten it just a bit more to stop the leaks. Usually, if you've tightened it properly, the only cause of leakage is a defective filter or gasket, which would have to be replaced before you drive.

11. Shut the engine down, allow the oil to drain into the pan, recheck the level, and add oil, if necessary.

Manual Transmission/Transaxle

LEVEL CHECK AND FLUID RECOMMENDATIONS

Both the manual transmission (Starion, Pickup and Montero) and the manual transaxles used on other models have filler plugs on the side of the unit. Be careful to identify the drain plug on the bottom and differentiate it from the filler plug on the side.

With the vehicle parked on a level surface and the wheels chocked, remove the filler plug. If the fluid runs out of the hole, the level is okay and you can immediately replace it. If fluid does not run out, you can stick your finger into the hole to check the level. As long as it is at or near the level of the bottom of the hole (no more than about ¼ inch below) the level is ok. If necessary, add API GL-4 fluid of viscosity of SAE 80W or SAE 75W-85W. You may want to purchase a syringe-like device for adding small quantities of such fluid from your local automotive parts store.

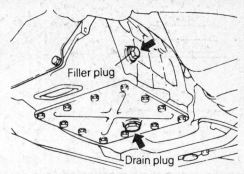

Manual transmission drain and fill plug locations on Starion

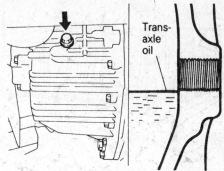

Checking oil level on manual transaxles

DRAIN AND REFILL

NOTE: Whenever you plan to change the transmission fluid, purchase a new drain plug gasket ahead of time. This item should be replaced at each fluid change to ensure a good drain plug seal.

1. Park the vehicle on a level surface.
2. Position a suitable drain pan under the transmission/transaxle and remove the fill plug.
3. Remove the drain plug and allow the oil to drain completely.
4. Wipe off the threads of the drain plug and install the plug with a new gasket.
5. Through the filler plug opening, add the proper amount of fluid (see Capacities Chart at end of this Section) until the level is just below the bottom of the opening. If a small amount runs out, that's OK.
6. After wiping the threads, install and tighten the filler plug.
7. Drive the vehicle to allow the oil to reach operating temperature and then check the drain and fill plugs for leaks.

Automatic Transmission/Transaxle

LEVEL CHECK AND FLUID RECOMMENDATIONS

The fluid level in these units is very sensitive to heat and must only be checked when the unit is at normal operating temperature. Drive the vehicle until a few miles after the engine has reached operating temperature. Then, with the engine idling, engage every gear selector position for a few seconds. Finally, put the gear selector in the neutral position and securely engage the handbrake.

Remove the dipstick, wipe it, reinstall it all the way, and remove it. Fluid level should be between the two marks on the stick. If the fluid level is at the lower mark, add one pint. Do not

1 GENERAL INFORMATION AND MAINTENANCE

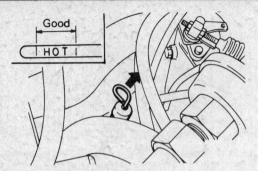

Checking the automatic transmission fluid on the Cordia

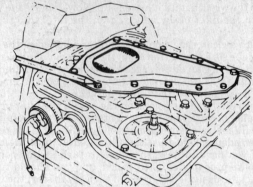

The Starion automatic transmission filter

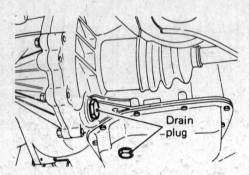

Draining the automatic transaxle fluid

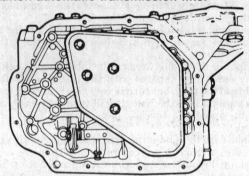

The automatic transaxle filter

overfill the transmission as this will cause foaming of the fluid and operating problems.

You should add Dexron®II automatic transmission fluid only. Shut the engine down and add the fluid with a small funnel through the dipstick tube. Mitsubishi recommends against the use of any transmission additives whatsoever.

DRAIN AND REFILL

It's best to change the fluid with the transmission hot. Drive the vehicle until the transmission is warmed up; a few miles after the engine temperature gauge reaches normal levels. Then follow these steps:

1. Support the vehicle securely on axle stands or ramps. It should be reasonably level.
2. There is no drain plug on most Mitsubishi automatic transmission oil pans. On 1983 Cordia/Tredia models, there *is* a drain plug. Remove it to drain the fluid. On transaxles only, remove the plug at the bottom of the differential case to allow the automatic transmission fluid to drain from that area. To remove the pan, loosen all the pan bolts just slightly and then place a large container under the oil pan. Tap one corner of the pan with a soft hammer to break the seal.
3. Once the pan is broken loose, support it, remove all the bolts, and then tilt it to one side to drain the fluid. on transaxles, there is an additional drain plug for fluid that accumulates in the housing. Remove this plug, located behind the pan on the side of the housing away from the engine.
4. Clean all the gasket surfaces thoroughly. Then, put the pan and gasket in position with boltholes lined up. Replace the bolts, tightening them only very gently with your fingers.
5. Then, tighten them diagonally in several stages to 4.4–5.7 ft. lb. on the Starion transmission and 7.5–8.5 ft. lb. on the transaxles used in other models. On front wheel drive models, torque the drain plug to 22–25 ft. lb. Pour fluid in cautiously until it reaches the lower mark on the dipstick. Refer to the "Capacities Chart" at the end of this Section for the amount you will need. Start the engine, put the gear selector in each of the positions for several seconds, and then go back to "Park". Check the fluid level again and make sure it's above the lower mark. Drive the vehicle until the transmission is hot and add fluid to the full mark.

NOTE: On 1983–85 Starion, the transmission bands are adjusted along with each fluid and filter change. See Section 7 for band adjustments on these models.

FILTER SERVICE

1. Remove the transmission/transaxle oil pan as described above.
2. Remove the attaching bolts and remove the filter assembly. On the Starion, the strainer forms the bottom portion of the valve body and is attached by 11 bolts. On all other models with transaxle, the filter is held in place by four bolts. Strainers may be cleaned in a safe solvent and air dried. Foam type filters should be replaced.
3. Install the filter or strainer and torque the bolts alternately (diagonally) is several stages to 25–34 inch lb. on the Starion and 48–60 inch lb. on other models.

Transfer Case

The 4–wheel drive family of vehicles use a transfer case between the transmission and the driveshaft. This case contains a series of gears and chains which must be properly lubricated. Although the transfer case is connected to the transmission, the two units have separate oil supplies. Do not make the mistake of thinking that draining or filling one unit takes care of both.

The transfer case has a drain plug at the lower edge and a fill plug about halfway up the case. Fluid level is checked with the vehicle parked on a level surface and the wheels chocked. Remove the filler plug. If the fluid runs out of the hole, the level is okay and you can immediately replace it. If fluid does not run

GENERAL INFORMATION AND MAINTENANCE 1

out, you can stick your finger into the hole to check the level. As long as it is at or near the level of the bottom of the hole (no more than about ¼ inch below) the level is ok. If necessary, add API GL-4 fluid of viscosity SAE 80W or SAE 75W–85W. This fluid specification does not change if the vehicle is equipped with automatic transmission. To drain the fluid:

1. Park the vehicle on a level surface.
2. Position a suitable drain pan under the transfer case drain plug and remove the fill plug.
3. Remove the drain plug and allow the oil to drain completely.
4. Wipe off the threads of the drain plug and install the plug with a new gasket.
5. Through the filler plug opening, add fluid (API GL-4, SAE 80W or SAE 75W–85W) until the level is just below the bottom of the opening. If a small amount runs out, that's OK.
6. After wiping the threads, install and tighten the filler plug.
7. Drive the vehicle to allow the oil to reach operating temperature and then check the drain and fill plugs for leaks.

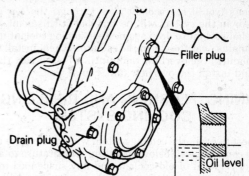

4-wheel drive transfer case drain and fill plug locations

Rear Drive Axle — Starion, Montero and Pickup
Front Axle — 4-Wheel Drive Trucks

LEVEL CHECK AND FLUID RECOMMENDATIONS

The fluid level must be checked every 30,000 miles on conventional differentials. On limited slip units, change the fluid every 15,000 miles. To check the fluid level, first make sure the wheels are chocked and then simply remove the plug from the side of the unit. If the fluid runs out or you can feel with your finger that it's not more than ¼ inch below the level of the hole, the level is satisfactory. Otherwise, add the fluid specified above with a syringe-like device you can buy at a local auto parts store.

For the limited slip transaxle, use the fluid recommended in the chart above. For the conventional axle, the API classification is GL-5 (as described in the chart above), and viscosity recommendations are as follows:

- Above −10°F (−23°C): SAE 90, SAE 85W–90, or SAE 80W–90
- −10°F (−23°C) to −30°F (−34°C): SAE 80W, SAE 80W–90
- Below −30°F (−34°C): SAE 75W

DRAIN AND REFILL

1. Make sure the wheels are chocked. Install a drain pan under the drain plug in the bottom of the differential.
2. Remove the fill plug from the side of the unit. Then, remove the drain plug, keeping it in your fingers.
3. When the fluid has drained completely, replace the drain plug, making sure the gasket is retained on the plug. Then, add fluid until it runs out the filler plug hole, and install the filler plug.

Cooling System

FLUID RECOMMENDATIONS

Mitsubishi recommends the use of a quality ethylene glycol coolant containing corrosion inhibitors, and approved for use with all the metals in the engine, including aluminum. Mitsubishi recommends against the use of additional cooling system additives as they may be incompatible with the coolant itself.

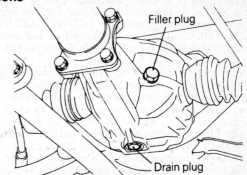

Rear axle filler and drain plug locations on Starion

LEVEL CHECK

Coolant level should always be checked, if possible, by simply looking at the level in the overflow tank. If there is a coolant leak, the level will drop in the tank and all the coolant will be drawn back into the radiator. Thus, if there is no coolant visible in the tank, the coolant level will have to be checked by removing the radiator cap. Also, if there are indications that the engine is overheating, it may be wise to check the level in the radiator, as a defective cap may permit the level to drop there even though there is an ample supply in the overflow system.

One of the reasons for the use of an overflow system is that it is often hazardous to remove the radiator cap. Under normal operating conditions, the temperature of the engine block exceeds the boiling point of the coolant. Thus, removing the cap on an engine that is still near operating temperature will usually result in a discharge of boiling hot coolant. The situation worsens substantially if the cooling system is dirty or the engine is overheating.

Thus the first rule to remember is: NEVER REMOVE THE RADIATOR CAP UNTIL THE ENGINE HAS COOLED SUBSTANTIALLY BELOW OPERATING TEMPERATURE AND THE TEMPERATURE GAUGE SHOWS THIS. Then, unless the engine is dead cold, USE A HEAVY RAG TO COVER THE CAP. Finally, TURN THE CAP SLOWLY TO THE FIRST NOTCH. This will release the pressure in the system. Give the pressure time to drop off, and then remove the cap. If the radiator cap is badly worn, the pressure may not be released. This is why it is necessary to use the rag, cool the engine first, and still proceed cautiously and slowly. If the cap reaches the position where it can be removed and hot coolant is suddenly expelled, the best procedure is to quickly reinstall the cap and then allow the engine to cool further.

Once the cap has been removed, you can add a 50–50 mix of antifreeze and water slowly to the radiator until the level comes up. Once the level is near the top of the radiator tank, have someone start the engine. Let the engine idle until the thermo-

1-35

1 GENERAL INFORMATION AND MAINTENANCE

stat opens (you'll see coolant flow through the top radiator tank). Air in the system will be expelled at this point, causing the level of coolant to drop again. Keep adding coolant until the level remains near the top of the radiator. Then, install the cap, add coolant to the overflow tank until it is well past the lower mark and install the overflow tank cap as well.

DRAINING, FLUSHING AND REFILLING THE COOLING SYSTEM

It's best to drain the cooling system when the engine is warm but has cooled to well below operating temperature to assist in complete removal of old coolant and any suspended material. Follow the procedure below to help ensure you will not be burned by hot coolant.

1. The first step is to loosen the drain cock or remove the drain plug located in the bottom radiator tank to begin the draining process and relieve pressure. DON'T START OUT BY REMOVING THE CAP! Just make sure you are well away from the direction coolant flow will take when you remove the plug, and then remove it.
2. Once coolant flow out of the bottom of the radiator has slowed, remove the radiator cap to vent the system. Then, remove the drain plug from the side of the engine block.
3. When all the coolant has drained, replace the plugs. If the hoses need attention, this would be an ideal time to relace them. Now, slowly fill the system with water until the water level reaches the top of the radiator tank. Start the engine and run it at idle. When the thermostat opens and water begins to flow through the top of the radiator, add more water as necessary until the engine is full. Now, shut the engine off and again carefully remove both drain plugs. Repeat the process of filling the system with water and draining it until drained water is clear. If the system cannot be cleaned out effectively this way, you may want to buy a reverse flushing kit at your local parts store and use it. An alternative is the use of a chemical cleaner; if you need to use one of these, just make sure it is compatible with the use of aluminum engine parts and that you follow the directions on the can carefully to ensure that you do not damage your engine.
4. Now, coat both drain plugs with sealer and install them snugly. Look up the coolant capacity in the "Capacities Chart" above. Then, see the chart on the side of the antifreeze container in order to calculate just how much antifreeze is required to protect the system down to the lowest expected temperature in your area. Pour the antifreeze in first. Then, follow up with clean water until the level reaches the top of the radiator tank. Finally, follow the steps at the end of the procedure above for filling the system with coolant after checking the level.

Brake And Clutch Master Cylinder

LEVEL CHECK AND FLUID RECOMMENDATIONS

The brake fluid reservoir on top of the master cylinder is clear so that you can check the fluid level without removing the top. If the level falls below the **A** mark, wipe off the cap and the outside of the reservoir and then remove the cap and refill the reservoir up to the **MAX** mark. Use brake fluid designed for use with vehicles equipped with disc brakes and meeting DOT 3 standards only.

As the vehicle ages, the presence of brake fluid in the reservoir will sometimes leave a residue that makes it difficult to read the fluid level reliably through the plastic. It is wise to occasionally check the fluid level by removing the cap to make sure you can accurately determine whether or not additional fluid is required.

Always be careful to keep dirt out of the system when removing the cap. It is normal for the level to drop gradually, as fluid

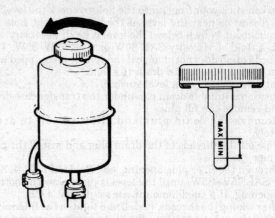

Checking power steering fluid level

takes up lost volume in the system occurring due to brake pad wear. A sudden loss of a significant amount of fluid should be investigated immediately. The system must be inspected thoroughly and all causes of leakage repaired.

NOTE: Follow these same procedures for checking the fluid level in the clutch master cylinder on those equipped with a hydraulic clutch. You should have the brake and hydraulic fluids changed as specified in the "Maintenance Intervals Chart". Use a fluid that meets DOT 3 standards.

Power Steering Pump

LEVEL CHECK AND FLUID RECOMMENDATIONS

Park the vehicle on a flat, level surface and leave the engine running. Turn the steering wheel several times to the left and right stops to raise the fluid to normal operating temperature. With the engine running, remove the cap from the power steering fluid reservoir. This unit is separate from the pump but connected to it by hoses. There are "MAX" and "MIN" marks on the dipstick connected to the cap. Add "DEXRON® II" automatic transmission fluid to the reservoir to bring the level up to the maximum.

At this time, you should inspect the hoses for signs of leakage or cracking. If either condition is noted, the defective hose or hoses should be replaced.

Manual Steering Gear

Mitsubishi vehicles with manual steering have a permanently lubricated steering box. Lubricant level checks or changes are not required.

Chassis Greasing

Ordinary greasing of the front suspension ball joints and steering linkage is not required on Mitsubishi vehicles as all such bearings are permanently lubricated. However, all grease seals should be inspected for leaks every two years and repaired as necessary. In addition, the transmission linkage, and parking brake cable should be lubricated periodically with NLGI Grade #2 grease.

Trucks and Monteros do have lubrication grease fittings at various locations within the steering and front suspension. These fittings should be lubricated with every oil change and after any off-pavement driving in wa-

GENERAL INFORMATION AND MAINTENANCE 1

ter or sand. You'll need a grease gun filled with Multipurpose grease SAE J310 (NLGI No. 2) and a large rag. Inexpensive grease guns are available almost everywhere and the newer styles use grease cartridges which eliminate the mess of loading the grease into the gun.

1. Elevate and safely support the vehicle on stands. Ramps can be used, but access is easier when the suspension hangs down.
2. Locate each grease nipple and wipe it clean of dust and grime. Fit the grease gun end onto each nipple. Apply the grease into the fitting until it is seen to come out of the dust seal cover-

ing the joint.

3. Remove the gun from the fitting and use the rag to wipe away excess grease at the fitting and the dust boot.
4. Continue with each fitting. Depending on the year and model, grease fittings may be found on the lower ball joint and steering joint (left and right), the upper ball joint (left and right), the tie rod ends (left and right) and the pivot for the pitman (steering) arm. Additionally, some trucks have fittings at both ends of the drive shaft running to the rear wheels.
5. Lower the vehicle to the ground.

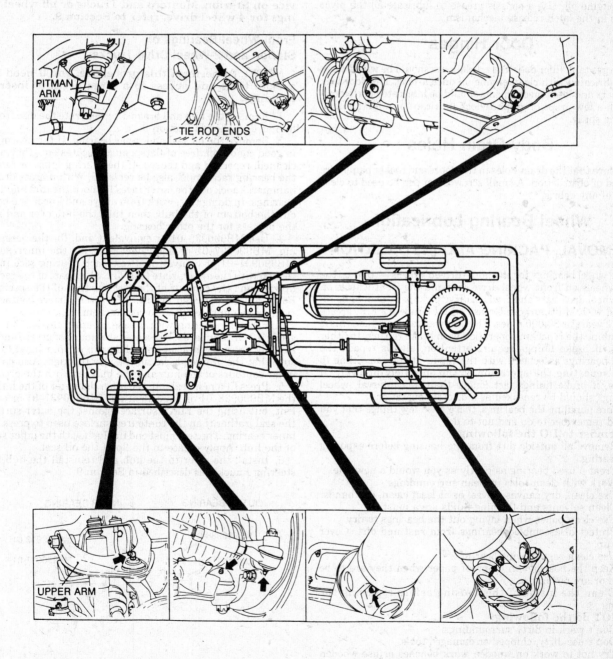

Lubrication points for Montero and trucks. Not all models have fittings at all points shown

1-37

1 GENERAL INFORMATION AND MAINTENANCE

OUTSIDE VEHICLE MAINTENANCE

Lock Cylinders

Apply graphite lubricant sparingly through the key slot. Insert the key and operate the lock several times to be sure that the lubricant is worked into the lock cylinder.

Hood Latch and Hinges

Clean the latch surfaces and apply clean engine oil to the latch pilot bolts and the spring anchor. Also lubricate the hood hinges with engine oil. Use a chassis grease to lubricate all the pivot points in the latch release mechanism.

Door Hinges

The gas tank filler door and truck doors should be wiped clean and lubricated with clean engine oil once a year. The door lock cylinders and latch mechanisms should be lubricated periodically with a few drops of graphite lock lubricant or a few shots of silicone spray.

Body Drain Holes

Be sure that the drain holes in the doors and rocker panels are cleared of obstruction. A small screwdriver can be used to clear them of any debris.

Wheel Bearing Lubrication

REMOVAL, PACKING AND INSTALLATION

The wheel bearings for the non–driven wheels – that is, the rear wheels on front wheel drive cars and the front wheels on rear wheel drive cars and 2-wheel drive pickups – should be repacked with Multipurpose Grease NLGI Grade #2 E.P. grease every 2 years or 30,000 miles. The best way to accomplish this is to combine the repacking operation with brake repairs. In other words, if brake linings require attention, always repack the wheel bearings associated with the repair at the same time to avoid repeating the operation at the specified interval. Of course, if brake linings last longer than this interval, wheel bearings should be repacked as a discrete operation.

Before handling the bearings, there are a few things that you should remember to do and not to do.

Remember to DO the following:
- Remove all outside dirt from the housing before exposing the bearing.
- Treat a used bearing as gently as you would a new one.
- Work with clean tools in clean surroundings.
- Use clean, dry canvas gloves, or at least clean, dry hands.
- Clean solvents and flushing fluids are a must.
- Use clean paper when laying out the bearings to dry.
- Protect disassembled bearings from rust and dirt. Cover them up.
- Use clean rags to wipe bearings.
- Keep the bearings in oil-proof paper when they are to be stored or are not in use.
- Clean the inside of the housing before replacing the bearing.

Do NOT do the following:
- Don't work in dirty surroundings.
- Don't use dirty, chipped or damaged tools.
- Try not to work on wooden work benches or use wooden mallets.
- Don't handle bearings with dirty or moist hands.
- Do not use gasoline for cleaning; use a safe solvent.
- Do not spin-dry bearings with compressed air. They will be damaged.
- Do not spin dirty bearings.
- Avoid using cotton waste or dirty cloths to wipe bearings.
- Try not to scratch or nick bearing surfaces.
- Do not allow the bearing to come in contact with dirt or rust at any time.

NOTE: Wheel bearing service in this Section covers only non–driven wheel bearings. For front wheel bearing service on front wheel drive cars or rear bearing service on Starion, Montero and Trucks or all wheel bearings for 4-wheel drive, refer to Section 9.

Front Wheel Bearings on Starion and 2-Wheel Drive Trucks

NOTE: To perform this procedure, you'll need three special tools and access to a large press to insert the bearings.

1. Remove the hub and brake disc; separate the disc from the hub as described in Section 9.
2. Remove the oil seal and inner bearing. If the bearings are in good condition, free of flats, gouges, scores etc., they may be cleaned, repacked and reused. If the bearing must be replaced, the bearing races must also be replaced. With a brass drift and hammer, knock out the outer races for both the inner and outer bearings. In doing this, work from above and knock the bearing out the bottom of the hub; then turn the hub over and repeat the process for the other bearing.
3. Use MB990938–01 or equivalent and, for the outer bearing, MB990927–01 or equivalent and, for the inner bearing, MB990931–01 or equivalent. Press each bearing race in from the top with the appropriate tool. The wider part of the race goes upward and the contour of the race fits that of the special tool. Races must be pressed in until the lower surface contacts the ridge in the hub designed to retain them.
4. Pack the bearing with grease meeting SAEJ310A NLGI grade #2 EP standards. Use a liberal amount of grease and occasionally press the bearing into the palm of your hand to make sure that the grease passes all the way through. Also pack the inner contours of the hub and the hub cap with the grease.
5. Press fit a new seal into the inner diameter of the hub with the MB990938–01 or equivalent and MB990931–01 or equivalent, but using the FLAT surface against the outer surface of the seal, rather than the contoured surface used to press in the inner bearing. The seal must end up flush with the inner surface of the hub. Apply grease to the lip of the oil seal.
6. Install the rotor to the hub and reinstall the hub to the steering knuckle as described in Section 9.

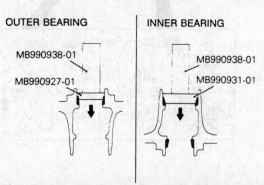

Special tools required for pressing the outer and inner bearings into the Starion front hub

GENERAL INFORMATION AND MAINTENANCE 1

Rear Wheel Bearings on Front Wheel Drive Cars with Rear Drum Brakes

1. Elevate and support the vehicle on jackstands. Make certain the parking brake is released and the rear wheels turn freely.
2. Remove the rear wheels.
3. Remove the small center cap, the cotter pin, the castellated nut and the lock nut.
4. Remove the drum, holding the outer bearing in place with a thumb as the drum comes off.
5. Remove the outer bearing. Turn the drum over and use a seal remover to remove the grease seal. After the seal is removed, the inner bearing may be lifted out.
6. If the bearings are in good condition, free of flats, gouges, scores etc., they may be cleaned, repacked and reused. If the bearings must be replaced, the bearing races must also be replaced. With a brass drift and hammer, knock out the outer races for both the inner and outer bearings. In doing this, work from above and knock the bearing out the bottom of the hub; then turn the hub over and repeat the process for the other bearing.
7. Use a bearing driver of the correct diameter and press each bearing race in from the top with the appropriate tool. The wider part of the race goes upward and the contour of the race fits that of the special tool. Races must be pressed in until the lower surface contacts the ridge in the hub designed to retain them.
8. Pack the bearing with grease meeting SAEJ310A NLGI grade #2 EP standards. Use a liberal amount of grease and occasionally press the bearing into the palm of your hand to make sure that the grease passes all the way through. Also pack the inner contours of the hub. Fill the small cap about half full of grease.
9. Install the inner bearing into the race. Press fit a new seal into the inner diameter of the hub with the a seal driver of the correct diameter. Use the FLAT surface against the outer surface of the seal. The seal must end up flush with the inner surface of the hub. Apply grease to the lip of the oil seal.
10. Install the outer bearing in the race. Hold it in position with a thumb and fit the drum and bearing assembly onto the axle.
11. Install the slotted washer onto the axle and install the lock nut. As the lock nut contacts the washer, make certain that the bearing is firmly seated in the race.
12. Tighten the lock nut to 20 Nm (14 ft. lbs). Turn the drum 2 or 3 turns in each direction to seat the bearing.
13. Release the locknut to 0 Nm; then retighten it to 10 Nm (7 ft. lbs). Turn the drum 2 or 3 times.
14. Set the locknut to its final torque of 10 Nm (7 ft. lbs) and install the castle nut and a new cotter pin.
15. If the holes of the castle nut do not align with the holes in the axle, simply reposition the castle nut; the holes are not evenly spaced and should be able to align with the axle. In the unlikely event that no holes align, the lock not may be loosened by no more than 15° of rotation.
16. Reinstall the rear wheels.
17. Lower the vehicle to the ground.

TRAILER TOWING

Trailer Weight

Trailer weight is the first, and most important, factor in determining whether or not your vehicle is suitable for towing the trailer you have in mind. The horsepower-to-weight ratio should be calculated. The basic standard is a ratio of 35:1. That is, 35 pounds of GVW for every horsepower. The maximum recommended weight of a trailer to be pulled by any Mitsubishi vehicle is 2000 lbs.; check the owner's manual for the correct maximum for your vehicle. Front wheel drive vehicles are severely limited in their ability to tow. Weight at the rear will cause the front to lift, reducing front wheel contact with the pavement. Since the front wheels do all the steering, all the propulsion and most of the braking, reduced tire contact is undesirable, particularly on slick or wet surfaces. Again, consult your owner's manual for limits and conditions specific to front wheel drive vehicles.

Hitch Weight

There are three kinds of hitches: bumper mounted, frame mounted, and load equalizing.

Bumper mounted hitches are those which attach solely to the vehicle's bumper. Many states prohibit towing with this type of hitch, when it attaches to the vehicle's stock bumper, since it subjects the bumper to stresses for which it was not designed. Aftermarket rear step bumpers, designed for trailer towing, are acceptable for use with bumper mounted hitches.

Frame mounted hitches can be of the type which bolts to two or more points on the frame, plus the bumper, or just to several points on the frame. Frame mounted hitches can also be of the tongue type, for Class I towing, or, of the receiver type, for classes II and III.

Load equalizing hitches are usually used for large trailers. Most equalizing hitches are welded in place and use equalizing bars and chains to level the vehicle after the trailer is hooked up.

The bolt-on hitches are the most common, since they are relatively easy to install.

Check the gross weight rating of your trailer. Tongue weight is usually figured as 10% of gross trailer weight. Therefore, a trailer with a maximum gross weight of 2,000 lb. will have a maximum tongue weight of 200 lb. Class I trailers fall into this category. Class II trailers are those with a gross weight rating of 2,000–3,500 lb., while Class III trailers fall into the 3,500–6,000 lb. category. Class IV trailers are those over 6,000 lb. and are for use with fifth wheel trucks, only.

1 GENERAL INFORMATION AND MAINTENANCE

When you've determined the hitch that you'll need, follow the manufacturer's installation instructions, exactly, especially when it comes to fastener torques. The hitch will subjected to a lot of stress and good hitches come with hardened bolts. Never substitute an inferior bolt for a hardened bolt.

Wiring

Wiring the vehicle for towing is fairly easy. There are a number of good wiring kits available and these should be used, rather than trying to design your own. All trailers will need brake lights and turn signals as well as tail lights and side marker lights. Most states require extra marker lights for overwide trailers. Also, most states have recently required back-up lights for trailers, and most trailer manufacturers have been building trailers with back-up lights for several years.

Additionally, some Class I, most Class II and just about all Class III trailers will have electric brakes. Electric brakes are strongly recommended on all trailers over 1000 lbs. Most states require trailer brakes above a certain weight.

Add to this number an accessories wire, to operate trailer internal equipment or to charge the trailer's battery, and you can have as many as seven wires in the harness.

Determine the equipment on your trailer and buy the wiring kit necessary. The kit will contain all the wires needed, plus a plug adapter set which included the female plug, mounted on the bumper or hitch, and the male plug, wired into, or plugged into the trailer harness.

When installing the kit, follow the manufacturer's instructions. The color coding of the wires is standard throughout the industry.

One final point: the best kits are those with a spring loaded cover on the vehicle mounted socket. This cover prevent dirt and moisture from corroding the terminals. Never let the vehicle socket hang loosely. Always mount it securely to the bumper or hitch.

Cooling

ENGINE

One of the most common, if not THE most common, problems associated with trailer towing is engine overheating.

With factory installed trailer towing packages, a heavy duty cooling system is usually included. Heavy duty cooling systems are available as optional equipment on most vehicles, with or without a trailer package. If you have one of these extra-capacity systems, you shouldn't have any overheating problems.

If you have a standard cooling system, without an expansion tank, you'll definitely need to get an aftermarket expansion tank kit, preferably one with at least a 2 quart capacity. These kits are easily installed on the radiator's overflow hose, and come with a pressure cap designed for expansion tanks.

Another helpful accessory is a Flex Fan. These fan are large diameter units are designed to provide more airflow at low speeds, with blades that have deeply cupped surfaces. The blades then flex, or flatten out, at high speed, when less cooling air is needed. These fans are far lighter in weight than stock fans, requiring less horsepower to drive them. Also, they are far quieter than stock fans.

If you do decide to replace your stock fan with a flex fan, note that if your vehicle has a fan clutch, a spacer between the flex fan and water pump hub will be needed.

Aftermarket engine oil coolers are helpful for prolonging engine oil life and reducing overall engine temperatures. Both of these factors increase engine life.

While not absolutely necessary in towing Class I and some Class II trailers, they are recommended for heavier Class II and all Class III towing.

Engine oil cooler systems consist of an adapter, screwed on in place of the oil filter, a remote filter mounting and a multi-tube, finned heat exchanger, which is mounted in front of the radiator or air conditioning condenser.

TRANSMISSION

An automatic transmission is usually recommended for trailer towing. Modern automatics have proven reliable and, of course, easy to operate, in trailer towing.

The increased load of a trailer, however, causes an increase in the temperature of the automatic transmission fluid. Heat is the worst enemy of an automatic transmission. As the temperature of the fluid increases, the life of the fluid decreases.

It is essential, therefore, that you install an automatic transmission cooler.

The cooler, which consists of a multi-tube, finned heat exchanger, is usually installed in front of the radiator or air conditioning compressor, and hooked inline with the transmission cooler tank inlet line. Follow the cooler manufacturer's installation instructions.

Select a cooler of at least adequate capacity, based upon the combined gross weights of the vehicle and trailer.

Cooler manufacturers recommend that you use an aftermarket cooler in addition to, and not instead of, the present cooling tank in your vehicle's radiator. If you do want to use it in place of the radiator cooling tank, get a cooler at least two sizes larger than normally necessary.

NOTE: A transmission cooler can, sometimes, cause slow or harsh shifting in the transmission during cold weather, until the fluid has a chance to come up to normal operating temperature. Some coolers can be purchased with or retrofitted with a temperature bypass valve which will allow fluid flow through the cooler only when the fluid has reached operating temperature, or above.

GENERAL INFORMATION AND MAINTENANCE 1

JUMP STARTING A DEAD BATTERY

The chemical reaction in a battery produces explosive hydrogen gas. This is the safe way to jump start a dead battery, reducing the chances of an accidental spark that could cause an explosion.

Jump Starting Precautions

1. Be sure both batteries are of the same voltage.
2. Be sure both batteries are of the same polarity (have the same grounded terminal).
3. Be sure the vehicles are not touching.
4. Be sure the vent cap holes are not obstructed.
5. Do not smoke or allow sparks around the battery.
6. In cold weather, check for frozen electrolyte in the battery. Do not jump start a frozen battery.
7. Do not allow electrolyte on your skin or clothing.
8. Be sure the electrolyte is not frozen.

CAUTION: Make certin that the ignition key, in the vehicle with the dead battery, is in the OFF position. Connecting cables to vehicles with on-board computers will result in computer destruction if the key is not in the OFF position.

Jump Starting Procedure

1. Determine voltages of the two batteries; they must be the same.
2. Bring the starting vehicle close (they must not touch) so that the batteries can be reached easily.
3. Turn off all accessories and both engines. Put both vehicles in Neutral or Park and set the handbrake.
4. Cover the cell caps with a rag—do not cover terminals.
5. If the terminals on the run-down battery are heavily corroded, clean them.
6. Identify the positive and negative posts on both batteries and connect the cables in the order shown.
7. Start the engine of the starting vehicle and run it at fast idle. Try to start the car with the dead battery. Crank it for no more than 10 seconds at a time and let it cool for 20 seconds in between tries.
8. If it doesn't start in 3 tries, there is something else wrong.
9. Disconnect the cables in the reverse order.
10. Replace the cell covers and dispose of the rags.

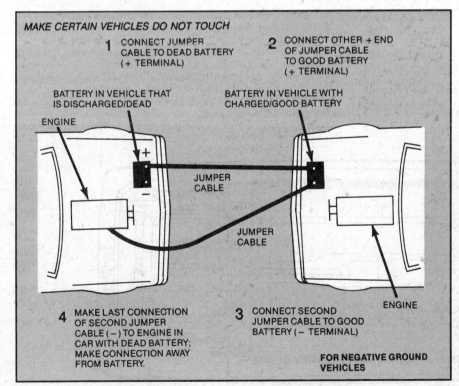

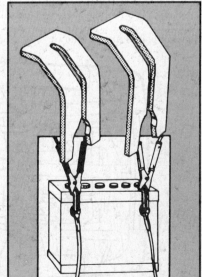

Side terminal batteries occasionally pose a problem when connecting jumper cables. There frequently isn't enough room to clamp the cables without touching sheet metal. Side terminal adaptors are available to alleviate this problem and should be removed after use

1-41

1 GENERAL INFORMATION AND MAINTENANCE

TOWING

Towing when the steering wheel can be unlocked with the ignition key should be accomplished by hooking cables to the towing hooks under the front of the chassis. As long as vehicles with manual transmissions/transaxles can be put in neutral, (4-wheel drive: Select neutral and 2H; unlock freewheeling hubs) there is no mileage limit to the distance they may be towed or limit as to maximum speed. Automatic transmission/trasaxle-equipped vehicles must not be towed faster than 19 mph or for a greater distance than 15 miles, as certain transmission parts will not be lubricated. To tow these vehicles a greater distance, raise the drive wheels off the ground or transport the vehicle on a flatbed.

If the ignition key is not available, the vehicle should be towed with the front wheels off the ground because of the locked steering wheel.

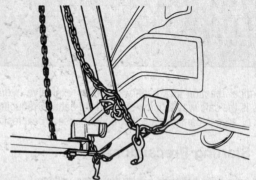

Towing the Cordia and Tredia with the rear wheels off the ground

JACKING

Passenger Cars

All the cars covered in this manual use a scissors type jack which supports the body when jacking is required. Follow these steps:
1. Block the wheel opposite to the one being removed in front of and to the rear of the tire. Apply the handbrake and make sure the engine is off. Make sure the jack will be supported on a hard surface.
2. Jacking points are as follows:

- Cordia/Tredia — in the front, place the jack under the bottom edge of the towing hook. At the rear, center the jack between the two marks on the lower body sheet metal seam.
- Starion — at both front and rear, center the jack between the two marks on the body sheet metal seam.
- On the Galant and Mirage, center the jack under the sheet metal seam right near the front or rear end of that seam.
- On Sigma front, center the jack on the portion of the frame member closest to the grill.
- On Sigma rear, center the jack on the support across from the catalytic converter.

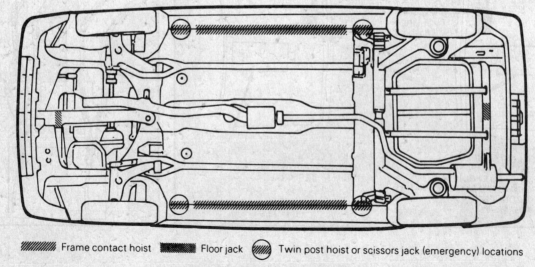

Frame contact hoist　　Floor jack　　Twin post hoist or scissors jack (emergency) locations

Lifting points for the Mirage

GENERAL INFORMATION AND MAINTENANCE 1

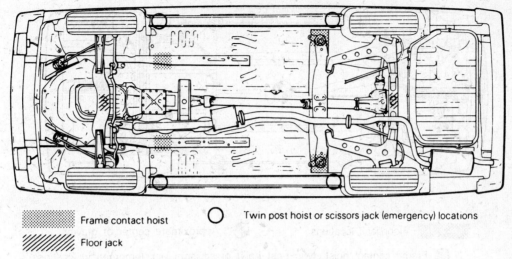

▓ Frame contact hoist ○ Twin post hoist or scissors jack (emergency) locations
▨ Floor jack

Lifting points for the Starion

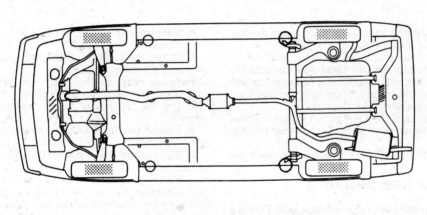

■ Drive on hoist ▨ Floor jack ○ Twin post hoist or scissors jack (emergency) locations

Lifting points for the Cordia and Tredia

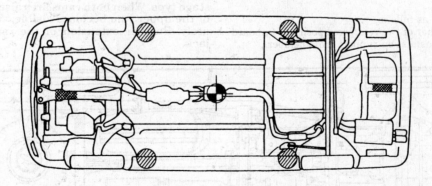

▓ Floor jack locations ⊕ Approximate center of gravity
◍ Frame contact hoist, twin post hoist or scissors jack (emergency) locations

Lifting points for Galant

1-43

1 GENERAL INFORMATION AND MAINTENANCE

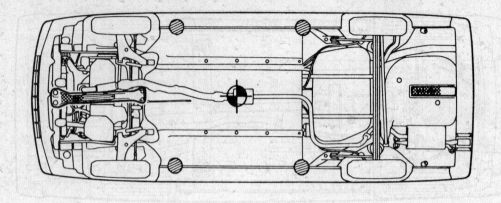

■ Floor jack locations　　⊕ Approximate center of gravity
● Frame contact hoist, twin post hoist or scissors jack (emergency) locations

Lifting points for Sigma

In all cases, the top of the jack must be swiveled so the seam or towing hook will fit down into the slot on the jack.

3. Once the jack is securely positioned, raise the jack until it supports the car securely but the tire tread still contacts the ground. Then loosen the lug nuts and, finally, raise the car just until the tire tread clears the ground. Make sure you don't get under the car while it is supported by the jack.

4. Install the new wheel, and then install and slightly tighten all the lug nuts. Now, lower the jack until the tire tread touches the ground and the
lug nuts can be fully tightened. Finally, lower and remove the jack.

Pickup and Montero

The truck and Montero use a cartridge or bottle jack. This is a jack with a heavier lifting capacity. Do not attempt the lift the vehicle with a scissors jack borrowed from a car.

1. Block the wheel opposite to the one being removed in front of and to the rear of the tire. Apply the handbrake and make sure the engine is off. Make sure the jack will be supported on a hard surface.

2. Jacking points are as follows:
• Montero, front: either of two reinforced positions (inboard or outboard) on the transmission support crossmember.
• Montero rear: on the rear axle, just inboard of the shock absorber mount.
• Pickup front: on the frame rail, just behind the transmission support crossmember.
• Pickup rear: on the frame rail just ahead of the spring mount (preferred) or on the rear axle, inboard of the spring.

3. Pry off the center wheel cap by using the wheel nut wrench.

4. Loosen the lug nuts with the wrench. Don't remove them yet.

5. Tighten the relief valve on the jack by using the slotted end of the jack handle.

6. Fit the jack handle into the holder pump up and down to partially raise the jack ram. Push the jack into position at the jacking point nearest the flat tire. Make certain the groove in the top of the ram aligns with the contact point.

7. Continue elevating the jack until the flat tire is clear of the ground.

WARNING: The jack is hydraulic and the ram is a two-stage type. When both rams are raised and the stop mark of the upper ram becomes visible, stop jacking immediately. Further extension of the ram may damage the jack.

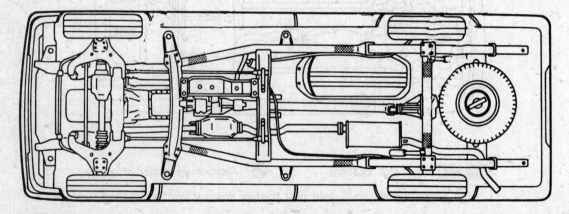

Jacking points, trucks

GENERAL INFORMATION AND MAINTENANCE 1

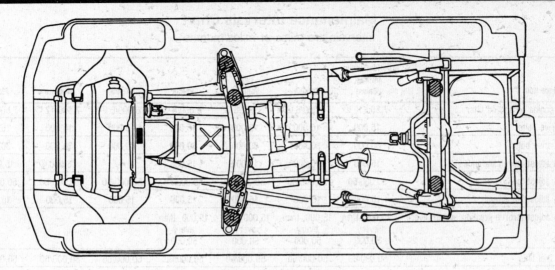

Montero jacking points

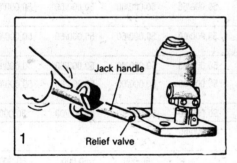
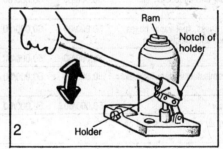
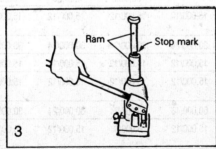
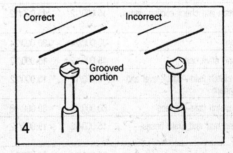

When using a bottle jack, tighten the relief valve (1), raise the ram by pumping the holder (2), don't exceed the stop marks (3) and align the groove with the jack point (4)

8. Install the new wheel, and then install and slightly tighten all the lug nuts.
9. Remove the lug wrench from the jack holder and place it on the relief valve.
10. SLOWLY loosen the valve, allowing the vehicle to settle to the ground. Do not release the jack suddenly and allow the vehicle to crash to the ground; damage to wheels, tires and suspension may result.
11. Tighten the lug nuts and install center cap or hub cap.
12. Remove the jack. Occasionally, the upper ram will not retract enough allow the jack to come out. If this happens, get in the vehicle and rock the body from side to side with some enthusiasm; the motion will compress the ram the needed fractions of an inch.
13. Securely stow the jack and the flat tire in the vehicle.
14. Remove the wheel chock.

1-45

1 GENERAL INFORMATION AND MAINTENANCE

Maintenance Intervals Chart
Intervals are miles or miles/months

Maintenance Item	Corida/Tredia, Mirage, Galant	Starion	Turbo Vehicles	Precis	Sigma	Montero	Pickup
Change engine oil and oil filter	7,500/12 ①	3,000/6 ①	3,000/6 ①	6,000/6 ①	6,000/6 ①	7,500/12 ①	7,500/12 ①
Check drive belts	15,000	15,000	15,000	15,000	15,000	15,000	15,000
Replace drive belts	30,000	30,000	30,000	30,000	30,000	30,000	30,000
Check & adjust valve clearance	15,000	15,000	15,000	15,000	—	15,000 ②	15,000 ②
Check & adjust ignition timing	50,000/60	50,000/60	50,000/60	50,000/60	50,000/60	50,000/60	50,000/60
Check & adjust idle speed	15,000	15,000	15,000	15,000	15,000	15,000	15,000
Check & adjust throttle position system every	15,000, then every 50,000	15,000, then every 50,000	15,000, then every 50,000	15,000, then every 50,000	—	—	—
Replace fuel filter	50,000/60	50,000/60	50,000/60	50,000/60	50,000/60	50,000/60	50,000/60
Inspect fuel system for leaks	50,000/60	50,000/60	50,000/60	50,000/60	50,000/60	50,000/60	50,000/60
Replace air cleaner	30,000	30,000	30,000	30,000	30,000	30,000	30,000
Replace spark plugs	30,000	30,000	30,000	30,000	30,000	30,000	30,000
Replace ignition wires	50,000/60	50,000/60	50,000/60	50,000/60	50,000/60	50,000/60	50,000/60
Replace vacuum, PCV, and secondary air hoses	50,000/60	50,000/60	50,000/60	50,000/60	50,000/60	50,000/60	50,000/60
Replace fuel system hoses, fuel vapor hoses, cooling system hoses, and fuel filler cap	50,000/60	50,000/60	50,000/60	50,000/60	50,000/60	50,000/60	50,000/60
Clean PCV system	50,000/60	50,000/60	50,000/60	50,000/60	50,000/60	50,000/60	50,000/60
Inspect evaporative emission system for leaks and replace canister	50,000/60	50,000/60	50,000/60	50,000/60	50,000/60	50,000/60	50,000/60
Replace oxygen sensor	50,000/60	50,000/60	50,000/60	50,000/60	50,000/60	80,000	50,000
Replace turbocharger air intake hoses and oil hoses	—	—	50,000/60	—	—	—	—
Replace engine timing belt	60,000			60,000	60,000	60,000	60,000 ④
Check coolant condition and check cooling sstem for leaks	15,000/12	15,000/12	15,000/12	15,000/12	15,000/12	15,000/12	15,000/12
Change engine coolant	30,000/24	30,000/24	30,000/24	30,000/24	30,000/24	30,000/24	30,000/24
Check A/C system and drive belt	15,000/12	15,000/12	15,000/12	15,000/12	15,000/12	15,000/12	15,000/12
Brake and hydraulic clutch fluid—check level and check for leaks in system	15,000/12	15,000/12	15,000/12	15,000/12	15,000/12	15,000/12	15,000/12
Brake and hydraulic clutch fluid—replace	60,000/48	60,000/48	60,000/48	30,000/24	30,000/24	30,000/24	60,000/48
Brake linings front and rear and brake hoses—inspect	15,000/12	15,000/12	15,000/12	6,000/6	15,000/12	15,000/12	15,000/12
Ball joint, steering linkage seals, and driveshaft boots (front wheel drive only)—inspect	30,000/24	30,000/24	30,000/24	30,000/24	30,000/24	30,000/24	30,000/24
Rear wheel bearings (front wheel drive only)—lubricate	30,000/24	—	30,000/24	15,000/12	30,000/24		
Front wheel bearings (rear drive only)—lubricate	—	30,000/24	30,000/24	—		30,000/24	30,000/24
Power steering fluid level, hoses, and belt—inspect	15,000/12	15,000/12	15,000/12	15,000/12	15,000/12	15,000/12	15,000/12
Automatic transaxle—change fluid	30,000	—	30,000	30,000	30,000	30,000 ③	30,000
Manual transmission (rear drive only)—check fluid level		30,000	30,000	—		30,000 ③	30,000/24
Rear drive axle—conventional—check fluid level		30,000	30,000	—	—	30,000	30,000
Rear drive axle—nonslip—change fluid		15,000	15,000	—	—	30,000	15,000/12

GENERAL INFORMATION AND MAINTENANCE 1

Maintenance Intervals Chart (continued)
Intervals are miles or miles/months

Maintenance Item	Cordia/Tredia, Mirage, Galant	Starion	Turbo Vehicles	Precis	Sigma	Montero	Pickup
Exhaust system—check all connections and inspect for excessive corrosion	15,000/12	15,000/12	15,000/12	15,000/12	15,000/12	15,000/12	15,000/12

① Under severe service conditions (short trips at very cold temperatures, driving in heavy dust, or towing a trailer), change the engine oil and filter every 3 months or 3,000 miles, whichever occurs first. Service the air cleaner filter, PCV system, and spark plugs at more frequent intervals.
② Jet valve only—adjust as required
③ Also drain & fill transfer case.
④ 2.0L engine (G63B)

CAPACITIES—AUTOMOBILES

Year	Model	Engine Displacement (cc)	Engine Crankcase (qts.) with Filter	Engine Crankcase (qts.) without Filter	Transmission (pts.) 4-Spd	Transmission (pts.) 5-Spd	Transmission (pts.) Auto.	Drive Axle (pts.)	Fuel Tank (gal.)	Cooling System (qts.)
1983	Cordia	1795	4.5	4.0	4.4	4.4	12.2	—	13.2	7.4
	Tredia	1795	4.5	4.0	4.4	4.4	12.2	—	13.2	7.4
	Starion	2555	5.0	4.5	4.8	—	14.8	2.7	19.8	9.7
1984	Cordia/Tredia	1795	4.5	4.0	4.4	4.4	12.2	—	13.2	7.4
	Cordia/Tredia	1997	4.5	4.0	4.4	4.4	12.2	—	13.2	7.4
	Starion	2555	5.0	4.5	4.8	—	14.8	2.7	19.8	9.7
1985	Cordia/Tredia	1795	4.5	4.0	4.4	4.4	12.2	—	13.2	7.4
	Cordia/Tredia	1997	4.5	4.0	4.4	4.4	12.2	—	13.2	7.4
	Starion	2555	5.5	5.0	4.8	—	14.8	2.7	19.8	9.7
	Mirage	1468	3.7	3.2	—	4.4	12.4	—	11.9	5.3
	Mirage	1597	4.5	4.0	—	4.8	12.4	—	11.9	5.3
	Galant	2350	4.5	4.0	—	—	12.4	—	15.9	7.4
1986	Cordia/Tredia	1795	4.5	4.0	4.4	4.4	12.2	—	13.2	7.4
	Cordia/Tredia	1997	4.5	4.0	4.4	4.4	12.2	—	13.2	7.4
	Starion	2555	5.0	4.5	4.8	—	14.8	2.7	19.8	9.7
	Mirage	1468	3.7	3.2	—	4.4	12.4	—	11.9	5.3
	Mirage	1597	4.5	4.0	—	4.8	12.4	—	11.9	5.3
	Galant	2350	4.5	4.0	—	—	12.4	—	15.9	7.4
1987	Cordia/Tredia	1795	4.5	4.0	4.4	4.4	12.2	—	13.2	7.4
	Cordia/Tredia	1997	4.5	4.0	4.4	4.4	12.2	—	13.2	7.4
	Starion	2555	5.0	4.5	4.8	—	14.8	2.7	19.8	9.7
	Mirage	1468	3.7	3.2	—	4.4	12.4	—	11.9	5.3
	Mirage	1597	4.5	4.0	—	4.8	12.4	—	11.9	5.3
	Galant	2350	4.5	4.0	—	—	12.4	—	15.9	7.4
	Precis	1468	3.7	3.2	—	4.4	—	—	11.9	5.3
1988	Cordia/Tredia	1795	4.5	4.0	4.4	4.4	12.2	—	13.2	7.4
	Cordia/Tredia	1997	4.5	4.0	4.4	4.4	12.2	—	13.2	7.4
	Starion	2555	5.0	4.5	4.8	—	14.8	2.7	19.8	9.7
	Mirage	1468	3.7	3.2	—	4.4	12.4	—	11.9	5.3
	Mirage	1597	4.5	4.0	—	4.8	12.4	—	11.9	5.3
	Galant	2350	4.5	4.0	—	—	12.4	—	15.9	7.4
	Galant	2972	4.5	4.0	—	5.3	12.3	—	15.9	9.7
	Precis	1468	3.7	3.2	—	4.4	—	—	11.9	5.3

1 GENERAL INFORMATION AND MAINTENANCE

CAPACITIES—AUTOMOBILES

Year	Model	Engine Displacement (cc)	Engine Crankcase (qts.) with Filter	Engine Crankcase (qts.) without Filter	Transmission (pts.) 4-Spd	Transmission (pts.) 5-Spd	Transmission (pts.) Auto.	Drive Axle (pts.)	Fuel Tank (gal.)	Cooling System (qts.)
1989	Starion	2555	5.0	4.5	4.8	—	14.8	2.7	19.8	9.7
	Mirage	1468	3.7	3.2	—	4.4	12.4	—	11.9	5.3
	Mirage	1597	3.7	3.2	—	4.8	12.4	—	11.9	5.3
	Galant	1997	①	①	—	3.8	12.9	—	15.9	7.6
	Precis	1468	3.7	3.2	—	4.4	—	—	11.9	5.3
	Sigma	2972	4.5	4.0	—	5.3	12.3	—	15.9	9.7

① SOHC engine—4 qts. w/o filter.
  4.5 qts. w/filter.
② DOHC engine—5 qts. w/o filter.
  5.5 qts. w/filter.

CAPACITIES—MONTERO & PICKUP

Year	Model	No. Cylinder Displacement cu. in. (liter)	Engine Crankcase (qts.) with Filter	Engine Crankcase (qts.) without Filter	Transmission (pts.) Manual	Transmission (pts.) Auto	Transfer Case (4WD)	Drive Axle (pts.)	Fuel Tank (gal.)	Cooling System (qts.)
1983	Pickups	4-122 (2.0)	4.2	3.7	4.9 ⑩	14.4	4.6	3.2	15.1	9.5
		4-143 (2.3)D	5.0	4.4	4.9 ⑪	14.4	4.6	3.2 ⑥	18.0	8.5
		4-156 (2.6)	5.2	4.7	4.9 ⑪	14.4	4.6	3.2 ⑥	18.0	9.5
	Montero	4-156 (2.6)	6.1	5.6	4.6	—	4.6	3.8 ⑥	15.9	8.5
1984	Pickups	4-122 (2.0)	4.2	3.7	4.9 ⑩	14.4	4.6	3.2	15.1	9.5
		4-143 (2.3)D	5.9	5.3	4.9 ⑪	14.4	4.6	3.2 ⑥	18.0	8.5
		4-156 (2.6)	5.2 ⑨	4.7	4.9 ⑪	14.4	4.6	3.2 ⑥	18.0	9.5
	Montero	4-156 (2.6)	6.1	5.6	4.6	14.4	4.6	3.8 ⑥	15.9	8.5
1985	Pickups	4-122 (2.0)	4.0	3.5	4.9 ⑩	14.4	—	3.2	15.1	9.5
		4-143 (2.3)D	5.9	5.3	4.9 ⑪	14.4	4.6	3.2 ⑥	18.0	8.5
		4-156 (2.6)	5.2 ⑨	4.7	4.9 ⑪	14.4	4.6	3.2 ⑥	18.0	9.5
	Montero	4-156 (2.6)	6.1	5.6	4.6	14.4	4.6	3.8 ⑥	15.9	8.5
1986	Pickups	4-122 (2.0)	4.2	3.7	4.9 ⑩	14.4	—	3.2	15.1	9.5
		4-156 (2.6)	5.2 ⑨	4.7	4.9 ⑪	14.4	4.6	3.2 ⑥	18.0	9.5
	Montero	4-156 (2.6)	6.1	5.6	4.6	14.4	4.6	3.8 ⑥	15.9	8.5
1987	Pickups	4-122 (2.0)	4.2	3.7	4.9	14.4	—	2.7	11.4 ⑧	8.3
		4-156 (2.6)	4.5 ⑨	4.0	4.7	14.4 ④	4.7	2.7 ⑥⑦	15.7 ②	8.3
	Montero	4-156 (2.6)	5.3	4.8	4.7	15.2	4.7	2.3 ⑬⑥	15.9	8.5
1988	Pickups	4-122 (2.0)	3.5	3.0	4.9	14.4	—	3.2	13.7 ①	8.3
		4-156 (2.6)	4.0 ③	3.5 ③	4.7	14.4 ④	4.7	3.2 ⑥	15.7 ②	8.3
	Montero	4-156 (2.6)	5.0	4.5	4.7	15.2	4.7	3.8 ⑥	15.9	8.5
1989	Pickups	4-122 (2.0)	3.5	3.0	4.9	14.4	—	3.2	13.7 ①	8.3
		4-156 (2.6)	4.0 ③	3.5 ③	4.7	14.4 ④	4.7	3.2 ⑥	15.7 ②	8.3
	Montero	4-156 (2.6)	5.5	5.0	4.7	—	4.7	3.8	15.9	8.5
		6-181 (3.0)	5.5	5.0	5.3	15.2	4.7	5.5	19.8 ⑫	9.5

D—Diesel
① Standard Body w/2WD; Long Body 2WD=18.2 Gal.
② Standard Body w/4WD; Long Body 4WD=19.8 gal.
③ 2WD; 4WD=5.5q w/filter, 5.0 without.
④ 2WD; 4WD=15.2 pts.
⑤ Rear Axle; Front Axle 3.2 pts.
⑥ Rear Axle; Front Axle 2.3 pts.
⑦ Limited Slip Rear, 3.2 pts.
⑧ Std. Body w/2WD, Long Body w/2WD 18.2
⑨ 4WD—6.1q w/filter, 5.6q without
⑩ 5 speed, 2WD, 4 speed 2WD, 4.4 pts.
⑪ 2WD-4WD=4.6 pts.
⑫ 2 door; 4 door=24.3 gal.
⑬ Normal rear, limited slip—3.8 pts.

Engine Performance and Tune-Up 2

QUICK REFERENCE INDEX

Electronic ignition	2-11
Firing Orders	2-11
Idle speed and mixture adjustment	2-22
Ignition Timing	2-13
Tune-up Charts	2-2
Valve lash adjustment	2-17

GENERAL INDEX

Carburetor Adjustments	2-22	Spark plugs	2-9	Procedures	2-2
Distributor	2-11	Spark plug wires	2-10	Spark plugs and wires	2-9
Electronic Ignition	2-11	Specifications Charts	2-2	Specifications	2-2
Firing orders	2-11	Timing	2-13	Troubleshooting	2-4
Idle speed and mixture adjustment	2-22	Tune-up		Valve lash adjustment	2-17, 19
		Idle speed and mixture	2-22	Wiring	
Ignition timing	2-13	Ignition timing	2-13	Spark plug	2-10

2–1

2 ENGINE PERFORMANCE AND TUNE-UP

TUNE-UP PROCEDURES

TUNE-UP SPECIFICATIONS

Year	Engine	Spark Plugs Type	Spark Plugs Gap (in.)	Distributor Dwell (deg.)	Distributor Gap ⑮ (in.)	Ignition Timing (deg.) Man. Trans.	Ignition Timing (deg.) Auto. Trans.	Valve Clearance (in.) In.	Valve Clearance (in.) Exh.	Idle Speed Man. Trans.	Idle Speed Auto. Trans.
1983	1,795	①	0.039–0.043	N/A	0.008	5B	5B	0.006 ②	0.010	650 ③	750
	2,555 ④	⑤	0.039–0.043	N/A	0.008	10B	10B	0.006 ②	0.010	800	N/A
1984	1,795 ④	⑥	0.035–0.039 ⑦	N/A	0.008	7B	7B	0.006	0.010 ②	700	700
	1,997	⑧	0.035–0.039 ⑦	N/A	0.031	5B	5B	0.006	0.010 ②	700	700
	2,555	⑤	0.039–0.043	N/A	0.008	10B	10B	0.006	0.010 ②	850	850
1985	1,468	⑫	0.039–0.043	N/A	0.031	3B	3B	0.006	0.010 ②	700	750
	1,597	⑬	0.039–0.043 ⑩	N/A	0.008	8B	8B	0.006	0.010 ②	700	700
	1,795 ④	⑨	0.039–0.043 ⑩	N/A	0.008	10B	10B	0.006	0.010 ②	700	700
	1,997	⑪	0.039–0.043 ⑩	N/A	0.031	5B	5B	0.010 ④	—	700	700
	2,350	⑪	0.035–0.039 ⑦	N/A	0.031	N/A	5B	0.010 ④	—	N/A	750
	2,555	⑪	0.039–0.043	N/A	0.008	10B	10B	0.006	0.010 ②	850	850
1986	1,468	⑫	0.039–0.043	N/A	0.031	5B	5B	0.006	0.010 ②	600–800	650–850
	1,597	⑫	0.039–0.043	N/A	0.008	5B	5B	0.006	0.010 ②	600–800	600–800
	1,795 ④	⑨	0.039–0.043 ⑩	N/A	0.008	7B	7B	0.006	0.010	600–800	600–800
	1,997	⑨	0.039–0.043 ⑩	N/A	0.008	5B	5B	0.010 ⑭	—	600–800	600–800
	2,350	⑪	0.039–0.043	N/A	0.031	5B	5B	0.010 ⑭	—	650–850	650–850
	2,555	⑪	0.039–0.043	N/A	0.008	10B	10B	0.010 ⑭	—	750–950	750–950
1987	1,468	⑫	0.035–0.039 ⑦	N/A	0.003	5B	5B	0.006	0.010 ②	600–800	650–850
	1,597	⑫	0.035–0.039 ⑦	N/A	0.008	10B	10B	0.006	0.010 ②	600–800	600–800
	1,795 ④	⑨	0.039–0.043 ⑩	N/A	—	7B	7B	0.006	0.010	600–800	600–800
	1,997	⑨	0.039–	N/A	0.008–	5B	5B	0.006	0.010 ②	600–	600

ENGINE PERFORMANCE AND TUNE-UP 2

TUNE-UP SPECIFICATIONS

Year	Engine	Spark Plugs Type	Spark Plugs Gap (in.)	Distributor Dwell (deg.)	Distributor Gap ⑮ (in.)	Ignition Timing (deg.) Man. Trans.	Ignition Timing (deg.) Auto. Trans.	Valve Clearance (in.) In.	Valve Clearance (in.) Exh.	Idle Speed Man. Trans.	Idle Speed Auto. Trans.
1987			0.043 ⑩		0.015					800	800
	2,350	⑪	0.039–0.043	N/A	—	5B	5B	0.010 ⑭	—	650–850	650–850
	2,555	①	0.039–0.043	N/A	—	10B	10B	0.010 ⑭	—	750–950	750–950
1988	1,468	⑫	0.035–0.039 ⑦	N/A	0.003	5B	5B	0.006	0.010	600–800	600–800
	1,597 ④	⑫	0.035–0.039 ⑦	N/A	0.008	10B	10B	0.006	0.010	600–800	650–850
	1,795 ④	⑯	0.035–0.039	N/A	—	8B	8B	0.010 ⑭	—	600–800	600–800
	1,997	⑰	0.035–0.039	N/A	—	8B	8B	0.010 ⑭	—	600–800	600–800
	2,555	⑱	0.035–0.039 ⑩	N/A	—	10B	10B	0.010	—	750–950	750–950
	2,972	⑲	0.039–0.043	N/A	—	5B	5B	⑳	⑳	600–700	600–700
1989	1,468	⑱	0.039–0.043	N/A	—	5B ㉒	5B ㉒	0.010 ⑭	—	650–850	650–850
	1,597 ④	㉑	0.028–0.031	N/A	—	8B	8B	⑳	⑳	650–850	650–850
	1,997	㉔	0.039–0.043	N/A	—	5B	5B	⑳	⑳	600–800	600–800
	1,997 ㉓	㉔	0.039–0.043	N/A	—	5B	5B	⑳	⑳	650–850	650–850
	2,555	㉕	0.039–0.043	N/A	—	10B	10B	0.010 ⑭	—	750–950	750–950
	2,972	⑲	0.039–0.043	N/A	—	5B	5B	⑳	⑳	600–800	600–800

Information on underhood labels takes precedence over chart data.
N/A—Not Applicable
① NGK BUR6EA-11, or Nippondenso W20EPR-S11
② Jet valve clearance is the same as this valve
③ Applies to 5-speed. 4x2-speed—700
④ Turbo engine
⑤ NGK BUR6EA-11 or Nippondenso W20EPR-S11
⑥ NGK BPR7ES-11 or Nippondenso W22EPR-U10
⑦ Applies to Nippondenso plug; NGK plug—0.039–0.043
⑧ NGK BPR6ES-11 or Nippondenso W20EPR-U10
⑨ NGK BUR7EA-11 or Nippondenso W22EPR-S11
⑩ Applies to Nipondenso plug; NGK—0.035–0.039
⑪ NGK EP6ES-11 or Nippondenso W20EP-U10
⑫ Nippondenso W20EP-U10
⑬ Nippondenso W20EP-U10 or NGK BUR7EA-11

B—Before Top Dead Center
⑭ Applies to jet valve only—intake and exhaust valves do not require adjustment
⑮ Reluctor gap
⑯ Nippondenso W22EPR-S11
⑰ Nippondenso W20EPR-S11
⑱ Nippondenso W22EP-U10
⑲ Nippondense P16PR-11
⑳ Uses hydraulic automatic lash adjusters. No adjustment is necessary.
㉑ NGK BUR7EZ-11
㉒ 3B: California
㉓ DOHC engine
㉔ Nippondenso W20EPR-11 or NGK BPR6ES-11
㉕ Nippondenso W22EPR-11 on NGK BPR7ES-11

2 ENGINE PERFORMANCE AND TUNE-UP

TROUBLESHOOTING BASIC POINT-TYPE IGNITION SYSTEM PROBLEMS

PROBLEM

ENGINE CRANKS, BUT WILL NOT START

Turn on lights—try starter. Note action of lights

- **Lights dim slightly**
  - Battery or starter and battery connections OK if cranking speed is good
  - Battery good, but engine will still not run
  - Remove spark plug wire and hold ¼" from engine while cranking

- **Lights dim considerably**
  - Battery weak or defective. Check for corroded or loose terminals

No spark
1. Points not closing
2. Points not opening
3. Points dirty, pitted, or burned
4. Broken primary wire or loose connection.
5. Shorted condenser
6. Grounded contact arm
7. Short or ground in primary circuit
8. High tension wire from coil to distributor defective
9. Defective coil or condenser
10. Cracked/burned rotor or cracked distributor cap
11. Wet coil, distributor or spark plug wires
12. Defective spark plugs

Weak spark
1. Dirty, pitted or burned points
2. Poor electrical connections
3. Defective plug wires
4. Defective condenser
5. Defective coil
6. Defective rotor
7. Cracked distributor cap or burned contacts
8. Wet coil, distributor or high tension wires

Good spark
Problem is not in ignition system. Check fuel supply.

ENGINE RUNS, BUT RUNS ROUGH

With engine running, remove one spark plug lead at a time to locate weak or misfiring cylinder

- **Weak or misfiring cylinder located**
  - Check condition of spark plug against chart in this chapter to determine cause of misfire—replace spark plug

- **No noticeable plug misfire**
  - Possible cause of misfiring may be:
    1. Plugs worn out
    2. Plug gap too wide
    3. Defective coil or condenser
    4. Breaker points worn out
    5. Spark advanced too far
    6. Incorrect point gap
    7. Loose primary circuit connections
    8. Cracked distributor cap
    9. Vacuum advance defective
    10. Defective rotor
    11. Defective plug wires

ENGINE PERFORMANCE AND TUNE-UP 2

Troubleshooting Engine Performance

Problem	Cause	Solution
Hard starting (engine cranks normally)	• Binding linkage, choke valve or choke piston	• Repair as necessary
	• Restricted choke vacuum diaphragm	• Clean passages
	• Improper fuel level	• Adjust float level
	• Dirty, worn or faulty needle valve and seat	• Repair as necessary
	• Float sticking	• Repair as necessary
	• Faulty fuel pump	• Replace fuel pump
	• Incorrect choke cover adjustment	• Adjust choke cover
	• Inadequate choke unloader adjustment	• Adjust choke unloader
	• Faulty ignition coil	• Test and replace as necessary
	• Improper spark plug gap	• Adjust gap
	• Incorrect ignition timing	• Adjust timing
	• Incorrect valve timing	• Check valve timing; repair as necessary
Rough idle or stalling	• Incorrect curb or fast idle speed	• Adjust curb or fast idle speed
	• Incorrect ignition timing	• Adjust timing to specification
	• Improper feedback system operation	• Refer to Chapter 4
	• Improper fast idle cam adjustment	• Adjust fast idle cam
	• Faulty EGR valve operation	• Test EGR system and replace as necessary
	• Faulty PCV valve air flow	• Test PCV valve and replace as necessary
	• Choke binding	• Locate and eliminate binding condition
	• Faulty TAC vacuum motor or valve	• Repair as necessary
	• Air leak into manifold vacuum	• Inspect manifold vacuum connections and repair as necessary
	• Improper fuel level	• Adjust fuel level
	• Faulty distributor rotor or cap	• Replace rotor or cap
	• Improperly seated valves	• Test cylinder compression, repair as necessary
	• Incorrect ignition wiring	• Inspect wiring and correct as necessary
	• Faulty ignition coil	• Test coil and replace as necessary
	• Restricted air vent or idle passages	• Clean passages
	• Restricted air cleaner	• Clean or replace air cleaner filler element
	• Faulty choke vacuum diaphragm	• Repair as necessary
Faulty low-speed operation	• Restricted idle transfer slots	• Clean transfer slots
	• Restricted idle air vents and passages	• Clean air vents and passages
	• Restricted air cleaner	• Clean or replace air cleaner filter element
	• Improper fuel level	• Adjust fuel level
	• Faulty spark plugs	• Clean or replace spark plugs

2-5

2 ENGINE PERFORMANCE AND TUNE-UP

Troubleshooting Engine Performance

Problem	Cause	Solution
Faulty low-speed operation (cont.)	• Dirty, corroded, or loose ignition secondary circuit wire connections	• Clean or tighten secondary circuit wire connections
	• Improper feedback system operation	• Refer to Chapter 4
	• Faulty ignition coil high voltage wire	• Replace ignition coil high voltage wire
	• Faulty distributor cap	• Replace cap
Faulty acceleration	• Improper accelerator pump stroke	• Adjust accelerator pump stroke
	• Incorrect ignition timing	• Adjust timing
	• Inoperative pump discharge check ball or needle	• Clean or replace as necessary
	• Worn or damaged pump diaphragm or piston	• Replace diaphragm or piston
	• Leaking carburetor main body cover gasket	• Replace gasket
	• Engine cold and choke set too lean	• Adjust choke cover
	• Improper metering rod adjustment (BBD Model carburetor)	• Adjust metering rod
	• Faulty spark plug(s)	• Clean or replace spark plug(s)
	• Improperly seated valves	• Test cylinder compression, repair as necessary
	• Faulty ignition coil	• Test coil and replace as necessary
	• Improper feedback system operation	• Refer to Chapter 4
Faulty high speed operation	• Incorrect ignition timing	• Adjust timing
	• Faulty distributor centrifugal advance mechanism	• Check centrifugal advance mechanism and repair as necessary
	• Faulty distributor vacuum advance mechanism	• Check vacuum advance mechanism and repair as necessary
	• Low fuel pump volume	• Replace fuel pump
	• Wrong spark plug air gap or wrong plug	• Adjust air gap or install correct plug
	• Faulty choke operation	• Adjust choke cover
	• Partially restricted exhaust manifold, exhaust pipe, catalytic converter, muffler, or tailpipe	• Eliminate restriction
	• Restricted vacuum passages	• Clean passages
	• Improper size or restricted main jet	• Clean or replace as necessary
	• Restricted air cleaner	• Clean or replace filter element as necessary
	• Faulty distributor rotor or cap	• Replace rotor or cap
	• Faulty ignition coil	• Test coil and replace as necessary
	• Improperly seated valve(s)	• Test cylinder compression, repair as necessary
	• Faulty valve spring(s)	• Inspect and test valve spring tension, replace as necessary
	• Incorrect valve timing	• Check valve timing and repair as necessary

ENGINE PERFORMANCE AND TUNE-UP 2

Troubleshooting Engine Performance (cont.)

Problem	Cause	Solution
Faulty high speed operation (cont.)	• Intake manifold restricted	• Remove restriction or replace manifold
	• Worn distributor shaft	• Replace shaft
	• Improper feedback system operation	• Refer to Chapter 4
Misfire at all speeds	• Faulty spark plug(s)	• Clean or replace spark plug(s)
	• Faulty spark plug wire(s)	• Replace as necessary
	• Faulty distributor cap or rotor	• Replace cap or rotor
	• Faulty ignition coil	• Test coil and replace as necessary
	• Primary ignition circuit shorted or open intermittently	• Troubleshoot primary circuit and repair as necessary
	• Improperly seated valve(s)	• Test cylinder compression, repair as necessary
	• Faulty hydraulic tappet(s)	• Clean or replace tappet(s)
	• Improper feedback system operation	• Refer to Chapter 4
	• Faulty valve spring(s)	• Inspect and test valve spring tension, repair as necessary
	• Worn camshaft lobes	• Replace camshaft
	• Air leak into manifold	• Check manifold vacuum and repair as necessary
	• Improper carburetor adjustment	• Adjust carburetor
	• Fuel pump volume or pressure low	• Replace fuel pump
	• Blown cylinder head gasket	• Replace gasket
	• Intake or exhaust manifold passage(s) restricted	• Pass chain through passage(s) and repair as necessary
	• Incorrect trigger wheel installed in distributor	• Install correct trigger wheel
Power not up to normal	• Incorrect ignition timing	• Adjust timing
	• Faulty distributor rotor	• Replace rotor
	• Trigger wheel loose on shaft	• Reposition or replace trigger wheel
	• Incorrect spark plug gap	• Adjust gap
	• Faulty fuel pump	• Replace fuel pump
	• Incorrect valve timing	• Check valve timing and repair as necessary
	• Faulty ignition coil	• Test coil and replace as necessary
	• Faulty ignition wires	• Test wires and replace as necessary
	• Improperly seated valves	• Test cylinder compression and repair as necessary
	• Blown cylinder head gasket	• Replace gasket
	• Leaking piston rings	• Test compression and repair as necessary
	• Worn distributor shaft	• Replace shaft
	• Improper feedback system operation	• Refer to Chapter 4
Intake backfire	• Improper ignition timing	• Adjust timing
	• Faulty accelerator pump discharge	• Repair as necessary
	• Defective EGR CTO valve	• Replace EGR CTO valve
	• Defective TAC vacuum motor or valve	• Repair as necessary

2 ENGINE PERFORMANCE AND TUNE-UP

Troubleshooting Engine Performance (cont.)

Problem	Cause	Solution
Intake backfire (cont.)	• Lean air/fuel mixture	• Check float level or manifold vacuum for air leak. Remove sediment from bowl
Exhaust backfire	• Air leak into manifold vacuum	• Check manifold vacuum and repair as necessary
	• Faulty air injection diverter valve	• Test diverter valve and replace as necessary
	• Exhaust leak	• Locate and eliminate leak
Ping or spark knock	• Incorrect ignition timing	• Adjust timing
	• Distributor centrifugal or vacuum advance malfunction	• Inspect advance mechanism and repair as necessary
	• Excessive combustion chamber deposits	• Remove with combustion chamber cleaner
	• Air leak into manifold vacuum	• Check manifold vacuum and repair as necessary
	• Excessively high compression	• Test compression and repair as necessary
	• Fuel octane rating excessively low	• Try alternate fuel source
	• Sharp edges in combustion chamber	• Grind smooth
	• EGR valve not functioning properly	• Test EGR system and replace as necessary
Surging (at cruising to top speeds)	• Low carburetor fuel level	• Adjust fuel level
	• Low fuel pump pressure or volume	• Replace fuel pump
	• Metering rod(s) not adjusted properly (BBD Model Carburetor)	• Adjust metering rod
	• Improper PCV valve air flow	• Test PCV valve and replace as necessary
	• Air leak into manifold vacuum	• Check manifold vacuum and repair as necessary
	• Incorrect spark advance	• Test and replace as necessary
	• Restricted main jet(s)	• Clean main jet(s)
	• Undersize main jet(s)	• Replace main jet(s)
	• Restricted air vents	• Clean air vents
	• Restricted fuel filter	• Replace fuel filter
	• Restricted air cleaner	• Clean or replace air cleaner filter element
	• EGR valve not functioning properly	• Test EGR system and replace as necessary
	• Improper feedback system operation	• Refer to Chapter 4

An engine tune-up is a service designed to restore the maximum capability of power, performance, economy and reliability in an engine, and, at the same time, assure the owner of a complete check and more lasting results in efficiency and trouble-free performance. Engine tune-up becomes increasingly important each year, to ensure that pollutant levels are in compliance with federal emissions standards.

It is advisable to follow a definite and thorough tune-up procedure. Tune-up consists of three separate steps: Analysis, the process of determining whether normal wear is responsible for performance loss, and whether parts require replacement or service; Parts Replacement or Service; and Adjustment, where engine adjustments are returned to the original factory specifications.

The extent of an engine tune-up is usually determined by the length of time since the previous service, although the type of driving and general mechanical conditioning of the engine must be considered. Specific maintenance should also be performed at regular intervals, depending on operating conditions.

Troubleshooting is a logical sequence of procedures designed

ENGINE PERFORMANCE AND TUNE-UP 2

to lead the owner or service man to the particular cause of trouble. Service usually comprises two areas; diagnosis and repair. While the apparent cause of trouble, in many cases, is worn or damaged parts, performance problems are less obvious. The first job is to locate the problem and cause. Once the problem has been isolated, refer to the appropriate section for repair, removal or adjustment procedures.

It is advisable to read the entire Section before beginning a tune-up, although those who are more familiar with tune-up procedures may wish to go directly to the instructions.

It should be noted that the diesel engine does not require tune-ups in the conventional sense. The diesel does not use spark ignition to fire the fuel (it uses compression ignition) so it has no spark plugs, distributor, or ignition wires. On the other hand diesels are much more sensitive to clogged air and fuel filters and should have their engine oil and filter changed more frequently as well. Diesel owners simply need to redefine "tune-up" to be the inspection and replacement of fluids and filters on a regular basis.

Spark Plugs

A typical spark plug consists of a metal shell surrounding a ceramic insulator. A metal electrode extends downward through the center of the insulator and protrudes a small distance. Located at the end of the plug and attached to the side of the outer metal shell is the side electrode. The side electrode bends in at a 90° angle so that its tip is even with, and parallel to, the tip of the center electrode. The distance between these two electrodes (measured in thousandths of an inch) is called the spark plug gap. The spark plug in no way produces a spark but merely provides a gap across which the current can arc. The coil produces anywhere from 20,000 to 40,000 volts which travels to the distributor where it is distributed through the spark plug wires to the spark plugs. The current passes along the center electrode and jumps the gap to the side electrode, and, in so doing, ignited the air/fuel mixture in the combustion chamber.

SPARK PLUG HEAT RANGE

Spark plug heat range is the ability of the plug to dissipate heat. The longer the insulator (or the farther it extends into the engine), the hotter the plug will operate; the shorter the insulator the cooler it will operate. A plug that absorbs little heat and remains too cool will quickly accumulate deposits of oil and carbon since it is not hot enough to burn them off. This leads to plug fouling and consequently to misfiring. A plug that absorbs too much heat will have no deposits, but, due to the excessive heat, the electrodes will burn away quickly and in some instances, preignition may result. Preignition takes place when plug tips get so hot that they glow sufficiently to ignite the fuel/air mixture before the actual spark occurs. This early ignition will usually cause a pinging during low speeds and heavy loads.

The general rule of thumb for choosing the correct heat range when picking a spark plug is: if most of your driving is *long distance*, high speed travel, use a colder plug; if most of your driving is *stop and go*, use a hotter plug. Original equipment plugs are compromise plugs, but most people never have occasion to change their plugs from the factory-recommended heat range.

REPLACING SPARK PLUGS

A set of spark plugs usually requires replacement after about 10,000 miles on cars with conventional ignition systems and after about 20,000 to 30,000 miles on cars with electronic ignition, depending on your style of driving. In normal operation, plug gap increased about 0.025mm (0.001 in.) for every 1,000–2,500 miles. As the gap increases, the plug's voltage requirement also increases. It requires a greater voltage to jump the wider gap and about two or three times as much voltage to fire a plug at high speeds than at idle.

When you're removing spark plugs, you should work on one at a time. Don't start by removing the plug wires all at once, because unless you number them, they may become mixed up. Take a minute before you begin and number the wires with tape. The best location for numbering is near where the wires come off the plug. Sometimes the wires are already numbered.

1. Twist the spark plug boot and remove the boot and wire from the plug. Do not pull on the wire itself as this will ruin the wire.

2. If possible, use a brush or rag to clean the area around the spark plug (a shop-vac is ideal for this). Make sure that all the dirt is removed so that none will enter the cylinder after the plug is removed.

3. Remove the spark plug using the proper size socket and assortment of universals and extensions as required. Turn the socket counterclockwise to remove the plug. Be sure to hold the socket straight on the plug to avoid breaking the plug, or rounding off the hex on the plug.

4. Once the plug is out, check its condition. This is crucial since plug readings are vital signs of engine condition.

The electrode end of a spark plug is a good indicator of the internal condition of your engine. If a spark plug is fouled, causing the engine to misfire, the problem will have to be found and corrected. Often, reading the plugs will lead you to the cause of the problem.

There are several reasons why a spark plug will foul and you can learn which reason by just looking at the plug. The two most common problems are oil fouling and pre-ignition/detonation.

Oil fouling is easily noticed as dark, wet oily deposits on the plug's electrodes. Oil fouling is caused by internal engine problems, the most common of which are worn valve seals or guides and worn or damaged piston rings. These problems can be corrected only by engine repairs.

Pre-ignition or detonation problems are characterized by extensive burning and/or damage to the plug's electrodes. The problem is caused by incorrect ignition timing or faulty spark control. Check the timing and/or diagnose the spark control system.

NOTE: A small amount of light tan or rust red colored deposits at the electrode end of the plug is normal. These plugs need not be renewed unless they are severely worn.

5. Use a round wire feeler gauge to check the plug gap. The correct size gauge should pass through the electrode gap with a slight drag. If you're in doubt, try one size smaller and one larger. The smaller gauge should go through easily while the larger one shouldn't go through at all. If the gap is incorrect, use the electrode bending tool on the end of the gauge to adjust the gap. When adjusting the gap, always bend the side electrode. The center electrode is non-adjustable.

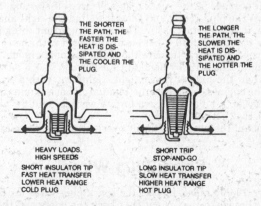

Spark plug heat range

2 ENGINE PERFORMANCE AND TUNE-UP

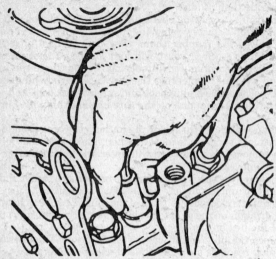

Twist and pull on the rubber boot to remove the spark plug wires. Never pull on the wire itself

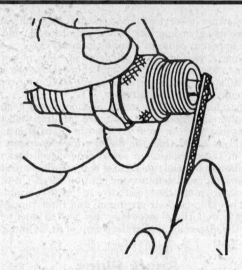

Plugs that are in good condition can be filed and reused

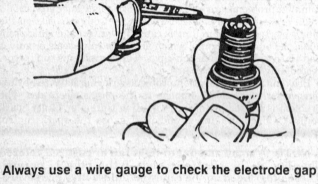

Always use a wire gauge to check the electrode gap

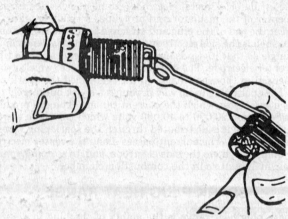

Adjust the electrode gap by bending the side electrode

6. Lightly coat the threads of the plug with high temperature anti-seize compound. To start the plug, make sure it is passing straight into the spark plug hole and turn it very gently. If it resists turning, it is probably beginning to cross-thread and you should turn it back out and start again, at a more favorable angle. The threads in the spark plug hole are self-starting, and if you turn the plug gently and allow it to find the right angle on its own, it should start easily. The most important thing is not to force it to turn until it has clearly engaged with the threads and turned in several turns without the use of a wrench. Turn the plug in clockwise by hand until it is snug.

7. When the plug is finger tight, tighten it with a wrench. Since the engines covered by this manual all have aluminum cylinder heads, it's best to use a torque wrench, and torque the plug to 15–21 ft. lbs.

8. Install the plug boot firmly over the plug. Proceed to the next plug.

Spark Plug Wires

Every 10,000 miles, inspect the spark plug wires for burns, cuts, or breaks in the insulation. Check the boots and the nipples on the distributor cap. Replace any damaged wiring.

Every 30,000 miles or so, the resistance of the wires should be checked with an ohmmeter. Wires with excessive resistance will cause misfiring, and may make the engine difficult to start in damp weather. Generally, the useful life of the cables is 45,000–60,000 miles.

To check resistance, remove the distributor cap, leaving the wires in place. Connect one lead of an ohmmeter to an electrode within the cap. Connect the other lead to the corresponding spark plug terminal (remove it from the spark plug for this test). Replace any wire which shows a resistance over 30,000Ω. Generally speaking, however, resistance should not be over 25,000Ω, and 30,000Ω must be considered the outer limit of acceptability.

It should be remembered that resistance is also a function of length. The longer the wire, the greater the resistance. Thus, if the wires on your truck are longer than the factory originals, the resistance will be higher, possibly outside these limits.

When installing new wires, replace them one at a time to avoid mixups. Start by replacing the longest one first. Install the boot firmly over the spark plug. Route the wire over the same path as the original. Insert the nipple firmly onto the tower on the distributor cap, then install the cap cover and latches to secure the wires.

ENGINE PERFORMANCE AND TUNE-UP 2

FIRING ORDERS

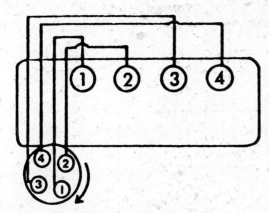

1795, 1997 and 2350cc engine firing order

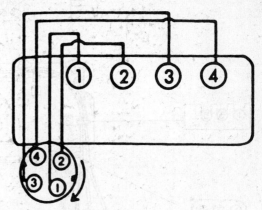

1468cc engine firing order

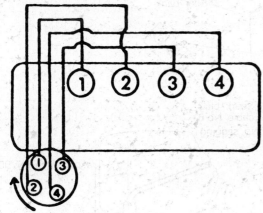

2555cc engine firing order

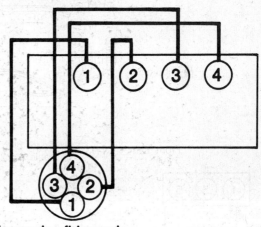

1597cc engine firing order

NOTE: To avoid confusion, always replace spark plug wires one at a time.

Should spark plug wires become mixed up for any reason, you can use the firing order illustrations shown here to install the plug wires properly. Start with the No. 1 plug wire, locating it properly in the cap and then installing it on the front cylinder or No. 1 cylinder in the illustration. Note that the engine does not fire in numerical order, but in the firing order of 1-3-4-2. Install the No. 3 spark plug next in order proceeding around the distributor cap in the direction of rotation shown, and then connect it to No. 3 cylinder. Continue in the same way for the wires feeding plugs in cylinders 4 and 2. For the 2,972cc engine, start with the No. 1 plug and proceed all the way to No. 6.

Electronic Ignition

RELUCTOR GAP CHECK

1,597, 1,795, and 2,555cc Engines

On these engines, the reluctor gap can only be checked. Doing so is not a matter of routine maintenance, but usually is necessary only if the distributor is overhauled. However, too tight a fit between the signal rotor and the pickup stator could cause rotation of the distributor to damage the parts involved or produce an incorrect signal. You might want to check the fit as described below and replace the stator and rotor to correct deficiencies.

In some cases, severe wear or incorrect original manufacturing tolerances in the distributor bearings could cause wear of these two pieces that resembles damage due to too tight a fit. However, in this case, the appropriate gap would exist between the rotor and stator and there would be excessive play in the distributor shaft.

1. Remove the distributor cap. Then, remove the two bolts that hold the rotor in place (on the distributor for the 1,795cc engine, merely pull the rotor off).

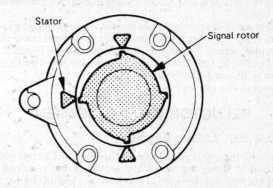

The reluctor gap on the distributor used on the 1597, 1795 and 2555 engines

2 ENGINE PERFORMANCE AND TUNE-UP

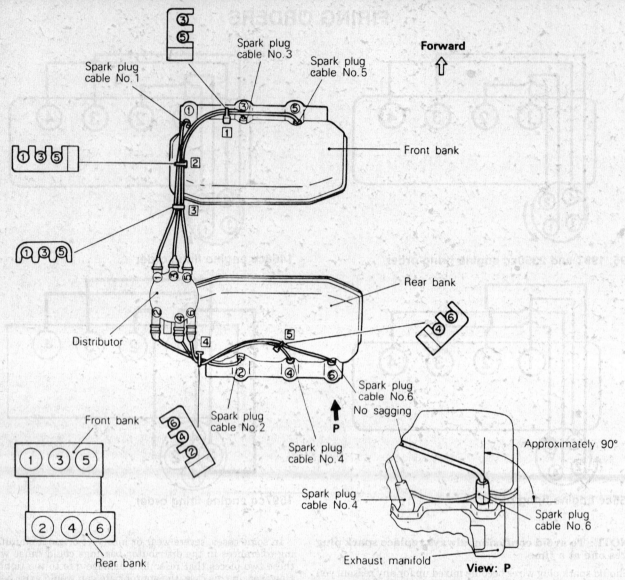

2972cc engine firing order: 1-2-3-4-5-6

2. Rotate the engine using a wrench on the front pulley until, on the distributor on the 1,795cc engine the two vertical stator pieces line up with two of the rotor lobes; on the other engines, the three vertical stator pieces should line up with three of the rotor lobes.
3. Get a non-magnetic (brass, plastic, or wood) feeler gauge of 0.2mm (0.008 in.) thickness, and insert it straight between the stator and rotor. As long as the gauge can be inserted and moved easily, the parts are okay; the gap may be wider than the 0.2mm (0.008 in.) specification.
4. Install the cap and rotor.

RELUCTOR GAP ADJUSTMENT

1,468, 1,997, 2,350cc Engines

The reluctor gap is adjustable on the Hitachi Electric distributor used in these engines. While service is not required as a part of normal maintenance, if you have worked on the distributor or if you suspect the reluctor gap might be incorrect because of ignition problems, you can check and adjust it.

1. Get a feeler gauge of non-magnetic material (brass, plastic

or wood) of 0.8mm (0.031 in.) thickness. Remove the distributor cap and rotor.
2. Rotate the engine (you can use a large socket wrench on the bolt that attaches the front pulley—but only turn the engine

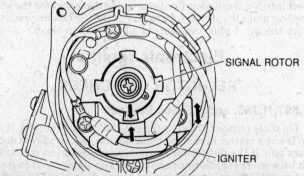

Adjusting the reluctor gap on the 1468, 1997 and 2350cc engines

2-12

ENGINE PERFORMANCE AND TUNE-UP 2

clockwise) so that one of the prongs of the rotor is directly across from the igniter pickup.

3. Insert or attempt to insert the feeler gauge between the prong and pickup. If the gap is correct, there will be a very slight drag. If the gauge fits loosely or cannot be inserted, loosen both mounting screws and then (if necessary widen the gap) and insert the gauge. Slowly close the gap by pivoting the igniter assembly on the left screw and rotating it at the right side, where it's slotted. When the gauge is just touching, tighten first the right side screw and then the screw on the left. Recheck the gap and readjust as necessary. Reinstall the cap and rotor.

Ignition Timing

It is wise to check the ignition timing at every tune-up, although timing varies little with electronic ignition systems. Mitsubishi permits a tolerance of 2° either side of the timing setting. Most engines run at their best and with maximum resistance to detonation if the timing is as close as possible to the setting.

CHECKING AND ADJUSTMENT

All 1983–86 Car Models
1983–89 Pickups with Gasoline Engine
1983–89 Montero with 2.6L Engine
1987 Galant
1987 Mirage
1987 Starion
1987–89 Precis
1987–88 Cordia and Tredia

1. Drive the car until the engine is hot; the temperature gauge indicates normal operating temperature. This is necessary as parts dimensions, and therefore timing may change slightly with temperature.
2. Leave the engine idling, apply the handbrake and put the transmission in Neutral (manual) or Park (automatic). Turn off all accessories.
3. Install a tachometer, connecting the red lead to the (−) terminal of the coil and the black lead to a clean ground (the battery minus terminal works well) or as otherwise detailed by the manufacturer of the tach. Verify that the engine idle speed is correct. If not adjust it, because incorrect idle speed will change the timing. See the idle adjustment procedure later it this Section.

NOTE: On 1986 Galant, to connect the tachometer, you will need special tool MD998439. This special tool is a wiring harness that connects between the ignition coil primary terminal and the connector that dangles from the base of the distributor. A view of this connector is shown in the idle speed adjustment section later in this Section.

4. Stop the engine, and connect the timing light high voltage pickup into the No.1 spark plug circuit (unless you have an inductive type of timing light, see the instructions that came with it). To do this, carefully pull the front spark plug wire off by the rubber boot at the spark plug end. Then, connect the timing light pickup to the end of the plug and install the other end of the pickup into the plug wire. On inductive timing lights, you'll have a device that looks like a clothespin which must simply be clamped around the wire. You won't have to disconnect the wire at all on these lights but you will have to pay attention to which way the pick–up faces when its clamped on the wire.
5. On all powered timing lights, connect the red lead to the battery (+) terminal and the black lead to the (−) terminal.
6. If the front pulley and timing mark are dirty, wipe them clean with a rag. If they are hard to see, you might want to put a small drop of white paint on both the timing scale and groove in the pulley. You may have to turn the engine over using a wrench on the bolt in the front of the crankshaft pulley to do this. Check the timing setting in degrees Before Top Dead Center in the chart above and verify that this is the correct setting by checking on the engine compartment sticker. Make sure you know where the correct setting is on the front cover. Then, make sure all wires are clear of any rotating parts and the light is in a secure place. Make sure you haven't left any rags or tools near the engine.
7. On the 1986–87 Mirage, 1987–88 Cordia and Tredia, 1987 Starion, 1987–89 Pickup or Montero, 1.5L engine (Mirage), 1987 Starion or 1.8L engine (Cordia and Tredia), and are performing the timing procedure at high altitudes (more than 3,900 ft. above sea level): Disconnect the pressure sensor electrical connector located just across from the distributor wires at the top of the cap before stopping the engine. The sensor is a box bolted to the fender with a vacuum hose connected to the bottom of the unit.

On the 1.6L engine (Mirage), 2.0L engine (Cordia, Tredia and Truck) or the 2.6L truck and Montero engine, disconnect the vacuum hose with the white stripes that is connected to the lowermost portion of the distributor vacuum advance. Plug the end of the hose.

8. Start the engine and allow it to idle. Point the timing light at the mark on the front cover and read the timing by noting the position of the groove in the front pulley in relation to the timing mark or scale on the front cover. If the timing is incorrect, loosen the distributor mounting bolt or nut. Turn the distributor slightly clockwise to retard the timing or counterclockwise to advance it (advance means turning to a setting representing more degrees Before Top Dead Center). Read the timing light as you turn the distributor. When the reading is correct, tighten the distributor mounting bolt back up and then verify that the setting has not changed. If necessary readjust the position of the distributor until the setting is correct after the bolt or nut is tight.
9. If so equipped, reconnect the white striped vacuum hose or the pressure sensor connector. On 1987–88 1.8L Cordia and Tredia, disconnect the negative battery cable for 20 seconds then reconnect it. Recheck the timing. On the 1.6L and 1.8L engines and Starion, at high altitudes the timing will advance to 15° BTDC after the hose is connected back to the distributor advance. This is the actual ignition timing. For the 2.0L truck, the actual timing should be 13° BTDC and for the 2.6L truck and Montero, 12° BTDC.
10. Turn the engine off, disconnect the timing light and tachometer and, if necessary, reconnect the ignition wire.

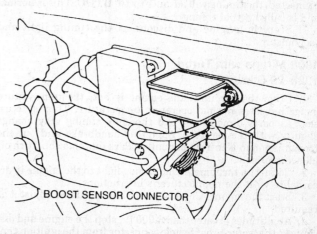

Boost sensor location

2-13

2 ENGINE PERFORMANCE AND TUNE-UP

1988 Mirage with 1.5L Engine

1. Drive the car until the engine is hot; the temperature gauge indicates normal operating temperature. This is necessary as parts dimensions, and therefore timing may change slightly with temperature. Apply the handbrake and put the transmission in Neutral (manual) or Park (automatic). Turn off all accessories.
2. Stop the engine and connect a tachometer and timing light in accordance with the manufacturer's instructions.
3. Start the engine. Check the curb idle speed and adjust as necessary as described in this Section.
4. If your checking the ignition timing at altitudes greater than 3,900 ft. disconect the yellow striped hose located at the sub-vacuum chamber on the distributor advance unit. Plug the end of the hose.
5. Illuminate the timing marks on the crankshaft pulley with the timing light and check the timing. If the timing is not within specification, loosen the distributor mounting nut and move the distributor as required to adjust the timing. When the timing is correct, tighten the nut.
6. At high altitudes, reconnect the vacuum hose to the sub-advance chamber and check the timing again. When the hose is connected, the timing will advance to **10° BTDC**. This is normal and is called actual timing.
7. Stop the engine and disconnect the timing light and the tachometer.

1988 Mirage with 1.6L Engine

1. Drive the car until the engine is hot; the temperature gauge indicates normal operating temperature. This is necessary as parts dimensions, and therefore timing may change slightly with temperature. Apply the handbrake and put the transmission in Neutral (manual) or Park (automatic). Turn off all accessories.
2. Stop the engine and connect a tachometer and timing light in accordance with the manufacturer's instructions.
3. Start the engine. Check the curb idle speed and adjust it as required.
4. If your checking the ignition timing at altitudes greater than 3,900 ft. disconnect the yellow striped hose located at the sub-vacuum chamber on the distributor advance unit. Plug end of the hose.
5. Illuminate the timing marks on the crankshaft pulley with the timing light and check the timing. If the timing is not within specification, loosen the distributor mounting nut and move the distributor as required to adjust the timing. When the timing is correct, tighten the nut.
6. At high altitudes, reconnect the vacuum hose to the sub-advance chamber and check the timing again. When the hose is connected, the timing will advance to **10° BTDC**. This is normal and is called actual timing.
7. Stop the engine and disconnect the timing light and tachometer.

1988 Mirage with Turbo
1988–89 Starion

1. Drive the car until the engine is hot; the temperature gauge indicates normal operating temperature. This is necessary as parts dimensions, and therefore timing may change slightly with temperature. Apply the handbrake and put the transmission in Neutral (manual) or Park (automatic). Turn off all accessories.
2. Connect a tachometer and timing light to the engine in accordance with the manuacturer's instructions.
3. Start the engine. Check the curb idle speed and adjust it as required.
4. At altitudes greater than 3,900 ft., stop the engine and disconnect the water-proof female connector from the ignition timing connector (see illustration). Connect a jumper wire with an

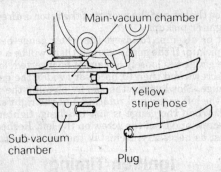

On 1988 Mirage with 1.5L engine, the yellow striped hose must be disconnected from the sub-vacuum advance to check timing at high altitudes

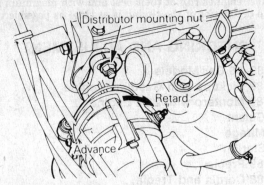

Rotating the distributor to adjust ignition timing

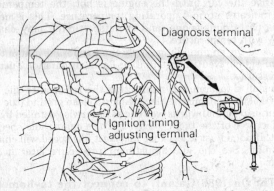

Grounding the ignition timing connector on 1988 Mirage with turbo engine

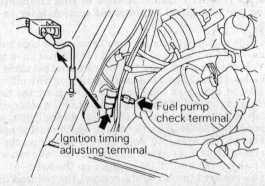

Grounding the ignition timing connector on 1988 Mirage with turbo engine

ENGINE PERFORMANCE AND TUNE-UP 2

alligator clip on the end to the ignition timing adjusting terminal to ground it. Make sure that you ground the right terminal as there is a diagnosis terminal in the connector also.

5. Illuminate the timing marks on the crankshaft pulley with the timing light and check the base timing. If the base timing is not within specification, loosen the distributor mounting nut and move the distributor as required to adjust the timing. When the timing is correct, tighten the nut being careful not to disturb the position of the distributor.

6. At altitudes greater than 3,900 ft. above sea level, stop the engine and remove the wire that was connected in Step 4. Start the engine and check the timing again. With the ground wire removed, the timing will advance to 15° BTDC. This is normal to ensure more efficient combustion and is called *actual* timing.

7. Stop the engine and disconnect the timing light and tachometer.

1989 Mirage with 1.5L Engine

NOTE: To adjust the timimg on these vehicles, you will need a jumper wire approximately two feet long with alligator clips on both ends. You will also need a paper clip. Make sure that you have these items on hand before attempting to check the timing.

1. Drive the car until the engine is hot; the temperature gauge indicates normal operating temperature. This is necessary as parts dimensions, and therefore timing may change slightly with temperature. Apply the handbrake and put the transmission in Neutral (manual) or Park (automatic). Turn off all accessories.

2. Stop the engine and connect a timing light in accordance with the manufacturer's instructions.

3. Insert a paper clip into the connector on the side of the valve cover. DO NOT separate the terminals of the connector. Make sure the paper clip makes full contact with the terminal surface.

4. Connect a tachometer to the paper clip and check the idle speed. Adjust the idle speed as necessary.

5. Stop the engine and connect one end of the jumper wire to the ignition timing adjustment connector and connect the other end to an engine ground.

6. Start the engine and illuminate the timing marks on the crankshaft pulley with the timing light. If the timing is not as specified, loosen the distributor mounting nut and turn the distributor to the right to retard the timing and to the left to advance it.

7. Once the timing is set, tighten the nut being careful not to disturb the distributor.

8. Stop the engine and disconnect the jumper wire from the ignition timing connector and ground.

9. Start the engine and run it at idle speed.

10. Check the timing again with the timing light. The timing should be 10° BTDC. This is called *actual* ignition timing. If the timing is not exactly 10°BTDC, don't be alarmed. This value may vary depending on what mode the computer is in at the time the base adjustment was made. If you do not see a change, check the base timing again (Steps 5 and 6). If the base timing is correct, the engine is in time and no further adjustment is necessary.

NOTE: At altitudes more than 2,300 ft above sea level, the actual ignition timing may vary more than 10° BTDC to ensure efficient combustion.

1989 Mirage with Turbo
1989 Galant with Dual Overhead Cam Engine

These models use a crank angle sensor instead of a distributor to control ignition timing. For the timing adjustment procedure on these vehicles, you will need a jumper wire approximately two feet long with alligator clips on both ends. You will also need

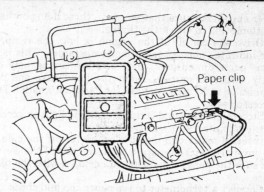

Tachometer hook-up on all 1989 Mirage and 1989 Galant with DOHC engine

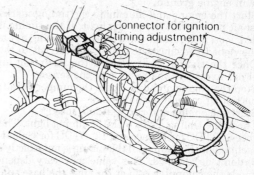

Grounding the ignition timing adjustment connector on all 1989 Mirage

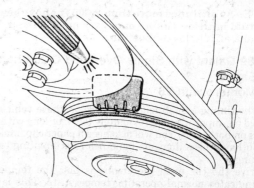

Timing marks on 1989 Mirage with 1.5L engine

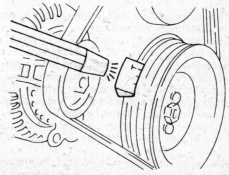

Timing marks on 1989 Mirage with turbo engine

2–15

2 ENGINE PERFORMANCE AND TUNE-UP

a paper clip. Make sure that you have these items on hand before attempting to check the timing.

1. Drive the car until the engine is hot; the temperature gauge indicates normal operating temperature. This is necessary as parts dimensions, and therefore timing may change slightly with temperature. Apply the handbrake and put the transmission in Neutral (manual) or Park (automatic). Turn off all accessories.
2. Stop the engine and connect a timing light in accordance with the manufacturer's instructions.
3. Insert a paper clip into the connector on the side of the valve cover. DO NOT separate the terminals of the connector. Make sure the paper clip makes full contact with the terminal surface.
4. Connect a tachometer to the paper clip and check the idle speed. Adjust the idle speed as necessary.
5. Stop the engine and connect one end of the jumper wire to the ignition timing adjustment connector and connect the other end to an engine ground.
6. Start the engine and illuminate the timing marks on the crankshaft pulley with the timing light. If the timing is not as specified, loosen the crank angle sensor mounting nut and turn it to the right to advance the timing and the left to retard it.
7. Once the timing is set, tighten the mounting nut being careful not to disturb the position of the sensor.
8. Stop the engine and disconnect the jumper wire from the ignition timing connector and ground.
9. Start the engine and run it at idle speed.
10. Check the timing again with the timing light. The timing should be 8° BTDC for Mirage and 12° BTDC for Galant. This is called *actual* ignition timing. If the timing is not exactly 8° or 12° BTDC, don't be alarmed. This value may vary depending on what mode the computer is in at the time the base adjustment was made. If you do not see a change, check the base timing again (Steps 5 and 6). If the base timing is correct, the engine is in time and no further adjustment is necessary.

NOTE: At altitudes more than 2,300 ft above sea level, the actual ignition timing may vary more than 8° BTDC to ensure efficient combustion.

1988–89 Galant with Single Overhead Cam Engine
1989 Sigma
1989 Montero with V6

For the timing adjustment procedure on these vehicles, you will need a jumper wire approximately two feet long with alligator clips on both ends. You will also need a paper clip. Make sure that you have these items on hand before attempting to check the timing.

1. Drive the car until the engine is hot; the temperature gauge indicates normal operating temperature. This is necessary as parts dimensions, and therefore timing may change slightly with temperature. Apply the handbrake and put the transmission in Neutral (manual) or Park (automatic). Turn off all accessories.
2. Stop the engine and connect a timing light in accordance with the manufacturer's instructions.
3. Trace the wire that runs from the primary side of the ignition coil to the noise filter. You will find a one pin harness connector between these two points. Insert a paper clip through the connector on either the female or the male side where the wire enters the connector. The paper clip must make full contact with the surface of the connector terminal and must be inserted at the proper angle or you will not be able to get it out.
4. Once the paper clip is in place, connect a tachometer to it.
5. Start the engine and check the idle speed. Adjust as necessary.
6. Stop the engine and connect one end of the jumper wire to the ignition timing adjustment connector and connect the other end to an engine ground.

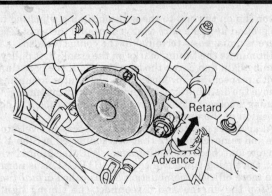

On 1989 Mirage turbo and Galant DOHC engine, ignition timing is adjusted by moving the crank angle sensor

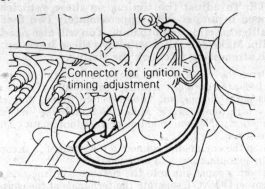

Grounding the ignition timing adjustment connector on all 1989 Galant with DOHC engine

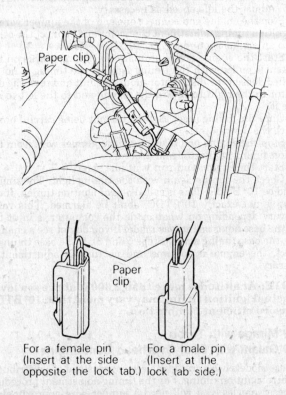

Making the tachometer connection with a paper clip

ENGINE PERFORMANCE AND TUNE-UP 2

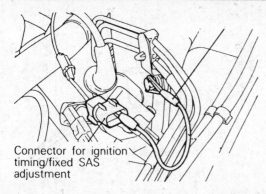

Grounding the ignition timing adjustment connector on 1988–89 Galant with SOHC engine and 1989 Sigma

7. Start the engine and illuminate the timing marks on the crankshaft pulley with the timing light. If the timing is not as specified, loosen the distributor mounting nut and turn the distributor to the right to retard the timing and to the left to advance it.
8. Once the timing is set, tighten the nut being careful not to disturb the distributor.
9. Stop the engine and disconnect the jumper wire from the ignition timing connector and ground.
10. Start the engine and run it at idle speed.
11. Check the timing again with the timing light. The timing should be 10° BTDC for the 1989 Galant and 15° BTDC for the 1988 Galant, 1989 Sigma and V6 Montero. This is called *actual* ignition timing. If the timing is not exactly 10° or 15° BTDC, don't be alarmed. This value may vary depending on what mode the computer is in at the time the base adjustment was made. If you do not see a change, check the base timing again (Steps 6 and 7). If the base timing is correct, the engine is in time and no further adjustment is necessary.

Diesel Injection Timing

Please refer to Section 5 – Fuel System for this procedure.

Intake and Exhaust Valve Lash Adjustment

Valve lash must be adjusted on all engines not equipped with hydraulic (oil-filled) automatic lash adjusters. Some of the engines covered in this guide have solid valve train systems requiring valve adjustments at the intervals stated in the Maintenance Intervals Chart in Section 1. Most engines are also equipped an additional set of valves called jet valves which require adjustment. These valves usually require special adjustment procedures which are addressed later in this Section. Where both procedures are combined, we will note it at the beginning of the procedure.

1983–86 Mirage
1987–89 Precis
1983–85 Starion
1983–86 Cordia and Tredia Except 2.0L Engine

NOTE: The 2.0L engine used on Cordia and Tredia is equipped with automatic hydraulic lash adjusters that do not require adjustment; however, jet valve clearance is adjustable. For the above models, intake, exhaust and jet valve adjustment procedures have been combined.

1. The first step is to warm the engine to operating temperature. Since the engine will cool as you work, it's best to actually drive the car at least several miles in order to heat the internal parts to maximum temperatures. Drive some distance after the temperature gauge has stabilized. Then, stop the car, turn off the engine, and block the wheels. Remove the cam cover as described in Section 3.
2. Torque the cylinder head bolts. Observe the sequence in the appropriate illustration (see Section 3). Turn each bolt in the sequence back just until it breaks loose, and then torque it to the following specification:
- Mirage: 58–61 ft. lbs.
- Cordia and Tredia: 73–79 ft. lbs.
- Galant: 73–79 ft. lbs.
- Starion: Bolts No. 1–10 – 73–79 ft. lbs. Bolts No. 11 – 11–15 ft. lbs.
- Precis: 58–61 ft. lbs.

After the first bolt in the sequence has been torqued, move on to the second one, repeating the procedure. Continue, in order, until all the bolts have been torqued.

3. Now, put the engine at Top Dead Center with No. 1 cylinder at the firing position. Turn the engine by using a wrench on the bolt in the front of the crankshaft until the **0** timing mark on the timing cover lines up with the notch in the front pulley. Observe the valve rockers for No. 1 cylinder. If both are in identical positions with the valves up, the engine is in the right position. If not, rotate the engine exactly 360° until the **0** timing mark is again aligned. Each jet valve is associated with an intake valve that is on the same rocker lever. In this position you'll be able to adjust all the valves marked **A** in the illustration, including associated jet valves (which are located on the rockers on the intake side only).
4. To adjust the appropriate jet valves, first loosen the regular (larger) intake valve adjusting stud right nearby by loosening the locknut and backing the stud off 2 turns. Now, loosen the jet valve (smaller) adjusting stud locknut, back the stud out

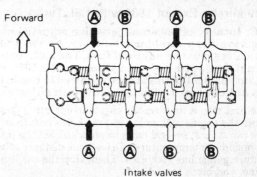

Valves marked 'A' are adjusted with the crankshaft at TDC No. 1 cylinder; turn the crankshaft exactly 360° from this position to adjust the 'B' valves

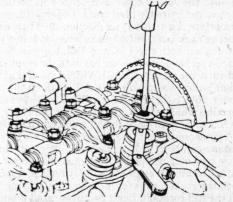

Adjusting the intake and exhaust valves (typical)

2-17

2 ENGINE PERFORMANCE AND TUNE-UP

slightly, and insert the feeler gauge between the jet valve and stud. Make sure the gauge lies flat on the top of the jet valve. Being careful not to twist the gauge or otherwise depress the jet valve spring, rotate the jet valve adjusting stud back in until it just touches the gauge. Now, tighten the locknut. Make sure the gauge still slides very easily between the stud and jet valve and that they both are still just touching the gauge. Readjust if necessary. Note that, especially with the jet valve, the clearance MUST NOT be too tight. Just make sure you are not clamping the gauge in between the stud and valve, but that the parts JUST TOUCH. Repeat entire the procedure for the other jet valves associated with rockers labeled **A**. Then, on engines not equipped with hydraulic lash adjusters, changing the thickness of the gauge if necessary, repeat the procedure for the intake valves labeled **A**.

5. Now, changing the thickness of the feeler gauge as necessary, repeat the basic adjustment procedure for exhaust valves labeled **A** on engines without hydraulic lash adjusters.

6. Turn the engine exactly 360°, until the timing marks are again aligned at **0** BTDC. First, perform Step 4 for all the jet valves on rockers labeled **B** (intake side only). On engines equipped with hydraulic lash adjusters, once jet valves on rockers on the intake side and labeled **B** are adjusted, the valve adjustment procedure is completed and you can proceed to the last step. On other engines, complete Step 4 for the regular intake valves labeled **B**. Finally, repeat Step 5 for exhaust valves labeled **B**.

7. Reinstall the cam cover as described in Section 3. Install new gaskets and seals wherever they are used, and to observe torque specifications for the cam cover bolts. Run the engine to check for oil leaks. Check/adjust the idle speed and ignition timing.

1987–89 Mirage Except 1989 with 1.6L Turbo

NOTE: Intake, exhaust and jet valve adjustment procedures have been combined; however, on the 1.5L engine jet valves must be adjusted first. The 1.6L turbo engine is not equipped with jet valves. On the 1989 turbo engine, no valve adjustments are required because these engines are equipped with hydraulic lash adjusters.

1. The first step is to warm the engine to operating temperature. Since the engine will cool as you work, it's best to actually drive the car at least several miles in order to heat the internal parts to maximum temperatures. Drive some distance after the temperature gauge has stabilized. Then, stop the car, turn off the engine, and block the wheels.

2. Remove the air cleaner making sure the hoses are properly labeled.

3. Remove the spark plugs to make turning the engine over easier.

4. Remove the rocker cover as described in Section 3.

5. Torque the cylinder head bolts. Observe the sequence in the appropriate illustration (see Section 3). Turn each bolt in the sequence back just until it breaks loose, and then torque it to 58–61 ft. lbs. on both the 1.5L and 1.6L engines.

6. Place a socket and a breaker bar onto the crankshaft pulley bolt and turn the engine CLOCKWISE until the notch on the crankshaft pulley is aligned with the **T** mark on the timing belt lower cover. Now either the No. 1 or No. 4 cylinders are at TDC. To determine which at TDC, grasp either the No. 1 and No. 4 rocker arms and try to jiggle them up and down. If you are able to jiggle either one, then that cylinder is at TDC. If you encounter resistance on both, the crankshaft pulley and timing cover marks are probably not properly aligned. To correct this, have an assistant rotate the engine in a CLOCKWISE direction while you check the rocker arms for movement by jiggling the rocker through the full rotation. Don't be discouraged if you don't get it the first time. It may take two or three!

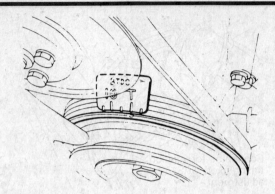

Timing marks on 1988 Mirage with 1.5L engine

Timing marks for 1988 Mirage with 1.6L turbo engine

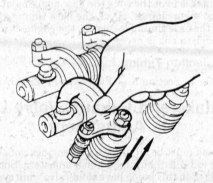

If the rocker arm can be jiggled up and down slightly, the piston is at TDC of the compression stroke

NEVER rotate the engine in a counterclockwise direction!

7. Depending on what cylinder is at TDC, measure the valve clearances at the points shown in the illustrations below using the proper size feeler gauge. Insert the feeler gauge between the rocker and the end of the valve stem and try to insert and withdraw the gauge several times. A slight drag should be felt as the feeler gauge moved in and out. Make sure that you use the proper illustration for each cylinder and refer to the Tune-Up Specifications chart for clearances. The clearances given in the chart are for a hot engine. For a cold engine, the clearances are as follows: 1.8mm (0.07 in.) for exhaust valves and 0.8mm (0.03 in.) for intake valves. On the 1.6L turbo engine, there are no jet valves.

ENGINE PERFORMANCE AND TUNE-UP 2

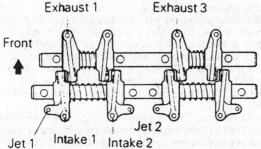

On 1987–89 Mirage 1.5L engines, adjust THESE valves when the No. 1 piston is at TDC

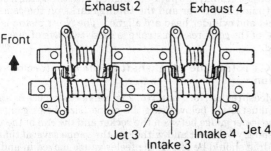

On 1987–89 Mirage 1.5L engines, adjust THESE valves when the No. 4 piston is at TDC

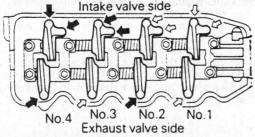

○ : When No. 1 piston is at top dead center on compression stroke
● : When No. 4 piston is at top dead center on compression stroke

Valve adjustment sequence on 1987–88 Mirage with 1.6L turbo engine

8. On the 1.5L engine, if the jet valve clearance is not as specified, loosen the rocker arm nut on the intake valve and loosen the intake valve adjusting screw two or three turns. Now, loosen the jet valve locknut and adjust the valve clearance by turning the adjusting screw until a slight drag is felt on the feeler gauge as you insert it and withdraw it between the rocker and the valve. Once the proper clearance is obtained, hold the screw and tighten the locknut. Check the clearance again to make sure that it didn't change when the locknut was tightened.

NOTE: When adjusting the jet valve clearance on the 1.5L engine, do not apply too much presure on the adjusting screw. The jet valve spring is very delicate.

9. To adjust the intake and exhaust valves, loosen the adjusting screw locknut and insert the proper size feeler gauge between the valve and the rocker. Adjust the valve clearance by turning the adjusting screw until a slight drag is felt on the feeler gauge as you insert it and withdraw it between the rocker and the valve. Once the proper clearance is obtained, hold the screw to prevent it from turning and tighten the locknut. Check the clearance again to make sure it didn't change when the locknut was tightened.

10. Once all the adjustments are complete, rotate the engine 360° until the notch on the pulley is re-aligned with the **T** mark on the lower timing belt cover to bring the remaining piston (either No. or No. 4) to TDC. Using the proper illustrations, repeat Steps 8 and 9 to check and adjust the remaining valves.

11. Install the rocker arm cover, sprak plugs and the air cleaner.

12. Check the timing and the idle speed and adjust as required.

1989 Sigma
1988–89 Galant
Montero V6
1989 Mirage with 1.6L with Turbo Engine

These models are equipped with hydraulic lash adjusters that require no adjustment. The best way to maintain hydraulic adjusters is through regular oil and filter changes following the intervals given in the Maintenance Intervals Chart in Section 1.

Jet Valve Adjustment

Almost every engine covered by this guide has an unusual third valve of very small size called a Jet Valve. Engines not equipped with jet valves are found on the 1988–89 Galant, 1989 Sigma and the 1.6L turbo engine on the 1989 Mirage 1.6L turbo engine. The Jet Valve must be adjusted on all engines, whether the engine uses hydraulic lash adjusters for the normal intake and exhaust valves or not. Thus, on most engines, there are three valves per cylinder that must be adjusted; on the engines equipped with hydraulic lash adjusters, only one valve must be adjusted per cylinder. Adjustment procedures for all engines are the same except for the omission of the adjustment procedure for the regular intake and exhaust valves on the engines with hydraulic lash adjusters. Specifications vary somewhat, however. The jet valve clearance may be the same as either the intake valve or the exhaust valve clearance. Make sure to clarify in your mind the locations and required adjustment dimensions for each valve before beginning work.

1983–88 Mirage
1987–89 Precis
1983–85 Starion
1983–86 Cordia and Tredia Except 2.0L Engine
1983–85 Truck with Gasoline Engine
1983–84 Montero

For these models refer to the above intake and exhaust valve adjustment procedure for Jet Valve Adjustment, as they are combined.

1986 Galant
1986 Starion
1986 Cordia and Tredia with 2.0L Engine
1986–89 Truck with Gasoline Engine
1985–89 Montero with 2.6L Engine

These are equipped with jet valves and hydraulic lash adjusters. The lash adjusters eliminate the need for intake and exhaust valve adjustments.

1. The first step is to warm the engine to operating tempera-

2 ENGINE PERFORMANCE AND TUNE-UP

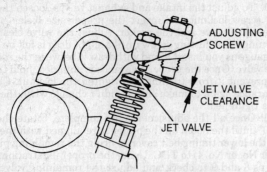

Adjust the jet valves at the location shown

ture. Since the engine will cool as you work, it's best to actually drive the car at least several miles in order to heat the internal parts to maximum temperatures. Drive some distance after the temperature gauge has stabilized. Then, stop the car, turn off the engine, and block the wheels.

2. Remove the spark plugs to make turning the engine over easier.
3. Remove the rocker cover as described in Section 3.
4. Torque the cylinder head bolts. Observe the sequence in the appropriate illustration (see Section 3). Turn each bolt in the sequence back just until it breaks loose, and then torque it to specification.
5. Engage the crankshaft pulley with a socket and breaker bar. Using the crankshaft pulley notch and the timing degree indicator plate, bring the No. 1 piston to TDC of the compresion stroke. To determine if it at TDC, grasp the rocker arm and try to jiggle it up and down. If you are able to jiggle it, then that cylinder is at TDC. If you encounter resistance, have an assistant rotate the engine while you check the rocker arms for movement by jiggling the rocker through the full rotation. You don't have to start at the No. 1 piston as long as you follow the firing order. Don't be discouraged if you don't get it the first time. It may take two or three!
6. Loosen the jet valve locknut and back off on the adjusting screw a small amount. Insert a 2.5mm (0.10 in.) feeler gauge between the rocker and the end of the jet valve stem check the clearance by moving the feeler gauge in and out. A slight drag should be felt. If the clearance is not as specified, turn the adjusting screw either in or out until the clearance is correct. Be careful not to apply to much pressure to the screw, because the valve spring is very weak. Hold the screw to prevent it from moving and tighten the locknut. Check the clearance again to make sure it didn't change when the locknut was tightened.
7. Depeding on what cylinder you started with, bring the next cylinder in the firing order to TDC and repeat the adjustment procedure as described in Steps 5 and 6. Repeat this procedure until all the valves are adjusted.
8. Install the rocker arm cover using a new gasket as required.
9. Install the spark plugs.
10. Check the timing and idle speed and adjust as necessary.

1987 Galant
1987–88 Cordia and Tredia

These are equipped with jet valves and hydraulic lash adjusters. The lash adjusters eliminate the need for intake and exhaust valve adjustments.

1. The first step is to warm the engine to operating temperature. Since the engine will cool as you work, it's best to actually drive the car at least several miles in order to heat the internal parts to maximum temperatures. Drive some distance after the temperature gauge has stabilized. Then, stop the car, turn off the engine, and block the wheels.

2. Remove the spark plugs to make turning the engine over easier.
3. Disconect the oxygen sensor connector and route the wire off to side and out of the way.
4. On Galant equipped with air conditioning, disconnect the high pressure hose from the engine mounting bracket.
5. Place a wooden block under the oil pan and support the block with a jack. Raise the engine. Remove the four nuts and two bolts that attach the engine support bracket. Remove the bracket and keep the engine supported with the jack.
6. Remove the timing belt upper front cover and set it off to the side.
7. On Cordia and Tredia with 1.8L engine, remove the air intake pipe. On the 2.0L engine, remove the air cleaner assembly.
8. Remove the rocker cover as described in Section 3.
9. Torque the cylinder head bolts. Observe the sequence in the appropriate illustration (see Section 3). Turn each bolt in the sequence back just until it breaks loose, and then torque it to specification.
10. Engage the crankshaft pulley with a socket and a breaker bar. Rotate the engine in a CLOCKWISE direction until the notch on the crankshaft pulley is aligned with the **T** mark on the lower timing belt cover and the timing marks on the camshaft sprocket and cylinder head are aligned. The No. 1 piston is now at TDC of the compression stroke is these two sets of marks are aligned.

DO NOT rotate the crankshaft in the counterclockwise rotation.

11. Measure the No. 1 and No. 2 jet valve clearances as shown in the illustration below using the proper size feeler gauge. Insert the feeler gauge between the rocker and the end of the valve stem and try to insert and withdraw the gauge several times. A slight drag should be felt as the feeler gauge moved in and out. Make sure that you use the proper illustration for each cylinder valve sequence and refer to the Tune-Up Specifications chart for clearances. The clearances given in the chart are for a hot engine. For a cold engine, the clearance 1.8mm (0.07 in.).
12. If the jet valve clearance is not as specified, loosen the jet valve locknut and adjust the valve clearance by adjusting the screw in or out until a slight drag is felt on the feeler gauge as you insert and withdraw it between the rocker and the end of the valve stem. Once the proper clearance is obtained, hold the screw to prevent it from turning and tighten the locknut. Check the clearance again to make sure that it didn't change when the locknut was tightened.

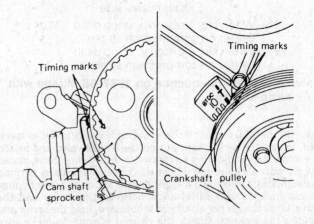

When these timing marks are alined, No 1 piston is at TDC for 1987 Galant and 1987–88 Cordia and Tredia

2-20

ENGINE PERFORMANCE AND TUNE-UP 2

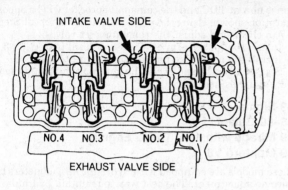

Adjust THESE valves when the No 1 piston is at TDC

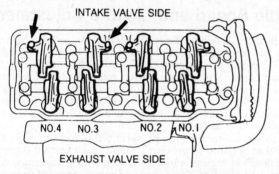

Adjust THESE valves when the No 4 piston is at TDC

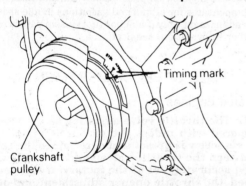

Starion timing marks

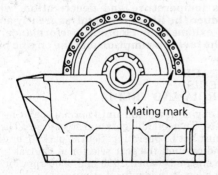

On 1987–89 Starion, when the camshaft sprocket timing mark is as shown, No 1 piston is at TDC When the timing mark is opposite the position shown, No. 4 piston is at TDC

NOTE: When adjusting the jet valve clearance, do not apply too much presure on the adjusting screw. The jet valve spring is very delicate.

13. Once all the adjustments on the No. 1 cylinder valves are complete, rotate the engine 360° until the notch on the pulley is re-aligned with the **T** mark on the lower timing belt cover. The No. 4 piston is now at TDC. Using the proper illustration, repeat Steps 10 and 11 to check and adjust the No. 3 and No. 4 jet valves.
14. Install the rocker arm cover as described in Section 3.
15. On Cordia and Tredia with 1.8L engine, install the air intake pipe. On the 2.0L engine, install the air cleaner assembly.
16. Install the upper timing belt cover making sure the surface is clean.
17. Attach the engine mounting bracket. With the engine still supported, torque the mounting bracket bolts to 36–47 ft. lbs. Lower the engine and remove the block of wood and the jack.
18. On Galant equipped with air conditioning, connect the high pressure hose to the engine mounting bracket.
19. Connect the oxygen sensor connector and install the spark plugs.
20. Start the engine and check the ignition timing and the idle speed. Adjust as necessary.

1987–89 Starion

1. The first step is to warm the engine to operating temperature. Since the engine will cool as you work, it's best to actually drive the car at least several miles in order to heat the internal parts to maximum temperatures. Drive some distance after the temperature gauge has stabilized. Then, stop the car, turn off the engine, and block the wheels.
2. Remove the spark plugs to make turning the engine over easier.
3. Remove the air intake pipe.
4. Remove the rocker arm cover as described in Section 3.
5. Engage the crankshaft pulley with a socket and a breaker bar. To set the No. 1 piston ar TDC, rotate the engine in a CLOCKWISE direction until the notch on the crankshaft pulley is aligned with the **T** mark on the timing chain cover. DO NOT rotate the crankshaft in the counterclockwise rotation. Then, make sure that the camshaft sprocket timing mark is in the position shown in the illustration.

NOTE: If the camshaft sprocket is opposite the side shown in the illustration then the No. 4 piston is at TDC.

6. Measure the jet valve clearances as shown in the illustration below using the proper size feeler gauge. Insert the feeler gauge between the rocker and the end of the valve stem and try to insert and withdraw the gauge several times. A slight drag should be felt as the feeler gauge moved in and out. Make sure that you use the proper illustration for each cylinder valve sequence and refer to the Tune-Up Specifications chart for clearances. The clearances given in the chart are for a hot engine. For a cold engine, the clearance 1.8mm (0.07 in.).
7. If the jet valve clearance is not as specified, loosen the jet valve locknut and adjust the valve clearance by adjusting the screw in or out until a slight drag is felt on the feeler gauge as you insert and withdraw it between the rocker and the end of the valve stem. Once the proper clearance is obtained, hold the screw to prevent it from turning and tighten the locknut. Check the clearance again to make sure it didn't change when you tightened the locknut.

NOTE: When adjusting the jet valve clearance, do not apply too much pressure on the adjusting screw. The jet valve spring is very delicate.

8. Once all the adjustments on the No. 1 cylinder valves are complete, rotate the engine 360° until the notch on the pulley is re-aligned with the **T** mark on the timing chain cover. The No. 4

2 ENGINE PERFORMANCE AND TUNE-UP

When No. 1 cylinder is at top dead center on compression stroke.

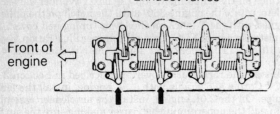

When No. 4 cylinder is at top dead center on compression stroke.

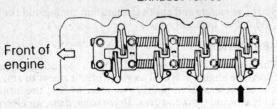

Jet valve adjustment sequence on 1987–89 Starion

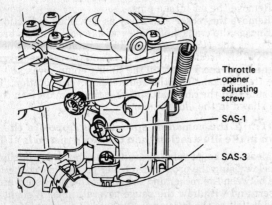

When adjusting the idle speed on carbureted engines, adjust only SAS-1

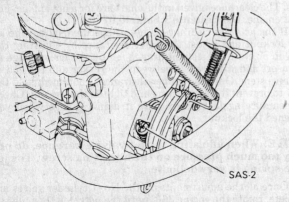

DON'T TOUCH THIS SCREW!

piston is now at TDC and the camshaft sprocket will be opposite the position shown. Using the proper illustration, repeat Steps 6 and 7 to check and adjust the remaining jet valves.
9. Install the rocker arm cover as described in Section 3.
10. Install the air intake pipe and the spark plugs.
11. Start the engine and check the ignition timing and the idle speed. Adjust as necessary.

1989 Sigma
1988–89 Galant
1989 Mirage with 1.6L Turbo Engine
1989 Montero V6

These models are equipped with hydraulic lash adjusters that require no adjustment. The best way to maintain hydraulic adjusters is through regular oil and filter changes.

Idle Speed and Mixture Adjustment

Idle speed is adjusted periodically to compensate for engine wear and to ensure smooth operation. Idle mixture adjustments are not required as a matter of routine, but only when major carburetor or throttle body work is required. The emission control system compensates as required to ensure a stable idle mixture. If you suspect trouble because of a rough idle, you can have the idle mixture checked by going to a shop or diagnostic center that has a CO meter.

Also, most Mitsubishis with carburetors have an idle-up solenoid that operates to prevent stalling under certain conditions. This does not require adjustment as a matter of routine either, but may be adjusted if you suspect the system is not functioning properly. Fuel injected engines employ an Idle Speed Control which compensates for all normal variations in idle speed. This requires adjustment only if a major fuel system part is replaced. For further information, see Section 5.

IDLE SPEED ADJUSTMENT

Carbureted Engines

NOTE: The throttle valve adjusting screw should not be tampered with unless the carburetor has been rebuilt. This screw is preset and determines the relationship between the throttle valve and the free lever, and has been accurately set at the factory. If this setting is disturbed, the throttle opener adjustment and or dashpot adjustment cannot be done accurately. Also the improper setting (throttle valve opening) will increase the exhaust gas temperature and deceleration, which in turn will reduce the life of the catalyst greatly and deteriorate the exhaust gas cleaning performance. It will also effect the fuel consumption and the engine braking.

1985 GALANT
1985–87 MIRAGE
1987–89 PRECIS
1983–84 CORDIA AND TREDIA

1. Run the engine at fast idle until normal operating temperature is reached. Turn off all lights and accessories. The electric cooling fan must not be operating. Set the parking brake, block the wheels and position the gear selector in neutral.
2. Run the engine between 2000 and 3000 rpm for about 5 seconds, then allow it to idle for 2 minutes.
3. Check the idle speed adjustment screw SAS 1 on the carburetor linkage arm.

1985–88 CORDIA AND TREDIA
1983–89 TRUCK WITH GASOLINE ENGINE
1983–89 MONTERO WITH 2.6L ENGINE

ENGINE PERFORMANCE AND TUNE-UP 2

NOTE: 1987–88 cars with automatic transaxle are equipped with an idle speed control (ISC) system which controls the idle speed automatically which makes adjustment unnecessary. For more information on the ISC system, see Section 5. To check the idle speed, proceed as follows:

1. Turn all lights and accessories to the OFF position.
2. Connect a tachometer and timing light to the engine.
3. Place the transaxle in Neutral and disconnect the electric cooling fan.
4. Start and warm up the engine at idle speed for a few minutes. On 1987–88 vehicles, press the gas pedal once to disengage the fast idle.
5. Check the ignition timing and adjust as necessary.
6. Check the idle speed. If it does not meet specifications, adjust the idle speed using the idle speed adjusting screw, which is located closest to the primary throttle valve shaft and labeled **SAS 1** in the illustration.
6. After the idle speed is set, stop the engine and reconnect the cooling fan connector. Disconnnect the tachometer and the timing light.

NOTE: DO NOT touch the screw labeled SAS 2. This preset screw determines the relationship between the throttle valve and the free lever and has been preset at the factory. If this screw is disturbed, throttle opener and dashpot adjustments will also be disturbed.

Fuel Injected Engines

1983–88 CORDIA AND TREDIA WITH TURBO

The idle speed check procedure for these vehicles is the same as for carbureted engines above. The turbo engine is equipped with an idle speed control (ISC) system which controls the idle speed automatically making adjustment unnecessary. For more information on the ISC system, see Section 5.

1983 STARION

1. Run the engine until normal operating temperature is reached. Make sure all lights and accessories are turned off.
2. Apply the parking brake and block the wheels. Position the gear selector in neutral and stop the engine.
3. Attach a tachometer and timing light. Start the engine and increase the engine speed to 2000–3000 rpm several times, return to idle and check the ignition timing, adjust if necessary.
4. Remove the rubber cap covering the idle speed adjuster switch, leaving the cable connector connected. The idle adjuster switch is located on the throttle linkage. Adjust the idle speed.
5. If the idle adjustment screw must be turned more than 1 turn during adjustment, disconnect the connector from the speed adjust switch and plug it into the dummy terminal on the injector base. Adjust to correct idle speed and reconnect to the idle switch. Remove the tachometer and timing light.

1985–89 MIRAGE
1984–89 STARION
1986–89 GALANT AND SIGMA
1989 MONTERO V6

These vehicles are equipped with an idle speed control (ISC) system which controls the idle speed automatically, making adjustment unnecessary. For more information on the ISC system, see Section 5. To check the idle speed, proceed as follows:

NOTE: On the 1986 Galant, a special harness connector MD998439 is needed to make the tachometer connection. A view of this harness is shown below. The tachometer hook-up for all Mirage and the 1989 Galant with DOHC engine is shown in the ignition timing section earlier in this Section.

1. Turn all lights and accessories to the OFF position and place the transaxle in Neutral. If equipped with power steering,

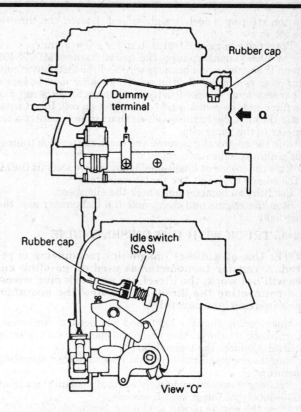

Idle speed adjustment on 1983 Starion

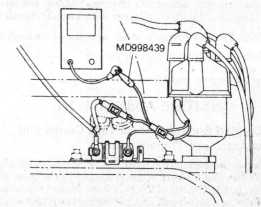

Special tachometer harness MD998439 used on 1986 Galant

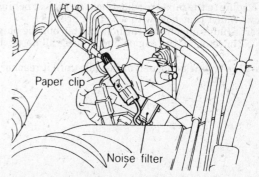

Making the tachometer connection on 1987–89 Galant (except DOHC engine) and Sigma

2-23

2 ENGINE PERFORMANCE AND TUNE-UP

place the steering wheel straight ahead. Loosen the throttle valve set screw.

2. Start the engine and warm it up for a few minutes.
3. On 1986 Galant, connect the special harness MD998439 between the ignition coil primary wire and the distributor connector as shown in the illustration. On all other models, locate the harness (engine speed detection) connector between the noise filter and the primary side of the ignition coil. Insert a paper clip through the harness side to short it and connect a tachometer to the paper clip.
4. Run the engine at idle speed and check the ignition timing. If not within specification, adjust it.
5. Race the engine at 2,000–3,000 rpm for five seconds the allow the engine to run at idle for two minutes.
6. Read the tachometer and check the idle speed.
7. Stop the engine and disconnect the tachometer and the timing light.

1983-85 TRUCK WITH 4D55 DIESEL ENGINE

NOTE: Use of a diesel (magnetic) tachometer is required. A regular tachometer as used on gasoline engines will not work; the diesel has no spark plug wires. When connecting the diesel tach, follow the manufacturer's instructions carefully.

1. Make certain that the lights and all electrical accessories are turned off. Set the parking brake and place the shift selector in neutral. Connect the tachometer.
2. Run the engine until the coolant rises to normal operating temperature.
3. Once the engine is completely warmed up, run the engine at 2000–3000 rpm for at least 5 seconds.
4. Allow the engine to run at idle for 2 minutes.
5. Read the idle speed on the tachometer. If the idle is not within specifications, adjust the idle speed screw. Loosen the locknut, then turn the adjusting screw with a small screwdriver.

WARNING: Do not disturb other screws or linkages.

6. When the proper idle speed is achieved, hold the screw in position and tighten the locknut.
7. Stop the engine and remove the tachometer.

MIXTURE ADJUSTMENT

All Carbureted Engines except 1985 Cordia and Tredia w/Automatic Transmission

1. Remove the carburetor from the engine as described in Section 5. The idle mixture screw is located in the base of the carburetor, just to the left of the PCV hose. Mount the carburetor, carefully, in a soft-jawed vise, protecting the gasket surface, and with the mixture adjusting screw facing upward.
2. Drill a 5/64 inch hole through the casting from the underside of the carburetor. Make sure that this hole intersects the passage leading to the mixture adjustment screw just behind the plug. Now, widen that hole with a 1/8 inch drill bit.
3. Insert a blunt punch into the hole and tap out the plug. Install the carburetor on the engine and connect all hoses, lines, etc.
4. Start the engine and run it at fast idle until it reaches nor-

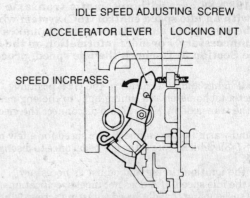

Diesel idle speed adjustment

mal operating temperature. Make sure that all accessories are OFF and the transaxle is in neutral. Turn the ignition switch OFF and disconnect the battery ground cable for about 5 seconds, then reconnect it. Disconnect the oxygen sensor.

5. Start the engine and run it for at least 5 seconds at 2,000–3,000 rpm. Then allow the engine to idle for about 2 minutes.
6. Connect a tachometer and allow the engine to operate at the specified curb idle speed. Adjust it, if necessary, to obtain this speed. Connect a CO meter to the exhaust pipe. A reading of 0.1–0.3% is necessary. Adjust the mixture screw to obtain the reading. If, during this adjustment, the idle speed is varied more than 100 rpm in either direction, reset the idle speed and readjust the CO until both specifications are met simultaneously. Shut off the engine, reconnect the oxygen sensor and install a new concealment plug.

1985 Cordia and Tredia with Automatic Transaxle

1. Refer to steps 1–3 of the above procedure.
2. Adjust the accelerator cable so that there is no tension on the throttle linkage.
3. Turn the ignition switch to ON for at least 18 seconds. Then, turn it OFF.
4. Disconnect the oxygen sensor and idle speed control.
5. Start the engine, run it to normal operating temperature and check the ignition timing. Adjust it if necessary.
6. Run the engine to 2,000–3,000 rpm and drop the throttle back, three times in quick succession. Allow the engine to idle for at least 30 seconds, then take a CO reading. CO should be 0.1–0.3%. If not, turn the mixture screw inward or outward until that range is obtained.
7. Using the ISC screw, adjust the engine speed to an idle of 700 rpm, unless the car has less than 300 miles on the odometer. In that case, the speed should be 650 rpm. Turn the SAS screw inward until an increase in rpm is noted. Then, back it out slowly until the point is reached at which it no longer changes idle speed. From that point, turn the screw inward 2/3 of a turn.
8. Check both CO and idle. If necessary, readjust both. It should not be necessary to readjust the SAS screw.
9. Adjust the accelerator cable to eliminate all play. Connect the ISC actuator and oxygen sensor.

3 Engine and Engine Overhaul

QUICK REFERENCE INDEX

Alternator and Regulator Specifications Chart	3-18	Exhaust System	3-272
Camshaft Specifications Chart	3-37	General Engine Specifications Chart	3-33
Crankshaft and Connecting Rod Specifications Chart	3-42	Piston Specifications Chart	3-40
Engine Electrical Systems	3-2	Piston Ring Specifications Chart	3-40
Engine Mechanical Systems	3-26	Starter Specifications Chart	3-25
Engine Torque Specifications Chart	3-45	Valve Specifications Chart	3-35

GENERAL INDEX

Air conditioning		Oil pan	3-237	Radiator	3-145
Compressor	3-143	Oil pump	3-239	Rear main oil seal	3-268
Condenser	3-151	Overhaul techniques	3-26	Regulator	3-17
Alternator		Piston pin	3-266	Ring gear	3-271
Alternator precautions	3-7	Pistons	3-263	Rings	3-265
Operation	3-7	Rear main seal	3-268	Rocker arms or shaft	3-93
Removal and installation	3-7	Removal and installation	3-48	Silent shafts	3-239
Specifications	3-18	Ring gear	3-271	Specifications Charts	
Troubleshooting	3-17	Rings	3-265	Alternator and regulator	3-18
Battery	3-17	Rocker arms and shafts	3-93	Camshaft	3-37, 39
Camshaft		Silent shafts	3-239	Crankshaft and connecting rod	3-42, 44
Inspection	3-262	Specifications	3-33	Fastener markings and	
Removal and Installation	3-255	Thermostat	3-107	torque standards	3-47
Catalytic converter	3-274	Timing belts and covers	3-202	General engine	3-33
Charging system	3-7	Timing chain	3-234	Piston and ring	3-40
Coil (ignition)	3-4	Tools	3-26	Starter	3-25
Combination manifold	3-138	Troubleshooting	3-28	Torque	3-45, 46
Compression testing	3-26	Turbocharger	3-138	Valves	3-35
Compressor	3-143	Valve (rocker) cover	3-91	Starter	
Condenser	3-151	Valve guides	3-202	Brush replacement	3-20
Connecting rods and bearings		Valves	3-199	Drive replacement	3-20
Service	3-263, 267	Valve seats	3-201	Overhaul	3-20
Specifications	3-42, 44	Valve springs	3-199	Removal and installation	3-19
Crankshaft		Water pump	3-154	Solenoid or relay replacement	3-19
Service	3-269	Exhaust Manifold	3-125	Specifications	3-25
Specifications	3-42, 44	Exhaust system	3-272	Troubleshooting	3-17, 24
Cylinder head		Fasteners	3-47	Stripped threads	3-26
Cleaning and inspection	3-197	Flywheel and ring gear	3-270	Tailpipe	3-274
Removal and installation	3-169	Ignition Coil	3-4	Thermostat	3-107
Resurfacing	3-198	Ignition Module	3-4	Timing belts and covers	3-202
Distributor	3-6	Intake manifold	3-109	Timing chain	3-234
Engine		Intercooler	3-142	Timing gear cover	3-202, 239
Camshaft	3-255	Main bearings	3-269	Tools	3-26
Combination manifold	3-138	Manifolds		Torque specifications	3-45, 46
Compression testing	3-26	Intake	3-109	Troubleshooting	
Connecting rods and bearings	3-263, 267	Exhaust	3-125	Battery and starting systems	3-3, 24
Crankshaft	3-269	Module (ignition)	3-4	Charging system	3-3, 17
Cylinder head	3-169	Muffler	3-274	Cooling system	3-31
Cylinders	3-263	Oil pan	3-237	Engine electrical systems	3-2
Exhaust manifold	3-125	Oil pump	3-239	Engine mechanical	3-2
Flywheel	3-270	Piston pin	3-266	Turbocharger	3-138
Front (timing) cover	3-202, 239	Pistons		Valve guides	3-202
Front seal	3-202, 239	Inspection	3-264	Valve seats	3-201
Intake manifold	3-109	Installation	3-265	Valve service	3-199
Intercooler	3-142	Piston pin	3-266	Valve specifications	3-35
Jet valve	3-200	Removal	3-263	Valve springs	3-199
Main bearings	3-269	Rings	3-265	Water pump	3-154

3 ENGINE AND ENGINE OVERHAUL

ENGINE ELECTRICAL

Troubleshooting the Engine Electrical System

For any electrical system to operate, it must make a complete circuit. This simply means that the power flow from the battery must make a complete circle. When an electrical component is operating, power flows from the battery to the components, passes through the component (load) causing it to function, and returns to the battery through the ground path of the circuit. This ground may be either another wire or the actual metal part of the car upon which the component is mounted.

Perhaps the easiest way to visualize this is to think of connecting a light bulb with two wires attached to it to the battery. If one of the two wires was attached to the negative (−) post of the battery and the other wire to the positive (+) post, the light bulb would light and the circuit would be complete. Electricity could follow a path from the battery to the bulb and back to the battery. Its not hard to see that with longer wires on our light bulb, it could be mounted anywhere on the car. Further, one wire could be fitted with a switch so that the light could be turned on and off at will. Various other items could be added to our primitive circuit to make the light flash, become brighter or dimmer under certain conditions or advise the user that it's burned out.

Some automotive components don't use a wire to battery--they ground to the metal of the car through their mounting points. The electrical current runs through the chassis of the vehicle and returns to the battery through the ground (−) cable; if you look, you'll see that the battery ground cable connects between the battery and the body of the car.

Every complete circuit must include a load--something to use the electricity coming from the source. If you were to connect a wire between the two terminals of the battery (DON'T do this) without the light bulb, the battery would attempt to deliver its entire power supply from one pole to another almost instantly. This is a short circuit. The electricity is taking a short-cut to get to ground and is not being used by any load in the circuit. This sudden and uncontrolled electrical flow can cause great damage to other components in the circuit and can develop a tremendous amount of heat. A short in an automotive wiring harness can develop sufficient heat to melt the insulation on all the surrounding wires and reduce a multi-wire cable to one sad lump of plastic and copper. Two common causes of shorts are broken insulation (thereby exposing the wire to contact with surrounding metal surfaces) or a failed switch (the pins inside the switch come out of place, touch each other and reroute the electricity).

Some electrical components which require a large amount of current to operate also have a relay in their circuit. Since these circuits carry a large amount of current (amperage or amps), the thickness of the wire in the circuit (wire gauge) is also greater. If this large wire were connected from the load to the control switch on the dash, the switch would have to carry the high amperage load and the dash would be twice as large to accommodate wiring harnesses as thick as your wrist. To prevent these problems, a relay is used. The large wires in the circuit are connected from the battery to one side of the relay and from the opposite side of the relay to the load. The relay is normally open, preventing current from passing through the circuit. An additional, smaller wire is connected from the relay to the control switch for the circuit. When the control switch is turned on, it grounds the smaller wire to the relay and completes its circuit. The main switch inside the relay closes, sending power to the component without routing the main power through the inside of the car. Some common circuits which may use relays are the horn, headlights, starter and rear window defogger systems.

It is possible for larger surges of current to pass through the electrical system of your car. If this surge of current were to reach the load in the circuit, it could burn it out or severely damage it. To prevent this, fuse and/or circuit breakers and/or fusible links are connected into the supply wires of the electrical system. These items are nothing more than a built-in weak spot in the system. It's much easier to go to a known location (the fusebox) to see why a circuit is inoperative than to disect 15 feet of wiring under the dashboard, looking for what happened.

When an electrical current of excessive power passes through the fuse, the fuse blows and breaks the circuit, preventing the passage of current and protecting the components.

A circuit breaker is basically a self-repairing fuse. It will open the circuit in the same fashion as a fuse, but when either the short is removed or the surge subsides, the circuit breaker resets itself and does not need replacement.

A fuse link (fusible link or main link) is a wire that acts as a fuse. It is normally connected between the starter relay and the main wiring harness under the hood. Since the starter is the highest electrical draw on the car, an internal short during starter use could direct about 130 amps into the wrong places. Consider the damage potential of introducing this current into a system whose wiring is rated at 15 amps and you'll understand the need for protection. Since this link is very early in the electrical path, it's the first place to look if nothing on the car works but the battery seems to be charged and is properly connected.

Electrical problems generally fall into one of three areas:
1. The component that is not functioning is not receiving current.
2. The component is receiving power but not using it or using it incorrectly (component failure).
3. The component is improperly grounded.

The circuit can be can be checked with a test light and a jumper wire. The test light is a device that looks like a pointed screwdriver with a wire on one end and a bulb in its handle. A jumper wire is simply a piece of wire with alligator clips on each end. If a component is not working, you must follow a systematic plan to determine which of the three causes is the villain.

1. Turn on the switch that controls the item not working.

NOTE: Some items only work when the ignition switch is turned on.

2. Disconnect the the power supply wire from the component.
3. Attach the ground wire on the test light to a good metal ground.
4. Touch the end probe of the test light to the power wire; if there is current in the wire, the light in the test light will come on. You have now established that current is getting to the component.
5. Turn the ignition or dash switch off and reconnect the wire to the component.

If the test light does not go on, then the problem is between the battery and the component. This includes all the switches, fuses, relays and the battery itself. Next place to look is the fusebox; check carefully either by eye or by using the test light across the fuse clips. The easiest way to check is to simply replace the fuse. If the fuse is blown, and upon replacement, immediately blows again, there is a short between the fuse and the component. This is generally (not always) a sign of an internal short in the component. Disconnect the power wire at the component again and replace the fuse; if the fuse holds, the component is the problem.

If all the fuses are good and the component is not receiving power, find the switch for the circuit. Bypass the switch with the jumper wire. This is done by connecting one end of the jumper to the power wire coming into the switch and the other end to the wire leaving the switch. If the component comes to life, the switch has failed.

ENGINE AND ENGINE OVERHAUL 3

WARNING: Never substitute the jumper for the component. The circuit needs the electrical load of the component. If you bypass it, you cause a short circuit.

Checking the ground for any circuit can mean tracing wires to the body, cleaning connections or tightening mounting bolts for the component itself. If the jumper wire can be connected to the case of the component or the ground connector, you can ground the other end to a piece of clean, solid metal on the car. Again, if the component starts working, you've found the problem.

It should be noted that generally the last place to look for an electrical problem is in the wiring itself. Unless the car has undergone unusual circumstances (major bodywork, flood damage, improper repairs, etc.) the wiring is not likely to change its condition. A systematic search through the fuse, the connectors and switches and the component itself will almost always yield an answer. Loose and/or corroded connectors--particularly in ground circuits--are becoming a larger problem in modern cars. The computers and on-board electronic (solid state) systems are highly sensitive to improper grounds and will change their function drastically if one occurs.

Remember that for any electrical circuit to work, ALL the connections must be clean and tight.

BATTERY AND STARTING SYSTEM

Basic Operating Principles

The battery is the first link in the chain of mechanisms which work together to provide cranking of the automobile engine. In most modern cars, the battery is a lead/acid electrochemical device consisting of six 2v subsections (cells) connected in series so the unit is capable of producing approximately 12v of electrical pressure. Each subsection consists of a series of positive and negative plates held a short distance apart in a solution of sulfuric acid and water.

The two types of plates are of dissimilar metals. This causes a chemical reaction to be set up, and it is this reaction which produces current flow from the battery when its positive and negative terminals are connected to an electrical appliance such as a lamp or motor. The continued transfer of electrons would eventually convert the sulfuric acid to water, and make the two plates identical in chemical composition. As electrical energy is removed from the battery, its voltage output tends to drop. Thus, measuring battery voltage and battery electrolyte composition are two ways of checking the ability of the unit to supply power. During the starting of the engine, electrical energy is removed from the battery. However, if the charging circuit is in good condition and the operating conditions are normal, the power removed from the battery will be replaced by the generator (or alternator) which will force electrons back through the battery, reversing the normal flow, and restoring the battery to its original chemical state.

The battery and starting motor are linked by very heavy electrical cables designed to minimize resistance to the flow of current. Generally, the major power supply cable that leaves the battery goes directly to the starter, while other electrical system needs are supplied by a smaller cable. During starter operation, power flows from the battery to the starter and is grounded through the car's frame and the battery's negative ground strap.

The starting motor is a specially designed, direct current electric motor capable of producing a very great amount of power for its size. One thing that allows the motor to produce a great deal of power is its tremendous rotating speed. It drives the engine through a tiny pinion gear (attached to the starter's armature), which drives the very large flywheel ring gear at a greatly reduced speed. Another factor allowing it to produce so much power is that only intermittent operation is required of it. Thus, little allowance for air circulation is required, and the windings can be built into a very small space.

The starter solenoid is a magnetic device which employs the small current supplied by the start circuit of the ignition switch. This magnetic action moves a plunger which mechanically engages the starter and closes the heavy switch connecting it to the battery. The starting switch circuit consists of the starting switch contained within the ignition switch, a transmission neutral safety switch or clutch pedal switch, and the wiring necessary to connect these in series with the starter solenoid or relay.

The pinion, a small gear, is mounted to a one-way drive clutch. This clutch is splined to the starter armature shaft. When the ignition switch is moved to the **START** position, the solenoid plunger slides the pinion toward the flywheel ring gear via a collar and spring. If the teeth on the pinion and flywheel match properly, the pinion will engage the flywheel immediately. If the gear teeth butt one another, the spring will be compressed and will force the gears to mesh as soon as the starter turns far enough to allow them to do so. As the solenoid plunger reaches the end of its travel, it closes the contacts that connect the battery and starter and then the engine is cranked.

As soon as the engine starts, the flywheel ring gear begins turning fast enough to drive the pinion at an extremely high rate of speed. At this point, the one-way clutch begins allowing the pinion to spin faster than the starter shaft so that the starter will not operate at excessive speed. (This overrun is similar to riding a bicycle downhill; the rear wheel is turning faster than the chain sprocket.) When the ignition switch is released from the starter position, the solenoid is de-energized, and a spring pulls the gear out of mesh, interrupting the current flow to the starter.

Some starters employ a separate relay, mounted away from the starter, to switch the motor and solenoid current on and off. The relay replaces the solenoid electrical switch, but does not eliminate the need for a solenoid mounted on the starter used to mechanically engage the starter drive gears. The relay is used to reduce the amount of current the starting switch must carry.

THE CHARGING SYSTEM

Basic Operating Principles

The automobile charging system provides electrical power for operation of the vehicle's ignition and starting systems and all the electrical accessories. The battery services as an electrical surge or storage tank, storing (in chemical form) the energy originally produced by the engine driven generator. The system also provides a means of regulating generator output to protect the battery from being overcharged and to avoid excessive voltage to the accessories.

The storage battery is a chemical device incorporating parallel lead plates in a tank containing a sulfuric acid/water solution. Adjacent plates are slightly dissimilar, and the chemical reaction of the two dissimilar plates produces electrical energy when the battery is connected to a load such as the starter motor. The chemical reaction is reversible, so that when the generator is producing a voltage (electrical pressure) greater than that produced by the battery, electricity is forced into the battery, and the battery is returned to its fully charged state.

The vehicle's generator is driven mechanically, through V-belts, by the engine crankshaft. It consists of two coils of fine wire, one stationary (the stator), and one movable (the rotor). The rotor may also be known as the armature, and consists of fine wire wrapped around an iron core which is mounted on a shaft. The electricity which flows through the two coils of wire (provided initially by the battery in some cases) creates an intense magnetic field around both rotor and stator, and the interaction between the two fields creates voltage, allowing the generator to power the accessories and charge the battery.

There are two types of generators: the earlier is the direct current (DC) type. The current produced by the DC generator is generated in the armature and carried off the spinning arma-

3 ENGINE AND ENGINE OVERHAUL

ture by stationary brushes contacting the commutator. The commutator is a series of smooth metal contact plates on the end of the armature. The commutator plates, which are separated from one another by a very short gap, are connected to the armature circuits so that current will flow in one direction only in the wires carrying the generator output. The generator stator consists of two stationary coils of wire which draw some of the output current from the generator to form a powerful magnetic field and create the interaction of fields which generates the voltage. The generator field is wired in series with the regulator.

Newer automobiles use alternating current generators or alternators, because they are more efficient, can be rotated at higher speeds, and have fewer brush problems. In an alternator, the field rotates while all the current produced passes only through the stator winding. The brushes bear against continuous slip rings rather than a commutator. This causes the current produced to periodically reverse the direction of its flow. Diodes (electrical one-way valves) block the flow of current from traveling in the wrong direction. A series of diodes is wired together to permit the alternating flow of the stator to be rectified back to 12 volts DC for use by the vehicles's electrical system.

The regulator consists of several circuits. Each circuit has a core, or magnetic coil of wire, which operates a switch. Each switch is connected to ground through one or more resistors. The coil of wire responds directly to system voltage. When the voltage reaches the required level, the magnetic field created by the winding of wire closes the switch and inserts a resistance into the generator field circuit, thus reducing the output. The contacts of the switch cycle open and close many times each second to precisely control voltage. On many newer cars, the regulating function is performed by solid-state (rather than mechanical) components. The regulator is often built in to the alternator; this system is termed an integrated or internal regulator.

While alternators are self-limiting as far as maximum current is concerned, DC generators employ a current regulating circuit which responds directly to the total amount of current flowing through the generator circuit rather than to the output voltage. The current regulator is similar to the voltage regulator except that all system current must flow through the energizing coil on its way to the various accessories.

Ignition Coil and Ballast Resistor

TESTING

1984–88 Cordia and Tredia, Pickup
1985–89 Mirage

1. Measure the resistance of the external resistor on the coil by connecting the probes of an ohmmeter across the two resistor connectors. Resistance should be as follows or the unit should be replaced.
- 1984–86 Cordia, Tredia, Pickup w/2.0L engine: 1.35Ω
- 1987–88 Cordia, Tredia, Pickup w/2.0L engine: 1.2–1.4Ω
- 1985–86 Mirage: 1.35Ω
- 1987–88 Mirage w/1.5L engine: 1.2–1.49Ω
- 1989 Mirage: specification does not apply

2. Measure the resistance across the coil primary circuit by connecting an ohmmeter between the (+) and (−) coil connectors. Resistance must be approximately as follows or the coil should be replaced.
- 1984–86 Cordia, Tredia, Pickup: 1.2Ω
- 1987–88 Cordia, Tredia, Pickup w/2.0L engine: 1.08–1.32Ω
- 1987–88 Cordia, Tredia, Pickup w/1.8L engine: 1.13–1.37Ω
- 1985–86 Mirage: 1.2Ω
- 1987–88 Mirage: 1.1–1.3Ω
- 1989 Mirage: 0.72–0.88Ω

3. Set your ohmmeter to the x1000 scale and measure the resistance between the connector inside the coil tower and the coil (+) terminal. This is secondary resistance and must be as follows, or the unit should be replaced.
- 1984–86 Cordia, Tredia, Pickup: 13.7kΩ
- 1987–88 Cordia, Tredia, Pickup w/2.0L engine: 11.7–15.7kΩ
- 1987–88 Cordia, Tredia, Pickup w/1.8L turbo engine: 9.4–12.6kΩ
- 1984–86 Mirage: 13.7kΩ
- 1987–88 Mirage w/1.5L engine: 11.6–15.8kΩ
- 1987–88 Mirage w/1.6L engine: 9.4–12.6kΩ
- 1989 Mirage: 10.3–13.9kΩ

4. Inspect the unit for oil leaks and cracks in the coil tower, and replace it if any defects are noted.

1983 Cordia and Tredia
1983–86 Starion, Pickup, Montero with G54B Engine

Follow the procedure above, but use the following primary and secondary resistances below. Note that there is no external resistor.

Primary
- 1983 Cordia and Tredia: 1.04–1.27Ω
- 1983–86 Starion, Pickup, Montero: 1.04–1.27Ω
- 1987–88 Starion, Pickup, Montero: 1.25Ω
- 1989 Starion, Pickup, Montero: 1.12–1.38Ω

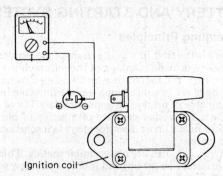

Checking primary coil resistance

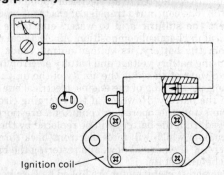

Checking secondary coil resistance

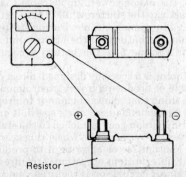

Checking resistance of external resistor

ENGINE AND ENGINE OVERHAUL 3

Secondary
- 1983 Cordia and Tredia: 7.1–9.6kΩ
- 1983–86 Starion, Pickup, Montero: 7.1–9.6kΩ
- 1987–88 Starion, Pickup, Montero: 11.0kΩ
- 1989 Starion, Pickup, Montero: 9.4–12.7kΩ

1985–87 Galant

Use the procedure given above, but substitute these resistance values:

Primary
- 1985–86 Galant: 1.2Ω
- 1987 Galant: 0.72–0.88Ω

Secondary
- 1985 Galant: 13.7kΩ
- 1986 Galant: 26.0kΩ
- 1987 Galant: 10.8–13.2kΩ

External
- 1985–86 Galant: 1.35Ω

SPARK TEST

1983 Starion, Cordia, Tredia, Pickup

1. Disconnect the high tension wire connecting the center of the distributor to the coil at the distributor. Hold the cable about 13mm (½ in.) away from the cylinder block. Hold the cable by the boot that normally seals the wire in the distributor using a suitable holding device such as spark plug wire pullers, and wear insulating gloves if you have them. Have someone crank the engine (make sure the parking brake is applied). The system should generate a fat blue spark if it is in proper operating condition. If there is spark, check spark plugs and ignition wires for defects, and then search for problems in the fuel system. If there is no spark, proceed with the next step.

2. Turn the ignition switch on and use a voltmeter to measure the voltage to the (−) terminal of the ignition coil (red lead to the coil terminal, black lead to a clean ground). If voltage is equal to battery voltage, proceed to Step 3. If it is 3 volts or less but the needle moves, proceed to 4, and if it is zero volts, proceed to 5.

3. Remove the ignition coil and substitute a known good coil. Again perform the test in Step 1. If there is now spark, the defect was in the coil. You may want to perform the coil tests below to avoid the need to buy a new coil.

4. The fault lies in the distributor pickup coil or igniter. Proceed to Step 5. in order to determine which.

5. First, measure the voltage to the (+) terminal of the coil. Make sure you actually put the positive meter probe on the coil connector stud. If there is 12 volts here, replace the coil. If there is no voltage or the voltage is low, disassemble and clean up the coil connection and reconnect it tightly; then retest. If necessary, search further for problems in the primary ignition wiring or ignition switch, and correct them.

6. Measure the resistance across the two prongs of the pickup coil with an ohmmeter. Resistance must be 920–1120 ohms. If the resistance is within this range, replace the ESC Igniter ; if resistance is beyond this range, replace the pickup coil.

1984–85 Starion, Pickup, Montero
1985 Mirage with 1.6L Turbo
1984–85 Cordia and Tredia with 1.8L Turbo

1. Disconnect the high tension wire connecting the center of the distributor to the coil at the distributor. Hold the cable about 13mm (½ in.) away from the cylinder block. Hold the cable with a suitable holding device such as a pair of spark plug wire pullers well behind the rubber boot that normally seals the wire in the distributor, and wear insulating gloves if you have them. Have someone crank the engine (make sure the parking brake is applied). The system should generate a fat blue spark if it is in proper operating condition. If there is spark, check spark plugs and ignition wires for defects, and then search for problems in the fuel system. If there is no spark, proceed with the next step.

2. Turn the ignition switch on and use a voltmeter to measure the voltage to the (−) terminal of the ignition coil (red lead to the coil terminal, black lead to a clean ground). If voltage is equal to battery voltage, proceed to Step 3. If it is 3 volts or less but the needle moves, proceed to 4, and if it is zero volts, proceed to 5.

3. Remove the ignition coil and substitute a known good coil. Again perform the test in Step 1. If there is now spark, the defect was in the coil. You may want to perform the coil tests below to avoid the need to buy a new coil.

4. The fault lies in the distributor igniter. Replace it.

5. First, measure the voltage to the (+) terminal of the coil. Make sure you actually put the positive meter probe on the coil connector stud. If there is 12 volts here, replace the coil. If there is no voltage or the voltage is low, disassemble and clean up the coil connection and reconnect it tightly; then retest. If necessary, search further for problems in the primary ignition wiring or ignition switch, and correct them.

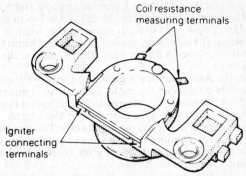

Measuring the pickup coil resistance on 1983 Cordia, Tredia and Starion

1984–85 Cordia, Tredia, Pickup w/2.0L Engine
1985 Mirage w/1.5L Engine
1985 Galant

1. Disconnect the high tension wire connecting the center of the distributor to the coil at the distributor. Hold the cable about 13mm (½ in.) away from the cylinder block. Hold the cable with a suitable holding device such as spark plug wire pullers well behind the rubber boot that normally seals the wire in the distributor, and wear insulating gloves if you have them. Have someone crank the engine (make sure the parking brake is applied). The system should generate a fat blue spark if it is in proper operating condition. If there is spark, check spark plugs and ignition wires for defects, and then search for problems in the fuel system. If there is no spark, proceed with the next step.

2. Turn the ignition switch on and use a voltmeter to read the voltage between the (+) terminal stud of the ignition coil and ground (red lead to the coil terminal, black lead to a good ground). If the voltage is equal to the battery voltage, (approx. 12 volts), proceed with the next test. If voltage is low or nonexistent, clean up the coil connection and tighten it. Proceed further as necessary to find the problem in the wiring or ignition switch.

3. With the ignition switch on, measure the voltage at the (−) terminal of the coil. If the voltage is equal to battery voltage (approx. 12 volts), or if voltage exists (the needle moves) but it is very low (about 1 volt), proceed to the next step. If there is no voltage, proceed to Step 5.

3 ENGINE AND ENGINE OVERHAUL

4. Install an ignition coil that is known to be good, and retest for spark. You may want to perform the coil test below to avoid the need to buy a new coil. If there is spark, the original coil must be replaced. If there is no spark, replace the pickup coil. You may want to proceed with the next step to confirm the results of your tests.

5. Measure the resistance of the pickup coil between the two wiring connectors directly on top of the coil with an ohmmeter. It must be 920–1120 ohms. If it is outside this range, replace it. If you got no voltage reading in Step 3, and the pickup coil checks out, you should now try a new ignition coil as described in Step 4. If these tests should fail to uncover the problem, you should check the wiring and connectors for corroded prongs, frayed insulation, or internal breakage.

Distributor

REMOVAL AND INSTALLATION

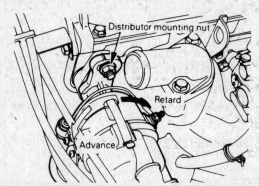

Typical location of the distributor mounting nut

1. Disconnect the battery ground cable. Disconnect the retaining clips and pull the distributor cap and seal off the distributor. Locate the cap and wires away from the distributor. Disconnect the vacuum advance line. Disconnect the distributor wiring connector.

2. Matchmark the location of the distributor on the cylinder block. A good place to do this is near where the mounting bolt passes through the slot. Then, matchmark the relationship between the tip of the distributor rotor and the distributor body.

3. Remove the mounting nut or bolt and pull the distributor carefully upward and out of the engine. Note the direction and degree to which the rotor turns as you pull it out. Mark the location of the rotor after it has turned, too.

4. If the engine has been rotated while the distributor is out, proceed to the next step. To install the distributor, position it so the distributor and block matchmarks are lined up. Now, position the rotor so it is lined up with the matchmark on the distributor body after the distributor was pulled part way out. Insert the distributor into the block until the gears at the bottom engage and then begin turning the rotor. If there is resistance, turn the rotor back and forth slightly so the gears mesh. Once the gears engage and inserting the distributor causes the rotor to turn, push the distributor in until it seats and the rotor is lined up with the first mark made on the body.

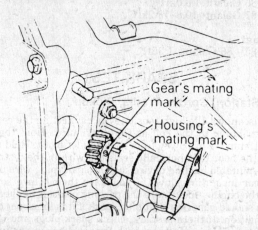

If the distributor has matchmarks already on it, set the engine to TDC No. 1 and align the housing mark with the gear mating mark

NOTE: **Later model distributors have alignment marks on the flange, housing and gear. Aligning these marks is more accurate than aligning the rotor with a homemade mark.**

5. If the engine has been rotated while the distributor was out, you'll have to first put the engine on No. 1 cylinder at Top Dead Center firing position. You can either remove the valve cover or No. 1 spark plug to determine engine position. Rotate the engine with a socket wrench on the nut at the center of the front pulley in the normal direction of rotation. Either feel for air being expelled forcefully through the spark plug hole or watch for the engine to rotate up to the Top Center mark without the valves moving (both valves will be closed or all the way up). If the valves are moving as you approach TDC or there is no air being expelled through the plug hole, turn the engine another full turn until you get the appropriate indication as the engine approaches TDC position.

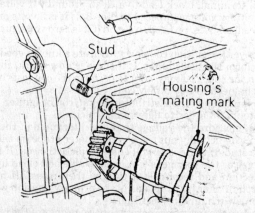

Align the groove or projection on the distributor flange wit the center of the mounting stud and install the distributor

6. Start the distributor into the block with the matchmarks between the distributor body and the block lined up. Turn the rotor slightly until the matchmarks on the bottom of the distributor body and the bottom of the distributor shaft near the gear are aligned. Then insert the distributor all the way into the block. If you have trouble getting the distributor and camshaft gears to mesh, turn the rotor back and forth very slightly until the distributor can be inserted easily. If the rotor is not lined up with the position of No. 1 plug terminal, you'll have to pull the distributor back out slightly, shift the position of the rotor appropriately, and then reinstall it.

7. Align the matchmarks between the distributor and block. Install the distributor mounting bolt and tighten it finger tight. Reconnect the vacuum advance line and distributor wiring connector, and reinstall the gasket and cap. Reconnect the negative battery cable. Adjust the ignition timing as described in Section 2. Tighten the distributor mounting bolt securely.

ENGINE AND ENGINE OVERHAUL 3

Alternator

The alternator charging system is a negative (−) ground system which consists of an alternator, a regulator, a charge indicator, a storage battery and wiring connecting the components, and fuse link wire.

The alternator is belt-driven from the engine. Energy is supplied from the alternator/regulator system to the rotating field through two brushes to two slip-rings. The slip-rings are mounted on the rotor shaft and are connected to the field coil. This energy supplied to the rotating field from the battery is called excitation current and is used to initially energize the field to begin the generation of electricity. Once the alternator starts to generate electricity, the excitation current comes from its own output rather than the battery.

The alternator produces power in the form of alternating current. The alternating current is rectified by 6 diodes into direct current. The direct current is used to charge the battery and power the rest of the electrical system.

When the ignition key is turned on, current flows from the battery, through the charging system indicator light on the instrument panel, to the voltage regulator, and to the alternator. Since the alternator is not producing any current, the alternator warning light comes on. When the engine is started, the alternator begins to produce current and turns the alternator light off. As the alternator turns and produces current, the current is divided in two ways: part to the battery to charge the battery and power the electrical components of the vehicle, and part is returned to the alternator to enable it to increase its output. In this situation, the alternator is receiving current from the battery and from itself. A voltage regulator is wired into the current supply to the alternator to prevent it from receiving too much current which would cause it to put out too much current. Conversely, if the voltage regulator does not allow the alternator to receive enough current, the battery will not be fully charged and will eventually go dead.

The battery is connected to the alternator at all times, whether the ignition key is turned on or not. If the battery were shorted to ground, the alternator would also be shorted. This would damage the alternator. To prevent this, a fuse link is installed in the wiring between the battery and the alternator. If the battery is shorted, the fuse link is melted, protecting the alternator.

ALTERNATOR PRECAUTIONS

To prevent damage to the alternator and regulator, the following precautions should be taken when working with the electrical system.

1. Never reverse the battery connections.
2. Booster batteries for starting must be connected properly: positive-to-positive and negative-to-negative. Remember that correct jumping procedure requires the final connection to be made to a good ground away from the failed battery. Look for a handy bracket or solid metal.
3. Disconnect the battery cables before using a fast charger; the charger will force current through the diodes in the opposite direction. This burns out the diodes.
4. Never use a fast charger as a booster for starting the vehicle.
5. Never disconnect the any component (voltage regulator, battery, alternator, etc) while the engine is running.
6. Avoid long soldering times when replacing diodes or transistors. Prolonged heat is damaging.
7. Do not use test lamps of more than 12 volts for checking diode continuity.
8. Do not short across or ground any of the terminals on the alternator.
9. The polarity of the battery, generator, and regulator must be matched and considered before making any electrical connections within the system.
10. Never operate the alternator on an open circuit. Make sure that all connections within the circuit are clean and tight.
11. Disconnect the battery terminals when performing any service on the electrical system. This will eliminate the possibility of accidental reversal of polarity.
12. Disconnect the battery ground cable if arc welding is to be done on any part of the car.

REMOVAL AND INSTALLATION

Cordia and Tredia
Pickup with G63B Engine

1. Turn off the ignition switch and disconnect both battery cables.
2. Loosen the support bolt and adjusting bolt, and then shift the alternator toward the engine so belt tension is relaxed. Remove the belt. On some models it will be necessary to loosen the power steering pump and belt and then remove the pump from its brackets.
3. Note locations of all connectors. Make a drawing, if necessary. Unplug electrical connectors and unscrew fastening nuts for terminal type connectors. Clean any dirty connections.
4. Remove the adjusting bolt. Remove the nut from the rear of the mounting bolt. Note that in the next step, shims located between the alternator and the rear of the engine front cover may fall out. Retain these shims for re-use. Supporting the alternator, slide the mounting bolt out and then remove the unit from the car.
5. To install the alternator, first position it so the mounting bolt can be inserted. If the same alternator is being re-used, insert the shims between the rear of the front cover and the rear hinge of the alternator body. Install the mounting bolt. If the alternator is a different one, you'll have to measure the clearance between the rear of the front cover and the section of alternator body that fits against it. Pull the alternator forward and use a flat feeler gauge. If the clearance exceeds 0.2mm (0.008 in.), insert shim(s) thick enough so there is a slight friction fit and they will not fall out when you let go of them. Put the shims between the front of the front cover and the rear of the front hinge on the alternator body. Make sure you have installed shims of adequate thickness (or that the clearance is within specification). Otherwise, the alternator body will be broken. Install the mounting bolt if it is not already in place, and then install the mounting bolt nut loosely.
6. Install the adjusting bolt, and then loosely install the nut at the rear of it. Install the belt and tighten it as described in the Section 1. Torque the adjusting bolt to 9–10 ft. lbs. and the mounting bolt and nut to 15–18 ft. lbs.
7. Reinstall the power steering pump and belt if it was removed. Adjust the belt to the proper tension. Tighten the power steering pump bolts to 18 ft. lbs.

Starion
Pickup and Montero with G54B Engine
WITHOUT AIR CONDITIONING

1. Turn off the ignition switch and disconnect both battery cables.
2. Loosen the support bolt and adjusting bolt, and then shift the alternator toward the engine so belt tension is relaxed. Remove the belt. On some models it will be necessary to loosen the power steering pump and belt and then remove the pump from its brackets.
3. Note locations of all connectors. Make a drawing, if necessary. Unplug electrical connectors and unscrew fastening nuts for terminal type connectors. Clean any dirty connections.
4. Remove the adjusting bolt. Remove the nut from the rear of the mounting bolt. Note that in the next step, shims located between the alternator and the rear of the engine front cover

3 ENGINE AND ENGINE OVERHAUL

may fall out. Retain these shims for re-use. Supporting the alternator, slide the mounting bolt out and then remove the unit from the car.

5. To install the alternator, first position it so the mounting bolt can be inserted. If the same alternator is being re-used, insert the shims between the rear of the front cover and the rear hinge of the alternator body. Install the mounting bolt. If the alternator is a different one, you'll have to measure the clearance between the rear of the front cover and the section of alternator body that fits against it. Pull the alternator forward and use a flat feeler gauge. If the clearance exceeds 0.2mm (0.008 in.), insert shim(s) thick enough so there is a slight friction fit and they will not fall out when you let go of them. Put the shims between the front of the front cover and the rear of the front hinge on the alternator body. Make sure you have installed shims of adequate thickness (or that the clearance is within specification). Otherwise, the alternator body will be broken. Install the mounting bolt if it is not already in place, and then install the mounting bolt nut loosely.

6. Install the adjusting bolt, and then loosely install the nut at the rear of it. Install the belt and tighten it as described in the Section 1. Torque the adjusting bolt to 9-10 ft. lbs. and the mounting bolt and nut to 15-18 ft. lbs.

7. Reinstall the power steering pump and belt if it was removed. Adjust the belt to the proper tension. Tighten the power steering pump bolts to 18 ft. lbs.

WITH AIR CONDITIONING

1. Disconnect the negative battery cable.
2. Safely discharge the air conditioning system.

CAUTION

Please re-read the Air Conditioning procedures in Section 1 so that the system may be properly discharged. Always wear eye protection and gloves when discharging the system. Observe NO SMOKING/NO OPEN FLAME rules.

3. Disconnect the discharge hose from the compressor. Cap the compressor port and plug the line immediately.
4. Disconnect the suction hose from the compressor. Cap the compressor port and plug the line immediately.
5. Disconnect the wiring connector to the compressor.
6. Loosen the belt tension adjuster and remove the belt from the compressor. Loosen and remove the lock bolt and the pivot bolt from the compressor. Support the compressor and remove it from the engine.
7. Loosen the upper and lower alternator bolts and pivot the alternator towards the engine. Release the belt tension for the alternator belt and remove the belt from the alternator.
8. Label and disconnect the wiring connections at the back of the alternator.
9. Remove the upper bolt. Remove the lower bolt while supporting the alternator. Be prepared to catch the bolt shims which will fall out when the bolt is removed. Remove the alternator from its mount.
10. Install the alternator in position and install the lower support bolt but do not install the nut. Install the upper bolt just enough to hold the bolt in place.
11. Push the alternator forward (towards the front of the car) and insert shims between the front leg of the alternator and the front case (mount). There should be enough shims to hold themselves in place when you let go. The alternator should still be movable with light pressure.
12. Install the shims between the alternator front leg and the front case. Tighten the upper and lower mounting bolts finger tight.
13. Install the alternator belt and adjust it to the correct tension. Tighten the upper mounting bolt to 10 ft. lbs. and the lower mounting bolt to 16 ft. lbs. Connect the wiring to the alternator.

14. Reinstall the compressor. Install the belt over the pulley (making sure the belt is still correctly mounted on the other pulleys) and gently snug the belt by moving the compressor on its mounts. As soon as there is a little tension on the belt, tighten the compressor mounting bolts.
15. Set the belt to its final tension by tightening the adjuster screw on the compressor. Do not pry on the compressor to adjust the belt.
16. Connect the compressor wiring harness.
17. Remove the plugs and caps from the A/C fittings. Connect the suction hose to its port on the compressor and then connect the discharge hose. Tighten the fitting finger tight, making very sure they are clean and correctly threaded. Tighten the suction hose to 24 ft. lbs. and tighten the discharge hose to 16 ft. lbs.
18. Evacuate and recharge the air conditioning system.
19. Double check the drive belts for proper tension and adjust as necessary. If a belt was replaced during repairs, it may need readjusting after 50-100 miles of driving.

1985-88 Mirage

WARNING: Perform this procedure on a cold motor only. If recently driven, allow the car to cool at least 3-4 hours before repairs.

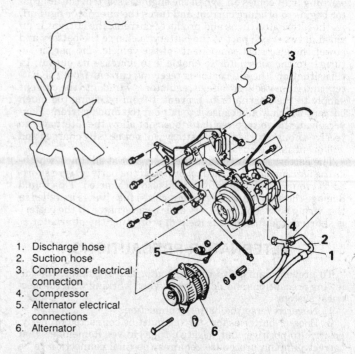

1. Discharge hose
2. Suction hose
3. Compressor electrical connection
4. Compressor
5. Alternator electrical connections
6. Alternator

Starion alternator and compressor

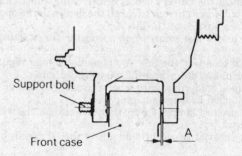

During reassembly, Starion alternator shims are placed at point A

ENGINE AND ENGINE OVERHAUL 3

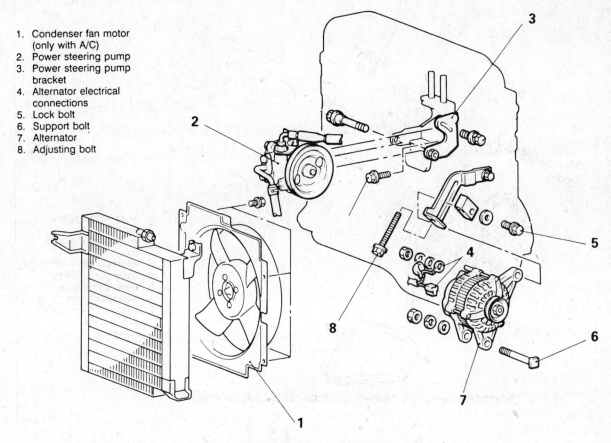

1. Condenser fan motor (only with A/C)
2. Power steering pump
3. Power steering pump bracket
4. Alternator electrical connections
5. Lock bolt
6. Support bolt
7. Alternator
8. Adjusting bolt

1985–88 Mirage: alternator removal

1. Disconnect the negative battery cable.
2. If equipped with air conditioning, disconnect the wiring connector running to the condenser (left side) fan. Remove the 4 bolts holding the fan and shroud assembly and remove it from the car.
3. Loosen the bolts holding the power steering pump to its bracket. Move the pump towards the engine and remove the belt. Remove the pump from the bracket and use a piece of stiff wire to hang the pump out of the way.

NOTE: Do not disconnect the pump hoses. Take care not to bend or twist the hoses when hanging up the pump.

4. Remove the power steering pump bracket.
5. Label and disconnect the electrical connections at the alternator
6. Loosen the adjusting bolt and remove the alternator belt.
7. Remove the upper (lock) bolt from the alternator.
8. Support the alternator and remove the lower (support) bolt. Remove the alternator from the engine.
9. Reinstall the alternator and install the support bolt and the lock bolt loosely.
10. Connect the wiring to the alternator.
11. Install the belt, making sure it is properly seated on all the pulleys.
12. Gently pull the alternator away from the engine, taking tension on the drive belt. When the belt is snug, hold the alternator in that position and tighten the lock bolt to 10 ft. lbs. and the support bolt to 16 ft. lbs. Use the belt adjusting bolt to adjust the belt to its final tension.

WARNING: Do not pry on the alternator with tools or prybars.

13. Install the power steering pump bracket to the engine and tighten its mounting bolts to 18 ft. lbs.
14. Place the power steering pump in the bracket and install its bolts loosely. Install the belt and move the pump outward to draw the correct tension on the belt. Tighten the upper pump bolt to 20 ft. lbs. and the lower support bolt to 16 ft. lbs.
15. Install the condenser fan and shroud assembly and tighten the 4 bolts. Connect the fan wiring to the harness.
16. Connect the negative battery cable.
17. Double check the drive belts for proper tension and adjust as necessary. If a belt was replaced during repairs, it may need readjusting after 50–100 miles of driving.

1989–90 Mirage with 1468cc Single Cam Engine

1. Disconnect the negative battery cable.
2. Elevate and safely support the vehicle. Remove the left side splash shield under the engine to allow access to the lower bolts.
3. Loosen the air conditioning compressor belt adjusting bolt and remove the belt.
4. Loosen the power steering pump bolts and remove the belt.
5. Loosen the upper and lower alternator mounting bolts, swing the alternator towards the engine and remove the belt.
6. Remove the 4 small bolts in the center of the outer water pump pulley and remove both pulleys. Remove the bolt holding the alternator brace to the engine and remove the alternator brace.
7. Disconnect the wiring from the back of the alternator.

3 ENGINE AND ENGINE OVERHAUL

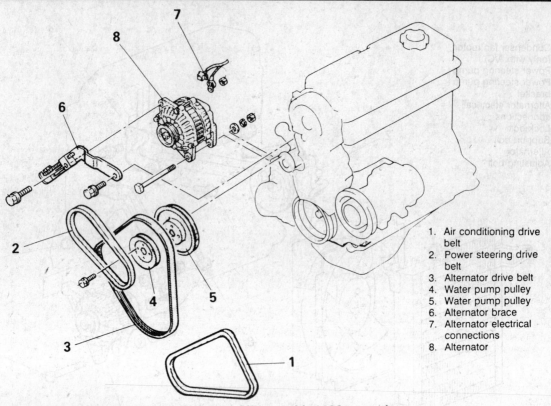

1. Air conditioning drive belt
2. Power steering drive belt
3. Alternator drive belt
4. Water pump pulley
5. Water pump pulley
6. Alternator brace
7. Alternator electrical connections
8. Alternator

Alternator removal: 1989–90 Mirage with 1498cc engine

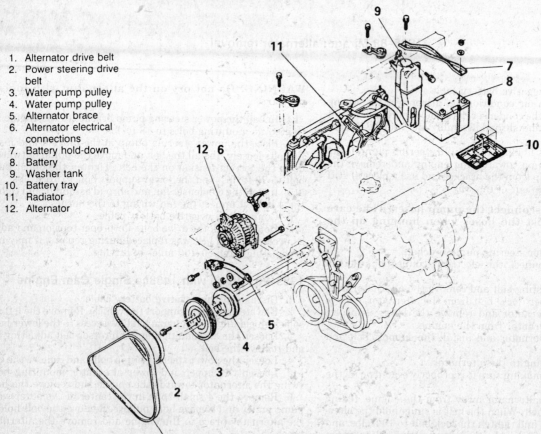

1. Alternator drive belt
2. Power steering drive belt
3. Water pump pulley
4. Water pump pulley
5. Alternator brace
6. Alternator electrical connections
7. Battery hold-down
8. Battery
9. Washer tank
10. Battery tray
11. Radiator
12. Alternator

Alternator removal: 1989–90 Mirage with 1595cc engine

ENGINE AND ENGINE OVERHAUL 3

8. Remove the lower support bolt and remove the alternator.
9. Reinstall the alternator and its support bolt. Tighten the bolt finger tight.
10. Install the alternator bracket. Tighten the bolt holding the bracket to the engine to 18 ft. lbs. Tighten the bolt holding the bracket to the alternator to 10 ft. lbs. Connect the wiring to the alternator.
11. Install the two water pump pulleys. Tighten the bolts alternately and evenly to 6 ft. lbs. Do not overtighten these bolts.
12. Install the alternator belt and adjust it to the proper tension.
13. Install the power steering drive belt and adjust it to the proper tension.
14. Install the compressor belt and use the adjuster bolt to set the correct belt tension.
15. Install the left side splash shield under the engine. Lower the vehicle to the ground.
16. Connect the negative battery cable.
17. Double check the drive belts for proper tension and adjust as necessary. If a belt was replaced during repairs, it may need readjusting after 50–100 (80–160 km) miles of driving.

1989–90 Mirage with 1595cc Twin Cam Engine

WARNING: Perform this procedure on a cold motor only. If recently driven, allow the car to cool at least 3–4 hours before repairs.

1. Disconnect the negative battery cable, then disconnect the positive battery cable. The battery will be removed later in the procedure.
2. Elevate and safely support the vehicle. Remove the left side splash shield under the engine to allow access to the lower bolts.
3. Loosen the upper and lower alternator mounting bolts, swing the alternator towards the engine and remove the belt.
4. Loosen the power steering pump bolts and remove the belt.
5. Remove the 4 small bolts in the center of the outer water pump pulley and remove both pulleys. Remove the bolt holding the alternator brace to the engine and remove the alternator brace.
6. Disconnect the wiring from the back of the alternator.
7. Remove the battery hold-down and remove the battery.
8. Remove the 2 bolts holding the washer solvent tank, remove the tank and remove the battery tray.
9. Remove the attaching bolts at the top of the radiator and remove the brackets. Label and disconnect the wiring running to the fan(s).
10. Without disconnecting any hoses, lift the radiator straight up and let it lean back towards the engine. The hoses will tolerate some bending but only move the radiator enough to gain clearance for the alternator.
11. Remove the bottom alternator bolt and remove the alternator. Once the alternator is clear of the car, position the radiator correctly to keep tension off the hoses.

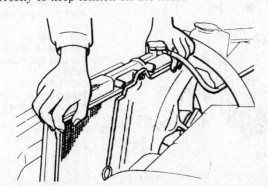

The Mirage radiator can be lifted and repositioned with the hoses attached

12. When reinstalling, carefully lift the radiator and install the alternator on its mount. Tighten the support bolt to 16 ft. lbs.
13. Push the radiator into its correct position, making sure it is seated at the bottom. Install the upper radiator supports and tighten them. Connect the fan wiring to the fan(s).
14. Install the battery tray and washer tank.
15. Install the battery and battery hold-down bracket. Do not connect the battery cables at this time.
16. Connect the alternator wiring.
17. Install the alternator brace. Tighten the bolt holding the brace to the engine to 18 ft. lbs. and the bolt holding the brace to the alternator to 10 ft. lbs.
18. Install the two water pump pulleys, tightening the 4 bolts alternately and evenly to 6 ft. lbs. Do not overtighten these bolts.
19. Install the power steering belt and tighten it to the correct tension.
20. Install the alternator drive belt and tighten it to the correct tension.
21. Install the left splash shield under the motor and lower the vehicle to the ground.
22. Connect the postive battery cable, then connect the negative battery cable.
23. Double check the drive belts for proper tension and adjust as necessary. If a belt was replaced during repairs, it may need readjusting after 50–100 miles of driving.

Galant, 1985–87

WARNING: Perform this procedure on a cold motor only. If recently driven, allow the car to cool at least 3–4 hours before repairs.

1. Disconnect the negative battery cable.
2. If equipped with air conditioning, disconnect the wiring connector running to the condenser (left side) fan. Remove the 4 bolts holding the fan and shroud assembly and remove it from the car.
3. Loosen the bolts holding the power steering pump to its bracket. Move the pump towards the engine and remove the belt. Remove the pump from the bracket and use a piece of stiff wire to hang the pump out of the way.

NOTE: Do not disconnect the pump hoses. Take care not to bend or twist the hoses when hanging up the pump.

4. Label and disconnect the electrical connections at the alternator
5. Loosen the lock bolt and remove the alternator belt.
6. Remove the upper (lock) bolt from the alternator.
7. Support the alternator and remove the nut and washers from the lower (support) bolt. Slide the bolt outward to remove it, noting that it will not come all the way out. The bracket that restricts the bolt head has a hole cut in it under a layer of tape. Remove the tape and then remove the bolt. Remember to catch the shims as the bolt is removed.
8. With the lower bolt removed, the alternator may be removed from the car.
9. Reinstall the alternator and install the support bolt and the lock bolt loosely.
10. Connect the wiring to the alternator. Place a piece of tape over the support bolt escape hole.
11. Push the alternator forward (towards the front of the car) and insert shims between the front leg of the alternator and the front case (mount). There should be enough shims to hold themselves in place when you let go. The alternator should still be movable with light pressure.
12. Install the shims between the alternator front leg and the front case. Tighten the upper and lower mounting bolts finger tight. Install the belt, making sure it is properly seated on all the pulleys.

3-11

3 ENGINE AND ENGINE OVERHAUL

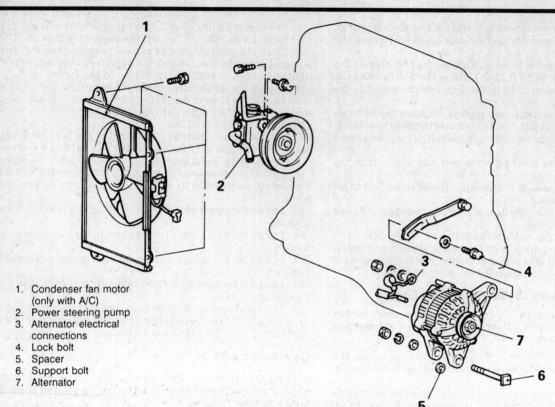

1. Condenser fan motor (only with A/C)
2. Power steering pump
3. Alternator electrical connections
4. Lock bolt
5. Spacer
6. Support bolt
7. Alternator

1985–87 Galant, alternator removal

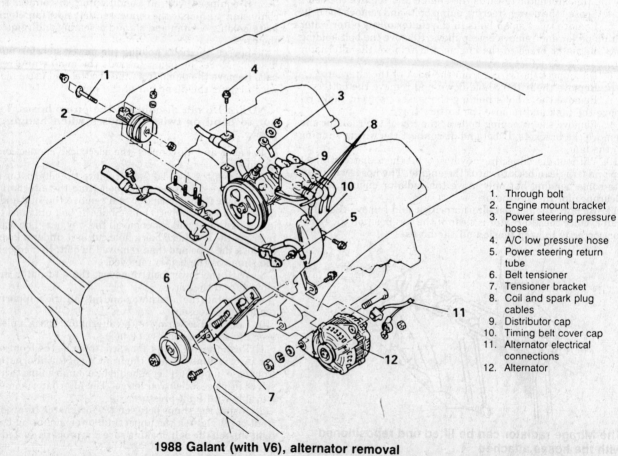

1. Through bolt
2. Engine mount bracket
3. Power steering pressure hose
4. A/C low pressure hose
5. Power steering return tube
6. Belt tensioner
7. Tensioner bracket
8. Coil and spark plug cables
9. Distributor cap
10. Timing belt cover cap
11. Alternator electrical connections
12. Alternator

1988 Galant (with V6), alternator removal

3-12

ENGINE AND ENGINE OVERHAUL 3

13. Gently pull the alternator away from the engine, taking tension on the drive belt. When the belt is snug, hold the alternator in that position and tighten the lock bolt to 10 ft. lbs. and the support bolt to 16 ft. lbs.
13. Place the power steering pump in the bracket and install its bolts loosely.
14. Install the belt and move the pump outward to draw the correct tension on the belt. Tighten the pump bolts to 20 ft. lbs.
15. Install the condenser fan and shroud assembly and tighten the 4 bolts. Connect the fan wiring to the harness.
16. Connect the negative battery cable.
17. Double check the drive belts for proper tension and adjust as necessary. If a belt was replaced during repairs, it may need readjusting after 50–100 miles of driving.

1988 Galant

1. Disconnect the negative battery cable.
2. Elevate and safely support the front of the car. Set the parking brake and block the rear wheels.
3. Remove the left front wheel.
4. Place a floor jack and a broad piece of wood under the engine oil pan. Make certain the wood is positioned to distribute the load evenly. Raise the jack just to the point of tension but not enough to raise the engine.
5. Remove the large through bolt from the left engine mount. Remove the 4 nuts holding the mount to the bracket and remove the mount.
6. Raise the jack until the maximum clearance is obtained between the engine and the left shock tower.

---- **CAUTION** ----

The car is elevated. Raise the jack only enough to move the engine; do not disturb the position of the car on the jackstands.

---

7. Remove the power steering pressure hose from the power steering pump. Be prepared to contain any spillage from the hose.
8. If equipped with air conditioning, remove the bracket bolts holding the low pressure hose to the power steering pump. Move the hose towards the center of the car.
9. Remove the bracket bolts holding the power steering return line and move the line out of the way.
10. Remove the drive belt tensioner pulley, then remove the tensioner bracket.
11. Label and disconnect the spark plug wires for cylinder Nos. 2, 4 and 6 (rear bank) from the distributor cap. Disconnect the coil wire.
12. Remove the distributor cap.
13. Remove the small cover cap from the timing belt cover.

WARNING: The timing belt will be exposed after the cap is removed. Great care must be taken to protect the belt from fluids and grease. It is recommended that a small protective shield such as a piece of cardboard be installed over the belt. If tape is used to secure the shield, make certain that no adhesive contacts the belt.

14. Remove the electrical connections at the alternator.
15. Remove the upper and lower bolts and remove the alternator. Be very careful not to damage neighboring components while removing the alternator.
16. When reinstalling, place the alternator on its bracket and install the upper and lower bolts. Tighten the bolts to 15 ft. lbs.
17. Connect the wiring to the alternator.
18. Remove the protective shield and install the timing belt cover cap, tightening the bolts to 96 inch lbs.
19. Install the distributor cap and connect the coil and spark plug wires.
20. Install the belt tensioner bracket, tightening the bolts to 30 ft. lbs.
21. Install the tensioner pulley, tightening the nut to 35 ft. lbs.

Install the belt onto the alternator and tensioner, making sure it is properly seated.
22. Reposition the power steering return tube and install the bracket bolts, tightening them to 8 ft. lbs.
23. Reposition the air conditioner low pressure hose and secure the bracket bolt.
24. Connect the power steering pressure hose to the pump and tighten the nut to 15 ft. lbs.
25. Install the engine mount onto its bracket; tighten the 4 nuts snug but not tight.
26. Lower the jack slowly until the holes for the through bolt align. Insert the through bolt and tighten just snug. Remove the jack from the engine.
27. Install the left front wheel and lower the vehicle to the ground.
28. Bounce the front end 2 or 3 times and allow the car to seek its normal "at rest" position. Final tighten the engine mount bracket nuts to 50 ft. lbs. Tighten the nut on the rear of the through bolt to 50 ft. lbs. and the nut on the front to 25 ft. lbs.

NOTE: These torque values are important. The engine mount absorbs vibration and must be correctly tightened to do its job.

29. Adjust the drive belt to the correct tension, using the belt adjuster.
30. Connect the negative battery cable.

1989–90 Galant

1. Disconnect the negative battery cable.
2. Elevate and safely support the vehicle. Remove the splash shield under the engine to allow access to the lower bolts.
3. Loosen the adjuster and remove the alternator drive belt.
4. Loosen the power steering pump bolts and remove the belt.
5. Remove the 4 small bolts in the center of the outer water pump pulley and remove both pulleys. Remove the bolt holding the alternator brace to the engine and remove the alternator brace.

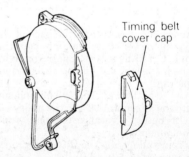

Always protect the timing belt when the cap is removed

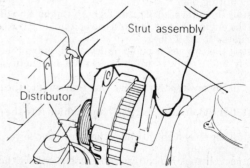

The alternator will fit between the engine and the strut tower. Be careful not to damage nearby components

3-13

3 ENGINE AND ENGINE OVERHAUL

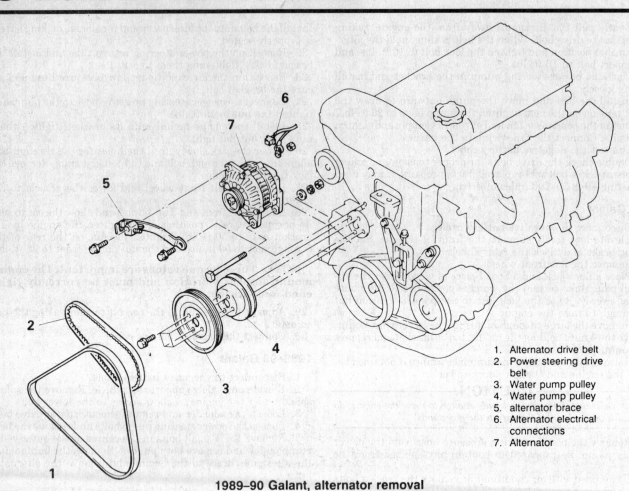

1. Alternator drive belt
2. Power steering drive belt
3. Water pump pulley
4. Water pump pulley
5. alternator brace
6. Alternator electrical connections
7. Alternator

1989–90 Galant, alternator removal

6. Disconnect the wiring from the back of the alternator.
7. Remove the power steering pump mounting bolt and support the pump. Carefully move the pump out of the way with the hoses attached. Place a rag on the rocker cover and rest the pump on the rag.
8. Remove the bottom alternator bolt and support the alternator. Rotate the alternator 90° (so the pulley end is up) and remove the alternator. Clearance is small; don't damage surrounding components.
9. When reinstalling, fit the alternator on its mount and tighten the support bolt to 16 ft. lbs.
10. Connect the alternator wiring.
11. Reinstall the power steering pump to its bracket and tighten the mounting bolt.
12. Install the alternator brace. Tighten the bolt holding the brace to the engine to 18 ft. lbs. and the bolt holding the brace to the alternator to 10 ft. lbs.
13. Install the two water pump pulleys, tightening the 4 bolts alternately and evenly to 6 ft. lbs. Do not overtighten these bolts.
14. Install the power steering belt and tighten it to the correct tension.
15. Install the alternator drive belt and tighten it to the correct tension.
16. Install the splash shield under the motor and lower the vehicle to the ground.
17. Connect the postive battery cable, then connect the negative battery cable.
18. Double check the drive belts for proper tension and adjust as necessary. If a belt was replaced during repairs, it may need readjusting after 50–100 miles of driving.

Precis

1. Turn off the ignition switch and disconnect both battery cables.
2. Loosen the support bolt and adjusting bolt, and then shift the alternator toward the engine so belt tension is relaxed. Remove the belt. On some models it will be necessary to loosen the power steering pump and belt and then remove the pump from its brackets.
3. Note locations of all connectors. Make a drawing, if necessary. Unplug electrical connectors and unscrew fastening nuts for terminal type connectors. Clean any dirty connections.
4. Remove the adjusting bolt. Remove the nut from the rear of the mounting bolt. Note that in the next step, shims located between the alternator and the rear of the engine front cover may fall out. Retain these shims for re-use. Supporting the al-

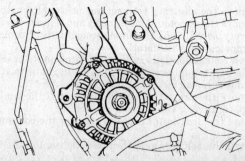

The alternator must be rotated 90° to be removed

ENGINE AND ENGINE OVERHAUL 3

ternator, slide the mounting bolt out and then remove the unit from the car.

5. To install the alternator, first position it so the mounting bolt can be inserted. If the same alternator is being re-used, insert the shims between the rear of the front cover and the rear hinge of the alternator body. Install the mounting bolt. If the alternator is a different one, you'll have to measure the clearance between the rear of the front cover and the section of alternator body that fits against it. Pull the alternator forward and use a flat feeler gauge. If the clearance exceeds 0.2mm (0.008 in.), insert shim(s) thick enough so there is a slight friction fit and they will not fall out when you let go of them. Put the shims between the front of the front cover and the rear of the front hinge on the alternator body. Make sure you have installed shims of adequate thickness (or that the clearance is within specification). Otherwise, the alternator body will be broken. Install the mounting bolt if it is not already in place, and then install the mounting bolt nut loosely.

6. Install the adjusting bolt, and then loosely install the nut at the rear of it. Install the belt and tighten it as described in the Section 1. Torque the adjusting bolt to 9–10 ft. lbs. and the mounting bolt and nut to 15–18 ft. lbs.

7. Reinstall the power steering pump and belt if it was removed. Adjust the belt to the proper tension. Tighten the power steering pump bolts to 18 ft. lbs.

Sigma V6
Pickup and Montero V6

1. Disconnect the negative battery cable.
2. Elevate and safely support the front of the car. Set the parking brake and block the rear wheels.
3. Remove the left front wheel.
4. Place a floor jack and a broad piece of wood under the engine oil pan. Make certain the wood is positioned to distribute the load evenly. Raise the jack just to the point of tension but not enough to raise the engine.
5. Remove the large through bolt from the left engine mount. Remove the 4 nuts holding the mount to the bracket and remove the mount.
6. Raise the jack until the maximum clearance is obtained between the engine and the left shock tower.

---- **CAUTION** ----
The car is elevated. Raise the jack only enough to move the engine; do not disturb the position of the car on the jackstands.

---

7. Remove the power steering pressure hose from the power steering pump. Be prepared to contain any spillage from the hose.
8. If equipped with air conditioning, remove the bracket bolts holding the low pressure hose to the power steering pump. Move the hose towards the center of the car.
9. Remove the bracket bolts holding the power steering return line and move the line out of the way.
10. Remove the drive belt tensioner pulley, then remove the tensioner bracket.
11. Label and disconnect the spark plug wires for cylinder Nos. 2,4 and 6 (rear bank) from the distributor cap. Disconnect the coil wire.
12. Remove the distributor cap.
13. Remove the small cover cap from the timing belt cover.

WARNING: The timing belt will be exposed after the cap is removed. Great care must be taken to protect the belt from fluids and grease. It is recommended that a small protective shield such as a piece of cardboard be installed over the belt. If tape is used to secure the shield, make certain that no adhesive contacts the belt.

14. Remove the electrical connections at the alternator.
15. Remove the upper and lower bolts and remove the alterna-

tor. Be very careful not to damage neighboring components while removing the alternator.
16. When reinstalling, place the alternator on its bracket and install the upper and lower bolts. Tighten the bolts to 15 ft. lbs.
17. Connect the wiring to the alternator.
18. Remove the protective shield and install the timing belt cover cap, tightening the bolts to 96 inch lbs.
19. Install the distributor cap and connect the coil and spark plug wires.
20. Install the belt tensioner bracket, tightening the bolts to 30 ft. lbs.
21. Install the tensioner pulley, tightening the nut to 35 ft. lbs. Install the belt onto the alternator and tensioner, making sure it is properly seated.
22. Reposition the power steering return tube and install the bracket bolts, tightening them to 8 ft. lbs.
23. Reposition the air conditioner low pressure hose and secure the bracket bolt.
24. Connect the power steering pressure hose to the pump and tighten the nut to 15 ft. lbs.
25. Install the engine mount onto its bracket; tighten the 4 nuts snug but not tight.
26. Lower the jack slowly until the holes for the through bolt align. Insert the through bolt and tighten just snug. Remove the jack from the engine.
27. Install the left front wheel and lower the vehicle to the ground.
28. Bounce the front end 2 or 3 times and allow the car to seek its normal "at rest" position. Final tighten the engine mount bracket nuts to 50 ft. lbs. Tighten the nut on the rear of the through bolt to 50 ft. lbs. and the nut on the front to 25 ft. lbs.

NOTE: These torque values are important. The engine mount absorbs vibration and must be correctly tightened to do its job.

29. Adjust the drive belt to the correct tension, using the belt adjuster.
30. Connect the negative battery cable.

Pickup and Montero
with the 4D55 Diesel

1. Make certain the ignition switch is **OFF**; disconnect the negative battery cable.
2. Disconnect the cable from terminal B of the alternator.
3. Remove the electrical connector form the back of the alternator.
4. Label and disconnect the oil hose, vacuum hose and return hose from the vacuum pump. Due to difficult access, it is usually easier to disconnect the oil return hose at the oil pan.
5. Remove the brace bolt and the support bolt nut; loosen the alternator.
6. Remove the drive belt from the alternator pulley.

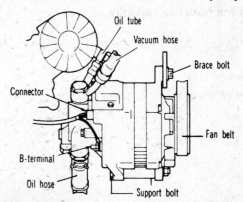

The 4D55 alternator

3-15

3 ENGINE AND ENGINE OVERHAUL

7. Support the alternator and pump. Remove the through bolt and remove the unit from the car.
8. Remove the vacuum housing bolts and then remove the housing.
9. Remove the rotor and vane from the shaft. Use great care not to damage the vane or O-ring. Replace any piece showing signs of damage or wear.
10. Use a roll of black PVC (electrical) tape to wind around the shaft splines. This prevents possible damage to the oil seal while the rotor assembly is removed.
11. At this point the alternator may be further disassembled (if desired) using normal procedures.
12. When reassembling, coat the O-ring with a light coat of white grease to hold it in place. Make certain it does not protrude from the groove.
13. Install the vane(s) with the rounded end outward. Never re-use a vane with a damaged or chipped end. Remove the black tape from the splines just before the shaft is inserted.
14. Install the housing carefully. Gently push the housing in the direction shown in the illustration; this minimizes clearance at critical points. The performance of the pump is affected by the installation of the housing.
15. Tighten the bolts evenly while keeping light pressure as shown.
16. If it is necessary to bench test the alternator after reassembly, make certain the pump is properly lubricated before turning the alternator. Take great care to insure the vacuum pump is never operated without lubrication.
17. Install the oil return hose to the vacuum pump and install the alternator in position. Once the unit is in place, it is impossible to attach the hose properly. Insert the through-bolt from the rear.
18. Install the drive belt onto the pulley; install the brace (upper) bolt.
19. Push the alternator towards the front of the engine and check for proper clearance between the alternator leg and the front case. If clearance is greater than 0.2mm (0.008 in) insert spacers to reduce clearance.

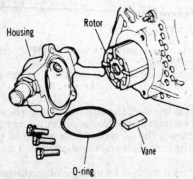

Remove the housing carefully

Always protect the shaft with tape

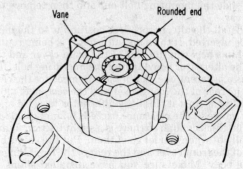

Correct installation of vanes

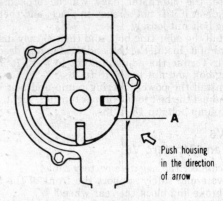

The housing must be pushed lightly in the direction shown

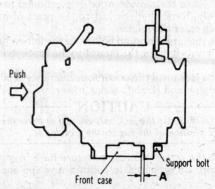

Do not tighten the support bolt if clearance A exceeds 0.2mm (0.008 in). Damage to the alternator will result

WARNING: If the support bolt is tightened without reducing the clearance, the alternator leg may be broken off.

20. Connect the oil delivery tube, the vacuum hose and the electrical connector.
21. Connect the wiring to terminal **B**.
22. Adjust the belt tension correctly.
23. Connect the negative battery terminal.

Voltage regulator

In all cases, the solid-state voltage regulator is contained within the alternator. This allows compact, efficient design and function. It does not allow ease of replacement without complete disassembly of the alternator. If replacement of the regulator is

ENGINE AND ENGINE OVERHAUL 3

Troubleshooting Basic Charging System Problems

Problem	Cause	Solution
Noisy alternator	• Loose mountings • Loose drive pulley • Worn bearings • Brush noise • Internal circuits shorted (High pitched whine)	• Tighten mounting bolts • Tighten pulley • Replace alternator • Replace alternator • Replace alternator
Squeal when starting engine or accelerating	• Glazed or loose belt	• Replace or adjust belt
Indicator light remains on or ammeter indicates discharge (engine running)	• Broken fan belt • Broken or disconnected wires • Internal alternator problems • Defective voltage regulator	• Install belt • Repair or connect wiring • Replace alternator • Replace voltage regulator
Car light bulbs continually burn out—battery needs water continually	• Alternator/regulator overcharging	• Replace voltage regulator/alternator
Car lights flare on acceleration	• Battery low • Internal alternator/regulator problems	• Charge or replace battery • Replace alternator/regulator
Low voltage output (alternator light flickers continually or ammeter needle wanders)	• Loose or worn belt • Dirty or corroded connections • Internal alternator/regulator problems	• Replace or adjust belt • Clean or replace connections • Replace alternator or regulator

indicated, either take the alternator to a shop specializing in automotive electrical repairs or replace the entire alternator as a unit.

Battery

REMOVAL AND INSTALLATION

1. Turn off the ignition switch. Disconnect the negative battery terminal, then disconnect the positive battery terminal. Be careful to keep the wrench from connecting the two battery terminals when loosening the nuts that tighten the battery connectors. If the connectors are difficult to pull off, either use a special puller designed for this purpose or carefully pry the halves of the connectors apart with a screwdriver. The connectors are very soft metal and are easily damaged.
2. Loosen and remove the retaining nuts, remove the battery mounting bracket, and remove the battery.

— CAUTION —
Battery fluid contains sulphuric acid, capable of causing skin and eye burns. Keep the battery upright at all times and wear heavy rubber gloves if any leakage is evident.

3. Clean the surface of the battery of grease and dust and inspect it for cracks while it is out of the car. If the case is cracked, the battery should be replaced. Clean the battery box or tray with a solution of baking soda and water to remove corrosion. Clean the top of the battery with the same solution, rinse it with clear water and dry it with clean rags.

NOTE: *If the battery is to be charged while out of the car, place it on wooden blocks or a workbench. Do not charge the battery while it is sitting on a concrete floor. The concrete will absorb heat and prevent the battery from charging.*

4. Reposition the battery and install the mounting brackets and nuts. Make sure the battery is mounted tightly to prevent vibration damage. Clean the battery terminal posts and connectors with a brush designed for this purpose. Reassemble the connectors (make sure red goes to (+) and black or ground goes to (−). Tighten the connectors securely. Then coat both connectors with petroleum grease.

Starter

DIAGNOSIS

Starter Won't Crank The Engine

1. Dead battery.
2. Open starter circuit, such as:
   a. Broken or loose battery cables.
   b. Inoperative starter motor solenoid.
   c. Broken or loose wire from ignition switch to solenoid.
   d. Poor solenoid or starter ground.
   e. Bad ignition switch.
3. Defective starter internal circuit, such as:
   a. Dirty or burnt commutator.
   b. Stuck, worn or broken brushes.
   c. Open or shorted armature.
   d. Open or grounded fields.

3-17

3 ENGINE AND ENGINE OVERHAUL

ALTERNATOR SPECIFICATIONS

Model	Year	Engine		Ampere Rating	Regulator	Regulated Voltage @ 68°F
Cordia/Tredia	1983	G62B		65A	Internal Electronic	14.1–14.7
	1985–87	G62B or G63B		65A	Internal Electronic	13.9–14.9
	1988	ALL		65A	Internal Electronic	14.2–15.2
Starion	1983–84	G54B		65A	Internal Electronic	14.1–14.7
	1985–86	G54B		65A	Internal Electronic	14.4–15.0
	1987	G54B	Intercooled	75A	Internal Electronic	13.9–14.9
			Non-Intercooled	65A		
	1988–89	G54B		75A	Internal Electronic	13.9–14.9
Mirage	1985–86	G15B	Manual	50A	Internal Electronic	14.4–15.0
			Automatic	55A		14.4–15.0
		G32B		65A	Internal Electronic	14.4–15.0
	1987–88	G15B	Manual	60A	Internal Electronic	14.2–15.2
			Automatic	65A		14.2–15.2
	1987–88	G32B	Manual	75A	Internal Electronic	14.2–15.2
	1989	4G15 or 4G61		75A	Internal Electronic	13.9–14.9
Galant	1985–86	G64B		65A	Internal Electronic	14.4–15.0
	1987	G64B		65A	Internal Electronic	13.9–14.9
	1988	6G72		75A	Internal Electronic	13.9–14.9
	1989	4G63		75A	Internal Electronic	13.9–14.9
Precis	1988	G15B	Manual	60A ①	Internal Electronic	14.4–15.0
			Automatic	55A ②		14.4–14.6
	1989	G15B	Manual	55A ③	Internal Electronic	14.4–15.0
			Automatic	60A ①		14.4–15.0
Sigma	1989	6G72		75A	Internal Electronic	13.9–14.9
Montero	1983	G54B		45	Internal Electronic	13.9–14.9
	1984–86	G54B	Manual	50	Internal Electronic	13.9–14.9
			Automatic	55		
	1987–88	G54B		50	Internal Electronic	13.9–14.9
	1989	G54B		50	Internal Electronic	13.9–14.9
	1989	6G72		75	Internal Electronic	13.9–14.9
Pickup	1983–85	G63B		45	Internal Electronic	1983–84: 13.9–14.9
		G54B		45	Internal Electronic	1985: 14.1–14.7
		4D55		50	Internal Electronic	
	1986–89	G63B		45	Internal Electronic	14.1–14.7
		G54B		45	Internal Electronic	14.1–14.7

① Melco
② Bosch
③ Mando

Note: Similar Amp ratings or voltage range does not mean alternators are interchangable. Always check part numbers and pulley size before changing alternators.

3–18

ENGINE AND ENGINE OVERHAUL 3

4. Starter motor mechanical faults, such as:
   a. Jammed armature end bearings.
   b. Bad bearings, allowing armature to rub fields.
   c. Bent shaft.
   d. Broken starter housing.
   e. Bad starter drive mechanism.
   f. Bad starter drive or flywheel-driven gear.
5. Engine hard or impossible to crank, such as:
   a. Hydrostatic lock, water in combustion chamber.
   b. Crankshaft seizing in bearings.
   c. Piston or ring seizing.
   d. Bent or broken connecting rod.
   e. Seizing of connecting rod bearings.
   f. Flywheel jammed or broken.

Starter Spins Freely, Won't Engage

1. Sticking or broken drive mechanism.
2. Damaged ring gear.

REMOVAL AND INSTALLATION

1. Disconnect the negative battery cable. Label and disconnect all wiring connectors at the starter.
2. Remove the two starter mounting bolts and remove the starter.
3. Clean the surfaces of the starter motor flange and the flywheel housing where the starter attaches.
4. Reinstall the starter and install the retaining bolts. Tighten the bolts to 16-23 ft. lbs.
5. Connect the wiring to the starter, making sure the terminals and connectors are clean and tight.
6. Connect the negative battery cable.

SOLENOID REPLACEMENT

Under some circumstances, the solenoid or magnetic switch may fail. This can be indicated by a failure of the starter to engage or produce adequate power. To test solenoid failure, disconnect the heavy starter motor wire at the **M** terminal on the solenoid. Run a jumper wire from the battery (+) terminal to the **S** terminal of the solenoid. If possible, use a remote starter control with a switch built into it. If this is not available, touch the wire to the **S** terminal briefly but do not leave the solenoid engaged for more than 10 seconds.

If the solenoid now engages in a positive manner, inspect the wiring and ignition switch for defects. (You have proven that the solenoid works if it gets the correct message.) If not, the solenoid should be replaced. You can test the solenoid switch itself by pulling the coil wire out of the distributor cap, having someone engage the starter, and then measuring the voltage at both the **B** and **M** terminals of the solenoid (**M** terminal wire connected). If voltage is close to battery voltage at the **B** terminal, but drops significantly at the **M** terminal with the starter turning, the solenoid switch is bad and the solenoid unit will have to be replaced.

1. Remove the starter from the car as described above. Disconnect the starter motor wire at the **M** terminal of the solenoid.
2. Remove the screw(s) from the front end of the solenoid. Disengage the solenoid plunger from the yoke inside the front of the starter and then remove the solenoid and the shims located between the solenoid and the starter front frame. Note the number and position of these shims--they are important and will be needed during reassembly. If you're replacing the solenoid, make sure you get extra shims.
3. Install the solenoid, making sure the plunger engages the drive yoke. Install the same number of shims.
4. Energize the solenoid by running jumper wires — including a switch if possible — from the (+) terminal of a 12 volt battery to the **S** terminal of the solenoid and from the (–) terminal

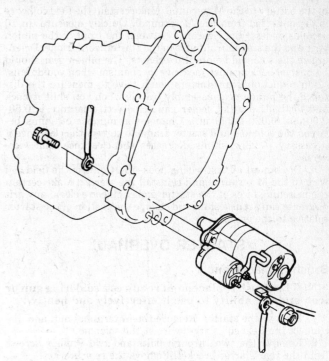

Typical starter installation. The starter is a fairly heavy component; be careful when lifting it out of the car

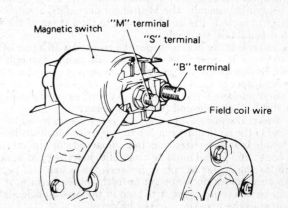

Solenoid terminals and wiring

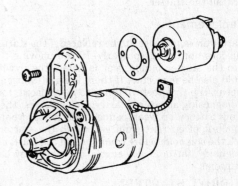

Solenoid and shim removal

3-19

3 ENGINE AND ENGINE OVERHAUL

of the battery to the **M** terminal. Make certain the field coil wire is disconnected from the **M** terminal. Quickly measure (in 10 seconds or less) the clearance between the front of the pinion gear and the stop in front of it in the starter front frame. De-energize the solenoid before it overheats. The pinion gear should be pushed back against the drive mechanism when you do this.

On reduction gear starters you'll have to measure the distance the pinion gear assembly travels when you shift it back and forth. Use a flat feeler gauge; correct clearance is 0.50-2.00mm (0.020–0.079 in.). Change the number of shims between the solenoid and starter frame to correct the clearance if necessary. Adding shims decreases the clearance, and viceversa.

5. Disconnect all test wiring hook-ups. Connect the field coil wire to the **M** terminal and reinstall the starter. Make certain the matching faces of the starter and engine are clean; any grit or grease can act as a shim and change the postion of the starter relative to the engine.

STARTER OVERHAUL

Brush Replacement

NOTE: Brush replacement requires a soldering gun or iron and the ability to use it effectively and neatly.

1. Remove the starter. Remove the **M** terminal nut, and disconnect the field coil (large) wire at the solenoid.
2. Remove the two through bolts and two Phillips screws from the rear starter bracket. Remove the rear bracket.
3. Pry the retaining springs back, slide the two brushes out of the brush holder and pull the brush holder off of the rear of the starter.
4. Inspect the brushes for excess wear. There is a manufacturer's symbol (usually the Mitsubishi diamonds) stamped on the side of each. If the brush is worn to the bottom of the emblem it should be replaced.
5. To replace the brush, it must be crushed with a pair of pliers to crack it where the wiring pigtail passes through the brush. Be careful not to damage the wiring pigtail in doing this. Use sandpaper to sand the end of the pigtail smooth. Also sand the outer surface of the last 6mm (0.25 in.) or so of the pigtail wire until it is bright and free of corrosion.
6. Insert the pigtail into the hole in the new brush until the flat end of the pigtail just reaches the opposite end of the hole in the brush. Insert the wire from the unmarked side of the brush. The brush and pigtail must be brought to just the right temperature for the solder to run in between the brush and pigtail. Make sure solder does not get onto the outer surface of the brush, as this could cause it to bind in the brush holder later.
7. Install the brushes into the holders.
8. Assemble the rear case of the starter and install bolts and screws. Connect the field coil wire to the **M** terminal of the solenoid.
9. Reinstall the starter.

Drive Replacement

The starter drive may need to be replaced if the starter motor turns but does not engage properly. If, in removing the drive, damage to the starter pinion gear is noted, the flywheel ring gear should also be inspected. If there is significant damage to the ring gear, the flywheel will have to be replaced, too.

When diagnosing apparent starter problems, test the battery first. A low battery or weak connections can prevent the required amount of current from getting to the starter motor. Also check the solenoid before condemning the starter drive itself; a solenoid that does not engage properly may cause the same symptoms.

DIRECT DRIVE STARTER

1. Remove the starter and remove the solenoid.

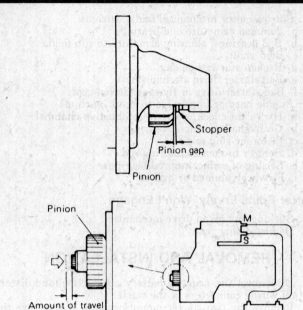

Measuring pinion gap or pinion travel (reduction type). To measure travel, the solenoid must be momentarily energized

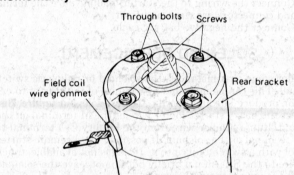

Disassembly of starter rear bracket or cover

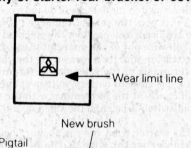

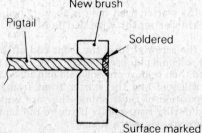

Correct installation of new starter brushes. Note that the solder is flush with the outer surface; any overflow or lump can cause binding when the brush is installed

ENGINE AND ENGINE OVERHAUL 3

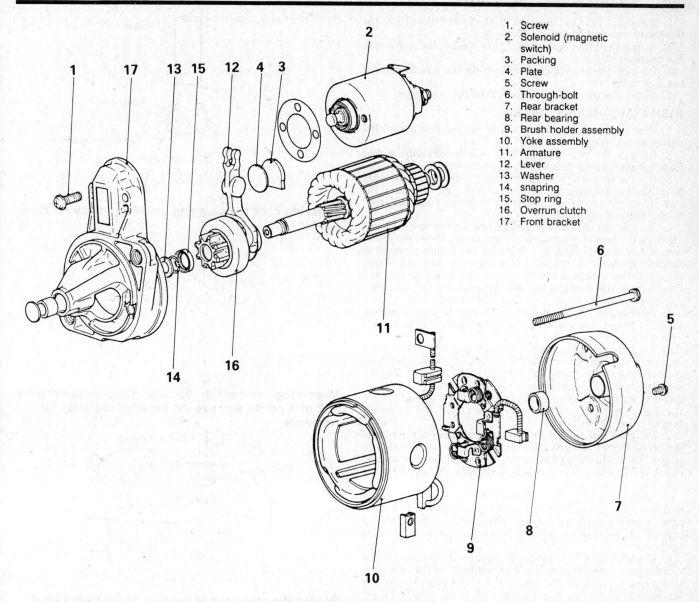

1. Screw
2. Solenoid (magnetic switch)
3. Packing
4. Plate
5. Screw
6. Through-bolt
7. Rear bracket
8. Rear bearing
9. Brush holder assembly
10. Yoke assembly
11. Armature
12. Lever
13. Washer
14. snapring
15. Stop ring
16. Overrun clutch
17. Front bracket

Typical direct drive starter

2. Remove the 2 through-bolts and 2 screws from the rear bracket. Remove the rear bracket.
3. Pry back the retaining rings and slide the 2 brushes out of the brush holder. Remove the brush holder and the yoke assembly.
4. Remove the washer from the rear of the armature. Remove the field coil assembly from the front frame. Remove the spring retainer, spring, and spring seat from the starter front frame.
5. Separate the armature from the front bracket by first pulling the armature back out of the front bearing and then shifting the armature so the starter drive is pulled out of the yoke. Make sure you don't lose the washer located in the front frame.
6. Invert the armature so the starter drive is on top and rest the rear of the armature on a solid surface. Use a deep well socket wrench that is just slightly larger than the diameter of the armature shaft to press the snap ring collar back. Install the socket over the top of the shaft and then press it downward or tap it very lightly to force the ring downward. Once the snap ring is exposed, use snap ring pliers to open it until it will slide upward, out of the groove and off the shaft. Pull the starter drive and snap ring collar upward and off the armature shaft.
7. With the starter disassembled, do not immerse parts in cleaning solvent. The yoke and field coil assembly will be damaged. Wipe the parts with a clean cloth. The overrun clutch is packed with lubricant which will be washed out by any solvent or fluid.
8. Use a micrometer or caliper to measure the diameter of the commutator and compare the measurement to the specifications chart. If below the limit, the commutator must be replaced.
9. To install, first coat the front of the armature shaft with a very light coating of a high temperature grease. Install the starter drive, snap ring collar, and snap ring. Make sure the snap ring seats in its groove. Then use a puller to pull the snap ring collar up and over the snap ring until the bottom of the collar touches the snap ring.
10. Place the washer in position in the front frame. Insert the armature through the lever and yoke. Make certain the armature is correctly seated in the bearing.
11. Install the spring seat, spring and spring retainer.

3-21

3 ENGINE AND ENGINE OVERHAUL

12. Install the field coil assembly to the front frame. Place the washer on the rear of the armature.
13. Install the brush holder and yoke assemblies and install the brushes.
14. Position the rear bracket and install the 2 screws and 2 through-bolts.
15. Install the solenoid and connect the field coil wire.

REDUCTION GEAR STARTER

1. Remove the starter from the car as described above. Remove the solenoid.
2. Remove the two through bolts and two Phillips screws from the rear starter bracket. Remove the rear bracket.
3. Pry the retaining springs back and slide the two brushes out of the brush holder. Pull the brush holder off the rear of the starter. Remove the field coil (yoke) assembly from the front frame. Remove the armature.
4. Remove the pinion shaft end cover from the center frame. Measure the clearance between the spacer and center cover and record it. If the pinion shaft is replaced, you'll have to insert or subtract spacer washers until the clearance is the same as that recorded. Use a screwdriver to remove the retaining clip and then remove the washers. Remove the retaining bolt and then separate the center frame from the front frame.
5. Remove the spring retainer and spring for the yoke from the front frame. Then remove the washer, reduction gear, shift yoke lever, and two lever supports.
6. Turn the front frame so the pinion gear is at the top and support it securely. Use a socket that fits tightly over the pinion shaft to force the snap ring collar (stop ring) downward. Tap the socket lightly at the top or use a press to do this. Use a screwdriver to work the snap ring out of its groove and remove it from the shaft. Remove the collar. Remove the pinion and the spring behind it from the shaft.
7. Pull the lever and pinion shaft assembly out of the rear of the front frame. Replace the pinion if its teeth are damaged (check the flywheel ring gear as well). Replace the overrunning clutch if the pinion gear is damaged or if the one-way action of the clutch is not precise.
8. With the starter disassembled, do not immerse parts in cleaning solvent. The yoke and field coil assembly will be damaged. Wipe the parts with a clean cloth. The overrun clutch is packed with lubricant which will be washed out by any solvent or fluid.
9. Use a micrometer or caliper to measure the diameter of the commutator and compare the measurement to the specifications chart. If below the limit, the commutator must be replaced.

To install:

10. Coat the front of the armature shaft with a very light coating of a high temperature grease. Install the pinion shaft, spring, gear and stop ring. Install the snap ring and use a puller to seat the stop ring over the snap ring. Make sure the snapring seats in its groove.
11. Place the washer in position in the front frame. Insert the armature through the lever and yoke. Make certain the armature is correctly seated in the bearing.
12. Install the lever and pinion shaft assembly with the reduction gear into the rear of the front frame. Make certain all the springs, spacers and washers are present and in the correct order. When the clearance is correct, install the cover and its small screw.
13. Fit the center frame onto the shaft and install the washer and retaining clip. Note that the clearance must be corrected by changing the thickness or number of washers if the overrunning clutch and pinion shaft assembly have been replaced. Double check the clearance and install or remove shims as necessary.
14. Install the armature into the yoke (field coil). Install the brush holder and install the brushes, making sure they are properly seated and do not bind in the holders.
15. Install the rear bracket and the small screws. Assemble the motor to the drive and install the through bolts.
16. Install the solenoid and connect the field coil wire to the M terminal.
17. Reinstall the starter.

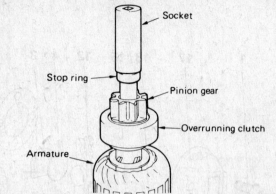

Pressing back the snapring collar on the direct drive starter

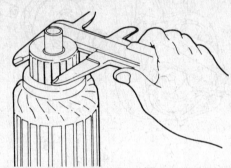

Measuring commutator diameter. Take measurements at 2 or 3 points and use the smallest diameter for reference

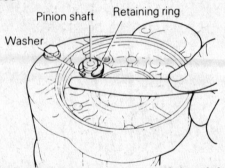

Record the clearance at the rear of the pinion shaft before disassembly

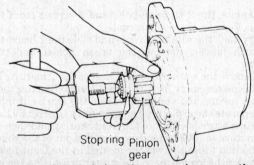

Use a small puller to bring the stop ring over the snapring

ENGINE AND ENGINE OVERHAUL 3

1. Screw
2. Solenoid (magnetic switch)
3. Screw
4. Through-bolt
5. Rear bracket
6. Brush holder
7. Yoke assembly
8. Armature
9. Front bearing
10. Rear bearing

11. Screw
12. Cover
13. Retaining ring
14. Washer
15. Screw
16. Center bracket
17. Spring seat
18. Lever spring
19. Adjusting washer(s)
20. Reduction gear
21. Lever
22. snapring
23. Stop ring
24. Pinion gear
25. Spring
26. Pinion shaft assembly
27. Front bracket

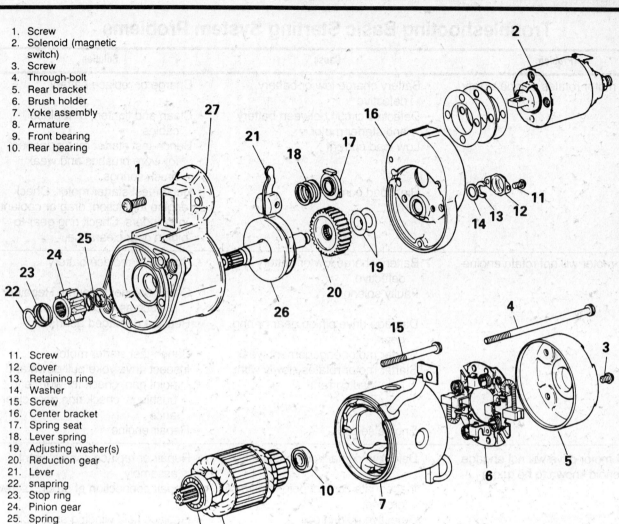

Typical reduction drive starter

1. Screw (2)
2. Front bracket
3. Lever
4. Spring, inner
5. Spring, outer
6. Packing
7. Center bracket
8. Washer set
9. Magnetic switch
10. snapring
11. Stop ring
12. Pinion
13. Overrunning clutch
14. Reduction gear
15. Adjusting washer
16. Washer

17. Retaining ring
18. Cover
19. Screw (2)
20. Screw
21. Front bearing
22. Armature
23. Rear Bearing
24. Conical spring washer
25. Yoke assembly
26. Brush spring
27. Brush holder assembly
28. Grommet
29. Rear bracket
30. Screw
31. Through bolt (2)

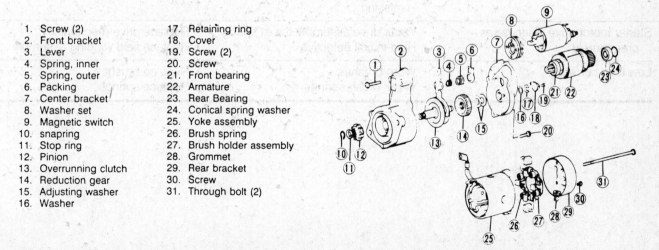

Diesel engine starter is similar to other truck starters but at a higher wattage

3-23

3 ENGINE AND ENGINE OVERHAUL

Troubleshooting Basic Starting System Problems

Problem	Cause	Solution
Starter motor rotates engine slowly	• Battery charge low or battery defective • Defective circuit between battery and starter motor • Low load current • High load current	• Charge or replace battery • Clean and tighten, or replace cables • Bench-test starter motor. Inspect for worn brushes and weak brush springs. • Bench-test starter motor. Check engine for friction, drag or coolant in cylinders. Check ring gear-to-pinion gear clearance.
Starter motor will not rotate engine	• Battery charge low or battery defective • Faulty solenoid • Damage drive pinion gear or ring gear • Starter motor engagement weak • Starter motor rotates slowly with high load current • Engine seized	• Charge or replace battery • Check solenoid ground. Repair or replace as necessary. • Replace damaged gear(s) • Bench-test starter motor • Inspect drive yoke pull-down and point gap, check for worn end bushings, check ring gear clearance • Repair engine
Starter motor drive will not engage (solenoid known to be good)	• Defective contact point assembly • Inadequate contact point assembly ground • Defective hold-in coil	• Repair or replace contact point assembly • Repair connection at ground screw • Replace field winding assembly
Starter motor drive will not disengage	• Starter motor loose on flywheel housing • Worn drive end busing • Damaged ring gear teeth • Drive yoke return spring broken or missing	• Tighten mounting bolts • Replace bushing • Replace ring gear or driveplate • Replace spring
Starter motor drive disengages prematurely	• Weak drive assembly thrust spring • Hold-in coil defective	• Replace drive mechanism • Replace field winding assembly
Low load current	• Worn brushes • Weak brush springs	• Replace brushes • Replace springs

3-24

ENGINE AND ENGINE OVERHAUL 3

STARTER SPECIFICATIONS CHART

Model	Year	Engine	Starter Drive	Nominal Watts @ 12.0V	Transmission	Volts	Max. Amps	Minimum RPM	Pinion Gap or Travel (in.)	Commutator minimum diameter (in.)
Cordia-Tredia	1983	G62B	D	700	Manual	11.5	60	6500	0.020–0.079	1.260
				900	Automatic					
	1984	G63B	D	700	Manual	11.5	60	6500	0.020–0.079	1.260
				900	Automatic					
		G62B	R	900	Manual	11.5	60	6500	0.020–0.079	1.157
				1200	Automatic	11.5	90	3300		
	1985–1988	G62B or G63B	D	900	Manual	11.5	60	6500	0.020–0.079	1.260
		G62B	R	1200	Automatic	11.5	90	3300	0.020–0.079	1.157
Starion	1983	G54B	D	900	All	11.5	60	6500	0.020–0.079	1.260
	1984–1986	G54B	D	900	Manual	11.5	60	6500	0.020–0.079	1.260
			R	1200	Automatic	11.5	100	3000		1.236
	1987–1989	G54B	R	1200	All	11.5	100	3000	0.020–0.079	1.236
Mirage	1985–1986	G15B or	R	900	Manual	11.5	60	3900	0.020–0.079	1.157
		G32B	D	900	Automatic	11.5	60	6500	0.020–0.079	1.260
	1987–1988	G15B or	D	700	Manual	11.5	60	6500	0.020–0.079	1.157
		G32B	D	900	Automatic	11.5	60	6600		
	1989–1990	4G15	D	700	Manual	11.5	60	6500	0.020–0.079	1.260
			D	900	Automatic	11.5	60	6600		
		4G61	R	1200	All	11.0	90	3000		1.157
Galant	1985–1987	G64B	R	1200	All	11.5	90	3300	0.020–0.079	1.157
	1988	6G72	R	1200	All	11.0	90	3000	0.020–0.079	1.157
	1989	4G63	D	900	SOHC-Manual	11.5	60	6600	0.020–0.079	1.260
			R	1200	All others	11.0	90	3000	0.020–0.079	1.157
Precis	1988–1989	G15B	D	700	Manual	11.5	60	6500	0.020–0.079	1.260
			D	900	Automatic	11.5	60	6600		
Sigma	1988–1989	6G72	R	1200	All	11.0	90	3000	0.020–0.079	1.157
Montero	1983	G54B	D	900	All	11.5	60	6500	0.020–0.079	1.484
	1984–1986	G54B	D	900	Manual	11.5	60	6500	0.020–0.079	1.484
			R	1200	Automatic	11.5	100	3000		1.220
	1987–1988	G54B	R	1200	All	11.0	90	3000	0.020–0.079	1.134
	1989	G54B or 6G62	R	1200	All	11.0	90	3000	0.020–0.079	1.134
Pickup	1983–1984	G63B &	D	900	Manual	11.5	60	6500	0.020–0.079	1.484
		G54B	R	1200	Automatic	11.5	100	3000		1.220
		4D55	R	2000	All	11.0	130	4000		1.484
	1985–1987	G63B &	D	900	Manual	11.5	60	6600	0.020–0.079	1.484
		G54B	R	1200	Automatic	11.5	90	3300		1.220
		4D55 ①	R	2000	All	11.0	130	4500		1.484
	1988–1989	G63B	D	900	Manual	11.5	60	6600	0.020–0.079	1.220
			R	1200	Automatic	11.0	90	3000		1.134
		G54B	R	1200	All	11.0	90	3000		

SOHC = Single Overhead Cam   D = Direct   R = Reduction
① Through 1985 only
Note: Similar test specifications for starters does not necessarily indicate interchangeable parts.

3 ENGINE AND ENGINE OVERHAUL

ENGINE MECHANICAL

Most engine overhaul procedures are fairly standard. In addition to specific parts replacement procedures and complete specifications for your individual engine, this Section also is a guide to accepted rebuilding procedures. Examples of standard rebuilding practice are shown and should be used along with specific details concerning your particular engine.

Competent and accurate machine shop services will ensure maximum performance, reliability and engine life.

In most instances it is more profitable for the do-it-yourself mechanic to remove, clean and inspect the component, buy the necessary parts and deliver these to a shop for actual machine work.

On the other hand, much of the rebuilding work (crankshaft, block, bearings, piston rods, and other components) is well within the scope of the do-it-yourself mechanic. Patience, proper tools, and common sense coupled with a basic understanding of the motor can yield satisfying and economical results.

TOOLS

The tools required for an engine overhaul or parts replacement will depend on the depth of your involvement. With a few exceptions, they will be the tools found in a mechanic's tool kit (see Section 1). More in-depth work will require any or all of the following:
- A dial indicator (reading in thousandths) mounted on a universal base
- Micrometers and telescope gauges
- Jaw and screw-type pullers
- Gasket scrapers; the best are wood or plastic
- Valve spring compressor
- Ring groove cleaner
- Piston ring expander and compressor
- Ridge reamer
- Cylinder hone or glaze breaker
- Plastigage®
- Engine stand

The use of most of these tools is illustrated in this Section. Many can be rented for a one-time use from a local parts jobber or tool supply house specializing in automotive work.

Occasionally, the use of special tools is called for. See the information on Special Tools and Safety Notice in the front of this book before substituting another tool.

INSPECTION TECHNIQUES

Procedures and specifications are given in this Section for inspecting, leaning and assessing the wear limits of most major components. Other procedures such as Magnaflux® and Zyglo® can be used to locate material flaws and stress cracks. Magnaflux® is a magnetic process applicable only to ferrous (iron and steel) materials. The Zyglo® process coats the material with a fluorescent dye penetrant and can be used on any material. Checks for suspected surface cracks can be more readily made using spot check dye. The dye is sprayed onto the suspected area, wiped off and the area sprayed with a developer. Cracks will show up brightly.

OVERHAUL TIPS

Aluminum has become extremely popular for use in engines, due to its low weight. Observe the following precautions when handling aluminum parts:
- Never hot tank aluminum parts (the caustic hot tank solution will eat the aluminum.).
- Remove all aluminum parts (identification tag, etc.) from engine parts prior to the tanking.
- Always coat threads lightly with engine oil or anti-seize compounds before installation to prevent seizure.
- Never overtighten bolts or spark plugs especially in aluminum threads.

Stripped threads in any component can be repaired using any of several commercial repair kits (Heli-Coil®, Microdot®, Keenserts®, etc.).

When assembling the engine, any parts that will be in frictional contact must be prelubed to provide lubrication at initial start-up. Any product specifically formulated for this purpose can be used, but engine oil is not recommended as a prelube.

When semi-permanent (locked, but removable) installation of bolts or nuts is desired, threads should be cleaned and coated with Loctite® or other similar, commercial non-hardening sealant.

REPAIRING DAMAGED THREADS

Several methods of repairing damaged threads are available. Heli-Coil® (shown here), Keenserts® and Microdot® are among the most widely used. All involve basically the same principle—drilling out stripped threads, tapping the hole and installing a prewound insert—making welding, plugging and oversize fasteners unnecessary.

Two types of thread repair inserts are usually supplied: a standard type for most Inch Coarse, Inch Fine, Metric Course and Metric Fine thread sizes and a spark plug type to fit most spark plug port sizes. Consult the individual manufacturer's catalog to determine exact applications. Typical thread repair kits will contain a selection of prewound threaded inserts, a tap (corresponding to the outside diameter threads of the insert) and an installation tool. Spark plug inserts usually differ because they require a tap equipped with pilot threads and a combined reamer/tap section. Most manufacturers also supply blister-packed thread repair inserts separately in addition to a master kit containing a variety of taps and inserts plus installation tools.

Before effecting a repair to a threaded hole, remove any snapped, broken or damaged bolts or studs. Penetrating oil can be used to free frozen threads. The offending item can be removed with locking pliers or with a screw or stud extractor. After the hole is clear, the thread can be repaired, as shown in the series of accompanying illustrations.

Checking Engine Compression

A noticeable lack of engine power, excessive oil consumption and/or poor fuel mileage measured over an extended period are all indicators of internal engine wear. Worn piston rings, scored or worn cylinder bores, leaking head gaskets, sticking or burnt valves and worn valve seats are all possible culprits here. A check of each cylinder's compression will help you locate the problems.

As mentioned in the Tools and Equipment section of Section 1, a screw-in type compression gauge is more accurate that the type you simply hold against the spark plug hole, although it takes slightly longer to use. It's worth it to obtain a more accurate reading. Follow the procedures below.

Gasoline Engines

1. Warm up the engine to normal operating temperature.
2. Remove all the spark plugs.
3. Disconnect the high tension lead from the ignition coil.
4. Fully open the throttle either by operating the carburetor throttle linkage by hand or by having an assistant floor the accelerator pedal.

ENGINE AND ENGINE OVERHAUL 3

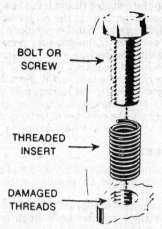

Damaged bolt holes can be repaired with thread repair inserts

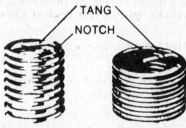

Standard thread repair insert (left) and spark plug thread insert (right)

Drill out the damaged threads with specified drill. Drill completely through the hole or to the bottom of a blind hole

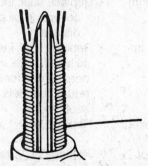

With the tap supplied, tap the hole to receive the thread insert. Keep the tap well oiled and back it out frequently to avoid clogging the threads

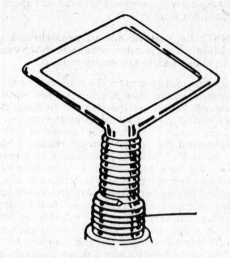

Screw the threaded insert onto the installation tool until the tang engages the slot. Screw the insert into the tapped hole until it is ¼–½ turn below the top surface. After installation break off the tang with a hammer and punch

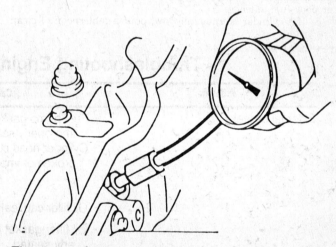

The screw-in type compression gauge is more accurate

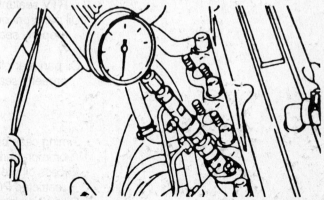

Diesel engines require a special compression gauge adaptor

3-27

3 ENGINE AND ENGINE OVERHAUL

5. Screw the compression gauge into the no.1 spark plug hole until the fitting is snug.

WARNING: Be careful not to crossthread the plug hole. On aluminum cylinder heads use extra care, as the threads in these heads are easily ruined.

6. Ask an assistant to depress the accelerator pedal fully on both carbureted and fuel injected vehicles. Then, while you read the compression gauge, ask the assistant to crank the engine two or three times in short bursts using the ignition switch.

7. Read the compression gauge at the end of each series of cranks, and record the highest of these readings. Repeat this procedure for each of the engine's cylinders. As a general rule, new motors will have compression on the order of 150-170 pounds per square inch (psi). This number will decrease with age and wear. The number of pounds of pressure that your test shows is not as important as the evenness between all the cylinders. Many cars run very well with all cylinders at 105 psi. The lower number simply shows a general deterioration internally. This car probably burns a little oil and may be a bit harder to start, but, based on these numbers, doesn't warrant an engine tear-down yet.

Compare the highest reading of all the cylinders. Any variation of more than 10% should be considered a sign of potential trouble. For example, if your compression readings for cylinders 1 through 4 were: 135 psi, 125 psi, 90 psi and 125 psi, it would be fair to say that cylinder number three is not working efficiently and is almost certainly the cause of your oil burning, rough idle or poor fuel mileage.

8. If a cylinder is unusually low, pour a tablespoon of clean engine oil into the cylinder through the spark plug hole and repeat the compression test. If the compression comes up after adding the oil, it appears that the cylinder's piston rings or bore are damaged or worn. If the pressure remains low, the valves may not be seating properly (a valve job is needed), or the head gasket may be blown near that cylinder. If compression in any two adjacent cylinders is low, and if the addition of oil doesn't help the compression, there is leakage past the head gasket. Oil and coolant in the combustion chamber can result from this problem. There may be evidence of water droplets on the engine dipstick when a head gasket has blown.

Diesel Engines

Checking cylinder compression on diesel engines is basically the same procedure as on gasoline engines except for the following:

1. A special compression gauge adaptor suitable for diesel engines (because these engines have much greater compression pressures) must be used.
2. Remove the injector tubes and remove the injectors from each cylinder.

NOTE: Don't forget to remove the washer underneath each injector; otherwise, it may get lost when the engine is cranked.

3. When fitting the compression gauge adaptor to the cylinder head, make sure the bleeder of the gauge (if equipped) is closed.
4. When reinstalling the injector assemblies, install new washers underneath each injector.

Troubleshooting Engine Mechanical Problems

Problem	Cause	Solution
External oil leaks	• Fuel pump gasket broken or improperly seated	• Replace gasket
	• Cylinder head cover RTV sealant broken or improperly seated	• Replace sealant; inspect cylinder head cover sealant flange and cylinder head sealant surface for distortion and cracks
	• Oil filler cap leaking or missing	• Replace cap
	• Oil filter gasket broken or improperly seated	• Replace oil filter
	• Oil pan side gasket broken, improperly seated or opening in RTV sealant	• Replace gasket or repair opening in sealant; inspect oil pan gasket flange for distortion
	• Oil pan front oil seal broken or improperly seated	• Replace seal; inspect timing case cover and oil pan seal flange for distortion
	• Oil pan rear oil seal broken or improperly seated	• Replace seal; inspect oil pan rear oil seal flange; inspect rear main bearing cap for cracks, plugged oil return channels, or distortion in seal groove
	• Timing case cover oil seal broken or improperly seated	• Replace seal
	• Excess oil pressure because of restricted PCV valve	• Replace PCV valve
	• Oil pan drain plug loose or has stripped threads	• Repair as necessary and tighten

ENGINE AND ENGINE OVERHAUL 3

Troubleshooting Engine Mechanical Problems (cont.)

Problem	Cause	Solution
External oil leaks	• Rear oil gallery plug loose	• Use appropriate sealant on gallery plug and tighten
	• Rear camshaft plug loose or improperly seated	• Seat camshaft plug or replace and seal, as necessary
	• Distributor base gasket damaged	• Replace gasket
Excessive oil consumption	• Oil level too high	• Drain oil to specified level
	• Oil with wrong viscosity being used	• Replace with specified oil
	• PCV valve stuck closed	• Replace PCV valve
	• Valve stem oil deflectors (or seals) are damaged, missing, or incorrect type	• Replace valve stem oil deflectors
	• Valve stems or valve guides worn	• Measure stem-to-guide clearance and repair as necessary
	• Poorly fitted or missing valve cover baffles	• Replace valve cover
	• Piston rings broken or missing	• Replace broken or missing rings
	• Scuffed piston	• Replace piston
	• Incorrect piston ring gap	• Measure ring gap, repair as necessary
	• Piston rings sticking or excessively loose in grooves	• Measure ring side clearance, repair as necessary
	• Compression rings installed upside down	• Repair as necessary
	• Cylinder walls worn, scored, or glazed	• Repair as necessary
	• Piston ring gaps not properly staggered	• Repair as necessary
	• Excessive main or connecting rod bearing clearance	• Measure bearing clearance, repair as necessary
No oil pressure	• Low oil level	• Add oil to correct level
	• Oil pressure gauge, warning lamp or sending unit inaccurate	• Replace oil pressure gauge or warning lamp
	• Oil pump malfunction	• Replace oil pump
	• Oil pressure relief valve sticking	• Remove and inspect oil pressure relief valve assembly
	• Oil passages on pressure side of pump obstructed	• Inspect oil passages for obstruction
	• Oil pickup screen or tube obstructed	• Inspect oil pickup for obstruction
	• Loose oil inlet tube	• Tighten or seal inlet tube
Low oil pressure	• Low oil level	• Add oil to correct level
	• Inaccurate gauge, warning lamp or sending unit	• Replace oil pressure gauge or warning lamp
	• Oil excessively thin because of dilution, poor quality, or improper grade	• Drain and refill crankcase with recommended oil
	• Excessive oil temperature	• Correct cause of overheating engine

3-29

3 ENGINE AND ENGINE OVERHAUL

Troubleshooting Engine Mechanical Problems (cont.)

Problem	Cause	Solution
Low oil pressure	• Oil pressure relief spring weak or sticking • Oil inlet tube and screen assembly has restriction or air leak • Excessive oil pump clearance • Excessive main, rod, or camshaft bearing clearance	• Remove and inspect oil pressure relief valve assembly • Remove and inspect oil inlet tube and screen assembly. (Fill inlet tube with lacquer thinner to locate leaks.) • Measure clearances • Measure bearing clearances, repair as necessary
High oil pressure	• Improper oil viscosity • Oil pressure gauge or sending unit inaccurate • Oil pressure relief valve sticking closed	• Drain and refill crankcase with correct viscosity oil • Replace oil pressure gauge • Remove and inspect oil pressure relief valve assembly
Main bearing noise	• Insufficient oil supply • Main bearing clearance excessive • Bearing insert missing • Crankshaft end play excessive • Improperly tightened main bearing cap bolts • Loose flywheel or drive plate • Loose or damaged vibration damper	• Inspect for low oil level and low oil pressure • Measure main bearing clearance, repair as necessary • Replace missing insert • Measure end play, repair as necessary • Tighten bolts with specified torque • Tighten flywheel or drive plate attaching bolts • Repair as necessary
Connecting rod bearing noise	• Insufficient oil supply • Carbon build-up on piston • Bearing clearance excessive or bearing missing • Crankshaft connecting rod journal out-of-round • Misaligned connecting rod or cap • Connecting rod bolts tightened improperly	• Inspect for low oil level and low oil pressure • Remove carbon from piston crown • Measure clearance, repair as necessary • Measure journal dimensions, repair or replace as necessary • Repair as necessary • Tighten bolts with specified torque
Piston noise	• Piston-to-cylinder wall clearance excessive (scuffed piston) • Cylinder walls excessively tapered or out-of-round • Piston ring broken • Loose or seized piston pin • Connecting rods misaligned • Piston ring side clearance excessively loose or tight	• Measure clearance and examine piston • Measure cylinder wall dimensions, rebore cylinder • Replace all rings on piston • Measure piston-to-pin clearance, repair as necessary • Measure rod alignment, straighten or replace • Measure ring side clearance, repair as necessary

ENGINE AND ENGINE OVERHAUL 3

Troubleshooting Engine Mechanical Problems (cont.)

Problem	Cause	Solution
Piston noise	• Carbon build-up on piston is excessive	• Remove carbon from piston
Valve actuating component noise	• Insufficient oil supply	• Check for: (a) Low oil level (b) Low oil pressure (c) Plugged push rods (d) Wrong hydraulic tappets (e) Restricted oil gallery (f) Excessive tappet to bore clearance
	• Push rods worn or bent	• Replace worn or bent push rods
	• Rocker arms or pivots worn	• Replace worn rocker arms or pivots
	• Foreign objects or chips in hydraulic tappets	• Clean tappets
	• Excessive tappet leak-down	• Replace valve tappet
	• Tappet face worn	• Replace tappet; inspect corresponding cam lobe for wear
	• Broken or cocked valve springs	• Properly seat cocked springs; replace broken springs
	• Stem-to-guide clearance excessive	• Measure stem-to-guide clearance, repair as required
	• Valve bent	• Replace valve
	• Loose rocker arms	• Tighten bolts with specified torque
	• Valve seat runout excessive	• Regrind valve seat/valves
	• Missing valve lock	• Install valve lock
	• Push rod rubbing or contacting cylinder head	• Remove cylinder head and remove obstruction in head
	• Excessive engine oil (four-cylinder engine)	• Correct oil level

Troubleshooting the Cooling System

Problem	Cause	Solution
High temperature gauge indication—overheating	• Coolant level low	• Replenish coolant
	• Fan belt loose	• Adjust fan belt tension
	• Radiator hose(s) collapsed	• Replace hose(s)
	• Radiator airflow blocked	• Remove restriction (bug screen, fog lamps, etc.)
	• Faulty radiator cap	• Replace radiator cap
	• Ignition timing incorrect	• Adjust ignition timing
	• Idle speed low	• Adjust idle speed
	• Air trapped in cooling system	• Purge air
	• Heavy traffic driving	• Operate at fast idle in neutral intermittently to cool engine
	• Incorrect cooling system component(s) installed	• Install proper component(s)
	• Faulty thermostat	• Replace thermostat
	• Water pump shaft broken or impeller loose	• Replace water pump

3 ENGINE AND ENGINE OVERHAUL

Troubleshooting the Cooling System (cont.)

Problem	Cause	Solution
High temperature gauge indication—overheating	• Radiator tubes clogged • Cooling system clogged • Casting flash in cooling passages • Brakes dragging • Excessive engine friction • Antifreeze concentration over 68% • Missing air seals • Faulty gauge or sending unit • Loss of coolant flow caused by leakage or foaming • Viscous fan drive failed	• Flush radiator • Flush system • Repair or replace as necessary. Flash may be visible by removing cooling system components or removing core plugs. • Repair brakes • Repair engine • Lower antifreeze concentration percentage • Replace air seals • Repair or replace faulty component • Repair or replace leaking component, replace coolant • Replace unit
Low temperature indication—undercooling	• Thermostat stuck open • Faulty gauge or sending unit	• Replace thermostat • Repair or replace faulty component
Coolant loss—boilover	• Overfilled cooling system • Quick shutdown after hard (hot) run • Air in system resulting in occasional "burping" of coolant • Insufficient antifreeze allowing coolant boiling point to be too low • Antifreeze deteriorated because of age or contamination • Leaks due to loose hose clamps, loose nuts, bolts, drain plugs, faulty hoses, or defective radiator • Faulty head gasket • Cracked head, manifold, or block • Faulty radiator cap	• Reduce coolant level to proper specification • Allow engine to run at fast idle prior to shutdown • Purge system • Add antifreeze to raise boiling point • Replace coolant • Pressure test system to locate source of leak(s) then repair as necessary • Replace head gasket • Replace as necessary • Replace cap
Coolant entry into crankcase or cylinder(s)	• Faulty head gasket • Crack in head, manifold or block	• Replace head gasket • Replace as necessary
Coolant recovery system inoperative	• Coolant level low • Leak in system • Pressure cap not tight or seal missing, or leaking • Pressure cap defective • Overflow tube clogged or leaking • Recovery bottle vent restricted	• Replenish coolant to FULL mark • Pressure test to isolate leak and repair as necessary • Repair as necessary • Replace cap • Repair as necessary • Remove restriction

ENGINE AND ENGINE OVERHAUL 3

Troubleshooting the Cooling System (cont.)

Problem	Cause	Solution
Noise	• Fan contacting shroud	• Reposition shroud and inspect engine mounts
	• Loose water pump impeller	• Replace pump
	• Glazed fan belt	• Apply silicone or replace belt
	• Loose fan belt	• Adjust fan belt tension
	• Rough surface on drive pulley	• Replace pulley
	• Water pump bearing worn	• Remove belt to isolate. Replace pump.
	• Belt alignment	• Check pulley alignment. Repair as necessary.
No coolant flow through heater core	• Restricted return inlet in water pump	• Remove restriction
	• Heater hose collapsed or restricted	• Remove restriction or replace hose
	• Restricted heater core	• Remove restriction or replace core
	• Restricted outlet in thermostat housing	• Remove flash or restriction
	• Intake manifold bypass hole in cylinder head restricted	• Remove restriction
	• Faulty heater control valve	• Replace valve
	• Intake manifold coolant passage restricted	• Remove restriction or replace intake manifold

NOTE: *Immediately after shutdown, the engine enters a condition known as heat soak. This is caused by the cooling system being inoperative while engine temperature is still high. If coolant temperature rises above boiling point, expansion and pressure may push some coolant out of the radiator overflow tube. If this does not occur frequently it is considered normal.*

GENERAL ENGINE SPECIFICATIONS

Year	Model	Engine	Fuel System Type	Net Horsepower @ rpm	Net Torque @ rpm (ft. lbs.)	Bore × Stroke (in.)	Compression Ratio	Oil Pressure @ rpm
1983	Cordia	G62B	Carb.	82 @ 5000	93 @ 3000	3.17 × 3.46	7.5:1	63 ①
	Tredia	G62B	Carb.	82 @ 5000	93 @ 3000	3.17 × 3.46	7.5:1	63 ①
	Starion	G54B	ECI ②	145 @ 5000	185 @ 2500	3.59 × 3.86	7.0:1	63 ①
	Montero	G54B	Carb.	105 @ 5000	139 @ 2500	3.59 × 3.86	8.2:1	57 ①
	Pickup	G63B	Carb.	93 @ 5200	108 @ 3000	3.35 × 3.46	8.5:1	64 ①
	Pickup	G54B	Carb.	105 @ 5000	139 @ 2500	3.59 × 3.86	8.2:1	57 ①
	Pickup	4D55	TD ⑥	84 @ 4200	136 @ 2500	3.59 × 3.54	21.0:1	78 ①
1984	Cordia	G62B	ECI ②	120 @ 5500	110 @ 2600	3.17 × 3.46	7.5:1	63 ①
	Cordia	G63B	Carb.	110 @ 5000	117 @ 3000	3.35 × 3.46	8.5:1	63 ①
	Tredia	G62B	ECI ②	120 @ 5500	110 @ 2600	3.17 × 3.46	7.5:1	63 ①
	Tredia	G63B	Carb.	110 @ 5000	117 @ 3000	3.35 × 3.46	8.5:1	63 ①
	Starion	G54B	ECI ②	145 @ 5000	185 @ 2500	3.59 × 3.86	7.0:1	63 ①
	Montero	G54B	Carb.	105 @ 5000	139 @ 2500	3.59 × 3.86	8.2:1	57
	Pickup	G63B	Carb.	93 @ 5200	108 @ 3000	3.35 × 3.46	8.5:1	64 ①
	Pickup	G54B	Carb.	105 @ 5000	139 @ 2500	3.59 × 3.86	8.2:1	57 ①
	Pickup	4D55	TD	84 @ 4200	136 @ 2500	3.59 × 3.54	21.0:1	78 ①

3 ENGINE AND ENGINE OVERHAUL

GENERAL ENGINE SPECIFICATIONS

Year	Model	Engine	Fuel System Type	Net Horsepower @ rpm	Net Torque @ rpm (ft. lbs.)	Bore × Stroke (in.)	Compression Ratio	Oil Pressure @ rpm
1985	Cordia	G62B	ECI ②	116 @ 5500	129 @ 3000	3.17 × 3.46	7.5:1	63 ①
	Cordia	G63B	Carb.	88 @ 5000	108 @ 3500	3.35 × 3.46	8.5:1	63 ①
	Tredia	G62B	ECI ②	116 @ 5500	129 @ 3000	3.17 × 3.46	7.5:1	63 ①
	Tredia	G63B	Carb.	88 @ 5000	108 @ 3500	3.35 × 3.46	8.5:1	63 ①
	Starion	G54B	ECI ②	145 @ 5000	185 @ 2500	3.59 × 3.86	7.0:1	63 ①
	Mirage	G15B	Carb.	68 @ 5500	82 @ 3500	2.97 × 3.23	9.4:1	63 ①
	Mirage	G32B	ECI ②	102 @ 5500	122 @ 3000	3.03 × 3.39	7.6:1	63 ①
	Galant	G64B	ECI ②	101 @ 5000	131 @ 2500	3.41 × 3.94	8.5:1	63 ①
	Montero	G54B	Carb.	108 @ 2500	142 @ 2500	3.59 × 3.86	8.7:1	78 ①
	Pickup	G63B	Carb.	93 @ 5200	108 @ 3000	3.35 × 3.46	8.5:1	78 ①
	Pickup	G54B	Carb.	105 @ 5000	139 @ 2500	3.59 × 3.86	8.2:1	57 ①
	Pickup	4D55	TD	84 @ 4200	136 @ 2500	3.59 × 3.54	21.0:1	78 ①
1986	Cordia	G62B	ECI ②	116 @ 5500	129 @ 3000	3.17 × 3.46	7.5:1	63 ①
	Cordia	G63B	Carb.	88 @ 5000	108 @ 3500	3.35 × 3.46	8.5:1	63 ①
	Tredia	G62B	ECI ②	116 @ 5500	129 @ 3000	3.17 × 3.46	7.5:1	63 ①
	Tredia	G63B	Carb.	88 @ 5000	108 @ 3500	3.35 × 3.46	8.5:1	63 ①
	Starion	G54B	ECI ②	145 @ 5000	185 @ 2500	3.59 × 3.86	7.0:1	63 ①
	Mirage	G15B	Carb.	68 @ 5000	82 @ 3500	2.97 × 3.23	9.4:1	63 ①
	Mirage	G32B	ECI ②	102 @ 5500	122 @ 3000	3.03 × 3.39	7.6:1	63 ①
	Galant	G64B	MPI ③	110 @ 4500	138 @ 3500	3.41 × 3.94	8.5:1	63 ①
	Montero	G54B	Carb.	108 @ 2500	142 @ 2500	3.59 × 3.86	8.7:1	78 ①
	Pickup	G63B	Carb.	90 @ 5000	108 @ 3500	3.35 × 3.46	8.5:1	78 ①
	Pickup	G54B	Carb.	105 @ 5000	139 @ 2500	3.59 × 3.86	8.2:1	57 ①
1987	Cordia	G62B	ECI ②	116 @ 5500	129 @ 3000	3.17 × 3.46	7.5:1	63 ①
	Cordia	G63B	Carb.	88 @ 5000	108 @ 3500	3.35 × 3.46	8.5:1	63 ①
	Tredia	G62B	ECI ②	116 @ 5500	129 @ 3000	3.17 × 3.46	7.5:1	63 ①
	Tredia	G63B	Carb.	88 @ 5000	108 @ 3500	3.35 × 3.46	8.5:1	63 ①
	Starion	G54B	ECI ②	145 @ 5000	185 @ 2500	3.59 × 3.86	7.0:1	63 ①
	Mirage	G15B	Carb.	68 @ 5000	82 @ 3500	2.97 × 3.23	9.4:1	63 ①
	Mirage	G32B	ECI ②	102 @ 5500	122 @ 3000	3.03 × 3.39	7.6:1	63 ①
	Galant	G64B	MPI ③	110 @ 4500	138 @ 3500	3.41 × 3.94	8.5:1	63 ①
	Precis	G15B	Carb.	68 @ 5000	82 @ 3500	2.97 × 3.23	9.4:1	63 ①
	Montero	G54B	Carb.	109 @ 5000	142 @ 3000	3.59 × 3.86	8.7:1	56 @ 2000
	Pickup	G63B	Carb.	90 @ 5500	109 @ 3500	3.35 × 3.46	8.5:1	49 @ 2000
	Pickup	G54B	Carb.	109 @ 5000	142 @ 3000	3.59 × 3.86	8.7:1	56 @ 2000
1988	Cordia	G62B	ECI ②	116 @ 5500	129 @ 3000	3.17 × 3.46	7.5:1	63 ①
	Cordia	G63B	Carb.	88 @ 5000	108 @ 3500	3.35 × 3.46	8.5:1	63 ①
	Tredia	G62B	ECI ②	116 @ 5500	129 @ 3000	3.17 × 3.46	7.5:1	63 ①
	Tredia	G63B	Carb.	88 @ 5000	108 @ 3500	3.35 × 3.46	8.5:1	63 ①
	Starion	G54B	ECI ②	145 @ 5000	185 @ 2500	3.59 × 3.86	7.0:1	63 ①
	Mirage	G15B	Carb.	68 @ 5000	82 @ 3500	2.97 × 3.23	9.4:1	63 ①
	Mirage	G32B	ECI ②	102 @ 5500	122 @ 3000	3.03 × 3.39	7.6:1	63 ①
	Galant	G64B	MPI ③	110 @ 4500	138 @ 3500	3.41 × 3.94	8.5:1	63 ①
	Galant	6G72	MPI ③	142 @ 5000	168 @ 2500	3.59 × 2.99	8.9:1	63 ①

ENGINE AND ENGINE OVERHAUL 3

GENERAL ENGINE SPECIFICATIONS

Year	Model	Engine	Fuel System Type	Net Horsepower @ rpm	Net Torque @ rpm (ft. lbs.)	Bore × Stroke (in.)	Compression Ratio	Oil Pressure @ rpm
1988	Precis	G15B	Carb.	68 @ 5500	82 @ 3500	2.97 × 3.23	9.4:1	63 ①
	Montero	G54B	Carb.	109 @ 5000	193 @ 3000	3.59 × 3.86	8.7:1	56 @ 2000
	Pickup	G63B	Carb.	90 @ 5500	109 @ 3500	3.35 × 3.46	8.5:1	49 @ 2000
	Pickup	G54B	Carb.	109 @ 5000	142 @ 3000	3.59 × 3.86	8.7:1	56 @ 2000
1989	Starion	G54B	ECI ②	145 @ 5000	185 @ 2500	3.59 × 3.86	7.0:1	63 ①
	Mirage	4G61	Carb.	68 @ 5000	82 @ 3500	2.97 × 3.23	9.4:1	63 ①
	Mirage	4G61	ECI ②	102 @ 5500	122 @ 3000	3.03 × 3.39	7.6:1	63 ①
	Galant	4G63 ④	MPI ③	120 @ 5000	116 @ 4500	3.35 × 3.46	8.5:1	11.4 @ 750
	Galant	4G63 ⑤	MPI ③	135 @ 6000	125 @ 5000	3.35 × 3.46	9.0:1	11.4 @ 750
	Sigma	6G72	MPI ③	142 @ 5000	168 @ 2500	3.59 × 2.99	8.9:1	63 ①
	Precis	G15B	Carb.	68 @ 5000	82 @ 3500	2.97 × 3.23	9.4:1	63 ①
	Montero	G54B	Carb.	109 @ 5000	142 @ 3000	3.59 × 3.86	8.7:1	56 @ 2000
	Montero	6G72	MPI	143 @ 5000	168 @ 2500	3.58 × 2.99	8.9:1	40 @ 2000
	Pickup	G63B	Carb.	90 @ 5500	109 @ 3500	3.35 × 3.46	8.5:1	49 @ 2000
	Pickup	G54B	Carb.	109 @ 5000	142 @ 3000	3.59 × 3.86	8.7:1	56 @ 2000

① Relief valve opening pressure
② Electronic controlled injection
③ Multi-point injection
④ Single overhead camshaft
⑤ Double overhead camshaft
⑥ Turbodiesel

VALVE SPECIFICATIONS

Year	Engine	Seat Angle (deg.)	Face Angle (deg.)	Spring Test Pressure (lbs. @ in.)	Spring Installed Height (in.)	Stem-to-Guide Clearance (in.) Intake	Stem-to-Guide Clearance (in.) Exhaust	Stem Diameter (in.) Intake	Stem Diameter (in.) Exhaust
1983	G62B	45	45	62 @ 1.591	1.591	0.0010–0.0022	0.0020–0.0035	0.315	0.315
	G54B	45	45	62 @ 1.591	1.591	0.0010–0.0024	0.0020–0.0035	0.315	0.315
	4D55	45	45	61 @ 1.591	1.591	0.0012–0.0024	0.0020–0.0035	0.315	0.315
1984	G62B	45	45	62 @ 1.591	1.591	0.0010–0.0022	0.0020–0.0035	0.315	0.315
	G63B	45	45	62 @ 1.591	1.591	0.0010–0.0022	0.0020–0.0035	0.315	0.315
	G54B	45	45	62 @ 1.591	1.591	0.0012–0.0024	0.0020–0.0035	0.315	0.315
	4D55	45	45	61 @ 1.591	1.591	0.0012–0.0024	0.0020–0.0035	0.315	0.315
1985	G62B	45	45	62 @ 1.591	1.591	0.0010–0.0022	0.0020–0.0035	0.315	0.315
	G63B	45	45	62 @ 1.591	1.591	0.0010–0.0022	0.0020–0.0035	0.315	0.315
	G54B	45	45	62 @ 1.591	1.591	0.0012–0.0024	0.0020–0.0035	0.315	0.315
	G15B	45	45	53 @ 1.469	1.417	0.0012–0.0024	0.0020–0.0035	0.315	0.315

3 ENGINE AND ENGINE OVERHAUL

VALVE SPECIFICATIONS

Year	Engine	Seat Angle (deg.)	Face Angle (deg.)	Spring Test Pressure (lbs. @ in.)	Spring Installed Height (in.)	Stem-to-Guide Clearance (in.) Intake	Stem-to-Guide Clearance (in.) Exhaust	Stem Diameter (in.) Intake	Stem Diameter (in.) Exhaust
1985	G32B	45	45	62 @ 1.469	1.469	0.0012–0.0024	0.0020–0.0035	0.315	0.315
	G64B	45	45	72 @ 1.591	1.591	0.0012–0.0024	0.0024–0.0035	0.322	0.315
	4D55	45	45	61 @ 1.591	1.591	0.0012–0.0024	0.0020–0.0035	0.315	0.315
1986	G62B	45	45	62 @ 1.591	1.591	0.0010–0.0022	0.0020–0.0035	0.315	0.315
	G63B	45	45	72 @ 1.591	1.591	0.0010–0.0022	0.0020–0.0035	0.315	0.315
	G54B	45	45	72 @ 1.591	1.591	0.0012–0.0024	0.0020–0.0035	0.315	0.315
	G15B	45	45	53 @ 1.469	1.417	0.0008–0.0020	0.0020–0.0035	0.315	0.315
	G32B	45	45	62 @ 1.469	1.469	0.0012–0.0024	0.0020–0.0035	0.315	0.315
	G64B	45	45	72 @ 1.591	1.591	0.0012–0.0024	0.0020–0.0035	0.322	0.315
1987	G62B	45	45	62 @ 1.591	1.591	0.0010–0.0022	0.0020–0.0035	0.315	0.315
	G63B	45	45	72 @ 1.591	1.591	0.0010–0.0022	0.0020–0.0035	0.315	0.315
	G54B	45	45	72 @ 1.591	1.591	0.0012–0.0024	0.0020–0.0035	0.315	0.315
	G15B	45	45	53 @ 1.469	1.417	0.0008–0.0020	0.0020–0.0035	0.315	0.315
	G32B	45	45	62 @ 1.469	1.469	0.0012–0.0024	0.0020–0.0035	0.315	0.315
	G64B	45	45	72 @ 1.591	1.591	0.0012–0.0024	0.0020–0.0035	0.322	0.315
1988	G62B	45	45	62 @ 1.591	1.591	0.0010–0.0022	0.0020–0.0035	0.315	0.315
	G63B	45	45	72 @ 1.591	1.591	0.0010–0.0022	0.0020–0.0035	0.315	0.315
	G54B	45	45	72 @ 1.591	1.591	0.0012–0.0024	0.0020–0.0035	0.315	0.315
	G15B	45	45	53 @ 1.469	1.417	0.0008–0.0020	0.0020–0.0035	0.315	0.315
	G32B	45	45	62 @ 1.469	1.469	0.0012–0.0024	0.0020–0.0035	0.315	0.315
	G64B	45	45	72 @ 1.591	1.591	0.0012–0.0024	0.0020–0.0035	0.322	0.315
	6G72	44	45	74 @ 1.591	1.591	0.0012–0.0024	0.0020–0.0035	0.314	0.313
1989	4G63	45	45	72 @ 1.591	1.591	0.0010–0.0022	0.0020–0.0035	0.315	0.315
	G54B	45	45	72 @ 1.591	1.591	0.0012–0.0024	0.0020–0.0035	0.315	0.315

ENGINE AND ENGINE OVERHAUL 3

VALVE SPECIFICATIONS

Year	Engine	Seat Angle (deg.)	Face Angle (deg.)	Spring Test Pressure (lbs. @ in.)	Spring Installed Height (in.)	Stem-to-Guide Clearance (in.) Intake	Stem-to-Guide Clearance (in.) Exhaust	Stem Diameter (in.) Intake	Stem Diameter (in.) Exhaust
1989	G15B	45	45	53 @ 1.469	1.417	0.0008–0.0020	0.0020–0.0035	0.315	0.315
	4G61	45	45	62 @ 1.469	1.469	0.0012–0.0024	0.0020–0.0035	0.315	0.315
	6G72	44	45	74 @ 1.591	1.591	0.0012–0.0024	0.0020–0.0035	0.314	0.313

CAMSHAFT SPECIFICATIONS—AUTOMOBILES

All measurements in inches.

Year	Engine	Journal Diameter	Bearing Clearance	Lobe Height Int.	Lobe Height Exh.	Fuel Pump Lobe	Maximum Lobe Wear	End Play Min./Max.
1983	G62B	1.339	0.002–0.004	1.661	1.661	1.496	0.020	0.004–0.008
	G54B	1.339	0.002–0.004	1.673	1.673	1.457	0.020	0.004–0.008
1984	G62B	1.339	0.002–0.004	1.661	1.661	1.496	0.020	0.004–0.008
	G63B	1.339	0.002–0.004	1.661	1.661	1.496	0.020	0.004–0.008
	G54B	1.339	0.002–0.004	1.673	1.673	1.457	0.020	0.004–0.008
1985	G62B	1.339	0.002–0.004	1.661	1.661	1.496	0.020	0.004–0.008
	G63B	1.339	0.002–0.004	1.657	1.657	1.496	0.020	0.004–0.008
	G54B	1.339	0.002–0.004	1.673	1.673	1.457	0.020	0.004–0.008
	G15B	1.8110	0.002–0.0035	1.5000	1.5039	—	0.020	0.002–0.008
	G32B	1.3363	0.002–0.0035	1.4331	1.4331	—	0.020	0.002–0.006
	G64B	1.339	0.002–0.004	1.665	1.665	—	0.020	0.004–0.008
1986	G62B	1.339	0.002–0.004	1.661	1.661	1.496	0.020	0.004–0.008
	G63B	1.339	0.002–0.004	1.657	1.657	1.496	0.020	0.004–0.008
	G54B	1.339	0.002–0.004	1.669	1.669	1.4567	0.020	0.004–0.008
	G15B	1.8110	0.0015–0.0031	1.5000	1.5039	—	0.020	0.002–0.008
	G32B	1.3363	0.002–0.0035	1.4331	1.4331	—	0.020	0.002–0.006
	G64B	1.339	0.002–0.004	1.669	1.669	—	0.020	0.004–0.008

3 ENGINE AND ENGINE OVERHAUL

CAMSHAFT SPECIFICATIONS—AUTOMOBILES
All measurements in inches.

Year	Engine	Journal Diameter	Bearing Clearance	Lobe Height Int.	Lobe Height Exh.	Fuel Pump Lobe	Maximum Lobe Wear	End Play Min./Max.
1987	G62B	1.339	0.002–0.004	1.661	1.661	1.496	0.020	0.004–0.008
	G63B	1.339	0.002–0.004	1.657	1.657	1.496	0.020	0.004–0.008
	G54B	1.3386	0.0012–0.0020	1.6693	1.6693	1.4567	0.0187	0.004–0.008
	G15B	1.8110	0.0015–0.0031	1.4992	1.5413	—	0.0197	0.0020–0.0079
	G32B	1.3363	0.0020–0.0035	1.4315	1.4335	—	0.0197	0.0020–0.0059
	G64B	1.3386	0.002–0.004	1.6693	1.6693	—	0.0197	0.004–0.008
1988	G62B	1.340	0.0020–0.0035	1.6565	1.6565	—	0.020	0.004–0.008
	G63B	1.3386	0.0020–0.0035	1.6565	1.6565	—	0.020	0.004–0.008
	G54B	1.3386	0.0020–0.0035	1.6705	1.6705	1.4567	0.0197	0.004–0.008
	G15B	1.8110	0.0015–0.0031	1.4992	1.5413	—	0.0197	0.0020–0.0079
	G32B	1.3363	0.0020–0.0035	1.4315	1.4335	—	0.0197	0.0020–0.0059
	6G72	1.3363	0.0020–0.0035	1.6240	1.6240	—	0.0197	0.004–0.008
1989	4G63 ①	1.3363	0.0020–0.0035	1.7529	1.7529	—	0.0197	0.004–0.008
	②	1.0221	0.0020–0.0035	1.3974	1.3858	—	0.0197	0.004–0.008
	G54B	1.3386	0.0020–0.0035	1.6705	1.6705	1.4567	0.0197	0.004–0.008
	G15B	1.8110	0.0015–0.0031	1.500	1.504	—	0.020	0.0020–0.0079
	4G61	1.020	0.0020–0.0035	1.3858	1.3743	—	0.0197	0.004–0.008
	6G72	1.340	0.0020–0.0035	1.6240	1.6240	—	0.0197	0.004–0.008
	4G15	1.8110	0.0015–0.0031	1.5318	1.5344	—	0.0197	0.0020–0.0079

① SOHC
② DOHC

ENGINE AND ENGINE OVERHAUL 3

CAMSHAFT SPECIFICATIONS—MONTERO & PICKUP
All measurements in inches.

Year	Engine	Journal Diameter	Bearing Clearance	Lobe Height Int.	Lobe Height Exh.	Fuel Pump Lobe	Maximum Lobe Wear	End Play Min./Max.
1983	G63B	1.339	0.002–0.004	1.661	1.661	1.496	0.020	0.004–0.008
	G54B	1.339	0.002–0.004	1.673	1.673	1.457	0.020	0.004–0.008
	4D55	1.181	0.002–0.004	1.461	1.461	—	0.020	0.004–0.008
1984	G63B	1.339	0.002–0.004	1.661	1.661	1.496	0.020	0.004–0.008
	G54B	1.339	0.002–0.004	1.673	1.673	1.457	0.020	0.004–0.008
	4D55	1.181	0.002–0.004	1.461	1.461	—	0.020	0.004–0.008
1985	G63B	1.339	0.002–0.004	1.657	1.657	1.496	0.020	0.004–0.008
	G54B	1.339	0.002–0.004	① 1.673 ② 1.669	1.673 1.669	1.457	0.020	0.004–0.008
	4D55	1.181	0.002–0.004	1.461	1.461	—	0.020	0.004–0.008
1986	G63B	1.339	0.002–0.004	1.657	1.657	1.496	0.020	0.004–0.008
	G54B	1.339	0.002–0.004	① 1.673 ② 1.669	1.673 1.669	1.457	0.020	0.004–0.008
1987	G63B	1.3386	0.002–0.0035	1.656	1.6567	1.496	0.020	0.004–0.008
	G54B	1.3386	0.0012–0.0020	1.669	1.669	1.456	0.020	0.004–0.008
1988	G63B	1.340	0.002–0.0035	1.656	1.656	—	0.0195	0.004–0.008
	G54B	1.340	0.002–0.0035	1.6705	1.6705	1.460	0.0197	0.004–0.008
1989	G63B	1.340	0.002–0.0035	1.656	1.656	1.500	0.0195	0.004–0.008
	G54B	1.340	0.002–0.0035	1.6705	1.6705	1.460	0.0197	0.004–0.008
	6G72	1.340	0.002–0.0035	1.624	1.624	—	0.0197	0.004–0.008

① Pickup
② Montero

3 ENGINE AND ENGINE OVERHAUL

PISTON AND RING SPECIFICATIONS—AUTOMOBILES
All measurements are given in inches.

Year	Engine	Piston Clearance	Ring Gap			Ring Side Clearance*	
			Top Compression	Bottom Compression	Oil Control	Top Compression	Bottom Compression
1983	G62B	0.0008–0.0016	0.0100–0.0180	0.0080–0.0160	0.0080–0.0200	0.002–0.004	0.001–0.002
	G54B	0.0008–0.0016	0.0120–0.0200	0.0100–0.0160	0.0120–0.0310	0.002–0.004	0.001–0.002
1984	G62B	0.0008–0.0016	0.0100–0.0180	0.0080–0.0160	0.0080–0.0200	0.002–0.004	0.001–0.002
	G63B	0.0008–0.0016	0.0100–0.0180	0.0080–0.0160	0.0080–0.0200	0.002–0.004	0.001–0.002
	G54B	0.0008–0.0016	0.0120–0.0200	0.0100–0.0160	0.0120–0.0310	0.002–0.004	0.001–0.002
1985	G62B	0.0008–0.0016	0.0100–0.0180	0.0080–0.0160	0.0080–0.0200	0.002–0.004	0.001–0.002
	G63B	0.0008–0.0016	0.0100–0.0180	0.0080–0.0160	0.0080–0.0200	0.002–0.004	0.001–0.002
	G54B	0.0008–0.0016	0.0120–0.0200	0.0100–0.0160	0.0120–0.0310	0.002–0.004	0.001–0.002
	G15B	0.0008–0.0016	0.0080–0.0160	0.0080–0.0160	0.0080–0.0280	0.0012–0.0028	0.0008–0.0024
	G32B	0.0008–0.0016	0.0080–0.0160	0.0080–0.0160	0.0080–0.0280	0.0012–0.0028	0.0008–0.0024
	G64B	0.0008–0.0016	0.0100–0.0180	0.0080–0.0160	0.0080–0.0280	0.002–0.004	0.001–0.002
1986	G62B	0.0008–0.0016	0.0100–0.0180	0.0080–0.0160	0.0080–0.0200	0.002–0.004	0.001–0.002
	G63B	0.0008–0.0016	0.0100–0.0180	0.0080–0.0160	0.0080–0.0200	0.002–0.004	0.001–0.002
	G54B	0.0008–0.0016	0.0120–0.0200	0.0100–0.0160	0.0120–0.0310	0.002–0.004	0.001–0.002
	G15B	0.0008–0.0016	0.0080–0.0160	0.0080–0.0160	0.0080–0.0280	0.0012–0.0028	0.0008–0.0024
	G32B	0.0008–0.0016	0.0080–0.0160	0.0080–0.0160	0.0080–0.0280	0.0012–0.0028	0.0008–0.0024
	G64B	0.0008–0.0016	0.0100–0.0180	0.0080–0.0160	0.0080–0.0280	0.002–0.004	0.001–0.002
1987	G62B	0.0008–0.0016	0.0100–0.0180	0.0080–0.0160	0.0080–0.0200	0.002–0.004	0.001–0.002
	G63B	0.0008–0.0016	0.0100–0.0180	0.0080–0.0160	0.0080–0.0200	0.002–0.004	0.001–0.002
	G54B	0.0008–0.0016	0.0120–0.0200	0.0100–0.0160	0.0120–0.0310	0.002–0.004	0.001–0.002
	G15B	0.0008–0.0016	0.0080–0.0160	0.0080–0.0160	0.0080–0.0280	0.0012–0.0028	0.0008–0.0024
	G32B	0.0008–0.0016	0.0080–0.0160	0.0080–0.0160	0.0080–0.0280	0.0012–0.0028	0.0008–0.0024
	G64B	0.0008–0.0016	0.0100–0.0180	0.0080–0.0160	0.0080–0.0280	0.0012–0.0028	0.0008–0.0024

ENGINE AND ENGINE OVERHAUL 3

PISTON AND RING SPECIFICATIONS—AUTOMOBILES

All measurements are given in inches.

Year	Engine	Piston Clearance	Ring Gap			Ring Side Clearance*	
			Top Compression	Bottom Compression	Oil Control	Top Compression	Bottom Compression
1988	G62B	0.0008–0.0016	0.0100–0.0180	0.0080–0.0160	0.0080–0.0200	0.002–0.004	0.001–0.002
	G63B	0.0008–0.0016	0.0100–0.0180	0.0080–0.0160	0.0080–0.0200	0.002–0.004	0.001–0.002
	G54B	0.0008–0.0016	0.0120–0.0200	0.0100–0.0160	0.0120–0.0310	0.002–0.004	0.001–0.002
	G15B	0.0008–0.0016	0.0080–0.0160	0.0080–0.0160	0.0080–0.0280	0.0012–0.0028	0.0008–0.0024
	G32B	0.0008–0.0016	0.0080–0.0160	0.0080–0.0160	0.0080–0.0280	0.0012–0.0028	0.0008–0.0024
	G64B	0.0008–0.0016	0.0100–0.0180	0.0080–0.0160	0.0080–0.0280	0.0012–0.0028	0.0008–0.0024
	6G72	0.0008–0.0016	0.0118–0.0177	0.0098–0.0157	0.0079–0.0276	0.0012–0.0035	0.0008–0.0024
1989	4G63	0.0008–0.0016	0.0100–0.0180	0.0080–0.0160	0.0080–0.0200	0.002–0.004	0.001–0.002
	G54B	0.0008–0.0016	0.0120–0.0200	0.0100–0.0160	0.0120–0.0310	0.002–0.004	0.001–0.002
	G15B	0.0008–0.0016	0.0080–0.0160	0.0080–0.0160	0.0080–0.0280	0.0012–0.0028	0.0008–0.0024
	4G61	0.0008–0.0016	0.0080–0.0160	0.0080–0.0160	0.0080–0.0280	0.0012–0.0028	0.0008–0.0024
	6G72	0.0008–0.0016	0.0118–0.0177	0.0098–0.0157	0.0079–0.0276	0.0012–0.0035	0.0008–0.0024

*Oil Control: Snug

PISTON AND RING SPECIFICATIONS—MONTERO & PICKUP

All measurements are given in inches.

Year	Engine	Piston Clearance	Ring Gap			Ring Side Clearance		
			Top Compression	Bottom Compression	Oil Control	Top Compression	Bottom Compression	Oil Control
1983	G63B	0.0008–0.0016	0.0100–0.0180	0.0080–0.0160	0.0080–0.0200	0.002–0.004	0.001–0.002	Snug
	4D55	0.0016–0.0024	0.0100–0.0160	0.0100–0.0160	0.0100–0.0180	0.001–0.002	0.001–0.003	0.001–0.003
	G54B	0.0008–0.0016	0.0100–0.0150	0.0100–0.0180	0.0120–0.0240	0.002–0.004	0.001–0.002	Snug
1984	G63B	0.0008–0.0016	0.0100–0.0180	0.0080–0.0160	0.0080–0.0200	0.002–0.004	0.001–0.002	Snug
	4D55	0.0016–0.0024	0.0100–0.0160	0.0100–0.0160	0.0100–0.0180	0.001–0.002	0.001–0.003	0.001–0.003
	G54B	0.0008–0.0016	0.0120–0.0180	0.0100–0.0150	0.0120–0.0240	0.002–0.004	0.001–0.002	Snug

3 ENGINE AND ENGINE OVERHAUL

PISTON AND RING SPECIFICATIONS—MONTERO & PICKUP

All measurements are given in inches.

Year	Engine	Piston Clearance	Ring Gap			Ring Side Clearance		
			Top Compression	Bottom Compression	Oil Control	Top Compression	Bottom Compression	Oil Control
1985	G63B	0.0008–0.0016	0.0100–0.0180	0.0080–0.0160	0.0080–0.0280	0.002–0.004	0.001–0.002	Snug
	4D55	0.0016–0.0024	0.0100–0.0160	0.0100–0.0160	0.0100–0.0180	0.001–0.002	0.001–0.003 ①	0.001–0.003
	G54B	0.0008–0.0016	0.0120–0.0180	0.0100–0.0150	0.0120–0.0240	0.002–0.004	0.001–0.002	Snug
1986	G63B	0.0008–0.0016	0.0100–0.0180	0.0080–0.0160	0.0080–0.0280	0.002–0.004	0.001–0.002	Snug
	G54B	0.0008–0.0016	0.0120–0.0180	0.0100–0.0150	0.0120–0.0240	0.002–0.004	0.001–0.002	Snug
1987	G63B	0.0004–0.0012	0.0100–0.0160	0.0080–0.0140	0.0080–0.0280	0.001–0.003	0.0008–0.0020	Snug
	G54B	0.0008–0.0016	0.0120–0.0180	0.0100–0.0160	0.0120–0.0310	0.0020–0.0035	0.0008–0.0024	Snug
1988	G63B	0.0004–0.0012	0.0100–0.0160	0.0080–0.0140	0.0080–0.0280	0.001–0.003	0.0008–0.0024	Snug
	G54B	0.0008–0.0016	0.0120–0.0180	0.0100–0.0160	0.0120–0.0310	0.0020–0.0035	0.0008–0.0024	Snug
1988	G63B	0.0008–0.0016	0.0100–0.0150	0.0080–0.0140	0.0080–0.0280	0.001–0.003	0.0008–0.0024	Snug
	G54B	0.0008–0.0016	0.0120–0.0180	0.0100–0.0160	0.0120–0.0310	0.0020–0.0035	0.0008–0.0024	Snug
	6G72	0.0008–0.0016	0.0120–0.0180	0.0100–0.0160	0.0080–0.0270	0.0012–0.0035	0.0008–0.0024	Snug

CRANKSHAFT AND CONNECTING ROD SPECIFICATIONS—AUTOMOBILES

All measurements are given in inches.

Year	Engine	Crankshaft				Connecting Rod		
		Main Brg. Journal Dia.	Main Brg. Oil Clearance	Shaft End-play	Thrust on No.	Journal Diameter	Oil Clearance	Side Clearance
1983	G62B	2.244	0.0008–0.0020	0.0020–0.0071	3	1.772	0.0008–0.0020	0.004–0.010
	G54B	2.362	0.0008–0.0020	0.0020–0.0071	3	2.087	0.0008–0.0024	0.004–0.010
1984	G62B	2.244	0.0008–0.0020	0.0020–0.0071	3	1.772	0.0008–0.0020	0.004–0.010
	G63B	2.244	0.0008–0.0020	0.0020–0.0071	3	1.772	0.0008–0.0020	0.004–0.010
	G54B	2.362	0.0008–0.0020	0.0020–0.0071	3	2.087	0.0008–0.0024	0.004–0.010
1985	G62B	2.244	0.0008–0.0020	0.0020–0.0071	3	1.772	0.0008–0.0020	0.004–0.010
	G63B	2.244	0.0008–0.0020	0.0020–0.0071	3	1.772	0.0008–0.0020	0.004–0.010
	G54B	2.362	0.0008–0.0020	0.0020–0.0071	3	2.087	0.0008–0.0024	0.004–0.010

ENGINE AND ENGINE OVERHAUL 3

CRANKSHAFT AND CONNECTING ROD SPECIFICATIONS—AUTOMOBILES

All measurements are given in inches.

Year	Engine	Crankshaft				Connecting Rod		
		Main Brg. Journal Dia.	Main Brg. Oil Clearance	Shaft End-play	Thrust on No.	Journal Diameter	Oil Clearance	Side Clearance
1985	G15B	1.889	0.0008–0.0020	0.0020–0.0071	3	1.653	0.0004–0.0024	0.004–0.010
	G32B	2.244	0.0008–0.0020	0.0020–0.0071	3	1.772	0.0004–0.0024	0.004–0.010
	G64B	2.244	0.0008–0.0020	0.0020–0.0071	3	2.087	0.0008–0.0024	0.004–0.010
1986	G62B	2.244	0.0008–0.0020	0.0020–0.0071	3	1.772	0.0008–0.0020	0.004–0.010
	G63B	2.244	0.0008–0.0020	0.0020–0.0071	3	1.772	0.0008–0.0020	0.004–0.010
	G54B	2.362	0.0008–0.0020	0.0020–0.0071	3	2.087	0.0008–0.0024	0.004–0.010
	G15B	1.889	0.0008–0.0020	0.0020–0.0071	3	1.653	0.0004–0.0024	0.004–0.010
	G32B	2.244	0.0008–0.0020	0.0020–0.0071	3	1.772	0.0004–0.0024	0.004–0.010
	G64B	2.244	0.0008–0.0020	0.0020–0.0071	3	2.087	0.0008–0.0024	0.004–0.010
1987	G62B	2.244	0.0008–0.0020	0.0020–0.0071	3	1.772	0.0008–0.0020	0.004–0.010
	G63B	2.244	0.0008–0.0020	0.0020–0.0071	3	1.772	0.0008–0.0020	0.004–0.010
	G54B	2.362	0.0008–0.0020	0.0020–0.0071	3	2.087	0.0008–0.0024	0.004–0.010
	G15B	1.889	0.0008–0.0020	0.0020–0.0071	3	1.653	0.0004–0.0024	0.004–0.010
	G32B	2.244	0.0008–0.0020	0.0020–0.0071	3	1.772	0.0004–0.0024	0.004–0.010
	G64B	2.244	0.0008–0.0020	0.0020–0.0071	3	2.087	0.0008–0.0024	0.004–0.010
1988	G62B	2.244	0.0008–0.0020	0.0020–0.0071	3	1.772	0.0008–0.0020	0.004–0.010
	G63B	2.244	0.0008–0.0020	0.0020–0.0071	3	1.772	0.0008–0.0020	0.004–0.010
	G54B	2.362	0.0008–0.0020	0.0020–0.0071	3	2.087	0.0008–0.0024	0.004–0.010
	G15B	1.889	0.0008–0.0020	0.0020–0.0071	3	1.653	0.0004–0.0024	0.004–0.010
	G32B	2.244	0.0008–0.0020	0.0020–0.0071	3	1.772	0.0004–0.0024	0.004–0.010
	G64B	2.244	0.0008–0.0020	0.0020–0.0071	3	2.087	0.0008–0.0024	0.004–0.010
	6G72	2.362	0.0008–0.0019	0.0020–0.0098	3	1.969	0.0006–0.0018	0.008–0.016

3 ENGINE AND ENGINE OVERHAUL

CRANKSHAFT AND CONNECTING ROD SPECIFICATIONS—AUTOMOBILES

All measurements are given in inches.

Year	Engine	Crankshaft				Connecting Rod		
		Main Brg. Journal Dia.	Main Brg. Oil Clearance	Shaft End-play	Thrust on No.	Journal Diameter	Oil Clearance	Side Clearance
1989–90	4G63	2.244	0.0008–0.0020	0.0020–0.0071	3	1.772	0.0008–0.0020	0.004–0.010
	G54B	2.362	0.0008–0.0020	0.0020–0.0071	3	2.087	0.0008–0.0024	0.004–0.010
	G15B	1.889	0.0008–0.0020	0.0020–0.0071	3	1.653	0.0004–0.0024	0.004–0.010
	4G61	2.244	0.0008–0.0020	0.0020–0.0071	3	1.772	0.0004–0.0024	0.004–0.010
	6G72	2.362	0.0008–0.0019	0.0020–0.0098	3	1.969	0.0006–0.0018	0.008–0.016

CRANKSHAFT AND CONNECTING ROD SPECIFICATIONS—MONTERO & PICKUP

All measurements are given in inches.

Year	Engine	Crankshaft				Connecting Rod		
		Main Brg. Journal Dia.	Main Brg. Oil Clearance	Shaft End-play	Thrust on No.	Journal Diameter	Oil Clearance	Side Clearance
1983	G63B	2.2440	0.0008–0.0020	0.002–0.007	3	1.7720	0.0008–0.0020	0.004–0.010
	4D55	2.5980	0.0008–0.0020	0.0008–0.0020	3	2.0866	0.0008–0.0020	0.004–0.010
	G54B	2.3622	0.0008–0.0020	0.002–0.007	3	2.0866	0.0008–0.0020	0.004–0.010
1984	G63B	2.2440	0.0008–0.0020	0.002–0.007	3	1.7720	0.0008–0.0020	0.004–0.010
	4D55	2.5980	0.0008–0.0020	0.0008–0.0020	3	2.0866	0.0008–0.0020	0.004–0.010
	G54B	2.3622	0.0008–0.0020	0.002–0.007	3	2.0866	0.0008–0.0020	0.004–0.010
1985	G63B	2.2440	0.0008–0.0020	0.002–0.007	3	1.7720	0.0008–0.0020	0.004–0.010
	4D55	2.5980	0.0008–0.0020	0.0008–0.0020	3	2.0866	0.0008–0.0020	0.004–0.010
	G54B	2.3622	0.0008–0.0020	0.002–0.007	3	2.0866	0.0008–0.0020	0.004–0.010
1986	G63B	2.2440	0.0008–0.0020	0.002–0.007	3	1.7720	0.0008–0.0020	0.004–0.010
	4D55	2.5980	0.0008–0.0020	0.0008–0.0020	3	2.0866	0.0008–0.0020	0.004–0.010
	G54B	2.3622	0.0008–0.0020	0.002–0.007	3	2.0866	0.0008–0.0020	0.004–0.010
1987	G63B	2.2440	0.0008–0.0020	0.002–0.007	3	1.7720	0.0008–0.0020	0.004–0.010
	G54B	2.3622	0.0008–0.0020	0.002–0.007	3	2.0866	0.0008–0.0020	0.004–0.010

ENGINE AND ENGINE OVERHAUL 3

CRANKSHAFT AND CONNECTING ROD SPECIFICATIONS—MONTERO & PICKUP

All measurements are given in inches.

Year	Engine	Crankshaft				Connecting Rod		
		Main Brg. Journal Dia.	Main Brg. Oil Clearance	Shaft End-play	Thrust on No.	Journal Diameter	Oil Clearance	Side Clearance
1988	G63B	2.2440	0.0008–0.0020	0.002–0.007	3	1.7720	0.0008–0.0020	0.004–0.010
	G54B	2.3622	0.0008–0.0020	0.002–0.007	3	2.0866	0.0008–0.0020	0.004–0.010
1989	G63B	2.2440	0.0008–0.0020	0.002–0.007	3	1.7720	0.0008–0.0020	0.0039–0.0098
	G54B	2.3622	0.0008–0.0020	0.0020–0.0071	3	2.090	0.0007–0.0022	0.0039–0.0098
	6G72	2.3622	0.0008–0.0019	0.0020–0.0098	3	1.970	0.0008–0.0019	0.0039–0.0098

COMPRESSOR-RELATED TORQUE SPECIFICATIONS

All values in ft. lbs.

Engine	Compressor Bracket to Engine Bolts	Compressor to Bracket Bolts	Idler Pulley Bracket Mounting Bolts	Compressor Through-Bolt	Alternator Bolts	High Pressure Line to Compressor (Discharge)	Low Pressure Line to Compressor (Suction)
Cordia/Tredia Mirage except 4G61 Precis	34	34	14	—	—	18	18
Galant G64B	34	14 ① 26	14	18	—	16	16
Mirage 4G61 Galant 4G63	36	18	18 ② 14	—	18	16	16
Starion	34 18 ①	18	14	—	—	24	16
Galant V-6 Sigma	34	34	14	—	—	16	16
Montero 2.6L	34	34	14	18	—	16	24
3.0L	34	34	14	18	—	18	18
Pickup	34	34	14	18	—	18	18

① Shorter bolts
② Mirage 4G61 only

3–45

3 ENGINE AND ENGINE OVERHAUL

TORQUE SPECIFICATIONS
All readings in ft. lbs.

Year	Engine	Cylinder Head Bolts ①	Main Bearing Bolts	Rod Bearing Bolts	Crankshaft Pulley Bolts	Flywheel Bolts	Manifold Intake	Manifold Exhaust
1983	G62B	73–79	38	37	80–94	94–101	11–14	11–14
	G54B	73–79	55–61	33	80–94	94–101	11–14	11–14
	4D55	76–83 ②	55–61	34	123–137	94–101	11–14	11–14
1984	G62B	73–79	38	37	80–94	94–101	11–14	11–14
	G63B	73–79	38	37	80–94	94–101	11–14	11–14
	G54B	73–79	55–61	33	80–94	94–101	11–14	11–14
	4D55	76–83 ②	55–61	34	123–137	94–101	11–14	11–14
1985	G62B	73–79	38	37	80–94	94–101	11–14	11–14
	G63B	73–79	38	37	80–94	94–101	11–14	11–14
	G54B	73–79	55–61	33	80–94	94–101	11–14	11–14
	G15B	58–61	38	24	51–72	94–101	11–14	11–14
	G32B	58–61	38	24	80–93	94–101	11–14	11–14
	G64B	73–79	38	33	80–94	94–101	11–14	11–14
	4D55	76–83 ②	55–61	34	123–137	94–101	11–14	11–14
1986	G62B	73–79	38	37	80–94	94–101	11–14	11–14
	G63B	73–79	38	37	80–94	94–101	11–14	11–14
	G54B	73–79	55–61	33	80–94	94–101	11–14	11–14
	G15B	58–61	38	24	51–72	94–101	11–14	11–14
	G32B	58–61	38	24	80–93	94–101	11–14	11–14
	G64B	73–79	38	33	80–94	94–101	11–14	11–14
1987	G62B	73–79	38	37	80–94	94–101	11–14	11–14
	G63B	73–79	38	37	80–94	94–101	11–14	11–14
	G54B	73–79	55–61	33	80–94	94–101	11–14	11–14
	G15B	58–61	38	24	51–72	94–101	11–14	11–14
	G32B	58–61	38	24	80–93	94–101	11–14	11–14
	G64B	73–79	38	33	80–94	94–101	11–14	11–14
1988	G62B	73–79	38	37	80–94	94–101	11–14	11–14
	G63B	73–79	38	37	80–94	94–101	11–14	11–14
	G54B	73–79	55–61	33	80–94	94–101	11–14	11–14
	G15B	58–61	38	24	51–72	94–101	11–14	11–14
	G32B	58–61	38	24	80–93	94–101	11–14	11–14
	G64B	73–79	38	33	80–94	94–101	11–14	11–14
	G672	73–79	55–61	38	109–115	53–55	11–14	11–16
1989–90	G64B	73–79	55–61	33	80–94	94–101	11–14	11–14
	4G61	58–61	38	24	51–72	94–101	11–14	11–14
	4G61	58–61	38	24	80–93	94–101	11–14	11–14
	4G63-SOHC	72–80	36–40	36–38	80–94	94–101	13–18	18–22
	4G63-DOHC	72–80	47–51	36–38	80–94	94–101	18–22	18–22
	6G72	73–79	55–61	38	109–115	53–55	11–14	11–16

① All figures are hot torque except as noted. See text for cold specs.
② Cold specification.

ENGINE AND ENGINE OVERHAUL 3

Standard Torque Specifications and Fastener Markings

In the absence of specific torques, the following chart can be used as a guide to the maximum safe torque of a particular size/grade of fastener.
- There is no torque difference for fine or coarse threads.
- Torque values are based on clean, dry threads. Reduce the value by 10% if threads are oiled prior to assembly.
- The torque required for aluminum components or fasteners is considerably less.

U.S. Bolts

SAE Grade Number / Number of lines always 2 less than the grade number. / Bolt Size (Inches)—(Thread)	1 or 2			5			6 or 7		
	Maximum Torque			Maximum Torque			Maximum Torque		
	Ft./Lbs.	Kgm	Nm	Ft./Lbs.	Kgm	Nm	Ft./Lbs.	Kgm	Nm
¼ — 20	5	0.7	6.8	8	1.1	10.8	10	1.4	13.5
— 28	6	0.8	8.1	10	1.4	13.6			
5/16 — 18	11	1.5	14.9	17	2.3	23.0	19	2.6	25.8
— 24	13	1.8	17.6	19	2.6	25.7			
⅜ — 16	18	2.5	24.4	31	4.3	42.0	34	4.7	46.0
— 24	20	2.75	27.1	35	4.8	47.5			
7/16 — 14	28	3.8	37.0	49	6.8	66.4	55	7.6	74.5
— 20	30	4.2	40.7	55	7.6	74.5			
½ — 13	39	5.4	52.8	75	10.4	101.7	85	11.75	115.2
— 20	41	5.7	55.6	85	11.7	115.2			
9/16 — 12	51	7.0	69.2	110	15.2	149.1	120	16.6	162.7
— 18	55	7.6	74.5	120	16.6	162.7			
⅝ — 11	83	11.5	112.5	150	20.7	203.3	167	23.0	226.5
— 18	95	13.1	128.8	170	23.5	230.5			
¾ — 10	105	14.5	142.3	270	37.3	366.0	280	38.7	379.6
— 16	115	15.9	155.9	295	40.8	400.0			
⅞ — 9	160	22.1	216.9	395	54.6	535.5	440	60.9	596.5
— 14	175	24.2	237.2	435	60.1	589.7			
1 — 8	236	32.5	318.6	590	81.6	799.9	660	91.3	894.8
— 14	250	34.6	338.9	660	91.3	849.8			

Metric Bolts

Relative Strength Marking / Bolt Markings / Bolt Size Thread Size x Pitch (mm)	4.6, 4.8			8.8		
	Maximum Torque			Maximum Torque		
	Ft./Lbs.	Kgm	Nm	Ft./Lbs.	Kgm	Nm
6 x 1.0	2-3	.2-.4	3-4	3-6	.4-.8	5-8
8 x 1.25	6-8	.8-1	8-12	9-14	1.2-1.9	13-19
10 x 1.25	12-17	1.5-2.3	16-23	20-29	2.7-4.0	27-39
12 x 1.25	21-32	2.9-4.4	29-43	35-53	4.8-7.3	47-72
14 x 1.5	35-52	4.8-7.1	48-70	57-85	7.8-11.7	77-110
16 x 1.5	51-77	7.0-10.6	67-100	90-120	12.4-16.5	130-160
18 x 1.5	74-110	10.2-15.1	100-150	130-170	17.9-23.4	180-230
20 x 1.5	110-140	15.1-19.3	150-190	190-240	26.2-46.9	160-320
22 x 1.5	150-190	22.0-26.2	200-260	250-320	34.5-44.1	340-430
24 x 1.5	190-240	26.2-46.9	260-320	310-410	42.7-56.5	420-550

3 ENGINE AND ENGINE OVERHAUL

ENGINE

REMOVAL AND INSTALLATION

Cordia and Tredia

NOTE: The engine and transaxle are removed as a unit, using a crane-type lift. All wires and hoses should be labeled at the time of removal. The amount of time saved during reassembly makes the extra effort well worthwhile.

1. Matchmark and then unbolt and remove the front hood. Disconnect both battery connectors and remove the battery.
2. Drain the engine coolant and transaxle fluid. Drain and remove the coolant reserve jug and the windshield washer jug.

---- CAUTION ----
When draining the coolant, keep in mind that cats and dogs are attracted by the ethylene glycol antifreeze, and are quite likely to drink any that is left in an uncovered container or in puddles on the ground. This will prove fatal in sufficient quantity. Always drain the coolant into a sealable container. Coolant should be reused unless it is contaminated or several years old.

3. Safely discharge the air conditioning system. See Section 1.
4. Disconnect the power steering fluid return hose (the hose that runs from the reservoir to the steering box, not the pump) at the reservoir, and drain the fluid into a clean container.
5. Disconnect the upper and lower radiator hoses at both ends and remove them. If the car has air conditioning, disconnect those hoses as close as possible to the condenser unit in front of the radiator. On 1987–88 models, disconnect at the compressor, and cap the openings securely. Remove the radiator or radiator and condenser assembly.
6. Remove the battery tray. Disconnect both heater hoses from the side of the engine.
7. Remove the air cleaner. On turbocharged vehicles, disconnect the turbocharger intake hose.
8. Disconnect the brake booster vacuum hose. On turbocharged vehicles only, disconnect the oil cooler hoses at the engine, avoiding spilling the oil and then capping the openings. On vehicles with an automatic transmission, disconnect the transmission cooler hoses, avoiding spilling the oil, and then capping the openings.
9. On vehicles with a manual transmission, disconnect the clutch cable on the side of the transaxle; on vehicles with an automatic, disconnect the shift control cable on the side of the unit. On all vehicles, disconnect the speedometer cable from the transaxle.
10. Disconnect the accelerator cable at the side of the engine. Disconnect the engine ground strap at the right front fender. On vehicles with an air conditioner, disconnect the hoses at the compressor and cap all openings securely.
11. Disconnect the power steering hoses at the side of the pump. Cap all openings. Disconnect the coil low and high tension wires. Label the low tension wires for reassembly to the proper terminals. Disconnect the battery negative cable from the side of the engine.
12. Label and disconnect the alternator connectors. Disconnect the oil pressure sending unit wire.
13. Remove the mounting screws for the vacuum unit and solenoid valve, and disconnect the electrical connector. Move the unit aside.
14. Disconnect the two smaller vacuum hoses from the purge control valve, remove the mounting screw, and move the unit aside.
15. Loosen the clamps and disconnect the vacuum hoses going to the evaporative emissions canister.
16. Disconnect the fuel return hose from the carburetor (normally aspirated engines) or injection mixer (turbo engines). Disconnect the fuel supply hose at the fuel filter.
17. Raise the vehicle and support it securely by the body or a crossmember. Disconnect the exhaust pipe at the manifold. Fasten the exhaust pipe with wire to keep it from falling.
18. On manual transmission vehicles, disconnect the shift control rod and extension and remove them.
19. Disconnect the left and right side strut bars and stabilizer bars where they connect to the lower control arms. Then remove the bolts fastening the control arms on both sides to the rearward crossmember.
20. Disconnect the lower arm ball joint at the steering knuckle on both sides. Disconnect the strut bar and stabilizer bar at the lower control arms. Now, using a prybar inserted between the transaxle case and driveshaft, carefully pry the driveshaft out of the transaxle on each side. Plug the openings in the transaxle to prevent dirt from getting in. Carefully lower all parts to the crossmember. Discard the retaining clips for the driveshafts. They must be replaced.

Disconnecting the clutch cable on Cordia and Tredia with manual transmission

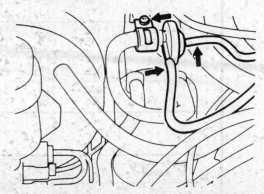

Disconnecting the purge control valve; Cordia and Tredia

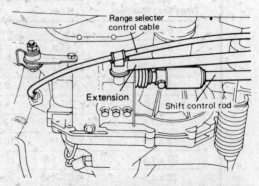

Disconnect the shift control rod and extension on Cordia and Tredia with manual transmission

3–48

ENGINE AND ENGINE OVERHAUL 3

21. Attach a cable securely supported by a lift and pulley arrangement to each engine lifting point. Put tension on all the cables to support the engine securely.

―――― CAUTION ――――
The vehicle is on jackstands. Raise the hoist only enough to tension the cables; do not disturb the position of the car on the stands.

23. Remove the nut from the left side engine mount insulator. Remove the four front roll bracket mounting bolts located on the side of the front crossmember.
24. Remove the mounting bolt from the rear roll insulator. Remove the nuts attaching the left engine mount insulator to the fender.
25. From inside the right fender shield, detach the protective cap and then remove the transaxle insulator bracket mounting bolts. Remove the bolts connecting the transaxle mount insulator.
26. Remove the bolts to the shift control selector. Remove the wiring connector going to the transaxle. Disconnect vacuum hoses.
27. Remove the transaxle insulator bracket. Increase the tension on the lifting cables so the engine weight is supported entirely by the cables and none of the weight is on the mounts. Remove the bolts passing through the insulators of the rear roll stop and left mounting bracket.
28. Check to make sure that all items are disconnected from the engine/transaxle assembly. Press downward on the transaxle to guide the assembly and lift it carefully out of the vehicle.
29. Mount the engine securely on an engine support stand or support it on wooden blocks. Do not allow it to rest on the oil pan; do not allow it to rest on its side.
30. When reinstalling, the transaxle must be mounted to the engine and the assembly installed as a unit. Support the assembly from the engine hoist and carefully move it into place within the car. Remember that the transaxle must be tilted downward to enter the body.
31. Install the insulator bolts for the rear roll stooper and the left mount bracket.
32. Adjust the engine hoist so that the weight is supported, but the mount holes align. Install the transaxle insulator bracket.
33. Install the transaxle mount insulator.
34. Install the transaxle insulator bracket mounting bolts. Install the cap within the right wheel well.
35. Mount the left mount insulator to the fender and tighten the nuts to 25 ft. lbs.
36. Install the rear roll insulator mounting bolt and tighten it to 25 ft. lbs.
37. Install the front roll bracket mounting bolts, tightening them to 33 ft. lbs.
38. Install the left mount insulator nut and tighten to 50 ft. lbs.

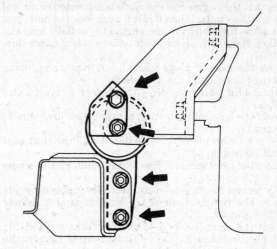

Remove the four front roll bracket mounting bolts on the side of the No. 1 crossmember

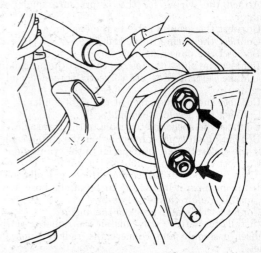

Remove the nuts attaching the left mount insulator to the fender

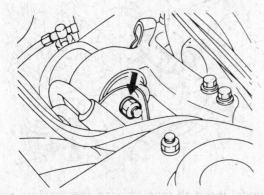

Remove the left mount insulator nut (Cordia and Tredia)

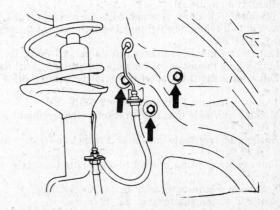

Remove the bolts connecting the transaxle mount insulator and the bracket to the fender

3-49

3 ENGINE AND ENGINE OVERHAUL

39. Check that the engine and transaxle is supported or pinned in place by all its mounts. When this is true, loosen the hoist just enough to allow the weight of the engine to partially load the mounts. Keep some tension on the hoist as a safety factor during reassembly.
40. Remove the caps or plugs from the transaxle and, using new retaining clips, install the driveshafts.
41. Install the bolts holding the lower arms to the crossmember.
42. Connect the left and right strut bars and stabilizer bars to the lower arms.
43. On vehicles with manual transaxles, install the shift control rod and extension.
44. Connect the front exhaust pipe to the manifold. Use a new gasket.
45. Slowly loosen the engine hoist until the engine is totally supported by the vehicle. Remove the hoist apparatus. Lower the car to the ground.
46. Connect the main fuel hose to the fuel filter. connect the fuel return hose to the carburetor; if turbocharged, connect the fuel return hose to the injection mixer.
47. Reinstall the canister and connect the hoses properly.
48. Install the purge control valve and connect its hoses.
49. Install the vacuum control unit and solenoid valve and attach the connector.
50. Attach the wiring for the oil pressure sender and the alternator.
51. Connect the ground cable to the engine.
52. Connect the wiring to the coil.
53. Connect the power steering hoses.
54. Remove the plugs or caps and connect the air conditioning hoses. Make certain they are correctly threaded and do not overtighten.
55. Connect the ground strap at the right fender.
56. Connect the accelerator cable.
57. Connect the speedometer cable to the transaxle.
58. Depending on type, connect either the clutch cable (manual) or the control cable (automatic) to the transaxle.
59. Connect the oil cooler lines to the automatic transaxle. If the engine is turbocharged, connect the engine oil cooler lines.
60. If turbocharged, install the turbo intake hose. On all engines, install the air cleaner assembly.
61. Connect the heater hoses to the engine.
62. Install the battery tray. Install the radiator or radiator/condenser assembly and connect the radiator hoses to the engine.
63. Install the washer tank and the radiator reserve tank.
64. Install the battery. Connect only the positive cable until ready to start the engine.
65. Refill the engine coolant, the power steering fluid and the transaxle oil.
66. Refill the engine oil.
67. Double check all mounts and bolts for correct torque. Engine mount bolts should be tightened with the load of the engine applied, as follows:
 • Left engine mount insulator nut (large): 43-58 ft. lbs.
 • Left engine mount insulator nuts (small): 22-29 ft. lbs.
 • Left engine mount bracket-to-engine bolts/nuts: 36-47 ft. lbs.
 • Transaxle mount insulator nut: 43-58 ft. lbs.
 • Manual transaxle insulator bracket-to-fender shield bolts: 40-43 ft. lbs.
 • Automatic transaxle insulator bracket-to-fender shield bolts: 22-29 ft. lbs.
 • Transaxle mounting bracket-to-automatic transaxle nuts: 43-58 ft. lbs.
 • Transaxle mounting bracket bolts: 22-29 ft. lbs.
 • Rear roll stop insulator nut: 22-29 ft. lbs.
 • Rear roll stop-to-rear crossmember nuts: 43-58 ft. lbs.
 • Rear roll stop bracket-to-rear roll stop stay bolt: 22-29 ft. lbs.
 • Front roll stop insulator nut: 36-47 ft. lbs.
 • Front roll bracket-to-front crossmember nuts: 29-36 ft. lbs.

68. Double check all installation items, paying particular attention to loose hoses or hanging wires, untightened nuts, poor routing of hoses and wires (too tight or rubbing) and tools left in the engine area.
69. Connect the negative battery cable and start the engine. Check carefully for leaks. Check all gauges for proper readings. Adjust clutch and shift linkages. Adjust the accelerator cable.
70. Recharge the air conditioning system.
71. With the help of an assistant, install the hood and adjust it for proper fit and latching.

Starion, 1983-1986

NOTE: All wires and hoses should be labeled at the time of removal. The amount of time saved during reassembly makes the extra effort well worthwhile.

1. Matchmark and then unbolt and remove the front hood. Drain the cooling system by unscrewing the drain cock in the bottom of the radiator and the plug in the side of the block.

CAUTION

When draining the coolant, keep in mind that cats and dogs are attracted by the ethylene glycol antifreeze, and are quite likely to drink any that is left in an unaaaered container or in puddles on the ground. This will prove fatal in sufficient quantity. Always drain the coolant into a sealable container. Coolant should be reused unless it is contaminated or several years old.

---

2. Disconnect the accelerator cable at the injection mixer. Disconnect both battery cables.
3. Disconnect the two heater hoses at the block and the brake booster vacuum hose at the intake manifold.
4. Safely relive pressure within the fuel system. Disconnect the fuel hoses at the injection system. One uses a hose clamp and the other a bolted flanged fitting. If the engine is hot, be careful not to spill fuel.
5. Disconnect the high tension wire at the center of the distributor. Disconnect the temperature sensor wire nearby. Disconnect the intake manifold ground cable connector. First label the wires for reassembly, and then disconnect the starter motor wiring harness.
6. Remove the power steering pump. Unbolt and disconnect the two engine oil cooler hoses at the oil filter adapter. Plug all openings to prevent the entry of dirt and leakage of oil.

NOTE: If equipped with air conditioning, disconnect the wire at the compressor. Unbolt the compressor from its mounts and move it out of the way, supported with stiff wire. It is not necessary to discharge the system or remove hoses.

Disconnect the high tension wire, temperature sensor connector and intake manifold ground connector on the Starion

ENGINE AND ENGINE OVERHAUL 3

7. Remove the radiator assembly.
8. Label and disconnect the alternator wiring.
9. Disconnect the engine ground cable. Disconnect the two plugs for the electronic injection (ECI) wiring harness.
10. Disconnect the vacuum hose from the boost sensor, located on the firewall.
11. Remove the rear catalytic converter. This requires unbolting the flanges at the rear of the front catalytic converter and at the rear of the rear converter.
12. Unscrew and disconnect the speedometer cable at the transmission. Disconnect the wiring for the oil pressure gauge sending unit on the block.
13. On automatic transmission-equipped vehicles, disconnect both oil cooler hoses and plug all openings. On all cars, disconnect the back-up light switch harness at the plug located under the transmission.
14. Remove the driveshaft.
15. Remove the clutch slave cylinder and suspend it out of the way with a piece of wire.
16. Put the gearshift in neutral. Unbolt the gearshift lever assembly by going in under the car and unbolting the unit. It's located on the top rear of the transmission.
17. Securely support the engine via a chain or cable arrangement and a crane hoist. Use both hooks located at the top of the block. Support the transmission with a jack to take load off the rear mounts.
18. Remove the front and rear engine mounting nuts and bolts, and the rear crossmember. Raise the assembly slightly and remove all the front support brackets and insulators. Gradually lower the transmission jack and pull the engine and transmission assembly out by raising the front of the engine so the transmission will clear the firewall.
19. Mount the engine securely on an engine support stand or support it on wooden blocks. Do not allow it to rest on the oil pan; do not allow it to rest on its side.
20. When reinstalling, the transaxle must be mounted to the engine and the assembly installed as a unit. Support the assembly from the engine hoist and carefully move it into place within the car. Remember that the transaxle must be tilted downward to enter the body.
21. Install the gearshift lever assembly.
22. Install the engine mounting bolts and brackets. Make sure all holes are properly aligned and that mounts are not distorted. On both front insulators, make sure the locating boss and hole in the insulator are in alignment. Torque values are as follows:
- Crossmember mounting bolts: 7.2 ft. lbs.
- Bolts holding front mounting brackets to the front crossmember: 22–29 ft. lbs.
- Nuts holding front mounts, and the bolt attaching the rear mount to the rear crossmember: 9.4–14 ft.-lbs.
- Nuts at the top of the rear mount (attaching it to the transmission): 14–17 ft. lbs.

23. On the rear crossmember insulators, after bolts are torqued, turn them until a flat will line up with lockwasher tabs and then bend the tabs up against the flats to keep the bolts from turning.
24. Connect the wiring to the oil pressure sender.
25. Install the driveshaft.
26. Connect the reverse light harness and the speedometer cable to the transmission.
27. Install the rear catalytic converter. Use new gaskets.
28. Connect the boost sensor hose and connect the ECI wiring harnesses.
29. Connect the alternator wiring and the engine ground strap.
30. Install the radiator and its hoses.
31. Connect the engine oil cooler hoses.
32. Install the power steering pump. If equipped with air conditioning, reinstall the compressor onto its bracket and connect its wiring.

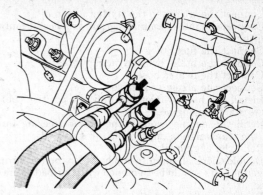
Disconnect the engine oil cooler lines on the Starion

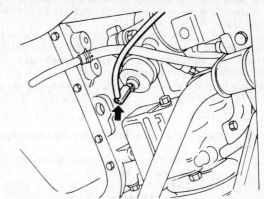
Disconnect the Starion oil pressure sending unit

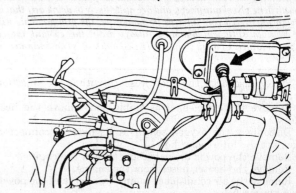
Removing the boost sensor hose

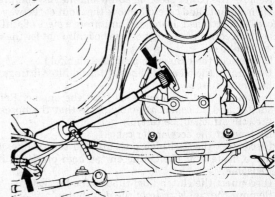
Location of the Starion speedometer cable and reverse light connector

3 ENGINE AND ENGINE OVERHAUL

33. Connect the wiring to the starter.
34. Connect the intake manifold ground cable and connect the coil wiring.
35. Connect the fuel hoses.
36. Connect the brake booster vacuum hose.
37. Connect the heater hoses.
38. Connect the accelerator cable.
39. Install the battery. Connect only the positive cable until ready to start the engine.
40. Refill the engine coolant, the power steering fluid and check the transmission oil, topping off if needed.
41. Refill the engine oil.
42. Double check all mounts and bolts for correct torque. Engine mount bolts should be tightened with the load of the engine applied.
43. Double check all installation items, paying particular attention to loose hoses or hanging wires, untightened nuts, poor routing of hoses and wires (too tight or rubbing) and tools left in the engine area.
44. Connect the negative battery cable and start the engine. Check carefully for leaks. Check all gauges for proper readings. Adjust clutch and shift linkages. Adjust the accelerator cable.
45. Recharge the air conditioning system.
46. With the help of an assistant, install the hood and adjust it for proper fit and latching.

Starion, 1987–1989

1. Matchmark and remove the hood. Disconnect the negative battery cable.
2. Drain the engine oil and transmission oil. Drain the clutch system if equipped with manual shift.
3. Drain the cooling system.

CAUTION

When draining the coolant, keep in mind that cats and dogs are attracted by the ethylene glycol antifreeze, and are quite likely to drink any that is left in an uncovered container or in puddles on the ground. This will prove fatal in sufficient quantity. Always drain the coolant into a sealable container. Coolant should be reused unless it is contaminated or several years old.

---

4. Safely discharge the fuel pressure within the injection lines.
5. Remove the air cleaner assembly and remove the heat shield(s).
6. Disconnect the oxygen sensor at its harness connector. Disconnect the intercooler hoses.
7. Remove the power brake booster hose at the engine.
8. Remove the heater hoses from the engine.
9. Remove the air conditioner compressor belt and the power steering pump belt.
10. Remove the power steering pump and the A/C compressor from their mounts and carefully move them out of the way with the hoses still attached. Hang each unit from a piece of stiff wire attached to a convenient point on the body. Do not let the units hang by the hoses.
11. Disconnect the exhaust pipe at the manifold.
12. Disconnect the oil cooler lines at the oil filter fittings. Cap the lines immediately.
13. Disconnect the radiator hoses from the engine. Remove the radiator and oil cooler assembly. Note that the hoses and lines are still attached; be careful.
14. Disconnect the accelerator cable.
15. Disconnect the fuel high pressure and return lines from the engine. Note that the O-ring on the high pressure line is NOT reusable.
16. Disconnect the clutch fluid tube at the bracket.
17. Remove the coil wire from the distributor.
18. Disconnect the speedometer cable from the transmission.
19. Disconnect the vacuum hoses at the firewall.
20. Label and disconnect all wiring running between the body and the engine. Disconnect each connector at the engine or transmission.

NOTE: All wires and hoses should be labeled at the time of removal. The amount of time saved during reassembly makes the extra effort well worthwhile.

21. The following items must be disconnected:
    a. Coolant temperature gauge sender
    b. Coolant temperature sensor
    c. Coolant temperature switch
    d. Secondary air solenoid valve
    e. EGR solenoid valve
    f. Injector harness
    g. Throttle position sensor
    h. ISC servo
    i. Motor position sensor
    j. Distributor signal generator
    k. Ground cable
    l. Alternator harness
    m. Oil pressure gauge unit
    n. Starter motor
    o. Detonation sensor
    p. Reverse light switch
    q. On cars with automatic transmissions, disconnect the overdrive cancel solenoid, the downshift solenoid, and the inhibitor switch harness. It is critical that these wires be labeled and reinstalled correctly.
22. Disconnect and remove the drive shaft.
23. Remove the gear shift assembly.
24. Install the hoist equipment and draw tension on the cables. Do not attempt to lift the engine.
25. Remove the nuts holding the engine mount brackets to the engine mounts. At the transmission mount, remove the four bolts holding the bracket to the frame.
26. Carefully elevate the hoist and guide the engine and transmission forward, up and out of the body. Securly mount the engine on a stand or support it on wooden block; do not allow it to rest on the oil pan or lie on its side.
27. Install the engine and transmission into the car. Align all the mounts and brackets. Install the four bolts at the transmission mount and tighten them to 84 inch lbs. Install the nuts on the front engine mounts and tighten them to 14 ft. lbs.
28. Install the gear shift lever assembly.
29. Install the driveshaft, tightening the rear flange bolts to 39 ft. lbs.
30. Remove the engine hoist apparatus.
31. Connect the wiring connectors to the engine, making certain each connection is clean and tight.
32. Connect the vacuum hoses and the speedometer cable. Connect the coil wire to the distributor.
33. Install the clutch fluid tube, tightening the fitting to 11 ft. lbs.
34. Using a new O-ring, attach the high pressure and return fuel lines.
35. Connect the accelerator cable.
36. Install the radiator and oil cooler with their lines. Attach the hoses to the engine and connect the oil cooler lines at the filter fitting. Tighten the bolts to 29 ft. lbs.
37. Connect the exhaust system, remembering to install the springs. Tighten the bolts to 26 ft. lbs.
38. Reposition the A/C compressor and power steering pump. Tighten the compressor mounting bolts to 18 ft. lbs. and the power steering pump mounting bolt to 20 ft. lbs. Tighten the pump adjusting bolt (upper) to 13 ft. lbs.
39. Install the drive belts for the air conditioning and power steering.
40. Connect the heater hoses.
41. Connect the brake booster vacuum hose to the engine.
42. Install the intercooler hoses to the turbocharger. Do not overtighten the clamps.

ENGINE AND ENGINE OVERHAUL 3

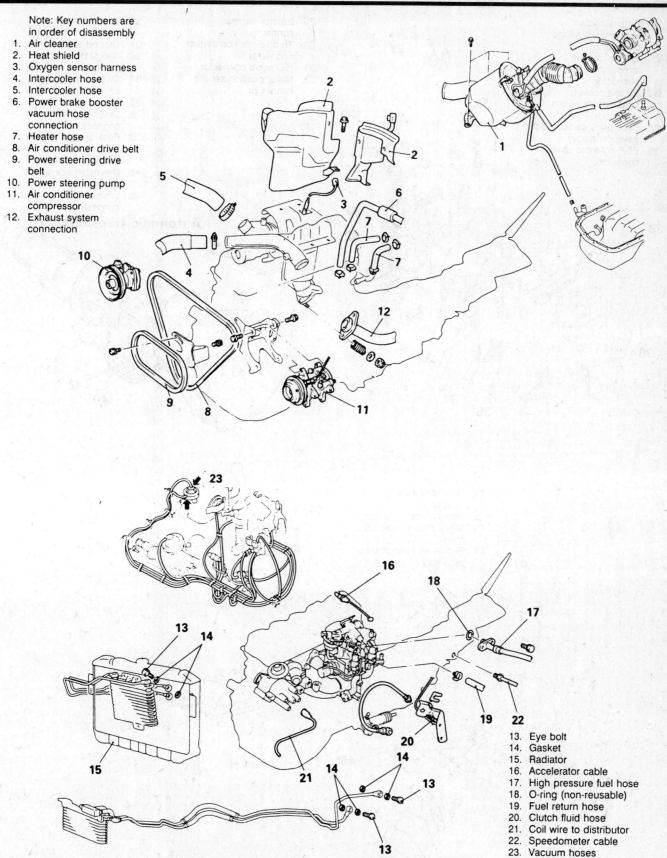

1987–89 Starion engine removal, external components

3 ENGINE AND ENGINE OVERHAUL

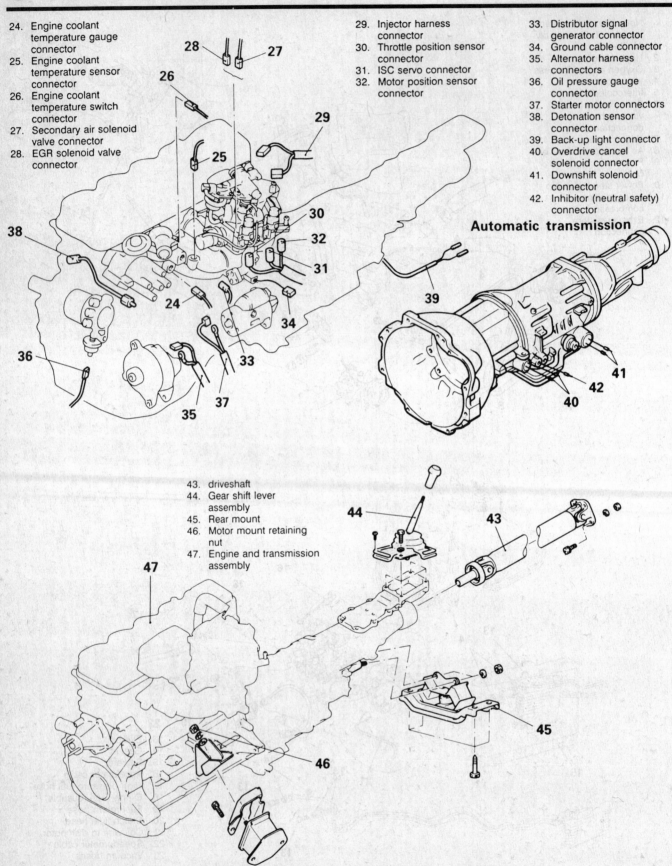

24. Engine coolant temperature gauge connector
25. Engine coolant temperature sensor connector
26. Engine coolant temperature switch connector
27. Secondary air solenoid valve connector
28. EGR solenoid valve connector
29. Injector harness connector
30. Throttle position sensor connector
31. ISC servo connector
32. Motor position sensor connector
33. Distributor signal generator connector
34. Ground cable connector
35. Alternator harness connectors
36. Oil pressure gauge connector
37. Starter motor connectors
38. Detonation sensor connector
39. Back-up light connector
40. Overdrive cancel solenoid connector
41. Downshift solenoid connector
42. Inhibitor (neutral safety) connector
43. driveshaft
44. Gear shift lever assembly
45. Rear mount
46. Motor mount retaining nut
47. Engine and transmission assembly

1987–89 Starion engine removal, external components

ENGINE AND ENGINE OVERHAUL 3

43. Connect the wiring for the oxygen sensor.
44. Install the heat shield(s) and install the air cleaner assembly.
45. Refill the clutch fluid and bleed the system.
46. Refill the engine coolant.
47. Refill the transmission oil.
48. Refill the engine oil.
49. Adjust the drive belts to the correct tension.
50. Double check all installation items, paying particular attention to loose hoses or hanging wires, untightened nuts, poor routing of hoses and wires (too tight or rubbing) and tools left in the engine area.
51. Start the engine and check carefully for leaks. Adjust the engine specifications as necessary. Adjust the accelerator cable and clutch freeplay. Make certain that each gauge operates correctly. Check the shifter assembly for correct function.
52. With the help of an assistant, reinstall the hood and adjust it for correct fit and latching.

1985-88 Mirage with G15B and G32B Engines

WARNING: Perform these operations only on a cold motor. If the car has been recently driven, allow it to cool at least 3-4 hours before beginning work. Overnight cooling is recommended.

1. Elevate and safely support the car as needed to gain access to components. Depending on tools, arm length and agility, each operation may be easier from above or below. Drain the engine coolant.

—— CAUTION ——
When draining the coolant, keep in mind that cats and dogs are attracted by the ethylene glycol antifreeze, and are quite likely to drink any that is left in an uncovered container or in puddles on the ground. This will prove fatal in sufficient quantity. Always drain the coolant into a sealable container. Coolant should be reused unless it is contaminated or several years old.

2. Drain the engine oil and the transmission oil. For G32B engines with manual transaxles, drain the clutch fluid.

—— CAUTION ——
Used motor oil may cause skin cancer if repeatedly left in contact with the skin for prolonged periods. Although this is unlikely unless you handle oil on a daily basis, it is wise to thoroughly wash your hands with soap and water immediately after handling used motor oil.

3. Remove the splash shield under the engine.
4. Matchmark the hood to the hinges and remove the hood.
5. Disconnect the battery, negative cable first. Remove the battery and the battery tray.
6. On vehicles equipped with automatic transaxle, disconnect the oil cooler hoses at the transaxle. Be prepared to deal with fluid spillage from the lines. Immediately plug the lines and cap the fittings on the transaxle the prevent entry of dirt and debris.
7. Disconnect the cooling hoses from the engine and remove the radiator. Note that the hoses and lines are still attached to the radiator; be careful during removal not to crimp or damage any lines. For G32B engines, disconnect the eye bolts (banjo fittings) for the oil cooler lines at the block.
8. Disconnect the heater hoses at the firewall.
9. Disconnect the accelerator cable from the engine.
10. Remove the main fuel hose and the fuel return hoses. Contain spillage immediately.

—— CAUTION ——
On the G32B engine, the fuel system is under pressure. Release pressure slowly and contain spillage. Observe no smoking/no open flame precautions. Have a Class B-C (dry powder) fire extinguisher within arm's reach at all times.

11. Disconnect the brake booster vacuum hose connection at the engine.
12. For cars with the G15B engine and a manual transaxle, disconnect the clutch cable and the selector cable. Note that the spring clips are not reusable; they must be replaced during reassembly. For cars with the G32B engine, disconnect the clutch fluid tube either at the clutch cylinder or at the first joint.
   For cars with automatic transaxles, disconnect the control cable and the throttle control cable.
13. Disconnect the back-up light connectors and, if automatic, disconnect the inhibitor switch connector at the rear of the transaxle.

NOTE: All wires and hoses should be labled at the time of removal. The amount of time saved during reassembly makes the extra effort well worthwhile.

14. Disconnect the speedometer cable.
15. Disconnect the starter motor wiring and the engine ground cable. Remove the coil wire from the distributor.
16. If equipped with air conditioning, loosen the belt tensioner pulley and remove the A/C drive belt. Do not loosen any hoses or discharge the system. Disconnect the wire from the compressor. Remove the four compressor mounting bolts and move the compressor away from the motor. Support it out of the way with a piece of stiff wire.
17. Loosen the power steering pump and remove the belt. Remove the pump mounting bolts and move the pump, with the hoses attached, out of the way. Hang the pump from a piece of stiff wire.
18. Disconnect the wiring from the alternator.
19. Label and disconnect all wiring running to the motor and any vacuum lines running to the firewall. The following electrical components must be disconnected:
   - oil pressure gauge sender
   - engine temperature gauge and sensor
   - coolant temperature switch (automatic trans only)
   - solenoid valve
   - feedback solenoid valve (G15B)
   - throttle position sensor
   - oxygen sensor connector
   - cold mixture heater connector
   - On the G32B fuel injected engine, also disconnect the idle speed control (ISC) servo motor, ISC position sensor, detonation sensor, distributor signal generator, EGR solenoid valve and the secondary air control solenoid valve.
20. Remove the self-locking nuts from the exhaust pipe to manifold joint. Separate the joint and remove the gasket. Suspend the pipe from a piece of wire.
21. Elevate and safely suport the car on jackstands. Remove the front wheels. Disconnect the stabilizer bar from the lower control arms.
22. Remove the cotter pin from the lower suspension ball joint. Loosen the nut but do NOT remove it. It should be unscrewed to the last point at which it is still fully threaded on the shaft. Install a ball joint separator (MB 991113 or equivalent) and slowly press the joint apart. Do not use a "pickle fork" or similar tool--the joint may be damaged. Repeat the procedure on the opposite side.
23. After each joint is separated (the nut will restrain a sudden separation), remove the nut and separate the joint.
24. Repeat the previous two steps and separate the tie rod from the steering knuckle. Insert a small pry bar into the gap between the transaxle case and the driveshaft. Do not insert it too far or an internal oil seal will be damaged.
25. Pry the driveshafts free of the transaxle. It will be necessary to swing the steering knuckle/hub/brake assembly to the outside of the body to gain clearance as the driveshaft comes free. Immediately suspend each axle with a piece of wire. Protect the boots and splined end of the shaft with rags or covers. Plug the holes in the transaxle case.

3-55

3 ENGINE AND ENGINE OVERHAUL

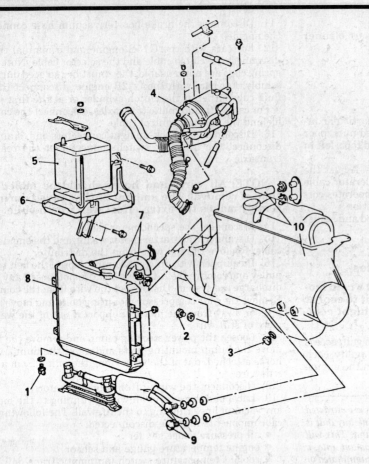

Note: Key numbers are in approximate order of disassembly

1. Radiator drain plug
2. Transaxle drain plug
3. Engine oil drain plug
4. Air cleaner
5. Battery
6. Battery tray
7. Auto. trans. oil cooler hoses
8. Radiator
9. Eye bolt
10. Heater hoses

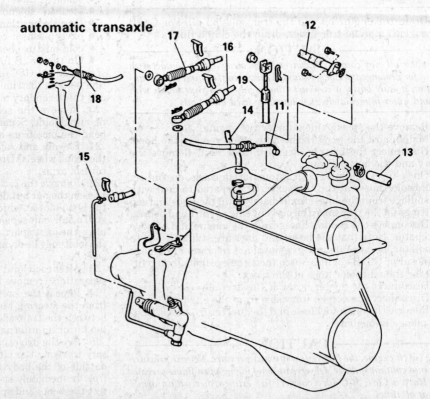

11. Accelerator cable
12. High pressure fuel hose
13. Fuel return hose
14. Brake booster vacuum hose
15. Clutch fluid tube (cable on G15B)
16. Shift cable (manual trans.)
17. Select cable (manual trans.)
18. Control cable (auto. trans.)
19. Throttle control cable (auto. trans.)

1985–89 Mirage engine removal, external components. G32B shown, G15B similar

ENGINE AND ENGINE OVERHAUL 3

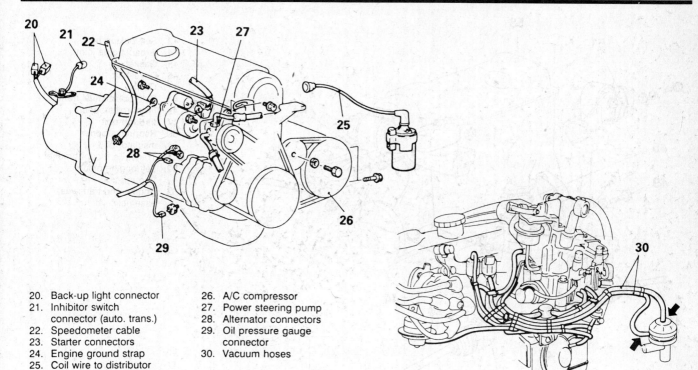

20. Back-up light connector
21. Inhibitor switch connector (auto. trans.)
22. Speedometer cable
23. Starter connectors
24. Engine ground strap
25. Coil wire to distributor
26. A/C compressor
27. Power steering pump
28. Alternator connectors
29. Oil pressure gauge connector
30. Vacuum hoses

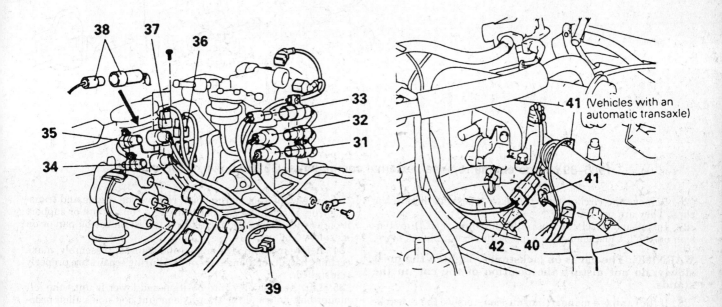

31. Throttle position sensor connector
32. Idle speed control servo motor connector
33. ISC position sensor connector
34. Detonation sensor connector
35. Distributor signal generator connector
36. EGR solenoid valve connector
37. Secondary air control solenoid valve connector
38. Oxygen sensor
39. Cold mixture heater connector
40. Coolant temperature sensor connector
41. Coolant temperature gauge connector
42. Coolant temperature switch connector

1985–89 Mirage engine removal, external components. G32B shown, G15B similar

3-57

3 ENGINE AND ENGINE OVERHAUL

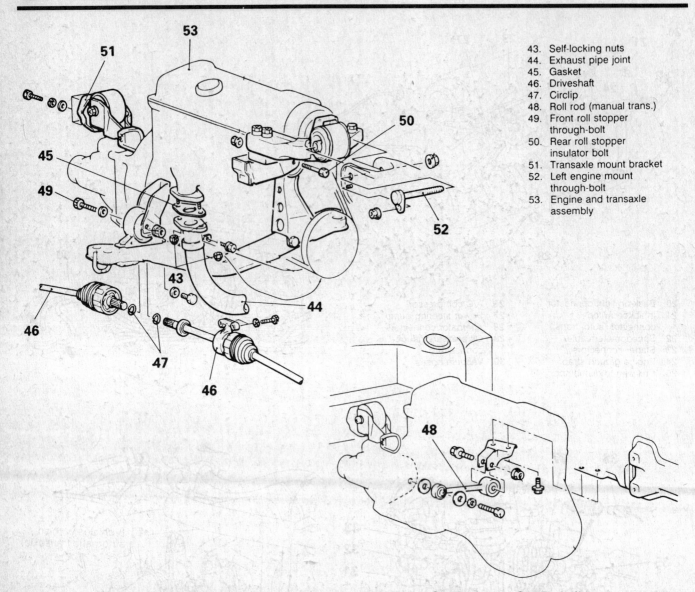

43. Self-locking nuts
44. Exhaust pipe joint
45. Gasket
46. Driveshaft
47. Circlip
48. Roll rod (manual trans.)
49. Front roll stopper through-bolt
50. Rear roll stopper insulator bolt
51. Transaxle mount bracket
52. Left engine mount through-bolt
53. Engine and transaxle assembly

1985–89 Mirage engine removal, external components. G32B shown, G15B similar

26. Remove the circlips from the drive axles and discard the clips. They are not reusable.
27. Install and attach the hoist or lifting equipment. Draw tension on the lift just enough to support the motor but no more.

WARNING: The car is on jackstands. Tighten the hoist slowly; do not disturb the position of the car on the stands.

28. If the car has a manual transmission, remove the rear roll-rod (anti-torque rod) and its bracket. Use tape or a marker to mark the bottom of the rod so that it may be installed in the same position. On cars with automatic transaxles, remove the through-bolt for the front roll stopper.
29. Remove the rear roll stopper through-bolt.
30. Double check the engine hoist, making sure it is securely braced and that there is proper tension on the cables or chains.
31. Remove the cover from the inside of the right fender inner shield. Remove the transaxle mount retaining bolts.
32. Remove the left engine mount through-bolt.
33. Double check the entire engine area for any wires, cables, hoses or components running to the body. Press down on the transaxle while raising the hoist; remove the engine and transaxle from the vehicle. Support the engine on a stand or support it on wooden blocks. Do not allow it to rest on the oil pan or on its side.
34. Before installing the engine and transaxle assembly, make certain that all components removed during repairs are properly reinstalled.
35. Lift the assembly into position and lower it until the left mount bolt hole and the transaxle mount bolt holes align properly. Install the bolts snug but do not overtighten; they will be set to final torque later in the procedure.
36. Install the rear roll stopper bolt and tighten snug.
37. Install either the front roll stopper bolt (automatic) or the rod assembly and its mount. Make certain it is positioned correctly. Tighten all the bolts snug.
38. Slowly loosen the hoist and let the engine weight bear fully on the mounts. Once the engine bears on the mounts, the hoist equipment may be removed.
39. Install new circlips within the transaxle housing and install the driveshafts. Make certain each axle engages the clip with a positive click.

ENGINE AND ENGINE OVERHAUL 3

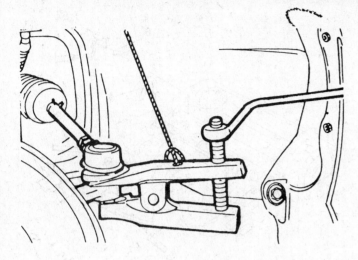

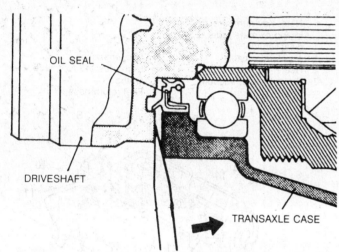

If using Mitsubishi's special ball joint tool (above), tie the safety line to a nearby component. When removing the driveshaft (below), don't insert the prybar too far

40. Move the suspension into position and attach the tie rod end to the knuckle. Tighten the nut to 18 ft. lbs. and install a new cotter pin. Repeat on the opposite side.
41. Connect the lower ball joint to the knuckle. Use a new self-locking nut if one was removed and tighten it to 48 ft. lbs. Install a new cotter pin if one was removed. Repeat on the opposite side.
42. Install a new gasket and attach the exhaust system to the engine. Use new self locking bolts; tighten the nuts and bolt to 18 ft. lbs.
43. Install the wheels and lower the car to the ground.
44. Carefully apply final torque to each of the motor mount and roll stopper nuts and bolts. The torque values are:
- Left mount through-bolt: 72 ft. lbs.
- Left mount through-bolt lock nut: 38 ft. lbs.
- Transaxle mount bracket bolts: 25 ft. lbs.
- Rear roll stopper insulator bolt: 38 ft. lbs.
- Front roll stopper through-bolt: 38 ft. lbs.
- Manual trans. roll rod bracket to firewall bolts: 38 ft. lbs.
- Roll rod through bolts: 47 ft. lbs.
45. Install the cover within the right front fender inner panel.

46. Connect the:
- oil pressure gauge sender
- engine temperature gauge and sensor
- coolant temperature switch (automatic trans only)
- solenoid valve
- feedback solenoid valve (G15B)
- throttle position sensor
- oxygen sensor connector
- cold mixture heater connector
- On the G32B fuel injected engine, also connect the idle speed control (ISC) servo motor, ISC position sensor, detonation sensor, distributor signal generator, EGR solenoid valve and the secondary air control solenoid valve.

NOTE: Elevate and safely support the car as needed to gain access to components. Depending on tools, arm length and agility, each operation may be easier from above or below.

47. Connect the wiring to the alternator.
48. Reinstall the power steering pump and drive belt, tightening the adjusting bolt to 21 ft. lbs.
49. Install the A/C compressor and belt, tightening the mounting bolts.
50. Connect the coil wire to the distributor, the ground strap to the engine and the wiring to the starter.
51. Reinstall the speedometer cable.
52. Connect the reverse light wiring and the inhibitor switch (auotmatic only) wiring to the transaxle.
53. Connect either the control cable and throttle cable (automatic trans) or the select cable and the shift cable (manual trans). In either case, the hairpin shaped lock springs must be replaced.
54. Install the clutch cable and adjust the clutch on manual shift cars with the G15B engine. Install the G32B clutch fluid hose.
55. Connect the brake booster vacuum line.
56. Connect the fuel hoses.
57. Connect the accelerator cable.
58. Connect the heater hoses. For cars with the G32B engine, connect the oil cooler hoses to the engine and tighten the eye bolts to 28 ft. lbs.
59. Install the radiator and connect the coolant hoses to the motor. Use new clamps. Connect the oil cooler lines to the automatic transaxle.
60. Install the battery tray and battery. Do not connect the negative battery cable until ready to start the engine.
61. Check the engine oil drain plug and secure it if necessary. Install the proper amount of engine oil.
62. Check the transaxle drain plug, tightening it if needed, and install the proper amount of transmission oil. On G32B engines with manual transmissions, refill the clutch fluid.
63. Check the radiator and engine drain cocks, closing them if necessary and refill the coolant system.
64. Double check all installation items, paying particular attention to loose hoses or hanging wires, untightened nuts, poor routing of hoses and wires (too tight or rubbing) and tools left in the engine area.
65. Connect the negative battery cable. Start the engine and check for leaks.
66. Attend to all leaks immediately, remembering that fluids and metal surfaces may be hot. Adjust the drive belts to the correct tension. Adjust all cables (transmission, throttle, shift selector) and check the fluid levels. Check the operation of all gauges and dashboard lights. Bleed the hydraulic clutch on the G32B with manual transaxle.
67. With the help of an assistant, install the hood and align it for proper body fit and latching.
68. In a safe location at low speed, road test the car for correct operation of steering brakes, transaxle, clutch and speedometer.
69. Install the splash shield under the engine.

3 ENGINE AND ENGINE OVERHAUL

1989 Mirage with 4G15 (SOHC) and 4G61 (DOHC) Engines

The transaxle must be removed from the car before removing the engine; they will not come out as a unit. Please refer to Section 7 for transaxle removal information. Location of components differs slightly between the two engines but the order of removal is the same.

WARNING: Perform these operations only on a cold motor. If the car has been recently driven, allow it to cool at least 3-4 hours before beginning work. Overnight cooling is recommended.

1. Elevate and safely support the car as needed to gain access to components. Depending on tools, arm length and agility, each operation may be easier from above or below. Disconnect the negative battery cable. Drain the engine coolant.

--- **CAUTION** ---

When draining the coolant, keep in mind that cats and dogs are attracted by the ethylene glycol antifreeze, and are quite likely to drink any that is left in an uncovered container or in puddles on the ground. This will prove fatal in sufficient quantity. Always drain the coolant into a sealable container. Coolant should be reused unless it is contaminated or several years old.

2. Drain the engine oil and the transmission oil.

--- **CAUTION** ---

Used motor oil may cause skin cancer if repeatedly left in contact with the skin for prolonged periods. Although this is unlikely unless you handle oil on a daily basis, it is wise to thoroughly wash your hands with soap and water immediately after handling used motor oil.

3. Safely relieve the pressure within the fuel injection system.

--- **CAUTION** ---

The fuel system is under pressure. Release pressure slowly and contain spillage. Observe no smoking/no open flame precautions. Have a Class B-C (dry powder) fire extinguisher within arm's reach at all times.

4. Matchmark the hood to the hinges and remove the hood.
5. Remove the transaxle assembly.
6. Remove the radiator, disconnecting the hoses at the engine.
7. Disconnect the accelerator cable and remove the bracket.
8. Disconnect the heater hoses.
9. Disconnect the brake booster vacuum hose at the engine.
10. Label and disconnect the vacuum hoses running to the firewall.
11. Disconnect the high pressure fuel line and discard the O-ring; it is not reusable.
12. Remove the fuel return hose.
13. Disconnect the electrical connectors to the engine components. All wires and connectors should be labled at the time of removal. The amount of time saved during reassembly makes the extra effort well worthwhile. Disconnect the:
- oxygen sensor
- coolant temperature gauge
- coolant temperature sensor
- thermo switch (auto. trans. only)
- idle speed control
- EGR temperature sensor (Calif. only)
- each injector, power transistor
- ignition coil lead, condenser
- throttle position sensor
- motor position sensor
- distributor connector
- control harness
- alternator
- oil pressure switch

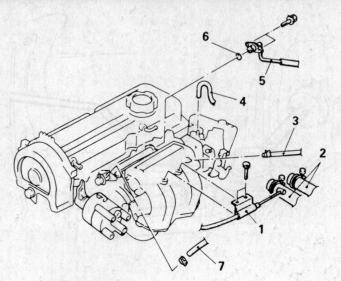

Note: Key numbers are in approximate order of disassembly
1. Accelerator cable
2. Heater hoses
3. Brake booster vacuum hose
4. Vacuum hoses
5. High pressure fuel hose
6. O-ring
7. Fuel return hose

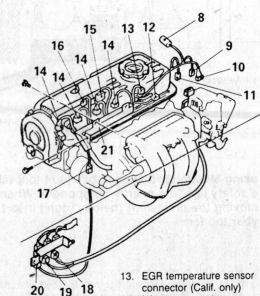

8. Oxygen sensor connector
9. Coolant temperature gauge connector
10. Coolant temperature sensor connector
11. Thermo switch connector (auto. trans.)
12. Idle speed control connector
13. EGR temperature sensor connector (Calif. only)
14. Injector connectors
15. Power transistor connector
16. Ignition coil connector.
17. Condenser connector
18. Throttle position sensor connector
19. Motor position connector
20. Distributor connector
21. Control harness

4G15 engine removal, external components

- On the 4G61 engine, the multi-wire connectors for the power transistor and the coil must also be removed
- Remove retaining brackets and mounts as necessary to clear the wiring away from the engine

ENGINE AND ENGINE OVERHAUL 3

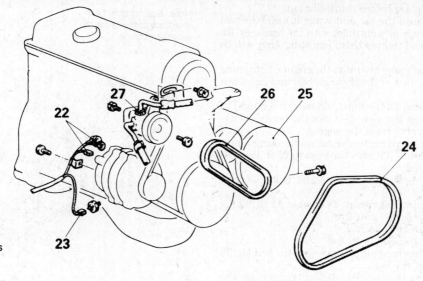

22. Alternator connectors
23. Oil pressure switch connector
24. A/C drive belt
25. A/C compressor
26. Power steering drive belt
27. Power steering pump
28. Self-locking nuts
29. Gasket
30. Upper engine mount assembly
31. Engine assembly

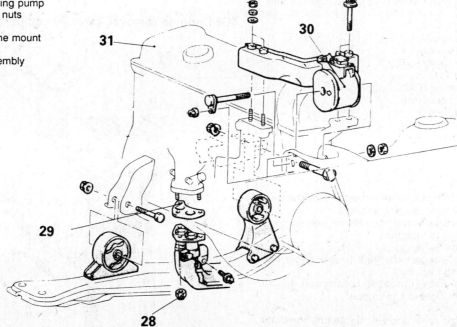

4G15 engine removal, external components

14. Loosen the adjuster and remove the air conditioning drive belt. Remove the compressor from its mount and hang it from stiff wire out of the way. Note that the hoses are still attached; do not loosen them or discharge the system.
15. Loosen the power steering drive belt and remove it. Remove the bolts holding the pump to its bracket and hang the pump out of the way. Do not disconnect the hoses and do not allow the pump to hang by the hoses.
16. Remove the self locking nuts and bolt at the exhaust system joint just below the manifold. Remove and discard the gasket and the two nuts.
17. Elevate the car and support it safely. Install the engine hoist equipment and make certain the attaching points on the engine are secure. Draw tension on the hoist just enough to support the engine's weight but no more. Do not disturb the place-

ment of the car on the stands.
18. Remove the nuts and bolts holding the upper (left side) engine mount to the engine. Remove the throught-bolt and remove the mount assembly.
19. Remove the through-bolt from the rear (firewall side) lower engine mount.
20. Remove the through-bolt from the front lower engine mount.
21. Double check for any remaining cables, wires or hoses running to the engine. Elevate the hoist and remove the engine from the car. Immediately place it on an engine stand or support it with wooden blocks. Do not allow it to rest on the oil pan or lie on its side. Never leave an engine hanging from a hoist.
22. After repairs, make certain the engine is fully reassembled before installation. All components removed with the engine out

3-61

3 ENGINE AND ENGINE OVERHAUL

of the car should be in place before reinstallation.

23. Install the engine into the car and lower it until the bolt holes for the front and rear mounts align with the brackets. Install the through-bolts and tighten them just snug; they will be final tightened later.
24. Install the upper (left side) mount to the engine, tightening the nuts and bolts snug. Install the through-bolt and tighten it snug.
25. Slowly release tension on the hoist, allowing the weight of the engine to bear fully on the mounts. Once the hoist is slack, remove the lifting apparatus from the engine.
26. Connect the exhaust system to the manifold, using a new gasket and new locking nuts. Tighten the nuts to 25 ft. lbs. and the small bolt to 18 ft. lbs.
27. Final tighten the engine mount nuts and bolts. Correct torque values are:
- Nuts and bolts holding left mount to engine: 41 ft. lbs.
- Left mount through-bolt: 72 ft. lbs.
- Rear mount through-bolt: 38 ft. lbs.
- Front mount through-bolt: 38 ft. lbs.
28. Install the power steering pump, tightening the bolts to 36 ft. lbs. Install and adjust the belt.
29. Install the air conditioning compressor, tightening the mounting bolts to 18 ft. lbs. Install the belt and adjust it.
30. Connect the wiring and harness connectors to the engine. Make certain each terminal is clean and the connector is firmly seated to its mate. Do not route wires near hot surfaces or moving parts. Connect the:
- oxygen sensor
- coolant temperature gauge
- coolant temperature sensor
- thermo switch (auto. trans. only)
- idle speed control
- EGR temperature sensor (Calif. only)
- each injector, power transistor
- ignition coil lead, condenser
- throttle position sensor
- motor position sensor
- distributor connector
- control harness
- alternator
- oil pressure switch
- On the 4G61 engine, the multi-wire connectors for the power transistor and the coil must also be removed
- Remove retaining brackets and mounts as necessary to clear the wiring away from the engine
31. Install the fuel return hose.
32. Using a new O-ring, connect the high pressure fuel line and tighten the bolts to 48 inch lbs.
33. Connect the vacuum lines running to the firewall. Install the brake booster vacuum hose to the engine.
34. Connect the heater hoses.
35. Install the accelerator cable bracket, tightening the bolts to 48 inch lbs., and connect the accelerator cable.
36. Install the radiator and connect the hoses.
37. Install the transaxle.
38. Check the engine oil drain plug and secure it if necessary. Install the proper amount of engine oil.
39. Check the transaxle drain plug, tightening it if needed, and install the proper amount of transmission oil.
40. Check the radiator and engine drain cocks, closing them if necessary and refill the coolant system.
41. Double check all installation items, paying particular attention to loose hoses or hanging wires, untightened nuts, poor routing of hoses and wires (too tight or rubbing) and tools left in the engine area.
42. Connect the negative battery cable. Start the engine and check for leaks.
43. Attend to all leaks immediately, remembering that fluids and metal surfaces may be hot. Adjust the drive belts to the cor-

Note: Key numbers are in approximate order of disassembly

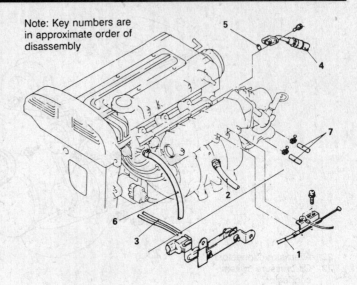

1. Accelerator cable
2. Brake booster vacuum hose
3. Vacuum hoses
4. High pressure fuel line
5. O-ring
6. Fuel return line
7. Heater hoses

4G61 engine removal, external components

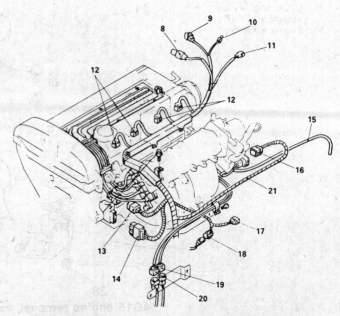

8. Oxygen sensor connector
9. Coolant temperature sensor connector
10. Coolant temperature gauge connector
11. Coolant temperature switch connector (auto. trans.)
12. Injector connectors
13. Ignition coil connectors
14. Power transistor connector
15. Vacuum hoses
16. Idle speed control motor connector
17. EGR temperature sensor connector (Calif. only)
18. Detonation sensor connector
19. Throttle position sensor
20. Crankshaft angle sensor connector
21. Control wiring harness

4G61 engine removal, external components

3-62

ENGINE AND ENGINE OVERHAUL 3

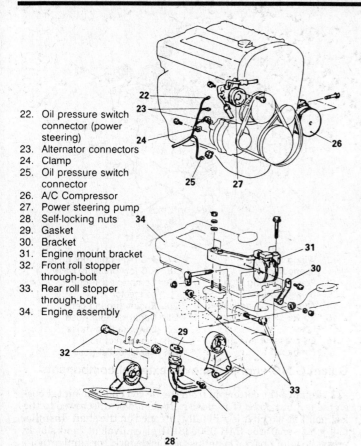

22. Oil pressure switch connector (power steering)
23. Alternator connectors
24. Clamp
25. Oil pressure switch connector
26. A/C Compressor
27. Power steering pump
28. Self-locking nuts
29. Gasket
30. Bracket
31. Engine mount bracket
32. Front roll stopper through-bolt
33. Rear roll stopper through-bolt
34. Engine assembly

4G61 engine removal, external components

4. Safely relieve the pressure in the fuel system.

— CAUTION —

The fuel system is under pressure. Release pressure slowly and contain spillage. Observe no smoking/no open flame precautions. Have a Class B-C (dry powder) fire extinguisher within arm's reach at all times.

5. Safely discharge the refrigerant within the air conditioning system. See Section 1.
6. If the car is equipped with the Electronic Controlled Suspension (ECS), the air compressor and reservoir tank must be removed. Please refer to Section 8 for information on this system.
7. Disconnect the battery, negative cable first, and remove it. Remove the battery tray.
8. Remove the air filter body and remove the air intake hoses. Disconnect the small breather hose from the valve cover.
9. Disconnect the brake booster vacuum hose from the intake manifold.
10. Label and disconnect the vacuum hoses running between the engine and the body panels.
11. Disconnect the high pressure fuel hose; discard the washer and O-ring.
12. Remove the accelerator cable and its brackets.
13. Remove the fuel return hose.
14. Disconnect the injector wiring and remove the bolts holding the wire guide channel. Label and disconnect the other connectors running from the harness. Work carefully; there will be no margin of error during reassembly.
15. Disconnect the power steering hose from the top of the pump. Contain spillage and immediately plug or cap the hose. Move it to a location out of the way and support there with string or wire.

rect tension. Adjust all cables (transmission, throttle, shift selector) and check the fluid levels. Check the operation of all gauges and dashboard lights.

44. With the help of an assistant, install the hood and align it for proper body fit and latching.
45. In a safe location at low speed, road test the car for correct operation of steering brakes, transaxle, clutch and speedometer.

Galant 1985–87 with G64B Engine

NOTE: Elevate and safely support the car as needed to gain access to components. Depending on tools, arm length and agility, each operation may be easier from above or below.

1. Matchmark and remove the hood.
2. Drain the engine coolant.

— CAUTION —

When draining the coolant, keep in mind that cats and dogs are attracted by the ethylene glycol antifreeze, and are quite likely to drink any that is left in an uncovered container or in puddles on the ground. This will prove fatal in sufficient quantity. Always drain the coolant into a sealable container. Coolant should be reused unless it is contaminated or several years old.

3. Drain the engine oil, the transaxle oil and, if manual transmission, the clutch fluid.

— CAUTION —

Used motor oil may cause skin cancer if repeatedly left in contact with the skin for prolonged periods. Although this is unlikely unless you handle oil on a daily basis, it is wise to thoroughly wash your hands with soap and water immediately after handling used motor oil.

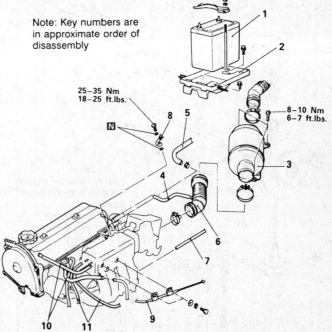

Note: Key numbers are in approximate order of disassembly

1. Battery
2. Battery tray
3. Air filter assembly
4. Breather hose
5. Brake booster vacuum hose
6. Air intake hose
7. Vacuum hoses
8. High pressure fuel hose
9. Accelerator cable
10. Fuel return hose
11. Vacuum hoses

Galant G64B engine removal, external components

3-63

3 ENGINE AND ENGINE OVERHAUL

16. Remove the connectors from the alternator.
17. Disconnect the return hose from the power steering pump.
18. Disconnect the oil pressure switch.
19. Remove the high pressure hose from the compressor, capping it immediately, then remove the low pressure hose. Cap this line and cap both compressor ports. Disconnect the wiring from the compressor.
20. Disconnect the engine ground cable.
21. Remove from the engine the upper and lower radiator hoses and disconnect both heater hoses.
22. If equipped with manual transaxle, disconnect the select cable, the shift cable and the clutch fluid hose. Note that the spring pins holding the cables are not reusable; they must be discarded and replaced during reassembly.
23. If equipped with automatic transaxle, disconnect the control cable and both oil cooler hoses at the transaxle.
24. Disconnect the front wiring harness and the engine wiring harness connectors. Remove the positive battery cable and fusible link box.
25. Disconnect the speedometer cable at the transaxle.
26. Elevate and safely suport the car on jackstands. Remove the front wheels. Disconnect the stabilizer bar from the lower control arms.

25. Select cable
26. Shift cable
27. Clutch fluid hose
28. Control hose
29. Oil cooler feed hose (auto. trans.)
30. Oil cooler return hose (auto. trans.)
31. Front wiring harness and engine wiring connectors

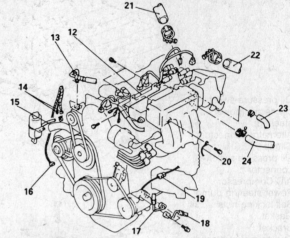

12. Wiring harness connectors
13. Power steering high pressure hose
14. Alternator connectors
15. Power steering return hose
16. Oil pressure switch connector
17. A/C high pressure hose
18. A/C low pressure hose
19. Compressor wiring connector
20. Engine ground cable
21. Upper radiator hose
22. Lower radiator hose
23. Heater inlet hose
24. Heater outlet hose

Galant G64B engine removal, external components

27. Remove the cotter pin from the tie rod end ball joint. Loosen the nut but do NOT remove it. It should be unscrewed to the last point at which it is still fully threaded on the shaft. Install a ball joint separator (MB 990635-01 or equivalent) and slowly press the joint apart. Do not use a "pickle fork" or similar tool-- the joint may be damaged. Repeat the procedure on the opposite side.
28. After each joint is separated (the nut will restrain a sudden separation), remove the nut and separate the joint.
29. Disconnect the stabilizer bar from the lower arm on each side. Take careful note of the order of nuts, washers, rubber bushings and spacers. Draw a diagram if necessary before disassembly.
30. Repeat Steps 27 and 28 to separate the lower arm ball joint from the steering knuckle. Insert a small pry bar into the gap between the transaxle case and the driveshaft. Do not insert it too far or an internal oil seal will be damaged.
31. Pry the driveshafts free of the transaxle. Don't pull on the driveshaft; the joint will be damaged. It will be necessary to swing the steering knuckle/hub/brake assembly to the outside of the body to gain clearance as the driveshaft comes free. Immediately suspend each axle with a piece of wire. Protect the boots and splined end of the shaft with rags or covers. Plug the holes in the transaxle case.
32. Remove the circlips from the drive axles and discard the clips. They are not reusable.
33. Install and attach the hoist or lifting equipment. Draw tension on the lift just enough to support the weight of the motor but no more. This releases the tension on the motor mount bolts, making removal easier.

WARNING: The car is on jackstands. Tighten the hoist slowly; do not disturb the position of the car on the stands.

34. Remove the through-bolt from the rear roll stopper.
35. Disconnect the connector for the oxygen sensor on top of the left motor mount. Remove the bracket holding the A/C high pressure hose and move the hose out of the way.
36. Disconnect the front exhaust pipe from the manifold discard the gasket.

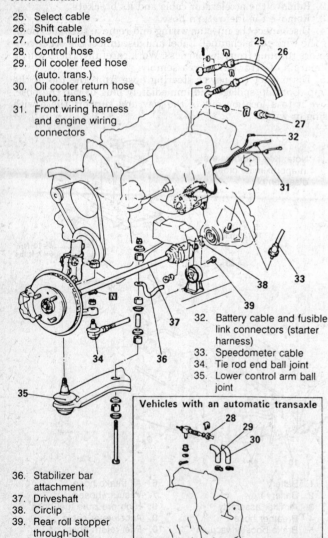

32. Battery cable and fusible link connectors (starter harness)
33. Speedometer cable
34. Tie rod end ball joint
35. Lower control arm ball joint
36. Stabilizer bar attachment
37. Driveshaft
38. Circlip
39. Rear roll stopper through-bolt

Vehicles with an automatic transaxle
28
29
30

Galant G64B engine removal, external components

ENGINE AND ENGINE OVERHAUL 3

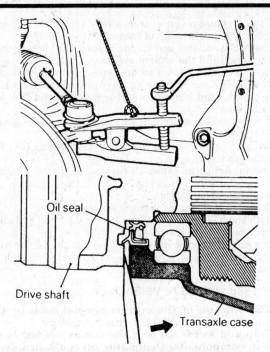

If using Mitsubishi's special ball joint tool (above), tie the safety line to a nearby component. When removing the driveshaft (below), don't insert the prybar too far

37. Remove the bolt holding the small heat shield and remove the shield from the front roll stopper mount.
38. Remove the through-bolt from front roll stopper.
39. Inside the right front fender inner liner, remove the caps from the transaxle mount bolts.
40. Double check the hoist equipment, making certain it is securely attached to the engine and is stable on its mounts. Check for any remaining wires, cables, or hoses running between the engine and the body work.
41. Remove the transaxle mount bolts. Most of the weight of the engine will now be on the hoist lines; adjust tension as necessary to keep the engine in place.
42. Remove the large through-bolt from the left (upper) engine mount. Note that two small safety nuts must be removed from the through-bolt head as well.
43. Elevate the hoist and remove the engine and transaxle assembly. It will be necessary to push down on the transaxle while lifting the engine.
44. Immediately attach the engine to a stand or support it on wooden blocks. Do not allow it to rest on the oil pan or lie on its side. Never leave an engine hanging from a hoist.
45. After repairs, make certain the engine and transaxle are fully reassembled before installation. All components removed with the engine out of the car should be in place before reinstallation.
46. Install the engine into the car and lower it until the bolt holes for the mounts align with the brackets. Install the through-bolts and the transaxle mount bolts, tightening them just snug; they will be final tightened later.
47. Connect the exhaust system to the manifold, using a new gasket and new locking nuts. Tighten the nuts and the small bolt to 18 ft. lbs.

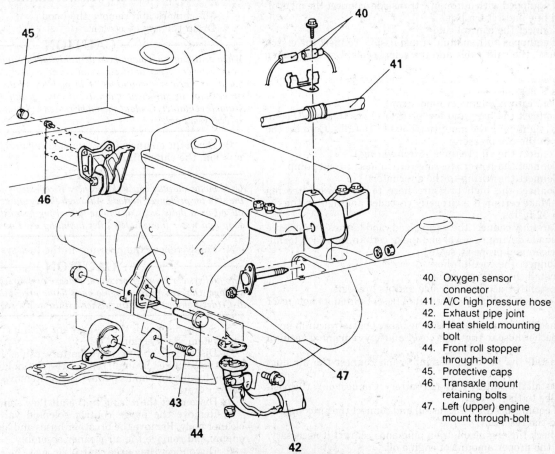

40. Oxygen sensor connector
41. A/C high pressure hose
42. Exhaust pipe joint
43. Heat shield mounting bolt
44. Front roll stopper through-bolt
45. Protective caps
46. Transaxle mount retaining bolts
47. Left (upper) engine mount through-bolt

Galant G64B engine removal, external components

3-65

3 ENGINE AND ENGINE OVERHAUL

48. Slowly release tension on the hoist, allowing the weight of the engine to bear fully on the mounts. Once the hoist is slack, remove the lifting apparatus from the engine.
49. Final tighten the mounting and through bolts. Use the following torque values:
   - Left engine mount, through-bolt: 50 ft. lbs.
   - Left engine mount, small nuts at head of through-bolt: 25 ft. lbs.
   - Transaxle mount bolts: 33 ft. lbs.
   - Rear roll stopper through-bolt: 28 ft. lbs.
   - Front roll stopper through-bolt: 40 ft. lbs.
50. Install the small heat shield at the front roll stopper bracket and install the protective caps on the transaxle mount bolts.
51. Install the A/C high pressure hose to its bracket on the uppper engine mount; connect the oxygen sensor wiring and secure the connector in the clip.
52. Install new circlips within the transaxle housing and install the driveshafts. Make certain each axle engages the clip with a positive click.
53. Connect the stabilizer bar to both lower arms. Make certain the hardware and spacers are assembled correctly. Use new self-locking nuts.
54. Connect the lower ball joint to the knuckle. Use a new self-locking nut if one was removed and tighten it to 48 ft. lbs. Install a new cotter pin if one was removed. Repeat on the opposite side.
55. Attach the tie rod end to the knuckle. Tighten the nut to 18 ft. lbs. and install a new cotter pin. Repeat on the opposite side.
56. Connect the speedometer cable.
57. Attach the positive battery cable and the fusible link wiring (starter harness).
58. Connect the front wiring harness and the engine wiring harness to their connectors.
59. If equipped with automatic transaxle, connect the oil cooler feed line and return line.
60. Connect the control cable.
61. If equipped with manual transmission, connect the clutch fluid hose, the shift cable and the select cable. Use new spring clips.
62. Install the heater hoses and radiator hoses, using new clamps if possible.
63. Connect the engine ground strap.
64. Connect the high and low pressure lines to the A/C compressor, tightening the mounting bolts to 15 ft.lbs. Connect the wiring to the compressor.
65. Connect the oil pressure switch wiring.
66. Connect the power steering return hose to the pump.
67. Connect the wiring to the alternator.
68. Connect the high pressure hose to the power steering pump. Make certain it is properly threaded and tighten the fitting to 32 ft. lbs.
69. Carefully connect the wiring leads and harnesses. Observe labels made during removal and make certain that each terminal is clean and properly seated.
70. Connect the vacuum hoses.
71. Install the fuel return lines.
72. Install the accelerator cable and its brackets.
73. Connect the high pressure fuel hose; tighten its bolt to 21 ft. lbs.
74. Install the air intake hose, the brake booster vacuum hose, the breather hose to the valve cover and any remaining vacuum hoses.
75. Install the air filter assembly and connect the air duct hoses.
76. Install the battery tray and battery. Connect only the positive calbe to the battery.
77. If equipped with ECS, install and connect the air compressor and the reservoir.
78. Check the engine oil drain plug and secure it if necessary. Install the proper amount of engine oil.
79. Check the transaxle drain plug, tightening it if needed, and install the proper amount of transmission oil. Refill the clutch fluid if equipped with manual transaxle.
80. Check the radiator and engine drain cocks, closing them if necessary and refill the coolant system.
81. Double check all installation items, paying particular attention to loose hoses or hanging wires, untightened nuts, poor routing of hoses and wires (too tight or rubbing) and tools left in the engine area.
82. Connect the negative battery cable. Start the engine and check for leaks.
83. Attend to all leaks immediately, remembering that fluids and metal surfaces may be hot. Adjust the drive belts to the correct tension. Adjust all cables (transmission, throttle, shift selector) and check the fluid levels. Check the operation of all gauges and dashboard lights.
84. Recharge the air conditioning system.
85. With the help of an assistant, install the hood and align it for proper body fit and latching.
86. In a safe location at low speed, road test the car for correct operation of steering, brakes, transaxle, clutch and speedometer.

Galant 1988 with 6G72 V6 Engine

NOTE: The use of the correct special tools or their equivalent is REQUIRED for this procedure.

Elevate and safely support the car as needed to gain access to components. Depending on tools, arm length and agility, each operation may be easier from above or below.

All wires and hoses should be labled at the time of removal. The amount of time saved during reassembly makes the extra effort well worthwhile.

1. Matchmark and remove the hood.
2. Drain the engine coolant.

CAUTION

When draining the coolant, keep in mind that cats and dogs are attracted by the ethylene glycol antifreeze, and are quite likely to drink any that is left in an uncovered container or in puddles on the ground. This will prove fatal in sufficient quantity. Always drain the coolant into a sealable container. Coolant should be reused unless it is contaminated or several years old.

---

3. Drain the engine oil, the transaxle oil and, if manual transmission, the clutch fluid.

CAUTION

Used motor oil may cause skin cancer if repeatedly left in contact with the skin for prolonged periods. Although this is unlikely unless you handle oil on a daily basis, it is wise to thoroughly wash your hands with soap and water immediately after handling used motor oil.

---

4. Safely relieve the pressure in the fuel system.

CAUTION

The fuel system is under pressure. Release pressure slowly and contain spillage. Observe no smoking/no open flame precautions. Have a Class B-C (dry powder) fire extinguisher within arm's reach at all times.

---

5. Safely discharge the refrigerant within the air conditioning system. See Section 1.
6. Disconnect the negative battery cable. Remove the cap over the positive terminal, remove the nut holding the terminal to the fusible link and separate them. Remove the battery and battery tray.
7. Disconnect the wiring to the air flow sensor.
8. Remove the purge control solenoid valve from the air cleaner body. Remove the breather hoses and air ducts from the engine and remove the air cleaner assembly.
9. Disconnect the purge control hose at the canister. Disconnect the brake booster vacuum hose.

ENGINE AND ENGINE OVERHAUL 3

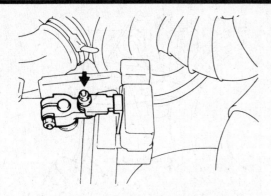

Disconnecting the positive battery terminal

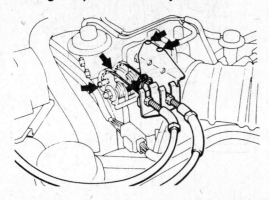

Removing the throttle and cruise control cables with the mounting bracket

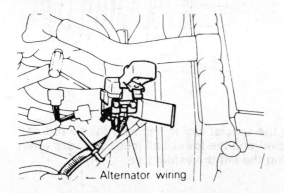

Alternator wiring

Detach the wiring inside the fusible link box

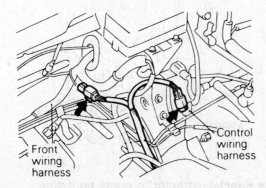

Front wiring harness — Control wiring harness

Harness connectors

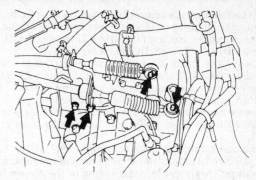

Removing manual transmission control cables and bracket

10. At the throttle body, identify the screws holding the throttle cable and cruise control bracket. Remove these screws without removing the individual cables. Lift the entire bracket with the cable intact away from the throttle body.
11. Disconnect the wiring to the idle switch and the motor position sensor.
12. Remove the high pressure fuel hose and the fuel return hose. Remove the clamp holding the two hoses to the throttle body and remove the hoses from the engine.
13. Open the cover of the fusible link box and disconnect the alternator wiring.
14. Remove the connector holding the battery ground strap to the body.
15. Disconnect the front wiring harness and the engine control wiring harness.
16. If equipped with manual transaxle, remove the retaining hardware holding the control cables to the transaxle. Unbolt the bracket holding the cables, then remove the bracket and cable assembly away from the transaxle. Discard the spring clips; they are non-reusable.
17. If equipped with automatic transaxle, disconnect the inhibitor switch and the pulse generator connectors. Remove the retaining hardware and disconnect the transaxle control cable. Discard the spring clip; it is not reusable.
18. If equipped with ECS (Electric Controlled Suspension), the compressor for the system must be removed. Please refer to Section 8 for information on this system.
19. If equipped with manual transaxle, remove the bracket holding the clutch hydraulic lines. Disconnect the clutch cylinder (slave cylinder) from the transaxle and carefully move the cylinder and lines out of the way. Disconnect the speedometer cable.
20. If equipped with automatic transaxle, disconnect the speedometer cable first, then label and disconnect the oil cooler lines at the transaxle. Contain spillage and plug or cap the lines and ports immediately to prevent entry of dirt.
21. Disconnect the heater hoses at the pipe fittings.
22. Disconnect the engine ground strap at the engine. Label and disconnect the engine control wiring harness connectors. Remove the nut holding the harness clamp and move harness out of the way.
23. Disconnect the high and low pressure hoses from the air conditioning compressor. Plug the hoses and compressor ports immediately. Remove the compressor wiring.

--- **CAUTION** ---
When disconnecting the lines, the residual coolant may splash. Wear gloves and eye protection

24. Disconnect the condenser wiring and remove the coil wire from the distributor.
25. Disconnect the power steering pressure hose and then the return hose from the power steering pump. Contain spillage. Plug or cap both the hoses and the pump port immediately.

3-67

3 ENGINE AND ENGINE OVERHAUL

26. At the radiator, disconnect the overflow tube from the jug and the upper and lower radiator hoses from the engine.
27. Disconnect the electric fan wiring. If equipped with automatic transaxle, disconnect the oil cooler lines.
28. Remove the radiator and the rubber bushings under it. Handle the radiator carefully and protect it while out of the car.
29. For cars with automatic transaxles, remove the oil cooler lines which ran from the transaxle to the radiator.
30. Elevate and safely support the car. Remove the lower inner panel covers in the wheel wells: 2 for automatics, 1 for manual transaxles.
31. Disconnect the stabilizer bar (sway bar) at the control arms. Take careful note of the arrangement of the nuts, spacers and bushings; they must be reassembled in the correct order.
32. Remove the cotter pin from the tie rod end ball joint. Loosen the nut but do NOT remove it. It should be unscrewed to the last point at which it is still fully threaded on the shaft. Install a ball joint separator (MB 991113 or eqivalent) and slowly press the joint apart. Do not use a "pickle fork" or similar tool--the joint may be damaged. Repeat the procedure on the opposite side.
33. After each joint is separated (the nut will restrain a sudden separation), remove the nut and separate the joint.
34. Repeat Steps 32 and 33 to separate the lower arm ball joint from the steering knuckle.
35. Insert a small pry bar into the gap between the transaxle case and the left driveshaft. Pry the driveshaft out of the case while swinging the suspension and hub assembly away (outward). The left driveshaft will stay attached to the hub. When the driveshaft is free, immediately cover the splined end with rags and support the shaft from a piece of stiff wire.
36. At the outer end of the right driveshaft, remove the cotter pin (discard it) and remove the large nut on the end of the shaft. This nut may be very tight; be careful not to disturb the position of the car on the stands during removal.
37. Install an extractor tool (Mitsubishi tool NB 990241-01 or equivalent) and force the shaft from the hub. Do not hammer on the end of the shaft.
38. Insert a large pry bar between the case and the inner shaft joint and release the shaft from the transaxle. Remove the circlip, discard it and protect both ends of the shaft with rags. Plug the holes in the transaxle case and put the right driveshaft in a safe location.
39. Disconnect the rubber hangers holding the exhaust system by sliding them off the body mounts.
40. Disconnect the wiring running to the oxygen sensor.
41. Remove the self-locking nuts holding the exhaust pipes to the manifolds. Separate the pipe and remove the gaskets. Use wire to suspend the pipe so it is not damaged.
42. Remove the distributor cap, install the hoisting equipment and make sure it is securely fastened to the engine. Draw tension on the hoist just enough to support the weight of the motor without elevating it.

CAUTION

The car is supported on stands. Tension the hoist slowly and do not disturb the position of the car on the supports.

---

43. Remove the through-bolt from the front roll stopper.
44. Remove the bolt holding the engine damper (looks like a small shock absorber, which it is) to the engine.
45. Remove the through-bolt from the rear roll stopper assembly.
46. Remove the large through-bolt from the left side engine mount. Note the two smaller nuts holding the head of the large bolt in place. Remove the 4 nuts holding the engine mount to the engine and remove the mount.
47. Remove the protective caps inside the right front fender liner. Loosen and remove the 4 transaxle mount bolts. Work carefully; do not allow the bolts to fall down into the fender liner.

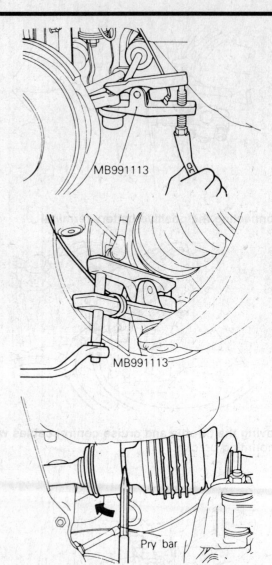

Use the special tool to separate the tie rod end (above) and the lower ball joint. Use care when prying the left driveshaft free

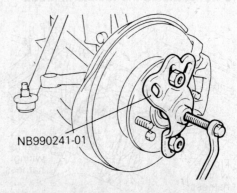

Use the special extractor to press tight side driveshaft free of the hub and bearing assembly

3-68

ENGINE AND ENGINE OVERHAUL 3

48. Double check for any remaining cables, hoses, wiring or lines still running to the engine and remove them. Check that any item which was moved out of the way is truly in a safe location.
49. Elevate the hoist slowly and raise the engine and transaxle assembly out of the car. Take your time; clearances may be close and the unit is heavy. As soon as the engine is clear, mount it on a stand or support it on wooden blocks. Do not allow it to rest on the oil pan or lie on its side. Never leave an engine hanging from a hoist.
50. After repairs, make certain the engine and transaxle are fully reassembled before installation. All components removed with the engine out of the car should be in place before reinstallation.
51. Install the engine into the car and lower it until the bolt holes for the mounts align with the brackets. Install the transaxle mount bolts, tightening them just snug. These bolts, along with the other mounting bolts, will be final tightened later.
52. Install the left mount bracket to the engine, tightening the nuts to 50 ft. lbs.
53. Install the through-bolt for the left mount and tighten the bolt snug. Install the two small nuts at the bolt head and tighten them to 26 ft. lbs.
54. Install the rear roll stopper through-bolt and tighten it snug.
55. Connect the engine damper to the engine and tighten it snug.
56. Use a new nut and install the through-bolt for the front roll stopper, tightening it just snug.
57. With the engine pinned in place, slowly relax tension on the hoist and allow the engine and transaxle to fully load the mounts. Remove the hoist apparatus from the engine. Reinstall the distributor cap.
58. Use new gaskets and new self-locking nuts, connect the exhaust pipe to the manifolds. Tighten the nuts to 25 ft. lbs.
59. Final tighten the engine mounts, damper and roll stopper bolts. The correct torque values are:
- Transaxle mount bolts: 32 ft. lbs.
- Left engine mount through-bolt: 50 ft. lbs.
- Rear roll stopper through-bolt: 25 ft. lbs.
- Engine damper bolt: 25 ft. lbs.
- Front roll stopper through-bolt: 39 ft. lbs.
60. Connect the oxygen sensor wiring at the exhaust system and reinstall the rubber exhaust hangers on the body mounts.
61. Install a new circlip in the transmission and install the right driveshaft. Fit the outer end through the hub assembly.

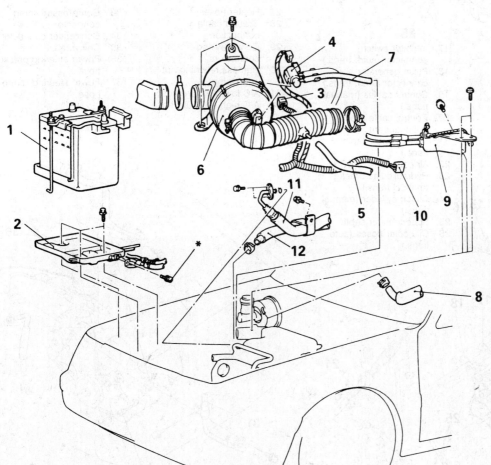

Note: Key numbers are in approximate order of disassembly
1. Battery
2. Battery tray
3. Air flow sensor connector
4. Purge control solenoid valve
5. Breather hose and air intake hoses
6. Air cleaner body
7. Purge hose
8. Brake booster vacuum hose
9. Accelerator and cruise control cable and bracket
10. Idle switch and motor position sensor connector
11. High pressure fuel hose
12. Fuel return hose

Galant 6G72 engine removal and external components

3 ENGINE AND ENGINE OVERHAUL

Manual transaxle

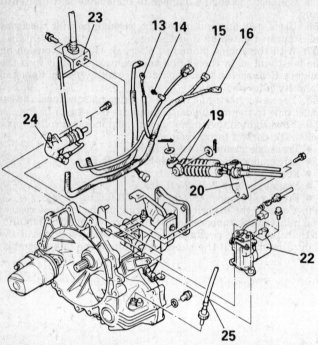
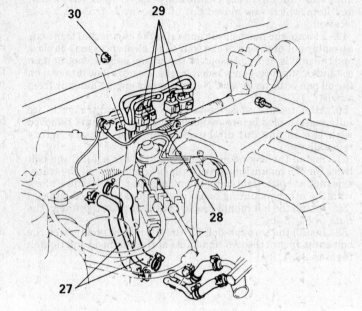

Automatic transaxle

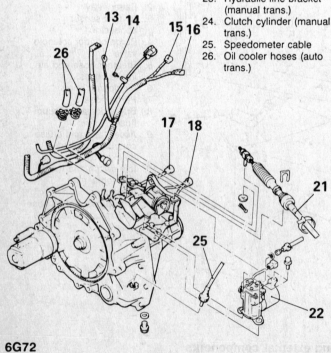
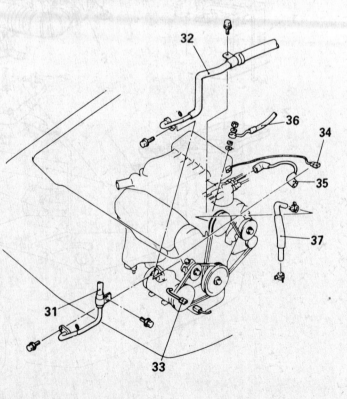

13. Fusible link and engine wire harness connectors
14. Battery to body ground connection
15. Front wiring harness connector
16. Control wiring harness connector
17. Inhibitor switch connector (auto. trans.)
18. Pulse generator connector (auto. trans.)
19. Control cables (manual trans.)
20. Control cable bracket (manual trans.)
21. Control cable (auto trans.)
22. Air compressor (ECS)
23. Hydraulic line bracket (manual trans.)
24. Clutch cylinder (manual trans.)
25. Speedometer cable
26. Oil cooler hoses (auto trans.)
27. Heater hoses
28. Ground cable connection
29. Control wiring connectors
30. Harness retaining nut and clamp
31. A/C low pressure hose
32. A/C high pressure hose
33. Compressor wiring connector
34. Condenser connector
35. Coil wire
36. Power steering pressure hose
37. Power steering return hose

6G72

6G72

3-70

ENGINE AND ENGINE OVERHAUL 3

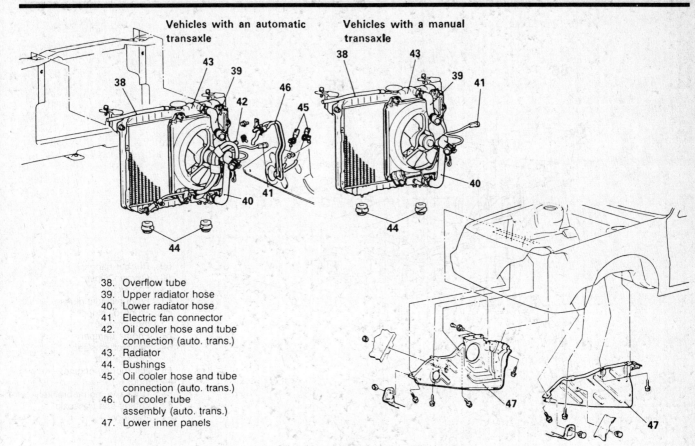

38. Overflow tube
39. Upper radiator hose
40. Lower radiator hose
41. Electric fan connector
42. Oil cooler hose and tube connection (auto. trans.)
43. Radiator
44. Bushings
45. Oil cooler hose and tube connection (auto. trans.)
46. Oil cooler tube assembly (auto. trans.)
47. Lower inner panels

6G72

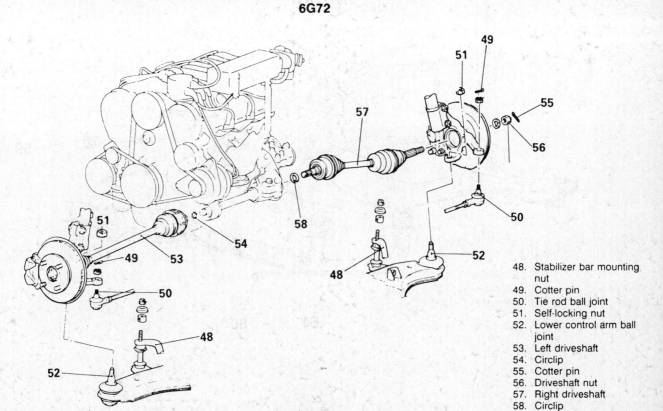

48. Stabilizer bar mounting nut
49. Cotter pin
50. Tie rod ball joint
51. Self-locking nut
52. Lower control arm ball joint
53. Left driveshaft
54. Circlip
55. Cotter pin
56. Driveshaft nut
57. Right driveshaft
58. Circlip

6G72

3-71

3 ENGINE AND ENGINE OVERHAUL

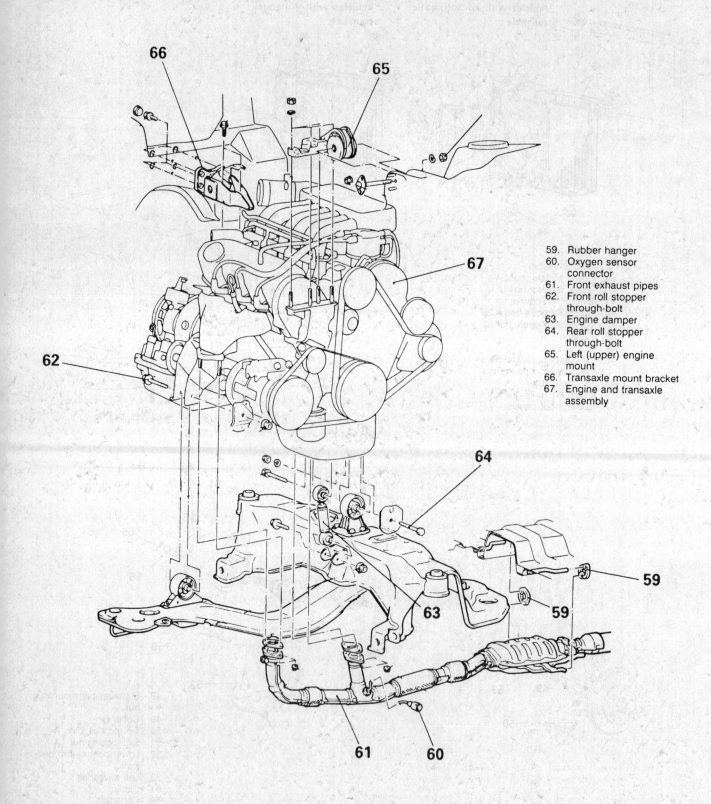

59. Rubber hanger
60. Oxygen sensor connector
61. Front exhaust pipes
62. Front roll stopper through-bolt
63. Engine damper
64. Rear roll stopper through-bolt
65. Left (upper) engine mount
66. Transaxle mount bracket
67. Engine and transaxle assembly

ENGINE AND ENGINE OVERHAUL 3

Make sure the washer and hub nut are installed in the correct direction. Use the nut to help pull the shaft into place. Tighten the nut comfortably tight but do not attempt to develop final torque; that must be done with the car on the ground.

62. Install a new circlip and install the left driveshaft.
63. Connect the lower control arm ball joint to the knuckle. Use a new nut and tighten it to 46 ft. lbs.
64. Connect the tie rod end to the knuckle, tighten the nut to 21 ft. lbs. and install a new cotter pin.
65. Connect the stabilizer bar to the control arms, installing the hardware in the correct order and using a new self-locking nut on each side. Tighten the nut until the threads project 15-18mm (0.6–0.7 in.) above the nut; this develops the proper compression on the bushings.
66. At this point it is recommended to install the wheels and lower the car to the ground, even though it may need to be raised again later. With the car on the ground, tighten the right driveshaft nut. Torque specification is 144–188 ft. lbs. Try for the mid-point, about 166 ft. lbs. Install a new cotter pin. If the cotter pin holes do not line up, the nut may be tightened to a maximum of 188 ft. lbs. Install the cotter pin in the first set of matching holes and bend it securely.

NOTE: The car may be raised and supported as necessary during the remaining steps.

67. Reinstall the lower inner panel covers.
68. For automatic transaxles, reinstall the oil cooler lines, attaching them to the hoses.
69. Install the lower bushings and the radiator.
70. Connect the oil cooler lines to the radiator lines (auto trans.).
71. Connect the electric fan wiring. Install the radiator hoses to the engine, using new clamps if needed. Connect the overflow tube.
72. Connect the power steering hoses to the pump. Tighten the pressure hose fitting to 32 ft. lbs.
73. Connect the coil wire and the condenser wire.
74. Connect the wiring to the compressor. Install the high and low pressure lines to the compressor and tighten them to 15 ft. lbs.
75. Attach the engine wiring harnesses to their connectors and install the retaining bolts.
76. Connect the engine ground cable.
77. Install the heater hoses, using new clamps if necessary.
78. Connect the oil cooler hoses to the automatic transaxle.
79. Connect the speedometer cable.
80. If equipped with ECS, install and connect the air compressor and lines.
81. Install the control cable and connect the inhibitor switch and pulse generator wiring for automatic transaxles. On manual transaxles, install the control cable bracket and connect the cables. Use new spring clips or cotter pins.
82. Connect the engine wiring harness and front wiring harness to their connectors.
83. Connect the negative battery cable to the body ground point, making sure the surfaces are clean and the bolt tight. Connect the starter harness to the fusible link box.
84. Install the high pressure fuel hose and the fuel return line to their ports. Use a new O-ring on the high pressure line. Install the clamp bolt holding the hoses in place.
85. Connect the wiring for the idle switch and the motor position sensor.
86. Install the bracket holding the accelerator and cruise control cables and connect the cables.
87. Connect the brake booster vacuum hose, the canister purge hose and install the air cleaner assembly.
88. Install the breather hose and connect the air intake ducts. Attach the purge control solenoid to the air cleaner.
89. Connect the air flow sensor wiring connector.
90. Install the battery tray and battery. Connect only the positive battery cable.

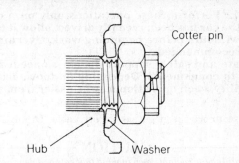

Right driveshaft end-nut and washer

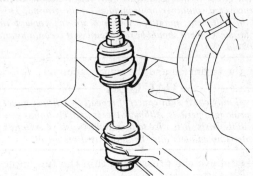

Correct projection of stabilizer bolt threads: 15–18mm (0.6–0.7 in.)

91. Check the engine oil drain plug and secure it if necessary. Install the proper amount of engine oil.
92. Check the transaxle drain plug, tightening it if needed, and install the proper amount of transmission oil. Refill the clutch fluid if equipped with manual transaxle. Bleed the clutch system.
93. Check the radiator and engine drain cocks, closing them if necessary and refill the coolant system.
94. Cfheck the fluid level in the power steering reservoir and refill as needed.
95. Double check all installation items, paying particular attention to loose hoses or hanging wires, untightened nuts, poor routing of hoses and wires (too tight or rubbing) and tools left in the engine area.
96. Connect the negative battery cable. Start the engine and check for leaks.
97. Attend to all leaks immediately, remembering that fluids and metal surfaces may be hot. Adjust the drive belts to the correct tension. Adjust all cables (transmission, throttle, shift selector) and check the fluid levels. Check the operation of all gauges and dashboard lights. Bleed the power steering fluid.
98. Recharge the air conditioning system.
99. With the help of an assistant, install the hood and align it for proper body fit and latching.
100. In a safe location at low speed, road test the car for correct operation of steering, brakes, transaxle, clutch and speedometer.

1989 Galant with 4G63 SOHC and DOHC Engines

The transaxle must be removed from the car before removing the engine; they will not come out as a unit. Please refer to Section 7 for transaxle removal information.

Component location may differ between the two engines; removal and installation sequence is the same.

3 ENGINE AND ENGINE OVERHAUL

NOTE: Perform these operations only on a cold motor. If the car has been recently driven, allow it to cool at least 3-4 hours before beginning work. Overnight cooling is recommended.

Elevate and safely support the car as needed to gain access to components. Depending on tools, arm length and agility, each operation may be easier from above or below.

1. Disconnect the negative battery cable. Drain the engine coolant.

CAUTION

When draining the coolant, keep in mind that cats and dogs are attracted by the ethylene glycol antifreeze, and are quite likely to drink any that is left in an uncovered container or in puddles on the ground. This will prove fatal in sufficient quantity. Always drain the coolant into a sealable container. Coolant should be reused unless it is contaminated or several years old.

2. Drain the engine oil and the transmission oil.

CAUTION

Used motor oil may cause skin cancer if repeatedly left in contact with the skin for prolonged periods. Although this is unlikely unless you handle oil on a daily basis, it is wise to thoroughly wash your hands with soap and water immediately after handling used motor oil.

3. Safely relieve the pressure within the fuel injection system.

CAUTION

The fuel system is under pressure. Release pressure slowly and contain spillage. Observe no smoking/no open flame precautions. Have a Class B-C (dry powder) fire extinguisher within arm's reach at all times.

4. Matchmark the hood to the hinges and remove the hood.
5. Remove the transaxle assembly.
6. Remove the radiator, disconnecting the hoses at the engine.
7. Disconnect the accelerator cable and remove the bracket.
8. Disconnect the heater hoses.
9. Disconnect the brake booster vacuum hose at the engine.
10. Label and disconnect the vacuum hoses running to the firewall.
11. Disconnect the high pressure fuel line and discard the O-ring; it is not reusable.
12. Remove the fuel return hose.
13. Disconnect the electrical connectors to the engine components. All wires and connectors should be labeled at the time of removal. The amount of time saved during reassembly makes the extra effort well worthwhile. Disconnect the:
    - oxygen sensor
    - coolant temperature gauge
    - coolant temperature sensor
    - idle speed control
    - EGR temperature sensor (Calif. only)
    - each injector
    - power transistor
    - condenser
    - throttle position sensor (TPS)
    - motor position sensor (MPS)
    - distributor connector
    - control harness
    - alternator
    - oil pressure switch
    - On the DOHC engine, the connectors for the coolant temperature switch, crankshaft angle sensor, body ground, power steering oil pressure switch must be removed.
14. Loosen the power steering drive belt and remove it. Remove the bolts holding the pump to its bracket and hang the pump out of the way. Do not disconnect the hoses and do not al-

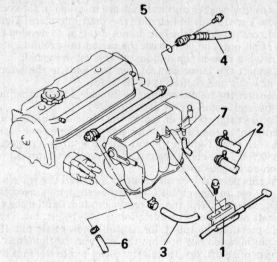

Note: Key numbers are in approximate order of disassembly
1. Accelerator cable
2. Heater hoses
3. Brake booster vacuum hose
4. High pressure fuel line
5. O-ring
6. Fuel return hose
7. Vacuum hose
8. Oxygen sensor connector
9. Coolant temperature gauge connector
10. Coolant temperature sensor connector
11. ISC connector
12. Injector connectors
13. EGR temperature sensor (Calif. only)
14. Power transistor connector
15. Condenser connector
16. Throttle position sensor connector
17. Motor position sensor connector
18. Distributor connector
19. Control wiring harness

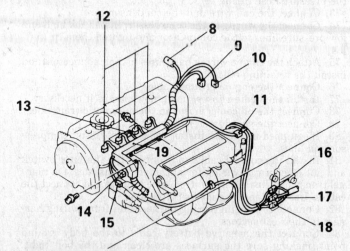

4G63 SOHC engine removal and external components

low the pump to hang by the hoses.

15. Loosen the adjuster and remove the air conditioning drive belt. Remove the compressor from its mount and hang it from stiff wire out of the way. Note that the hoses are still attached; do not loosen them or discharge the system.
16. Remove the self locking nuts and bolt at the exhaust system joint just below the manifold. Separate the exhaust pipes; discard the gasket and the two nuts.
17. Remove the clamp holding the power steering and air con-

ENGINE AND ENGINE OVERHAUL 3

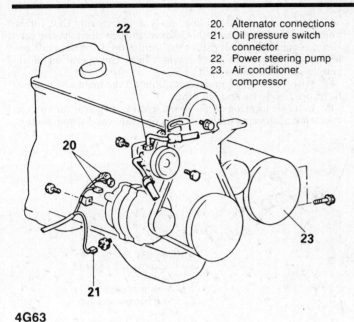

20. Alternator connections
21. Oil pressure switch connector
22. Power steering pump
23. Air conditioner compressor

4G63

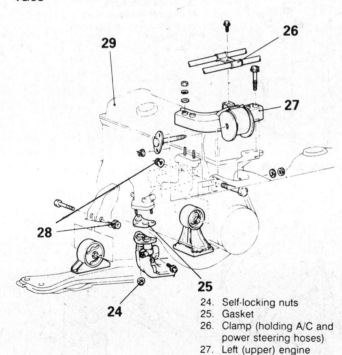

24. Self-locking nuts
25. Gasket
26. Clamp (holding A/C and power steering hoses)
27. Left (upper) engine mount bracket
28. Self-locking nuts for roll stopper through-bolts
29. Engine assembly

4G63

ditioning hoses to the top of the left engine mount. Move the hoses out of the way.

18. Elevate the car and support it safely. Install the engine hoist equipment and make certain the attaching points on the engine are secure. Draw tension on the hoist just enough to support the engine's weight but no more. Do not disturb the placement of the car on the stands.
19. Remove the through-bolt from the rear (firewall side) roll stopper. Remove the through-bolt from the front engine roll stopper.

20. Remove the nuts and bolts holding the upper (left side) engine mount to the engine. Remove the throught-bolt and remove the mount assembly. For DOHC engines, also remove the support bracket below the mount.
21. Double check for any remaining cables, wires or hoses running to the engine. Elevate the hoist and remove the engine from the car. Immediately place it on an engine stand or support it with wooden blocks. Do not allow it to rest on the oil pan or lie on its side. Never leave an engine hanging from a hoist.
22. After repairs, make certain the engine is fully reassembled before installation. All components removed with the engine out of the car should be in place before reinstallation.
23. Install the engine into the car and lower it until the bolt holes for the mounts and roll stoppers align with the brackets. Install the through-bolts and new self-locking nuts, tightening them just snug; they will be final tightened later.
24. Install the upper (left side) mount to the engine, tightening the nuts and bolts snug. Install the through-bolt and tighten it snug.
25. Slowly release tension on the hoist, allowing the weight of the engine to bear fully on the mounts. Once the hoist is slack, remove the lifting apparatus from the engine.
26. Connect the exhaust system to the manifold, using a new gasket and new locking nuts. Tighten the nuts and the small bolt to 25 ft. lbs.
27. Final tighten the engine mount nuts and bolts. Correct torque values are:
- Nuts and bolts holding left mount to engine: 41 ft. lbs.
- Left mount through-bolt: 50 ft. lbs.
- Small nuts on head of through-bolt: 25 ft. lbs.
- Left mount support nuts and bolts (DOHC): 15 ft. lbs.
- Rear roll stopper through-bolt: 33 ft. lbs.
- Front roll stopper through-bolt: 41 ft. lbs.

28. Install the air conditioning compressor, tightening the mounting bolts to 18 ft. lbs. Install the belt and adjust it.
29. Install the power steering pump, tightening the bolts to 36 ft. lbs. Install and adjust the belt.
30. Connect the wiring and harness connectors to the engine. Make certain each terminal is clean and the connector is firmly seated to its mate. Do not route wires near hot surfaces or moving parts. Connect the:
- oxygen sensor
- coolant temperature gauge
- coolant temperature sensor
- idle speed control
- EGR temperature sensor (Calif. only)
- each injector
- power transistor
- condenser
- throttle position sensor (TPS)
- motor position sensor (MPS)
- distributor connector
- control harness
- alternator
- oil pressure switch
- On the DOHC engine, the connectors for the coolant temperature switch, crankshaft angle sensor, body ground, power steering oil pressure switch must be removed.

31. Install the fuel return hose.
32. Using a new O-ring, connect the high pressure fuel line and tighten the bolts to 48 inch lbs.
33. Connect the vacuum lines running to the firewall. Install the brake booster vacuum hose to the engine.
34. Connect the heater hoses.
35. Install the accelerator cable bracket, tightening the bolts to 48 inch lbs., and connect the accelerator cable.
36. Install the radiator and connect the hoses.
37. Install the transaxle.
38. Check the engine oil drain plug and secure it if necessary. Install the proper amount of engine oil.

3-75

3 ENGINE AND ENGINE OVERHAUL

39. Check the transaxle drain plug, tightening it if needed, and install the proper amount of transmission oil.
40. Check the radiator and engine drain cocks, closing them if necessary and refill the coolant system.
41. Double check all installation items, paying particular attention to loose hoses or hanging wires, untightened nuts, poor routing of hoses and wires (too tight or rubbing) and tools left in the engine area.
42. Connect the negative battery cable. Start the engine and check for leaks.
43. Attend to all leaks immediately, remembering that fluids and metal surfaces may be hot. Adjust the drive belts to the correct tension. Adjust all cables (transmission, throttle, shift selector) and check the fluid levels. Check the operation of all gauges and dashboard lights.
44. With the help of an assistant, install the hood and align it for proper body fit and latching.
45. In a safe location at low speed, road test the car for correct operation of steering brakes, transaxle, clutch and speedometer.

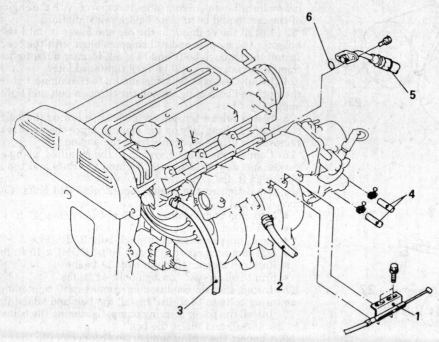

Note: Key numbers are in approximate order of disassembly
1. Accelerator cable
2. Brake booster vacuum hose
3. Fuel return hose
4. Heater hoses
5. High pressure fuel hose
6. O-ring

4G63 DOHC engine removal and external components

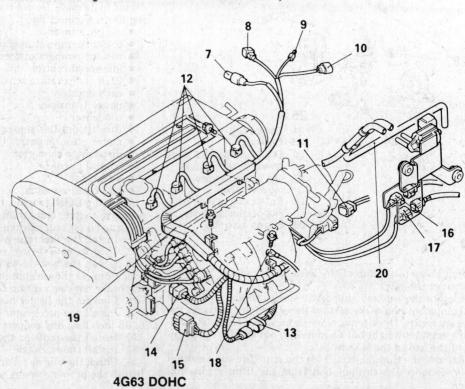

7. Oxygen sensor connector
8. Coolant temperature sensor connector
9. Coolant temperature gauge connector
10. Coolant temperature switch for A/C
11. ISC motor connector
12. Injector connectors
13. EGR temperature sensor connector (Calif. only)
14. Ignition coil lead
15. Power transistor connector
16. Throttle position sensor
17. Crank angle sensor connector
18. Ground cable connector
19. Engine control wiring harness
20. Vacuum hose connections

4G63 DOHC

ENGINE AND ENGINE OVERHAUL 3

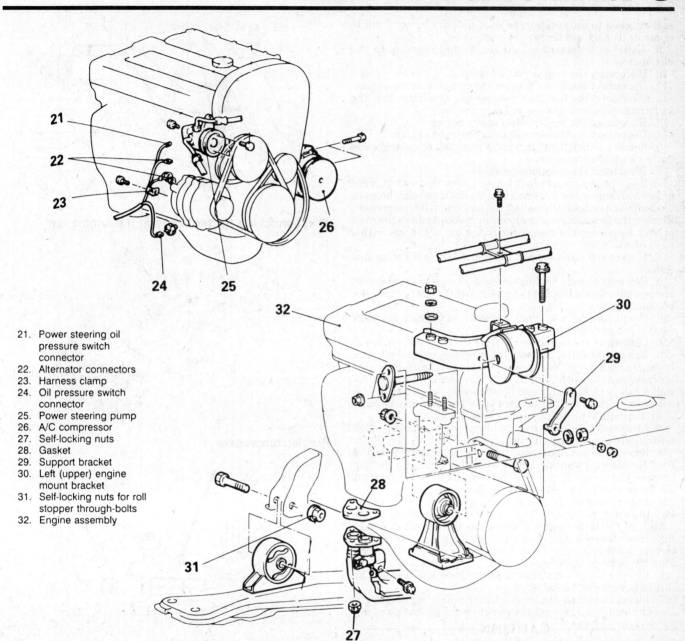

21. Power steering oil pressure switch connector
22. Alternator connectors
23. Harness clamp
24. Oil pressure switch connector
25. Power steering pump
26. A/C compressor
27. Self-locking nuts
28. Gasket
29. Support bracket
30. Left (upper) engine mount bracket
31. Self-locking nuts for roll stopper through-bolts
32. Engine assembly

4G63 DOHC

Precis

NOTE: Perform these operations only on a cold motor. If the car has been recently driven, allow it to cool at least 3–4 hours before beginning work. Overnight cooling is recommended.

Elevate and safely support the car as needed to gain access to components. Depending on tools, arm length and agility, each operation may be easier from above or below.

1. Disconnect the battery, negative cable first and remove the battery.
2. Remove the air cleaner.
3. Label and disconnect the wiring for the reverse lights and engine harness.
4. If equipped with manual transaxle, disconnect the select valve connector.
5. Detach the connectors for the alternator harness and the oil pressure gauge wiring.
6. Drain the engine coolant.

--- **CAUTION** ---

When draining the coolant, keep in mind that cats and dogs are attracted by the ethylene glycol antifreeze, and are quite likely to drink any that is left in an uncovered container or in puddles on the ground. This will prove fatal in sufficient quantity. Always drain the coolant into a sealable container. Coolant should be reused unless it is contaminated or several years old.

7. If equipped with automatic transaxle, label and disconnect the oil cooler hoses. Contain spillage and plug the lines and ports.
8. Disconnect the upper and lower radiator hoses from the engine. Remove the radiator. Note that the hoses and lines are

3-77

3 ENGINE AND ENGINE OVERHAUL

still attached to the radiator; be careful during removal not to crimp or damage any lines.

9. Label and disconnect all the wiring running to the distributor.
10. Disconnect the engine ground strap.
11. Disconnect the brake booster vacuum hose at the engine.
12. Disconnect the fuel inlet, return and vapor hoses from the carburetor. Contain spillage and plug the lines.
13. Disconnect the heater hoses at the engine.
14. Disconnect the accelerator cable from the engine.
15. Remove either the clutch cable (manual) or control cable (auto.) from the transaxle.
16. Disconnect the speedometer cable.
17. If equipped with air conditioning, loosen the belt tensioner pulley and remove the A/C drive belt. Do not loosen any hoses or discharge the system. Disconnect the wire from the compressor. Remove the four compressor mounting bolts and move the compressor away from the motor. Support it out of the way with a piece of stiff wire.
18. Elevate and safely support the vehicle. Drain the transmission oil.
19. Remove the nuts holding the exhaust pipe to the manifold and separate them. Discard the gasket and the nuts — they are not reusable.
20. If equipped with manual transaxle, disconnect the shift control rod and the extension rod.
21. Disconnect the stabilizer bar from the lower control arms.
22. Remove the cotter pin from the lower suspension ball joint. Loosen the nut but do NOT remove it. It should be unscrewed to the last point at which it is still fully threaded on the shaft. Install a ball joint separator and slowly press the joint apart. Do not use a "pickle fork" or similar tool — the joint may be damaged. Repeat the procedure on the opposite side.
23. After each joint is separated (the nut will restrain a sudden separation), remove the nut and separate the joint.
24. Repeat the previous two steps and separate the tie rod from the steering knuckle. Insert a small pry bar into the gap between the transaxle case and the driveshaft. Do not insert it too far or an internal oil seal will be damaged.
25. Pry the driveshafts free of the transaxle. It will be necessary to swing the steering knuckle/hub/brake assembly to the outside of the body to gain clearance as the driveshaft comes free. Immediately suspend each axle with a piece of wire. Protect the boots and splined end of the shaft with rags or covers. Plug the holes in the transaxle case.
26. Remove the circlips from the drive axles and discard the clips. They are not reusable.
27. Install and attach the hoist or lifting equipment. Draw tension on the lift just enough to support the motor but no more.

— **CAUTION** —
The car is on jackstands. Tighten the hoist slowly; do not disturb the position of the car on the stands.

28. Remove the front roll stopper through-bolt.
29. Remove the rear roll stopper mounting bolts from the crossmember. Leave the through-bolt in place.
30. Double check the engine hoist, making sure it is securely braced and that there is proper tension on the cables or chains.
31. Remove the left engine mount through-bolt. Remove the nuts and bolts holding the mount to the engine.
32. Double check the entire engine area for any wires, cables, hoses or components running to the body.
33. Remove the cover from the inside of the right fender inner shield. Remove the transaxle mount retaining bolts.
34. Press down on the transaxle while raising the hoist; remove the engine and transaxle from the vehicle. Support the engine on a stand or support it on wooden blocks. Do not allow it to rest on the oil pan or on its side.
35. Before installing the engine and transaxle assembly, make certain that all components removed during repairs are properly

Disconnect the heater hoses at the engine end

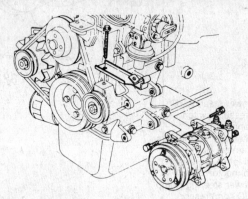

Precis compressor

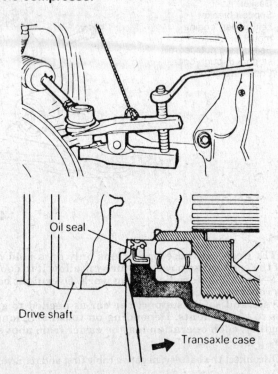

If using Mitsubishi's special ball joint tool (above), tie the safety line to a nearby component. When removing the driveshaft (below), don't insert the prybar too far

3-78

ENGINE AND ENGINE OVERHAUL 3

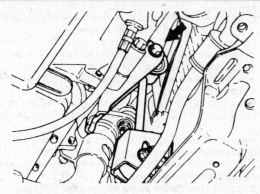

Upper through-bolt for Precis front roll stopper

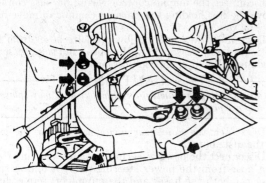

Nuts and bolts holding left engine mount to Precis engine

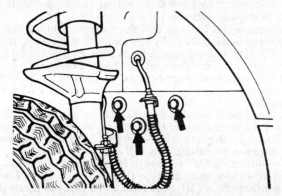

Gain access to the transaxle mount bolts through the right inner fender

reinstalled. Lift the assembly into position and lower it until the roll stopper holes and the transaxle mount bolt holes align properly. Maintain tension on the hoist and install the bolts and nuts holding the engine mount to the engine, tightening them to 50 ft. lbs.

36. Install the bolts holding the rear roll stopper to the crossmember and tighten them to 26 ft. lbs.

37. Install the through-bolts for the front roll stopper and the upper mount and install the bolts for the right side transaxle mount; tighten all the bolts snug but do not overtighten. Final tightening will occur later in the installation.

38. Slowly loosen the hoist and let the engine weight bear fully on the mounts. Once the engine bears on the mounts, the hoist equipment may be removed.

39. Install new circlips within the transaxle housing and install the driveshafts. Make certain each axle engages the clip with a positive click.

40. Move the suspension into position and attach the tie rod end to the knuckle. Tighten the nut to 18 ft. lbs. and install a new cotter pin. Repeat on the opposite side.

41. Connect the lower ball joint to the knuckle. Use a new self-locking nut if one was removed and tighten it to 78 ft. lbs. Install a new cotter pin. Repeat on the opposite side.

42. Attach the shift control rod and extension rod to the manual transmission.

43. Install a new gasket and attach the exhaust system to the engine. Use new self locking bolts; tighten the nuts and bolt to 18 ft. lbs.

44. Install the wheels and lower the car to the ground. Carefully apply final torque to each of the motor mount and roll stopper nuts and bolts. The torque values are:
- Left mount through-bolt: 50 ft. lbs.
- Left mount through-bolt lock nuts: 25 ft. lbs.
- Nuts and bolts holding left mount to engine: 41 ft. lbs.
- Transaxle mount bracket bolts: 25 ft. lbs.
- Rear roll stopper through-bolt: 26 ft. lbs.
- Front roll stopper through-bolt: 26 ft. lbs.

45. Install the air conditioner compressor and the belt.

NOTE: Elevate and safely support the car as needed to gain access to components. Depending on tools, arm length and agility, each operation may be easier from above or below.

46. Connect the speedometer cable.
47. Either connect the control cable to the automatic transaxle or connect the clutch cable to the manual transaxle.
48. Connect the accelerator cable.
49. connect the heater hoses, using new clamps if necessary.
50. Connect the fuel hoses to the carburetor.
51. connect the brake booster vacuum hose to the engine and connect the engine ground strap.
52. Connect the distributor wiring, making sure each cable is firmly seated.
53. Install the radiator and connect the hoses to the engine, using new clamps as needed.
54. Connect the automatic transaxle oil cooler hoses to the transaxle.
55. Connect the electrical connectors to the engine and transaxle.
56. Install the air cleaner and install the battery. Connect only the positive battery cable until ready to start the engine.
57. Check the engine oil drain plug and secure it if necessary. Install the proper amount of engine oil.
58. Check the transaxle drain plug, tightening it if needed, and install the proper amount of transmission oil.
58. Check the radiator and engine drain cocks, closing them if necessary and refill the coolant system.
59. Double check all installation items, paying particular attention to loose hoses or hanging wires, untightened nuts, poor routing of hoses and wires (too tight or rubbing) and tools left in the engine area.
60. Connect the negative battery cable. Start the engine and check for leaks.
61. Attend to all leaks immediately, remembering that fluids and metal surfaces may be hot. Adjust the drive belts to the correct tension. Adjust all cables (transmission, throttle, shift selector) and check the fluid levels. Check the operation of all gauges and dashboard lights.
62. With the help of an assistant, install the hood and align it for proper body fit and latching.
63. In a safe location at low speed, road test the car for correct operation of steering, brakes, transaxle, clutch, speedometer and dashboard indicators.

3-79

3 ENGINE AND ENGINE OVERHAUL

Sigma V6 (2972cc)

NOTE: The use of the correct special tools or their equivalent is REQUIRED for this procedure.

Elevate and safely support the car as needed to gain access to components. Depending on tools, arm length and agility, each operation may be easier from above or below.

All wires and hoses should be labled at the time of removal. The amount of time saved during reassembly makes the extra effort well worthwhile.

1. Matchmark and remove the hood.
2. Drain the engine coolant.

CAUTION

When draining the coolant, keep in mind that cats and dogs are attracted by the ethylene glycol antifreeze, and are quite likely to drink any that is left in an uncovered container or in puddles on the ground. This will prove fatal in sufficient quantity. Always drain the coolant into a sealable container. Coolant should be reused unless it is contaminated or several years old.

---

3. Drain the engine oil, the transaxle oil and, if manual transmission, the clutch fluid.

CAUTION

Used motor oil may cause skin cancer if repeatedly left in contact with the skin for prolonged periods. Although this is unlikely unless you handle oil on a daily basis, it is wise to thoroughly wash your hands with soap and water immediately after handling used motor oil.

---

4. Safely relieve the pressure in the fuel system.

CAUTION

The fuel system is under pressure. Release pressure slowly and contain spillage. Observe no smoking/no open flame precautions. Have a Class B-C (dry powder) fire extinguisher within arm's reach at all times.

---

5. Safely discharge the refrigerant within the air conditioning system. See Section 1.
6. Disconnect the negative battery cable. Remove the cap over the positive terminal, remove the nut holding the terminal to the fusible link and separate them. Remove the battery and battery tray.
7. Disconnect the wiring to the air flow sensor.
8. Remove the purge control solenoid valve from the air cleaner body. Remove the breather hoses and air ducts from the engine and remove the air cleaner assembly.
9. Disconnect the purge control hose at the canister. Disconnect the brake booster vacuum hose.
10. At the throttle body, identify the screws holding the throttle cable and cruise control bracket. Remove these screws without removing the individual cables. Lift the entire bracket with the cable intact away from the throttle body.
11. Disconnect the wiring to the idle switch and the motor position sensor.
12. Remove the high pressure fuel hose and the fuel return hose. Remove the clamp holding the two hoses to the throttle body and remove the hoses from the engine.
13. Open the cover of the fusible link box and disconnect the alternator wiring.
14. Remove the connector holding the battery ground strap to the body.
15. Disconnect the front wiring harness and the engine control wiring harness.
16. If equipped with manual transaxle, remove the retaining hardware holding the control cables to the transaxle. Unbolt the bracket holding the cables, then remove the bracket and cable assembly away from the transaxle. Discard the spring clips; they are non-reusable.
17. If equipped with automatic transaxle, disconnect the inhibitor switch and the pulse generator connectors. Remove the retaining hardware and disconnect the transaxle control cable. Discard the spring clip; it is not reusable.
18. If equipped with ECS (Electric Controlled Suspension), the compressor for the system must be removed. Please refer to Section 8 for information on this system.
19. If equipped with manual transaxle, remove the bracket holding the clutch hydraulic lines. Disconnect the clutch cylinder (slave cylinder) from the transaxle and carefully move the cylinder and lines out of the way. Disconnect the speedometer cable.
20. If equipped with automatic transaxle, disconnect the speedometer cable first, then label and disconnect the oil cooler lines at the transaxle. Contain spillage and plug or cap the lines and ports immediately to prevent entry of dirt.
21. Disconnect the heater hoses at the pipe fittings.
22. Disconnect the engine ground strap at the engine. Label and disconnect the engine control wiring harness connectors. Remove the nut holding the harness clamp and move harness out of the way.
23. Disconnect the high and low pressure hoses from the air conditioning compressor. Plug the hoses and compressor ports immediately. Remove the compressor wiring.

CAUTION

When disconnecting the lines, the residual coolant may splash. Wear gloves and eye protection

---

24. Disconnect the condenser wiring and remove the coil wire from the distributor.
25. Disconnect the power steering pressure hose and then the return hose from the power steering pump. Contain spillage. Plug or cap both the hoses and the pump port immediately.
26. At the radiator, disconnect the overflow tube from the jug and the upper and lower radiator hoses from the engine.
27. Disconnect the electric fan wiring. If equipped with automatic transaxle, disconnect the oil cooler lines.
28. Remove the radiator and the rubber bushings under it. Handle the radiator carefully and protect it while out of the car.
29. For cars with automatic transaxles, remove the oil cooler lines which ran from the transaxle to the radiator.
30. Elevate and safely support the car. Remove the lower inner panel covers in the wheel wells: 2 for automatics, 1 for manual transaxles.
31. Disconnect the stabilizer bar (sway bar) at the control arms. Take careful note of the arrangement of the nuts, spacers and bushings; they must be reassembled in the correct order.
32. Remove the cotter pin from the tie rod end ball joint. Loosen the nut but do NOT remove it. It should be unscrewed to the last point at which it is still fully threaded on the shaft. Install a ball joint separator (MB 991113 or eqivalent) and slowly press the joint apart. Do not use a "pickle fork" or similar tool--the joint may be damaged. Repeat the procedure on the opposite side.
33. After each joint is separated (the nut will restrain a sudden separation), remove the nut and separate the joint.
34. Repeat Steps 32 and 33 to separate the lower arm ball joint from the steering knuckle.
35. Insert a small pry bar into the gap between the transaxle case and the left driveshaft. Pry the driveshaft out of the case while swinging the suspension and hub assembly away (outward). The left driveshaft will stay attached to the hub. When the driveshaft is free, immediately cover the splined end with rags and support the shaft from a piece of stiff wire.
36. At the outer end of the right driveshaft, remove the cotter pin (discard it) and remove the large nut on the end of the shaft. This nut may be very tight; be careful not to disturb the position of the car on the stands during removal.
37. Install an extractor tool (Mitsubishi tool MB 990241-01 or equivalent) and force the shaft from the hub. Do not hammer on the end of the shaft.
38. Insert a large pry bar between the case and the inner shaft

ENGINE AND ENGINE OVERHAUL 3

joint and release the shaft from the transaxle. Remove the circlip, discard it and protect both ends of the shaft with rags. Plug the holes in the transaxle case and put the right driveshaft in a safe location.

39. Disconnect the rubber hangers holding the exhaust system by sliding them off the body mounts.
40. Disconnect the wiring running to the oxygen sensor.
41. Remove the self-locking nuts holding the exhaust pipes to the manifolds. Separate the pipe and remove the gaskets. Use wire to suspend the pipe so it is not damaged.
42. Remove the distributor cap, install the hoisting equipment and make sure it is securely fastened to the engine. Draw tension on the hoist just enough to support the weight of the motor without elevating it.

CAUTION
The car is supported on stands. Tension the hoist slowly and do not disturb the position of the car on the supports.

43. Remove the through-bolt from the front roll stopper.
44. Remove the bolt holding the engine damper (looks like a small shock absorber, which it is) to the engine.
45. Remove the through-bolt from the rear roll stopper assembly.
46. Remove the large through-bolt from the left side engine mount. Note the two smaller nuts holding the head of the large bolt in place. Remove the 4 nuts holding the engine mount to the engine and remove the mount.
47. Remove the protective caps inside the right front fender liner. Loosen and remove the 4 transaxle mount bolts. Work carefully; do not allow the bolts to fall down into the fender liner.
48. Double check for any remaining cables, hoses, wiring or lines still running to the engine and remove them. Check that any item which was moved out of the way is truly in a safe location.
49. Elevate the hoist slowly and raise the engine and transaxle assembly out of the car. Take your time; clearances may be close and the unit is heavy. As soon as the engine is clear, mount it on a stand or support it on wooden blocks. Do not allow it to rest on the oil pan or lie on its side. Never leave an engine hanging from a hoist.
50. After repairs, make certain the engine and transaxle are fully reassembled before installation. All components removed with the engine out of the car should be in place before reinstallation.
51. Install the engine into the car and lower it until the bolt holes for the mounts align with the brackets. Install the transaxle mount bolts, tightening them just snug. These bolts, along with the other mounting bolts, will be final tightened later.
52. Install the left mount bracket to the engine, tightening the nuts to 50 ft. lbs.
53. Install the through-bolt for the left mount and tighten the bolt snug. Install the two small nuts at the bolt head and tighten them to 26 ft. lbs.
54. Install the rear roll stopper through-bolt and tighten it snug.
55. Connect the engine damper to the engine and tighten it snug.
56. Use a new nut and install the through-bolt for the front roll stopper, tightening it just snug.
57. With the engine pinned in place, slowly relax tension on the hoist and allow the engine and transaxle to fully load the mounts. Remove the hoist apparatus from the engine. Reinstall the distributor cap.
58. Use new gaskets and new self-locking nuts, connect the exhaust pipe to the manifolds. Tighten the nuts to 25 ft. lbs.
59. Final tighten the engine mounts, damper and roll stopper bolts. The correct torque values are:
- Transaxle mount bolts: 32 ft. lbs.
- Left engine mount through-bolt: 50 ft. lbs.
- Rear roll stopper through-bolt: 25 ft. lbs.
- Engine damper bolt: 25 ft. lbs.
- Front roll stopper through-bolt: 39 ft. lbs.

60. Connect the oxygen sensor wiring at the exhaust system and reinstall the rubber exhaust hangers on the body mounts.
61. Install a new circlip in the transmission and install the right driveshaft. Fit the outer end through the hub assembly. Make sure the washer and hub nut are installed in the correct direction. Use the nut to help pull the shaft into place. Tighten the nut comfortably tight but do not attempt to develop final torque; that must be done with the car on the ground.
62. Install a new circlip and install the left driveshaft.
63. Connect the lower control arm ball joint to the knuckle. Use a new nut and tighten it to 46 ft. lbs.
64. Connect the tie rod end to the knuckle, tighten the nut to 21 ft. lbs. and install a new cotter pin.
65. Connect the stabilizer bar to the control arms, installing the hardware in the correct order and using a new self-locking nut on each side. Tighten the nut until the threads project 15-18mm (0.6-0.7 in.) above the nut; this develops the proper compression on the bushings.
66. At this point it is recommended to install the wheels and lower the car to the ground, even though it may need to be raised again later. With the car on the ground, tighten the right driveshaft nut. Torque specification is 144-188 ft. lbs. Try for the mid-point, about 166 ft. lbs. Install a new cotter pin. If the cotter pin holes do not line up, the nut may be tightened to a maximum of 188 ft. lbs. Install the cotter pin in the first set of matching holes and bend it securely.

NOTE: The car may be raised and supported as necessary during the remaining steps.

67. Reinstall the lower inner panel covers.
68. For automatic transaxles, reinstall the oil cooler lines, attaching them to the hoses.
69. Install the lower bushings and the radiator.
70. Connect the oil cooler lines to the radiator lines (auto trans.).
71. Connect the electric fan wiring. Install the radiator hoses to the engine, using new clamps if needed. Connect the overflow tube.
72. Connect the power steering hoses to the pump. Tighten the pressure hose fitting to 32 ft. lbs.
73. Connect the coil wire and the condenser wire.
74. Connect the wiring to the compressor. Install the high and low pressure lines to the compressor and tighten them to 15 ft. lbs.
75. Attach the engine wiring harnesses to their connectors and install the retaining bolts.
76. Connect the engine ground cable.
77. Install the heater hoses, using new clamps if necessary.
78. Connect the oil cooler hoses to the automatic transaxle.
79. Connect the speedometer cable.
80. If equipped with ECS, install and connect the air compressor and lines.
81. Install the control cable and connect the inhibitor switch and pulse generator wiring for automatic transaxles. On manual transaxles, install the control cable bracket and connect the cables. Use new spring clips or cotter pins.
82. Connect the engine wiring harness and front wiring harness to their connectors.
83. Connect the negative battery cable to the body ground point, making sure the surfaces are clean and the bolt tight. Connect the starter harness to the fusible link box.
84. Install the high pressure fuel hose and the fuel return line to their ports. Use a new O-ring on the high pressure line. Install the clamp bolt holding the hoses in place.
85. Connect the wiring for the idle switch and the motor position sensor.
86. Install the bracket holding the accelerator and cruise con-

3-81

3 ENGINE AND ENGINE OVERHAUL

trol cables and connect the cables.
87. Connect the brake booster vacuum hose, the canister purge hose and install the air cleaner assembly.
88. Install the breather hose and connect the air intake ducts. Attach the purge control solenoid to the air cleaner.
89. Connect the air flow sensor wiring connector.
90. Install the battery tray and battery. Connect only the positive battery cable.
91. Check the engine oil drain plug and secure it if necessary. Install the proper amount of engine oil.
92. Check the transaxle drain plug, tightening it if needed, and install the proper amount of transmission oil. Refill the clutch fluid if equipped with manual transaxle. Bleed the clutch system.
93. Check the radiator and engine drain cocks, closing them if necessary and refill the coolant system.
94. Check the fluid level in the power steering reservoir and refill as needed.
95. Double check all installation items, paying particular attention to loose hoses or hanging wires, untightened nuts, poor routing of hoses and wires (too tight or rubbing) and tools left in the engine area.
96. Connect the negative battery cable. Start the engine and check for leaks.
97. Attend to all leaks immediately, remembering that fluids and metal surfaces may be hot. Adjust the drive belts to the correct tension. Adjust all cables (transmission, throttle, shift selector) and check the fluid levels. Check the operation of all gauges and dashboard lights. Bleed the power steering fluid.
98. Recharge the air conditioning system.
99. With the help of an assistant, install the hood and align it for proper body fit and latching.
100. In a safe location at low speed, road test the car for correct operation of steering, brakes, transaxle, clutch and speedometer.

2-wheel Drive Pickup

1. Matchmark and remove the hood. Disconnect the negative battery cable at the battery.
2. Elevate and safely support the truck on jackstands. Remove the undercovers and shields below the engine.
3. Drain the engine coolant.

---------- CAUTION ----------
When draining the coolant, keep in mind that cats and dogs are attracted by the ethylene glycol antifreeze, and are quite likely to drink any that is left in an uncovered container or in puddles on the ground. This will prove fatal in sufficient quantity. Always drain the coolant into a sealable container. Coolant should be reused unless it is contaminated or several years old.

4. Drain the engine oil and the transmission oil.

---------- CAUTION ----------
Used motor oil may cause skin cancer if repeatedly left in contact with the skin for prolonged periods. Although this is unlikely unless you handle oil on a daily basis, it is wise to thoroughly wash your hands with soap and water immediately after handling used motor oil.

5. Disconnect the negative battery cable at the engine block.
6. Remove the air cleaner assembly.
7. If the vehicle is equipped with air conditioning, remove the upper and lower fan shrouds from the radiator.
8. Disconnect the upper and lower hose from the radiator and remove the radiator.
9. Disconnect the coil wire from the distributor cap.
10. Label and disconnect the water temperature sensor (with automatic trans.), the water temperature gauge sender and the starter motor connector.
11. Disconnect the positive battery cable from the starter.
12. Disconnect the transmission harness, the oxygen sensor connection, the wiring to the alternator and the oil pressure sender wire.
13. Disconnect the accelerator cable from the carburetor.
14. Carefully disconnect both hoses from the power steering pump. Do not damage the hose and mop up any spilled fluid.
15. Safely discharge the refrigerant within the air conditioning system. See Section 1.
16. Disconnect both air conditioning hoses from the compressor. Carefully move the hoses out of the way but do not crimp or crease the hoses.
17. Disconnect the brake booster vacuum hose.
18. Disconnect the exhaust pipe from the bottom of the exhaust manifold. Use wire or string to support the pipe to the side. Do not let the pipe hang of its own weight.
19. Carefully label and disconnect the fuel lines, breather and vacuum lines and the canister lines as far from the carburetor as possible; disconnecting everything at the carburetor makes reassembly confusing.
20. Inside the vehicle, remove the shifter assembly (manual trans.) or disconnect the automatic shift selector linkage.

NOTE: When removing the manual shifter, remove the mounting bolts from the stopper plate and remove the lever assembly with the stopper plate.

21. Remove the driveshaft. Before disconnecting the rear flange from the axle, make matchmarks on both the shaft and axle flange so that the shaft may be reinstalled in its original position. If the driveshaft is the 3-joint type, disconnect the center carrier from the floorpan of the vehicle. After the flange is free at the rear, slide the shaft out of the transmission housing. Immediately plug or cover the transmission; do not allow dirt or foreign matter to enter the transmission case.
22. Install the hoisting equipment and make sure it is securely fastened to the engine. Draw tension on the hoist just enough to support the weight of the motor without elevating it.

---------- CAUTION ----------
The truck is supported on stands. Tension the hoist slowly and do not disturb the position of the truck on the supports.

23. Position a floor jack under the transmission (No.2) crossmember and adjust it to just touch to crossmember for support.
24. Remove the outer bolts holding the crossmember to the vehicle. Carefully remove the two center bolts holding the crossmember to the transmission case. Lower the jack slowly to remove the crossmember from the vehicle.
25. Double check for any remaining cables, hoses, wiring or lines still running to the engine and/or transmission and remove them. Check that any item which was moved out of the way is truly in a safe location.
26. Place the floor jack with a large block of wood under the transmission case for support. The jack will need to be adjusted and eventually removed during the engine removal.
27. Double check the security of the lifting apparatus. Make certain the cables are tensioned properly. Remove the nut, lockwasher and spacer from each motor mount.
28. Elevate the hoist slowly and raise the engine and transmission assembly out of the truck. The transmission will need to be lowered and the assembly lifted out on an angle. Take your time; clearances may be close and the unit is heavy. As soon as the engine is clear, mount it on a stand or support it on wooden blocks. Do not allow it to rest on the oil pan or lie on its side. Never leave an engine hanging from a hoist. The transmission should be disconnected from the engine as soon as convenient.

To install:
29. After repairs, make certain the engine and transaxle are fully reassembled before installation. All components removed with the engine out of the truck should be in place before reinstallation.
30. Install the engine into the truck and lower it until the bolt holes for the mounts align with the brackets. Use the floor jack

ENGINE AND ENGINE OVERHAUL 3

1. Air cleaner
2. Radiator upper shroud (Vehicles with an air conditioner)
3. Radiator lower shroud (Vehicles with an air conditioner)
4. Radiator
5. Battery ground cable
6. High tension cable
7. Water temperature sensor (Vehicles with an automatic transmission)
8. Water temperature gauge unit
9. Starter motor connector
10. Battery positive cable
11. Transmission harness connector
12. Oxygen sensor connector
13. Alternator connector
14. Oil pressure switch connector
15. Accelerator cable
16. Power steering oil pipe
17. Air conditioner compressor pipe

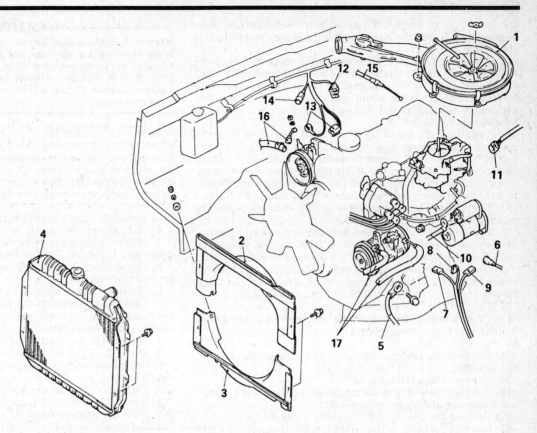

Engine removal, 2-wheel drive truck with 2.0L engine

18. Gear shift assembly
19. Rear propeller shaft
20. No. 2 crossmember
21. Engine and transmission assembly

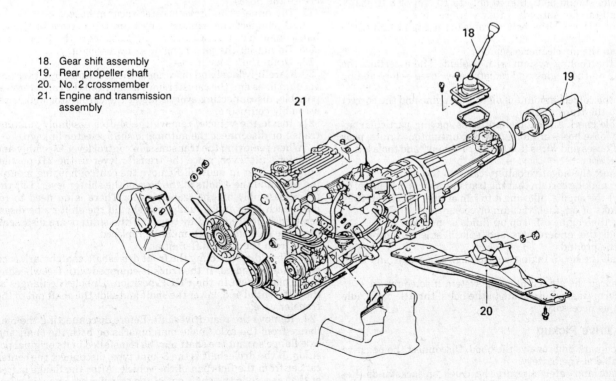

Engine removal, 2-wheel drive truck with 2.0L engine

3-83

3 ENGINE AND ENGINE OVERHAUL

and wood block to support and guide the transmission into place as necessary. Install the mount nuts, tightening them to 19 Nm (14 ft. lbs).

31. Using the floor jack as necessary for support, install the No.2 crossmember. Tighten the body bolts to 55 Nm (41 ft. lbs.) and the nuts holding the crossmember to the transmission to 24 Nm (17 ft. lbs). When the mounts and crossmember are securely fastened, the floor jack and the engine lifting apparatus may be removed from the vehicle.

32. Install the driveshaft. Remove the plug or cover from the transmission and insert the driveshaft. If the driveshaft is of the 3-joint type, connect the center support and tighten the nuts to 35 Nm (26 ft. lbs). Turn the differential flange until the mark made during removal aligns with the mark on the shaft flange. Install the retaining bolts and tighten to 55 Nm or 41 ft. lbs.

33. Reinstall the manual shifter assembly with new gaskets or reconnect the automatic shift linkage.

34. Connect the fuel, vacuum, breather and canister lines to their correct locations. Make certain lines are not crimped or split. Route the lines clear of any hot surfaces or moving parts.

35. Use a new gasket with new nuts, connect the exhaust pipe to the manifold. Tighten the nuts to 50 Nm (37 ft. lbs.) and tighten the small bracket bolt to 25 Nm or 19 ft. lbs.

36. Connect the brake booster vacuum hose.

37. Using a new O-ring inside each line, connect the air conditioning lines to the compressor. Tighten the fittings carefully to 25 Nm or 18 ft. lbs.

38. Connect the power steering hoses to the pump. Tighten the retaining nut to 20 Nm (15 ft. lbs.).

39. Connect the accelerator cable.

40. Connect the wiring to the transmission, the oxygen sensor, the alternator and oil pressure sender.

41. Connect the positive battery cable to the starter.

42. Connect the other wiring to the starter. Connect the water temperature sender and gauge wiring.

43. Connect the coil wire to the distributor.

44. Install the radiator and connect the hoses. The radiator retaining bolts should be tightened only to 10 Nm or 8 ft. lbs.

45. Install the fan shrouds if they were removed.

46. Connect the negative battery cable to the engine (not to the battery).

47. Install the air cleaner assembly.

48. Fill the cooling system with coolant. Make certain the draincocks on the engine and radiator are closed before adding the fluid.

49. Add the correct amount of oil to the engine and the correct amount to the transmission.

50. Double check all installation items, paying particular attention to loose hoses or hanging wires, untightened nuts, poor routing of hoses and wires (too tight or rubbing) and tools left in the engine area.

51. Connect the negative battery cable to the battery.

52. After making certain that the transmission is in Neutral or Park, start the engine, allowing it to run at idle. Check carefully for any leaks of oil, fuel, vacuum or coolant.

53. Shut the engine off. Top up fluids as necessary.

54. Install the undercovers and splash shields. Lower the truck to the ground.

55. Install the hood, taking care to align the seams and latch correctly.

56. Recharge the air conditioning system if so equipped.

57. Perform final adjustments to the belts, throttle cable, idle speed etc. as necessary.

4-wheel Drive Pickup

1. Matchmark and remove the hood. Disconnect the negative battery cable at the battery.

2. Elevate and safely support the truck on jackstands. Remove the undercovers and shields below the engine.

3. Drain the engine coolant.

CAUTION

When draining the coolant, keep in mind that cats and dogs are attracted by the ethylene glycol antifreeze, and are quite likely to drink any that is left in an uncovered container or in puddles on the ground. This will prove fatal in sufficient quantity. Always drain the coolant into a sealable container. Coolant should be reused unless it is contaminated or several years old.

---

4. Drain the engine oil and the transmission oil.

CAUTION

Used motor oil may cause skin cancer if repeatedly left in contact with the skin for prolonged periods. Although this is unlikely unless you handle oil on a daily basis, it is wise to thoroughly wash your hands with soap and water immediately after handling used motor oil.

---

5. Disconnect the negative battery cable at the engine block.

6. Remove the air cleaner assembly.

7. If the vehicle is equipped with air conditioning, remove the upper and lower fan shrouds from the radiator.

8. Disconnect the upper and lower hose from the radiator and remove the radiator.

9. Disconnect the coil wire from the distributor cap.

10. Label and disconnect the water temperature sensor (with automatic trans.), the water temperature gauge sender and the starter motor connector.

11. Disconnect the positive battery cable from the starter.

12. Disconnect the transmission harness, the oxygen sensor connection, the wiring to the alternator and the oil pressure sender wire.

13. Disconnect the accelerator cable from the carburetor.

14. Carefully disconnect both hoses from the power steering pump. Do not damage the hose and mop up any spilled fluid.

15. Safely discharge the refrigerant within the air conditioning system. See Section 1.

16. Disconnect both air conditioning hoses from the compressor. Carefully move the hoses out of the way but do not crimp or crease the hoses.

17. Disconnect the brake booster vacuum hose.

18. Disconnect the exhaust pipe from the bottom of the exhaust manifold. Use wire or string to support the pipe to the side. Do not let the pipe hang of its own weight.

19. Drain the transfer case.

20. Carefully label and disconnect the fuel lines, breather and vacuum lines and the canister lines as far from the carburetor as possible; disconnecting everything at the carburetor makes reassembly confusing.

21. Inside the vehicle, remove the shifter assembly (manual trans.) or disconnect the automatic shift selector linkage.

When removing the transmission control lever assembly and transfer shift lever, place the transfer lever in the **2H** postion and the shifter in neutral. Remove the control housing mounting bolts and the 4 bolts at the base of the shifter lever. Lift the entire assembly off the transmission. There is no need to remove each lever individually, nor should the shifters be disassembled if not required by repair. If the shifters are disassembled, the gaskets in each must be replaced.

22. Drain the front differential.

23. When disconnecting the front driveshaft, matchmark each end before removal. If the truck is equipped with freewheeling hubs, place them in the **FREE** position. Unbolt the flanges at the differential end, lower the shaft and slide the shaft out of the transfer case.

24. Remove the rear driveshaft. Before disconnecting the rear flange from the axle, make matchmarks on both the shaft and axle flange so that the shaft may be reinstalled in its original position. If the driveshaft is the 3-joint type, disconnect the center carrier from the floorpan of the vehicle. After the flange is free at the rear, slide the shaft out of the transfer case housing. Immediately plug or cover the transfer case; do not allow dirt or foreign matter to enter the transfer case.

ENGINE AND ENGINE OVERHAUL 3

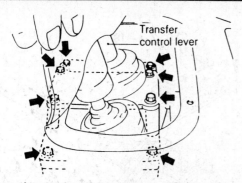

Removing the 4-wheel drive manual shifter assembly

25. Disconnect the small support mount on the left side of the transfer case.
26. Install the hoisting equipment and make sure it is securely fastened to the engine. Draw tension on the hoist just enough to support the weight of the motor without elevating it.
   The added weight and size of the transfer case will change the balance of the unit. Take this into consideration when installing the hoist equipment. A second floor jack with a piece of lumber should be used to support the transfer case before the crossmember is removed.

CAUTION
The truck is supported on stands. Tension the hoist slowly and do not disturb the position of the truck on the supports.

---

27. Position a floor jack under the transmission (No.2) crossmember and adjust it to just touch to crossmember for support.
28. Remove the outer bolts holding the crossmember to the vehicle. Carefully remove the two center bolts holding the crossmember to the transmission case. Lower the jack slowly to remove the crossmember from the vehicle.
29. Double check for any remaining cables, hoses, wiring or lines still running to the engine and/or transmission and remove them. Check that any item which was moved out of the way is truly in a safe location.
30. Place the floor jack with a large block of wood under the transmission case for support. The jack will need to be adjusted and eventually removed during the engine removal.
31. Double check the security of the lifting apparatus. Make certain the cables are tensioned properly. Remove the nut, lockwasher and spacer from each motor mount.

WARNING: When hoisting the engine and transmission, remember that the transfer case adds width on the left side; clearances will be reduced and weight increased.

32. Elevate the hoist slowly and raise the engine and transmission assembly out of the truck. The transmission will need to be lowered and the assembly lifted out on an angle. Take your time; clearances may be close and the unit is heavy. As soon as the engine is clear, mount it on a stand or support it on wooden blocks. Do not allow it to rest on the oil pan or lie on its side. Never leave an engine hanging from a hoist. The transmission should be disconnected from the engine as soon as convenient.

To install:
33. After repairs, make certain the engine and transaxle are fully reassembled before installation. All components removed with the engine out of the truck should be in place before reinstallation.
34. Install the engine into the truck and lower it until the bolt holes for the mounts align with the brackets. Use the floor jack and wood block to support and guide the transmission into place as necessary. Install the mount nuts, tightening them to 19 Nm (14 ft. lbs)
   When reinstalling, use the jack to support the transmission as necessary. Tighten the crossmember mounting bolts to 65 Nm (48 ft. lbs)
35. Using the floor jack as necessary for support, install the No.2 crossmember. Tighten the body bolts to 55 Nm (41 ft. lbs.) and the nuts holding the crossmember to the transmission to 24 Nm (17 ft. lbs). When the mounts and crossmember are securely fastened, the floor jack and the engine lifting apparatus may be removed from the vehicle.
36. Install the front driveshaft after crossmember No. 2 and the engine mounts are secured. Tighten the flange bolts to 55 Nm (41 ft.lbs).
37. Install the rear driveshaft. Remove the plug or cover from the transfer case and insert the driveshaft. If the driveshaft is of the 3-joint type, connect the center support and tighten the nuts to 35 Nm (26 ft. lbs). Turn the differential flange until the mark made during removal aligns with the mark on the shaft flange. Install the retaining bolts and tighten to 55 Nm or 41 ft. lbs.
38. Tighten the support on the left side of the transfer case to 20 Nm (15 ft. lbs.)
39. When reinstalling the shifter and transfer levers, the gasket at the base of the control housing must be replaced. The control housing retaining bolts and the bolts at the base of the shifter should be tightened to 20 Nm or 15 ft. lbs.
40. Reconnect the automatic shift linkage.
41. Connect the fuel, vacuum, breather and canister lines to their correct locations. Make certain lines are not crimped or split. Route the lines clear of any hot surfaces or moving parts.
42. Use a new gasket with new nuts, connect the exhaust pipe to the manifold. Tighten the nuts to 50 Nm (37 ft. lbs.) and tighten the small bracket bolt to 25 Nm or 19 ft. lbs.
43. Connect the brake booster vacuum hose.
44. Using a new O-ring inside each line, connect the air conditioning lines to the compressor. Tighten the fittings carefully to 25 Nm or 18 ft. lbs.
45. Connect the power steering hoses to the pump. Tighten the retaining nut to 20 Nm (15 ft. lbs.).
46. Connect the accelerator cable.
47. Connect the wiring to the transmission, the oxygen sensor, the alternator and oil pressure sender.
48. Connect the positive battery cable to the starter.
49. Connect the other wiring to the starter. Connect the water temperature sender and gauge wiring.
50. Connect the coil wire to the distributor.
51. Install the radiator and connect the hoses. The radiator retaining bolts should be tightened only to 10 Nm or 8 ft. lbs.
52. Install the fan shrouds if they were removed.
53. Connect the negative battery cable to the engine (not to the battery).
54. Install the air cleaner assembly.
55. The transfer case and front axle housings must be refilled with the correct amount of the proper type of oil.
56. Fill the cooling system with coolant. Make certain the draincocks on the engine and radiator are closed before adding the fluid.
57. Add the correct amount of oil to the engine and the correct amount to the transmission.
58. Double check all installation items, paying particular attention to loose hoses or hanging wires, untightened nuts, poor routing of hoses and wires (too tight or rubbing) and tools left in the engine area.
59. Connect the negative battery cable to the battery.
60. After making certain that the transmission is in Neutral or Park, start the engine, allowing it to run at idle. Check carefully for any leaks of oil, fuel, vacuum or coolant.
61. Shut the engine off. Top up fluids as necessary.
62. Install the undercovers and splash shields. Lower the truck to the ground.
63. Install the hood, taking care to align the seams and latch correctly.
64. Recharge the air conditioning system if so equipped.

3-85

3 ENGINE AND ENGINE OVERHAUL

18. Gear shift assembly
19. Rear propeller shaft
20. No. 2 crossmember
21. Engine and transmission assembly
22. Front propeller shaft

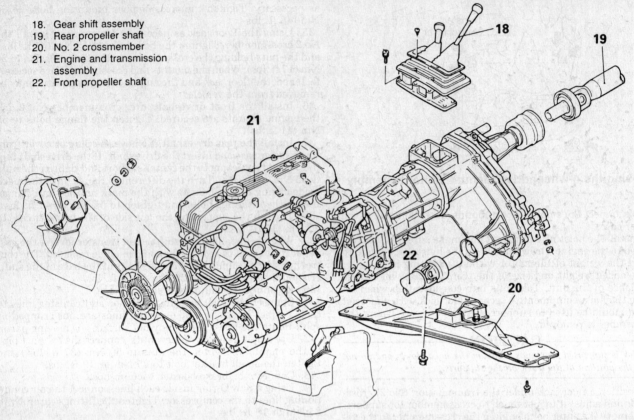

Take note of the driveline and shifter differences when removing the engine from a 4-wheel drive truck

65. Perform final adjustments to the belts, throttle cable, idle speed etc. as necessary.

Montero with 2.6L Engine

NOTE: The transmission, transfer case and front driveshaft must be removed before the engine can be removed. Please refer to Section 7 – Drive Train for detailed instructions on the removal of these components.

1. Matchmark and remove the hood. Disconnect the negative battery cable at the battery.
2. Elevate and safely support the truck on jackstands. Remove the undercovers and shields below the engine and transmission.
3. Drain the engine coolant.

CAUTION

When draining the coolant, keep in mind that cats and dogs are attracted by the ethylene glycol antifreeze, and are quite likely to drink any that is left in an uncovered container or in puddles on the ground. This will prove fatal in sufficient quantity. Always drain the coolant into a sealable container. Coolant should be reused unless it is contaminated or several years old.

---

4. Drain the engine oil and the transmission oil. Drain the transfer case and the front axle lubricant.

CAUTION

Used motor oil may cause skin cancer if repeatedly left in contact with the skin for prolonged periods. Although this is unlikely unless you handle oil on a daily basis, it is wise to thoroughly wash your hands with soap and water immediately after handling used motor oil.

---

5. Remove the air cleaner assembly.
6. Remove the transmission assembly, following procedures outlined in Section 7.
7. Remove the upper and lower fan shrouds from the radiator.
8. Disconnect the upper and lower hose from the radiator and remove the radiator.
9. Disconnect the accelerator cable.
10. Disconnect the heater hoses at the engine.
11. Disconnect the brake booster vacuum hose.
12. Disconnect the coolant by-pass hose, the small coolant hose and the upper radiator hose from the engine.
13. Disconnect the large multi-plug for the engine control harness.
14. Disconnect the engine coolant temperature sensor wire; if the vehicle is air conditioned, disconnect the coolant temperature switch wire.
15. Disconnect the vacuum hose and the coolant temperature gauge sender.
16. Disconnect the oxygen sensor (carefully) and the oil pressure sending unit wire.
17. Loosen the power steering pump and remove the belt.
18. Remove the power steering pump with the hoses attached and undisturbed; place or hang the pump out of the way. Do not disconnect the hoses.
19. Loosen the air conditioning compressor and remove the belt.
20. Remove the air conditioning compressor with the hoses attached and undisturbed; place or hang the unit out of the way. Do not loosen or disconnect the hoses; do not allow the hoses to become crimped or crushed when the unit is placed aside.

WARNING: Do not allow the compressor to hang by the hoses.

21. Disconnect the fuel hoses running to the carburetor. Label or identify them for correct reinstallation.

ENGINE AND ENGINE OVERHAUL 3

22. Disconnect the 3 wiring connectors at the alternator.
23. Remove the ground cable from the engine block.
24. Disconnect the exhaust pipe from the bottom of the exhaust manifold. Use wire or string to support the pipe to the side. Do not let the pipe hang of its own weight.
25. Carefully label and disconnect any breather, vacuum and canister lines running to the firewall or body.
26. Remove the clamp holding the fuel lines to the block.
27. Install the hoisting equipment and make sure it is securely fastened to the engine. Draw tension on the hoist just enough to support the weight of the motor without elevating it.

CAUTION
The truck is supported on stands. Tension the hoist slowly and do not disturb the position of the truck on the supports.

28. Remove the heat shield from the right motor mount.

29. Double check for any remaining cables, hoses, wiring or lines still running to the engine and remove them. Check that any item which was moved out of the way is truly in a safe location.
30. Double check the security of the lifting apparatus. Make certain the cables or chains are tensioned properly. Remove the bolts holding the motor mounts to the vehicle frame.
31. Elevate the hoist slowly and raise the engine and transmission assembly out of the truck. The assembly may need to be lifted out on an angle. Take your time; clearances may be close and the unit is heavy. As soon as the engine is clear, mount it on a stand or support it on wooden blocks. Do not allow it to rest on the oil pan or lie on its side. Never leave an engine hanging from a hoist.
32. After repairs, make certain the engine is fully reassembled before installation. All components removed when the engine

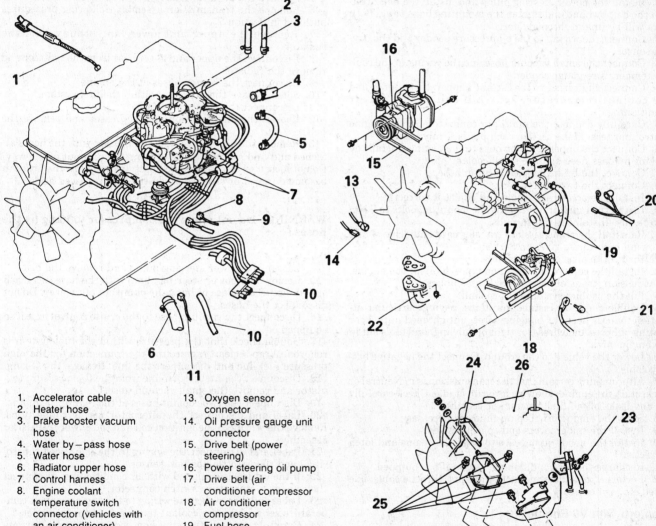

1. Accelerator cable
2. Heater hose
3. Brake booster vacuum hose
4. Water by–pass hose
5. Water hose
6. Radiator upper hose
7. Control harness
8. Engine coolant temperature switch connector (vehicles with an air conditioner)
9. Engine coolant temperature sensor connector
10. Vacuum hose
11. Engine coolant temperature gauge unit connector
13. Oxygen sensor connector
14. Oil pressure gauge unit connector
15. Drive belt (power steering)
16. Power steering oil pump
17. Drive belt (air conditioner compressor)
18. Air conditioner compressor
19. Fuel hose
20. Alternator connector
21. Ground cable
22. Front exhaust pipe
23. Fuel hose clamp
24. Heat protector
25. Engine mounting front insulator attaching bolt
26. Engine assembly

Removal of Montero 2.6 liter engine

3–87

3 ENGINE AND ENGINE OVERHAUL

was out of the truck should be in place before reinstallation.

33. Install the engine into the truck and lower it until the bolt holes for the mounts align with the frame rails. Install the bolts and tighten them to 35 Nm or 26 ft. lbs. Install the heat shield on the right mount.
34. Clamp the fuel lines into place.
35. Connect the various vacuum lines and breather hoses running to the body and firewall.
36. Use a new gasket and new bolts to connect the exhaust pipe to the manifold. Tighten the nuts to 25 Nm (19 ft. lbs).
37. Connect the ground cable to the engine.
38. Connect the wiring to the alternator.
39. Connect the fuel hoses to the carburetor. Install them correctly or the engine won't run.
40. Install the air conditioning compressor and install the belt. Tension the belt by hand and tighten the compressor mounting bolts snug.
41. Install the power steering pump and install the belt. Tension the belt by hand and tighten the mounting bolts snug. Both belts will be final tightened later.
42. Connect the wiring to the oil pressure sender and the oxygen sensor.
43. Connect the small vacuum hose and the wiring to the coolant temperature gauge sender.
44. Connect the wiring to the coolant temperature sender and the coolant temperature switch if equipped with air conditioning.
45. Carefully connect the multi-pin connector for the engine control harness. Make certain each pin is firmly seated.
46. Connect the upper radiator hose to the engine. Install the coolant by-pass hose and the small coolant hose.
47. Connect the brake booster vacuum hose.
48. Connect the two heater hoses.
49. Install the accelerator cable and adjust it correctly.
50. Install the radiator, tightening the mounting bolts to 7 Nm or 5 ft. lbs. Install the fan shrouds.
51. Reinstall the transmission, transfer case and drive shafts following procedures given in Section 7.
52. Install the air cleaner assembly.
53. Install the correct amounts and types of oil for the engine, transmission, transfer case and axle assemblies.
54. Fill the cooling system with coolant.
55. Double check all installation items, paying particular attention to loose hoses or hanging wires, untightened nuts, poor routing of hoses and wires (too tight or rubbing) and tools left in the engine area.
56. Lower the vehicle to the ground. Connect the negative battery cable.
57. After making certain that the transmission is in Neutral or Park, start the engine, allowing it to run at idle. Check carefully for any leaks of oil, fuel, vacuum or coolant.
58. Shut the engine off. Top up fluids as necessary.
59. Install the undercovers and splash shields.
60. Install the hood, taking care to align the seams and latch correctly.
61. Recharge the air conditioning system if so equipped.
62. Perform final adjustments to the belts, throttle cable, idle speed etc. as necessary.

Montero with V6 Engine

NOTE: The transmission, transfer case and front drive shaft must be removed before the engine can be removed. Please refer to Section 7 – Drive Train for detailed instructions on the removal of these components.

1. Matchmark and remove the hood. Disconnect the negative battery cable at the battery.
2. Elevate and safely support the truck on jackstands. Remove the undercovers and shields below the engine and transmission.
3. Drain the engine coolant.

— **CAUTION** —
When draining the coolant, keep in mind that cats and dogs are attracted by the ethylene glycol antifreeze, and are quite likely to drink any that is left in an uncovered container or in puddles on the ground. This will prove fatal in sufficient quantity. Always drain the coolant into a sealable container. Coolant should be reused unless it is contaminated or several years old.

4. Drain the engine oil and the transmission oil. Drain the transfer case and the front axle lubricant.

— **CAUTION** —
Used motor oil may cause skin cancer if repeatedly left in contact with the skin for prolonged periods. Although this is unlikely unless you handle oil on a daily basis, it is wise to thoroughly wash your hands with soap and water immediately after handling used motor oil.

5. Remove the air cleaner assembly and its ductwork.
6. Remove the transmission assembly, following procedures outlined in Section 7.
7. Remove the upper and lower fan shrouds from the radiator.
8. Disconnect the upper and lower hose from the radiator and remove the radiator.
9. Disconnect the heater hoses at the engine.
10. Safely relieve the pressure within the fuel system.
11. Disconnect the accelerator cable.
12. Loosen the air conditioning compressor and remove the belt.
13. Remove the air conditioning compressor with the hoses attached and undisturbed; place or hang the unit out of the way. Do not loosen or disconnect the hoses; do not allow the hoses to become crimped or crushed when the unit is placed aside.

WARNING: Do not allow the compressor to hang by the hoses.

14. Loosen the power steering pump and remove the belt.
15. Remove the power steering pump with the hoses attached and undisturbed; place or hang the pump out of the way. Do not disconnect the hoses.
16. Disconnect the vacuum hose to the cruise control unit if so equipped.
17. Double check that the pressure within the fuel system is relieved. Wrap a clean rag around the connection for the high pressure fuel line and disconnect the line. Remove the O-ring.
18. Disconnect the fuel return line from the fuel pressure regulator and remove the small vacuum hose.
19. Disconnect the two coolant hoses at the rear of the motor.
20. Label and disconnect the alternator wiring connections. Remove the wiring to the ignition coil and the power transistor assembly.
21. Label and disconnect the wiring to the idle speed control (ISC) and the throttle position sensor.
22. If the vehicle is equipped with air conditioning, disconnect the coolant temperature switch connector.
23. Label and disconnect the wiring to the engine coolant temperature sensor and the coolant temperature gauge sender.
24. Carefully remove the emission control vacuum harness and disconnect the engine ground cable from the block.
25. Disconnect the vacuum hose for the brake vacuum booster.
26. Disconnect the engine control wiring harnesses.
27. Disconnect the firewall ground cable from the block.
28. Disconnect the wiring to the EGR temperature sensor and the oil pressure gauge sender.
29. Install the hoisting equipment and make sure it is securely fastened to the engine. Draw tension on the hoist just enough to support the weight of the motor without elevating it.

ENGINE AND ENGINE OVERHAUL 3

CAUTION
The truck is supported on stands. Tension the hoist slowly and do not disturb the position of the truck on the supports.

30. Remove the heat shields from each motor mount.
31. Double check for any remaining cables, hoses, wiring or lines still running to the engine and remove them. Check that any item which was moved out of the way is truly in a safe location.
32. Double check the security of the lifting apparatus. Make certain the cables or chains are tensioned properly. Remove the bolts holding the motor mounts to the frame.
33. Elevate the hoist slowly and raise the engine out of the truck. The assembly may need to be lifted out on an angle. Take your time; clearances may be close and the unit is heavy. As soon as the engine is clear, mount it on a stand or support it on wooden blocks. Do not allow it to rest on the oil pan or lie on its side. Never leave an engine hanging from a hoist.
34. After repairs, make certain the engine is fully reassembled before installation. All components removed when the engine was out of the truck should be in place before reinstallation.
35. Install the engine into the truck and lower it until the bolt holes for the mounts align. Install the bolts and tighten them to 35 Nm (26 ft. lbs). Install the ground wire on the left mount and install the heat shields; tighten the nuts to 12 Nm or 9 ft. lbs.
36. Connect the wiring to the oil pressure sender and the EGR temperature sensor. Connect the firewall ground cable.
37. Connect the engine control harness connectors and attach the vacuum hose for the vacuum brake booster.
38. Connect the ground cable for the engine block and attach the vacuum harness for the emission system.
39. Connect the wiring for the coolant temperature gauge sender, the coolant temperature sensor and the coolant temperature switch if equipped with air conditioning.
40. Connect the wiring for the ISC and throttle position sensor.
41. Connect the wiring for the ignition coil, the power transistor and the alternator.
42. Install the two coolant hoses at the rear of the block.
43. Connect the vacuum hose and the fuel return line to the pressure regulator.
44. Use a new O-ring and connect the fuel delivery line to the regulator. Tighten the connection only to 3 Nm or 2 ft. lbs (24 inch lbs).
45. Install the cruise control vacuum hose if so equipped.
46. Loosely install the power steering pump and install the belt. Tension the belt correctly and tighten the pump mounting bolts to 40 Nm (30 ft. lbs).
47. Install the air conditioning compressor and install the belt. Tension the belt correctly.
48. Install and adjust the accelerator cable. The retaining bolts should be tightened only to 5 Nm (3.5 ft. lbs or 42 inch lbs).
49. Install the radiator, tightening the mounting bolts to 9 Nm or 7 ft. lbs. Install the fan shrouds.
50. Reinstall the transmission, transfer case and drive shafts following procedures given in Section 7.
51. Install the air cleaner assembly and the ductwork.
52. Install the correct amounts and types of oil for the engine,

1. Air cleaner duct
2. Accelerator cable
3. Compressor belt
4. Compressor
5. Power steering belt
6. Power steering pump
7. Cruise control vacuum hose
8. High pressure fuel line
9. O-ring
10. Fuel return hose
11. Vacuum hose
12. Coolant hose
13. Coolant hose
14. Alternator connectors
15. Ignition coil and power transistor
16. ISC connector
17. Throttle position sensor connector
18. Coolant temperature switch connector (with A/C)
19. Coolant temperature sensor connector
20. Coolant temperature gauge connector
21. Emission vacuum harness
22. Ground cable
23. Brake booster vacuum hose
24. Control wiring harness connectors
25. Ground cable
26. EGR temperature sensor connector
27. Oil pressure gauge sender connector
28. Ground cable
29. Heat shield
30. Engine mount bolts
31. Engine assembly

Montero V6 (3.0L) engine removal

Montero V6 (3.0L) engine removal

3 ENGINE AND ENGINE OVERHAUL

transmission, transfer case and axle assemblies.
53. Fill the cooling system with coolant.
54. Double check all installation items, paying particular attention to loose hoses or hanging wires, untightened nuts, poor routing of hoses and wires (too tight or rubbing) and tools left in the engine area.
55. Lower the vehicle to the ground. Connect the negative battery cable.
56. After making certain that the transmission is in Neutral or Park, start the engine, allowing it to run at idle. Check carefully for any leaks of oil, fuel, vacuum or coolant.
57. Shut the engine off. Top up fluids as necessary.
58. Install the undercovers and splash shields.
59. Install the hood, taking care to align the seams and latch correctly.
60. Recharge the air conditioning system if so equipped.

Pickup with 4D55 Diesel Engine

NOTE: All wires and hoses should be labled at the time of removal. The amount of time saved during reassembly makes the extra effort well worthwhile.

1. Disconnect the negative battery cable
2. Drain the engine oil.

--- **CAUTION** ---

Used motor oil may cause skin cancer if repeatedly left in contact with the skin for prolonged periods. Although this is unlikely unless you handle oil on a daily basis, it is wise to thoroughly wash your hands with soap and water immediately after handling used motor oil.

3. Drain the coolant.

WARNING: Housepets and small animals are attracted to the odor and taste of engine coolant (antifreeze). It is a highly poisonous mixture of chemicals; special care must be taken to protect open containers and spillage.

4. Matchmark the hinges and remove the hood.
5. Remove the air cleaner duct and remove the heater hoses. Disconnect the throttle cable.
6. Disconnect the fuel lines (carefully!) and disconnect the water level sensor connector.
7. Remove the fuel filter.
8. Remove the power steering pump, if so equipped.
9. Disconnect the glow plug cable and disconnect the gauge unit wiring harness.
10. Disconnect the engine ground strap.
11. Disconnect the starter motor wiring harness at the starter.
12. Remove the clutch release (slave) cylinder from the bell housing. Support it out of the way from a piece of stiff wire. It is not necessary to remove the fluid hose.
13. Disconnect the hoses to the engine oil cooler.
14. Disconnect the brake booster vacuum hose.
15. Disconnect the alternator wiring harness from the alternator. Label the wires carefully.
16. Depending on equipment, remove the wiring to either the oil pressure switch or the oil pressure gauge sending unit.
17. Disconnect the radiator hoses from the engine.
18. Remove the radiator.
19. Disconnect the exhaust pipe at the first joint under the truck. Support the remainder of the system with wire. Remove the bolts holding the front pipe to the turbocharger and remove the front pipe.
20. Remove the undercovers. On 4-wheel drive vehicles, remove the protective skid plates at the engine and transfer case.
21. Disconnect the speedometer cable.
22. Disconnect the reverse light switch at the transmission. If 4-wheel drive, disconnect the 4wd indicator switch harness also.
23. Remove the driveshafts.
24. Remove the gearshift lever assembly.
25. Use a floor jack to support the transmission. Remove the rear insulator (mount) from the transmission.
26. Remove the crossmember.
27. On 4 wheel drive models, the transfer case must be supported by a second floor jack or adjustable stand. Remove the transfer case mounting bracket and the insulator.
28. Remove the plate from the side frame. Remove the mounting bracket from the transfer case.
29. Position and secure a chain or overhead engine hoist. Attach it to the lift points and draw tension on the chain or cable to support the engine without lifting it.
30. Remove the engine mount nuts from the front engine mounts. Double check completely around the motor (and below it) for any wire, cable, hose, or linkage still running to the body. At this point, the engine should be completely isolated from the vehicle.
31. Elevate the hoist, raising the engine and transmission as a unit. Push downward on the rear of the transmission; remove the engine/trans diagonally.
32. Once clear of the car, the unit should be carefully lowered onto wooden blocks. Once the transmission is removed, the engine should be mounted to a secure stand, allowing it to be worked on.
33. When reinstalling, position the engine/transmission unit with the hoist and lower it into the vehicle. The front mounts should engage and the transmission and/or transfer case should

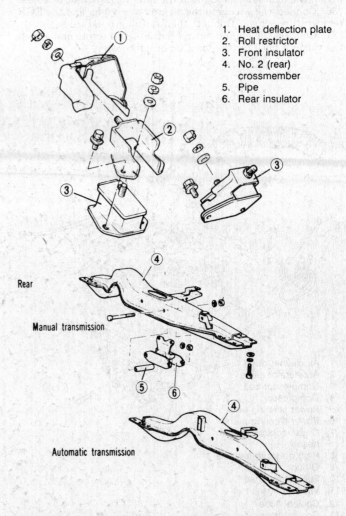

1. Heat deflection plate
2. Roll restrictor
3. Front insulator
4. No. 2 (rear) crossmember
5. Pipe
6. Rear insulator

Engine and transmission mounts, 2-wheel drive trucks with 4D55 diesel

ENGINE AND ENGINE OVERHAUL 3

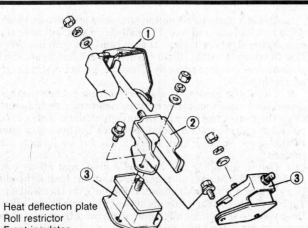

1. Heat deflection plate
2. Roll restrictor
3. Front insulator
4. No. 2 (rear) crossmember
5. Rear insulator
6. Transfer case mounting bracket
7. Transfer case support insulator
8. Plate
9. Stopper
10. Pipe

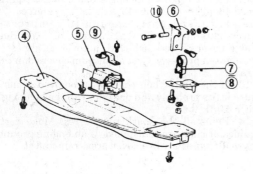

Engine and transmission mounts, 4-wheel drive trucks with 4D55 diesel

rest on the jacks.

34. Install the nuts for the front mounts, tightening them to 17 Nm (13 ft lbs.) Resist the temptation to overtighten these nuts.
35. On 4-wheel drive vehicles, install the mounting bracket to the transfer case and tighten the bolts and nuts to 36 Nm (27 ft. lbs).
36. If 4-wheel drive, install the transfer case mounting bracket and insulator.
37. Install the rear crossmember and the transmission mount (insulator). Tighten the bolts holding the crossmember to the body to 48 Nm (36 ft. lbs.) The bolts holding the transmission insulator to the crossmember should be tightened to 18 Nm (13.5 ft. lbs).
38. When the transmission (and transfer case) is properly attached to the supports, the jacks may be removed. The engine hoist chain may also be removed once the engine and transmission is firmly mounted to the vehicle.
39. Install the gearshift lever assembly.
40. Install the driveshaft(s).
41. Connect the wiring to the transmission. Make certain the connectors are tight and the wiring is correctly routed out of the way of moving parts.
42. Install the speedometer cable.
43. Install the skid plates and/or undercovers.
44. Use new gaskets and install the front exhaust pipe. It will be easier to attach the pipe to the system and then to the turbocharger. Tighten the pipe-to-turbocharger nuts to 20 Nm (15 ft.lbs). Tighten the pipe-to-pipe joint under the vehicle to 25 Nm (18.5 ft. lbs.)
45. Install the radiator assembly. Connect the hoses to the engine. Use new clamps as necessary.
46. Connect the wiring harnesses to the alternator, the oil pressure switch and or the oil pressure gauge sender.
47. Connect the brake booster vacuum hose.
48. Connect the engine oil cooler lines.
49. Install the clutch release cylinder. Adjust the clutch.
50. Install the starter motor wiring harness; connect the engine ground harness to the engine block.
51. Install the power steering pump. Tighten the bolts to 18 Nm (13.5 ft. lbs.).
52. Connect the fuel lines. Install the fuel filter and connect the water level sensor connector.
53. Connect the glow system harness and the gauge unit harness.
54. Connect the throttle cable and adjust it.
55. Connect the heater hoses. Install the air cleaner ductwork.
56. Double check all installation items, paying particular attention to loose hoses or hanging wires, untightened nuts, poor routing of hoses and wires (too tight or rubbing) and tools left in the engine area.
57. Fill the cooling system with the correct amount of coolant.
58. Install the proper amount of engine oil. If not already done, replace the oil filter.
59. Bleed the fuel system.
60. Connect the negative battery cable.
61. Start the engine, following the correct starting procedures. The engine may crank longer than usual; this is normal.
62. Allow the engine to run at idle. While it is warming up, check carefully for any leaks of oil, fuel, vacuum or coolant. Shut the engine off and attend to any leaks immediately. Allow the engine to cool completely before working on the cooling system.
63. Reinstall the hood and align it with the marks made during removal. Close the hood and check the alignment of the seams.

Valve Cover (Rocker Arm Cover)

Usually made of either pressed steel or cast aluminum, the valve cover serves to keep dirt and debris out of the top of the motor and contain the oil pumped to the valve train and camshaft. Removing the valve cover is usually a simple procedure with few surprises.

REMOVAL AND INSTALLATION

Gasoline Engines

1. If the engine is carbureted, remove the air cleaner assembly. Certain injected engines also require removal of the air cleaner for access.
2. Disconnect the breather hoses running to the cover. Check for any other hose or line running over the cover. It can usually be repositioned by removing a clamp or bracket — don't disconnect any hoses if not absolutely necessary.
3. Disconnect the spark plug cables and place them to the side. On DOHC engines, remove the plug connectors very carefully — they are easily damaged.
4. Look at the end of the valve cover closest to the pulleys. Some covers overlap the upper timing belt cover. If this is so, remove the retaining bolts holding the valve cover.
5. If the upper timing belt cover overlaps the valve cover, remove the timing belt cover, then remove the valve cover bolts. Cover the belt with a thick rag to protect it.
6. Gently lift the valve cover up and off the head. If it is stuck, release it by tapping it with a wooden tool handle or plastic mallet. Do not — ever — pry it up with a screwdriver; this will distort the cover and guarantee an oil leak after reassembly.

3 ENGINE AND ENGINE OVERHAUL

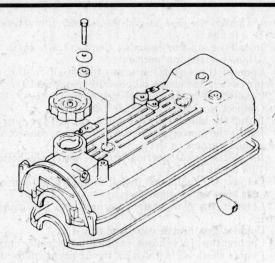

Single cam valve cover, typical

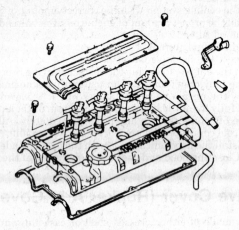

Twin cam valve cover, typical

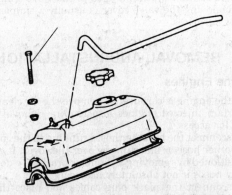

V6 valve cover. Right bank shown, left similar

WARNING: Lift the valve cover and gasket away from the timing belt. Do not allow any oil or foreign matter to fall on the belt. It can be damaged by petroleum based products.

7. At the end farthest from the pulleys, remove the half-circle seal from either the head or the valve cover. The V6 engine does not use this seal.

8. Using a parts cleaning solvent and a stiff brush, clean the valve cover inside and out. The half-circle seal will have the remains of sealant on it; clean the seal and the head or cover surface to which it mates. If these surfaces are not clean, the new sealant won't adhere properly. Clean the gasket mating surfaces on the head.

9. Check the gasket. The square-section rubber gaskets are generally reusable, but they must be clean and free of distortion. Once the gasket has been on and off a few times, it should be replaced as a preventative measure. Check the groove or channel in the cover, making sure there is no dirt or foreign matter present.

10. Before installing the half-circle plug, coat all its contact surfaces with a bead of oil-proof gasket sealer. Install the plug in place. Place a very small amount of sealant on the head surface 10mm (0.4 in.) on each side of the seal. this small area will allow the gasket to seal under all temperature conditions as it expands and contracts. For V6 engines, apply a small amount of sealer to the valve cover gasket at the vertical corners.

11. Fit the gasket into the channel in the cover and place it into position on the head. Make sure the gasket stays in place and that the cover lines up properly on all edges.

12. Install the rubber bushings, the metal washers and the retaining bolts through the cover. The trick to a leak proof installation is even tightening of the bolts. Bolt torque should not exceed 48–60 inch lbs; 36–48 inch lbs. is preferred. The cover must be held just tightly enough to compress the gasket. Overtightening is a leading cause of oil leaks due to deformed valve covers; once distorted, a cover is almost impossible to straighten properly. Cast aluminum covers don't distort — they crack.

13. If the timing belt cover was removed, reinstall it, making sure it fits properly and does not rub the belt. Tighten the bolts to 7–10 ft. lbs.

14. Reinstall the spark plug wires. Make certain any wire guides or clips are reinstalled. If an engine ground wire was removed from the valve cover area, reinstall it.

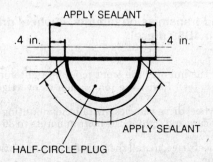

Apply sealant to the half-circle plug as shown

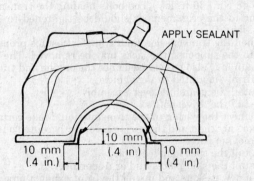

Installing V6 valve cover

ENGINE AND ENGINE OVERHAUL 3

15. Connect the breather hoses and attach the clamps for any other hoses removed. Install the air cleaner assembly as necessary.

16. It is recommended that the car not be started for 1-2 hours, giving the sealant time to set. While the manufacturer of the sealer may disagree, it seems reasonable to allow a tight bond to develop before subjecting it to a hot, oily environment.

17. After the engine has been run for 20-30 minutes, shut it off and check the valve cover gasket area for oil leaks. Do not attempt to cure a leak by tightening the bolts; it won't work. If leaks are present, the cover must be removed and the cause identified.

Diesel Engines

1. Disconnect the breather hose from the valve cover.
2. Remove the two retaining bolts. Catch the washer and seal on each bolt as it is removed.
3. Lift the valve cover up and away from the head. If it does not come loose easily, tap it on a corner with a rubber or plastic mallet.
4. Remove the gasket and check it carefully. If it is not deformed, stretched, nicked or damaged, it may be reused.
5. Clean the inside of the valve cover thoroughly. The amount of sludge present in the cover and on the head is a "report card" on the frequency of your oil changes. If your oil and filter changes are frequent, there should be little or no sludge in the valve train.
6. When reinstalling, fit the gasket into the cover, but do not use any sealer or adhesive. Put the cover in place on the engine.
7. Install the two bolts, making sure the washer and seal are in place on each.
8. Tighten the bolts to 5 Nm (4 ft. lbs) ONLY. This is just tight enough to hold the cover firmly and develop a small amount of compression on the rubber gasket. Overtightening deforms the cover; the gasket cannot seal properly and oil leaks develop.

Rocker Arms and Shafts

REMOVAL AND INSTALLATION

Cordia and Tredia
Pickup with G63B Engine

NOTE: On models which have hydraulic lash adjusters (1997cc), you'll need eight special holders, Mitsubishi Special Tool No. MD 998443 to retain the hydraulic lash adjusters when disassembled.

1. Remove the rocker cover and the upper timing belt cover. Loosen the camshaft sprocket bolt until you can turn it with your fingers.

2. Turn the engine over until the camshaft sprocket timing mark lines up with the timing mark on the cylinder head. The Top Dead Center mark on the front crankshaft pulley must line up with the timing scale on the front cover.

3. Remove the camshaft sprocket bolt, and without allowing tension on the belt to be lost, place the sprocket in the sprocket holder of the front cover or lower timing belt cover. Make sure you don't lose tension on the belt as this will require a lot of additional labor! Make sure, also, that you do not turn the crankshaft throughout the work. On models with hydraulic lash adjusters, put the special clips on the eight hydraulic adjusters at the outer ends of all eight rocker arms. Note that these clips go over the lash adjusters that actuate the large intake valves, not on the small adjusting screw for the smaller jet valves.

4. Loosen but do not remove the camshaft bearing cap bolts (not the outer or cylinder head bolts!), rotating them outward in steps, a little at a time, moving from bolt to bolt. Remove the bolts and then, holding the ends so the assembly stays together,

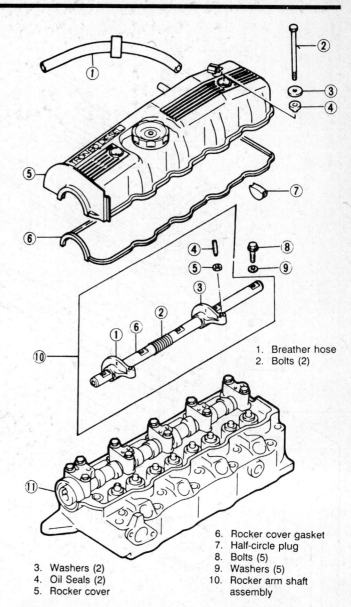

1. Breather hose
2. Bolts (2)
3. Washers (2)
4. Oil Seals (2)
5. Rocker cover
6. Rocker cover gasket
7. Half-circle plug
8. Bolts (5)
9. Washers (5)
10. Rocker arm shaft assembly

Diesel valve cover and rocker shaft assembly

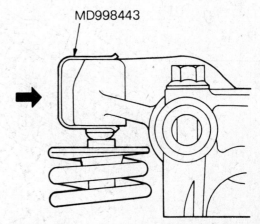

The special tool is used to retain hydraulic lash adjusters

3-93

3 ENGINE AND ENGINE OVERHAUL

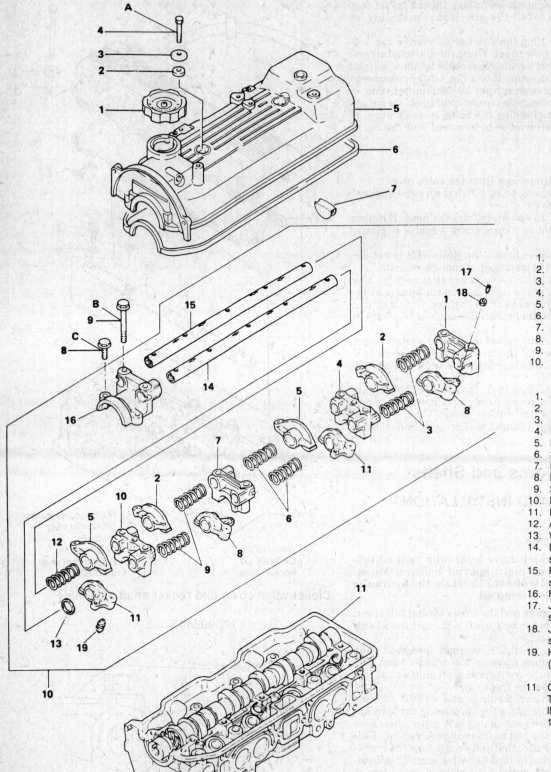

(Numbers are in order of assembly)
1. Oil filler cap
2. Bolt (2)
3. Washer (2)
4. Oil seal (2
5. Rocker cover
6. Rocker cover gasket
7. Semi–circular packing
8. Flange bolt (2)
9. Flange bolt (1)
10. Rocker arm and shaft assembly

1. Rear bearing cap
2. Rocker arm "D" (2)
3. Spring (2)
4. Bearing cap No. 4
5. Rocker arm "C" (2)
6. Spring (2)
7. Bearing cap No. 3
8. Rocker arm "B"
9. Spring (2)
10. Bearing cap No. 2
11. Rocker arm "A"
12. Arm
13. Wave washer
14. Left side rocker arm shaft
15. Right side rocker arm shaft
16. Front bearing cap
17. Jet valve adjusting screw (4)
18. Jet valve adjusting screw locknut (4)
19. Hydraulic lash adjuster (8)

11. Camshaft
Torques: A–3.7–5.0 ft. lbs; B–14–15 ft. lbs; C–15–19 ft. lbs.

Exploded view of valve gear on Cordia and Tredia G63B engines

ENGINE AND ENGINE OVERHAUL 3

remove the rocker shaft assembly from the cylinder head.

5. Place the assembly on a clean work bench. Remove the bolts (which retain the bearing caps in position on the shafts) two at a time along with the associated cap, washers, springs and rockers. Continue until all parts are disassembled. Keep all parts in original order. All parts that will be re-used must be in original positions.

6. Inspection of parts should include verifying that the pads that follow the cam lobes are not worn excessively and that the oil holes are clear. Replace the rockers and/or shafts if rockers are loose on the shafts. If any of the hydraulic lash adjusters have been diagnosed as defective (valves tapping or not closing fully) they can be replaced by removing clips, allowing the adjuster to fall out of the rocker, and replacing the adjuster with another. Install the clip to keep the new adjuster from dropping out of the arm.

7. Assemble the parts of the rocker assembly as follows:
   a. Install the left and right side rocker shafts into the front bearing cap. Notches in the ends of the shaft must be upward.
   b. Install the bolts for the front cap to retain the shafts in place. Note that the left rocker shaft is longer than the right rocker shaft.
   c. Install the wave washer onto the left rocker shaft with the bulge forward.

8. Coat the inner surfaces of the rockers and the upper bearing surfaces of the bearing caps with clean engine oil and assemble rockers, springs, and the remaining bearing caps in order. The intake rockers are the only ones with the jet valve actuators. Note that the rockers are labeled for cylinders 1–3 and 2–4 because the direction the jet valve actuator faces changes. Use mounting bolts to hold the caps in place after each is assembled. When the assembly is complete, install it onto the head and start all mounting bolts into the head and tighten finger tight.

9. Tighten the attaching bolts for the rocker assembly to 14–15 ft. lbs., working from the center outward.

10. Without removing tension from the timing chain or belt, lift the sprocket out of the holder and position it against the front of the cam. Make sure the locating tang on the sprocket goes into the hole in the front of the cam. Install the bolt. Tighten it to 65 ft. lbs.

11. Adjust the valves.

12. Apply sealant to the top surface of the semi-circular seals in the head and install the valve cover. Install the upper timing belt cover.

Starion
Pickup and Montero with G54B Engine

WARNING: The use of the correct special tools or their equivalent is REQUIRED for this procedure.

1. Remove the valve cover.
2. Loosen the large center bolt holding the camshaft sprocket to the cam shaft; don't remove it, just loosen it.
3. Turn the crankshaft pulley clockwise until the timing marks align on both the cylinder head and the crank pulley. This sets the engine at TDC No.1 cylinder. Once this position is established, the engine MUST NOT be rotated during repairs.
4. Remove the center bolt from the sprocket and remove the small distributor drive gear.
5. Pull the cam sprocket with the chain attached out from the camshaft and place it on top of the camshaft sprocket holder. Do not allow the chain to come off the sprocket.
6. Install the retaining clips (tool MD 998443) over the auto lash adjusters.
7. Loosen but do not remove the camshaft bearing cap bolts (not the outer or cylinder head bolts!). Start at an inner or center bolt and work outward in a circular pattern, loosening each a little at a time, moving from bolt to bolt.
8. Remove the bolts after all are loosened. Hold the ends so the assembly stays together and remove the rocker shaft assem-

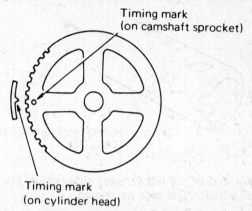

Before removing the cam sprocket, align the timing marks on the cylinder head and sprocket

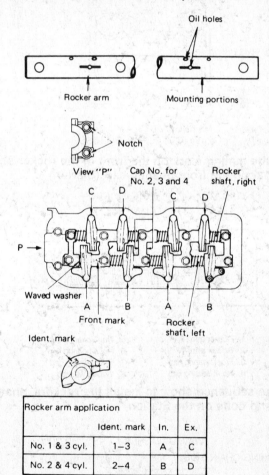

Proper installation sequence for rocker arms on Cordia and Tredia

Rocker arm application			
	Ident. mark	In.	Ex.
No. 1 & 3 cyl.	1–3	A	C
No. 2 & 4 cyl.	2–4	B	D

bly from the cylinder head.

9. Place the assembly on a clean work bench. Remove the bolts (which retain the bearing caps in position on the shafts) two at a time along with the associated cap, washers, springs and rockers. Continue until all parts are disassembled. Keep all parts in original order. All parts that will be re-used must be in original positions.

10. Inspection of parts should include verifying that the pads

3-95

3 ENGINE AND ENGINE OVERHAUL

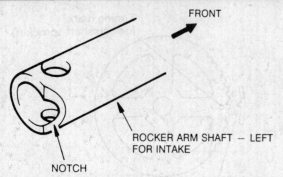

The rear end of the left (intake) side rocker shaft has a notch which must face as shown

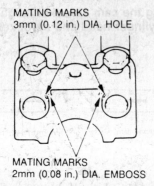

Align the mating mark on the front of the rocker shaft with that on the front bearing cap (Starion)

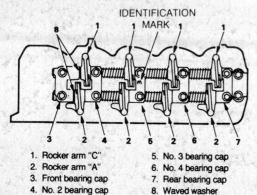

1. Rocker arm "C"
2. Rocker arm "A"
3. Front bearing cap
4. No. 2 bearing cap
5. No. 3 bearing cap
6. No. 4 bearing cap
7. Rear bearing cap
8. Waved washer

Use the sequence show to install the rockers, shafts, caps and bolts on the Starion

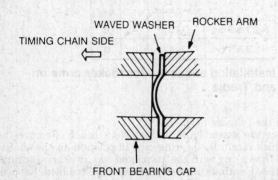

Correct placement of wave washer on Starion rocker shafts

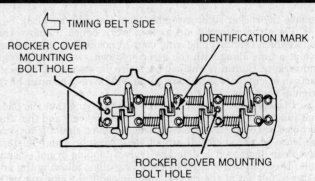

Check the identification marks before reassembling the rocker caps

that follow the cam lobes are not worn excessively and that the oil holes are clear. Replace the rockers and/or shafts if rockers are loose on the shafts. If any of the hydraulic lash adjusters have been diagnosed as defective (valves tapping or not closing fully) they can be replaced by removing clips, allowing the adjuster to fall out of the rocker, and replacing the adjuster with another. Install the clip to keep the new adjuster from dropping out of the arm.

11. Install the left and right side rocker shafts into the front bearing cap. The rear end of the left (intake) rocker arm shaft has a notch. Align the mating marks on the front of each rocker shaft to the mating marks on the front bearing cap. Insert the bolts to hold the shafts in the cap.

12. Install the wave washer so that the rounded side bulges toward the timing chain.

13. Coat the inner surfaces of the rockers and the upper bearing surfaces of the bearing caps with clean engine oil and assemble rockers, springs, and the remaining bearing caps in order. The intake rockers are the only ones with the jet valve actuators. Note that the rockers are labeled for cylinders 2, 3 and 4. While similar in shape, they must be reinstalled in their original position. Use mounting bolts to hold the caps in place after each is assembled. When the assembly is complete, install it onto the head and start all mounting bolts into the head and tighten finger tight.

14. Tighten the attaching bolts for the rocker assembly to 15 ft. lbs., working from the center outward. Remove the special clips from the rocker arms.

15. Without removing tension from the timing chain, lift the sprocket out of the holder and position it against the front of the cam. Make sure the locating tang on the sprocket goes into the hole in the front of the cam.

16. Install the distributor drive gear, making certain it is properly seated. Install the sprocket bolt and tighten it finger tight.

17. Tighten the center sprocket bolt to 40 ft. lbs.

18. Adjust the valve clearances.

19. Apply sealant to the flat face of the half-circle plug in the head. Install the valve cover and gasket.

Mirage with 1597cc Engine

1. Disconnect the hoses that run across the rocker cover and then remove the air cleaner. Label and then disconnect the spark plug high tension wires. Remove the upper front timing belt cover.

2. Turn the crankshaft until No. 1 piston is at TDC of its compression stroke. Align the timing mark on the upper cover at the rear of the timing belt with the mark on the camshaft sprocket. Use a marker to mark the relationship between the timing belt and the mark on the cam sprocket.

3. Loosen and remove the bolt fastening the sprocket to the camshaft, holding the sprocket in position. Don't let the belt slip off! Rest the sprocket on the sprocket holder provided on the lower front cover. If necessary, slip a short piece of used timing

ENGINE AND ENGINE OVERHAUL 3

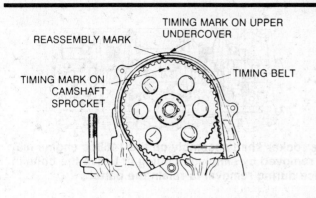

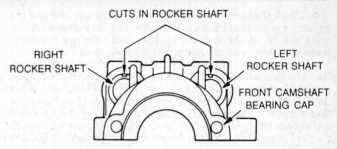

On the Mirage, install the rocker shafts into the front bearing cap with the cuts in the shafts located as shown

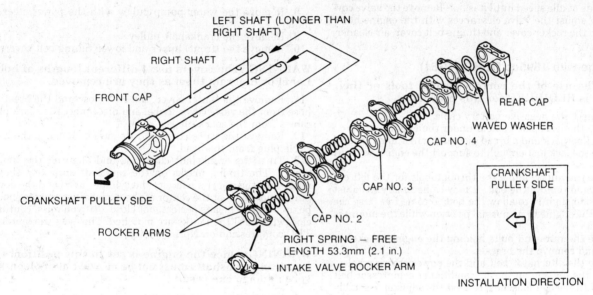

Exploded view of the rocker shaft and arms on the G32B Mirage engine

belt or other thin, flexible material between the holder and the sprocket to keep tension and avoid losing belt timing. Be sure the crankshaft does NOT turn throughout this work.

4. Remove the upper cover located behind the timing belt. Remove the rocker cover. Loosen the camshaft bearing cap bolts without pulling them out of the caps and remove the caps, rockers, and shafts as an assembly. Place on a clean work bench.

5. Remove the bolts from the camshaft bearing caps and remove the caps, rocker arms, and springs from the shaft, keeping all parts in exact order. Check the rocker arm face contacting the cam lobe and the adjusting screw that contacts the valve stem for excess wear. Inspect the fit of the rockers on the shaft. Replace adjusting screws, rockers, and/or shafts that show excessive wear.

6. To reassemble, first lubricate all wear surfaces with clean engine oil and then insert the two rocker shafts into the front bearing cap with the cuts at the top/front of the caps at the tops. Note that the longer shaft goes on the left side (facing the crankshaft pulley). The intake rockers only have the jet valve actuators, and that the waved washers are installed behind the last set of rockers with the bulge at the center of the washer facing the crankshaft pulley.

7. After each cap goes on and the holes are lined up, install the bolts to keep it in place. If the camshaft front oil seal has been damaged it must be replaced. If it needs replacement, remove it while the cam bearing caps are off the engine and then replace it after they are reinstalled. Refer to the Camshaft removal and installation section later in this Section.

8. Lubricate the wear surfaces of the cam bearing caps, and then install them. Tighten the bolts to 15 ft. lbs.

9. Install the timing belt rear cover. Pull the camshaft sprocket upward and install it to the camshaft. Turn the camshaft slightly if necessary to make the dowel pin fit into the hole in the sprocket. Make sure that the mating marks made during disassembly are still aligned; otherwise the camshaft timing will be incorrect. Install the sprocket attaching bolt, torquing it to 51 ft. lbs.

10. Install the timing belt upper cover and spark plug high tension wire supports. Set the valve clearances.

11. Apply sealant to the top of the front bearing cap and rear of the head where the rocker cover seals. Install the rocker cover. Reconnect the spark plug wires, and install the air cleaner and PCV and evaporative emissions hoses. Run the engine at idle speed until it is fully warmed up.

12. Remove the rocker cover and set the valves with the engine hot.

Mirage with 1468cc Engine

1. Remove the air cleaner.
2. Remove the upper timing belt cover. Remove the rocker cover.
3. Loosen the rocker shaft mounting bolts, but do not remove them. Remove each rocker shaft, rocker arms and springs as an assembly. Disassemble the whole assembly by progressively removing each bolt, and then the associated springs and rockers, keeping all parts in the exact order of disassembly.

3 ENGINE AND ENGINE OVERHAUL

4. Check the rocker arm face contacting the cam lobe and the adjusting screw that contacts the valve stem for excess wear. Inspect the fit of the rockers on the shaft. Replace adjusting screws, rockers, and/or shafts that show excessive wear.
5. Assemble all the parts, noting the differences between intake and exhaust parts. The intake rocker shaft is much longer; the intake rocker shaft springs are over 76mm (3 in.) long, while those for the exhaust side are less than 50mm (2 in.) long; intake rockers have the extra adjusting screw for the jet valve; rockers are labeled "1–3" and "2–4" for the cylinder with which they are associated. Tighten the rocker shaft mounting bolts to 17 ft. lbs.
6. Adjust the valve clearances. Install the rocker cover with a new gasket if necessary. Install the air cleaner and PCV valve.
7. Remembering that there is no timing belt cover in place, run the engine at idle speed until it is hot. Remove the valve cover again and adjust the valve clearances with the engine hot.
8. Replace the rocker cover and timing belt cover, air cleaner, and PCV valve.

1989 Mirage with 1595cc DOHC (4G61)

NOTE: The use of the correct special tools or their equivalent is REQUIRED for this procedure.

1. Disconnect the negative battery cable.
2. Remove the left splash shield under the engine.
3. Place a floor jack and a broad piece of lumber under the oil pan. Elevate the jack just enough to support the engine without raising it.
4. Remove the nuts holding the through-bolt for the left engine bracket and remove the bolt. It may be necessary to adjust the jack tension slightly to allow the bolt to come free. Use the jack to keep the engine in its normal position while the mount is disconnected.
5. Remove the nuts and bolts holding the engine bracket to the engine and remove the bracket.
6. Remove the alternator belt and the power steering belt.
7. Loosen the tension on the air conditioner compressor belt tensioner. With the belt loosened, remove the adjuster assembly and remove the belt.

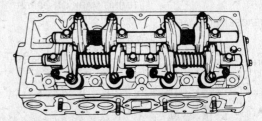

The rocker shaft assembly on the 1468cc engine may be removed by simply unbolting it. Leave the bolts in place during removal to retain the parts

8. Remove the water pump pulley with the power steering pulley.
9. Remove the crankshaft pulley.
10. Remove the (front) upper and lower timing belt covers.

WARNING: The covers use 4 different lengths of bolts. Label or diagram them as they are removed.

11. Remove the center cover, the PCV hose and the breather hose from the valve cover. Label and disconnect the spark plug cables from the spark plugs.
12. Remove the valve cover and its gasket. Remove the half-circle plug from the head.
13. Turn the crankshaft clockwise and align all the timing marks. The timing marks on the camshaft sprockets should align with the upper surface of the cylinder head and the dowel pins (guide pins) at the center of the camshaft sprockets should be up. Aligned means perfect, not close — if you miss, continue clockwise until the marks are matched. This sets the motor at TDC/compression on No.1 cylinder.

WARNING: Once the engine is set to this position, the crank and camshafts must not be moved out of place. Severe damage can result.

14. Remove the auto-tensioner.

IDENTIFICATION MARK	INSTALLATION POSITION
1–3	NOS. 1 and 3 CYLINDERS (POSITIONS A AND C IN THE ILLUSTRATION BELOW)
2–4	NOS. 2 and 4 CYLINDERS (POSITIONS B AND D IN THE ILLUSTRATION BELOW)

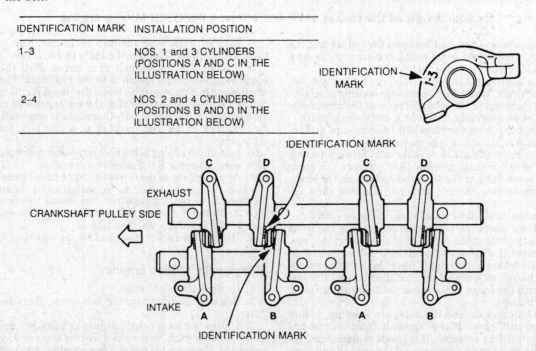

Installed positions of the rockers on the Mirage G15B engine

ENGINE AND ENGINE OVERHAUL 3

15. Before removing the timing belt, mark it with an arrow to show the direction of rotation. If the belt is to be reused, it must be reinstalled so that it rotates in the same direction as before.
16. Remove the timing belt, keeping it free of grease, oil and fluids. Do not crease or crimp the belt. After the belt is clear, remove the tensioner pulley.
17. Remove the throttle body bracket.
18. Carefully remove the crankshaft angle sensor from the rear end of the left (intake) camshaft.
19. Remove the exhaust camshaft sprocket, the intake camshaft sprocket and oil seals at the front of the camshafts. Use a wrench to counterhold the camshaft at the hexagonal flats ONLY to prevent turning. Do not try to wedge the two sprockets; they are easily damaged.
20. Beginning at the front camshaft bearing caps, loosen the cap bolts a little at a time, progressing from front to rear in 2 or 3 passes. Remove the bearing caps and keep them in strict order both front/rear and left/right. If the caps are difficult to remove, use a small plastic mallet to tap gently on the rear of the camshaft. The vibration will free the caps.
21. Remove the camshafts and put them in a safe location, protecting them from dust and grit.
22. Remove the lifters individually and place them in a labeled containers to maintain strict order. They are not interchangable.
23. Inspect the rollers and wear surfaces for any sign of scoring or metal wear. Check each roller for smooth rotation. Replace lifters as needed.

NOTE: Check the camshaft lobes for the same conditions. If either shaft is worn, it must be replaced along with the lifters for that cam.

24. Before reassembly, coat the camshafts liberally with clean motor oil, covering both the lobes and the journals (where the bearings hold it).
25. Install the lifters, making sure each one is correctly mounted on the lash adjuster and the valve stem end.
26. Install the camshafts on the cylinder head. Remember that the intake (left) cam has the slotted end to drive the crank angle sensor.
27. After each camshaft is in place, turn it so that the small dowel pin at the sprocket end is up (12 o'clock). The exhaust cam must be turned an additional 3° clockwise.
28. Check the markings on the bearing caps. Nos. 2 through 5 are marked **L** (left, or intake) or **R** (right; exhaust) with a position number. The front caps (No. 1) are only marked **L** or **R**. Install the bolts finger snug — no more.
29. Tighten the bearing caps in the correct order, making 2 or 3 passes through the pattern to achieve the final torque of 15 ft. lbs. The secret is to draw the caps down very evenly, eliminating any drag or binding on the cam.
30. Apply fresh motor oil to each new camshaft oil seal. Using special tools MD 998307 and MD 998306-01, press fit the oil seal into the cylinder head over the end of the camshaft.
31. Install the camshaft sprockets, paying attention to which is exhaust and intake. Counterhold the camshaft as before and tighten each sprocket bolt to 65 ft. lbs. Maker certain the guide dowels line up with the pulley before tightening the bolt.
32. Align the notch or punch mark on the end of the crank angle sensor blade with the punched mark on the sensor case. This alignment is critical to the correct operation of the fuel injection system.
33. Install the crank angle sensor to the cylinder head, making sure the blade fits properly into the slot in the camshaft. When tightening the nuts, make certain the sensor does not turn.
34. Install the throttle body stay, tightening the nuts to 14 ft. lbs.
35. Check the sprockets and tensioner for wear. The sprocket teeth should be well defined, not rounded and the valleys between the teeth should be clean. The tensioner and idler pulleys

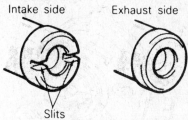

Don't confuse the rocker shafts during reinstallation

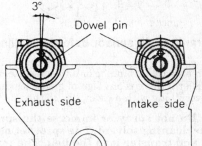

Correct camshaft positions after reinstallation

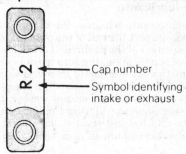

Pay attention to the markings on the bearing caps — they are not interchangable

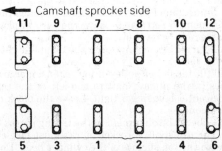

Tighten the Mirage camshaft bearings in this order

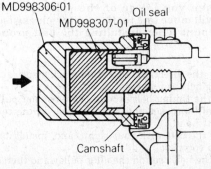

Installing the camshaft oil seals with the special guide and driver

3 ENGINE AND ENGINE OVERHAUL

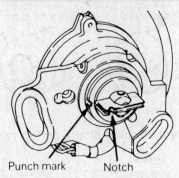

Punch mark   Notch

Align the crank angle sensor before installing it

should spin freely with no binding or unusual noise. Replace the tensioner or idler if there is any sign of grease leaking from the seal. Clean everything with a clean, dry cloth.

WARNING: Do not spray or immerse the sprockets or tensioner in cleaning solvent. The sprocket may absorb the solvent and transfer it to the belt. The tensioner is internally lubricated and the solvent will dilute or dissolve the lubricant.

36. Inspect the auto-tensioner for wear or leakage around the seals. Closely inspect the end of the pushrod for any wear. Measure the projection of the pushrod. It should be 12mm (0.47 in.). If it is out of specification, the auto tensioner must be replaced.

37. Mount the auto-tensioner in a vise with protected jaws. Keep the unit level at all times, and, if the plug at the bottom projects, protect it with a common washer. Smoothly tighten the vise to compress the adjuster tip; if the the rod is easily retracted, the unit has lost tension and should be replaced. You should feel a fair amount of resistance when compressing the rod.

38. As the rod is contracted into the body of the auto-tensioner, watch for the hole in the pushrod to align with the hole in the body. When this occurs, pin the pushrod in place with a piece of stiff wire 1.5mm diameter. It will be easier to install the tensioner with the pushrod retracted. Leave the wire in place and remove the assembly from the vise.

39. Install the auto-tensioner onto the engine, tightening the bolts to 18 ft. lbs.

40. Install the tensioner pulley onto the tensioner arm. Locate the pinholes in the pulley shaft to the left of the center bolt. Tighten the center bolt finger tight. Leave the blocking wire in the auto-tensioner.

41. Double check the alignment of the timing marks for all the sprockets. The camshaft sprocket marks must face each other and align with the top surface of the cylinder head. Don't forget to check the oil pump sprocket alignment as well as the crankshaft sprocket.

NOTE: When you let go of the exhaust camshaft sprocket, it will move one tooth counter-clockwise. Take this into account when installing the belt around the sprockets.

42. Observing the direction of rotation mark made earlier, install the timing belt as follows:
    a. Install the timing belt around the tensioner pulley and the crankshaft sprocket and hold the belt on the tensioner with your left hand.
    b. Pulling the belt with your right hand, install it around the oil pump sprocket.
    c. Install the belt around the idler pulley and then around the intake camshaft sprocket.
    d. Double check the exhaust camshaft sprocket alignment mark. It has probably moved one tooth; if so, rotate it clockwise to align the timing mark with the top of the cylinder head and install the belt over the sprocket. The belt is a snug fit but can be fitted with the fingers; do not use tools to force the belt onto the sprocket.
    e. Raise the tensioner pulley against the belt to prevent sagging and temporarily tighten the center bolt.

43. Turn the crankshaft ¼ turn (90°) counter-clockwise, then turn it clockwise and align the timing marks.

44. Loosen the center bolt on the tensioner pulley and attach special tool MD 998752 and a torque wrench. (This is a purpose-built tool; substitutes will be hard to find.) Apply a torque of 1.88–2.03 ft. lbs. to the tensioner pulley; this establishes the correct loading against the belt.

NOTE: You must use a torque wrench with a 0–3 ft. lb. scale. Normal scales will not read to the accuracy needed. If the body interferes with the tools, elevate the jack slightly to raise the engine above the interference point.

45. Hold the tensioner pulley in place with the torque wrench and tighten the center bolt of the tensioner pulley to 36 ft. lbs. Remove the torque wrench assembly.

46. On the left side of the engine, located in the rear lower timing belt cover, is a rubber plug. Remove the plug and install special tool MD 998738 (another one difficult to substitute). Screw the tool in until it makes contact with the tensioner arm. After the point of contact, screw it in some more to compress the arm slightly.

47. Remove the wire holding the tip of the auto-tensioner. The tip will push outward and engage the tensioner arm. Remove the screw tool.

48. Rotate the crankshaft 2 complete turns clockwise. Allow everything to sit as is for about 15 minutes — longer if the temperature is cool — then check the distance the tip of the auto-tensioner protrudes. (It is almost impossible with the engine in the car; if you can measure the distance, it should be 3.8-4.5mm (0.15–0.18 in.). The alternate method is given in the next step.

49. Reinstall the special screw tool into the left support bracket until the end just makes contact with the tensioner arm. Starting from that position, turn the tool against the arm, counting the number of full turns until the tensioner arm contacts the auto-tensioner body. At this point, the arm has compressed the tensioner tip through its projection distance. The correct number of turns to this point is 2.5 to 3 turns. If you took the correct number of turns to bottom the arm, the projection was correct to begin with and you may now unwind the screw tool EXACTLY the number of turns you turned it in.

If your number of turns was too high or too low when the arm bottomed, you have a problem with the tensioner and should replace it.

50. Remove the special screw tool and install the rubber plug into the access hole in the timing belt cover.

51. Install the half-circle plug with the proper sealant, then install the valve cover and gasket.

52. Install the spark plug wires, the breather and PCV hoses and the center cover.

53. Install the lower and upper timing belt covers, paying attention to the correct placement of the four different sizes of bolts.

54. Install the crankshaft pulley, tightening the bolts to 18 ft. lbs.

55. Install the water pump pulley(s) and tighten the bolts to 72 inch lbs.

56. Install air conditioning drive belt and install the tensioner bracket. Tighten the bracket bolts to 18 ft. lbs.

57. Install the power steering belt and the alternator belt.

58. Install the engine mount bracket to the engine. Tighten the mounting nuts and bolts to 42 ft. lbs.

59. Adjust the jack (if necessary) so that the engine mount bushing aligns with the bodywork bracket. Install the through-bolt and tighten the nuts snug.

ENGINE AND ENGINE OVERHAUL 3

60. Slowly release tension on the floor jack so that the weight of the engine bears fully on the mount. Tighten the through-bolt to 72 ft. lbs. and the small safety nut to 38 ft. lbs. Don't forget to install the small support bracket for the engine mount. Tighten its bolt and nut to 16 ft. lbs.
61. Double check all installation items, paying particular attention to loose hoses or hanging wires, untightened nuts, poor routing of hoses and wires (too tight or rubbing) and tools left in the engine area.
62. Connect the negative battery cable.
63. Start the engine and let it idle, listening for any unusual noises from the area of the timing belt. Possible causes of noise are the belt rubbing against the covers or a sprocket flange, the belt being too loose and slapping, or a tensioner binding. Do not accelerate the engine if abnormal noises are heard from the timing belt train — severe damage can result.
64. Reinstall the splash shield under the car. Final adjustment of the drive belts may be needed.

Galant with G64B Engine

WARNING: The use of the correct special tools or their equivalent is REQUIRED for this procedure.

1. Remove the air cleaner.
2. Remove the rocker cover.
3. Loosen the rocker shaft mounting bolts, but do not remove them. Install the special retaining clips (MB 998443-01) to hold the auto-lash adjusters in place.
4. Remove each rocker shaft, rocker arms and springs as an assembly. Disassemble the whole assembly by progressively removing each bolt, and then the associated springs and rockers, keeping all parts in the exact order of disassembly.
5. Check the rocker arm face contacting the cam lobe and the adjusting screw that contacts the valve stem for excess wear. Inspect the fit of the rockers on the shaft. Replace adjusting screws, rockers, and/or shafts that show excessive wear.
6. Assemble all the parts. The rocker shafts are inserted into the front bearing cap with the notches facing up. Install the wave washer with the bulge facing the pulley end of the motor. Bearing caps Nos. 2, 3, and 4 are very similar in appearance but must not be interchanged. Observe the identifying numbers and letters on each rocker and install them in their correct positions.
7. Install the assembly onto the head and tighten the bolts finger tight. Tighten the bearing cap bolts to 15 ft. lbs. in two steps, working from the center bearing outward. Remove the retaining clips from the rockers.
8. Adjust the valve clearances. Install the rocker cover with a new gasket if necessary. Install the air cleaner, PCV and breather hoses.

Galant and Montero with 6G72 V6 Engine

NOTE: If only the rocker assembly is to be removed, continue with the procedure below. If the camshaft or other components are to be removed, the timing belt and camshaft pulley must be removed before this procedure.

The use of the correct special tools or their equivalent is REQUIRED for this procedure.

1. Remove the valve cover for the affected side.
2. Install the special clips (MB 998443-01) to hold the auto-adjusters in place. Carefully loosen the retaining bolts at each bearing cap. Keep light downward pressure on the rockers to hold the camshaft in place.
3. When all the bolts are loose, remove them from the front (pulley end) to the rear while keeping the same light pressure on the assembly. Have an assistant hold the cam firmly in place as you remove the rocker assembly. The tension of the timing belt will try to pop the cam out of its journals; this must NOT be allowed to happen or your work instantly triples in difficulty. If this should happen, you'll need to reinstall the camshaft and

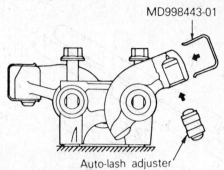
Install the clips to hold the auto-adjuster in place

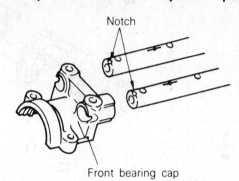
Install the rocker shafts with the notches up

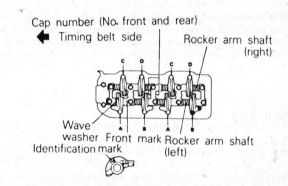

Assembled cylinders	Identification marks	Intake	Exhaust
1 and 3	1-3	A	C
2 and 4	2-4	B	D

Galant G64B rocker identification and placement

timing belt. Instructions for each are found later in this Section.
4. As soon as the rocker assembly is clear of the car, remove the rearmost bearing cap and install it in its original position on the cam. Tighten the bolts just snug enough to hold it in place. If, during the inspection process, this bearing cap must be removed for cleaning or replacement, install No. 3 bearing cap before removing No. 4.

NOTE: Take care during removal that no oil, grease or dirt comes into contact with the timing belt.

5. Disassemble the rocker assembly, checking each component for wear, scoring, or plugged oil passages. Check the roller for correct and smooth rotation. Inspect the inner diameter of each rocker for any scoring or enlargement. If wear is found inside the rocker, replace it and inspect the shaft for damage.

3 ENGINE AND ENGINE OVERHAUL

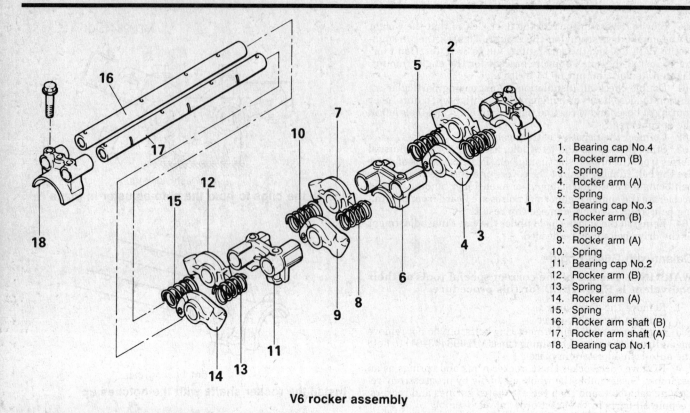

1. Bearing cap No.4
2. Rocker arm (B)
3. Spring
4. Rocker arm (A)
5. Spring
6. Bearing cap No.3
7. Rocker arm (B)
8. Spring
9. Rocker arm (A)
10. Spring
11. Bearing cap No.2
12. Rocker arm (B)
13. Spring
14. Rocker arm (A)
15. Spring
16. Rocker arm shaft (B)
17. Rocker arm shaft (A)
18. Bearing cap No.1

V6 rocker assembly

6. Before reassembly, coat the contact faces liberally with clean motor oil. Observe the numbers on the bearing caps so that they are replaced in the correct location. Reassemble the rocker shafts into the front bearing cap so that the notches face outward. As you continue to assemble the springs, rockers and bearing caps, remember that the arrows on the on the bearing caps must point in the same direction as the arrow on the head. This is particularly important if the rockers have been removed from both heads.

7. When the rocker arm assembly is complete (except for the rear bearing cap, which is holding the cam down), have your trusted assistant hold the cam while the rear cap is removed. Assemble the last cap onto the rocker assembly and carefully fit the assembly onto the head. Remember that the cam must be held in place during this exchange.

8. Install the retaining bolts finger tight, making sure each rocker is correctly aligned with the camshaft and valve stem. Once the rockers are securely in place, the special retaining clips may be removed.

9. Starting at the center bearing cap and working outward in two passes, tighten the bearing cap bolts to 15 ft. lbs.

10. Apply sealer to the correct locations and reinstall the valve cover.

Galant with 4G63 SOHC ENGINE

WARNING: The use of the correct special tools or their equivalent is REQUIRED for this procedure.

1. Remove the valve cover.
2. Turn the crankshaft clockwise until No. 1 piston is at TDC of its compression stroke. Align the timing mark on the upper cover at the rear of the timing belt with the mark on the camshaft sprocket. Use a marker to mark the relationship between the timing belt and the mark on the cam sprocket.
3. Loosen and remove the bolt fastening the sprocket to the camshaft, holding the sprocket in position. Don't let the belt slip off! Rest the sprocket on the sprocket holder provided on the lower front cover. If necessary, slip a short piece of used timing belt or other thin, flexible material between the holder and the sprocket to keep tension and avoid losing belt timing. Be sure the crankshaft does NOT turn throughout this work. An alternate scheme is to hold the sprocket and belt upward with a wire tied to something overhead. This is a bit risky; be careful not to deform the belt or allow the sprocket to come out from under the belt.

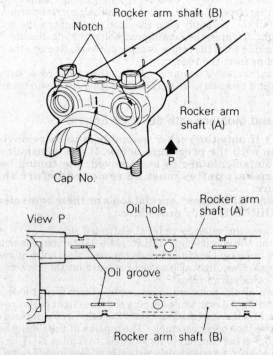

Correct reassembly of V6 rocker shafts

ENGINE AND ENGINE OVERHAUL 3

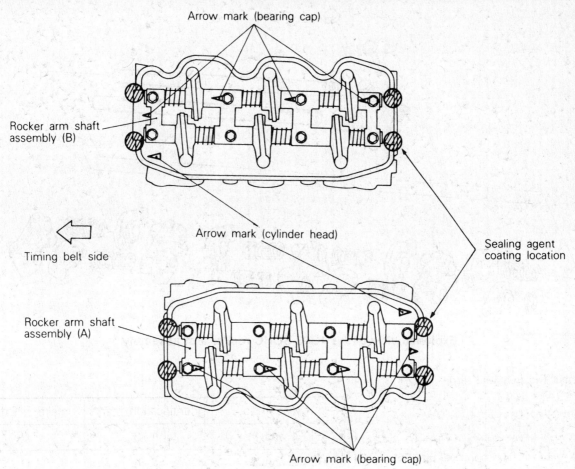

Arrows on the bearing caps must agree with the arrow on the head

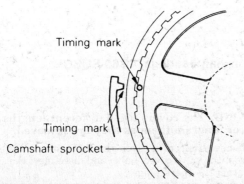

align the timing mark before removing the 4G63 camshaft pulley

4. Install the special clips (MB 998443-01) to hold the auto-adjusters in place. Loosen the camshaft bearing cap bolts without pulling them out of the caps and remove the caps, rockers, and shafts as an assembly. Place on a clean work bench.

5. Remove the bolts from the camshaft bearing caps and remove the caps, rocker arms, and springs from the shaft, keeping all parts in exact order. Check the rocker arm face contacting the cam lobe and the adjuster that contacts the valve stem for excess wear. Inspect the fit of the rockers on the shaft. Replace adjusters, rockers, and/or shafts that show excessive wear.

6. To reassemble, first lubricate all wear surfaces with clean engine oil and then insert the two rocker shafts into the front bearing cap with the notches facing up. Pay close attention to the identifying marks on each rocker and bearing cap. Install the wave washer with the bulge towards the pulley end of the motor.

7. After each cap goes on and the holes are lined up, install the bolts to keep it in place. If the camshaft front oil seal has been damaged it must be replaced. If it needs replacement, remove it while the cam bearing caps are off the engine and then replace it after they are reinstalled. Refer to the camshaft removal and installation section later in this Section.

8. Lubricate the wear surfaces of the cam bearing caps, and then install them. Beginning at the center bearing cap and working outward in two passes, tighten the bolts to 15 ft. lbs.

9. Pull the camshaft sprocket into place and install it to the camshaft. Turn the camshaft slightly if necessary to make the dowel pin fit into the hole in the sprocket. Make sure that the mating marks made during disassembly are still aligned; otherwise the camshaft timing will be incorrect. Install the sprocket attaching bolt, torquing it to 65 ft. lbs.

10. Set the valve clearances.

11. Install the timing belt upper cover and spark plug high tension wire supports.

12. Apply sealant to the top of the front bearing cap and rear of the head where the rocker cover seals. Install the rocker cover. Reconnect the spark plug wires, and install the air cleaner, PCV and evaporative emissions hoses as necessary.

Galant with 4G63 DOHC Engine

NOTE: The use of the correct special tools or their equivalent is REQUIRED for this procedure.

3 ENGINE AND ENGINE OVERHAUL

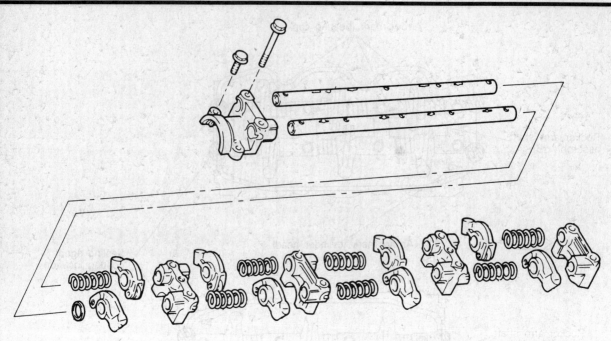

Exploded view of the 4G63 SOHC rocker arm assembly

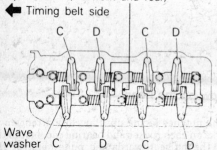

Cylinder No.	Identification marks	Intake	Exhaust
1 and 3	1–3		C
2 and 4	2–4		D

Positions and markings for rockers and camshaft bearings, 4G63 SOHC

1. Disconnect the negative battery cable.
2. Remove the left splash shield under the engine. Remove the bracket bolt holding the two hoses to the top of the left engine mount.
3. Place a floor jack and a broad piece of lumber under the oil pan. Elevate the jack just enough to support the engine without raising it.
4. Remove the nuts holding the through-bolt for the left engine bracket and remove the bolt. It may be necessary to adjust the jack tension slightly to allow the bolt to come free. Use the jack to keep the engine in its normal position while the mount is disconnected.
5. Remove the nuts and bolts holding the engine bracket to the engine and remove the bracket.
6. Remove the alternator belt and the power steering belt.
7. Loosen the tension on the air conditioner compressor belt tensioner. With the belt loosened, remove the adjuster assembly and remove the belt.
8. Remove the water pump pulley with the power steering pulley.
9. Remove the crankshaft pulley.
10. Remove the (front) upper and lower timing belt covers.

WARNING: The covers use 4 different lengths of bolts. Label or diagram them as they are removed.

11. Remove the center cover, the PCV hose and the breather hose from the valve cover. Label and disconnect the spark plug cables from the spark plugs.
12. Remove the valve cover and its gasket. Remove the half-circle plug from the head.
13. Turn the crankshaft clockwise and align all the timing marks. The timing marks on the camshaft sprockets should align with the upper surface of the cylinder head and the dowel pins (guide pins) at the center of the camshaft sprockets should be up. Aligned means perfect, not close — if you miss, continue clockwise until the marks are matched. This sets the motor at TDC/compression on No.1 cylinder.

WARNING: Once the engine is set to this position, the crank and camshafts must not be moved out of place. Severe damage can result.

14. Remove the auto-tensioner.
15. Before removing the timing belt, mark it with an arrow to show the direction of rotation. If the belt is to be reused, it must

ENGINE AND ENGINE OVERHAUL 3

be reinstalled so that it rotates in the same direction as before.

16. Remove the timing belt, keeping it free of grease, oil and fluids. Do not crease or crimp the belt. After the belt is clear, remove the tensioner pulley.
17. Carefully remove the crankshaft angle sensor from the rear end of the left (intake) camshaft.
18. Remove the exhaust camshaft sprocket, the intake camshaft sprocket and oil seals at the front of the camshafts. Use a wrench to counterhold the camshaft at the hexagonal flats ONLY to prevent turning. Do not try to wedge the two sprockets; they are easily damaged.
19. Beginning at the front camshaft bearing caps, loosen the cap bolts a little at a time, progressing from front to rear in 2 or 3 passes. Remove the bearing caps and keep them in strict order both front/rear and left/right. If the caps are difficult to remove, use a small plastic mallet to tap gently on the rear of the camshaft. The vibration will free the caps.
20. Remove the camshafts and put them in a safe location, protecting them from dust and grit.
21. Remove the lifters individually and place them in a labeled containers to maintain strict order. They are not interchangable.
22. Inspect the rollers and wear surfaces for any sign of scoring or metal wear. Check each roller for smooth rotation. Replace lifters as needed.

NOTE: Check the camshaft lobes for the same conditions. If either shaft is worn, it must be replaced along with the lifters for that cam.

23. Before reassembly, coat the camshafts liberally with clean motor oil, covering both the lobes and the journals (where the bearings hold it).
24. Install the lifters, making sure each one is correctly mounted on the lash adjuster and the valve stem end.
25. Install the camshafts on the cylinder head. Remember that the intake (left) cam has the slotted end to drive the crank angle sensor.
26. After each camshaft is in place, turn it so that the small dowel pin at the sprocket end is up (12 o'clock). The exhaust cam must be turned an additional 3° clockwise.
27. Check the markings on the bearing caps. Nos. 2 through 5 are marked **L** (left, or intake) or **R** (right; exhaust) with a position number. The front caps (No. 1) are only marked **L** or **R**. Install the bolts finger snug — no more.
28. Tighten the bearing caps in the correct order, making 2 or 3 passes through the pattern to achieve the final torque of 15 ft. lbs. The secret is to draw the caps down very evenly, eliminating any drag or binding on the cam.
29. Apply fresh motor oil to each new camshaft oil seal. Using special tools MD 998307 and MD 998306-01, press fit the oil seal into the cylinder head over the end of the camshaft.
30. Install the camshaft sprockets, paying attention to which is exhaust and intake. Counterhold the camshaft as before and tighten each sprocket bolt to 65 ft. lbs. Maker certain the guide dowels line up with the pulley before tightening the bolt.
31. Align the notch or punch mark on the end of the crank angle sensor blade with the punched mark on the sensor case. This alignment is critical to the correct operation of the fuel injection system.
32. Install the crank angle sensor to the cylinder head, making sure the blade fits properly into the slot in the camshaft. When tightening the nuts, make certain the sensor does not turn.
33. Install the throttle body stay, tightening the nuts to 14 ft. lbs.
34. Check the sprockets and tensioner for wear. The sprocket teeth should be well defined, not rounded and the valleys between the teeth should be clean. The tensioners should spin freely with no binding or unusual noise. Replace the tensioner if there is any sign of grease leaking from the seal. Clean everything with a clean, dry cloth.

WARNING: Do not spray or immerse the sprockets or tensioners in cleaning solvent. The sprocket may absorb the solvent and transfer it to the belt. The tensioners are internally lubricated and the solvent will dilute or dissolve the lubricant.

35. Inspect the auto-tensioner for wear or leakage around the seals. Closely inspect the end of the pushrod for any wear. Measure the projection of the pushrod. It should be 12mm (0.47 in.). If it is out of specification, the auto tensioner must be replaced.
36. Mount the auto-tensioner in a vise with protected jaws. Keep the unit level at all times, and, if the plug at the bottom projects, protect it with a common washer. Smoothly tighten the vise to compress the adjuster tip; if the rod is easily retracted, the unit has lost tension and should be replaced. You should feel a fair amount of resistance when compressing the rod.
37. As the rod is contracted into the body of the auto-tensioner, watch for the hole in the pushrod to align with the hole in the body. When this occurs, pin the pushrod in place with a piece of stiff wire 1.5mm. It will be easier to install the tensioner with the pushrod retracted. Leave the wire in place and remove the assembly from the vise.
38. To reinstall, make certain the crankshaft sprocket and the silent shaft sprocket are aligned with the timing marks. Slip the belt over the crank sprocket and then over the silent shaft sprocket, keeping tension on the upper side. Make certain there is no slack in the tension side of the belt.
39. Install the tensioner and temporarily position it so that the center of the pulley is to the left and above the center of the installation bolt.
40. Apply finger pressure on the tensioner, forcing it into the belt. Make certain the opposite side of the belt is taut. Tighten the installation bolt to hold the tensioner in place.
41. Push down on the center of the tension side of the belt (side opposite the tensioner pulley) and check the distance the belt deflects. Correct deflection is 6mm (¼ inch).
42. Install the flange on the crankshaft sprocket, making sure it is positioned correctly.
43. Install the outer crankshaft sprocket and/or the oil pump sprocket if either was removed.
44. Install the auto-tensioner onto the engine, tightening the bolts to 18 ft. lbs.
45. Install the tensioner pulley onto the tensioner arm. Locate the pinholes in the pulley shaft to the left of the center bolt. Tighten the center bolt finger tight. Leave the blocking wire in the auto-tensioner.
46. Double check the alignment of the timing marks for all the sprockets. The camshaft sprocket marks must face each other and align with the top surface of the cylinder head. Don't forget to check the oil pump sprocket alignment as well as the crankshaft sprocket.

NOTE: When you let go of the exhaust camshaft sprocket, it will move one tooth counter-clockwise. Take this into account when installing the belt around the sprockets.

47. Observing the direction of rotation mark made earlier, install the timing belt as follows:
   a. Install the timing belt around the tensioner pulley and the crankshaft sprocket and hold the belt on the tensioner with your left hand.
   b. Pulling the belt with your right hand, install it around the oil pump sprocket.
   c. Install the belt around the idler pulley and then around the intake camshaft sprocket.
   d. Double check the exhaust camshaft sprocket alignment mark. It has probably moved one tooth; if so, rotate it clockwise to align the timing mark with the top of the cylinder head and install the belt over the sprocket. The belt is a snug fit but can be fitted with the fingers; do not use tools to force the belt

3-105

3 ENGINE AND ENGINE OVERHAUL

onto the sprocket.

    e. Raise the tensioner pulley against the belt to prevent sagging and temporarily tighten the center bolt.

48. Turn the crankshaft ¼ turn (90°) counter-clockwise, then turn it clockwise and align the timing marks.

49. Loosen the center bolt on the tensioner pulley and attach special tool MD 998752 and a torque wrench. (This is a purpose-built tool; substitutes will be hard to find.) Apply a torque of 1.88–2.03 ft. lbs. to the tensioner pulley; this establishes the correct loading against the belt.

NOTE: You must use a torque wrench with a 0–3 ft. lb. scale. Normal scales will not read to the accuracy needed. If the body interferes with the tools, elevate the jack slightly to raise the engine above the interference point.

50. Hold the tensioner pulley in place with the torque wrench and tighten the center bolt of the tensioner pulley to 36 ft. lbs. Remove the torque wrench assembly.

51. On the left side of the engine, located in the rear lower timing belt cover, is a rubber plug. Remove the plug and install special tool MD 998738 (another one difficult to substitute). Screw the tool in until it makes contact with the tensioner arm. After the point of contact, screw it in some more to compress the arm slightly.

52. Remove the wire holding the tip of the auto-tensioner. The tip will push outward and engage the tensioner arm. Remove the screw tool.

53. Rotate the crankshaft 2 complete turns clockwise. Allow everything to sit as is for about 15 minutes — longer if the temperature is cool — then check the distance the tip of the auto-tensioner protrudes. (It is almost impossible with the engine in the car; if you can measure the distance, it should be 3.8–4.5mm (0.15–0.18 in.). The alternate method is given in the next step.

54. Reinstall the special screw tool into the left support bracket until the end just makes contact with the tensioner arm. Starting from that position, turn the tool against the arm, counting the number of full turns until the tensioner arm contacts the auto-tensioner body. At this point, the arm has compressed the tensioner tip through its projection distance. The correct number of turns to this point is 2.5 to 3 turns. If you took the correct number of turns to bottom the arm, the projection was correct to begin with and you may now unwind the screw tool EXACTLY the number of turns you turned it in.

If your number of turns was too high or too low when the arm bottomed, you have a problem with the tensioner and should replace it.

55. Remove the special screw tool and install the rubber plug into the access hole in the timing belt cover.

56. Install the half-circle plug with the proper sealant, then install the valve cover and gasket.

57. Install the spark plug wires, the breather and PCV hoses and the center cover.

58. Install the lower and upper timing belt covers, paying attention to the correct placement of the four different sizes of bolts.

59. Install the crankshaft pulley, tightening the bolts to 18 ft. lbs.

60. Install the water pump pulley(s) and tighten the bolts to 6 ft. lbs.

61. Install air conditioning drive belt and install the tensioner bracket. Tighten the bracket bolts to 18 ft. lbs.

62. Install the power steering belt and the alternator belt.

63. Install the engine mount bracket to the engine. Tighten the mounting nuts and bolts to 42 ft. lbs.

64. Adjust the jack (if necessary) so that the engine mount bushing aligns with the bodywork bracket. Install the through-bolt and tighten the nuts snug.

65. Slowly release tension on the floor jack so that the weight of the engine bears fully on the mount. Tighten the through-bolt to 72 ft. lbs. and the small safety nut to 38 ft. lbs. Don't forget to install the small support bracket for the engine mount. Tighten its bolt and nut to 16 ft. lbs.

66. Double check all installation items, paying particular attention to loose hoses or hanging wires, untightened nuts, poor routing of hoses and wires (too tight or rubbing) and tools left in the engine area.

67. Connect the negative battery cable.

68. Start the engine and let it idle, listening for any unusual noises from the area of the timing belt. Possible causes of noise are the belt rubbing against the covers or a sprocket flange, the belt being too loose and slapping, or a tensioner binding. Do not accelerate the engine if abnormal noises are heard from the timing belt train — severe damage can result.

68. Reinstall the splash shield under the car. Final adjustment of the drive belts may be needed.

Precis

1. Remove the air cleaner.
2. Remove the upper timing belt cover. Remove the rocker cover.
3. Loosen the rocker shaft mounting bolts, but do not remove them. Remove each rocker shaft, rocker arms and springs as an assembly. Disassemble the whole assembly by progressively removing each bolt, and then the associated springs and rockers, keeping all parts in the exact order of disassembly.
4. Check the rocker arm face contacting the cam lobe and the adjusting screw that contacts the valve stem for excess wear. Inspect the fit of the rockers on the shaft. Replace adjusting screws, rockers, and/or shafts that show excessive wear.
5. Assemble all the parts, noting the differences between intake and exhaust parts. The intake rocker shaft is much longer; the intake rocker shaft springs are over 76mm (3 in.) long, while those for the exhaust side are less than 50mm (2 in.) long; intake rockers have the extra adjusting screw for the jet valve; rockers are labeled "1-3" and "2-4" for the cylinder with which they are associated. Tighten the rocker shaft mounting bolts to 17 ft. lbs.
6. Adjust the valve clearances. Install the rocker cover with a new gasket if necessary. Install the air cleaner and PCV valve.
7. Remembering that there is no timing belt cover in place, run the engine at idle speed until it is hot. Remove the valve cover again and adjust the valve clearances with the engine hot.
8. Replace the rocker cover and timing belt cover, air cleaner, and PCV valve.

Sigma V6

NOTE: If only the rocker assembly is to be removed, continue with the procedure below. If the camshaft or other components are to be removed, the timing belt and camshaft pulley must be removed before this procedure.

The use of the correct special tools or their equivalent is REQUIRED for this procedure.

1. Remove the valve cover for the affected side.
2. Install the special clips (MB 998443-01) to hold the auto-adjusters in place. Carefully loosen the retaining bolts at each bearing cap. Keep light downward pressure on the rockers to hold the camshaft in place.
3. When all the bolts are loose, remove them from the front (pulley end) to the rear while keeping the same light pressure on the assembly. Have an assistant hold the cam firmly in place as you remove the rocker assembly. The tension of the timing belt will try to pop the cam out of its journals; this must NOT be allowed to happen or your work instantly triples in difficulty. If this should happen, you'll need to reinstall the camshaft and timing belt. Instructions for each are found later in this Section.
4. As soon as the rocker assembly is clear of the car, remove the rearmost bearing cap and install it in its original position on the cam. Tighten the bolts just snug enough to hold it in place.

3-106

ENGINE AND ENGINE OVERHAUL 3

If, during the inspection process, this bearing cap must be removed for cleaning or replacement, install No. 3 bearing cap before removing No. 4.

NOTE: Take care during removal that no oil, grease or dirt comes into contact with the timing belt.

5. Disassemble the rocker assembly, checking each component for wear, scoring, or plugged oil passages. Check the roller for correct and smooth rotation. Inspect the inner diameter of each rocker for any scoring or enlargement. If wear is found inside the rocker, replace it and inspect the shaft for damage.
6. Before reassembly, coat the contact faces liberally with clean motor oil. Observe the numbers on the bearing caps so that they are replaced in the correct location. Reassemble the rocker shafts into the front bearing cap so that the notches face outward. As you continue to assemble the springs, rockers and bearing caps, remember that the arrows on the on the bearing caps must point in the same direction as the arrow on the head. This is particularly important if the rockers have been removed from both heads.
7. When the rocker arm assembly is complete (except for the rear bearing cap, which is holding the cam down), have your trusted assistant hold the cam while the rear cap is removed. Assemble the last cap onto the rocker assembly and carefully fit the assembly onto the head. Remember that the cam must be held in place during this exchange.
8. Install the retaining bolts finger tight, making sure each rocker is correctly aligned with the camshaft and valve stem. Once the rockers are securely in place, the special retaining clips may be removed.
9. Starting at the center bearing cap and working outward in two passes, tighten the bearing cap bolts to 15 ft. lbs.
10. Apply sealer to the correct locations and reinstall the valve cover.

Pickup and Montero
with the 4D55 Diesel Engine

1. Remove the valve cover.
2. Remove the timing belt covers.
3. Using a wrench, turn the crankshaft in a clockwise direction until all the timing marks align. This positions No.1 piston at TDC/compression.
4. Remove the 5 flange bolts and washers holding the rocker rail to the head.
5. Lift the rocker assembly with the rockers attached away from the engine.
6. If the assembly is to be disassembled, take note that the intake and exhaust rockers are different. Each is labled **I** or **E**; they must be reassembled in the correct positions.
7. Inspect the arms and shafts after a thorough cleaning. The face of the cam follower must be smooth and free of grooves. The end of the adjuster screw must be round (spherical), not mushroomed or flattened. The oil passages in the shaft must be open.
8. Reassemble the rockers onto the shaft, making certain each component is in the correct location. Coat the entire shaft and rockers with a light coat of clean engine oil. Pay particular attention to coating the face of each cam follower.
9. Install the assembly onto the head. Install the 5 retaining bolts and tighten them alternately and evenly to 37 Nm. (27.5 ft. lbs.)
10. If the adjusting screws have been changed or replaced, make a preliminary valve adjustment now. Adjust the clearance to 0.015–0.020 in.; this is a rough setting, not a final one.
11. Install the valve cover.
12. Start the engine.

WARNING: The timing belt covers are still removed. Do not allow tools or clothing to get near the moving belt and pulleys.

13. After the engine is warmed to normal temperature, shut it off and remove the valve cover. Adjust the valves to their final setting of 0.010 in.
14. Install the timing belt covers and install the valve cover.

Thermostat

REMOVAL AND INSTALLATION

To operate at peak efficiency, an engine must maintain its internal temperatures within certain upper and lower limits. The cooling system circulates fluid around the combustion cylinders and conducts this heated fluid to the radiator, where the heat is exchanged into the airflow created by the fan and the motion of the car.

While most people realize that an engine running too hot (overheated) is a sign of trouble, few know that an engine can run too cool as well. If the proper internal temperatures are not achieved, fuel is not burned efficiently and the lubricating oil does not reach its best working temperature. While a too cold condition is rarely disabling, it can cause a variety of problems which can be mistaken for tune-up or electrical causes.

The thermostat controls the flow of coolant within the system. It reacts to the heat of the coolant and allows more fluid (or less) to circulate. Depending on the amount of fluid being circulated, more or less heat is drawn away from the inside of the engine. While we are beyond the days of having to install different thermostats for summer and winter driving, it is wise to check the function of the thermostat periodically. Special use of the car such as trailer towing or carrying heavy loads may require the installation of a thermostat with different temperature characteristics.

The thermostat requires replacement if the engine runs too cold, as indicated by low operating temperature on the gauge, low heater output, poor gas mileage and a slight lack of performance. If the engine is overheating, there are a number of different possible causes, including not only low coolant due to leakage, but improper engine tuning and/or a dirty cooling system or partially plugged radiator.

If the engine is overheating, it's best to test the thermostat before getting into more complicated diagnosis. You should also test the thermostat if the water temperature gauge is reading low and you're not sure whether it's a thermostat or a gauge problem. Any time the engine has suffered an overheating period, the thermostat should be changed; high temperature can damage it.

WARNING: Perform this procedure on a cold engine only. If the car has been recently driven, allow it to cool for at least 3–4 hours. Allowing the engine to cool overnight is preferred.

All Gasoline Engines

1. To replace the thermostat, remove the air cleaner (if necessary to gain access) and then drain the cooling system below the level of the tubes in the top tank of the radiator.

--- **CAUTION** ---

When draining the coolant, keep in mind that cats and dogs are attracted by the ethylene glycol antifreeze, and are quite likely to drink any that is left in an uncovered container or in puddles on the ground. This will prove fatal in sufficient quantity. Always drain the coolant into a sealable container. Coolant should be reused unless it is contaminated or several years old.

2. You can usually remove the thermostat housing, located at the intake manifold end of the top radiator hose, without disconnecting the hose. It may be necessary to remove other small hoses and/or wiring connectors from the housing. Remove the two mounting bolts and gently lift the housing off the manifold.

3-107

3 ENGINE AND ENGINE OVERHAUL

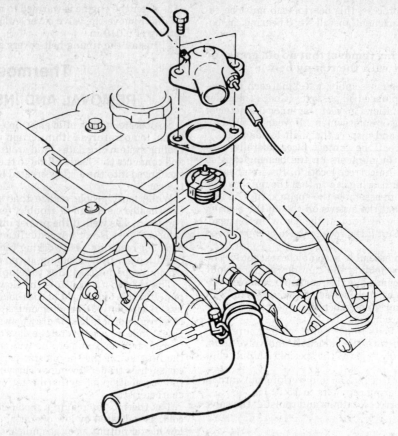

Typical thermostat installation. Note that the gasket goes above the thermostat

Once the housing is removed, you may wish to remove it from the end of the hose, making reinstallation easier.

3. If the thermostat is to be tested, suspend it well off the bottom of a pan of water on the stove. Immerse a thermometer that reads up to about 250°F (121°C) in the water. Heat the water. Note the temperature at which the thermostat valve begins to open and continue heating until the water boils. The valve should begin opening around 190°F (88°C), the exact temperature should be stamped on the thermostat, and be wide open as the water boils. The valve must open at least 8mm (0.31 in.) and should start opening within just a few degrees of the specified temperature; otherwise, replace it.

4. Scrape both gasket surfaces thoroughly, including the indentation in the intake manifold. Install the thermostat with the wax pellet and spring downward. Some models have a ribbed section inside the manifold to keep you from installing the unit upside down. Make sure the unit seats in the indentation so there is a flush surface for the gasket to seal against.

5. Coat a new gasket with sealer on both sides and install it to the manifold with bolt holes matching up. Note that the gasket must be installed above the thermostat.

6. Install the housing and two bolts, turning them gently and going back and forth until they are just snug. You are dealing with cast aluminum which is very brittle — DO NOT overtighten these bolts!

7. Connect the hoses and wiring as necessary and install the air cleaner if it was removed.

8. Refill the system with coolant. Operate the engine just above idle until the thermostat opens. Refill the system as necessary. Check for leaks. If any leak is found around the thermostat housing, do not attempt to tighten the bolts. After the engine cools the housing must be removed and scraped, then reinstalled with a fresh gasket and fresh sealer.

Diesel Engine

The thermostat for the diesel engine is is located on the water intake side of the cooling system, between the lower radiator hose and the block. The thermostat is equipped with a bypass valve. To remove the thermostat:

1. Drain the coolant to below the level of the thermostat.

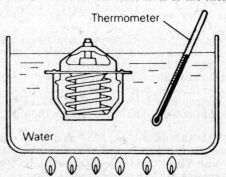

Testing the thermostat

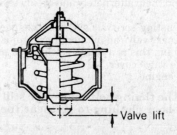

ENGINE AND ENGINE OVERHAUL 3

CAUTION

When draining the coolant, keep in mind that cats and dogs are attracted by the ethylene glycol antifreeze, and are quite likely to drink any that is left in an uncovered container or in puddles on the ground. This will prove fatal in sufficient quantity. Always drain the coolant into a sealable container. Coolant should be reused unless it is contaminated or several years old.

2. Remove the lower radiator hose from the water inlet fitting.
3. Remove the water inlet fitting and remove the thermostat.
4. When reinstalling, make certain the thermostat flange is correctly seated against the dimpled area of the inlet fitting.
5. Apply sealant to both sides of a new inlet fitting gasket and install the gasket against the water pump.
6. Install the inlet fitting; carefully tighten the two bolts.
7. Install the lower radiator hose and secure the clamp.
8. Refill the coolant.
9. Start the engine, checking carefully for leaks. Allow the engine to warm up fully. Watch the temperature gauge for any indication of improper warm-up (underheat or overheat).

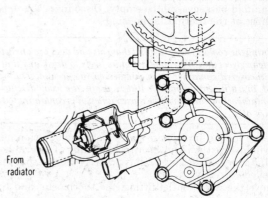

Location of the diesel thermostat

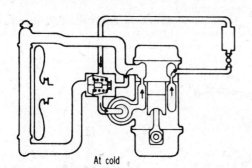
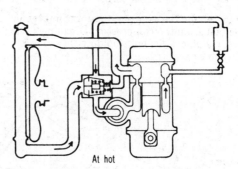

Coolant circulation, 4D55

Intake Manifold

REMOVAL AND INSTALLATION

Cordia and Tredia
Pickup with the G63B Engine

1. Disconnect the negative battery cable.
2. On fuel injected vehicles, safely relieve the residual fuel pressure within the system. The fuel system is under pressure. Release pressure slowly and contain spillage. Observe no smoking/no open flame precautions. Have a Class B-C (dry powder) fire extinguisher within arm's reach at all times.
3. Drain most of the coolant out of the cooling system so the intake manifold passages will be empty. Disconnect the upper radiator hose at the thermostat housing.

CAUTION

When draining the coolant, keep in mind that cats and dogs are attracted by the ethylene glycol antifreeze, and are quite likely to drink any that is left in an uncovered container or in puddles on the ground. This will prove fatal in sufficient quantity. Always drain the coolant into a sealable container. Coolant should be reused unless it is contaminated or several years old.

4. Remove either the carburetor or injection mixer. This will involve the removal or repositioning of several hoses, lines and wires. Make sure each is properly labeled so it may be reinstalled correctly.
5. Disconnect any remaining vacuum lines and on some models, EGR system piping.

6. Remove all the intake manifold retaining nuts and washers. These may be difficult to break loose. Use rust penetrant freely and keep the wrench square on the bolt to prevent breakage. Gently tap the manifold loose from the gaskets, and remove it.
7. If a new manifold is being installed, transfer parts such as the thermostat and gasket, and any vacuum fittings. Thoroughly clean the sealing surfaces of the cylinder head and, if it's being re-used, the manifold. Check the cylinder head and manifold sealing surfaces for flatness with a straightedge. Correct a warped surface by replacing the manifold or having the cylinder head surface machined. Install new gasket(s) against the head. Make sure all bolt holes and passages for the cooling water and intake ports align properly.
8. Position the manifold against the gasket and support it there while installing washers and nuts finger tight. Tighten the bolts evenly and in at least two passes to 12 ft. lbs.; do not overtighten them.
9. Install the carburetor or injection mixer and all hoses, wires and vacuum lines.
10. Refill the coolant. Connect the negative battery cable. Start the engine and check carefully for leaks of either coolant or vacuum.

Starion
Pickup and Montero with the G54B Engine

1. Disconnect the negative battery cable.
2. Safely relieve the residual fuel pressure within the system. The fuel system is under pressure. Release pressure slowly and contain spillage. Observe no smoking/no open flame precautions. Have a Class B-C (dry powder) fire extinguisher within

3-109

3 ENGINE AND ENGINE OVERHAUL

arm's reach at all times.

3. Drain most of the coolant out of the cooling system so the intake manifold passages will be empty. Disconnect the upper radiator hose at the thermostat housing.

CAUTION
When draining the coolant, keep in mind that cats and dogs are attracted by the ethylene glycol antifreeze, and are quite likely to drink any that is left in an uncovered container or in puddles on the ground. This will prove fatal in sufficient quantity. Always drain the coolant into a sealable container. Coolant should be reused unless it is contaminated or several years old.

4. Remove the air intake hose and remove the injection mixer. This will involve the removal or repositioning of several hoses, lines and wires. Make sure each is properly labeled so it may be reinstalled correctly.
5. Remove the water outlet fitting, the thermostat and the gasket.
6. Remove the secondary air cleaner assembly and the secondary air pipe.
7. Disconnect the bolt holding the engine oil dipstick tube and remove the tube and dipstick. Don't lose the O-ring on the tube.
8. Loosen the adjuster for the air conditioning compressor belt and remove the belt from the compressor. Disconnect the wire to the compressor. Remove the bolts holding the compressor to its mounting bracket and move the compressor (with the lines attached) out of the way. Support from a piece of stiff wire.

NOTE: Do not loosen any lines or discharge the system.

9. Remove the compressor bracket from the engine.
10. Disconnect the heater hose at the manifold.
11. disconnect the brake booster vacuum hose at the manifold.
12. Disconnect the small water hose at the rear of the manifold.
13. Disconnect the wiring and retainers running to the manifold. Check carefully for concealed wires underneath.
14. Remove the EGR valve. Discard the gasket; it is not reusable.
15. Label and disconnect the vacuum hoses from the thermo valve at the thermostat housing.
16. Loosen the nuts and bolt holding the manifold to the engine. Support the manifold and remove it.
17. Thoroughly clean the sealing surfaces of the cylinder head and the manifold. Check the cylinder head and manifold sealing surfaces for flatness with a straightedge. Correct a warped surface by replacing the manifold or having the cylinder head surface machined.

If the manifold is to be replaced, each fitting and sensor must be removed and transferred to the new unit. Note that when the thermo-vacuum valve is transferred, it must be sealed with a

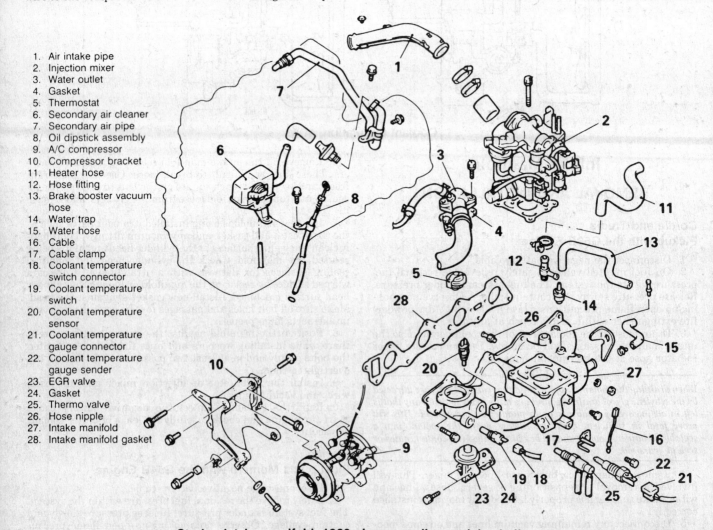

1. Air intake pipe
2. Injection mixer
3. Water outlet
4. Gasket
5. Thermostat
6. Secondary air cleaner
7. Secondary air pipe
8. Oil dipstick assembly
9. A/C compressor
10. Compressor bracket
11. Heater hose
12. Hose fitting
13. Brake booster vacuum hose
14. Water trap
15. Water hose
16. Cable
17. Cable clamp
18. Coolant temperature switch connector
19. Coolant temperature switch
20. Coolant temperature sensor
21. Coolant temperature gauge connector
22. Coolant temperature gauge sender
23. EGR valve
24. Gasket
25. Thermo valve
26. Hose nipple
27. Intake manifold
28. Intake manifold gasket

Starion intake manifold. 1989 shown, earlier models similar

ENGINE AND ENGINE OVERHAUL 3

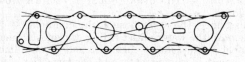

Measure the intake manifold faces with a straightedge and a feeler gauge. Maximum allowable warpage is 0.03mm (0.0012 in.)

waterproof sealer such as 3M 4171 or equivalent. The new manifold should be completely equipped with its supply of sensors, vacuum ports and fittings before installation.

18. Install new gasket(s) against the head. Make sure all bolt holes and passages for the cooling water and intake ports align properly.
19. Install the manifold and retaining nuts and bolt. Tighten the hardware finger tight.
20. Working from the center of the manifold outward, tighten the nuts and bolt to 12 ft. lbs. in at least 2 stages.
21. Install the EGR valve with a new gasket.
22. Connect the vacuum hoses to the thermo-vacuum valve. Connect the wiring to each sensor or switch on the manifold. Make certain that the wires are connected to the correct unit. Install any cable clamps which were removed.
23. Install the small water hose to the back of the manifold.
24. Connect the brake booster vacuum hose.
25. Install the compressor bracket to the engine, tightening the bolts to 33 ft. lbs.
26. Install the compressor, tightening the bolts to 18 ft. lbs. Connect the wire to the compressor and install the belt. Tighten the belt adjuster to draw the correct tension on the belt.
27. Install the oil dipstick and tube. Make certain the tube is secure in the engine.
28. Install the secondary air tube and secondary air cleaner assembly.
29. Install the thermostat with a new gasket and sealer; install the water outlet. Tighten the bolts no more than 9 ft. lbs.
30. Install the injection mixer. Tighten the retaining bolts to 13 ft. lbs.
31. Connect the air intake ducting. Refill the coolant and connect the negative battery cable.
32. Start the engine and check for leaks of coolant, oil or vacuum. Confirm the proper operation of all electrical lights and gauges on the dashboard. Turn on the air conditioning and confirm the compressor's operation.

Mirage with the G15B Engine

1. Disconnect the negative battery cable.
2. Drain most of the coolant out of the cooling system so the intake manifold passages will be empty. Disconnect the upper radiator hose at the thermostat housing.

------- CAUTION -------

When draining the coolant, keep in mind that cats and dogs are attracted by the ethylene glycol antifreeze, and are quite likely to drink any that is left in an uncovered container or in puddles on the ground. This will prove fatal in sufficient quantity. Always drain the coolant into a sealable container. Coolant should be reused unless it is contaminated or several years old.

3. Remove the air cleaner assembly.
4. Remove the carburetor and gasket.
5. Disconnect the brake booster vacuum hose.
6. Remove the upper radiator hose from the thermostat housing.
7. Remove the water outlet, the thermostat and its gasket.
8. Disconnect the coolant hoses from the intake manifold.
9. Label and disconnect the electrical connections and the vacuum hoses.
10. Remove the EGR valve. Remove the gasket and discard it; it is not reusable.
11. Label and disconnect the spark plug and coil wiring from the distributor cap.
12. Remove the distributor cap and mark the position of the rotor in relation to the case. This will insure correct reinstallation.
13. Remove the distributor.
14. Remove the manifold support bracket.
15. Remove the engine lifting hook. It is mounted under one of the manifold retaining nuts.
16. Remove the retaining nuts and carefully remove the manifold. As soon as the manifold is free of the engine, remove the vacuum pipe assembly from the bottom of the manifold.
17. Remove the gaskets from the engine and clean the mating faces thoroughly. Examine the manifold for any damage or cracking. Check each vacuum port and water passage for any clogging or plugging. Use a straightedge and feeler gauge to check the manifold and head surfaces for warpage. Maximum allowable warpage on the head face is 0.3mm (0.012 in.). If the manifold is to be replaced, transfer the vacuum fittings, water ports and other hardware to the new unit. The manifold should be completely assembled before installation.
18. Before reinstallation, attach the vacuum pipe assembly to the bottom of the manifold. Install the new gaskets and fit the manifold into position on the studs. Install the manifold support bracket so that the end with the word **UP** is in contact with the manifold. Tighten the bracket bolts snug.
19. Install the nuts and washers which retain the manifold snugly. Don't forget the engine lifting hook.
20. After the manifold and its bracket are in place, tighten the support bracket bolts to 18 ft. lbs. Working from the center out, make two complete passes and tighten the manifold retaining nuts in two steps to 12 ft. lbs.
21. Reinstall the distributor, observing the alignment marks made earlier. Install the distributor cap and connect the spark plug and coil wires.
22. Install the EGR valve with a new gasket.
23. Connect the water hoses, electrical connections and vacuum hoses.
24. Install the thermostat and water outlet fitting with a new gasket and connect the upper radiator hose.
25. Connect the brake booster vacuum hose. Install the PCV hose to the manifold.
26. Install the base gasket and install the carburetor. Connect all the vacuum hoses and fuel lines. Connect the accelerator cable.
27. Install the air cleaner assembly.
28. Double check all installation items, paying particular attention to loose hoses or hanging wires, untightened nuts, poor routing of hoses and wires (too tight or rubbing) and tools left in the engine area.
29. Refill the coolant to the proper level.
30. Connect the negative battery cable and start the engine. Check carefully for any leaks of fluid or vacuum. Adjust the engine timing as necessary and adjust the throttle cable.

Mirage with the G32B Engine

1. Disconnect the negative battery cable.
2. Safely relieve the residual fuel pressure within the system. The fuel system is under pressure. Release pressure slowly and contain spillage. Observe no smoking/no open flame precautions. Have a Class B-C (dry powder) fire extinguisher within arm's reach at all times.
3. Drain most of the coolant out of the cooling system so the intake manifold passages will be empty. Disconnect the upper radiator hose at the thermostat housing.

3-111

3 ENGINE AND ENGINE OVERHAUL

> **CAUTION**
>
> When draining the coolant, keep in mind that cats and dogs are attracted by the ethylene glycol antifreeze, and are quite likely to drink any that is left in an uncovered container or in puddles on the ground. This will prove fatal in sufficient quantity. Always drain the coolant into a sealable container. Coolant should be reused unless it is contaminated or several years old.

4. Remove the air intake hose and remove the injection mixer. This will involve the removal or repositioning of several hoses, lines and wires. Make sure each is properly labeled so it may be reinstalled correctly.
5. Remove the solenoid valve assembly.
6. Remove the EGR valve and its gasket.
7. Disconnect the brake booster vacuum hose at the manifold.
8. Disconnect the water hose at the manifold.
9. Remove the water outlet fitting, thermostat and gasket.
10. Label and disconnect each wire and vacuum line running to the manifold components and senders. Don't forget the ground strap on the lower part of the manifold body.
11. Remove the water by-pass and heater hoses.
12. Remove the secondary air cleaner assembly.
13. Remove the intake manifold support bracket.
14. Remove the engine lifting hook. It is mounted under one of the manifold retaining nuts.
15. Remove the retaining nuts and bolt and carefully remove the manifold.

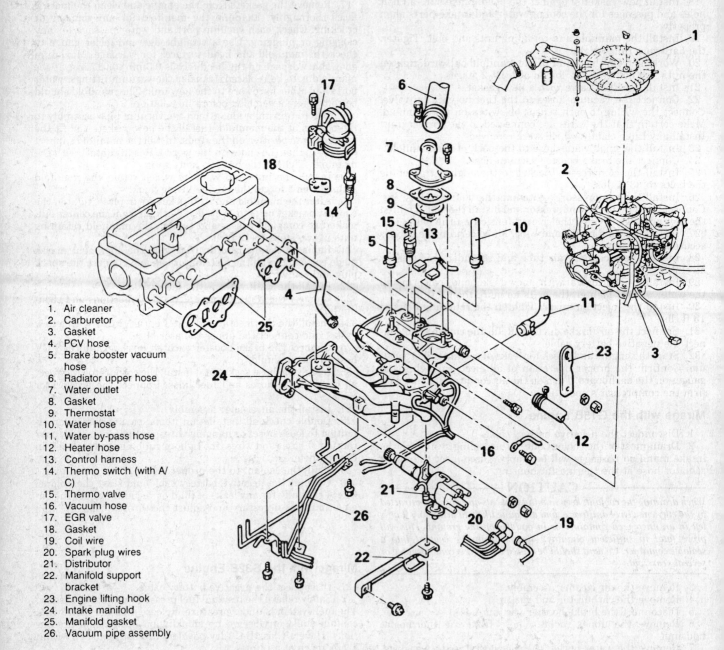

1. Air cleaner
2. Carburetor
3. Gasket
4. PCV hose
5. Brake booster vacuum hose
6. Radiator upper hose
7. Water outlet
8. Gasket
9. Thermostat
10. Water hose
11. Water by-pass hose
12. Heater hose
13. Control harness
14. Thermo switch (with A/C)
15. Thermo valve
16. Vacuum hose
17. EGR valve
18. Gasket
19. Coil wire
20. Spark plug wires
21. Distributor
22. Manifold support bracket
23. Engine lifting hook
24. Intake manifold
25. Manifold gasket
26. Vacuum pipe assembly

Intake manifold assembly, Mirage with G15B engine

ENGINE AND ENGINE OVERHAUL 3

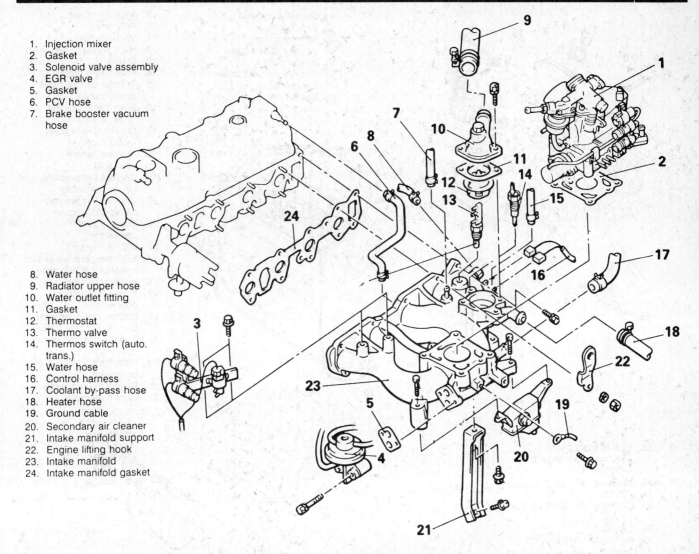

1. Injection mixer
2. Gasket
3. Solenoid valve assembly
4. EGR valve
5. Gasket
6. PCV hose
7. Brake booster vacuum hose
8. Water hose
9. Radiator upper hose
10. Water outlet fitting
11. Gasket
12. Thermostat
13. Thermo valve
14. Thermos switch (auto. trans.)
15. Water hose
16. Control harness
17. Coolant by-pass hose
18. Heater hose
19. Ground cable
20. Secondary air cleaner
21. Intake manifold support
22. Engine lifting hook
23. Intake manifold
24. Intake manifold gasket

Intake manifold assembly, Mirage with G32B engine

16. Remove the gaskets from the engine and clean the mating faces thoroughly. Examine the manifold for any damage or cracking. Check each vacuum port and water passage for any clogging or plugging. Use a straightedge and feeler gauge to check the manifold and head surfaces for warpage. Maximum allowable warpage on the head face is 0.3mm (0.012 in.). If the manifold is to be replaced, transfer the vacuum fittings, water ports and other hardware to the new unit. The manifold should be completely assembled before installation.
17. Install the new gasket and fit the manifold into position on the studs. Install the manifold support bracket and tighten the bracket bolts snug.
18. Install the nuts and washers which retain the manifold snugly. Don't forget the engine lifting hook.
19. After the manifold and its bracket are in place, tighten the support bracket bolts to 15 ft. lbs. Working from the center out, make two complete passes and tighten the manifold retaining nuts and bolt in two steps to 12 ft. lbs.
19. Install the secondary air cleaner assembly.
20. Connect the ground strap to the manifold. Connect the heater hose and the water by-pass hose.
21. Observing the labels made during disassembly, connect each vacuum line and wiring connector to its proper component.
22. Install the thermostat with a new gasket and install the water outlet fitting. Tighten the bolts to 8 ft. lbs. and install the upper radiator hose.
23. Connect the brake booster vacuum hose.
24. Connect the PCV hose.
25. Install the EGR valve with a new gasket.
26. Install the solenoid valve assembly. Tighten the bracket bolt to 10 ft. lbs.
27. Install a new gasket and install the injection mixer assembly.
28. Double check all installation items, paying particular attention to loose hoses or hanging wires, untightened nuts, poor routing of hoses and wires (too tight or rubbing) and tools left in the engine area.
29. Refill the coolant to the correct level.
30. Connect the negative battery cable; start the engine and check carefully for any fluid or vacuum leaks. Adjust the engine specifications as necessary. Adjust the throttle cable.

Mirage with the 4G15 SOHC Engine

1. Disconnect the negative battery cable.
2. Safely relieve the residual fuel pressure within the system. The fuel system is under pressure. Release pressure slowly and contain spillage. Observe no smoking/no open flame precau-

3 ENGINE AND ENGINE OVERHAUL

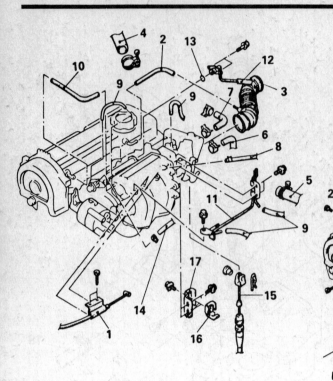

1. Accelerator cable
2. Breather hose
3. Air intake hose
4. Upper radiator hose
5. Heater hose
6. Coolant by-pass hose
7. Coolant hose
8. Brake booster vacuum hose
9. Vacuum hoses
10. PCV hose
11. Vacuum pipe
12. High pressure fuel hose
13. O-ring
14. Fuel return hose
15. Throttle control cable (auto. trans.)
16. Inner cable bracket (auto. trans.)
17. Outer cable bracket (auto. trans.)
18. Oxygen sensor connector
19. Coolant temperature gauge connector
20. Coolant temperature sensor connector
21. Thermo switch connector
22. ISC connector
23. EGR temperature sensor connector (Calif. only)
24. Injector connector
25. Power transistor connector
26. Ignition coil connector
27. Condenser connector
28. Throttle position sensor connector
29. Motor position sensor connector
30. Distributor connector
31. Control harness

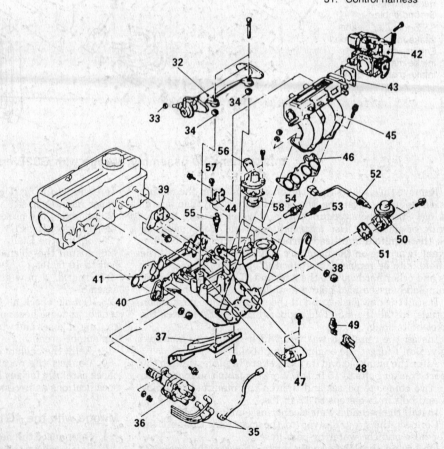

32. Fuel delivery pipe, injectors and pressure regulator
33. Insulator
34. Insulator
35. Coil wire and spark plug cables
36. Distributor
37. Intake manifold support bracket
38. Engine lifting hook
39. Power transistor bracket
40. Intake manifold
41. Manifold gasket
42. Throttle body assembly
43. Gasket
44. Air intake plenum support bracket
45. Air intake plenum
46. Plenum gasket
47. Ignition coil
48. Vacuum hose
49. Thermo valve
50. EGR valve
51. EGR gasket
52. EGR temperature sensor (Calif. only)
53. Coolant temperature gauge unit
54. Coolant temperature sensor
55. Thermo switch (auto trans)
56. Water outlet fitting
57. Gasket
58. Thermostat

Mirage intake manifold, 4G15 SOHC engine

ENGINE AND ENGINE OVERHAUL 3

tions. Have a Class B-C (dry powder) fire extinguisher within arm's reach at all times.

3. Drain most of the coolant out of the cooling system so the intake manifold passages will be empty. Disconnect the upper radiator hose at the thermostat housing.

—————————— **CAUTION** ——————————
When draining the coolant, keep in mind that cats and dogs are attracted by the ethylene glycol antifreeze, and are quite likely to drink any that is left in an uncovered container or in puddles on the ground. This will prove fatal in sufficient quantity. Always drain the coolant into a sealable container. Coolant should be reused unless it is contaminated or several years old.

4. Disconnect the accelerator cable.
5. Remove the breather hose and the air intake hose.
6. Disconnect the heater hose, the coolant by-pass hose and the small water hose.
7. Disconnect the brake booster vacuum hose.
8. Label and disconnect the vacuum hoses, the PCV hose and the metal vacuum pipe.
9. Disconnect the high pressure fuel hose and discard the O-ring; it is not reusable. Remove the fuel return hose.

WARNING: Although the pressure was relieved earlier, cover the joint with a rag during removal; some pressure may remain.

10. On cars equipped with automatic transmissions, disconnect the throttle control cable, and remove the inner and outer cable brackets.
11. Label and disconnect each wire harness and lead running to the manifold and injection components. Some of the wiring will need to disconnected away from the manifold; follow the wires to the first connector and disconnect it.
12. Remove the delivery pipe (fuel rail), the injectors and the pressure regulator as a unit.

WARNING: Store this assembly in a clean, protected location. The injectors must be kept totally free of dirt. Do not drop the injectors during removal and handling.

13. Remove the insulators from the fuel rail.
14. Label and disconnect the coil wire and spark plug wires from the distributor cap. Remove the distributor, making a mark on the shaft and gear showing the alignment. Alternately, remove the cap and mark the position of the rotor on the outer case.
15. Remove the intake manifold support and remove the engine lifting hook.
16. Remove the power transistor bracket.
17. Unbolt and remove the intake manifold assembly and remove the gasket.
18. Remove the throttle body assembly and its gasket. Discard the gasket.
19. Remove the air intake plenum support bracket; unbolt the plenum and remove it with its gasket.
20. Remove the gaskets from the engine and clean the mating faces thoroughly. Examine the manifold and the plenum for any damage or cracking. Check each vacuum port and water passage for any clogging or plugging. Use a straightedge and feeler gauge to check the manifold and head surfaces for warpage. Maximum allowable warpage on the head face is 0.3mm (0.012 in.).

If the manifold is to be replaced, transfer the vacuum fittings, water ports, thermostat, EGR valve and other hardware to the new unit. When any of the coolant sensors are transferred, the threads must be coated with an adhesive sealant such as 3M 4171 or eqivalent. The manifold should be completely assembled before installation.

21. Install a new gasket between the plenum and the manifold. Assemble the two and tighten the bolts to 12 ft. lbs. Install the air plenum support bracket, tightening its bolts to 12 ft. lbs.
22. Using a new gasket, reinstall the throttle body assembly. Tighten the mounting bolts evenly to 8 ft. lbs.
23. Install a new manifold gasket and place the manifold in position on the engine. Replace the engine lifting hook and the power transistor bracket and tighten all the mounting nuts and bolts snug. Install the manifold support bracket, again bringing the bolts just snug.
24. Working from the center outward, and in at least two passes, tighten the manifold retaining bolts and nuts to 12 ft. lbs. Tighten the manifold bracket bolts to 15 ft. lbs. after the manifold bolts are properly tightened.
25. Install the distributor observing the alignment marks made during disassembly. Install the cap if it was removed and connect the coil and spark plug wiring.
26. Install the insulators which sit below the fuel rail. Install a new insulator (O-ring) on the fuel rail nipple.
27. Install the fuel rail (delivery pipe), injectors and pressure regulator as a unit. tighten the mounting bolts to 8 ft. lbs. Take care to protect the injectors from dropping or impact during installation.
28. Connect the vacuum lines and electrical connectors to their respective components. Make certain each is firmly seated and correctly installed.
29. On cars equipped with automatic transaxles, install the inner and outer cable brackets and tighten the bolts to 10 ft. lbs. Install the throttle control cable.
30. Connect the fuel return line. Install a new O-ring on the high pressure fuel line and connect it to the fuel rail.
31. Connect the vacuum hoses, the PCV hose and the brake booster vacuum hose.
32. Install the small coolant hose, the by-pass hose and the heater hose.
33. Connect the upper radiator hose to the thermostat housing.
34. Install the air intake hose and connect the breather hose from the valve cover to the air intake.
35. Connect the accelerator cable.
36. Double check all installation items, paying particular attention to loose hoses or hanging wires, untightened nuts, poor routing of hoses and wires (too tight or rubbing) and tools left in the engine area.
37. Refill the coolant system to the proper level.
38. Connect the negative battery cable and start the engine. Check carefully for leaks of either coolant or vacuum. Check the operation of all electrical components, including dash gauges and warning lights.
39. Perform final adjustments to the engine specifications. Adjust the throttle control cable and/or the accelerator cable as necessary.

Mirage with the 4G61 DOHC Engine

1. Disconnect the negative battery cable.
2. Safely relieve the residual fuel pressure within the system. The fuel system is under pressure. Release pressure slowly and contain spillage. Observe no smoking/no open flame precautions. Have a Class B-C (dry powder) fire extinguisher within arm's reach at all times.
3. Drain most of the coolant out of the cooling system so the intake manifold passages will be empty. Disconnect the upper radiator hose at the thermostat housing.

—————————— **CAUTION** ——————————
When draining the coolant, keep in mind that cats and dogs are attracted by the ethylene glycol antifreeze, and are quite likely to drink any that is left in an uncovered container or in puddles on the ground. This will prove fatal in sufficient quantity. Always drain the coolant into a sealable container. Coolant should be reused unless it is contaminated or several years old.

4. Disconnect the accelerator cable.
5. Remove the breather hose and the air intake hose. Also re-

3-115

3 ENGINE AND ENGINE OVERHAUL

move the throttle body support bracket.

6. Disconnect the heater hose, the coolant by-pass hose and the small water hose.
7. Disconnect the brake booster vacuum hose.
8. Label and disconnect the vacuum hoses and the PCV hose.
9. Disconnect the high pressure fuel hose and discard the O-ring; it is not reusable. Remove the fuel return hose.

WARNING: Although the pressure was relieved earlier, cover the joint with a rag during removal; some pressure may remain.

10. Label and disconnect each wire harness and lead running to the manifold and injection components. Some of the wiring will need to disconnected away from the manifold; follow the wires to the first connector and disconnect it.
11. Remove the delivery pipe (fuel rail), the injectors and the pressure regulator as a unit.

WARNING: Store this assembly in a clean, protected location. The injectors must be kept totally free of dirt. Do not drop the injectors during removal and handling.

12. Remove the insulators from the fuel rail.
13. There is no mechanical distributor. The coil and power transistor are bolted to the manifold. They can be removed with the manifold but it is safer to remove them and put them in a protected location.
14. Remove the intake manifold support and remove the tension rod bracket.
15. Remove the power transistor bracket.
16. Unbolt and remove the intake manifold assembly and remove the gasket.
17. Remove the throttle body assembly and its gasket. Discard the gasket.
18. Remove the air intake plenum support bracket.
19. Remove the gaskets from the engine and clean the mating faces thoroughly. Examine the manifold and the plenum for any damage or cracking. Check each vacuum port and water passage for any clogging or plugging. Use a straightedge and feeler gauge to check the manifold and head surfaces for warpage. Maximum allowable warpage on the head face is 0.3mm (0.012 in.).

If the manifold is to be replaced, transfer the vacuum fittings, water ports, thermostat, EGR valve and other hardware to the new unit. When any of the coolant sensors are transferred, the threads must be coated with an adhesive sealant such as 3M 4171 or eqivalent. The manifold should be completely assembled before installation.

20. Install the air plenum support bracket, tightening its bolts to 12 ft. lbs.
21. Using a new gasket, reinstall the throttle body assembly. Tighten the mounting bolts evenly to 8 ft. lbs.
22. Install a new manifold gasket and place the manifold in position on the engine. Replace the tension rod bracket. Torque the nuts to 30 ft. lbs. Install the power transistor bracket and tighten all the mounting nuts and bolts snug. Install the manifold support bracket, again bringing the bolts just snug.
23. Working from the center outward, and in at least two passes, tighten the manifold retaining bolts and nuts to 12 ft. lbs. Tighten the manifold bracket bolts to 20 ft. lbs. after the manifold bolts are properly tightened.
24. are bolted to the manifold. They can be removed with the manifold but it is safer to remove them and put them in a protected location. Install the coil and power transistor Tighten the power transistor retaining bolts to 8 ft. lbs. and the ignition coil bolts to 18 ft. lbs.
25. Install the insulators which sit below the fuel rail. Install a new insulator (O-ring) on the fuel rail nipple.
26. Install the fuel rail (delivery pipe), injectors and pressure regulator as a unit. tighten the mounting bolts to 8 ft. lbs. Take care to protect the injectors from dropping or impact during installation.
27. Connect the vacuum lines and electrical connectors to their respective components. Make certain each is firmly seated and correctly installed.
28. Connect the fuel return line. Install a new O-ring on the high pressure fuel line and connect it to the fuel rail.
29. Connect the vacuum hoses, the PCV hose and the brake booster vacuum hose.
30. Install the small coolant hose, the by-pass hose and the heater hose.
31. Connect the upper radiator hose to the thermostat housing. Install the throttle body support bracket. Tighten the nuts to 13 ft. lbs.
32. Install the air intake hose and connect the breather hose from the valve cover to the air intake.
33. Connect the accelerator cable.
34. Double check all installation items, paying particular attention to loose hoses or hanging wires, untightened nuts, poor routing of hoses and wires (too tight or rubbing) and tools left in the engine area.
35. Refill the coolant system to the proper level.
36. Connect the negative battery cable and start the engine. Check carefully for leaks of either coolant or vacuum. Check the operation of all electrical components, including dash gauges and warning lights.
37. Perform final adjustments to the engine specifications. Adjust the throttle control cable and/or the accelerator cable as necessary.

Galant with the G64B Engine

1. Disconnect the negative battery cable.
2. Safely relieve the residual fuel pressure within the system. The fuel system is under pressure. Release pressure slowly and contain spillage. Observe no smoking/no open flame precautions. Have a Class B-C (dry powder) fire extinguisher within arm's reach at all times.
3. Drain most of the coolant out of the cooling system so the intake manifold passages will be empty. Disconnect the upper radiator hose at the thermostat housing.

— CAUTION —
When draining the coolant, keep in mind that cats and dogs are attracted by the ethylene glycol antifreeze, and are quite likely to drink any that is left in an uncovered container or in puddles on the ground. This will prove fatal in sufficient quantity. Always drain the coolant into a sealable container. Coolant should be reused unless it is contaminated or several years old.

---

4. Disconnect the upper radiator hose, the coolant by-pass hose, the water hose and the heater hose from the manifold.
5. Label and disconnect the bolt holding the control harness, the ground cable, the coil wire and the spark plug cables.
6. Disconnect the accelerator cable and remove the bracket holding it to the manifold.
7. label and disconnect the PCV hose, the vacuum hoses at the manifold, the brake booster vacuum line and the air intake duct.
8. Remove the distributor, making a mark on the shaft and gear showing the alignment. Alternately, remove the cap and mark the position of the rotor on the outer case.
9. Disconnect the high pressure fuel line and discard the washer; it is not reusable. Disconnect the fuel return line.
10. Remove the bolt holding the lower end of the manifold support brace to the block. The brace will come off with the manifold.
11. Remove the nut holding the engine lifting hook and remove the hook.
12. Continue removing the retaining nuts and bolts and remove the intake manifold. Remove the gasket and clean the mating faces.
13. Examine the manifold for any damage or cracking. Check each vacuum port and water passage for any clogging or plug-

ENGINE AND ENGINE OVERHAUL 3

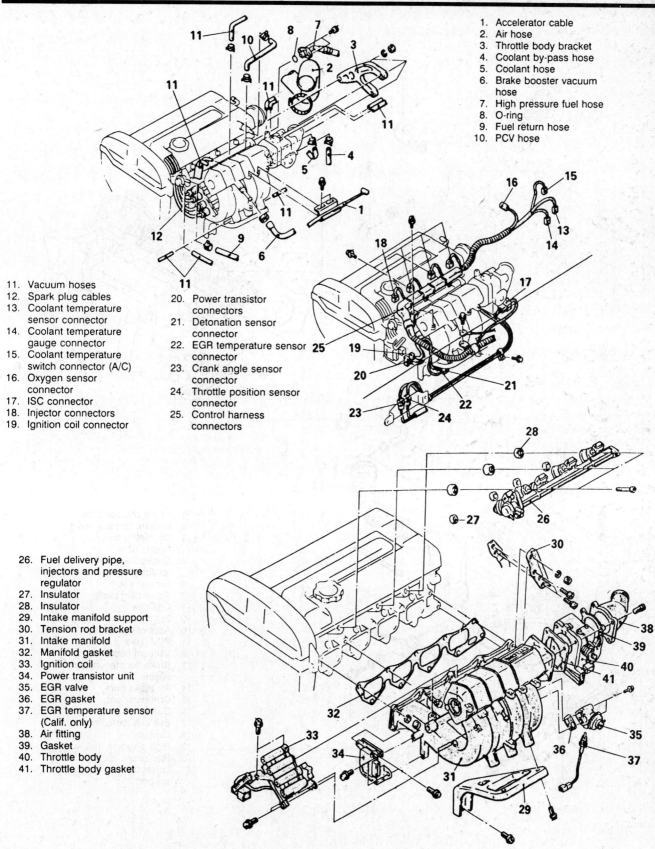

1. Accelerator cable
2. Air hose
3. Throttle body bracket
4. Coolant by-pass hose
5. Coolant hose
6. Brake booster vacuum hose
7. High pressure fuel hose
8. O-ring
9. Fuel return hose
10. PCV hose

11. Vacuum hoses
12. Spark plug cables
13. Coolant temperature sensor connector
14. Coolant temperature gauge connector
15. Coolant temperature switch connector (A/C)
16. Oxygen sensor connector
17. ISC connector
18. Injector connectors
19. Ignition coil connector
20. Power transistor connectors
21. Detonation sensor connector
22. EGR temperature sensor connector
23. Crank angle sensor connector
24. Throttle position sensor connector
25. Control harness connectors

26. Fuel delivery pipe, injectors and pressure regulator
27. Insulator
28. Insulator
29. Intake manifold support
30. Tension rod bracket
31. Intake manifold
32. Manifold gasket
33. Ignition coil
34. Power transistor unit
35. EGR valve
36. EGR gasket
37. EGR temperature sensor (Calif. only)
38. Air fitting
39. Gasket
40. Throttle body
41. Throttle body gasket

4G61 Mirage DOHC intake manifold removal

3-117

3 ENGINE AND ENGINE OVERHAUL

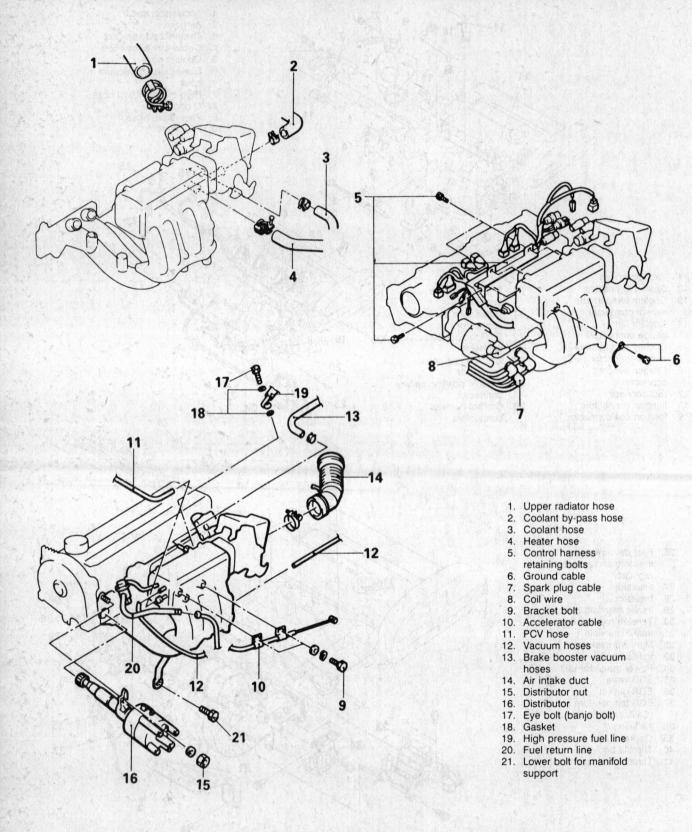

1. Upper radiator hose
2. Coolant by-pass hose
3. Coolant hose
4. Heater hose
5. Control harness retaining bolts
6. Ground cable
7. Spark plug cable
8. Coil wire
9. Bracket bolt
10. Accelerator cable
11. PCV hose
12. Vacuum hoses
13. Brake booster vacuum hoses
14. Air intake duct
15. Distributor nut
16. Distributor
17. Eye bolt (banjo bolt)
18. Gasket
19. High pressure fuel line
20. Fuel return line
21. Lower bolt for manifold support

Galant G64B intake manifold

ENGINE AND ENGINE OVERHAUL 3

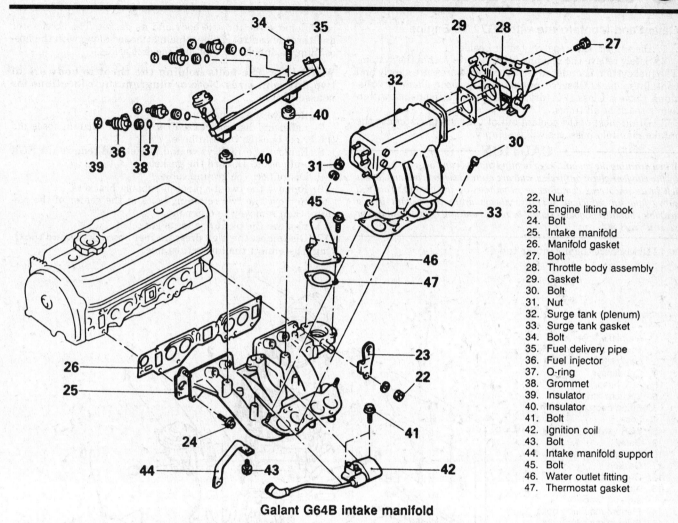

22. Nut
23. Engine lifting hook
24. Bolt
25. Intake manifold
26. Manifold gasket
27. Bolt
28. Throttle body assembly
29. Gasket
30. Bolt
31. Nut
32. Surge tank (plenum)
33. Surge tank gasket
34. Bolt
35. Fuel delivery pipe
36. Fuel injector
37. O-ring
38. Grommet
39. Insulator
40. Insulator
41. Bolt
42. Ignition coil
43. Bolt
44. Intake manifold support
45. Bolt
46. Water outlet fitting
47. Thermostat gasket

Galant G64B intake manifold

ging. Use a straightedge and feeler gauge to check the manifold and head surfaces for warpage. Maximum allowable warpage on the head face is 0.3mm (0.012 in.).

If the manifold is to be replaced, transfer the vacuum fittings, water ports, thermostat, EGR valve and other hardware to the new unit. Note that the throttle body, fuel injector rail, and surge tank (intake plenum) came off with the unit. These must all be transferred to the new unit with the use of new gaskets and O-rings. Tighten the fuel rail bolts and the throttle body bolts to 96 inch lbs. and the surge tank-to-manifold bolts to 12 ft. lbs.

When any of the coolant sensors are transferred, the threads must be coated with an adhesive sealant such as 3M 4171 or eqivalent. The manifold should be completely assembled before installation.

14. Place a new gasket in position and install the manifold. Tighten the nuts and bolts finger tight. Don't forget to install the lifting hook. Install the lower bolt for the manifold support bracket and tighten snug. Make certain it is in contact with the engine boss behind it.
15. Starting from the center and working outward in two stages, tighten the manifold nuts and bolts to 12 ft. lbs. Tighten the lower support bolt to 15 ft. lbs. after the manifold bolts are set. If the support bracket was removed from the manifold, the upper bolt should be checked and tightened to 15 ft. lbs.
16. Connect the fuel return hose; install a new washer and connect the high pressure fuel line. Tighten the bolt to 22 ft. lbs.
17. Install the distributor, observing alignment marks made earlier.
18. Connect the air intake hose, the brake booster vacuum hose, the small vacuum hoses running to the manifold and the PCV hose.
19. Install the accelerator cable brackets and connect the cable.
20. Install and connect the coil and spark plug wiring, the ground cable and the control harness and its mounting bolts.
21. Connect the heater hose, coolant hose, coolant by-pass hose and the upper radiator hose to the intake manifold.
22. Fill the system with coolant.
23. Double check all installation items, paying particular attention to loose hoses or hanging wires, untightened nuts, poor routing of hoses and wires (too tight or rubbing) and tools left in the engine area.
24. Connect the negative battery cable and start the engine. Check carefully for leaks of either coolant or vacuum. Check the operation of all electrical components, including dash gauges and warning lights.
25. Perform final adjustments to the engine specifications. Adjust the accelerator cable as necessary.

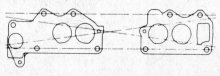

Check the manifold face for warpage or distortion

3-119

3 ENGINE AND ENGINE OVERHAUL

Galant and Montero the with 6G72 V6 Engine

1. Disconnect the negative battery cable.
2. Safely relieve the residual fuel pressure within the system. The fuel system is under pressure. Release pressure slowly and contain spillage. Observe no smoking/no open flame precautions. Have a Class B-C (dry powder) fire extinguisher within arm's reach at all times.
3. Drain most of the coolant out of the cooling system so the intake manifold passages will be empty.

---------------------- **CAUTION** ----------------------

When draining the coolant, keep in mind that cats and dogs are attracted by the ethylene glycol antifreeze, and are quite likely to drink any that is left in an uncovered container or in puddles on the ground. This will prove fatal in sufficient quantity. Always drain the coolant into a sealable container. Coolant should be reused unless it is contaminated or several years old.

4. Disconnect the air intake tube.
5. Remove the throttle body and its gasket. The accelerator and throttle control cables remain attached along with the hoses. Simply unbolt the unit and move it aside.

WARNING: The bolts holding the throttle body are of two different sizes. Note or diagram their locations for reassembly.

6. Remove the condenser.
7. Label and disconnect the PCV hose, the vacuum hoses and the brake booster vacuum hose.
8. Remove the EGR valve and gasket and remove the EGR pipe and gasket. Discard the gaskets.
9. Disconnect the ground cable.
10. Remove the two air plenum support brackets.
11. Remove the two retaining bolts in the center of the plenum, then remove the outer nuts.
12. Remove the plenum and its gasket.
13. Disconnect the high pressure fuel hose and discard the O-ring. Disconnect the fuel return line.

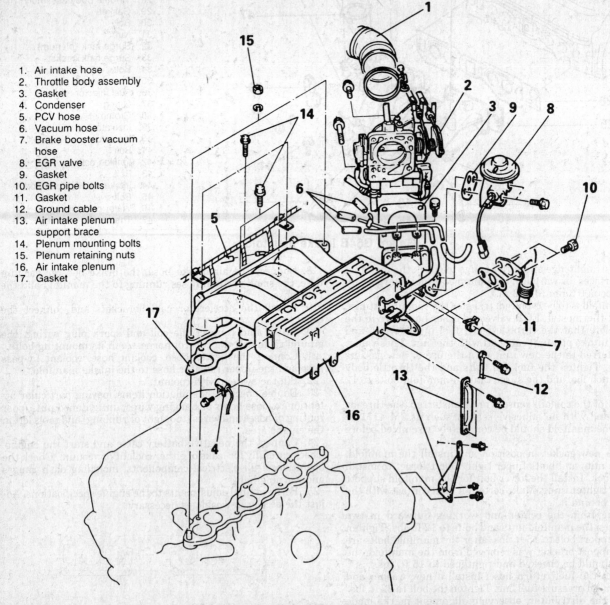

1. Air intake hose
2. Throttle body assembly
3. Gasket
4. Condenser
5. PCV hose
6. Vacuum hose
7. Brake booster vacuum hose
8. EGR valve
9. Gasket
10. EGR pipe bolts
11. Gasket
12. Ground cable
13. Air intake plenum support brace
14. Plenum mounting bolts
15. Plenum retaining nuts
16. Air intake plenum
17. Gasket

Galant and Sigma V6 intake plenum removal

3-120

ENGINE AND ENGINE OVERHAUL 3

1. Air intake plenum
2. High pressure fuel line
3. O-ring
4. Fuel return line
5. Vacuum hoses
6. Delivery pipe, injectors and pressure regulator
7. Upper radiator hose
8. Heater hose
9. Coolant hose
10. Water outlet fitting
11. Gasket
12. Wiring harness connector(s)
13. Intake manifold
14. Insulators
15. Intake manifold gasket

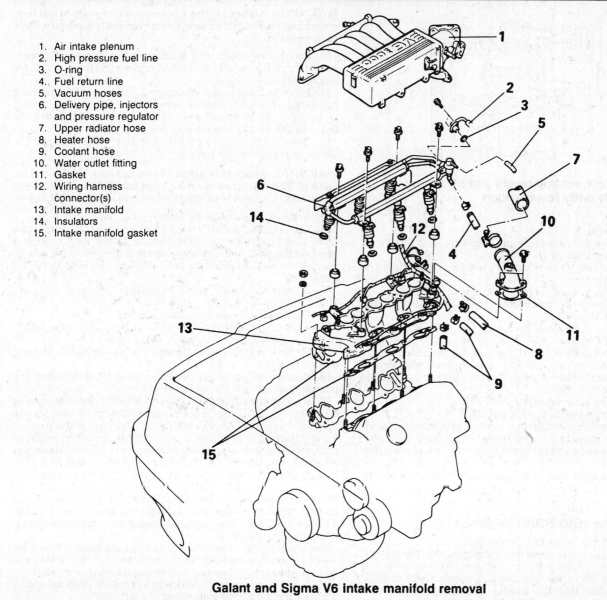

Galant and Sigma V6 intake manifold removal

14. Disconnect the vacuum line at the fuel delivery rail.
15. Remove the fuel delivery rail with the injectors and pressure regulator. Store them in a safe location. Do not drop or jar the injectors and keep them absolutely free of dirt and foreign matter.
16. Disconnect the upper radiator hose, the heater hose and the nearby coolant hose.
17. Remove the water outlet fitting and the gasket.
18. Disconnect the wiring harness connectors.
19. Remove the retaining nuts and remove the intake manifold. Remove the insulators and the manifold gaskets.
20. Examine the manifold and plenum for any damage or cracking. Check each vacuum port and water passage for any clogging or plugging. Use a straightedge and feeler gauge to check the manifold and head surfaces for warpage. Maximum allowable warpage on the head face is 0.3mm (0.012 in.).
    If the manifold and/or plenum is to be replaced, transfer the vacuum fittings, water ports, thermostat, EGR valve and other hardware to the new unit. The manifold should be completely assembled before installation. Clean the mating faces of any gasket remains or foreign matter.

21. Before reinstallation, place new gaskets in position. Make certain the adhesive side is towards the manifold, not towards the head. Do NOT add any gasket sealer to this gasket.
22. Install new insulators into the manifold.
23. Place the manifold in postion and install the nuts snug. When all are equally snug, work from the center outward in a circular pattern, using two passes to tighten the nuts to 12 ft. lbs.
24. Connect the wiring harness connectors and route the wiring properly.
25. Install the water outlet fitting with a new gasket.
26. Connect the coolant hose, the heater hose and the upper radiator hose.
27. Install the fuel delivery pipe and injectors, taking care not to drop the injectors or subject them to impact. Tighten the retaining bolts to 8 ft. lbs.
28. Connect the vacuum hoses.
29. Connect the fuel return line. Install a new O-ring and connect the high pressure fuel line.
30. Install a new plenum gasket (use no sealant) and install the plenum. Make certain the printed surface of the gasket faces

3 ENGINE AND ENGINE OVERHAUL

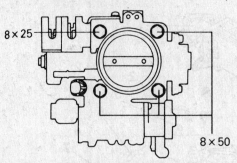

Observe the correct bolt length and placement during throttle body reinstallation

the intake plenum.

31. Tighten the nuts on the outer edges to 13 ft. lbs., then tighten the inner bolts to 13 ft. lbs.
32. Install the two plenum support brackets, tightening the bolts to 10 ft. lbs.
33. Connect the ground cable to the plenum.
34. Install a new gasket and connect the EGR pipe. Install the EGR valve with a new gasket.
35. Connect the brake booster vacuum line, the small vacuum hoses and the PCV hose.
36. Install the condenser on the side of the plenum.
37. Install the throttle body with a new gasket. Remember that the bolts are different sizes. Position them correctly and tighten them to 8 ft. lbs.
38. Connect the air intake hose.
39. Fill the cooling system with coolant.
40. Double check all installation items, paying particular attention to loose hoses or hanging wires, untightened nuts, poor routing of hoses and wires (too tight or rubbing) and tools left in the engine area.
41. Start the engine and check for leaks of either coolant or vacuum. Adjust the throttle cable as needed.

Galant with the 4G63 SOHC Engine

1. Disconnect the negative battery cable.
2. Safely relieve the residual fuel pressure within the system. The fuel system is under pressure. Release pressure slowly and contain spillage. Observe no smoking/no open flame precautions. Have a Class B-C (dry powder) fire extinguisher within arm's reach at all times.
3. Drain most of the coolant out of the cooling system so the intake manifold passages will be empty. Disconnect the upper radiator hose at the thermostat housing.

CAUTION

When draining the coolant, keep in mind that cats and dogs are attracted by the ethylene glycol antifreeze, and are quite likely to drink any that is left in an uncovered container or in puddles on the ground. This will prove fatal in sufficient quantity. Always drain the coolant into a sealable container. Coolant should be reused unless it is contaminated or several years old.

---

4. Disconnect the accelerator cable.
5. Remove the breather hose and the air intake hose.
6. Disconnect the heater hose, the coolant by-pass hose and the small water hose.
7. Disconnect the brake booster vacuum hose.
8. Label and disconnect the vacuum hoses, the PCV hose and the metal vacuum pipe.
9. Disconnect the high pressure fuel hose and discard the O-ring; it is not reusable. Remove the fuel return hose.

WARNING: Although the pressure was relieved earlier, cover the joint with a rag during removal; some pressure may remain.

10. On cars equipped with automatic transmissions, disconnect the throttle control cable, and remove the inner and outer cable brackets.
11. Label and disconnect each wire harness and lead running to the manifold and injection components. Some of the wiring will need to disconnected away from the manifold; follow the wires to the first connector and disconnect it.
12. Remove the delivery pipe (fuel rail), the injectors and the pressure regulator as a unit.

WARNING: Store this assembly in a clean, protected location. The injectors must be kept totally free of dirt. Do not drop the injectors during removal and handling.

13. Remove the insulators from the fuel rail.
14. Label and disconnect the coil wire and spark plug wires from the distributor cap. Remove the distributor, making a mark on the shaft and gear showing the alignment. Alternately, remove the cap and mark the position of the rotor on the outer case.
15. Remove the intake manifold support and remove the engine lifting hook.
16. Remove the power transistor bracket.
17. Unbolt and remove the intake manifold assembly and remove the gasket.
18. Remove the throttle body assembly and its gasket. Discard the gasket.
19. Remove the air intake plenum support bracket; unbolt the plenum and remove it with its gasket.
20. Remove the gaskets from the engine and clean the mating faces thoroughly. Examine the manifold and the plenum for any damage or cracking. Check each vacuum port and water passage for any clogging or plugging. Use a straightedge and feeler gauge to check the manifold and head surfaces for warpage. Maximum allowable warpage on the head face is 0.3mm (0.012 in.).

If the manifold is to be replaced, transfer the vacuum fittings, water ports, thermostat, EGR valve and other hardware to the new unit. When any of the coolant sensors are transferred, the threads must be coated with an adhesive sealant such as 3M 4171 or eqivalent. The manifold should be completely assembled before installation.

21. Install a new gasket between the plenum and the manifold. Assemble the two and tighten the bolts to 12 ft. lbs. Install the air plenum support bracket, tightening its bolts to 12 ft. lbs.
22. Using a new gasket, reinstall the throttle body assembly. Tighten the mounting bolts evenly to 8 ft. lbs.
23. Install a new manifold gasket and place the manifold in position on the engine. Replace the engine lifting hook and the power transistor bracket and tighten all the mounting nuts and bolts snug. Install the manifold support bracket, again bringing the bolts just snug.
24. Working from the center outward, and in at least two passes, tighten the manifold retaining bolts and nuts to 12 ft. lbs. Tighten the manifold bracket bolts to 15 ft. lbs. after the manifold bolts are properly tightened.
25. Install the distributor observing the alignment marks made during disassembly. Install the cap if it was removed and connect the coil and spark plug wiring.
26. Install the insulators which sit below the fuel rail. Install a new insulator (O-ring) on the fuel rail nipple.
27. Install the fuel rail (delivery pipe), injectors and pressure regulator as a unit. tighten the mounting bolts to 8 ft. lbs. Take care to protect the injectors from dropping or impact during installation.
28. Connect the vacuum lines and electrical connectors to their respective components. Make certain each is firmly seated and correctly installed.
29. On cars equipped with automatic transaxles, install the in-

ENGINE AND ENGINE OVERHAUL 3

ner and outer cable brackets and tighten the bolts to 10 ft. lbs. Install the throttle control cable.

30. Connect the fuel return line. Install a new O-ring on the high pressure fuel line and connect it to the fuel rail.
31. Connect the vacuum hoses, the PCV hose and the brake booster vacuum hose.
32. Install the small coolant hose, the by-pass hose and the heater hose.
33. Connect the upper radiator hose to the thermostat housing.
34. Install the air intake hose and connect the breather hose from the valve cover to the air intake.
35. Connect the accelerator cable.
36. Double check all installation items, paying particular attention to loose hoses or hanging wires, untightened nuts, poor routing of hoses and wires (too tight or rubbing) and tools left in the engine area.
37. Refill the coolant system to the proper level.
38. Connect the negative battery cable and start the engine. Check carefully for leaks of either coolant or vacuum. Check the operation of all electrical components, including dash gauges and warning lights.
39. Perform final adjustments to the engine specifications. Adjust the throttle control cable and/or the accelerator cable as necessary.

Galant with the 4G63 DOHC Engine

1. Disconnect the negative battery cable.
2. Safely relieve the residual fuel pressure within the system. The fuel system is under pressure. Release pressure slowly and contain spillage. Observe no smoking/no open flame precautions. Have a Class B-C (dry powder) fire extinguisher within arm's reach at all times.
3. Drain most of the coolant out of the cooling system so the intake manifold passages will be empty. Disconnect the upper radiator hose at the thermostat housing.

CAUTION

When draining the coolant, keep in mind that cats and dogs are attracted by the ethylene glycol antifreeze, and are quite likely to drink any that is left in an uncovered container or in puddles on the ground. This will prove fatal in sufficient quantity. Always drain the coolant into a sealable container. Coolant should be reused unless it is contaminated or several years old.

---

4. Disconnect the accelerator cable.
5. Remove the breather hose and the air intake hose. Also remove the throttle body support bracket.
6. Disconnect the heater hose, the coolant by-pass hose and the small water hose.
7. Disconnect the brake booster vacuum hose.
8. Label and disconnect the vacuum hoses and the PCV hose.
9. Disconnect the high pressure fuel hose and discard the O-ring; it is not reusable. Remove the fuel return hose.

WARNING: Although the pressure was relieved earlier, cover the joint with a rag during removal; some pressure may remain.

10. Label and disconnect each wire harness and lead running to the manifold and injection components. Some of the wiring will need to disconnected away from the manifold; follow the wires to the first connector and disconnect it.
11. Remove the delivery pipe (fuel rail), the injectors and the pressure regulator as a unit.

WARNING: Store this assembly in a clean, protected location. The injectors must be kept totally free of dirt. Do not drop the injectors during removal and handling.

12. Remove the insulators from the fuel rail.
13. There is no mechanical distributor. The coil and power transistor are bolted to the manifold. They can be removed with the manifold but it is safer to remove them and put them in a protected location.
14. Remove the intake manifold support and remove the tension rod bracket.
15. Remove the power transistor bracket.
16. Unbolt and remove the intake manifold assembly and remove the gasket.
17. Remove the throttle body assembly and its gasket. Discard the gasket.
18. Remove the air intake plenum support bracket.
19. Remove the gaskets from the engine and clean the mating faces thoroughly. Examine the manifold and the plenum for any damage or cracking. Check each vacuum port and water passage for any clogging or plugging. Use a straightedge and feeler gauge to check the manifold and head surfaces for warpage. Maximum allowable warpage on the head face is 0.3mm (0.012 in.).

If the manifold is to be replaced, transfer the vacuum fittings, water ports, thermostat, EGR valve and other hardware to the new unit. When any of the coolant sensors are transferred, the threads must be coated with an adhesive sealant such as 3M 4171 or eqivalent. The manifold should be completely assembled before installation.

20. Install the air plenum support bracket, tightening its bolts to 12 ft. lbs.
21. Using a new gasket, reinstall the throttle body assembly. Tighten the mounting bolts evenly to 8 ft. lbs.
22. Install a new manifold gasket and place the manifold in position on the engine. Replace the tension rod bracket. Torque the nuts to 30 ft. lbs. Install the power transistor bracket and tighten all the mounting nuts and bolts snug. Install the manifold support bracket, again bringing the bolts just snug.
23. Working from the center outward, and in at least two passes, tighten the manifold retaining bolts and nuts to 12 ft. lbs. Tighten the manifold bracket bolts to 20 ft. lbs. after the manifold bolts are properly tightened.
24. are bolted to the manifold. They can be removed with the manifold but it is safer to remove them and put them in a protected location. Install the coil and power transistor Tighten the power transistor retaining bolts to 8 ft. lbs. and the ignition coil bolts to 18 ft. lbs.
25. Install the insulators which sit below the fuel rail. Install a new insulator (O-ring) on the fuel rail nipple.
26. Install the fuel rail (delivery pipe), injectors and pressure regulator as a unit. tighten the mounting bolts to 8 ft. lbs. Take care to protect the injectors from dropping or impact during installation.
27. Connect the vacuum lines and electrical connectors to their respective components. Make certain each is firmly seated and correctly installed.
28. Connect the fuel return line. Install a new O-ring on the high pressure fuel line and connect it to the fuel rail.
29. Connect the vacuum hoses, the PCV hose and the brake booster vacuum hose.
30. Install the small coolant hose, the by-pass hose and the heater hose.
31. Connect the upper radiator hose to the thermostat housing. Install the throttle body support bracket. Tighten the nuts to 13 ft. lbs.
32. Install the air intake hose and connect the breather hose from the valve cover to the air intake.
33. Connect the accelerator cable.
34. Double check all installation items, paying particular attention to loose hoses or hanging wires, untightened nuts, poor routing of hoses and wires (too tight or rubbing) and tools left in the engine area.
35. Refill the coolant system to the proper level.
36. Connect the negative battery cable and start the engine. Check carefully for leaks of either coolant or vacuum. Check the operation of all electrical components, including dash gauges and warning lights.

3 ENGINE AND ENGINE OVERHAUL

37. Perform final adjustments to the engine specifications. Adjust the throttle control cable and/or the accelerator cable as necessary.

Precis

1. Disconnect the negative battery cable.
2. Drain most of the coolant out of the cooling system so the intake manifold passages will be empty. Disconnect the upper radiator hose at the thermostat housing.

------- CAUTION -------
When draining the coolant, keep in mind that cats and dogs are attracted by the ethylene glycol antifreeze, and are quite likely to drink any that is left in an uncovered container or in puddles on the ground. This will prove fatal in sufficient quantity. Always drain the coolant into a sealable container. Coolant should be reused unless it is contaminated or several years old.

3. Remove the air cleaner assembly.
4. Remove the carburetor and gasket.
5. Disconnect the brake booster vacuum hose.
6. Remove the upper radiator hose from the thermostat housing.
7. Remove the water outlet, the thermostat and its gasket.
8. Disconnect the coolant hoses from the intake manifold.
9. Label and disconnect the electrical connections and the vacuum hoses.
10. Remove the EGR valve. Remove the gasket and discard it; it is not reusable.
11. Label and disconnect the spark plug and coil wiring from the distributor cap.
12. Remove the distributor cap and mark the position of the rotor in relation to the case. This will insure correct reinstallation.
13. Remove the distributor.
14. Remove the manifold support bracket.
15. Remove the engine lifting hook. It is mounted under one of the manifold retaining nuts.
16. Remove the retaining nuts and carefully remove the manifold. As soon as the manifold is free of the engine, remove the vacuum pipe assembly from the bottom of the manifold.
17. Remove the gaskets from the engine and clean the mating faces thoroughly. Examine the manifold for any damage or cracking. Check each vacuum port and water passage for any clogging or plugging. Use a straightedge and feeler gauge to check the manifold and head surfaces for warpage. Maximum allowable warpage on the head face is 0.3mm (0.012 in.). If the manifold is to be replaced, transfer the vacuum fittings, water ports and other hardware to the new unit. The manifold should be completely assembled before installation.
18. Before reinstallation, attach the vacuum pipe assembly to the bottom of the manifold. Install the new gaskets and fit the manifold into position on the studs. Install the manifold support bracket so that the end with the word **UP** is in contact with the manifold. Tighten the bracket bolts snug.
19. Install the nuts and washers which retain the manifold snugly. Don't forget the engine lifting hook.
20. After the manifold and its bracket are in place, tighten the support bracket bolts to 18 ft. lbs. Working from the center out, make two complete passes and tighten the manifold retaining nuts in two steps to 12 ft. lbs.
21. Reinstall the distributor, observing the alignment marks made earlier. Install the distributor cap and connect the spark plug and coil wires.
22. Install the EGR valve with a new gasket.
23. Connect the water hoses, electrical connections and vacuum hoses.
24. Install the thermostat and water outlet fitting with a new gasket and connect the upper radiator hose.
25. Connect the brake booster vacuum hose. Install the PCV hose to the manifold.
26. Install the base gasket and install the carburetor. Connect all the vacuum hoses and fuel lines. Connect the accelerator cable.
27. Install the air cleaner assembly.
28. Double check all installation items, paying particular attention to loose hoses or hanging wires, untightened nuts, poor routing of hoses and wires (too tight or rubbing) and tools left in the engine area.
29. Refill the coolant to the proper level.
30. Connect the negative battery cable and start the engine. Check carefully for any leaks of fluid or vacuum. Adjust the engine timing as necessary and adjust the throttle cable.

Sigma

1. Disconnect the negative battery cable.
2. Safely relieve the residual fuel pressure within the system. The fuel system is under pressure. Release pressure slowly and contain spillage. Observe no smoking/no open flame precautions. Have a Class B-C (dry powder) fire extinguisher within arm's reach at all times.
3. Drain most of the coolant out of the cooling system so the intake manifold passages will be empty.

------- CAUTION -------
When draining the coolant, keep in mind that cats and dogs are attracted by the ethylene glycol antifreeze, and are quite likely to drink any that is left in an uncovered container or in puddles on the ground. This will prove fatal in sufficient quantity. Always drain the coolant into a sealable container. Coolant should be reused unless it is contaminated or several years old.

4. Disconnect the air intake tube.
5. Remove the throttle body and its gasket.

WARNING: The bolts holding the throttle body are of two different sizes. Note or diagram their locations for reassembly.

The accelerator and throttle control cables remain attached along with the hoses. Simply unbolt the unit and move it aside.

6. Remove the condenser.
7. Label and disconnect the PCV hose, the vacuum hoses and the brake booster vacuum hose.
8. Remove the EGR valve and gasket and remove the EGR pipe and gasket. Discard the gaskets.
9. Disconnect the ground cable.
10. Remove the two air plenum support brackets.
11. Remove the two retaining bolts in the center of the plenum, then remove the outer nuts.
12. Remove the plenum and its gasket.
13. Disconnect the high pressure fuel hose and discard the O-ring. Disconnect the fuel return line.
14. Disconnect the vacuum line at the fuel delivery rail.
15. Remove the fuel delivery rail with the injectors and pressure regulator. Store them in a safe location. Do not drop or jar the injectors and keep them absolutely free of dirt and foreign matter.
16. Disconnect the upper radiator hose, the heater hose and the nearby coolant hose.
17. Remove the water outlet fitting and the gasket.
18. Disconnect the wiring harness connectors.
19. Remove the retaining nuts and remove the intake manifold. Remove the insulators and the manifold gaskets.
20. Examine the manifold and plenum for any damage or cracking. Check each vacuum port and water passage for any clogging or plugging. Use a straightedge and feeler gauge to check the manifold and head surfaces for warpage. Maximum allowable warpage on the head face is 0.3mm (0.012 in.).

If the manifold and/or plenum is to be replaced, transfer the vacuum fittings, water ports, thermostat, EGR valve and other

ENGINE AND ENGINE OVERHAUL 3

hardware to the new unit. The manifold should be completely assembled before installation. Clean the mating faces of any gasket remains or foreign matter.

21. Before reinstallation, place new gaskets in position. Make certain the adhesive side is towards the manifold, not towards the head. Do NOT add any gasket sealer to this gasket.
22. Install new insulators into the manifold.
23. Place the manifold in postion and install the nuts snug. When all are equally snug, work from the center outward in a circular pattern, using two passes to tighten the nuts to 12 ft. lbs.
24. Connect the wiring harness connectors and route the wiring properly.
25. Install the water outlet fitting with a new gasket.
26. Connect the coolant hose, the heater hose and the upper radiator hose.
27. Install the fuel delivery pipe and injectors, taking care not to drop the injectors or subject them to impact. Tighten the retaining bolts to 8 ft. lbs.
28. Connect the vacuum hoses.
29. Connect the fuel return line. Install a new O-ring and connect the high pressure fuel line.
30. Install a new plenum gasket (use no sealant) and install the plenum. Make certain the printed surface of the gasket faces the intake plenum.
31. Tighten the nuts on the outer edges to 13 ft. lbs., then tighten the inner bolts to 13 ft. lbs.
32. Install the two plenum support brackets, tightening the bolts to 10 ft. lbs.
33. Connect the ground cable to the plenum.
34. Install a new gasket and connect the EGR pipe. Install the EGR valve with a new gasket.
35. Connect the brake booster vacuum line, the small vacuum hoses and the PCV hose.
36. Install the condenser on the side of the plenum.
37. Install the throttle body with a new gasket. Remember that the bolts are different sizes. Position them correctly and tighten them to 8 ft. lbs.
38. Connect the air intake hose.
39. Fill the cooling system with coolant.
40. Double check all installation items, paying particular attention to loose hoses or hanging wires, untightened nuts, poor routing of hoses and wires (too tight or rubbing) and tools left in the engine area.
41. Start the engine and check for leaks of either coolant or vacuum. Adjust the throttle cable as needed.

Exhaust Manifold

REMOVAL AND INSTALLATION

Cordia and Tredia
Pickup with the G63B Engine

1. Remove the air cleaner.
2. Remove the heat stove and/or heat shield on the exhaust manifold, if so equipped.
3. If the engine is turbocharged, see the procedure later in this Section and remove the turbocharger.
4. With the manifold cold, soak all the manifold nuts and studs with a liquid penetrant. The hardware will be difficult to loosen.
5. If the engine is not turbocharged, disconnect the exhaust pipe or primary catalytic converter at the exhaust manifold.
6. Disconnect and remove the oxygen sensor. If there is a secondary air line connected to the exhaust manifold, disconnect it.
7. Suspend the manifold and remove all attaching nuts and washers. Pull the manifold off the head, if necessary rocking it to break it loose.
8. Thoroughly clean the sealing surfaces on the cylinder head and manifold. Replace any nuts, washers, or studs that are excessively rusted or may have been damaged during removal. Studs may sometimes be removed by installing two nuts and twisting them in opposite directions to lock them, and then using your wrench on the inner nut. Use a straightedge to check manifold and cylinder head sealing surfaces for flatness. Correct problems by replacing the manifold or having the cylinder head surface machined.
9. Install new gaskets in such a way that all bolt holes and ports are lined up. Make sure all the nuts turn freely, oiling them lightly if necessary. Also make sure all the studs are screwed all the way into the block.
10. Place the manifold in position and support it while you install all the washers and nuts hand tight.
11. Tighten the nuts and bolts alternately and in several stages to 12 ft. lbs.
12. If the engine is turbocharged, install the turbocharger as described in the procedure later in this Section.
13. Install the piping, heat stoves, and shields.
14. Using a new gasket and new nuts, attach the exhaust pipe or primary catalytic converter. Tighten the nuts to 25 ft. lbs. and the small bracket bolt to 18 ft. lbs.
15. Operate the engine and check carefully for leaks.

Starion
Pickup and Montero with the G54B Engine

WARNING: Perform this procedure only on a cold motor. The manifold and turbocharger will retain heat for a period of hours after the engine is stopped.

1. Remove the turbocharger, following procedures listed later in this Section.
2. Soak all the manifold retaining nuts and bolts with rust penetrant. Tap the end of each one with a metal object and allow the penetrant to work for about 30 minutes.
3. Support the manifold and remove all attaching nuts and washers. Pull the manifold off the head, if necessary rocking it to break it loose.
4. Thoroughly clean the sealing surfaces on the cylinder head and manifold. Replace any nuts, washers, or studs that are excessively rusted or may have been damaged during removal. Studs may sometimes be removed by installing two nuts and twisting them in opposite directions to lock them, and then using your wrench on the inner nut.
5. Use a straightedge to check manifold and cylinder head sealing surfaces for flatness. Maximum acceptable head warpage is 0.3mm (0.012 in.). Minor warpage may be corrected by resurfacing the head and/or replacing the manifold.
6. Install a new gasket so that all bolt holes and ports are lined up. Make sure all the nuts turn freely, oiling them lightly if necessary. Also make sure all the studs are screwed all the way into the block.
7. Place the manifold in position and support it while you install all the washers and nuts hand tight.
8. Tighten the nuts and bolts alternately and in several stages to 12 ft. lbs.
9. Install the turbocharger.
10. Operate the engine and check carefully for leaks.

Mirage with the G15B Engine

CAUTION

Perform this operation only on a cold motor. The exhaust components and other metal parts will retain heat for a period of hours after the engine is shut off.

1. Loosen and remove the power steering pump belt. Remove the pump from its brackets and move it out of the way with its hoses attached. Suspend it from a piece of stiff wire if necessary.
2. Remove the power steering pump bracket from the engine.

3-125

3 ENGINE AND ENGINE OVERHAUL

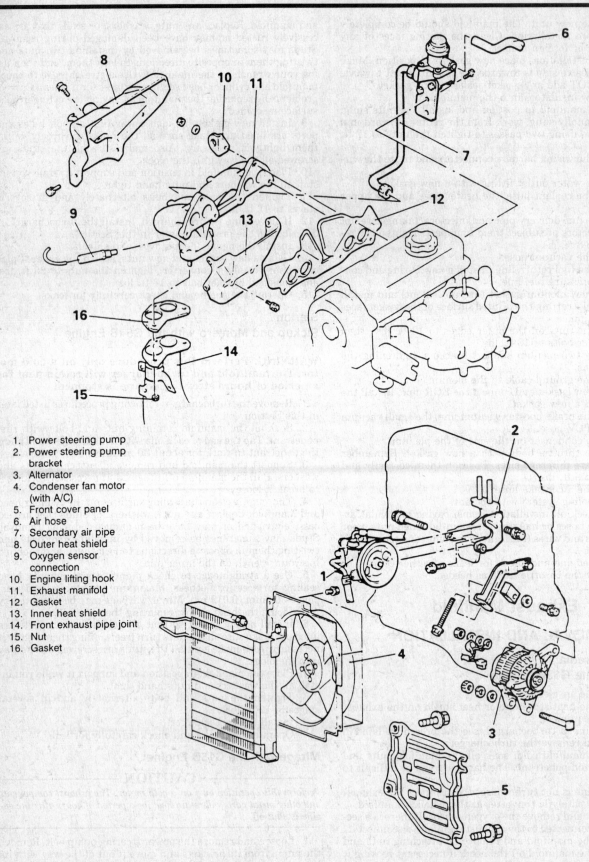

1. Power steering pump
2. Power steering pump bracket
3. Alternator
4. Condenser fan motor (with A/C)
5. Front cover panel
6. Air hose
7. Secondary air pipe
8. Outer heat shield
9. Oxygen sensor connection
10. Engine lifting hook
11. Exhaust manifold
12. Gasket
13. Inner heat shield
14. Front exhaust pipe joint
15. Nut
16. Gasket

G15B exhaust manifold removal

ENGINE AND ENGINE OVERHAUL 3

3. Remove the alternator.
4. On air conditioned vehicles, remove the condenser fan with the shroud. Don't forget to disconnect the fan wiring before removal.
5. Remove the front end cover panel.
6. Label and disconnect the air hose running to the secondary air pipe. disconnect the pipe from the exhaust manifold, remove the mounting bolt and remove the air pipe assembly.
7. Remove the protective heat shield from the outside of the manifold.
8. Disconnect the oxygen sensor.
9. Remove the upper manifold nut holding the engine lifting hook and remove the hook.
10. Continue carefully removing the manifold nuts and remove the manifold from the studs. Remove the gasket.
11. Either have a helper hold the manifold or place wood blocks under the car to support the exhaust system. Don't let the exhaust system support the weight of the manifold. Remove the inner heat shield from the exhaust manifold.
12. Disconnect the manifold from the exhaust pipe. Lift the manifold clear of the engine and remove the exhaust pipe gasket.
13. Check the manifold carefully for cracks and damage. Use a straightedge and feeler gauges to measure the cylinder head face for distortion. Maximum allowable warpage is 0.3mm (0.012 in.).
14. Install the new gasket in position, making sure it is correctly aligned with all the ports and studs.
15. Attach the manifold to the exhaust system using new nuts and a new gasket. Tighten the nuts to 18 ft. lbs.
16. Install the inner heat shield to the manifold and tighten the bolts to 10 ft. lbs.
17. Move the manifold into position against the head and install the nuts finger tight. don't forget to reinstall the engine lifting hook on the correct stud.
18. In an alterating pattern and in at least two steps, tighten the manifold nuts to 12 ft. lbs.
19. Connect the oxygen sensor.
20. Install the outer heat shield on the manifold and tighten the bolts to 10 ft. lbs.
21. Install the secondary air pipe by connecting it loosely to the exhaust manifold, then loosely connecting the bracket to the head. Once it is held in its correct position, tighten the manifold nut to 60 ft. lbs. and the mounting bolt to 8 ft. lbs. Failure to follow this procedure may result in stripped threads or improper sealing at the connection.
22. Connect the air hose to the secondary air pipe. Make certain that the white mark on the hose is up (facing the hood) and that the hose is pushed onto the valve about an 12mm (1 inch). It is important that this hose not be crushed, crimped or twisted.
23. Install the front end cover panel. On air conditioned vehicles, install the condenser fan.
24. Install the alternator.
25. Install the power steering pump mounting bracket, tightening the bolts to 21 ft. lbs.
26. Install the power steering pump and tighten the bolts to 18 ft. lbs.
27. Check all the drive belts and adjust to the correct tension if necessary. Start the engine and check for exhaust leaks.

Mirage with the G32B Engine

CAUTION
Perform this operation only on a cold motor. The exhaust components and other metal parts will retain heat for a period of hours after the engine is shut off.

1. Disconnect the breather hose, remove the hose clamp and remove the air intake duct. Remove the O-ring.

2. Loosen and remove the power steering pump belt. Remove the pump from its brackets and move it out of the way with its hoses attached. Suspend it from a piece of stiff wire if necessary.
3. Remove the power steering pump bracket from the engine.
4. Remove the alternator.
5. On air conditioned vehicles, remove the condenser fan with the shroud. Don't forget to disconnect the fan wiring before removal.
6. Remove the front end cover panel.
7. Remove the engine oil dipstick. Remove the bolt holding the dipstick tube and remove the tube from the engine. Discard the O-ring on the bottom of the tube.
8. Label and disconnect the air hose running to the secondary air pipe. Disconnect the pipe from the exhaust manifold, remove the mounting bolt and remove the air pipe assembly.
9. Remove the protective heat shield from the outside of the manifold.
10. Remove the upper manifold nut holding the engine lifting hook and remove the hook.
11. Continue carefully removing the manifold nuts and remove the manifold from the studs. Remove the gasket.
12. Either have a helper hold the manifold or place wood blocks under the car to support the exhaust system. Don't let the exhaust system support the weight of the manifold.
13. Disconnect the manifold from the exhaust pipe. Lift the manifold clear of the engine and remove the exhaust pipe gasket.
14. Check the manifold carefully for cracks and damage. Use a straightedge and feeler gauges to measure the cylinder head face for distortion. Maximum allowable warpage is 0.3mm (0.012 in.).
15. Install the new gasket in position, making sure it is correctly aligned with all the ports and studs.
16. Attach the manifold to the exhaust system using new nuts and a new gasket. Tighten the nuts to 25 ft. lbs. The long bolt

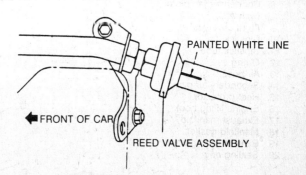

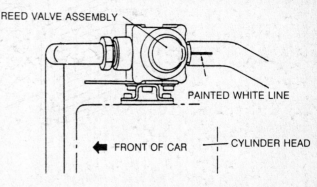

The white painted identification mark must face upwards to prevent the air hose from kinking. G15B shown above, G32B below

3-127

3 ENGINE AND ENGINE OVERHAUL

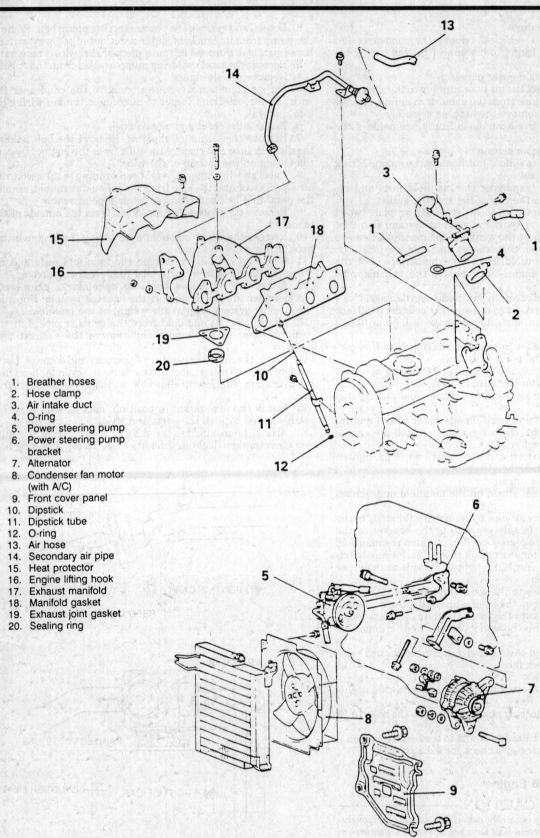

1. Breather hoses
2. Hose clamp
3. Air intake duct
4. O-ring
5. Power steering pump
6. Power steering pump bracket
7. Alternator
8. Condenser fan motor (with A/C)
9. Front cover panel
10. Dipstick
11. Dipstick tube
12. O-ring
13. Air hose
14. Secondary air pipe
15. Heat protector
16. Engine lifting hook
17. Exhaust manifold
18. Manifold gasket
19. Exhaust joint gasket
20. Sealing ring

G32B exhaust manifold removal

3-128

ENGINE AND ENGINE OVERHAUL 3

which runs through the manifold into the exhaust pipe joint should be tightened to 42 ft. lbs.

17. Move the manifold into position against the head and install the nuts finger tight. Don't forget to reinstall the engine lifting hook on the correct stud.
18. In an alternating pattern and in at least two steps, tighten the manifold nuts to 20 ft. lbs.
19. Install the outer heat shield on the manifold and tighten the bolts to 10 ft. lbs.
20. Install the secondary air pipe by connecting it loosely to the exhaust manifold, then loosely connecting the brackets to their mounts. Once it is held in its correct position, tighten the manifold nut to 60 ft. lbs. and the bracket mounting bolt to 8 ft. lbs. Failure to follow this procedure may result in stripped threads or improper sealing at the connection.
21. Connect the air hose to the secondary air pipe. Make certian that the white mark on the hose is up (facing the hood) and that the hose is pushed onto the valve about 12mm (1 inch). It is important that this hose not be crushed, crimped or twisted.
22. Install a new O-ring on the dipstick tube and coat the O-ring and lower tube with engine oil. Install the tube, making sure it is firmly seated in the block. Don't damage the O-ring during installation. Install the dipstick.
23. Install the front end cover panel. On air conditioned vehicles, install the condenser fan and connect the wiring.
24. Install the alternator.
25. Install the power steering pump mounting bracket, tightening the bolts to 21 ft. lbs.
26. Install the power steering pump and tighten the bolts to 18 ft. lbs.
27. Install the air intake duct. Tighten the small bolt holding the duct to the valve cover first, then tighten the fittings holding the duct to the turbocharger.
28. Install the hose clamp and connect the breather hoses.
29. Check all the drive belts and adjust to the correct tension if necessary. Start the engine and check for exhaust leaks.

Mirage with the 4G15 SOHC Engine

CAUTION
Perform this operation only on a cold motor. The exhaust components and other metal parts will retain heat for a period of hours after the engine is shut off.

WARNING: The use of the correct special tools or their equivalent is REQUIRED for this procedure.

1. Remove the self-locking nuts holding the exhaust system to the bottom of the manifold. Remove the exhaust bracket bolt.
2. Separate the exhaust system from the manifold and either hang it from a piece of wire or support it from underneath with wooden blocks. Remove the gasket and discard it.
3. Remove the outer heat shield from the manifold.
4. Using the special socket (MB998748 or equivalent), carefully remove the oxygen sensor from the manifold. Place the sensor in a protected location and keep the tip completely free of any dirt or oil.
5. Remove the nut holding the engine lifting hook and remove the hook.
6. Continue removing the nuts and remove the exhaust manifold. Remove the gasket.
7. Remove the inner heat shield from the manifold.
8. Check the manifold carefully for cracks and damage. Use a straightedge and feeler gauges to measure the cylinder head face for distortion. Maximum allowable warpage is 0.3mm (0.012 in.)
9. Install the new gasket in position, making sure it is correctly aligned with all the ports and studs.

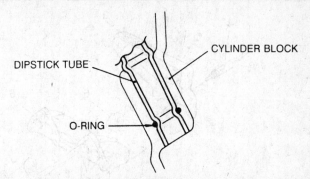

Correct installation of G32B oil dipstick tube

10. Install the inner heat shield, tioghtening the bolts to 22 ft. lbs.
11. Place the manifold over the studs and install all the nuts finger tight. Don't forget the engine lift hook.
12. In an alternating pattern and in at least two steps, tighten the manifold nuts to 12 ft. lbs.
13. Carefully install the oxygen sensor and tighten it to 33 ft. lbs. Make sure the wiring is connected.
14. Install the outer heat shield on the manifold. Tighten the bolts to 22 ft. lbs.
15. Using a new gasket and new self-locking nuts, connect the exhaust system to the bottom of the manifold. Tighten the nuts to 25 ft. lbs. and the exhaust bracket bolt to 18 ft. lbs.

Mirage with the 4G61 DOHC Engine

CAUTION
Perform this operation only on a cold engine. The exhaust components and other metal parts will retain heat for a period of hours after the engine is shut off.

WARNING: The use of the correct special tools or their equivalent is REQUIRED for this procedure.

1. Disconnect the negative battery cable.
2. Drain the cooling system.

CAUTION
When draining the coolant, keep in mind that cats and dogs are attracted by the ethylene glycol antifreeze, and are quite likely to drink any that is left in an uncovered container or in puddles on the ground. This will prove fatal in sufficient quantity. Always drain the coolant into a sealable container. Coolant should be reused unless it is contaminated or several years old.

3. Disconnect the upper and lower coolant hoses from the engine and remove the radiator with the hoses attached. Leave the electric fan assembly attached to the radiator.
4. Using the special socket (MB998748 or equivalent), carefully remove the oxygen sensor. Place the sensor in a protected location and keep the tip completely free of any dirt or oil.
5. Disconnect the air intake hose from the turbocharger.
6. Label and remove the small vacuum hoses from the turbocharger.
7. Disconnect air hose **A** from the turbocharger.
8. Remove the heat shield from the top of the manifold, then remove the heat shield from the front and side of the turbocharger.
9. Remove the two nuts holding the engine lifting bracket and remove the bracket.
10. Disconnect the eye bolt (banjo bolt) above the turbocharger. Remove and discard the two washers (gaskets) from the joint.
11. Label and remove the small coolant hose from the turbocharger. Carefully unscrew and remove the connection for wa-

3 ENGINE AND ENGINE OVERHAUL

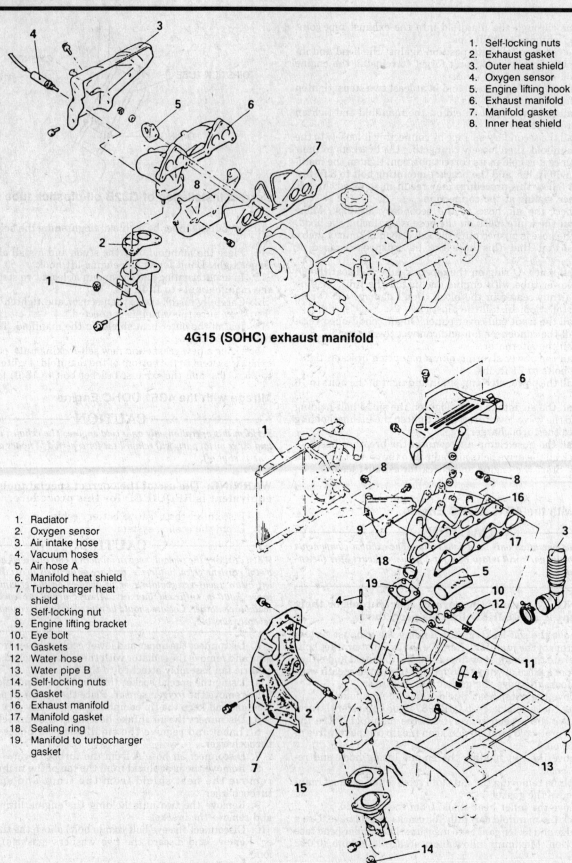

1. Self-locking nuts
2. Exhaust gasket
3. Outer heat shield
4. Oxygen sensor
5. Engine lifting hook
6. Exhaust manifold
7. Manifold gasket
8. Inner heat shield

4G15 (SOHC) exhaust manifold

1. Radiator
2. Oxygen sensor
3. Air intake hose
4. Vacuum hoses
5. Air hose A
6. Manifold heat shield
7. Turbocharger heat shield
8. Self-locking nut
9. Engine lifting bracket
10. Eye bolt
11. Gaskets
12. Water hose
13. Water pipe B
14. Self-locking nuts
15. Gasket
16. Exhaust manifold
17. Manifold gasket
18. Sealing ring
19. Manifold to turbocharger gasket

4G61 exhaust manifold disassembly

3-130

ENGINE AND ENGINE OVERHAUL 3

ter pipe B at the turbocharger. Use a special line wrench if available; do not damage the fitting during removal.

12. Remove the exhaust pipe bracket bolt (just below the turbo charger) and remove the self-locking nuts holding the exhaust pipe to the turbocharger.
13. Gently separate the pipe from the turbocharger and remove the gasket from the joint.

NOTE: The turbocharger is loose but not completely free. Handle it carefully and do not place excess strain on lines and hoses.

14. Remove the exhaust manifold retaining nuts and bolts, including the vertical ones holding the manifold to the turbocharger.
15. Remove the manifold, its gasket and sealing ring and support the turbocharger with a piece of stiff wire or string.
16. Check the manifold carefully for cracks and damage. Use a straightedge and feeler gauges to measure the cylinder head face for distortion. Maximum allowable warpage is 0.3mm (0.012 in.).
17. Install the new manifold gasket in position, making sure it is correctly aligned with all the ports and studs.
18. Install a new gasket on the top of the turbocharger and install the sealing ring in the bottom of the manifold. Place the manifold on the turbocharger, install the 3 through bolts and the nut and draw them up snug. Evenly tighten the 3 bolts and the nut to 44 ft. lbs.
19. Install a new gasket and connect the exhaust pipe to the bottom of the turbocharger. Use new self-locking nuts and tighten them to 18 ft. lbs.
20. Apply a light coat of machine oil (or other lightweight oil) to the edge of flare on water pipe B. Carefully position the pipe and tighten the fitting to 34 ft. lbs. Connect the water hose to the turbocharger.
21. If not already done for convenience, fit the manifold onto the head studs, making sure the gasket is not disturbed. Install new self-locking nuts and tighten each snug; don't forget the engine lifting bracket. Install the lower exhaust pipe bracket bolt and tighten it to 18 ft. lbs.
22. In an alternating pattern and in at least two steps, tighten the manifold nuts to 20 ft. lbs.
23. Use new washers and connect the eye bolt for the oil line. Tighten the fitting to 12 ft. lbs.
24. Install the heat shield over the turbocharger and then install the heat shield above the manifold. Tighten the heat shield retaining bolts to 10 ft. lbs.
25. Connect air hose **A** to the turbocharger and connect the smaller vacuum hoses. Install the air intake duct.
26. Carefully install the oxygen sensor, tightening it to 34 ft. lbs., and connect its wiring.
27. Install the radiator and connect the hoses to the engine. Remember to connect the wiring for the electric fan.
28. Fill the cooling system and radiator with the proper coolant.
29. Start the engine and check carefully for leaks of exhaust, vacuum, coolant or oil. The intake air and exhaust leaks will be the most difficult to identify; they must be eliminated if the turbocharger is to work correctly. If leak repair is needed, remember that fluids and surfaces will become hot as the engine runs.

Galant with the G64B Engine

--- **CAUTION** ---

Perform this operation only on a cold motor. The exhaust components and other metal parts will retain heat for a period of hours after the engine is shut off.

1. Remove the bolt holding the heat shield and remove the shield.
2. Disconnect the exhaust pipe from the bottom of the mani-

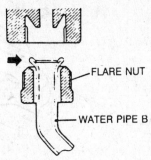

Apply light oil to the flange of water pipe "B" before reinstalling

fold and discard the gasket.

3. Disconnect and carefully remove the oxygen sensor from the exhaust manifold. Protect the tip of the sensor from grease or dirt and place it in a safe location.
4. Remove the self-locking nuts holding the manifold and remove the engine lifting bracket.
5. Remove the exhaust manifold and its gasket. Inspect the manifold and head surfaces for cracks or distortion. Use a straightedge and feeler gauge to check the surfaces for warpage. Maximum acceptable warpage is 0.3mm (0.012 in.).
6. Install the new gasket and place the manifold on the head. Install the nuts finger tight, remembering to replace the engine lifting hook. When all the nuts are snug, work from the center nuts outward in a circular pattern, tightening the nuts to 18 ft. lbs. in at least two steps.
7. Carefully reinstall the oxygen sensor and tighten it to 34 ft. lbs. Connect the wiring.
8. Install a new gasket and, using new nuts, connect the exhaust pipe to the bottom of the manifold. Tighten the nuts to 18 ft. lbs.
9. Install the heat shield and tighten its bolt to 10 ft. lbs.
10. Start the engine and check for exhaust leaks.

Galant with the 6G72 V6 Engine

--- **CAUTION** ---

Perform this operation only on a cold motor. The exhaust components and other metal parts will retain heat for a period of hours after the engine is shut off.

FRONT MANIFOLD

1. Disconnect the oxygen sensor wiring. Leave the sensor in the exhaust pipe.
2. Remove the self-locking nuts holding both exhaust pipes to the front and rear manifolds. Remove the gaskets and discard them. Either hang the exhaust pipe from stiff wire or support it from underneath with wooden blocks.
3. Remove the heat shield from the manifold.
4. Remove the retainig nuts and remove the exhaust manifold.
5. Remove the dipstick. Remove the dipstick tube retaining bolt and remove the dipstick tube. Take care not to damage the O-ring when removing the tube.
6. Remove the manifold gasket.
7. Inspect the manifold and head surfaces for cracks or distortion. Use a straightedge and feeler gauge to check the surfaces for warpage. Maximum acceptable warpage is 0.3mm (0.012 in.).

To install:

8. Place a new gasket against the head. Apply a light coat of engine oil to the O-ring on the dipstick tube and install the tube, making certain the tube is properly seated and the O-ring is not damaged during installation. Install the dipstick.

3 ENGINE AND ENGINE OVERHAUL

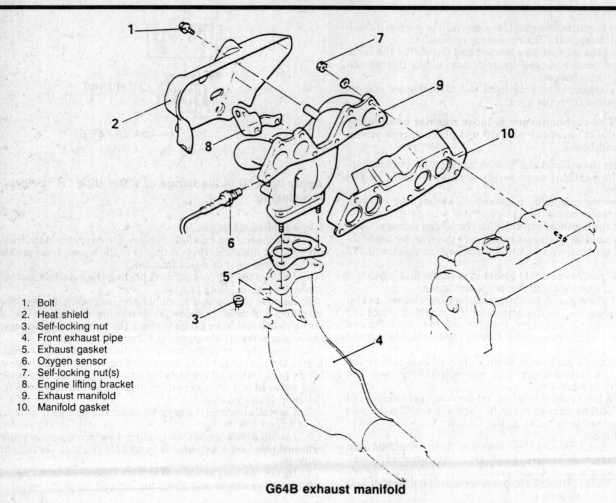

1. Bolt
2. Heat shield
3. Self-locking nut
4. Front exhaust pipe
5. Exhaust gasket
6. Oxygen sensor
7. Self-locking nut(s)
8. Engine lifting bracket
9. Exhaust manifold
10. Manifold gasket

G64B exhaust manifold

9. Install the manifold. Tighten the all the nuts snug. In two passes and working from the inner nuts outward, tighten the retaining nuts to 14 ft. lbs.
10. Install the heat shield, tightening the bolts to 10 ft. lbs.
11. Using new gaskets and new self-locking nuts, connect the exhaust system to both the front and rear manifolds. Tighten the retaining nuts to 26 ft. lbs.
12. Connect the oxygen sensor wiring.
13. Start the engine and check for exhaust leaks.

REAR MANIFOLD

1. Disconnect the negative battery cable. Disconnect the oxygen sensor wiring. Leave the sensor in the exhaust pipe.
2. Remove the self-locking nuts holding both exhaust pipes to the front and rear manifolds. Remove the gaskets and discard them. Either hang the exhaust pipe from stiff wire or support it from underneath with wooden blocks.
3. Disconnect the air intake tube.
4. Remove the throttle body and its gasket.

WARNING: The bolts holding the throttle body are of two different sizes. Note or diagram their locations for reassembly.
The accelerator and throttle control cables remain attached along with the hoses. Simply unbolt the unit and move it aside.

5. Remove the condenser.
6. Label and disconnect the PCV hose, the vacuum hoses and the brake booster vacuum hose.
7. Remove the EGR valve and gasket and remove the EGR pipe and gasket. Discard the gaskets.

8. Disconnect the ground cable.
9. Remove the two air plenum support brackets.
10. Remove the two retaining bolts in the center of the plenum, then remove the outer nuts.
11. Remove the plenum and its gasket.
12. Remove the heat shield from the rear manifold.
13. Disconnect the upper end of the EGR pipe. The pipe should be removed with the manifold. If a new manifold is to be installed, transfer the pipe on the workbench.
14. Remove the retaining nuts and remove the manifold and its gasket.
15. Inspect the manifold and head surfaces for cracks or distortion. Use a straightedge and feeler gauge to check the surfaces for warpage. Maximum acceptable warpage is 0.3mm (0.012 in.).

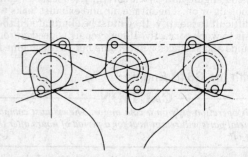

Check the face of the manifold in all dimensions for warpage

3-132

ENGINE AND ENGINE OVERHAUL 3

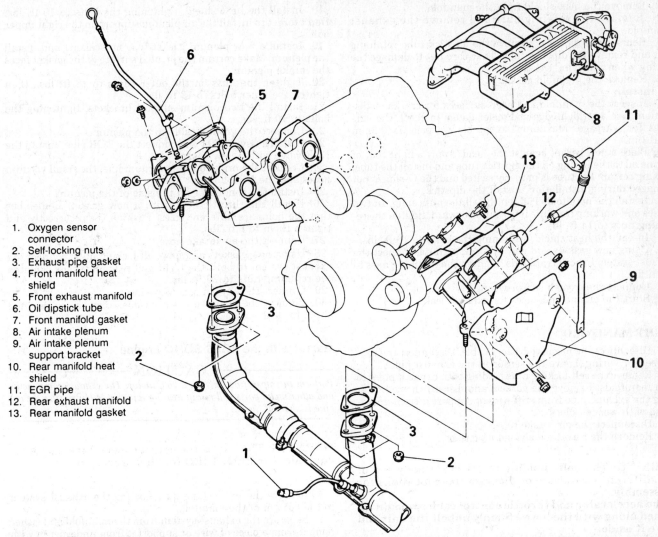

1. Oxygen sensor connector
2. Self-locking nuts
3. Exhaust pipe gasket
4. Front manifold heat shield
5. Front exhaust manifold
6. Oil dipstick tube
7. Front manifold gasket
8. Air intake plenum
9. Air intake plenum support bracket
10. Rear manifold heat shield
11. EGR pipe
12. Rear exhaust manifold
13. Rear manifold gasket

Either V6 exhaust manifold may be removed individually, but the exhaust system must be disconnected from both manifolds

To install:
16. Install a new gasket and place the manifold in position. Tighten the nuts snug. In two passes and working from the inner nuts outward, tighten the retaining nuts to 14 ft. lbs.
17. Replace the gasket for the EGR pipe with a new one and install the pipe to the EGR valve.
18. Install the heat shield, tightening the bolts to 10 ft. lbs. Don't forget to install the air plenum support on the right upper bolt.
19. Install a new plenum gasket (use no sealant) and install the plenum. Make certain the printed surface of the gasket faces the intake plenum.
20. Tighten the nuts on the outer edges to 13 ft. lbs., then tighten the inner bolts to 13 ft. lbs.
21. Install the two plenum support brackets, tightening the bolts to 10 ft. lbs.
22. Connect the ground cable to the plenum.
23. Install a new gasket and connect the EGR pipe. Install the EGR valve with a new gasket.
24. Connect the brake booster vacuum line, the small vacuum hoses and the PCV hose.
25. Install the condenser on the side of the plenum.
26. Install the throttle body with a new gasket. Remember that the bolts are different sizes. Position them correctly and tighten them to 8 ft. lbs.
27. Connect the air intake hose.
28. Using new gaskets and new self-locking nuts, connect the exhaust system to both the front and rear manifolds. Tighten the retaining nuts to 26 ft. lbs.
29. Connect the oxygen sensor wiring.
30. Connect the negative battery cable; start the engine and check for leaks.

Montero with the 6G72 V6 Engine

---- **CAUTION** ----
Perform this operation only on a cold motor. The exhaust components and other metal parts will retain heat for a period of hours after the engine is shut off.

LEFT MANIFOLD
1. Disconnect the oxygen sensor wiring. Leave the sensor in the exhaust pipe.
2. Remove the self-locking nuts holding both exhaust pipes to the manifolds. Remove the gaskets and discard them. Either hang the exhaust pipe from stiff wire or support it from underneath with wooden blocks.

3-133

ENGINE AND ENGINE OVERHAUL

3. Remove the heat shield from the manifold.
4. Remove the retainig nuts and remove the exhaust manifold.
5. Remove the dipstick. Remove the dipstick tube retaining bolt and remove the dipstick tube. Take care not to damage the O-ring when removing the tube.
6. Remove the manifold gasket.

To install:

7. Inspect the manifold and head surfaces for cracks or distortion. Use a straightedge and feeler gauge to check the surfaces for warpage. Maximum acceptable warpage is 0.3mm (0.012 in.).
8. Place a new gasket against the head. Apply a light coat of engine oil to the O-ring on the dipstick tube and install the tube, making certain the tube is properly seated and the O-ring is not damaged during installation. Install the dipstick.
9. Install the manifold. Tighten the all the nuts snug. In two passes and working from the inner nuts outward, tighten the retaining nuts to 14 ft. lbs.
10. Install the heat shield, tightening the bolts to 10 ft. lbs.
11. Using new gaskets and new self-locking nuts, connect the exhaust system to both manifolds. Tighten the retaining nuts to 26 ft. lbs.
12. Connect the oxygen sensor wiring.
13. Start the engine and check for exhaust leaks.

RIGHT MANIFOLD

1. Disconnect the negative battery cable. Disconnect the oxygen sensor wiring. Leave the sensor in the exhaust pipe.
2. Remove the self-locking nuts holding both exhaust pipes to both manifolds. Remove the gaskets and discard them. Either hang the exhaust pipe from stiff wire or support it from underneath with wooden blocks.
3. Disconnect the air intake tube.
4. Remove the throttle body and its gasket.

WARNING: The bolts holding the throttle body are of two different sizes. Note or diagram their locations for reassembly.

The accelerator and throttle control cables remain attached along with the hoses. Simply unbolt the unit and move it aside.

5. Remove the condenser.
6. Label and disconnect the PCV hose, the vacuum hoses and the brake booster vacuum hose.
7. Remove the EGR valve and gasket and remove the EGR pipe and gasket. Discard the gaskets.
8. Disconnect the ground cable.
9. Remove the two air plenum support brackets.
10. Remove the two retaining bolts in the center of the plenum, then remove the outer nuts.
11. Remove the plenum and its gasket.
12. Remove the heat shield from the manifold.
13. Disconnect the upper end of the EGR pipe. The pipe should be removed with the manifold. If a new manifold is to be installed, transfer the pipe on the workbench.
14. Remove the retaining nuts and remove the manifold and its gasket.

To install:

15. Inspect the manifold and head surfaces for cracks or distortion. Use a straightedge and feeler gauge to check the surfaces for warpage. Maximum acceptable warpage is 0.3mm (0.012 in.).
16. Install a new gasket and place the manifold in position. Tighten the nuts snug. In two passes and working from the inner nuts outward, tighten the retaining nuts to 14 ft. lbs.
17. Replace the gasket for the EGR pipe with a new one and install the pipe to the EGR valve.

18. Install the heat shield, tightening the bolts to 10 ft. lbs. Don't forget to install the air plenum support on the right upper bolt.
19. Install a new plenum gasket (use no sealant) and install the plenum. Make certain the printed surface of the gasket faces the intake plenum.
20. Tighten the nuts on the outer edges to 13 ft. lbs., then tighten the inner bolts to 13 ft. lbs.
21. Install the two plenum support brackets, tightening the bolts to 10 ft. lbs.
22. Connect the ground cable to the plenum.
23. Install a new gasket and connect the EGR pipe. Install the EGR valve with a new gasket.
24. Connect the brake booster vacuum line, the small vacuum hoses and the PCV hose.
25. Install the condenser on the side of the plenum.
26. Install the throttle body with a new gasket. Remember that the bolts are different sizes. Position them correctly and tighten them to 8 ft. lbs.
27. Connect the air intake hose.
28. Using new gaskets and new self-locking nuts, connect the exhaust system to both the front and rear manifolds. Tighten the retaining nuts to 26 ft. lbs.
29. Connect the oxygen sensor wiring.
30. Connect the negative battery cable; start the engine and check for leaks.

Galant with the 4G63 SOHC Engine

— **CAUTION** —

Perform this operation only on a cold motor. The exhaust components and other metal parts will retain heat for a period of hours after the engine is shut off.

WARNING: The use of the correct special tools or their equivalent is REQUIRED for this procedure.

1. Remove the self-locking nuts holding the exhaust system to the bottom of the manifold.
2. Separate the exhaust system from the manifold and either hang it from a piece of wire or support it from underneath with wooden blocks. Remove the gasket and discard it.
3. Remove the outer heat shield from the manifold.
4. Using the special socket (MB998703 or equivalent), carefully remove the oxygen sensor from the manifold. Place the sensor in a protected location and keep the tip completely free of any dirt or oil.
5. Remove the nuts holding the engine lifting hook and remove the hook.
6. Continue removing the nuts and remove the exhaust manifold. Remove the gasket.
7. Check the manifold carefully for cracks and damage. Use a straightedge and feeler gauges to measure the cylinder head face for distortion. Maximum allowable warpage is 0.3mm (0.012 in.).
8. Install the new gasket in position, making sure it is correctly aligned with all the ports and studs.
9. Place the manifold over the studs and install all the nuts finger tight. Don't forget the engine lift hook.
10. In an alternating pattern and in at least two steps, tighten the manifold nuts to 12 ft. lbs.
11. Carefully install the oxygen sensor and tighten it to 33 ft. lbs. Make sure the wiring is connected.
12. Install the outer heat shield on the manifold. Tighten the bolts to 10 ft. lbs.
13. Using a new gasket and new self-locking nuts, connect the exhaust system to the bottom of the manifold. Tighten the nuts to 25 ft. lbs. and the exhaust bracket bolt to 18 ft. lbs.

ENGINE AND ENGINE OVERHAUL 3

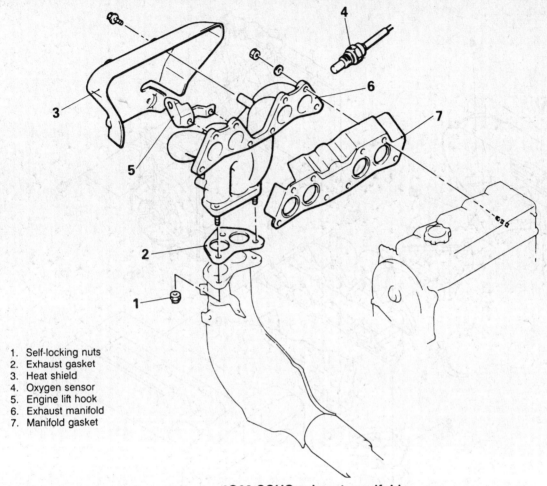

1. Self-locking nuts
2. Exhaust gasket
3. Heat shield
4. Oxygen sensor
5. Engine lift hook
6. Exhaust manifold
7. Manifold gasket

4G63 SOHC exhaust manifold

Galant with the 4G63 DOHC Engine

── **CAUTION** ──

Perform this operation only on a cold motor. The exhaust components and other metal parts will retain heat for a period of hours after the engine is shut off.

WARNING: The use of the correct special tools or their equivalent is REQUIRED for this procedure.

1. If the vehicle is air conditioned, unplug the condenser fan (left side fan) and remove the fan assembly.
2. Remove the self-locking nuts holding the exhaust pipe to the manifold. Remove the small exhaust bracket bolt and separate the pipe from the manifold. Discard the gasket. Support the exhaust system with wire or wooden blocks.
3. Remove the heat shield from the manifold. Don't forget the bolt on the lower left corner.
4. Using the special socket (MB998703 or equivalent), carefully remove the oxygen sensor from the manifold. Place the sensor in a protected location and keep the tip completely free of any dirt or oil.
5. Remove the nuts holding the engine lifting hook and remove the hook.
6. Continue removing the nuts and remove the exhaust manifold. Remove the gasket. Remove the inner heat shield from the manifold.
7. Check the manifold carefully for cracks and damage. Use a straightedge and feeler gauges to measure the cylinder head face for distortion. Maximum allowable warpage is 0.3mm (0.012 in.).
8. Install the new gasket in position, making sure it is correctly aligned with all the ports and studs.
9. Install the inner heat shield onto the manifold. Tighten the mounting bolts to 10 ft. lbs.
10. Place the manifold over the studs and install all the nuts finger tight. Don't forget the engine lift hook.
11. In an alternating pattern and in at least two steps, tighten the manifold nuts to 20 ft. lbs.
12. Carefully install the oxygen sensor and tighten it to 33 ft. lbs. Make sure the wiring is connected.
13. Install the outer heat shield on the manifold. Tighten the bolts to 10 ft. lbs.
14. Using a new gasket and new self-locking nuts, connect the exhaust system to the bottom of the manifold. Tighten the nuts and the bracket bolt to 20 ft. lbs.
15. Reinstall the condenser fan on air conditioned vehicles and connect the wiring harness.
16. Start the engine. Check for leaks and check the operation of the condenser fan.

Precis

── **CAUTION** ──

Perform this operation only on a cold motor. The exhaust components and other metal parts will retain heat for a period of hours after the engine is shut off.

3 ENGINE AND ENGINE OVERHAUL

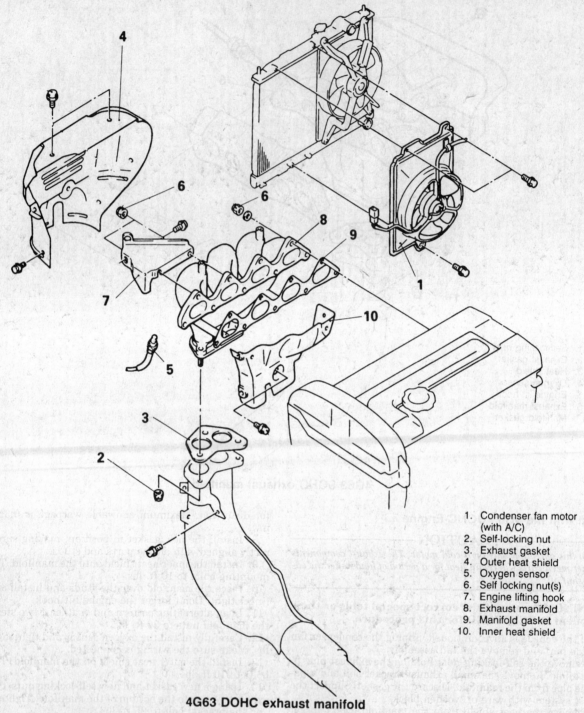

1. Condenser fan motor (with A/C)
2. Self-locking nut
3. Exhaust gasket
4. Outer heat shield
5. Oxygen sensor
6. Self locking nut(s)
7. Engine lifting hook
8. Exhaust manifold
9. Manifold gasket
10. Inner heat shield

4G63 DOHC exhaust manifold

1. Loosen and remove the power steering pump belt. Remove the pump from its brackets and move it out of the way with its hoses attached. Suspend it from a piece of stiff wire if necessary.
2. Remove the power steering pump bracket from the engine.
3. Remove the alternator.
4. On air conditioned vehicles, remove the condenser fan with the shroud. Don't forget to disconnect the fan wiring before removal.
5. Remove the front end cover panel.
6. Label and disconnect the air hose running to the secondary air pipe. disconnect the pipe from the exhaust manifold, remove the mounting bolt and remove the air pipe assembly.

7. Remove the protective heat shield from the outside of the manifold.
8. Disconnect the oxygen sensor.
9. Remove the upper manifold nut holding the engine lifting hook and remove the hook.
10. Continue carefully removing the manifold nuts and remove the manifold from the studs. Remove the gasket.
11. Either have a helper hold the manifold or place wood blocks under the car to support the exhaust system. Don't let the exhaust system support the weight of the manifold. Remove the inner heat shield from the exhaust manifold.
12. Disconnect the manifold from the exhaust pipe. Lift the

ENGINE AND ENGINE OVERHAUL

manifold clear of the engine and remove the exhaust pipe gasket.

13. Check the manifold carefully for cracks and damage. Use a straightedge and feeler gauges to measure the cylinder head face for distortion. Maximum allowable warpage is 0.3mm (0.012 in.).
14. Install the new gasket in position, making sure it is correctly aligned with all the ports and studs.
15. Attach the manifold to the exhaust system using new nuts and a new gasket. Tighten the nuts to 18 ft. lbs.
16. Install the inner heat shield to the manifold and tighten the bolts to 10 ft. lbs.
17. Move the manifold into position against the head and install the nuts finger tight. don't forget to reinstall the engine lifting hook on the correct stud.
18. In an alterating pattern and in at least two steps, tighten the manifold nuts to 12 ft. lbs.
19. Connect the oxygen sensor.
20. Install the outer heat shield on the manifold and tighten the bolts to 10 ft. lbs.
21. Install the secondary air pipe by connecting it loosely to the exhaust manifold, then loosely connecting the bracket to the head. Once it is held in its correct position, tighten the manifold nut to 60 ft. lbs. and the mounting bolt to 8 ft. lbs. Failure to follow this procedure may result in stripped threads or improper sealing at the connection.
22. Connect the air hose to the secondary air pipe. Make certain that the white mark on the hose is up (facing the hood) and that the hose is pushed onto the valve about an 12mm (1 inch). It is important that this hose not be crushed, crimped or twisted.
23. Install the front end cover panel. On air conditioned vehicles, install the condenser fan.
24. Install the alternator.
25. Install the power steering pump mounting bracket, tightening the bolts to 21 ft. lbs.
26. Install the power steering pump and tighten the bolts to 18 ft. lbs.
27. Check all the drive belts and adjust to the correct tension if necessary. Start the engine and check for exhaust leaks.

Sigma

CAUTION

Perform this operation only on a cold motor. The exhaust components and other metal parts will retain heat for a period of hours after the engine is shut off.

FRONT MANIFOLD

1. Disconnect the oxygen sensor wiring. Leave the sensor in the exhaust pipe.
2. Remove the self-locking nuts holding both exhaust pipes to the front and rear manifolds. Remove the gaskets and discard them. Either hang the exhaust pipe from stiff wire or support it from underneath with wooden blocks.
3. Remove the heat shield from the manifold.
4. Remove the retainig nuts and remove the exhaust manifold.
5. Remove the dipstick. Remove the dipstick tube retaining bolt and remove the dipstick tube. Take care not to damage the O-ring when removing the tube.
6. Remove the manifold gasket.
7. Inspect the manifold and head surfaces for cracks or distortion. Use a straightedge and feeler gauge to check the surfaces for warpage. Maximum acceptable warpage is 0.3mm (0.012 in.).

To install:
8. Place a new gasket against the head. Apply a light coat of engine oil to the O-ring on the dipstick tube and install the tube, making certain the tube is properly seated and the O-ring is not damaged during installation. Install the dipstick.
9. Install the manifold. Tighten the all the nuts snug. In two passes and working from the inner nuts outward, tighten the retaining nuts to 14 ft. lbs.
10. Install the heat shield, tightening the bolts to 10 ft. lbs.
11. Using new gaskets and new self-locking nuts, connect the exhaust system to both the front and rear manifolds. Tighten the retaining nuts to 26 ft. lbs.
12. Connect the oxygen sensor wiring.
13. Start the engine and check for exhaust leaks.

REAR MANIFOLD

1. Disconnect the negative battery cable. Disconnect the oxygen sensor wiring. Leave the sensor in the exhaust pipe.
2. Remove the self-locking nuts holding both exhaust pipes to the front and rear manifolds. Remove the gaskets and discard them. Either hang the exhaust pipe from stiff wire or support it from underneath with wooden blocks.
3. Disconnect the air intake tube.
4. Remove the throttle body and its gasket.

WARNING: The bolts holding the throttle body are of two different sizes. Note or diagram their locations for reassembly.

The accelerator and throttle control cables remain attached along with the hoses. Simply unbolt the unit and move it aside.

5. Remove the condenser.
6. Label and disconnect the PCV hose, the vacuum hoses and the brake booster vacuum hose.
7. Remove the EGR valve and gasket and remove the EGR pipe and gasket. Discard the gaskets.
8. Disconnect the ground cable.
9. Remove the two air plenum support brackets.
10. Remove the two retaining bolts in the center of the plenum, then remove the outer nuts.
11. Remove the plenum and its gasket.
12. Remove the heat shield from the rear manifold.
13. Disconnect the upper end of the EGR pipe. The pipe should be removed with the manifold. If a new manifold is to be installed, transfer the pipe on the workbench.
14. Remove the retaining nuts and remove the manifold and its gasket.
15. Inspect the manifold and head surfaces for cracks or distortion. Use a straightedge and feeler gauge to check the surfaces for warpage. Maximum acceptable warpage is 0.3mm (0.012 in.).

To install:
16. Install a new gasket and place the manifold in position. Tighten the nuts snug. In two passes and working from the inner nuts outward, tighten the retaining nuts to 14 ft. lbs.
17. Replace the gasket for the EGR pipe with a new one and install the pipe to the EGR valve.
18. Install the heat shield, tightening the bolts to 10 ft. lbs. Don't forget to install the air plenum support on the right upper bolt.
19. Install a new plenum gasket (use no sealant) and install the plenum. Make certain the printed surface of the gasket faces the intake plenum.
20. Tighten the nuts on the outer edges to 13 ft. lbs., then tighten the inner bolts to 13 ft. lbs.
21. Install the two plenum support brackets, tightening the bolts to 10 ft. lbs.
22. Connect the ground cable to the plenum.
23. Install a new gasket and connect the EGR pipe. Install the EGR valve with a new gasket.
24. Connect the brake booster vacuum line, the small vacuum hoses and the PCV hose.
25. Install the condenser on the side of the plenum.
26. Install the throttle body with a new gasket. Remember

3 ENGINE AND ENGINE OVERHAUL

that the bolts are different sizes. Position them correctly and tighten them to 8 ft. lbs.

27. Connect the air intake hose.

28. Using new gaskets and new self-locking nuts, connect the exhaust system to both the front and rear manifolds. Tighten the retaining nuts to 26 ft. lbs.

29. Connect the oxygen sensor wiring.

30. Connect the negative battery cable; start the engine and check for leaks.

Combination Manifold

NOTE: Although they are individual pieces, the diesel engine intake and exhaust manifolds share the same gasket. Whenever one is removed, the other must also be removed. The gasket must always be replaced.

REMOVAL AND INSTALLATION

Pickup with the 4D55 Diesel Engine

1. Perform this work only on a cold engine. Make sure all surfaces are cool to the touch before beginning.
2. Disconnect the exhaust pipe from the turbocharger and remove the gasket. Remove the heat shield.
3. Remove the bolts holding the inlet fitting to the inlet manifold. Loosen the hose clamps and remove the fitting; remove the rubber connecting hose.
4. Loosen the hose clamps from the oil return line and remove it.
5. Carefully disconnect the oil supply pipe from the turbocharger.
6. Remove the bolts holding the turbocharger to the exhaust flange. Remove the turbocharger and discard the gasket.
7. Immediately after removing the turbocharger, plug the intake and exhaust ports and the oil ports with clean, lint-free cloths or crumpled paper. The turbocharger MUST be protected from dirt and grit; place the unit in a protected location away from the work area.
8. Disconnect the intake manifold retaining bolts and remove the manifold.
9. Disconnect the exhaust manifold retaining bolts and remove the manifold. Remove the gasket(s) and discard. Make certain the mating surfaces are completely free of gasket material.

To install:

10. Install the new gasket(s) and install the manifolds (exhaust first, then intake) into position.
11. Start the retaining nuts and bolts by hand; tighten them to 16 Nm (12 ft. lbs.)
12. Remove the plugs from the turbocharger passages and install the turbocharger with a new gasket onto the exhaust manifold. Tighten the nuts evenly.
13. Connect the oil return line; make certain the clamp is secure.
14. Pour clean engine oil into the oil supply port to pre-lubricate the turbocharger. Install the oil supply line and tighten it carefully.
15. Install the rubber connecting hose on the inlet fitting. Use a new gasket and attach the inlet fitting to the intake manifold.
16. Use a new gasket and install the exhaust pipe to the bottom of the turbocharger. Install the heat shield.
17. Start the engine, listening carefully for any sign of unusual noise from either the manifolds or the turbocharger. Watch the oil pressure indicator for any sign of incorrect pressure.

Turbocharger

A turbocharger is an exhaust-driven turbine which drives a compressor wheel on the other end of the same shaft. The turbine is located in the exhaust flow, generally just below the exhaust manifold. The compressor is located in the intake air path, usually between the air cleaner and the intake manifold. Even though the exhaust and intake air channels are connected by the turbocharger shaft, the exhaust is kept separate from the intake air at all times. (Think of the old water wheel turning the grinding wheel at the mill; the force is transferred but the water never touched the grain.)

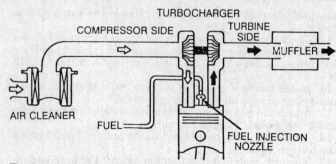

Typical turbocharging schematic

By compressing the intake air, more air is squeezed into the cylinder carrying more oxygen for better combustion. Since more air is introduced, more fuel can be introduced, yielding more power. Turbocharging is one way of coaxing more power out of relatively small engines. A larger engine of the same output would add additional weight, thus reducing overall performance. A turbocharger system is relatively light in comparison, and the additional weight of strengthened components is still well below the weight of a larger motor.

It is possible to get too much of a good thing. Turbocharging is self-perpetuating; that is, as boost (air compression) builds, the exhaust volume builds and the turbine turns faster, providing more boost and so on. If left alone, the turbocharger would build pressure well beyond the operating ability of the engine. To prevent these costly and spectacular failures, boost is held to a reasonable level by a wastegate. Usually located in the output elbow area, the wastegate is a pressure valve which activates at a pre-determined level of pressure. When it opens, it simply allows exhaust flow to bypass the turbine, thus limiting its speed. NEVER attempt to change the setting of the wastegate. If the wastgate or actuator is suspected of faulty operation, replace it.

As the intake air is compressed in the turbocharger, it becomes heated and expands. This expanding air flow is less dense so less air is forced into the engine, partially defeating the purpose of the turbocharger. To overcome this condition, some engines are fitted with an intercooler to remove heat from the air charge. A properly designed intercooler system can reduce air temperature by 90° or more. The intercooler is simply a heat exchanger located between the turbocharger and the intake manifold.

The compressed air charge is directed through ductwork to the intercooler where it is cooled and then on to the intake manifold. The system works in the same fashion as the radiator for the cooling system except that air is being cooled instead of fluid. In some cases the intercooler even looks like a small radiator. The cooled air charge once again becomes dense, introducing more air into the engine and providing more power, greater economy and quiter performance.

Since turbine speeds routinely reach 140,000 rpm, adequate lubrication is absolutely vital. Turbochargers are lubricated by engine oil. Since all parts of the rotating assemblies are protected by a film of oil, no metal-to-metal contact occurs. If a supply of clean, fresh oil is maintained, bearing life should be indefinite. All clearances in the turbocharger are closely controlled and carefully machined. Any dirt in the oil will seriously affect the life of the unit. Oil and filter changes should occur at frequent intervals. The oil filter should ALWAYS be changed with the engine oil. ALWAYS use an oil of the recommended viscosity

ENGINE AND ENGINE OVERHAUL 3

for your particular engine. Check the owner's manual or underhood label for the correct oil. Additionally, periodically check with your dealer for the latest recommendations. New petroleum technology constantly changes and improves available motor oils; the best oil when you bought the car may be old news two years later.

While the turbocharging system requires no special care and feeding (other than good maintenance habits), some general rules do apply to engine operation.

- After starting the engine, make sure there is sufficient oil pressure before accelerating or applying load. Run the engine at low RPM for several minutes to allow the oil to circulate and warm up a bit.
- When starting in cold weather, allow the engine to warm up a minute or two before driving. The very thick oil must be allowed to thin and begin to work.
- Before stopping the engine after driving (particularly after hard driving or high RPM operation) allow the engine to idle for a short length of time. This equalizes temperatures within the engine and cools the oil. Sudden shut-off of the engine will trap hot oil within the very hot turbocharger. The trapped oil undergoes "coking", a process of hardening and cooking out suspended solids. These carbon bits can plug oil passages and ruin bearings.
- If the engine should stall during normal operation on a warmed-up engine, restart it immediately. This prevents the rapid rise of temperatures within the turbocharger and possible mechanical damage to the turbine and compressor.
- If it is necessary to transport a turbocharged engine removed from the car, plug the air passages with rags or tape them shut. This will prevent entry of dirt and dust which can damage the bearings. Additionally, this will prevent the turbine from turning on its bearings while no oil is being delivered.
- Be sensitive to what the engine tells you as you drive. Impaired performance, noise, vibration or smoking can be a sign of impending failure. Periodically check for loose or restricted hoses and fittings. Replace the air filter at frequent intervals.

NOTE: After the engine is shut off, the turbocharger may whine as it runs down. Don't confuse this air whine with bearing failure noise, usually a more mechanical high-pitched sound.

REMOVAL AND INSTALLATION

CAUTION
In all cases, allow the engine to cool at least 3–4 hours before attempting any work on the turbocharger. The exhaust and turbocharger retain heat long after the engine is shut off.

WARNING: Do not disassemble the turbocharger unit after removal. The tolerances within are beyond the ability of even the best home mechanic. It can be checked without disassembly.

When reinstalling the turbocharger, observe the torque specifications exactly. Extreme heat and different expansion rates can crack overtightened components.

Cordia and Tredia

1. Remove the heat shields from the turbocharger and exhaust manifold.
2. Disconnect the oil return line from the turbocharger to the oil pan. Be prepared to contain spillage.
3. Remove the oil supply line from the top of the turbocharger to the oil filter fitting. Be careful when disconnecting the eye bolt (banjo bolt) fittings; they are easily damaged.
4. Carefully disconnect the air intake hoses.
5. Remove the turbocharger retaining nuts and bolts. Re-

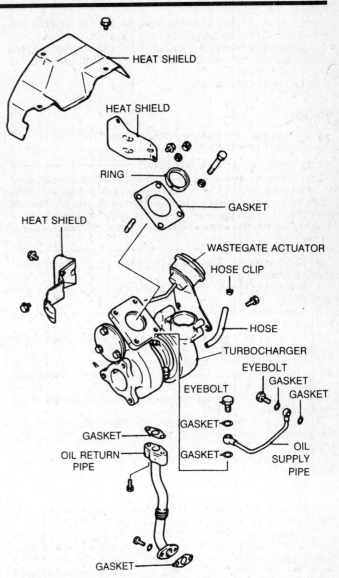

Cordia and Tredia turbocharger external components

move the unit from the exhaust manifold. Handle the unit carefully and keep it free of any foreign matter.

6. Inspect the case for cracks or damage. Check the turbine and compressor wheels for damage. Make sure the wheels turn smoothly without contacting the casing. Gently check the shaft between the wheels for any looseness or wobble. If any play is found, the unit must be replaced.
7. When reinstalling the turbocharger, always replace the gaskets. Both the manifold gasket and the turbocharger gasket (stainless steel) are subject to damage from the extreme heat present in the system.
8. Install the turbocharger. Tighten the bolts holding the turbocharger to the exhaust manifold to 42 ft. lbs. and make certain the locking ring is correctly seated. Tighten the bolts between the turbocharger and the front catalytic converter to 25 ft. lbs.
9. Install the air intake ducts and make sure they are properly seated.
10. Pour a small amount of clean motor oil into the top of the turbocharger. Install the oil supply line between the oil filter bracket and the top of the turbocharger. Always use new gaskets (washers) on either side of the banjo fitting. Tighten the

3-139

3 ENGINE AND ENGINE OVERHAUL

upper eye bolt (at the oil filter bracket) to 12 ft. lbs. and the lower bolt (on the turbocharger) to 22 ft. lbs.

11. Use new gaskets and install the oil return pipe between the oil pan and the bottom of the turbocharger. Tighten the upper and lower bolts to 6 ft. lbs.

12. Install the heat shields.

Starion

1. Disconnect the negative battery cable.
2. Drain the engine coolant.

── **CAUTION** ──
When draining the coolant, keep in mind that cats and dogs are attracted by the ethylene glycol antifreeze, and are quite likely to drink any that is left in an uncovered container or in puddles on the ground. This will prove fatal in sufficient quantity. Always drain the coolant into a sealable container. Coolant should be reused unless it is contaminated or several years old.

3. Remove the large heat shield from the manifold and turbocharger area.
4. Disconnect the secondary air pipe and remove the smaller heat shield.
5. Carefully remove the oxygen sensor. Place it in a protected location and keep the tip absolutely free of grease and dirt.
6. If the engine does not have an intercooler, disconnect and remove the intake air hose and its O-ring.
7. If the engine is intercooled, disconnect air hose A (running from the turbocharger to the intercooler) and the boost hose (running from the intercooler to the engine).
8. Disconnect the air intake hose (from the air cleaner to the turbocharger).
9. Disconnect the catalytic converter from the exhaust system, then disconnect it from the turbocharger. Discard the gaskets.
10. Disconnect the oil supply pipe at both the engine and turbocharger. Make certain that NO dirt or foreign matter enters the oil line or ports. Don't overlook the small bracket in the middle of the pipe.
11. Remove the two water hoses from the turbocharger. Use care in loosening the eye bolts. Discard the small washers.
12. The oil return pipe runs from the turbocharger to a flexible hose. Loosen the clamp where the pipe meets the hose and separate the pipe and hose. If this cannot be done easily, don't deform the hose. Remove the bolt holding the oil return pipe bracket to the engine.
13. Disconnect the turbocharger from the exhaust manifold. Lift the turbocharger clear of the engine. If the oil return pipe is still connected to the flexible hose, some upward tension will separated them. Remove the turbocharger with the oil return pipe attached; it may be removed on the workbench if necessary. The gasket and sealing ring must be removed at the manifold joint.
14. Inspect the case for cracks or damage. Check the turbine and compressor wheels for damage. Make sure the wheels turn smoothly without contacting the casing. Gently check the shaft between the wheels for any looseness or wobble. If any play is found, the unit must be replaced. Check the oil supply and return pipes for any clogging, collapse or deformation. Keep everything very clean.
15. Before reinstalling the turbocharger, the oil return pipe and a new gasket must be connected to it with the bolts tightened to 72 inch lbs. Use a new gasket and install the turbocharger and sealing ring to the manifold. Tighten the nuts to 44 ft. lbs.
16. Connect the oil return pipe to the flexible hose and secure the clamp on the joint.
17. Using new washers, connect the water lines to the turbocharger, tightening the eye bolts to 30 ft. lbs.
18. Attach the oil supply line to the engine and tighten the fit-

1. Heat shield
2. Secondary air pipe
3. Heat shield
4. Oxygen sensor
5. Air intake pipe (without intercooler)
6. O-ring
7. Air hose A (with intercooler)
8. Boost hose (with intercooler)
9. Air intake hose (with intercooler)
10. Catalytic converter
11. Gasket
12. Oil supply pipe
13. Nut
14. Gasket
15. Water pipe
16. Water pipe
17. Oil hose
18. Turbocharger
19. Oil return pipe
20. Gasket
21. Gasket
22. Sealing ring

Starion turbocharger removal

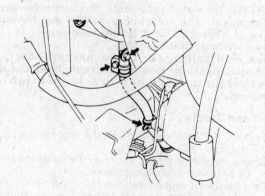

Remove the Starion oil return pipe bracket and disconnect the clamp at the hose

ENGINE AND ENGINE OVERHAUL 3

ting to 15 ft. lbs. Before attaching the upper fitting at the turbocharger, pour a small amount of clean motor oil into the oil passage in the turbocharger.

19. Using new gaskets, install the catalytic converter. Tighten the bolts at the turbocharger snug with a wrench, then tighten the bolts at the exhaust pipe snug with a wrench. When both are in place, tighten the bolts between the converter and turbocharger to 44 ft. lbs. and the converter–to–exhaust bolts to 26 ft. lbs.
20. Install the air intake hose from the air cleaner.
21. On intercooled vehicles, connect the boost hose and air hose A to their proper positions. Do not overtighten the hose clamps; 36–48 inch lbs. is enough.
22. On vehicles without an intercooler, install the air intake hose with a new O-ring and tighten the retaining bolt to 8 ft. lbs.
23. Install the oxygen sensor.
24. Use a new gasket and install the secondary air pipe, tightening the nuts to 20 ft. lbs.
25. Install the heat shields, tightening the larger bolts to 12 ft. lbs. and the smaller ones to 6 ft. lbs.

Mirage with G32B Engine

1. Remove the exhaust manifold. This is a separate operation with several steps. Please refer to the Exhaust Manifold section of this Section.
2. With the manifold removed, disconnect the air intake hose from the turbocharger.
3. Disconnect the oxygen sensor wiring.
4. Disconnect the water hose from water pipe A and disconnect water pipe B from the turbocharger.
5. Remove the eye bolt from each end of the oil supply line and disconnect the small bracket supporting the line. Remove the line.
6. Remove the bolts holding the oil return line to the engine. Remove the gasket and discard it.
7. Remove the nuts and bolts holding the exhaust pipe to the bottom of the turbocharger. Separate the pipe and discard the gasket. Remove the small heat shield from the turbocharger.
8. Remove the turbocharger assembly. Note that it will be coming off with various lines still attached. If the turbocharger is to be replaced, these items must be transferred to the new unit. If removed only for inspection, the lines and sensors should be left in place.
9. Inspect the case for cracks or damage. Check the turbine and compressor wheels for damage. Make sure the wheels turn smoothly without contacting the casing. Gently check the shaft between the wheels for any looseness or wobble. If any play is found, the unit must be replaced. Check the oil supply and return pipes for any clogging, collapse or deformation. Keep everything very clean.
10. To reinstall, mount the turbocharger onto the exhaust pipe using new self-locking nuts and a new gasket. Tighten the nuts to 26 ft. lbs. Install the heat shield, tightening its bolt to 10 ft. lbs.
11. Install a new gasket and connect the oil return pipe to the engine. Tighten the bolts to 6 ft. lbs. If the pipe was removed from the turbocharger, those mounting bolts should also be tightened to 6 ft. lbs.
12. Using new washers, connect water pipe B to the turbocharger, tightening the eye bolt to 24 ft. lbs. If the other end of this tube was disconnected, apply a thin coating of machine oil to the flange of the tube before reinstalling it. Tighten the threaded fitting to 34 ft. lbs.
13. Install the oil supply tube. Tighten the engine fitting to 12 ft. lbs. Pour a small amount of clean, fresh motor oil into the oil port on the turbocharger, then connect the line and tighten the fitting to 12 ft. lbs.
14. Connect water hose A to its pipe. If the pipe was disconnected from the turbocharger, it should be reinstalled with new gaskets and tightened to 24 ft. lbs.

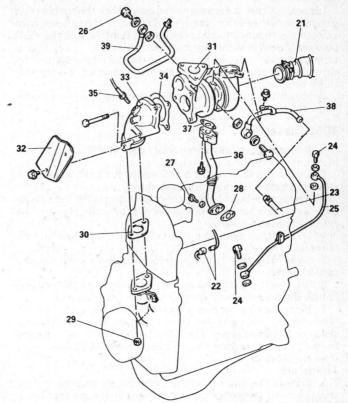

21. Air intake hose
22. Oxygen sensor wiring connector
23. Water hose connection
24. Eye bolts
25. Oil supply line
26. Eye bolt
27. Bolt
28. Gasket
29. Self-locking nut
30. Gasket
31. Turbocharger assembly
32. Heat shield
33. Exhaust fitting (elbow)
34. Gasket
35. Oxygen sensor
36. Oil return pipe
37. Gasket
38. Water pipe A
39. Water pipe B

G32B turbocharger and related components

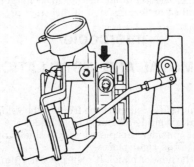

Pour a small amount of clean engine oil into the turbocharger before reinstalling the oil supply pipe

15. Connect the wiring connector for the oxygen sensor.
16. Connect the air intake hose.
17. Reinstall the exhaust manifold.

Mirage with 4G61 DOHC Engine

Refer to the Exhaust Manifold section in this Section. After the manifold is removed, the turbocharger is easily removed by disconnecting the remaining ductwork and lines.

3 ENGINE AND ENGINE OVERHAUL

Inspect the case for cracks or damage. Check the turbine and compressor wheels for damage. Make sure the wheels turn smoothly without contacting the casing. Gently check the shaft between the wheels for any looseness or wobble. If any play is found, the unit must be replaced. Check the oil supply and return pipes for any clogging, collapse or deformation. Keep everything very clean.

Reinstall the unit following the directions for the exhaust manifold.

4D55 Diesel

1. Perform this work only on a cold engine. Make sure all surfaces are cool to the touch before beginning.
2. Disconnect the exhaust pipe from the turbocharger and remove the gasket. Remove the heat shield.
3. Remove the bolts holding the inlet fitting to the inlet manifold. Loosen the hose clamps and remove the fitting; remove the rubber connecting hose.
4. Loosen the hose clamps from the oil return line and remove it.
5. Carefully disconnect the oil supply pipe from the turbocharger.
6. Remove the bolts holding the turbocharger to the exhaust flange. Remove the turbocharger and discard the gasket.
7. Immediately after removing the turbocharger, plug the intake and exhaust ports and the oil ports with clean, lint-free cloths or crumpled paper. The turbocharger MUST be protected from dirt and grit; place the unit in a protected location away from the work area.

To install:

8. Remove the plugs from the turbocharger pasages and install the turbocharger with a new gasket onto the exhaust manifold. Tighten the nuts evenly.
9. Connect the oil return line; make certain the clamp is secure.
10. Pour clean engine oil into the oil supply port to pre-lubricate the turbocharger. Install the oil supply line and tighten it carefully.
11. Install the rubber connecting hose on the inlet fitting. Use a new gasket and attach the inlet fitting to the intake manifold.
12. Use a new gasket and install the exhaust pipe to the bottom of the turbocharger. Install the heat shield.
13. Start the engine, listening carefully for any sign of unusual noise from either the manifolds or the turbocharger. Watch the oil pressure indicator for any sign of incorrect pressure.

Intercooler

REMOVAL AND INSTALLATION

Starion

1. Elevate and safely support the front of the vehicle. Remove the lower front air guide assembly from under the front of the car. You might consider this piece a splash shield, but it's there to force air into the right places.
2. Remove the header panel. It's the piece in front of the hood and between the headlights. It is held in place by 4 screws; two can be reached through the grille at the outer edges and the other two are found under the hood.
3. Remove air hose **A** and then air hose **D**.

NOTE: It is strongly recommended that the air hoses and ducts be labled or diagrammed before removing them. The six or seven pieces can present quite a puzzle at reassembly.

4. Label and disconnect the vacuum harness hoses (from the air cleaner) at the turbocharger. Earlier cars may not have this harness.
5. Disconnect the air intake hose at the turbocharger.

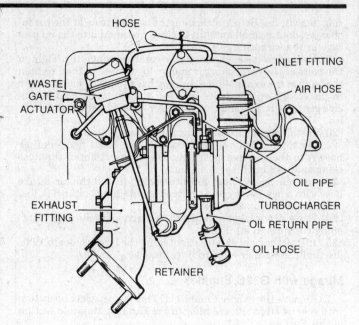

Turbocharger fittings and installation

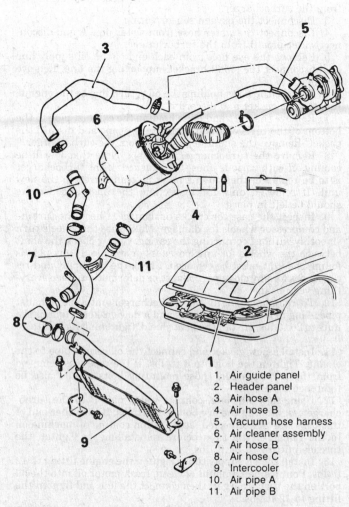

1. Air guide panel
2. Header panel
3. Air hose A
4. Air hose B
5. Vacuum hose harness
6. Air cleaner assembly
7. Air hose B
8. Air hose C
9. Intercooler
10. Air pipe A
11. Air pipe B

Starion intercooler. 1989 model shown; earlier cars may differ slightly

ENGINE AND ENGINE OVERHAUL 3

6. Remove the air cleaner assembly.
7. Remove air hose **B**, then remove air hose **C**.
8. Remove the bolts holding the right side intercooler bracket to the bodywork.
9. Remove the bolts holding the intercooler to the left and right brackets.
10. Carefully remove the intercooler from below. Note that some air pipes are still connected. Remove air pipe A and B from the unit after it is removed. Check the intercooler fins for bending, damage or foreign matter which could block the air flow. Check the hoses for cracking or damage.
11. When reinstalling the intercooler, place it into the car and tighten the retaining bolts, including the right mount bolts. Make sure the unit goes back in with air pipes A and B attached.

NOTE: The air hoses and pipes are marked for correct alignment. Look for a straight line either raised or indented in the plastic and align it with the painted dot on the hoses. This insures correct routing of the duct system. The correct tightness for the hose clamps connecting the pieces is 36 inch lbs.

12. Install air hoses **B** and **C**.
13. Install the air cleaner and connect the vacuum harness to the turbocharger. Make certain the hoses are correctly connected. Attach the air intake hose to the turbocharger.
14. Connect air hoses **A** and **D**.
15. Install the header panel, taking care to align it correctly with the surrounding body work.

Air Conditioning Compressor

REMOVAL AND INSTALLATION

———————— CAUTION ————————
Please re-read the air conditioning part of Section 1, so that the system can be discahrged safely. Always wear eye protection and gloves when discharging the system. Don't smoke when handling refrigerant!.

All Cordia and Tredia
Pickup with G63B Engine
Mirage with SOHC 4G15 Engine
Mirage with G15B Engine
Mirage with G32B Engine
Galant with G64B Engine
All Precis

1. Disconnect the negative battery cable.
2. Safely discharge the air conditioning system. See Section 1.
3. Disconnect the distributor cap from the distributor and move the cap (with the wires attached) to a safe location.
4. Loosen the belt tension adjuster (idler pulley) and remove the belt. It should not be necessary to remove this pulley if only the compressor is to be removed.
5. Label and disconnect the wiring running from the compressor. Some models have more than one wire connector. Check carefully.
6. Disconnect the discharge hose and the suction hose from the compressor. As soon as the lines are free, remove the small O-rings inside the hoses and discard them. Immediately plug or cover each open line and port to prevent the entry of any dirt.
7. Identify the upper bolts holding the compressor to the engine bracket. On some models, these bolts are near the hose ports; on others the bolts are either straight through the compressor body or parallel to the body (pivot bolt). Remove the upper bolts.
8. Support the compressor and remove the lower mounting bolts. Be prepared to hold the compressor as the last bolt comes free. Note the mounting bolts may be of different lengths; they

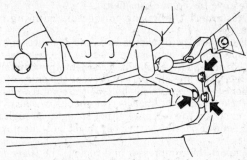

Disconnect the right bracket from the bodywork and the intercooler from the bracket

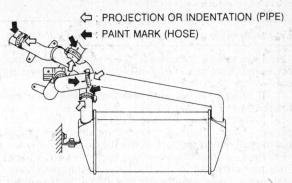

For proper hose routing, observe the alignment marks during reassembly

must be reinstalled correctly.

———————— CAUTION ————————
The compressor is a heavy component; be ready for the weight.

9. Remove the compressor, keeping it roughly level as it comes out. Excessive tilting will allow the compressor oil to run out.
10. If related repairs are needed, the compressor mounting bracket(s) and adjuster pulley and/or bracket may be removed.
11. Reinstall the compressor to its bracket and install the upper and lower bolts finger tight. Once all are snug, tighten the mounting bolts to the torque specification shown in the chart.
12. Replace the O-rings in the suction and discharge hoses with new ones and install the hoses to the compressor. Make certain the hoses are properly seated and the threaded connectors are properly engaged. Tighten the fittings to 18 ft. lbs.
13. Proceed with reassembly following the preliminary procedure for your car in reverse order. Consult the chart for torque specifications.
14. Evacuate the system and recharge it according to procedures in Section 1.
15. Adjust the compressor belt and other drive belts as necessary.

Mirage with 4G61 Engine
Galant with 4G63 SOHC and DOHC Engines

1. Disconnect the negative battery cable.
2. Safely discharge the air conditioning system. See Section 1.
3. Loosen the alternator and remove the alternator drive belt.
4. Loosen the adjuster bolt to relieve tension on the compressor drive belt.
5. Remove the bolts holding the adjuster pulley bracket to the engine and compressor bracket. Remove the entire adjuster assembly on its bracket.

3-143

3 ENGINE AND ENGINE OVERHAUL

6. Remove the compressor belt.
7. Disconnect the wiring connectors running from the compressor.
8. Disconnect the discharge hose and the suction hose from the compressor. As soon as the lines are free, remove the small O-rings inside the hoses and discard them. Immediately plug or cover each open line and port to prevent the entry of any dirt.
9. Identify the upper bolts holding the compressor to the engine bracket. On some models, these bolts are near the hose ports; on others the bolts are either straight through the compressor body or parallel to the body (pivot bolt). Remove the upper bolts.
10. Support the compressor and remove the lower mounting bolts. Be prepared to hold the compressor as the last bolt comes free. Note the mounting bolts may be of different lengths; they must be reinstalled correctly.

CAUTION
The compressor is a heavy component; be ready for the weight!

11. Remove the compressor, keeping it roughly level as it comes out. Excessive tilting will allow the compressor oil to run out.
12. If related repairs are needed, the compressor mounting bracket(s) and adjuster pulley and/or bracket may be removed.
13. Reinstall the compressor to its bracket and install the upper and lower bolts finger tight. Once all are snug, tighten the mounting bolts to the torque specification shown in the chart.
14. Replace the O-rings in the suction and discharge hoses with new ones and install the hoses to the compressor. Make certain the hoses are properly seated and the threaded connectors are properly engaged. Tighten the fittings to 18 ft. lbs.
15. Proceed with reassembly following the preliminary procedure for your car in reverse order. Consult the chart for torque specifications.
16. Evacuate the system and recharge it according to procedures in Section 1.
17. Adjust the compressor belt and other drive belts as necessary.

Starion
Pickup and Montero with G54B Engine

1. Disconnect the negative battery cable.
2. Safely discharge the air conditioning system. See Section 1.
3. Remove the ignition coil and move it to a safe location. The wiring may be left intact. If you choose to remove the wiring, be certain to label each wire.
4. Loosen the belt tension adjuster (idler pulley) and remove the belt. It should not be necessary to remove this pulley if only the compressor is to be removed.
5. Label and disconnect the wiring running from the compressor. Some models have more than one wire connector. Check carefully.
6. Disconnect the discharge hose and the suction hose from the compressor. As soon as the lines are free, remove the small O-rings inside the hoses and discard them. Immediately plug or cover each open line and port to prevent the entry of any dirt.
7. Identify the upper bolts holding the compressor to the engine bracket. On some models, these bolts are near the hose ports; on others the bolts are either straight through the compressor body or parallel to the body (pivot bolt). Remove the upper bolts.
8. Support the compressor and remove the lower mounting bolts. Be prepared to hold the compressor as the last bolt comes free. Note the mounting bolts may be of different lengths; they must be reinstalled correctly.

CAUTION
The compressor is a heavy component; be ready for the weight.

9. Remove the compressor, keeping it roughly level as it comes out. Excessive tilting will allow the compressor oil to run out.
10. If related repairs are needed, the compressor mounting bracket(s) and adjuster pulley and/or bracket may be removed.
11. Reinstall the compressor to its bracket and install the upper and lower bolts finger tight. Once all are snug, tighten the mounting bolts to the torque specification shown in the chart.
12. Replace the O-rings in the suction and discharge hoses with new ones and install the hoses to the compressor. Make certain the hoses are properly seated and the threaded connectors are properly engaged. Tighten the fittings to 18 ft. lbs.
13. Proceed with reassembly following the preliminary procedure for your car in reverse order. Consult the chart for torque specifications.
14. Evacuate the system and recharge it according to procedures in Section 1.
15. Adjust the compressor belt and other drive belts as necessary.

Galant, Sigma and Montero with V6 6G72

1. Disconnect the negative battery cable.
2. Remove the air cleaner assembly.
3. Safely discharge the air conditioning system. See Section 1.
4. Loosen the belt tension adjuster (idler pulley) and remove the belt.
5. Label and disconnect the wiring running from the compressor. Some models have more than one wire connector. Check carefully.
6. Disconnect the discharge hose and the suction hose from the compressor. As soon as the lines are free, remove the small O-rings inside the hoses and discard them. Immediately plug or cover each open line and port to prevent the entry of any dirt.
7. Identify the upper bolts holding the compressor to the engine bracket. On some models, these bolts are near the hose ports; on others the bolts are either straight through the compressor body or parallel to the body (pivot bolt). Remove the upper bolts.
8. Support the compressor and remove the lower mounting bolts. Be prepared to hold the compressor as the last bolt comes free. Note the mounting bolts may be of different lengths; they must be reinstalled correctly.

CAUTION
The compressor is a heavy component; be ready for the weight!

9. Remove the compressor, keeping it roughly level as it comes out. Excessive tilting will allow the compressor oil to run out.
10. If related repairs are needed, the compressor mounting bracket(s) and adjuster pulley and/or bracket may be removed.
11. Reinstall the compressor to its bracket and install the upper and lower bolts finger tight. Once all are snug, tighten the mounting bolts to the torque specification shown in the chart.
12. Replace the O-rings in the suction and discharge hoses with new ones and install the hoses to the compressor. Make certain the hoses are properly seated and the threaded connectors are properly engaged. Tighten the fittings to 18 ft. lbs.
13. Proceed with reassembly following the preliminary procedure for your car in reverse order. Consult the chart for torque specifications.
14. Evacuate the system and recharge it according to procedures in Section 1.
15. Adjust the compressor belt and other drive belts as necessary.

Pickup with 4D55 Diesel

1. Disconnect the negative battery cable.
2. Remove the air cleaner assembly.
3. Safely discharge the air conditioning system. See Section 1.

ENGINE AND ENGINE OVERHAUL 3

4. Loosen the belt tension adjuster (idler pulley) and remove the belt.
5. Label and disconnect the wiring running from the compressor. Some models may have more than one wire connector. Check carefully.
6. Loosen the tensioner (idler) pulley and remove the compressor drive belt.
7. Disconnect the high and low pressure lines.
8. Remove the compressor mounting bolts. Support the compressor securely during removal; it is heavy.
9. To reinstall, place the compressor in position and install the mounting bolts.
10. Install the high and low pressure lines to their correct ports. Apply a thin coat of clean engine oil to the threads and to the O-ring before installation.
11. Install the belt; adjust and tighten the idler pulley.
12. Connect the wiring harness to the compressor.
13. Connect the negative battery cable.
14. Evacuate and recharge the system, keeping a close watch for leaks.

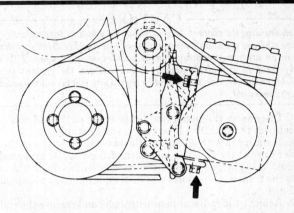

Location of 4D55 compressor mounting bolts

Radiator

The radiator is nothing more than a large heat exchanger. It is mounted so that the airflow at the front of the car is forced through the fins of the unit, carrying heat away from the engine coolant flowing through the unit. The fan(s) supplement the airflow by drawing in cool air, thus providing cooling even when the car is not moving.

Because of the need for good air flow, modern radiators and the fans have shrouding or ducting to guide the air through the fins. This duct work, including undercar covers and shields, must be in place for proper cooling. Leaving the shrouds and covers off can reduce cooling efficiency and reduce driveability.

Periodically, check the radiator surfaces for blockage by leaves, insects, or mud. Most debris can be removed by hand, and the force of a water hose can be useful in dislodging other items. When cleaning the radiator fins, don't use anything metallic or sharp; the fins are very thin and are easily bent or punctured. Generally, the only times a radiator must be removed are either for repair of a leak or to allow access to other components.

It should be noted that most radiators are mounted to rubber bushings rather than directly to the bodywork. This allows the unit to serve as a vibration damper while the engine is running. The mounts and bushings must be properly reinstalled. Replace the rubber bushings if they show signs of wear or lack of flexibility.

REMOVAL AND INSTALLATION

— **CAUTION** —
Always perform this work on a fully cool engine. The coolant will retain heat and pressure for several hours after use. Never remove the radiator cap or hoses if the radiator feels hot or warm to the touch. Liquid under pressure can turn to steam instantly when released; severe scalding can result.

Before beginning any work, make certain the ignition is OFF and unplug the connector(s) to the electric fan(s).

Cordia and Tredia

1. Set the heater temperature control to HOT inside the car.
2. Using a large capacity container, loosen the radiator drain plug and drain the cooling system. It will drain quicker with the radiator cap removed.

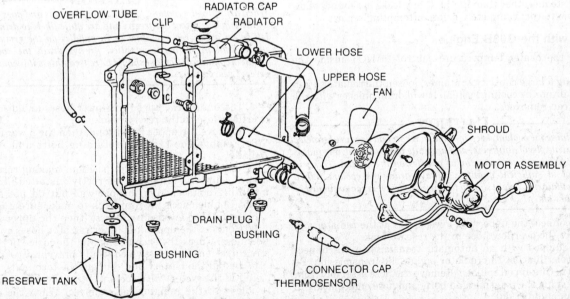

Cordia and Tredia radiator and related components. The reserve tank is on the opposite side if the engine is not turbocharged

3 ENGINE AND ENGINE OVERHAUL

CAUTION

When draining the coolant, keep in mind that cats and dogs are attracted by the ethylene glycol antifreeze, and are quite likely to drink any that is left in an uncovered container or in puddles on the ground. This will prove fatal in sufficient quantity. Always drain the coolant into a sealable container. Coolant should be reused unless it is contaminated or several years old.

---

3. Disconnect the upper and lower hoses at the radiator and disconnect the overflow hose to the reserve tank.
4. If equipped with an automatic transaxle, disconnect the oil cooler lines at the radiator and at the transaxle. Be prepared to contain oil spillage. Remove the hose assembly from the car; plug the transmission ports and hose ends quickly to keep oil in and dirt out.
5. Remove the radiator mounting bolts and remove the radiator with the fan and motor attached.
6. Reinstall the radiator in position, making certain all the mounts and bushings are correctly installed. Tighten the mounting bolts enough to hold well but do not overtighten them. Double check the draincock to make sure it is closed.
7. Reassemble the automatic transaxle oil cooler lines and install the retaining bracket. Make certain the hoses are properly routed and firmly attached at both ends.
8. Connect the upper and lower radiator hoses and the overflow hose. Check the each fitting and the inside of the hose for any corrosion or debris which would prevent a good seal.
9. Fill the system with coolant. Connect the fan wiring.
10. Start the engine and allow it to idle with the radiator cap removed. When the engine has warmed enough, the thermostat will open and water flow will be visible within the radiator. Fill the coolant to the bottom of the radiator neck and install the cap.

CAUTION

Do not lean over the radiator neck while the engine is running. Hot coolant may be splashed out by air trapped within the system. Keep hands, tools and clothing away from the fans which may engage at any time as the coolant heats up.

---

11. Allow the engine to warm up fully and check that the fans cycle on and off correctly.
12. Shut the engine off and check the hose connections carefully for leaks. With the engine fully warm, a small leak may be emitting steam rather than liquid. If any leaks are found, allow the engine to cool completely before attempting repairs.

Pickup with the G63B Engine

1. Set the heater temperature control to HOT inside the truck.
2. Using a large capacity container, loosen the radiator drain plug and drain the cooling system. It will drain quicker with the radiator cap removed.

CAUTION

When draining the coolant, keep in mind that cats and dogs are attracted by the ethylene glycol antifreeze, and are quite likely to drink any that is left in an uncovered container or in puddles on the ground. This will prove fatal in sufficient quantity. Always drain the coolant into a sealable container. Coolant should be reused unless it is contaminated or several years old.

---

3. Disconnect the upper and lower hoses at the radiator and disconnect the overflow hose to the reserve tank.
4. If equipped with an automatic transmission, disconnect the oil cooler lines at the radiator and at the transmission. Be prepared to contain oil spillage. Remove the hose assembly from the truck; plug the transmission ports and hose ends quickly to keep oil in and dirt out.
5. Remove the radiator mounting bolts and lift out the radiator.

To install:
6. Position the radiator in the truck, making certain all the mounts and bushings are correctly installed. Tighten the mounting bolts enough to hold well but do not overtighten them. Double check the draincock to make sure it is closed.
7. Reassemble the automatic transmission oil cooler lines and install the retaining bracket. Make certain the hoses are properly routed and firmly attached at both ends.
8. Connect the upper and lower radiator hoses and the overflow hose. Check each fitting and the inside of the hose for any corrosion or debris which would prevent a good seal.
9. Fill the system with coolant.
10. Start the engine and allow it to idle with the radiator cap removed. When the engine has warmed enough, the thermostat will open and water flow will be visible within the radiator. Fill the coolant to the bottom of the radiator neck and install the cap.

CAUTION

Do not lean over the radiator neck while the engine is running. Hot coolant may be splashed out by air trapped within the system. Keep hands, tools and clothing away from the fans which may engage at any time as the coolant heats up.

---

11. Shut the engine off and check the hose connections carefully for leaks. With the engine fully warm, a small leak may be emitting steam rather than liquid. If any leaks are found, allow the engine to cool completely before attempting repairs.

Starion, Without Intercooler
Pickup and Montero with G54B Engine

NOTE: The fan is belt driven; there is no electrical connector for it.

1. Set the heater temperature control to HOT inside the vehicle.
2. Disconnect and remove the battery.
3. Using a large capacity container, loosen the radiator drain plug and drain the cooling system. It will drain quicker with the radiator cap removed.

CAUTION

When draining the coolant, keep in mind that cats and dogs are attracted by the ethylene glycol antifreeze, and are quite likely to drink any that is left in an uncovered container or in puddles on the ground. This will prove fatal in sufficient quantity. Always drain the coolant into a sealable container. Coolant should be reused unless it is contaminated or several years old.

---

4. Disconnect the upper and lower hoses and disconnect the overflow hose to the reserve tank.
5. Remove the upper fan shroud, then the lower fan shroud.
6. Remove the radiator retaining bolts and remove the radiator.
7. Reinstall the radiator in position, making certain all the mounts and bushings are correctly installed. Tighten the mounting bolts enough to hold well but do not overtighten them. Double check the draincock to make sure it is closed.
8. Install the lower shroud first, then the upper shroud.
9. Connect the upper and lower radiator hoses and the overflow hose. Check the each fitting and the inside of the hose for any corrosion or debris which would prevent a good seal.
10. Install the battery and connect the terminals.
11. Fill the system with coolant. Start the engine and allow it to idle with the radiator cap removed. When the engine has warmed enough, the thermostat will open and water flow will be visible within the radiator. Fill the coolant to the bottom of the radiator neck and install the cap.

ENGINE AND ENGINE OVERHAUL 3

CAUTION

Do not lean over the radiator neck while the engine is running. Hot coolant may be splashed out by air trapped within the system. Keep hands, tools and clothing away from the fan.

12. Shut the engine off and check the hose connections carefully for leaks. With the engine fully warm, a small leak may be emitting steam rather than liquid. If any leaks are found, allow the engine to cool completely before attempting repairs.

Starion, with Intercooler

NOTE: **The electric fan connectors have waterproof caps over them. Make sure they are reinstalled.**

1. Set the heater temperature control to HOT inside the car.
2. Disconnect and remove the battery.
3. Using a large capacity container, loosen the radiator drain plug and drain the cooling system. It will drain quicker with the radiator cap removed.

CAUTION

When draining the coolant, keep in mind that cats and dogs are attracted by the ethylene glycol antifreeze, and are quite likely to drink any that is left in an uncovered container or in puddles on the ground. This will prove fatal in sufficient quantity. Always drain the coolant into a sealable container. Coolant should be reused unless it is contaminated or several years old.

4. Disconnect the upper and lower hoses and disconnect the overflow hose to the reserve tank.
5. Disconnect the wiring to the thermosensors at the bottom of the radiator.
6. Remove the radiator retaining bolts and remove the radiator.
7. If the fan and shroud assemblies have been removed, they must be reinstalled before installing the radiator. Make certain the ground wires for the fan motors are properly attached under the mounting bolts. The mounting bolts for the fans should be tightened to 10 ft. lbs. If the thermosensors were removed, they

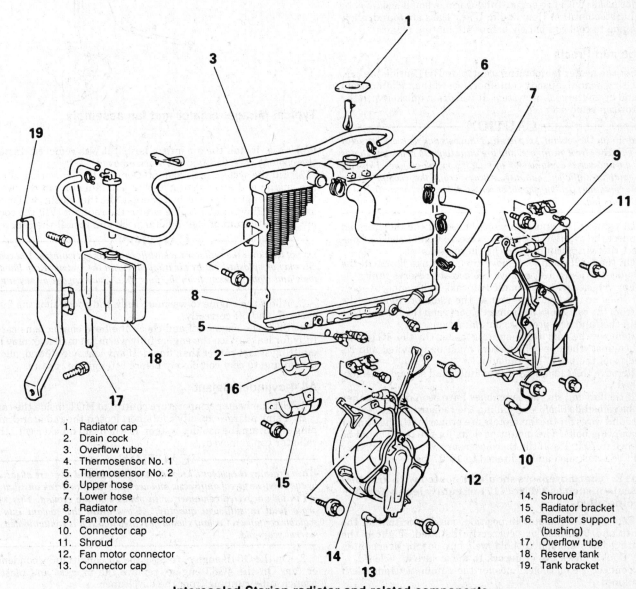

1. Radiator cap
2. Drain cock
3. Overflow tube
4. Thermosensor No. 1
5. Thermosensor No. 2
6. Upper hose
7. Lower hose
8. Radiator
9. Fan motor connector
10. Connector cap
11. Shroud
12. Fan motor connector
13. Connector cap
14. Shroud
15. Radiator bracket
16. Radiator support (bushing)
17. Overflow tube
18. Reserve tank
19. Tank bracket

Intercooled Starion radiator and related components

3-147

3 ENGINE AND ENGINE OVERHAUL

should be reinstalled and tightened to 10 ft. lbs.

8. Reinstall the radiator in position, making certain all the mounts and bushings are correctly installed. Tighten the mounting bolts enough to hold well but do not overtighten them. Double check the draincock to make sure it is closed.
9. Connect the thermosensor wiring.
10. Connect the upper and lower radiator hoses and the overflow hose. Check the each fitting and the inside of the hose for any corrosion or debris which would prevent a good seal.
11. Install the battery and connect the terminals.
12. Fill the system with coolant. Connect the fan wiring. Start the engine and allow it to idle with the radiator cap removed. When the engine has warmed enough, the thermostat will open and water flow will be visible within the radiator. Fill the coolant to the bottom of the radiator neck and install the cap.

CAUTION

Do not lean over the radiator neck while the engine is running. Hot coolant may be splashed out by air trapped within the system. Keep hands, tools and clothing away from the fans which may engage at any time.

13. Shut the engine off and check the hose connections carefully for leaks. With the engine fully warm, a small leak may be emitting steam rather than liquid. If any leaks are found, allow the engine to cool completely before attempting repairs.

Mirage and Precis

1. Set the heater temperature control to HOT inside the car.
2. Using a large capacity container, loosen the radiator drain plug and drain the cooling system. It will drain quicker with the radiator cap removed.

CAUTION

When draining the coolant, keep in mind that cats and dogs are attracted by the ethylene glycol antifreeze, and are quite likely to drink any that is left in an uncovered container or in puddles on the ground. This will prove fatal in sufficient quantity. Always drain the coolant into a sealable container. Coolant should be reused unless it is contaminated or several years old.

3. Disconnect the upper and lower hoses at the radiator and disconnect the overflow hose to the reserve tank. Label and disconnect the wiring running to the thermosensors.
4. On the G32B engine, remove the air intake hose. On the 4G15 and 4G61 engines, remove the coolant reserve tank.
5. For vehicles with automatic transaxles, disconnect the oil cooler lines at the transaxle and at the radiator. Remove the lines from the car. Plug the transaxle ports and the hose ends to contain the fluid and prevent contamination.
6. Remove the radiator mounting bolts. On the 4G15 and 4G61 engines, the upper insulators must be unbolted and removed from the top of the radiator.
7. Remove the radiator with the fan(s) and shroud assembly attached.
8. If the fan and shroud assemblies have been removed, they must be reinstalled before installing the radiator. Make certain the ground wires for the fan motors are properly attached under the mounting bolts. The mounting bolts for the fans should be tightened to 10 ft. lbs. If the thermosensors were removed, they should be reinstalled and tightened to 10 ft. lbs.

NOTE: Thermosensors should be coated with an adhesive sealant such as 3M No. 4171 or equivalent before being reinstalled.

9. Reinstall the radiator in position, making certain all the mounts and bushings are correctly installed. Tighten the mounting bolts enough to hold well but do not overtighten them. Double check the draincock to make sure it is closed.
10. Connect the oil cooler lines for the automatic transaxle if so equipped.
11. Connect the upper and lower radiator hoses and the over-

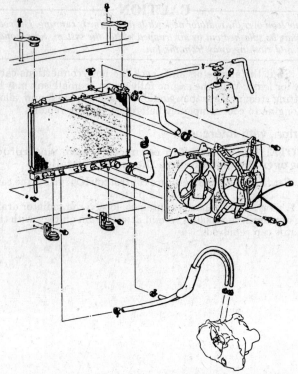

Typical Mirage radiator and fan assembly

flow hose. Install the air intake hose if it was removed. Install the coolant reserve tank if it was removed.
12. Fill the system with coolant. Connect the fan wiring. Start the engine and allow it to idle with the radiator cap removed. When the engine has warmed enough, the thermostat will open and water flow will be visible within the radiator. Fill the coolant to the bottom of the radiator neck and install the cap.

CAUTION

Do not lean over the radiator neck while the engine is running. Hot coolant may be splashed out by air trapped within the system. Keep hands, tools and clothing away from the fans which may engage at any time.

13. Allow the engine to warm up fully and check that the fans cycle on and off correctly.
14. Shut the engine off and check the hose connections carefully for leaks. With the engine fully warm, a small leak may be emitting steam rather than liquid. If any leaks are found, allow the engine to cool completely before attempting repairs.

All 4-cylinder Galant

1. Set the heater temperature control to HOT inside the car.
2. Using a large capacity container, loosen the radiator drain plug and drain the cooling system. It will drain quicker with the radiator cap removed.

CAUTION

When draining the coolant, keep in mind that cats and dogs are attracted by the ethylene glycol antifreeze, and are quite likely to drink any that is left in an uncovered container or in puddles on the ground. This will prove fatal in sufficient quantity. Always drain the coolant into a sealable container. Coolant should be reused unless it is contaminated or several years old.

3. On the G64B engine, remove the air conditioning condenser fan. On the 4G63 engine, remove the bracket and plastic branch tube running from the air cleaner.
4. Disconnect the overflow tube and remove the coolant re-

3-148

ENGINE AND ENGINE OVERHAUL 3

serve tank. Tank shape, location and mounting will vary by model.

5. Disconnect the upper radiator hose.

NOTE: It is recommended that each clamp be matchmarked to the hose. Observe the marks and reinstall the clamps exactly when reinstalling the radiator.

6. Label and disconnect the wiring to the thermosensors and the electric fan assemblies.
7. For vehicles with automatic transaxles, disconnect the oil cooler lines at the transaxle and at the radiator. Remove the lines from the car. Plug the transaxle ports and the hose ends to contain the fluid and prevent contamination.
8. On G64B engines, remove the cooling fan assembly.
9. Remove the lower radiator hose.
10. Remove the bolts holding the upper mounting brackets to the bodywork. Lift the radiator clear of the car with the brackets attached. 4G63 radiators come out with the fans still attached as well.
11. If the fan and shroud assemblies were removed with the radiator, they must be reinstalled before installing the radiator. The mounting bolts for the fans should be tightened to 10 ft. lbs. If the thermosensors were removed, they should be reinstalled and tightened to 10 ft. lbs.
12. Reinstall the radiator in position, making certain all the mounts and bushings are correctly installed. Tighten the mounting bolts to 10 ft. lbs. Double check the draincock to make sure it is closed.
13. Install the G64B fan assembly.
14. Connect the oil cooler lines and attach the brackets.
15. Connect the wiring to the electrical components, making sure each is correctly located and securely fastened.
16. Connect the upper and lower radiator hoses and the overflow hose. Install the coolant reserve tank.
17. For G64B engines, install the condenser fan assembly. For 4G63 engines, install the branch tube and its bracket.
18. Fill the system with coolant. Connect the fan wiring. Start the engine and allow it to idle with the radiator cap removed. When the engine has warmed enough, the thermostat will open and water flow will be visible within the radiator. Fill the coolant to the bottom of the radiator neck and install the cap.

CAUTION

Do not lean over the radiator neck while the engine is running. Hot coolant may be splashed out by air trapped within the system. Keep hands, tools and clothing away from the fans which may engage at any time.

19. Allow the engine to warm up fully and check that the fans cycle on and off correctly.
20. Shut the engine off and check the hose connections carefully for leaks. With the engine fully warm, a small leak may be emitting steam rather than liquid. If any leaks are found, allow the engine to cool completely before attempting repairs.

Galant V6 and Sigma

1. Set the heater temperature control to HOT inside the vehicle.
2. Using a large capacity container, loosen the radiator drain plug and drain the cooling system. It will drain quicker with the radiator cap removed.

CAUTION

When draining the coolant, keep in mind that cats and dogs are attracted by the ethylene glycol antifreeze, and are quite likely to drink any that is left in an uncovered container or in puddles on the ground. This will prove fatal in sufficient quantity. Always drain the coolant into a sealable container. Coolant should be reused unless it is contaminated or several years old.

3. Disconnect and remove the battery.

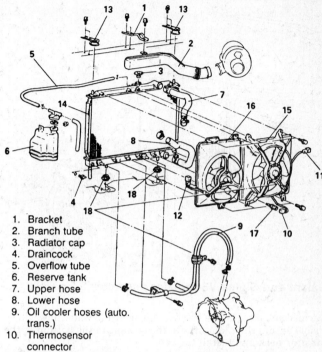

1. Bracket
2. Branch tube
3. Radiator cap
4. Draincock
5. Overflow tube
6. Reserve tank
7. Upper hose
8. Lower hose
9. Oil cooler hoses (auto. trans.)
10. Thermosensor connector
11. Fan motor connector
12. Condenser fan motor connector (with A/C)
13. Upper insulators
14. Radiator
15. Fan motor assembly
16. Condenser fan motor assembly
17. Thermosensor
18. Lower insulators

Galant radiator assembly. 4G63 shown

4. If equipped with air conditioning, remove the air conditioning condenser fan.
5. Disconnect the overflow tube and remove the coolant reserve tank and its bracket.
6. Disconnect the upper radiator hose.

NOTE: It is recommended that each clamp be matchmarked to the hose. Observe the marks and reinstall the clamps exactly when reinstalling the radiator.

7. Label and disconnect the wiring to the thermosensors and the electric fan.
8. If equipped with automatic transaxle, disconnect the oil cooler lines at the radiator. Remove the retaining bracket bolts and move the lines out of the way. Plug the lines to prevent leakage.
9. Remove the radiator fan assembly.
10. Remove the lower radiator hose.
11. Remove the bolts holding the upper radiator brackets to the bodywork. Lift the radiator out of the car. The brackets may be removed on the workbench if so desired. Before installation, make certain the radiator is completely assembled with the thermosensors and brackets if they were removed.
12. Place the radiator in position and tighten the upper mount bolts. Connect the lower radiator hose.
13. Install the oil cooler hoses for the automatic transaxle.
14. Install the radiator fan assembly. Connect the wiring for the fan and thermosensors.
15. Install the bracket and coolant reserve tank; connect the overflow tube.
16. Install the condenser fan and connect the wiring.
17. Install the battery and connect the terminals.
18. Fill the system with coolant. Start the engine and allow it to idle with the radiator cap removed. When the engine has

3-149

3 ENGINE AND ENGINE OVERHAUL

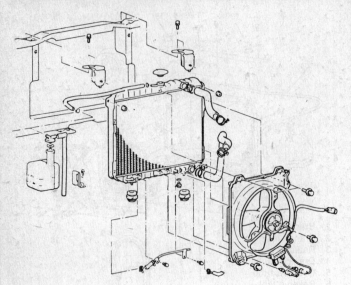

Galant and Sigma V6 radiator assembly

flow hose. Check the each fitting and the inside of the hose for any corrosion or debris which would prevent a good seal.
12. Install the battery and connect the terminals.
13. Fill the system with coolant. Start the engine and allow it to idle with the radiator cap removed. When the engine has warmed enough, the thermostat will open and water flow will be visible within the radiator. Fill the coolant to the bottom of the radiator neck and install the cap.

CAUTION
Do not lean over the radiator neck while the engine is running. Hot coolant may be splashed out by air trapped within the system. Keep hands, tools and clothing away from the fan.

---

14. Shut the engine off and check the hose connections carefully for leaks. With the engine fully warm, a small leak may be emitting steam rather than liquid. If any leaks are found, allow the engine to cool completely before attempting repairs.

Pickup with 4D55 Diesel

WARNING: Perform this work only on a cold engine. Minimum cooling time is 3–4 hours; overnight cold is preferred.

1. Remove the splash shield from under the vehicle.
2. Drain the coolant from the radiator.

WARNING: Housepets and small animals are attracted to the odor and taste of engine coolant (antifreeze). It is a highly poisonous mixture of chemicals; special care must be taken to protect open containers and spillage.

3. Disconnect the upper and lower radiator hoses from the engine. Disconnect the hose running to the coolant reserve tank.

warmed enough, the thermostat will open and water flow will be visible within the radiator. Fill the coolant to the bottom of the radiator neck and install the cap.

CAUTION
Do not lean over the radiator neck while the engine is running. Hot coolant may be splashed out by air trapped within the system. Keep hands, tools and clothing away from the fans which may engage at any time.

---

19. Allow the engine to warm up fully and check that the fans cycle on and off correctly.
20. Shut the engine off and check the hose connections carefully for leaks. With the engine fully warm, a small leak may be emitting steam rather than liquid. If any leaks are found, allow the engine to cool completely before attempting repairs.

Montero V6

1. Set the heater temperature control to HOT inside the vehicle.
2. Disconnect and remove the battery.
3. Using a large capacity container, loosen the radiator drain plug and drain the cooling system. It will drain quicker with the radiator cap removed.

CAUTION
When draining the coolant, keep in mind that cats and dogs are attracted by the ethylene glycol antifreeze, and are quite likely to drink any that is left in an uncovered container or in puddles on the ground. This will prove fatal in sufficient quantity. Always drain the coolant into a sealable container. Coolant should be reused unless it is contaminated or several years old.

---

4. Disconnect the upper and lower hoses and disconnect the overflow hose to the reserve tank.
5. Remove the upper fan shroud, then the lower fan shroud.
6. Disconnect the automatic transmission cooler lines.
7. Remove the radiator retaining bolts and remove the radiator.

To install:

8. Reinstall the radiator in position, making certain all the mounts and bushings are correctly installed. Tighten the mounting bolts enough to hold well but do not overtighten them. Double check the draincock to make sure it is closed.
9. Connect the automatic transmission cooler lines.
10. Install the lower shroud first, then the upper shroud.
11. Connect the upper and lower radiator hoses and the over-

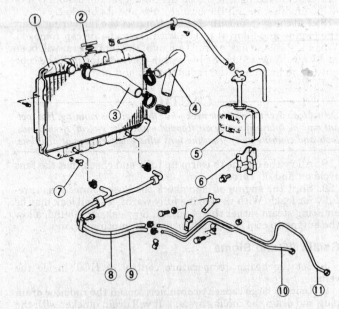

1. Exhaust rocker arms (4)
2. Rocker shaft springs (4)
3. Intake rocker arms (4)
4. Rocker arm adjusting screws
5. Nut
6. Rocker arm shaft
11. Cylinder head

Typical truck radiator components. Rear wheel drive with automatic transmission shown; others similar

3-150

ENGINE AND ENGINE OVERHAUL 3

4. If equipped with automatic transmission, disconnect the oil cooler lines from the radiator. Plug the lines to keep out dirt as soon as they are removed.
5. Remove the radiator shroud attaching bolts and remove the radiator shroud.
6. Remove the radiator mounting bolts and remove the radiator.
7. Reinstall the radiator and tighten the mounting bolts.
8. Install the shroud.
9. If equipped with automatic transmission, connect the oil cooler lines.
10. Connect the upper and lower radiator hoses. Use new clamps if necessary and make certain they are correctly positioned. Connect the coolant reservoir hose.
11. Double check that the radiator draincock is tightly closed. Fill the system with coolant.
12. Install the radiator cap. Start the engine and check carefully for leaks.

Condenser

Removing the condenser is not difficult, although some bolts may require deft wrench-work to gain access. Most home mechanics encounter air conditioning troubles after the condenser is reinstalled and the system fails to work properly. Remember some simple rules to avoid air conditioning problems.

- Dirt and moisture are the enemy. Any line which is disconnected should be immediately plugged or capped with a tight-fitting seal.
- Have everything you need at hand before beginning the work. Time saved usually equals a cleaner system.
- Handle lines and hoses carefully; any bend or kink reduces system capacity. Never attempt to straighten a bent line — replace it.
- If the line fitting has two wrench faces, counterhold one while turning the other. The lines are lightweight metal and bend easily.
- Even though the system is discharged, loosen fittings slowly and listen for any remaining pressure; allow it to bleed slowly before removing the line.
- Most line joints or fittings contain a small rubber O-ring. Always replace it before rejoining the lines. Make sure the new O-ring is free of nicks or scratches.
- Keep all tools and charging equipment clean. Keep removed components in a clean, protected location.
- Fittings must be reconnected carefully and tightened to the correct torque. Too tight damages threads, too loose will leak.

REMOVAL AND INSTALLATION

All Cars

1. Disconnect the negative battery cable. On the Mirage 4G15 and 4G61 engines, remove the battery and battery tray. The 4G63 engine in the Galant requires removal of the plastic duct tube from the air cleaner.
2. Safely discharge the air conditioning system. See Section 1.
3. If the condenser fan is located in front of the condenser, the fan must be removed. In most cases, this requires removing the grille for access. During the fan removal be aware of other lines, hoses or brackets attached to the fan framework. These may be loosened and repositioned as needed.
4. Label and disconnect both lines running to the condenser. If the line seems to be in the way of removal, disconnect the other end and remove the line. Do not attempt to bend the lines out of the way.
5. Locate and remove the condenser retaining bolts. Carefully lift the condenser out of the car.

NOTE: **In some cases, the radiator must be loosened to gain clearance. Remove the upper radiator retaining brackets, lift the radiator slightly (the hoses, etc. are still attached) and lean it back towards the engine.**

6. Install the condenser, making sure the rubber cushions below it are correctly placed.
7. If the radiator was moved for clearance, reposition it and install the upper mounts. Install the condenser retaining bolts.
8. Install new O-rings at every joint that was disconnected. Position the lines carefully and start the threaded fittings with your fingers. The threads should match perfectly and draw tension smoothly. Don't force anything.
9. Tighten each fitting to the correct torque. The small bolts holding flange-type fittings should be tightened only to 36–48 inch lbs. Threaded pipe unions have different tightnesses depending on their position: Discharge pipe fittings (from compressor to condenser) should be tightened to 14–16 ft. lbs.; liquid lines from the condenser to the receiver/dryer should be tightened to 10 ft. lbs. Make certain that any brackets or clamps holding lines or hoses are reinstalled and secure.
10. Install the condenser fan if it was removed. Don't forget to remount any other lines or wiring clipped to the fan frame. Connect the fan wiring to the harness. Reinstall the grille.
11. Reinstall the battery and/or the duct tube if it was removed.
12. Connect the charging equipment. Evacuate the system and check for leaks. Recharge the system according to procedures in Section 1.

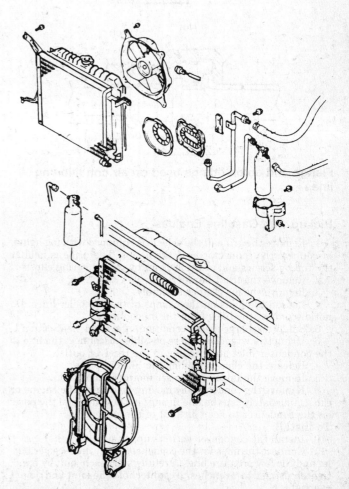

Typical condenser assemblies

3 ENGINE AND ENGINE OVERHAUL

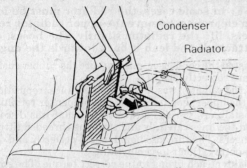

It may be necessary to move the radiator to remove the condenser

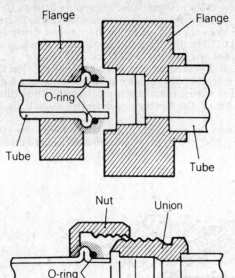

Flange and union fittings used on air conditioning lines

Pickup with Gasoline Engines

1. Remove the grille. If the vehicle is equipped with fog lights, disconnect the connector(s) to the lamps. The grille is held at the top by screws and at the bottom by plastic spring clips.
2. Remove the front bumper.
3. Remove the air cleaner assembly.
4. Loosen the bolts on the clamps of the oil cooler lines. Do not disconnect the lines from the oil cooler.
5. Safely discharge the air conditioning system. See Section 1.
6. Using two wrenches where possible, disconnect the line to the condenser inlet and the line to the receiver outlet.
7. Release the clamps holding the lines in place.
8. Remove the upper condenser mounting bolts.
9. Remove the condenser by moving it towards the engine or cab. Immediately cap the line fittings in the truck and the ports on the condenser to keep dirt out of the system.

To install:

10. Install the condenser with its upper support bolts.
11. Connect the lines for the condenser inlet, the receiver outlet and the low pressure line. Carefully start each joint by hand, tighten using two wrenches to counterhold the joint and do not overtighten.
12. Install the retaining clamps for the lines and hoses.
13. Install the air cleaner.
14. Install the front bumper.
15. Install the grille. Reconnect the fog lights if they were disconnected.
16. Evacuate and recharge the system.

Pickup with 4D55 Diesel Engine

1. Remove the grille. If the vehicle is equipped with fog lights, disconnect the connector(s) to the lamps. The grille is held at the top by screws and at the bottom by plastic spring clips.
2. Remove the front bumper.
3. Loosen but do not remove the bolts holding the hood latch support.
4. Remove the air cleaner assembly.
5. Loosen the bolts on the clamps of the oil cooler lines. Do not disconnect the lines from the oil cooler.
6. Safely discharge the air conditioning system. See Section 1.
7. Using two wrenches where possible, disconnect the line to the condenser inlet and the line to the receiver outlet.
8. Disconnect the low pressure line running across the front of the condenser.
9. Release the clamps holding the lines in place.
10. Remove the upper condenser mounting bolts.
11. Move the oil cooler forward about 4 inches and remove the condenser by moving it towards the engine or cab. Immediately cap the line fittings in the car and the ports on the condenser to

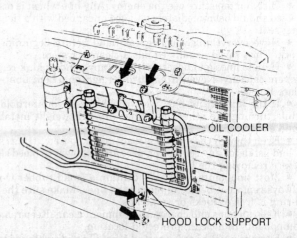

Loosen the hood lock support piece to allow the oil cooler to move out of the way

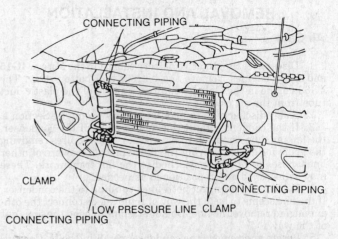

Truck and diesel condenser and related components

ENGINE AND ENGINE OVERHAUL 3

keep dirt out of the system.
To install:
12. Move the oil cooler forward and install the condenser with its upper support bolts.
13. Connect the lines for the condenser inlet, the receiver outlet and the low pressure line. Carefully start each joint by hand, tighten using two wrenches to counterhold the joint and do not overtighten.
14. Install the retaining clamps for the lines and hoses.
15. Position the oil cooler and tighten the clamps holding the oil lines.
16. Install the air cleaner.
17. Tighten the hood latch support mounting bolts.
18. Install the front bumper.
19. Install the grille. Reconnect the fog lights if they were disconnected.
20. Evacuate and recharge the system.

Montero with Single Air Conditioner

1. Safely discharge the air conditioning system. See Section 1.
2. Remove the grille. Use care not to damage the plastic.
3. Carefully, and with the use of two wrenches, disconnect the lines in front of and running to the condenser. On the V6 there are three line junctions on the right side of the condenser and and four junctions on the left side.
   On the 4-cylinder engine, disconnect the lines at the condenser ports and disconnect the receiver/dryer.
4. Remove the receiver/dryer mounting bolts(s) and remove the receiver/drier.
5. Unplug the electric condenser fan, remove the mounting bolts for the fan and remove the fan.
6. Remove the condenser mounting bolt on the side opposite the receiver/dryer and carefully lift the condenser out of the vehicle.
7. Install the unit and tighten the right retaining bolt.
8. Install the fan assembly and tighten the retaining bolts. Connect the fan wiring harness.
9. Install the receiver/dryer and its mounting bolt(s); at least one of these bolts serves as the other condenser mounting bolt.
10. Connect each line and hose by hand, making certain the O-rings are intact and not damaged during installation. Tighten each fitting by hand first, then using two wrenches, tighten them to the specified torque. Do not overtighten these fittings.
11. Evacuate and recharge the system; check carefully for leaks.
12. Reinstall the grille.

Montero with Dual Air Conditioners

1. Safely discharge the air conditioning system. See Section 1.
2. Remove the grille. Use care not to damage the plastic.
3. Carefully, and with the use of two wrenches, disconnect the lines in front of and running to the condenser. The presence of the second condenser adds two more lines to the maze; disconnect the lines at the joints closest to each condenser. Disconnect both ends of any line in the way; do not bend or force the lines out of the way.
4. Remove the receiver/dryer mounting bolts(s) and remove the receiver/drier.
5. Unplug the large electric condenser fan, remove the mounting bolts for the fan and remove the fan.
6. Remove the condenser mounting bolt on the side opposite the receiver/dryer and carefully lift the main condenser out of the vehicle.
7. Unplug the connector for the smaller auxilary fan. Unbolt the sub-condenser from its brackets and remove it.
8. Remove the fan assembly from the sub-condenser.
9. To reinstall, attach the small fan to the condenser before placing the unit in the vehicle. Install the bracket bolts.
10. Install the main condenser and tighten the right retaining bolt.

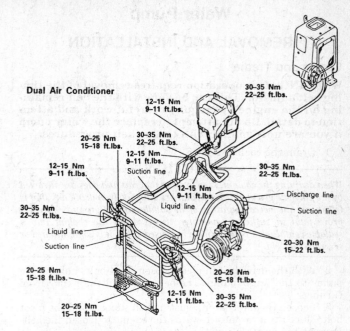

Montero single air conditioner system; line routing and tightening values

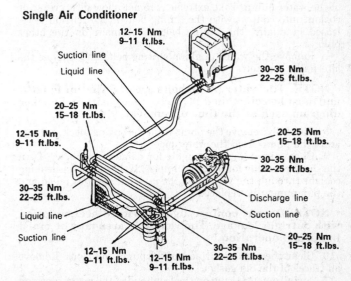

Montero dual air conditioner system; line routing and tightening values

11. Install the large fan assembly and tighten the retaining bolts. Connect the fan wiring harness.
12. Install the receiver/dryer and its mounting bolt(s); at least one of these bolts serves as the other condenser mounting bolt.
13. Connect each line and hose by hand, making certain the O-rings are intact and not damaged during installation. Tighten each fitting by hand first, then using two wrenches, tighten them to the specified torque. Do not overtighten these fittings.
14. Evacuate and recharge the system; check carefully for leaks.
15. Reinstall the grille.

3-153

3 ENGINE AND ENGINE OVERHAUL

Water Pump

REMOVAL AND INSTALLATION

Cordia and Tredia

WARNING: This operation requires removal of the timing belt adjuster. This can be done with the belt remaining on the engine but requires careful work and attention to detail. Do not attempt to replace the water pump if you are not familiar with timing belt procedures.

1. Drain the cooling system.

— CAUTION —

When draining the coolant, keep in mind that cats and dogs are attracted by the ethylene glycol antifreeze, and are quite likely to drink any that is left in an uncovered container or in puddles on the ground. This will prove fatal in sufficient quantity. Always drain the coolant into a sealable container. Coolant should be reused unless it is contaminated or several years old.

2. With the drive belt in place, loosen the 4 bolts holding the pulley to the water pump. The belt will keep the pulley from turning.
3. Loosen the alternator adjusting bolt (upper) and the pivot bolt. Move the alternator towards the engine to loosen the belt. Remove the belt.
4. Remove the pulley bolts and the pulley.
5. Remove the timing belt covers.
6. Remove the timing belt adjuster which is partially in front of the water pump. Take extreme care not allow dirt, grease or coolant into contact with the timing belt. If the tension is released gradually, the timing belt can remain on the other pulleys.
7. Remove the water pump mounting bolts and remove the alternator bracket.

NOTE: The water pump bolts are of different lengths and must be reinstalled in the correct position. Label or diagram each at the time of removal.

8. Carefully remove the pump from the engine block and separate the pump from the water pipe.
9. Check the pump thoroughly for damage or cracks. Turn the shaft, checking for binding or noise. If the pump was leaking coolant through the vent hole, the internal seal has failed and the pump must be replaced.

NOTE: Do not confuse light moisture accumulation with outright leakage. The vent hole area may be moist in normal operation.

10. Clean the mating surfaces of the pump and block. Remove all traces of the old gasket.
11. Install a new O-ring on the front end of the water pipe and wet the O-ring with water or coolant. Do not apply grease or oil.
12. Install a new water pump gasket, making sure all the bolt holes line up. Do not apply any sealer to the gasket.
13. Install the water pump retaining bolts in their correct holes (don't forget the alternator brace) and tighten each finger tight. Tighten the bolts in two steps to their final torque: Bolt heads marked **4** to 10 ft. lbs. and bolt heads marked **7** to 18 ft. lbs.
14. Install the timing belt adjuster and adjust the timing belt tension as described later in this Section.
15. Install the timing belt covers. Install the water pump pulley and tighten the bolts snug. Install the belt and tension the alternator to correctly adjust the belt. Tighten the alternator in position.
16. Tighten the water pump pulley bolts to 6 ft. lbs.
17. Double check the draincock, closing it if necessary, and refill the cooling system.

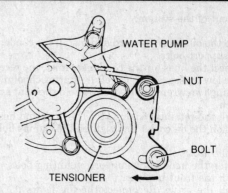

Removing the timing belt tensioner

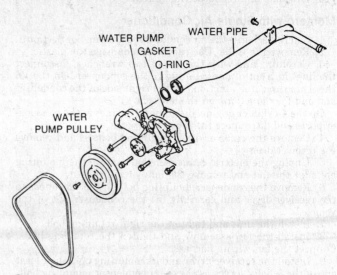

Cordia and Tredia water pump

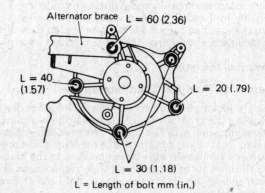

Bolt length and placement on Cordia and Tredia water pump

18. Start the engine and check carefully for leaks. Allow the engine to warm up with the radiator cap removed. After the thermostat opens, adjust the coolant level to the bottom of the radiator neck and install the radiator cap.

— CAUTION —

Do not lean over the radiator neck while the engine is running. Hot coolant may be splashed out by air trapped within the system. Keep hands, tools and clothing away from the fans which may engage at any time.

3-154

ENGINE AND ENGINE OVERHAUL 3

Starion
Pickup and Montero with G54B Engine

1. Disconnect and remove the battery.
2. Drain the cooling system.

― CAUTION ―

When draining the coolant, keep in mind that cats and dogs are attracted by the ethylene glycol antifreeze, and are quite likely to drink any that is left in an uncovered container or in puddles on the ground. This will prove fatal in sufficient quantity. Always drain the coolant into a sealable container. Coolant should be reused unless it is contaminated or several years old.

3. Remove the lower radiator hose from the water pump.
4. If the engine does not have an intercooler, remove the upper fan shroud. Carefully unbolt and remove the fan with the clutch assembly and store the fan in an upright position.
5. Loosen the adjuster and remove the air conditioning drive belt. Loosen the alternator and power steering pump and remove their drive belts.
6. Remove the water pump pulley from the studs.
7. Disconnect the heater hose from the water pump.
8. Remove the mounting bolts holding the pump to the engine. Note that the bolts are of different lengths; label or diagram them as they are removed.
9. Remove the water pump and gasket. Clean the mating surfaces thoroughly and remove all traces of the old gasket.
10. Check the pump thoroughly for damage or cracks. Turn the shaft, checking for binding or noise. If the pump was leaking coolant through the vent hole, the internal seal has failed and the pump must be replaced.

NOTE: Do not confuse light moisture accumulation with outright leakage. The vent hole area may be moist in normal operation.

11. If the fan was removed (without intercooler), inspect the fan blades for cracking or bending. Check the area of each bolt hole for cracks.
12. Install a new gasket to the engine, making sure all the bolt holes are aligned. Install the water pump and tighten the bolts finger tight. Don't forget to install the alternator bracket. When all the bolts are snug, tighten them to the correct torque as shown in the chart.
13. Connect the heater hose to the water pump.
14. Install the water pump pulley. On intercooled engines, tighten the pulley bolts to 8 ft. lbs.
15. Install the drive belts beginning with the power steering belt, then the alternator belt and the A/C compressor belt. Seat the belts on all the pulleys but do not tighten them yet.
16. For engines without intercooler, install the fan and clutch assembly. Tighten the nuts to 8 ft. lbs. If the fan was removed from the clutch for any reason, the mounting bolts should be tightened to 6 ft. lbs. Install the upper fan shroud.

WARNING: The shroud retaining bolts must not exceed 16mm (0.63 in.) in length. Longer bolts will interfere with the radiator and cause leaks.

17. Adjust the drive belts to the proper tension and tighten the adjusting fittings for each component.
18. Install the lower radiator hose.
19. Install the battery.
20. Double check the draincock, closing it if necessary, and refill the cooling system.
21. Start the engine and check carefully for leaks. Allow the engine to warm up with the radiator cap removed. After the thermostat opens, adjust the coolant level to the bottom of the radiator neck and install the radiator cap.

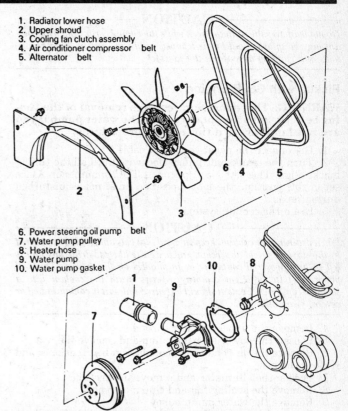

1. Radiator lower hose
2. Upper shroud
3. Cooling fan clutch assembly
4. Air conditioner compressor belt
5. Alternator belt
6. Power steering oil pump belt
7. Water pump pulley
8. Heater hose
9. Water pump
10. Water pump gasket

Starion water pump as found on engines without intercooler. Intercooled engines do not have belt driven fans but pump assembly is similar

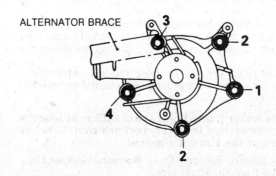

ALTERNATOR BRACE

No.	d x ℓ mm (in.)	Tightening torque Nm (ft.lbs.)
1	8 x 23 (0.90)	10 – 12 (7 – 9)
2	8 x 28 (1.10)	
3	8 x 88 (3.46)	15 – 21 (11 – 15)
4	8 x 78 (3.07)	10 – 12 (7 – 9)

Starion water pump bolt placement and torque values. Data applies to engines with or without intercooler

3 ENGINE AND ENGINE OVERHAUL

CAUTION
Do not lean over the radiator neck while the engine is running. Hot coolant may be splashed out by air trapped within the system. Keep hands, tools and clothing away from the fan(s).

Pickup with G63B Engine

WARNING: This operation requires removal of the timing belt. Do not attempt to replace the water pump if you are not familiar with timing belt procedures.

1. Disconnect the negative battery cable.
2. Turn the crankshaft pulley clockwise until all the timing marks align. This sets No. 1 piston at TDC/compression. Once set in this position, the motor position must not be disturbed during repairs.
3. Drain the cooling system.

CAUTION
When draining the coolant, keep in mind that cats and dogs are attracted by the ethylene glycol antifreeze, and are quite likely to drink any that is left in an uncovered container or in puddles on the ground. This will prove fatal in sufficient quantity. Always drain the coolant into a sealable container. Coolant should be reused unless it is contaminated or several years old.

---

4. Remove the upper radiator shroud.
5. Loosen the power steering pump and remove the belt.
6. Remove the air conditioner compressor belt tensioner and remove the belt.
7. Loosen the alternator and remove the drive belt.
8. Remove the cooling fan and clutch assembly.
9. Remove the water pump pulley.
10. Remove the center bolt holding the crankshaft pulleys. The bolt will be tight.
11. Remove the smaller pulley bolt and remove the power steering pulley and the crankshaft pulley from the crankshaft.
12. Remove the upper and lower timing belt covers.
13. Confirm that all the timing marks align and that the engine is at TDC/compression on No. 1 cylinder.
14. Loosen and remove the timing belt adjuster.
15. Use chalk or a crayon to mark the direction of rotation on the timing belt. Carefully remove the timing belt from the pulleys.
16. Disconnect the lower radiator hose from the water pump.
17. Remove the water pump mounting bolts and remove the alternator bracket.

NOTE: The water pump bolts are of different lengths and must be reinstalled in the correct position. Label or diagram each at the time of removal.

18. Carefully remove the pump from the engine block and separate the pump from the water pipe.
19. Check the pump thoroughly for damage or cracks. Turn the shaft, checking for binding or noise. If the pump was leaking coolant through the vent hole, the internal seal has failed and the pump must be replaced.

NOTE: Do not confuse light moisture accumulation with outright leakage. The vent hole area may be moist in normal operation.

20. Clean the mating surfaces of the pump and block. Remove all traces of the old gasket.
21. Install a new O-ring on the front end of the water pipe and wet the O-ring with water or coolant. Do not apply grease or oil.
22. Install a new water pump gasket, making sure all the bolt holes line up. Do not apply any sealer to the gasket.
23. Install the water pump retaining bolts in their correct holes (don't forget the alternator brace) and tighten each finger tight. Tighten the bolts in two steps to their final torque: Bolt heads marked **4** to 14 Nm (10 ft. lbs.) and bolt heads marked **7** to

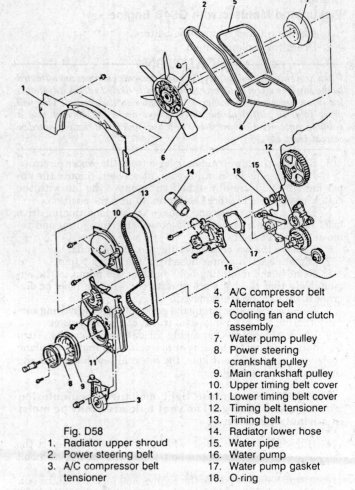

Fig. D58
1. Radiator upper shroud
2. Power steering belt
3. A/C compressor belt tensioner
4. A/C compressor belt
5. Alternator belt
6. Cooling fan and clutch assembly
7. Water pump pulley
8. Power steering crankshaft pulley
9. Main crankshaft pulley
10. Upper timing belt cover
11. Lower timing belt cover
12. Timing belt tensioner
13. Timing belt
14. Radiator lower hose
15. Water pipe
16. Water pump
17. Water pump gasket
18. O-ring

Water pump and related components; pickup with 2.0L engine

24 Nm or 18 ft. lbs.
25. Install the water pipe, making sure the O-ring stays in position.
26. Connect the lower radiator hose to the water pump. Use a new hose clamp if necessary.
27. Install the timing belt onto the crankshaft sprocket, the oil pump sprocket and the camshaft sprocket in that order. Make certain there is no slack between the crankshaft sprocket and the oil pump or between the oil pump sprocket and the camshaft sprocket.
28. Loosen the tensioner mounting bolts. Allow the tensioner to move against the belt under its spring tension; do not force or push on the tensioner.
29. Turn the crankshaft in its normal clockwise direction until the camshaft pulley has moved 2 teeth from the timing mark.

WARNING: This rotation is to create the correct tension on the belt. Do not rotate the engine counterclockwise and do not press on the belt.

30. Carefully observe the belt and the way it fits on each sprocket. It may have lifted from the camshaft sprocket, particularly in the "10 o'clock" position. If this is the case, GENTLY push on the tensioner with your fingers (counterclockwise or downward) to take up the slack. Do not push on the tensioner any more than needed to seat the belt on the sprocket.
31. Tighten the lower bolt on the tensioner first, then the pivot

ENGINE AND ENGINE OVERHAUL 3

or upper bolt. The order is important; if done incorrectly, the belt will not be under the correct tension.

32. Install the lower timing belt cover and then the upper cover. Tighten the bolts to 11 Nm or 8 ft. lbs.
33. Install the crankshaft pulley and the power steering pulley. Tighten the small pulley bolt to 25 Nm (19 ft. lbs.) and the crankshaft center bolt to 120 Nm or 88 ft. lbs.
34. Install the pulley and fan assembly onto the water pump. Tighten the nuts to 11 Nm (8 ft. lbs).
35. Install the alternator drive belt and adjust it to the proper tension.
36. Install the compressor drive belt. Install the belt tensioner and adjust the belt tension.
37. Install the power steering pump drive belt and adjust it to the proper tension.
38. Install the upper radiator shroud.
39. Double check the draincock, closing it if necessary, and refill the cooling system.
40. Connect the negative battery cable.
41. Start the engine and check carefully for leaks. Allow the engine to warm up with the radiator cap removed. After the thermostat opens, adjust the coolant level to the bottom of the radiator neck and install the radiator cap.

— CAUTION —
Do not lean over the radiator neck while the engine is running. Hot coolant may be splashed out by air trapped within the system. Keep hands, tools and clothing away from the fans which may engage at any time.

42. Shut the engine off and perform any necessary adjustments to the drive belts.

Mirage — all Engines
Precis — G15B Engine

WARNING: Extensive disassembly is required involving bolts of different lengths. Develop a system of identifying bolts by location and component so that reassembly is made easier. Extreme damage can be caused by forcing a long bolt into a shorter passage.

1. Disconnect the negative battery cable.
2. Remove the left splash shield under the engine.
3. Drain the engine coolant.

— CAUTION —
When draining the coolant, keep in mind that cats and dogs are attracted by the ethylene glycol antifreeze, and are quite likely to drink any that is left in an uncovered container or in puddles on the ground. This will prove fatal in sufficient quantity. Always drain the coolant into a sealable container. Coolant should be reused unless it is contaminated or several years old.

4. On the G15B engine, remove the air cleaner assembly and the air intake duct.
5. Loosen the appropriate component or adjuster and remove (in order) the air conditioning compressor drive belt, the power steering belt and the ribbed belt driving the water pump.

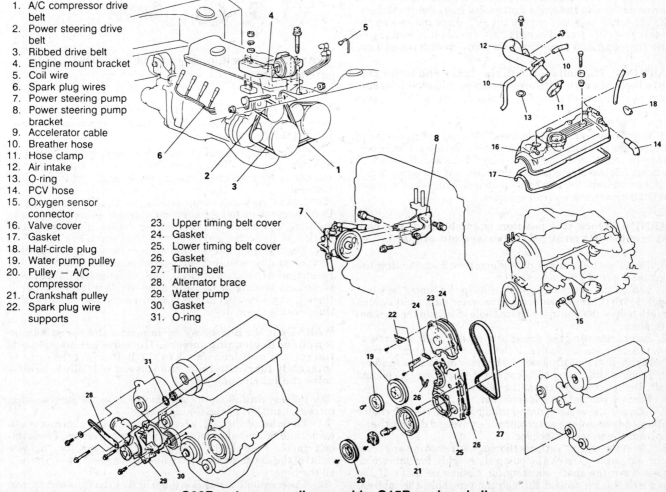

1. A/C compressor drive belt
2. Power steering drive belt
3. Ribbed drive belt
4. Engine mount bracket
5. Coil wire
6. Spark plug wires
7. Power steering pump
8. Power steering pump bracket
9. Accelerator cable
10. Breather hose
11. Hose clamp
12. Air intake
13. O-ring
14. PCV hose
15. Oxygen sensor connector
16. Valve cover
17. Gasket
18. Half-circle plug
19. Water pump pulley
20. Pulley — A/C compressor
21. Crankshaft pulley
22. Spark plug wire supports
23. Upper timing belt cover
24. Gasket
25. Lower timing belt cover
26. Gasket
27. Timing belt
28. Alternator brace
29. Water pump
30. Gasket
31. O-ring

G32B water pump disassembly. G15B engine similar

3 ENGINE AND ENGINE OVERHAUL

6. Place a floor jack and a broad piece of lumber under the oil pan. Elevate the jack just enough to support the engine without raising it.
7. Remove the nuts holding the through-bolt for the left engine bracket and remove the bolt. It may be necessary to adjust the jack tension slightly to allow the bolt to come free. Use the jack to keep the engine in its normal position while the mount is disconnected.
8. Remove the nuts and bolts holding the engine bracket to the engine and remove the bracket.
9. Label and remove the coil wire and the spark plug wires from the distributor cap and move them out of the way. On 4G61 (DOHC) engines, remove the spark plug wire cover and remove the leads from the plugs.
10. Remove the power steering pump from its bracket and move it, with the hoses attached, out of the way. Hang the pump from a piece of stiff wire or string; do not allow the pump to hang by the hoses.
11. Remove the power steering pump bracket from the engine.
12. On G32B engines, disconnect the accelerator cable.
13. Except on 4G15 and 4G61 engines, disconnect the wiring connector for the oxygen sensor (located under the distributor).
14. Except on 4G15 engines, label and disconnect each air tube or vacuum line running to the valve cover. On G32B engines, disconnect the air intake tube as well.
15. For all engines except 4G15, remove the valve cover and its gasket. Remove the half-circle plug from the G32B valve cover.
16. Remove the water pump pulleys.
17. On the G15B engine remove the upper and lower timing belt covers with the gaskets. On all other engines, remove the A/C drive pulley and the crankshaft pulley from the crankshaft.
18. On G15B engines, remove the A/C drive pulley and the crankshaft pulley from the crankshaft. On all other engines, remove the upper and lower timing belt covers with the gaskets.

WARNING: The bolts for both the upper and lower timing belt covers are of different lengths. Label or diagram them at the time of removal; replacement in the correct location is required.

19. Rotate the crankshaft clockwise until the timing marks on the crank pulley align at **0** on the scale (on the lower timing cover) and the mark on the camshaft pulley aligns with the mark on the head. Aligned means perfect, not close — if you miss, continue clockwise until the marks are matched. This sets the motor at TDC/compression on No.1 cylinder.

WARNING: Once this position is established, the cam and crankshafts must NOT be moved out of place.

20. Remove the crankshaft pulley(s) without moving the crankshaft out of position.
21. Loosen (do not remove) both bolts in the timing belt tensioner. Move the tensioner toward the water pump and tighten the lock bolt — the one in the slotted hole — to hold the tensioner in place.
22. Before removing the timing belt, mark it with an arrow to show the direction of rotation. If the belt is to be reused, it must be reinstalled so that it rotates in the same direction as before.
23. Remove the timing belt, keeping it free of grease, oil and fluids. Do not crease or crimp the belt.
24. Remove the timing belt tensioner.
25. Remove the water pump retaining bolts, the alternator bracket and remove the water pump. Remove and discard the O-ring from the water intake pipe.
26. Clean the mating surfaces thoroughly, removing any trace of the old gasket. Check the pump thoroughly for damage or cracks. Turn the shaft, checking for binding or noise. If the pump was leaking coolant through the vent hole, the internal seal has failed and the pump must be replaced.

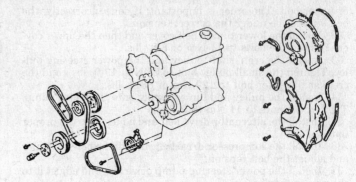

4G15 belt, pulley and timing cover assembly

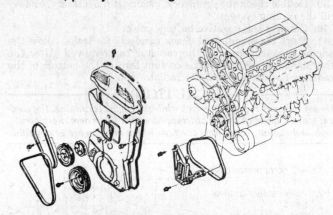

4G61 (DOHC) belt, pulley and timing cover arrangement

NOTE: Do not confuse light moisture accumulation with outright leakage. The vent hole area may be moist in normal operation.

27. Install a new gasket to the engine, making sure all the bolt holes are aligned. Place a new O-ring on the water pipe and moisten it with water or coolant. Never apply grease or oil to this seal.
28. Install the water pump and tighten the bolts finger tight. Don't forget to install the alternator bracket. When all the bolts are snug, tighten them to the correct torque: Bolts with head mark **4** to 10 ft. lbs. and bolts with head mark **7** to 18 ft. lbs.
29. Check the sprockets and tensioner for wear. The sprocket teeth should be well defined, not rounded and the valleys between the teeth should be clean. The tensioner should spin freely with no binding or unusual noise. Replace the tensioner if there is any sign of grease leaking from the seal. Clean everything with a clean, dry cloth.

WARNING: Do not spray or immerse the sprockets or tensioner in cleaning solvent. The sprocket may absorb the solvent and transfer it to the belt. The tensioner is internally lubricated and the solvent will dilute or dissolve the lubricant.

30. Before reinstalling the belt, double check the sprockets for correct alignment of the timing marks.
31. To reinstall the belt, observe the directional arrow made earlier and fit the belt under the crankshaft sprocket. Track the belt up and over the right side (your right, not the engine's right) of the camshaft sprocket while keeping tension on the belt all the time.
32. The camshaft sprocket is quite likely to have moved during the belt installation. Gently turn the cam sprocket counter-

ENGINE AND ENGINE OVERHAUL 3

clockwise to tension the belt; when the belt is taut, the timing marks on the head and cam sprocket should align perfectly. If they do not, remove the belt and reinstall it.

33. Install the crankshaft pulley temporarily. It is needed to hold the belt in place.
36. Loosen the lock bolt in the tensioner. Allow the tensioner to move against the belt under its own spring tension; don't pry it or force it.
37. Tighten the lock bolt (in the slotted hole) in the tensioner, then tighten the pivot bolt for the tensioner. This order is important; if done in the wrong order, the belt will be over-tightened.
38. Turn the crankshaft one full revolution clockwise. Turn it smoothly and continuously and realign the timing mark. Loosen the pivot bolt, then the lock bolt on the tensioner. This allows the tensioner to pick-up any remaining slack in the belt.
39. Tighten the lock nut to 18 ft. lbs. and then tighten the pivot bolt to 18 ft. lbs.
40. Hold the tensioner and timing belt by hand and give the belt slight thumb pressure at a point level with the tensioner. The belt should deflect enough to allow the top of a tooth to cover about ¼ of the head of the lock bolt.
41. Finish the installation of the crank pulley(s). The large center bolt should be tightened to 62 ft. lbs. and the smaller bolts (for the air conditioning pulley) should be tightened to 10 ft. lbs. if they were removed.
42. Install the lower timing belt cover with its gasket, making sure the gasket doesn't shift during installation. tighten the bolts to 8 ft. lbs.
43. On G15B engines, install the crankshaft pulley and tighten the bolt to 62 ft. lbs. Install the A/C compressor pulley and tighten its bolts to 10 ft. lbs. On all other engines, install the upper and lower timing belt covers, making sure each gasket is properly seated. Don't forget the small supports which carry the spark plug wires. Tighten the cover bolts to 8 ft. lbs.
44. On G15B engines, install the upper and lower timing belt covers, making sure each gasket is in place. Tighten the bolts to 8 ft. lbs. For all other engines, install the crankshaft pulley and tighten the bolt. Correct torque for G32B is 88 ft. lbs.; 4G15 is 62 ft. lbs. and 4G61 is 18 ft. lbs. for the bolts holding the pulley to the sprocket. Install the A/C drive pulley, tightening the bolts to 12 ft. lbs.
45. Install the water pump pulleys, tightening the bolts to 6 ft. lbs.
46. Reinstall the valve cover and gasket. On G32B engines, install the half-circle plug before mounting the cover.
47. Connect the air hoses and vacuum lines to the valve cover. Connect the G32B air intake tube.
48. Connect the wiring for the oxygen sensor if it was disconnected.
49. Connect the G32B accelerator cable.
50. Install the power steering pump bracket to the engine block. Tighten the mounting bolts to 18 ft. lbs.
51. Install the power steering pump to the bracket, tightening the bolts just enough to hold the pump in place.
52. Connect the coil and spark plug cables to the distributor cap and/or the spark plugs. Remember to route the wiring correctly on the wire supports.
53. Install the motor mount bracket to the engine and tighten the nuts and bolts to 42 ft. lbs.
54. Align the mount bushing with the body bracket and install the through-bolt. Tighten both the large nut and the smaller safety nut snug.
55. Remove the jack from under the engine. With the weight of the engine on the mount, tighten the large nut on the through-bolt to 72 ft. lbs. and the smaller safety nut to 38 ft. lbs.
56. Install the ribbed water pump drive belt, making sure it is correctly seated on each pulley. Install the power steering belt and the compressor drive belt. Adjust each belt to the correct tension and tighten the fittings or adjuster on each component.

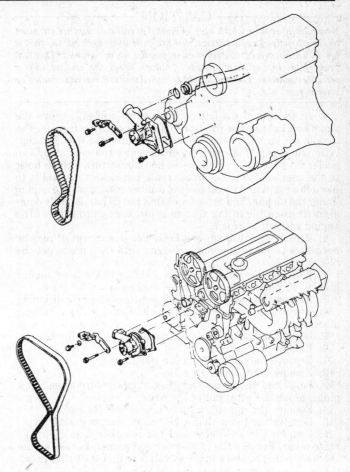

Water pumps: 4G15 above, 4G61 below

57. Install the G15B air cleaner assembly and intake duct.
58. Double check all installation items, paying particular attention to loose hoses or hanging wires, untightened nuts, poor routing of hoses and wires (too tight or rubbing) and tools left in the engine area.
59. Double check the draincock, closing it if necessary, and refill the cooling system.
60. Connect the negative battery cable.
61. Start the engine and check carefully for leaks. Allow the engine to warm up with the radiator cap removed. After the thermostat opens, adjust the coolant level to the bottom of the radiator neck and install the radiator cap.

--- CAUTION ---
Do not lean over the radiator neck while the engine is running. Hot coolant may be splashed out by air trapped within the system. Keep hands, tools and clothing away from the fan(s).

62. Reinstall the splash shield under the car. Final adjustment of the drive belts and accelerator cable may be needed.

Galant with G64B Engine

WARNING: Extensive disassembly is required involving bolts of different lengths. Develop a system of identifying bolts by location and component so that reassembly is made easier. Extreme damage can be caused by forcing a long bolt into a shorter passage.

1. Disconnect the negative battery cable.
2. Remove the left splash shield under the engine.
3. Drain the engine coolant.

3-159

3 ENGINE AND ENGINE OVERHAUL

CAUTION

When draining the coolant, keep in mind that cats and dogs are attracted by the ethylene glycol antifreeze, and are quite likely to drink any that is left in an uncovered container or in puddles on the ground. This will prove fatal in sufficient quantity. Always drain the coolant into a sealable container. Coolant should be reused unless it is contaminated or several years old.

4. At the top of the left engine mount bracket, remove the small bolt holding the wiring clip. Disconnect the wiring for the oxygen sensor and release the high pressure hose from the clip.
5. The engine must be supported before the next step. The preferred method is to remove the hood and attach a chain hoist to the engine lift hooks. An alternate but riskier method is to use a floor jack and broad piece of lumber to support the engine under the oil pan. You are warned that the oil pan is easily damaged. Tension the lifting apparatus just enough to support the engine without raising it.
6. Remove the through-bolt for the engine mount. It may be necessary to make small adjustments with the jack to get the bolt free.
7. Remove the nuts and bolts holding the engine mount bracket to the engine and remove the bracket.
8. Release the tension on the adjuster for the air conditioning compressor belt and remove the belt. Remove the tensioner assembly.
9. Loosen the power steering pump and remove the belt.
10. Remove the water pump belt.
11. Remove the pulleys from the water pump.
12. Remove the crankshaft pulley.
13. Label and disconnect the spark plug wires from the spark plugs. Move the wires out of the way.
14. Remove the upper timing belt cover and its gasket.
15. Remove the lower timing belt cover and its gasket.
16. Remove the PCV hose and the breather hose from the valve cover. Remove the valve cover and gasket. Loosen all the valve adjusting screws and the jet valves so that the tip of the adjuster projects from 0-1mm (0–0.04 in.).
17. Use a socket wrench on the projecting crankshaft bolt to turn the engine clockwise (only!) and align all the timing marks on the sprockets and cases. It may be necessary to wipe off the area to see the marks clearly. Do not use spray cleaners around the timing belt.
When the marks all align exactly, the engine is set to TDC/compression on No. 1 cylinder. From this point onward, the engine position MUST NOT be changed.
18. If the timing belt is to be reused, make a chalk or crayon arrow on the belt showing the direction of rotation so that it may be reinstalled correctly.
19. Loosen the bolts holding the tensioner and pivot the tensioner towards the water pump. Temporarily tighten the bolts to hold the tensioner in its slack position.
20. Carefully slide the belt off the sprockets. Place the belt in a clean, dry, protected location away from the work area.
21. Remove the tensioner for the silent shaft belt. (Tensioner B).
22. Remove the crankshaft sprocket retaining bolt. Remove the sprocket and flange.
23. Remove the silent shaft belt (timing belt B).
24. Remove the retaining the bolts for the water pump, taking note of the different lengths and positions. Remove the pump and gasket. Clean the mating surfaces thoroughly, removing all traces of the old gasket. Discard the O-ring from the water pipe. Check the pump thoroughly for damage or cracks. Turn the shaft, checking for binding or noise. If the pump was leaking coolant through the vent hole, the internal seal has failed and the pump must be replaced.

NOTE: Do not confuse light moisture accumulation with outright leakage. The vent hole area may be moist in normal operation.

1. Left side shield
2. Bolt
3. Oxygen sensor connector
4. Bracket
5. High pressure hose
6. Engine mount bracket
7. A/C compressor belt
8. Tensioner assembly
9. Power steering belt
10. Water pump belt
11. Water pump pulley
12. Crankshaft pulley (damper)
13. Spark plug wires
14. Timing belt upper cover
15. Gasket
16. Timing belt lower cover
17. Gasket

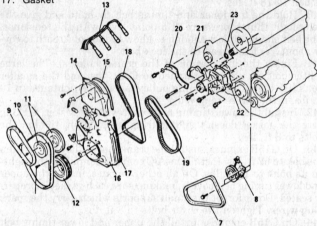

18. Timing belt
19. Timing belt (B)
20. Alternator brace
21. Water pump
22. Gasket
23. O-ring

G64B water pump and related components

25. Install a new gasket to the engine, making sure all the bolt holes are aligned. Place a new O-ring on the water pipe and moisten it with water or coolant. Never apply grease or oil to this seal.
26. Install the water pump and tighten the bolts finger tight. Don't forget to install the alternator bracket. When all the bolts are snug, tighten them to the correct torque. The two longest bolts — mounted in roughly the 12 o'clock and 9 o'clock positions — should be tightened to 18 ft. lbs. The other bolts are tightened to 10 ft. lbs.
27. Check the sprockets and tensioner for wear. The sprocket teeth should be well defined, not rounded and the valleys between the teeth should be clean. The tensioners should spin freely with no binding or unusual noise. Replace the tensioner if there is any sign of grease leaking from the seal. Clean everything with a clean, dry cloth.

WARNING: Do not spray or immerse the sprockets or tensioners in cleaning solvent. The sprocket may absorb the solvent and transfer it to the belt. The tensioners are internally lubricated and the solvent will dilute or dissolve the lubricant.

28. To reinstall, make certain the crankshaft sprocket and the silent shaft sprocket are aligned with the timing marks. Slip the belt over the crank sprocket and then over the silent shaft sprocket, keeping tension on the upper side. Make certain there is no slack in the tension side of the belt.
29. Install the tensioner and temporarily position it so that the

ENGINE AND ENGINE OVERHAUL 3

center of the pulley is to the left and above the center of the installation bolt.

30. Apply finger pressure on the tensioner, forcing it into the belt. Make certain the opposite side of the belt is taut. Tighten the installation bolt to hold the tensioner in place.

31. Push down on the center of the tension side of the belt (side opposite the tensioner pulley) and check the distance the belt deflects. Correct deflection is 6mm (¼ inch).

32. Install the flange on the crankshaft sprocket, making sure it is positioned correctly.

33. Install the timing belt sprocket onto the crankshaft and tighten the bolt to 88 ft. lbs.

34. If the timing belt tensioner was removed, reinstall it and connect the spring by hooking it under the water pump casting. If not already done, move the tensioner towards the water pump and temporarily tighten the lock bolt to hold it in place.

35. Double check that the timing marks on all the sprockets are still aligned, including the oil pump sprocket.

36. Observe the directional arrow made during removal and install the timing belt by placing on the crankshaft sprocket, then the oil pump sprocket and then over the camshaft. Maintain the belt taut on the tension side while installing it.

37. Loosen the adjuster lock bolt and allow the spring to move the pulley against the belt. Do not push or pry on the adjuster.

38. Recheck that each sprocket is still aligned with its timing marks.

39. Rotate the crankshaft clockwise until the camshaft sprocket has moved 2 teeth away from its mark. This applies tension to the belt. Do NOT rotate the crank counter-clockwise and do not push on the belt to check the tension.

40. With your fingers, apply gentle upward pressure on the tensioner and check the placement of the belt on the cam sprocket. The belt may tend to lift or float in the left side of the sprocket — roughly between 7 o'clock and 12 o'clock as you view the sprocket.

41. When the belt is correctly seated, tighten the tensioner lock bolt to 36 ft. lbs, then tighten the pivot nut to 36 ft. lbs. The order of tightening is important; if not followed, the belt could be over-tightened.

42. Check the deflection of the belt. At the middle of the right (tension) side of the belt, deflect the belt outward with your finger, toward the timing case. The distance between the belt and the line of the cover seal should be about 13mm (½ inch).

43. Adjust the valves to the correct clearances. Install the valve cover and gasket, applying sealant as needed to the cover and the half-circle plug. Connect the breather and PCV hoses.

44. Install the timing belt covers, lower one first; make certain the gaskets are properly seated and do not shift during installation. Tighten the cover bolts to 8 ft. lbs.

45. Connect the spark plug cables to the spark plugs and route the wires correctly.

46. Install the crankshaft pulley, tightening the bolts to 18 ft. lbs.

47. Install the water pump pulleys, tightening the bolts evenly to 6 ft. lbs.

48. Install the water pump belt, making certain the ribs align properly with each pulley. Install the power steering pump belt.

49. Install the belt tensioner assembly for the compressor belt and install the belt.

50. Adjust the tension on the belts and tighten the components and adjusters.

51. Install the engine mount bracket on the engine. Tighten the nuts and bolts to 42 ft. lbs.

52. Align the mount bushing with the bodywork bracket and install the through-bolt. Install the large nut and the smaller safety nut snug.

53. Carefully release tension on the engine support apparatus. When the weight of the engine is fully on the mount, tighten the through-bolt nut to 50 ft. lbs. and the smaller safety nut to 26 ft. lbs.

54. Install the high-pressure hose into its clamp. Connect the oxygen sensor wiring and install it into its retaining clip.

55. Install the left side splash shield.

56. Double check all installation items, paying particular attention to loose hoses or hanging wires, untightened nuts, poor routing of hoses and wires (too tight or rubbing) and tools left in the engine area.

57. Double check the draincock, closing it if necessary, and refill the cooling system.

58. Connect the negative battery cable.

59. Start the engine and check carefully for leaks. Allow the engine to warm up with the radiator cap removed. After the thermostat opens, adjust the coolant level to the bottom of the radiator neck and install the radiator cap.

— CAUTION —

Do not lean over the radiator neck while the engine is running. Hot coolant may be splashed out by air trapped within the system. Keep hands, tools and clothing away from the fan(s).

60. Shut the engine off and perform final adjustments of the drive belts as needed.

Galant and Sigma with V6 6G72 Engine

WARNING: Extensive disassembly is required involving bolts of different lengths. Develop a system of identifying bolts by location and component so that reassembly is made easier. Extreme damage can be caused by forcing a long bolt into a shorter passage.

The use of the correct special tools or their equivalent is REQUIRED for this procedure. The crankshaft sprocket must be removed and installed with the use of tools MB 990767-01 and MB 998715 or their equivalents. Attempts to use other tools or methods may cause damage to the sprocket and/or crankshaft.

1. Disconnect the negative battery cable.
2. Drain the engine coolant.

— CAUTION —

When draining the coolant, keep in mind that cats and dogs are attracted by the ethylene glycol antifreeze, and are quite likely to drink any that is left in an uncovered container or in puddles on the ground. This will prove fatal in sufficient quantity. Always drain the coolant into a sealable container. Coolant should be reused unless it is contaminated or several years old.

3. Remove the bolt holding the high-pressure hose and position the hose out of the way.

4. Disconnect the electrical connector at the power steering pump. Carefully disconnect the power steering pressure from the pump.

5. Place a floor jack and a broad piece of lumber under the oil pan. Raise the jack until the engine is supported but not raised.

6. Remove the through-bolt for the left engine mount.

7. Remove the four nuts holding the upper part of the engine mount to the lower bracket.

8. Remove the air conditioning compressor drive belt, then remove the belt tensioner assembly.

9. Remove the water pump drive belt.

10. Remove the power steering pump and position it out of the way.

11. Remove the belt tensioner assembly and its bracket.

12. Remove the upper outer timing belt cover (B) for the front bank of cylinders. Remove the gaskets and keep them with the cover.

NOTE: The bolts holding the belt covers are of different lengths and must be correctly placed during reinstallation. Label or diagram them as they are removed.

3-161

3 ENGINE AND ENGINE OVERHAUL

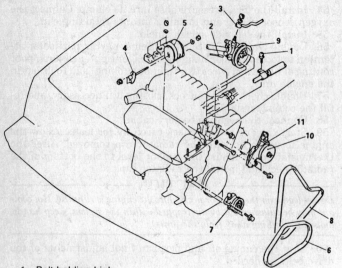

1. Bolt holding high pressure hose
2. Power steering oil pump wiring connector
3. Power steering pressure hose
4. Engine mount through-bolt
5. Engine mount bracket
6. A/C compressor belt
7. A/C belt tensioner assembly
8. Water pump belt (ribbed)
9. Power steering pump
10. Belt tensioner pulley
11. Tensioner bracket

Removal of surrounding components to gain access to the 6G72 water pump. Timing belt, covers and pulleys shown in Timing Belt section of this chapter

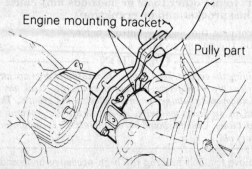

There's only one way to remove the V6 water pump

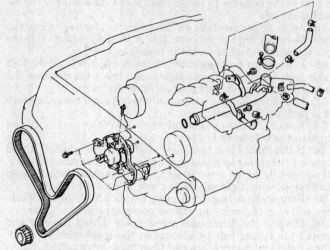

V6 water pump and related components

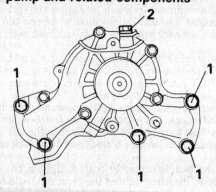

No.	d × ℓ mm (in.)	Torque Nm (ft.lbs.)
1	8×25 (.31×.98)	20–27
2	8×14 (.31×.55)	(14–19)

Correct bolt placement and torque for 6G72 water pump

13. Remove the lower portion of the engine support bracket. Remove the bolts in the order shown in the figure. Use spray lubricant to assist in removing the reamer bolt. Keep in mind that the reamer bolt may be heat-seized on the bracket. Remove it slowly.
14. Remove the small end cap from the rear bank belt cover, then remove the upper outer belt cover. Make certain all the gaskets are removed and placed with the covers.
15. Remove the left side splash shield. Turn the crankshaft clockwise until all the timing marks align. This will set the engine to TDC/compression on No. 1 cylinder.
16. Install the special counterholding tools and carefully remove the crankshaft pulley without moving the engine out of position.
17. Remove the front flange from the crank sprocket.
18. Remove the lower outer timing belt cover with its gaskets.
19. Loosen the timing belt tensioner bolt and turn the tensioner counterclockwise along the elongated hole. This will relax the tension on the belt.

20. If the timing belt is to be reused, make a chalk or crayon arrow on the belt showing the direction of rotation so that it may be reinstalled correctly.
21. Carefully slide the belt off the sprockets. Place the belt in a clean, dry, protected location away from the work area. if the tensioner is to be removed, disconnect the spring and remove the retaining bolt.
22. Remove the crankshaft sprocket.
23. Remove the water pump retaining bolts, noting their length and placement, and loosen the pump. When the pump is loose, move it so the pulley part of the pump fits into the body-work part of the engine mount; remove the pump upward. This is the only way the pump comes out.
24. Remove the O-ring from the water pipe. Clean the mating surfaces of the pump and block thoroughly, removing all traces of the old gasket. Discard the O-ring from the water pipe.
25. Check the pump for damage or cracks. Turn the shaft, checking for binding or noise. If the pump was leaking coolant through the vent hole, the internal seal has failed and the pump

ENGINE AND ENGINE OVERHAUL

must be replaced.

NOTE: Do not confuse light moisture accumulation with outright leakage. The vent hole area may be moist in normal operation.

26. Install a new gasket to the engine, making sure all the bolt holes are aligned. Place a new O-ring on the water pipe and moisten it with water or coolant. Never apply grease or oil to this seal.
27. Install the water pump and tighten the bolts finger tight. Remember that the shortest bolt mounts vertically into the pump. When all the bolts are snug, tighten them evenly to 18 ft. lbs.
28. Install the crankshaft sprocket.
29. If the tensioner was removed, it must be reinstalled. After bolting it loosely in place, connect the spring onto the water pump pin. Make certain the spring faces in the correct direction on the tensioner. Turn the tensioner to the extreme counterclockwise position on the elongated hole and tighten the bolt just enough to hold the tensioner in this position.
30. Double check the alignment of the timing marks on the camshaft and crankshaft sprockets.
31. Observing the direction of rotation marks made earlier, install the belt onto the crankshaft sprocket and then onto the rear bank camshaft sprocket. Maintain tension on the belt between the sprockets.
32. Continue installing the belt onto the water pump, the front bank cam sprocket and the tensioner.
33. With your fingers, apply gentle counterclockwise force to the rear camshaft sprocket. When the belt is taut on the tension side, the timing marks should align perfectly.
34. Install the flange on the crankshaft sprocket.
35. Loosen the bolt holding the tensioner one or two turns and allow the spring tension to draw the tensioner against the belt.
36. Using special tool MB 998716 or equivalent adapter, turn the crankshaft two complete revolutions clockwise. Turn the crank smoothly and re-align the timing marks at the end of the second revolution. This allows the tensioner to compensate for the normal amount of slack in the belt.
37. With the timing marks aligned, tighten the tensioner bolts to 18 ft. lbs.
38. Using a belt tension gauge, measure the belt tension at a point halfway between the crankshaft sprocket and the rear camshaft sprocket. Correct value is 57-84 lbs.
39. Making certain all the gaskets are correctly in place, install the lower timing belt cover and tighten the bolts to 8 ft. lbs.
40. Install the crankshaft pulley. Use the special tools to counterhold it and tighten the bolt to 112 ft. lbs.
41. Install the left splash shield.
42. Install the upper outer cover on the rear bank camshaft sprocket and install the smaller end cap. Tighten the bolts to 8 ft. lbs.
43. Install the lower portion of the engine support bracket. Install the bolts in the correct order. Of the 3 mounting bolts arranged vertically on the right side of the mount, the upper two are tightened to 50 ft. lbs; the bottom one is tightened to 78 ft. lbs. The two smaller bolts in the side of the mount should be tightened to 30 ft. lbs. Use spray lubricant to help with the reamer bolt, and tighten it slowly. Access may be easier if the engine is elevated slightly with the floor jack.
44. Install the cover and gaskets for the front bank camshaft sprocket.
45. Install the tensioner and bracket for the ribbed belt.
46. Install the power steering pump onto its bracket.
47. Install the ribbed belt and adjust it as needed. Make certain it is properly seated on all the pulleys.
48. Install the tensioner and bracket for the air conditioner drive belt and install the belt, adjusting it as necessary.
49. Install the upper section of the engine mounting bracket to the lower section. Tighten the nuts to 50 ft. lbs.
50. Adjust the jack so that the bushing of the engine mount aligns with the bodywork bracket. Install the through-bolt and tighten the nuts snug.
51. Carefully lower the jack, allowing the full weight of the engine to bear on the mount. Tighten the nut on the through-bolt to 50 ft. lbs. and the smaller safety nut to 26 ft. lbs.
52. Install the pressure hose to the power steering pump and tighten the nut to 32 ft. lbs.
53. Connect the wiring to the power steering pump and rebolt the bracket for the high pressure hose.
54. Double check all installation items, paying particular attention to loose hoses or hanging wires, untightened nuts, poor routing of hoses and wires (too tight or rubbing) and tools left in the engine area.
55. Connect the negative battery cable. Start the engine and let it idle, listening for any unusual noises from the area of the timing belt. Possible causes of noise are the belt rubbing against the covers or a sprocket flange, the belt being too loose and slapping, or a tensioner binding. Do not accelerate the engine if abnormal noises are heard from the timing belt train — severe damage can result.
56. Shut the engine off and perform final adjustments of the drive belts and/or engine specifications as needed.
57. Double check the draincock, closing it if necessary, and refill the cooling system.
58. Start the engine and check carefully for leaks of coolant, vacuum or power steering oil. Allow the engine to warm up with the radiator cap removed. After the thermostat opens, adjust the coolant level to the bottom of the radiator neck and install the radiator cap.

--- **CAUTION** ---

Do not lean over the radiator neck while the engine is running. Hot coolant may be splashed out by air trapped within the system. Keep hands, tools and clothing away from the fan(s).

Galant with 4G63 SOHC Engine

WARNING: Extensive disassembly is required involving bolts of different lengths. Develop a system of identifying bolts by location and component so that reassembly is made easier. Extreme damage can be caused by forcing a long bolt into a shorter passage.

1. Disconnect the negative battery cable.
2. Remove the left splash shield under the engine.
3. Drain the engine coolant.

--- **CAUTION** ---

When draining the coolant, keep in mind that cats and dogs are attracted by the ethylene glycol antifreeze, and are quite likely to drink any that is left in an uncovered container or in puddles on the ground. This will prove fatal in sufficient quantity. Always drain the coolant into a sealable container. Coolant should be reused unless it is contaminated or several years old.

4. Remove the air cleaner assembly and the air intake duct.
5. Loosen the appropriate component or adjuster and remove (in order) the air conditioning compressor drive belt, the power steering belt and the ribbed belt driving the water pump.
6. Place a floor jack and a broad piece of lumber under the oil pan. Elevate the jack just enough to support the engine without raising it.
7. Remove the nuts holding the through-bolt for the left engine bracket and remove the bolt. It may be necessary to adjust the jack tension slightly to allow the bolt to come free. Use the jack to keep the engine in its normal position while the mount is disconnected.
8. Remove the nuts and bolts holding the engine bracket to the engine and remove the bracket.
9. Label and remove the coil wire and the spark plug wires from the distributor cap and move them out of the way.

3-163

3 ENGINE AND ENGINE OVERHAUL

10. Remove the power steering pump from its bracket and move it, with the hoses attached, out of the way. Hang the pump from a piece of stiff wire or string; do not allow the pump to hang by the hoses.
11. Remove the power steering pump bracket from the engine.
12. Remove the water pump pulleys.
13. Remove the A/C drive pulley and the crankshaft pulley from the crankshaft.
14. Remove the upper timing belt cover and its gaskets. Rotate the crankshaft clockwise until the timing marks on the crank pulley align at 0 on the scale (on the lower timing cover) and the mark on the camshaft pulley aligns with the indicator on the head. Aligned means perfect, not close — if you miss, continue clockwise until the marks are matched. This sets the motor at TDC/compression on No.1 cylinder.

WARNING: Once the engine is set to this position, the crank and camshafts must not be moved out of place. Severe damage can result.

15. Remove the air conditioner pulley from the crankshaft and remove the crankshaft pulley. Remember that the pulley (and therefore, the crankshaft) must be held steady while the bolt is loosened. This can be made easier by wrapping a scrapped belt tightly around the pulley. Mitsubishi has a special tool (MD 998747) for this trick, but other variations work almost as well. DO NOT use any of the drive belts which will be reinstalled; the belt used to hold the pulley will be damaged or weakened.
16. Remove the lower timing belt cover and gaskets. Note the different bolt lengths and placements.
17. Loosen (do not remove) both bolts in the timing belt tensioner. Move the tensioner toward the water pump and tighten the lock bolt — the one in the slotted hole — to hold the tensioner in place.
18. Before removing the timing belt, mark it with an arrow to show the direction of rotation. If the belt is to be reused, it must be reinstalled so that it rotates in the same direction as before.
19. Remove the timing belt, keeping it free of grease, oil and fluids. Do not crease or crimp the belt.
20. Remove the crankshaft sprocket and the flange.
21. Loosen the adjuster for the silent shaft belt (timing belt B). Carefully remove the belt. The sprockets and tensioners may be removed if desired.
22. Remove the water pump retaining bolts, the alternator bracket and remove the water pump. Remove and discard the O-ring from the water intake pipe.
23. Clean the mating surfaces thoroughly, removing any trace of the old gasket. Check the pump thoroughly for damage or cracks. Turn the shaft, checking for binding or noise. If the pump was leaking coolant through the vent hole, the internal seal has failed and the pump must be replaced.

NOTE: Do not confuse light moisture accumulation with outright leakage. The vent hole area may be moist in normal operation.

24. Install a new gasket to the engine, making sure all the bolt holes are aligned. Place a new O-ring on the water pipe and moisten it with water or coolant. Never apply grease or oil to this seal.
25. Install the water pump and tighten the bolts finger tight. Don't forget to install the alternator bracket. When all the bolts are snug, tighten them to the correct torque: Bolts with head mark **4** to 10 ft. lbs. and bolts with head mark **7** to 18 ft. lbs.
26. Check the sprockets and tensioner for wear. The sprocket teeth should be well defined, not rounded and the valleys between the teeth should be clean. The tensioners should spin freely with no binding or unusual noise. Replace the tensioner if there is any sign of grease leaking from the seal. Clean everything with a clean, dry cloth.

WARNING: Do not spray or immerse the sprockets or tensioners in cleaning solvent. The sprocket may absorb the solvent and transfer it to the belt. The tensioners are internally lubricated and the solvent will dilute or dissolve the lubricant.

27. To reinstall, make certain the crankshaft sprocket and the silent shaft sprocket are aligned with the timing marks. Slip the belt over the crank sprocket and then over the silent shaft sprocket, keeping tension on the upper side. Make certain there is no slack in the tension side of the belt.
28. Install the tensioner and temporarily position it so that the center of the pulley is to the left and above the center of the installation bolt.
29. Apply finger pressure on the tensioner, forcing it into the belt. Make certain the opposite side of the belt is taut. Tighten the installation bolt to hold the tensioner in place.
30. Push down on the center of the tension side of the belt (side opposite the tensioner pulley) and check the distance the belt deflects. Correct deflection is 6mm (¼ inch).
31. Install the flange on the crankshaft sprocket, making sure it is positioned correctly.
32. Install the outer crankshaft sprocket and/or the oil pump sprocket if either was removed.
33. Install the timing belt tensioner if it was removed and position it towards the water pump. Tighten the lock bolt to hold it in this position.
34. Make certain that all the timing marks for the camshaft sprocket, the crankshaft sprocket and the oil pump sprocket are correctly aligned.
35. Once the oil pump sprocket is aligned, remove the plug in the left side of the block. Insert a Phillips screwdriver with a shaft diameter of 8mm (0.31 in.) into the plug hole. It should enter 58mm (2.3 in.) or more. Do not remove the screwdriver until the timing belt is completely installed. If the screwdriver shaft only enters about 25mm (1 inch), turn the oil pump sprocket one rotation and align the timing mark again; insert the screwdriver to block the silent shaft from turning.
36. Observe the directional arrow made during removal and install the timing belt by placing on the crankshaft sprocket, then the oil pump sprocket and then over the camshaft. Maintain the belt taut on the tension side while installing it.
37. Loosen the adjuster lock bolt and allow the spring to move the pulley against the belt. Do not push or pry on the adjuster.
38. Recheck that each sprocket is still aligned with its timing marks.
39. Remove the screwdriver from the silent shaft. Rotate the crankshaft clockwise until the camshaft sprocket has moved 2 teeth away from its mark. This applies tension to the belt. Do NOT rotate the crank counter-clockwise and do not push on the belt to check the tension.
40. With your fingers, apply gentle upward pressure on the tensioner and check the placement of the belt on the cam sprocket. The belt may tend to lift or float in the left side of the sprocket — roughly between 7 o'clock and 12 o'clock as you view the sprocket.
41. When the belt is correctly seated, tighten the tensioner lock bolt to 36 ft. lbs., then tighten the pivot nut to 36 ft. lbs. The order of tightening is important; if not followed, the belt could be over-tightened.
42. Check the deflection of the belt. At the middle of the right (tension) side of the belt, deflect the belt outward with your finger, toward the timing case. The distance between the belt and the line of the cover seal should be about 13mm (½ inch).
43. Tighten the crankshaft sprocket bolt to 88 ft. lbs.
44. Install the lower timing belt cover, paying attention to the correct placement of the bolts. Tighten the bolts to 8 ft. lbs.
45. Install the upper timing belt cover and gaskets, tightening the bolts to 8 ft. lbs. Remember to install the guides for the spark plug wires.

ENGINE AND ENGINE OVERHAUL 3

46. Install the crankshaft damper pulley. Tighten the bolts to 18 ft. lbs.
47. Install the water pump pulleys, tightening the bolts to 6 ft. lbs.
48. Reinstall the valve cover and gasket.
49. Connect the air hoses and vacuum lines to the valve cover.
50. Connect the wiring for the oxygen sensor if it was disconnected.
51. Install the power steering pump bracket to the engine block. Tighten the mounting bolts to 18 ft. lbs.
52. Install the power steering pump to the bracket, tightening the bolts just enough to hold the pump in place.
53. Connect the coil and spark plug cables to the distributor cap and/or the spark plugs. Remember to route the wiring correctly on the wire supports.
54. Install the motor mount bracket to the engine and tighten the nuts and bolts to 42 ft. lbs.
55. Align the mount bushing with the body bracket and install the through-bolt. Tighten both the large nut and the smaller safety nut snug.
56. Remove the jack from under the engine. With the weight of the engine on the mount, tighten the large nut on the through-bolt to 72 ft. lbs. and the smaller safety nut to 38 ft. lbs.
57. Install the ribbed water pump drive belt, making sure it is correctly seated on each pulley.
58. Install the power steering belt and the compressor drive belt. Adjust each belt to the correct tension and tighten the fittings or adjuster on each component.
59. Install the air cleaner assembly and intake duct.
60. Double check all installation items, paying particular attention to loose hoses or hanging wires, untightened nuts, poor routing of hoses and wires (too tight or rubbing) and tools left in the engine area.
61. Double check the draincock, closing it if necessary, and refill the cooling system.
62. Connect the negative battery cable. Start the engine and let it idle, listening for any unusual noises from the area of the timing belt. Possible causes of noise are the belt rubbing against the covers or a sprocket flange, the belt being too loose and slapping, or a tensioner binding. Do not accelerate the engine if abnormal noises are heard from the timing belt train — severe damage can result.
63. Check carefully for leaks. Allow the engine to warm up with the radiator cap removed. After the thermostat opens, adjust the coolant level to the bottom of the radiator neck and install the radiator cap.

―――――――― CAUTION ――――――――
Do not lean over the radiator neck while the engine is running. Hot coolant may be splashed out by air trapped within the system. Keep hands, tools and clothing away from the fan(s).

64. Reinstall the splash shield under the car. Final adjustment of the drive belts and accelerator cable may be needed.

Galant with 4G63 DOHC Engine

WARNING: Extensive disassembly is required involving bolts of different lengths. Develop a system of identifying bolts by location and component so that reassembly is made easier. Extreme damage can be caused by forcing a long bolt into a shorter passage.

1. Disconnect the negative battery cable.
2. Remove the left splash shield under the engine.
3. Drain the engine coolant.

―――――――― CAUTION ――――――――
When draining the coolant, keep in mind that cats and dogs are attracted by the ethylene glycol antifreeze, and are quite likely to drink any that is left in an uncovered container or in puddles on the ground. This will prove fatal in sufficient quantity. Always drain the coolant into a sealable container. Coolant should be reused unless it is contaminated or several years old.

4. Loosen the appropriate component or adjuster and remove (in order) the air conditioning compressor drive belt, the power steering belt and the ribbed belt driving the water pump.
5. Place a floor jack and a broad piece of lumber under the oil pan. Elevate the jack just enough to support the engine without raising it.
6. Remove the nuts holding the through-bolt for the left engine bracket and remove the bolt. It may be necessary to adjust the jack tension slightly to allow the bolt to come free. Use the jack to keep the engine in its normal position while the mount is disconnected.
7. Remove the nuts and bolts holding the engine bracket to the engine and remove the bracket.
8. Label and remove the coil wire and the spark plug wires from the distributor cap and move them out of the way.
9. Remove the spark plug wire cover and remove the leads from the plugs.
10. Remove the power steering pump from its bracket and move it, with the hoses attached, out of the way. Hang the pump from a piece of stiff wire or string; do not allow the pump to hang by the hoses.
11. Remove the power steering pump bracket from the engine.
12. Label and disconnect each air tube or vacuum line running to the valve cover.
13. Remove the valve cover and its gasket.
14. Remove the water pump pulleys.
15. Remove the A/C drive pulley and the crankshaft pulley from the crankshaft.
16. Remove the (front) upper and lower timing belt covers.

WARNING: The covers use 4 different lengths of bolts. Label or diagram them as they are removed.

17. Remove the center cover, the PCV hose and the breather hose from the valve cover. Label and disconnect the spark plug cables from the spark plugs.
18. Remove the valve cover and its gasket. Remove the half-circle plug from the head.
19. Turn the crankshaft clockwise and align all the timing marks. The timing marks on the camshaft sprockets should align with the upper surface of the cylinder head and the dowel pins (guide pins) at the center of the camshaft sprockets should be up. Aligned means perfect, not close — if you miss, continue clockwise until the marks are matched. This sets the motor at TDC/compression on No. 1 cylinder.

WARNING: Once the engine is set to this position, the crank and camshafts must not be moved out of place. Severe damage can result.

20. Remove the auto-tensioner.
21. Before removing the timing belt, mark it with an arrow to show the direction of rotation. If the belt is to be reused, it must be reinstalled so that it rotates in the same direction as before.
22. Remove the timing belt, keeping it free of grease, oil and fluids. Do not crease or crimp the belt. After the belt is clear, remove the tensioner pulley.
23. Remove the crankshaft sprocket and the flange.
24. Loosen the adjuster for the silent shaft belt (timing belt B). Carefully remove the belt. The sprockets and tensioners may be removed if desired.
25. Remove the water pump retaining bolts, the alternator bracket and remove the water pump. Remove and discard the O-ring from the water intake pipe.
26. Clean the mating surfaces thoroughly, removing any trace of the old gasket. Check the pump thoroughly for damage or cracks. Turn the shaft, checking for binding or noise. If the pump was leaking coolant through the vent hole, the internal seal has failed and the pump must be replaced.

3-165

3 ENGINE AND ENGINE OVERHAUL

NOTE: Do not confuse light moisture accumulation with outright leakage. The vent hole area may be moist in normal operation.

27. Install a new gasket to the engine, making sure all the bolt holes are aligned. Place a new O-ring on the water pipe and moisten it with water or coolant. Never apply grease or oil to this seal.
28. Install the water pump and tighten the bolts finger tight. Don't forget to install the alternator bracket. When all the bolts are snug, tighten them to the correct torque: Bolts with head mark **4** to 10 ft. lbs. and bolts with head mark **7** to 18 ft. lbs.
29. Check the sprockets and tensioner for wear. The sprocket teeth should be well defined, not rounded and the valleys between the teeth should be clean. The tensioners should spin freely with no binding or unusual noise. Replace the tensioner if there is any sign of grease leaking from the seal. Clean everything with a clean, dry cloth.

WARNING: Do not spray or immerse the sprockets or tensioners in cleaning solvent. The sprocket may absorb the solvent and transfer it to the belt. The tensioners are internally lubricated and the solvent will dilute or dissolve the lubricant.

30. Inspect the auto-tensioner for wear or leakage around the seals. Closely inspect the end of the pushrod for any wear. Measure the projection of the pushrod. It should be 12mm (0.47 in.). If it is out of specification, the auto tensioner must be replaced.
31. Mount the auto-tensioner in a vise with protected jaws. Keep the unit level at all times, and, if the plug at the bottom projects, protect it with a common washer. Smoothly tighten the vise to compress the adjuster tip; if the the rod is easily retracted, the unit has lost tension and should be replaced. You should feel a fair amount of resistance when compressing the rod.
32. As the rod is contracted into the body of the auto-tensioner, watch for the hole in the pushrod to align with the hole in the body. When this occurs, pin the pushrod in place with a piece of stiff wire 1.5mm. It will be easier to install the tensioner with the pushrod retracted. Leave the wire in place and remove the assembly from the vise.
33. To reinstall, make certain the crankshaft sprocket and the silent shaft sprocket are aligned with the timing marks. Slip the belt over the crank sprocket and then over the silent shaft sprocket, keeping tension on the upper side. Make certain there is no slack in the tension side of the belt.
34. Install the tensioner and temporarily position it so that the center of the pulley is to the left and above the center of the installation bolt.
35. Apply finger pressure on the tensioner, forcing it into the belt. Make certain the opposite side of the belt is taut. Tighten the installation bolt to hold the tensioner in place.
36. Push down on the center of the tension side of the belt (side opposite the tensioner pulley) and check the distance the belt deflects. Correct deflection is 6mm (¼ inch).
37. Install the flange on the crankshaft sprocket, making sure it is positioned correctly.
38. Install the outer crankshaft sprocket and/or the oil pump sprocket if either was removed.
39. Install the auto-tensioner onto the engine, tightening the bolts to 18 ft. lbs.
40. Install the tensioner pulley onto the tensioner arm. Locate the pinholes in the pulley shaft to the left of the center bolt. Tighten the center bolt finger tight. Leave the blocking wire in the auto-tensioner.
41. Double check the alignment of the timing marks for all the sprockets. The camshaft sprocket marks must face each other and align with the top surface of the cylinder head. Don't forget to check the oil pump sprocket alignment as well as the crankshaft sprocket.

NOTE: When you let go of the exhaust camshaft sprocket, it will move one tooth counter-clockwise. Take this into account when installing the belt around the sprockets.

42. Observing the direction of rotation mark made earlier, install the timing belt as follows:
    a. Install the timing belt around the tensioner pulley and the crankshaft sprocket and hold the belt on the tensioner with your left hand.
    b. Pulling the belt with your right hand, install it around the oil pump sprocket.
    c. Install the belt around the idler pulley and then around the intake camshaft sprocket.
    d. Double check the exhaust camshaft sprocket alignment mark. It has probably moved one tooth; if so, rotate it clockwise to align the timing mark with the top of the cylinder head and install the belt over the sprocket. The belt is a snug fit but can be fitted with the fingers; do not use tools to force the belt onto the sprocket.
    e. Raise the tensioner pulley against the belt to prevent sagging and temporarily tighten the center bolt.
43. Turn the crankshaft ¼ turn (90°) counterclockwise, then turn it clockwise and align the timing marks.
44. Loosen the center bolt on the tensioner pulley and attach special tool MD 998752 and a torque wrench. (This is a purpose-built tool; substitutes will be hard to find.) Apply a torque of 22–24 inch lbs. to the tensioner pulley; this establishes the correct loading against the belt.

NOTE: You must use a torque wrench with a 0–3 ft. lb. scale. Normal scales will not read to the accuracy needed. If the body interferes with the tools, elevate the jack slightly to raise the engine above the interference point.

45. Hold the tensioner pulley in place with the torque wrench and tighten the center bolt of the tensioner pulley to 36 ft. lbs. Remove the torque wrench assembly.
46. On the left side of the engine, located in the rear lower timing belt cover, is a rubber plug. Remove the plug and install special tool MD 998738 (another one difficult to substitute). Screw the tool in until it makes contact with the tensioner arm. After the point of contact, screw it in some more to compress the arm slightly.
47. Remove the wire holding the tip of the auto-tensioner. The tip will push outward and engage the tensioner arm. Remove the screw tool.
48. Rotate the crankshaft 2 complete turns clockwise. Allow everything to sit as is for about 15 minutes — longer if the temperature is cool — then check the distance the tip of the auto-tensioner protrudes. (It is almost impossible with the engine in the car; if you can measure the distance, it should be 3.8-4.5mm (0.15-0.18 in.). The alternate method is given in the next step.).
49. Reinstall the special screw tool into the left support bracket until the end just makes contact with the tensioner arm. Starting from that position, turn the tool against the arm, counting the number of full turns until the tensioner arm contacts the auto-tensioner body. At this point, the arm has compressed the tensioner tip through its projection distance. The correct number of turns to this point is 2.5 to 3 turns. If you took the correct number of turns to bottom the arm, the projection was correct to begin with and you may now unwind the screw tool EXACTLY the number of turns you turned it in.

If your number of turns was too high or too low when the arm bottomed, you have a problem with the tensioner and should replace it.

50. Remove the special screw tool and install the rubber plug into the access hole in the timing belt cover.
51. Install the half-circle plug with the proper sealant, then install the valve cover and gasket.
52. Install the spark plug wires, the breather and PCV hoses and the center cover.

ENGINE AND ENGINE OVERHAUL 3

53. Install the lower and upper timing belt covers, paying attention to the correct placement of the four different sizes of bolts.
54. Install the crankshaft pulley and tighten the bolt. Correct torque for is 18 ft. lbs. for the bolts holding the pulley to the sprocket. Install the A/C drive pulley, tightening the bolts to 12 ft. lbs.
55. Install the water pump pulleys, tightening the bolts to 6 ft. lbs.
56. Reinstall the valve cover and gasket.
57. Connect the air hoses and vacuum lines to the valve cover.
58. Connect the wiring for the oxygen sensor if it was disconnected.
59. Install the power steering pump bracket to the engine block. Tighten the mounting bolts to 18 ft. lbs.
60. Install the power steering pump to the bracket, tightening the bolts just enough to hold the pump in place.
61. Connect the coil and spark plug cables to the distributor cap and/or the spark plugs. Remember to route the wiring correctly on the wire supports.
62. Install the motor mount bracket to the engine and tighten the nuts and bolts to 42 ft. lbs.
63. Align the mount bushing with the body bracket and install the through-bolt. Tighten both the large nut and the smaller safety nut snug.
64. Remove the jack from under the engine. With the weight of the engine on the mount, tighten the large nut on the through-bolt to 72 ft. lbs. and the smaller safety nut to 38 ft. lbs.
65. Install the ribbed water pump drive belt, making sure it is correctly seated on each pulley.
66. Install the power steering belt and the compressor drive belt. Adjust each belt to the correct tension and tighten the fittings or adjuster on each component.
67. Double check all installation items, paying particular attention to loose hoses or hanging wires, untightened nuts, poor routing of hoses and wires (too tight or rubbing) and tools left in the engine area.
68. Double check the draincock, closing it if necessary, and refill the cooling system.
69. Connect the negative battery cable.
70. Start the engine and let it idle, listening for any unusual noises from the area of the timing belt. Possible causes of noise are the belt rubbing against the covers or a sprocket flange, the belt being too loose and slapping, or a tensioner binding. Do not accelerate the engine if abnormal noises are heard from the timing belt train — severe damage can result.
71. Check carefully for leaks. Allow the engine to warm up with the radiator cap removed. After the thermostat opens, adjust the coolant level to the bottom of the radiator neck and install the radiator cap.

CAUTION
Do not lean over the radiator neck while the engine is running. Hot coolant may be splashed out by air trapped within the system. Keep hands, tools and clothing away from the fan(s).

72. Reinstall the splash shield under the car. Final adjustment of the drive belts and accelerator cable may be needed.

Montero with 6G72 V6 Engine

WARNING: Extensive disassembly is required involving bolts of different lengths. Develop a system of identifying bolts by location and component so that reassembly is made easier. Extreme damage can be caused by forcing a long bolt into a shorter passage.

The use of the correct special tools or their equivalent is REQUIRED for this procedure. The crankshaft sprocket must be removed and installed with the use of tools MB 990767-01 and MB 998715 or their equivalents. Attempts to use other tools or methods may cause damage to the sprocket and/or crankshaft.

1. Disconnect the negative battery cable.
2. Drain the engine coolant.

CAUTION
When draining the coolant, keep in mind that cats and dogs are attracted by the ethylene glycol antifreeze, and are quite likely to drink any that is left in an uncovered container or in puddles on the ground. This will prove fatal in sufficient quantity. Always drain the coolant into a sealable container. Coolant should be reused unless it is contaminated or several years old.

3. Remove the bolt holding the high-pressure hose and position the hose out of the way.
4. Disconnect the electrical connector at the power steering pump. Carefully disconnect the power steering pressure from the pump.
5. Place a floor jack and a broad piece of lumber under the oil pan. Raise the jack until the engine is supported but not raised.
6. Remove the through-bolt for the left engine mount.
7. Remove the four nuts holding the upper part of the engine mount to the lower bracket.
8. Remove the air conditioning compressor drive belt, then remove the belt tensioner assembly.
9. Remove the water pump drive belt.
10. Remove the power steering pump and position it out of the way.
11. Remove the engine fan and pulley.
12. Remove the belt tensioner assembly and its bracket.
13. Remove the upper outer timing belt cover (B) for the front bank of cylinders. Remove the gaskets and keep them with the cover.

NOTE: The bolts holding the belt covers are of different lengths and must be correctly placed during reinstallation. Label or diagram them as they are removed.

14. Remove the lower portion of the engine support bracket. Remove the bolts in the order shown in the figure. Use spray lubricant to assist in removing the reamer bolt. Keep in mind that the reamer bolt may be heat-seized on the bracket. Remove it slowly.
15. Remove the small end cap from the rear bank belt cover, then remove the upper outer belt cover. Make certain all the gaskets are removed and placed with the covers.
16. Remove the left side splash shield. Turn the crankshaft clockwise until all the timing marks align. This will set the engine to TDC/compression on No. 1 cylinder.
17. Install the special counterholding tools and carefully remove the crankshaft pulley without moving the engine out of position.
18. Remove the front flange from the crank sprocket.
19. Remove the lower outer timing belt cover with its gaskets.
20. Loosen the timing belt tensioner bolt and turn the tensioner counterclockwise along the elongated hole. This will relax the tension on the belt.
21. If the timing belt is to be reused, make a chalk or crayon arrow on the belt showing the direction of rotation so that it may be reinstalled correctly.
22. Carefully slide the belt off the sprockets. Place the belt in a clean, dry, protected location away from the work area. if the tensioner is to be removed, disconnect the spring and remove the retaining bolt.
23. Remove the crankshaft sprocket.
24. Remove the water pump retaining bolts, noting their length and placement, and loosen the pump. When the pump is loose, move it so the pulley part of the pump fits into the bodywork part of the engine mount; remove the pump upward. This is the only way the pump comes out.
25. Remove the O-ring from the water pipe. Clean the mating surfaces of the pump and block thoroughly, removing all traces of the old gasket. Discard the O-ring from the water pipe.

ENGINE AND ENGINE OVERHAUL

26. Check the pump for damage or cracks. Turn the shaft, checking for binding or noise. If the pump was leaking coolant through the vent hole, the internal seal has failed and the pump must be replaced.

NOTE: Do not confuse light moisture accumulation with outright leakage. The vent hole area may be moist in normal operation.

To install:

27. Install a new gasket to the engine, making sure all the bolt holes are aligned. Place a new O-ring on the water pipe and moisten it with water or coolant. Never apply grease or oil to this seal.
28. Install the water pump and tighten the bolts finger tight. Remember that the shortest bolt mounts vertically into the pump. When all the bolts are snug, tighten them evenly to 18 ft. lbs.
29. Install the crankshaft sprocket.
30. If the tensioner was removed, it must be reinstalled. After bolting it loosely in place, connect the spring onto the water pump pin. Make certain the spring faces in the correct direction on the tensioner. Turn the tensioner to the extreme counter-clockwise position on the elongated hole and tighten the bolt just enough to hold the tensioner in this position.
31. Double check the alignment of the timing marks on the camshaft and crankshaft sprockets.
32. Observing the direction of rotation marks made earlier, install the belt onto the crankshaft sprocket and then onto the rear bank camshaft sprocket. Maintain tension on the belt between the sprockets.
33. Continue installing the belt onto the water pump, the front bank cam sprocket and the tensioner.
34. With your fingers, apply gentle counterclockwise force to the rear camshaft sprocket. When the belt is taut on the tension side, the timing marks should align perfectly.
35. Install the flange on the crankshaft sprocket.
36. Loosen the bolt holding the tensioner one or two turns and allow the spring tension to draw the tensioner against the belt.
37. Using special tool MB 998716 or equivalent adapter, turn the crankshaft two complete revolutions clockwise. Turn the crank smoothly and re-align the timing marks at the end of the second revolution. This allows the tensioner to compensate for the normal amount of slack in the belt.
38. With the timing marks aligned, tighten the tensioner bolts to 18 ft. lbs.
39. Using a belt tension gauge, measure the belt tension at a point halfway between the crankshaft sprocket and the rear camshaft sprocket. Correct value is 57-84 lbs.
40. Making certain all the gaskets are correctly in place, install the lower timing belt cover and tighten the bolts to 8 ft. lbs.
41. Install the crankshaft pulley. Use the special tools to counterhold it and tighten the bolt to 112 ft. lbs.
42. Install the left splash shield.
43. Install the upper outer cover on the rear bank camshaft sprocket and install the smaller end cap. Tighten the bolts to 8 ft. lbs.
44. Install the lower portion of the engine support bracket. Install the bolts in the correct order. Of the 3 mounting bolts arranged vertically on the right side of the mount, the upper two are tightened to 50 ft. lbs; the bottom one is tightened to 78 ft. lbs. The two smaller bolts in the side of the mount should be tightened to 30 ft. lbs. Use spray lubricant to help with the reamer bolt, and tighten it slowly. Access may be easier if the engine is elevated slightly with the floor jack.
45. Install the cover and gaskets for the front bank camshaft sprocket.
46. Install the tensioner and bracket for the ribbed belt.
47. Install the power steering pump onto its bracket.
48. Install the ribbed belt and adjust it as needed. Make certain it is properly seated on all the pulleys.
49. Install the tensioner and bracket for the air conditioner drive belt and install the belt, adjusting it as necessary.
50. Install the upper section of the engine mounting bracket to the lower section. Tighten the nuts to 50 ft. lbs.
51. Adjust the jack so that the bushing of the engine mount aligns with the bodywork bracket. Install the through-bolt and tighten the nuts snug.
52. Carefully lower the jack, allowing the full weight of the engine to bear on the mount. Tighten the nut on the through-bolt to 50 ft. lbs. and the smaller safety nut to 26 ft. lbs.
53. Install the pressure hose to the power steering pump and tighten the nut to 32 ft. lbs.
54. Connect the wiring to the power steering pump and rebolt the bracket for the high pressure hose.
55. Install the engine fan and pulley.
56. Double check all installation items, paying particular attention to loose hoses or hanging wires, untightened nuts, poor routing of hoses and wires (too tight or rubbing) and tools left in the engine area.
57. Connect the negative battery cable. Start the engine and let it idle, listening for any unusual noises from the area of the timing belt. Possible causes of noise are the belt rubbing against the covers or a sprocket flange, the belt being too loose and slapping, or a tensioner binding. Do not accelerate the engine if abnormal noises are heard from the timing belt train — severe damage can result.
58. Shut the engine off and perform final adjustments of the drive belts and/or engine specifications as needed.
59. Double check the draincock, closing it if necessary, and refill the cooling system.
60. Start the engine and check carefully for leaks of coolant, vacuum or power steering oil. Allow the engine to warm up with the radiator cap removed. After the thermostat opens, adjust the coolant level to the bottom of the radiator neck and install the radiator cap.

CAUTION

Do not lean over the radiator neck while the engine is running. Hot coolant may be splashed out by air trapped within the system. Keep hands, tools and clothing away from the fan(s).

Pickup with 4D55 Diesel

1. Perform this procedure only on a completely cold engine. Minimum cooling time is 3-4 hours; overnight cold is preferred.
2. Disconnect the negative battery cable.
3. Drain the coolant into clean containers for reuse.

CAUTION

When draining the coolant, keep in mind that cats and dogs are attracted by the ethylene glycol antifreeze, and are quite likely to drink any that is left in an uncovered container or in puddles on the ground. This will prove fatal in sufficient quantity. Always drain the coolant into a sealable container. Coolant should be reused unless it is contaminated or several years old.

4. Disconnect the lower radiator hose from the water pump.
5. Loosen the bolts holding the fan, fan pulley and fan clutch.
6. Loosen the drive belt and remove it.
7. Remove the cooling fan, fan clutch and pulley.
8. Remove all the retaining bolts and remove the water pump. If it is stuck in place, it may be tapped with a plastic or rubber mallet.
9. Inspect the pump parts for any sign of wear or damage. Turn the shaft by hand, checking for binding, noise or uneven rotation.
10. Check the seal area for any sign of leakage. Any fluid leakage from the breather hole requires replacement of the pump.

NOTE: Very slight moisture or vapor around the hole is normal. Use common sense to determine outright leakage.

ENGINE AND ENGINE OVERHAUL 3

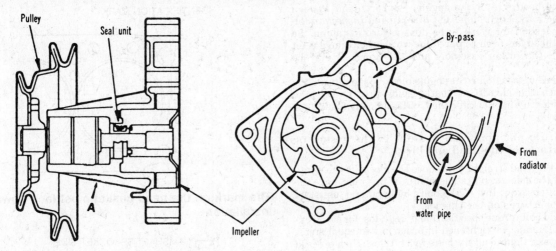

Diesel water pump; point "A" is the breather hole in the case

To install:
11. Make certain the mating surfaces are free of any trace of the old gasket.
12. Use a new gasket, install the pump and tighten the bolts evenly and alternately. Do not overtighten.
13. Install the fan clutch, pulley and cooling fan.
14. Install the drive belt and adjust it to the proper tension.
15. Connect the lower radiator hose; make certain the clamp is correctly placed and secure.
16. Check the draincock and make sure it is firmly closed. Refill the cooling system with coolant.
17. Connect the negative battery cable. Start the engine and check for leaks.

Cylinder Head and Gasket

REMOVAL AND INSTALLATION

NOTE: The use of the correct special tools or their equivalent is REQUIRED for this procedure. The head bolts used on some Mitsubishi engines require a special hex wrench; the bolts heads are round with internal facets. Use tool MD 998051-01 or equivalent.

Several external components must be removed before the head can be removed. The steps below are general in nature; specific steps for individual components may be found in the appropriate part or Section. Refer to these other procedures frequently; use particular care during the timing belt removal and installation. Extreme damage can result from improper work.

Cordia and Tredia

1. Disconnect the negative battery cable.
2. Drain the cooling system.

CAUTION

When draining the coolant, keep in mind that cats and dogs are attracted by the ethylene glycol antifreeze, and are quite likely to drink any that is left in an uncovered container or in puddles on the ground. This will prove fatal in sufficient quantity. Always drain the coolant into a sealable container. Coolant should be reused unless it is contaminated or several years old.

---

3. Remove the carburetor or injection mixer from the intake manifold.
4. Remove the intake manifold.
5. Remove the exhaust manifold.

6. Remove the valve cover, labeling all lines and wires during disassembly.
7. Remove the upper timing belt cover.
8. Turn the motor until the timing marks align and the valves for cylinder No.1 are closed (rockers are both loose). If the marks align, but the rockers are not loose, turn the crankshaft another 360°. This will put the engine at TDC compression on cylinder No.1.
9. Loosen and remove the bolt holding the upper timing belt sprocket to the camshaft. Hold the sprocket inside the belt and maintain the tension on the belt at all times. Carefully move the sprocket with the belt away from the head and support it on the pad in the lower cover. It may be necessary to use a small shim of soft material to keep the sprocket in position and under tension.

WARNING: If the belt tension is lost or the sprocket fall out of place, the timing belt must be reinstalled "from scratch", requiring the removal of the lower cover and other items.

10. Double check the head to insure no remaining wires or hoses are connected.
11. Loosen the head bolts in the order shown (this is important) and in three passes until all are finger loose. Remove the bolts.
12. Rock the head gently to break it loose; if tapping is necessary, do so with a rubber or wooden mallet at the corners of the head. DO NOT pry the head up by wedging tools between the head and the block.

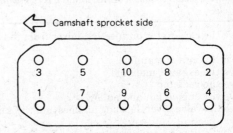

Bolt removal pattern, Cordia and Tredia

3 ENGINE AND ENGINE OVERHAUL

13. Lift the head free of the engine. Support it on wooden blocks on the workbench. Refer to *Cleaning and Inspection* following this section for work to be done before installing the head. If the head has been removed for work other than gasket replacement, the rocker assembly and the camshaft may be removed.
14. Place a new gasket onto the engine block. Make certain it is installed with the identifying mark facing up (towards the head). The identifying mark for the G62B engine is **62** and for the G63B is **63**.

WARNING: Do not apply any sealant to the gasket or mating surfaces of the head and block.

15. Install the head straight down onto the block. Try to eliminate most of the side-to-side adjustments as this may move the gasket out of position. Install the bolts by hand and just start each bolt 1 or 2 turns on the threads.
16. The head bolt torque specification is 68 ft. lbs. for a cold engine. The bolts must be tightened in order in two equal steps. On the first pass, tighten all the bolts to 34 ft. lbs.; then proceed through the order again, tightening each bolt to 68 ft. lbs.
17. Check that the cam and rocker position has not changed while the head was removed. Carefully install the camshaft sprocket with the timing belt to the camshaft. Tighten the retaining bolt to 65 ft. lbs.
18. Install the upper timing belt cover and the valve cover. This will protect the valve train and timing belt from contamination during reassembly of other components.
19. Install the exhaust manifold with a new gasket.
20. Install the intake manifold with a new gasket.
21. Install the carburetor or injection mixer.
22. Connect all wiring, hoses, and lines to the proper location. Double check all connectors and clamps.
23. Double check all installation items, paying particular attention to loose hoses or hanging wires, untightened nuts, poor routing of hoses and wires (too tight or rubbing) and tools left in the engine area.
24. Fill the cooling system with the proper coolant. Changing the engine oil and filter is recommended to eliminate any contaminants in the oil.
25. Connect the negative battery cable. Start the engine and check carefully for leaks of oil, vacuum, fuel or coolant.
26. Perform final engine adjustments as necessary.

Pickup with G63B Engine

1. Disconnect the negative battery cable.
2. Drain the cooling system.

--- **CAUTION** ---
When draining the coolant, keep in mind that cats and dogs are attracted by the ethylene glycol antifreeze, and are quite likely to drink any that is left in an uncovered container or in puddles on the ground. This will prove fatal in sufficient quantity. Always drain the coolant into a sealable container. Coolant should be reused unless it is contaminated or several years old.

3. Remove the carburetor or injection mixer from the intake manifold.
4. Remove the intake manifold.
5. Remove the exhaust manifold.
6. Remove the radiator.
7. Remove the engine fan and pulley.
8. Remove the valve cover, labeling all lines and wires during disassembly.
9. Remove the upper timing belt cover.
10. Turn the motor until the timing marks align and the valves for cylinder No.1 are closed (rockers are both loose). If the marks align, but the rockers are not loose, turn the crankshaft another 360°. This will put the engine at TDC compression on cylinder No.1.

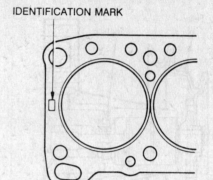

The mark on the head gasket must face upwards at installation

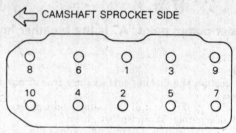

Bolt tightening pattern, Cordia and Tredia

11. Loosen and remove the bolt holding the upper timing belt sprocket to the camshaft. Hold the sprocket inside the belt and maintain the tension on the belt at all times. Carefully move the sprocket with the belt away from the head and support it on the pad in the lower cover. It may be necessary to use a small shim of soft material to keep the sprocket in position and under tension.

WARNING: If the belt tension is lost or the sprocket fall out of place, the timing belt must be reinstalled "from scratch", requiring the removal of the lower cover and other items.

12. Double check the head to insure no remaining wires or hoses are connected.
13. Loosen the head bolts in the order shown (this is important) and in three passes until all are finger loose. Remove the bolts.
14. Rock the head gently to break it loose; if tapping is necessary, do so with a rubber or wooden mallet at the corners of the head. DO NOT pry the head up by wedging tools between the head and the block.
15. Lift the head free of the engine. Support it on wooden blocks on the workbench. Refer to *Cleaning and Inspection* following this section for work to be done before installing the head. If the head has been removed for work other than gasket replacement, the rocker assembly and the camshaft may be removed.
16. Place a new gasket onto the engine block. Make certain it is installed with the identifying mark facing up (towards the head). The identifying mark is **63**.

WARNING: Do not apply any sealant to the gasket or mating surfaces of the head and block.

17. Install the head straight down onto the block. Try to eliminate most of the side-to-side adjustments as this may move the gasket out of position. Install the bolts by hand and just start each bolt 1 or 2 turns on the threads.
18. The head bolt torque specification is 68 ft. lbs. for a cold engine. The bolts must be tightened in order in two equal steps.

ENGINE AND ENGINE OVERHAUL 3

On the first pass, tighten all the bolts to 34 ft. lbs.; then proceed through the order again, tightening each bolt to 68 ft. lbs.

19. Check that the cam and rocker position has not changed while the head was removed. Carefully install the camshaft sprocket with the timing belt to the camshaft. Tighten the retaining bolt to 65 ft. lbs.
20. Install the upper timing belt cover and the valve cover. This will protect the valve train and timing belt from contamination during reassembly of other components.
21. Install the exhaust manifold with a new gasket.
22. Install the intake manifold with a new gasket.
23. Install the carburetor or injection mixer.
24. Connect all wiring, hoses, and lines to the proper location. Double check all connectors and clamps.
25. Install the engine fan and pulley.
26. Install the radiator.
27. Double check all installation items, paying particular attention to loose hoses or hanging wires, untightened nuts, poor routing of hoses and wires (too tight or rubbing) and tools left in the engine area.
28. Fill the cooling system with the proper coolant. Changing the engine oil and filter is recommended to eliminate any contaminants in the oil.
29. Connect the negative battery cable. Start the engine and check carefully for leaks of oil, vacuum, fuel or coolant.
30. Perform final engine adjustments as necessary.

Starion
Pickup and Montero with G54B Engine

1. Disconnect the negative battery cable.
2. Drain the coolant.

---- **CAUTION** ----

When draining the coolant, keep in mind that cats and dogs are attracted by the ethylene glycol antifreeze, and are quite likely to drink any that is left in an uncovered container or in puddles on the ground. This will prove fatal in sufficient quantity. Always drain the coolant into a sealable container. Coolant should be reused unless it is contaminated or several years old.

---

3. Remove the air conditioner compressor drive belt.
4. Remove the heat shield over the brake master cylinder.
5. Disconnect the oxygen sensor harness connector and remove the heat shield from the exhaust manifold and turbocharger.
6. Remove the air cleaner assembly.
7. Disconnect air hose A, air hose D and the boost hose.
8. Remove the upper radiator hose from the thermostat housing.
9. Remove the air intake duct and the secondary air cleaner.
10. Disconnect the PCV hose and accelerator cable from the valve cover. Remove the valve cover and gasket. Remove the half-circle plug from the head.
11. Turn the crankshaft clockwise to align the timing marks and set the engine to TDC/Compression for No. 1 cylinder.
12. Label and disconnect the coil and spark plug wires from the distributor.
13. Disconnect the high pressure fuel line, remove the O-ring and disconnect the fuel return line

---- **CAUTION** ----

The fuel system is under pressure. Release pressure slowly and contain spillage. Observe no smoking/no open flame precautions. Have a Class B-C (dry powder) fire extinguisher within arm's reach at all times.

---

14. Disconnect the vacuum hose to the power brake booster from the engine. Disconnect the heater hose and engine coolant hose at the rear of the engine.
15. Remove the dipstick tube and dipstick. Discard the O-ring.
16. Disconnect the oil supply pipe and the oil return pipe from the turbocharger.

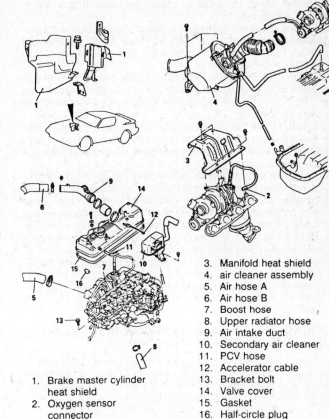

1. Brake master cylinder heat shield
2. Oxygen sensor connector
3. Manifold heat shield
4. Air cleaner assembly
5. Air hose A
6. Air hose B
7. Boost hose
8. Upper radiator hose
9. Air intake duct
10. Secondary air cleaner
11. PCV hose
12. Accelerator cable
13. Bracket bolt
14. Valve cover
15. Gasket
16. Half-circle plug

Preliminary steps for Starion head removal

Disconnecting Starion vacuum hoses

17. Disconnect the coolant line from the turbocharger.
18. Disconnect the rear (bottom) of the catalytic converter from the exhaust system.
19. Remove the distributor.
20. Label and disconnect the vacuum hoses at or near the head.
21. Label and disconnect the electrical connectors at the head and intake manifold. Remember the connectors on the rear of the head, too.
22. Remove the ground cable connection at the engine.
23. Remove the bolt holding the camshaft sprocket to the camshaft. Remove the distributor drive gear.
24. Pull the camshaft sprocket (with the timing chain attached) from the camshaft and place it on top of the camshaft sprocket holder.

WARNING: The crankshaft must not be rotated after the sprocket is removed from the camshaft. Do not allow the timing chain to come off the camshaft sprocket.

3-171

3 ENGINE AND ENGINE OVERHAUL

25. Double check the head to insure no remaining wires or hoses are connected.
26. Loosen the head bolts in the order shown (this is important) and in three passes until all are finger loose. Remove the bolts. Note that the two front bolts must each be loosened before bolt No.2 is loosened.
27. Rock the head gently to break it loose; if tapping is necessary, do so with a rubber or wooden mallet at the corners of the head. DO NOT pry the head up by wedging tools between the head and the block.
28. Lift the head free of the engine, remembering that the intake and exhaust manifolds (with the catalytic converter) are still attached to the head. This assembly is heavy; an assistant is helpful when lifting.

Support the head assembly on wooden blocks on the workbench. Refer to "Cleaning and Inspection" following this section for work to be done before installing the head. If the head has been removed for work other than gasket replacement, the manifolds, rocker assembly and the camshaft may be removed.

29. Before placing a new gasket onto the engine block, apply 3M ART sealant No. 8660 or equivalent to the top surface of each joint between the timing case and the cylinder block. Apply just enough sealant to fill the seam and make certain NO sealant enters the oil passages in the block. Place the new gasket on the block with the identifying mark ("54") facing up (towards the head).

WARNING: Do not apply any sealant to the gasket or mating surfaces of the head and block.

30. Install the head straight down onto the block. Try to eliminate most of the side-to-side adjustments as this may move the gasket out of position. Install the bolts by hand and just start each bolt 1 or 2 turns on the threads.
31. The head bolt torque specification is 68 ft. lbs. for a cold engine. The bolts must be tightened in the order shown in 3 steps. On the first pass, tighten all the bolts to about 20 ft. lbs. then proceed through the order tightening each bolt to about 45 ft. lbs. The final torque is achieved on the third pass.

WARNING: The two bolts into the timing case (No. 11 in the tightening diagram) are tightened ONLY to 14 ft. lbs. Achieve this torque in steps of 4, 9 and 13 ft. lbs. Do not overtighten these bolts.

32. Check that the cam and rocker position has not changed while the head was removed. Carefully install the camshaft sprocket with the timing chain to the camshaft. Install the distributor drive gear and tighten the retaining bolt to 40 ft. lbs.
33. Connect the engine ground strap and connect each electrical connector to its proper component.

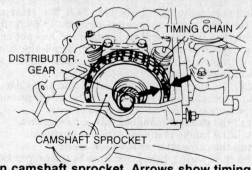

Starion camshaft sprocket. Arrows show timing marks to be aligned before removal

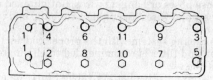

Starion head bolt removal pattern

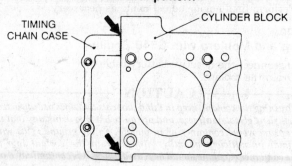

Apply sealant only to the areas shown

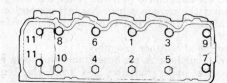

Starion head bolt installation pattern

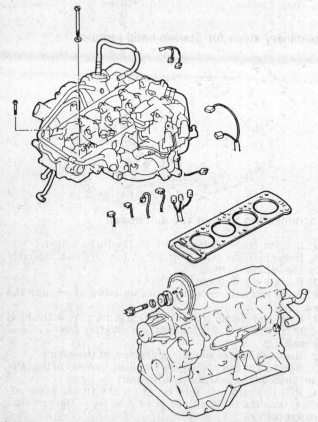

Starion head (with manifolds) and gasket. The large number of electrical connectors makes a strong case for labeling everything before disconnecting

ENGINE AND ENGINE OVERHAUL 3

34. Connect the vacuum hoses to the proper ports, making sure each is firmly seated and not crimped or crushed.
35. Install the distributor.
36. Connect the catalytic converter to the exhaust system; tighten the nuts to 26 ft. lbs.
37. Connect the coolant hose, the oil return hose and the oil supply hose to the turbocharger. The oil supply line bolts should be tightened to 16 ft. lbs.
38. Install a new O-ring on the dipstick tube and install the tube to the engine. Make sure the O-ring is not damaged during installation.
39. Connect the small coolant hose and the heater hose to the head or intake manifold. Install the brake booster vacuum hose.
40. Connect the fuel return hose. Install a new O-ring and connect the high pressure fuel hose.
41. Install the spark plug and coil wires.
42. Install the half-circle plug with sealant. Install the valve cover and gasket, tightening the bolts only to 48 inch lbs.
43. Install the accelerator cable and connect the PCV hose to the valve cover.
44. Install the secondary air cleaner and connect the air intake ducts.
45. Connect the boost hose and the air hoses (A and D) to the turbocharger.
46. Install the air cleaner assembly.
47. Install the manifold heat shield and connect the oxygen sensor wiring.
48. Install the heat shield on the brake master cylinder.
49. Install and adjust the air conditioning compressor belt.
50. Double check all installation items, paying particular attention to loose hoses or hanging wires, untightened nuts, poor routing of hoses and wires (too tight or rubbing) and tools left in the engine area.
51. Fill the cooling system with the proper coolant. Changing the engine oil and filter is recommended to eliminate any contaminants in the oil.
52. Connect the negative battery cable. Start the engine and check carefully for leaks of oil, vacuum, fuel or coolant.
53. Perform final engine adjustments as necessary.

Mirage and Precis with the G15B Engine

1. Disconnect the negative battery cable.
2. Remove the left side splash shield. Drain the engine coolant.

CAUTION
When draining the coolant, keep in mind that cats and dogs are attracted by the ethylene glycol antifreeze, and are quite likely to drink any that is left in an uncovered container or in puddles on the ground. This will prove fatal in sufficient quantity. Always drain the coolant into a sealable container. Coolant should be reused unless it is contaminated or several years old.

3. Remove the air cleaner assembly along with the various air ducts and hoses.
4. Disconnect the accelerator cable.
5. Remove the upper radiator hose, the heater hose and the small coolant hose from the head or intake manifold.
6. Disconnect the brake booster vacuum hose.
7. disconnect either the throttle control cable (automatic) or the clutch cable (manual trans).
8. Label and disconnect the electrical connectors running to the head and intake manifold, including the oxygen sensor connector.
9. Label and disconnect the spark plug wires and the coil wire from the distributor.
10. Position a floor jack and a broad piece of lumber under the engine. Elevate the jack just enough to support the engine without raising it.
11. Remove the through-bolt from the left side engine mount.

It may be necessary to adjust the jack slightly to allow the bolt to come free. When the bolt has been removed, disconnect the nuts and bolts holding the mounting bracket to the engine and remove the bracket.
12. Label and disconnect the vacuum hoses at the intake manifold.
13. Remove the valve cover and gasket.
14. Remove the heat shields from the exhaust manifold.
15. Remove the engine oil dipstick.
16. Remove the upper timing belt cover. Turn the engine clockwise to align the timing marks at TDC/compression for No. 1 cylinder.
17. Remove the bolt holding the camshaft sprocket to the camshaft. With the belt still on the sprocket, move the sprocket away from the cam and suspend the sprocket from a piece of wire attached to the hood. Do not allow the belt to come off the

26. Vacuum hose
27. Valve cover
28. Gasket
29. Exhaust manifold heat shield (upper)
30. Exhaust manifold heat shield (lower)
31. Dipstick
32. Upper timing belt cover
33. Gasket
34. Camshaft sprocket retaining bolt
35. Exhaust system retaining bolt
36. Self-locking nuts
37. Exhaust gasket
38. Head assembly
39. Head gasket

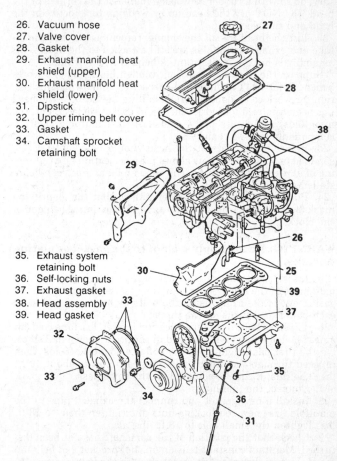

Head and related components, G15B engine

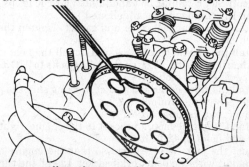

When suspending the sprocket, take great care not to damage the timing belt

3–173

3 ENGINE AND ENGINE OVERHAUL

sprocket and maintain constant upward tension on the belt. Do not allow slack in the timing belt.

WARNING: The crankshaft must not be turned while the sprocket is off the camshaft.

18. Remove the self-locking nuts and the small bolt holding the exhaust system to the bottom of the exhaust manifold. Remove and discard the exhaust gasket.
19. Label and disconnect the main fuel hose and the fuel return hose.
20. Double check the head to insure no remaining wires or hoses are connected.
21. Loosen the head bolts in the order shown (this is important) and in three passes until all are finger loose. Remove the bolts.
22. Rock the head gently to break it loose; if tapping is necessary, do so with a rubber or wooden mallet at the corners of the head. DO NOT pry the head up by wedging tools between the head and the block.
23. Lift the head free of the engine, remembering that the intake and exhaust manifolds are still attached to the head. This assembly is heavy; an assistant is helpful when lifting.

Support the head assembly on wooden blocks on the workbench. Refer to "Cleaning and Inspection" following this section for work to be done before installing the head. If the head has been removed for work other than gasket replacement, the manifolds, rocker assembly, camshaft, distributor and other components may be removed.

23. Before reinstallation, the head should be completely assembled on the bench. This allows proper location and tightening of all the external items. The head and manifolds can be installed as a unit, saving time and effort.
24. Place a new gasket on the engine so that the identifying mark faces up (towards the head) and is at the timing belt end of the block.

WARNING: Do not apply sealant to the gasket or mating surfaces.

25. Install the head straight down onto the block. Try to eliminate most of the side-to-side adjustments as this may move the gasket out of position. Install the bolts by hand and just start each bolt 1 or 2 turns on the threads.
26. The head bolt torque specification is 52 ft. lbs. for a cold engine. The bolts must be tightened in the order shown in 3 steps. On the first pass, tighten all the bolts to about 20 ft. lbs. then proceed through the order tightening each bolt to about 40 ft. lbs. The final torque is achieved on the third pass.
27. Connect the fuel hoses.
28. Install a new gasket and connect the exhaust pipe to the manifold. Use new self-locking nuts and tighten them to 18 ft. lbs. Tighten the small bolt to 18 ft. lbs. also.
29. Check that the position of the camshaft has not been disturbed. Maintain constant tension on the sprocket and belt and move the sprocket onto the camshaft with the belt in place. Install the sprocket retaining bolt and tighten it to 50 ft. lbs. Make certain all the timing marks are aligned.
30. Install the timing belt cover and gasket. Tighten the bolts to 96 inch lbs.
31. Install the engine oil dipstick and install the heat shields on the exhaust manifold. Tighten the upper bolts to 9 ft. lbs. and the lower bolts to 72 inch lbs.
32. Install the valve cover and gasket, making sure the gasket is properly seated all around.
33. Connect the vacuum hoses.
34. Install the engine mount bracket to the engine, tightening the nuts and bolts to 42 ft. lbs.
35. Adjust the jack (if necessary) so that the engine mount bushing aligns with the bodywork bracket. Install the through-bolt and tighten the nuts snug.
36. Slowly release tension on the floor jack so that the weight

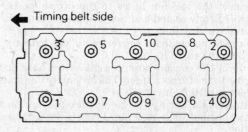

Bolt removal pattern, G15B head

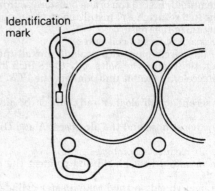

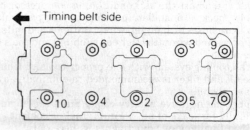

Bolt tightening pattern, G15B engine

of the engine bears fully on the mount. Tighten the through-bolt to 75 ft. lbs. and the small safety nut to 38 ft. lbs.
37. Install the spark plug and coil wires to the distributor. Connect the wiring leads running to the head and intake manifold.
38. Connect either the throttle control cable or the clutch cable.
39. Connect the brake booster vacuum hose.
40. Connect the heater hose, water hose and upper radiator hose.
41. Install the accelerator cable.
42. Install the air cleaner assembly with its ducts and hoses.
43. Double check all installation items, paying particular attention to loose hoses or hanging wires, untightened nuts, poor routing of hoses and wires (too tight or rubbing) and tools left in the engine area.
44. Fill the cooling system with coolant. Changing the engine oil and filter is recommended to eliminate pollutants in the oil.
45. Connect the negative battery cable. Start the engine and check for leaks of fluid or vacuum.
46. Adjust the clutch, throttle and/or accelerator cables as needed. Check and top up the coolant.
47. Install the left splash shield.

Mirage with the G32B Engine

1. Disconnect the negative battery cable.
2. Remove the left splash shield. Drain the engine coolant.

ENGINE AND ENGINE OVERHAUL 3

CAUTION

When draining the coolant, keep in mind that cats and dogs are attracted by the ethylene glycol antifreeze, and are quite likely to drink any that is left in an uncovered container or in puddles on the ground. This will prove fatal in sufficient quantity. Always drain the coolant into a sealable container. Coolant should be reused unless it is contaminated or several years old.

3. Safely relieve the pressure within the fuel system.

CAUTION

The fuel system is under pressure. Release pressure slowly and contain spillage. Observe no smoking/no open flame precautions. Have a Class B-C (dry powder) fire extinguisher within arm's reach at all times.

4. Disconnect the oil drain hose and the air flow sensor wiring connector from the air cleaner assembly. Remove the air cleaner assembly.

5. With the hoses attached, remove the power steering pump from its bracket and support on wire out of the way. If the vehicle is air conditioned, the compressor belt will need to be removed first. Do not allow the pump to hang by the hoses.
6. Remove the power steering pump bracket from the engine.
7. Disconnect the wiring, loosen the belt and remove the alternator.
8. if the vehicle is air conditioned, remove the condenser fan assembly. Don't remove the condenser — just the fan, motor and shroud.
9. Remove the engine oil dipstick and tube.
10. Disconnect the secondary air hose and remove the secondary air pipe.
11. Label and disconnect each electrical lead running to the head or intake manifold. Accurate labeling is critical for reassembly. Label and remove the coil and spark plug wires from the distributor.

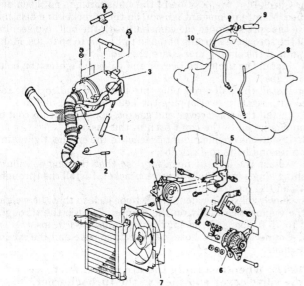

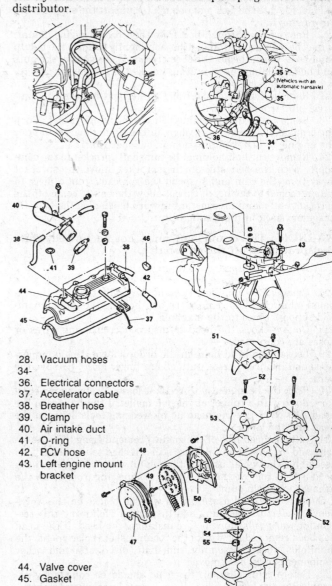

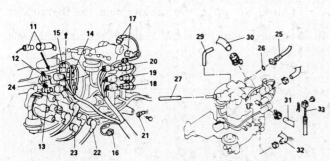

1. Oil drain hose
2. Air flow sensor connector
3. Air cleaner
4. Power steering pump
5. Power steering pump bracket
6. Alternator
7. Condenser fan and motor (with A/C)
8. Dipstick tube
9. Secondary air hose
10. Secondary air pipe
11. -
20. Electrical connectors
21. Engine ground strap
22. Coil wire
23. Spark plug cables
24. Solenoid valve bracket
25. High pressure fuel line
26. O-ring
27. Fuel return hose
28. Brake booster vacuum hose
30. Upper radiator hose
31. Heater hose
32. Coolant hose
33. Throttle control cable

28. Vacuum hoses
34. -
36. Electrical connectors
37. Accelerator cable
38. Breather hose
39. Clamp
40. Air intake duct
41. O-ring
42. PCV hose
43. Left engine mount bracket
44. Valve cover
45. Gasket
46. Half-circle plug
47. Upper timing belt cover
48. Gasket
49. Camshaft sprocket retaining bolt
50. Upper rear timing belt cover
51. Heat shield
52. Manifold through-bolt
53. Cylinder head assembly
54. Exhaust gasket
55. Sealing ring
56. Head gasket

G32B head removal

3 ENGINE AND ENGINE OVERHAUL

12. Remove the solenoid valve bracket (just above the distributor).
13. Disconnect the high pressure fuel line, remove the O-ring, and disconnect the fuel return line.
14. Label and disconnect the vacuum hoses and the brake booster vacuum line.
15. Disconnect the upper radiator hose, the heater hose and the small coolant line from the engine.
16. If equipped with automatic transmission, disconnect the throttle control cable.
17. If not already done, label and disconnect the water temperature connectors near the thermostat housing.
18. Disconnect the accelerator cable.
19. Remove the breather hose, the PCV hose and the air intake duct.
20. Position a floor jack and a broad piece of lumber under the engine. Elevate the jack just enough to support the engine without raising it.
21. Remove the through-bolt from the left side engine mount. It may be necessary to adjust the jack slightly to allow the bolt to come free. When the bolt has been removed, disconnect the nuts and bolts holding the mounting bracket to the engine and remove the bracket.
22. Remove the valve cover and gasket and remove the half-circle plug from the head.
23. Remove the upper timing belt cover and its gasket. Turn the crankshaft clockwise to align all the timing marks and put the engine at TDC compression on cylinder No. 1.
24. Remove the bolt holding the camshaft sprocket to the camshaft. With the belt still on the sprocket, move the sprocket away from the cam and suspend the sprocket from a piece of wire attached to the hood. Do not allow the belt to come off the sprocket and maintain constant upward tension on the belt. Do not allow slack in the timing belt.

WARNING: The crankshaft must not be turned while the sprocket is off the camshaft.

25. Remove the upper rear timing cover.
26. Remove the heat shield from the exhaust manifold.
27. Remove the long bolt running vertically through the exhaust manifold. Also remove the bolt holding the intake manifold support brace to the manifold.
28. Double check the head to insure no remaining wires or hoses are connected.
29. Loosen the head bolts in the order shown (this is important) and in three passes until all are finger loose. Remove the bolts.
30. Rock the head gently to break it loose; if tapping is necessary, do so with a rubber or wooden mallet at the corners of the head. DO NOT pry the head up by wedging tools between the head and the block.
31. Lift the head free of the engine, remembering that the intake and exhaust manifolds are still attached to the head. This assembly is heavy; an assistant is helpful when lifting. After the head is free, remove the gasket and sealing ring from below the exhaust manifold joint.

Support the head assembly on wooden blocks on the workbench. Refer to "Cleaning and Inspection" following this section for work to be done before installing the head. If the head has been removed for work other than gasket replacement, the manifolds, rocker assembly, camshaft, distributor and other components may be removed.

32. Before reinstallation, the head should be completely assembled on the bench. This allows proper location and tightening of all the external items. The head and manifolds can be installed as a unit, saving time and effort.
33. Place a new gasket on the engine so that the identifying mark faces up (towards the head) and is at the timing belt end of the block. Install a new exhaust gasket and position the sealing ring for the exhaust manifold.

WARNING: Do not apply sealant to the head gasket or mating surfaces of the head and engine.

34. Install the head straight down onto the block. Try to eliminate most of the side-to-side adjustments as this may move the gasket out of position. Install the bolts by hand and just start each bolt 1 or 2 turns on the threads.
35. The head bolt torque specification is 52 ft. lbs. for a cold engine. The bolts must be tightened in the order shown in 3 steps. On the first pass, tighten all the bolts to about 20 ft. lbs. then proceed through the order tightening each bolt to about 40 ft. lbs. The final torque is achieved on the third pass.
36. Install the through-bolt into the exhaust manifold and tighten it to 42 ft. lbs. Install the bolt for the intake manifold support brace and tighten it to 18 ft. lbs.
37. Install the heat shield on the exhaust manifold.
38. Install the upper rear timing belt cover, tightening the bolts to 8 ft. lbs.
39. Check that the position of the camshaft has not been disturbed. Maintain constant tension on the sprocket and belt and move the sprocket onto the camshaft with the belt in place. Install the sprocket retaining bolt and tighten it to 66 ft. lbs. Make certain all the timing marks are aligned.
40. Install the timing belt cover. Tighten the bolts to 8 ft. lbs.
41. Install the half-circle plug into the head and coat the top edge with sealant, overlapping the head by 10mm (0.4 in.).
42. Install the valve cover and gasket, applying a thin coat of clean engine oil to the front arch in the gasket.
43. Install the engine mount bracket to the engine, tightening the nuts and bolts to 42 ft. lbs.
44. Adjust the jack (if necessary) so that the engine mount bushing aligns with the bodywork bracket. Install the through-bolt and tighten the nuts snug.
45. Slowly release tension on the floor jack so that the weight of the engine bears fully on the mount. Tighten the through-bolt to 75 ft. lbs. and the small safety nut to 38 ft. lbs.
46. Connect the PCV hose, the breather hose and the air intake pipes.

NOTE: When installing the air intake duct, secure it to the valve cover and then to the turbocharger.

47. Connect the accelerator cable.
48. Connect the wiring connectors near the thermostat housing.
49. If equipped with automatic transaxle, install the throttle control cable.

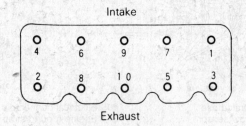

G32B head bolt removal pattern

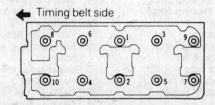

Head bolt tightening sequence, G32B

ENGINE AND ENGINE OVERHAUL 3

50. Install the coolant hose, the heater hose and the upper radiator hose to the engine.
51. Install the brake booster vacuum hose and connect the other vacuum lines.
52. Connect the fuel return hose; install a new O-ring and connect the high pressure fuel line.
53. Connect the electrical connectors to the head and intake manifold and install the spark plug and coil wiring to the distributor.
54. Install the secondary air pipe and connect the secondary air hose.
55. Install the dipstick tube. Use a new O-ring and don't damage it during installation.
56. If air conditioned, install the condenser fan and connect the wiring.
57. Install the alternator and tension the belt properly.
58. Install the power steering pump bracket, tightening the bolts to 18 ft. lbs.
59. Install the power steering pump into the bracket, install the belt and tighten the bolts to 18 ft. lbs. when the belt is correctly adjusted. Install the air conditioning compressor belt (and adjust it) if it was removed.
60. Install the air cleaner assembly and connect its sensor wiring and ductwork.
61. Connect the oil drain hose.
62. Double check all installation items, paying particular attention to loose hoses or hanging wires, untightened nuts, poor routing of hoses and wires (too tight or rubbing) and tools left in the engine area.
63. Fill the cooling system with coolant. An oil and filter change is recommended to eliminate pollutants in the oil.
64. Connect the negative battery cable. Start the engine and check for leaks of fuel, vacuum or oil. Check the operation of all engine electrical systems as well as dashboard gauges and lights.
65. Perform necessary adjustments to the drive belts, accelerator cable and throttle control cable. Adjust the coolant level after the engine has cooled off.
66. Install the left splash shield.

Mirage with the 4G15 Engine

1. Disconnect the negative battery cable.
2. Drain the engine coolant.

---- **CAUTION** ----

When draining the coolant, keep in mind that cats and dogs are attracted by the ethylene glycol antifreeze, and are quite likely to drink any that is left in an uncovered container or in puddles on the ground. This will prove fatal in sufficient quantity. Always drain the coolant into a sealable container. Coolant should be reused unless it is contaminated or several years old.

3. Safely relieve the pressure within the fuel system.

---- **CAUTION** ----

The fuel system is under pressure. Release pressure slowly and contain spillage. Observe no smoking/no open flame precautions. Have a Class B-C (dry powder) fire extinguisher within arm's reach at all times.

4. Disconnect the accelerator cable.
5. Disconnect the air intake hose and breather tube from the head or intake manifold.
6. Remove the upper radiator hose, the heater hose and the coolant by-pass hose at the engine.
7. Disconnect the brake booster vacuum hose. Label and disconnect the other vacuum hoses.
8. Disconnect the high pressure fuel line and remove the O-ring. Disconnect the fuel return line.
9. Label and remove the spark plug wires from the distributor.
10. If equipped with automatic transaxle, disconnect the throttle control cable.

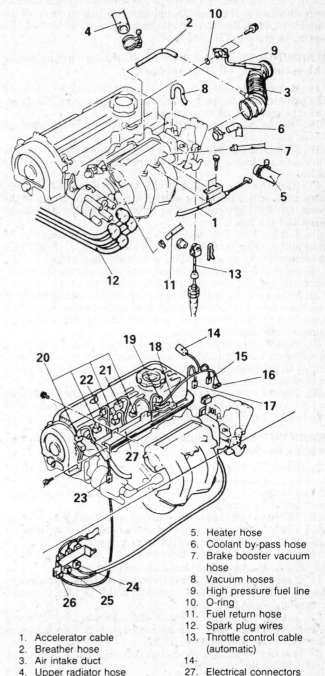

1. Accelerator cable
2. Breather hose
3. Air intake duct
4. Upper radiator hose
5. Heater hose
6. Coolant by-pass hose
7. Brake booster vacuum hose
8. Vacuum hoses
9. High pressure fuel line
10. O-ring
11. Fuel return hose
12. Spark plug wires
13. Throttle control cable (automatic)
14-
27. Electrical connectors

Removal of external components for access to 4G15 head

11. Label and disconnect the electrical connectors on the head and intake manifold. Note that this includes the wiring to the fuel injectors and the distributor. Be aware that some of the harnesses run from the engine to the firewall and must be disconnected at the firewall. Label everything carefully.
12. Remove the upper timing belt cover and its gasket. Turn the crankshaft clockwise to align all the timing marks and put the engine at TDC compression on cylinder No. 1.
13. Remove the bolt holding the camshaft sprocket to the camshaft. With the belt still on the sprocket, move the sprocket

3-177

3 ENGINE AND ENGINE OVERHAUL

away from the cam and suspend the sprocket from a piece of wire attached to the hood. Do not allow the belt to come off the sprocket and maintain constant upward tension on the belt. Do not allow slack in the timing belt.

WARNING: The crankshaft must not be turned while the sprocket is off the camshaft.

14. Remove the self-locking nuts and retaining bolt holding the exhaust pipe to the bottom of the exhaust manifold. Separate the pipe and remove the gasket.
15. Double check the head to insure no remaining wires or hoses are connected.
16. Loosen the head bolts in the order shown (this is important) and in three passes until all are finger loose. Remove the bolts.
17. Rock the head gently to break it loose; if tapping is necessary, do so with a rubber or wooden mallet at the corners of the head. DO NOT pry the head up by wedging tools between the head and the block.
18. Lift the head free of the engine, remembering that the intake and exhaust manifolds are still attached to the head. This assembly is heavy; an assistant is helpful when lifting.

Support the head assembly on wooden blocks on the workbench. Refer to "Cleaning and Inspection" following this section for work to be done before installing the head. If the head has been removed for work other than gasket replacement, the manifolds, rocker assembly, camshaft, distributor and other components may be removed.

19. Before reinstallation, the head should be completely assembled on the bench. This allows proper location and tightening of all the external items. The head and manifolds can be installed as a unit, saving time and effort.
20. Place a new gasket on the engine so that the identifying mark faces up (towards the head) and is at the timing belt end of the block. Install a new exhaust gasket.

WARNING: Do not apply sealant to the head gasket or mating surfaces.

21. Install the head straight down onto the block. Try to eliminate most of the side–to–side adjustments as this may move the gasket out of position. Install the bolts by hand and just start each bolt 1 or 2 turns on the threads.
22. The head bolt torque specification is 52 ft. lbs. for a cold engine. The bolts must be tightened in the order shown in 3 steps. On the first pass, tighten all the bolts to about 20 ft. lbs. then proceed through the order tightening each bolt to about 40 ft. lbs. The final torque is achieved on the third pass.
23. Using new self-locking nuts, connect the exhaust system to the manifold and tighten the nuts to 26 ft. lbs. Tighten the small retaining bolt to 18 ft. lbs.
24. Check the position of the camshaft has not changed during repairs. Carefully install the sprocket and belt onto the camshaft and tighten the bolt to 50 ft. lbs.
25. Install the valve cover and gasket and install the upper timing belt cover.
26. Reconnect each electrical connection, making sure the wires are routed properly and the connectors are firmly seated.
27. Install the throttle control cable if equipped with automatic transaxle.
28. Connect the spark plug cables to the distributor.
29. Connect the fuel return hose, install a new O-ring and connect the high pressure fuel line.
30. Connect the vacuum hoses and the brake booster vacuum hose.
31. Install the coolant by-pass hose, the heater hose and the upper radiator hose.
32. Install the air intake and breather hoses.
33. Connect the accelerator cable.
34. Double check all installation items, paying particular attention to loose hoses or hanging wires, untightened nuts, poor

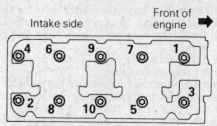

4G15 head bolt removal pattern

routing of hoses and wires (too tight or rubbing) and tools left in the engine area.

35. Fill the cooling system with coolant. Changing the oil and filter is recommended to eliminate pollutants in the oil.
36. Connect the negative battery cable. Start the engine and check for leaks of fuel, vacuum or oil. Check the operation of all engine electrical systems as well as dashboard gauges and lights.
37. Perform necessary adjustments to the drive belts, accelerator cable and throttle control cable. Adjust the coolant level after the engine has cooled off.

Mirage with the 4G61 Engine

WARNING: The use of the correct special tools or their equivalent is REQUIRED for this procedure. The timing belt will be removed. Special tools are required for the correct reinstallation and tensioning of the belt.

1. Disconnect the negative battery cable.
2. Remove the splash shield from under the engine. Drain the engine coolant.

CAUTION

When draining the coolant, keep in mind that cats and dogs are attracted by the ethylene glycol antifreeze, and are quite likely to drink any that is left in an uncovered container or in puddles on the ground. This will prove fatal in sufficient quantity. Always drain the coolant into a sealable container. Coolant should be reused unless it is contaminated or several years old.

3. Remove the radiator.
4. Disconnect the accelerator cable.
5. Unplug the connector for the air flow sensor at the air cleaner.
6. Remove the breather hose and the purge hose from the air intake hose. Disconnect the large air ducts to the turbocharger and the intake manifold (hoses A and E), disconnect the air by-pass hose and remove the air intake hose.
7. Remove the air cleaner assembly.
8. Remove the PCV hose from the valve cover.
9. Disconnect the coolant by-pass hose and the heater hose.
10. Label and disconnect the vacuum hoses at the intake plenum.
11. Disconnect the brake booster vacuum hose.
12. Remove the high pressure fuel hose and remove the O-ring.

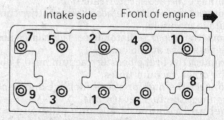

4G15 head bolt tightening sequence

ENGINE AND ENGINE OVERHAUL 3

CAUTION

The fuel system is under pressure. Release pressure slowly and contain spillage. Observe no smoking/no open flame precautions. Have a Class B-C (dry powder) fire extinguisher within arm's reach at all times.

13. Remove the fuel return hose.
14. Label and disconnect the the electrical connectors running to the head and intake manifold. This includes removing the spark plug wires completely (remove the center panel from the valve cover) and disconnecting the injector wiring. Assorted clips and brackets will need to be removed to get the wiring harnesses clear of the head and manifolds.
15. Position a floor jack and a broad piece of lumber under the engine. Elevate the jack just enough to support the engine without raising it.
16. Remove the through-bolt from the left side engine mount. It may be necessary to adjust the jack slightly to allow the bolt to come free. When the bolt has been removed, disconnect the nuts and bolts holding the mounting bracket to the engine and remove the bracket.
17. Remove the alternator belt and the power steering belt.
18. Loosen the tension on the air conditioner compressor belt tensioner. With the belt loosened, remove the adjuster assembly and remove the belt.
19. Remove the water pump pulley with the power steering pulley.
20. Remove the crankshaft pulley.
21. Remove the (front) upper and lower timing belt covers.

WARNING: The covers use 4 different lengths of bolts. Label or diagram them as they are removed.

22. Remove the center cover, the PCV hose and the breather hose from the valve cover. Label and disconnect the spark plug cables from the spark plugs.
23. Remove the valve cover and its gasket. Remove the half-circle plug from the head.
24. Turn the crankshaft clockwise and align all the timing marks. The timing marks on the camshaft sprockets should align with the upper surface of the cylinder head and the dowel pins (guide pins) at the center of the camshaft sprockets should be up. Aligned means perfect, not close — if you miss, continue clockwise until the marks are matched. This sets the motor at TDC/compression on No.1 cylinder.

WARNING: Once the engine is set to this position, the crank and camshafts must not be moved out of place. Severe damage can result.

25. Remove the auto-tensioner.
26. Before removing the timing belt, mark it with an arrow to show the direction of rotation. If the belt is to be reused, it must be reinstalled so that it rotates in the same direction as before.
27. Remove the timing belt, keeping it free of grease, oil and fluids. Do not crease or crimp the belt. After the belt is clear, remove the tensioner pulley.
28. Remove the exhaust manifold heat shield.
29. Disconnect the eye bolt holding coolant pipe B to the turbocharger. Remove the gaskets and discard them.
30. Disconnect the exhaust system from the bottom of the turbocharger. Discard the self-locking nuts.
31. Disconnect the oil return pipe at the turbocharger. Discard the gasket.
32. Disconnect the tension rod bolted to the right rear corner of the head.
33. Disconnect the intake manifold support bracket from the manifold.
34. Double check the head to insure no remaining wires or hoses are connected.
35. Loosen the head bolts in the order shown (this is important) and in three passes until all are finger loose. Remove the bolts.

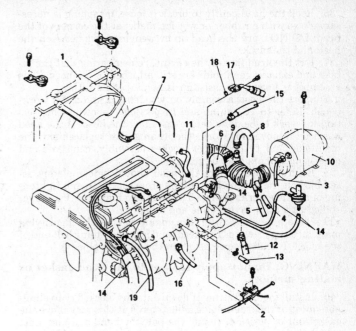
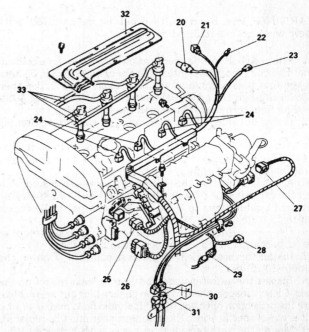

1. Radiator
2. Accelerator
3. Air flow sensor connector
4. Breather hose
5. Purge hose
6. Air duct E
7. Air duct A
8. Air by-pass hose
9. Air intake hose
10. Air cleaner assembly
11. PCV hose
12. Coolant by-pass hose
13. Heater hose
14. Vacuum hose
15. Vacuum hose
16. Brake booster vacuum hose
17. High pressure fuel hose
18. O-ring
19. Fuel return hose
20. -
31. Electrical connectors
32. Center cover
33. Spark plug wires

Removal of external components for access to 4G61 head

3-179

3 ENGINE AND ENGINE OVERHAUL

36. Rock the head gently to break it loose; if tapping is necessary, do so with a rubber or wooden mallet at the corners of the head. DO NOT pry the head up by wedging tools between the head and the block.
37. Lift the head free of the engine, remembering that the intake and exhaust manifolds are still attached to the head. This assembly is heavy; an assistant is helpful when lifting.

Support the head assembly on wooden blocks on the workbench. Refer to "Cleaning and Inspection" following this section for work to be done before installing the head. If the head has been removed for work other than gasket replacement, the manifolds and turbocharger, rocker assembly, camshaft and other components may be removed.

38. Before reinstallation, the head should be completely assembled on the bench. This allows proper location and tightening of all the external items. The head and manifolds can be installed as a unit, saving time and effort.
39. Place a new gasket on the engine so that the identifying mark faces up (towards the head) and is at the timing belt end of the block. Install a new gasket on the exhaust pipe.

WARNING: Do not apply sealant to the head gasket or mating surfaces.

40. Install the head straight down onto the block. Try to eliminate most of the side-to-side adjustments as this may move the gasket out of position. Install the bolts by hand and just start each bolt 1 or 2 turns on the threads.

WARNING: The head bolts must be reinstalled with their washers positioned correctly.

41. The head bolt torque specification is 76 ft. lbs. for a cold engine. The bolts must be tightened in the order shown in 3 steps. On the first pass, tighten all the bolts to about 25 ft. lbs. then proceed through the order tightening each bolt to about 50 ft. lbs. The final torque is achieved on the third pass.
42. Install or connect the intake manifold support, tightening the bolts to 20 ft. lbs.
43. Attach the tension rod, tightening the bolts to 34 ft. lbs.
44. Install a new gasket and connect the oil return line to the turbocharger; bolt torque is 8 ft. lbs.
45. After checking that the new gasket is correctly placed, connect the exhaust pipe to the turbocharger. Use new self-locking nuts and tighten them to 26 ft. lbs. Tighten the small bracket bolt to 18 ft. lbs.
46. Use new washers (gaskets) and connect the coolant pipe to the turbocharger. Tighten the eyebolt to 24 ft. lbs.
47. Install the heat shield on exhaust manifold. Tighten the bolts to 10 ft. lbs.
48. Inspect the auto-tensioner for wear or leakage around the seals. Closely inspect the end of the pushrod for any wear. Measure the projection of the pushrod. It should be 12mm (0.47 in.). If it is out of specification, the auto tensioner must be replaced.
49. Mount the auto-tensioner in a vise with protected jaws. Keep the unit level at all times, and, if the plug at the bottom projects, protect it with a common washer. Smoothly tighten the vise to compress the adjuster tip; if the the rod is easily retracted, the unit has lost tension and should be replaced. You should feel a fair amount of resistance when compressing the rod.
50. As the rod is contracted into the body of the auto-tensioner, watch for the hole in the pushrod to align with the hole in the body. When this occurs, pin the pushrod in place with a piece of stiff wire 1.5mm diameter. It will be easier to install the tensioner with the pushrod retracted. Leave the wire in place and remove the assembly from the vise.
51. Install the auto-tensioner onto the engine, tightening the bolts to 18 ft. lbs.
52. Install the tensioner pulley onto the tensioner arm. Locate the pinholes in the pulley shaft to the left of the center bolt.

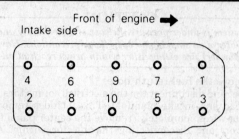

4G61 head bolt removal sequence

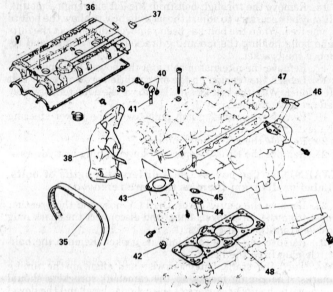

35. Timing belt
36. Valve cover
37. Half-circle plug
38. Heat shield
39. Eye bolt (banjo bolt)
40. Gaskets
41. Coolant pipe B

42. Self-locking nuts
43. Exhaust gasket
44. Oil return pipe
45. Gasket
46. Tension rod
47. Cylinder head assembly
48. Head gasket

4G61 head and gasket

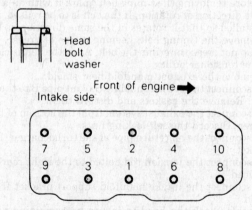

Head bolt installation sequence. Note that the washer must be reinstalled and positioned correctly

ENGINE AND ENGINE OVERHAUL

Tighten the center bolt finger tight. Leave the blocking wire in the auto-tensioner.

53. Double check the alignment of the timing marks for all the sprockets. The camshaft sprocket marks must face each other and align with the top surface of the cylinder head. Don't forget to check the oil pump sprocket alignment as well as the crankshaft sprocket.

NOTE: When you let go of the exhaust camshaft sprocket, it will move one tooth counter-clockwise. Take this into account when installing the belt around the sprockets.

54. Observing the direction of rotation mark made earlier, install the timing belt as follows:
   a. Install the timing belt around the tensioner pulley and the crankshaft sprocket and hold the belt on the tensioner with your left hand.
   b. Pulling the belt with your right hand, install it around the oil pump sprocket.
   c. Install the belt around the idler pulley and then around the intake camshaft sprocket.
   d. Double check the exhaust camshaft sprocket alignment mark. It has probably moved one tooth; if so, rotate it clockwise to align the timing mark with the top of the cylinder head and install the belt over the sprocket. The belt is a snug fit but can be fitted with the fingers; do not use tools to force the belt onto the sprocket.
   e. Raise the tensioner pulley against the belt to prevent sagging and temporarily tighten the center bolt.
55. Turn the crankshaft ¼ turn (90°) counter-clockwise, then turn it clockwise and align the timing marks.
56. Loosen the center bolt on the tensioner pulley and attach special tool MD 998752 and a torque wrench. (This is a purpose-built tool; substitutes will be hard to find.) Apply a torque of 1.88–2.03 ft. lbs. to the tensioner pulley; this establishes the correct loading against the belt.

NOTE: You must use a torque wrench with a 0–3 ft. lb. scale. Normal scales will not read to the accuracy needed. If the body interferes with the tools, elevate the jack slightly to raise the engine above the interference point.

57. Hold the tensioner pulley in place with the torque wrench and tighten the center bolt of the tensioner pulley to 36 ft. lbs. Remove the torque wrench assembly.
58. On the left side of the engine, located in the rear lower timing belt cover, is a rubber plug. Remove the plug and install special tool MD 998738 (another one difficult to substitute). Screw the tool in until it makes contact with the tensioner arm. After the point of contact, screw it in some more to compress the arm slightly.
59. Remove the wire holding the tip of the auto-tensioner. The tip will push outward and engage the tensioner arm. Remove the screw tool.
60. Rotate the crankshaft 2 complete turns clockwise. Allow everything to sit as is for about 15 minutes — longer if the temperature is cool — then check the distance the tip of the auto-tensioner protrudes. (It is almost impossible with the engine in the vehicle; if you can measure the distance, it should be 3.8–4.5mm (0.15–0.18 in.). The alternate method is given in the next step.
61. Reinstall the special screw tool into the left support bracket until the end just makes contact with the tensioner arm. Starting from that position, turn the tool against the arm, counting the number of full turns until the tensioner arm contacts the auto-tensioner body. At this point, the arm has compressed the tensioner tip through its projection distance. The correct number of turns to this point is 2.5 to 3 turns. If you took the correct number of turns to bottom the arm, the projection was correct to begin with and you may now unwind the screw tool EXACTLY the number of turns you turned it in.

If your number of turns was too high or too low when the arm bottomed, you have a problem with the tensioner and should replace it.

62. Remove the special screw tool and install the rubber plug into the access hole in the timing belt cover.
63. Install the half-circle plug with the proper sealant, then install the valve cover and gasket.
64. Install the spark plug wires, the breather and PCV hoses and the center cover.
65. Install the lower and upper timing belt covers, paying attention to the correct placement of the four different sizes of bolts.
66. Install the crankshaft pulley. Tighten the bolts to 18 ft. lbs.
67. Install the water pump pulleys, tightening the bolts to 6 ft. lbs.
68. Fit the air conditioning compressor belt into place and install the tensioner assembly. Tighten the mounting bolts to 18 ft. lbs.
69. Install the power steering belt and the alternator belt.
70. Install the engine mount bracket to the engine. Tighten the mounting nuts and bolts to 42 ft. lbs.
71. Adjust the jack (if necessary) so that the engine mount bushing aligns with the bodywork bracket. Install the through-bolt and tighten the nuts snug.
72. Slowly release tension on the floor jack so that the weight of the engine bears fully on the mount. Tighten the through-bolt to 72 ft. lbs. and the small safety nut to 38 ft. lbs.
73. Install and connect the wiring harnesses and leads. Make certain the cables are properly routed and that each connector is firmly seated with its mate.
74. Install the fuel return hose. Use a new O-ring and install the high pressure fuel line, tightening the bolts to 4 ft. lbs.

NOTE: Apply a light coat of oil to the O-ring and insert the hose into the delivery pipe carefully to avoid damage to the O-ring.

75. Connect the vacuum hoses and connect the brake booster vacuum hose.
76. Install the heater hose and the coolant by-pass hose.
77. Connect the PCV hose.
78. Install the air cleaner assembly and the related ductwork and tubing, including the tubes to and from the turbocharger.
79. Connect the air flow sensor wiring to the air cleaner.
80. Connect the accelerator cable and adjust it.
81. Install the radiator.
82. Double check all installation items, paying particular attention to loose hoses or hanging wires, untightened nuts, poor routing of hoses and wires (too tight or rubbing) and tools left in the engine area.
83. Fill the cooling system with coolant. Changing the oil and filter is recommended to eliminate pollutants in the oil.

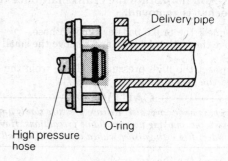

Correct installation of the O-ring on the high pressure fuel line. Apply a light coating of gasoline to the shaded area before connecting the lines

3 ENGINE AND ENGINE OVERHAUL

84. Connect the negative battery cable. Start the engine and check for leaks of fuel, vacuum or oil. Check the operation of all engine electrical systems as well as dashboard gauges and lights.
85. Perform necessary adjustments to the drive belts, accelerator cable and engine specifications. Adjust the coolant level after the engine has cooled off.

Galant, Sigma with 2350cc (G64B) Engine

1. Disconnect the negative battery cable.
2. Drain the engine coolant.

CAUTION

When draining the coolant, keep in mind that cats and dogs are attracted by the ethylene glycol antifreeze, and are quite likely to drink any that is left in an uncovered container or in puddles on the ground. This will prove fatal in sufficient quantity. Always drain the coolant into a sealable container. Coolant should be reused unless it is contaminated or several years old.

---

3. Disconnect the high pressure hose from its clamp on top of the left engine mount. Disconnect the oxygen sensor connector and remove the clip.
4. Position a floor jack and a broad piece of lumber under the engine. Elevate the jack just enough to support the engine without raising it.
5. Remove the through-bolt from the left side engine mount. It may be necessary to adjust the jack slightly to allow the bolt to come free. When the bolt has been removed, disconnect the nuts and bolts holding the mounting bracket to the engine and remove the bracket.
6. Remove the upper radiator hose from the engine. Disconnect the two heater hoses at the head.
7. Label and disconnect the control wiring harness connectors.
8. Remove the engine ground cable from the air intake plenum.
9. Label and remove the spark plug cables from the distributor cap.
10. Disconnect the accelerator cable.
11. Disconnect the PCV hose and the breather hose from the valve cover.
12. Remove the valve cover and gasket. Remove the half-circle plug from the head.
13. Remove the upper timing belt cover; don't lose the spark plug wire guide.
14. Turn the crankshaft clockwise to align the timing marks and position the engine on TDC/compression for cylinder No. 1. Remove the bolt holding the camshaft sprocket to the camshaft. Remove the washer.
15. Remove the camshaft sprocket — with the belt attached — and place it on the lower timing belt cover. Do not allow the belt to come off the sprocket.

WARNING: Do not rotate the crankshaft once the camshaft sprocket is removed.

16. Disconnect the brake booster vacuum hose.
17. Remove the air intake hose and remove the small vacuum hose.
18. Disconnect the high pressure fuel line from the fuel rail and discard the gaskets.

CAUTION

The fuel system is under pressure. Release pressure slowly and contain spillage. Observe no smoking/no open flame precautions. Have a Class B-C (dry powder) fire extinguisher within arm's reach at all times.

---

19. Disconnect the fuel return hose.
20. Remove the two self-locking nuts and the small bracket bolt; separate the exhaust pipe from the bottom of the exhaust manifold. Discard the gasket.
21. Disconnect the intake manifold support brace from the en-

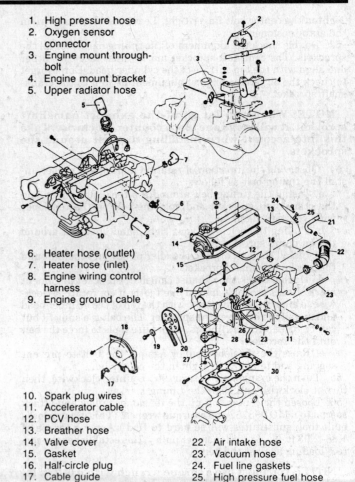

1. High pressure hose
2. Oxygen sensor connector
3. Engine mount through-bolt
4. Engine mount bracket
5. Upper radiator hose
6. Heater hose (outlet)
7. Heater hose (inlet)
8. Engine wiring control harness
9. Engine ground cable
10. Spark plug wires
11. Accelerator cable
12. PCV hose
13. Breather hose
14. Valve cover
15. Gasket
16. Half-circle plug
17. Cable guide
18. Upper timing belt cover
19. Washer
20. Camshaft sprocket
21. Brake booster vacuum hose
22. Air intake hose
23. Vacuum hose
24. Fuel line gaskets
25. High pressure fuel hose
26. Fuel return hose
27. Front exhaust pipe
28. Exhaust gasket
29. Cylinder head assembly
30. Head gasket

Head, gasket and related components, G64B engine

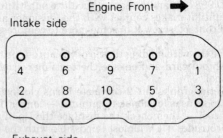

G64B head bolt removal pattern

gine block. Loosen the head bolts in the order shown (this is important) and in three passes until all are finger loose. Remove the bolts.

22. Rock the head gently to break it loose; if tapping is necessary, do so with a rubber or wooden mallet at the corners of the head. DO NOT pry the head up by wedging tools between the head and the block.
23. Lift the head free of the engine, remembering that the intake and exhaust manifolds are still attached. This assembly is

3-182

ENGINE AND ENGINE OVERHAUL 3

heavy; an assistant is helpful when lifting.

Support the head assembly on wooden blocks on the workbench. Refer to "Cleaning and Inspection" following this section for work to be done before installing the head. If the head has been removed for work other than gasket replacement, the manifolds, rocker assembly, camshaft and other components may be removed.

24. Before reinstallation, the head should be completely assembled on the bench. This allows proper location and tightening of all the external items. The head and manifolds can be installed as a unit, saving time and effort.

25. Place a new gasket on the engine so that the identifying mark faces up (towards the head) and is at the timing belt end of the block. Install a new gasket on the exhaust pipe.

WARNING: Do not apply sealant to the head gasket or mating surfaces.

26. Install the head straight down onto the block. Try to eliminate most of the side-to-side adjustments as this may move the gasket out of position. Install the bolts by hand and just start each bolt 1 or 2 turns on the threads.

27. The head bolt torque specification is 68 ft. lbs. for a cold engine. The bolts must be tightened in the order shown in 3 steps. On the first pass, tighten all the bolts to about 22 ft. lbs. then proceed through the order tightening each bolt to about 45 ft. lbs. The final torque is achieved on the third pass. Install the intake manifold support brace to the engine block and tighten the bolt to 16 ft. lbs.

28. Making sure the gasket is still in place, connect the exhaust pipe to the base of the exhaust manifold. Use new self-locking nuts; tighten the nuts and the small bracket bolt to 18 ft. lbs.

29. Connect the fuel return hose. Install new gaskets and connect the high pressure fuel line and tighten the eye bolt to 22 ft. lbs.

30. Reconnect the small vacuum hoses, the brake booster vacuum hose and the air intake duct.

31. Carefully install the camshaft sprocket and timing belt onto the camshaft. Make certain that the cam has not changed position during the removal of the head. Install the small washer and the retaining bolt, tightening the bolt to 66 ft. lbs.

32. Install the upper timing case cover. Tighten the bolts to 8 ft. lbs. and don't forget the spark plug wire guide.

33. Install the half-circle plug into the head and coat the flat surface of the plug with a thin layer of sealant. Install the valve cover and gasket.

34. Connect the breather hose, PCV hose and the accelerator cable.

35. Install the spark plug wires, making certain they are correctly routed and firmly seated.

36. Connect the engine ground cable. Connect the engine control harness and properly attach the brackets holding it.

37. Install the heater hoses (2) and the upper radiator hose. Make sure the hoses are not crushed or restricted.

38. Install the engine mount bracket to the engine. Tighten the mounting nuts and bolts to 42 ft. lbs.

39. Adjust the jack (if necessary) so that the engine mount bushing aligns with the bodywork bracket. Install the through-bolt and tighten the nuts snug.

40. Slowly release tension on the floor jack so that the weight of the engine bears fully on the mount. Tighten the through-bolt to 52 ft. lbs. and the small safety nut to 26 ft. lbs.

41. Connect the oxygen sensor wiring and install its clip. Position the high pressure hose and secure its bracket.

42. Double check all installation items, paying particular attention to loose hoses or hanging wires, untightened nuts, poor routing of hoses and wires (too tight or rubbing) and tools left in the engine area.

43. Fill the cooling system with coolant. Changing the oil and filter is recommended to eliminate pollutants in the oil.

44. Connect the negative battery cable. Start the engine and check for leaks of fuel, vacuum or oil. Check the operation of all engine electrical systems as well as dashboard gauges and lights.

45. Perform necessary adjustments to the accelerator cable and engine specifications. Adjust the coolant level after the engine has cooled off.

Galant, Sigma with 6G72 V6 Engine

WARNING: The use of the correct special tools or their equivalent is REQUIRED for this procedure. The camshaft sprocket must be removed and installed with the use of tools MB 990767-01 and MB 998715 or their equivalents. Attempts to use other tools or methods may cause damage to the sprocket and/or camshaft.

FRONT HEAD

1. Disconnect the negative battery cable.
2. Drain the cooling system.

CAUTION

When draining the coolant, keep in mind that cats and dogs are attracted by the ethylene glycol antifreeze, and are quite likely to drink any that is left in an uncovered container or in puddles on the ground. This will prove fatal in sufficient quantity. Always drain the coolant into a sealable container. Coolant should be reused unless it is contaminated or several years old.

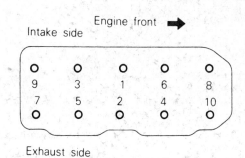

Head bolt tightening sequence, G64B

3. Remove the air intake plenum.
4. Disconnect the upper radiator hose from the thermostat housing.
5. Label and disconnect the various connectors along the engine control harness. Position the harness and wiring leads out of the way.
6. Remove the intake manifold with the injectors and fuel rail attached.

CAUTION

The fuel system is under pressure. Release pressure slowly and contain spillage. Observe no smoking/no open flame precautions. Have a Class B-C (dry powder) fire extinguisher within arm's reach at all times.

7. Label and disconnect the wiring to the front spark plugs (Nos. 1, 3, and 5).
8. Disconnect the oxygen sensor wiring.
9. Remove the self-locking nuts holding each (front and rear) exhaust pipe to the exhaust manifolds. Remove the small bracket bolts and separate the exhaust system from the engine.
10. Remove the heat shield from the exhaust manifold and remove the manifold from the head.
11. Remove the dipstick and tube from the engine block. Remove the exhaust manifold gasket.
12. Remove the bolt holding the high pressure hose to the engine.
13. Disconnect the pressure switch wiring from the power steering pump.

3-183

3 ENGINE AND ENGINE OVERHAUL

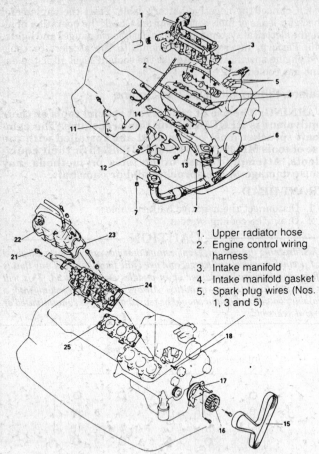

1. Upper radiator hose
2. Engine control wiring harness
3. Intake manifold
4. Intake manifold gasket
5. Spark plug wires (Nos. 1, 3 and 5)
6. Oxygen sensor connector
7. Self-locking nuts
11. Heat shield
12. Front exhaust manifold
13. Dipstick tube
14. Exhaust manifold gasket
15. Timing belt
16. Camshaft sprocket
17. Upper rear timing belt cover
18. Power steering pump bracket bolts
21. Bracket bolts
22. Valve cover
23. Valve cover gasket
24. Cylinder head assembly
25. Head gasket

Removal of V6 front cylinder head. Timing belt components and covers not shown

14. Disconnect the high pressure line from the power steering pump.
15. Place a floor jack and a broad piece of lumber under the oil pan. Raise the jack until the engine is supported but not raised.
16. Remove the through-bolt for the left engine mount.
17. Remove the four nuts holding the upper part of the engine mount to the lower bracket.
18. Remove the air conditioning compressor drive belt, then remove the belt tensioner assembly.
19. Remove the water pump drive belt.
20. Remove the power steering pump and position it out of the way.
21. Remove the belt tensioner assembly and its bracket.
22. Remove the upper outer timing belt cover (B) for the front bank of cylinders. Remove the gaskets and keep them with the cover.

NOTE: The bolts holding the belt covers are of different lengths and must be correctly placed during reinstallation. Label or diagram them as they are removed.

23. Remove the lower portion of the engine support bracket. Remove the bolts in the order shown in the figure. Use spray lubricant to assist in removing the reamer bolt. Keep in mind that the reamer bolt may be heat-seized on the bracket. Remove it slowly.
24. Remove the small end cap from the rear bank belt cover, then remove the upper outer belt cover. Make certain all the gaskets are removed and placed with the covers.
25. Remove the left side splash shield. Turn the crankshaft clockwise until all the timing marks align. This will set the engine to TDC/compression on No. 1 cylinder.
26. Install the special counterholding tools and carefully remove the crankshaft pulley without moving the engine out of position.
27. Remove the front flange from the crank sprocket.
28. Remove the lower outer timing belt cover with its gaskets.
29. Loosen the timing belt tensioner bolt and turn the tensioner counterclockwise along the elongated hole. This will relax the tension on the belt.
30. If the timing belt is to be reused, make a chalk or crayon arrow on the belt showing the direction of rotation so that it may be reinstalled correctly.
31. Carefully slide the belt off the sprockets. Place the belt in a clean, dry, protected location away from the work area. if the tensioner is to be removed, disconnect the spring and remove the retaining bolt.
32. Remove the valve cover and gasket.
33. Use the special hex wrench (MB 998051-01) and loosen the head bolts in the order shown in 2 or 3 passes. When all are finger loose, remove the bolts.
34. Rock the head gently to break it loose; if tapping is necessary, do so with a rubber or wooden mallet at the corners of the head. DO NOT pry the head up by wedging tools between the head and the block.
35. Lift the head free of the engine. Support the head assembly on wooden blocks on the workbench. Refer to *Cleaning and Inspection* following this section for work to be done before installing the head. If the head has been removed for work other than gasket replacement, the rocker assembly and camshaft or other components may be removed.

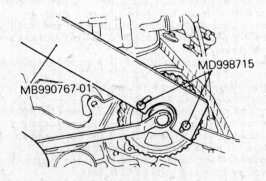

Use the counterhold tools to prevent the camshaft from turning

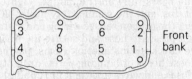

Removal sequence for front head bolts, 6G72 V6

ENGINE AND ENGINE OVERHAUL 3

36. Before reinstallation, the head should be completely assembled on the bench. This allows proper location and tightening of all the external items.

To install:

37. The head gaskets are different for each head. The gasket for this head is marked **R** (right); the gasket should be installed with the identifying mark at the timing belt end and facing up (towards the head).

WARNING: Do not apply sealant to the head gasket or mating surfaces.

38. Install the head straight down onto the block. Try to eliminate most of the side–to–side adjustments as this may move the gasket out of position. Install the bolts and the special washers by hand and just start each bolt 1 or 2 turns on the threads.

WARNING: The washers must be installed correctly. The rounded shoulder of the washer denotes the face in contact with the bolt. The flat face contacts the head.

39. Correct tightening of the head bolts requires 3 steps:
    a. Following the tightening sequence, draw each bolt to 62 ft. lbs.
    b. Follow the loosening sequence and loosen each bolt in order until it is completely loose.
    c. Follow the tightening sequence and tighten each bolt to 70 ft. lbs.
40. Install the valve cover and gasket.
41. Install the power steering pump mounting bracket, tightening the bolts to 42 ft. lbs. Install the small bracket on the rear of the head and tighten the bolts to 10 ft. lbs.
42. Check the sprockets and tensioner for wear. The sprocket teeth should be well defined, not rounded and the valleys between the teeth should be clean. The tensioners should spin freely with no binding or unusual noise. Replace the tensioner if there is any sign of grease leaking from the seal. Clean everything with a clean, dry cloth.

WARNING: Do not spray or immerse the sprockets or tensioners in cleaning solvent. The sprocket may absorb the solvent and transfer it to the belt. The tensioners are internally lubricated and the solvent will dilute or dissolve the lubricant.

43. If the tensioner was removed, it must be reinstalled. After bolting it loosely in place, connect the spring onto the water pump pin. Make certain the spring faces in the correct direction on the tensioner. Turn the tensioner to the extreme counterclockwise position on the elongated hole and tighten the bolt just enough to hold the tensioner in this position.
44. Double check the alignment of the timing marks on the camshaft and crankshaft sprockets.
45. Observing the direction of rotation marks made earlier, install the belt onto the crankshaft sprocket and then onto the rear bank camshaft sprocket. Maintain tension on the belt between the sprockets.
46. Continue installing the belt onto the water pump, the front bank cam sprocket and the tensioner.
47. With your fingers, apply gentle counterclockwise force to the rear camshaft sprocket. When the belt is taut on the tension side, the timing marks should align perfectly.
48. Install the flange on the crankshaft sprocket.
49. Loosen the bolt holding the tensioner one or two turns and allow the spring tension to draw the tensioner against the belt.
50. Using special tool MB 998716 or equivalent adapter, turn the crankshaft two complete revolutions clockwise. Turn the crank smoothly and re–align the timing marks at the end of the second revolution. This allows the tensioner to compensate for the normal amount of slack in the belt.
51. With the timing marks aligned, tighten the tensioner bolts to 18 ft. lbs.
52. Using a belt tension gauge, measure the belt tension at a point halfway between the crankshaft sprocket and the rear camshaft sprocket. Correct value is 57–84 lbs.
53. Making certain all the gaskets are correctly in place, install the lower timing belt cover and tighten the bolts to 8 ft. lbs.
54. Install the crankshaft pulley. Use the special tools to counterhold it and tighten the bolt to 112 ft. lbs.
55. Install the left splash shield.
56. Install the upper outer cover on the rear bank camshaft sprocket and install the smaller end cap. Tighten the bolts to 8 ft. lbs.
57. Install the lower portion of the engine support bracket. Install the bolts in the correct order. Of the 3 mounting bolts arranged vertically on the right side of the mount, the upper two are tightened to 50 ft. lbs; the bottom one is tightened to 78 ft. lbs. The two smaller bolts in the side of the mount should be tightened to 30 ft. lbs. Use spray lubricant to help with the reamer bolt, and tighten it slowly. Access may be easier if the engine is elevated slightly with the floor jack.
58. Install the cover and gaskets for the front bank camshaft sprocket.
59. Install the tensioner and bracket for the ribbed belt.
60. Install the power steering pump onto its bracket.
61. Install the ribbed belt and adjust it as needed. Make certain it is properly seated on all the pulleys.
62. Install the tensioner and bracket for the air conditioner drive belt and install the belt, adjusting it as necessary.
63. Install the upper section of the engine mounting bracket to the lower section. Tighten the nuts to 50 ft. lbs.
64. Adjust the jack so that the bushing of the engine mount aligns with the bodywork bracket. Install the through-bolt and tighten the nuts snug.
65. Carefully lower the jack, allowing the full weight of the engine to bear on the mount. Tighten the nut on the through-bolt to 50 ft. lbs. and the smaller safety nut to 26 ft. lbs.
66. Install the pressure hose to the power steering pump and tighten the nut to 32 ft. lbs.
67. Connect the wiring to the power steering pump and rebolt the bracket for the high pressure hose.
68. Install a new exhaust manifold gasket. Install the dipstick

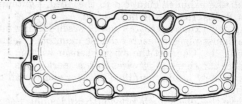

Identifying mark on front head gasket

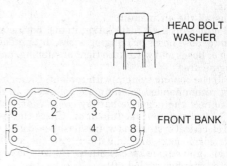

Head bolt tightening sequence. Note the correct installation of the washers

3 ENGINE AND ENGINE OVERHAUL

and dipstick tube, using a new O-ring on the tube and coating it with oil before installation.

69. Install the exhaust manifold to the head, tightening the nuts to 14 ft. lbs. Install the heat shield.
70. Using new exhaust gaskets and new self-locking nuts; attach the exhaust pipes to each exhaust manifold. Tighten the nuts to 26 ft. lbs. Connect the oxygen sensor wiring.
71. Connect the spark plug wires to the spark plugs.
72. Position new intake manifold gaskets and install the intake manifold.
73. Observe the labels made previously and connect each lead of the engine control harness to its proper component.
74. Connect the upper radiator hose to the thermostat housing.
75. Install the air plenum.
76. Double check all installation items, paying particular attention to loose hoses or hanging wires, untightened nuts, poor routing of hoses and wires (too tight or rubbing) and tools left in the engine area.
77. Fill the cooling system with coolant. Changing the oil and filter is recommended to eliminate pollutants in the oil.
78. Connect the negative battery cable. Start the engine and check for leaks of fuel, vacuum or oil. Check the operation of all engine electrical systems as well as dashboard gauges and lights.
79. Perform necessary adjustments to the accelerator cable, drive belts and engine specifications. Adjust the coolant level after the engine has cooled off.

REAR HEAD

1. Disconnect the negative battery cable.
2. Drain the cooling system.

CAUTION

When draining the coolant, keep in mind that cats and dogs are attracted by the ethylene glycol antifreeze, and are quite likely to drink any that is left in an uncovered container or in puddles on the ground. This will prove fatal in sufficient quantity. Always drain the coolant into a sealable container. Coolant should be reused unless it is contaminated or several years old.

---

3. Remove the air intake plenum.
4. Disconnect the upper radiator hose from the thermostat housing.
5. Label and disconnect the various connectors along the engine control harness. Position the harness and wiring leads out of the way.
6. Remove the intake manifold with the injectors and fuel rail attached.

CAUTION

The fuel system is under pressure. Release pressure slowly and contain spillage. Observe no smoking/no open flame precautions. Have a Class B-C (dry powder) fire extinguisher within arm's reach at all times.

---

7. Label and disconnect the wiring to the rear spark plugs (Nos. 2, 4 and 6).
8. Disconnect the oxygen sensor wiring.
9. Remove the self-locking nuts holding each (front and rear) exhaust pipe to the exhaust manifolds. Remove the small bracket bolts and separate the exhaust system from the engine.
10. Remove the distributor.
11. Remove the air intake plenum support.
12. Remove the bracket bolt holding the two hoses to the rear of the head.
13. Remove the heat shield from the rear exhaust manifold; remove the manifold and its gasket.
14. Remove the bolt holding the high pressure hose to the engine.
15. Disconnect the pressure switch wiring from the power steering pump.
16. Disconnect the high pressure line from the power steering pump.
17. Place a floor jack and a broad piece of lumber under the oil pan. Raise the jack until the engine is supported but not raised.
18. Remove the through-bolt for the left engine mount.
19. Remove the four nuts holding the upper part of the engine mount to the lower bracket.
20. Remove the air conditioning compressor drive belt, then remove the belt tensioner assembly.
21. Remove the water pump drive belt.
22. Remove the power steering pump and position it out of the way.
23. Remove the belt tensioner assembly and its bracket.
24. Remove the upper outer timing belt cover (B) for the front bank of cylinders. Remove the gaskets and keep them with the cover.

NOTE: The bolts holding the belt covers are of different lengths and must be correctly placed during reinstallation. Label or diagram them as they are removed.

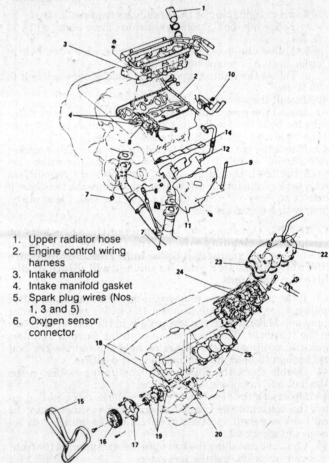

1. Upper radiator hose
2. Engine control wiring harness
3. Intake manifold
4. Intake manifold gasket
5. Spark plug wires (Nos. 1, 3 and 5)
6. Oxygen sensor connector
7. Self-locking nuts
8. Distributor
9. Air intake plenum support
10. Bracket bolt
11. Heat shield
12. Rear exhaust manifold
14. Exhaust manifold gasket
15. Timing belt
16. Camshaft sprocket
17. Upper rear timing belt cover
18. Bracket bolts
19. Distributor adapter assembly
20. Alternator brace upper bolt
22. Valve cover
23. Valve cover gasket
24. Cylinder head assembly
25. Head gasket

Removal of V6 rear cylinder head. Timing belt components and covers not shown

ENGINE AND ENGINE OVERHAUL 3

25. Remove the lower portion of the engine support bracket. Remove the bolts in the order shown in the figure. Use spray lubricant to assist in removing the reamer bolt. Keep in mind that the reamer bolt may be heat–seized on the bracket. Remove it slowly.
26. Remove the small end cap from the rear bank belt cover, then remove the upper outer belt cover. Make certain all the gaskets are removed and placed with the covers.
27. Remove the left side splash shield. Turn the crankshaft clockwise until all the timing marks align. This will set the engine to TDC/compression on No. 1 cylinder.
28. Install the special counterholding tools and carefully remove the crankshaft pulley without moving the engine out of position.
29. Remove the front flange from the crank sprocket.
30. Remove the lower outer timing belt cover with its gaskets.
31. Loosen the timing belt tensioner bolt and turn the tensioner counterclockwise along the elongated hole. This will relax the tension on the belt.
32. If the timing belt is to be reused, make a chalk or crayon arrow on the belt showing the direction of rotation so that it may be reinstalled correctly.
33. Carefully slide the belt off the sprockets. Place the belt in a clean, dry, protected location away from the work area. if the tensioner is to be removed, disconnect the spring and remove the retaining bolt.
34. Install the counterholding tools and remove the camshaft sprocket. Remove the upper rear timing belt cover.
35. Remove the bolts holding the small bracket to the front of the head. Unbolt and remove the distributor adapter, oil seal and O-ring.

WARNING: Use care not to damage the O-ring during removal. If it becomes damaged or cracked, replace it.

36. Remove the upper bolt from the alternator brace.
37. Remove the valve cover and gasket.
38. Use the special hex wrench (MB 998051-01) and loosen the head bolts in the order shown in 2 or 3 passes. When all are finger loose, remove the bolts.
39. Rock the head gently to break it loose; if tapping is necessary, do so with a rubber or wooden mallet at the corners of the head. DO NOT pry the head up by wedging tools between the head and the block.
40. Lift the head free of the engine. Support the head assembly on wooden blocks on the workbench. Refer to *Cleaning and Inspection* following this section for work to be done before installing the head. If the head has been removed for work other than gasket replacement, the rocker assembly and camshaft or other components may be removed.

To install:
41. Before reinstallation, the head should be completely assembled on the bench. This allows proper location and tightening of all the external items.
42. The head gaskets are different for each head. The gasket for this head is marked **L** (left); the gasket should be installed with the identifying mark at the timing belt end and facing up (towards the head).

WARNING: Do not apply sealant to the head gasket or mating surfaces.

43. Install the head straight down onto the block. Try to eliminate most of the side–to–side adjustments as this may move the gasket out of position. Install the bolts and the special washers by hand and just start each bolt 1 or 2 turns on the threads.

WARNING: The washers must be installed correctly. The rounded shoulder of the washer denotes the face in contact with the bolt. The flat face contacts the head.

44. Correct tightening of the head bolts requires 3 steps:

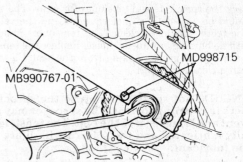

Use the counterhold tools to prevent the camshaft from turning

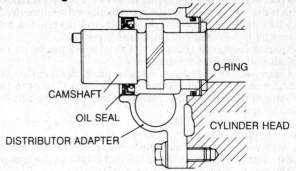

Distributor adapter. Use great care not to damage the O-ring or seal during removal and installation

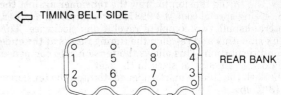

Removal sequence for rear head bolts, 6G72 V6

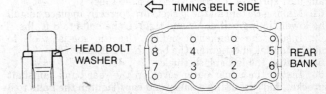

Bolt installation sequence, V6 rear head. Note the correct position of the washers

    a. Following the tightening sequence, draw each bolt to 62 ft. lbs.
    b. Follow the loosening sequence and loosen each bolt in order until it is completely loose.
    c. Follow the tightening sequence and tighten each bolt to 70 ft. lbs.
45. Install the valve cover and gasket.
46. Install the upper bolt for the alternator brace and tighten it to 14 ft. lbs.
47. Carefully install the distributor adapter assembly with the seal and O-ring. Tighten the retaining bolts to 10 ft. lbs.
48. Install the small bracket bolts to the front of the head, tightening them to 16 ft. lbs.
49. Install the upper rear timing belt cover. Install the camshaft sprocket (use the counterholding tools) and tighten the retaining bolt to 66 ft. lbs.

3 ENGINE AND ENGINE OVERHAUL

50. Check the sprockets and tensioner for wear. The sprocket teeth should be well defined, not rounded and the valleys between the teeth should be clean. The tensioners should spin freely with no binding or unusual noise. Replace the tensioner if there is any sign of grease leaking from the seal. Clean everything with a clean, dry cloth.

WARNING: Do not spray or immerse the sprockets or tensioners in cleaning solvent. The sprocket may absorb the solvent and transfer it to the belt. The tensioners are internally lubricated and the solvent will dilute or dissolve the lubricant.

51. If the tensioner was removed, it must be reinstalled. After bolting it loosely in place, connect the spring onto the water pump pin. Make certain the spring faces in the correct direction on the tensioner. Turn the tensioner to the extreme counterclockwise position on the elongated hole and tighten the bolt just enough to hold the tensioner in this position.
52. Double check the alignment of the timing marks on the camshaft and crankshaft sprockets.
53. Observing the direction of rotation marks made earlier, install the belt onto the crankshaft sprocket and then onto the rear bank camshaft sprocket. Maintain tension on the belt between the sprockets.
54. Continue installing the belt onto the water pump, the front bank cam sprocket and the tensioner.
55. With your fingers, apply gentle counterclockwise force to the rear camshaft sprocket. When the belt is taut on the tension side, the timing marks should align perfectly.
56. Install the flange on the crankshaft sprocket.
57. Loosen the bolt holding the tensioner one or two turns and allow the spring tension to draw the tensioner against the belt.
58. Using special tool MB 998716 or equivalent adapter, turn the crankshaft two complete revolutions clockwise. Turn the crank smoothly and re-align the timing marks at the end of the second revolution. This allows the tensioner to compensate for the normal amount of slack in the belt.
59. With the timing marks aligned, tighten the tensioner bolts to 18 ft. lbs.
60. Using a belt tension gauge, measure the belt tension at a point halfway between the crankshaft sprocket and the rear camshaft sprocket. Correct value is 57–84 lbs.
61. Making certain all the gaskets are correctly in place, install the lower timing belt cover and tighten the bolts to 8 ft. lbs.
62. Install the crankshaft pulley. Use the special tools to counterhold it and tighten the bolt to 112 ft. lbs.
63. Install the left splash shield.
64. Install the upper outer cover on the rear bank camshaft sprocket and install the smaller end cap. Tighten the bolts to 8 ft. lbs.
65. Install the lower portion of the engine support bracket. Install the bolts in the correct order. Of the 3 mounting bolts arranged vertically on the right side of the mount, the upper two are tightened to 50 ft. lbs; the bottom one is tightened to 78 ft. lbs. The two smaller bolts in the side of the mount should be tightened to 30 ft. lbs. Use spray lubricant to help with the reamer bolt, and tighten it slowly. Access may be easier if the engine is elevated slightly with the floor jack.
66. Install the cover and gaskets for the front bank camshaft sprocket.
67. Install the tensioner and bracket for the ribbed belt.
68. Install the power steering pump onto its bracket.
69. Install the ribbed belt and adjust it as needed. Make certain it is properly seated on all the pulleys.
70. Install the tensioner and bracket for the air conditioner drive belt and install the belt, adjusting it as necessary.
71. Install the upper section of the engine mounting bracket to the lower section. Tighten the nuts to 50 ft. lbs.
72. Adjust the jack so that the bushing of the engine mount aligns with the bodywork bracket. Install the through-bolt and tighten the nuts snug.
73. Carefully lower the jack, allowing the full weight of the engine to bear on the mount. Tighten the nut on the through-bolt to 50 ft. lbs. and the smaller safety nut to 26 ft. lbs.
74. Install the pressure hose to the power steering pump and tighten the nut to 32 ft. lbs.
75. Connect the wiring to the power steering pump and rebolt the bracket for the high pressure hose.
76. Install a new exhaust manifold gasket and install the rear exhaust manifold to the head. tighten the manifold retaining nuts to 14 ft. lbs.
77. Install the heat shield on the rear exhaust manifold, tightening the bolts to 10 ft. lbs. Don't forget the intake plenum support.
78. Reinstall the bracket bolt holding the two hoses to the rear of the head.
79. Install the distributor.
80. Using new exhaust gaskets and new self-locking nuts; attach the exhaust pipes to each exhaust manifold. Tighten the nuts to 26 ft. lbs. Connect the oxygen sensor wiring.
81. Connect the spark plug wires to the spark plugs.
82. Position new intake manifold gaskets and install the intake manifold.
83. Observe the labels made previously and connect each lead of the engine control harness to its proper component.
84. Connect the upper radiator hose to the thermostat housing.
85. Install the air plenum.
86. Double check all installation items, paying particular attention to loose hoses or hanging wires, untightened nuts, poor routing of hoses and wires (too tight or rubbing) and tools left in the engine area.
87. Fill the cooling system with coolant. Changing the oil and filter is recommended to eliminate pollutants in the oil.
88. Connect the negative battery cable. Start the engine and check for leaks of fuel, vacuum or oil. Check the operation of all engine electrical systems as well as dashboard gauges and lights.
89. Perform necessary adjustments to the accelerator cable, drive belts and engine specifications. Adjust the coolant level after the engine has cooled off.

Montero with 6G72 V6 Engine

WARNING: The use of the correct special tools or their equivalent is REQUIRED for this procedure. The camshaft sprocket must be removed and installed with the use of tools MB 990767-01 and MB 998715 or their equivalents. Attempts to use other tools or methods may cause damage to the sprocket and/or camshaft.

RIGHT HEAD

1. Disconnect the negative battery cable.
2. Drain the cooling system.

CAUTION

When draining the coolant, keep in mind that cats and dogs are attracted by the ethylene glycol antifreeze, and are quite likely to drink any that is left in an uncovered container or in puddles on the ground. This will prove fatal in sufficient quantity. Always drain the coolant into a sealable container. Coolant should be reused unless it is contaminated or several years old.

---

3. Remove the air intake plenum.
4. Disconnect the upper radiator hose from the thermostat housing.
5. Label and disconnect the various connectors along the engine control harness. Position the harness and wiring leads out of the way.
6. Remove the intake manifold with the injectors and fuel rail attached.

ENGINE AND ENGINE OVERHAUL 3

CAUTION

The fuel system is under pressure. Release pressure slowly and contain spillage. Observe no smoking/no open flame precautions. Have a Class B-C (dry powder) fire extinguisher within arm's reach at all times.

---

7. Label and disconnect the wiring to the front spark plugs (Nos. 1, 3, and 5).
8. Disconnect the oxygen sensor wiring.
9. Remove the self-locking nuts holding each exhaust pipe to the exhaust manifolds. Remove the small bracket bolts and separate the exhaust system from the engine.
10. Remove the heat shield from the exhaust manifold and remove the manifold from the head.
11. Remove the dipstick and tube from the engine block. Remove the exhaust manifold gasket.
12. Remove the bolt holding the high pressure hose to the engine.
13. Disconnect the pressure switch wiring from the power steering pump.
14. Disconnect the high pressure line from the power steering pump.
15. Place a floor jack and a broad piece of lumber under the oil pan. Raise the jack until the engine is supported but not raised.
16. Remove the through-bolt for the right engine mount.
17. Remove the four nuts holding the upper part of the engine mount to the lower bracket.
18. Remove the air conditioning compressor drive belt, then remove the belt tensioner assembly.
19. Remove the engine fan and pulley.
20. Remove the water pump drive belt.
21. Remove the power steering pump and position it out of the way.
22. Remove the belt tensioner assembly and its bracket.
23. Remove the upper outer timing belt cover (B) for the front bank of cylinders. Remove the gaskets and keep them with the cover.

NOTE: The bolts holding the belt covers are of different lengths and must be correctly placed during reinstallation. Label or diagram them as they are removed.

24. Remove the lower portion of the engine support bracket. Remove the bolts in the order shown in the figure. Use spray lubricant to assist in removing the reamer bolt. Keep in mind that the reamer bolt may be heat-seized on the bracket. Remove it slowly.
25. Remove the small end cap from the rear bank belt cover, then remove the upper outer belt cover. Make certain all the gaskets are removed and placed with the covers.
26. Remove the left side splash shield. Turn the crankshaft clockwise until all the timing marks align. This will set the engine to TDC/compression on No. 1 cylinder.
27. Install the special counterholding tools and carefully remove the crankshaft pulley without moving the engine out of position.
28. Remove the front flange from the crank sprocket.
29. Remove the lower outer timing belt cover with its gaskets.
30. Loosen the timing belt tensioner bolt and turn the tensioner counterclockwise along the elongated hole. This will relax the tension on the belt.
31. If the timing belt is to be reused, make a chalk or crayon arrow on the belt showing the direction of rotation so that it may be reinstalled correctly.
32. Carefully slide the belt off the sprockets. Place the belt in a clean, dry, protected location away from the work area. if the tensioner is to be removed, disconnect the spring and remove the retaining bolt.
33. Remove the valve cover and gasket.
34. Use the special hex wrench (MB 998051-01) and loosen the head bolts in the order shown in 2 or 3 passes. When all are finger loose, remove the bolts.
35. Rock the head gently to break it loose; if tapping is necessary, do so with a rubber or wooden mallet at the corners of the head. DO NOT pry the head up by wedging tools between the head and the block.
36. Lift the head free of the engine. Support the head assembly on wooden blocks on the workbench. Refer to *Cleaning and Inspection* following this section for work to be done before installing the head. If the head has been removed for work other than gasket replacement, the rocker assembly and camshaft or other components may be removed.

To install:
37. Before reinstallation, the head should be completely assembled on the bench. This allows proper location and tightening of all the external items.
38. The head gaskets are different for each head. The gasket for this head is marked **R** (right); the gasket should be installed with the identifying mark at the timing belt end and facing up (towards the head).

WARNING: Do not apply sealant to the head gasket or mating surfaces.

39. Install the head straight down onto the block. Try to eliminate most of the side-to-side adjustments as this may move the gasket out of position. Install the bolts and the special washers by hand and just start each bolt 1 or 2 turns on the threads.

WARNING: The washers must be installed correctly. The rounded shoulder of the washer denotes the face in contact with the bolt. The flat face contacts the head.

40. Correct tightening of the head bolts requires 3 steps:
   a. Following the tightening sequence, draw each bolt to 62 ft. lbs.
   b. Follow the loosening sequence and loosen each bolt in order until it is completely loose.
   c. Follow the tightening sequence and tighten each bolt to 70 ft. lbs.
41. Install the valve cover and gasket.
42. Install the power steering pump mounting bracket, tightening the bolts to 43 ft. lbs. Install the small bracket on the rear of the head and tighten the bolts to 10 ft. lbs.
44. Check the sprockets and tensioner for wear. The sprocket teeth should be well defined, not rounded and the valleys between the teeth should be clean. The tensioners should spin freely with no binding or unusual noise. Replace the tensioner if there is any sign of grease leaking from the seal. Clean everything with a clean, dry cloth.

WARNING: Do not spray or immerse the sprockets or tensioners in cleaning solvent. The sprocket may absorb the solvent and transfer it to the belt. The tensioners are internally lubricated and the solvent will dilute or dissolve the lubricant.

45. If the tensioner was removed, it must be reinstalled. After bolting it loosely in place, connect the spring onto the water pump pin. Make certain the spring faces in the correct direction on the tensioner. Turn the tensioner to the extreme counterclockwise position on the elongated hole and tighten the bolt just enough to hold the tensioner in this position.
46. Double check the alignment of the timing marks on the camshaft and crankshaft sprockets.
47. Observing the direction of rotation marks made earlier, install the belt onto the crankshaft sprocket and then onto the rear bank camshaft sprocket. Maintain tension on the belt between the sprockets.
48. Continue installing the belt onto the water pump, the front bank cam sprocket and the tensioner.
49. With your fingers, apply gentle counterclockwise force to the rear camshaft sprocket. When the belt is taut on the tension side, the timing marks should align perfectly.

3-189

ENGINE AND ENGINE OVERHAUL

50. Install the flange on the crankshaft sprocket.
51. Loosen the bolt holding the tensioner one or two turns and allow the spring tension to draw the tensioner against the belt.
52. Using special tool MB 998716 or equivalent adapter, turn the crankshaft two complete revolutions clockwise. Turn the crank smoothly and re–align the timing marks at the end of the second revolution. This allows the tensioner to compensate for the normal amount of slack in the belt.
53. With the timing marks aligned, tighten the tensioner bolts to 18 ft. lbs.
54. Using a belt tension gauge, measure the belt tension at a point halfway between the crankshaft sprocket and the rear camshaft sprocket. Correct value is 57–84 lbs.
55. Making certain all the gaskets are correctly in place, install the lower timing belt cover and tighten the bolts to 8 ft. lbs.
56. Install the crankshaft pulley. Use the special tools to counterhold it and tighten the bolt to 112 ft. lbs.
57. Install the left splash shield.
58. Install the upper outer cover on the rear bank camshaft sprocket and install the smaller end cap. Tighten the bolts to 8 ft. lbs.
59. Install the lower portion of the engine support bracket. Install the bolts in the correct order. Of the 3 mounting bolts arranged vertically on the right side of the mount, the upper two are tightened to 50 ft. lbs; the bottom one is tightened to 78 ft. lbs. The two smaller bolts in the side of the mount should be tightened to 30 ft. lbs. Use spray lubricant to help with the reamer bolt, and tighten it slowly. Access may be easier if the engine is elevated slightly with the floor jack.
60. Install the cover and gaskets for the front bank camshaft sprocket.
61. Install the tensioner and bracket for the ribbed belt.
62. Install the power steering pump onto its bracket.
63. Install the ribbed belt and adjust it as needed. Make certain it is properly seated on all the pulleys.
64. Install the tensioner and bracket for the air conditioner drive belt and install the belt, adjusting it as necessary.
65. Install the upper section of the engine mounting bracket to the lower section. Tighten the nuts to 50 ft. lbs.
66. Adjust the jack so that the bushing of the engine mount aligns with the bodywork bracket. Install the through-bolt and tighten the nuts snug.
67. Carefully lower the jack, allowing the full weight of the engine to bear on the mount. Tighten the nut on the through-bolt to 50 ft. lbs. and the smaller safety nut to 26 ft. lbs.
68. Install the pressure hose to the power steering pump and tighten the nut to 32 ft. lbs.
69. Connect the wiring to the power steering pump and rebolt the bracket for the high pressure hose.
70. Install the engine fan and pulley.
71. Install a new exhaust manifold gasket. Install the dipstick and dipstick tube, using a new O-ring on the tube and coating it with oil before installation.
72. Install the exhaust manifold to the head, tightening the nuts to 14 ft. lbs. Install the heat shield.
73. Using new exhaust gaskets and new self-locking nuts; attach the exhaust pipes to each exhaust manifold. Tighten the nuts to 26 ft. lbs. Connect the oxygen sensor wiring.
74. Connect the spark plug wires to the spark plugs.
75. Position new intake manifold gaskets and install the intake manifold.
76. Observe the labels made previously and connect each lead of the engine control harness to its proper component.
77. Connect the upper radiator hose to the thermostat housing.
78. Install the air plenum.
79. Double check all installation items, paying particular attention to loose hoses or hanging wires, untightened nuts, poor routing of hoses and wires (too tight or rubbing) and tools left in the engine area.
80. Fill the cooling system with coolant. Changing the oil and filter is recommended to eliminate pollutants in the oil.
81. Connect the negative battery cable. Start the engine and check for leaks of fuel, vacuum or oil. Check the operation of all engine electrical systems as well as dashboard gauges and lights.
82. Perform necessary adjustments to the accelerator cable, drive belts and engine specifications. Adjust the coolant level after the engine has cooled off.

LEFT HEAD

1. Disconnect the negative battery cable.
2. Drain the cooling system.

--- **CAUTION** ---

When draining the coolant, keep in mind that cats and dogs are attracted by the ethylene glycol antifreeze, and are quite likely to drink any that is left in an uncovered container or in puddles on the ground. This will prove fatal in sufficient quantity. Always drain the coolant into a sealable container. Coolant should be reused unless it is contaminated or several years old.

3. Remove the air intake plenum.
4. Disconnect the upper radiator hose from the thermostat housing.
5. Label and disconnect the various connectors along the engine control harness. Position the harness and wiring leads out of the way.
6. Remove the intake manifold with the injectors and fuel rail attached.

--- **CAUTION** ---

The fuel system is under pressure. Release pressure slowly and contain spillage. Observe no smoking/no open flame precautions. Have a Class B-C (dry powder) fire extinguisher within arm's reach at all times.

7. Label and disconnect the wiring to the rear spark plugs (Nos. 2, 4 and 6).
8. Disconnect the oxygen sensor wiring.
9. Remove the self-locking nuts holding each (front and rear) exhaust pipe to the exhaust manifolds. Remove the small bracket bolts and separate the exhaust system from the engine.
10. Remove the distributor.
11. Remove the air intake plenum support.
12. Remove the bracket bolt holding the two hoses to the rear of the head.
13. Remove the heat shield from the rear exhaust manifold; remove the manifold and its gasket.
14. Remove the engine fan and pulley.
15. Remove the bolt holding the high pressure hose to the engine.
16. Disconnect the pressure switch wiring from the power steering pump.
17. Disconnect the high pressure line from the power steering pump.
18. Place a floor jack and a broad piece of lumber under the oil pan. Raise the jack until the engine is supported but not raised.
19. Remove the through-bolt for the left engine mount.
20. Remove the four nuts holding the upper part of the engine mount to the lower bracket.
21. Remove the air conditioning compressor drive belt, then remove the belt tensioner assembly.
22. Remove the water pump drive belt.
23. Remove the power steering pump and position it out of the way.
24. Remove the belt tensioner assembly and its bracket.
25. Remove the upper outer timing belt cover (B) for the front bank of cylinders. Remove the gaskets and keep them with the cover.

NOTE: The bolts holding the belt covers are of different lengths and must be correctly placed during reinstallation. Label or diagram them as they are removed.

ENGINE AND ENGINE OVERHAUL 3

26. Remove the lower portion of the engine support bracket. Remove the bolts in the order shown in the figure. Use spray lubricant to assist in removing the reamer bolt. Keep in mind that the reamer bolt may be heat-seized on the bracket. Remove it slowly.
27. Remove the small end cap from the rear bank belt cover, then remove the upper outer belt cover. Make certain all the gaskets are removed and placed with the covers.
28. Remove the left side splash shield. Turn the crankshaft clockwise until all the timing marks align. This will set the engine to TDC/compression on No. 1 cylinder.
29. Install the special counterholding tools and carefully remove the crankshaft pulley without moving the engine out of position.
30. Remove the front flange from the crank sprocket.
31. Remove the lower outer timing belt cover with its gaskets.
32. Loosen the timing belt tensioner bolt and turn the tensioner counterclockwise along the elongated hole. This will relax the tension on the belt.
33. If the timing belt is to be reused, make a chalk or crayon arrow on the belt showing the direction of rotation so that it may be reinstalled correctly.
34. Carefully slide the belt off the sprockets. Place the belt in a clean, dry, protected location away from the work area. if the tensioner is to be removed, disconnect the spring and remove the retaining bolt.
35. Install the counterholding tools and remove the camshaft sprocket. Remove the upper rear timing belt cover.
36. Remove the bolts holding the small bracket to the front of the head. Unbolt and remove the distributor adapter, oil seal and O-ring.

WARNING: Use care not to damage the O-ring during removal. If it becomes damaged or cracked, replace it.

37. Remove the upper bolt from the alternator brace.
38. Remove the valve cover and gasket.
39. Use the special hex wrench (MB 998051-01) and loosen the head bolts in the order shown in 2 or 3 passes. When all are finger loose, remove the bolts.
40. Rock the head gently to break it loose; if tapping is necessary, do so with a rubber or wooden mallet at the corners of the head. DO NOT pry the head up by wedging tools between the head and the block.
41. Lift the head free of the engine. Support the head assembly on wooden blocks on the workbench. Refer to *Cleaning and Inspection* following this section for work to be done before installing the head. If the head has been removed for work other than gasket replacement, the rocker assembly and camshaft or other components may be removed.

To install:
42. Before reinstallation, the head should be completely assembled on the bench. This allows proper location and tightening of all the external items.
43. The head gaskets are different for each head. The gasket for this head is marked **L** (left); the gasket should be installed with the identifying mark at the timing belt end and facing up (towards the head).

WARNING: Do not apply sealant to the head gasket or mating surfaces.

44. Install the head straight down onto the block. Try to eliminate most of the side-to-side adjustments as this may move the gasket out of position. Install the bolts and the special washers by hand and just start each bolt 1 or 2 turns on the threads.

WARNING: The washers must be installed correctly. The rounded shoulder of the washer denotes the face in contact with the bolt. The flat face contacts the head.

45. Correct tightening of the head bolts requires 3 steps:
  a. Following the tightening sequence, draw each bolt to 62 ft. lbs.
  b. Follow the loosening sequence and loosen each bolt in order until it is completely loose.
  c. Follow the tightening sequence and tighten each bolt to 70 ft. lbs.
46. Install the valve cover and gasket.
47. Install the upper bolt for the alternator brace and tighten it to 14 ft. lbs.
48. Carefully install the distributor adapter assembly with the seal and O-ring. Tighten the retaining bolts to 10 ft. lbs.
49. Install the small bracket bolts to the front of the head, tightening them to 16 ft. lbs.
50. Install the upper rear timing belt cover. Install the camshaft sprocket (use the counterholding tools) and tighten the retaining bolt to 66 ft. lbs.
51. Check the sprockets and tensioner for wear. The sprocket teeth should be well defined, not rounded and the valleys between the teeth should be clean. The tensioners should spin freely with no binding or unusual noise. Replace the tensioner if there is any sign of grease leaking from the seal. Clean everything with a clean, dry cloth.

WARNING: Do not spray or immerse the sprockets or tensioners in cleaning solvent. The sprocket may absorb the solvent and transfer it to the belt. The tensioners are internally lubricated and the solvent will dilute or dissolve the lubricant.

52. If the tensioner was removed, it must be reinstalled. After bolting it loosely in place, connect the spring onto the water pump pin. Make certain the spring faces in the correct direction on the tensioner. Turn the tensioner to the extreme counterclockwise position on the elongated hole and tighten the bolt just enough to hold the tensioner in this position.
53. Double check the alignment of the timing marks on the camshaft and crankshaft sprockets.
54. Observing the direction of rotation marks made earlier, install the belt onto the crankshaft sprocket and then onto the rear bank camshaft sprocket. Maintain tension on the belt between the sprockets.
55. Continue installing the belt onto the water pump, the front bank cam sprocket and the tensioner.
56. With your fingers, apply gentle counterclockwise force to the rear camshaft sprocket. When the belt is taut on the tension side, the timing marks should align perfectly.
57. Install the flange on the crankshaft sprocket.
58. Loosen the bolt holding the tensioner one or two turns and allow the spring tension to draw the tensioner against the belt.
59. Using special tool MB 998716 or equivalent adapter, turn the crankshaft two complete revolutions clockwise. Turn the crank smoothly and re-align the timing marks at the end of the second revolution. This allows the tensioner to compensate for the normal amount of slack in the belt.
60. With the timing marks aligned, tighten the tensioner bolts to 18 ft. lbs.
61. Using a belt tension gauge, measure the belt tension at a point halfway between the crankshaft sprocket and the rear camshaft sprocket. Correct value is 57–84 lbs.
62. Making certain all the gaskets are correctly in place, install the lower timing belt cover and tighten the bolts to 8 ft. lbs.
63. Install the crankshaft pulley. Use the special tools to counterhold it and tighten the bolt to 112 ft. lbs.
64. Install the left splash shield.
65. Install the upper outer cover on the rear bank camshaft sprocket and install the smaller end cap. Tighten the bolts to 8 ft. lbs.
66. Install the lower portion of the engine support bracket. Install the bolts in the correct order. Of the 3 mounting bolts arranged vertically on the right side of the mount, the upper two are tightened to 50 ft. lbs; the bottom one is tightened to 78 ft. lbs. The two smaller bolts in the side of the mount should be

3 ENGINE AND ENGINE OVERHAUL

tightened to 30 ft. lbs. Use spray lubricant to help with the reamer bolt, and tighten it slowly. Access may be easier if the engine is elevated slightly with the floor jack.

67. Install the cover and gaskets for the front bank camshaft sprocket.
68. Install the tensioner and bracket for the ribbed belt.
69. Install the power steering pump onto its bracket.
70. Install the ribbed belt and adjust it as needed. Make certain it is properly seated on all the pulleys.
71. Install the tensioner and bracket for the air conditioner drive belt and install the belt, adjusting it as necessary.
72. Install the upper section of the engine mounting bracket to the lower section. Tighten the nuts to 50 ft. lbs.
73. Adjust the jack so that the bushing of the engine mount aligns with the bodywork bracket. Install the through-bolt and tighten the nuts snug.
74. Carefully lower the jack, allowing the full weight of the engine to bear on the mount. Tighten the nut on the through-bolt to 50 ft. lbs. and the smaller safety nut to 26 ft. lbs.
75. Install the pressure hose to the power steering pump and tighten the nut to 32 ft. lbs.
76. Connect the wiring to the power steering pump and rebolt the bracket for the high pressure hose.
77. Install the engine fan and pulley.
78. Install a new exhaust manifold gasket and install the rear exhaust manifold to the head. tighten the manifold retaining nuts to 14 ft. lbs.
79. Install the heat shield on the rear exhaust manifold, tightening the bolts to 10 ft. lbs. Don't forget the intake plenum support.
80. Reinstall the bracket bolt holding the two hoses to the rear of the head.
81. Install the distributor.
82. Using new exhaust gaskets and new self-locking nuts; attach the exhaust pipes to each exhaust manifold. Tighten the nuts to 26 ft. lbs. Connect the oxygen sensor wiring.
83. Connect the spark plug wires to the spark plugs.
84. Position new intake manifold gaskets and install the intake manifold.
85. Observe the labels made previously and connect each lead of the engine control harness to its proper component.
86. Connect the upper radiator hose to the thermostat housing.
87. Install the air plenum.
88. Double check all installation items, paying particular attention to loose hoses or hanging wires, untightened nuts, poor routing of hoses and wires (too tight or rubbing) and tools left in the engine area.
89. Fill the cooling system with coolant. Changing the oil and filter is recommended to eliminate pollutants in the oil.
90. Connect the negative battery cable. Start the engine and check for leaks of fuel, vacuum or oil. Check the operation of all engine electrical systems as well as dashboard gauges and lights.
91. Perform necessary adjustments to the accelerator cable, drive belts and engine specifications. Adjust the coolant level after the engine has cooled off.

Galant and Sigma with 1997cc (4G63) SOHC Engine

1. Disconnect the negative battery cable.
2. Drain the cooling system.

---- **CAUTION** ----

When draining the coolant, keep in mind that cats and dogs are attracted by the ethylene glycol antifreeze, and are quite likely to drink any that is left in an uncovered container or in puddles on the ground. This will prove fatal in sufficient quantity. Always drain the coolant into a sealable container. Coolant should be reused unless it is contaminated or several years old.

3. Remove the air intake hose.
4. Disconnect the accelerator cable and remove the bracket.

5. Disconnect the high pressure fuel line and remove the O-ring.

---- **CAUTION** ----

The fuel system is under pressure. Release pressure slowly and contain spillage. Observe no smoking/no open flame precautions. Have a Class B-C (dry powder) fire extinguisher within arm's reach at all times.

6. Disconnect the upper radiator hose, the coolant by-pass hose and the heater hose from the head and/or intake manifold.
7. Disconnect the brake booster vacuum hose.
8. Remove the fuel return hose.
9. Label and disconnect the vacuum hose(s) running to the manifold. Disconnect the PCV hose at the valve cover.
10. Label and disconnect the spark plug wires from the distributor cap.
11. Label and disconnect each electrical connector, including the distributor lead and the injector connectors. Note that some of the wiring must be disconnected at the firewall. When all the connectors are loose, remove the bracket bolts holding the control wiring harness in place and move the harness to an out-of-the-way location.
12. Remove the clamp holding the power steering and air con-

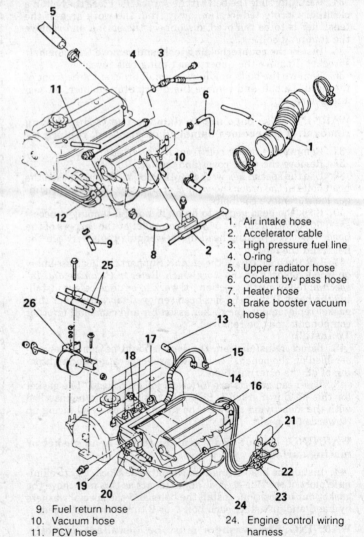

1. Air intake hose
2. Accelerator cable
3. High pressure fuel line
4. O-ring
5. Upper radiator hose
6. Coolant by-pass hose
7. Heater hose
8. Brake booster vacuum hose
9. Fuel return hose
10. Vacuum hose
11. PCV hose
12. Spark plug wires
13.
23. Electrical connectors
24. Engine control wiring harness
25. Clamp for hoses
26. Engine mounting bracket

4G63 SOHC head and related components

ENGINE AND ENGINE OVERHAUL 3

27. Valve cover
28. Half-circle plug
29. Upper timing belt cover
30. Camshaft sprocket
31. Self-locking nuts
32. Exhaust gasket
33. Cylinder head assembly
34. Head gasket

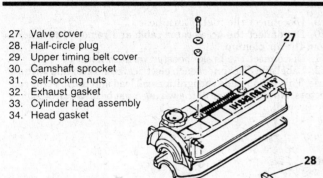

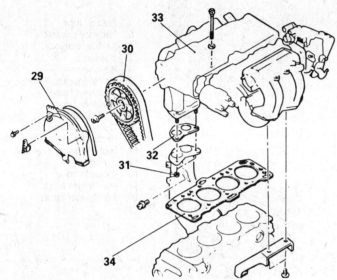

4G63 SOHC head and related components

ditioning hoses to the top of the left engine mount bracket. Move the hoses out of the way but don't disconnect either hose from its system.

13. Position a floor jack and a broad piece of lumber under the engine. Elevate the jack just enough to support the engine without raising it.
14. Remove the through-bolt from the left side engine mount. It may be necessary to adjust the jack slightly to allow the bolt to come free. When the bolt has been removed, disconnect the nuts and bolts holding the mounting bracket to the engine and remove the bracket.
15. Remove the valve cover and gasket. Remove the half-circle plug from the head.
16. Remove the upper timing belt cover. Turn the crankshaft clockwise until all the timing marks align, setting the engine to TDC/compression for No. 1 cylinder.
17. Remove the bolt holding the camshaft sprocket to the camshaft. Remove the camshaft sprocket — with the belt attached — and place it on the lower timing belt cover. Do not allow the belt to come off the sprocket.

WARNING: Do not rotate the crankshaft once the camshaft sprocket is removed.

18. Remove the self-locking nuts and the small retaining bolt holding the exhaust pipe to the bottom of the exhaust manifold. Separate the pipe from the manifold and remove the gasket.
19. Remove the bolts holding the support brace to the bottom of the intake manifold.
20. Use the special hex wrench (MB 998051-01) and loosen the head bolts in the order shown in 2 or 3 passes. When all are finger loose, remove the bolts.

Head bolt removal pattern, 4G63 SOHC

21. Rock the head gently to break it loose; if tapping is necessary, do so with a rubber or wooden mallet at the corners of the head. DO NOT pry the head up by wedging tools between the head and the block.
22. Lift the head free of the engine. It is coming off with both manifolds and the intake plenum attached; the help of an assistant is recommended for lifting.

Support the head assembly on wooden blocks on the workbench. Refer to *Cleaning and Inspection* following this section for work to be done before installing the head. If the head has been removed for work other than gasket replacement, the rocker assembly and camshaft or other components may be removed.

23. Before reinstallation, the head should be completely assembled on the bench. This allows proper location and tightening of all the external items.
24. Place a new gasket on the engine so that the identifying mark faces up (towards the head) and is at the timing belt end of the block. Install a new gasket on the exhaust pipe.

WARNING: Do not apply sealant to the head gasket or mating surfaces.

25. Install the head straight down onto the block. Try to eliminate most of the side-to-side adjustments as this may move the gasket out of position. Install the bolts by hand and just start each bolt 1 or 2 turns on the threads.
26. The head bolt torque specification is 68 ft. lbs. for a cold engine. The bolts must be tightened in the order shown in 3 steps. On the first pass, tighten all the bolts to about 22 ft. lbs. then proceed through the order tightening each bolt to about 45 ft. lbs. The final torque is achieved on the third pass. Install the intake manifold support brace to the manifold and tighten the bolts to 16 ft. lbs.
27. Making sure the gasket is still in place, connect the exhaust pipe to the base of the exhaust manifold. Use new self-locking nuts; tighten the nuts and the small bracket bolt to 26 ft. lbs.
28. Make sure the camshaft has not changed position during repairs. Carefully install the camshaft sprocket and belt onto the camshaft. Tighten the retaining bolt to 66 ft. lbs.
29. Install the upper timing belt cover, tightening the bolts to 8 ft. lbs.

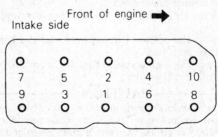

4G63 head bolt tightening sequence

3-193

3 ENGINE AND ENGINE OVERHAUL

30. Apply sealant to the contact surfaces of the half-circle plug and install the plug in the head. Install the valve cover and gasket.
31. Install the engine mount bracket to the engine. Tighten the mounting nuts and bolts to 42 ft. lbs.
32. Adjust the jack (if necessary) so that the engine mount bushing aligns with the bodywork bracket. Install the through-bolt and tighten the nuts snug.
33. Slowly release tension on the floor jack so that the weight of the engine bears fully on the mount. Tighten the through-bolt to 52 ft. lbs. and the small safety nut to 26 ft. lbs.
34. Install the bracket holding the power steering hose and air conditioning hose to the top of the engine mount.
35. Position the control wiring harness and install the retaining bolts. Connect each electrical connector to its proper location, making sure the wires are properly routed and firmly connected.
36. Install the spark plug wires in the distributor cap.
37. Connect the PCV hose and the vacuum hose(s).
38. Connect the fuel return line. Connect the brake booster vacuum hose.
39. Install the heater hose, the coolant by-pass hose and the upper radiator hose. Pay close attention to the position and routing of these hoses and insure that they are not crimped or constricted. Install the clamps in the same location as before removal.
40. Install a new O-ring on the high pressure fuel line and lubricate it with a coating of gasoline. Carefully connect the high pressure fuel line to the fuel rail, taking care not to damage the O-ring. Tighten the bolts only to 4 ft. lbs.
41. Connect the accelerator cable and adjust it as necessary.
42. Install the air intake hose.
43. Fill the cooling system with coolant. Changing the oil and filter is recommended to eliminate pollutants in the oil.
44. Connect the negative battery cable. Start the engine and check for leaks of fuel, vacuum or oil. Check the operation of all engine electrical systems as well as dashboard gauges and lights.
45. Perform necessary adjustments to the accelerator cable, drive belts and engine specifications. Adjust the coolant level after the engine has cooled off.

Galant and Sigma with 1997cc (4G63) DOHC Engine

1. Disconnect the negative battery cable.
2. Drain the engine coolant.

CAUTION

When draining the coolant, keep in mind that cats and dogs are attracted by the ethylene glycol antifreeze, and are quite likely to drink any that is left in an uncovered container or in puddles on the ground. This will prove fatal in sufficient quantity. Always drain the coolant into a sealable container. Coolant should be reused unless it is contaminated or several years old.

---

3. Remove the branch tube and the radiator assembly.
4. Disconnect the air flow wiring from the air cleaner. Remove the resonator assembly (next to the air cleaner) and remove the air cleaner assembly.
5. Remove the air intake hose, the breather hose and the PCV hose.
6. Disconnect the coolant by-pass hose and the heater hose from the intake manifold.
7. Disconnect the high pressure fuel line from the fuel delivery rail and remove the O-ring.

CAUTION

The fuel system is under pressure. Release pressure slowly and contain spillage. Observe no smoking/no open flame precautions. Have a Class B-C (dry powder) fire extinguisher within arm's reach at all times.

---

8. Remove the small vacuum hose at the front of the intake manifold.
9. Disconnect the fuel return hose.
10. Disconnect the accelerator cable and remove the bracket from the air plenum.
11. Disconnect the brake booster vacuum hose.
12. Label and disconnect each electrical connector and vacuum hose. This will include the ignition coil lead and the injector connectors. Note that some of the wiring must be disconnected near the firewall.

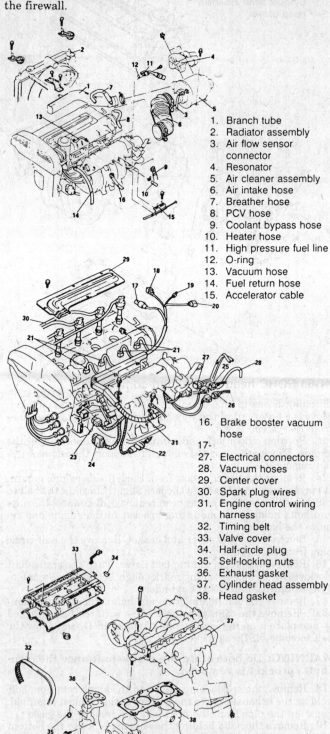

1. Branch tube
2. Radiator assembly
3. Air flow sensor connector
4. Resonator
5. Air cleaner assembly
6. Air intake hose
7. Breather hose
8. PCV hose
9. Coolant bypass hose
10. Heater hose
11. High pressure fuel line
12. O-ring
13. Vacuum hose
14. Fuel return hose
15. Accelerator cable
16. Brake booster vacuum hose
17-
27. Electrical connectors
28. Vacuum hoses
29. Center cover
30. Spark plug wires
31. Engine control wiring harness
32. Timing belt
33. Valve cover
34. Half-circle plug
35. Self-locking nuts
36. Exhaust gasket
37. Cylinder head assembly
38. Head gasket

Removing the 4G63 DOHC head. Timing belt components are not shown

ENGINE AND ENGINE OVERHAUL 3

13. Remove the center cover and disconnect the spark plug cables from both the spark plugs and the ignition coil. Label each wire before removal.
14. When all the connectors are loose, remove the bracket bolts holding the control wiring harness in place and move the harness to an out-of-the-way location. Assorted clips and brackets may need to be removed to get the wiring harnesses clear of the head and manifolds.
15. Position a floor jack and a broad piece of lumber under the engine. Elevate the jack just enough to support the engine without raising it.
16. Remove the through-bolt from the left side engine mount. It may be necessary to adjust the jack slightly to allow the bolt to come free. When the bolt has been removed, disconnect the nuts and bolts holding the mounting bracket to the engine and remove the bracket.
17. Loosen the appropriate component or adjuster and remove (in order) the air conditioning compressor drive belt, the power steering belt and the alternator belt. Remove the air conditioning belt tension adjuster and bracket.
18. Remove the power steering and water pump pulleys.
19. Remove the upper timing belt cover and its gaskets. Rotate the crankshaft clockwise until the timing marks on the crank pulley align at 0 on the scale (on the lower timing cover) and the mark on the camshaft pulley aligns with the indicator on the head. Aligned means perfect, not close — if you miss, continue clockwise until the marks are matched. This sets the motor at TDC/compression on No.1 cylinder.

WARNING: Once the engine is set to this position, the crank and camshafts must not be moved out of place. Severe damage can result.

20. Remove the air conditioner pulley from the crankshaft and remove the crankshaft pulley. Remember that the pulley (and therefore, the crankshaft) must be held steady while the bolt is loosened. This can be made easier by wrapping a scrapped belt tightly around the pulley. Mitsubishi has a special tool (MD 998747) for this trick, but other variations work almost as well. DO NOT use any of the drive belts which will be reinstalled; the belt used to hold the pulley will be damaged or weakened.
21. Remove the lower timing belt cover and gaskets. Note the different bolt lengths and placements.
22. Loosen (do not remove) both bolts in the timing belt tensioner. Move the tensioner toward the water pump and tighten the lock bolt — the one in the slotted hole — to hold the tensioner in place.
23. Before removing the timing belt, mark it with an arrow to show the direction of rotation. If the belt is to be reused, it must be reinstalled so that it rotates in the same direction as before.
24. Remove the timing belt, keeping it free of grease, oil and fluids. Do not crease or crimp the belt.
25. Remove the valve cover and gasket. Remove the half-circle plug from the head.
26. Remove the self-locking nuts and small bracket bolt holding the exhaust system to the bottom of the exhaust manifold. Separate the pipe from the manifold and remove the gasket.
27. Remove the bolts holding the intake manifold support to the intake manifold.
28. Use the special hex wrench (MB 998051-01) and loosen the head bolts in the order shown in 2 or 3 passes. When all are finger loose, remove the bolts.
29. Rock the head gently to break it loose; if tapping is necessary, do so with a rubber or wooden mallet at the corners of the head. DO NOT pry the head up by wedging tools between the head and the block.
30. Lift the head free of the engine. It is coming off with both manifolds and the intake plenum attached; the help of an assistant is recommended for lifting.

Support the head assembly on wooden blocks on the workbench. Refer to *Cleaning and Inspection* following this section for work to be done before installing the head. If the head has been removed for work other than gasket replacement, the rocker assembly and camshaft or other components may be removed.

To install:

31. Before reinstallation, the head should be completely assembled on the bench. This allows proper location and tightening of all the external items.
32. Place a new gasket on the engine so that the identifying mark faces up (towards the head) and is at the timing belt end of the block. Install a new gasket on the exhaust pipe.

WARNING: Do not apply sealant to the head gasket or mating surfaces.

33. Install the head straight down onto the block. Try to eliminate most of the side-to-side adjustments as this may move the gasket out of position. Install the bolts by hand and just start each bolt 1 or 2 turns on the threads.
34. The head bolt torque specification is 68 ft. lbs. for a cold engine. The bolts must be tightened in the order shown in 3 steps. On the first pass, tighten all the bolts to about 22 ft. lbs. then proceed through the order tightening each bolt to about 45 ft. lbs. The final torque is achieved on the third pass. Install the intake manifold support brace to the manifold and tighten the bolts to 20 ft. lbs.
35. Making sure the gasket is still in place, connect the exhaust pipe to the base of the exhaust manifold. Use new self-locking nuts; tighten the nuts and the small bracket bolt to 26 ft. lbs.
36. Coat the contact surfaces of the half-circle plug with sealer and install it in the head. Install the valve cover and gasket.
37. Check the sprockets and tensioner for wear. The sprocket teeth should be well defined, not rounded and the valleys between the teeth should be clean. The tensioners should spin freely with no binding or unusual noise. Replace the tensioner if there is any sign of grease leaking from the seal. Clean everything with a clean, dry cloth.

WARNING: Do not spray or immerse the sprockets or tensioners in cleaning solvent. The sprocket may absorb the solvent and transfer it to the belt. The tensioners are internally lubricated and the solvent will dilute or dissolve the lubricant.

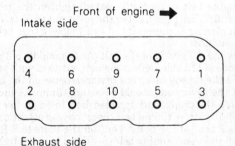

Removal sequence for 4G63 DOHC head bolts

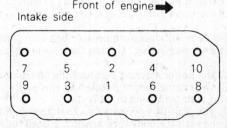

Head bolt tightening sequence, 4G63 DOHC

3 ENGINE AND ENGINE OVERHAUL

38. Install the tensioner and temporarily position it so that the center of the pulley is to the left and above the center of the installation bolt.
39. Apply finger pressure on the tensioner, forcing it into the belt. Make certain the opposite side of the belt is taut. Tighten the installation bolt to hold the tensioner in place.
40. Push down on the center of the tension side of the belt (side opposite the tensioner pulley) and check the distance the belt deflects. Correct deflection is 6mm (¼ inch).
41. Install the flange on the crankshaft sprocket, making sure it is positioned correctly.
42. Install the timing belt tensioner if it was removed and position it towards the water pump. Tighten the lock bolt to hold it in this position.
43. Make certain that all the timing marks for the camshaft sprocket, the crankshaft sprocket and the oil pump sprocket are correctly aligned.
44. Once the oil pump sprocket is aligned, remove the plug in the left side of the block. Insert a Phillips screwdriver with a shaft diameter of 8mm (0.31 in.) into the plug hole. It should enter 58mm (2.3 in.) or more. Do not remove the screwdriver until the timing belt is completely installed. If the screwdriver shaft only enters about 25mm (1 inch), turn the oil pump sprocket one rotation and align the timing mark again; insert the screwdriver to block the silent shaft from turning.
45. Observe the directional arrow made during removal and install the timing belt by placing on the crankshaft sprocket, then the oil pump sprocket and then over the camshaft. Maintain the belt taut on the tension side while installing it.
46. Loosen the adjuster lock bolt and allow the spring to move the pulley against the belt. Do not push or pry on the adjuster.
47. Recheck that each sprocket is still aligned with its timing marks.
48. Remove the screwdriver from the silent shaft. Rotate the crankshaft clockwise until the camshaft sprocket has moved 2 teeth away from its mark. This applies tension to the belt. Do NOT rotate the crank counter-clockwise and do not push on the belt to check the tension.
49. With your fingers, apply gentle upward pressure on the tensioner and check the placement of the belt on the cam sprocket. The belt may tend to lift or float in the left side of the sprocket — roughly between 7 o'clock and 12 o'clock as you view the sprocket.
50. When the belt is correctly seated, tighten the tensioner lock bolt to 36 ft. lbs, then tighten the pivot nut to 36 ft. lbs. The order of tightening is important; if not followed, the belt could be over-tightened.
51. Check the deflection of the belt. At the middle of the right (tension) side of the belt, deflect the belt outward with your finger, toward the timing case. The distance between the belt and the line of the cover seal should be about 13mm (½ inch).
52. Tighten the crankshaft sprocket bolt to 88 ft. lbs.
53. Install the lower timing belt cover, paying attention to the correct placement of the bolts. Tighten the bolts to 8 ft. lbs.
54. Install the upper timing belt cover and gaskets, tightening the bolts to 8 ft. lbs. Remember to install the guides for the spark plug wires.
55. Install the crankshaft pulley. Tighten the bolts to 18 ft. lbs.
56. Install the water pump pulleys, tightening the bolts to 6 ft. lbs.
57. Fit the air conditioning compressor belt into place and install the tensioner assembly. Tighten the mounting bolts to 18 ft. lbs.
58. Install the power steering belt and the alternator belt.
59. Install the engine mount bracket to the engine. Tighten the mounting nuts and bolts to 42 ft. lbs.
60. Adjust the jack (if necessary) so that the engine mount bushing aligns with the bodywork bracket. Install the through-bolt and tighten the nuts snug.
61. Slowly release tension on the floor jack so that the weight of the engine bears fully on the mount. Tighten the through-bolt to 50 ft. lbs. and the small safety nut to 26 ft. lbs.
62. Position the engine control harness and install the retaining bolts and brackets.
63. Connect the spark plug wires to the plugs and the coil. Install the center cover.
64. Observing the labels made earlier, connect each wire and vacuum hose to its correct location. Make sure each is properly routed and firmly attached.
65. Connect the brake booster vacuum hose.
66. Connect the accelerator cable and attach the bracket. Tighten the bracket bolts only to 4 ft. lbs.
67. Connect the fuel return line. Install a new O-ring on the high pressure fuel line, coat it lightly with gasoline and install the line. Tighten the retaining bolts to 4 ft. lbs.
68. Install the heater hose and the water by-pass hose.
69. Connect the PCV hose, the breather hose and the air intake hose.
70. Install the air cleaner and resonator. Attach the air intake tube and connect the wiring for the air flow sensor.
71. Install the radiator assembly and the branch tube.
72. Double check all installation items, paying particular attention to loose hoses or hanging wires, untightened nuts, poor routing of hoses and wires (too tight or rubbing) and tools left in the engine area.
73. Fill the cooling system with coolant. Changing the oil and filter is recommended to eliminate pollutants in the oil.
74. Connect the negative battery cable. Start the engine and check for leaks of fuel, vacuum or oil. Check the operation of all engine electrical systems as well as dashboard gauges and lights.
75. Perform necessary adjustments to the accelerator cable, drive belts and engine specifications. Adjust the coolant level after the engine has cooled off.

Pickup with the 4D55 Diesel Engine

WARNING: Perform this work only on a completely cold engine!

1. Disconnect the negative battery cable.
2. Drain the coolant into clean containers for reuse.

--- **CAUTION** ---

When draining the coolant, keep in mind that cats and dogs are attracted by the ethylene glycol antifreeze, and are quite likely to drink any that is left in an uncovered container or in puddles on the ground. This will prove fatal in sufficient quantity. Always drain the coolant into a sealable container. Coolant should be reused unless it is contaminated or several years old.

3. Remove the upper radiator hose from the engine.
4. Label and disconnect the breather hoses running to the valve cover and across the head.
5. Disconnect the heater hose at the intake manifold.
6. Disconnect the wiring to the temperature sensor.
7. Disconnect and remove the inlet piping and the oil lines running to the turbocharger. Make certain that both the lines and the turbocharger ports are plugged or covered as soon as they are disconnected. NO foreign matter is permitted in the turbocharger system.
8. Carefully disconnect the fuel injection lines. See Section 5.
9. Disconnect the glow plug connections.
10. Remove the intake and exhaust manifolds.
11. Remove the upper and lower timing belt covers. Turn the crankshaft and align all the timing marks; this positions the motor at TDC/compression for No. 1 cylinder.
12. Mark the direction of rotation on the timing belts with chalk or a felt marker.
13. Remove the center bolt holding the crankshaft pulley and remove the pulley. Do not turn the engine out of position during this removal.
14. Remove the lower timing belt cover.

ENGINE AND ENGINE OVERHAUL 3

15. Slightly loosen the retaining (lock) bolts for the timing belt tensioner. Move the tensioner towards the water pump and tighten the bolts; this holds the tensioner in the slack position.
16. Remove the camshaft sprocket and the injection pump sprocket. Do not let either shaft turn out of place during the removal. Remove the timing belt without forcing it or kinking it.
17. Remove the fuel injection pump.

WARNING: Do not move or disturb the position of the motor once the timing belt is removed!

18. Remove the valve cover and gasket.
19. Loosen the head bolts in the correct sequence. Loosen each bolt gradually and in two or three passes. Sudden loosening of any bolt may cause head damage by warping.
20. Remove the head by lifting it off the engine; do not attempt to slide it to the side.
21. Remove the gasket. Clean the mating surfaces of all gasket remains. If the head is to be off the engine for any length of time, plug the cylinders with clean cloths or crumpled newspaper. Place the head on wood blocks on the work bench for examination or further work. The head may be cleaned, inspected or disassembled following instructions given later in this section.

To install:
22. Before reinstallation, double check the head and block faces for any trace of gasket material. Remove any plugs or paper from the head and/or cylinders.
23. Place a new head gasket in position on the block. Pay close attention to the alignment of the gasket with the holes and passages in the block. The gaskets generally do not have any identifying marks for up or front. An incorrectly installed gasket may block an oil or coolant passage; neither the engine nor your bankbook will tolerate this error.
24. Place the head onto the block carefully; do not disturb the position of the gasket. Start the head bolts in each hole by hand and turn them in 1 or 2 turns.
25. Following the correct bolt–tightening pattern, tighten each bolt to 54 Nm (40 ft. lbs.). This is one half the final tension.
26. Repeat the bolt tightening sequence, tightening each bolt to 108 Nm (80 ft. lbs.) The acceptable range is 103–113 Nm (76–83 ft. lbs.)
27. Install the crankshaft pulley; tighten the bolt to 70 Nm (52 ft. lbs.).
28. Install the fuel injection pump.
29. Move the timing belt tensioner all the way toward the water pump and temporarily tighten it in this position.
30. Install the timing belt onto the injection pump sprocket and camshaft sprocket, making certain NOT to turn the crankshaft sprocket during the procedure. Make certain there is no slack in the belt between the camshaft sprocket and the injection pump sprocket.

NOTE: The injection pump sprocket tends to rotate during this installation. Hold it in place while installing the timing belt.

31. With the belt in place, check the alignment of the timing marks on both sprockets. Loosen the tensioner mounting bolts which will apply tension to the belt.
32. Tighten the belt tensioner nut and bolt. Tighten the slot side bolt first, then the fulcrum side. If it's done the other way, the adjuster will turn and the belt will be too tight.
33. Check the alignment of the timing marks.
34. Using a wrench on the crankshaft bolt, turn the engine clockwise enough to move the camshaft sprocket 2 teeth past the timing mark. Use the wrench to reverse (counterclockwise) direction and turn the engine back to the original position with the timing marks aligned.
35. Use your finger to press down on the belt midway between the cam sprocket and the injection pump sprocket. Deflection in the belt should be 4–5 mm (0.16–0.20 in.).
36. Install the valve cover.

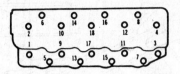

Diesel cylinder head bolt removal sequence

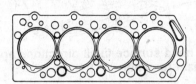

Diesel cylinder head gasket, top side

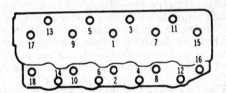

Diesel head bolt installation torque sequence

37. Install the upper and lower timing covers.
38. Install the intake and exhaust manifolds. Always use new gaskets.
39. Connect the fuel injection lines.
40. Connect the wiring for the glow plug circuit.
41. Install the turbocharger if it was removed from the manifolds. Remove any plugs from the turbocharger passages; prime the unit by pouring a clean engine oil into the turbocharger oil inlet port.
42. Connect the oil drain hose and the oil inlet pipe to the turbocharger.
43. Connect the turbocharger air inlet fitting and rubber hose (duct).
44. Connect the temperature sensor lead. Connect the heater hose to the intake manifold.
45. Connect the breather and vacuum hoses as necessary.
46. Install the upper radiator hose to the engine and make certain the clamp is properly located and secure.
47. Refill the cooling system with coolant.
48. Connect the negative battery cable; start the engine and check for leaks of fuel, oil or coolant.
49. Check the injection timing, adjusting it if needed.

CYLINDER HEAD CLEANING AND INSPECTION

Immediately after the head is removed from the engine, the old gasket material should be removed from the head and block. Stuff the cylinders with newspaper or rags to prevent accumulation of foreign matter, but don't press down on the pistons too hard. If a piston moves, the crankshaft will turn and the timing relationship will be lost. (This is particularly important if the belt is still attached to the camshaft gear.)

Remove as much of the old gasket as possible from the block with your fingers. The rest must be scraped with a tool which will not damage the metal by scoring or scratching. A hardwood or plastic scraper is highly recommended; the edge of an old credit card is a good substitute. Take extreme care not to get any

3 ENGINE AND ENGINE OVERHAUL

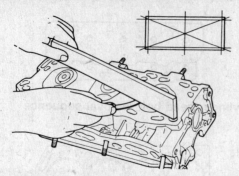

Check the head surface in all directions and take several measurements

of the scrapings into the water and oil passages or the bolt holes in the head. This common error carries very expensive penalties.

Once the head is on the work bench, the manifolds and other external components should be removed for all but the simplest of cleaning. The head is easier to manipulate and removed items are less likely to be damaged. Unless the head is to be fully stripped, the cam and rockers may stay in place, but take measures to protect the valve train from dust and impact during repairs.

The head surface should be scraped with the non-marring scraper, again observing the protection of oil and water passages. The combustion chamber and valves will have carbon deposits in them. Attempt to remove as much of the deposit as possible with the scraper. The remaining deposits will need to be removed with a rotary wire brush mounted in an electric drill. Make sure that the brush is actually removing the deposit and not just polishing it into invisibility.

After the head surface is completely clean and free of any debris, place a quality straight-edge across the gasket surface of the cylinder head. Using feeler gauges, determine the clearance at the center of the straight-edge and at other locations on either side. Measure along the diagonals of the head, along the longitudinal center line and across the head at several locations. Keep a written record of the thickest feeler which will pass between the straight-edge and the head at every location.

Acceptable warpage is at or below 0.05mm (0.0020 in.) at EVERY location measured. If all the measurements are at or below this specification, the head may be reused as is. If any one measurement is greater than this limit, the head must be resurfaced or replaced. Maximum allowable warpage is 0.20mm (0.0080 in.). Warpage exceeding this requires replacement of the head. Measure the engine block surface (deck) in the same fashion. Look for clearance under the straight-edge to be 0.05mm (0.0020 in.) or less. Any reading over this value requires replacement or resurfacing of the block.

RESURFACING

If the head is to be resurfaced (planed), it must be completely stripped of all components and taken to a machine shop. At this point, it is worthwhile considering having the head crack-checked by the shop. Commercial methods allow much more sophisticated checking than can be done by eye. Additionally, having the head cleaned in a solvent bath will remove deposits from all the passages within the head.

WARNING: Do not allow the head to be immersed in "hot tank" cleaners designed for iron heads. The chemicals will attack the aluminum alloy in the Mitsubishi head and make the head unusable. Specify to the shop that the head contains aluminum.

Resurfacing the head involves removing metal from the face of the head, much as a wooden surface is leveled with a plane. This can remove the high and low spots and give a true surface to mount the gasket onto the engine, clamping the gasket evenly across its surface. The two critical dimensions in resurfacing the head are the amount of metal removed (usually expressed in thousandths of a millimeter) and the overall thickness of the head.

The minimum head thickness must NOT be exceeded under any circumstance. If the amount of resurfacing required would cause the head to become too thin, the head is unusable and must be discarded. The maximum thickness allowed to be removed from a Mitsubishi head is 0.20mm (0.008 in.). This figure is valid for a head which has never been cut (original thickness) and would reduce it to minimum thickness. If the head has been resurfaced previously, the cutting limit is the present height of the head minus the minimum thickness as follows:

- G54B, G62B, G63B, G64B: 90.0mm – 89.8mm
- G15B: 107.0mm – 106.8mm
- G32B: 88.5mm – 88.3mm
- 4G61, 4G63 DOHC: 131.9–132.1mm – 0.20mm
- 4G15: 106.9–107.1mm – 0.20mm
- 4G63 SOHC: 89.9–90.1mm – 0.20mm
- 6G72: 83.9–84.1mm – 0.20mm

An additional reminder: when resurfacing a V6 head, the profile of the intake manifold mating surface will change (relative to the manifold) after the head is cut. The intake manifold face

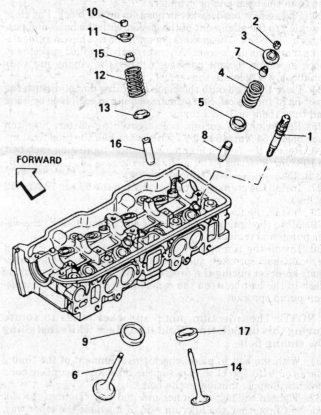

1. Jet valve assembly
2. Retainer
3. Spring retainer
4. Valve spring
5. Spring seat
6. Intake valve
7. Valve stem seal
8. Intake valve guide
9. Intake valve seat
10. Retainer
11. Spring retainer
12. Valve spring
13. Spring seat
14. Exhaust valve
15. Valve stem seal
16. Exhaust valve guide
17. Exhaust valve seat

Typical head and valve assembly. The jet valve is not found on all engines

ENGINE AND ENGINE OVERHAUL 3

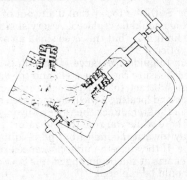

Use a spring compressor when removing or installing valve springs and retainers

will have to be resurfaced to compensate for this. Most machine shops will do this automatically; remind them anyway.

After the head is reconditioned, build it back up on the workbench by installing any and all removed components. It is much easier to install a head complete with cam, rockers and manifolds than to attempt final assembly with the head on the block. Keep the head clean and free of grit. Refer to the applicable procedures of this Section for details and torque specifications for individual components (camshaft, manifolds, etc.)

Valves and Valve Springs

NOTE: On all engines, the head must be removed from the car and the camshaft(s) and rockers removed from the head.

Valves and components must be reinstalled in their original positions. Develop a system to store and identify parts as they are removed. Identify components by location and application (Example: Cyl. 2 exhaust) to avoid confusion. Keep the work area clean and store small parts immediately.

The use of the correct special tools or their equivalents may be required for some procedures. Other specialty tools, such as spring compressors, micrometers, and reamers may not be in your tool box.

REMOVAL

The early Mitsubishi engines — G15B, G32B, G54B and G64B — use a small "jet valve" next to the main intake valve. This small additional valve directs a portion of the air/fuel mixture at the spark plug, allowing hotter and more complete combustion. The jet valve must be removed before removing the larger valves.

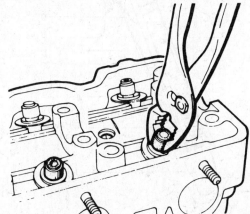

Remove the valve stem seal with a pair of pliers

The jet valve assembly is screwed into the head. The use of special tool (socket) MD 998310 is required to remove and install the valve. When using the tool, make certain that the wrench is not tilted with respect to the center of the valve. If the wrench is tilted, the excess pressure on the spring retainer may bend the valve stem, causing it to bind during operation. Remove the assembly and set it aside.

On all heads, install a standard valve spring compressor and tighten it to compress the spring. Remove the valve spring retainers (collets) from the valve stem. A small magnet can be very helpful for this. Slowly loosen the spring compressor and remove the valve spring retainer, the spring and the lower spring washer (spring seat). The valve can now be withdrawn from the combustion side of the head. Use a pair of pliers to remove the valve stem seal from each position. Discard the seals immediately; they cannot be reused.

Each valve should be cleaned to remove the carbon deposits from the valve head and stem. This may be done either with chemical cleaners (wear eye protection and gloves) or with the wire brush and power drill used to clean the head. Do not use sandpaper or other abrasives to clean the valves.

INSPECTION

Inspect each valve for burning of the head or deformation of the lower stem. Severe corrosion, pitting, burning, or cracking of these parts mean the valve must be replaced. Inspect the top of the stem for pitting. If the pitting is light and the valve is otherwise in good condition, pitting may be removed by light application of an oil stone.

Measure the stem diameter at several points along its length. A common wear point is just above the head of the valve where the stem enters the valve guide. If this area is worn, the valve must be replaced. Visually check the top of the stem for mushrooming or deformation and replace the valve if any such damage is found.

Inspect the valve margin for adequate thickness. This is the vertical section of the valve head below the seating surface. The margin is just that — a thickness of metal giving strength to the valve head. The minimum margin must be maintained, even after the valve is resurfaced. Minimum margins for valves are given below.

- G15B: Int. 0.50mm; Exh. 1.00mm
- G32B: Int. 1.00mm; Exh. 1.00mm
- G54B: Int. 0.70mm; Exh. 1.50mm
- G62B: Int. 0.50mm; Exh. 0.50mm
- G63B: Int. 0.50mm; Exh. 0.50mm
- G64B: Int. 1.55mm; Exh. 2.30mm
- 4G15: Int. 0.50mm; Exh. 1.00mm
- 4G61: Int. 0.70mm; Exh. 1.00mm
- 4G63/SOHC: Int. 0.70mm; Exh. 1.50mm
- 4G63/DOHC: Int. 0.70mm; Exh. 1.00mm
- 6G72: Int. 1.55mm; Exh. 1.50mm

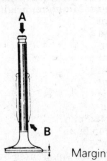

Check for deformation or pitting at "A" and wear at point "B". Do not reuse the valve if the margin is at or below minimum

3 ENGINE AND ENGINE OVERHAUL

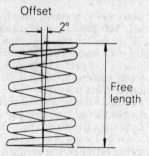

check the free length and squareness of the spring

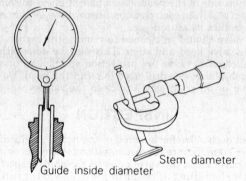

One method of determining stem-to-guide clearance

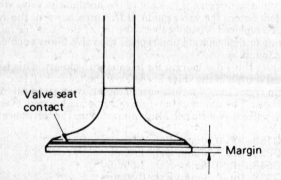

Valve margin and contact area

ber on the error. Suffice it to say that if any out of square is detected, the spring should be replaced. A very slight deviation is acceptable (the limit is 2°) but this is so small an angle that you may not even notice it.

Measure stem-to-guide clearance. Depending on the tools available, either measure the valve stem and the valve guide individually, then subtract to find the clearance or install the valve in its guide and locating a securely mounted dial indicator against the top of the stem. Rock the valve away from the indicator and hold it there while you zero the indicator. Then rock the valve all the way toward the indicator and hold it there as you take the reading. If the stem-to-guide clearance exceeds specifications (in the charts at the beginning of this Section) the valve and/or guide should be replaced.

The valve face and seat surfaces must have consistent contact near the center of each. Coat the surface of the seat with a metal dye (such as prussian blue) or a thin coat of lapping compound. Insert the valve all the way so the marking material is transferred to the valve face. If a consistent mark is found toward the center of the valve face all around, the valve and seat may be reused as is, assuming there are no other problems with the valve. If the seating is questionable or obviously bad, the valve and seat should be machined to the proper dimensions and angles by an automotive machine shop.

It then must be lapped in (polished) by coating the seating surfaces with a lapping compound and rotating the valve with a special suction cup tool until the surfaces mate perfectly; lapping compound must then be thoroughly cleaned from both surfaces.

Jet Valve Service

While the jet valve is replaced as an assembly rather than being serviced, you can disassemble it to inspect it and replace the valve stem seal. You should use tool MD 998309 or equivalent.

1. Disassemble the jet valve with the special tool used as a spring compressor. The forked portion of the tool fits over the ridge at the center of the assembly while the hole in the upper section fits over the valve stem. Compress the spring enough to remove the keepers and remove them; then slowly release the pressure.

2. Disassemble each jet valve assembly so all parts are kept

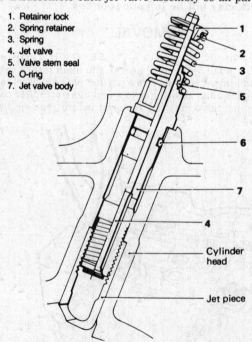

1. Retainer lock
2. Spring retainer
3. Spring
4. Jet valve
5. Valve stem seal
6. O-ring
7. Jet valve body

The jet valve assembly

Measure the height of valve springs. This height (or free length) is an indicator of the strength and condition of the spring. A worn spring will compress but not return to its full length. This may allow the valve to stay partially open, causing driveability problems. Minimum heights are shown below.
- G15B: 43.6mm
- G32B: 45.0mm
- G54B: 49.0mm
- G62B: 46.5mm
- G63B: 48.8mm
- G64B: 48.8mm
- 4G15: 43.6mm
- 4G61: 45.0mm
- 4G63/SOHC: 48.8mm
- 4G63/DOHC: 45.0mm
- 6G72: 49.5mm

Check each spring for squareness by placing the side of the spring next to a
T-square and rotating the spring until you find maximum out-of-square. Squareness is measured in degrees and without very accurate measuring equipment, you won't be able to put a num-

ENGINE AND ENGINE OVERHAUL 3

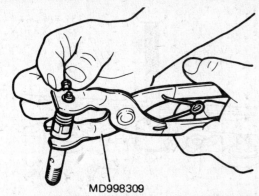

Disassembling the jet valve with the jet valve spring pliers

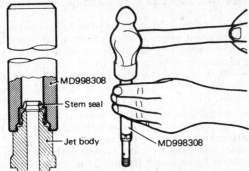

Replacing the jet valve stem seal with the special installer

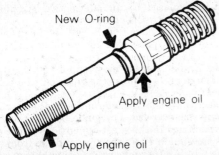

Jet valve assembly before installation

together. Check the seat and valve faces for any signs of burning or cracks. Make sure the valve slides in the jet body smoothly and without play. If there are any signs of burning or excessive valve stem wear, the jet valve assembly must be replaced. Inspect the spring for cracks of any kind, and replace defective springs.

3. Coat the valve stem with clean engine oil and insert it into the jet valve body. Sit the assembly on a flat surface, place a new seal squarely over the valve stem, and install the seal with the special installation tool, tapping downward on it gently with a hammer. The tool numbers are: Cordia/Tredia and Galant — MD 998308 or equivalent; for Starion and Mirage — MD 998301-01. Make sure the valve stem still moves up and down smoothly after the seal is installed; if movement is not smooth, the seal has been damaged or is not installed squarely.

4. Install the spring and spring retainer, compress the spring with the special tool and install the keepers, making sure they remain squarely in the valve stem groove until tension is removed and the retainer covers them. If they do not lock in place securely, the valve spring could pop out either as you are working on the engine, or later, when the engine is operating.

5. Coat a new O-ring with clean engine oil and coat the threaded area and sealing areas of the jet body. Install the new O-ring into the groove provided for it. Then, insert each jet valve squarely into the hole in the cylinder head and start the threads carefully. Use the special socket and a torque wrench to tighten the assembly to 13–15 ft. lbs., keeping the wrench square to the valve stem at all times.

INSTALLATION

Before reassembly, check that all lapping compound has been cleaned from the head. Remember that each piece being reused must be reinstalled in its original position.

1. Position the spring seat on the head. The seat will be forced into its final position in the next step.
2. You'll need a special tool to install the valve stem seals and insure proper installation of the spring seats. The tool delivers square installation of the seal so the engine will not consume excessive amounts of oil. The necessary tools are:
- Cordia, Tredia and Starion to 1984: MD 998377
- Starion in 1985: MD 998377-01
- 1468 engines used in Mirage: MD 998302
- 1597 Mirage: MD 998005
- Galant: MD 998115

3. Install the tool squarely over the seal and valve guide and lightly tap the seal and spring seat in place.
4. Thoroughly coat the valve stem with clean engine oil. Insert the valve slowly from underneath, guiding it gently through the new seal. Make sure the valve will slide up and down smoothly, indicating the seal is in proper position.
5. Install the valve springs with the color-coded end facing upward. Install the retainer on top of the spring. Compress the retainer with the valve spring compressor until the retainer sits below the groove in the valve stem.
6. Install the keepers, making sure they stay properly positioned in the stem groove, and release the tension on the compressor slowly. Note that if the keepers do not sit squarely in the grooves, the valve spring tension could be released suddenly either as you work, or later as the car is in operation. Once the keepers have locked securely, remove the valve spring compressor.
7. Repeat the sequence for each intake and exhaust valve.

Valve Seats

REMOVAL AND INSTALLATION

Valve Seats

As in the case of valves, seats may corrected for inadequate seal width and trueness with machining. After the seat is ma-

Using the special tool reduces the chance of improper installation of valve seals

3-201

3 ENGINE AND ENGINE OVERHAUL

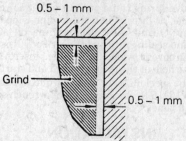

Grind away most of the valve seat before removal

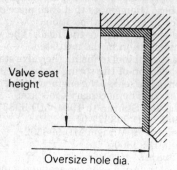

Once the new seat insert is selected, accurate measurement is critical to proper installation

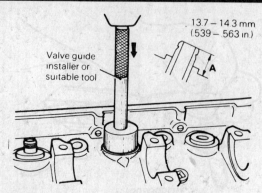

Pressing in the new guide to the proper dimension of the Mirage 1597cc engine

chined the valve must be lapped in. If the valve seat insert shows severe damage or wear it must be replaced.

The seats should be replaced by an experienced machine shop. The procedure requires precise measuring and milling of the head. The new seat is installed through the use of extreme temperatures which require the use of special safety and handling equipment.

Before removal, the seat thickness is removed through drilling or reaming, leaving about 0.5mm around the outside. The seat can then be collapsed out of the head. After cleaning the area, the opening in the head is precisely enlarged to accept the new, oversize valve seat. The hole in the head must conform exactly to the outer diameter and depth of the seat insert. These dimensions vary for each engine; select your parts carefully.

On all but the G62B and G63B engines, the seat is installed after the head is heated to 482°F (250°C). The room temperature seat is pressed (not hammered) into the heated head. As the head cools, it grips the seat very tightly. G62B and G63B engines also require pressing the seat into the head, but the head is at normal temperatures and the seat has to be cooled in liquid nitrogen, contracting it to fit. Remember that as the seat insert is cooled, it becomes very brittle; handle it carefully or it may crack during installation.

After the seat is installed, the metal must be allowed to return to normal temperatures slowly. Resist any temptation to hurry the heating or cooling; sudden temperature change can cause the metal to crack. Simply let the head sit — usually on the press or workbench — for a few hours.

After the head has returned to normal temperature, the new seat insert must be machined to the correct angles. The valve is then inserted and the valve and seat lapped with compound to assure perfect contact.

Valve Guides

The valve guides are so named because they guide the stem of the valve during its motion. By properly locating the stem of the valve, the head and face of the valve are kept in proper relationship to the head. Occasionally, the valve guides will wear causing poor valve sealing and poor engine performance. If this problem is suspected, the valve stem-to-guide clearance must be checked.

If the wear is minimal, it may not be necessary to replace the guides. A competant machine shop can knurl the guides. This process raises a spiral ridge on the inside of the guide (without removing it), thereby eliminating the wobble in the stem. This is not a cure-all; it can only be done to correct minimal wear.

Excessive wear of the valve guides is indicated by excessive stem-to-guide clearance. You can try inserting a new valve into the old guide and measuring stem-to-guide clearance again. If this brings tolerances well within specifications, it may be considered satisfactory to just to replace the valve.

If this does not bring the clearance to within specifications, the guide must be replaced. Provided the old valve is in good condition, it may be checked in the new guide. However, if the engine is receiving a complete overhaul, maximum service life will be achieved by replacing both the guide and valve. Valve guides should be replaced by an automotive machine shop equipped with the proper tools and presses.

The guide must be pressed out of the cylinder head using special tools and a hydraulic press. The tool number varies by specific motor (due to the diameter of the guide) and equivalent tools should be available from most suppliers. Use the tool backed up by a large press to press the guide out the bottom of the cylinder head.

Do NOT replace the guide with one of the same size, but use a guide that is one size oversize. The guide bore in the cylinder head must be machined to the exact dimensions of the outside diameter of the new guide. Press the new guide in, from the top, using the same tool. All these operations should be performed with the parts at room temperature. For each motor, the amount of projection of the valve guide is critical. Use of the correct special tool will ensure the guide is pressed in the correct distance.

The new or re-used valves should operate freely inside the new guides. A new guide may require reaming to achieve the correct stem-to-guide clearance. Once the guide is correctly mated to the valve, the valve face and seat must be lapped and then checked for correct seating. This must be done because the new guide may change the exact angle of the valve stem.

Timing Belts and Covers

REMOVAL AND INSTALLATION

NOTE: Elevate and safely support the car as necessary for access to various components. Depending on tools used and part location, some components may be easier to reach from underneath. When working over the fender, always use a fender cover to protect the paintwork.

ENGINE AND ENGINE OVERHAUL 3

If the timing belt is to be reused, extreme care must be taken to protect it from oil, fluids and grease. These chemicals can cause the belt to deteriorate and possibly break during engine operation. When handling the belt, do not crimp or fold it; sharp bends can break the fibers within the belt.

Cordia and Tredia with G62B and G63B Engines
Pickup with G63B Engine

NOTE: These engines have two timing belts. One connects the crankshaft and camshaft, the second one drives the right silent shaft.

1. With the car in neutral or park, make certain the parking brake is set and the wheels are blocked. Disconnect the negative battery cable.
2. Remove the water pump drive belt and water pump pulley. If the car is equipped with air conditioning, loosen the compressor belt tensioner and remove the compressor belt.
3. Remove the bolts holding the crankshaft pulley(s) and remove the pulley(s).
4. Remove the upper and lower timing belt covers and their gaskets.
5. Use a socket wrench on the projecting crankshaft bolt to turn the engine clockwise (only!) and align all the timing marks on the sprockets and cases. It may be necessary to wipe off the area to see the marks clearly. Do not use spray cleaners around the timing belt.

When the marks all align exactly, the engine is set to TDC/

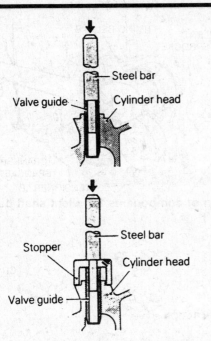

Valve guide removal (upper) and installation (below) using Mitsubishi special tools. Tool numbers vary with each engine. Note the stopper to allow correct projection during installation

G62B and G63B timing belts and components

3-203

3 ENGINE AND ENGINE OVERHAUL

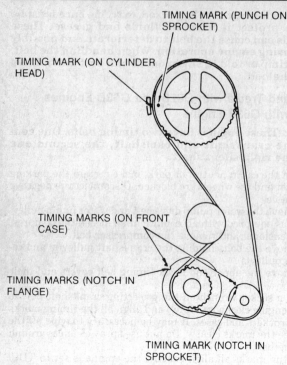

Location and correct alignment of timing marks for the timing belt (camshaft belt)

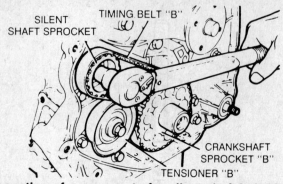

Location of components for silent shaft belt (timing belt "B")

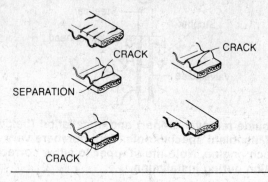

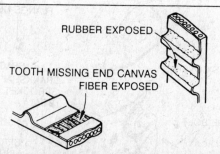

Common timing belt faults. Replace the belt if any of these conditions exist

compression on No. 1 cylinder. From this point onward, the engine position MUST NOT be changed.

6. If the timing belt is to be reused, make a chalk or crayon arrow on the belt showing the direction of rotation so that it may be reinstalled correctly.
7. Loosen the bolts holding the tensioner and pivot the tensioner towards the water pump. Temporarily tighten the bolts to hold the tensioner in its slack position.
8. Carefully slide the belt off the sprockets. Place the belt in a clean, dry, protected location away from the work area.
9. Remove the camshaft sprocket bolt and remove the sprocket.
10. Remove the crankshaft sprocket retaining bolt. Remove the sprocket and flange.
11. Remove the plug from the left side of the cylinder block and insert a screwdriver to keep the left silent shaft in place. The screwdriver should have a shaft diameter of 8mm (0.3 in.) and a shaft length of at least 60mm (2.4 in.).
12. Remove the bolt holding the oil pump sprocket and remove the sprocket.
13. Loosen the mounting bolt for the right silent shaft sprocket until it can be turned with your fingers. Do not remove it.
14. Remove the tensioner for the silent shaft belt. (Tensioner B).
15. Remove the silent shaft belt (timing belt B). Remove the crankshaft pulley for the silent shaft belt if so desired.

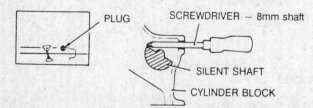

Correct placement of the screwdriver to block the left silent shaft

WARNING: After the timing belt has been removed, do not attempt to loosen the silent shaft bolt by holding the sprocket with pliers. If the sprocket is to be removed, hold the pulley with your fingers.

16. Inspect the timing belts in detail for any flaw or wear. If the belt is not virtually perfect, replace it. A case can be made for replacing the belt every time it is removed, particularly on high-mileage engines. Some of the conditions to look for are:

3-204

ENGINE AND ENGINE OVERHAUL 3

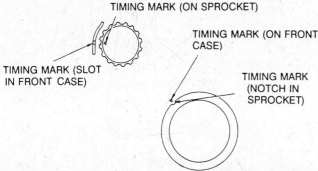

Install the spacer on the right silent shaft correctly

Align the timing marks before installing the silent shaft belt

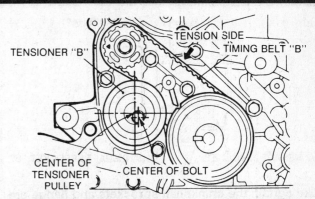

Correct installation of tensioner "B" for the silent shaft belt. The center of the pulley is offset to the left of the bolt

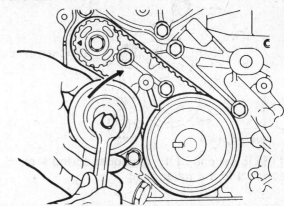

Hold the tensioner while tightening the bolt

- Hardened back surface; non-elastic and glossy; hard to mark with a fingernail.
- Cracking on back of belt, bottom of teeth or side of belt.
- Missing teeth or teeth lifting from belt.
- Side of belt worn or fuzzy. Normal belt should have clean sides as if cut with a sharp knife.
- Wear on teeth as shown by distinct color change or worn rubber.
- Separation of inner coating from backing.
- Any uneven wear patterns on the teeth of the belt. Wear pattern should be even across each tooth and not differ from one tooth to another.

17. Check the sprockets and tensioner for wear. The sprocket teeth should be well defined, not rounded and the valleys between the teeth should be clean. The tensioners should spin freely with no binding or unusual noise. Replace the tensioner if there is any sign of grease leaking from the seal. Clean everything with a clean, dry cloth.

WARNING: Do not spray or immerse the sprockets or tensioners in cleaning solvent. The sprocket may absorb the solvent and transfer it to the belt. The tensioners are internally lubricated and the solvent will dilute or dissolve the lubricant.

18. To reassemble, install the sprocket for the silent shaft belt onto the crankshaft. Make certain it is installed correctly.
19. If the sprocket for the right silent shaft was removed, coat the spacer with a light coating of clean engine oil and install the spacer to the shaft. Be sure to install it in the correct direction. Install the silent shaft sprocket and tighten the bolt finger tight.
20. Double check the timing marks on the silent shaft sprocket and crankshaft sprocket. Carefully align the marks if necessary.
21. Install the silent shaft belt, observing the direction of rotation mark made earlier. Handle the belt carefully and do not use metal tools to guide or force the belt into place. When installing the belt, make sure the tension side has no slack in it.

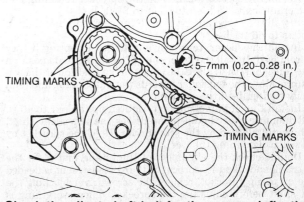

Check the silent shaft belt for the proper deflection

22. Install the tensioner (B) with the center of the pulley located to the left side of the mounting bolt and with the pulley flange to the front of the engine.
23. Lift the tensioner with your hand so that the belt becomes taut. Hold the tensioner in this position and tighten its bolt. Use care that only the bolt and not the tensioner shaft is turned during tightening. The bolt should be tightened to 13 ft. lbs.
24. Tighten the right silent shaft bolt to 26 ft. lbs.
25. Check that the timing marks are still aligned. Push down on the center of the tension side of the belt with your finger. Correct belt deflection is 5–7mm or approximately ¼ inch. If the deflection is not correct, the tensioner must be released fully and the belt re-tensioned.
26. Install the flange and the crankshaft sprocket onto the crankshaft. Make certain the flange is installed in the correct di-

3 ENGINE AND ENGINE OVERHAUL

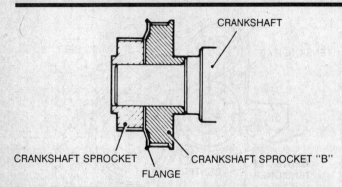

Make certain the crankshaft sprockets and flange are correctly assembled

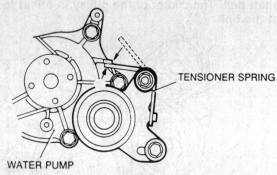

Install the lower end of the tensioner spring first, then attach it to the water pump

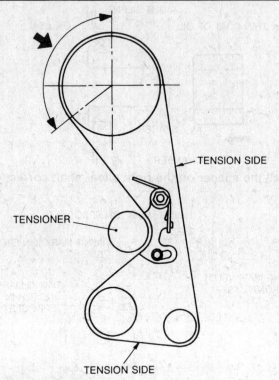

Check the cam sprocket in the area shown for any sign of the belt lifting after it is tensioned

rection. If it is put on incorrectly, the belt will wear and break.
27. Install the special washer and the sprocket retaining bolt to the crankshaft. Tighten the bolt to 88 ft. lbs.
28. Install the camshaft sprocket to the camshaft and tighten the bolts to 66 ft. lbs.
29. Install the spacer, tensioner and tensioner spring if they were removed. Install the lower end of the spring to its position on the tensioner, then place the upper end in position at the water pump. Move the tensioner towards the water pump and temporarily tighten it in this position.
31. Install the oil pump sprocket, tightening the nut to 40 ft. lbs. Remember that the oil pump drives the left silent shaft, which is still blocked by the screwdriver. Hold the sprocket by hand when tightening the nut.
32. Double check the alignment of the timing marks for the cam, crank and oil pump sprockets. If any adjustment is needed to the oil pump sprocket, remove the screwdriver blocking the silent shaft before adjusting the sprocket. Once everything is aligned, replace the screwdriver in the left side of the block and leave it there until installation of the timing belt is complete.

NOTE: If the screwdriver can only be inserted about 25mm (1 inch) or less, the shaft is out of position. Turn the oil pump sprocket through one full turn clockwise; the screwdriver should then go in about 63mm (2½ inches).

33. Install the timing belt onto the crankshaft sprocket, the oil pump sprocket and the cam sprocket in that order. Keep the belt taut between sprockets. If reusing an old belt, make certain that the direction of rotation arrow is properly oriented.
34. Loosen the tensioner mounting nut and bolt. The spring will move the tensioner against the belt and tension it.
35. Check the belt as it passes over the camshaft sprocket. The belt may tend to lift in the area to the left of the sprocket. (Roughly 7 o'clock to 12 o'clock when viewed from the pulley end.) Make certain the belt is well seated and not rubbing on any flanges or nearby surfaces.
36. At the tensioner, tighten the bolt in the slotted hole first, then tighten the nut on the pivot. If this order is not followed, the belt will become too tight and break.
37. Once again, check all the timing marks for alignment. Nothing should have changed; check anyway.
38. Remove the screwdriver blocking the left side silent shaft. Using a socket on the crankshaft bolt, turn the crankshaft smoothly one full turn (360°). Do NOT turn the engine backwards.
39. Loosen the tensioner nut and bolt. The spring will allow the tensioner to tighten a little bit more because of the slack picked up during engine rotation. Tighten the bolt, then the nut to 36 ft. lbs.
40. Check the deflection of the belt. At the middle of the right (tension) side of the belt, deflect the belt outward with your finger, toward the timing case. The distance between the belt and the line of the cover seal should be about 14mm (9/16 in.).

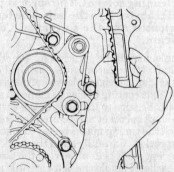

The pinch test for checking belt deflection. Distance between the arrows should be about 14mm (9/16 in.)

ENGINE AND ENGINE OVERHAUL 3

16. Power steering pulley
17. Water pump pulley
18. Valve cover
19. Gasket
20. Upper timing belt cover
21. Gasket
22. Lower timing belt cover
23. Gasket
24. Damper pulley (A/C pulley)
25. Crankshaft pulley
26. Timing belt

G15B timing belt and covers

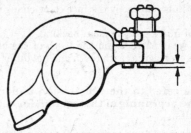

Back off the valve adjusters until the distance shown is almost zero

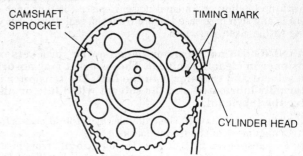

Aligning the cam sprocket

41. Install the lower and upper timing belt covers. Make certain the gaskets are properly seated in the covers and that they don't come loose during installation. Tighten the bolts to 8 ft. lbs.
42. Install the crankshaft pulley(s) and tighten the retaining bolts to 18 ft. lbs.
43. Install the water pump pulley, tightening the bolts to 6 ft. lbs, and install the belt. Adjust the belt to the correct tension.
44. Connect the negative battery cable. Start the engine and let it idle, listening for any unusual noises from the area of the timing belt. Possible causes of noise are the belt rubbing against the covers or a sprocket flange, the belt being too loose and slapping, or a tensioner binding. Do not accelerate the engine if abnormal noises are heard from the timing belt train — severe damage can result.

Mirage and Precis with G15B Engine

1. Disconnect the negative battery cable.
2. Remove the left splash shield under the engine.
3. Remove the air cleaner assembly, the air intake duct and the hot air duct.
4. Disconnect the accelerator cable.
5. Label and remove the coil wire and the spark plug wires from the distributor. Label and disconnect the oxygen sensor connector on the side of the valve cover. Move all the wires out of the way.
6. Loosen the appropriate component or adjuster and remove (in order) the air conditioning compressor drive belt, the power steering belt and the ribbed belt driving the water pump.
7. Place a floor jack and a broad piece of lumber under the oil pan. Elevate the jack just enough to support the engine without raising it.
8. Remove the nuts holding the through-bolt for the left engine bracket and remove the bolt. It may be necessary to adjust the jack tension slightly to allow the bolt to come free. Use the jack to keep the engine in its normal position while the mount is disconnected.
9. Remove the nuts and bolts holding the engine bracket to the engine and remove the bracket.

10. Remove the water pump pulley. If the car is equipped with power steering, remove the smaller pulley first.
11. Remove the valve cover and gasket. At the valve adjusters, back off each adjusting screw until the ends of the screws project only 1.0mm (0.04 in.) or less from the rocker.
12. Remove the upper timing belt cover and its gasket.

WARNING: The bolts for both the upper and lower timing belt covers are of different lengths. Label or diagram them at the time of removal; replacement in the correct location is required.

13. Rotate the crankshaft clockwise until the timing marks on the crank pulley align at 0 on the scale (on the lower timing cover) and the mark on the camshaft pulley aligns with the mark on the head. Aligned means perfect, not close — if you miss, continue clockwise until the marks are matched. This sets the motor at TDC/compression on No.1 cylinder.

WARNING: Once this position is established, the cam and crankshafts must NOT be moved out of place.

14. Remove the lower timing belt cover and its gasket.
15. Remove the crankshaft pulley(s) without moving the crankshaft out of position.
16. Loosen (do not remove) both bolts in the timing belt tensioner. Move the tensioner toward the water pump and tighten the lock bolt — the one in the slotted hole — to hold the tensioner in place.
17. Before removing the timing belt, mark it with an arrow to show the direction of rotation. If the belt is to be reused, it must be reinstalled so that it rotates in the same direction as before.
18. Remove the timing belt, keeping it free of grease, oil and fluids. Do not crease or crimp the belt.
19. Inspect the timing belt in detail for any flaw or wear. If the belt is not virtually perfect, replace it. A case can be made for replacing the belt every time it is removed, particularly on high-mileage engines. Some of the conditions to look for are:
• Hardened back surface; non-elastic and glossy; hard to mark with a fingernail.
• Cracking on back of belt, bottom of teeth or side of belt.
• Missing teeth or teeth lifting from belt.
• Side of belt worn or fuzzy. Normal belt should have clean sides as if cut with a sharp knife.

3-207

3 ENGINE AND ENGINE OVERHAUL

- Wear on teeth as shown by distinct color change or worn rubber.
- Separation of inner coating from backing.
- Any uneven wear patterns on the teeth of the belt. Wear pattern should be even across each tooth and not differ from one tooth to another.

NOTE: Please refer to the illustration under Cordia and Tredia at the beginning of this section for samples of timing belt faults.

20. Check the sprockets and tensioner for wear. The sprocket teeth should be well defined, not rounded and the valleys between the teeth should be clean. The tensioner should spin freely with no binding or unusual noise. Replace the tensioner if there is any sign of grease leaking from the seal. Clean everything with a clean, dry cloth.

WARNING: Do not spray or immerse the sprockets or tensioner in cleaning solvent. The sprocket may absorb the solvent and transfer it to the belt. The tensioner is internally lubricated and the solvent will dilute or dissolve the lubricant.

21. Before reinstalling the belt, double check the sprockets for correct alignment of the timing marks.
22. To reinstall the belt, observe the directional arrow made earlier and fit the belt under the crankshaft sprocket. Track the belt up and over the right side (your right, not the engine's right) of the camshaft sprocket while keeping tension on the belt all the time.
23. The camshaft sprocket is quite likely to have moved during the belt installation. Gently turn the cam sprocket counterclockwise to tension the belt; when the belt is taut, the timing marks on the head and cam sprocket should align perfectly. If they do not, remove the belt and reinstall it.
24. Install the crankshaft pulley temporarily. It is needed to hold the belt in place.
25. Loosen the lock bolt in the tensioner. Allow the tensioner to move against the belt under its own spring tension; don't pry it or force it.

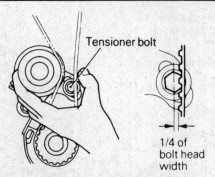

Checking belt deflection after final adjustment

26. Tighten the lock bolt (in the slotted hole) in the tensioner, then tighten the pivot bolt for the tensioner. This order is important; if done in the wrong order, the belt will be over-tightened.
27. Turn the crankshaft one full revolution clockwise. Turn it smoothly and continuously and realign the timing mark. Loosen the pivot bolt, then the lock bolt on the tensioner. This allows the tensioner to pick-up any remaining slack in the belt.
28. Tighten the lock nut to 18 ft. lbs. and then tighten the pivot bolt to 18 ft. lbs.
29. Hold the tensioner and timing belt by hand and give the belt slight thumb pressure at a point level with the tensioner. The belt should deflect enough to allow the top of a tooth to cover about ¼ of the head of the lock bolt.
30. Finish the installation of the crank pulley(s). The large center bolt should be tightened to 62 ft. lbs. and the smaller bolts (for the air conditioning pulley) should be tightened to 10 ft. lbs. if they were removed.
31. Install the lower timing belt cover with its gasket, making sure the gasket doesn't shift during installation. tighten the bolts to 8 ft. lbs.
32. Install the upper timing belt cover with its gaskets and tighten the bolts to 8 ft. lbs.
33. Adjust the valve clearance. Follow the procedure outlined in Section 2.
34. Install the valve cover and gasket.
35. Install the water pump pulley (and power steering pulley if removed), tightening the bolts to 6 ft. lbs.

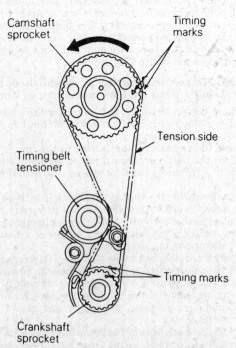

Arrow shows direction of belt installation, NOT direction of engine rotation

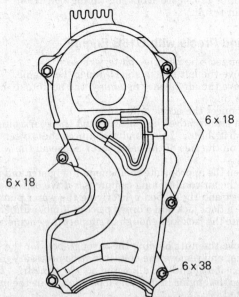

Correct placement of G15B timing belt cover bolts. All measurements shown are millimeters

3-208

ENGINE AND ENGINE OVERHAUL

36. Install the motor mount bracket to the engine and tighten the nuts and bolts to 42 ft. lbs.
37. Align the mount bushing with the body bracket and install the through-bolt. Tighten both the large nut and the smaller safety nut snug.
38. Remove the jack from under the engine. With the weight of the engine on the mount, tighten the large nut on the through-bolt to 72 ft. lbs. and the smaller safety nut to 38 ft. lbs.
39. Install the coil and spark plug wires. Connect the oxygen sensor wiring. Make sure all the wiring is properly held by the clips and wire guides.
40. Install the ribbed water pump drive belt, making sure it is correctly seated on each pulley. Install the power steering belt and the compressor drive belt. Adjust each belt to the correct tension and tighten the fittings or adjuster on each component.
41. Connect the accelerator cable and adjust it.
42. Install the air cleaner assembly and the ducts.
43. Double check all installation items, paying particular attention to loose hoses or hanging wires, untightened nuts, poor routing of hoses and wires (too tight or rubbing) and tools left in the engine area.
44. Connect the negative battery cable.
45. Start the engine and let it idle, listening for any unusual noises from the area of the timing belt. Possible causes of noise are the belt rubbing against the covers or a sprocket flange, the belt being too loose and slapping, or a tensioner binding. Do not accelerate the engine if abnormal noises are heard from the timing belt train — severe damage can result.
46. Reinstall the splash shield under the car. Final adjustment of the drive belts and accelerator cable may be needed.

Mirage with G32B Engine

1. Disconnect the negative battery cable.
2. Remove the left splash shield under the engine.
3. Disconnect the air intake duct, the breather hose and the O-ring from the engine.
4. Disconnect the accelerator cable.
5. Label and remove the coil wire and the spark plug wires from the distributor and from the spark plugs.
6. Loosen the appropriate component or adjuster and remove (in order) the air conditioning compressor drive belt, the power steering belt and the ribbed belt driving the water pump.
7. Place a floor jack and a broad piece of lumber under the oil pan. Elevate the jack just enough to support the engine without raising it.
8. Remove the nuts holding the through-bolt for the left engine bracket and remove the bolt. It may be necessary to adjust the jack tension slightly to allow the bolt to come free. Use the jack to keep the engine in its normal position while the mount is disconnected.
9. Remove the nuts and bolts holding the engine bracket to the engine and remove the bracket.
10. Remove the water pump pulley. If the car is equipped with power steering, remove the smaller pulley first.
11. Disconnect the PCV hose and remove the valve cover and gasket. At the valve adjusters, back off each adjusting screw until the ends of the screws project only 1.0mm (0.04 in.) or less from the rocker.
12. Remove the upper timing belt cover and its gasket.

WARNING: The bolts for both the upper and lower timing belt covers are of different lengths. Label or diagram them at the time of removal; replacement in the correct location is required.

13. Rotate the crankshaft clockwise until the timing marks on the crank pulley align at 0 on the scale (on the lower timing cover) and the mark on the camshaft pulley aligns with the mark on the rear upper cover. Aligned means perfect, not close. If you miss, continue clockwise until the marks are matched. This sets the motor at TDC/compression on No.1 cylinder.

WARNING: Once this position is established, the cam and crankshafts must NOT be moved out of place.

14. Remove the crankshaft pulley(s) without moving the crankshaft out of position.
15. Remove the lower timing belt cover and its gasket.
16. Loosen (do not remove) the nut and the bolt in the timing belt tensioner. Move the tensioner toward the water pump and tighten the nut to hold the tensioner in place.
17. Before removing the timing belt, mark it with an arrow to show the direction of rotation. If the belt is to be reused, it must be reinstalled so that it rotates in the same direction as before.
18. Remove the timing belt, keeping it free of grease, oil and fluids. Do not crease or crimp the belt.
19. Inspect the timing belt in detail for any flaw or wear. If the belt is not virtually perfect, replace it. A case can be made for re-

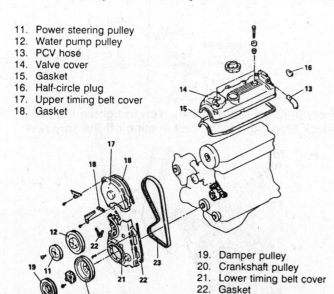

11. Power steering pulley
12. Water pump pulley
13. PCV hose
14. Valve cover
15. Gasket
16. Half-circle plug
17. Upper timing belt cover
18. Gasket
19. Damper pulley
20. Crankshaft pulley
21. Lower timing belt cover
22. Gasket
23. Timing belt

G32B timing belt and covers

Back off the valve adjusters until the distance shown is almost zero

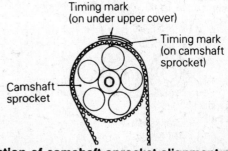

Location of camshaft sprocket alignment marks

3-209

3 ENGINE AND ENGINE OVERHAUL

placing the belt every time it is removed, particularly on high-mileage engines. Some of the conditions to look for are:
- Hardened back surface; non-elastic and glossy; hard to mark with a fingernail.
- Cracking on back of belt, bottom of teeth or side of belt.
- Missing teeth or teeth lifting from belt.
- Side of belt worn or fuzzy. Normal belt should have clean sides as if cut with a sharp knife.
- Wear on teeth as shown by distinct color change or worn rubber.
- Separation of inner coating from backing.
- Any uneven wear patterns on the teeth of the belt. Wear pattern should be even across each tooth and not differ from one tooth to another.

NOTE: Please refer to the illustration under Cordia and Tredia at the beginning of this section for samples of timing belt faults.

20. Check the sprockets and tensioner for wear. The sprocket teeth should be well defined, not rounded and the valleys between the teeth should be clean. The tensioner should spin freely with no binding or unusual noise. Replace the tensioner if there is any sign of grease leaking from the seal. Clean everything with a clean, dry cloth.

WARNING: Do not spray or immerse the sprockets or tensioner in cleaning solvent. The sprocket may absorb the solvent and transfer it to the belt. The tensioner is internally lubricated and the solvent will dilute or dissolve the lubricant.

21. Before reinstalling the belt, double check the sprockets for correct alignment of the timing marks.
22. To reinstall the belt, observe the directional arrow made earlier and fit the belt under the crankshaft sprocket. Track the belt up and over the right side (your right, not the engine's right) of the oil pump sprocket and the camshaft sprocket while keeping tension on the belt all the time.
23. The camshaft sprocket is quite likely to have moved during the belt installation. Gently turn the cam sprocket counter-

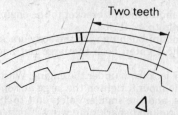

When adjusting the belt tension, turn the crankshaft clockwise enough to move the camshaft sprocket two teeth

clockwise to tension the belt; when the belt is taut, the upper timing marks for the cam sprocket should align perfectly. If they do not, remove the belt and reinstall it.

24. Install the crankshaft pulley temporarily. It is needed to hold the belt in place.
25. Loosen the lock bolt in the tensioner. Allow the tensioner to move against the belt under its own spring tension; don't pry it or force it.
26. Turn the crankshaft clockwise so that the camshaft sprocket timing mark moves 2 teeth away from the mark on the cover. This is necessary to draw the proper tension on the timing belt.

WARNING: Do not turn the crankshaft counterclockwise and do not push on the belt to check its tension.

27. Now apply finger force on the tensioner, pushing it towards the belt. This will seat the belt squarely on the cam sprocket and overcome the belt's tendency to float during adjustment.
28. Tighten the tensioner nut in the tensioner, then tighten the pivot bolt for the tensioner. This order is important; if done in the wrong order, the belt will be over-tightened.
29. Hold the center of the timing belt tension side (between the cam sprocket and the oil pump sprocket) between a thumb

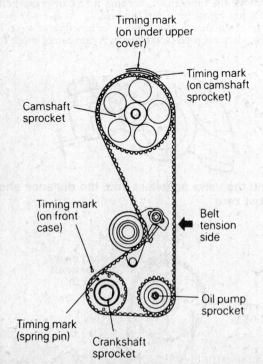

G32B Sprocket alignment marks

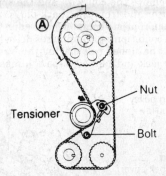

Apply light pressure at the arrow to tighten the belt. Check area "A" for the belt floating off the sprocket

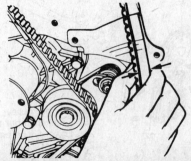

The pinch test checks proper belt deflection. Look for about 6mm (¼ in.) between the arrows

ENGINE AND ENGINE OVERHAUL 3

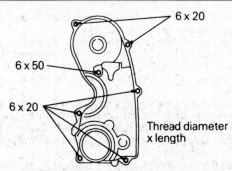

Upper and lower timing cover bolt placement. All measurements shown are in millimeters

and forefinger and gently squeeze the belt outward. The belt should deflect enough to allow 6mm (¼ inch) clearance between the back of the belt and the edge of the belt cover.

30. Carefully remove the crankshaft pulley. Install the lower timing belt cover and gasket, making sure the gaskets stay in place during installation. Observe correct bolt placement and tighten them to 8 ft. lbs.
31. Install the crankshaft pulley(s), tightening the bolts to 12 ft. lbs.
32. Install the upper timing belt cover with its gaskets and tighten the bolts to 8 ft. lbs.
33. Adjust the valve clearance. Follow the procedure outlined in Section 2.
34. Install the valve cover and gasket. Connect the PCV hose.
35. Install the water pump pulley (and power steering pulley if removed), tightening the bolts to 6 ft. lbs.
36. Install the motor mount bracket to the engine and tighten the nuts and bolts to 42 ft. lbs.
37. Align the mount bushing with the body bracket and install the through-bolt. Tighten both the large nut and the smaller safety nut snug.
38. Remove the jack from under the engine. With the weight of the engine on the mount, tighten the large nut on the through-bolt to 72 ft. lbs. and the smaller safety nut to 38 ft. lbs.
39. Install the water pump drive belt, making sure it is correctly seated on each pulley. Install the power steering belt and the compressor drive belt. Adjust each belt to the correct tension and tighten the fittings or adjuster on each component.
40. Install the coil and spark plug wires. Connect the oxygen sensor wiring. Make sure all the wiring is properly held by the clips and wire guides.
41. Using a new O-ring, reconnect the air intake duct and connect the breather hoses. Tighten the retaining nuts to 7 ft. lbs.
42. Connect the accelerator cable and adjust it.
43. Double check all installation items, paying particular attention to loose hoses or hanging wires, untightened nuts, poor routing of hoses and wires (too tight or rubbing) and tools left in the engine area.
44. Connect the negative battery cable.
45. Start the engine and let it idle, listening for any unusual noises from the area of the timing belt. Possible causes of noise are the belt rubbing against the covers or a sprocket flange, the belt being too loose and slapping, or a tensioner binding. Do not accelerate the engine if abnormal noises are heard from the timing belt train — severe damage can result.
46. Reinstall the splash shield under the car. Final adjustment of the drive belts and accelerator cable may be needed.

Mirage with 4G15 Engine

1. Disconnect the negative battery cable.
2. Remove the splash shield under the engine.
3. Place a floor jack and a broad piece of lumber under the oil pan. Elevate the jack just enough to support the engine without raising it.

4. Remove the nuts holding the through-bolt for the left engine bracket and remove the bolt. It may be necessary to adjust the jack tension slightly to allow the bolt to come free. Use the jack to keep the engine in its normal position while the mount is disconnected.
5. Remove the nuts and bolts holding the engine bracket to the engine and remove the bracket.
6. Loosen the appropriate component or adjuster and remove (in order) the air conditioning compressor drive belt, the power steering belt and the alternator belt. Remove the air conditioning belt tension adjuster and bracket.
7. Remove the power steering and water pump pulleys.
8. Remove the upper timing belt cover and its gaskets. Rotate the crankshaft clockwise until the timing marks on the crank pulley align at 0 on the scale (on the lower timing cover) and the mark on the camshaft pulley aligns with the indicator on the head. Aligned means perfect, not close — if you miss, continue clockwise until the marks are matched. This sets the motor at TDC/compression on No.1 cylinder.

WARNING: Once the engine is set to this position, the crank and camshafts must not be moved out of place. Severe damage can result.

9. Remove the air conditioner pulley from the crankshaft and remove the crankshaft pulley. Remember that the pulley (and therefore, the crankshaft) must be held steady while the bolt is loosened. This can be made easier by wrapping a scrapped belt tightly around the pulley. Mitsubishi has a special tool (MD 998747) for this trick, but other variations work almost as well.

10. Upper timing belt cover
11. Lower timing belt cover
12. Tensioner spacer
13. Tensioner spring
14. Tensioner
15. Timing belt
16. Camshaft sprocket
17. Crankshaft sprocket
18. Flange

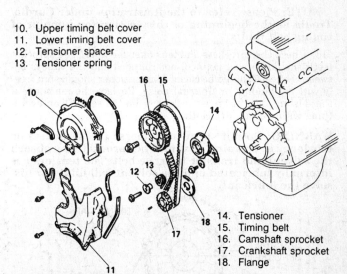

4G15 timing belt, sprockets and covers Camshaft alignment marks

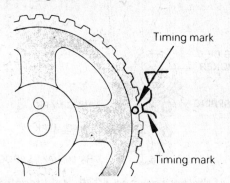

Camshaft alignment marks

3-211

3 ENGINE AND ENGINE OVERHAUL

DO NOT use any of the drive belts which will be reinstalled; the belt used to hold the pulley will be damaged or weakened.

10. Remove the lower timing belt cover and gaskets. Note the different bolt lengths and placements.
11. Loosen (do not remove) both bolts in the timing belt tensioner. Move the tensioner toward the water pump and tighten the lock bolt — the one in the slotted hole — to hold the tensioner in place.
12. Before removing the timing belt, mark it with an arrow to show the direction of rotation. If the belt is to be reused, it must be reinstalled so that it rotates in the same direction as before.
13. Remove the timing belt, keeping it free of grease, oil and fluids. Do not crease or crimp the belt.
14. Inspect the timing belt in detail for any flaw or wear. If the belt is not virtually perfect, replace it. A case can be made for replacing the belt every time it is removed, particularly on high-mileage engines. Some of the conditions to look for are:
- Hardened back surface; non-elastic and glossy; hard to mark with a fingernail.
- Cracking on back of belt, bottom of teeth or side of belt.
- Missing teeth or teeth lifting from belt.
- Side of belt worn or fuzzy. Normal belt should have clean sides as if cut with a sharp knife.
- Wear on teeth as shown by distinct color change or worn rubber.
- Separation of inner coating from backing.
- Any uneven wear patterns on the teeth of the belt. Wear pattern should be even across each tooth and not differ from one tooth to another.

NOTE: Please refer to the illustration under Cordia/Tredia at the beginning of this section for samples of timing belt faults.

15. Check the sprockets and tensioner for wear. The sprocket teeth should be well defined, not rounded and the valleys between the teeth should be clean. The tensioner should spin freely with no binding or unusual noise. Replace the tensioner if there is any sign of grease leaking from the seal. Clean everything with a clean, dry cloth.

WARNING: Do not spray or immerse the sprockets or tensioner in cleaning solvent. The sprocket may absorb the solvent and transfer it to the belt. The tensioner is internally lubricated and the solvent will dilute or dissolve the lubricant.

16. Before reinstalling the belt, double check the sprockets for correct alignment of the timing marks.
17. To reinstall the belt, observe the directional arrow made earlier and fit the belt under the crankshaft sprocket. Track the belt up and over the right side (your right, not the engine's right) of the camshaft sprocket while keeping tension on the belt all the time.
18. The camshaft sprocket may have moved during the belt installation. Gently turn the cam sprocket counter-clockwise to tension the belt; when the belt is taut, the timing marks on the head and cam sprocket should align perfectly. If they do not, remove the belt and reinstall it.
19. Install the crankshaft pulley temporarily. It is needed to hold the belt in place.
20. Loosen the lock bolt in the tensioner. Allow the tensioner to move against the belt under its own spring tension; don't pry it or force it.
21. Tighten the lock bolt (in the slotted hole) in the tensioner, then tighten the pivot bolt for the tensioner. This order is important; if done in the wrong order, the belt will be over-tightened.
22. Turn the crankshaft one full revolution clockwise. Turn it smoothly and continuously and realign the timing mark. Loosen the pivot bolt, then the lock bolt on the tensioner. This allows the tensioner to pick-up any remaining slack in the belt.
23. Tighten the lock nut to 18 ft. lbs, then tighten the pivot bolt to 18 ft. lbs.
24. Hold the tensioner and timing belt by hand and give the belt slight thumb pressure at a point level with the tensioner. The belt should deflect enough to allow the top of a tooth to cover about ¼ of the head of the lock bolt.
25. Install the lower and upper timing belt covers, making sure the gaskets are in place. Observe the correct placement of the bolts and tighten the cover bolts to 8 ft. lbs.
26. Finish the installation of the crank pulley(s). The large center bolt should be tightened to 62 ft. lbs. (you can use the belt trick here, too) and the smaller bolts for the air conditioning pulley should be tightened to 10 ft. lbs. if they were removed.
27. Install the water pump pulley and power steering pulley, tightening the bolts to 6 ft. lbs.
28. Install the alternator drive belt, making sure it is correctly seated on each pulley. Install the power steering belt and the compressor drive belt. Adjust each belt to the correct tension and tighten the fittings or adjuster on each component.
29. Install the motor mount bracket to the engine and tighten the nuts and bolts to 42 ft. lbs.
30. Align the mount bushing with the body bracket and install the through-bolt. Tighten both the large nut and the smaller safety nut snug.
31. Remove the jack from under the engine. With the weight of the engine on the mount, tighten the large nut on the through-bolt to 72 ft. lbs. and the smaller safety nut to 38 ft. lbs.
32. Double check all installation items, paying particular attention to loose hoses or hanging wires, untightened nuts, poor

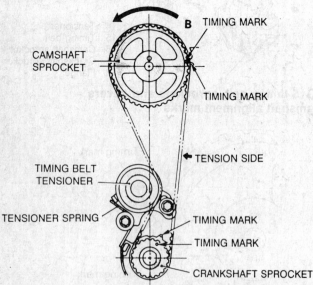

Alignment of timing marks on 4G15. Hold tension in direction "B" when installing the belt

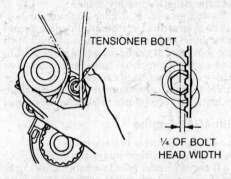

Checking belt deflection after final adjustment

ENGINE AND ENGINE OVERHAUL 3

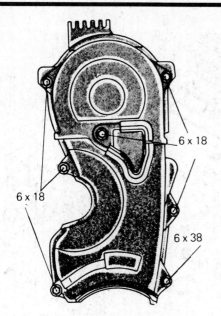

Correct placement of 4G15 timing cover bolts. Measurements shown are in millimeters

routing of hoses and wires (too tight or rubbing) and tools left in the engine area.

33. Connect the negative battery cable.
34. Start the engine and let it idle, listening for any unusual noises from the area of the timing belt. Possible causes of noise are the belt rubbing against the covers or a sprocket flange, the belt being too loose and slapping, or a tensioner binding. Do not accelerate the engine if abnormal noises are heard from the timing belt train — severe damage can result.
35. Reinstall the splash shield under the car. Final adjustment of the drive belts may be needed.

Mirage with 4G61 Engine

WARNING: The use of the correct special tools or their equivalent is REQUIRED for this procedure.

1. Disconnect the negative battery cable.
2. Remove the left splash shield under the engine.
3. Place a floor jack and a broad piece of lumber under the oil pan. Elevate the jack just enough to support the engine without raising it.
4. Remove the nuts holding the through-bolt for the left engine bracket and remove the bolt. It may be necessary to adjust the jack tension slightly to allow the bolt to come free. Use the jack to keep the engine in its normal position while the mount is disconnected.
5. Remove the nuts and bolts holding the engine bracket to the engine and remove the bracket.
6. Remove the alternator belt and the power steering belt.
7. Loosen the tension on the air conditioner compressor belt tensioner. With the belt loosened, remove the adjuster assembly and remove the belt.
8. Remove the water pump pulley with the power steering pulley.
9. Remove the crankshaft pulley.
10. Remove the (front) upper and lower timing belt covers.

WARNING: The covers use 4 different lengths of bolts. Label or diagram them as they are removed.

11. Remove the center cover, the PCV hose and the breather hose from the valve cover. Label and disconnect the spark plug cables from the spark plugs.
12. Remove the valve cover and its gasket. Remove the half-circle plug from the head.
13. Turn the crankshaft clockwise and align all the timing marks. The timing marks on the camshaft sprockets should align with the upper surface of the cylinder head and the dowel pins (guide pins) at the center of the camshaft sprockets should be up. Aligned means perfect, not close — if you miss, continue clockwise until the marks are matched. This sets the motor at TDC/compression on No.1 cylinder.

WARNING: Once the engine is set to this position, the crank and camshafts must not be moved out of place. Severe damage can result.

14. Remove the auto-tensioner.
15. Before removing the timing belt, mark it with an arrow to show the direction of rotation. If the belt is to be reused, it must be reinstalled so that it rotates in the same direction as before.
16. Remove the timing belt, keeping it free of grease, oil and fluids. Do not crease or crimp the belt. After the belt is clear, remove the tensioner pulley.
17. Inspect the timing belt in detail for any flaw or wear. If the belt is not virtually perfect, replace it. A case can be made for replacing the belt every time it is removed, particularly on high-mileage engines. Some of the conditions to look for are:

- Hardened back surface; non-elastic and glossy; hard to mark with a fingernail.
- Cracking on back of belt, bottom of teeth or side of belt.
- Missing teeth or teeth lifting from belt.
- Side of belt worn or fuzzy. Normal belt should have clean sides as if cut with a sharp knife.
- Wear on teeth as shown by distinct color change or worn rubber.
- Separation of inner coating from backing.
- Any uneven wear patterns on the teeth of the belt. Wear pattern should be even across each tooth and not differ from one tooth to another.

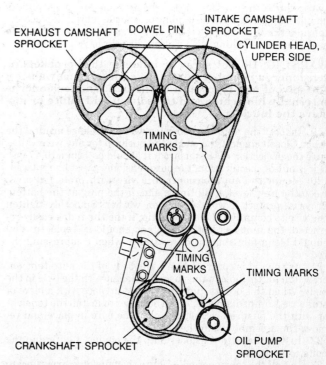

Align all the timing marks before removing the belt

3-213

3 ENGINE AND ENGINE OVERHAUL

3. Alternator belt
4. Power steering belt
5. Tensioner pulley bracket
6. A/C belt
7. Water pump pulley
8. Crankshaft pulley
9. Front upper timing belt cover
10. Front lower timing belt cover
11. Center cover
12. Breather hose
13. PCV hose
14. Spark plug wires
15. Valve cover
16. Half-circle plug
17. Access hole plug
18. Auto tensioner
19. Timing belt
20. Tensioner pulley
21. Tensioner arm
22. Idler pulley
23. Camshaft sprockets
24. Oil pump sprocket
25. Crankshaft sprocket bolt
26. Special washer
27. Crankshaft sprocket
28. Flange
29. Spacer
30. Rear timing belt cover
31. Rear timing belt cover
32. Rear timing belt cover

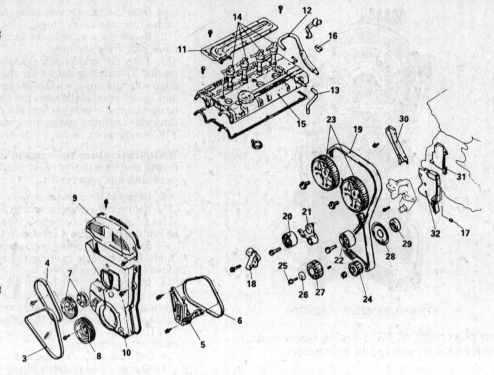

4G61 timing belt, sprockets, tensioners and covers

NOTE: Please refer to the illustration under Cordia/Tredia at the beginning of this section for samples of timing belt faults.

18. Check the sprockets and tensioner for wear. The sprocket teeth should be well defined, not rounded and the valleys between the teeth should be clean. The tensioner and idler pulleys should spin freely with no binding or unusual noise. Replace the tensioner or idler if there is any sign of grease leaking from the seal. Clean everything with a clean, dry cloth.

WARNING: Do not spray or immerse the sprockets or tensioner in cleaning solvent. The sprocket may absorb the solvent and transfer it to the belt. The tensioner is internally lubricated and the solvent will dilute or dissolve the lubricant.

19. Inspect the auto-tensioner for wear or leakage around the seals. Closely inspect the end of the pushrod for any wear. Measure the projection of the pushrod. It should be 12mm (0.47 in.). If it is out of specification, the auto tensioner must be replaced.
20. Mount the auto-tensioner in a vise with protected jaws. Keep the unit level at all times, and, if the plug at the bottom projects, protect it with a common washer. Smoothly tighten the vise to compress the adjuster tip; if the rod is easily retracted, the unit has lost tension and should be replaced. You should feel a fair amount of resistance when compressing the rod.
21. As the rod is contracted into the body of the auto-tensioner, watch for the hole in the pushrod to align with the hole in the body. When this occurs, pin the pushrod in place with a piece of stiff wire 1.5mm diameter. It will be easier to install the tensioner with the pushrod retracted. Leave the wire in place and remove the assembly from the vise.
22. Install the auto-tensioner onto the engine, tightening the bolts to 18 ft. lbs.
23. Install the tensioner pulley onto the tensioner arm. Locate the pinholes in the pulley shaft to the left of the center bolt.

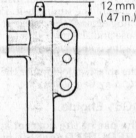

Correct projection for the tip of the auto-tensioner

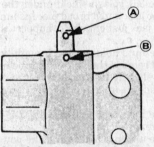

Compress the tip of the auto-tensioner until holes "A" and "B" align. Insert a piece of stiff wire to keep the tip retracted

Tighten the center bolt finger tight. Leave the blocking wire in the auto-tensioner.

24. Double check the alignment of the timing marks for all the sprockets. The camshaft sprocket marks must face each other and align with the top surface of the cylinder head. Don't forget to check the oil pump sprocket alignment as well as the crankshaft sprocket.

ENGINE AND ENGINE OVERHAUL 3

NOTE: When you let go of the exhaust camshaft sprocket, it will move one tooth counter-clockwise. Take this into account when installing the belt around the sprockets.

25. Observing the direction of rotation mark made earlier, install the timing belt as follows:
   a. Install the timing belt around the tensioner pulley and the crankshaft sprocket and hold the belt on the tensioner with your left hand.
   b. Pulling the belt with your right hand, install it around the oil pump sprocket.
   c. Install the belt around the idler pulley and then around the intake camshaft sprocket.
   d. Double check the exhaust camshaft sprocket alignment mark. It has probably moved one tooth; if so, rotate it clockwise to align the timing mark with the top of the cylinder head and install the belt over the sprocket. The belt is a snug fit but can be fitted with the fingers; do not use tools to force the belt onto the sprocket.
   e. Raise the tensioner pulley against the belt to prevent sagging and temporarily tighten the center bolt.

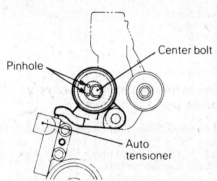

Align the tensioner pulley so the pinholes are to the left of the center bolt. The special pulley tool mounts to the pinholes

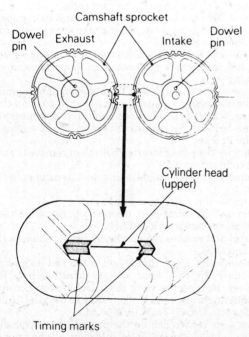

Aligning the camshaft pulleys

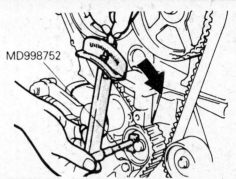

The special adapter is required to properly set the pulley tension

26. Turn the crankshaft ¼ turn (90°) counter-clockwise, then turn it clockwise and align the timing marks.
27. Loosen the center bolt on the tensioner pulley and attach special tool MD 998752 and a torque wrench. (This is a purpose-built tool; substitutes will be hard to find.) Apply a torque of 1.88–2.03 ft. lbs. to the tensioner pulley; this establishes the correct loading against the belt.

NOTE: You must use a torque wrench with a 0–36 inch lb. scale. Normal scales will not read to the accuracy needed. If the body interferes with the tools, elevate the jack slightly to raise the engine above the interference point.

28. Hold the tensioner pulley in place with the torque wrench and tighten the center bolt of the tensioner pulley to 36 ft. lbs. Remove the torque wrench assembly.
29. On the left side of the engine, located in the rear lower timing belt cover, is a rubber plug. Remove the plug and install special tool MD 998738 (another one difficult to substitute). Screw the tool in until it makes contact with the tensioner arm. After the point of contact, screw it in some more to compress the arm slightly.
30. Remove the wire holding the tip of the auto-tensioner. The tip will push outward and engage the tensioner arm. Remove the screw tool.
31. Rotate the crankshaft 2 complete turns clockwise. Allow everything to sit as is for about 15 minutes — longer if the temperature is cool — then check the distance the tip of the auto-tensioner protrudes. (It is almost impossible with the engine in the car; if you can measure the distance, it should be 3.8-4.5mm (0.15-0.18 in.). The alternate method is given in the next step.
32. Reinstall the special screw tool into the left support bracket until the end just makes contact with the tensioner arm. Starting from that position, turn the tool against the arm, counting the number of full turns until the tensioner arm contacts the auto-tensioner body. At this point, the arm has compressed the tensioner tip through its projection distance. The correct number of turns to this point is 2.5 to 3 turns. If you took the correct number of turns to bottom the arm, the projection was correct to begin with and you may now unwind the screw tool EXACTLY the number of turns you turned it in.

If your number of turns was too high or too low when the arm bottomed, you have a problem with the tensioner and should replace it.

33. Remove the special screw tool and install the rubber plug into the access hole in the timing belt cover.
34. Install the half-circle plug with the proper sealant, then install the valve cover and gasket.
35. Install the spark plug wires, the breather and PCV hoses and the center cover.
36. Install the lower and upper timing belt covers, paying attention to the correct placement of the four different sizes of bolts.

3–215

3 ENGINE AND ENGINE OVERHAUL

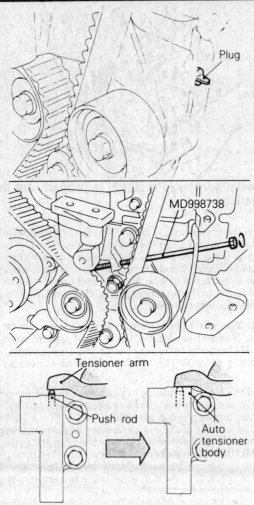

Remove the plug, insert the special screw tool and compress the tip of the auto-tensioner pushrod

37. Install the crankshaft pulley, tightening the bolts to 18 ft. lbs.
38. Install the water pump pulley(s) and tighten the bolts to 72 inch lbs.
39. Install air conditioning drive belt and install the tensioner bracket. Tighten the bracket bolts to 18 ft. lbs.
40. Install the power steering belt and the alternator belt.
41. Install the engine mount bracket to the engine. Tighten the mounting nuts and bolts to 42 ft. lbs.
42. Adjust the jack (if necessary) so that the engine mount bushing aligns with the bodywork bracket. Install the through-bolt and tighten the nuts snug.
43. Slowly release tension on the floor jack so that the weight of the engine bears fully on the mount. Tighten the through-bolt to 72 ft. lbs. and the small safety nut to 38 ft. lbs. Don't forget to install the small support bracket for the engine mount. Tighten its bolt and nut to 16 ft. lbs.
44. Double check all installation items, paying particular attention to loose hoses or hanging wires, untightened nuts, poor routing of hoses and wires (too tight or rubbing) and tools left in the engine area.
45. Connect the negative battery cable.
46. Start the engine and let it idle, listening for any unusual noises from the area of the timing belt. Possible causes of noise are the belt rubbing against the covers or a sprocket flange, the belt being too loose and slapping, or a tensioner binding. Do not accelerate the engine if abnormal noises are heard from the timing belt train — severe damage can result.
47. Reinstall the splash shield under the car. Final adjustment of the drive belts may be needed.

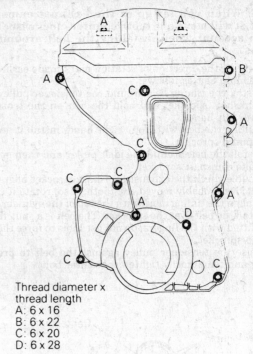

Bolt placement for the timing belt covers. Measurements are in millimeters

Galant with G64B Engine

NOTE: These engines have two timing belts. One connects the crankshaft and camshaft, the second one drives the right silent shaft.

1. Disconnect the negative battery cable.
2. Disconnect the high pressure hose from its clamp on top of the left engine mount. Disconnect the oxygen sensor connector and remove the clip.
3. Position a floor jack and a broad piece of lumber under the engine. Elevate the jack just enough to support the engine without raising it.
4. Remove the through-bolt from the left side engine mount. It may be necessary to adjust the jack slightly to allow the bolt to come free. When the bolt has been removed, disconnect the nuts and bolts holding the mounting bracket to the engine and remove the bracket.
5. Remove the power steering belt, then remove the air conditioning compressor belt.
6. Remove the air conditioning belt tensioner pulley and bracket.
7. Remove the water pump pulleys.
8. Remove the crankshaft (damper) pulley from the crankshaft, then remove the ribbed belt.
9. Label and remove the spark plug wires. Remove the cable guides and clips.
10. Remove the upper timing belt cover and its gasket.
11. Remove the PCV hose and the breather hose from the valve cover. Remove the valve cover and gasket. Loosen all the valve adjusting screws and the jet valves so that the tip of the adjuster projects from 0-1mm (0-0.04 in.).
12. Use a socket wrench on the projecting crankshaft bolt to turn the engine clockwise (only!) and align all the timing marks on the sprockets and cases. It may be necessary to wipe off the

ENGINE AND ENGINE OVERHAUL 3

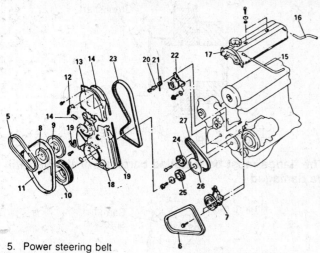

5. Power steering belt
6. Air conditioning belt
7. Tensioner pulley and bracket
8. Water pump pulley
9. Water pump pulley
10. Damper (A/C) pulley
11. Ribbed belt
12. Cable support
13. Upper timing belt cover
14. Gasket
15. PCV hose
16. Breather hose
17. Valve cover
18. Lower timing belt cover
19. Gasket
20. Spacer
21. Spring
22. Tensioner
23. Timing belt
24. Timing belt B tensioner
25. Crankshaft sprocket
26. Flange
27. Timing belt B (silent shaft belt)

G64B timing belts, covers and sprockets

area to see the marks clearly. Do not use spray cleaners around the timing belt.

When the marks all align exactly, the engine is set to TDC/compression on No. 1 cylinder. From this point onward, the engine position MUST NOT be changed.

13. If the timing belt is to be reused, make a chalk or crayon arrow on the belt showing the direction of rotation so that it may be reinstalled correctly.
14. Loosen the bolts holding the tensioner and pivot the tensioner towards the water pump. Temporarily tighten the bolts to hold the tensioner in its slack position.
15. Carefully slide the belt off the sprockets. Place the belt in a clean, dry, protected location away from the work area.
16. Remove the tensioner for the silent shaft belt. (Tensioner B).
17. Remove the crankshaft sprocket retaining bolt. Remove the sprocket and flange.
18. Remove the silent shaft belt (timing belt B).
19. Inspect the timing belts in detail for any flaw or wear. If the belt is not virtually perfect, replace it. A case can be made for

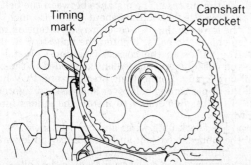

Alignment of the camshaft sprocket timing mark

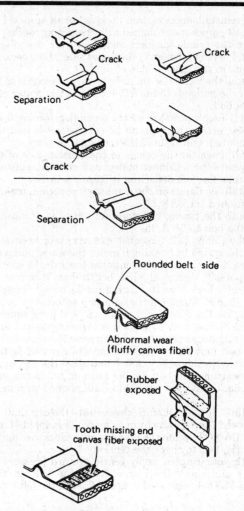

Common timing belt faults. Replace the belt if any of these conditions exist

replacing the belt every time it is removed, particularly on high-mileage engines. Some of the conditions to look for are:
• Hardened back surface; non-elastic and glossy; hard to mark with a fingernail.
• Cracking on back of belt, bottom of teeth or side of belt.
• Missing teeth or teeth lifting from belt.
• Side of belt worn or fuzzy. Normal belt should have clean sides as if cut with a sharp knife.
• Wear on teeth as shown by distinct color change or worn rubber.
• Separation of inner coating from backing.
• Any uneven wear patterns on the teeth of the belt. Wear pattern should be even across each tooth and not differ from one tooth to another.

20. Check the sprockets and tensioner for wear. The sprocket teeth should be well defined, not rounded and the valleys between the teeth should be clean. The tensioners should spin freely with no binding or unusual noise. Replace the tensioner if there is any sign of grease leaking from the seal. Clean everything with a clean, dry cloth.

WARNING: Do not spray or immerse the sprockets or tensioners in cleaning solvent. The sprocket may absorb the solvent and transfer it to the belt. The tensioners are internally lubricated and the solvent will dilute or dissolve the lubricant.

3-217

3 ENGINE AND ENGINE OVERHAUL

21. To reinstall, make certain the crankshaft sprocket and the silent shaft sprocket are aligned with the timing marks. Slip the belt over the crank sprocket and then over the silent shaft sprocket, keeping tension on the upper side. Make certain there is no slack in the tension side of the belt.
22. Install the tensioner and temporarily position it so that the center of the pulley is to the left and above the center of the installation bolt.
23. Apply finger pressure on the tensioner, forcing it into the belt. Make certain the opposite side of the belt is taut. Tighten the installation bolt to hold the tensioner in place.
24. Push down on the center of the tension side of the belt (side opposite the tensioner pulley) and check the distance the belt deflects. Correct deflection is 6mm (¼ inch).
25. Install the flange on the crankshaft sprocket, making sure it is positioned correctly.
26. Install the timing belt sprocket onto the crankshaft and tighten the bolt to 88 ft. lbs.
27. If the timing belt tensioner was removed, reinstall it and connect the spring by hooking it under the water pump casting. If not already done, move the tensioner towards the water pump and temporarily tighten the lock bolt to hold it in place.
28. Double check that the timing marks on all the sprockets are still aligned, including the oil pump sprocket.
29. Observe the directional arrow made during removal and install the timing belt by placing on the crankshaft sprocket, then the oil pump sprocket and then over the camshaft. Maintain the belt taut on the tension side while installing it.
30. Loosen the adjuster lock bolt and allow the spring to move the pulley against the belt. Do not push or pry on the adjuster.
31. Recheck that each sprocket is still aligned with its timing marks.
32. Rotate the crankshaft clockwise until the camshaft sprocket has moved 2 teeth away from its mark. This applies tension to the belt. Do NOT rotate the crank counterclockwise and do not push on the belt to check the tension.
33. With your fingers, apply gentle upward pressure on the tensioner and check the placement of the belt on the cam sprocket. The belt may tend to lift or float in the left side of the

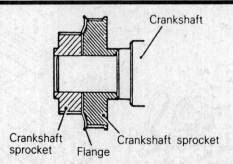

The flange must be installed correctly or the belt will be damaged

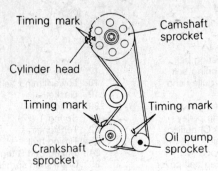

Timing mark alignment, G64B

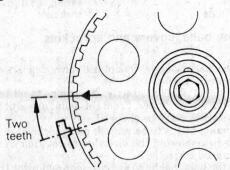

Rotate the crankshaft until the camshaft sprocket has moved 2 teeth

sprocket — roughly between 7 o'clock and 12 o'clock as you view the sprocket.
34. When the belt is correctly seated, tighten the tensioner lock bolt to 36 ft. lbs, then tighten the pivot nut to 36 ft. lbs. The order of tightening is important; if not followed, the belt could be over–tightened.
35. Check the deflection of the belt. At the middle of the right (tension) side of the belt, deflect the belt outward with your finger, toward the timing case. The distance between the belt and the line of the cover seal should be about 13mm (½ inch).
36. Install the lower timing belt cover, paying attention to the correct placement of the bolts. Tighten the bolts to 8 ft. lbs.
37. Adjust the valves to the correct clearances. Install the valve cover and gasket, applying sealant as needed to the cover and the half-circle plug. Connect the breather and PCV hoses.
38. Install the upper timing belt cover and gaskets, tightening the bolts to 8 ft. lbs. Remember to install the guides for the spark plug wires.
39. Reinstall the spark plug wires and connect them.
40. Install the ribbed drive belt and the damper pulley on the crankshaft. Tighten the bolts to 18 ft. lbs.
41. Install the water pump pulleys and tighten the bolts to 6 ft. lbs.

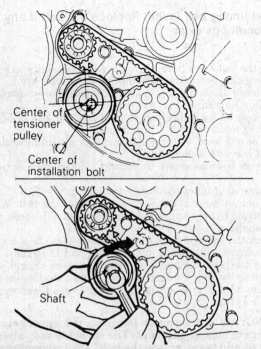

Position the tensioner properly, then move it inward to tighten the belt

ENGINE AND ENGINE OVERHAUL

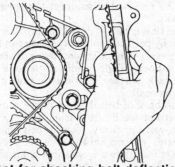

The pinch test for checking belt deflection. Distance between the arrows should be about 14mm (9/16 in.)

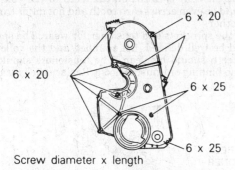

Correct bolt placement is important when reassembling the timing belt covers

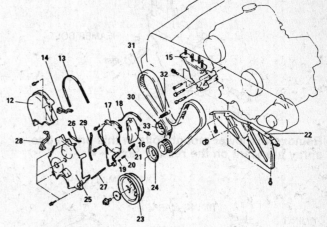

12. Upper outer timing belt cover—front bank
13. Gasket D
14. Gasket X
15. Engine mount bracket, lower section
16. Timing belt cover cap
17. Upper outer timing belt cover—rear bank
18. Gasket E
19. Gasket M
20. Gasket F
21. Gasket Z
22. Splash shield
23. Crankshaft pulley
24. Front flange
25. Lower outer timing belt cover
26. Gasket C
27. Gasket B
28. Gasket W
29. Gasket A
30. Timing belt tensioner bolt
31. Timing belt
32. Tensioner spring
33. Timing belt tensioner

V6 Timing belt and covers. Keep track of all the cover gaskets; they must be reinstalled with the covers

42. Install the tensioner pulley and the A/C drive belt; adjust the belt to the correct tension.
43. Install the power steering belt and adjust it.
44. Install the engine mount bracket to the engine. Tighten the mounting nuts and bolts to 42 ft. lbs.
45. Adjust the jack (if necessary) so that the engine mount bushing aligns with the bodywork bracket. Install the through-bolt and tighten the nuts snug.
46. Slowly release tension on the floor jack so that the weight of the engine bears fully on the mount. Tighten the through-bolt to 52 ft. lbs. and the small safety nut to 26 ft. lbs.
47. Connect the oxygen sensor wiring and install its clip. Position the high pressure hose and secure its bracket.
48. Double check all installation items, paying particular attention to loose hoses or hanging wires, untightened nuts, poor routing of hoses and wires (too tight or rubbing) and tools left in the engine area.
49. Connect the negative battery cable. Start the engine and let it idle, listening for any unusual noises from the area of the timing belt. Possible causes of noise are the belt rubbing against the covers or a sprocket flange, the belt being too loose and slapping, or a tensioner binding. Do not accelerate the engine if abnormal noises are heard from the timing belt train — severe damage can result.
50. Final adjustment of the drive belts may be needed.

Galant and Sigma with 6G72 V6 Engine

WARNING: The use of the correct special tools or their equivalent is REQUIRED for this procedure. The crankshaft sprocket must be removed and installed with the use of tools MB 990767-01 and MB 998715 or their equivalents. Attempts to use other tools or methods may cause damage to the sprocket and/or crankshaft.

1. Disconnect the negative battery cable.
2. Remove the bolt holding the high pressure hose to the engine.
3. Disconnect the pressure switch wiring from the power steering pump.
4. Disconnect the high pressure line from the power steering pump.
5. Place a floor jack and a broad piece of lumber under the oil pan. Raise the jack until the engine is supported but not raised.
6. Remove the through-bolt for the left engine mount.
7. Remove the four nuts holding the upper part of the engine mount to the lower bracket.
8. Remove the air conditioning compressor drive belt, then remove the belt tensioner assembly.
9. Remove the water pump drive belt.
10. Remove the power steering pump and position it out of the way.
11. Remove the belt tensioner assembly and its bracket.
12. Remove the upper outer timing belt cover (B) for the front bank of cylinders. Remove the gaskets and keep them with the cover.

NOTE: The bolts holding the belt covers are of different lengths and must be correctly placed during reinstallation. Label or diagram them as they are removed.

13. Remove the lower portion of the engine support bracket. Remove the bolts in the order shown in the figure. Use spray lubricant to assist in removing the reamer bolt. Keep in mind that the reamer bolt may be heat-seized on the bracket. Remove it slowly.
14. Remove the small end cap from the rear bank belt cover, then remove the upper outer belt cover. Make certain all the gaskets are removed and placed with the covers.
15. Remove the left side splash shield. Turn the crankshaft

3-219

3 ENGINE AND ENGINE OVERHAUL

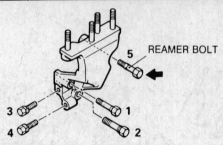

Remove the lower mount bolts in this order. Use spray lubricant on the reamer bolt

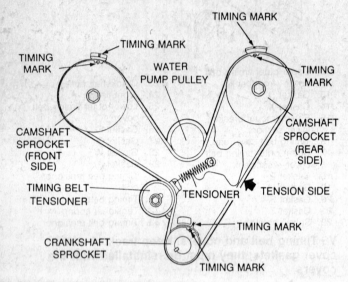

Aligning the timing marks

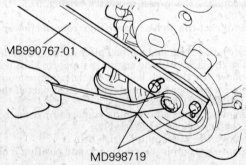

Use the special tools to hold the crankshaft pulley during removal and installation

clockwise until all the timing marks align. This will set the engine to TDC/compression on No. 1 cylinder.
16. Install the special counterholding tools and carefully remove the crankshaft pulley without moving the engine out of position.
17. Remove the front flange from the crank sprocket.
18. Remove the lower outer timing belt cover with its gaskets.
19. Loosen the timing belt tensioner bolt and turn the tensioner counterclockwise along the elongated hole. This will relax the tension on the belt.
20. If the timing belt is to be reused, make a chalk or crayon arrow on the belt showing the direction of rotation so that it may be reinstalled correctly.
21. Carefully slide the belt off the sprockets. Place the belt in a clean, dry, protected location away from the work area. If the tensioner is to be removed, disconnect the spring and remove the retaining bolt.
22. Inspect the timing belt in detail for any flaw or wear. If the belt is not virtually perfect, replace it. A case can be made for replacing the belt every time it is removed, particularly on high-mileage engines. Some of the conditions to look for are:
• Hardened back surface; non-elastic and glossy; hard to mark with a fingernail.
• Cracking on back of belt, bottom of teeth or side of belt.
• Missing teeth or teeth lifting from belt.
• Side of belt worn or fuzzy. Normal belt should have clean sides as if cut with a sharp knife.
• Wear on teeth as shown by distinct color change or worn rubber.
• Separation of inner coating from backing.
• Any uneven wear patterns on the teeth of the belt. Wear pattern should be even across each tooth and not differ from one tooth to another.
23. Check the sprockets and tensioner for wear. The sprocket teeth should be well defined, not rounded and the valleys between the teeth should be clean. The tensioners should spin freely with no binding or unusual noise. Replace the tensioner if

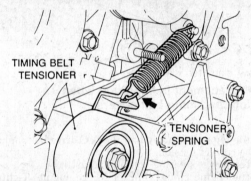

Make sure the spring is correctly installed or damage may result

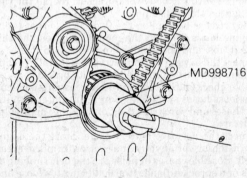

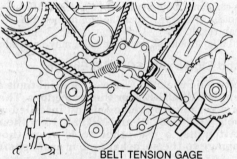

Special tools for turning the crankshaft and measuring belt tension

ENGINE AND ENGINE OVERHAUL

there is any sign of grease leaking from the seal. Clean everything with a clean, dry cloth.

WARNING: Do not spray or immerse the sprockets or tensioners in cleaning solvent. The sprocket may absorb the solvent and transfer it to the belt. The tensioners are internally lubricated and the solvent will dilute or dissolve the lubricant.

24. If the tensioner was removed, it must be reinstalled. After bolting it loosely in place, connect the spring onto the water pump pin. Make certain the spring faces in the correct direction on the tensioner. Turn the tensioner to the extreme counterclockwise position on the elongated hole and tighten the bolt just enough to hold the tensioner in this position.
25. Double check the alignment of the timing marks on the camshaft and crankshaft sprockets.
26. Observing the direction of rotation marks made earlier, install the belt onto the crankshaft sprocket and then onto the rear bank camshaft sprocket. Maintain tension on the belt between the sprockets.
27. Continue installing the belt onto the water pump, the front bank cam sprocket and the tensioner.
28. With your fingers, apply gentle counterclockwise force to the rear camshaft sprocket. When the belt is taut on the tension side, the timing marks should align perfectly.
29. Install the flange on the crankshaft sprocket.
30. Loosen the bolt holding the tensioner one or two turns and allow the spring tension to draw the tensioner against the belt.
31. Using special tool MB 998716 or equivalent adapter, turn the crankshaft two complete revolutions clockwise. Turn the crank smoothly and re-align the timing marks at the end of the second revolution. This allows the tensioner to compensate for the normal amount of slack in the belt.
32. With the timing marks aligned, tighten the tensioner bolts to 18 ft. lbs.
33. Using a belt tension gauge, measure the belt tension at a point halfway between the crankshaft sprocket and the rear camshaft sprocket. Correct value is 57-84 lbs.

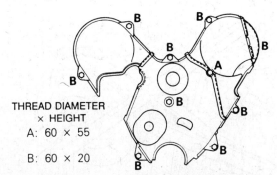

Correct bolt placement for V6 timing belt covers

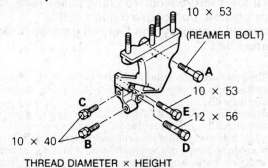

Installation order for lower engine mount bracket bolts

34. Making certain all the gaskets are correctly in place, install the lower timing belt cover and tighten the bolts to 8 ft. lbs.
35. Install the crankshaft pulley. Use the special tools to counterhold it and tighten the bolt to 112 ft. lbs.
36. Install the left splash shield.
37. Install the upper outer cover on the rear bank camshaft sprocket and install the smaller end cap. Tighten the bolts to 8 ft. lbs.
38. Install the lower portion of the engine support bracket. Install the bolts in the correct order. Of the 3 mounting bolts arranged vertically on the right side of the mount, the upper two are tightened to 50 ft. lbs; the bottom one is tightened to 78 ft. lbs. The two smaller bolts in the side of the mount should be tightened to 30 ft. lbs. Use spray lubricant to help with the reamer bolt, and tighten it slowly. Access may be easier if the engine is elevated slightly with the floor jack.
39. Install the cover and gaskets for the front bank camshaft sprocket.
40. Install the tensioner and bracket for the ribbed belt.
41. Install the power steering pump onto its bracket.
42. Install the ribbed belt and adjust it as needed. Make certain it is properly seated on all the pulleys.
43. Install the tensioner and bracket for the air conditioner drive belt and install the belt, adjusting it as necessary.
44. Install the upper section of the engine mounting bracket to the lower section. Tighten the nuts to 50 ft. lbs.
45. Adjust the jack so that the bushing of the engine mount aligns with the bodywork bracket. Install the through-bolt and tighten the nuts snug.
46. Carefully lower the jack, allowing the full weight of the engine to bear on the mount. Tighten the nut on the through-bolt to 50 ft. lbs. and the smaller safety nut to 26 ft. lbs.
47. Install the pressure hose to the power steering pump and tighten the nut to 32 ft. lbs.
48. Connect the wiring to the power steering pump and rebolt the bracket for the high pressure hose.
49. Double check all installation items, paying particular attention to loose hoses or hanging wires, untightened nuts, poor routing of hoses and wires (too tight or rubbing) and tools left in the engine area.
50. Connect the negative battery cable. Start the engine and let it idle, listening for any unusual noises from the area of the timing belt. Possible causes of noise are the belt rubbing against the covers or a sprocket flange, the belt being too loose and slapping, or a tensioner binding. Do not accelerate the engine if abnormal noises are heard from the timing belt train — severe damage can result.
51. Final adjustment of the drive belts may be needed.

Montero with V6 (6G72) Engine

WARNING: The use of the correct special tools or their equivalent is REQUIRED for this procedure. The crankshaft pulley must be removed and installed with the use of tools MB 998716-01 and MB 998747 or their equivalents. Attempts to use other tools or methods may cause damage to the pulley and/or crankshaft.

1. Disconnect the negative battery cable.
2. Drain the cooling system.

CAUTION

When draining the coolant, keep in mind that cats and dogs are attracted by the ethylene glycol antifreeze, and are quite likely to drink any that is left in an uncovered container or in puddles on the ground. This will prove fatal in sufficient quantity. Always drain the coolant into a sealable container. Coolant should be reused unless it is contaminated or several years old.

---

3. Disconnect the upper radiator hose at the radiator.
4. Remove the upper fan shroud from the radiator.
5. Remove the cooling fan and clutch assembly.

3-221

3 ENGINE AND ENGINE OVERHAUL

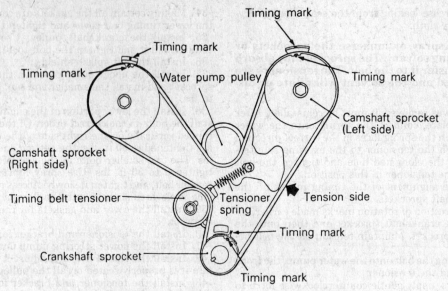

V6 Montero timing belt and sprocket layout

6. Loosen the necessary components and remove the drive belts from the air conditioning compressor, the power steering pump and the water pump/alternator.
7. Remove the fan pulley from the water pump.
8. Remove the power steering pump from the engine. Hang it out of the way from string or stiff wire. Do not disconnect any lines or hoses; just move the whole pump with the lines attached.
9. Remove the power steering pump bracket and mount.

WARNING: The bolts are of different lengths. Label or diagram each bolt and its location; correct reassembly is required.

10. Remove the tension pulley bracket (idler pulley) located just behind the power steering pump bracket.
11. Remove the air conditioning compressor. Hang it out of the way without kinking or twisting the lines.

WARNING: Do not disconnect any lines from the compressor. Do not allow the compressor to hang by the lines; support it securely.

12. Remove the compressor bracket.
13. Remove the cooling fan bracket assembly.
14. Remove the upper timing belt covers and their gaskets; keep the gaskets with the covers.

WARNING: Bolts are of three different lengths; label or diagram their location during removal.

15. Remove the lower timing belt cover.
16. Use a wrench on the crankshaft bolt to turn the engine clockwise until all the timing marks align. This positions the engine at TDC/compression for No. 1 piston. Once positioned, the engine must not be moved out of place.
17. Install the special counterholding tools and remove the crankshaft pulley. The large center bolt will be tight; do not turn the motor during removal.
18. Remove the front flange from the crankshaft sprocket.
19. Loosen the timing belt tensioner bolt and turn the timing belt tensioner counterclockwise along the elongated hole; this will relax the belt tension.
20. If the timing belt is to be reused, mark the direction of rotation on the belt with chalk or crayon. The belt must be reinstalled in its original position.
21. Carefully slide the belt off the sprockets. Place the belt in a

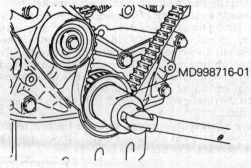
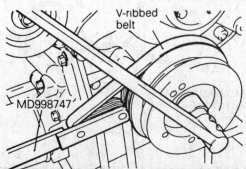

Special tools for V6 Montero crank pulley removal and installation. Do not use a good belt to counterhold the pulley; the tool will damage it. Use a discarded belt

clean, dry, protected location away from the work area. if the tensioner is to be removed, disconnect the spring and remove the retaining bolt.

22. Inspect the timing belt in detail for any flaw or wear. If the belt is not virtually perfect, replace it. A case can be made for replacing the belt every time it is removed, particularly on high-mileage engines. Some of the conditions to look for are:
- Hardened back surface; non-elastic and glossy; hard to mark with a fingernail.
- Cracking on back of belt, bottom of teeth or side of belt.
- Missing teeth or teeth lifting from belt.

ENGINE AND ENGINE OVERHAUL 3

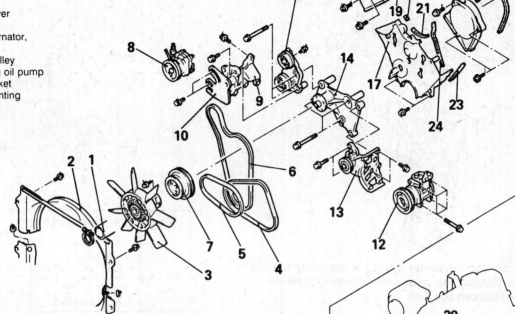

1. Connection for radiator upper hose
2. Radiator upper shroud
3. Cooling fan clutch assembly
4. Drive belt (air conditioner)
5. Drive belt (power steering)
6. Drive belt (alternator, water pump)
7. Cooling fan pulley
8. Power steering oil pump
9. Oil pump bracket
10. Oil pump mounting bracket
11. Tension pulley bracket
12. Compressor
13. Compressor bracket
14. Cooling fan bracket assembly
15. Timing belt upper cover outer (B)
16. Timing belt upper cover outer (A)
17. Timing belt lower cover outer
18. Gasket K
19. Gasket J
20. Gasket L
21. Gasket I
22. Gasket J
23. Gasket G
24. Gasket H
25. Crankshaft pulley
26. Front flange
27. Timing belt tensioner bolt
28. Timing belt
29. Tensioner spring
30. Timing belt tensioner

Montero V6 timing belt and related components

• Side of belt worn or fuzzy. Normal belt should have clean sides as if cut with a sharp knife.
• Wear on teeth as shown by distinct color change or worn rubber.
• Separation of inner coating from backing.
• Any uneven wear patterns on the teeth of the belt. Wear pattern should be even across each tooth and not differ from one tooth to another.

23. Check the sprockets and tensioner for wear. The sprocket teeth should be well defined, not rounded and the valleys between the teeth should be clean. The tensioners should spin freely with no binding or unusual noise. Replace the tensioner if there is any sign of grease leaking from the seal. Clean everything with a clean, dry cloth.

WARNING: Do not spray or immerse the sprockets or tensioners in cleaning solvent. The sprocket may absorb the solvent and transfer it to the belt. The tensioners are internally lubricated and the solvent will dilute or dissolve the lubricant.

24. If the tensioner was removed, it must be reinstalled. After bolting it loosely in place, connect the spring onto the pin. Make certain the spring faces in the correct direction on the tensioner. Turn the tensioner to the extreme counterclockwise position on the elongated hole and tighten the bolt just enough to hold the tensioner in this position.
25. Double check the alignment of the timing marks on the camshaft and crankshaft sprockets.
26. Observing the direction of rotation marks made earlier, install the belt onto the crankshaft sprocket and then onto the left bank camshaft sprocket. Maintain tension on the belt between the sprockets.
27. Continue installing the belt onto the water pump, the right bank cam sprocket and the tensioner.
28. With your fingers, apply gentle counterclockwise force to the left camshaft sprocket. When the belt is taut on the tension side, the timing marks should align perfectly.
29. Install the flange on the crankshaft sprocket.
30. Loosen the bolt holding the tensioner one or two turns and allow the spring tension to draw the tensioner against the belt.

3-223

3 ENGINE AND ENGINE OVERHAUL

Make certain the tensioner spring is correctly placed (A-upper). Moving the tensioner counterclockwise (lower) will loosen the belt

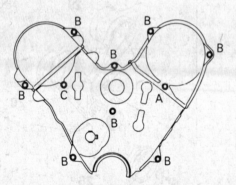

Thread diameter x height mm
A : 60 x 60
B : 60 x 20
C : 60 x 55

Correct bolt location by length, Montero V6 timing belt covers

31. Using special tool MB 998716–01 or equivalent adapter, turn the crankshaft two complete revolutions clockwise. Turn the crank smoothly and re–align the timing marks at the end of the second revolution. This allows the tensioner to compensate for the normal amount of slack in the belt.
32. With the timing marks aligned, tighten the tensioner bolt to 25 Nm (18 ft. lbs).
33. Install the crankshaft pulley. Use the counterholding tool to maintain the engine position and tighten the center bolt to 145 Nm or 112 ft. lbs.
34. Inspect each gasket for the timing covers. The gaskets should be clean and pliable. Replace any which are distorted, cracked or broken. Coat the channel in each timing cover with a

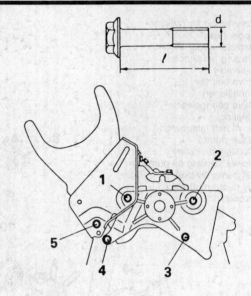

No.	d x l mm (in.)	Torque Nm (ft.lbs.)
1	10 x 95 (.39 x 3.74)	33–50 (24–36)
2	10 x 95 (.39 x 3.74)	
3	10 x 110 (.39 x 4.33)	
4	12 x 100 (.47 x 3.93)	65–85 (47–61)
5	10 x 85 (.39 x 3.34)	33–50 (24–36)

Bolt length and torque value for Montero V6 fan bracket and tension pulley bracket

light coat of adhesive such as 3M® EC 870 or equivalent. Press the seals squarely into their channels.
35. Install the lower timing cover. Then install the upper timing covers, making sure each is properly seated. Tighten the bolts to 11 Nm or 8 ft. lbs.
36. Install the cooling fan bracket.
37. Install the compressor bracket and install the compressor.
38. Install the idler pulley bracket assembly.
39. Install the mount and bracket for the power steering pump. Tighten the small bolts to 40 Nm (30 ft. lbs).
40. Install the power steering pump.
41. Install the cooling fan pulley.
42. Install the drive belts: alternator/water pump, power steering and A/C compressor in that order. Adjust each belt to the correct tension.
43. Install the fan and clutch assembly, tightening the nuts to 11 Nm or 8 ft. lbs.
44. Install the upper radiator shroud. Tighten the upper bolts to only 5 Nm (4 ft. lbs.) and the lower bolts to 9 Nm (7 ft. lbs.)
45. Connect the upper radiator hose.
46. Fill the cooling system with coolant.
47. Double check all installation items, paying particular attention to loose hoses or hanging wires, untightened nuts, poor routing of hoses and wires (too tight or rubbing) and tools left in the engine area.
48. Connect the negative battery cable. Start the engine and let it idle, listening for any unusual noises from the area of the

ENGINE AND ENGINE OVERHAUL

timing belt. Possible causes of noise are the belt rubbing against the covers or a sprocket flange, the belt being too loose and slapping, or a tensioner binding. Do not accelerate the engine if abnormal noises are heard from the timing belt train — severe damage can result.

49. Final adjustment of the drive belts may be needed.

Galant with 4G63 SOHC Engine

1. Disconnect the negative battery cable. Remove the bolt holding the two hoses to the top of the left engine mount.
2. Remove the splash shield under the engine.
3. Place a floor jack and a broad piece of lumber under the oil pan. Elevate the jack just enough to support the engine without raising it.
4. Remove the nuts holding the through-bolt for the left engine bracket and remove the bolt. It may be necessary to adjust the jack tension slightly to allow the bolt to come free. Use the jack to keep the engine in its normal position while the mount is disconnected.
5. Remove the nuts and bolts holding the engine bracket to the engine and remove the bracket.
6. Loosen the appropriate component or adjuster and remove (in order) the air conditioning compressor drive belt, the power steering belt and the alternator belt. Remove the air conditioning belt tension adjuster and bracket.
7. Remove the power steering and water pump pulleys.
8. Remove the upper timing belt cover and its gaskets. Rotate the crankshaft clockwise until the timing marks on the crank pulley align at 0 on the scale (on the lower timing cover) and the mark on the camshaft pulley aligns with the indicator on the head. Aligned means perfect, not close — if you miss, continue clockwise until the marks are matched. This sets the motor at TDC/compression on No.1 cylinder.

WARNING: Once the engine is set to this position, the crank and camshafts must not be moved out of place. Severe damage can result.

9. Remove the air conditioner pulley from the crankshaft and remove the crankshaft pulley. Remember that the pulley (and therefore, the crankshaft) must be held steady while the bolt is

14. Crankshaft sprocket bolt
15. Special washer
16. Tensioner spacer
17. Tensioner spring
18. Timing belt tensioner
19. Spacer
20. Timing belt
21. Camshaft sprocket
22. Oil Pump sprocket
23. Crankshaft sprocket
24. Flange
25. Tensioner B
26. Timing belt B (silent shaft belt)
27. Right silent shaft sprocket
28. Spacer
29. Crankshaft sprocket B
30. Key
31. Timing belt rear cover

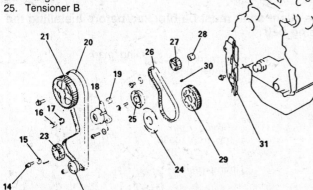

4G63 SOHC timing belts and sprockets

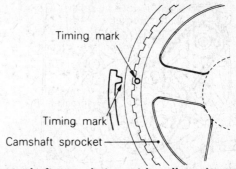

The camshaft sprocket must be aligned exactly

loosened. This can be made easier by wrapping a scrapped belt tightly around the pulley. Mitsubishi has a special tool (MD 998747) for this trick, but other variations work almost as well. DO NOT use any of the drive belts which will be reinstalled; the belt used to hold the pulley will be damaged or weakened.

10. Remove the lower timing belt cover and gaskets. Note the different bolt lengths and placements.
11. Loosen (do not remove) both bolts in the timing belt tensioner. Move the tensioner toward the water pump and tighten the lock bolt — the one in the slotted hole — to hold the tensioner in place.
12. Before removing the timing belt, mark it with an arrow to show the direction of rotation. If the belt is to be reused, it must be reinstalled so that it rotates in the same direction as before.
13. Remove the timing belt, keeping it free of grease, oil and fluids. Do not crease or crimp the belt.
14. Remove the crankshaft sprocket and the flange.
15. Loosen the adjuster for the silent shaft belt (timing belt B). Carefully remove the belt. The sprockets and tensioners may be removed if desired.

WARNING: If the oil pump sprocket is to be removed, remove the plug in the left side of the engine and insert a Phillips screwdriver to hold the silent shaft in place. The screwdriver must enter at least 50mm (2 in.) to hold the shaft. If the screwdriver enters less than 25mm (1 inch), turn the sprocket one rotation and realign the timing marks.

16. Inspect the timing belts in detail for any flaw or wear. If the belt is not virtually perfect, replace it. A case can be made for replacing the belt every time it is removed, particularly on high-mileage engines. Some of the conditions to look for are:
 • Hardened back surface; non-elastic and glossy; hard to mark with a fingernail.
 • Cracking on back of belt, bottom of teeth or side of belt.
 • Missing teeth or teeth lifting from belt.
 • Side of belt worn or fuzzy. Normal belt should have clean sides as if cut with a sharp knife.
 • Wear on teeth as shown by distinct color change or worn rubber.
 • Separation of inner coating from backing.
 • Any uneven wear patterns on the teeth of the belt. Wear pattern should be even across each tooth and not differ from one tooth to another.

NOTE: Please refer to the illustration under Cordia/Tredia at the beginning of this section for samples of timing belt faults.

17. Check the sprockets and tensioner for wear. The sprocket teeth should be well defined, not rounded and the valleys between the teeth should be clean. The tensioners should spin freely with no binding or unusual noise. Replace the tensioner if there is any sign of grease leaking from the seal. Clean everything with a clean, dry cloth.

3-225

3 ENGINE AND ENGINE OVERHAUL

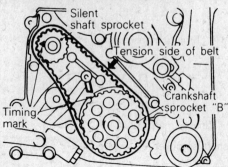

Align the timing marks first, and keep the tension side taut when installing the silent shaft belt

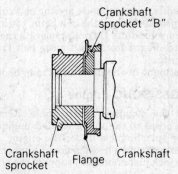

Correct installation of crankshaft sprockets and flange

WARNING: Do not spray or immerse the sprockets or tensioners in cleaning solvent. The sprocket may absorb the solvent and transfer it to the belt. The tensioners are internally lubricated and the solvent will dilute or dissolve the lubricant.

18. To reinstall, make certain the crankshaft sprocket and the silent shaft sprocket are aligned with the timing marks. Slip the belt over the crank sprocket and then over the silent shaft sprocket, keeping tension on the upper side. Make certain there is no slack in the tension side of the belt.
19. Install the tensioner and temporarily position it so that the center of the pulley is to the left and above the center of the installation bolt.
20. Apply finger pressure on the tensioner, forcing it into the belt. Make certain the opposite side of the belt is taut. Tighten the installation bolt to hold the tensioner in place.
21. Push down on the center of the tension side of the belt (side opposite the tensioner pulley) and check the distance the belt deflects. Correct deflection is 6mm (¼ inch).
22. Install the flange on the crankshaft sprocket, making sure it is positioned correctly.

23. Install the outer crankshaft sprocket and/or the oil pump sprocket if either was removed.
24. Install the timing belt tensioner if it was removed and position it towards the water pump. Tighten the lock bolt to hold it in this position.
25. Make certain that all the timing marks for the camshaft sprocket, the crankshaft sprocket and the oil pump sprocket are correctly aligned.
26. Once the oil pump sprocket is aligned, remove the plug in the left side of the block. Insert a Phillips screwdriver with a shaft diameter of 8mm (0.31 in.) into the plug hole. It should enter 58mm (2.3 in.) or more. Do not remove the screwdriver until the timing belt is completely installed. If the screwdriver shaft only enters about 25mm (1 inch), turn the oil pump sprocket one rotation and align the timing mark again; insert the screwdriver to block the silent shaft from turning.
27. Observe the directional arrow made during removal and install the timing belt by placing on the crankshaft sprocket, then the oil pump sprocket and then over the camshaft. Maintain the belt taut on the tension side while installing it.
28. Loosen the adjuster lock bolt and allow the spring to move

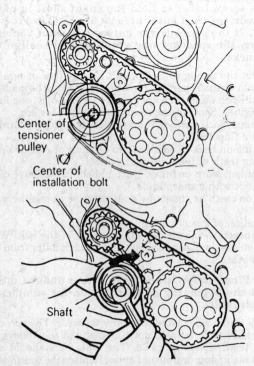

Position the tensioner properly, then move it inward to tighten the belt

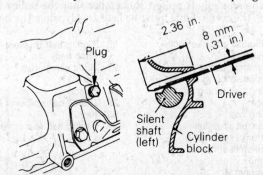

The silent shaft must be blocked before installing the timing belt

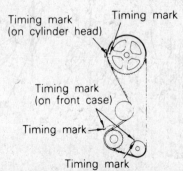

Aligning the 4G63 SOHC timing marks

ENGINE AND ENGINE OVERHAUL 3

the pulley against the belt. Do not push or pry on the adjuster.

29. Recheck that each sprocket is still aligned with its timing marks.

30. Remove the screwdriver from the silent shaft. Rotate the crankshaft clockwise until the camshaft sprocket has moved 2 teeth away from its mark. This applies tension to the belt. Do NOT rotate the crank counter-clockwise and do not push on the belt to check the tension.

31. With your fingers, apply gentle upward pressure on the tensioner and check the placement of the belt on the cam sprocket. The belt may tend to lift or float in the left side of the sprocket — roughly between 7 o'clock and 12 o'clock as you view the sprocket.

32. When the belt is correctly seated, tighten the tensioner lock bolt to 36 ft. lbs, then tighten the pivot nut to 36 ft. lbs. The order of tightening is important; if not followed, the belt could be over-tightened.

33. Check the deflection of the belt. At the middle of the right (tension) side of the belt, deflect the belt outward with your finger, toward the timing case. The distance between the belt and the line of the cover seal should be about 13mm (½ inch).

34. Tighten the crankshaft sprocket bolt to 88 ft. lbs.

35. Install the lower timing belt cover, paying attention to the correct placement of the bolts. Tighten the bolts to 8 ft. lbs.

36. Install the upper timing belt cover and gaskets, tightening the bolts to 8 ft. lbs. Remember to install the guides for the spark plug wires.

37. Reinstall the spark plug wires and connect them.

38. Install the crankshaft damper pulley. Tighten the bolts to 18 ft. lbs.

39. Install the water pump pulleys and tighten the bolts to 6 ft. lbs.

40. Install the tensioner pulley and the A/C drive belt; adjust the belt to the correct tension.

41. Install the power steering belt and adjust it.

42. Install the engine mount bracket to the engine. Tighten the mounting nuts and bolts to 42 ft. lbs.

43. Adjust the jack (if necessary) so that the engine mount bushing aligns with the bodywork bracket. Install the through-bolt and tighten the nuts snug.

44. Slowly release tension on the floor jack so that the weight of the engine bears fully on the mount. Tighten the through-bolt to 52 ft. lbs. and the small safety nut to 26 ft. lbs.

45. Position the high pressure hose and secure its bracket.

46. Double check all installation items, paying particular attention to loose hoses or hanging wires, untightened nuts, poor routing of hoses and wires (too tight or rubbing) and tools left in the engine area.

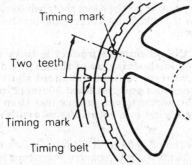

Rotating the engine until the cam sprocket moves 2 teeth places the correct pressure on the belt and tensioner

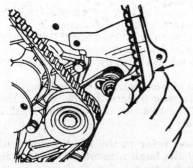

The pinch test checks proper belt deflection. Look for about 6mm (¼ in.) between the arrows

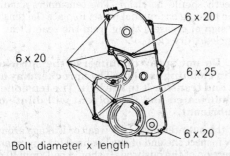

Correct placement of the timing belt cover bolts

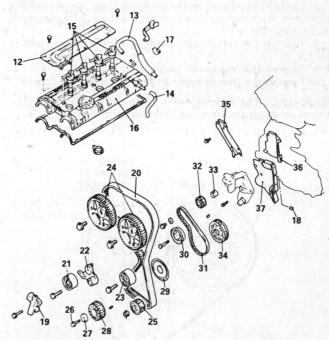

12. Center cover
13. Breather hose
14. PCV hose
15. Spark plug cables
16. Valve cover
17. Half-circle plug
18. Access plug
19. Auto tensioner
20. Timing belt
21. Tensioner pulley
22. Tensioner arm
23. Idler pulley
24. Camshaft sprocket
25. Oil pump sprocket
26. Crankshaft sprocket bolt
27. Special washer
28. Crankshaft sprocket
29. Flange
30. Tensioner B
31. Timing belt B (silent shaft belt)
32. Silent shaft sprocket
33. Spacer
34. Crankshaft sprocket B
35. Timing belt right rear cover
36. Timing belt left rear upper cover
37. Timing belt left rear lower cover

4G63 DOHC timing belts and sprockets

3-227

3 ENGINE AND ENGINE OVERHAUL

47. Connect the negative battery cable. Start the engine and let it idle, listening for any unusual noises from the area of the timing belt. Possible causes of noise are the belt rubbing against the covers or a sprocket flange, the belt being too loose and slapping, or a tensioner binding. Do not accelerate the engine if abnormal noises are heard from the timing belt train — severe damage can result.
48. Final adjustment of the drive belts may be needed. Reinstall the splash shield.

Galant with 4G63 DOHC Engine

1. Disconnect the negative battery cable.
2. Remove the left splash shield under the engine. Remove the bracket bolt holding the two hoses to the top of the left engine mount.
3. Place a floor jack and a broad piece of lumber under the oil pan. Elevate the jack just enough to support the engine without raising it.
4. Remove the nuts holding the through-bolt for the left engine bracket and remove the bolt. It may be necessary to adjust the jack tension slightly to allow the bolt to come free. Use the jack to keep the engine in its normal position while the mount is disconnected.
5. Remove the nuts and bolts holding the engine bracket to the engine and remove the bracket.
6. Remove the alternator belt and the power steering belt.
7. Loosen the tension on the air conditioner compressor belt tensioner. With the belt loosened, remove the adjuster assembly and remove the belt.
8. Remove the water pump pulley with the power steering pulley.
9. Remove the crankshaft pulley.
10. Remove the (front) upper and lower timing belt covers.

WARNING: The covers use 4 different lengths of bolts. Label or diagram them as they are removed.

11. Remove the center cover, the PCV hose and the breather hose from the valve cover. Label and disconnect the spark plug cables from the spark plugs.
12. Remove the valve cover and its gasket. Remove the half-circle plug from the head.
13. Turn the crankshaft clockwise and align all the timing marks. The timing marks on the camshaft sprockets should align with the upper surface of the cylinder head and the dowel pins (guide pins) at the center of the camshaft sprockets should be up. Aligned means perfect, not close — if you miss, continue clockwise until the marks are matched. This sets the motor at TDC/compression on No.1 cylinder.

WARNING: Once the engine is set to this position, the crank and camshafts must not be moved out of place. Severe damage can result.

14. Remove the auto-tensioner.
15. Before removing the timing belt, mark it with an arrow to show the direction of rotation. If the belt is to be reused, it must be reinstalled so that it rotates in the same direction as before.
16. Remove the timing belt, keeping it free of grease, oil and fluids. Do not crease or crimp the belt. After the belt is clear, remove the tensioner pulley.
17. Remove the crankshaft sprocket and the flange.
18. Loosen the adjuster for the silent shaft belt (timing belt B). Carefully remove the belt. The sprockets and tensioners may be removed if desired.

WARNING: If the oil pump sprocket is to be removed, remove the plug in the left side of the engine and insert a Phillips screwdriver to hold the silent shaft in place. The screwdriver must enter at least 50mm (2 in.) to hold the shaft. If the screwdriver enters less than 25mm (1 in.), turn the sprocket one rotation and realign the timing marks.

19. Inspect the timing belts in detail for any flaw or wear. If the belt is not virtually perfect, replace it. A case can be made for replacing the belt every time it is removed, particularly on high-mileage engines. Some of the conditions to look for are:
   • Hardened back surface; non-elastic and glossy; hard to mark with a fingernail.
   • Cracking on back of belt, bottom of teeth or side of belt.
   • Missing teeth or teeth lifting from belt.
   • Side of belt worn or fuzzy. Normal belt should have clean sides as if cut with a sharp knife.
   • Wear on teeth as shown by distinct color change or worn rubber.
   • Separation of inner coating from backing.
   • Any uneven wear patterns on the teeth of the belt. Wear pattern should be even across each tooth and not differ from one tooth to another.

NOTE: Please refer to the illustration under Cordia and Tredia at the beginning of this section for samples of timing belt faults.

20. Check the sprockets and tensioner for wear. The sprocket teeth should be well defined, not rounded and the valleys between the teeth should be clean. The tensioners should spin freely with no binding or unusual noise. Replace the tensioner if there is any sign of grease leaking from the seal. Clean everything with a clean, dry cloth.

WARNING: Do not spray or immerse the sprockets or tensioners in cleaning solvent. The sprocket may absorb the solvent and transfer it to the belt. The tensioners are internally lubricated and the solvent will dilute or dissolve the lubricant.

21. Inspect the auto-tensioner for wear or leakage around the seals. Closely inspect the end of the pushrod for any wear. Measure the projection of the pushrod. It should be 12mm (0.47 in.). If it is out of specification, the auto tensioner must be replaced.

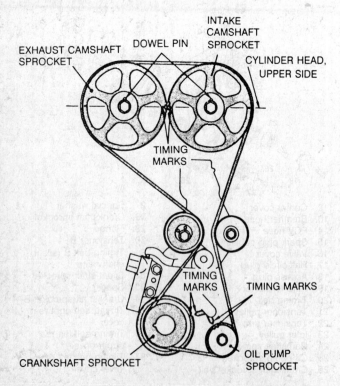

Align all the timing marks before removing the belt

ENGINE AND ENGINE OVERHAUL 3

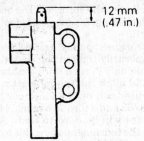

Correct projection for the tip of the auto-tensioner

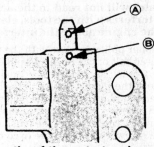

Compress the tip of the auto-tensioner until holes "A" and "B" align. Insert a piece of stiff wire to keep the tip retracted

22. Mount the auto-tensioner in a vise with protected jaws. Keep the unit level at all times, and, if the plug at the bottom projects, protect it with a common washer. Smoothly tighten the vise to compress the adjuster tip; if the the rod is easily retracted, the unit has lost tension and should be replaced. You should feel a fair amount of resistance when compressing the rod.
23. As the rod is contracted into the body of the auto-tensioner, watch for the hole in the pushrod to align with the hole in the body. When this occurs, pin the pushrod in place with a piece of stiff wire 1.5mm. It will be easier to install the tensioner with the pushrod retracted. Leave the wire in place and remove the assembly from the vise.
24. To reinstall, make certain the crankshaft sprocket and the silent shaft sprocket are aligned with the timing marks. Slip the belt over the crank sprocket and then over the silent shaft sprocket, keeping tension on the upper side. Make certain there is no slack in the tension side of the belt.
25. Install the tensioner and temporarily position it so that the center of the pulley is to the left and above the center of the installation bolt.
26. Apply finger pressure on the tensioner, forcing it into the belt. Make certain the opposite side of the belt is taut. Tighten the installation bolt to hold the tensioner in place.
27. Push down on the center of the tension side of the belt (side opposite the tensioner pulley) and check the distance the belt deflects. Correct deflection is 6mm (¼ inch).
28. Install the flange on the crankshaft sprocket, making sure it is positioned correctly.
29. Install the outer crankshaft sprocket and/or the oil pump sprocket if either was removed.
30. Install the auto-tensioner onto the engine, tightening the bolts to 18 ft. lbs.
31. Install the tensioner pulley onto the tensioner arm. Locate the pinholes in the pulley shaft to the left of the center bolt. Tighten the center bolt finger tight. Leave the blocking wire in the auto-tensioner.
32. Double check the alignment of the timing marks for all the sprockets. The camshaft sprocket marks must face each other and align with the top surface of the cylinder head. Don't forget

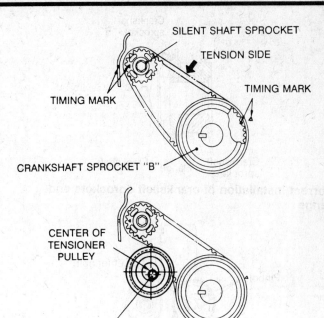

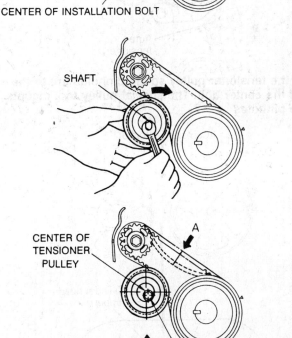

4 steps for installing the silent shaft belt. Check belt deflection at point A

to check the oil pump sprocket alignment as well as the crankshaft sprocket.

NOTE: When you let go of the exhaust camshaft sprocket, it will move one tooth counter-clockwise. Take this into account when installing the belt around the sprockets.

33. Observing the direction of rotation mark made earlier, install the timing belt as follows:
    a. Install the timing belt around the tensioner pulley and the crankshaft sprocket and hold the belt on the tensioner with your left hand.

3-229

3 ENGINE AND ENGINE OVERHAUL

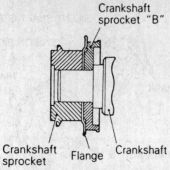

Correct installation of crankshaft sprockets and flange

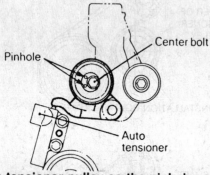

Align the tensioner pulley so the pinholes are to the left of the center bolt. The special pulley tool mounts to the pinholes

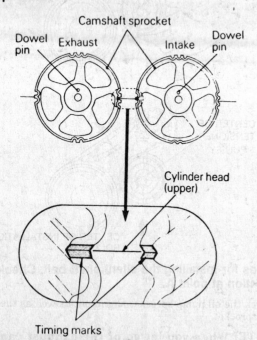

Aligning the camshaft pulleys

and install the belt over the sprocket. The belt is a snug fit but can be fitted with the fingers; do not use tools to force the belt onto the sprocket.

e. Raise the tensioner pulley against the belt to prevent sagging and temporarily tighten the center bolt.

34. Turn the crankshaft ¼ turn (90°) counter-clockwise, then turn it clockwise and align the timing marks.

35. Loosen the center bolt on the tensioner pulley and attach special tool MD 998752 and a torque wrench. (This is a purpose-built tool; substitutes will be hard to find.) Apply a torque of 1.88–2.03 ft. lbs. to the tensioner pulley; this establishes the correct loading against the belt.

NOTE: You must use a torque wrench with a 0–3 ft. lb. scale. Normal scales will not read to the accuracy needed. If the body interferes with the tools, elevate the jack slightly to raise the engine above the interference point.

36. Hold the tensioner pulley in place with the torque wrench and tighten the center bolt of the tensioner pulley to 36 ft. lbs. Remove the torque wrench assembly.

37. On the left side of the engine, located in the rear lower timing belt cover, is a rubber plug. Remove the plug and install special tool MD 998738 (another one difficult to substitute). Screw the tool in until it makes contact with the tensioner arm. After the point of contact, screw it in some more to compress the arm slightly.

38. Remove the wire holding the tip of the auto-tensioner. The tip will push outward and engage the tensioner arm. Remove the screw tool.

39. Rotate the crankshaft 2 complete turns clockwise. Allow everything to sit as is for about 15 minutes — longer if the temperature is cool — then check the distance the tip of the auto-tensioner protrudes. (It is almost impossible with the engine in the car; if you can measure the distance, it should be 3.8-4.5mm (0.15–0.18 in.). The alternate method is given in the next step.)

40. Reinstall the special screw tool into the left support bracket until the end just makes contact with the tensioner arm. Starting from that position, turn the tool against the arm, counting the number of full turns until the tensioner arm contacts the auto-tensioner body. At this point, the arm has compressed the tensioner tip through its projection distance. The correct number of turns to this point is 2.5 to 3 turns. If you took the correct number of turns to bottom the arm, the projection was correct to begin with and you may now unwind the screw tool EXACTLY the number of turns you turned it in.

If your number of turns was too high or too low when the arm bottomed, you have a problem with the tensioner and should replace it.

41. Remove the special screw tool and install the rubber plug into the access hole in the timing belt cover.

42. Install the half-circle plug with the proper sealant, then install the valve cover and gasket.

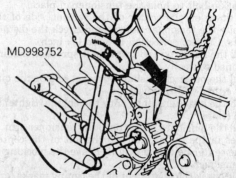

The special adapter is required to properly set the pulley tension

b. Pulling the belt with your right hand, install it around the oil pump sprocket.

c. Install the belt around the idler pulley and then around the intake camshaft sprocket.

d. Double check the exhaust camshaft sprocket alignment mark. It has probably moved one tooth; if so, rotate it clockwise to align the timing mark with the top of the cylinder head

ENGINE AND ENGINE OVERHAUL 3

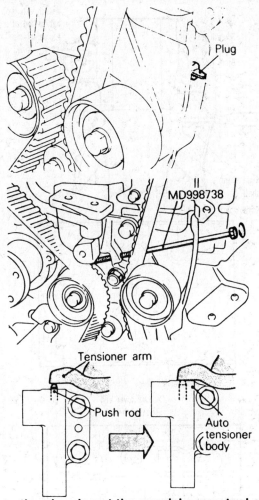

Remove the plug, insert the special screw tool and compress the tip of the auto-tensioner pushrod

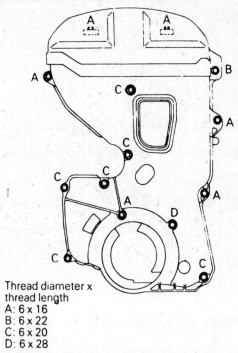

Bolt placement for the timing belt covers. Measurements are in millimeters

Thread diameter x thread length
A: 6 x 16
B: 6 x 22
C: 6 x 20
D: 6 x 28

43. Install the spark plug wires, the breather and PCV hoses and the center cover.
44. Install the lower and upper timing belt covers, paying attention to the correct placement of the four different sizes of bolts.
45. Install the crankshaft pulley, tightening the bolts to 18 ft. lbs.
46. Install the water pump pulley(s) and tighten the bolts to 6 ft. lbs.
47. Install air conditioning drive belt and install the tensioner bracket. Tighten the bracket bolts to 18 ft. lbs.
48. Install the power steering belt and the alternator belt.
49. Install the engine mount bracket to the engine. Tighten the mounting nuts and bolts to 42 ft. lbs.
50. Adjust the jack (if necessary) so that the engine mount bushing aligns with the bodywork bracket. Install the through-bolt and tighten the nuts snug.
51. Slowly release tension on the floor jack so that the weight of the engine bears fully on the mount. Tighten the through-bolt to 72 ft. lbs. and the small safety nut to 38 ft. lbs. Don't forget to install the small support bracket for the engine mount. Tighten its bolt and nut to 16 ft. lbs.
52. Double check all installation items, paying particular attention to loose hoses or hanging wires, untightened nuts, poor routing of hoses and wires (too tight or rubbing) and tools left in the engine area.
53. Connect the negative battery cable.
54. Start the engine and let it idle, listening for any unusual noises from the area of the timing belt. Possible causes of noise are the belt rubbing against the covers or a sprocket flange, the belt being too loose and slapping, or a tensioner binding. Do not accelerate the engine if abnormal noises are heard from the timing belt train — severe damage can result.
55. Reinstall the splash shield under the car. Final adjustment of the drive belts may be needed.

Pickup with the 4D55 Diesel Engine

1. Rotate the engine by hand (clockwise) until all the timing marks align. This positions No. 1 cylinder at TDC/compression.
2. Remove the upper timing belt cover. Note that the bolts are of different lengths; diagram their location for proper reinstallation.
3. If either or both of the timing belts are to be reused, mark the direction of rotation on the belt with chalk or a felt marker.
4. Remove the center bolt holding the crankshaft pulley and remove the pulley. Do not turn the engine out of position during this removal.
5. Remove the lower timing belt cover.
6. Slightly loosen the retaining (lock) bolts for the timing belt tensioner. Move the tensioner towards the water pump and tighten the bolts; this holds the tensioner in the slack position.

NOTE: If the silent shaft timing belt is not being removed, DO NOT loosen the tensioner for the silent shaft belt.

7. Remove the camshaft sprocket and the injection pump sprocket. Do not let either shaft turn out of place during the removal. Remove the timing belt without forcing it or kinking it.
8. If the silent shaft belt is to be removed, the silent shafts must be locked in place. On the right side, remove the rubber plug from the silent shaft case and insert a screwdriver or punch; the shaft should have a diameter of 8mm or 0.32 in. On the left side, remove the cover panel from the silent shaft case and insert a broad round bar; a short extension for a socket wrench is an ideal tool.

3 ENGINE AND ENGINE OVERHAUL

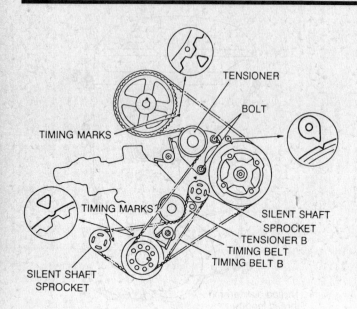

Timing marks and timing belt routing, 4D55

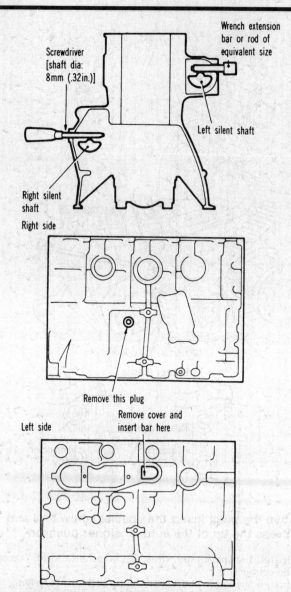

The silent shafts must be held in position

9. Remove the silent shaft sprockets; label each sprocket so it may be reinstalled in its original position. Remove the timing belt.
10. Check the belt carefully for damage, wear or deterioration. Check the teeth carefully for any cracking at the base or separation from the backing. Do not use detergents or solvents to clean the belt or the sprockets. Clean them only with dry cloths or brushes. Pay particular attention to the edges of the belt; they should not be frayed or worn.
11. To reinstall, install the crankshaft sprocket for the silent shaft belt, the flange and the timing belt sprocket. Make certain the sprockets and flange are installed in the correct positions; if either sprocket is installed backwards, the belt may jump off or become damaged.
12. Install the crankshaft pulley; tighten the bolt to 70 Nm (52 ft. lbs.).
13. Install the injection pump sprocket and flange.
14. Install the spacer on the left silent shaft with its tapered end inward or toward the oil seal. If the spacer is not placed correctly, the oil seal will be damaged.
15. Install the silent shaft sprockets and flanges, tightening the bolts to 38 Nm (28 ft. lbs.). The shafts must be held in place with the tools used during disassembly.
16. The tensioner for the silent shaft belt should still be in the slack position. Align the timing mark of the crankshaft sprocket if necessary.
17. Align the timing marks of each silent shaft sprocket.
18. Install the silent shaft timing belt (belt B), making certain that there is no slack between the crankshaft pulley and the right (upper) silent shaft. This side of the belt is called the tension side.

WARNING: If the belt is being reused, it must be reinstalled in the same direction of rotation. Observe the marks made during disassembly.

19. Use a finger to push down on the slack side of the belt—between the tensioner and the right side silent shaft—and check that the timing marks on the sprockets remain in alignment.
20. Keep your finger on the belt. Loosen the mounting nut and bolt on the tensioner and allow the tensioner to move against the belt under its own spring tension.

WARNING: Do not assist the tensioner by pushing on it.

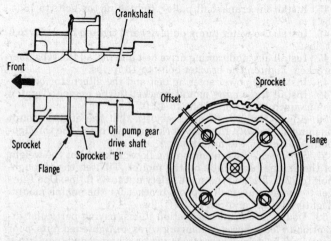

The crankshaft sprockets and flange must be correctly installed

ENGINE AND ENGINE OVERHAUL 3

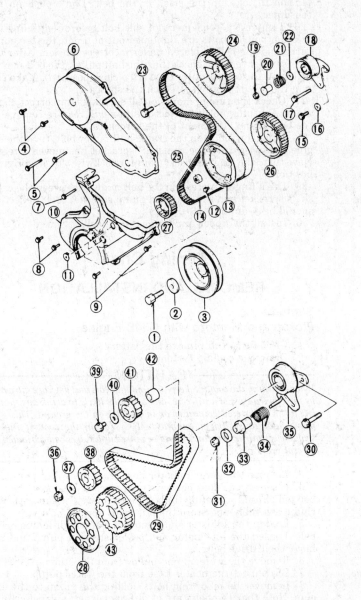

1. Crank pulley bolt
2. Special washer
3. Dumper pulley
4. Flange bolt (2)
5. Flange bolt (2)
6. Timing belt front upper cover
7. Flange bolt
8. Flange bolt (2)
9. Flange bolt (3)
10. Timing belt front lower cover
11. Access cover
12. Flange bolt (4)
13. Flange
14. Timing belt
15. Flange bolt
16. Washer
17. Flange bolt
18. Timing belt tensioner assembly
19. Flange nut
20. Tensioner spacer
21. Tensioner spring
22. Plain washer
23. Flange bolt
24. Camshaft sprocket
25. Nut
26. Injection pump sprocket
27. Crankshaft sprocket
28. Flange
29. Timing belt "B"
30. Flange bolt
31. Flange nut
32. Gasket
33. Tensioner spacer
34. Tensioner spring "B"
35. Timing belt tensioner assembly "B"
36. Flange nut
37. Washer
38. Counterbalance shaft sprocket
39. Flange bolt
40. Washer
41. Counterbalance shaft sprocket
42. Spacer
43. Crankshaft sprocket "B"

4D55 timing belts, sprockets and related components

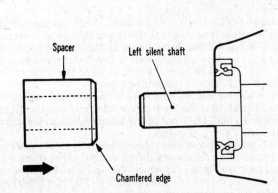

Install the spacer correctly

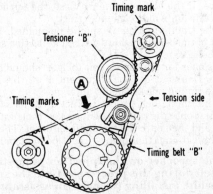

Arrangement of silent shaft timing belt. Press at point "A" when tensioning the belt

3-233

3 ENGINE AND ENGINE OVERHAUL

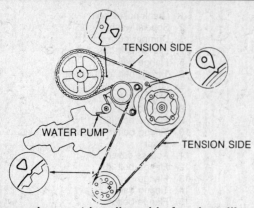

Timing marks must be aligned before installing the timing belt

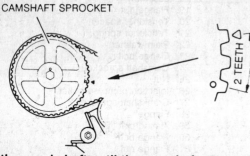

Turn the crankshaft until the camshaft moves 2 teeth

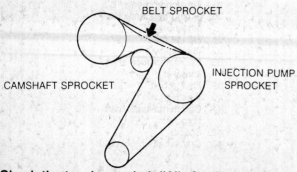

Check the tension on belt "A" after installation

21. Tighten the tensioner nut, then the bolt to hold the tensioner in place. If the bolt is tightened first, the tensioner will turn and the tension will become incorrect.
22. Check the timing marks; all should be aligned. Press on the mid-point of the belt's tension side; belt deflection should be 4–5 mm (0.16–0.20 in.).
23. The tensioner for the timing belt (belt A) should be in the slack position. Double check the alignment of all three pulleys for belt A; the timing marks should all align.
24. Install the timing belt (in the correct direction if reusing the old belt) first onto the crankshaft sprocket, then onto the injection pump and then onto the camshaft sprocket. The tension side—crankshaft to injection pump—must be kept taut.

NOTE: The sprocket on the injection pump tends to rotate by itself during this procedure. Hold the sprocket in position while installing the belt. An assistant can be useful during this three–handed installation.

25. Loosen the tensioner mounting bolt and allow the spring tension to take up the slack in the belt. Don't push on the tensioner.
26. Tighten the tensioner slot side bolt before tightening the fulcrum bolt. If bolts are tightened out of order, the tensioner turns and the belt is placed under incorrect tension. After the belt is under tension, check the camshaft pulley; the belt may be lifting on the upper, outer edge (10 o'clock position). Make certain the belt is firmly seated on the sprocket.
27. Check the timing marks on all sprockets for correct alignment. Anything not in correct position will require removal of the belt(s) and alignment of the sprockets.
28. Using a wrench on the crankshaft pulley bolt, turn the engine clockwise until the camshaft sprocket has moved 2 teeth. Now turn the engine back (counterclockwise) to its original position.
29. Use a finger to press on the belt midway between the camshaft sprocket and the injection pump sprocket. Belt deflection should be 4–5mm (0.16–0.20 in).
30. Install the upper and lower timing belt covers.

Timing Chain

REMOVAL AND INSTALLATION

Starion
Pickup and Montero with G54B Engine

1. Disconnect and remove the battery.
2. Drain the cooling system.

--- **CAUTION** ---
When draining the coolant, keep in mind that cats and dogs are attracted by the ethylene glycol antifreeze, and are quite likely to drink any that is left in an uncovered container or in puddles on the ground. This will prove fatal in sufficient quantity. Always drain the coolant into a sealable container. Coolant should be reused unless it is contaminated or several years old.

3. Remove the lower radiator hose from the water pump.
4. If the engine does not have an intercooler, remove the upper fan shroud. Carefully unbolt and remove the fan with the clutch assembly and store the fan in an upright position.
5. Loosen the adjuster and remove the air conditioning drive belt. Loosen the alternator and power steering pump and remove their drive belts.
6. Remove the water pump pulley from the studs.
7. Disconnect the heater hose from the water pump.
8. Remove the mounting bolts holding the pump to the engine. Note that the bolts are of different lengths; label or diagram them as they are removed.
9. Remove the water pump and gasket.
10. Remove the crankshaft bolt. Use a puller to remove the crankshaft pulley.
11. Remove the valve cover.
12. Remove ONLY the two small front bolts from the cylinder head; these screw into and seal the top of the timing cover.
13. Remove the oil pan bolts (front and side) that screw into the timing cover. Use a moderately sharp instrument to separate the oil pan gasket from the underside of the front cover.
14. Unbolt and remove the timing cover.
15. Clean all the gasket surfaces. If the oil pan gasket was damaged in removing the front cover, carefully cut the oil pan gasket off flush with the front of the block on both sides and remove the cutoff piece of gasket.
16. The crankshaft seal should be replaced any time the timing cover is removed. Carefully pry the oil seal out of the cover without scratching the bore into which the seal fits. Install a new seal with an seal driver or installer such as MD 998376-01 and MB 990938-01 or their equivalents.
17. Turn the crankshaft clockwise (only!) until the timing

ENGINE AND ENGINE OVERHAUL 3

1. Special washer
2. Crankshaft pulley
3. Timing chain case
4. Gasket
5. Chain guide access cover
6. Gasket
7. Crankshaft seal
8. Chain guide B
9. Chain guide A
10. Chain guide C
11. Silent shaft chain (B)
12. Crankshaft sprocket (B)
13. Oil pump sprocket
14. Left silent shaft sprocket
15. Spacer
16. Tensioner (foot)
17. Rubber seal
18. Tensioner spring
19. Distributor drive gear
20. Spring pin (roll pin)
21. Camshaft sprocket
22. Timing chain (camshaft chain)
23. Crankshaft sprocket
24. Loose side chain guide
25. Tension side chain guide
26. Sprocket holder

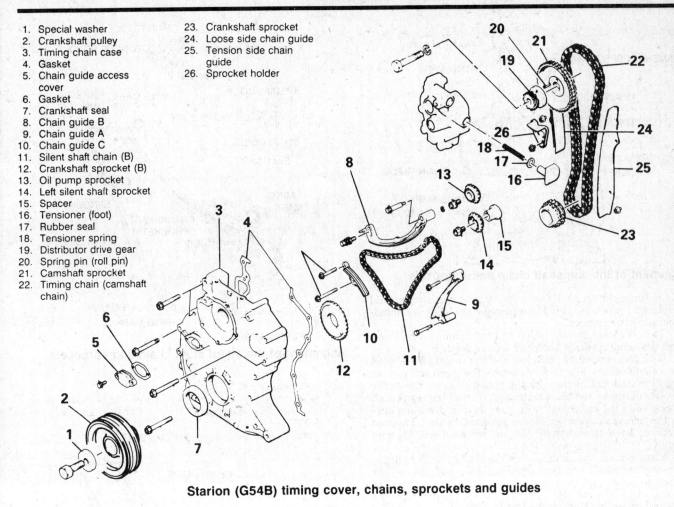

Starion (G54B) timing cover, chains, sprockets and guides

marks align. (Both valves for cylinder No. 1 are closed and the rockers are loose.).

18. Remove the 3 chain guides bearing on the silent shaft chain. Label or diagram each bolt as it is removed. The bolts are of different lengths and styles; they must be replaced in the correct locations.

18. Remove the chain from the silent shaft and oil pump sprockets and then from the crankshaft sprocket. If the chain will not come off the sprockets, don't pry on it. Unbolt the silent shaft and oil pump sprockets and carefully remove them with the chain. If the chain is removed individually, remove the sprockets afterward. Some sprockets use a small key on the shaft; don't lose it.

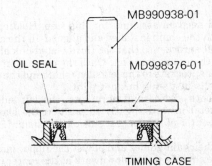

Use a seal driver of the correct diameter to replace the crankshaft seal

NOTE: The two sprockets are identical, that the oil pump drive sprocket is installed with the concave side toward the engine while the left silent shaft sprocket has the concave side out.

19. Remove the crankshaft sprocket for the silent shaft chain.
20. The timing chain tensioner maintains constant spring pressure on the chain. To prevent it from popping out of the oil pump it must be fastened in place. Run a piece of wire around the plunger and the left side of the oil pump.
21. Remove the camshaft sprocket bolt and pull the distributor drive gear and the sprocket off the camshaft. Separate the chain from the sprockets and remove it. An alternate method is to loosen the crankshaft and camshaft sprockets and remove them simultaneously with the chain attached.
22. If not already done, remove the the sprocket from the crankshaft.
23. Remove the chain guides, keeping close track of all the small bolts, washers and other hardware.
24. Check the chain carefully for any sign of play in the rollers, wear or damage. Replace the chain if any wear exists. Timing chains are most often replaced because of a gradual increase in length and lack of proper tension.

Inspect the tensioner plunger and replace it if it shows a deep grooving where the chain has ridden against it. If the plunger needs replacement, remove the wire holding it in place and allow the spring to gradually push it out of the oil pump body. Replace the rubber seal in the oil pump body and the spring behind it when you replace the plunger. Make sure the thinner part of the plunger faces downward. Wire the new follower in place just as the old one was.

3 ENGINE AND ENGINE OVERHAUL

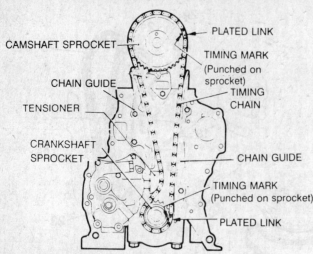

Alignment of the camshaft chain and sprockets

If the timing chain guides show heavy grooving, they should be replaced. Sprockets should be replaced if the teeth are rounded, deformed or cracked.

25. To reassemble, install the left and right chain guide. Don't forget the small sprocket holder; it will be needed.
26. Both the crankshaft sprocket and the camshaft sprocket have a small dot on their faces. Assemble the chain and sprockets so that each dot aligns with the plated links on the chain. Both sets of marks and links must align. Install the crankshaft sprocket onto the crankshaft with the chain in place and support the camshaft sprocket on the sprocket holder. This can sometimes be a three-handed job; an assistant can be very helpful.
27. Make certain the chain is riding correctly on the guides. Have the camshaft sprocket retaining bolt and the proper wrench at hand. Lift the cam sprocket into place against the cam and install the bolt. Make certain the sprocket engages the guide pin correctly. Tighten the bolt to 40 ft. lbs.
28. Remove the wire holding the tensioner plunger and allow it to tension the chain.
29. On the work bench, assemble the silent shaft sprocket, the oil pump sprocket and the silent shaft chain. Again, each sprocket is marked with a dot which must be aligned with the plated links on the chain.

WARNING: The sprocket for the silent shaft mounts with the concave side facing out (away from the engine block) and the sprocket for the oil pump mounts with the concave side facing the engine block.

30. Install the silent shaft drive sprocket onto the crankshaft (sprocket B).

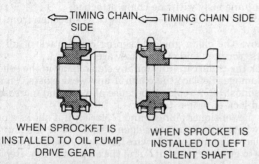

Correct installation of the silent shaft chain sprockets is required. The pulleys mount differently depending on location, but are otherwise identical

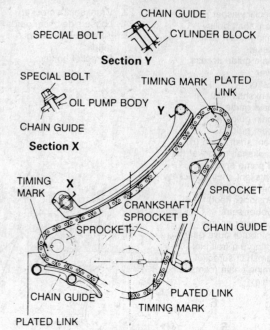

Alignment of the silent shaft chain and sprockets

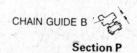

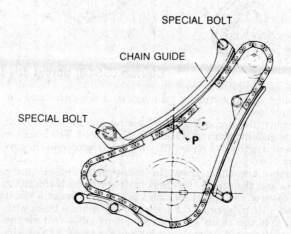

Adjust the upper chain guide while pulling gently on the chain at point "P"

31. You will need an assistant for this step. Holding the two sprockets with the chain installed and aligned, fit the chain over the crankshaft sprocket so that the third plated link aligns with the mark on the crank sprocket. Once the chain is in place, install the other sprockets to the left silent shaft and the oil pump. Install the bolts just snug but not tight.
32. Rotate both silent shaft sprockets (the left side and oil pump sprockets) inward slightly to position the slack in the chain in the center of the longest side of the chain. (Point P in the figure).
32. Install the chain guides and adjust the upper guide (B) so that when the chain is pulled downward at the center (Point P), the clearance between the chain and the guide is 0.20-0.80mm (0.008-0.031 in.). Tighten the guide locking bolt to 14 ft. lbs.
33. Before replacing the timing chain cover, check all the gas-

ENGINE AND ENGINE OVERHAUL 3

ket surfaces for cleanliness and inspect the cover for cracks or damage.

34. Using a new gasket, cut an exact replacement for the section of oil pan gasket damaged during removal. Insert this piece of gasket onto the front of the pan in the exact position of the old piece. Use liquid sealer on the joint between the two sections of gasket on both sides.
35. Using new gaskets carefully positioned, install the chain cover and tighten the bolts to 10 ft. lbs.
36. Lightly coat the outside diameter of the crankshaft pulley boss with clean engine oil. Install the pulley onto the crankshaft, followed by the washer and bolt. Tighten the bolt to force the pulley all the way on. Tighten the bolt to 88 ft. lbs.
37. Install the two head bolts through the head and into the timing chain cover. Tighten the bolts to 14 ft. lbs.
38. Install the oil pan bolts, tightening them evenly to 4 ft. lbs.
39. Install the water pump with a new gasket. Don't forget to install the alternator bracket.
40. Connect the heater hose to the water pump.
41. Install the water pump pulley.
42. Install the drive belts beginning with the power steering belt, then the alternator belt and the A/C compressor belt.
43. Install the fan and clutch assembly if it was removed. Install the upper fan shroud.
44. Adjust the drive belts to the proper tension and tighten the adjusting fittings for each component.
45. Install the lower radiator hose.
46. Install the battery.
47. Double check the draincock, closing it if necessary, and refill the cooling system.
48. Connect the negative battery cable and start the engine. Check carefully for leaks, particularly around the water pump and oil pan seams.
49. Listen for any unusual noises from the area of the timing chain. Possible causes of noise are the chain rubbing against the cover or a guide being too loose or too tight. Do not accelerate the engine if abnormal noises are heard from the timing chain case — severe damage can result.
50. After making needed adjustments, shut the engine off, allow it to cool and set the coolant to the correct level.

Oil Pan
REMOVAL AND INSTALLATION

WARNING: The use of the correct special tools or their equivalent is REQUIRED for this procedure.

--- CAUTION ---
Used motor oil may cause skin cancer if repeatedly left in contact with the skin for prolonged periods. Although this is unlikely unless you handle oil on a daily basis, it is wise to thoroughly wash your hands with soap and water immediately after handling used motor oil.

The oil pan must be pulled downward as much as 152mm (6 in.) to clear the oil pickup. In many cases, this requires that the engine mounts be disconnected and the engine raised to clear a crossmember underneath the shallower section of the pan. Survey the area to determine clearance. Some front wheel drive, 4 cylinder engines will allow the pan to be removed without much drama.

If raising the engine is needed, support the car securely high enough to allow easy access to the underside. Refer to the engine removal and installation procedures and disconnect all the hoses and wires that would prevent the engine from being lifted the necessary distance.

WARNING: Raising the engine will require the use of overhead or hoist type equipment attached to the top of the engine by chains. Do not attempt to raise the engine with a jack underneath. Damage or injury can result.

Typical oil pan. The pick-up screen will vary in shape and location

Support the engine on the hoist and remove the through-bolts from the mounts. Elevate the engine only enough to gain the needed clearance and keep a close watch for interference with other components as the engine is raised.

On the V6, the starter, transaxle mounts, and rear bell housing cover will need to be removed. Disconnect the oxygen sensor wiring and disconnect both exhaust pipes at the manifolds.

1. Drain the engine oil. When the pan is empty, reinstall the drain plug finger tight.
2. If the engine is turbocharged, disconnect the oil return tube and/or the oil drain tube from the pan.
3. Remove the bolts holding the oil pan. The length may vary by location. Keep the bolts in order or identify them so that they may be correctly installed.
4. Most engines use a bead of sealer on the oil pan rather than a paper gasket. After the bolts are removed, tap the corners of the pan with a rubber or plastic mallet. If the pan pops loose, it may be removed on the spot.

In most cases, the pan will remain firmly glued to the block. Insert the special gasket slicing tool (MB 998727 or equivalent) by tapping it into place with a mallet or hammer. Once in place, drive the tool along the pan to release the sealer and the pan.

WARNING: DO NOT attempt to loosen the pan with a screwdriver or chisel. The flange of the pan will deform and cause oil leaks after reassembly.

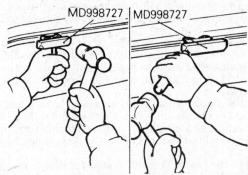

Use the special tool for separating the pan from the block. Other tools may cause damage

3-237

3 ENGINE AND ENGINE OVERHAUL

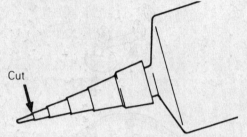

Cut the tip of the sealant tube to allow very narrow flow

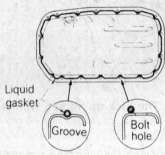

Correct application of sealant to the oil pan, except Starion and V6 engines

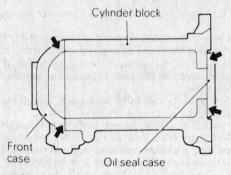

Apply sealant only to the indicated areas on the Starion block

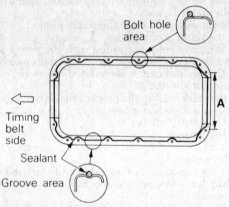

Do not apply sealant to area "A" on V6 engines. The area must be protected from sealant seepage

The V6 pan bolts must be tightened in this order. The alternating pattern can be used for any pan

5. When the pan is removed, both the flange surface of the pan and the matching surface of the block must be completely cleaned of any remaining gasket or sealer. When cleaning with a scraper, take great care not to gouge or mar the metal surfaces.

6. With the pan removed, the oil pick-up screen and tube may be removed by unbolting it from the block. If removed, it should be thoroughly cleaned in a solvent bath and air dried before reinstallation. The gasket surfaces must be cleaned and the old gasket discarded.

7. Clean the oil pan with solvent, taking great care to remove all particulate matter and sludge before air drying. The pan must be reinstalled free of any matter which could be circulated through the engine. Remove the drain plug, replace the washer and reinstall the plug, tightening it to the correct torque.

8. If the pick-up was removed, reinstall it with a new gasket and tighten the bolts to 13 ft. lbs.

9. On all engines except the Starion, Galant/Sigma V6 and diesel truck, apply a thin bead of sealant (MZ 100168 or equivalent) all around the pan. Cut the tip of the sealant tube to allow a bead about 0.16 in (4mm) to flow. The sealant goes in the groove in the pan flange. Note that the bead should flow to the inside of the bolt holes or between the bolt and the pan.

The V6 engines also receive the bead of sealer but the area between the two rear-most bolt holes must not be sealed. Care must be taken to prevent oozing of sealant into the rear valley of the pan.

The Starion engine and the diesel truck engine use a paper gasket which does not require sealant. Sealant such as 3M ART 8660 or its equivalent must be applied to the block at the four seams where the front and rear cases join. Apply the sealer only to these areas.

10. Once the sealant is applied, the pan must be installed within 10 minutes or the sealant becomes unusable. Fit the pan into place and install the bolts finger tight only, making certain the bolts are replaced in the correct holes as determined by their length.

11. Tighten the pan bolts evenly and alternating from side to side to 5 ft. lbs. only. The pan bolts for the V6 must be tightened in the pattern shown; the pattern may be used as a guide for the correct tightening of other pans.

WARNING: The tightness of the bolts does not seal the pan, the gasket or sealer does. DO NOT OVERTIGHTEN.

12. Since the oil has been drained, replacement of the oil filter is recommended before refilling the engine oil.

13. Connect the oil drain and/or oil return lines to the pan if they were removed.

14. Install the correct amount of engine oil. Make sure the drain plug is tight before pouring the oil.

15. If the engine is supported by a hoist, lower it into position and secure the mounts and through-bolts as necessary. Refer to the *Engine Removal and Installation* section for the correct torque values.

On the V6 engine, use new gaskets and connect the exhaust system to the manifolds. Install the bell housing cover, the transaxle mounts and the starter.

16. Lower the car to the ground. If possible, the sealant should be allowed to cure for a period of hours before the engine is started. Once started, let the engine warm up at idle and then shut it off.

ENGINE AND ENGINE OVERHAUL 3

17. Check the pan carefully around the seam for any sign of leakage. If any leakage is found, do not attempt to cure it by tightening the pan bolts. This usually aggravates the leak rather than curing it. The pan will need to be removed and resealed.

Oil Pump and Silent Shafts

REMOVAL AND INSTALLATION

NOTE: This repair is much easier if the engine is removed from the car. Expect difficulty if attempting this with the engine in the car.

The use of the correct special tools or their equivalent may be REQUIRED for this procedure.

Cordia and Tredia
Pickup with G63B Engines
Galant with G64B Engine

1. With the car in neutral or park, make certain the parking brake is set and the wheels are blocked. Disconnect the negative battery cable.
2. Remove the water pump drive belt and water pump pulley. If the car is equipped with air conditioning, loosen the compressor belt tensioner and remove the compressor belt.
3. Remove the bolts holding the crankshaft pulley(s) and remove the pulley(s).
4. Remove the upper and lower timing belt covers and their gaskets.
5. Use a socket wrench on the projecting crankshaft bolt to turn the engine clockwise (only!) and align all the timing marks on the sprockets and cases. It may be necessary to wipe off the area to see the marks clearly. Do not use spray cleaners around the timing belt.

When the marks all align exactly, the engine is set to TDC/compression on No. 1 cylinder. From this point onward, the engine position MUST NOT be changed.

6. If the timing belt is to be reused, make a chalk or crayon arrow on the belt showing the direction of rotation so that it may be reinstalled correctly.
7. Loosen the bolts holding the tensioner and pivot the tensioner towards the water pump. Temporarily tighten the bolts to hold the tensioner in its slack position.
8. Carefully slide the belt off the sprockets. Place the belt in a clean, dry, protected location away from the work area.
9. Remove the camshaft sprocket bolt and remove the sprocket.
10. Remove the crankshaft sprocket retaining bolt. Remove the sprocket and flange.
11. Remove the oil filter. Drain the engine oil.

— CAUTION —
Used motor oil may cause skin cancer if repeatedly left in contact with the skin for prolonged periods. Although this is unlikely unless you handle oil on a daily basis, it is wise to thoroughly wash your hands with soap and water immediately after handling used motor oil.

12. Remove the oil pan following procedures outlined previously.
13. Remove the oil pick up screen and gasket.
14. Remove the plug from the left side of the cylinder block and insert a screwdriver to keep the left silent shaft in place. The screwdriver should have a shaft diameter of 8mm (0.3 in.) and a shaft length of at least 60mm (2.4 in.).

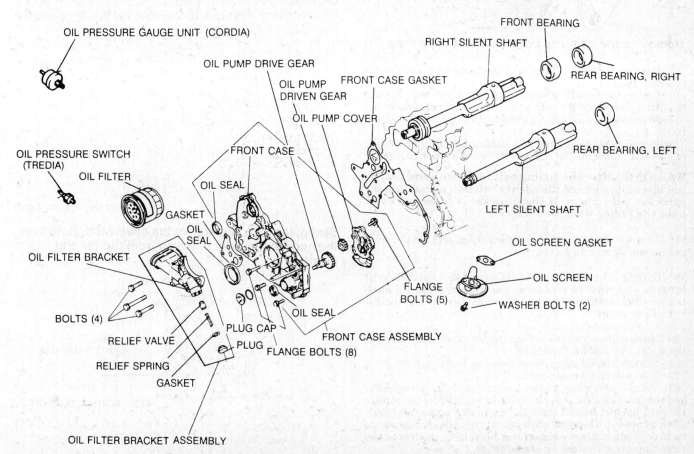

Cordia and Tredia oil pump assembly, front case and silent shafts

3 ENGINE AND ENGINE OVERHAUL

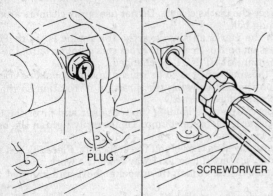

Use a screwdriver to block the left (firewall side) silent shaft

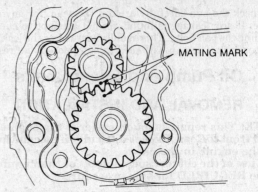

Carefully align the marks on the oil pump gears

check for smooth operation. Check the free length of the spring; correct uncompressed length is 46.6mm (1.835 in.).

25. Apply engine clean engine oil to the oil pump gears and install them in the front case. Align the mating marks.
26. Install the oil pump cover to the front case and tighten the bolts to 16 ft. lbs.
27. Carefully install the left and right silent shafts into the engine; the bearings can be damaged by careless installation.
28. Install a seal guide tool such as MD 998285-01 or its equivalent to the crankshaft. Note that the tapered end should face away from the block. Coat the surface of the guide with clean engine oil. This guide will help in installing a new seal in the front case. If the seal is already in place, the guide will protect the seal during installation.
29. Install a new front case gasket, carefully aligning all the bolt holes.
30. Install the front case to the block and lightly tighten the

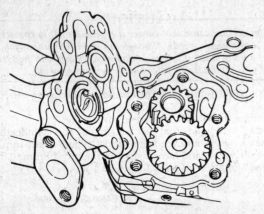

Removing the oil pump cover from the front case

15. Remove the bolt holding the oil pump sprocket and remove the sprocket.
16. Loosen the mounting bolt for the right silent shaft sprocket until it can be turned with your fingers. Do not remove it.
17. Remove the tensioner for the silent shaft belt. (Tensioner B).
18. Remove the silent shaft belt (timing belt B). Remove the crankshaft pulley for the silent shaft belt if so desired.

WARNING: After the timing belt has been removed, do not attempt to loosen the silent shaft bolt by holding the sprocket with pliers. If the sprocket is to be removed, hold the pulley with your fingers.

19. Remove the front case mounting bolts; remove the front case assembly and its gasket.

WARNING: The front case bolts are of several different lengths and must be replaced exactly as removed. Labeling or diagramming their placement during removal is very important.

20. Remove the screwdriver from the left silent shaft. Remove the silent shafts from the engine.
21. Remove the oil pump cover from the front case.
22. Remove the oil pump gears.
23. Visually check the contact surfaces of the oil pump cover and the front case for wear. If excessive wear is present, replace the components. Inspect the oil passages for clogging and clean them as needed. The silent shafts should be inspected for any of wear or seizure. Closely inspect the oil seals on the front case, replacing any with worn or damaged lips.
24. Insert the relief plunger into the oil filter bracket and

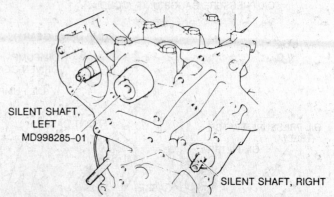

Seal guide tool in place on the crankshaft. Note that the engine shown is removed from the car and inverted

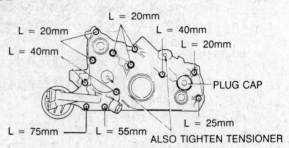

Correct bolt placement by length for front case

ENGINE AND ENGINE OVERHAUL 3

eight bolts. When the case is properly seated, the seal guide may be removed.

31. Insert a screwdriver through the left side access hole and block the left silent shaft.
32. Install the silent shaft end bolt and tighten it to 26 ft. lbs.
33. Install the oil filter bracket and gasket. Tighten the front case mounting bolts to 17 ft. lbs. and the oil filter bracket bolts to 13 ft. lbs.
34. Install the plug cap at the end of the left silent shaft.
35. Coat the relief plunger and spring in clean engine oil and install them into the oil filter bracket. Install the relief plug and gasket, tightening the plug to 34 ft. lbs.
36. Install the oil screen and gasket.
37. Install the oil pan.
38. Install a new oil filter.
39. Check that the timing marks are still aligned. Push down on the center of the tension side of the belt with your finger. Correct belt deflection is 5–7mm or approximately ¼ inch. If the deflection is not correct, the tensioner must be released fully and the belt re-tensioned.
40. Install the flange and the crankshaft sprocket onto the crankshaft. Make certain the flange is installed in the correct direction. If it is put on incorrectly, the belt will wear and break.
41. Install the special washer and the sprocket retaining bolt to the crankshaft. Tighten the bolt to 88 ft. lbs.
42. Install the camshaft sprocket to the camshaft and tighten the bolts to 66 ft. lbs.
43. Install the spacer, tensioner and tensioner spring if they were removed. Install the lower end of the spring to its position on the tensioner, then place the upper end in position at the water pump. Move the tensioner towards the water pump and temporarily tighten it in this position.
44. Install the oil pump sprocket, tightening the nut to 40 ft. lbs. Remember that the oil pump drives the left silent shaft, which is still blocked by the screwdriver. Hold the sprocket by hand when tightening the nut.
45. Double check the alignment of the timing marks for the cam, crank and oil pump sprockets. If any adjustment is needed to the oil pump sprocket, remove the screwdriver blocking the silent shaft before adjusting the sprocket. Once everything is aligned, replace the screwdriver in the left side of the block and leave it there until installation of the timing belt is complete.

NOTE: If the screwdriver can only be inserted about 25mm (1 inch) or less, the shaft is out of position. Turn the oil pump sprocket through one full turn clockwise; the screwdriver should then go in about 63mm (2½ inches).

46. Install the timing belt onto the crankshaft sprocket, the oil pump sprocket and the cam sprocket in that order. Keep the belt taut between sprockets. If reusing an old belt, make certain that the direction of rotation arrow is properly oriented.
47. Loosen the tensioner mounting nut and bolt. The spring will move the tensioner against the belt and tension it.
48. Check the belt as it passes over the camshaft sprocket. The belt may tend to lift in the area to the left of the sprocket. (Roughly 7 o'clock to 12 o'clock when viewed from the pulley end.) Make certain the belt is well seated and not rubbing on any flanges or nearby surfaces.
49. At the tensioner, tighten the bolt in the slotted hole first, then tighten the nut on the pivot. If this order is not followed, the belt will become too tight and break.
50. Once again, check all the timing marks for alignment. Nothing should have changed; check anyway.
51. Remove the screwdriver blocking the left side silent shaft. Using a socket on the crankshaft bolt, turn the crankshaft smoothly one full turn (360°). Do NOT turn the engine backwards.
52. Loosen the tensioner nut and bolt. The spring will allow the tensioner to tighten a little bit more because of the slack picked up during engine rotation. Tighten the bolt, then the nut to 36 ft. lbs.
53. Check the deflection of the belt. At the middle of the right (tension) side of the belt, deflect the belt outward with your finger, toward the timing case. The distance between the belt and the line of the cover seal should be about 14mm (9/16 in.).
54. Install the lower and upper timing belt covers. Make certain the gaskets are properly seated in the covers and that they don't come loose during installation. Tighten the bolts to 8 ft. lbs.
55. Install the crankshaft pulley(s) and tighten the retaining bolts to 18 ft. lbs.
56. Install the water pump pulley, tightening the bolts to 6 ft. lbs., and install the belt. Adjust the belt to the correct tension.
57. Install the correct amount of motor oil.
58. Connect the negative battery cable. Start the engine and let it idle, listening for any unusual noises from the area of the timing belt. Possible causes of noise are the belt rubbing against the covers or a sprocket flange, the belt being too loose and slapping, or a tensioner binding. Do not accelerate the engine if abnormal noises are heard from the timing belt train — severe damage can result.

Starion, Montero, and Pickup with G54B Engine

1. Remove the timing chains and sprockets, following procedures outlined previously in this Section.
2. Drain the engine oil and remove the oil filter.

---- CAUTION ----
Used motor oil may cause skin cancer if repeatedly left in contact with the skin for prolonged periods. Although this is unlikely unless you handle oil on a daily basis, it is wise to thoroughly wash your hands with soap and water immediately after handling used motor oil.

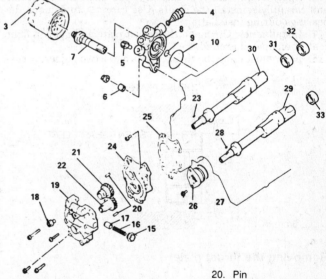

3. Oil filter
4. Oil cooler bypass valve
5. Oil pipe joint
6. Bushing
7. Oil filter bracket bolt
8. Oil filter bracket
9. O-ring
10. O-ring
15. Plug
16. Relief spring
17. Relief plunger
18. Flange bolt
19. Oil pump body
20. Pin
21. Driven gear
22. Drive gear
23. Woodruff key
24. Oil pump cover
25. Oil pump gasket
26. Thrust plate
27. O-ring
28. Woodruff key
29. Left silent shaft
30. Right silent shaft
31. Front bearing
32. Rear bearing
33. Rear bearing

Starion oil pump and silent shafts

3–241

3 ENGINE AND ENGINE OVERHAUL

3. Remove the oil pan following procedures outlined previously.
4. Disconnect the oil cooler bypass valve and the oil pipe joint at the filter bracket.
5. Remove the bracket bolt and nut. Remove the bushing and remove the oil filter bracket from the engine. Remove both O-rings from the filter bracket.
6. Remove the oil pan and gasket, following procedures outlined previously in this Section.
7. Remove the oil pick-up screen and gasket.
8. Remove the plug, relief spring and plunger from the side of the oil pump body.
9. Remove the flange bolts and remove the oil pump body. Don't lose the small pins in the case.
10. Remove the driven gear and drive gear. Don't lose the woodruff key in the end of the silent shaft.
11. Remove the oil pump cover and its gasket.
12. If the left (firewall side) silent shaft is to be removed, the thrust plate must be extracted first. Install two 8mm diameter bolts into the threaded holes of the flange and turn the bolts to remove the thrust plate by forcing it outward.
13. If desired, remove the right silent shaft.
14. Visually check the contact surfaces of the oil pump cover and the front case for wear. Use a straight edge and a feeler gauge to measure the clearance between the gears and the case. Maximum allowable clearance is 0.15mm (0.0060 in.) If excessive wear is present, replace the the gears or case as needed.

Inspect the oil passages for clogging and clean them as needed. The silent shafts should be inspected for any of wear or seizure.

15. Insert the relief plunger into the front case and check for smooth operation. Check the free length of the spring; correct uncompressed length is 46.6mm.
16. Apply a light coat of engine oil to the silent shaft journals and carefully reinstall the shafts. Use care not to damage the bearings during installation.
17. Install a new O-ring on the thrust plate and apply a light coat of engine oil around the O-ring.
18. Before installing the thrust plate, make two guides by cut-

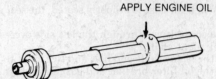

The silent shaft journals must be lightly coated with oil before reinstallation

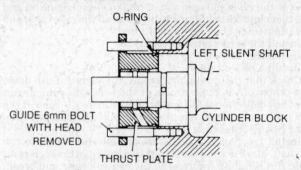

Use the homemade guides to locate the thrust plate correctly

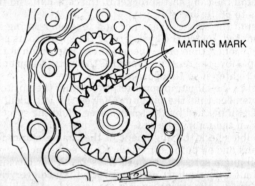

Carefully align the marks on the oil pump gears

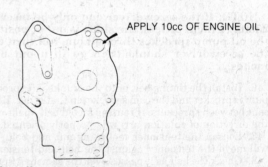

Pre-lube the oil pump before installation

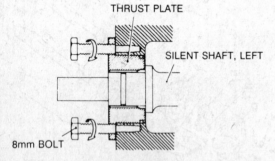

Removing the thrust plate

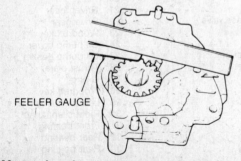

Measuring the gear-to-case clearance

ting off the heads of two 6mm bolts. The bolts should be 50mm long. Install the bolts in the threaded holes.

19. Install the thrust plate into the cylinder block along the guides. Without the guide bolts, it will be almost impossible to align the bolt holes. Install the bolts and tighten them to 8 ft. lbs.
20. Apply engine clean engine oil to the oil pump gears and install them in the oil pump body. Align the mating marks. If the marks are not aligned, the silent shaft will be out of phase and engine vibration will occur.
21. Place a new oil pump gasket in position and install the cover, tightening the bolts to 8 ft. lbs.
22. Hold the oil pump body in hand in the same relative posi-

ENGINE AND ENGINE OVERHAUL 3

tion as it will be on the engine and put approx. 10cc of clean engine oil into the delivery port.

23. Install the pump body. Tighten the large flange bolts to 48 ft. lbs. The smaller screws and bolts should be tightened only to 7 ft. lbs.
24. Assemble the relief plunger and spring into the pump body and tighten the plug to 28 ft. lbs.
25. Install the oil pick-up screen with a new gasket. Tighten the bolts to 13 ft. lbs.
26. Install the oil pan and gasket.
27. Replace the O-rings and install the oil filter bracket. Tighten the large bracket bolt to 33 ft. lbs. and the nut to 22 ft. lbs. Don't forget the bushing.
28. Install a new oil filter.
29. Install the correct amount of engine oil.
30. Reinstall the timing chains and sprockets.

Mirage and Precis with G15B Engine
Mirage with 4G15 Engine

1. Disconnect the negative battery cable.
2. Remove the left splash shield under the engine.
3. Remove the air cleaner assembly, the air intake duct and the hot air duct.
4. Disconnect the accelerator cable.
5. Label and remove the coil wire and the spark plug wires from the distributor. Label and disconnect the oxygen sensor connector on the side of the valve cover. Move all the wires out of the way.
6. Loosen the appropriate component or adjuster and remove (in order) the air conditioning compressor drive belt, the power steering belt and the ribbed belt driving the water pump.
7. Place a floor jack and a broad piece of lumber under the oil pan. Elevate the jack just enough to support the engine without raising it.
8. Remove the nuts holding the through-bolt for the left engine bracket and remove the bolt. It may be necessary to adjust the jack tension slightly to allow the bolt to come free. Use the jack to keep the engine in its normal position while the mount is disconnected.
9. Remove the nuts and bolts holding the engine bracket to the engine and remove the bracket.
10. Remove the water pump pulley. If the car is equipped with power steering, remove the smaller pulley first.
11. Remove the valve cover and gasket. At the valve adjusters, back off each adjusting screw until the ends of the screws project only 1.0mm (0.04 in.) or less from the rocker.
12. Remove the upper timing belt cover and its gasket.

WARNING: The bolts for both the upper and lower timing belt covers are of different lengths. Label or diagram them at the time of removal; replacement in the correct location is required.

13. Rotate the crankshaft clockwise until the timing marks on the crank pulley align at 0 on the scale (on the lower timing cover) and the mark on the camshaft pulley aligns with the mark on the head. Aligned means perfect, not close — if you miss, continue clockwise until the marks are matched. This sets the motor at TDC/compression on No.1 cylinder.

WARNING: Once this position is established, the cam and crankshafts must NOT be moved out of place.

14. Remove the lower timing belt cover and its gasket.
15. Remove the crankshaft pulley(s) without moving the crankshaft out of position.
16. Loosen (do not remove) both bolts in the timing belt tensioner. Move the tensioner toward the water pump and tighten the lock bolt — the one in the slotted hole — to hold the tensioner in place.
17. Before removing the timing belt, mark it with an arrow to show the direction of rotation. If the belt is to be reused, it must

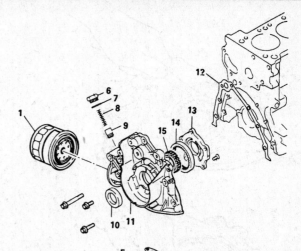

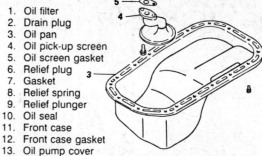

1. Oil filter
2. Drain plug
3. Oil pan
4. Oil pick-up screen
5. Oil screen gasket
6. Relief plug
7. Gasket
8. Relief spring
9. Relief plunger
10. Oil seal
11. Front case
12. Front case gasket
13. Oil pump cover
14. Outer gear
15. Inner gear

G15B oil pump assembly

be reinstalled so that it rotates in the same direction as before.
18. Remove the timing belt, keeping it free of grease, oil and fluids. Do not crease or crimp the belt.
19. Drain the engine oil and remove the oil filter.

CAUTION
Used motor oil may cause skin cancer if repeatedly left in contact with the skin for prolonged periods. Although this is unlikely unless you handle oil on a daily basis, it is wise to thoroughly wash your hands with soap and water immediately after handling used motor oil.

---

20. Remove the oil pan following procedures outlined previously.
21. Remove the oil pick-up screen and gasket
22. Remove the relief plug, gasket, spring and plunger from the oil filter bracket.
23. Remove the oil seal from the front case. Note that this seal must NOT be reused; replacement is mandatory during reassembly.
24. Remove the front case and gasket from the engine block.

WARNING: The bolts are of different lengths. Label or diagram their locations as they are removed.

25. Remove the oil pump cover.
26. Remove the inner and outer gears from the casing. The outer gear has no identifying mark as to which side faces the block — mark it with a felt pen before removal. It must be reinstalled in its original position.
27. Visually check the contact surfaces of the oil pump cover and the front case for wear. Inspect the gear tooth surfaces for wear or damage. After cleaning, reassemble the gears and crescent in the case. With a feeler gauge, check the clearance between the outer gear and the case. Maximum acceptable clearance is 0.2mm (0.0079 in.). Check the clearance between the

3-243

3 ENGINE AND ENGINE OVERHAUL

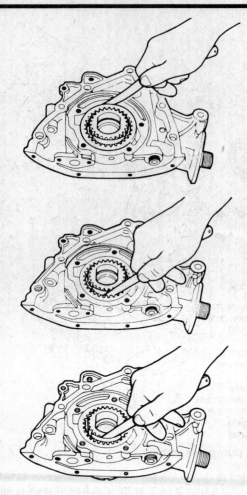

Measuring gear clearances. Refer to text for maximum acceptable clearances

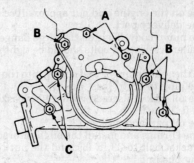

Bolt lengths (A): 30 mm
(B): 20 mm
(C): 60 mm

Correct bolt placement when reinstalling the front case

teeth of the inner gear and the crescent, looking for a clearance no greater than 0.34mm (0.00134 in.). Measure the clearance between the teeth of the outer gear and the crescent. Maximum clearance should be 0.44mm (0.0173 in.) or less. If any clearance exceeds the maximum, the gears and/or case must be replaced.

28. Insert the relief plunger into the front case and check for smooth operation. Check the free length of the spring; correct uncompressed length is 46.6mm (1.8346 in.).
29. Double check that the outer gear is correctly installed according to the marks made during removal. Coat the gears in clean engine oil.
30. Install the oil pump cover, tightening the bolts to 9 ft. lbs.
31. Install a new front case gasket and install the front case onto the cylinder block. Tighten the mounting bolts to 9 ft. lbs. Make certain the bolts are installed in the correct positions.
32. Install special tool MD 998305-01 or equivalent to the front of the crankshaft to guide the front seal into place. Apply clean oil to the outer surface of the tool and slide the new oil seal along the special tool by hand until it touches the front case.
33. Install the oil seal in the front case using a seal installer such as MD 998304-01 or its equivalent. Tap the installer until the seal is properly seated in the case.
34. Install the relief plunger, spring, gasket and relief plug into the case body. Tighten the plug to 34 ft. lbs.
35. Install the oil pick-up screen with a new gasket. Tighten the bolts to 13 ft. lbs.
36. Install the oil pan.

37. Install a new oil filter and install the proper amount of engine oil.
38. Check the sprockets and tensioner for wear. The sprocket teeth should be well defined, not rounded and the valleys between the teeth should be clean. The tensioner should spin freely with no binding or unusual noise. Replace the tensioner if there is any sign of grease leaking from the seal. Clean everything with a clean, dry cloth.

WARNING: Do not spray or immerse the sprockets or tensioner in cleaning solvent. The sprocket may absorb the solvent and transfer it to the belt. The tensioner is internally lubricated and the solvent will dilute or dissolve the lubricant.

39. Before reinstalling the belt, double check the sprockets for correct alignment of the timing marks.
40. To reinstall the belt, observe the directional arrow made earlier and fit the belt under the crankshaft sprocket. Track the belt up and over the right side (your right, not the engine's right) of the camshaft sprocket while keeping tension on the belt all the time.
41. The camshaft sprocket is quite likely to have moved during the belt installation. Gently turn the cam sprocket counterclockwise to tension the belt; when the belt is taut, the timing marks on the head and cam sprocket should align perfectly. If they do not, remove the belt and reinstall it.
42. Install the crankshaft pulley temporarily. It is needed to hold the belt in place.
43. Loosen the lock bolt in the tensioner. Allow the tensioner to move against the belt under its own spring tension; don't pry it or force it.
44. Tighten the lock bolt (in the slotted hole) in the tensioner, then tighten the pivot bolt for the tensioner. This order is important; if done in the wrong order, the belt will be over-tightened.

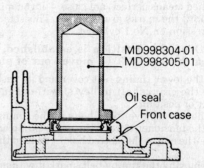

Use the proper tools to install the front seal

ENGINE AND ENGINE OVERHAUL 3

45. Turn the crankshaft one full revolution clockwise. Turn it smoothly and continuously and realign the timing mark. Loosen the pivot bolt, then the lock bolt on the tensioner. This allows the tensioner to pick-up any remaining slack in the belt.
46. Tighten the lock nut to 18 ft. lbs. and then tighten the pivot bolt to 18 ft. lbs.
47. Hold the tensioner and timing belt by hand and give the belt slight thumb pressure at a point level with the tensioner. The belt should deflect enough to allow the top of a tooth to cover about ¼ of the head of the lock bolt.
48. Finish the installation of the crank pulley(s). The large center bolt should be tightened to 62 ft. lbs. and the smaller bolts (for the air conditioning pulley) should be tightened to 10 ft. lbs. if they were removed.
49. Install the lower timing belt cover with its gasket, making sure the gasket doesn't shift during installation. tighten the bolts to 8 ft. lbs.
50. Install the upper timing belt cover with its gaskets and tighten the bolts to 8 ft. lbs.
51. Adjust the valve clearance. Follow the procedure outlined in Section 2.
52. Install the valve cover and gasket.
53. Install the water pump pulley (and power steering pulley if removed), tightening the bolts to 6 ft. lbs.
54. Install the motor mount bracket to the engine and tighten the nuts and bolts to 42 ft. lbs.
55. Align the mount bushing with the body bracket and install the through-bolt. Tighten both the large nut and the smaller safety nut snug.
56. Remove the jack from under the engine. With the weight of the engine on the mount, tighten the large nut on the through-bolt to 72 ft. lbs. and the smaller safety nut to 38 ft. lbs.
57. Install the coil and spark plug wires. Connect the oxygen sensor wiring. Make sure all the wiring is properly held by the clips and wire guides.
58. Install the ribbed water pump drive belt, making sure it is correctly seated on each pulley. Install the power steering belt and the compressor drive belt. Adjust each belt to the correct tension and tighten the fittings or adjuster on each component.
59. Connect the accelerator cable and adjust it.
60. Install the air cleaner assembly and the ducts.
61. Double check all installation items, paying particular attention to loose hoses or hanging wires, untightened nuts, poor routing of hoses and wires (too tight or rubbing) and tools left in the engine area.
62. Connect the negative battery cable.
63. Start the engine and let it idle, listening for any unusual noises from the area of the timing belt. Possible causes of noise are the belt rubbing against the covers or a sprocket flange, the belt being too loose and slapping, or a tensioner binding. Do not accelerate the engine if abnormal noises are heard from the timing belt train — severe damage can result.
64. Reinstall the splash shield under the car. Final adjustment of the drive belts and accelerator cable may be needed.

Mirgage with G32B Engine

1. Disconnect the negative battery cable.
2. Remove the left splash shield under the engine.
3. Disconnect the air intake duct, the breather hose and the O-ring from the engine.
4. Disconnect the accelerator cable.
5. Label and remove the coil wire and the spark plug wires from the distributor and from the spark plugs.
6. Loosen the appropriate component or adjuster and remove (in order) the air conditioning compressor drive belt, the power steering belt and the ribbed belt driving the water pump.
7. Place a floor jack and a broad piece of lumber under the oil pan. Elevate the jack just enough to support the engine without raising it.
8. Remove the nuts holding the through-bolt for the left engine bracket and remove the bolt. It may be necessary to adjust the jack tension slightly to allow the bolt to come free. Use the jack to keep the engine in its normal position while the mount is disconnected.
9. Remove the nuts and bolts holding the engine bracket to the engine and remove the bracket.
10. Remove the water pump pulley. If the car is equipped with power steering, remove the smaller pulley first.
11. Disconnect the PCV hose and remove the valve cover and gasket. At the valve adjusters, back off each adjusting screw until the ends of the screws project only 1.0mm (0.04 in.) or less from the rocker.
12. Remove the upper timing belt cover and its gasket.

WARNING: The bolts for both the upper and lower timing belt covers are of different lengths. Label or diagram them at the time of removal; replacement in the correct location is required.

13. Rotate the crankshaft clockwise until the timing marks on the crank pulley align at 0 on the scale (on the lower timing cover) and the mark on the camshaft pulley aligns with the mark on the rear upper cover. Aligned means perfect, not close. If you miss, continue clockwise until the marks are matched. This sets the motor at TDC/compression on No.1 cylinder.

WARNING: Once this position is established, the cam and crankshafts must NOT be moved out of place.

14. Remove the crankshaft pulley(s) without moving the crankshaft out of position.
15. Remove the lower timing belt cover and its gasket.
16. Loosen (do not remove) the nut and the bolt in the timing belt tensioner. Move the tensioner toward the water pump and tighten the nut to hold the tensioner in place.
17. Before removing the timing belt, mark it with an arrow to show the direction of rotation. If the belt is to be reused, it must be reinstalled so that it rotates in the same direction as before.
18. Remove the timing belt, keeping it free of grease, oil and fluids. Do not crease or crimp the belt.
19. Remove the oil filter and drain the oil. Remove the oil pressure switch by using the correct sized socket.

CAUTION

Used motor oil may cause skin cancer if repeatedly left in contact with the skin for prolonged periods. Although this is unlikely unless you handle oil on a daily basis, it is wise to thoroughly wash your hands with soap and water immediately after handling used motor oil.

---

20. Remove the oil filter bracket and gasket.
21. Remove the oil pan.
22. Remove the oil pick-up screen and its gasket.
23. Carefully remove the oil pump cover. Check the seal carefully for wear and remove it if damaged or worn.
24. Remove the oil pump rotor assembly and discard the gasket.
25. Remove the plug, relief spring and plunger from the front case.
26. Remove the front case and gasket. Discard the crankshaft oil seal from the front case; it must be replaced after each removal from the crankshaft.

WARNING: The bolts are of different lengths. Label or diagram their locations as they are removed.

27. Check the front case and pump components for any sign of wear or cracking. Install the rotor into the pump cover and check the clearances as shown in the illustration. Maximum allowable clearances are:
 - Side clearance: 0.2mm (0.0079 in.)
 - Tip clearance: 0.18mm (0.0071 in.)
 - Body clearance: 0.2mm (0.0079 in.)
28. Insert the relief plunger into the front case and check for

3–245

ENGINE AND ENGINE OVERHAUL

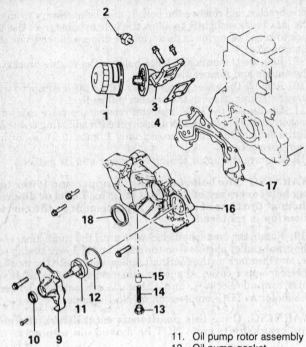

1. Oil filter
2. Oil pressure switch
3. Oil filter bracket
4. Gasket
9. Oil pump cover
10. Oil seal
11. Oil pump rotor assembly
12. Oil pump gasket
13. Plug
14. Relief spring
15. Relief plunger
16. Front case
17. Front case gasket
18. Crankshaft oil seal

G32B oil pump and front case assembly

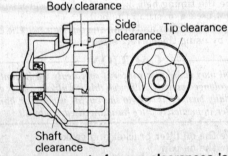

Correct measurement of pump clearances is important

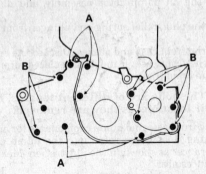

The bolts must be correctly installed in the front case. Bolts "A" are 8 × 35mm, bolts "B" are 8 × 40mm

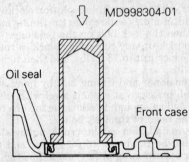

Use a seal driver to install the case seal

smooth operation. Check the free length of the spring; correct uncompressed length is 46.6mm (1.8346 in.).

29. Install a new crankshaft seal into the front case using tool MD 998304–01 or equivalent seal installer. Make certain the seal is installed in the correct direction.
30. Install tool MD 998285–01 on the crankshaft with the tapered end facing away from the engine block. Coat the outside of the guide tool with clean engine oil. Install the front case onto the block by fitting the seal over the guide. Observe correct placement of the case bolts, install them and tighten them to 12 ft. lbs. Remove the guide tool.
31. Install the relief plunger, spring and plug into the front case. Tighten the plug to 34 ft. lbs.
32. Install a new oil pump cover gasket in the front case groove so that the rounded edge faces the oil pump cover.
33. Apply clean engine oil to the entire surface of the inner and outer rotor. Install the rotors in place with the marks facing each other.
34. Before installing the oil pump cover, install a new oil seal using an appropriate flat plate to guide the seal into place. Make certain the seal faces in the correct direction.
35. Install the oil pump cover, tightening the bolts to 72 inch lbs.
36. Install the oil pick–up screen with a new gasket and tighten the nuts to 14 ft. lbs.
37. Install the oil pan.

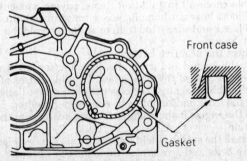

Install the pump gasket with the rounded edge facing out of the groove

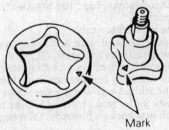

Oil pump alignment marks

ENGINE AND ENGINE OVERHAUL 3

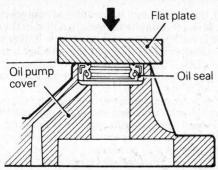

Installing the cover seal requires a flat plate to seat the seal evenly

38. Using a new gasket, install the oil filter bracket and tighten the bolts to 13 ft. lbs.
39. Coat the threads of the oil pressure switch with a sealant such as 3M ART 8660 or equivalent, but avoid getting sealant in the tip of the switch. Install the switch with the correct socket. Tighten it to 7 ft. lbs. but DO NOT overtighten the switch.
40. Install a new oil filter.
41. Install the proper amount of engine oil.
42. Check the sprockets and tensioner for wear. The sprocket teeth should be well defined, not rounded and the valleys between the teeth should be clean. The tensioner should spin freely with no binding or unusual noise. Replace the tensioner if there is any sign of grease leaking from the seal. Clean everything with a clean, dry cloth.

WARNING: Do not spray or immerse the sprockets or tensioner in cleaning solvent. The sprocket may absorb the solvent and transfer it to the belt. The tensioner is internally lubricated and the solvent will dilute or dissolve the lubricant.

43. Before reinstalling the belt, double check the sprockets for correct alignment of the timing marks.
44. To reinstall the belt, observe the directional arrow made earlier and fit the belt under the crankshaft sprocket. Track the belt up and over the right side (your right, not the engine's right) of the oil pump sprocket and the camshaft sprocket while keeping tension on the belt all the time.
45. The camshaft sprocket is quite likely to have moved during the belt installation. Gently turn the cam sprocket counterclockwise to tension the belt; when the belt is taut, the upper timing marks for the cam sprocket should align perfectly. If they do not, remove the belt and reinstall it.
46. Install the crankshaft pulley temporarily. It is needed to hold the belt in place.
47. Loosen the lock bolt in the tensioner. Allow the tensioner to move against the belt under its own spring tension; don't pry it or force it.
48. Turn the crankshaft clockwise so that the camshaft sprocket timing mark moves 2 teeth away from the mark on the cover. This is necessary to draw the proper tension on the timing belt.

WARNING: Do not turn the crankshaft counterclockwise and do not push on the belt to check its tension.

49. Now apply finger force on the tensioner, pushing it towards the belt. This will seat the belt squarely on the cam sprocket and overcome the belt's tendency to float during adjustment.
50. Tighten the tensioner nut in the tensioner, then tighten the pivot bolt for the tensioner. This order is important; if done in the wrong order, the belt will be over-tightened.
51. Hold the center of the timing belt tension side (between the cam sprocket and the oil pump sprocket) between a thumb and forefinger and gently squeeze the belt outward. The belt should deflect enough to allow 6mm (¼ inch) clearance between the back of the belt and the edge of the belt cover.
52. Carefully remove the crankshaft pulley. Install the lower timing belt cover and gasket, making sure the gaskets stay in place during installation. Observe correct bolt placement and tighten them to 8 ft. lbs.
53. Install the crankshaft pulley(s), tightening the bolts to 12 ft. lbs.
54. Install the upper timing belt cover with its gaskets and tighten the bolts to 8 ft. lbs.
55. Adjust the valve clearance. Follow the procedure outlined in Section 2.
56. Install the valve cover and gasket. Connect the PCV hose.
57. Install the water pump pulley (and power steering pulley if removed), tightening the bolts to 6 ft. lbs.
58. Install the motor mount bracket to the engine and tighten the nuts and bolts to 42 ft. lbs.
59. Align the mount bushing with the body bracket and install the through-bolt. Tighten both the large nut and the smaller safety nut snug.
60. Remove the jack from under the engine. With the weight of the engine on the mount, tighten the large nut on the through-bolt to 72 ft. lbs. and the smaller safety nut to 38 ft. lbs.
61. Install the water pump drive belt, making sure it is correctly seated on each pulley. Install the power steering belt and the compressor drive belt. Adjust each belt to the correct tension and tighten the fittings or adjuster on each component.
62. Install the coil and spark plug wires. Connect the oxygen sensor wiring. Make sure all the wiring is properly held by the clips and wire guides.
63. Using a new O-ring, reconnect the air intake duct and connect the breather hoses. Tighten the retaining nuts to 7 ft. lbs.
64. Connect the accelerator cable and adjust it.
65. Double check all installation items, paying particular attention to loose hoses or hanging wires, untightened nuts, poor routing of hoses and wires (too tight or rubbing) and tools left in the engine area.
66. Connect the negative battery cable.
67. Start the engine and let it idle, listening for any unusual noises from the area of the timing belt. Possible causes of noise are the belt rubbing against the covers or a sprocket flange, the belt being too loose and slapping, or a tensioner binding. Do not accelerate the engine if abnormal noises are heard from the timing belt train — severe damage can result.
68. Reinstall the splash shield under the car. Final adjustment of the drive belts and accelerator cable may be needed.

Mirage with 4G61 Engine

1. With the car in neutral or park, make certain the parking brake is set and the wheels are blocked. Disconnect the negative battery cable.
2. Remove the water pump drive belt and water pump pulley. If the car is equipped with air conditioning, loosen the compressor belt tensioner and remove the compressor belt.
3. Remove the bolts holding the crankshaft pulley(s) and remove the pulley(s).
4. Remove the upper and lower timing belt covers and their gaskets.
5. Use a socket wrench on the projecting crankshaft bolt to turn the engine clockwise (only!) and align all the timing marks on the sprockets and cases. It may be necessary to wipe off the area to see the marks clearly. Do not use spray cleaners around the timing belt.

When the marks all align exactly, the engine is set to TDC/compression on No. 1 cylinder. From this point onward, the engine position MUST NOT be changed.

6. If the timing belt is to be reused, make a chalk or crayon arrow on the belt showing the direction of rotation so that it may be reinstalled correctly.

3 ENGINE AND ENGINE OVERHAUL

7. Loosen the bolts holding the tensioner and pivot the tensioner towards the water pump. Temporarily tighten the bolts to hold the tensioner in its slack position.
8. Carefully slide the belt off the sprockets. Place the belt in a clean, dry, protected location away from the work area.
9. Remove the camshaft sprocket bolt and remove the sprocket.
10. Remove the crankshaft sprocket retaining bolt. Remove the sprocket and flange.
11. Remove the oil filter. Drain the engine oil.

--- CAUTION ---

Used motor oil may cause skin cancer if repeatedly left in contact with the skin for prolonged periods. Although this is unlikely unless you handle oil on a daily basis, it is wise to thoroughly wash your hands with soap and water immediately after handling used motor oil.

12. Remove the oil pan following procedures outlined previously.
13. Remove the oil pick up screen and gasket.
14. Remove the bolt holding the oil pump sprocket and remove the sprocket.
15. Remove the front case mounting bolts; remove the front case assembly and its gasket.

WARNING: The front case bolts are of several different lengths and must be replaced exactly as removed. Labeling or diagramming their placement during removal is very important.

16. Remove the oil pump cover from the front case.
17. Remove the oil pump gears.
18. Visually check the contact surfaces of the oil pump cover and the front case for wear. If excessive wear is present, replace the components. Inspect the oil passages for clogging and clean them as needed. Closely inspect the oil seals on the front case, replacing any with worn or damaged lips.
19. Insert the relief plunger into the oil filter bracket and check for smooth operation. Check the free length of the spring; correct uncompressed length is 46.6mm (1.835 in.).
20. Using a straight edge and feeler gauge to check the gear to case clearances, look for a maximum clearance of 0.25mm (0.0098 in.).
21. Apply engine clean engine oil to the oil pump gears and install them in the front case. Align the mating marks.
22. Install the oil pump cover to the front case and tighten the bolts to 16 ft. lbs.
23. Install a seal guide tool such as MD 998285-01 or its equivalent to the crankshaft. Note that the tapered end should face away from the block. Coat the surface of the guide with clean engine oil. This guide will help in installing a new seal in the front case. If the seal is already in place, the guide will protect the seal during installation.
24. Install a new front case gasket, carefully aligning all the bolt holes.
25. Install the front case to the block and lightly tighten the eight bolts. When the case is properly seated, the seal guide may be removed.
26. Install the oil filter bracket and gasket. Tighten the front case mounting bolts to 17 ft. lbs. and the oil filter bracket bolts to 13 ft. lbs.
27. Coat the relief plunger and spring in clean engine oil and install them into the oil filter bracket. Install the relief plug and gasket, tightening the plug to 34 ft. lbs.
28. Install the oil screen and gasket.
29. Install the oil pan.
30. Install a new oil filter.
31. Check that the timing marks are still aligned. Push down on the center of the tension side of the belt with your finger. Correct belt deflection is 5–7mm or approximately ¼ inch. If the deflection is not correct, the tensioner must be released fully and the belt re-tensioned.
32. Install the flange and the crankshaft sprocket onto the crankshaft. Make certain the flange is installed in the correct direction. If it is put on incorrectly, the belt will wear and break.
33. Install the special washer and the sprocket retaining bolt to the crankshaft. Tighten the bolt to 88 ft. lbs.
34. Install the camshaft sprocket to the camshaft and tighten the bolts to 66 ft. lbs.
35. Install the spacer, tensioner and tensioner spring if they were removed. Install the lower end of the spring to its position on the tensioner, then place the upper end in position at the water pump. Move the tensioner towards the water pump and temporarily tighten it in this position.
36. Install the oil pump sprocket, tightening the nut to 40 ft. lbs. Remember that the oil pump drives the left silent shaft, which is still blocked by the screwdriver. Hold the sprocket by hand when tightening the nut.
37. Double check the alignment of the timing marks for the cam, crank and oil pump sprockets.
38. Install the timing belt onto the crankshaft sprocket, the oil pump sprocket and the cam sprocket in that order. Keep the belt taut between sprockets. If reusing an old belt, make certain that the direction of rotation arrow is properly oriented.
39. Loosen the tensioner mounting nut and bolt. The spring will move the tensioner against the belt and tension it.
40. Check the belt as it passes over the camshaft sprocket. The belt may tend to lift in the area to the left of the sprocket. (Roughly 7 o'clock to 12 o'clock when viewed from the pulley end.) Make certain the belt is well seated and not rubbing on any flanges or nearby surfaces.
41. At the tensioner, tighten the bolt in the slotted hole first, then tighten the nut on the pivot. If this order is not followed, the belt will become too tight and break.
42. Once again, check all the timing marks for alignment. Nothing should have changed; check anyway.
43. Loosen the tensioner nut and bolt. The spring will allow the tensioner to tighten a little bit more because of the slack picked up during engine rotation. Tighten the bolt, then the nut to 36 ft. lbs.
44. Check the deflection of the belt. At the middle of the right (tension) side of the belt, deflect the belt outward with your finger, toward the timing case. The distance between the belt and the line of the cover seal should be about 14mm (9/16 in.).
45. Install the lower and upper timing belt covers. Make certain the gaskets are properly seated in the covers and that they don't come loose during installation. Tighten the bolts to 8 ft. lbs.
46. Install the crankshaft pulley(s) and tighten the retaining bolts to 18 ft. lbs.
47. Install the water pump pulley, tightening the bolts to 6 ft. lbs, and install the belt. Adjust the belt to the correct tension.
48. Install the correct amount of motor oil.
49. Connect the negative battery cable. Start the engine and let it idle, listening for any unusual noises from the area of the timing belt. Possible causes of noise are the belt rubbing against the covers or a sprocket flange, the belt being too loose and slapping, or a tensioner binding. Do not accelerate the engine if abnormal noises are heard from the timing belt train — severe damage can result.

Galant, Sigma, and Montero with 6G72 V6 Engine

WARNING: The use of the correct special tools or their equivalent is REQUIRED for this procedure. The crankshaft sprocket must be removed and installed with the use of tools MB 990767-01 and MB 998715 or their equivalents. Attempts to use other tools or methods may cause damage to the sprocket and/or crankshaft.

1. Disconnect the negative battery cable.
2. Remove the bolt holding the high pressure hose to the engine.

ENGINE AND ENGINE OVERHAUL 3

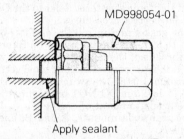

Use the special socket for turning the oil switch. Note the application of sealant to the threads before installation

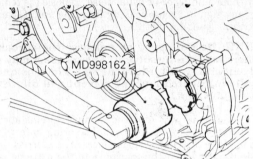

The plug cap requires a special wrench

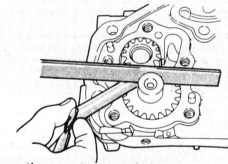

Measure the gear-to-case clearance

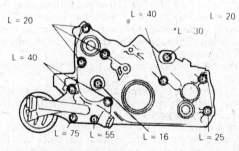

L = Bolt length below head

Correct bolt placement for the 4G61 front case. Note that the bolt marked with an asterisk is tightened differently from the others

3. Disconnect the pressure switch wiring from the power steering pump.
4. Disconnect the high pressure line from the power steering pump.
5. Place a floor jack and a broad piece of lumber under the oil pan. Raise the jack until the engine is supported but not raised.
6. Remove the through-bolt for the left engine mount.

7. Remove the four nuts holding the upper part of the engine mount to the lower bracket.
8. Remove the air conditioning compressor drive belt, then remove the belt tensioner assembly.
9. Remove the water pump drive belt.
10. Remove the power steering pump and position it out of the way.
11. Remove the belt tensioner assembly and its bracket.
12. Remove the upper outer timing belt cover (B) for the front bank of cylinders. Remove the gaskets and keep them with the cover.

NOTE: The bolts holding the belt covers are of different lengths and must be correctly placed during reinstallation. Label or diagram them as they are removed.

13. Remove the lower portion of the engine support bracket. Remove the bolts in the order shown in the figure. Use spray lubricant to assist in removing the reamer bolt. Keep in mind that the reamer bolt may be heat-seized on the bracket. Remove it slowly.
14. Remove the small end cap from the rear bank belt cover, then remove the upper outer belt cover. Make certain all the gaskets are removed and placed with the covers.
15. Remove the left side splash shield. Turn the crankshaft clockwise until all the timing marks align. This will set the engine to TDC/compression on No. 1 cylinder.
16. Install the special counterholding tools and carefully remove the crankshaft pulley without moving the engine out of position.
17. Remove the front flange from the crank sprocket.
18. Remove the lower outer timing belt cover with its gaskets.
19. Loosen the timing belt tensioner bolt and turn the tensioner counterclockwise along the elongated hole. This will relax the tension on the belt.
20. If the timing belt is to be reused, make a chalk or crayon arrow on the belt showing the direction of rotation so that it may be reinstalled correctly.
21. Carefully slide the belt off the sprockets. Place the belt in a clean, dry, protected location away from the work area. if the tensioner is to be removed, disconnect the spring and remove the retaining bolt.
22. Remove the crankshaft sprocket.
23. Support the engine/transaxle assembly from above with a hoist or crane. Remove the transaxle supports A and B.
24. Use the correct socket (MD 998054-01) to remove the oil pressure switch.

WARNING: The threads are coated with sealant. Do not break the switch during removal.

25. Drain the engine oil.

CAUTION

Used motor oil may cause skin cancer if repeatedly left in contact with the skin for prolonged periods. Although this is unlikely unless you handle oil on a daily basis, it is wise to thoroughly wash your hands with soap and water immediately after handling used motor oil.

26. Remove the oil filter, the oil filter bracket and its gasket.
27. Remove the oil pan, following instructions given earlier in this Section.
28. Remove the oil pick-up screen and its gasket.
29. Remove the plug from the front case; remove the relief spring and plunger.
30. Remove the crankshaft oil seal from the front case; discard the seal.
31. Remove the oil pump case and the gasket.

WARNING: The bolts are of different lengths. Label or diagram their locations as they are removed.

3-249

3 ENGINE AND ENGINE OVERHAUL

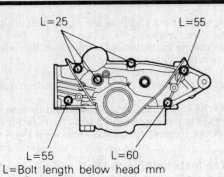

Front case bolt placement for the 6G72 V6 engine

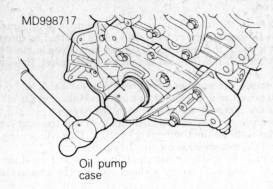

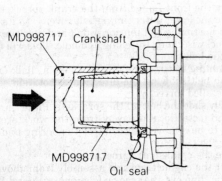

Use the correct tools to install the front crankshaft seal

32. Remove the oil pump cover from the case. Remove the oil pump gears.
33. Inspect the case for any sign of cracking or unusual wear. Check the teeth of the rotors (gears) for any sign of uneven wear or chipping. Assemble the gears into the pump and check for wear by measuring the side clearance with a straight edge and a feeler gauge. Maximum allowable clearance is 0.095mm (0.0037 in.).
34. Insert the relief plunger into the front case and check for smooth operation.
35. Install the oil pump cover and tighten the bolts to 6 ft. lbs.
36. Using a new gasket, install the front case to the engine and tighten the bolts to 9 ft. lbs.
37. The front crankshaft seal should be replaced using a seal guide and driver such as MD 998717 or equivalent. Make certain the seal is installed flush.
38. Install the relief plunger, relief spring and plug into the front case. Tighter the plug cap to 34 ft. lbs.
39. With a new gasket, install the oil pick-up screen, tightening the bolts to 13 ft. lbs.
40. Install the oil pan, observing correct application of sealant and correct bolt tightening pattern. Refer to the Oil Pan section of this Section for correct procedures.
41. Replace the gasket and install the oil filter bracket, tightening the bolts to 9 ft. lbs. Install a new oil filter.
42. Coat the threads of the oil pressure switch with a sealant such as 3M ART 8660 or equivalent, but avoid getting sealant in the tip of the switch. Install the switch with the correct socket. Tighten it to 7 ft. lbs. but DO NOT overtighten the switch.
43. Install the proper amount of engine oil.
44. Install the transaxle mounts, A and B, and tighten the bolts to 54 ft. lbs.
45. Install the crankshaft sprocket.
46. Check the sprockets and tensioner for wear. The sprocket teeth should be well defined, not rounded and the valleys between the teeth should be clean. The tensioners should spin freely with no binding or unusual noise. Replace the tensioner if there is any sign of grease leaking from the seal. Clean everything with a clean, dry cloth.

WARNING: Do not spray or immerse the sprockets or tensioners in cleaning solvent. The sprocket may absorb the solvent and transfer it to the belt. The tensioners are internally lubricated and the solvent will dilute or dissolve the lubricant.

47. If the tensioner was removed, it must be reinstalled. After bolting it loosely in place, connect the spring onto the water pump pin. Make certain the spring faces in the correct direction on the tensioner. Turn the tensioner to the extreme counterclockwise position on the elongated hole and tighten the bolt just enough to hold the tensioner in this position.
48. Double check the alignment of the timing marks on the camshaft and crankshaft sprockets.
49. Observing the direction of rotation marks made earlier, install the belt onto the crankshaft sprocket and then onto the rear bank camshaft sprocket. Maintain tension on the belt between the sprockets.
50. Continue installing the belt onto the water pump, the front bank cam sprocket and the tensioner.
51. With your fingers, apply gentle counterclockwise force to the rear camshaft sprocket. When the belt is taut on the tension side, the timing marks should align perfectly.
52. Install the flange on the crankshaft sprocket.
53. Loosen the bolt holding the tensioner one or two turns and allow the spring tension to draw the tensioner against the belt.
54. Using special tool MB 998716 or equivalent adapter, turn the crankshaft two complete revolutions clockwise. Turn the crank smoothly and re-align the timing marks at the end of the second revolution. This allows the tensioner to compensate for the normal amount of slack in the belt.
55. With the timing marks aligned, tighten the tensioner bolts to 18 ft. lbs.
56. Using a belt tension gauge, measure the belt tension at a point halfway between the crankshaft sprocket and the rear camshaft sprocket. Correct value is 57–84 lbs.
57. Making certain all the gaskets are correctly in place, install the lower timing belt cover and tighten the bolts to 8 ft. lbs.
58. Install the crankshaft pulley. Use the special tools to counterhold it and tighten the bolt to 112 ft. lbs.
59. Install the left splash shield.
60. Install the upper outer cover on the rear bank camshaft sprocket and install the smaller end cap. Tighten the bolts to 8 ft. lbs.
61. Install the lower portion of the engine support bracket. Install the bolts in the correct order. Of the 3 mounting bolts arranged vertically on the right side of the mount, the upper two are tightened to 50 ft. lbs; the bottom one is tightened to 78 ft. lbs. The two smaller bolts in the side of the mount should be tightened to 30 ft. lbs. Use spray lubricant to help with the reamer bolt, and tighten it slowly. Access may be easier if the engine is elevated slightly with the floor jack.

ENGINE AND ENGINE OVERHAUL 3

62. Install the cover and gaskets for the front bank camshaft sprocket.
63. Install the tensioner and bracket for the ribbed belt.
64. Install the power steering pump onto its bracket.
65. Install the ribbed belt and adjust it as needed. Make certain it is properly seated on all the pulleys.
66. Install the tensioner and bracket for the air conditioner drive belt and install the belt, adjusting it as necessary.
67. Install the upper section of the engine mounting bracket to the lower section. Tighten the nuts to 50 ft. lbs.
68. Adjust the jack so that the bushing of the engine mount aligns with the bodywork bracket. Install the through-bolt and tighten the nuts snug.
69. Carefully lower the jack, allowing the full weight of the engine to bear on the mount. Tighten the nut on the through-bolt to 50 ft. lbs. and the smaller safety nut to 26 ft. lbs.
70. Install the pressure hose to the power steering pump and tighten the nut to 32 ft. lbs.
71. Connect the wiring to the power steering pump and rebolt the bracket for the high pressure hose.
72. Double check all installation items, paying particular attention to loose hoses or hanging wires, untightened nuts, poor routing of hoses and wires (too tight or rubbing) and tools left in the engine area.
73. Connect the negative battery cable. Start the engine and let it idle, listening for any unusual noises from the area of the timing belt. Possible causes of noise are the belt rubbing against the covers or a sprocket flange, the belt being too loose and slapping, or a tensioner binding. Do not accelerate the engine if abnormal noises are heard from the timing belt train — severe damage can result.
74. Final adjustment of the drive belts may be needed.

Galant with 4G63 SOHC or DOHC Engine

1. Disconnect the negative battery cable. Remove the bolt holding the two hoses to the top of the left engine mount.
2. Remove the splash shield under the engine.
3. Place a floor jack and a broad piece of lumber under the oil pan. Elevate the jack just enough to support the engine without raising it.
4. Remove the nuts holding the through-bolt for the left engine bracket and remove the bolt. It may be necessary to adjust the jack tension slightly to allow the bolt to come free. Use the jack to keep the engine in its normal position while the mount is disconnected.
5. Remove the nuts and bolts holding the engine bracket to the engine and remove the bracket.
6. Loosen the appropriate component or adjuster and remove (in order) the air conditioning compressor drive belt, the power steering belt and the alternator belt. Remove the air conditioning belt tension adjuster and bracket.
7. Remove the power steering and water pump pulleys.
8. Remove the upper timing belt cover and its gaskets. Rotate the crankshaft clockwise until the timing marks on the crank pulley align at 0 on the scale (on the lower timing cover) and the mark on the camshaft pulley aligns with the indicator on the head. Aligned means perfect, not close — if you miss, continue clockwise until the marks are matched. This sets the motor at TDC/compression on No.1 cylinder.

WARNING: Once the engine is set to this position, the crank and camshafts must not be moved out of place. Severe damage can result.

9. Remove the air conditioner pulley from the crankshaft and remove the crankshaft pulley. Remember that the pulley (and therefore, the crankshaft) must be held steady while the bolt is loosened. This can be made easier by wrapping a scrapped belt tightly around the pulley. Mitsubishi has a special tool (MD 998747) for this trick, but other variations work almost as well.

DO NOT use any of the drive belts which will be reinstalled; the belt used to hold the pulley will be damaged or weakened.
10. Remove the lower timing belt cover and gaskets. Note the different bolt lengths and placements.
11. Loosen (do not remove) both bolts in the timing belt tensioner. Move the tensioner toward the water pump and tighten the lock bolt — the one in the slotted hole — to hold the tensioner in place.
12. Before removing the timing belt, mark it with an arrow to show the direction of rotation. If the belt is to be reused, it must be reinstalled so that it rotates in the same direction as before.
13. Remove the timing belt, keeping it free of grease, oil and fluids. Do not crease or crimp the belt.
14. Remove the oil filter. Drain the engine oil.

CAUTION

Used motor oil may cause skin cancer if repeatedly left in contact with the skin for prolonged periods. Although this is unlikely unless you handle oil on a daily basis, it is wise to thoroughly wash your hands with soap and water immediately after handling used motor oil.

---

15. Remove the oil pressure switch using the correct socket

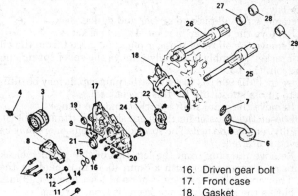

16. Driven gear bolt
17. Front case
18. Gasket
19. Oil seal
3. Oil filter
4. Oil pressure switch
6. Oil pick-up screen
7. Gasket
8. Oil filter bracket
9. Gasket
10. Relief plug
11. Gasket
12. Relief spring
13. Relief plunger
14. Plug cap
15. O-ring
20. Oil seal
21. Crankshaft front oil seal
22. Oil pump cover
23. Oil pump driven gear
24. Oil pump drive gear
25. Left silent shaft
26. Right silent shaft
27. Silent shaft front bearing
28. Right silent shaft rear bearing
29. Left silent shaft rear bearing

4G63 SOHC and DOHC oil pump, front case and silent shafts

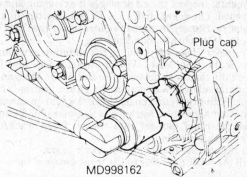

Use the special tool for turning the plug cap. Other tools may cause damage

3-251

3 ENGINE AND ENGINE OVERHAUL

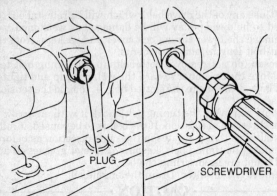

Use a screwdriver to block the left (firewall side) silent shaft

(Md 998054-01 or equivalent. The switch has sealant on the threads; take care not to break the switch during removal.

16. Remove the oil pan following procedures outlined previously.
17. Remove the oil pick up screen and gasket.
18. Remove the oil filter bracket and its gasket.
19. Remove the oil relief plunger plug and gasket from the oil filter bracket assembly. Carefully remove the relief spring and valve from the bore.
20. Use tool MD 998162 to remove the plug cap. It may be difficult to find a substitute for this special tool.
21. Remove the crankshaft sprocket and the flange.
22. Loosen the adjuster for the silent shaft belt (timing belt B). Carefully remove the belt. The sprockets and tensioners may be removed if desired.
23. Remove the plug from the left side of the cylinder block and insert a screwdriver with a 8mm (0.31 in.) shank in the hole. The screwdriver shaft must go in at least 58mm (2.3 in.). This blocks the silent shaft and keeps it from turning.
24. Remove the bolt holding the oil pump driven gear to the silent shaft and remove the gear.
25. Remove the front case mounting bolts; remove the front case assembly and its gasket.

WARNING: The front case bolts are of several different lengths and must be replaced exactly as removed. Labeling or diagramming their placement during removal is very important.

26. Remove the screwdriver from the left silent shaft. Remove the silent shafts from the engine.
27. Remove the oil pump cover from the front case.
28. Remove the oil pump gears.
29. Visually check the contact surfaces of the oil pump cover and the front case for wear. If excessive wear is present, replace the components. Inspect the oil passages for clogging and clean them as needed. The silent shafts should be inspected for any of wear or seizure. Closely inspect the oil seals on the front case, replacing any with worn or damaged lips.
30. Apply engine clean engine oil to the oil pump gears and install them in the front case. With a straight edge and feeler gauge, check the clearance between the side of the gears and the case. Maximum allowable clearance for either gear is 0.25mm (0.0098 in.). With the feeler gauge, check the gear-to-case clearance. Maximum allowable clearance is 0.25mm (0.0098 in.).
31. Insert the relief plunger into the oil filter bracket and check for smooth operation.
32. Align the mating marks on the pump gears.
33. Install the oil pump cover to the front case and tighten the bolts to 12 ft. lbs.
34. Carefully install the left and right silent shafts into the engine; the bearings can be damaged by careless installation.

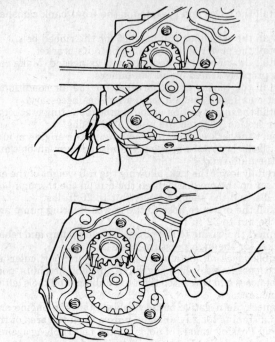

Inspecting pump gear clearances

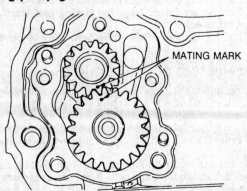

Carefully align the marks on the oil pump gears

35. Use a seal driver such as MD 998375-01 or equivalent to install a new crankshaft seal into the front case. This seal should be replaced every time the case is removed.
36. Install a seal guide tool such as MD 998285-01 or its equivalent to the crankshaft. Note that the tapered end should face away from the block. Coat the surface of the guide with clean engine oil. This guide will help in installing the front case. If the seal is already in place, the guide will protect the seal during installation.
37. Install a new front case gasket, carefully aligning all the bolt holes.
38. Install the front case to the block and lightly tighten the eight bolts. When the case is properly seated, the seal guide may be removed.
39. Insert a screwdriver through the left side access hole and block the left silent shaft.
40. Install the oil pump driven gear onto the left silent shaft and tighten the bolt to 26 ft. lbs.
41. Install a new O-ring and, with the special tool, install the plug cap, tightening it to 18 ft. lbs.
42. Coat the relief plunger and spring in clean engine oil and install them into the oil filter bracket. Install the relief plug and gasket, tightening the plug to 32 ft. lbs.

ENGINE AND ENGINE OVERHAUL 3

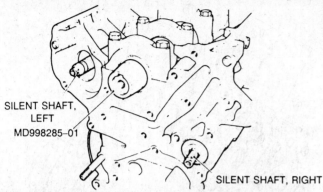

Seal guide tool in place on the crankshaft. Note that the engine shown is removed from the car and inverted

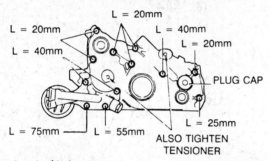

Correct bolt placement by length for front case

43. Install the oil filter bracket and gasket. Tighten the front case mounting bolts to 17 ft. lbs. and the oil filter bracket bolts to 13 ft. lbs.
44. Install the oil screen and gasket, tightening the bolts to 14 ft. lbs.
45. Install the oil pan.
46. Coat the threads of the oil pressure switch with sealant such as 3M ART 8660 or equivalent. Take care not to allow the sealant to ooze out the top of the threaded area and do not allow sealant to get on the tip of the switch. Use the special socket to install the switch and tighten it to 8 ft. lbs. Do not overtighten this switch.
47. Install a new oil filter.
48. Install the correct amount of motor oil.
49. Make certain the crankshaft sprocket and the silent shaft sprocket are aligned with the timing marks. Slip the belt over the crank sprocket and then over the silent shaft sprocket, keeping tension on the upper side. Make certain there is no slack in the tension side of the belt.
50. Install the tensioner and temporarily position it so that the center of the pulley is to the left and above the center of the installation bolt.
51. Apply finger pressure on the tensioner, forcing it into the belt. Make certain the opposite side of the belt is taut. Tighten the installation bolt to hold the tensioner in place.
52. Push down on the center of the tension side of the belt (side opposite the tensioner pulley) and check the distance the belt deflects. Correct deflection is 6mm (¼ inch).
53. Install the flange on the crankshaft sprocket, making sure it is positioned correctly.
54. Install the outer crankshaft sprocket and/or the oil pump sprocket if either was removed.
55. Install the timing belt tensioner if it was removed and position it towards the water pump. Tighten the lock bolt to hold it in this position.
56. Make certain that all the timing marks for the camshaft sprocket, the crankshaft sprocket and the oil pump sprocket are correctly aligned.
57. Once the oil pump sprocket is aligned, remove the plug in the left side of the block. Insert a Phillips screwdriver with a shaft diameter of 8mm (0.31 in.) into the plug hole. It should enter 58mm (2.3 in.) or more. Do not remove the screwdriver until the timing belt is completely installed. If the screwdriver shaft only enters about 25mm (1 inch), turn the oil pump sprocket one rotation and align the timing mark again; insert the screwdriver to block the silent shaft from turning.
58. Observe the directional arrow made during removal and install the timing belt by placing on the crankshaft sprocket, then the oil pump sprocket and then over the camshaft. Maintain the belt taut on the tension side while installing it.
59. Loosen the adjuster lock bolt and allow the spring to move the pulley against the belt. Do not push or pry on the adjuster.
60. Recheck that each sprocket is still aligned with its timing marks.
61. Remove the screwdriver from the silent shaft. Rotate the crankshaft clockwise until the camshaft sprocket has moved 2 teeth away from its mark. This applies tension to the belt. Do NOT rotate the crank counter-clockwise and do not push on the belt to check the tension.
62. With your fingers, apply gentle upward pressure on the tensioner and check the placement of the belt on the cam sprocket. The belt may tend to lift or float in the left side of the sprocket — roughly between 7 o'clock and 12 o'clock as you view the sprocket.
63. When the belt is correctly seated, tighten the tensioner lock bolt to 36 ft. lbs, then tighten the pivot nut to 36 ft. lbs. The order of tightening is important; if not followed, the belt could be over-tightened.
64. Check the deflection of the belt. At the middle of the right (tension) side of the belt, deflect the belt outward with your finger, toward the timing case. The distance between the belt and the line of the cover seal should be about 13mm (½ inch).
65. Tighten the crankshaft sprocket bolt to 88 ft. lbs.
66. Install the lower timing belt cover, paying attention to the correct placement of the bolts. Tighten the bolts to 8 ft. lbs.
67. Install the upper timing belt cover and gaskets, tightening the bolts to 8 ft. lbs. Remember to install the guides for the spark plug wires.
68. Reinstall the spark plug wires and connect them.
69. Install the crankshaft damper pulley. Tighten the bolts to 18 ft. lbs.
70. Install the water pump pulleys and tighten the bolts to 6 ft. lbs.
71. Install the tensioner pulley and the A/C drive belt; adjust the belt to the correct tension.
72. Install the power steering belt and adjust it.
73. Install the engine mount bracket to the engine. Tighten the mounting nuts and bolts to 42 ft. lbs.
74. Adjust the jack (if necessary) so that the engine mount bushing aligns with the bodywork bracket. Install the through-bolt and tighten the nuts snug.
75. Slowly release tension on the floor jack so that the weight of the engine bears fully on the mount. Tighten the through-bolt to 52 ft. lbs. and the small safety nut to 26 ft. lbs.
76. Position the high pressure hose and secure its bracket.
77. Double check all installation items, paying particular attention to loose hoses or hanging wires, untightened nuts, poor routing of hoses and wires (too tight or rubbing) and tools left in the engine area.
78. Connect the negative battery cable. Start the engine and let it idle, listening for any unusual noises from the area of the timing belt. Possible causes of noise are the belt rubbing against the covers or a sprocket flange, the belt being too loose and slapping, or a tensioner binding. Do not accelerate the engine if abnormal noises are heard from the timing belt train — severe damage can result.

3 ENGINE AND ENGINE OVERHAUL

79. Final adjustment of the drive belts may be needed. Reinstall the splash shield.

Pickup with the 4D55 Diesel Engine

NOTE: The oil pump on the diesel engine is located behind the front lower case assembly and is driven directly by the crankshaft. Removal requires disassembly of the lower front case, a job best performed with the engine out of the truck. Expect difficulty if attempting this with the engine in the truck. The use of the correct special tools or their equivalent may be REQUIRED for this procedure.

1. Elevate and safely support the vehicle.
2. Drain the engine oil.

―――――― CAUTION ――――――
Used motor oil may cause skin cancer if repeatedly left in contact with the skin for prolonged periods. Although this is unlikely unless you handle oil on a daily basis, it is wise to thoroughly wash your hands with soap and water immediately after handling used motor oil.

3. Remove the oil pan.
4. Rotate the engine by hand (clockwise) until all the timing marks align. This positions No. 1 cylinder at TDC/compression.
5. Remove the upper timing belt cover. Note that the bolts are of different lengths; diagram their location for proper reinstallation.
6. If either or both of the timing belts are to be reused, mark the direction of rotation on the belt with chalk or a felt marker.
7. Remove the center bolt holding the crankshaft pulley and remove the pulley. Do not turn the engine out of position during this removal.
8. Remove the lower timing belt cover.
9. Slightly loosen the retaining (lock) bolts for the timing belt tensioner. Move the tensioner towards the water pump and tighten the bolts; this holds the tensioner in the slack position.

NOTE: If the silent shaft timing belt is not being removed, DO NOT loosen the tensioner for the silent shaft belt.

10. Remove the camshaft sprocket and the injection pump sprocket. Do not let either shaft turn out of place during the removal. Remove the timing belt without forcing it or kinking it.
11. If the silent shaft belt is to be removed, the silent shafts must be locked in place. On the right side, remove the rubber plug from the silent shaft case and insert a screwdriver or punch; the shaft should have a diameter of 8mm or 0.32 in. On the left side, remove the cover panel from the silent shaft case and insert a broad round bar; a short extension for a socket wrench is an ideal tool.
12. Remove the silent shaft sprockets; label each sprocket so it may be reinstalled in its original position. Remove the timing belt.
13. Check the belt carefully for damage, wear or deterioration. Check the teeth carefully for any cracking at the base or separation from the backing. Do not use detergents or solvents to clean the belt or the sprockets. Clean them only with dry cloths or brushes. Pay particular attention to the edges of the belt; they should not be frayed or worn.
14. Remove the upper front case. The oil seal in the case should be carefully removed and discarded. This seal must be replaced any time the case is removed.
15. Remove the lower front case.
16. Before removing the inner and outer drive gears, make alignment marks on both gears for reference during reassembly.
17. Remove the oil pump.
18. Remove the oil pump cover from the front case.
19. Remove the oil pump gears.
20. Visually check the contact surfaces of the oil pump cover and the front case for wear. If excessive wear is present, replace

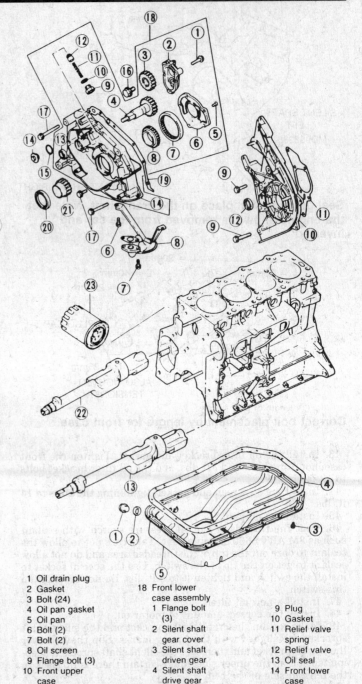

1 Oil drain plug
2 Gasket
3 Bolt (24)
4 Oil pan gasket
5 Oil pan
6 Bolt (2)
7 Bolt (2)
8 Oil screen
9 Flange bolt (3)
10 Front upper case
11 Front upper case gasket
12 Oil seal
13 Left silent shaft
14 Plug cap
15 O-ring
16 Flange bolt
17 Flange bolt (7)
18 Front lower case assembly
1 Flange bolt (3)
2 Silent shaft gear cover
3 Silent shaft driven gear
4 Silent shaft drive gear
5 Machine screw (5)
6 Oil pump cover
7 Oil pump outer gear
8 Oil pump inner gear
9 Plug
10 Gasket
11 Relief valve spring
12 Relief valve
13 Oil seal
14 Front lower case
19 Front lower case gasket
20 Front oil seal
21 Oil pump gear drive shaft
22 Right silent shaft
23 Oil filter

Diesel engine front cover assembly showing oil pump and silent shafts

the components. Inspect the oil passages for clogging and clean them as needed. The silent shafts should be inspected for any of wear or seizure. Closely inspect the oil seals on the front case, replacing any with worn or damaged lips.

21. If the silent shafts are to be removed, do so by lifting each

3-254

ENGINE AND ENGINE OVERHAUL 3

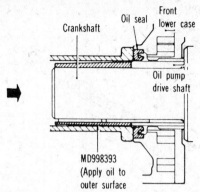

Installing diesel front cover oil seal

out of its housing. Handle the shafts carefully and do not damage the bearing surfaces.

22. When reinstalling the silent shafts, coat each one liberally with clean engine oil. Tighten the chamber cover bolts to 5 Nm (3.5 ft. lbs or 42 INCH lbs.).
23. Install the oil pump gears in the same direction as before. Take great care to align the the marks.
24. Coat the gears with a light coat of clean engine oil before installing the oil pump cover.
25. Install the oil pump cover to the front case and tighten the bolts to 16 ft. lbs.
26. Install a seal guide tool such as MD 998393 or its equivalent to the crankshaft. Note that the tapered end should face away from the block. Coat the surface of the guide with clean engine oil. This guide will help in installing a new seal in the front case. If the seal is already in place, the guide will protect the seal during installation.
27. Install a new front case gasket, carefully aligning all the bolt holes.
28. Install the upper and lower front cases to the block and lightly tighten the bolts. When the case is properly seated, the seal guide may be removed.
29. To reinstall, install the crankshaft sprocket for the silent shaft belt, the flange and the timing belt sprocket. Make certain the sprockets and flange are installed in the correct positions; if either sprocket is installed backwards, the belt may jump off or become damaged.
30. Install the crankshaft pulley; tighten the bolt to 70 Nm (52 ft. lbs.).
31. Install the injection pump sprocket and flange.
32. Install the spacer on the left silent shaft with its tapered end inward or toward the oil seal. If the spacer is not placed correctly, the oil seal will be damaged.
33. Install the silent shaft sprockets and flanges, tightening the bolts to 38 Nm (28 ft. lbs.). The shafts must be held in place with the tools used during disassembly.
34. The tensioner for the silent shaft belt should still be in the slack position. Align the timing mark of the crankshaft sprocket if necessary.
35. Align the timing marks of each silent shaft sprocket.
36. Install the silent shaft timing belt (belt B), making certain that there is no slack between the crankshaft pulley and the right (upper) silent shaft. This side of the belt is called the tension side.

WARNING: If the belt is being reused, it must be reinstalled in the same direction of rotation. Observe the marks made during disassembly.

37. Use a finger to push down on the slack side of the belt — between the tensioner and the right side silent shaft — and check that the timing marks on the sprockets remain in alignment.
38. Keep your finger on the belt. Loosen the mounting nut and bolt on the tensioner and allow the tensioner to move against the belt under its own spring tension.

WARNING: Do not assist the tensioner by pushing on it.

39. Tighten the tensioner nut, then the bolt to hold the tensioner in place. If the bolt is tightened first, the tensioner will turn and the tension will become incorrect.
40. Check the timing marks; all should be aligned. Press on the mid-point of the belt's tension side; belt deflection should be 4–5 mm (0.16–0.20 in.).
41. The tensioner for the timing belt (belt A) should be in the slack position. Double check the alignment of all three pulleys for belt A; the timing marks should all align.
42. Install the timing belt (in the correct direction if reusing the old belt) first onto the crankshaft sprocket, then onto the injection pump and then onto the camshaft sprocket. The tension side — crankshaft to injection pump — must be kept taut.

NOTE: The sprocket on the injection pump tends to rotate by itself during this procedure. Hold the sprocket in position while installing the belt. An assistant can be useful during this three-handed installation.

43. Loosen the tensioner mounting bolt and allow the spring tension to take up the slack in the belt. Don't push on the tensioner.
44. Tighten the tensioner slot side bolt before tightening the fulcrum bolt. If bolts are tightened out of order, the tensioner turns and the belt is placed under incorrect tension. After the belt is under tension, check the camshaft pulley; the belt may be lifting on the upper, outer edge (10 o'clock position). Make certain the belt is firmly seated on the sprocket.
45. Check the timing marks on all sprockets for correct alignment. Anything not in correct position will require removal of the belt(s) and alignment of the sprockets.
46. Using a wrench on the crankshaft pulley bolt, turn the engine clockwise until the camshaft sprocket has moved 2 teeth. Now turn the engine back (counterclockwise) to its original position.
47. Use a finger to press on the belt midway between the camshaft sprocket and the injection pump sprocket. Belt deflection should be 4–5mm (0.16–0.20 in).
48. Install the upper and lower timing belt covers.

Camshafts and Bearings
REMOVAL AND INSTALLATION

WARNING: The camshaft is brittle. Do not drop it or subject to impact; it can break into pieces.

G62B, G63B and G64B Engines

1. Disconnect the negative battery cable.
2. Remove the distributor.
3. Remove the valve cover.
4. Remove the rocker arm assembly following procedures given in "Rocker Arms and Shafts Removal and Installation" earlier in this Section.
5. Lift the camshaft off the cylinder head. Place it on a padded surface in a protected area. Refer to the Inspection section following for correct procedures.
6. To reinstall, thoroughly lubricate the bearing journals on the camshaft and the bearing seats in the head with clean engine oil. Place the camshaft in position on the head.
7. Liberally coat the bearing caps with clean engine oil. Install the rocker shaft and bearing cap assembly, following instructions given in "Rocker Arms and Shafts Removal and Installation".
8. Reinstall the camshaft sprocket and timing belt.
9. Adjust the valve clearance.
10. Install the valve cover.

3-255

3 ENGINE AND ENGINE OVERHAUL

11. Install the distributor. Connect the negative battery cable.

G54B Engine

1. Disconnect the negative battery cable.
2. Remove the distributor.
3. Remove the valve cover.
4. Remove the rocker arm assembly following procedures given in "Rocker Arms and Shafts Removal and Installation" earlier in this Section.
5. Remove the rear bearing cap bolts and the cap. These are not associated with the rocker shaft assembly.
6. Lift the camshaft off the cylinder head. Place it on a padded surface in a protected area. Refer to the Inspection section following for correct procedures.
7. To reinstall, thoroughly lubricate the bearing journals on the camshaft and the bearing seats in the head with clean engine oil. Place the camshaft in position on the head. Be careful not to damage the camshaft journals during installation.
8. Apply a sealer to the outside diameter of the circular seal for the rear bearing and install it in the head with one side directly in contact with the rear of the camshaft. The packing will end up under the rearmost portion of the rear bearing cap.
9. Install and torque the rocker shaft/bearing cap assembly as described in the Rocker Arm Shaft Removal and Installation procedure. Include the rear bearing cap, using the same torque.
10. Reinstall the cam sprocket and chain to the camshaft.
11. Inspect the semi-circular seal that goes in the front of the timing chain cover. Seal the top with an adhesive such as 3M Adhesive 8001 or equivalent.
12. Adjust the valve clearances.
13. Install the rocker cover.
14. Install the distributor. Connect the negative battery cable.
15. Start the engine and allow it to idle until the temperature gauge indicates normal operating temperature.
16. Remove the valve cover again and adjust the valves with the engine hot.

G15B, 4G15 and 4G63 SOHC Engine

1. Disconnect the negative battery cable.
2. Remove the rocker cover.
3. Remove the timing belt cover.
4. Remove the distributor.
5. At the timing belt tensioner, loosen the two bolts, move the timing belt tensioner toward the water pump as far as it will go and then retighten the timing belt tensioner adjusting bolt.
6. Disengage the timing belt from the camshaft sprocket and then unbolt and remove the sprocket. The timing belt may be left engaged with the crankshaft sprocket, and tensioner.

NOTE: You have effectively removed the timing belt and will need to follow the timing belt reinstallation procedures given earlier in this Section.

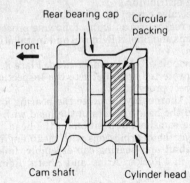

Installing the rear camshaft packing on the Starion engine

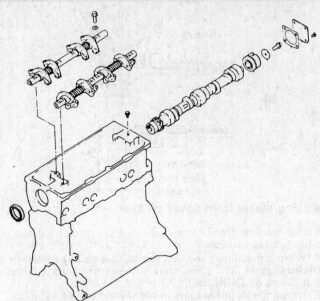

G15B camshaft with endplate and thrust case

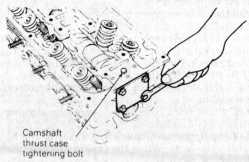

Removing the camshaft rear thrust case cover and thrust bolt from the rear of the head. This must be done before the camshaft can be removed

7. Remove the rocker shaft assembly as described previously in *Rocker Arm Shafts Removal and Installation*.
8. Remove the small, square cover that sits directly behind the camshaft on the transaxle side of the head. Remove the camshaft thrust case tightening bolt that sits on the top of the head near the square cover.
9. Very carefully, slide the entire camshaft out of the head through the hole in the transaxle side of the head. Make certain the cam lobes do not strike the bearing bores in the head. Refer to the Inspection section following for correct procedure to check the camshaft.
10. Lubricate all journal and thrust surfaces with clean engine oil. Carefully insert the camshaft into the engine, again keeping the cam lobes from touching the bearing bores. Make sure the camshaft goes in with the threaded hole in the top of the thrust case straight upward. Align the bolt hole in the thrust case and the cylinder head surface once the camshaft is all the way inside the head.
11. Install the thrust case bolt and tighten firmly.
12. Install the rear cover with a new gasket and install and tighten the four bolts.
13. Coat the external surface of the front oil seal lip with engine oil. With a special installer and guide (MD 998306-01 and MD 998307-01 or equivalent), drive the new front camshaft oil seal into place between the cam and head at the forward end. Make sure the seal seats fully.

ENGINE AND ENGINE OVERHAUL 3

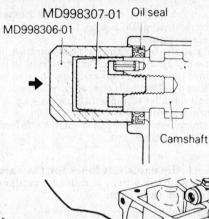

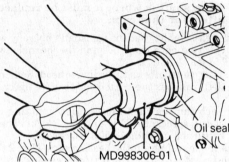

Using the special tools to install the oil seal. Note that the inner guide keeps the seal centered during installation

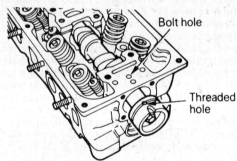

The hole in the thrust case must align with the bolt hole in the head

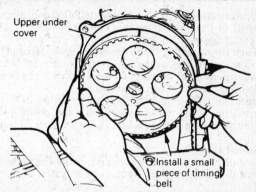

A small piece of old timing belt may be slipped under the upper camshaft sprocket to maintain timing belt tension during removal. G32B engine shown

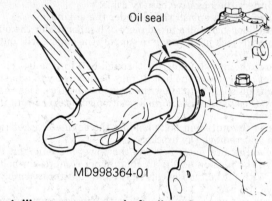

Installing a new camshaft oil seal

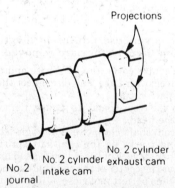

If the dowel pin hole for the camshaft sprocket cannot be aligned with the pin on the front of the camshaft, turn the cam using the two projections shown. Rest a small tool against the projection and tap the end (of the tool) lightly with a hammer

14. Install the camshaft sprocket and torque the bolt to 50 ft. lbs.
15. Reconnect the timing belt and tension the belt as described in the Timing Belt section.
16. Reinstall the rocker shaft assembly.
17. Adjust the valves in the correct sequence.
18. Install the timing belt covers.
19. Install the valve cover. Connect the negative battery cable.

G32B Engine

1. Disconnect the negative battery cable.
2. Remove the distributor.
3. Remove the valve cover and the upper timing cover. Turn the crankshaft clockwise until the timing mark on the rear timing belt cover aligns with the mark on the camshaft sprocket. It's a good idea to mark the timing belt itself to align with the marks on the sprocket and rear timing belt cover to make precise reassembly easier.
4. Remove the camshaft sprocket from the camshaft.
5. Remove the rocker arms and shaft assemblies, following procedures listed in "Rocker Arms and Shafts Removal and Installation".
6. Pull the camshaft front oil seal off the front of the camshaft. Remove the camshaft. Place it in a protected area. Refer to the following Inspection section for inspection procedures.
7. Before reinstalling, thoroughly lubricate the camshaft bearing journals, the bearing saddles in the cylinder head, and the inner surfaces of the caps with clean engine oil.
8. Install the camshaft onto the cylinder head, being careful not to damage any of the camshaft journals. Install the rocker arm and shaft assembly to the head, torquing the bolts to 15 ft. lbs.
9. Coat the outside diameter of the front end of the camshaft

3-257

ENGINE AND ENGINE OVERHAUL

with clean engine oil. With a seal installation tool such as MD 998354-01 or equivalent, tap in a new front seal.

10. Install the rear timing belt cover.
11. Turn the camshaft so the dowel pin on the front of the cam lines up with the hole in the sprocket. If you need to turn the cam, you can do so by exerting force on either of the two projections behind the No. 2 exhaust valve.
12. Reconnect the camshaft sprocket by lifting it off the rest and installing it to the camshaft with the dowel pin going through the hole in the sprocket. Tighten the retaining bolt to 50 ft. lbs.
13. Adjust the valves.
14. Install the timing belt cover and install the valve cover.
15. Install the distributor. Connect the negative battery cable.

4G61 and 4G63 DOHC Engines

NOTE: The use of the correct special tools or their equivalent is REQUIRED for this procedure.

1. Disconnect the negative battery cable.
2. Remove the left splash shield under the engine.
3. Place a floor jack and a broad piece of lumber under the oil pan. Elevate the jack just enough to support the engine without raising it.
4. Remove the nuts holding the through-bolt for the left engine bracket and remove the bolt. It may be necessary to adjust the jack tension slightly to allow the bolt to come free. Use the jack to keep the engine in its normal position while the mount is disconnected.
5. Remove the nuts and bolts holding the engine bracket to the engine and remove the bracket.
6. Remove the alternator belt and the power steering belt.
7. Loosen the tension on the air conditioner compressor belt tensioner. With the belt loosened, remove the adjuster assembly and remove the belt.
8. Remove the water pump pulley with the power steering pulley.
9. Remove the crankshaft pulley.
10. Remove the (front) upper and lower timing belt covers.

WARNING: The covers use 4 different lengths of bolts. Label or diagram them as they are removed.

11. Remove the center cover, the PCV hose and the breather hose from the valve cover. Label and disconnect the spark plug cables from the spark plugs.
12. Remove the valve cover and its gasket. Remove the half-circle plug from the head.
13. Turn the crankshaft clockwise and align all the timing marks. The timing marks on the camshaft sprockets should align with the upper surface of the cylinder head and the dowel pins (guide pins) at the center of the camshaft sprockets should be up. Aligned means perfect, not close — if you miss, continue clockwise until the marks are matched. This sets the motor at TDC/compression on No.1 cylinder.

WARNING: Once the engine is set to this position, the crank and camshafts must not be moved out of place. Severe damage can result.

14. Remove the auto-tensioner.
15. Before removing the timing belt, mark it with an arrow to show the direction of rotation. If the belt is to be reused, it must be reinstalled so that it rotates in the same direction as before.
16. Remove the timing belt, keeping it free of grease, oil and fluids. Do not crease or crimp the belt. After the belt is clear, remove the tensioner pulley.
17. Remove the throttle body bracket.
18. Carefully remove the crankshaft angle sensor from the rear end of the left (intake) camshaft.
19. Remove the exhaust camshaft sprocket, the intake camshaft sprocket and oil seals at the front of the camshafts. Use a wrench to counterhold the camshaft at the hexagonal flats ONLY to prevent turning. Do not try to wedge the two sprockets; they are easily damaged.
20. Beginning at the front camshaft bearing caps, loosen the cap bolts a little at a time, progressing from front to rear in 2 or 3 passes. Remove the bearing caps and keep them in strict order both front/rear and left/right. If the caps are difficult to remove, use a small plastic mallet to tap gently on the rear of the camshaft. The vibration will free the caps.
21. Remove the camshafts and put them in a safe location, protecting them from dust and grit.

To install:

NOTE: Check the camshaft lobes for the same conditions. If either shaft is worn, it must be replaced along with the lifters for that cam.

23. Before reassembly, coat the camshafts liberally with clean motor oil, covering both the lobes and the journals (where the bearings hold it).
24. Install the camshafts on the cylinder head. Remember that the intake (left) cam has the slotted end to drive the crank angle sensor.
25. After each camshaft is in place, turn it so that the small dowel pin at the sprocket end is up (12 o'clock). The exhaust cam must be turned an additional 3° clockwise.
26. Check the markings on the bearing caps. Nos. 2 through 5 are marked **L** (left, or intake) or **R** (right; exhaust) with a position number. The front caps (No. 1) are only marked **L** or **R**. Install the bolts finger snug — no more.
27. Tighten the bearing caps in the correct order, making 2 or 3 passes through the pattern to achieve the final torque of 15 ft. lbs. The secret is to draw the caps down very evenly, eliminating any drag or binding on the cam.
28. Apply fresh motor oil to each new camshaft oil seal. Using special tools MD 998307 and MD 998306-01, press fit the oil seal into the cylinder head over the end of the camshaft.
29. Install the camshaft sprockets, paying attention to which is exhaust and intake. Counterhold the camshaft as before and tighten each sprocket bolt to 65 ft. lbs. Maker certain the guide dowels line up with the pulley before tightening the bolt.
30. Align the notch or punch mark on the end of the crank angle sensor blade with the punched mark on the sensor case. This alignment is critical to the correct operation of the fuel injection system.
31. Install the crank angle sensor to the cylinder head, making sure the blade fits properly into the slot in the camshaft. When tightening the nuts, make certain the sensor does not turn.
32. Install the throttle body stay, tightening the nuts to 14 ft. lbs.
33. Check the sprockets and tensioner for wear. The sprocket teeth should be well defined, not rounded and the valleys between the teeth should be clean. The tensioner and idler pulleys should spin freely with no binding or unusual noise. Replace the tensioner or idler if there is any sign of grease leaking from the seal. Clean everything with a clean, dry cloth.

WARNING: Do not spray or immerse the sprockets or tensioner in cleaning solvent. The sprocket may absorb the solvent and transfer it to the belt. The tensioner is internally lubricated and the solvent will dilute or dissolve the lubricant.

34. Inspect the auto-tensioner for wear or leakage around the seals. Closely inspect the end of the pushrod for any wear. Measure the projection of the pushrod. It should be 12mm (0.47 in.). If it is out of specification, the auto tensioner must be replaced.
35. Mount the auto-tensioner in a vise with protected jaws. Keep the unit level at all times, and, if the plug at the bottom projects, protect it with a common washer. Smoothly tighten the vise to compress the adjuster tip; if the the rod is easily retracted, the unit has lost tension and should be replaced. You

ENGINE AND ENGINE OVERHAUL 3

should feel a fair amount of resistance when compressing the rod.

36. As the rod is contracted into the body of the auto-tensioner, watch for the hole in the pushrod to align with the hole in the body. When this occurs, pin the pushrod in place with a piece of stiff wire 1.5mm diameter. It will be easier to install the tensioner with the pushrod retracted. Leave the wire in place and remove the assembly from the vise.
37. Install the auto-tensioner onto the engine, tightening the bolts to 18 ft. lbs.
38. Install the tensioner pulley onto the tensioner arm. Locate the pinholes in the pulley shaft to the left of the center bolt. Tighten the center bolt finger tight. Leave the blocking wire in the auto-tensioner.
39. Double check the alignment of the timing marks for all the sprockets. The camshaft sprocket marks must face each other and align with the top surface of the cylinder head. Don't forget to check the oil pump sprocket alignment as well as the crankshaft sprocket.

NOTE: When you let go of the exhaust camshaft sprocket, it will move one tooth counter-clockwise. Take this into account when installing the belt around the sprockets.

40. Observing the direction of rotation mark made earlier, install the timing belt as follows:
    a. Install the timing belt around the tensioner pulley and the crankshaft sprocket and hold the belt on the tensioner with your left hand.
    b. Pulling the belt with your right hand, install it around the oil pump sprocket.
    c. Install the belt around the idler pulley and then around the intake camshaft sprocket.
    d. Double check the exhaust camshaft sprocket alignment mark. It has probably moved one tooth; if so, rotate it clockwise to align the timing mark with the top of the cylinder head and install the belt over the sprocket. The belt is a snug fit but can be fitted with the fingers; do not use tools to force the belt onto the sprocket.
    e. Raise the tensioner pulley against the belt to prevent sagging and temporarily tighten the center bolt.
41. Turn the crankshaft ¼ turn (90°) counter-clockwise, then turn it clockwise and align the timing marks.
42. Loosen the center bolt on the tensioner pulley and attach special tool MD 998752 and a torque wrench. (This is a purpose-built tool; substitutes will be hard to find.) Apply a torque of 1.88–2.03 ft. lbs. to the tensioner pulley; this establishes the correct loading against the belt.

NOTE: You must use a torque wrench with a 0–3 ft. lb. scale. Normal scales will not read to the accuracy needed. If the body interferes with the tools, elevate the jack slightly to raise the engine above the interference point.

43. Hold the tensioner pulley in place with the torque wrench and tighten the center bolt of the tensioner pulley to 36 ft. lbs. Remove the torque wrench assembly.
44. On the left side of the engine, located in the rear lower timing belt cover, is a rubber plug. Remove the plug and install special tool MD 998738 (another one difficult to substitute). Screw the tool in until it makes contact with the tensioner arm. After the point of contact, screw it in some more to compress the arm slightly.
45. Remove the wire holding the tip of the auto-tensioner. The tip will push outward and engage the tensioner arm. Remove the screw tool.
46. Rotate the crankshaft 2 complete turns clockwise. Allow everything to sit as is for about 15 minutes — longer if the temperature is cool — then check the distance the tip of the auto-tensioner protrudes. (It is almost impossible with the engine in the car; if you can measure the distance, it should be 3.8-4.5mm (0.15–0.18 in.). The alternate method is given in the next step.
47. Reinstall the special screw tool into the left support bracket until the end just makes contact with the tensioner arm. Starting from that position, turn the tool against the arm, counting the number of full turns until the tensioner arm contacts the auto-tensioner body. At this point, the arm has compressed the tensioner tip through its projection distance. The correct number of turns to this point is 2.5 to 3 turns. If you took the correct number of turns to bottom the arm, the projection was correct to begin with and you may now unwind the screw tool EXACTLY the number of turns you turned it in.

If your number of turns was too high or too low when the arm bottomed, you have a problem with the tensioner and should replace it.

48. Remove the special screw tool and install the rubber plug into the access hole in the timing belt cover.
49. Install the half-circle plug with the proper sealant, then install the valve cover and gasket.
50. Install the spark plug wires, the breather and PCV hoses and the center cover.
51. Install the lower and upper timing belt covers, paying attention to the correct placement of the four different sizes of bolts.
52. Install the crankshaft pulley, tightening the bolts to 18 ft. lbs.
53. Install the water pump pulley(s) and tighten the bolts to 72 inch lbs.
54. Install air conditioning drive belt and install the tensioner bracket. Tighten the bracket bolts to 18 ft. lbs.
55. Install the power steering belt and the alternator belt.
56. Install the engine mount bracket to the engine. Tighten the mounting nuts and bolts to 42 ft. lbs.
57. Adjust the jack (if necessary) so that the engine mount bushing aligns with the bodywork bracket. Install the through-bolt and tighten the nuts snug.
58. Slowly release tension on the floor jack so that the weight of the engine bears fully on the mount. Tighten the through-bolt to 72 ft. lbs. and the small safety nut to 38 ft. lbs. Don't forget to install the small support bracket for the engine mount. Tighten its bolt and nut to 16 ft. lbs.
59. Double check all installation items, paying particular attention to loose hoses or hanging wires, untightened nuts, poor routing of hoses and wires (too tight or rubbing) and tools left in the engine area.
60. Connect the negative battery cable.
61. Start the engine and let it idle, listening for any unusual noises from the area of the timing belt. Possible causes of noise are the belt rubbing against the covers or a sprocket flange, the belt being too loose and slapping, or a tensioner binding. Do not accelerate the engine if abnormal noises are heard from the timing belt train — severe damage can result.
62. Reinstall the splash shield under the car. Final adjustment of the drive belts may be needed.

6G72 V6 Engine

NOTE: This procedure applies to either bank of the V6. Each camshaft may be removed individually.

The use of the correct special tools or their equivalent is REQUIRED for this procedure. The crankshaft sprocket must be removed and installed with the use of tools MB 990767-01 and MB 998715 or their equivalents. Attempts to use other tools or methods may cause damage to the sprocket and/or crankshaft.

1. Disconnect the negative battery cable.
2. Remove the bolt holding the high pressure hose to the engine.
3. Disconnect the pressure switch wiring from the power steering pump.
4. Disconnect the high pressure line from the power steering pump.

3-259

3 ENGINE AND ENGINE OVERHAUL

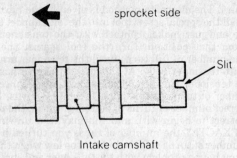

The slit drives the crank angle sensor

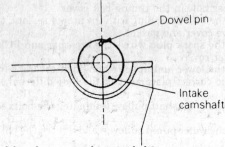

The guide pins must be straight up

5. Place a floor jack and a broad piece of lumber under the oil pan. Raise the jack until the engine is supported but not raised.
6. Remove the through-bolt for the left engine mount.
7. Remove the four nuts holding the upper part of the engine mount to the lower bracket.
8. Remove the air conditioning compressor drive belt, then remove the belt tensioner assembly.
9. Remove the water pump drive belt.
10. Remove the power steering pump and position it out of the way.
11. Remove the belt tensioner assembly and its bracket.
12. Remove the upper outer timing belt cover (B) for the front bank of cylinders. Remove the gaskets and keep them with the cover.

NOTE: The bolts holding the belt covers are of different lengths and must be correctly placed during reinstallation. Label or diagram them as they are removed.

13. Remove the lower portion of the engine support bracket. Remove the bolts in the order shown in the figure. Use spray lubricant to assist in removing the reamer bolt. Keep in mind that the reamer bolt may be heat-seized on the bracket. Remove it slowly.
14. Remove the small end cap from the rear bank belt cover, then remove the upper outer belt cover. Make certain all the gaskets are removed and placed with the covers.
15. Remove the left side splash shield. Turn the crankshaft clockwise until all the timing marks align. This will set the engine to TDC/compression on No. 1 cylinder.
16. Install the special counterholding tools and carefully remove the crankshaft pulley without moving the engine out of position.
17. Remove the front flange from the crank sprocket.
18. Remove the lower outer timing belt cover with its gaskets.
19. Loosen the timing belt tensioner bolt and turn the tensioner counterclockwise along the elongated hole. This will relax the tension on the belt.
20. If the timing belt is to be reused, make a chalk or crayon arrow on the belt showing the direction of rotation so that it may be reinstalled correctly.

21. Carefully slide the belt off the sprockets. Place the belt in a clean, dry, protected location away from the work area. if the tensioner is to be removed, disconnect the spring and remove the retaining bolt.
22. Remove the rocker cover.
23. Remove the rear camshaft seal from the end of the cam. Discard the seal.
24. Remove the camshaft oil seal from the sprocket end of the camshaft and discard the seal.
25. If working on the left (firewall side) head, remove the distributor and the distributor adapter with its O-ring.
26. Following the instructions found in the Rocker Arms and Shafts section of this Section, remove the rocker arm and shaft assembly.
27. Carefully remove the camshaft(s) and place in a protected location. If both camshafts are removed, label them plainly so that they are not interchanged during reassembly. (The left side cam has the distributor drive gear, but label them anyway.) Refer to the following Inspection section for inspection procedures.
28. Before reinstalling, thoroughly lubricate the camshaft bearing journals, the bearing saddles in the cylinder head, and the inner surfaces of the caps with clean engine oil.
29. Install the camshaft onto the cylinder head, being careful not to damage any of the camshaft journals. Install the rocker arm and shaft assembly to the head, torquing the bolts to 14–15 ft. lbs.
30. If working on the left head, reinstall the distributor adapter with a new O-ring.
31. Apply a thin coat of engine oil to the circumference of the camshaft (sprocket end) oil seal lip. Use a seal installer such as MD 998713 or equivalent to install a new front seal.
32. Install a new rear camshaft seal with the use of a seal driver such as MD 998714 or equivalent.
33. Install the camshaft sprocket, tightening the bolt to 65 ft. lbs.
34. Check the sprockets and tensioner for wear. The sprocket teeth should be well defined, not rounded and the valleys between the teeth should be clean. The tensioners should spin freely with no binding or unusual noise. Replace the tensioner if there is any sign of grease leaking from the seal. Clean everything with a clean, dry cloth.

WARNING: Do not spray or immerse the sprockets or tensioners in cleaning solvent. The sprocket may absorb the solvent and transfer it to the belt. The tensioners are internally lubricated and the solvent will dilute or dissolve the lubricant.

35. If the tensioner was removed, it must be reinstalled. After bolting it loosely in place, connect the spring onto the water pump pin. Make certain the spring faces in the correct direction on the tensioner. Turn the tensioner to the extreme counterclockwise position on the elongated hole and tighten the bolt just enough to hold the tensioner in this position.
36. Double check the alignment of the timing marks on the camshaft and crankshaft sprockets.
37. Observing the direction of rotation marks made earlier, install the belt onto the crankshaft sprocket and then onto the rear bank camshaft sprocket. Maintain tension on the belt between the sprockets.
38. Continue installing the belt onto the water pump, the front bank cam sprocket and the tensioner.
39. With your fingers, apply gentle counterclockwise force to the rear camshaft sprocket. When the belt is taut on the tension side, the timing marks should align perfectly.
40. Install the flange on the crankshaft sprocket.
41. Loosen the bolt holding the tensioner one or two turns and allow the spring tension to draw the tensioner against the belt.
42. Using special tool MB 998716 or equivalent adapter, turn the crankshaft two complete revolutions clockwise. Turn the crank smoothly and re-align the timing marks at the end of the

ENGINE AND ENGINE OVERHAUL 3

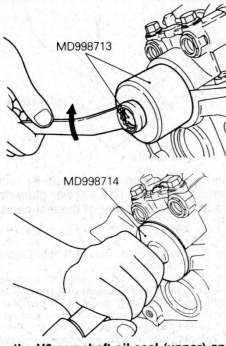

Installing the V6 camshaft oil seal (upper) and rear camshaft seal (lower) with the correct seal installation tools

second revolution. This allows the tensioner to compensate for the normal amount of slack in the belt.

43. With the timing marks aligned, tighten the tensioner bolts to 18 ft. lbs.
44. Using a belt tension gauge, measure the belt tension at a point halfway between the crankshaft sprocket and the rear camshaft sprocket. Correct value is 57-84 lbs.
45. Making certain all the gaskets are correctly in place, install the lower timing belt cover and tighten the bolts to 8 ft. lbs.
46. Install the crankshaft pulley. Use the special tools to counterhold it and tighten the bolt to 112 ft. lbs.
47. Install the left splash shield.
48. Install the upper outer cover on the rear bank camshaft sprocket and install the smaller end cap. Tighten the bolts to 8 ft. lbs.
49. Install the lower portion of the engine support bracket. Install the bolts in the correct order. Of the 3 mounting bolts arranged vertically on the right side of the mount, the upper two are tightened to 50 ft. lbs; the bottom one is tightened to 78 ft. lbs. The two smaller bolts in the side of the mount should be tightened to 30 ft. lbs. Use spray lubricant to help with the reamer bolt, and tighten it slowly. Access may be easier if the engine is elevated slightly with the floor jack.
50. Install the cover and gaskets for the front bank camshaft sprocket.
51. Install the tensioner and bracket for the ribbed belt.
52. Install the power steering pump onto its bracket.
53. Install the ribbed belt and adjust it as needed. Make certain it is properly seated on all the pulleys.
54. Install the tensioner and bracket for the air conditioner drive belt and install the belt, adjusting it as necessary.
55. Install the upper section of the engine mounting bracket to the lower section. Tighten the nuts to 50 ft. lbs.
56. Adjust the jack so that the bushing of the engine mount aligns with the bodywork bracket. Install the through-bolt and tighten the nuts snug.
57. Carefully lower the jack, allowing the full weight of the engine to bear on the mount. Tighten the nut on the through-bolt to 50 ft. lbs. and the smaller safety nut to 26 ft. lbs.

58. Install the pressure hose to the power steering pump and tighten the nut to 32 ft. lbs.
59. Connect the wiring to the power steering pump and rebolt the bracket for the high pressure hose.
60. Double check all installation items, paying particular attention to loose hoses or hanging wires, untightened nuts, poor routing of hoses and wires (too tight or rubbing) and tools left in the engine area.
61. Connect the negative battery cable. Start the engine and let it idle, listening for any unusual noises from the area of the timing belt. Possible causes of noise are the belt rubbing against the covers or a sprocket flange, the belt being too loose and slapping, or a tensioner binding. Do not accelerate the engine if abnormal noises are heard from the timing belt train — severe damage can result.
62. Final adjustment of the drive belts may be needed.

Pickup with the 4D55 Diesel Engine

1. Label and disconnect the breather hose and any vacuum hoses running around or across the valve cover.
2. Remove the valve cover.
3. Remove the upper timing cover.
4. Using a wrench on the center bolt, turn the crankshaft clockwise until all the timing marks line up. This positions the No.1 piston on TDC/compression. Both valves for No.1 cylinder are closed.
5. Without disturbing the engine position, loosen the center bolt on the camshaft sprocket.
6. Loosen the retaining bolts on the timing belt adjuster. Move the adjuster towards the water pump and temporarily tighten the bolts, holding the adjustor in the slack position.

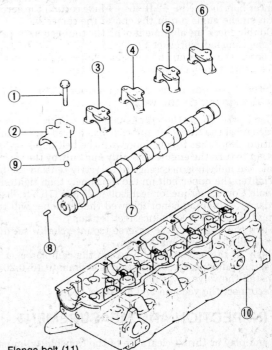

1. Flange bolt (11)
2. Camshaft bearing cap, front
3. Camshaft bearing cap, No. 2
4. Camshaft bearing cap, No. 3
5. Camshaft bearing cap, No. 5
6. Camshaft bearing cap, rear
7. Camshaft
8. Spring pin
9. Knock bushing
10. Cylinder head

Diesel camshaft and bearing caps

3 ENGINE AND ENGINE OVERHAUL

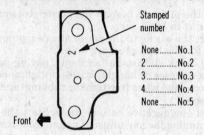

Reinstall the camshaft caps correctly

7. Remove the sprocket with the timing belt attached. Keep upward tension on the sprocket and belt. The sprocket may be supported from a piece of wire attached to the hood or overhead fixture.
8. Remove the rocker assembly, following directions outlined earlier in this Section.
9. Remove the camshaft bearing caps, keeping them in order.
10. Lift the camshaft off the cylinder head. Place it on a padded surface in a protected area. Refer to the Inspection section following for correct procedures.
11. To reinstall, thoroughly lubricate the bearing journals on the camshaft and the bearing seats in the head with clean engine oil. Place the camshaft in position on the head.
12. Liberally coat the bearing caps with clean engine oil. Install the bearing caps. Tighten the bolts evenly in two or three steps; final torque should be 20 Nm (15 ft. lbs.)
13. Reinstall the camshaft sprocket and timing belt. Tighten the center bolt to 70 Nm (52 ft. lbs.). Make certain the sprocket lines up on the guide pin in the end of the camshaft.
14. Double check the alignment of all the timing marks, particularly the camshaft sprocket.
15. Loosen the bolts on the tensioner, allowing the tensioner to move against the belt under its spring tension.

WARNING: Do not assist the tensioner by pushing on it – let the spring do the work.

16. Use a wrench on the crankshaft bolt to turn the engine clockwise until the camshaft sprocket has moved two teeth out of position. Note that this is done with the belt adjustor loose. Be certain to turn the engine smoothly and only by the specified amount; the motion is necessary to correctly tension the belt.
17. Tighten the upper bolt on the tensioner, then tighten the lower bolt. Correct torque for each bolt is 25 Nm (18.5 ft. lbs.). If the order of tightening is not followed the tensioner will turn, allowing the belt to receive incorrect tension.
18. Turn the engine in the reverse (counterclockwise) direction until all the timing mark align.
19. Push on the belt midway between the cam sprocket and the injection pump; correct deflection is 4–5mm (0.16–0.20 in.)
20. Install the valve cover and gasket.
21. Reconnect the vacuum and breather hoses.

INSPECTION AND MEASUREMENT

The end play or thrust clearance of the camshaft(s) must be measured with the camshaft installed in the head. In most cases it may be checked before removal or after reinstallation. To check the end play, mount a dial indicator accurate to ten one-thousandths (four decimal places) on the end of the block, so that the tip bears on the end of the camshaft. The timing belt must be removed. On some motors, it will be necessary to remove the sprockets for unobstructed access to the camshaft. Set the scale on the dial indicator to zero. Using a screwdriver or similar tool, gently lever the camshaft fore-and-aft in its mounts. Record the amount of deflection shown on the gauge

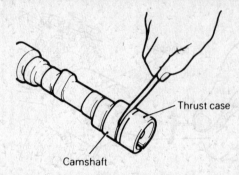

If the thrust case is removable, it must be reinstalled to measure the end play. Use a feeler gauge between the case surface and the rear of the last camshaft journal

and compare this number to the Camshaft Specifications Chart at the beginning of this Section.

If the camshaft has a removable thrust case, the case must be reinstalled to the end of the camshaft and the bolt tightened. Check the end play with a feeler gauge. If the end play is excessive, replace the thrust case with a new one and recheck the clearance. If the end play is still excessive, replace the camshaft.

Excessive end-play may indicate either a worn camshaft or a worn head; the worn cam is most likely and much cheaper to replace. Chances are good that if the cam is worn in this dimension (axial), substantial wear will show up in other measurements.

Using a micrometer or Vernier caliper, measure the diameter of all the journals and the height of all the lobes. Record the readings and compare them to the Camshaft Specifications Chart in the beginning of this Section. Any measurement beyond the stated limits indicates wear and the camshaft must be replaced.

Lobe wear is generally accompanied by scoring or visible metal damage on the lobes. Overhead camshaft engines are very sensitive to proper lubrication with clean, fresh oil. A worn cam may be your report card for poor maintenance intervals and late oil changes.

If a new cam is required, order new rockers to accompany it so that there are two new surfaces in contact. On the twin-cam motors either camshaft (with rockers) may be replaced individually.

The clearance between the camshaft and its journals (bearings) must also be measured. The amount of space between the camshaft and its bearings is the oil clearance; less space means

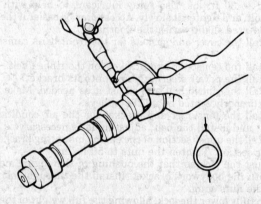

Measure the camshaft dimensions to determine wear

ENGINE AND ENGINE OVERHAUL 3

less oil and more wear. Excessive play will allow the camshaft to hammer the bearing and cause wear. Clean the camshaft, the journals and the bearing caps of any remaining oil and place the camshaft in position on the head. Lay a piece of compressable gauging material (Plastigage® or similar) on top of each journal on the cam.

Install the bearing caps in their correct order and install the rocker assembly. Tighten the bolts in three passes to the correct torque.

WARNING: Do not turn the camshaft with the gauging material installed.

Remove the rocker assembly and measure the gauging material at its widest point by comparing it to the scale provided with the package. Compare these measurements to the Camshaft Specifications Chart at the beginning of this Section. Any measurement beyond specifications indicates wear. If you have already measured the cam (or replaced it) and determined it to be usable, excess bearing clearance indicates the need for a new head due to wear of the journals.

Remove the camshaft from the head and remove all traces of the gauging material. Check carefully for any small pieces clinging to contact faces.

After a careful cleaning, coat the camshaft and bearing surfaces with clean engine oil and continue reassembly.

Pistons and Connecting Rods

NOTE: This procedure requires removal of the head and oil pan. It is much easier to perform this work with the engine removed from the vehicle and mounted on a stand.

These procedures require certain hand tools which may not be in your tool box. A cylinder ridge reamer, a numbered punch set, piston ring expander, snap-ring tools and piston installation tool (ring compressor) are all necessary for correct piston and rod repair. These tools are commonly available from retail tool suppliers; you may be able to rent them from larger automotive supply houses.

REMOVAL

1. If the pistons and rods are being removed as part of a complete tear-down, follow the procedures described in *Crankshaft and Main Bearings* until the crankshaft is removed. If you are removing the pistons and connecting rods without removing the crankshaft and main bearings, proceed from this point.
2. Remove the cylinder head, following correct procedures listed earlier in this Section.
3. Remove the oil pan, following correct procedures listed earlier in this Section.
4. The connecting rods are marked to indicate which surface faces front, but the bearing caps should be matchmarked with numbers 1–4 (front to rear) before disassembly. Use a marking punch and a small hammer; install the number over the seam so that each piece will be re-used in its original location.
5. Remove the connecting rod cap bolts, pull the caps off the rods, and place them on a bench in order.
6. Inspect the upper portions of the cylinder (near the head) for a ridge formed by ring wear. If there is a ridge, it must be removed by first shifting the piston down in the cylinder and then covering the piston top completely with a clean rag. Use a ridge reamer to remove metal at the lip until the cylinder is smooth. If this is not done, the rings will be damaged during removal of the piston.
7. Once the ridges have been removed, the pistons and rods may be pushed upward and out of the cylinders. Place pieces of rubber tubing over the rod bolts to protect the cylinder walls.

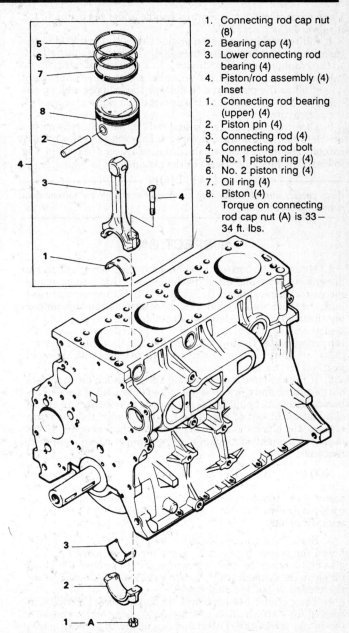

1. Connecting rod cap nut (8)
2. Bearing cap (4)
3. Lower connecting rod bearing (4)
4. Piston/rod assembly (4)
   Inset
1. Connecting rod bearing (upper) (4)
2. Piston pin (4)
3. Connecting rod (4)
4. Connecting rod bolt (4)
5. No. 1 piston ring (4)
6. No. 2 piston ring (4)
7. Oil ring (4)
8. Piston (4)
   Torque on connecting rod cap nut (A) is 33–34 ft. lbs.

Exploded view of Starion pistons, connecting rods and associated parts

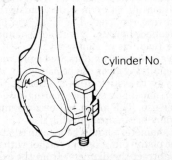

Mark the side of the piston rod and rod cap with the cylinder number. Note that by overlapping the seam, only 1 mark need be made

3-263

3 ENGINE AND ENGINE OVERHAUL

Use a piece of wood or a hammer handle under the piston to tap it upward.

If you're working under an engine that's still installed in the car with the crankshaft still in position, turn the crankshaft until the crankpin for each cylinder is in a convenient position. Be careful not to subject the piston and/or rod to heavy impact and do not allow the piston rod to damage the cylinder wall on the way out. The slightest nick in the metal can cause problems after reassembly.

8. Clean the pistons, rings and rods in parts solvent with a bristle brush. Do not use a wire brush, even to remove heavy carbon. The metal may be damaged. Use a piece of a broken ring to clean the lands (grooves) in the piston.

---------- CAUTION ----------
Wear goggles and gloves during cleaning. Do not spatter solvent onto painted surfaces.

INSPECTION

1. Measure the bore of the cylinder at three levels and in two dimensions (fore-and-aft and side-to-side). That's six measurements for each cylinder. By comparing the three vertical readings, the taper of the cylinder can be determined and by comparing the front-rear and left-right readings the out-of-round can be determined. The block should be measured: at the level of the top piston ring at the top of piston travel; in the center of the cylinder; and at the bottom. Compare your readings with the specifications in the chart.

2. If the cylinder bore is within specifications for taper and out-of-round, and the wall is not scored or scuffed, it need not be bored. If not, it should be bored oversize as necessary to ensure elimination of out-of-round and taper. Under these circumstances, the block should be taken to a machine shop for proper boring by a qualified machinist using the specialized equipment required.

NOTE: If the cylinder is bored, oversize pistons and rings must be installed. Since all pistons must be the same size (for correct balance within the engine) ALL cylinders must be rebored if any one is out of specification.

3. Even if the cylinders need not be bored, they should be fine honed for proper break-in by a qualified machine shop. A deglazing tool may be used in a power drill to remove the glossy finish on the cylinder walls. Use only the smooth stone type, not the beaded or bottle-brush type.

4. The cylinder head top deck (gasket surface) should be inspected for warpage. Run a straightedge along all four edges of the block, across the center, and diagonally. If you can pass a feeler gauge of 0.1mm (0.004 in.) under the straightedge, the top surface of the block should be machined.

5. The rings should be removed from the pistons with a ring expander. Keep all rings in order and with the piston from which they were removed. The rings and piston ring grooves should be cleaned thoroughly with solvent and a brush as deposits will alter readings of ring wear.

6. Before any measurements are begun, visually examine the piston (a magnifying glass can be handy) for any signs of cracks--particularly in the skirt area--or scratches in the metal. Anything other than light surface scoring disqualifies the piston from further use. The metal will become unevenly heated and the piston may break apart during use.

7. Piston diameter should be measured at the skirt, at right angles to the piston pin. Compare either with specified piston diameter or subtract the diameter from the cylinder bore dimension to get clearance, depending upon the information in the specifications.

If clearance is excessive, the piston should be replaced. If a new piston still does not produce piston-to-wall clearance within

Inspecting the cylinder block. Inspect each cylinder at the locations and in the directions shown

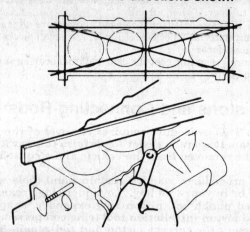

Inspect the top deck of the block in the directions shown

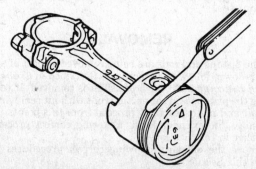

Measuring piston ring side clearance

specifications, use an oversize piston and bore out the cylinder accordingly.

8. Compression ring side clearance should be measured by using a ring expander to put cleaned rings back in their original positions on the pistons. Measure side clearance on one side by attempting to slide a feeler gauge of the thickness specified between the ring and the edge of the ring groove. If the gauge will not pass into the groove, the ring may be re-used. If the gauge will pass, but a gauge of slightly greater thickness representing the wear limit will not, the piston may be re-used, but new rings must be installed.

9. Ring end gap must be measured for all three rings in the cylinder by using a piston top (upside down) to press the ring squarely into the top of the cylinder. The rings must be at least 15mm (0.59 in.) from the bottom of the bore. Use a feeler gauge

ENGINE AND ENGINE OVERHAUL 3

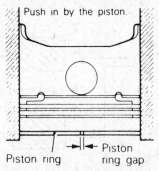

Use an inverted piston to seat a ring in the cylinder. Measure the end gap with a feeler gauge

to measure the end gap and compare it with specifications. If the gap is too great, the ring should be checked with a gauge of 1.0mm (0.039 in.) to see if it is within the wear limit.

If cylinder bore wear is very slight, you may use new rings to bring the end gap to specification without boring the cylinder. Measure the gap with the ring located near the minimum dimension at the bottom of the cylinder, not nearer the top where wear is greatest.

10. The connecting rods must be free from wear, cracking and bending. Visually examine the rod, particularly at its upper and lower ends. Look for any sign of metal stretching or wear. The piston pin should fit cleanly and tightly through the upper end, allowing no sideplay or wobble. The bottom end should also be an exact ½ circle, with no deformity of shape. The bolts must be firmly mounted and parallel.

The rods may be taken to a machine shop for exact measurement of twist or bend. This is generally easier and cheaper than purchasing a seldom used rod-alignment tool.

INSTALLATION

1. Remember that if you are installing an oversize piston, you must also use new rings that are also of the correct oversize.
2. Install the rings on the piston, *lowest ring first*. Generally the side rails for the oil control ring can be installed by hand with care. The other rings require the use of the ring expander. There is a high risk of ring breakage or piston damage if the rings are installed without the expander. The correct spacing of the ring end gaps is critical to oil control.

No two gaps should align; they should be evenly spaced around the piston with the gap in the oil ring expander facing the front of the piston (aligned with the mark on the top of the piston). Once the rings are installed, the pistons must be handled carefully and protected from dirt and impact.

3. Install the number two compression ring next and then the top compression ring using a ring expander. Note that these rings have the same thickness but different cross-sections; make sure positioning is correct. Make sure all markings face upward and that the gaps are all staggered. Gaps must also not be in line with either the piston pin or thrust faces of the piston.

4. All the pistons, rods and caps must be reinstalled in the correct cylinder. Make certain that all labels and stamped numbers are present and legible. Double check the piston rings; make certain that the ring gaps DO NOT line up, but are evenly spaced around the piston at about 120° intervals. Double check the bearing insert at the bottom of the rod for proper mounting. Reinstall the protective rubber hose pieces on the bolts.

5. Liberally coat the cylinder walls and the crankshaft journals with clean, fresh engine oil. Also apply oil to the bearing surfaces on the connecting rod and the cap.

6. Identify the **Front** mark on each piston and rod and position the piston loosely in its cylinder with the marks facing the front (pulley end) of the motor.

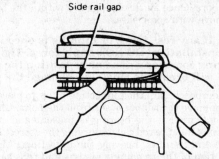

Installing the oil ring side rail without damaging it

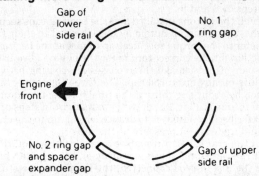

Stagger the ring gaps to reduce oil consumption and increase compression

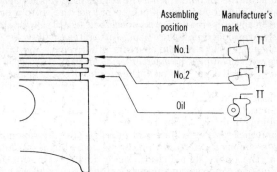

Diesel piston rings must be installed correctly

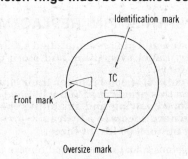

4D55 diesel piston markings

WARNING: Failure to observe the "Front" marking and its correct placement can lead to sudden and catastrophic engine failure.

7. Install the ring compressor (piston installation tool) around one piston and tighten it gently until the rings are compressed almost completely.
8. Gently push down on the piston top with a wooden hammer handle or similar soft-faced tool and drive the piston into

3-265

3 ENGINE AND ENGINE OVERHAUL

the cylinder bore. Once all three rings are within the bore, the piston will move with some ease.

WARNING: If any resistance or binding is encountered during the installation, DO NOT apply force. Tighten or adjust the ring compressor and/or reposition the piston. Brute force will break the ring(s) or damage the piston.

9. From underneath, pull the connecting rod into place on the crankshaft. Remove the rubber hoses from the bolts. Check the rod cap to confirm that the bearing is present and correctly mounted, then install the rod cap (observing the correct number and position) and its nuts. Leaving the nuts finger tight will make installation of the remaining pistons and rods easier.
10. Assemble the remaining pistons in the same fashion, repeating steps 7, 8 and 9.
11. With all the pistons installed and the bearing caps secured finger tight, the retaining nuts may be tightened to their final setting. Refer to the torque specifications chart at the beginning of this Section for the correct torque for the engine in your car. For each pair of nuts, make three passes alternating between the two nuts on any given rod cap. The three tightening steps should each be about one third of the final torque; for example, if the final torque is 36 ft. lbs, draw the nuts tight in steps of 12, 24 and 36 ft. lbs. The intent is to draw each cap up to the crank straight and under even pressure at the nuts.
12. Turn the crankshaft through several clockwise rotations, making sure everything moves smoothly and there is no binding. With the piston rods connected, the crank may be stiff to turn. Try to turn it in a smooth continuous motion so that any binding or stiff spots may be felt.
13. Reinstall the oil pan. Even if the engine is to remain apart for other repairs, install the oil pan to protect the bottom end and tighten the bolts to the correct specification; this eliminates one easily overlooked mistake during future reassembly.
14. If the engine is to remain apart for other repairs, pack the cylinders with crumpled newspaper or clean rags (to keep out dust and grit) and cover the top of the motor with a large rag. If the engine is on a stand, the whole block can be protected with a large plastic trash bag.
15. If no further work is to be performed, continue reassembly by installing the head, timing belt, etc.
16. When the engine is restarted after reassembly, the exhaust will be very smoky as the oil within the cylinders burns off. This is normal; the smoke should clear quickly during warm up. Depending on the condition of the spark plugs, it may be wise to check for any oil fouling after the engine is shut off.

PISTON PIN (WRIST PIN) REPLACEMENT

NOTE: The piston and pin are a matched set and must be kept together. Label everything and store parts in identified containers.

The use of the correct special tools or their equivalent is REQUIRED for this procedure. You will need tool set MD 998184-01 for removing and installing the piston pin. You will also need access to a hydraulic press, capable of delivering up to 4000 lbs. of force.

1. Remove the pistons from the engine and remove the rings from the pistons.
2. Set up the tool set with the pin pusher on the press bed. Make certain the press plates are securely fastened to the press frame.
3. Fit the piston and rod assembly onto the body of the special tool, with the front marks on the rod and piston facing up. The numbers or letters on the side of the rod serve as the front mark for the rod; the piston usually has an arrowhead ➤ pointing to the front. The V6 pistons are marked **L** and **R**; if the piston came from the left (firewall) side, the **L** designates the front. If from the right bank, use the **R** as the front marker.
4. Position the piston so that the lip of the tool insert fits be-

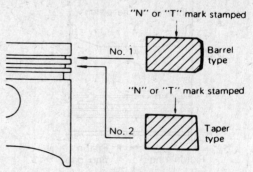

Install the rings with the marks upward. The barrel-faced chrome plated ring goes in the No. 1 position while the taper type goes in position No. 2

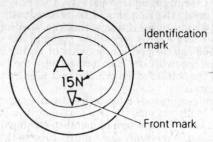

Piston identification marks. The V6 pistons use "L" and "R" depending on which bank contained the piston

tween the connecting rod boss and the inside of the piston. The boss should contact as much of the tool insert as possible to support it.
5. Adjust the support at the rear of the connecting rod until the rod is horizontal (level) to the press bed. Misalignment of the pin and receiving tube may occur if the rod is crooked.
6. Position the pin pusher on the pin and use the press to drive the pin out. As the pin is removed, it must pass through the receiving tube. Adjust the alignment as necessary.
7. Remove the rod and piston from the press. Check the pin for obvious wear or damage. A quick check of piston pin wear is to push the pin into the piston (without the rod) with your thumb. The pin should fit with slight resistance. If the pin is too loose or has any play, either the pin or piston is worn; both should be replaced.
8. To reinstall, set up the special press tools with the stop plug threaded about halfway into the bottom of the receiving tube.
9. Select the largest piston pin guide that will pass through the piston and rod. Install the spring, spacer and guide into the receiving tube.
10. With the connecting rod removed from the piston, insert the piston pin into the piston so that it is centered in the piston. Accurately measure the amount of protrusion (or recess) between the pin and piston and make certain the measurement is the same on the opposite side. Record the measurement.
11. Place the connecting rod and piston over the special tool on the press with the front marks facing up. The spring loaded piston guide will pass through the piston and rod and align them.
12. Coat the piston pin with fresh motor oil and install it into the piston. Place the pin pusher on the piston pin and use the press to insert it through the connecting rod and opposite side of the piston.
13. Continue pressing until the pin projection (or recess) is equal to that measured in Step 10. apply pressure to the pin and adjust the stop plug until it comes into contact with the spacer.

ENGINE AND ENGINE OVERHAUL 3

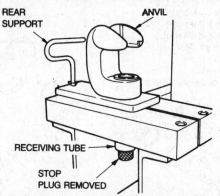

Assemble the piston pin tools onto the bed of a hydraulic press

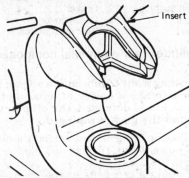

The insert will support the rod during removal and installation

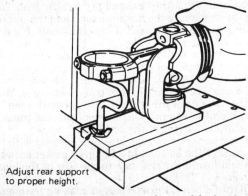

Adjust the rear support so the rod is horizontal and parallel to the press bed

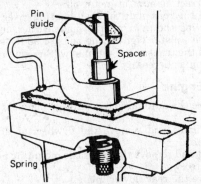

Install the spring, spacer and guide into the receiving tube

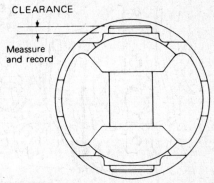

The clearance measurement is important; the pin must be centered after installation

14. Remove the piston assembly from the press and check the piston pin for centering. If it is not centered, adjust the stop plug up or down to obtain the correct position. The pin stop is now set for any remaining pistons.

WARNING: Installation pressure on the press ranges from about 1100 lbs. to about 3600 lbs. If the pin cannot be installed within this range, replace the rod and/or pin. Pressure extremes will break the piston or rod.

ROD BEARING REPLACEMENT

1. Crankshaft crankpins (the part of the crank on which the bearing rides) should be inspected for clogged oil holes and scoring. If scoring is evident, the crankpins should be machined undersize and the proper undersize (thicker) bearings installed in all the rods.
2. If the crankpins and rod bearing shells appear to show minimal wear and normal wear patterns, they still must be checked for proper fit with Plastigage. Clean all the surfaces of oil and then cut plastic measuring media (Plastigage® or similar) inserts to a length that will fit the width of the bearing. Apply the insert to an area of the crankpin away from an oil hole and install the rod cap. Tighten the bolts to specification. Make sure you don't turn the crankshaft.
3. Remove the connecting rod bolts and then remove the caps. Read the width of the Plastigage at its widest part by comparing it with the scale on the package to get actual bearing clearance. If the bearing clearance is within specification, the bearings may be reused.
4. You must also measure side clearance between the connecting rod side surface and the crank "cheek" with the rod assembled and torqued. Excessive wear here indicates the rod and cap must be replaced to restore clearance.

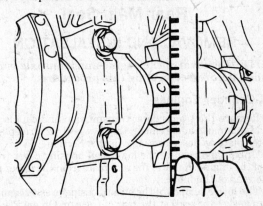

Reading the width of the mark left by Plastigage®

3 ENGINE AND ENGINE OVERHAUL

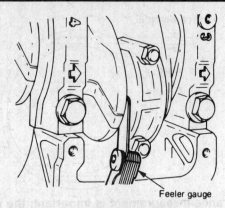

Checking connecting rod side clearance

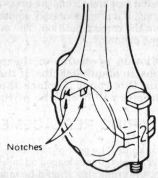

Make sure the notches in the connecting rod and cap align as shown and that notches in the bearings align with these notches

5. When the bearing shells are replaced, make sure bearing backs and inner surfaces of the rod and cap are dry and that bearings and crankpins are well lubricated with clean engine oil. It is important to insure that all bearings are installed with the notches (in the bearings, rod and cap) aligned; this will keep the bearing from spinning out of position.

Notches in the rod and cap should be on the same side. This is a way to make sure rod and cap are properly assembled.

6. Torque the rod bolts to specification, going back and forth in several steps.

Rear Main Seal

REMOVAL AND INSTALLATION

NOTE: Removal of the engine and transaxle is recommended before performing this procedure.

All 4-Cylinder Engines

1. Remove the transaxle or transmission and clutch.
2. Matchmark and remove the flywheel or driveplate and adapter plate as described later in this Section.
3. Unbolt and remove the lower bell housing cover from the rear of the engine. Remove the rear plate from the upper portion of the rear of the block.
4. The lower surface of the oil seal casing presses against the oil pan gasket (or seal) at the rear of the pan. On engines with a gasket, carefully separate the gasket from the bottom of the seal case with a moderately sharp instrument.

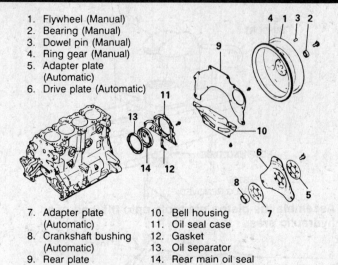

1. Flywheel (Manual)
2. Bearing (Manual)
3. Dowel pin (Manual)
4. Ring gear (Manual)
5. Adapter plate (Automatic)
6. Drive plate (Automatic)
7. Adapter plate (Automatic)
8. Crankshaft bushing (Automatic)
9. Rear plate
10. Bell housing
11. Oil seal case
12. Gasket
13. Oil separator
14. Rear main oil seal

Typical 4-cylinder rear main seal components

NOTE: You may want to loosen the oil pan bolts slightly at the rear to make it easier to separate the two surfaces. If the gasket is damaged, the oil pan will have to be removed and the gasket replaced.

5. On vehicles employing sealant around the oil pan, unbolt and remove the pan. Clean the pan and block surfaces of all traces of sealant.
6. Remove the oil seal case bolts, and pull the case straight off the rear of the crankshaft. Remove the case gasket.
7. Remove the seal retainer or oil separator (not all engines use an oil separator) from the case and pry out the seal. Take care not to gouge or damage the metal surrounding the seal. Inspect the sealing surface at the rear of the crankshaft. If a deep groove is worn into the surface, the crankshaft will have to be replaced. Lubricate the sealing surface with clean engine oil.
8. Using a seal installer of the correct size (examples: MD 998376-01 and MD 990938-01 or equivalents; 998376 for diesel), install the new seal into the bore of rear oil seal case. Make certain that the flat side of the seal will face outward when the case is installed on the engine. The inside of the seal must be flush with the inside surface of the seal case.
9. Install the retainer or oil separator over the seal with the small hole located at the bottom. Install a new gasket onto the block surface and install the seal case to the rear of the block. Tighten the bolts to 8 ft. lbs.
10. Reinstall the oil pan if it was removed, making certain the sealant is correctly applied. If the pan was only loosened, retighten the bolts to the correct torque. If the oil was drained, install the correct amount of motor oil.
11. Install the rear plate and bell housing cover.
12. Observing the matchmarks made earlier, install the flywheel or drive plate. Tighten the bolts to 98 ft. lbs.
13. Install the transaxle. Continue with engine reassembly or reinstallation as necessary.

6G72 V6 Engine

1. Remove the automatic transaxle or manual transaxle and clutch.
2. Matchmark the flywheel or driveplate and adapter plate.
3. Unbolt and remove the transaxle mounting plate from the rear of the block.
4. Remove the rear (lower) plate.
5. Unbolt the oil seal case and then pull it straight back and off the crankshaft. Remove the case gasket. Pry the old seal out of the case.

ENGINE AND ENGINE OVERHAUL 3

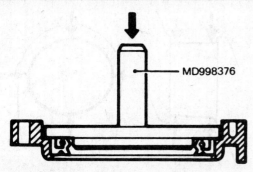

The new oil seal is installed into the case with the correct seal installation tool. Case shown is common to Cordia, Tredia, Galant and Starion

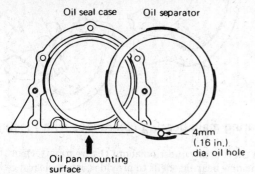

When pushing the oil separator into the case, make certain the oil hole is in the bottom position for Cordia, Tredia, Galant and Starion

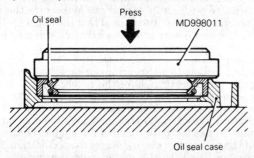

Using the correct tool to press a new seal into the Mirage seal case

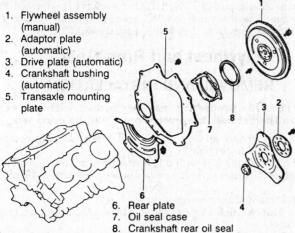

1. Flywheel assembly (manual)
2. Adaptor plate (automatic)
3. Drive plate (automatic)
4. Crankshaft bushing (automatic)
5. Transaxle mounting plate
6. Rear plate
7. Oil seal case
8. Crankshaft rear oil seal

V6 rear main seal components

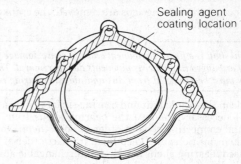

Apply sealant to the face of the V6 seal case

6. Inspect the sealing surface at the rear of the crankshaft. If a deep groove is worn into the surface, the crankshaft will have to be replaced. Press a new seal into the case with a seal installing tool such as MD 998718. The seal must be pressed in square until it bottoms in the case.

7. Apply liquid sealant such as 3M No. 8660 or its equivalent to the flange and bolt areas of the case. Apply a small amount of clean engine oil to the lip of the new seal.

8. Install the seal, gasket and seal case straight over the crankshaft surface. Install the case bolts, tightening them to 8 ft. lbs.

9. Install the rear plate, tightening the bolts to 7 ft. lbs.

10. Install the transaxle mounting plate, tightening the bolts to 8 ft. lbs.

11. Install the flywheel or drive plate and adapter. Tighten the mounting bolts to 54 ft. lbs.

12. Install the transaxle. Continue with engine reassembly or reinstallation as necessary.

Crankshaft and Main Bearings

REMOVAL

1. Remove the engine from the vehicle. Mount it on a stand which allows the engine to be rotated for easy access.
2. If not already done, remove the transaxle from the engine.
3. Remove the cylinder head and the front cover, timing belt or chains and sprockets as described earlier in this Section.
4. Remove the oil pan (if not removed already) and pick-up screen.
5. Remove the oil pump.
6. Remove the flywheel and pilot bearing or torque converter drive plate, adapter plate and crankshaft bushing.
7. Remove the bell housing cover and rear plate, on those models so equipped.
8. Remove the rear main seal housing and seal.
9. Check the crankshaft end play. Shift the crankshaft as far to the rear as possible. Install a dial indicator on the front of the engine with the stem directly on the front of the crankshaft. Zero the indicator and use a small pry bar to shift the crankshaft forward as far as it will go. The amount of motion will show on the dial indicator; read the figure and compare with specifications. Excess end play is a sign of wear.

NOTE: It's best to check the end play with the engine in its upright (normal) position.

10. Invert the engine, and remove the connecting rod caps, keeping them in exact order. Loosen the bolts slowly and evenly. It is recommended that each cap be marked with its cylinder number before disassembly.

11. Remove the main bearing caps (crankshaft caps), keeping them in order. The caps are marked with arrows indicating the front of the engine, and are numbered in order from front to rear.

3-269

3 ENGINE AND ENGINE OVERHAUL

12. Once all the bearing caps are removed, the crankshaft may be removed if necessary.

CAUTION

The crankshaft is a heavy component and can cause damage or injury if dropped or banged into other components during removal. The polished surfaces must be protected from dirt and abrasion during repairs.

---

13. Clean the crankshaft and bearing surfaces with a parts solvent. If necessary, remove the bearing shells from the caps, block and connecting rods in order to soak them. Note or diagram their positions and keep them in exact order at all times. Once used, bearing shells are NOT interchangable and must be reinstalled in their original location.

INSPECTION AND CLEANING

14. Inspect the crankshaft journals for scuffing, grooving or scoring. If crankshaft wear is uneven, the crank should be turned on a lathe by a competant machine shop. Each journal will be undersize and new, undersize (thicker) bearings must be installed.
15. Make sure all the oil passages are clean, if necessary cleaning them with solvent and a stiff bristle brush.
16. It is recommended that crankpins and main bearing journals be measured with a micrometer to check for wear, out-of-roundness, and taper. Out-of-round is checked by measuring the journal across 2 diameters which are 90° apart (north-south and east-west) and comparing the two readings. The difference between the two readings gives the out-of-round specification.

Taper is determined by measuring each journal at each end and comparing the numbers. If one reading is larger than the other, the journal tapers from large to small. Please refer to the charts at the beginning of this Section for the correct crankshaft specifications for your engine.

17. Inspect the bearing surfaces for burning, grooving, or scoring of any kind. Any but the slightest scratches disqualify a bearing from reuse; if there is any doubt, replace the bearings. (A good rule of thumb is to replace the bearing every time they are disassembled.) If the end play, as measured above, is excessive, new bearings must be installed.

NOTE: If new bearings fail to correct end play problems, the crankshaft will have to be machined undersize and different bearings installed.

INSTALLATION

18. Once the crankshaft has been cleaned, inspected or machined, the bearing clearance must be checked. this must be done in ALL cases, including reuse of old bearings or installation of thicker bearings. The bearing clearance (or oil clearance) is a critical dimension and MUST be proper before final assembly.
19. Install the crankshaft into the block (the block is inverted) and make sure all bearing surfaces and journals are clean and dry. Cut pieces of plastic measuring media (such as Plastigage® or similar product) to fit the width of the bearing. Lay the piece widthwise across the journal. The insert must not rest on top of oil holes and the bearing shell that clamps against it must not be a grooved one.
20. Install and torque all rod and main bearing caps to specification. Do NOT turn the crankshaft! Remove the caps and read the width of the measuring media at the widest point. This is done by comparing the width at its widest point to lines on the scale included with the package.
21. If the bearing clearance is to specification, old bearings may be re-used if otherwise in good condition. In cases of moderate, regular wear, it may be feasible to replace the bearings with new ones of original size and avoid machining the crankshaft, provided this gives the specified clearance.

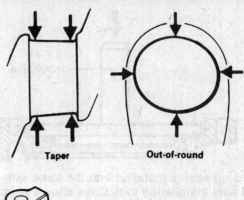

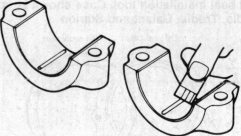

Measuring main bearing clearances with Plastigage®

22. If there is any question about the wear pattern or the ability of the new bearing shells to provide correct clearance, it is advisable to have the crank machined and install matching bearings. This approach to the problem will allow maximum service life and prevent you having to do this all again in six months.
23. Before final assembly, all bearing surfaces must be thoroughly lubricated with clean engine oil. Make sure the main bearings have the oil holes lined up and that the upper shells for bearings 1, 2, 4, and 5 have grooves so that oil will pass to the connecting rod journals.

The center main bearings, both top and bottom, are grooveless but include thrust bearing surfaces (flanges). Make sure connecting rod bearings have the notches in the bearing and cap together. Make sure all main bearing caps are installed in their original positions and directions. Main bearing caps are numbered in order from front to rear and have an arrow indicating the front (crankshaft pulley) end of the engine. Bearing cap bolts should be torqued evenly in 3 stages to specification. Work in sequence, starting with the center main bearing cap, then moving to No. 2, No. 4, front, and rear caps.

24. Reinstall the crankshaft seals, the oil pick-up screen and oil pan as soon as possible. Even if other work is to performed on the engine, the seals and pan will serve to protect the crank, rods and bearings from dirt and impact damage.

Flywheel and Ring Gear

REMOVAL AND INSTALLATION

NOTE: This procedure may performed with the engine in the vehicle, however, access will be cramped.

1. Remove the transaxle or transmission, following procedures outlined in Section 7.
2. For cars equipped with automatic transmissions, remove the torque converter carefully and examine the teeth on the ring gear.
3. For manual transmission cars:
   a. Matchmark the pressure plate assembly and the flywheel.
   b. Loosen the pressure plate retaining bolts a little at a time and in a criss-cross pattern. Support the pressure plate

ENGINE AND ENGINE OVERHAUL 3

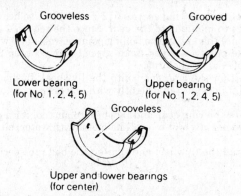

Locations of the various types of bearings, grooveless, grooved and end-thrust. Notches in the bearing caps correspond with tabs on the bearing. This is further insurance that each type of bearing will be properly assembled

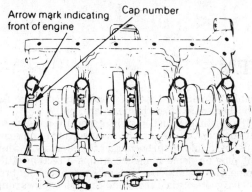

The bearing caps must be installed with the arrow facing the front of the engine and in proper numbered order

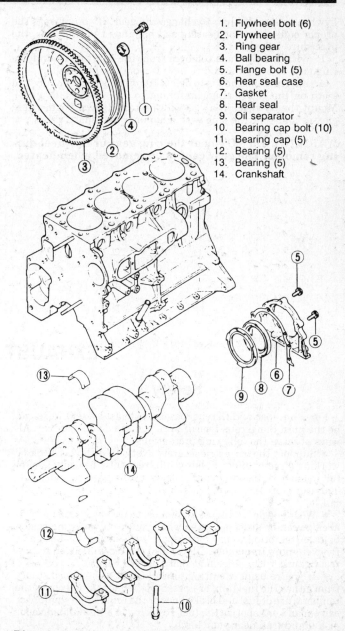

1. Flywheel bolt (6)
2. Flywheel
3. Ring gear
4. Ball bearing
5. Flange bolt (5)
6. Rear seal case
7. Gasket
8. Rear seal
9. Oil separator
10. Bearing cap bolt (10)
11. Bearing cap (5)
12. Bearing (5)
13. Bearing (5)
14. Crankshaft

Diesel engine crankshaft, flywheel and ring gear assembly

and clutch assembly as the last bolt is removed and lift them away from the flywheel.

― CAUTION ―
The clutch disc contains asbestos, which has been determined to be a cancer causing agent. Never clean the clutch assembly with compressed air! Avoid inhaling dust from the clutch during disassembly. When cleaning components, use commercially available brake cleaning fluids.

 c. Matchmark the flywheel and crankshaft. Loosen the retaining bolts evenly and in a criss-cross pattern. Support the flywheel during removal of the last bolts and remove the flywheel.
 4. Carefully inspect the teeth on the flywheel or torque converter driveplate for any signs of wearing or chipping. If anything beyond minimal contact wear is found, replace the unit.

NOTE: Since the flywheel is driven by the starter gear, you would be wise to inspect the starter drive if any wear is found on the flywheel teeth. A worn starter can cause damage to the flywheel.

 5. When reassembling, place the flywheel in position on the crankshaft and make sure the matchmarks align. Install the retaining bolts finger tight. The torque converter should be reinstalled on the transaxle, not on the engine.
 6. Tighten the flywheel bolts (manual trans) in a diagonal pattern and in three passes to 98 ft. lbs. for 4-cylinder engines or 54 ft. lbs. for V6 engines.

 7. Install the clutch and pressure plate assembly and tighten its mounting bolts properly.

NOTE: If the clutch appears worn or cracked in any way, replace it with a new disc, pressure plate and release bearing. The slight extra cost of the parts will prevent having to remove the transaxle again later.

 8. Reinstall the transaxle assembly.

RING GEAR REPLACEMENT

If the ring gear teeth on the driveplate (automatic trans) are damaged, the torque converter must be replaced. The ring gear cannot be separated or reinstalled individually.
If a the ring gear is damaged on the flywheel of a manual transmission car, it is usually cheaper and easier to buy a new

3 ENGINE AND ENGINE OVERHAUL

flywheel assembly than to change the gear. If you possess the proper equipment for heating and handling the ring gear, the procedure follows.

1. With the flywheel removed from the car, tap around the ring gear to loosen it and remove it from the flywheel.

An alternate method is to drill a 10mm hole between any two teeth on the gear, being careful not to drill into the flywheel. Mount the flywheel in a vise with protected jaws. Wear safety glasses and split the ring with a hammer and sharp chisel.

WARNING: Do NOT heat the ring gear or flywheel during removal. The gear cannot be removed when heated.

2. Heat the new ring gear to 572°F (300°C). Use heavy gloves and tongs to manipulate the gear. Since the required temperature is more than twice the boiling point of water and well above the flame point of paper, take special care to keep anything flammable out of the area.

3. Position the ring gear onto the flywheel. It should align easily. If it doesn't, tap it lightly with a brass punch to get it in postion.

4. Allow the ring gear and flywheel to air cool for a period of hours. Do not attempt to cool the metal with water, oil or other fluids.

5. Install the flywheel and tighten the bolts to specification.

EXHAUST SYSTEM

Safety

For a number of different reasons, exhaust system work can be the most dangerous type of work you can do on your car. Always observe the following precautions:

- Support the car extra securely. Not only will you often be working directly under it, but you'll frequently be using a lot of force, such as heavy hammer blows to dislodge rusted parts. This can cause an improperly supported car to shift and possibly fall.
- Wear goggles. Exhaust system parts are always rusty. Metal chips can be dislodged, even when you're only turning rusted bolts. Attempting to pry pipes apart with a chisel makes chips fly even more frequently. Gloves are also recommended to protect against rusty chips and sharp, jagged edges.
- If you're using a cutting torch, keep it at a great distance from either the fuel tank or lines. Stop frequently and check the temperature of fuel and brake lines or the tank. Even slight heat can expand or vaporize the fuel, resulting in accumulated vapor or a liquid leak near your torch.
- Watch where your hammer blows fall. You could easily tap a brake or fuel line when you hit an exhaust system part with a glancing blow. Inspect all lines and hoses in the work area before driving the car.

---- CAUTION ----

Be very careful when working on or near the catalytic converter. External temperatures can reach 1500°F (815°C) and more, causing severe burns. Removal or installation should be performed only on a cold exhaust system.

Special Tools

A number of special exhaust system tools can be rented from auto supply houses or local stores that rent special equipment. A common one is a tail pipe expander, designed to enable you to join pipes of identical diameter.

It may also be quite helpful to use solvents designed to loosen rusted bolts or flanges. Soaking rusted hardware the night before you do the job can speed the work of freeing rusted parts considerably. Remember that these solvents are often flammable. Apply them only after the parts are cool.

The Mitsubishi exhaust system generally consists of four pieces. At the front of the car, the front section of pipe connects the exhaust manifold or turbocharger to the catalytic converter. The front pipe on some models contains a section of flexible, braided pipe which allows the system to move without breaking. On turbocharged Cordia/Tredia and Starion engines, one catalytic converter is found just below the turbocharger (ahead of the front pipe) and a second one at the end of the front pipe. With the exception of the Starion, the catalytic converter (a sealed, non-servicable unit) can be easily unbolted from the system and replaced if necessary.

The diesel engine does not require a catalyst and no converter is found on diesel powered trucks. The gasoline engines in the truck have two catalytic converters. Although they are referred to as front and rear converters, they are attached to each other and are located immediately below the exhaust manifold. Replacement of the front (upper) catalyst requires removal of the exhaust manifold before removal of the converter.

INSPECTION

An intermediate or center pipe runs from the catalytic converter to the muffler at the rear of the car. Depending on the engine in the car, this pipe may contain a resonator or pre–muffler which serves to quiet and smooth out the exhaust flow before it enters the rear muffler. The resonator is welded into the pipe and is not removable. Should the resonator fail, the entire pipe must be replaced. On the G15B and G32B engines (used in Cordia/Tredia, Mirage and Precis) the intermediate pipe runs directly into the muffler with no joint.

The muffler, with its entry pipe and tailpipe, complete the system and serve to further quiet and cool the exhaust. Mitsubishi mufflers have the lead–in and exhaust pipes welded to the body of the muffler. If one of the pipes is damaged, the complete muffler assembly should be replaced.

The exhaust system is attached to the body by several welded hooks and flexible rubber hangers; these hangers absorb exhaust vibrations and isolate the system from the body of the car. A series of metal heat shields runs along the exhaust piping,

ENGINE AND ENGINE OVERHAUL 3

1. O-ring
2. Muffler
3. O-ring
4. Hanger
5. Center (intermediate) exhaust pipe
6. Hook
7. O-ring
8. Hanger
9. Rear catalytic converter
10. O-ring
11. Hanger
12. Bracket
13. Hanger
14. O-ring
15. Bracket
16. Front catalytic converter

1. Hanger
2. Muffler
3. Gasket
4. Tailpipe trim (extensions)
5. Hanger bracket
6. Self-locking nuts
7. O-ring
8. Hook
9. Bracket
10. Stopper
11. Hanger bracket
12. Hanger
13. Protector

14. Hanger bracket
15. Center exhaust pipe
16. Gasket
17. Catalytic converter
18. Gasket
19. Self-locking nuts
20. Hanger
21. Front exhaust pipe
22. Gasket
23. Heat shield

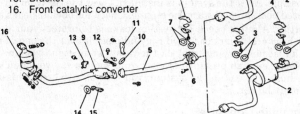

Starion exhaust system. Note the two catalytic converters in the system. Cordia and Tredia similar except the rear converter is bolted to the front pipe

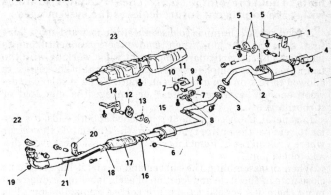

Typical 4-piece exhaust system. Heat shields and hangers will vary by model

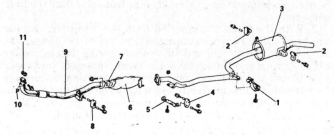

1. Band
2. Rubber hanger
3. Muffler
4. Rubber hanger
5. Hanger
6. Catalytic converter
7. Gasket
8. Rubber hanger
9. Front exhaust pipe
10. Nut
11. Gasket

Typical 3-piece exhaust system. The intermediate pipe is welded to the muffler and does not separate from it

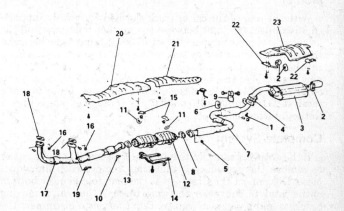

protecting the underbody from excess heat. These heat shields should be the first place to look when chasing a light metallic rattle or buzz under the car. Because the shields are exposed under the body work, they may become bent, loose or packed with road debris.

When inspecting or replacing exhaust system parts, make sure there is adequate clearance from all points on the body to avoid possible overheating of the floorpan. Check the complete system for broken, damaged, missing or poorly positioned parts. Rattles and vibrations in the exhaust system may be caused by misalignment of parts. When aligning the system, leave all the nuts and bolts loose until everything is in its proper place, then tighten the hardware working from the front to the rear. Remember that what appears to be proper clearance during repair may change as the car moves down the road. The motion of the engine, body and suspension must be considered when replacing parts.

1. Bolts
2. Rubber hanger
3. Muffler
4. Gasket
5. Self-locking nuts
6. Rubber hanger
7. Center exhaust pipe
8. Gasket
9. Damper (insulator)
10. Bolts
11. O-rings
12. Catalytic converter

13. Gasket
14. Hanger bracket
15. Hanger brackets
16. Self-locking nuts
17. Front exhaust pipe
18. Gasket
19. Oxygen sensor connection
20. Heat shield
21. Heat shield
22. Hanger brackets
23. Rear heat shield

V6 exhaust system. The double catalytic converters do not separate from each other and must be considered one unit

3 ENGINE AND ENGINE OVERHAUL

COMPONENT REMOVAL AND INSTALLATION

CAUTION

DO NOT perform exhaust repairs with the engine or exhaust hot. Allow the system to cool completely before attempting any work.

Exhaust systems are noted for sharp edges, flaking metal and rusted bolts. Gloves and eye protection are required. A healthy supply of penetrating oil and rags is highly recommended.

NOTE: ALWAYS use a new gasket at each pipe joint whenever the joint is disassembled. Use new nuts and bolts to hold the joint properly. These two low-cost items will serve to prevent future leaks as the system ages.

A good rule of thumb is to disconnect the forward-most joint first, then the rear joint, then remove the component from its hangers. This is particularly important when working with the front pipe or the complete system; if the weight of the pipe(s) is allowed to rest on the ground, the length of the system can develop enough leverage to break or damage the manifold or turbocharger.

In general, if any component is attached at both ends, it's easier to remove the entire system from the car and do the repair where you can see it. Remember to use new gaskets at every disassembled joint.

When disassembling the rubber and/or metal hangers, take close note of the style and location of each, labeling or diagramming them if necessary to insure correct reinstallation. The manufacturer selects the hangers and bushings to deliver the maximum support and isolation. If any of the hangers or hardware look to be in poor condition, replace the piece.

Muffler

With the rear of the car or truck elevated and firmly supported, disconnect the joint between the muffler inlet pipe and the intermediate pipe. Support the muffler and remove the hangers. Install the new muffler by supporting it from at least 2 hangers; this holds it in place but allows you to move it around. Install new gasket(s) at the joint and install new nuts and bolts, tightening them to 14 ft. lbs. if the muffler inlet pipe connects to the catalyst (G15B and G32B) or 22 ft. lbs. if the inlet pipe connects to the intermediate pipe.

Complete System

If the entire exhaust system is to be replaced, it is much easier to remove the system as a unit than remove each individual piece. Disconnect the first pipe at the manifold joint or bottom of the catalytic converter and work towards the rear, removing brackets and hangers as you go. Don't forget to disconnect the oxygen sensor connector on the V6 double pipe (Y-pipe) under the engine.

NOTE: Most exhaust pipes have a small bracket bolt just below the manifold or turbocharger joint. It must be removed as well as the joint nuts or bolts.

Remove the rear muffler hangers and slide the entire exhaust system out from under the car. The new system can then be bolted up on the workbench and easily checked for proper tightness and gasket integrity.

When installing the new assembly, suspend it from the flexible hangers first, then attach the fixed (solid) brackets. Check the clearance to the body and suspension and install the manifold joint bolts, tightening them to 22 ft. lbs. The small bracket bolt should be tightened to 18 ft. lbs. On the G54B (Starion) engine, don't forget to reinstall the springs under the nuts at the catalyst joint.

Constantly check for system to body interference; remember that the system is not rigidly mounted and must be free to move under the car. Minimum clearance to any body part should be 9.5mm (³⁄₈ inch), with 12-19mm (½-¾ inch) preferred where possible. After the system is installed, lower the car from its elevated position and start the engine. Check for obvious leaks and rattles. A quick check for system leaks is to protect your palm with several layers of folded rag and place it over the end of the exhaust pipe, trying to block the flow. An air tight system will develop enough pressure to push your hand away very quickly. If there is a leak in the system, you'll be able to hear a hissing under the car.

Catalytic Converter

1. The converter on most models has flanged connections on both ends. The second converter on Starions is flanged at the end of the front pipe, not at the converter itself. Remove the nuts and bolts at the rear joint, separate the flanges, and remove all the gaskets. Then disconnect the front joint or, on the Starion, disconnect the upper end of the front pipe from the first converter.

2. With the pipes disconnected from the converter, remove the brackets or hangers supporting the converter. hanger arrangements and attaching points vary by model. Pay attention to the location and placement of the various pieces of hardware.

3. Install the catalytic converter by supporting it from its hangers (or with your hand) and loosely connecting the front and rear joints to the pipes. Final tighten major parts after you've checked that all parts are properly positioned.

Turbocharger with Catalytic Converter

1. The Starion and turbocharged Cordia/Tredia use an auxiliary catalytic converter located right below the turbocharger and bolted to it. This is removed by supporting the pipe leading to the main converter and then removing flange bolts and separating the auxiliary converter at both flanges. On the Starion, it will be necessary to disconnect the oxygen sensor and secondary air pipe before removing the converter from the turbocharger.

2. During installation, note that the Starion uses springs under the nuts to guarantee a proper seal. Replace the springs if they are weak or excessively rusted. Install the catalyst to the exhaust port of the turbocharger, tightening the bolts to 44 ft. lbs. Connect the front pipe to the converter, , using new flange gaskets and tightening the bolts to 26 ft. lbs.

3. On the Starion, connect the oxygen sensor and install the secondary air pipe.

ENGINE AND ENGINE OVERHAUL 3

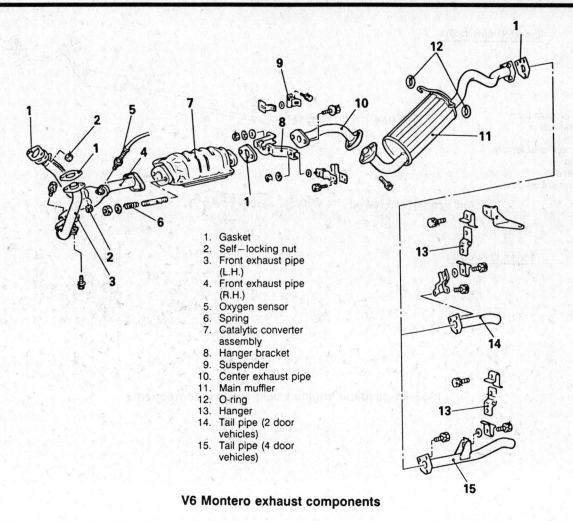

1. Gasket
2. Self-locking nut
3. Front exhaust pipe (L.H.)
4. Front exhaust pipe (R.H.)
5. Oxygen sensor
6. Spring
7. Catalytic converter assembly
8. Hanger bracket
9. Suspender
10. Center exhaust pipe
11. Main muffler
12. O-ring
13. Hanger
14. Tail pipe (2 door vehicles)
15. Tail pipe (4 door vehicles)

V6 Montero exhaust components

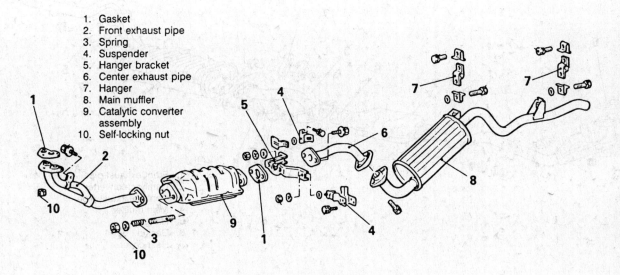

1. Gasket
2. Front exhaust pipe
3. Spring
4. Suspender
5. Hanger bracket
6. Center exhaust pipe
7. Hanger
8. Main muffler
9. Catalytic converter assembly
10. Self-locking nut

Exhaust components, Montero with 2.6L engine

3-275

3 ENGINE AND ENGINE OVERHAUL

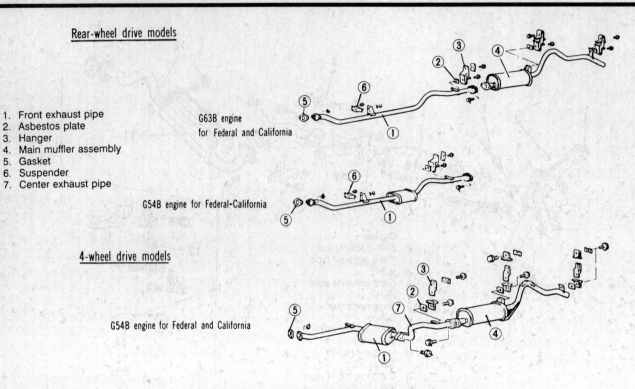

1. Front exhaust pipe
2. Asbestos plate
3. Hanger
4. Main muffler assembly
5. Gasket
6. Suspender
7. Center exhaust pipe

1983–86 gasoline engine truck; exhaust components

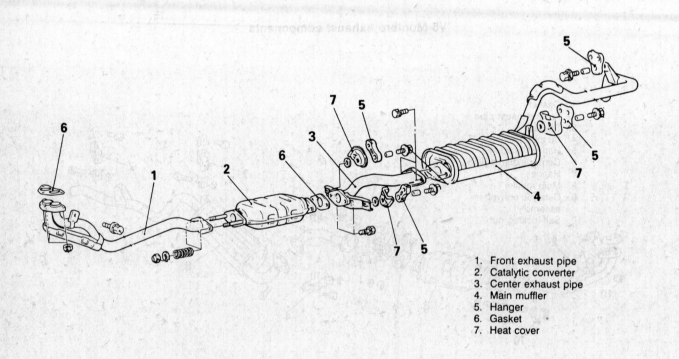

1. Front exhaust pipe
2. Catalytic converter
3. Center exhaust pipe
4. Main muffler
5. Hanger
6. Gasket
7. Heat cover

Truck exhaust, 1987–89

Emission Controls

QUICK REFERENCE INDEX

Diesel Engine Emission Control Systems	4-13
Gasoline Engine Emission Control Systems	4-2
Vacuum diagrams	4-15

GENERAL INDEX

Application chart	4-2	Gasoline Engine Emission controls		Evaporative canister	4-3
Carburetor	4-12	Evaporative canister	4-13	Exhaust emission controls	4-5
Crankcase ventilation valve	4-2	Exhaust Gas Recirculation (EGR) system	4-5	Exhaust Gas Recirculation (EGR) system	4-5
Diesel Engine Emission Controls		Feedback system	4-12	Maintenance reminder light	4-11
Crankcase ventilation system	4-13	Jet valve	4-12	Oxygen (O_2) sensor	4-9
Evaporative emission controls	4-13	Oxygen (O_2) sensor	4-9	PCV valve	4-2
Exhaust emission controls	4-14	PCV valve	4-2	Vacuum diagrams	4-15
EGR valve	4-5				

4 EMISSION CONTROLS

GASOLINE ENGINE EMISSION CONTROLS

Component location and vacuum routing diagrams are located at the end of this Section. Please refer to them before beginning any disassembly or testing.

There are three sources of automotive pollutants; crankcase fumes, exhaust gases, and gasoline evaporation. The pollutants formed from these substances fall into three categories: unburnt hydrocarbons (HC), carbon monoxide (CO), and oxides of nitrogen (NOx). The equipment used to limit these pollutants is called emission control equipment.

Due to varying state, federal, and provincial regulations, specific emission control equipment may vary by area of sale. The U.S. emission equipment is divided into two categories: California and 49 State. In this section, the term "California" applies only to cars originally built to be sold in California. Some California emissions equipment is not shared with equipment installed on cars built to be sold in the other 49 states. Models built to be sold in Canada also have specific emissions equipment, although in many cases the 49 State and Canadian equipment is the same.

Both carbureted and fuel injected cars require an assortment of systems and devices to control emissions. Newer cars rely more heavily on computer management of many of the engine controls. This eliminates the many of the vacuum hoses and linkages around the engine. In the lists that follow, remember that not every component is found on every car.

ECM CONTROLLED SYSTEMS
- Fuel Evaporative Control
- Carburetor Feedback System
- Deceleration Fuel Cutoff
- Ignition Timing Controls
- Cold Mixture Heater

NON-ECM CONTROLLED SYSTEMS
- Positive Crankcase Ventilation (PCV)
- Throttle Positioner
- Exhaust Gas Recirculation (EGR)
- Air Suction
- High Altitude Compensation
- Automatic Hot Air Intake
- Automatic Choke
- Choke Breaker
- Choke Opener

Positive Crankcase Ventilation (PCV) System

SYSTEM OPERATION

A closed positive crankcase ventilation system is used on all Mitsubishi models. This system cycles incompletely burned fuel (which works its way past the piston rings into the crankcase) back into the intake manifold for reburning with the fuel/air mixture. The oil filler cap is sealed and the air is drawn from the top of the crankcase into the intake manifold through a valve with a variable orifice.

This valve (commonly known as the PCV valve) regulates the flow of air into the manifold according to the amount of manifold vacuum. When the throttle plates are open fairly wide, the valve opens fully. However, at idle speed, when the manifold vacuum is at maximum, the PCV valve reduces the flow in order not to unnecessarily affect the small volume of mixture passing into the engine.

During most driving conditions, manifold vacuum is high and all of the vapor from the crankcase, plus a small amount of fresh air, is drawn into the manifold via the PCV valve. At full throttle, the increase in the volume of blow-by and the decrease in manifold vacuum make the flow via the PCV valve inadequate.

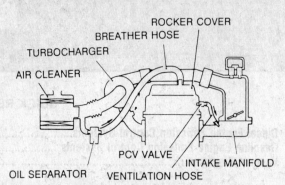

Typical PCV system for fuel injected engine

Under these conditions, excess vapors are drawn into the air cleaner and pass into the engine along with the fresh air.

A plugged valve or hose may cause a rough idle, stalling or low idle speed, oil leaks in the engine and/or sludging and oil deposits within the engine and air cleaner. A leaking valve or hose could cause an erratic idle or stalling.

TESTING AND TROUBLESHOOTING

The PCV is easily checked with the engine running at normal idle speed (warmed up). Remove the PCV valve from the valve cover, but leave it connected to its hose. (Most valves are pressed into the valve cover but some are threaded and screw into the cover.) Place your thumb over the end of the valve to check for vacuum. If there is no vacuum, check for plugged hoses or ports. If these are open, the valve is faulty. With the engine off, remove the PCV valve completely. Shake it end to end, listening for the rattle of the needle inside the valve. Generally, if no rattle is

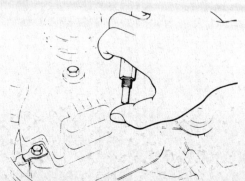

Check for suction in the PCV system. If vacuum is present, you should hear the ball rattle inside the valve

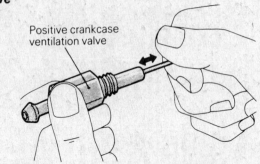

Checking the plunger on a threaded PCV valve

EMISSION CONTROLS 4

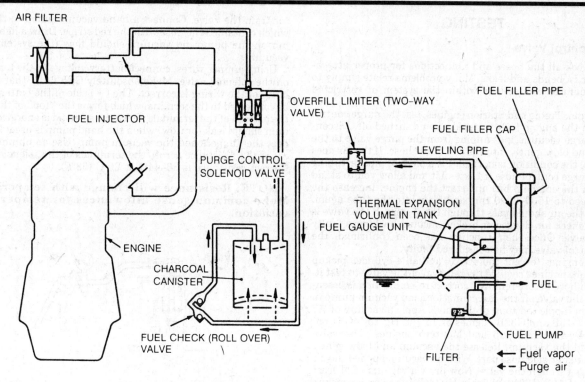

Schematic of typical evaporative emissions control system

heard, the needle is jammed (probably with oil sludge) and the valve should be replaced. If the valve is a threaded type, it may be necessary to use a thin probe inserted in the threaded end to check for plunger motion. If no motion is felt, replace the valve.

An engine which is operated without crankcase ventilation can be damaged very quickly. It is important to check and change the PCV valve at regular maintenance intervals.

REMOVAL AND INSTALLATION

Remove the PCV valve from the valve cover. Remove the hose from the valve. Take note of which end of the valve was in the manifold. This one-way valve must be reinstalled correctly or it will not function. While the valve is removed, the hoses should be checked for splits, kinks and blockages. Check the vacuum port (that the hoses connect to) for any clogging.

Remember that the correct function of the PCV system is based on a sealed engine. An air leak at the oil filler cap and/or around the oil pan can defeat the design of the system.

If any of the following operating problems occur, inspect the PCV system:

1. Rough idle, not explained by an ordinary vacuum leak, or fuel delivery problem.
2. Oil leaks past the valve cover, oil pan seals or even front and rear crankshaft seals not explainable by age, high mileage or lack of basic maintenance.
3. Excessive dirtiness of the air cleaner cartridge at low mileage.
4. Noticeable dirtiness in the engine oil due to fuel dilution well before normal oil change interval.

NOTE: An engine with badly worn piston rings and/or valve seals may produce so much blow-by that even a normally functioning PCV system cannot deal with it. A compression test should be performed if extreme wear is suspected.

Evaporative Emission Control System

— CAUTION —
Fuel vapors are EXTREMELY explosive. Observe no smoking/no open flame precautions. Have a "Type B-C" (dry chemical) fire extinguisher within arm's reach at all times and know how to use it.

OPERATION

The heart of this system is a charcoal canister located in the engine compartment. Fuel vapor that collects in the carburetor float bowl or gas tank is stored in the canister instead of being released to the atmosphere.

In order to restore the ability of the charcoal to hold fuel, fresh air is drawn through the canister under certain operating conditions, drawing the fuel vapors back out and introducing them into the combustion chambers.

At idle speed, or when the engine is cold, the addition of any fuel vapor to the correct mixture would cause excessive tailpipe emissions. For this reason, a port in the carburetor or fuel injection system throttle body allows the fuel vapors to be drawn out of the canister only after the throttle has been opened past the normal idle position. If there is no vacuum, the canister purge valve remains closed.

The flow of air and fuel are further restricted when the engine is cold by a thermal valve. This valve prevents the vacuum signal from going to the canister purge valve until the engine reaches a pre-determined temperature. On turbocharged vehicles, two vacuum signals are sent to the purge valve so that the system will work only above idle speed and below full throttle.

When the canister purge valve opens, air is drawn under slight vacuum from the air intake. If the engine exhibits operating problems during warm-up and basic fuel system and engine tune-up adjustments are correct, check the thermo valve for proper operation.

4 EMISSION CONTROLS

TESTING

Purge Control Valve

First check all the hoses and connections for proper attachment, cracks, bends and leaks. Many problems relate simply to poor mechanical connections within the system or restricted hoses.

For Cordia, Tredia and Starion engines, test the purge control valve with the engine fully warmed up and turned off. Disconnect the large vacuum line running from the purge valve to the canister and blow gently into the end of the hose. If the air does NOT pass, the condition is normal. Fit a piece of pipe or tubing into the purge control hose (to extend it and allow you to stand back from the engine a bit) and start the engine. Increase the engine speed to 1500-2000 rpm and blow into the tube again. This time the air should pass through the system. If the valve is not open, check for clogged or broken vacuum hoses or a faulty thermo valve. Once their function has been confirmed, the purge control valve may be considered faulty.

Mirage engines (except the 4G15), and all 4-cylinder pickup and Montero engines require the removal of the valve to test it. Label each hose before removal; correct re-connection is essential. With the valve off the car, connect a hand vacuum pump to the vacuum nipple (bottom) of the valve. Apply a vacuum of 7.7 in. Hg for G15B and G32B engines or 15.7 in Hg. for 4G61 engines or 4-cylinder pickup and Montero engines. The valve should hold the vacuum. Release the vacuum and blow gently into the canister side hose port. With no vacuum applied, no air should pass through the valve. Now draw a vacuum of at least 3.9 in. Hg. (G15B/G32B) or 8.0 in.Hg (4G61, 4-cylinder pickup and Montero engines) and blow into the port again. Air should flow through the valve.

The 4G15 and all the Galant engines, including the V6, and V6 Montero, control the canister through the purge control solenoid valve. This can be checked in place on the firewall. With the engine off, label and disconnect the two vacuum hoses running to the valve. One hose will have a red stripe on it; take note of which port it was connected to. Remove the electrical connector from the valve. Connect a hand vacuum pump to the port which contained the hose with the red stripe. Draw a light vacuum on the pump; no vacuum should flow (the system holds vacuum).

Using jumper wires, connect battery voltage to the terminals on the solenoid valve. Make absolutely CERTAIN that the polarity of the wiring is correct. The (+) side of the battery must be connected to the terminal which forms the "top" of the tee in the pattern of the terminals. Once the solenoid is energized, vacuum should leak (or flow) when the hand pump is used. Disconnect the jumpers and the vacuum pump. Use an ohmmeter to check the resistance across the terminals of the solenoid valve. Correct resistance is 36–44Ω at 20°C (68°F).

NOTE: Resistance will change with temperature. Make common sense allowances for temperature variation.

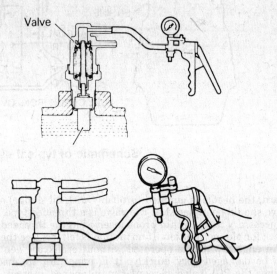

Checking the thermovalve. G15B valve shown below the more common 2-port type

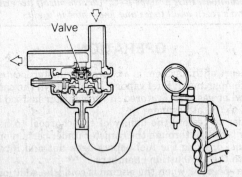

Checking the purge control valve

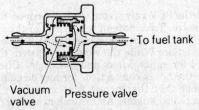

The two-way valve responds to pressure changes within the fuel vapor system

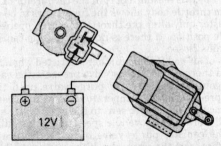

Use jumper wires to check the operation of the purge control solenoid valve. Make certain the terminals are connected as shown

The two-way valve should offer some resistance before allowing air to pass

EMISSION CONTROLS 4

Thermo Valve

The thermo valve found on all G62-3-4B, G15B and G32B engines and all 4-cylinder pickup and Montero engines is located on the intake manifold or cylinder head where its bottom will be immersed in coolant. (The later engines, beginning with the designations 4 or 6, control the purge system through the fuel injection computer.) The thermo valve blocks or passes vacuum to the purge control valve depending on the coolant temperature. Testing begins on an engine which has cooled overnight.

On all but the G15B engine, label and disconnect the two vacuum hoses on the thermo valve. On the G15B, remove only the hose closest to the engine. With the engine off, connect a hand-held vacuum pump to the upper port on the two port valves or the single port on the G15B valve. Apply vacuum and confirm that the valve leaks or does NOT hold vacuum. Start the engine and allow it to warm up. When the coolant has reached normal operating temperature, disconnect the appropriate hose as before and repeat the test; the valve should hold vacuum and not leak.

If it becomes necessary to replace the thermovalve, do so only on a cold motor. Fit the wrench only onto the faceted base of the valve, never on the plastic parts. When installing the new unit, apply sealer such as 3M No. 4171® or equivalent to the threads and tighten the new unit to 30 Nm (22 ft. lbs.)

Overfill Limiter (2-way Valve)

This device, sometimes mistaken for a fuel filter, is found in the vapor line running from the tank to the canister. Usually located at or near the tank, this valve is both a pressure– and suction–sensitive unit. Its purpose is to compensate for the pressure changes within the fuel tank. (Since the filler cap is tight enough to be considered sealed, the pressure must be equalized somewhere within the system.

When the pressure builds within the tank, such as on a very hot day or after a long period of driving, the valve releases the pressure and vapor into the charcoal canister, thereby venting the tank without raising emissions. Conversely, should the tank develop a vacuum, the valve will bleed some air (and vapor) from the canister into the tank.

CHILTON TIP: If you've heard a ghostly, high-pitched whining noise from the rear of your car on a summer day (even with the engine off), you're hearing this valve releasing pressure. Two ways to prevent the noise (for a while) are to either loosen the gas cap and then retighten it after the pressure equalizes or keep the tank $1/2-3/4$ full of fuel. Replacing the valve may change the sound but rarely eliminates it.

The control pressures within the valve are pre-set and not adjustable, but a quick check can be performed as follows:
1. Look at the valve and observe which end is toward the tank. Label or diagram the correct position.
2. Remove the valve from the vapor line. It may be necessary to remove other obvious components such as a parking brake cable bracket for access.
3. Lightly blow through either end of the valve. If air passes after some resistance, the valve is in good condition.
4. Install the valve into the line in the correct direction and secure the clamps. Make certain the lines are firmly seated on the ports before installing the clamps.

Fuel Check Valve (Roll-over Valve)

Usually mounted on the firewall on cars, or, on pickups and Monteros, near the tank, this simple valve is found in the vapor line coming from the tank to the charcoal canister. Normally the line carries vapor which is easily absorbed and held by the charcoal in the canister. If the car rolls over, the line would fill with liquid gasoline and exceed the canister's ability to absorb fuel. Once saturated, the canister would allow the liquid fuel to run

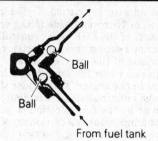

The fuel check valve is one of the simplest components on the car, but can be very valuable at the right time

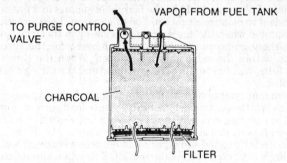

In the charcoal canister, fresh air is taken in only when the purge control system is open

out, possibly onto the hot surfaces of the engine. Since an engine fire is the last thing you want when your car is on its head (or any other time), the roll-over valve will block the vapor line and keep the fuel out of the canister.

This valve rarely, if ever, fails. If it must be checked, unbolt it from the firewall and remove the hoses. Shake it — if it rattles, it's OK. Since even the simplest job can be done incorrectly, make sure the valve is reinstalled right side up. Connect the hoses firmly and install the clamps.

Charcoal Canister

Since the canister cannot be tested on the workbench, it should be checked periodically for cracks, obstructions and proper hose connections. Check the maintenance schedule for your car; many models require replacement of the canister after a period of years or miles. Additionally, the canister should be considered suspect on a carbureted car after any incident of severe engine flooding. It is possible to deliver enough fuel vapor from the carburetor to overcome the capacity of the charcoal. This can result in the air/fuel mixture becoming too rich and causing driveability problems. If a high mileage carburetor is overhauled or replaced, a new charcoal canister should be installed as well.

When working around the canister, remember to label or diagram every hose before removal. Vacuum and vapor must flow correctly if the system is to work properly.

Exhaust Gas Recirculation (EGR) System

OPERATION

Exhaust Gas Recirculation is used to reduce peak flame temperatures in the combustion chamber. A small amount of exhaust gas is diverted from the exhaust manifold and re–entered into the intake manifold, where it mixes with the air/fuel charge

4-5

4 EMISSION CONTROLS

and enters the cylinder to be burned. Cooler combustion reduces the formation of Nitrogen Oxide (NO_2) emissions.

The system consists of the EGR valve, controlling the flow of exhaust gas and various vacuum and/or electric controls to keep the EGR from working at the incorrect time.

No EGR is required when the engine is cold due to lower flame temperatures in the engine. EGR under these conditions would produce rough running so EGR function is cut off either by a thermo valve or by the fuel injection computer (which is monitoring coolant temperature.) Additionally, EGR flow is cut off at warm idle to eliminate any roughness or stumble on initial acceleration.

Cooler combustion temperatures also result in slightly reduced power output. This isn't felt during normal, part-throttle driving and the emission benefits outweigh the slight loss. However, in a wide-open throttle situation a power reduction is not desirable; full power could be the margin of success in a passing or accident avoidance situation. For this reason, EGR function is eliminated when the engine goes on wide–open throttle. Normally, the vacuum to the EGR valve can overcome the spring tension within the valve and hold it open. When the throttle opens fully, vacuum to the EGR is reduced and the spring closes the valve.

A common symptom of EGR malfunction is light engine ping at part throttle, particularly noticable under load such as going uphill or carrying several passengers. An EGR valve which fails to close properly can also cause a rough or uneven idle. If the engine is correctly tuned and other common causes (vacuum leaks, bad plug wires, etc.) are eliminated, EGR function should considered as a potential cause when troubleshooting a rough idle.

Since the majority of EGR components do not require routine maintenance
and should not clog or corrode if unleaded gas is used, you should check all other reasonable causes of a problem before checking this system.

CHECKING THE EGR SYSTEM AND COMPONENTS

NOTE: A vacuum pump capable of producing about 10 in.Hg of vacuum will be needed to perform some of these tests.

EGR System
All 49 State Car Engines
1983–87 California Car Engines
4-Cylinder Montero and Pickup

1. Allow the engine to cool overnight. Since the EGR system works differently for warm and cold engines, a completely cold engine is required for testing.
2. Disconnect the vacuum hose with the green stripe from either the throttle body (fuel injected) or the base of the carburetor. Attach the end of the hose to vacuum pump.
3. Plug the port from which the hose was removed. Start the engine and attempt to draw a vacuum with the hand pump. On fuel injected engines, the system should NOT hold vacuum with the engine cold and running at idle. On cars with feedback carburetors (Cordia/Tredia and G15B Mirage), accelerate the engine to 3500 rpm and look for the vacuum to bleed off through the thermovalve.
4. Allow the engine to warm up to normal operating temperature. The coolant must be 80–85°C (175–185°F) before testing. Using the pump, draw a vacuum of 1.2–1.7 in.Hg. The system on a fuel injected engine should hold this vacuum with no noticable change in the idle quality. The carbureted system will leak vacuum at warm idle.
5. For carbureted engines, increase the engine speed to 3500 rpm. Slowly draw vacuum with the hand pump and observe the vacuum gauge on the pump. The system should leak vacuum

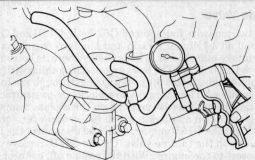

Using the "T" fitting to check California EGR systems

until the pump reaches about 1.5 in. Hg, at which time the vacuum should be held.

6. For fuel injected engines, allow the engine to run at warm idle as before. Draw a vacuum of 3.3 in.Hg (7.5–7.7 in.Hg for 4G15, 4G61 and 4G63 engines) and listen to the engine idle. As the correct vacuum is reached, the idle should roughen, become uneven, or possibly stall as the EGR valve opens and admits exhaust gas. The system should hold vacuum while the valve is open.

EGR System
1988–89 California Car Engines
Montero with V6 Engine

1. Allow the engine to cool overnight.
2. Disconnect the vacuum hose with the green stripe from the EGR valve. Use a "tee" (T) connector to connect the hand vacuum pump into the system and connect the hose back to the EGR.
3. Start the engine and observe the vacuum gauge on the hand pump. Press the accelerator suddenly to race the engine. On a cold engine, there should be no change in the vacuum; normal (atmospheric) pressure is maintained.
4. Allow the engine to warm up to normal operating temperature, generally 68–77°C (158–170°F). Repeat the sudden rpm test while watching the gauge on the pump. The vacuum should rise temporarily to about 3.9 in.Hg.
5. Disconnect the tee from the system and connect the vacuum pump directly to the EGR valve.
6. Draw a vacuum of 6–8 in.Hg while the engine is at warm idle. The quality of the idle should change noticably, becoming rough or even stalling as the EGR valve opens. The exact vacuum level at which this occurs varies by engine family but none opens higher than 7.7 in.Hg.

EGR Valve

1. Label and disconnect the hoses from the valve. Carefully loosen and remove the retaining bolts, remembering that they are probably heat–seized and rusty. Use penetrating oil freely.
2. Remove the valve and clean the gasket remains from both mating surfaces.
3. Inspect the valve for any sign of carbon deposits or other cause of binding or sticking. The valve must close and seal properly; the pintle area may be cleaned with solvent to remove soot and carbon.
4. Attach the vacuum pump to the vacuum port on the EGR valve. If the valve has two vacuum ports, pick one and plug the other.
5. Perform the vacuum holding test. Refer to the chart and draw the correct amount of vacuum, making sure it is held. If the correct vacuum cannot be maintained, the valve is leaking internally.
6. Release the vacuum but keep the pump attached to the valve. Devise a way to blow into the valve while drawing a vacu-

EMISSION CONTROLS 4

EGR SPECIFICATIONS CHART

Engine	Holds Vacuum ① at	Will Not Pass ① Air at	Air Passes ① through at	Installation ② Torque
G54B, G62B, G63B G15B, G32B	10	1.2	3.3	9.5 (7)
4G15, 4G61 4G63 SOHC & DOHC	19.7	1.7	7.6	4G15-12 (9) 4G61-18 (13) 4G63 SOHC/12 (9) 4G63 DOHC 18 (13)
G64B	9.7	1.0	3.3	23 (17)
6G72	9.7	0.8	3.3	23 (17)

① Values given in inches Hg.
② Values given in Nm (ft. lbs.)

um and reading the gauge on the pump. Draw a slight vacuum according to the chart and make sure that the valve is closed (your breath does not pass) at the specified vacuum.

7. Now increase the vacuum according to the chart and check that the valve passes air at the specified vacuum. If the valve is sticky or worn, it may not open properly. A weakened valve will open too soon. If either condition is encountered, replace the valve.

8. Install the valve with a new gasket. (Don't forget to remove the plug from the second vacuum port). Tighten the bolts according to the torque specification in the chart.

9. Connect the hoses and lines to their proper ports.

EGR Control Solenoid

1. Label and disconnect the vacuum hoses, taking note of the position of each.
2. Remove the wiring harness connector.
3. Connect the hand vacuum pump to the port which contained the vacuum hose with the red stripe.
4. Use jumper wires to bridge battery voltage to the terminals of the solenoid.

Draw a vacuum with the pump. When battery voltage is present, the solenoid should NOT hold vacuum. When the voltage is removed, it should be possible to draw and hold a vacuum. If either condition is not met, the unit must be replaced.

Secondary Air Supply System

The engines whose designations end with **B** (G54B, etc.) except the G64B use a second system to supply fresh air into the exhaust stream. This extra air is rich with oxygen and enhances the conversion process within the catalytic converter. At certain times, the pulses within the exhaust system create sufficient vacuum to open the reed inside the secondary air control valve and allows fresh air to pass into the exhaust stream. The fresh air may be either from the air cleaner or a separate intake system containing its own air cleaner. The entire system is controlled by the secondary air control solenoid, an electrically operated vacuum switch capable of disabling the system when it is not needed.

Although this system rarely fails, the reed valve may be checked by removing the hose from the air cleaner to the valve and unscrewing the valve from the line. The valve is a simple

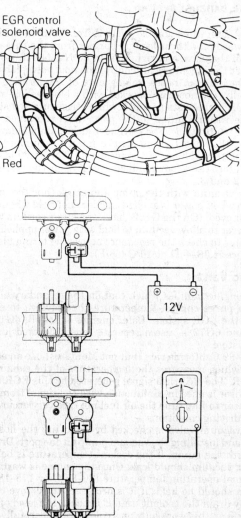

Three steps in checking the EGR control solenoid

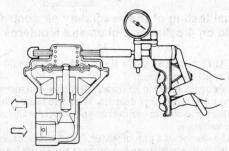

Checking the EGR valve after removal. Note that the second vacuum port is capped

4-7

4 EMISSION CONTROLS

The reed valve has one threaded fitting and one slip-on hose

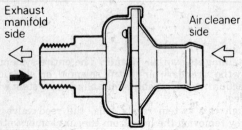

The reed valve insures a one-way flow of fresh air into the exhaust system

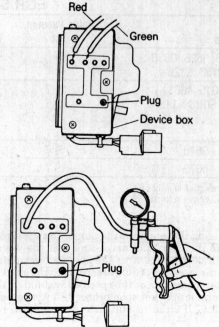

Vacuum testing the secondary air control solenoid on 4 cylinder pickup and Monteros

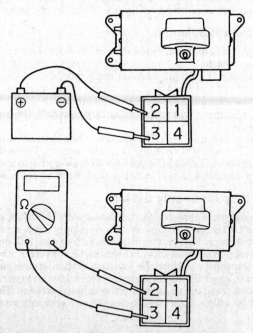

Electrical testing of the secondary air control solenoid on 4 cylinder pickup and Monteros

one-way gate; when you blow into it from both ends, air should pass towards the threaded end but not from the threaded end. Reinstall the valve and tighten it to 60 Nm (44 ft. lbs.)

To test the secondary air control solenoid valve, label and remove the vacuum hoses, taking note of the location of each. (Generally, the hoses are marked with a blue, yellow and white stripe OR a green stripe and a white stripe or a red stripe. Attach the vacuum pump to the port which contained the hose with the white stripe. For 4-cylinder pickups and Monteros, use the port that had the red striped hose. Use jumper wires to bridge battery voltage to the terminals. The 4-cylinder pickup and Montero use a four conductor plug; apply the voltage to terminals 2 and 3.

Draw vacuum with the pump and check that the unit holds vacuum when power is applied and does not hold when the power is removed. (On the G15B, block the other vacuum port with your finger to allow vacuum to hold with power applied.) Use an ohmmeter to check the resistance across the terminals; correct resistance is 36–44Ω at 10°C (68°F).

Thermo Valve

The engine families which contain the secondary air system use the purge control temperature thermovalve to enable or trigger the EGR system. Later engines — G64B, 4G15, 4G61, 4G63 and 6G72 — use a separate sensor devoted just to the EGR system.

1988–89 California cars (but not Monteros) use an additional sensor which measures the temperature of the exhaust gas at the EGR. The electrical signal generated by this EGR temperature sensor is used in conjunction with the signal from the oxygen sensor to fine tune the air/fuel mixture very accurately under all driving conditions.

The vacuum valve is checked by removing the hoses (label them!) and installing the vacuum pump on one port. Draw a vacuum with the pump: if the coolant temperature is below 50°C (122°F), vacuum should leak. Once the engine is warmed up to its normal operating temperature (80–85°C or 175–185°F) the vacuum should be held. If it is necessary to remove the valve, partially drain the coolant until it is below the level of the sensor. Perform this work only on a cold engine. Carefully unscrew the unit, applying wrench force only to the faceted part, never on the plastic. Before reinstalling, coat the threads with sealant (3M No. 4171 or equivalent); tighten the valve to 30 Nm (22 ft. lbs.) and refill the coolant.

The California electric EGR temperature sensor must be removed from the car before testing. With the motor cold, carefully disconnect the wiring connector and unscrew the sensor from the EGR valve.

Place the sensor in a pan of water. Use a thermometer to measure the water temperature as you heat the pan. Use an ohmmeter to measure the resistance at the terminals of the sensor as the temperature increases. At 50°C (122°F)) the resistance

4–8

EMISSION CONTROLS 4

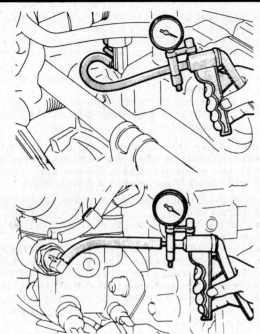

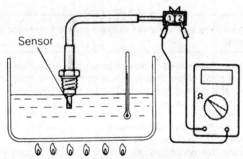

Varying locations for the thermovalve

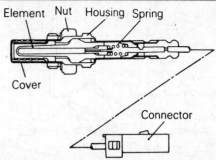

cut away view of an oxygen sensor. It requires no maintenance and must be tested on the car

Testing the California EGR temperature sensor

should be 60,000–83,000Ω. When the water reaches 100°C (212°F), the resistance should be 11,000–14,000Ω. The sensor should be replaced if there is significant deviation in the resistance. When reinstalling the sensor, tighten it to 11 Nm (8 ft. lbs.)

Oxygen Sensor

OPERATION

With the exception of the 1983 Cordia/Tredia (G62B), the 1983–84 pickup and the 1984 49–state Montero, all Mitsubishi engines use an oxygen sensor to aid in the control of the air/fuel mixture. The ideal mixture within the engine is 14.7 parts of air to one part of fuel. If this ratio can be maintained under all conditions, emissions will be kept to an absolute minimum. The trick is to inform the control computer (ECU or engine control unit) of any change in conditions so that it can react and make necessary changes. The oxygen sensor is one of many sensors which detect changes during driving.

Located in either the exhaust manifold or the exhaust pipe ahead of the catalytic converter, the oxygen sensor reads the amount of oxygen in the exhaust flow and generates a proportional electrical voltage. This voltage is transmitted to ECU which interprets it and sends necessary messages to fuel and air control components. Remember that the oxygen sensor is reading the result of combustion and reacting to it. If there is a problem in the air/fuel mixture entering the engine, the combustion will be imperfect and the oxygen sensor will generate a signal which shows the error. The signal does not necessarily indicate that the sensor has failed, only that it has detected a different oxygen concentration.

Since the oxygen sensor is the furthest "downstream" in the combustion process, it essential to check all other sensors and controls on the engine before assuming this sensor to be bad. Obviously, if the engine is running inefficiently, replacing the oxygen sensor won't cure the problem; the new sensor will continue to correctly read the imperfect exhaust content. About the only failure common to all oxygen sensors is loose or corroded connectors in the electrical wires. If a trouble code indicates an oxygen sensor malfunction, the first place to look is at the connector, making sure the pins are clean and fit tightly together. The low voltages flowing in this system can be changed or blocked by a high resistance (poor) connection.

TESTING

All Engines Except 4G61 and 4G63 DOHC

NOTE: An accurate digital voltmeter is required for this test.

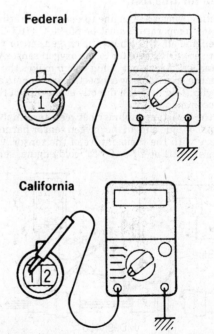

Testing the oxygen sensor for all but 4G61 and 4G63 DOHC engines

4-9

4 EMISSION CONTROLS

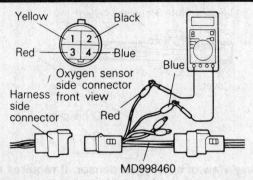

room temperature) and 30Ω or more at 400°C (752°F) (fully warm engine, sensor at working temperature).

Use common sense when reading the resistance. If you know the engine has cooled for 20 minutes while you hooked up the equipment, a reading of 25Ω probably reflects a sensor cooling off rather than a failed sensor. Wildly inaccurate readings may indicate a failed sensor.

4. Using jumper wires, connect terminal 3 (red clip) to the positive terminal of the battery. Connect terminal 4 (blue clip) to the negative terminal.

WARNING: Apply the voltage to the correct pins and avoid short-circuits or cross connections. Never apply voltage to the other terminals. Severe and expensive damage will result from improper connections.

5. Connect the digital voltmeter between terminals 1 (yellow clip) and 2 (black clip).
6. Place the meter where it can be seen from the driver's seat. Start the engine.
7. Race the engine to about 4000 rpm and observe the meter; it should show about 1 volt (600–1000 mV). If a faulty reading is obtained, first make certain the engine is fully warm. Recheck all the test connections and retest. If the reading is still inaccurate, it is probable that the oxygen sensor has failed.
8. Shut the engine off, remove the test equipment and reconnect the sensor harness.

Use the plug adapter to check the oxygen sensor on 4G61 and 4G63 DOHC engines. Damage may result if the adapter is not used

1. Before testing, warm the engine to normal operating temperature. Coolant temperature must be 80–85°C (175–185°F) or more.
2. Shut the engine off. Disconnect the oxygen sensor connector and connect the positive probe of the voltmeter to the sensor connector. If the connector has two terminals (California spec), use the terminal on the left.
3. Ground the negative probe of the meter to the body or the engine as convenient but do not ground it back to the sensor or connect it to the second terminal.
4. Place the meter where it can be seen from the driver's seat. Start the engine.
5. Race the engine to about 4000 rpm and observe the meter; it should show about 1 volt (600–1000 mV).
6. Shut the engine off, remove the test equipment and reconnect the sensor harness.

4G61 and 4G63 DOHC Engines

NOTES: 1. The use of special tool MD 998460 is highly recommended. This wiring adapter will help eliminate costly short-circuits during testing.
2. The use of an accurate digital volt/ohmmeter is required for this test.

1. Before testing, warm the engine to normal operating temperature. Coolant temperature must be 85–96°C (185–205°F.)
2. Shut the engine off. Disconnect the oxygen sensor connector and connect the wiring adapter to the oxygen sensor wiring. The test harness is a "break-out" plug which makes the terminal pins much more accessable by connecting each pin to a color-coded alligator clip. Trying to test directly at the pins of the connector is not recommended.
3. Measure the resistance (ohms) between terminal 3 (red clip) and terminal 4 (blue clip) of the oxygen sensor harness. Resistance will vary with the temperature of the sensor. Control values are: approx. 12Ω at 20°C (68°F) (cold engine, sensor at

REMOVAL AND INSTALLATION

CAUTION
Perform this work only after the exhaust system has cooled enough to avoid burns.

It is more common to remove the oxygen sensor for protection or access during other repairs than to replace it because of failure. Once the sensor is removed, it must be protected from impact and/or chemical contact. Never attempt to clean the tip with solvent and never allow the tip to contact grease, oil or other chemicals. The zirconia element in the tip will be polluted and the sensor will function poorly, if at all.

1. Locate the oxygen sensor. On non-turbocharged 4-cylinder engines, it will be located in the exhaust manifold, usually mounted either underneath or from the side. Turbocharged engines will have the sensor between the turbo and the first catalyst, usually mounted in the "bell" of the exhaust housing at the turbocharger. The V6 engine has the oxygen sensor mounted in the exhaust pipe, just beyond the **Y** where the left and right pipes connect into one.
2. Follow the wiring from the sensor to the first connector and disconnect it. Do not attempt to disconnect the wiring at the sensor.

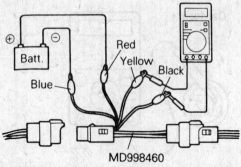

Voltage must be carefully applied or the sensor will be damaged

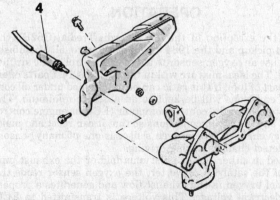

Typical oxygen sensor location for non-turbocharged 4-cylinder engine

4–10

EMISSION CONTROLS 4

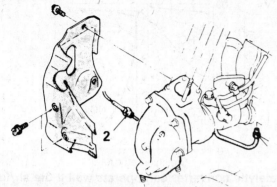

Typical location for oxygen sensor on turbocharged engine. The turbo need not be removed to remove the sensor

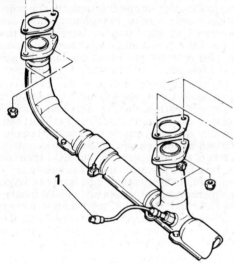

the V6 oxygen sensor is located where it can monitor the exhaust from both banks of cylinders

3. Particularly on non-turbocharged, 4-cylinder engines, the sensor may be obstructed by heat shields on the exhaust manifold. Remove them as necessary.
4. Install the proper size wrench on the flats of the sensor. 4G61 and 4G63 engines require a special socket for removal. Place the socket on the sensor and use a box wrench to turn the socket.
5. Keeping the wrench (or socket) square to the sensor while removing it. Do not allow the wrench to become crooked or to come off the flats. Remember that the sensor has been exposed to extreme temperature and corrosive exhaust gasses. It may be difficult to remove.
6. Once the sensor is removed, place it in a clean, protected location. For reinstallation, the threads of the sensor may be lightly coated with an anti-seize compound but extreme care must be taken to protect the tip and shield area of the sensor from even the slightest contamination.
7. Handle the oxygen sensor carefully, protecting it from impact, and install it in place. Start the threads by hand and hand tighten it as far as possible.
8. Use the wrench(es) to finish tightening the sensor to 45 Nm (33 ft. lbs.)
9. Install the heat shields if any were removed. Tighten the bolts to 16 Nm (12 ft. lbs.)
10. Connect the sensor wiring to the harness connector. Make certain the wiring is correctly run and out of the way of hot or moving components.

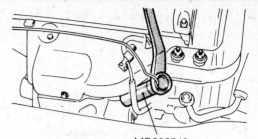

Without the adapter socket, it's almost impossible to extract the oxygen sensor from the 4G61 or 4G63 DOHC engines

Maintenance Reminder Light Check EGR Light

Pickups and Monteros with gasoline engines from 1985 incorporate a dashboard light to remind the owner of necessary maintenance intervals. Reading either "Check EGR" (1985–87) or "Maintenance Reqd", these lights are controlled by the elapsed mileage on the odometer. The light will illuminate first at 50,000 miles and then at 100,000 miles. 1989 vehicles will energize the light at 80,000 and 120,000 miles as well. The light is simply a reminder that system maintenance and inspection is due at these intervals; it is not a warning light in the usual sense.

RESETTING

When the light comes on, resist the temptation to simply turn it off. Leave it on until you have the time to perform at least basic maintenance. Chances are that your pickup or Montero is running pretty well at 50,000 miles. A good tune up with fresh filters will keep it that way. Check the vacuum hoses for cracks or damage. Check the EGR valve using the procedure outlined in this section. If this maintenance isn't possible, at least take the vehicle to a dealer or service station capable of running it on an exhaust analyzer. The content of the exhaust gas can tell you quite a bit about how efficiently the engine is burning fuel.

Once the needed maintenance has been performed, the reminder light may be turned off (which resets it for the next interval) by moving a small switch.

On 1985–86 pickups, the switch is located on the back of the instrument cluster at the top. Reaching it behind the dashboard can require contortions and creative language, but it can be done without removing the instrument cluster. 1987–89 pickups have the switch on the front of the instrument cluster, at the bottom right; the clear plastic window over the cluster must

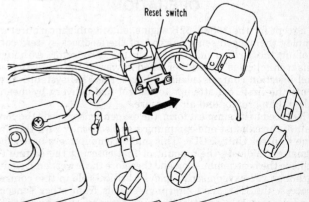

"Check EGR" reset switch, 1985–86 pickup

4 EMISSION CONTROLS

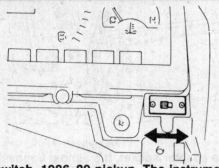

Reset switch, 1986–89 pickup. The instrument cluster lens must be removed for access

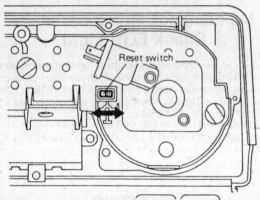

Maintenance reminder reset switch for 1985–89 Montero

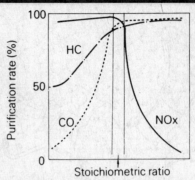

the catalytic converter can operate well if the air/fuel mixture in the engine is closely controlled

be removed to gain access. Carefully remove the four screws and the lens, reset the switch and reinstall the lens.

All 1985–1989 Monteros have the reset switch located on the back of the speedometer, roughly behind the 80 mph marking. Again, reaching it can be difficult but it can be done. When reaching up behind the dashboard, be careful not to dislodge any wiring harnesses or connectors; don't force or push any components out of the way.

Mitsubishi recommends that after the 100,000 mile (1989: 120,000 mile) service is performed, the light bulb be removed from the indicator. This does not relieve the owner from performing needed maintenance; it just turns the light off permanently. Removing the bulb or disabling the system before the specified mileage is a violation of the Federal Automobile Emissions Act.

Feedback Carburetor System (FBC)

OPERATION

Except for the 1983 G62B engine, all Mitsubishi carburetors employ the oxygen sensor as part of the feedback system controlling the air/fuel mixture within the carburetor. Although this technology is now rather dated in view of the sophisticated fuel injection controls available today, the FBC system represents the first real attempt to control fuel delivery by means other than gravity and air pressure.

Reduced to its simplest form, the oxygen sensor reads the content of the exhaust and communicates electrically with the Engine Control Unit (ECU). This micro computer evaluates the signal and adjusts the amount of fuel entering the engine to achieve the best combustion and therefore the lowest emissions. Although the oxygen sensor plays a critical role in this control process, the ECU also interprets signals from other sensors reading throttle position, engine temperature and other variables.

In ordinary operating conditions after engine warm up, the air/fuel ratio is within the usable range of the oxygen sensor. By attempting to keep the air and fuel mixed at the proportionally perfect (stoichiometric) ratio, the exhaust will be almost completely processed by the catalytic converter. If the exhaust changes drastically, pollutants may exceed the capacity of the converter. This constant process of checking the exhaust and adjusting the mixture is called closed loop operation.

There are times during normal operation, however, when the exhaust is outside these usable limits. Conditions of engine start-up, partially warmed driving, high load operation or sudden deceleration can each cause sufficient exhaust conditions to exceed the range of the oxygen sensor. For this reason, when the ECU is advised by the sensors of one of these conditions, the system enters open loop operation. The ECU then controls the carburetor based on pre-programmed values (often called default values) and disregards the signal from the oxygen sensor. These pre-programmed values are in the ROM (read only memory) of the ECU and, since they are installed at manufacture, cannot be altered by the oxygen sensor. They control the carburetor to allow the quickest warm up of a cold engine or the best compromise of performance and emissions under the existing condition.

The feedback carburetor and its sensors are covered in Section 5.

Jet Valve

OPERATION

Although a mechanical component of the valve train, the jet valve (See Section 3) is present to improve combustion and

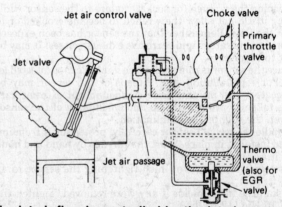

The jet air flow is controlled by the jet air control valve. Jet air does not flow until the coolant reaches a pre-determined temperature.

EMISSION CONTROLS 4

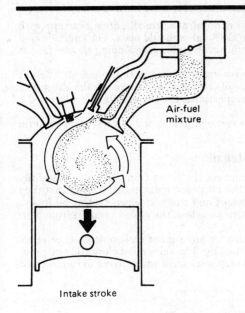
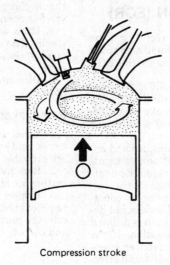
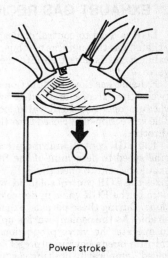

Intake stroke   Compression stroke   Power stroke

The jet valve admits additional air to promote complete combustion

thereby reduce emissions. The jet valve admits a super–lean mixture (almost all air with no fuel) into the cylinder at the same time the intake valve admits the normal air/fuel charge.

When the throttle position is almost closed, such as at idle or light throttle, a large pressure difference is created within the cylinder, drawing in large amounts of jet air. This rapid flow swirls in the cylinder, scavenging the remaining exhaust fumes near the spark plug and creating a good combustion environment. The swirl within the cylinder continues during the piston's compression stroke and aids in more complete combustion.

As the throttle opening is increased, the volume of jet air is proportionally reduced, but the greater volume of air entering through the intake valve is sufficient to promote good combustion.

As with most other emission control systems, there are times when the leaning effect of the jet valve is undesirable. A cold start, with the choke engaged, requires a rich mixture to allow the engine to run smoothly. The jet valve still opens with each intake stroke (it's driven by the camshaft) but the air passage is blocked by the Jet Air Volume Control valve. The valve is controlled by coolant temperature; when the engine warms to a pre–determined point, the temperature sensor activates the Jet Air Volume control valve and the air flows into the cylinder on each intake stroke. The temperature sensor controlling this system is the same one used to control the EGR valve.

DIESEL ENGINES EMISSION CONTROLS

The diesel engine in Mitsubishi trucks has inherently low air pollution characteristics due to the nature of its combustion system. Exhaust emissions — carbon monoxide, oxides of nitrogen and particulate matter — are controlled through careful internal engine design.

A portion of the combustion chamber consists of a swirl chamber into which atomized fuel is injected under high pressure. The fuel is instantly mixed with fresh air which is swirling quickly within the chamber. This forms an air/fuel mixture yielding good fuel economy and low air pollution characteristics.

Crankcase Ventilation System

A sealed crankcase ventilation system is used to prevent the blow-by gasses from escaping into the atmosphere. This system consists of a breather hose, an outlet pipe at the rocker cover, and an inlet pipe at the air intake hose.

Blow-by gasses generated in the crankcase are drawn into the air intake hose through the breather hose in the valve cover. Aside from keeping the hose and passages clean and open, the system requires no maintenance.

Evaporative Emission Controls

Diesel fuel does not pose an evaporant hazard; therefore no serviceable evaporative controls are installed on the diesel truck.

4–13

4 EMISSION CONTROLS

Exhaust Emission Controls

EXHAUST GAS RECIRCULATION (EGR)

EGR is used to control oxides of nitrogen by recycling some exhaust gas through the combustion process. A small amount of exhaust gas is admitted into the intake manifold by the EGR valve.

Exhaust gas flow is a function of the pressure differential between the exhaust and intake systems. It is controlled by the vacuum operated EGR valve located on the intake manifold. The valve is open at closed throttle (idle) and closed at wide open throttle.

The EGR control unit monitors coolant temperature and engine speed to determine if the EGR control unit output signal should be sent to open the EGR valve. The control unit de-energizes the EGR control solenoid, which in turn shuts off the vacuum to the EGR valve under cold start and warming-up conditions. During these periods, exhaust gas recirculation is not desirable. As the coolant warms up to normal, vacuum is allowed to operate the valve proportional to throttle position. If the throttle opens beyond a preset level, a vacuum reducer is engaged, temporarily limiting vacuum to the EGR valve and effectively turning it off during the wide open throttle period. The vacuum signal to the valve is also reduced at high altitude, reducing tailpipe smoke.

Testing

1. Start the engine and warm it to full operating temperature. This can be achieved quicker by driving the vehicle than by allowing it to idle.
2. Disconnect the vacuum hose from the EGR valve and install a vacuum pump and short piece of hose. Alternately, disconnect the EGR hose at the control solenoid and connect the hose directly to the vacuum pump.
3. With the engine running at warm idle, draw a vacuum with the hand pump; the EGR valve should open. The engine sound should change, possibly even stalling, depending on how far the valve is opened.
4. Draw and release vacuum several times and at different speeds. The response of the valve (and therefore the idle speed/quality) should be proportional the the vacuum signal from the pump.
5. Disconnect the test equipment and reconnect the vacuum hose.

Removal and Installation

1. Label and disconnect the hoses from the valve. Carefully loosen and remove the retaining bolts, remembering that they are probably heat-seized and rusty. Use penetrating oil freely.
2. Remove the valve and clean the gasket remains from both mating surfaces.
3. Inspect the valve for any sign of carbon deposits or other cause of binding or sticking. The valve must close and seal properly; the pintle area may be cleaned with solvent to remove soot and carbon.

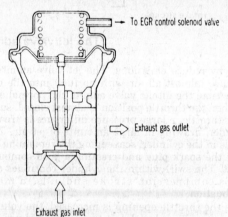

Diesel engine EGR valve. Check for soot and dirt holding the valve open

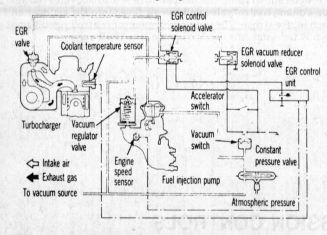

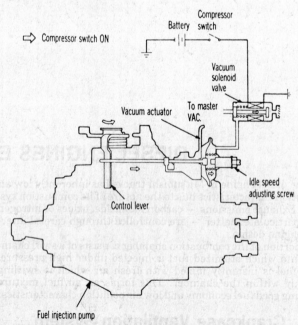

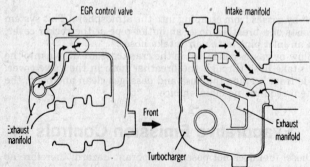

Diesel engine EGR system schematic

Schematic of the air conditioning compensation circuit.

EMISSION CONTROLS 4

4. After cleaning and inspection, recheck the motion of the valve with a hand vacuum pump. Watch the shaft area closely. A stuck or binding valve will not open properly; a weakened valve will open too soon. If either condition is encountered, replace the valve.

5. Install the valve with a new gasket. Tighten the bolts evenly to 7 Nm or 5 ft. lbs.

6. Connect the hoses and lines to their proper ports.

High Altitude Compensation System

The diesel trucks sold as Federal specification low altitude vehicles will meet the Federal high altitude requirements with no adjustments required. Trucks sold for principal use in high altitude areas are adjusted by the dealer in order to reduce levels of visible smoke and soot. The adjustment involves recalibrating the injection timing from 5° ATDC to 3° ATDC.

Air Conditioning Idle Speed Compensator

When the optional air conditioning unit is installed, the idle speed must be increased when the air conditioner compressor is engaged. The compressor adds sufficient load to the engine that uneven idle or stalling may occur if compensation is not provided.

This system consists of a vacuum actuator and a vacuum solenoid valve. When the compressor is switched on, the solenoid valve is energized, allowing some vacuum from the power brake booster assembly to be bled into the actuator. As a result, the control lever is slightly pulled by the actuator, admitting more fuel to the engine. Consequently, the engine runs at a faster idle speed.

Fuel Injection

The fuel injection system, the timing of fuel delivery into the engine and the quality of fuel used all contribute significantly to emission reduction. Correct diesel fuel — either 1-D, 2-D or winterized 2-D is to be used. Do not use heating fuel oil or diesel fuels intended for industrial or marine use. Buy fuel from a reputable dealer and avoid the use of additives. The diesel fuel system is discussed in detail in Section 5 of this book.

COMPONENT LOCATION AND VACUUM DIAGRAMS

For any model not shown in these diagrams or in case of conflict between the actual engine arrangement and a diagram, refer to the underhood label on the vehicle. Changes in equipment and vacuum routing may be made throughout the production of the vehicle; the underhood label will show the correct data for the car.

In the following diagrams, the symbol **C** within a circle indicates that the wire so noted connects to the computer (ECU).

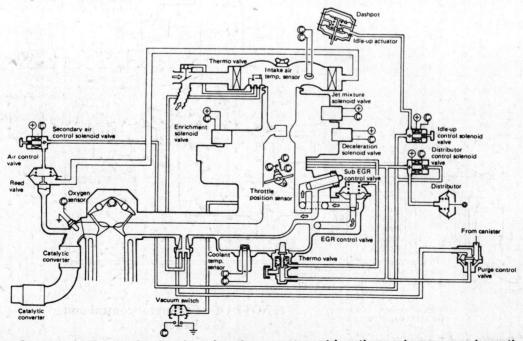

1984 Cordia and Tredia: feedback carburetor component location and vacuum schematic

4-15

4 EMISSION CONTROLS

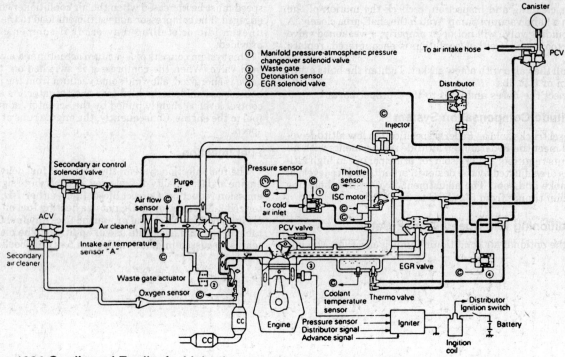

1984 Cordia and Tredia: fuel injection component location and vacuum schematic

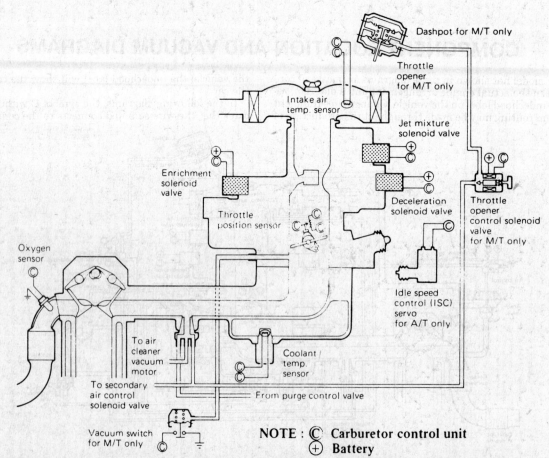

1985 Cordia and Tredia: feedback carburetor component location and vacuum schematic

EMISSION CONTROLS 4

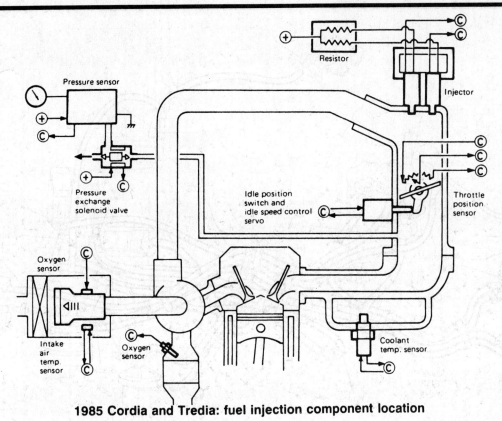

1985 Cordia and Tredia: fuel injection component location

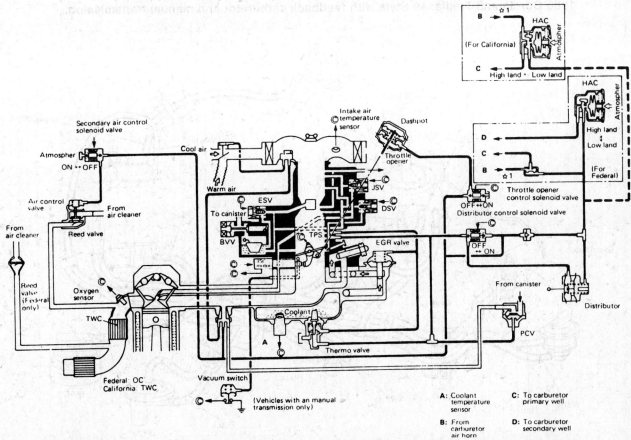

1986 Cordia and Tredia: feedback carburetor component location and vacuum schematic

4-17

4 EMISSION CONTROLS

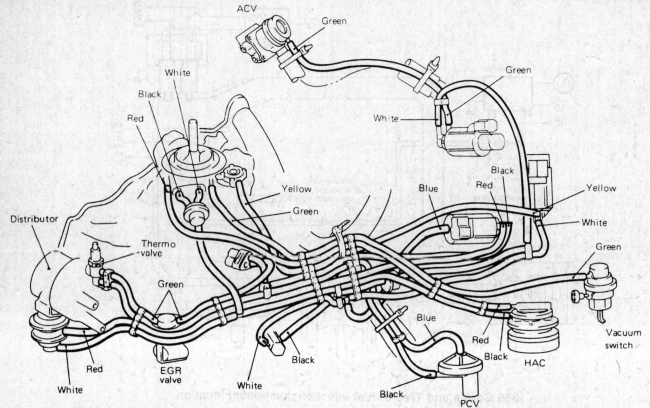

1986 Cordia and Tredia: 49 State with feedback carburetor and manual transmission.

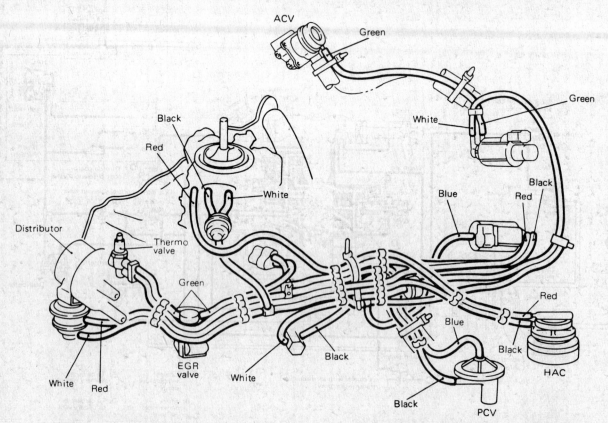

1986 Cordia and Tredia: 49 State with feedback carburetor and automatic transmission

EMISSION CONTROLS 4

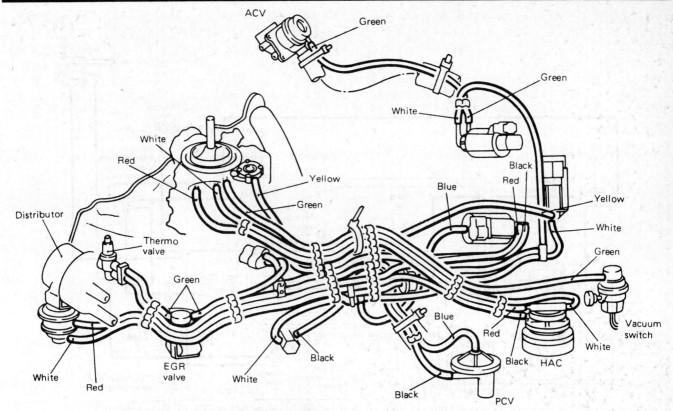

1986 Cordia and Tredia: California with feedback carburetor and manual transmission.

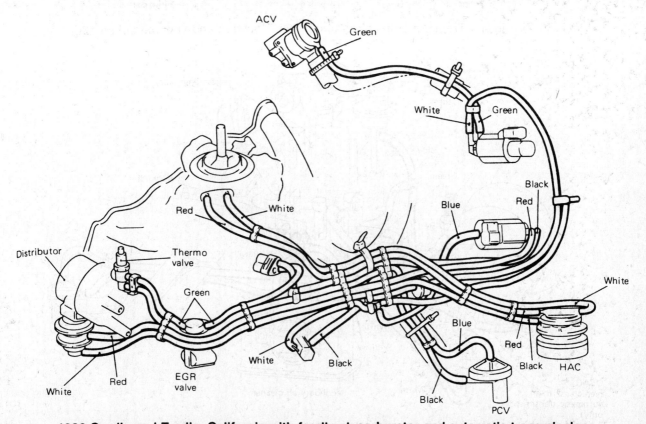

1986 Cordia and Tredia: California with feedback carburetor and automatic transmission.

4 EMISSION CONTROLS

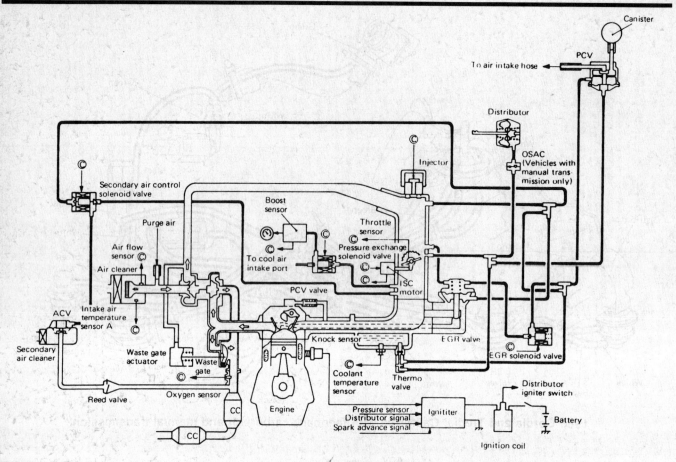

1986 Cordia and Tredia: fuel injection component location and vacuum schematic

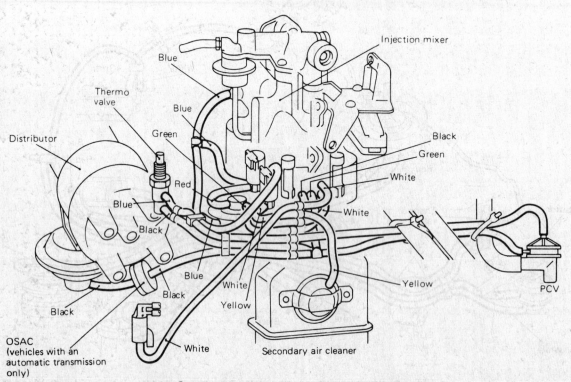

1986 Cordia and Tredia: vacuum hose routing

EMISSION CONTROLS 4

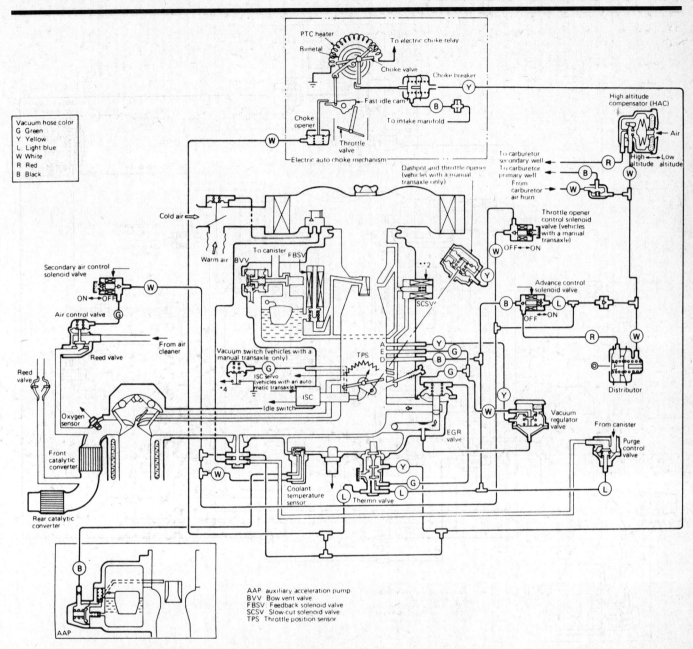

1987 Cordia and Tredia, 49 state with feedback carburetor, component location and vacuum schematic

4-21

4 EMISSION CONTROLS

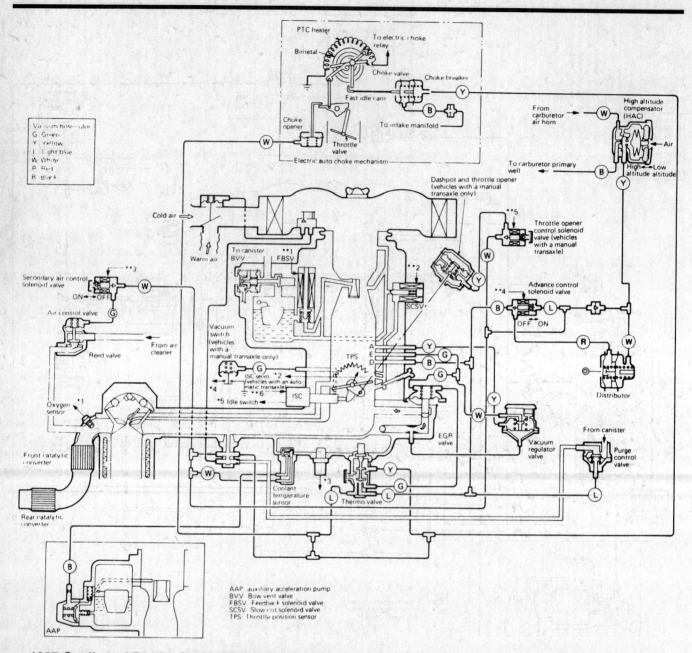

1987 Cordia and Tredia: California with feedback carburetor, component location and vacuum schematic

EMISSION CONTROLS 4

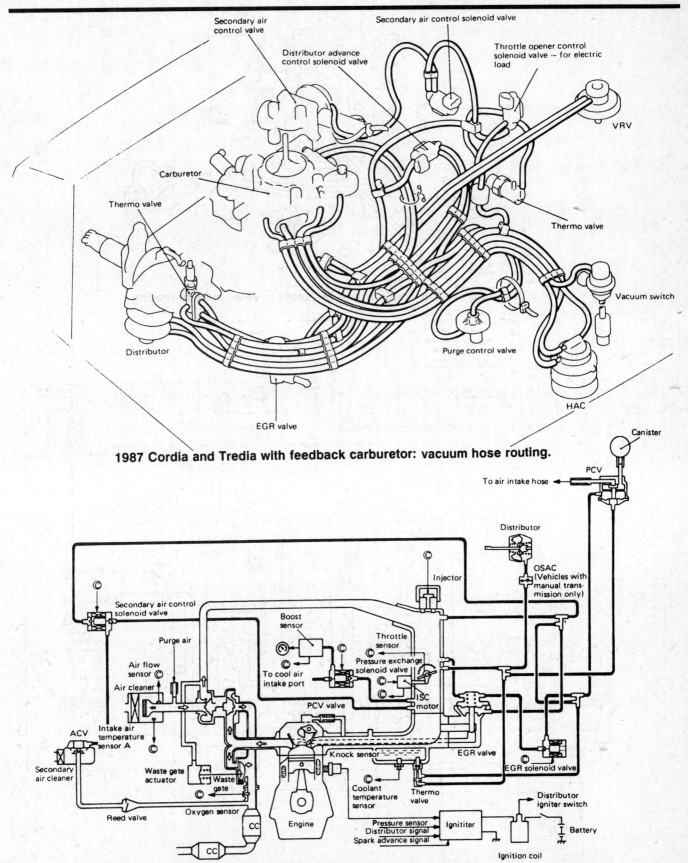

1987 Cordia and Tredia with feedback carburetor: vacuum hose routing.

1987 Cordia and Tredia with fuel injection: component location and vacuum schematic

4-23

4 EMISSION CONTROLS

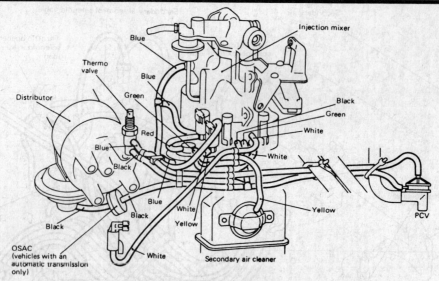

1987 Cordia and Tredia with fuel injection: vacuum hose routing

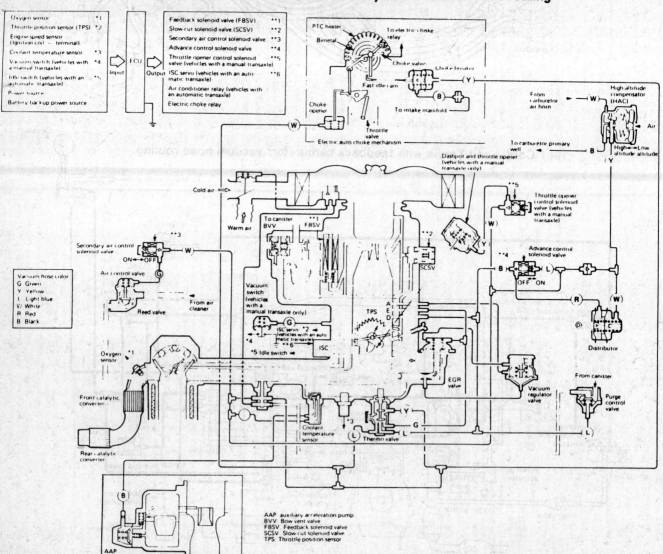

1988 Cordia and Tredia: California, with feedback carburetor. Component location and vacuum schematic

EMISSION CONTROLS 4

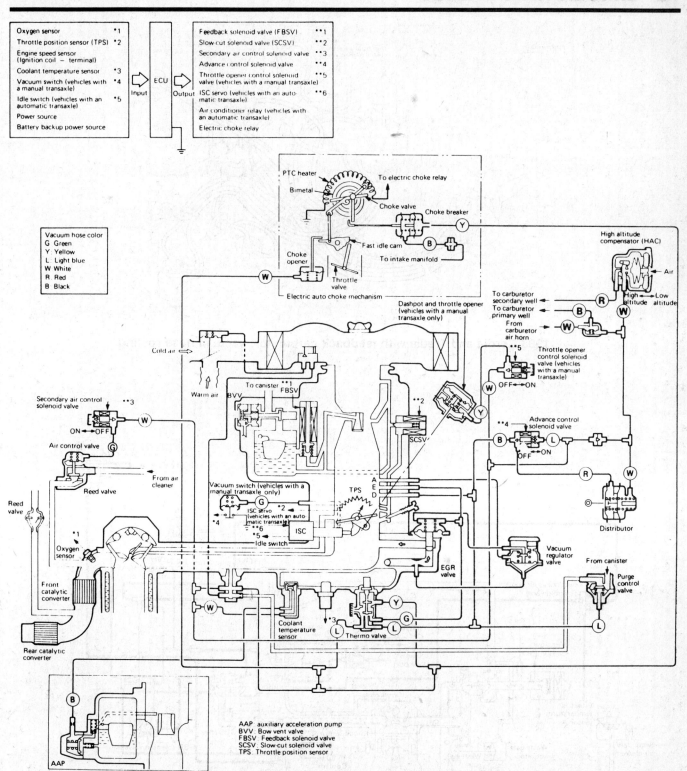

1988 Cordia and Tredia: 49 State with feedback carburetor. Component location and vacuum schematic

4 EMISSION CONTROLS

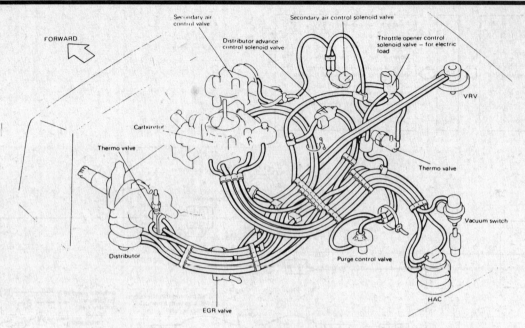

1988 Cordia and Tredia with feedback carburetor: vacuum hose routing

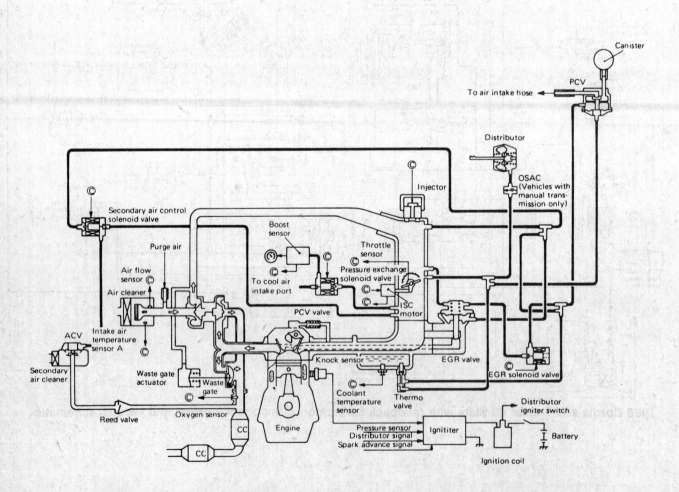

1988 Cordia and Tredia: fuel injection component location and vacuum schematic

EMISSION CONTROLS 4

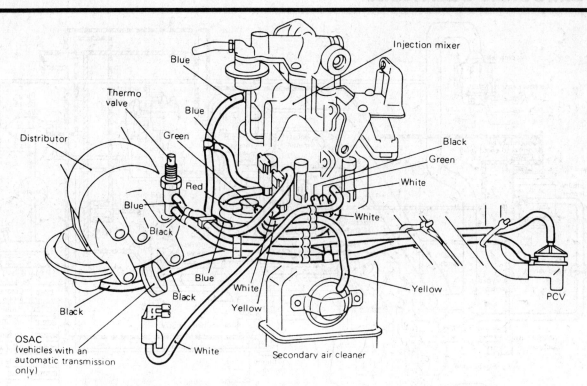

1988 Cordia and Tredia with fuel injection: vacuum hose routing

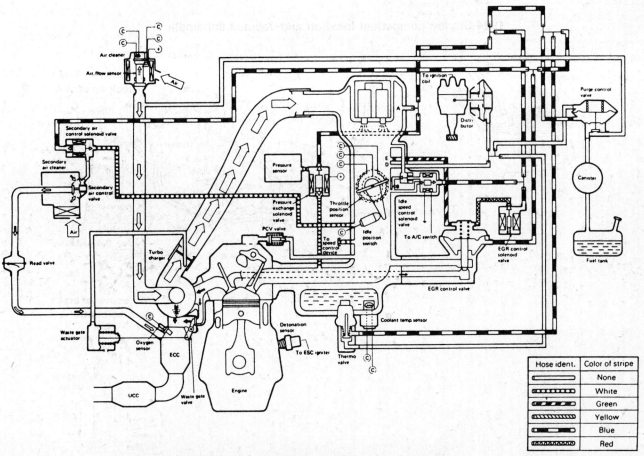

1983 Starion component location and vacuum schematic

4-27

4 EMISSION CONTROLS

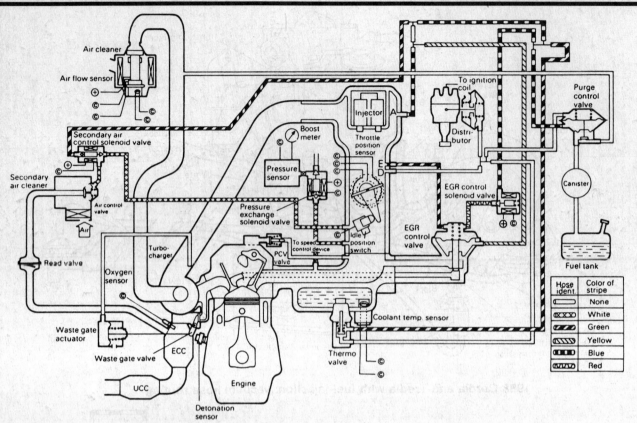

1984 Starion component location and vacuum schematic

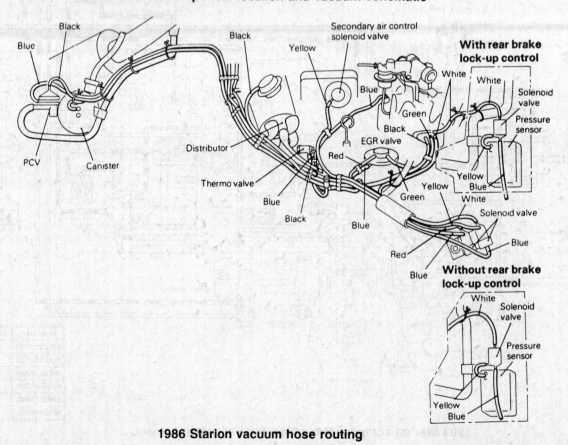

1986 Starion vacuum hose routing

EMISSION CONTROLS 4

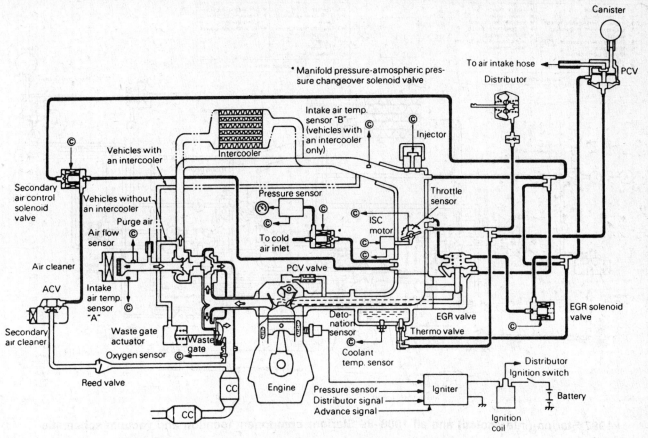

1986 Starion component location and vacuum schematic

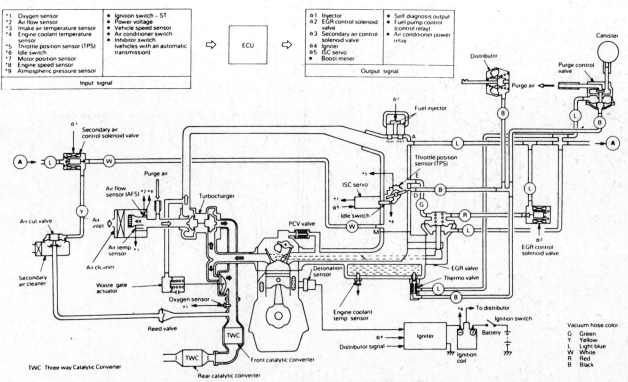

1987 Starion without intercooler: component location and vacuum schematic

4-29

4 EMISSION CONTROLS

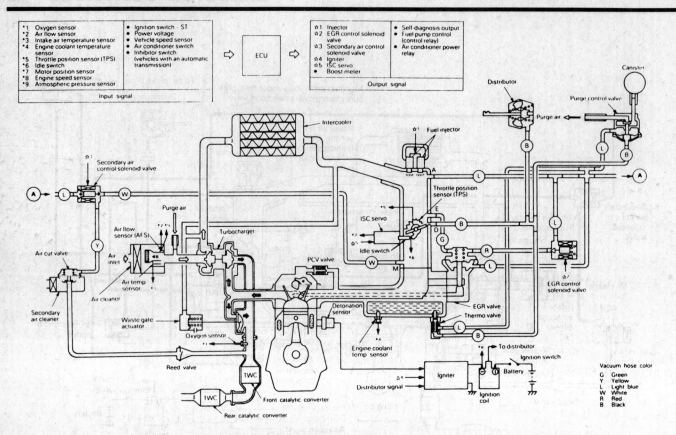

1987 Starion (intercooled) and all 1988–89 Starion: component location and vacuum schematic

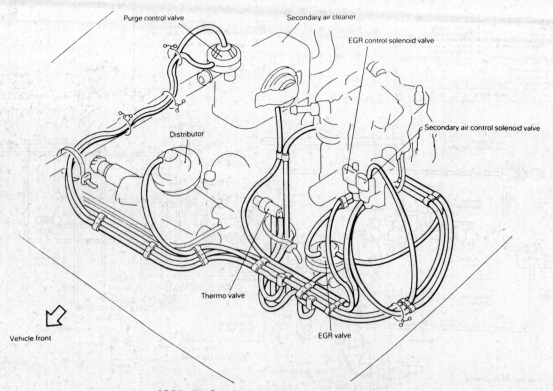

1987–89 Starion vacuum hose routing

4-30

EMISSION CONTROLS 4

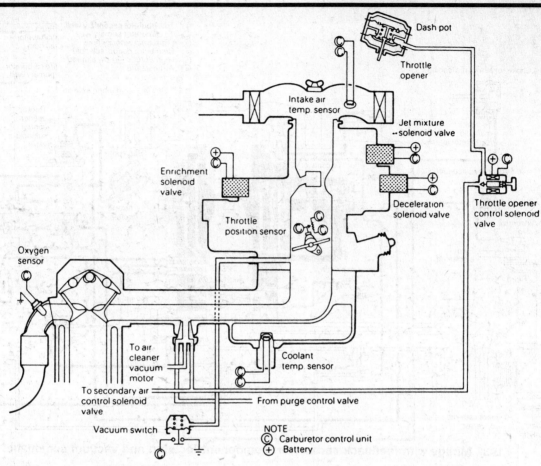

1985 Mirage with feedback carburetor: component location and vacuum schematic

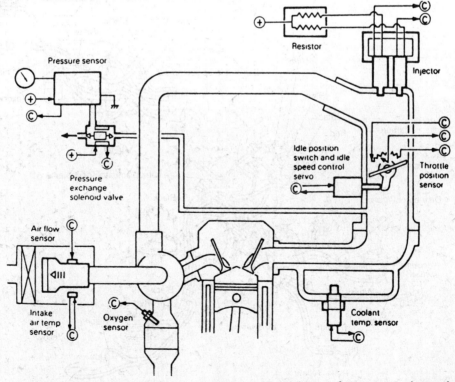

1985 Mirage with fuel injection: component location and vacuum schematic

4-31

4 EMISSION CONTROLS

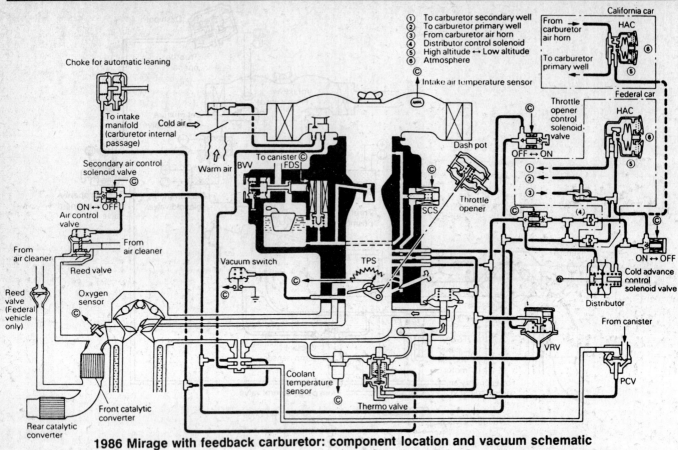

1986 Mirage with feedback carburetor: component location and vacuum schematic

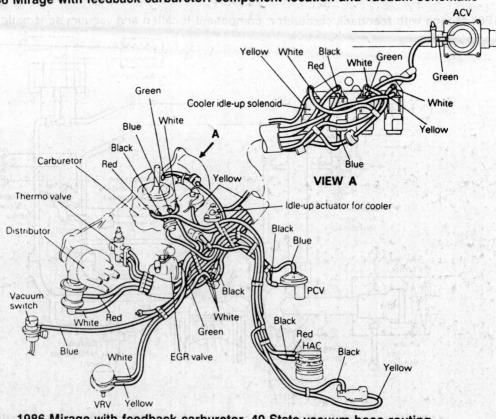

1986 Mirage with feedback carburetor, 49 State vacuum hose routing

4–32

EMISSION CONTROLS 4

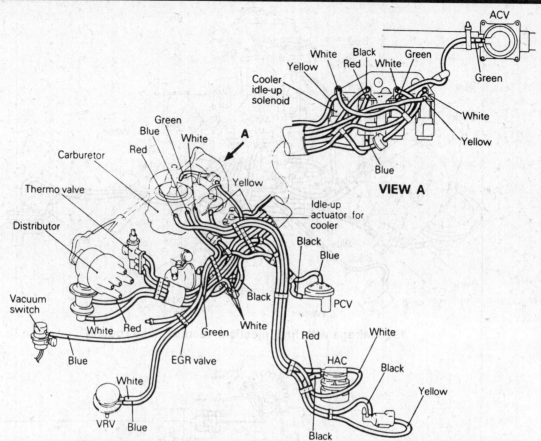

1986 Mirage with feedback carburetor, California vacuum hose routing

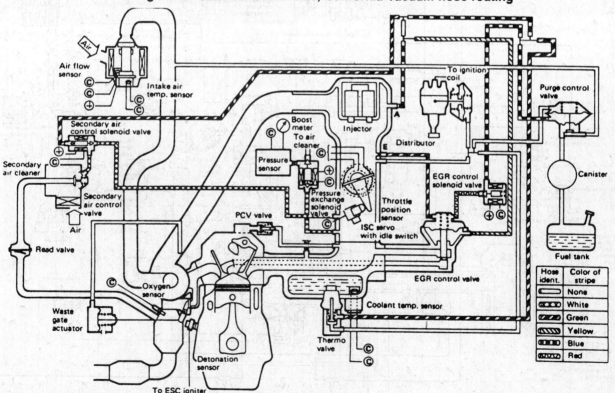

1986 Mirage with fuel injection, component location and vacuum schematic

4-33

4 EMISSION CONTROLS

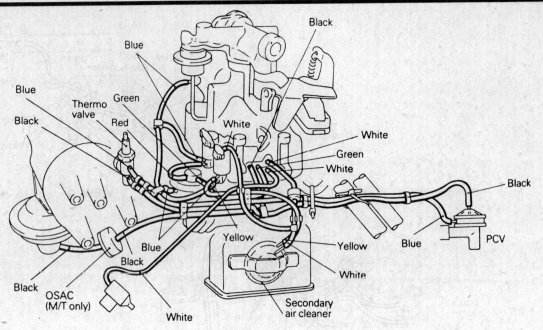

1986 Mirage with fuel injection: vacuum hose routing

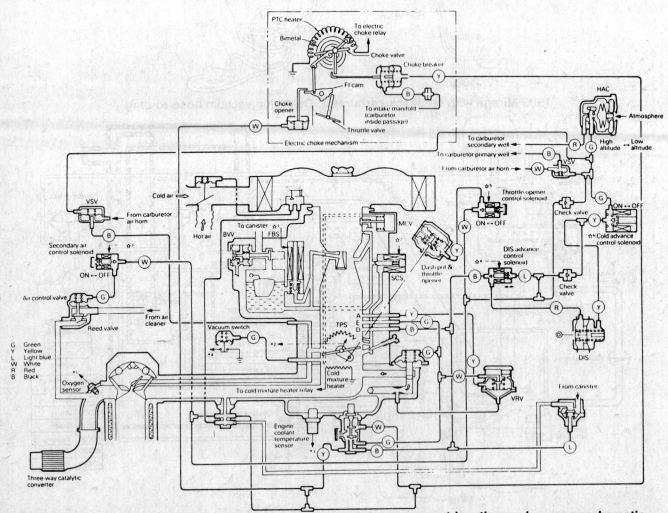

1987 Mirage and Precis with feedback carburetor: 49 state component location and vacuum schematic

EMISSION CONTROLS 4

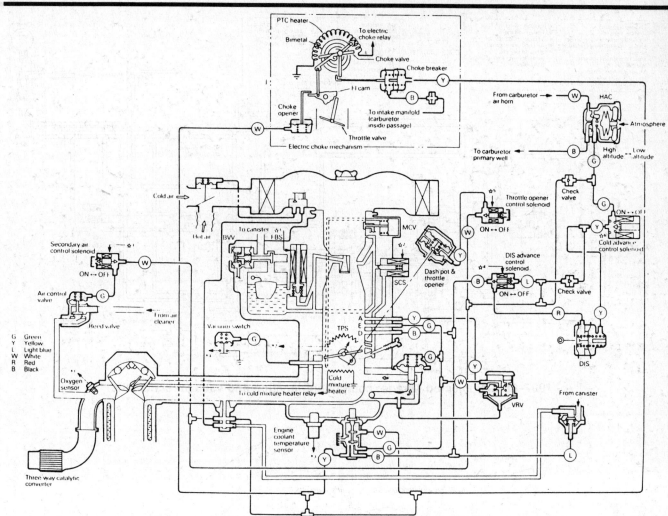

1987 Mirage and Precis with feedback carburetor: California component location and vacuum schematic

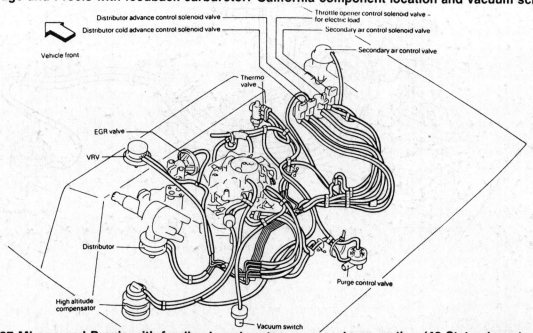

1987 Mirage and Precis with feedback carburetor: vacuum hose routing (49 State shown)

4-35

4 EMISSION CONTROLS

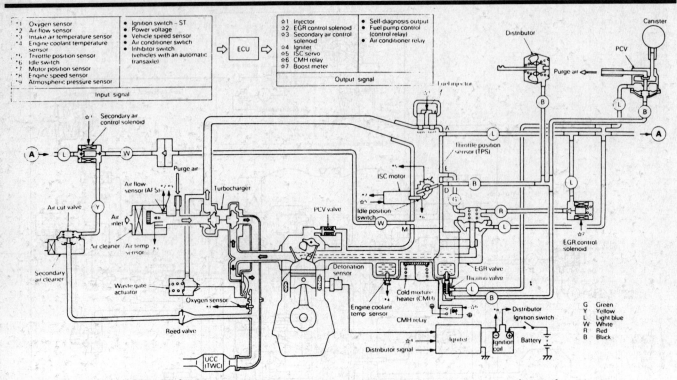

1987-88 fuel injected Mirage: component location and vacuum schematic

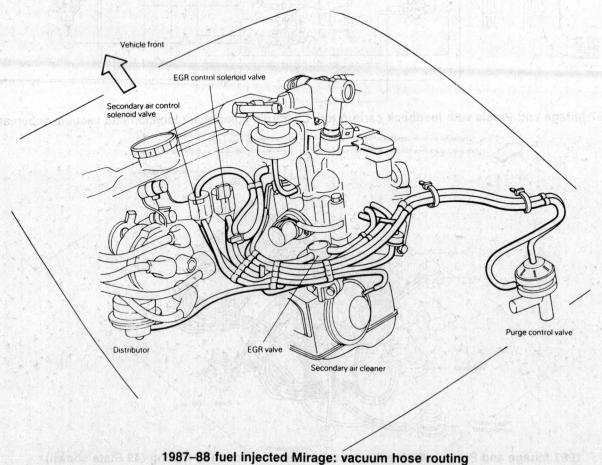

1987-88 fuel injected Mirage: vacuum hose routing

4-36

EMISSION CONTROLS 4

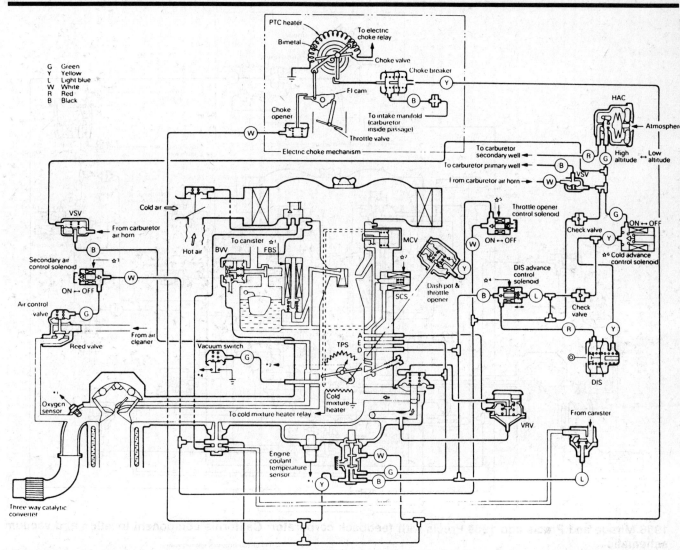

1988 Mirage and Precis and 1989 Precis with feedback carburetor: 49 state component location and vacuum schematic

4 EMISSION CONTROLS

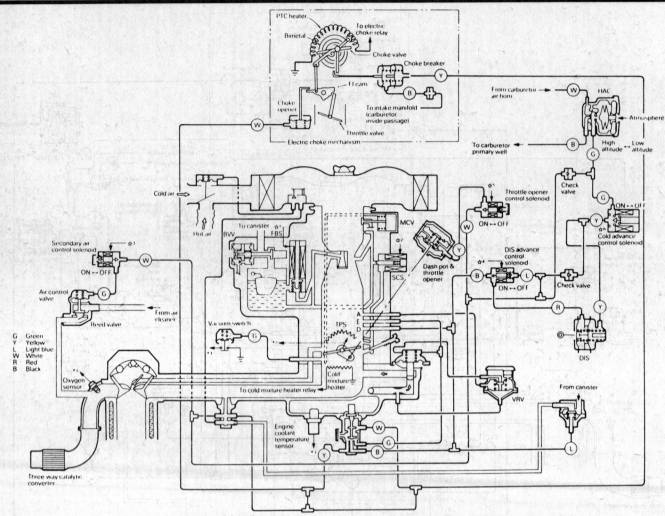

1988 Mirage and Precis and 1989 Precis with feedback carburetor: California component location and vacuum schematic

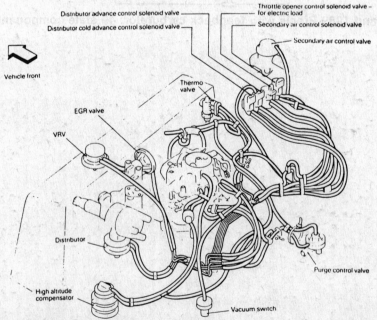

1988 Mirage and Precis and 1989 Precis with feedback carburetor: vacuum hose routing (49 State shown)

EMISSION CONTROLS 4

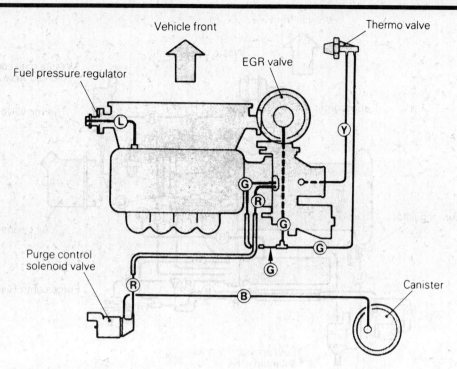

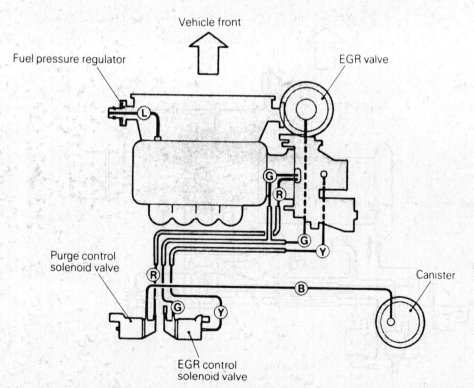

1989 Mirage (4G15) vacuum hose schematic: Upper, 49 State; California below

4-39

4 EMISSION CONTROLS

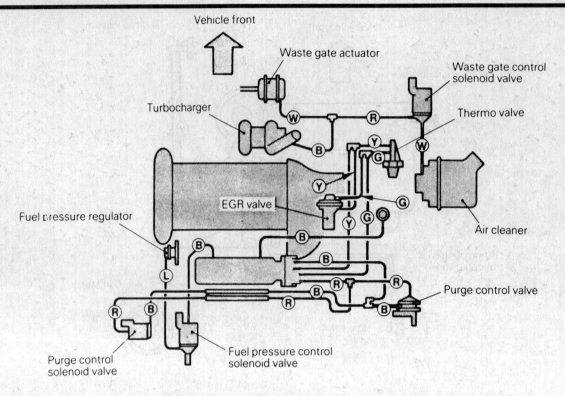

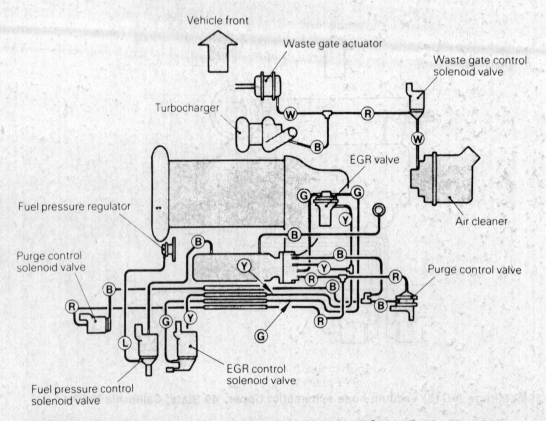

1989 Mirage (4G61) vacuum hose schematic: Upper, 49 State; California below

4-40

EMISSION CONTROLS 4

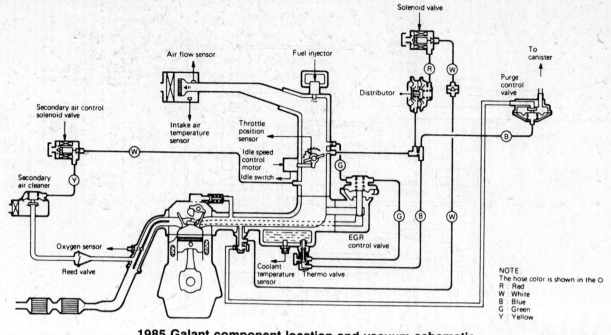

1985 Galant component location and vacuum schematic

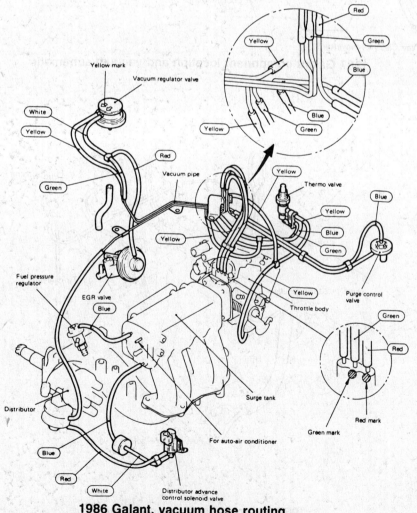

1986 Galant, vacuum hose routing

4-41

4 EMISSION CONTROLS

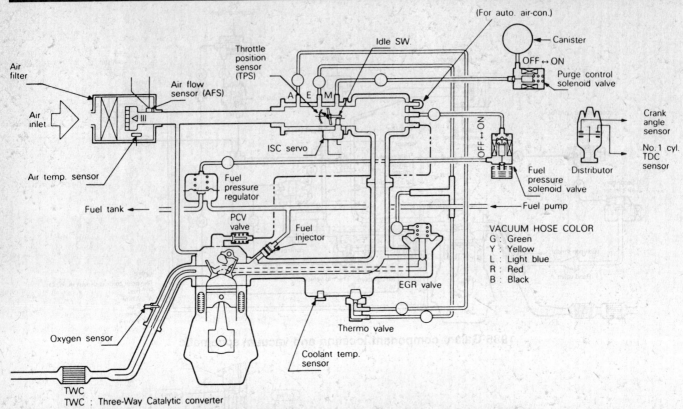

1987 Galant component location and vacuum schematic

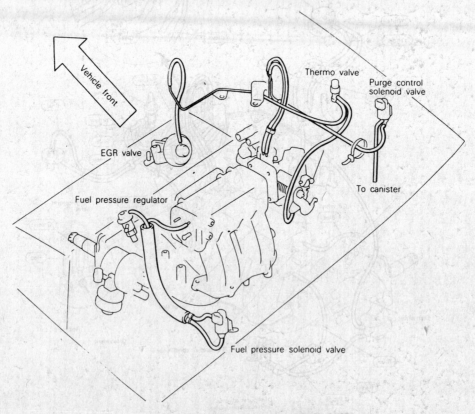

1987 Galant vacuum hose routing

4-42

EMISSION CONTROLS 4

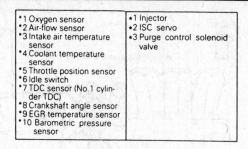

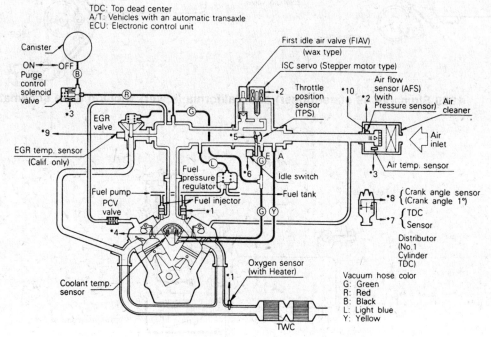

1988 Galant Sigma component location and vacuum schematic

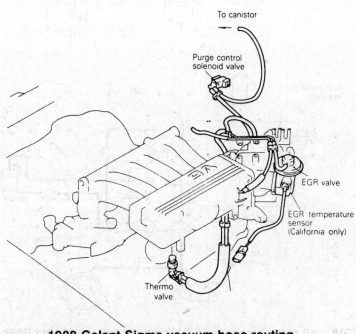

1988 Galant Sigma vacuum hose routing

4-43

4 EMISSION CONTROLS

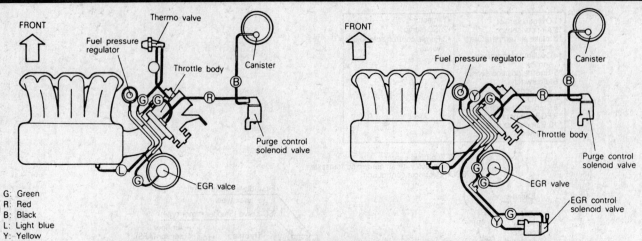

1989 Sigma V6 49 state (upper) and California (lower) vacuum hose schematic

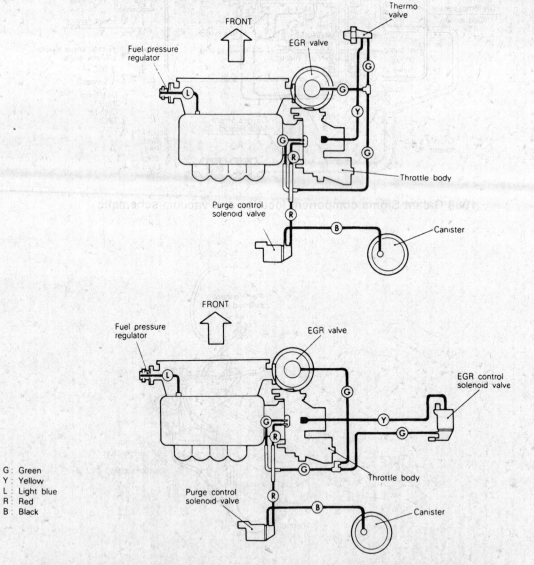

1989 Galant SOHC: 49 state (upper) and California (lower) vacuum hose schematic

4-44

EMISSION CONTROLS 4

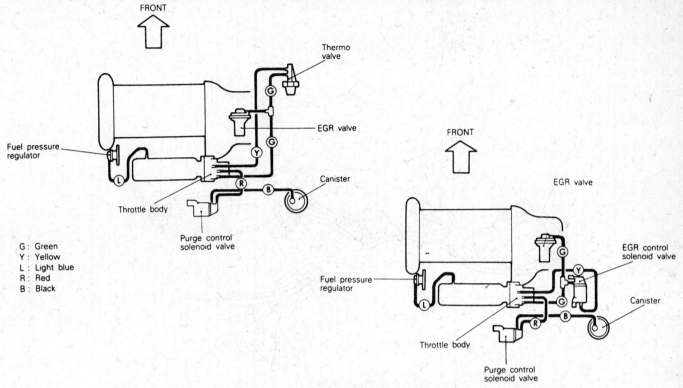

1989 Galant DOHC: 49 state (upper) and California (lower) vacuum hose schematic

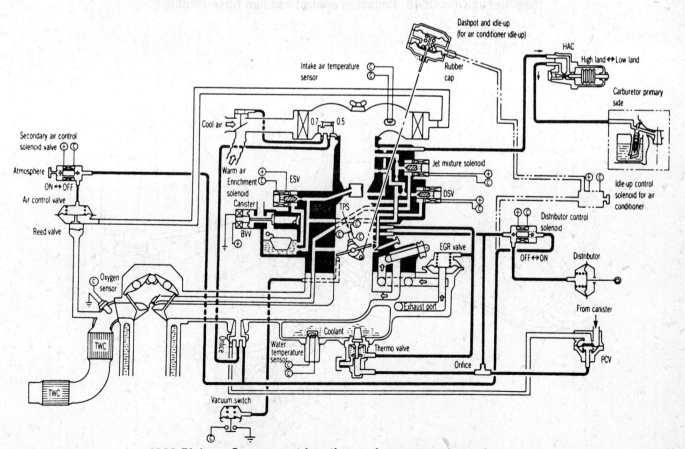

1986 Pickup: Component location and vacuum schematic

4-45

4 EMISSION CONTROLS

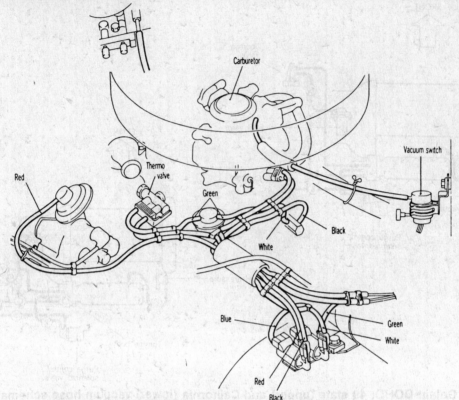

1986 Pickup with G54B: Emission control vacuum hose routing

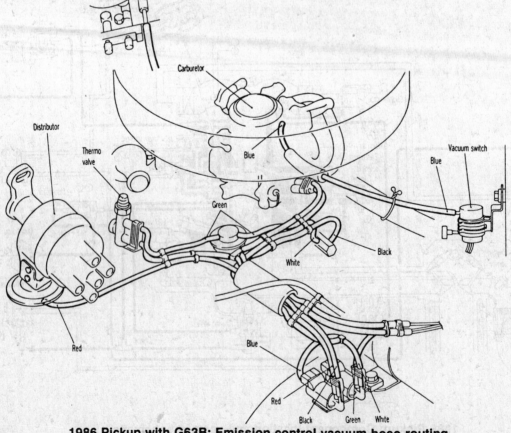

1986 Pickup with G63B: Emission control vacuum hose routing

Emission Controls 4

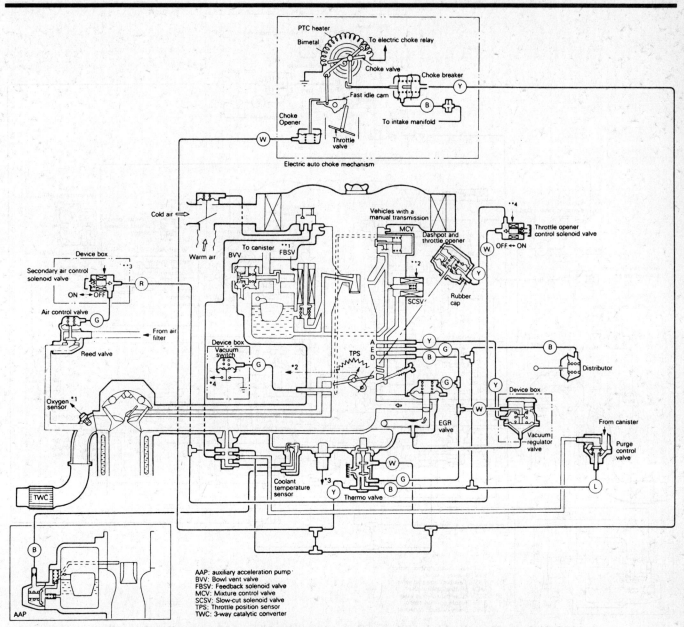

1987 Pickup, 49-state: Component location and vacuum schematic

4-47

4 EMISSION CONTROLS

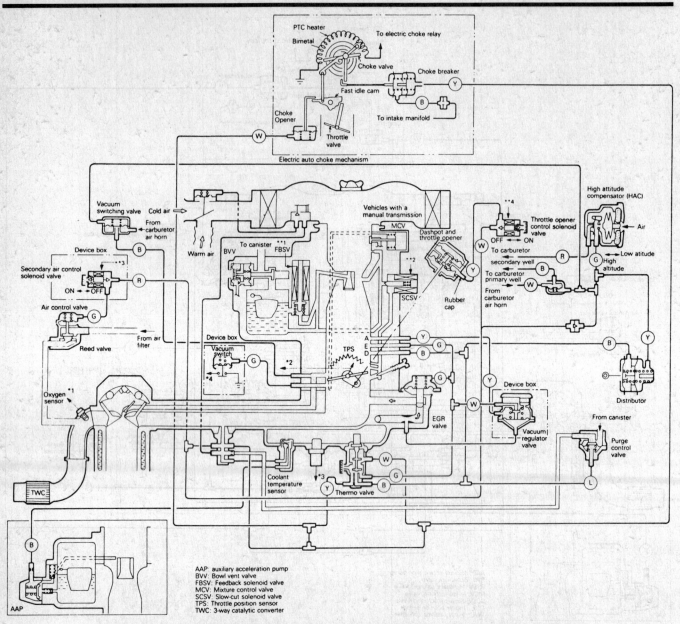

1987 Pickup, 49-state, high altitude: Component location and vacuum schematic

EMISSION CONTROLS 4

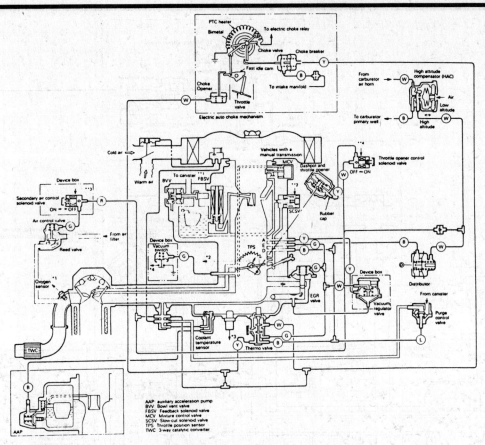

1987 Pickup, California: Component location and vacuum schematic

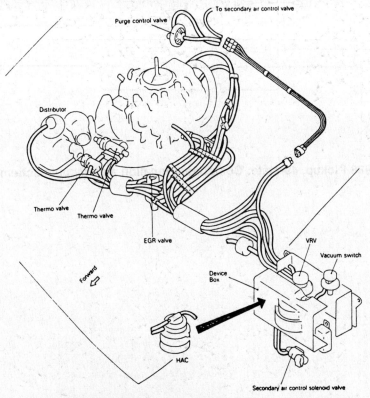

1987 Pickup: Emission control vacuum hose routing

4-49

4 EMISSION CONTROLS

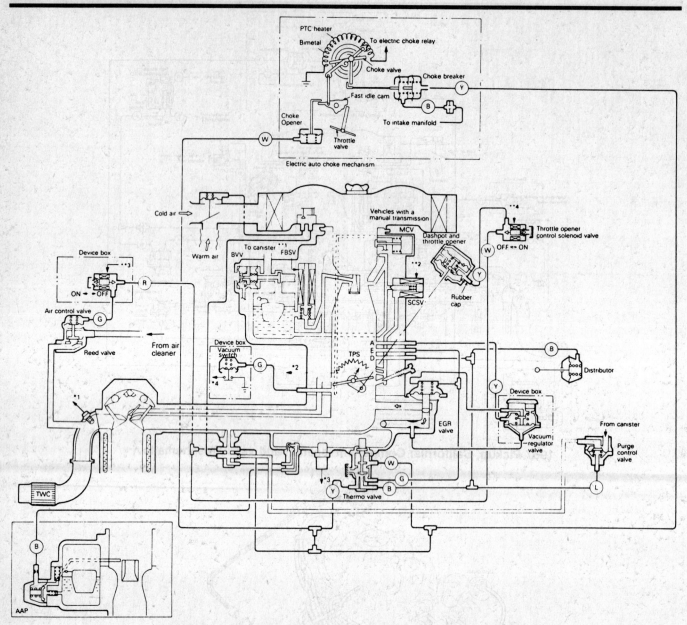

1988 Pickup, 49-state: Component location and vacuum schematic

EMISSION CONTROLS 4

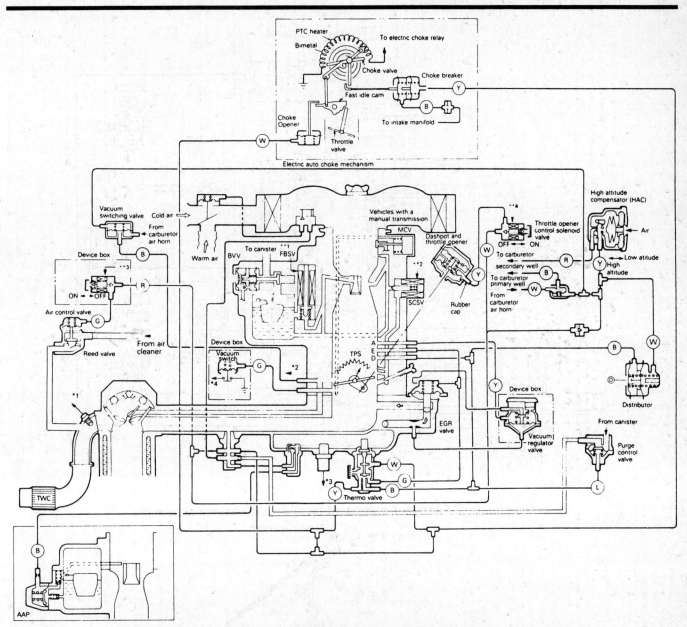

1988 Pickup, 49-state, high altitude: Component location and vacuum schematic

4 EMISSION CONTROLS

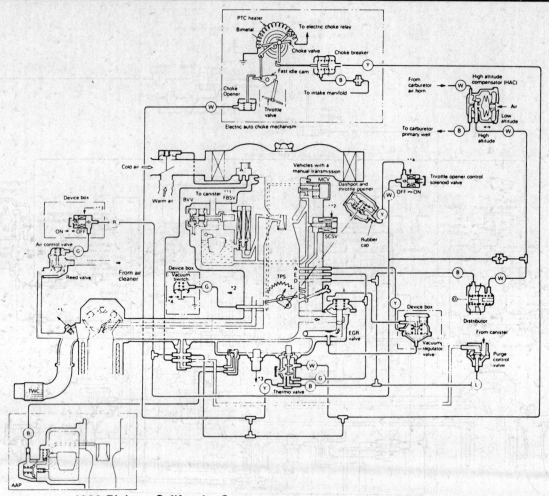

1988 Pickup, California: Component location and vacuum schematic

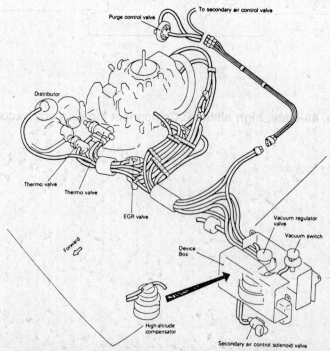

1988 Pickup: Emission control vacuum hose routing

EMISSION CONTROLS 4

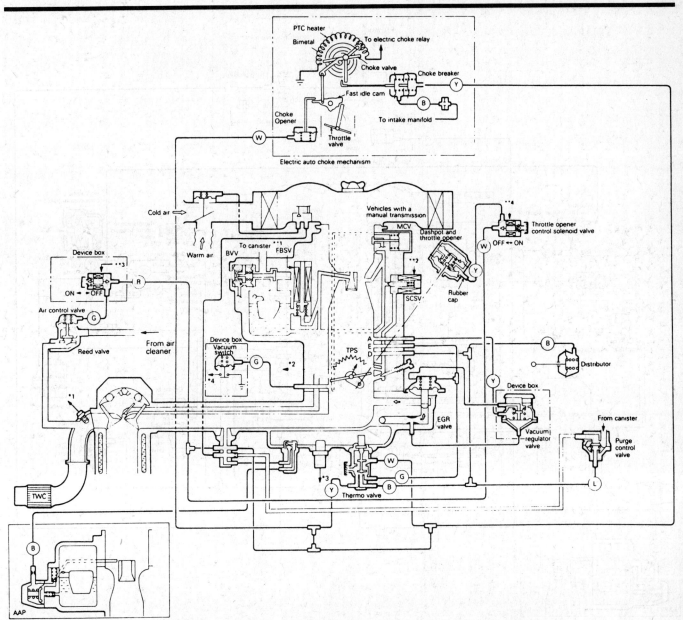

1989 Pickup, 49-state: Component location and vacuum schematic

4-53

4 EMISSION CONTROLS

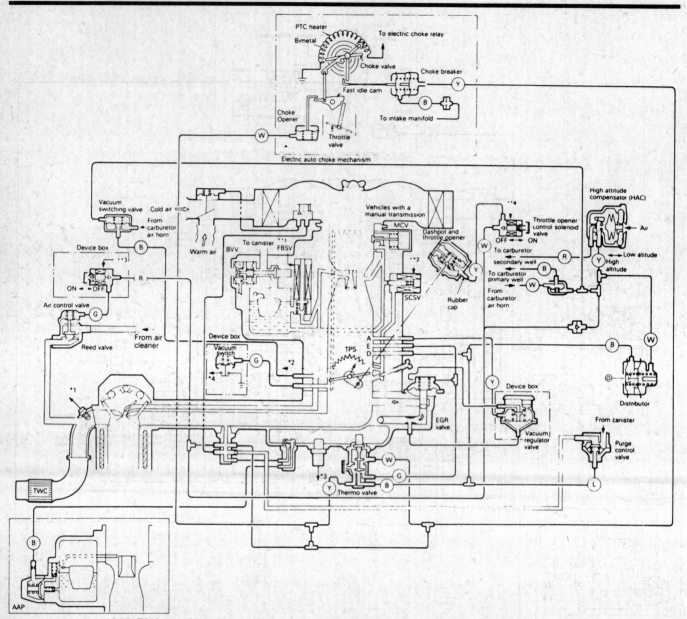

1989 Pickup, 49-state, high altitude: Component location and vacuum schematic

Emission Controls 4

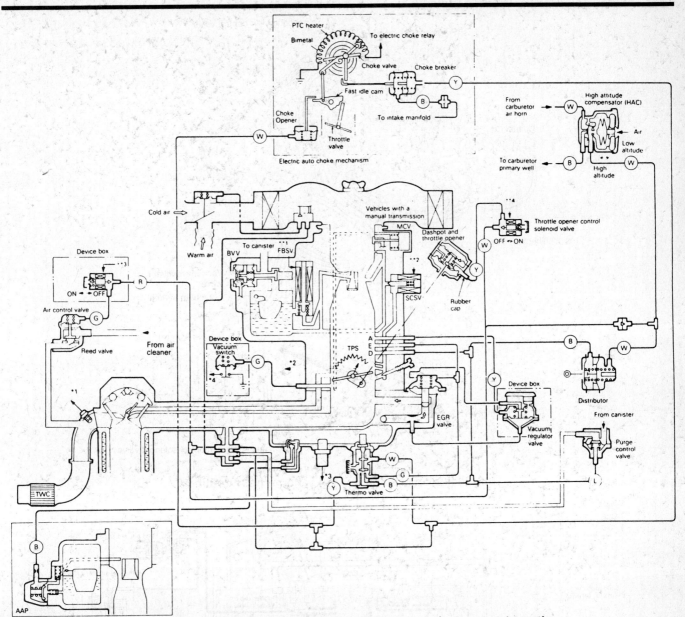

1989 Pickup, California: Component location and vacuum schematic

4-55

4 EMISSION CONTROLS

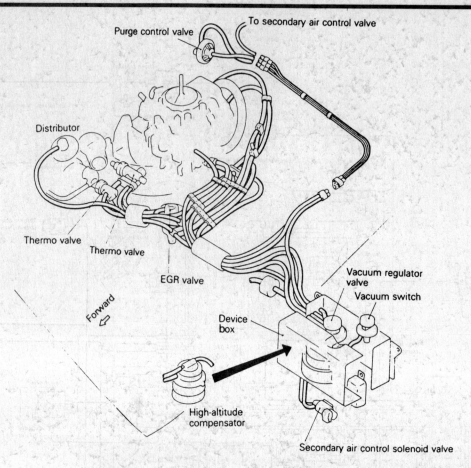

1989 Pickup: Emission control vacuum hose routing

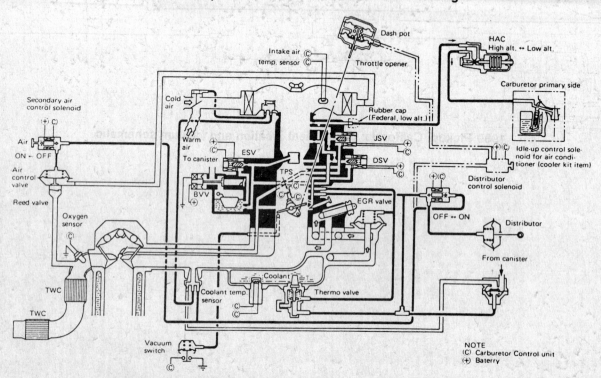

1986 Montero: Component location and vacuum schematic

EMISSION CONTROLS 4

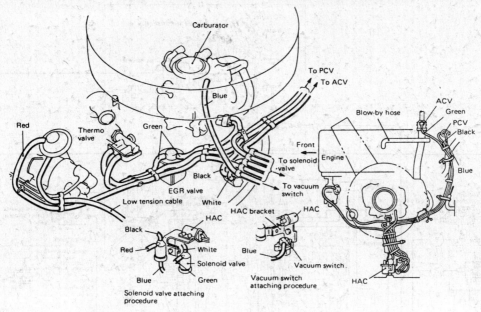

1986 Montero: Emission control vacuum hose routing

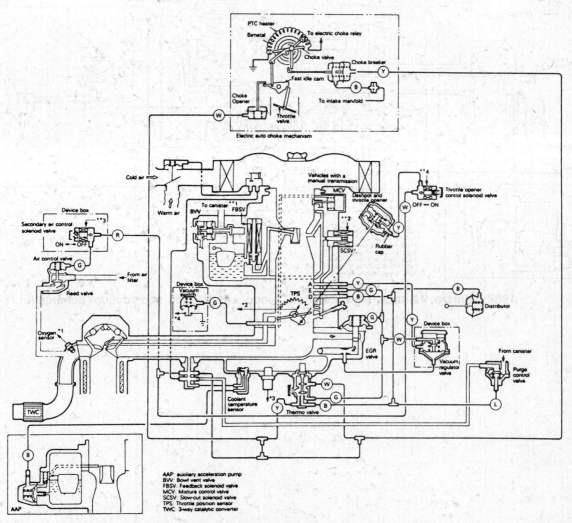

1987 Montero, 49-state: Component location and vacuum schematic

4-57

4 EMISSION CONTROLS

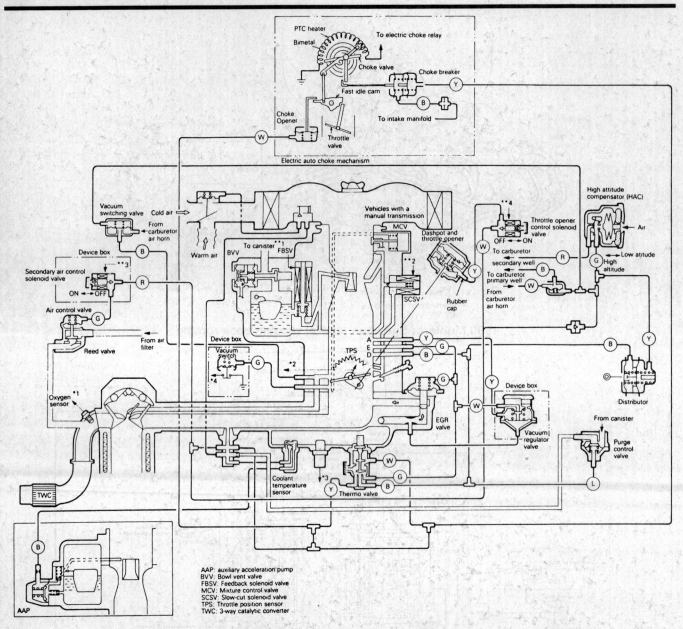

1987 Montero, 49-state, high altitude: Component location and vacuum schematic

EMISSION CONTROLS 4

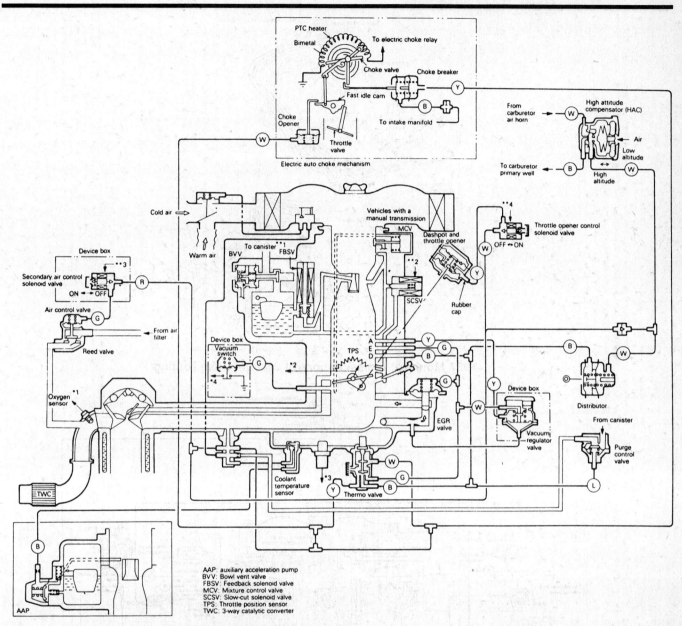

1987 Montero, California: Component location and vacuum schematic

4-59

4 EMISSION CONTROLS

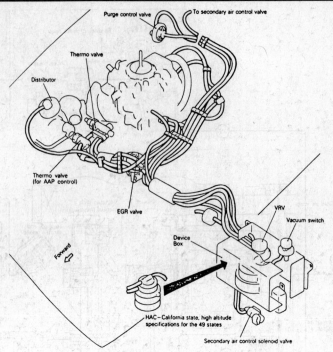

1987 Montero: Emission control vacuum hose routing

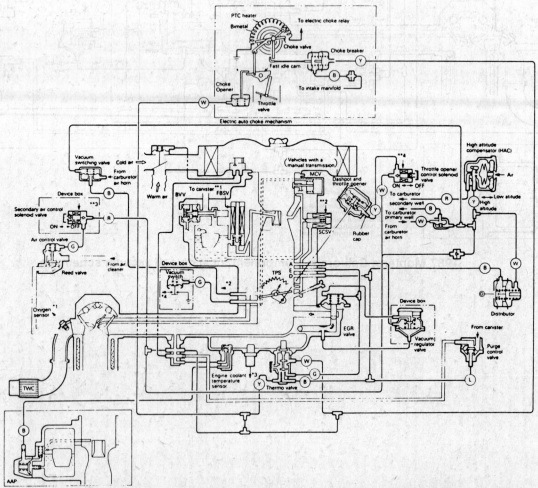

1988 Montero, 49-state: Component location and vacuum schematic

EMISSION CONTROLS 4

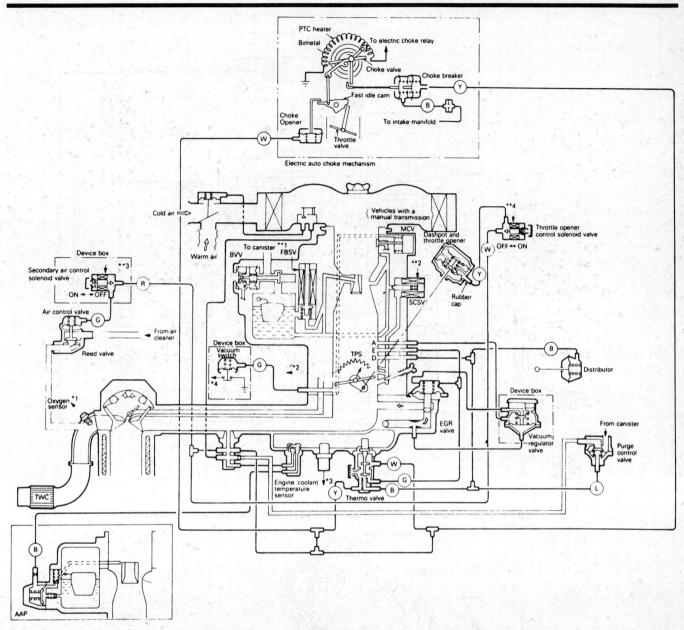

1988 Montero, 49-state, high altitude: Component location and vacuum schematic

4–61

4 EMISSION CONTROLS

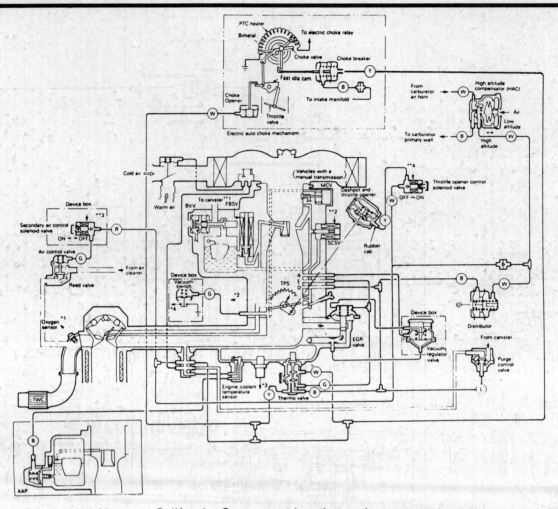

1988 Montero, California: Component location and vacuum schematic

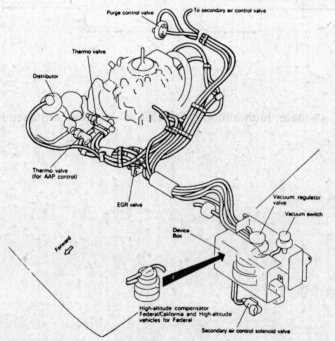

1988 Montero: Emission control vacuum hose routing

EMISSION CONTROLS 4

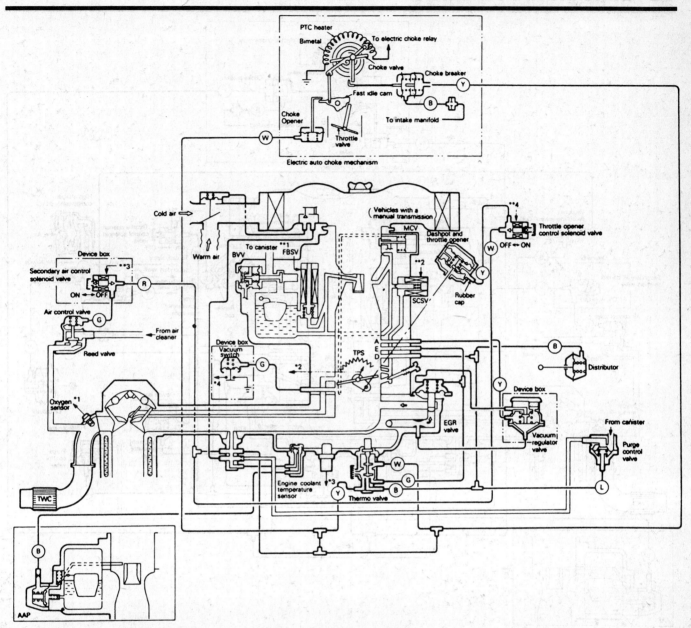

1989 Montero, 2.6L, 49-state: Component location and vacuum schematic

4 EMISSION CONTROLS

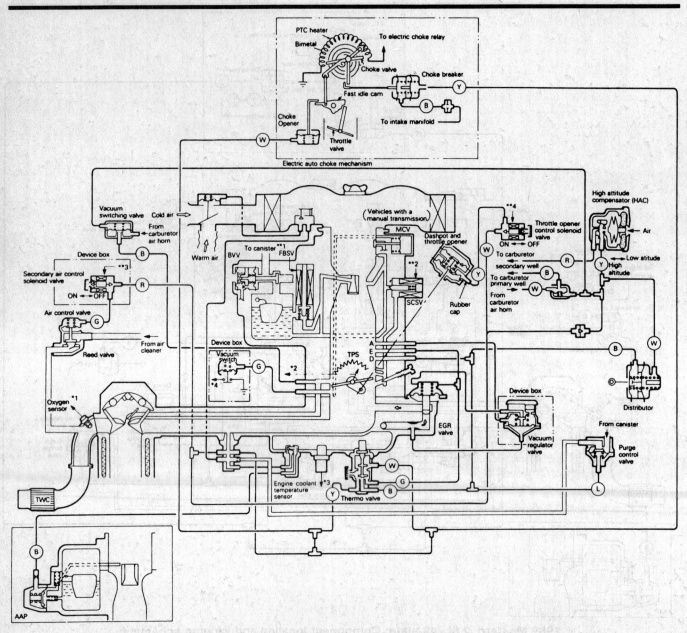

1989 Montero, 2.6L, 49-state, high altitude: Component location and vacuum schematic

EMISSION CONTROLS 4

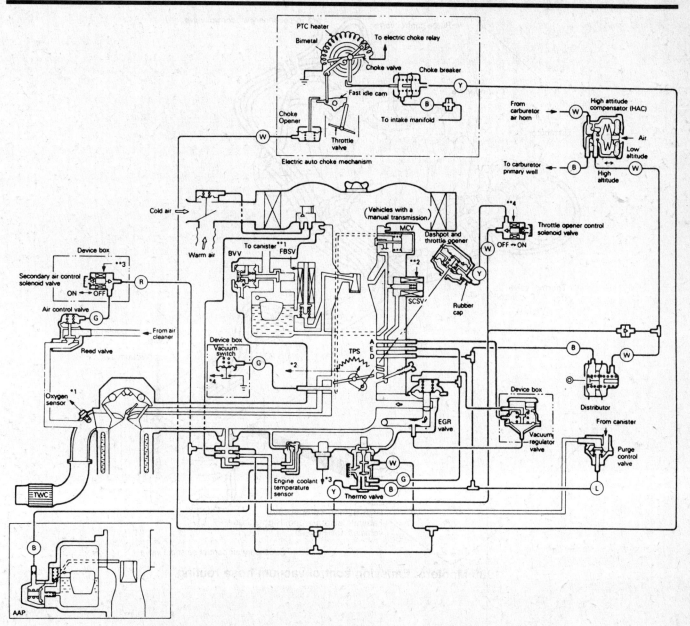

1989 Montero, 2.6L, California: Component location and vacuum schematic

4-65

4 EMISSION CONTROLS

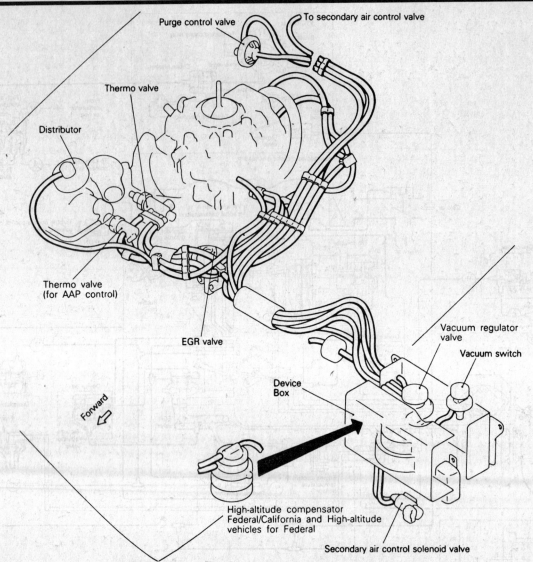

1989 Montero: Emission control vacuum hose routing

4-66

5 Fuel System

QUICK REFERENCE INDEX

Carburetor Specifications Chart	5-18
Carbureted Fuel System	5-2
Diesel Fuel System	5-48
Fuel Injection Systems	5-21, 5-29

GENERAL INDEX

Accelerator cable	5-16	Fuel pump		Fuel pump	5-29
Carburetor		Electric	5-22, 5-29	Injectors	5-34
Adjustments	5-16	Mechanical	5-2, 5-21	Throttle body	5-33
Overhaul	5-5	Fuel system		Troubleshooting	5-39
Removal and Installation	5-3	Carbureted	5-2	Specifications Charts	5-18
Specifications	5-18	Gasoline Fuel injection	5-21, 5-29	Throttle body	5-24
Diesel Fuel System		Troubleshooting	5-25	Throttle Body Electronic Fuel injection	
Description	5-48	Fuel tank	5-52	Fuel pump	
Fuel injectors	5-49	Glow plugs	5-51	Electric	5-22
Fuel lines	5-48	Idle speed adjustment	5-51	Mechanical	5-21
Glow plugs	5-51	Idle-up adjustment	5-18	Injectors	5-24
Idle speed adjustment	5-51	Injection pump	5-49	Operation	5-21
Injection pump	5-49	Injection timing	5-50	Throttle body	5-24, 5-33
Injection timing	5-50	Injectors, fuel	5-24, 5-34, 5-49	Troubleshooting	5-25
Float and fuel level adjustments	5-16	Multi-Point Fuel injection		Throttle opener	5-17
Fuel lines	5-48			Troubleshooting	5-25, 5-39

5–1

5 FUEL SYSTEM

Carbureted Fuel System

Mechanical Fuel Pump

REMOVAL AND INSTALLATION

Mechanical fuel pumps are found on the 1983 Cordia and Tredia, the 1984–88 Cordia and Tredia with the G63B (non-turbocharged) engine, the 1985–88 Mirage and Precis with the G15B engine, all pickup trucks through 1989 and all 4-cylinder Monteros.

The mechanical fuel pump is mounted on the side of the head, centered between the runners of the intake manifold. The intake manifold does not require removal to remove the pump, but access can be tricky. A selection of socket extensions, swivels and open end wrenches can be helpful.

A pushrod or lever arm runs from a special camshaft lobe to the pump rocker arm, transferring the motion to the pump diaphragm. As the diaphragm moves, fuel is alternately drawn into the pump and then sent to the carburetor.

CAUTION
Gasoline in either liquid or vapor state is EXTREMELY explosive. Take great care to contain spillage. Work in an open or well-ventilated area. Do not connect or disconnect electrical connectors while fuel hoses are removed or loosened. Observe no smoking/no open flame rules during repairs. Have a dry-chemical fire extinguisher (type B-C) within arm's reach at all times and know how to use it.

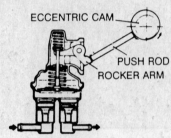

The mechanical fuel pump is driven from the camshaft through a pushrod

1. Disconnect the negative battery cable. Depending on the vehicle, it may be necessary to remove the air cleaner assembly.
2. Set the engine to TDC/compression for No. 1 cylinder. This is generally done by turning the crankshaft pulley bolt clockwise (only!) until the timing marks on the case align. Double check the positioning by removing the distributor cap and making sure the rotor points to the No.1 terminal.
3. Label or diagram the fuel lines running to the pump. Disconnect the three fuel lines by using a pair of pliers to shift clamps away from the nipples on the pump and then pulling the lines off with a twisting motion.
4. Remove the two mounting bolts from the head. Remove the pump, spacer, and two gaskets from the head. As you pull the pump off the head, catch the pushrod which is located just behind the pump. If the pump is the lever type, it will come off with the pump.

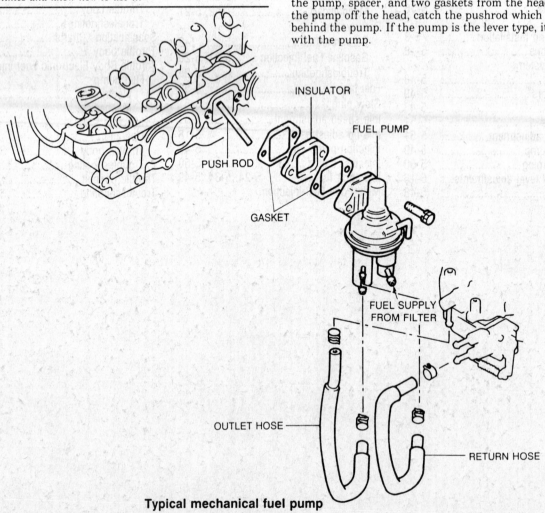

Typical mechanical fuel pump

FUEL SYSTEM 5

5. Clean the gasket surfaces of the insulator, pump and cylinder head. Insert the two bolts through the pump's mounting base. Slide a new gasket, the insulator, and a second new gasket into position over the two bolts. Turn the pump so its mounting surface faces the cylinder head.

6. Apply a light coat of clean engine oil to the pushrod. Place the pump pushrod against the cupped surface of the operating lever and angle it upward in the position it was in during removal. Hold the pushrod at that angle as you insert it into the bore in the head. Once the pushrod is in the bore in the cylinder head, you can release it your fingers and move the pump toward the head along the line of the pushrod.

If the pump is the lever-arm type, make certain the arm is installed above the camshaft. The cam will move the lever as it turns.

7. Start the two bolts into the bores in the head and tighten them finger tight. Tighten the mounting bolts alternately and evenly to 16 Nm (12 ft. lbs.).

8. Inspect the hoses for cracks (even hairline cracks can leak) and replace if necessary. Reconnect the fuel hoses. Make sure the hoses are installed all the way onto the nipples.

9. Work the clamps into position. Make sure the clamps are located well past the bulged portion of the nipples but do not sit at the extreme ends of the hoses.

10. Replace the distributor cap.

NOTE: If the fuel pump was replaced because of a ruptured diaphragm, there is a strong possibility of fuel contamination in the engine oil. Changing the oil and oil filter is strongly recommended after the fuel pump is replaced.

11. Install the air cleaner assembly if it was removed.

12. Connect the negative battery cable. Start the engine and check for leaks. Because the fuel lines have been partially emptied, the engine may crank for a longer than normal period before starting.

TESTING

On Car Testing

1. Disconnect the inlet line (coming from the filter) at the pump. Connect a vacuum gauge to the pump nipple. Remove the coil-to-distributor high tension cable at the coil.

2. Have an assistant crank the engine with the key as you watch the gauge. A vacuum should be produced in a regular cycle as the pump turns over. There should be no blowback of pressure (abrupt drop in vacuum or even positive pressure for a short time).

If there is blowback of pressure, the inlet valve on the pump is leaking and the unit must be replaced.

3. The vacuum shown with each pump stroke should be strong and constant. If the vacuum is low (or none at all), the diaphragm is leaking and the pump must be replaced.

Off Car Testing

1. Inspect the small breather hole or tube (above the diaphragm) which vents the pump's upper chamber. Leakage of fuel or oil here confirms that the diaphragm or oil seal is leaking.

2. Inspect the end of the pushrod and the contact surface on the pump operating lever. Replace the pushrod or pump if there is obvious wear. If the camshaft end of the pushrod is badly worn, remove the valve cover and inspect the camshaft eccentric (which operates the fuel pump) for excessive wear.

3. If the pump has a lever arm, check the contact pad (face) of the lever for wear or scoring. Inspect the camshaft is heavy scoring is present. Check the motion of the arm and the tension of the spring.

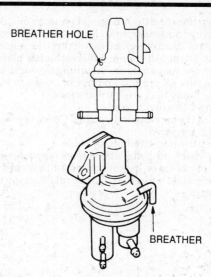

The diaphragm will vent to the atmosphere through a hole or tube in the upper case

Carburetor

REMOVAL AND INSTALLATION

CAUTION

Gasoline in either liquid or vapor state is EXTREMELY explosive. Take great care to contain spillage. Work in an open or well-ventilated area. Do not connect or disconnect electrical connectors while fuel hoses are removed or loosened. Observe no smoking/no open flame rules during repairs. Have a dry-chemical fire extinguisher (type B-C) within arm's reach at all times and know how to use it.

Mechanical Carburetor used on:
1983 Cordia and Tredia w/G62B Engine
1983–84 Pickup w/G54B or G63B Engine
1983 Montero
1984 Montero, exc. Calif.

1. Disconnect the negative battery cable.
2. Remove the air cleaner.
3. Place a container under the fuel line inlet fitting to contain spillage. Disconnect the fuel inlet line from the carburetor nipple.
3. Label or diagram each vacuum hose connection; disconnect the hoses from the carburetor ports.
4. Disconnect the throttle cable at the carburetor.
5. Remove the mounting bolts. Lift the carburetor off the engine and remove it to a workbench. Keep it level to avoid spilling of fuel from the float bowl.
6. Carefully drain the carburetor into a container with an airtight lid.
7. Before installation, inspect the mating surfaces of the carburetor and manifold. They should be clean and free of nicks, burrs or any pieces of gasket material. Clean the surfaces as necessary and remove any slight imperfections with crocus cloth.
8. Put a new carburetor gasket on the surface of the manifold.
9. Carefully locate the carburetor on top of the gasket with all holes lined up. Install the carburetor bolts, tightening them alternately and evenly in small increments. The gasket must be compressed evenly to prevent leakage.

5-3

5 FUEL SYSTEM

10. Connect the throttle linkage. Have someone depress the accelerator pedal and make sure the throttle blade opens all the way; it should be exactly vertical at full throttle. Adjust the cable as needed to obtain full opening of the throttle plate.
11. Connect the vacuum hoses according to your drawing or labeling. Make sure all are soft and free of cracks to make a good seal. Replace hoses that are hard or cracked.
12. Reconnect the fuel hoses.
13. Reinstall the air cleaner assembly. Check the filter element and replace it if needed.
14. Connect the negative battery cable.
15. Double check all installation items, paying particular attention to loose hoses or hanging wires, untightened nuts, poor routing of hoses and wires (too tight or rubbing) and tools left in the engine area.
16. Start the engine. It will require a longer cranking period and two or three pumps of the accelerator pedal before it starts.

— **CAUTION** —

Do NOT prime the engine by pouring fuel into the air horn of the carburetor. This outdated and foolish practice can result in severe injury or damage.

17. While the engine is warming up, check the work area carefully for any sign of fuel or vacuum leaks and attend to them immediately. Set the idle speed and make other necessary adjustments after the engine is running smoothly and is fully warmed up.

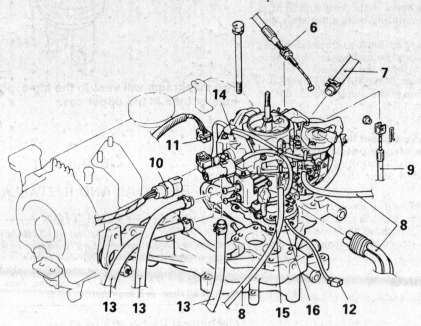

6. Accelerator cable
7. Coolant hose
8. Vacuum hoses
9. Throttle control cable (auto. trans.)
10. Throttle position sensor connector
11. Solenoid valve connector
12. Carburetor heater connector
13. Fuel hoses
14. Carburetor
15. Carburetor heater
16. Gasket

Typical feedback carburetor

Feedback Carburetor used on:
Cordia and Tredia with G63B Engine
Mirage and Precis with G15B Engine
1985–89 Pickup
1984 Montero for California
1985–89 4-cylinder Montero

NOTE: All wires and hoses should be labeled at the time of removal. The amount of time saved during reassembly makes the extra effort well worthwhile.

1. Disconnect the negative battery cable.
2. Drain the engine coolant to a point below the level of the intake manifold.

— **CAUTION** —

When draining the coolant, keep in mind that cats and dogs are attracted by the ethylene glycol antifreeze, and are quite likely to drink any that is left in an uncovered container or in puddles on the ground. This will prove fatal in sufficient quantity. Always drain the coolant into a sealable container. Coolant should be reused unless it is contaminated or several years old.

3. Remove the air cleaner assembly, disconnecting the various hoses and ducts as necessary.
4. Disconnect the accelerator cable from the carburetor.
5. Disconnect the coolant hose connection from the back of the carburetor.
6. Disconnect the vacuum hoses.
7. For vehicles with automatic transmissions, remove the lock pin and disconnect the throttle control cable linkage.
8. Disconnect the wiring to the throttle position sensor.
9. Disconnect the wiring to the solenoid valve.
10. Unplug the connector for the carburetor heater. Take care to separate the connectors without pulling on the wiring.
11. Use a small pan or jar to hold under the fuel hose connections. Disconnect them one at a time and catch any spilled fuel.
12. Remove the carburetor retaining bolts and lift the carburetor off the engine. Keep the carburetor level to avoid spillage. Once the carb is clear of the car, drain the remaining fuel into a container with an airtight lid.
13. Remove the carburetor base gasket (insulator). If desired, remove the heater element. Handle the heater carefully, protecting it from impact or pulling on the wire.
14. Before installation, inspect the mating surfaces of the carburetor and manifold. They should be clean and free of nicks, burrs or any pieces of gasket material. Clean the surfaces as necessary and remove any slight imperfections with crocus cloth.
15. Reinstall the heater element and put a new carburetor gasket on the surface of the manifold.
16. Carefully locate the carburetor on top of the gasket with all holes lined up. Install the carburetor bolts, tightening them al-

FUEL SYSTEM 5

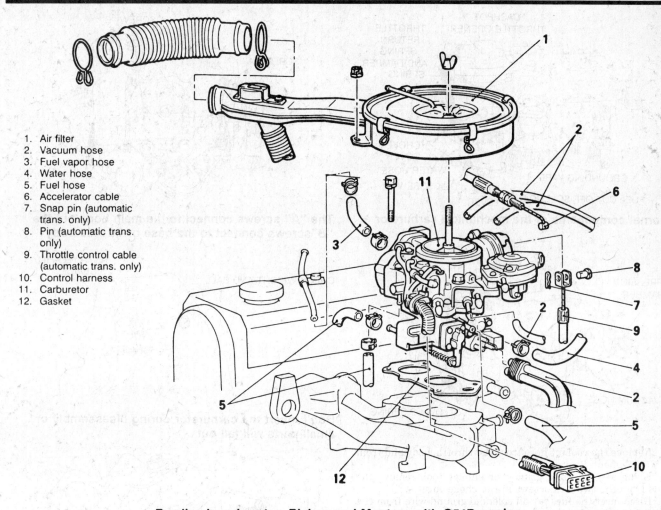

1. Air filter
2. Vacuum hose
3. Fuel vapor hose
4. Water hose
5. Fuel hose
6. Accelerator cable
7. Snap pin (automatic trans. only)
8. Pin (automatic trans. only)
9. Throttle control cable (automatic trans. only)
10. Control harness
11. Carburetor
12. Gasket

Feedback carburetor, Pickup and Montero with G54B engine

ternately and evenly in small increments. The gasket must be compressed evenly to prevent leakage. Tighten the bolts to 18 Nm (13 ft. lbs.).
17. Connect the wiring harness to the heater element, the solenoid valve and the throttle position sensor.
18. If equipped with automatic transmission, connect the throttle control cable and install the locking pin securely.
19. Connect the vacuum hoses, inspecting each one for any sign of cracking or hardening. If a hose is suspect, replace it.
20. Install the coolant hose and secure the clamp.
21. connect the accelerator cable.
22. Reinstall the air cleaner, connecting the hoses and ductwork as necessary. Make certain the air cleaner body is properly seated on the carburetor and that each hose is firmly attached.
23. Refill the cooling system with coolant.
24. Connect the negative battery cable.
25. Double check all installation items, paying particular attention to loose hoses or hanging wires, untightened nuts, poor routing of hoses and wires (too tight or rubbing) and tools left in the engine area.
26. Start the engine. It will require a longer cranking period and two or three pumps of the accelerator pedal before it starts.

── **CAUTION** ──
Do NOT prime the engine by pouring fuel into the air horn of the carburetor. This outdated and foolish practice can result in severe injury or damage.

27. While the engine is warming up, check the work area carefully for any sign of fuel, coolant or vacuum leaks and attend to them immediately. Set the idle speed and make other necessary adjustments to the accelerator or throttle control cables after the engine is running smoothly and is fully warmed up.

CARBURETOR OVERHAUL

Mechanical Carburetor used on:
1983 Cordia and Tredia w/G62B Engine
1983–84 Pickup w/G54B or G63B Engine
1983 Montero
1984 Montero, exc. Calif.

NOTE: The carburetor is in three main pieces: the top or float bowl cover, the main body and the throttle body or base. Separating these sections requires the removal and installation of many small parts and fittings. Do not disassemble anything unnecessarily. Some important components are not removed or adjusted during an overhaul.

1. Remove the carburetor from the car, following instructions given earlier in this section. Drain any remaining fuel into a container with an airtight lid. Place the carburetor on the workbench in a clean, dry area. Placing it on a large, lint-free cloth will help prevent parts from getting lost or rolling around.

5-5

5 FUEL SYSTEM

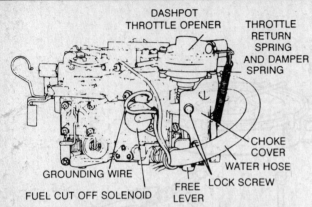

External components of the mechanical carburetor

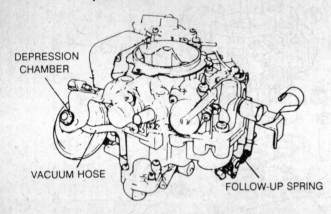

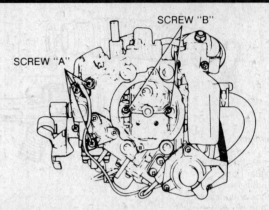

The "A" screws connect to the main body but the "B" screws connect to the base or throttle body

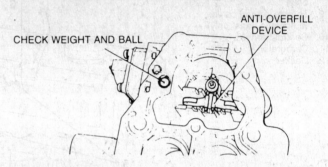

Don't invert the carburetor during disassembly or small parts will fall out

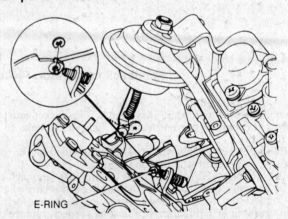

Disconnecting the choke unloader. Keep a close watch on the E-ring — it will vanish if dropped

2. Remove the coolant hose from the throttle body and from the wax element.
3. Using a small hand grinder or similar tool, remove the heads from the two lock screws in the choke cover.
4. Disconnect the fuel cut-off solenoid ground wire from the top of the carburetor.
5. Remove the throttle return spring and the damper spring.
6. Disconnect the vacuum hose running from the depression chamber to the throttle body.
7. Remove the accelerator pump rod from the throttle lever.
8. Remove the dashpot rod (for manual transaxle) or the throttle opener rod (automatic transaxle) from the free lever.
9. Remove the depression chamber rod from the secondary throttle lever.
10. Remove the six screws from the carburetor top. The four outer ones connect to the main body of the carburetor; the two bolts within the air passage connect to the throttle body.

NOTE: Many of the screws use Phillips-type heads. Use a screwdriver which fits the head exactly. An improper tool can damage the head of the screw and cause problems during reassembly.

11. Remove the main body with the top attached (the top cannot come free yet) by lifting straight up. Do not turn the carburetor upside down during the removal; if it is inverted, the accelerator pump check weight, ball and steel ball of the anti-overfill device will fall out.
12. Remove the E-clip from the lower end of the choke unloader rod and disconnect the rod from the lever.
13. Separate the top from the main body.
14. At this point, the carburetor is disassembled enough to perform common overhaul replacements. Do not disassemble any further components without good diagnostic reasons. In particular, do not disassemble the automatic choke system or attempt to remove the throttle plates; both systems require very precise alignment which is beyond the ability of the home mechanic.
15. Remove the float from the float arm by removing the pivot pin.
16. Inspect the float bowl for any sign of particulate dirt or solid matter. Carefully wipe the bowl clean. Shake the float, listening for any sign of liquid fuel inside. If the float has absorbed fuel, it must be replaced.
17. Remove the retaining screw and bracket holding the needle valve. Carefully remove the needle valve and inspect it for uneven wear or pitting. (A magnifying glass is very helpful in checking the tip.)

Check the seat for signs of pitting. Don't remove the seat without planning to replace it; if it looks OK, leave it alone. If

FUEL SYSTEM 5

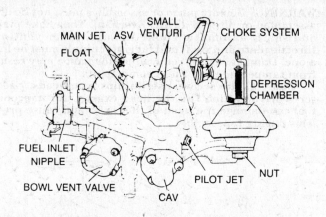

Don't remove anything without good reason. Some systems cannot be adjusted without expensive flow monitoring equipment

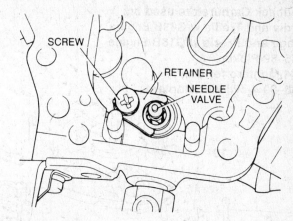

Remove the retainer to remove the needle valve

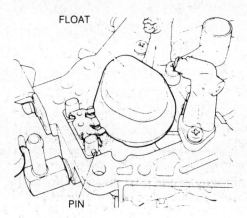

Replacing the float is common during most overhauls

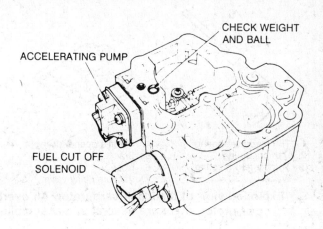

Remove these components from the mechanical carburetor before cleaning any passages or ports

the seat is to be removed, it must be carefully unscrewed with pliers. It will be difficult to loosen and care must be take not to damage or deform the seat. When the seat is removed, the spacing shim below it must be recovered and reinstalled. This shim determines float level adjustment.
18. Remove the accelerator pump and the fuel cut-off solenoid. Remove the check weight and ball.
19. Wearing eye protection and gloves, carefully clean the fuel and air passages with a spray cleaner and, if available, compressed air. A majority of carburetor problems are caused by very small bits of dirt lodging in the air or fuel passages. Clean everything thoroughly.
20. Inspect the motion of both the choke and throttle plates. They must move smoothly with absolutely no sign of binding or notching. Clean the linkages and plates as necessary, then apply a small amount of lubricant to the pivot points.
21. If any of the fuel jets are to be replaced due to wear or etching, they must be replaced with the identical item. Each jet has a number on the side of it to aid in identification. (The jets are selected based on precise airflow measurements during assembly. Installation of the wrong jet will send the wrong fuel mixture to the engine under almost all conditions).
22. Install the accelerator pump, the fuel cut-off solenoid and the check weight and ball.
23. Install the needle valve and its retainer.
24. Hold the float in position and install the pivot pin.
25. Carefully place the carb top onto the main body and install the four retaining screws holding the top to the body.

26. Install the choke unloader rod to the lever and install the E-ring to hold the rod in place.

NOTE: Work carefully. This E-ring has a habit of springing out of place and getting lost during installation.

27. Install the two screws through the air horn and tighten them.
28. Install the depression chamber rod to the secondary throttle lever.
29. Connect the dashpot or throttle opener rod to the free lever.
30. Install the accelerator pump rod to the throttle lever.
31. Install the vacuum hose between the depression chamber and the throttle body.
32. Install the throttle return spring and the damper spring.
33. Connect the ground wire for the fuel cut-off solenoid.
34. Install new screws to hold the choke cover in place.
35. Install the coolant hose from the throttle body to the wax element.
36. Move the carburetor linkage by hand, checking that motions are smooth and there is no binding in any of the mechanisms.
37. Reinstall the carburetor.

5-7

5 FUEL SYSTEM

Feedback Carburetors used on:
Cordia and Tredia w/G63B Engine
Mirage and Precis w/G15B Engine
1985–89 Pickup
1984 Montero for Calif.
1985–89 4-cylinder Montero

WARNING: Certain parts or assemblies must not be disassembled or altered during overhaul. The choke plate and shaft, automatic choke linkage, inner venturi, throttle plate and shaft and fuel inlet nipple must be left alone. Damage and or reduced performance may result from tampering with these components.

Many of the screws use Phillips-type heads. Use a screwdriver which fits the head exactly. An improper tool can damage the head of the screw and cause problems during reassembly.

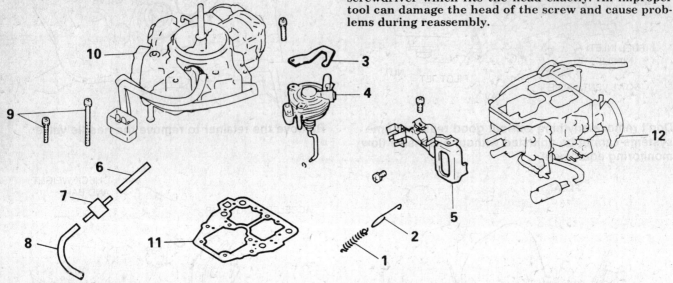

1. Throttle return spring
2. Damper spring
3. Throttle return spring bracket
4. Throttle opener/dashpot
5. Bracket
6. Hose
7. Vacuum delay valve
8. Hose
9. Screw
10. Float chamber cover (carb. top)
11. Gasket
12. Mixing body (main body) and throttle body (base)

Exploded view of feedback carburetor. An overhaul does not require the total disassembly of the carburetor; there are several sub-assemblies which should not be removed or adjusted

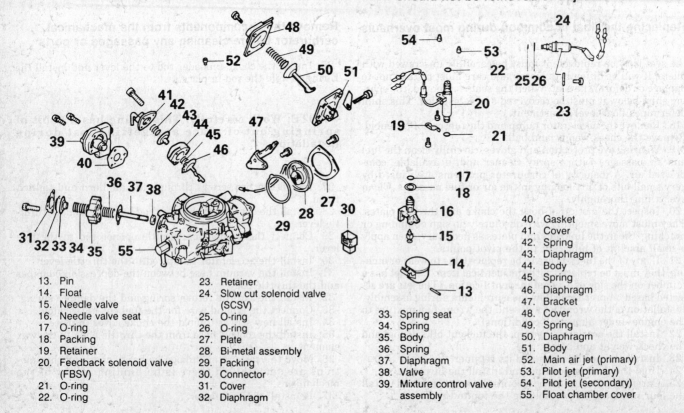

13. Pin
14. Float
15. Needle valve
16. Needle valve seat
17. O-ring
18. Packing
19. Retainer
20. Feedback solenoid valve (FBSV)
21. O-ring
22. O-ring
23. Retainer
24. Slow cut solenoid valve (SCSV)
25. O-ring
26. O-ring
27. Plate
28. Bi-metal assembly
29. Packing
30. Connector
31. Cover
32. Diaphragm
33. Spring seat
34. Spring
35. Body
36. Spring
37. Diaphragm
38. Valve
39. Mixture control valve assembly
40. Gasket
41. Cover
42. Spring
43. Diaphragm
44. Body
45. Spring
46. Diaphragm
47. Bracket
48. Cover
49. Spring
50. Diaphragm
51. Body
52. Main air jet (primary)
53. Pilot jet (primary)
54. Pilot jet (secondary)
55. Float chamber cover

5-8

FUEL SYSTEM 5

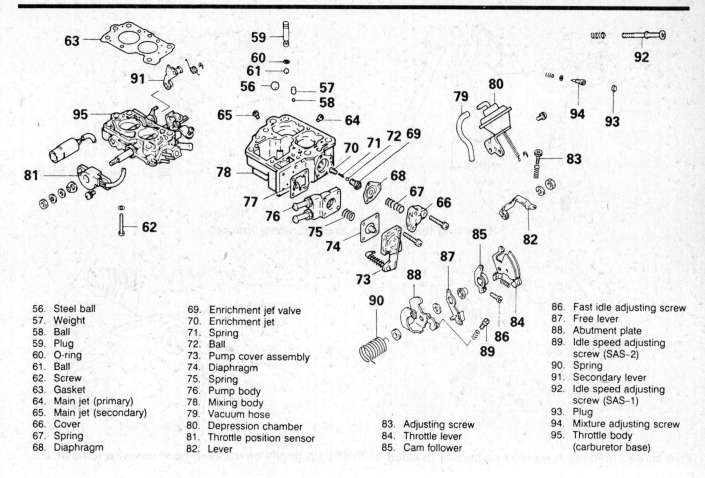

56. Steel ball	69. Enrichment jet valve	86. Fast idle adjusting screw
57. Weight	70. Enrichment jet	87. Free lever
58. Ball	71. Spring	88. Abutment plate
59. Plug	72. Ball	89. Idle speed adjusting
60. O-ring	73. Pump cover assembly	screw (SAS-2)
61. Ball	74. Diaphragm	90. Spring
62. Screw	75. Spring	91. Secondary lever
63. Gasket	76. Pump body	92. Idle speed adjusting
64. Main jet (primary)	78. Mixing body	screw (SAS-1)
65. Main jet (secondary)	79. Vacuum hose	93. Plug
66. Cover	80. Depression chamber	94. Mixture adjusting screw
67. Spring	81. Throttle position sensor	95. Throttle body
68. Diaphragm	82. Lever	(carburetor base)
	83. Adjusting screw	
	84. Throttle lever	
	85. Cam follower	

1. Remove the carburetor from the car, following instructions given earlier in this section. Drain any remaining fuel into a container with an airtight lid. Place the carburetor on the workbench in a clean, dry area. Placing it on a large, lint-free cloth will help prevent parts from getting lost or rolling around.
2. Remove the throttle return spring and the damper spring.
3. Remove the throttle opener (automatic transaxle) or the dashpot (manual transaxle) rod from the free lever and remove the opener or dashpot unit from the top of the carburetor.
4. If equipped with automatic transmission, remove the idle speed control (ISC) servo by removing the bracket screws. Put the ISC servo out of the way until reassembly.

WARNING: Do not attempt to test the servo with battery voltage. It runs on a lower voltage sent from the ECU. Applying battery voltage to this unit will destroy it.

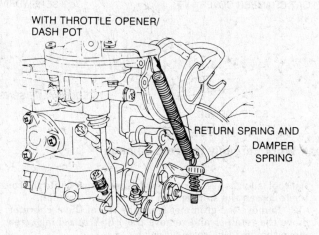

Removing the springs

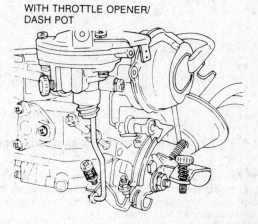

Don't deform the connecting rods when removing the dashpot or throttle opener

5-9

5 FUEL SYSTEM

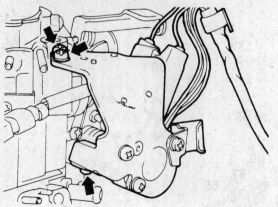

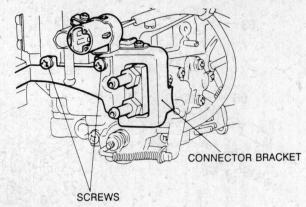

Removing the ISC servo and connector bracket

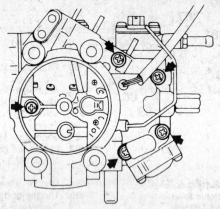

Five screws hold the top of the carburetor in place

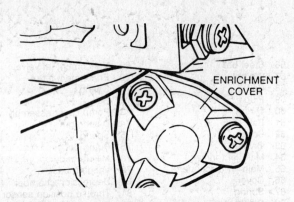

Pry up gently on the float bowl cover to remove it

5. Remove the connector bracket.
6. Remove the vacuum hose running from the base to the choke breaker. This vacuum line will have a delay valve in it.
7. Remove the five screws holding the top of the carburetor to the body and base.
8. The top of the carburetor will be firmly held to the mixing body by the gasket. Do not attempt to lift the top by hand. Use a screwdriver blade or similar thin, flat tool inserted between the top and the enrichment cover. Lightly pry the top upwards and lift the top slowly. Do not apply excessive force and don't try to rush the job.
9. Remove the float pivot pin and remove the float. Carefully remove the needle valve.

WARNING: Do not let the float drop and do not apply any force to the float. The needle valve controlled by the float will be damaged.

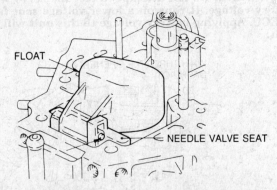

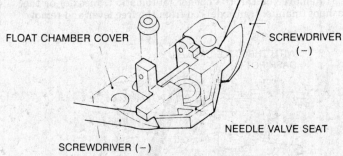

Removing the float, needle and valve seat. Work carefully to avoid damage

10. The valve seat may be removed by using two small, flat-bladed screwdrivers to gently lever the seat upwards and out of position. Use care not to damage the surrounding area or the seat mounts during this process.
11. Find the electrical connector for the feedback solenoid valve (FBSV). The terminals must be removed from the connector housing before the valve can be removed. Use a very thin, flat tool (such as a jewelry screwdriver) inserted into the connector to loosen the stopper and remove each terminal.
12. Remove the grommet from the top of the carburetor. Remove the retainer and remove the FBSV attaching screw; remove the feedback solenoid valve.
13. Remove the retainer and remove the slow cut solenoid valve (SCSV) from the carb top. Hold the solenoid by the body

FUEL SYSTEM 5

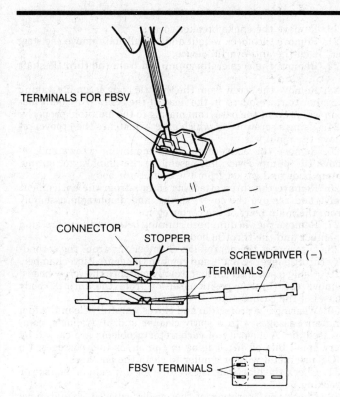

Removing the terminals from the connector body. Use a very small screwdriver to gently lift the stop tab within the plastic shell

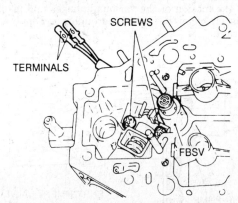

Remove the feedback solenoid valve after the terminals are free

Hold the SCSV only by the body; pulling on the wire will cause damage

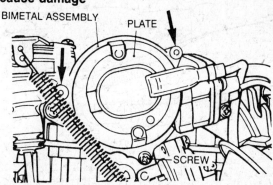

Removing the bi-metal choke coil. The rivets must be ground or drilled for removal. Don't forget the small screw at the bottom

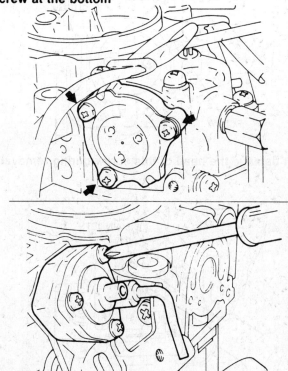

Removing the bowl vent valve (upper) and the choke breaker cover. Don't lose the small springs or diaphragms

and avoid pulling on the wiring.

14. Using the same small screwdriver trick, disconnect the SCSV terminals from the plastic connector body and remove the SCSV from the carburetor.
15. Use a hand grinder or similar tool to remove the heads of the rivets holding the cover of the choke assembly. Remove the small screw in the bottom of the cover.
16. Remove the packing (gasket), bi-metal assembly and plate.
17. Using a pin punch or similar tool, remove the remainder of the rivets from each hole. Take care not to damage the surrounding material.
18. Remove the bi-metal terminal from the wiring connector.
19. Remove the bowl vent valve.

NOTE: There are small springs within this unit. Take note of their location and placement and don't lose them.

5-11

5 FUEL SYSTEM

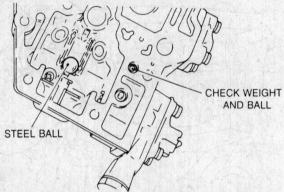

A small magnet can be very helpful when removing the steel balls

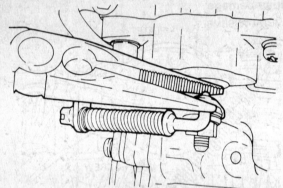

Removing the accelerator pump rod

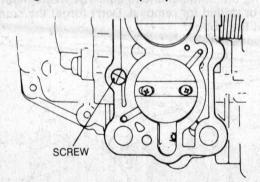

Don't damage the head of this screw during removal

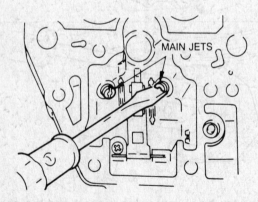

Carefully remove the main jets for cleaning or inspection. Do not damage the slotted heads

20. Remove the choke breaker cover.
21. Remove the check weight and its ball and remove the steel ball from the anti-overfill device.
22. Remove the accelerator pump rod from the throttle shaft lever.
23. Remove the screw from the throttle body assembly, taking care not to raise burrs on the head of the screw. Any deformation will prevent the base from mating to the manifold properly.
24. Using a screwdriver which exactly matches the groove, remove the main jets.
25. Remove the accelerator pump mounting screws and remove the pump cover link assembly, the diaphragm, spring, pump body and gasket from the carburetor body.
26. Remove the three attaching screws from the enrichment valve and remove the cover, spring, and diaphragm assembly from the main body of the carburetor.
27. Remove the vacuum hose running between the depression chamber and the throttle body.
28. Disconnect the depression chamber rod from the secondary throttle lever. Unbolt and remove the depression chamber.
29. Using a screwdriver that exactly matches the screw heads, remove the throttle position sensor from the throttle body (base) of the carburetor.
30. Wearing eye protection and gloves, carefully clean the fuel and air passages with a spray cleaner and, if available, compressed air. A majority of carburetor problems are caused by very small bits of dirt lodging in the air or fuel passages. Do NOT use metal wire or similar to clean the passages.
31. Check the diaphragms carefully for any sign of damage or cracking.
32. Check the operation of the needle valve; it should move lightly and smoothly. If any binding is felt, replace it.
33. Check the fuel inlet filter (above the needle valve) for clogging.
34. Check the float for cracks, deformation or internal leakage.
35. Inspect the motion of the various linkages and pivots. If any binding is felt, clean the system thoroughly and apply a light coat of lubricant.

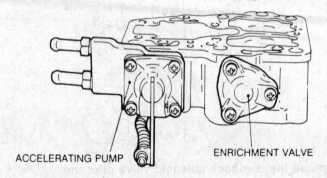

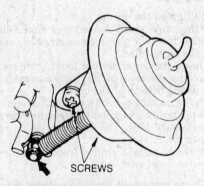

Removing the other external components, including the depression chamber (lower)

FUEL SYSTEM 5

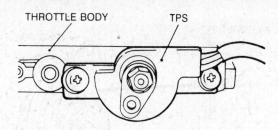

Carefully remove the throttle position sensor

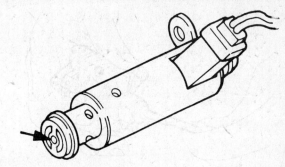

The tip of the feedback solenoid valve must be free of debris

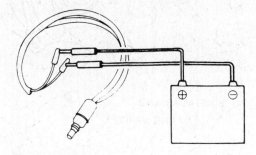

Test the two solenoids by applying battery voltage to the terminals. Each unit should operate with a distinct click

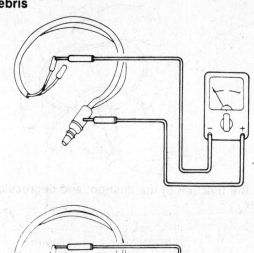

36. Check the operation of both solenoid valves. Apply battery voltage to the terminals; the solenoid should operate with a distinct click each time power is applied or removed. Inspect the tip of the FBSV to insure the jet is open and clean.
37. Use an ohmmeter to check the solenoids' resistance. The SCSV should have a resistance of 48–60Ω at 68°F (20°C); the FBSV should have 54–66Ω resistance at the same temperature.

NOTE: Resistance will increase or decrease as the temperature rises or falls. Make common sense allowances for the temperature in which you are testing the units.

38. Hold one lead of the ohmmeter against the case of each solenoid and touch the other lead to each terminal. There should NOT be continuity between the case and the terminals.
39. Use the ohmmeter to check the bi-metal assembly. Connect one lead to the wire terminal and the other lead to the body of the assembly. Correct resistance is approximately 6Ω at 68°F (20°C).
40. The dashpot should be inspected by pulling outward on the rod. Resistance should be felt; when released, the lever should return quickly to its original position.
41. The depression chamber is checked by pushing the rod all the way into the unit and then blocking the vacuum port firmly with a finger. Release the rod; if it stays in place (with the vacuum port blocked), the unit is good. If the rod returns to its original extended position, the diaphragm inside has failed.
42. Before reassembly, make certain that all parts are clean and dry. Any gasket or O-ring which was removed MUST be replaced with a new one during reinstallation.
43. If the jets are to be replaced, the new ones must be exact replacements. The jets have a number stamped on them for identification. Jet size is selected based on sophisticated air flow measurements during assembly of the carburetor; changing the jets will lead to extreme driveability and emission problems.
44. Install the throttle position sensor onto the throttle body and tighten the screws without causing damage.
45. Install the depression chamber. Connect the chamber rod to the secondary throttle lever.

Using the ohmmeter to test the solenoids. Note that in the upper test, the shell of the solenoid is insulated from the terminals; the meter should show NO continuity

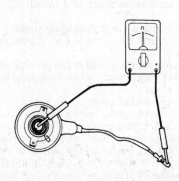

Checking the resistance of the bi-metal coil

5-13

5 FUEL SYSTEM

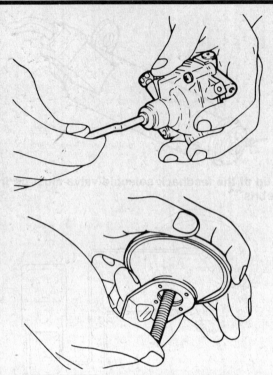

Inspect the function of the dashpot and depression chamber

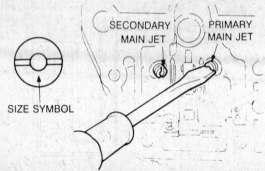

Sizes are marked on each jet; exact replacements must be used

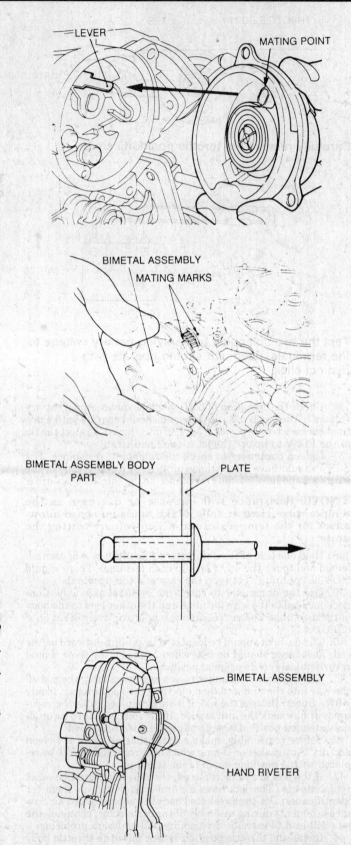

Four steps to the correct installation of the bi-metal choke unit

46. Install the vacuum hose from the depression chamber to the throttle body.
47. Assemble the diaphragm, spring and cover for the enrichment valve and install it (with its cover) onto the main body.
48. Assemble the accelerator pump and install it to the body with a new gasket.
49. Install the main jets without damaging them.
50. Install the screw into the throttle body assembly without damaging the head of the screw.
51. Install the accelerator pump rod to the throttle shaft lever.
52. Install the check weight and ball and the steel ball for the anti-overfill device.
53. Install the choke breaker cover.
54. Install the bowl vent valve, taking care to assemble the small springs and diaphragm correctly.
55. Loosely hold the bi-metal choke assembly in place. Route the terminal and wiring correctly to the plastic connector. Install the terminal in the connector by pushing it into the correct location. The stopper pin will engage the terminal automatically. A slight click may be heard or felt when the terminal is in position.

FUEL SYSTEM 5

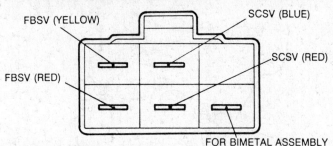

The terminals must be correctly installed in the connector body

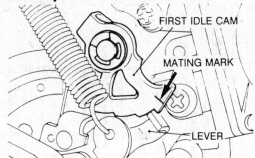

The throttle plate clearance must be checked after the high idle cam is correctly set

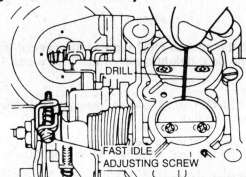

Use a drill or clearance gauge to check throttle plate clearance

56. To install the bi-metal assembly:
    a. Using a new gasket, place the cup on the top of the spiral spring in line with the choke lever. Fit the cap into place and use the small screw to hold it in place.
    b. Line up the mating marks on the case and body.
    c. Once aligned, install the rivets to hold the case in place. A hand rivetter is required; the use of nuts and bolts is NOT recommended.
57. Route the wiring for the FBSV and SCSV correctly. Install terminals into the correct connector port by pushing them in firmly.
58. Install the SCSV and its retainer into the top of the carburetor. Handle the unit only by the body and avoid pulling on the wire.
59. Install the FBSV into the carburetor. Install the retainer and attaching screw as well as a new grommet.
60. Install the seat and needle valve, making sure each is correctly placed and securely installed.
61. Install the float and pivot pin, taking great care not to put any undue force on the float. Refer to the *Carburetor Adjustment* section for float level adjustment procedures.
62. Install the carburetor top to the main body (with a new gasket) and install the five screws. Make sure each screw is tight without deforming the head.
63. Install the vacuum hose and delay valve running from the base of the carb to the choke breaker.
64. Install the connector bracket.
65. Install the ISC servo if it was removed (automatic transaxle only).
66. Install either the dashpot or throttle opener unit and connect the rod to the free lever.
67. Install the throttle return spring and damper spring.
68. Move the carburetor linkages by hand, checking that motions are smooth and there is no binding in any of the mechanisms.
69. With the carburetor correctly assembled, some adjustments must be made on the bench. Set the high idle cam to the second highest position and turn the carburetor upside down. Using a drill bit of known diameter or a clearance tool, check the clearance between the primary throttle plate and the throttle bore. Correct clearances are:
    - Cordia/Tredia with manual trans: 0.59mm (0.023 in.)
    - Cordia/Tredia with auto. trans.: 0.67mm (0.026 in.)
    - Mirage with manual transmission: 0.93mm (0.036 in.)
    - Mirage with automatic trans: 1.02mm (0.040 in.)

The clearance may be adjusted by turning the high-idle adjustment screw.

70. Check the choke unloader clearance by using your finger to lightly press and set the choke plate. When it is fully closed, move the throttle linkage to open the throttle plate(s) all the way; the throttle plates should be vertical in their bores. Measure the clearance between the choke plate and the choke bore. Correct clearance is 2mm (0.079 in) for Cordia/Tredia, Mirage, Pickups and Monteros.

If adjustment is needed, gently bend the throttle lever to achieve the correct clearance. Bending the lever upwards increases the clearance and bending it downward reduces the clearance. Bending it too much breaks off the tab and requires a new carburetor.

NOTE: Generally, the choke unloader clearance should not change after an overhaul. Adjusting this clearance greatly affects cold driveability; check the clearance after an overhaul, but don't adjust it if it isn't needed.

71. Install a new base gasket and reinstall the carburetor on the engine, following instructions given previously in this Section.

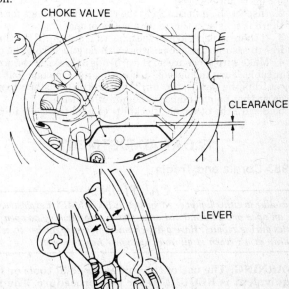

Check the choke plate clearance with the throttle wide open; make needed adjustments by carefully bending the lever

5-15

5 FUEL SYSTEM

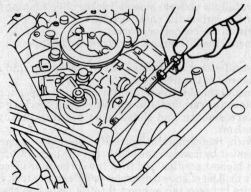

Adjusting accelerator cable play on the '83 Cordia and Tredia

Carburetor Adjustments

ACCELERATOR CABLE ADJUSTMENT

1983 Cordia and Tredia w/G62B Engine
All 4-Cylinder Pickup and Montero

1. Warm the engine up completely and make certain the high idle cam is not engaged.
2. Check the action of the accelerator pedal. There should be little to no free play between the normal "at rest" position and the point that engine speed begins to increase. Correct free play measurement is 0–20mm (0–0.8 in.).
3. If there is excessive free play, loosen the cable adjusting nuts located on the cable mount on the carburetor. With both nuts loosened, the cable may be adjusted to remove slack. Do not adjust the cable beyond the point of no free play; the engine idle speed will be changed.

Cordia and Tredia w/G63B Engine
Mirage and Precis w/G15B Engine

1. Warm the engine up completely and make certain the high idle cam is not engaged.
2. Inspect the inner cable at the carburetor linkage for slack. If there is no slack, the adjustment is proper.
3. If there is slack, loosen the adjusting nuts until the throttle is free to assume the idle position with no effect from the cable.
4. Make sure there are no sharp bends or kinks in the accelerator cable. Turn the outboard adjusting nut (the one that's farthest from the carburetor) until the throttle just starts to move. Stop at that point and back the nut off ½ turn.
5. Secure the other nut (locknut) to hold the cable in place.

FLOAT ADJUSTMENT

1983 Cordia and Tredia

CAUTION

Gasoline in either liquid or vapor state is EXTREMELY explosive. Work in an open or well-ventilated area. Observe no smoking/no open flame rules during repairs. Have a dry-chemical fire extinguisher (type B-C) within arm's reach at all times and know how to use it.

WARNING: The use of the correct special tools or their equivalent is REQUIRED for this procedure. The float level cannot be checked without tool MD998161.

1. The float level is checked with the carburetor installed on the engine and completely connected. Start and run the engine for a short period before testing. This will compensate for any

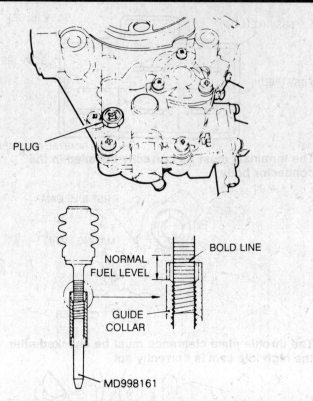

Remove the access plug and use the special tool to check the float height on the 1983 Cordia and Tredia

fuel loss from the float bowl due to evaporation or leakage. Shut the engine off before testing.
2. Remove the air cleaner assembly.
3. Remove the plug and washer from the fuel level check hole on top of the fuel bowl cover, just to the side of the fuel inlet fitting.
4. Hold the top of the special tool and turn the lower portion, the guide collar. The top of the threads inside the guide collar must be 4 lines (4mm or 0.16 in.) below the the bold line on the upper portion of the tool.
5. Insert the gauge into the check hole until the bottom of the guide collar touches the seat. Make sure the guage is inserted straight.
6. Push the top of the rubber head a little to confirm that fuel is being drawn up.
7. If there is no fuel in the gauge, remove it, turn the collar half a turn upward and reinsert the gauge. Continue this check and adjust procedure until fuel enters the tube.
8. When gas comes into the tube, remove it and turn the collar ½ turn downward, allowing you to see the fuel level and read it. It should be at the bold line. The acceptable range is from 2mm (0.08 in.) BELOW the bold line to 1mm (0.04 in.) ABOVE the line.
9. If the fuel level is correct, replace the access plug.
10. If the fuel level is not correct, the carburetor top and float must be removed. The carburetor need not be removed. Follow the detailed procedures listed within the *Carburetor Overhaul* section of this Section.
11. Remove the float seat retainer and then gently pull out the seat with a pair of pliers. Handle the pliers gently; the seat is easily deformed.
12. The float level is adjusted by replacing the shim below the seat. To raise the float level, use a thinner shim. To lower the float level, use a thicker shim. The proper replacement shims

FUEL SYSTEM 5

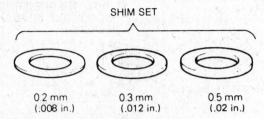

Sets of shims are available for adjusting float height. Shim thicknesses will vary by model

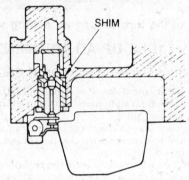

The shims fit under the seat of the needle valve. Careful disassembly is required

come in a kit and include thicknesses of 0.2mm, 0.3mm and 0.5mm.
13. Reassemble the seat, needle valve and float. Reinstall the carburetor top.
14. Start the engine, allowing it to idle briefly and shut it off.
15. Recheck the float adjustment.

1984–88 Cordia and Tredia
1985–88 Mirage and Precis
All 4-Cylinder Pickup and Montero

1. Remove the top of the carburetor and remove the gasket. The carburetor does not need to be removed for this procedure. Follow the directions within the *Carburetor Overhaul* procedure earlier in this Section.
2. Hold the carburetor top and float upside down. Use a float gauge or depth gauge to measure the distance from the bottom of the float (now the top, since it's upside down) to the inside of the carburetor top. The correct distance is 20mm ± 1mm (0.8 in. ± 0.04 in.).
3. If the dimension is not correct, the shim below the needle seat must be changed. Use a thicker shim to increase the measurement (or lower the float level) and a thinner shim to decrease the measurement (raise the float level).
4. The shim kit contains 3 shims: 0.3mm, 0.4mm and 0.5mm. The float measurement will change by 3 times the measurement of the shim installed or removed. Shims may be combined as necessary. Some arithmetic after measuring should provide the correct shim thickness on the first try.
5. To replace the shim, remove the float pin and the float. Remove the needle valve.
6. Use a pair of pliers to gently unscrew the seat at its widest point. Take great care not to damage the seat or the surrounding metal.
7. Slip the shim(s) over the narrow portion of the seat and then reinstall the seat. Gently tighten it without damage.
8. Reassemble the needle valve and float.
9. Re-measure the float height and replace the shim(s) if necessary to get the correct height.
10. Reinstall the carburetor top.

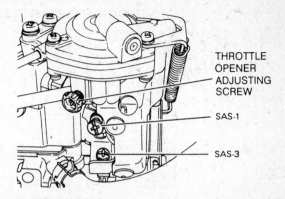

Location of the throttle opener adjusting screw. Don't adjust SAS 1 or 3

THROTTLE OPENER ADJUSTMENT

1984–88 Cordia and Tredia w/G63B Engine
Mirage and Precis w/G15B Engine

1. Check and adjust the normal idle speed as described in Section 2.
2. The engine should be idling at normal operating temperature with the transmission in neutral. Connect a tachometer to the engine.
3. Locate the throttle opener on the top of the carburetor. (The throttle return spring is attached to a bracket on top of the throttle opener). Follow the vacuum line leading out of the throttle opener to the throttle opener solenoid.
4. Connect a jumper wire to the positive battery terminal. Disconnect the electrical lead at the throttle opener solenoid and connect the other end of the jumper wire to the solenoid connector.
5. Open the throttle until the engine reaches about 2,000 rpm, and slowly release the throttle.
6. Check the engine speed on the tachometer. It should be:
- Mirage — 850 rpm
- 1984–88 Cordia/Tredia w/manual trans. — 700 rpm
- 1984–88 Cordia/Tredia w/automatic trans. — 750 rpm.

Adjust the throttle opener adjusting screw to achieve the correct rpm. The screw has finger grips on it and is located on the lower portion of the throttle opener. Check that the rpm remains at the correct level. Continue adjusting as necessary until the idle is stable.
7. Disconnect the jumper wire and reconnect the throttle opener solenoid wire.

1983–87 Pickup
1983–86 Montero

1. Disconnect the vacuum hose running to the throttle opener port.
2. Connect a hand vacuum pump to the nipple of the throttle opener. Install a tachometer to read engine speed. Make certain the idle speed for the engine is set correctly.
3. Run the engine at curb idle (fully warm). Apply 11.8 in.Hg with the hand pump; idle speed should increase to 900–950 rpm. If the idle does not increase, replace the dashpot/throttle opener assembly.
   a. Remove the throttle return spring from the throttle lever.
   b. Disconnect the throttle opener rod from the free lever.
   c. Remove the two attaching screws and remove the throttle opener/dashpot assembly.
   d. Install the new unit, connect the rod and install the spring.

5 FUEL SYSTEM

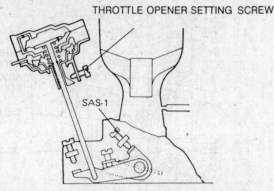

Adjusting the throttle opener, 1983–86 Pickup and Montero

4. Reconnect the vacuum hose to the nipple. Start the engine, run it at curb idle and turn the air conditioning on. Idle speed should increase to 900–950 rpm.
5. If the throttle opener needs adjustment, do so by turning the screw on the throttle opener/dashpot assembly. Don't turn any other screws to adjust the throttle opener; the curb idle or other important settings may be affected.

1988–89 Pickup
1987–89 Montero

1. Identify the vacuum control solenoid on the firewall. Label and carefully remove the two vacuum hoses running to the unit.
2. Disconnect the electrical connector from the solenoid.
3. Connect a hand vacuum pump to the solenoid port which held the vacuum hose with the white stripe.
4. Using jumper wires, connect battery voltage to one terminal of the solenoid and connect the other terminal to a good ground.
5. With battery voltage applied to the solenoid, draw vacuum with the hand pump. Use the chart to check the properties of the valve with proper combinations of vacuum and electricity applied or removed.
6. Remove the 12 volt jumpers. Use an ohmmeter to measure the resistance of the solenoid. Resistance should be 40–46Ω at 68°F (20°C).
7. If the solenoid does not behave correctly under ALL test conditions, it must be replaced.

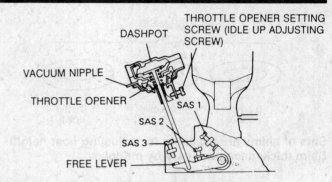

Location of the throttle opener setting screw

IDLE-UP ADJUSTMENT

1984–88 Cordia and Tredia w/G63B ENGINE AND A/C

1. The engine must be at normal operating temperature. Wheels must be in straight ahead position so there is no load on the power steering pump.
2. Disconnect the electric cooling fan connector. Apply the parking brake and make sure the transmission is in neutral. Turn the air conditioner on.
3. Connect a tachometer. Using the throttle opener adjusting screw, (located near the throttle return spring mount on top of the carburetor), adjust the engine speed to 850 rpm. Make sure the rpm stays steady at this level, readjusting it if necessary. Disconnect the tachometer and reconnect the cooling fan electrical connector.

—— **CAUTION** ——
The cooling fan may come on when connected! Keep hands, tools and clothing clear of the fan blades at all times.

Battery voltage	The other nipple of solenoid valve	Normal state
Applied	Open	Vacuum leaks
	Closed with finger	Vacuum is held
Not applied	Open	Vacuum is held

CARBURETOR CHART—AUTOMOBILES

Engine	Year	Application	Type	Model #	Throttle Bore mm (in.) Primary	Secondary
G62B	1983	CAL-M/T	M	32-35 DIDTA-110	32 (1.260)	35 (1.378)
		CAL-A/T, P/S	M	32-35 DIDTA-111	32 (1.260)	35 (1.378)
		CAL-A/T	M	32-35 DIDTA-119	32 (1.260)	35 (1.378)
		FED-5 MT	M	32-35 DIDTA-114	32 (1.260)	35 (1.378)
		FED-4x2 M/T	M	32-35 DIDTA-112	32 (1.260)	35 (1.378)
		FED-A/T P/S	M	32-35 DIDTA-113	32 (1.260)	35 (1.378)
		FED-A/T	M	32-35 DIDTA-120	32 (1.260)	35 (1.378)
G63B	1984	M/T	F	32-35 DIDTA-160	32 (1.260)	35 (1.378)
		A/T	F	32-35 DIDTA-161	32 (1.260)	35 (1.378)

FUEL SYSTEM 5

CARBURETOR CHART—AUTOMOBILES

Engine	Year	Application	Type	Model #	Throttle Bore mm (in.) Primary	Secondary
G63B	1985–86	FED-M/T	F	32-35 DIDTF-200	32 (1.260)	35 (1.378)
		FED-A/T	F	32-35 DIDTF-201	32 (1.260)	35 (1.378)
		CAL-M/T	F	32-35 DIDTF-195	32 (1.260)	35 (1.378)
		CAL-A/T	F	32-35 DIDTF-196	32 (1.260)	35 (1.378)
	1987–88	FED-M/T	F	32-35 DIDTF-415	32 (1.260)	35 (1.378)
		FED-A/T	F	32-35 DIDTF-416	32 (1.260)	35 (1.378)
		CAL-M/T	F	32-35 DIDTF-417	32 (1.260)	35 (1.378)
		CAL-A/T	F	32-35 DIDTF-418	32 (1.260)	35 (1.378)
G15B	1985	FED-M/T	F	28-32 DIDTF-407	28 (1.102)	32 (1.260)
		FED-A/T	F	28-32 DIDTF-408	28 (1.102)	32 (1.260)
		CAL-M/T	F	28-32 DIDTF-405	28 (1.102)	32 (1.260)
		CAL-A/T	F	28-32 DIDTF-406	28 (1.102)	32 (1.260)
	1986	FED-M/T	F	30-32 DIDTF-300	30 (1.181)	32 (1.260)
		FED-A/T	F	30-32 DIDTF-301	30 (1.181)	32 (1.260)
		CAL-M/T	F	30-32 DIDTF-302	30 (1.181)	32 (1.260)
		CAL-A/T	F	30-32 DIDTF-303	30 (1.181)	32 (1.260)
	1987–88 Mirage	FED-M/T	F	30-32 DIDEF-310	30 (1.181)	32 (1.260)
		FED-A/T	F	30-32 DIDEF-311	30 (1.181)	32 (1.260)
		CAL-M/T	F	30-32 DIDEF-312	30 (1.181)	32 (1.260)
		CAL-A/T	F	30-32 DIDEF-313	30 (1.181)	32 (1.260)
	1988–89 Precis	FED-M/T	F	30-32 DIDEF-410	30 (1.181)	32 (1.260)
		FED-A/T	F	30-32 DIDEF-411	30 (1.181)	32 (1.260)
		CAL-M/T	F	30-32 DIDEF-412	30 (1.181)	32 (1.260)
		CAL-A/T	F	30-32 DIDEF-413	30 (1.181)	32 (1.260)

NOTES: M/T—manual transmission, A/T—automatic transmission
5 MT—5 speed manual transmission, P/S—power steering
CAL—California, FED—Federal (49 state)
M—mechanical carburetor, F—feedback carburetor
HA—high altitude

CARBURETOR CHART—MONTERO

Engine	Year	Application	Type	Model #	Throttle Bore mm (in.) Primary	Secondary
Montero G54B	1983	All	M	32-35 DIDTA-106	32 (1.260)	35 (1.378)
	1984	Fed HA, M/T	M	32-35 DIDTA-170	32 (1.260)	35 (1.378)
		Fed HA, A/T	M	32-35 DIDTA-171	32 (1.260)	35 (1.378)
		Fed M/T	M	32-35 DIDTA-186	32 (1.260)	35 (1.378)
		Fed A/T	M	32-35 DIDTA-187	32 (1.260)	35 (1.378)
		Cal M/T	F	32-35 DIDTA-184	32 (1.260)	35 (1.378)
		Cal A/T	F	32-35 DIDTA-185	32 (1.260)	35 (1.378)
	1985–86	M/T	F	32-35 DIDTF-209	32 (1.260)	35 (1.378)
		A/T	F	32-35 DIDTF-210	32 (1.260)	35 (1.378)

5 FUEL SYSTEM

CARBURETOR CHART—MONTERO

Engine	Year	Application	Type	Model #	Throttle Bore mm (in.) Primary	Secondary
Montero G54B	1987	Fed M/T	F	32-35 DIDEF-410	32 (1.260)	35 (1.378)
		Fed A/T	F	32-35 DIDEF-411	32 (1.260)	35 (1.378)
		Cal M/T	F	32-35 DIDEF-412	32 (1.260)	35 (1.378)
		Cal A/T	F	32-35 DIDEF-413	32 (1.260)	35 (1.378)
	1988	Fed M/T	F	32-35 DIDEF-431	32 (1.260)	35 (1.378)
		Fed A/T	F	32-35 DIDEF-432	32 (1.260)	35 (1.378)
		Cal M/T	F	32-35 DIDEF-433	32 (1.260)	35 (1.378)
		Cal A/T	F	32-35 DIDEF-434	32 (1.260)	35 (1.378)
	1989	Fed	F	32-35 DIDEF-431	32 (1.260)	35 (1.378)
		Cal	F	32-35 DIDEF-441	32 (1.260)	35 (1.378)

CARBURETOR CHART—PICKUP

Engine	Year	Application	Type	Model No.	Throttle Bore mm (in.) Primary	Secondary
Pickup G63B	1983–84	Fed	M	—	32 (1.260)	35 (1.378)
		Cal	M	—	32 (1.260)	35 (1.378)
	1985–86	M/T	F	32-35 DIDTF-205	32 (1.260)	35 (1.378)
		A/T	F	32-35 DIDTF-206	32 (1.260)	35 (1.378)
	1987–89	Fed-M/T	F	32-35 DIDEF-400	32 (1.260)	35 (1.378)
		Fed-A/T	F	32-35 DIDEF-401	32 (1.260)	35 (1.378)
		Cal-M/T	F	32-35 DIDEF-402	32 (1.260)	35 (1.378)
		Cal-A/T	F	32-35 DIDEF-403	32 (1.260)	35 (1.378)
Pickup G54B	1983–84	Fed	M	—	32 (1.102)	35 (1.378)
		Cal	M	—	32 (1.102)	35 (1.378)
	1985–86	M/T	F	32-35 DIDTF-207	32 (1.102)	35 (1.378)
		A/T	F	32-35 DIDTF-208	32 (1.102)	35 (1.378)
	1987	Fed-M/T	F	32-35 DIDTF-404	32 (1.102)	35 (1.378)
		Cal-A/T	F	32-35 DIDTF-405	32 (1.102)	35 (1.378)
		Fed-M/T	F	32-35 DIDTF-406	32 (1.102)	35 (1.378)
		Cal-A/T	F	32-35 DIDTF-407	32 (1.102)	35 (1.378)
	1988	Fed	F	32-35 DIDEF-429	32 (1.102)	35 (1.378)
		Cal/Fed	F	32-35 DIDEF-435	32 (1.102)	35 (1.378)
	1989	Fed M/T	F	32-35 DIDEF-429	32 (1.102)	35 (1.378)
		Fed A/T	F	32-35 DIDEF-430	32 (1.102)	35 (1.378)
		Cal M/T	F	32-35 DIDEF-435	32 (1.102)	35 (1.378)
		Cal A/T	F	32-35 DIDEF-436	32 (1.102)	35 (1.378)

FUEL SYSTEM 5

ELECTRONICALLY CONTROLLED FUEL INJECTION SYSTEM (ECI)

Mechanical Fuel Pump

REMOVAL AND INSTALLATION

Found only on the 1984 Cordia and Tredia with the G63B (turbocharged) engine, the mechanical fuel pump is mounted on the side of the head, centered between the runners of the intake manifold. The intake manifold does not require removal to remove the pump, but access can be tricky. A selection of socket extensions, swivels and open end wrenches can be helpful.

A pushrod runs from a special camshaft lobe to the pump rocker arm, transferring the motion to the pump diaphragm. As the diaphragm moves, fuel is alternately drawn into the pump and then sent to the carburetor.

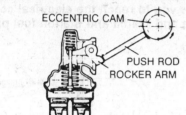

The mechanical fuel pump is driven from the camshaft through a pushrod

--- CAUTION ---

Gasoline in either liquid or vapor state is EXTREMELY explosive. Take great care to contain spillage. Work in an open or well-ventilated area. Do not connect or disconnect electrical connectors while fuel hoses are removed or loosened. Observe smoking/no open flame rules during repairs. Have a dry-chemical fire extinguisher (type B-C) within arm's reach at all times and know how to use it.

1. Disconnect the negative battery cable.
2. Set the engine to TDC/compression for No. 1 cylinder. This is generally done by turning the crankshaft pulley bolt clockwise (only!) until the timing marks on the case align. Double check the positioning by removing the distributor cap and making sure the rotor points to the No.1 terminal.
3. Label or diagram the fuel lines running to the pump. Disconnect the three fuel lines by using a pair of pliers to shift clamps away from the nipples on the pump and then pulling the lines off with a twisting motion.
4. Remove the two mounting bolts from the head. Remove the pump, spacer, and two gaskets from the head. As you pull the pump off the head, catch the pushrod which is located just behind the pump.
5. Clean the gasket surfaces of the insulator, pump and cylinder head. Insert the two bolts through the pump's mounting base. Slide a new gasket, the insulator, and a second new gasket into position over the two bolts. Turn the pump so its mounting surface faces the cylinder head.
6. Apply a light coat of clean engine oil to the pushrod. Place the pump pushrod against the cupped surface of the operating lever and angle it upward in the position it was in during removal. Hold the pushrod at that angle as you insert it into the bore in the head. Once the pushrod is in the bore in the cylinder head, you can release your fingers and move the pump toward the head along the line of the pushrod.
7. Start the two bolts into the bores in the head and tighten them finger tight. Tighten the mounting bolts alternately and evenly to 16 Nm (12 ft. lbs.).
8. Inspect the hoses for cracks (even hairline cracks can leak) and replace if necessary. Reconnect the fuel hoses. Make sure the hoses are installed all the way onto the nipples.
9. Work the clamps into position. Make sure the clamps are located well past the bulged portion of the nipples but do not sit at the extreme ends of the hoses.
10. Replace the distributor cap.

NOTE: If the fuel pump was replaced because of a ruptured diaphragm, there is a strong possibility of fuel contamination in the engine oil. Changing the oil and oil filter is strongly recommended after the fuel pump is replaced.

11. Connect the negative battery cable. Start the engine and check for leaks. Because the fuel lines have been partially emptied, the engine may crank for a longer than normal period before starting.

TESTING

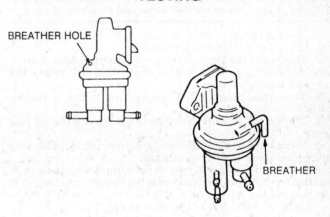

The diaphragm will vent to the atmosphere through a hole or tube in the upper case

On-Car Testing

1. Disconnect the inlet line (coming from the filter) at the pump. Connect a vacuum gauge to the pump nipple. Remove the coil-to-distributor high tension cable at the coil.
2. Have an assistant crank the engine with the key as you watch the gauge. A vacuum should be produced in a regular cycle as the pump turns over. There should be no blowback of pressure (abrupt drop in vacuum or even positive pressure for a short time).

If there is blowback of pressure, the inlet valve on the pump is leaking and the unit must be replaced.

3. The vacuum shown with each pump stroke should be strong and constant. If the vacuum is low (or none at all), the diaphragm is leaking and the pump must be replaced.

Off-Car Testing

1. Inspect the small breather hole or tube (above the diaphragm) which vents the pump's upper chamber. Leakage of fuel or oil here confirms that the diaphragm or oil seal is leaking.
2. Inspect the end of the pushrod and the contact surface on the pump operating lever. Replace the pushrod or pump if there is obvious wear. If the camshaft end of the pushrod is badly worn you should remove the valve cover and inspect the camshaft eccentric (which operates the fuel pump) for excessive wear.

5-21

5 FUEL SYSTEM

Electric Fuel Pump

Mechanical fuel pumps are generally not capable of delivering the high pressures required for fuel injection. Because of alternating motion of the pump arm, the mechanical pump is prone to wear and failure. An electric motor, driving a rotary or vane-type pump, provides a constant speed and pressure in the fuel line.

REMOVAL AND INSTALLATION

WARNING: In almost every case, replacing the Mitsubishi electric fuel pump requires removal of the fuel tank. Please refer to "Fuel Tank—Removal and Installation" at the end of this Section.

1985–88 Cordia and Tredia
All Starion

---- CAUTION ----
The fuel system is under pressure. Release pressure slowly and contain spillage. Observe no smoking/no open flame precautions. Have a Class B-C (dry powder) fire extinguisher within arm's reach at all times.

1. Remove the inner panel from the floor of the luggage area. Start the engine and allow it to idle.
2. Disconnect the electric fuel pump connector. Allow the engine to continue running until it stalls. This will relieve the pressure in the fuel lines.
3. Turn the ignition off. Disconnect the negative battery cable.
4. Elevate and support the car securely. Remove the left rear wheel.
5. Loosen and remove the fuel tank from the car, following procedures given later in this Section.
6. Label and disconnect the fuel lines, carefully noting their locations and remove the pump from the tank by unbolting it. If the pump is being replaced, switch the mounting clamp to the new pump and install it at the same angle.

NOTE: Even though the lines have been depressurized, wrap a towel or rag around each connection before removing the line. This will guard against any spray from residual pressure within the system.

7. Connect the lines to the pump and tank, making sure they are correctly placed and tightly installed.
8. Install the fuel tank. Install the left rear wheel.
9. Lower the car to the ground.
10. Connect the wiring connector to the fuel pump. Install the floor panel in the luggage area.
11. Connect the negative battery cable. Turn the ignition key to **ON** but do not start the engine. You should hear the pump running. Allow it to run for about 10 seconds to build line pressure, then start the engine.
12. After the engine has run smoothly for a minute or two, shut the ignition off and check the work area for any trace of leakage. Attend to any fuel leaks immediately, remembering that the system has repressurized and must be handled safely.

Mirage

1. Remove the rear seat cushion and move the carpet as necessary.
2. Start the engine. Disconnect the fuel pump wiring connector under the rear seat and allow the engine to run until it quits.
3. Turn the ignition key to the **OFF** position. Disconnect the negative battery cable.
4. If the pump is to be tested on the car, reconnect the pump wiring connector. If the pump is to be removed, leave the wiring disconnected.

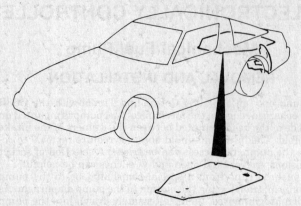

Removing the panel from the floor of the luggage area will allow you to reach the electrical connector for the Cordia and Tredia and Starion fuel pump

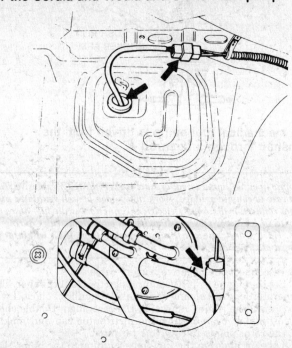

The wiring for the pump must be disconnected to reduce pressure in the fuel system

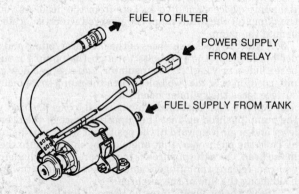

Typical electric fuel pump found on Cordia, Tredia and Starion

FUEL SYSTEM 5

5. Drain and remove the fuel tank, following instructions given later in this Section.

NOTE: All wires and hoses should be labeled at the time of removal. The amount of time saved during reassembly makes the extra effort well worthwhile.

6. With the tank removed from the car, remove the 5 nuts holding the pump to the tank. Carefully lift the pump straight up and out of the tank. Some pumps have filter at the foot of the pump which must be manipulated through the top of the tank. If you knock the filter off during removal, you'll have to fish it out of the tank before reinstallation.

CAUTION

The fuel pump may still contain liquid fuel. Drain it into a suitable container with an airtight lid before performing any tests or inspections.

---

7. Carefully install the pump into the tank and install the five retaining nuts onto the studs. Tighten the nuts evenly and in a criss-cross pattern to 16 Nm (12 ft. lbs.).

8. Reinstall the tank and connect the lines properly.

WARNING: When connecting the hoses to the pipes, make sure the hose extends up the pipe to the second bulge. Take great care when connecting the threaded high pressure lines. Start the threads by hand, then counterhold the hose side (to keep it from turning) and tighten the flare nut to 37 Nm (27 ft. lbs.).

9. Connect the pump wiring harness and install the rear seat cushion.

10. Connect the negative battery cable. Turn the ignition key to **ON** but do not start the engine. You should hear the pump running. Allow it to run for about 10 seconds to build line pressure, then start the engine.

11. After the engine has run smoothly for a minute or two, shut the ignition off and check the work area for any trace of leakage. Attend to any fuel leaks immediately, remembering that the system has repressurized and must be handled safely.

TESTING

When diagnosing engine problems, particularly a "no start" condition, the fuel pump function should be checked. Because of the difficulty in reaching the fuel pump connections, Mitsubishi automobiles have a conveniently located test connector under the hood. This connector bypasses all of the controls in the system (ignition switch, pump relay, etc.) and sends voltage directly to the pump. When power is applied to the pump you should be able to hear it running, although the in-tank pumps require you remove the fuel filler cap and listen at the filler.

To use the test connector, make certain the ignition is **OFF**. Great damage may be caused if the system is tested with the key on. Locate the test connector in the engine compartment—it is generally on a short lead out of a wiring harness and looks like a plug that isn't connected to anything. Once found, use a jumper wire to connect the test terminal to the positive battery terminal. Listen for the fuel pump; if it runs, you know the pump motor is good. While the pump is running, gently squeeze one of the fuel lines and confirm that the pump is delivering fuel pres-

1.
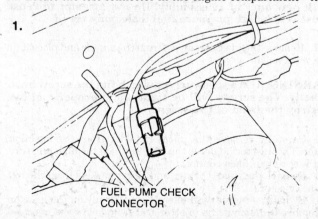
FUEL PUMP CHECK CONNECTOR

2.
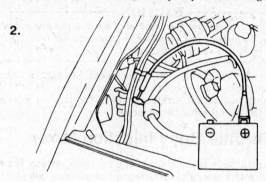

3.
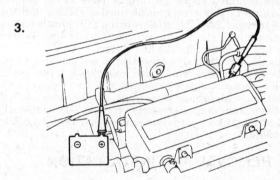

4.
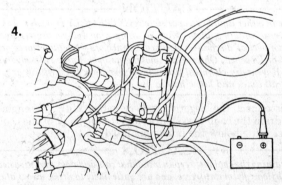

5.
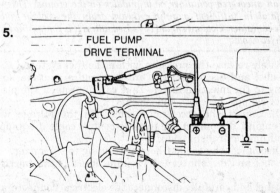
FUEL PUMP DRIVE TERMINAL

Location of fuel pump testing connectors. 1: Cordia and Tredia, 2: Starion, 3: Galant and Sigma, 4: Mirage and Precis except 5: 1989 Mirage

5-23

5 FUEL SYSTEM

sure. If it does not run, the pump itself is most likely defective OR the connector to the pump is faulty.

When the pump is removed from the tank, it may be checked by applying battery voltage (+) to the connector and grounding the housing. Only run the pump for one or two seconds during this test; the pump can be damaged by running without liquid.

CAUTION
Make certain the pump has been drained of residual fuel and is free of fuel vapors. Connecting and removing the test leads will cause sparks which can ignite any vapor in the area.

Additionally, the pump motor may be checked with an ohmmeter. Some resistance should be seen on the meter scale. If the meter shows no motion (infinite), the motor has an internal fault and the pump must be replaced.

Throttle Body Injection Mixer

The throttle body is the heart of the injection system. It contains the throttle plate and throttle position sensor. Two fuel injectors are mounted above the throttle plate in the throttle body. Because of this arrangement (mixing the air and fuel above the throttle), Mitsubishi refers to the ECI throttle body as the injection mixer. The designation injection mixer will be used throughout this section for the ECI system.

Usually the throttle body or injection mixer should not need removal except for replacement and even that is a rare event. These units contain and are connected to delicate components. When working with or near these units, care must be taken to protect them from impact or excessive force. The lightweight metal is easily broken; wires and sensors may be damaged by rough handling.

REMOVAL AND INSTALLATION

CAUTION
Gasoline in either liquid or vapor state is EXTREMELY explosive. Take great care to contain spillage. Work in an open or well-ventilated area. Do not connect or disconnect electrical connectors while fuel hoses are removed or loosened. Observe no smoking/no open flame rules during repairs. Have a dry-chemical fire extinguisher (type B-C) within arm's reach at all times and know how to use it.

1. Disconnect the negative battery cable. With the engine cooled, drain the cooling system at least to the point of the coolant level being below the intake manifold.

CAUTION
When draining the coolant, keep in mind that cats and dogs are attracted by the ethylene glycol antifreeze, and are quite likely to drink any that is left in an uncovered container or in puddles on the ground. This will prove fatal in sufficient quantity. Always drain the coolant into a sealable container. Coolant should be reused unless it is contaminated or several years old.

2. Safely release the pressure within the fuel system.
3. Label and disconnect the breather and air intake hoses running to the injection mixer. Take careful note of the placement of clamps, brackets and gaskets.
4. Disconnect the accelerator cable. On the Mirage with G32B engine, disconnect the throttle control cable if equipped with automatic transmission.
5. Disconnect the coolant hose(s) running to the mixer.
6. Label and disconnect the vacuum hoses at the injection mixer.
7. Label and carefully disconnect the electrical connectors at or near the injection mixer. Different models contain different numbers of connectors, but each must be unlocked before removal. Take note of any brackets or supports holding the wire harnesses or connectors.
8. Disconnect the high pressure fuel line at the mixer. Remove and discard the O-ring.

CAUTION
Wrap or protect the connection with a clean towel before removal. Some pressure may remain within the fuel system.

9. Disconnect the fuel return line.
10. Using the proper sized wrench, carefully remove the bolts holding the injection mixer to the manifold. Lift the mixer away from the engine and remove the gasket.
11. If the unit is being replaced, compare the old and the new ones. Look for any components which need to be transferred from the old unit. The throttle plate area may be cleaned with a spray cleaner, but must be completely dry before installation. Disassembly of the injection mixer is not recommended.
12. Before reinstalling, make sure that all remains of the old gasket are removed from both the manifold flange and the base of the mixer. Place a new gasket on the manifold and set the mixing unit in place.
13. Install the four bolts finger tight. Tighten them evenly, alternating from bolt to bolt, in small increments. The bolts must draw down evenly, creating an airtight seal against the gasket. Tighten the bolts to 18 Nm (13.5 ft. lbs.).
14. Connect the fuel return hose.
15. Before connecting the high pressure fuel line, install a new O-ring and coat the mating surfaces lightly with gasoline. Do not use oil or grease. Install the line and tighten the bolts to 6 Nm (48 INCH lbs.).
16. Carefully connect the electrical harnesses. Make certain each is properly seated and the locks engaged. Install any brackets or wire retainers which were removed.
17. Install the vacuum hoses.
18. Connect the water hose(s), making sure there are no kinks.
19. Install the accelerator cable and the throttle control cable if it was removed.
20. Reassemble the air intake and breather hoses and connect them to the injection mixer. Make certain that these hoses are correctly seated and that the clamps establish a good seal at each joint.
21. Refill the coolant.
22. Connect the negative battery cable.

Fuel Injectors

REMOVAL AND INSTALLATION

WARNING: The injectors are extremely sensitive to dirt and impact. They must be handled gently and protected at all times. The entire work area must be as clean as possible. Any particle of dirt entering the system can foul an injector or change its operation. Any gaskets or O-rings removed with the injector MUST be replaced with new ones at reassembly. Do not attempt to reuse these seals; high pressure fuel leaks may result.

1. Remove the injection mixer from the engine and place it on the workbench.

WARNING: Use a screwdriver which fits the screw head exactly. The screws will be tight; an improper tool can destroy the head of the screw.

2. Carefully remove the injector holder from the injection mixer. The injector holder is the unit to which the high pressure and fuel return lines connect.
3. Remove the small O-rings and larger gaskets from the top of the injectors.
4. By hand — never with pliers — pull firmly on the injector to remove it. Remove the injector seat from below the injector.

FUEL SYSTEM 5

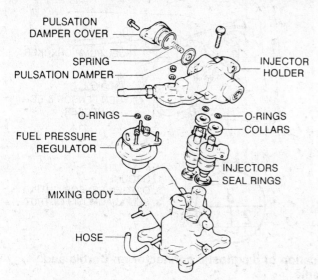

Typical ECI injector arrangement. Disassembly of the pulsation damper is not required

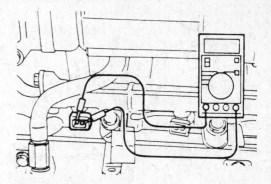

Use a good digital ohmmeter to accurately check injector resistance

NOTE: On the G54B and G32B engines, the injectors have differently colored electrical connectors. Before removal, note and record which color is in which position (front, rear, left, right). The injectors are not identical and must be reinstalled in their proper locations.

5. As soon as both injectors have been removed, use tape to seal the injector port against entry of dirt.
6. After the injectors have been tested, checked or replaced, remove the tape over the injector ports and install new seats. Note that the injector seats have a flat side and a round side; the flat side faces up (towards the injector).
7. Each injector should be fitted with a new collar and O-ring. The O-ring may be lightly coated with clean gasoline the make installation easier; do not use oil or grease to lubricate it.
8. Place the injector in its correct location and press it firmly into place.
9. Before the injector holder is reinstalled, examine the small screen filters carefully for any clogging or obstruction.
10. Carefully install the injector holder onto the injection mixer. Install the retaining screws and tighten them evenly to 4 Nm (36 INCH lbs.).
11. Reinstall the injection mixer.

FUEL INJECTOR TESTING

The simplest way to test the injectors is simply to listen to them with the engine running. Use either a stethoscope-type tool or the blade of a long screwdriver to touch each injector while the engine is idling. You should hear a distinct clicking as each injector opens and closes. Check that the operating sound increases as the engine speed is increased.

NOTE: The sounds of the other injector(s) may be heard, even though the one being checked is not operating. Listen to each injector to get a feel for normal sounds; the one with the abnormal sound is the problem.

Additionally, the resistance of the injector can be easily checked. Disconnect the negative battery cable and remove the electrical connector from the injector to be tested. Use an ohmmeter to check the resistance across the terminals of the injector. Correct resistance at 68°F (20°C) is 2–3Ω.

Slight variations are acceptable due to temperature conditions.

Bench testing of the injectors can only be done using expensive special equipment. Generally this equipment can be found at a dealership and sometimes at a well-equipped machine shop or performance shop. There is no provision for field testing the injectors by the owner/mechanic. DO NOT attempt to test the injector by removing it from the engine and making it spray into a jar.

Never attempt to check a removed injector by hooking it directly to the battery. The injector runs on a much smaller voltage and the 12 volts from the battery will destroy it internally. Since this happens at the speed of electricity, you don't get a second chance.

TESTING THE INJECTION SYSTEM AND SENSORS

As stated before, the heart of any fuel injection system is the computer or ECU. Besides reacting to the changing signals from various sensors and controlling various relays, switches and injectors, the ECU can serve an important diagnostic function. It "knows" what characteristics to look for in the signal from each unit. If any irregularity is detected (improper voltage, improper time duration, etc) the ECU notes the problem and assigns it a predetermined identifying number. If two or more faults are sensed, the codes are stored in numeric order. In some cases a light on the dashboard will come on to show a fault has occured. Even without a dash light, the fault code is stored in the computer until someone asks for it.

If you're going to read these codes, certain conditions must be observed. Since the codes are maintained by battery voltage, the battery must be fully charged. If the battery is low on charge, the codes may be lost. The codes will be lost (erased) if the battery cable is disconnected or if the ECU is disconnected before the codes are read. Additionally, the engine should be fully warmed up and driven a good distance before retrieving the codes; this allows the sensors and the oxygen sensor to enter their proper ranges.

Mitsubishi, like most manufacturers, has a diagnostic tool which plugs into the system between the wire harness and the ECU. This tester allows the operator to read out trouble codes and check the operation of various components during operation. Think of the tester as a window through which you can see the voltages change as they are transmitted to or from the ECU. If the normal values are known, a faulty item can be quickly found.

Unfortunately, these diagnostic tools are extremely expensive. The cost far exceeds the need of the owner/mechanic who is occasionally servicing one or two cars. If you suspect a fuel injection system problem that cannot be found through more common diagnostic means, take the car to either a dealer or a reputable diagnostic shop. The cost will be well repaid by the speed of diagnosis.

5-25

5 FUEL SYSTEM

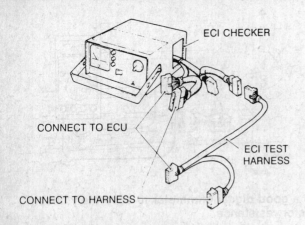

Mitsubishi's ECI checker allows very detailed inspection of the fuel injection system

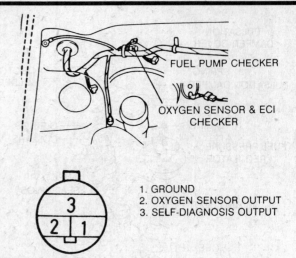

Location of diagnostic connector on Cordia and Tredia

The stored codes may also be read on an analog (dial type) voltmeter hooked into the system. Since the output from the ECU is electrical, the meter needle will deflect or sweep as the pulses are generated. Recording the number of sweeps and their time duration will yield the numeric code involved. (For example: Two ½-second sweeps followed by a 2 second pause followed by three ½-sweeps might indicate code 23 or code 2 and code 3, depending on the system.).

WARNING: The order of the fault codes does NOT indicate the order of occurence. Multiple codes are stored in numerical order, regardless of which occured first.

The code, when interpreted, points you to the unit which may be the problem. It must still be checked along with the attendant wiring, connectors and controls. A great number of fault codes are set because of loose or dirty connections in the wiring which fool the ECU into thinking the unit has failed.

The following section gives the diagnosis codes for each fuel injected engine, where to hook up the voltmeter and volt or ohmmeter testing for certain components. Note that not every component can be tested with a meter. Additional testing information for some components may also be found in the appropriate Engine or Emissions Sections in this book. In each case remember that you are only reading voltage used to transmit a code; the actual voltage running within the system can only be checked with the factory diagnostic unit.

NOTE: All resistances given are for 68°F (20°C). Remember that resistance will increase or decrease respectively as the temperature rises or falls. Use common sense in interpreting readings.

Cordia and Tredia w/G62B Engine

With the engine warm and running at idle, connect the voltmeter to the diagnostic (ECI checker) terminal. It is located under the hood on the right side of the firewall. The meter must be set to receive 12 volts DC (or more) and the operator must be ready to observe the meter as soon as it is connected. If the engine is switched off before reading the codes, certain codes will disappear due to the ECU resetting itself for the next operating cycle.

If all is well with the system and no fault codes are set, the meter will show a constant 12 volt signal with no interruptions. If trouble codes are to be transmitted, the needle will deflect from 0 to 12v in 0.4 second intervals. A two second pause is provided after each number group.

TESTING THE AIR FLOW SENSOR

1. With the engine off, disconnect the air flow sensor wiring at the air cleaner.
2. Remove the air cleaner cover and remove the filter element.
3. Carefully remove the airflow sensor.
4. Measure the resistance of the airflow sensor by connecting the positive probe of the meter to the No.4 terminal and the negative probe of the sensor to the No. 1 terminal of the sensor. Correct value is $2.65 k\Omega$.
5. Reassemble the airflow sensor and filter. Install the cover and connect the plug, making certain the terminals are securely connected.

TESTING THE IDLE SWITCH

1. With the ignition switch off, disconnect the ISC servo.
2. Check for continuity between terminal No.2 and the body of the injection mixer. There should be continuity with the throttle in the idle position. Move the throttle linkage by hand to open the throttle plate. The meter should show no continuity when the throttle moves off the idle position. If either condition is not met, replace the ISC servo unit.
3. Reconnect the ISC wiring.

TESTING THE THROTTLE POSITION SENSOR

1. With the engine off, disconnect the TPS connector.
2. Check the total resistance of the unit by connecting an ohmmeter across terminal Nos. 1 and 3. Correct resistance $3.5-6.5 k\Omega$.
3. Connect the meter across either terminal Nos. 1 and 2 or terminal Nos. 2 and 3. Move the throttle smoothly from idle to wide open and observe the meter; the resistance should change smoothly and progressively with the throttle motion.
4. Remove the meter and reconnect the TPS connector.

RESETTING THE SYSTEM

After recording the fault code and making repairs based on diagnosis, disconnect the negative battery cable for at least 15 seconds. Reconnect the cable and check for fault codes after running the engine. If the repair cured the problem, the original fault code should not reappear. The fault code will remain stored (even after repair) if the battery cable is not disconnected.

FUEL SYSTEM 5

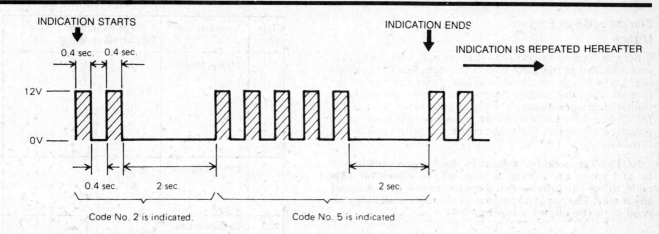

Mal-function No.	Diagnosis item	Self-diagnosis output pattern and output code
1	Oxygen sensor	12V / 0V
2	Ignition signal	12V / 0V
3	Air flow sensor	12V / 0V
4	Pressure sensor	12V / 0V

Mal-function No.	Diagnosis item	Self-diagnosis output pattern and output code
5	Throttle position sensor	12V / 0V
6	ISC motor position switch	12V / 0V
7	Coolant temperature sensor	12V / 0V

G62B fault codes. Note that the lowest numbered code will be displayed first regardless of order of occurrence

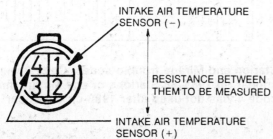

Correct polarity must be observed when testing the intake air temperature sensor

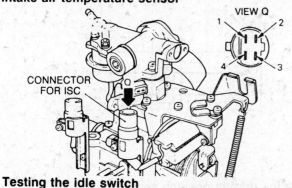

Testing the idle switch

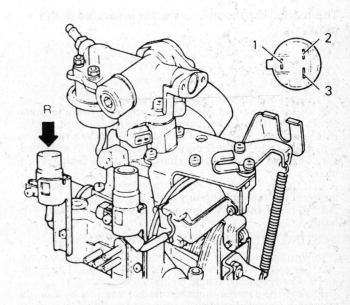

Checking the G62B throttle position sensor

5-27

5 FUEL SYSTEM

Starion w/G54B Engine
Mirage

For the Starion, connect the voltmeter to the diagnostic connector located in the glove box on the right side of the dashboard. The Mirage connector is located under the hood at the firewall. Turn the ignition switch to **ON**; the codes will begin transmitting immediately. The meter must be set to read 12v DC. If no codes have been stored, the system will generate an unbroken 12v signal; the meter needle will deflect to 12v and stay there.

NOTE: The code for a fault in the oxygen sensor will be lost when the engine is shut off. To check for this code, allow the fully warmed engine to idle and connect the meter. The oxygen sensor and all other codes may be read with the engine running.

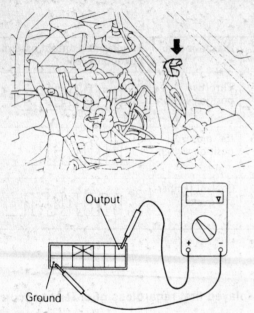

The Starion diagnostic connector is located in the glovebox

Code No.	Diagnosis item	Voltage waveform (abnormal code)
1	Oxygen sensor	
2	Ignition signal	
3	Air flow sensor	
4	Pressure sensor	
5	Throttle position sensor	
6	ISC motor position switch	
7	Water temperature sensor	
8	Car speed signal	

Starion and Mirage trouble codes. Code 8 will not appear on 1985–88 Starions or any Mirage model; code 4 was not used after 1986 on either car

TESTING THE INTAKE AIR TEMPERATURE SENSOR

1. Disconnect the air flow sensor connector.
2. Connect the terminals of the ohmmeter between terminal Nos. 2 and 4. Resistance will depend on air temperature. 2 reference points are: 32°F (0°C), 6kΩ and 68°F (20°C), 2.7kΩ.
3. Use a hand-held hair dryer to blow warm air over the sensor. The resistance should change with the temperature increase.
4. If the values are not close to target or do not change with temperature, the unit is faulty and must be replaced.
5. Remove the test equipment and reconnect the wire harness.

TESTING THE THROTTLE POSITION SENSOR

1. With the engine off, disconnect the TPS wiring connector.
2. Connect the ohmmeter across terminal Nos. 1 (ground) and 3. The resistance should be 3.5–6.5 kΩ.
3. Move the positive ohmmeter lead to terminal No.2 on the connector. Operate the throttle slowly and smoothly from idle to wide open; the resistance shown on the meter should change smoothly and in proportion to the throttle motion.

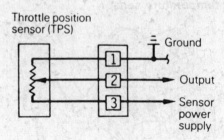

Testing the throttle position sensor

FUEL SYSTEM 5

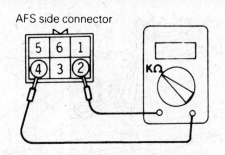
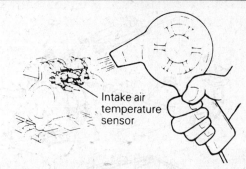

Testing the air intake temperature sensor

4. Even though the resistance may be correct, if it does not change smoothly with throttle motion, the unit is defective and must be replaced.

5. Remove the test equipment and connect the wiring harness.

RESETTING THE SYSTEM

After recording the fault code and making repairs based on diagnosis, disconnect the negative battery cable for at least 15 seconds. Reconnect the cable and check for fault codes after running the engine. If the repair cured the problem, the original fault code should not reappear. The fault code will remain stored (even after repair) if the battery cable is not disconnected.

MULTI-POINT FUEL INJECTION SYSTEM (MPI)

Electric Fuel Pump

Mechanical fuel pumps are generally not capable of delivering the high pressures required for fuel injection. Because of alternating motion of the pump arm, the mechanical pump is prone to wear and failure. An electric motor, driving a rotary or vane-type pump, provides a constant speed and pressure in the fuel line.

REMOVAL AND INSTALLATION

WARNING: In almost every case, replacing the Mitsubishi electric fuel pump requires removal of the fuel tank. Please refer to "Fuel Tank – Removal and Installation" at the end of this Section.

Mirage

1. Remove the rear seat cushion and move the carpet as necessary.
2. Start the engine. Disconnect the fuel pump wiring connector under the rear seat and allow the engine to run until it quits.
3. Turn the ignition key to the **OFF** position. Disconnect the negative battery cable.
4. If the pump is to be tested on the car, reconnect the pump wiring connector. If the pump is to be removed, leave the wiring disconnected.
5. Drain and remove the fuel tank, following instructions given later in this Section.

NOTE: All wires and hoses should be labeled at the time of removal. The amount of time saved during reassembly makes the extra effort well worthwhile.

6. With the tank removed from the car, remove the 5 nuts holding the pump to the tank. Carefully lift the pump straight up and out of the tank. Some pumps have filter at the foot of the pump which must be manipulated through the top of the tank. If you knock the filter off during removal, you'll have to fish it out of the tank before reinstallation.

The Mirage fuel pump connection is found under the rear seat cushion

5-29

5 FUEL SYSTEM

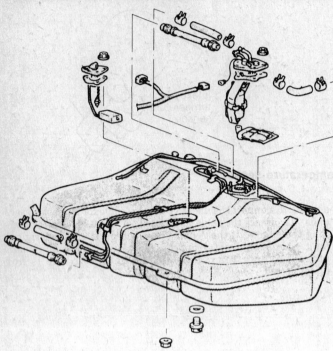

The Galant and Sigma fuel pump connector is on the front of the tank. Make certain the connector is reinstalled in the retaining clip when the work is finished

The Mirage in-tank electric fuel pump can only be removed after the tank is removed from the car. 1989 type shown, others similar

---- CAUTION ----
The fuel pump may still contain liquid fuel. Drain it into a suitable container with an airtight lid before performing any tests or inspections.

7. Carefully install the pump into the tank and install the five retaining nuts onto the studs. Tighten the nuts evenly and in a criss-cross pattern to 16 Nm (12 ft. lbs.).
8. Reinstall the tank and connect the lines properly.

WARNING: When connecting the hoses to the pipes, make sure the hose extends up the pipe to the second bulge. Take great care when connecting the threaded high pressure lines. Start the threads by hand, then counterhold the hose side (to keep it from turning) and tighten the flare nut to 37 Nm (27 ft. lbs.).

The vapor hoses must be pushed onto the pipes until they touch the second bulge

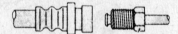

FLARE NUT

Use great care when tightening the high pressure fuel connections

9. Connect the pump wiring harness and install the rear seat cushion.
10. Connect the negative battery cable. Turn the ignition key to **ON** but do not start the engine. You should hear the pump running. Allow it to run for about 10 seconds to build line pressure, then start the engine.
11. After the engine has run smoothly for a minute or two, shut the ignition off and check the work area for any trace of leakage. Attend to any fuel leaks immediately, remembering that the system has repressurized and must be handled safely.

Galant and Sigma, all Engines

---- CAUTION ----
This procedure requires draining the fuel tank. It is to your advantage to perform the repair when the fuel tank is almost empty. A large container with an airtight lid is required for storing the drained fuel.

1. Disconnect the wiring connector for the fuel pump. The connector is held within a bracket on the front (leading edge) of the tank under the car.
2. Start the engine and allow it to run until it quits. Switch the ignition to **OFF**. Disconnect the negative battery cable. Elevate and safely support the car on stands.
3. Remove the fuel filler cap. Using a large container appropriate for the amount of fuel expected, drain the fuel tank. As soon as the tank is empty, reinstall the drain plug and tighten it to 20 Nm (15 ft. lbs.). Cover or cap the fuel container with an airtight lid.
4. Disconnect the high pressure fuel hose from the fuel line (pipe). Use a second wrench to counterhold the hose while turning the flare nut on the pipe. Do not allow the hose to twist.

---- CAUTION ----
Wrap the line connection in rags or a towel before disconnecting the joint. This will contain any spray from residual pressure within the lines.

5. The tank retaining straps are held to the body studs by self-locking bolts. Loosen the bolts just to the end of the studs, but do not remove them. This will allow the tank to be lowered without removal.
6. Disconnect the rear height sensor and its connection to the lateral rod (Panhard rod). Do not alter the length of the vertical link; just remove the bolt holding it at the bottom.
7. Remove the through-bolt holding the lateral rod to the body and lower the rod out of the way. The tank should now be free to slide all the way down on the straps and allow access to the pump mounting area.
8. Remove the ring of bolts holding the pump to the tank and remove the bracket bolt at the bottom of the pump.
9. Carefully remove the pump from the tank.

---- CAUTION ----
The fuel pump may still contain liquid fuel. Drain it into a suitable container with an airtight lid before performing any tests or inspections.

5-30

FUEL SYSTEM 5

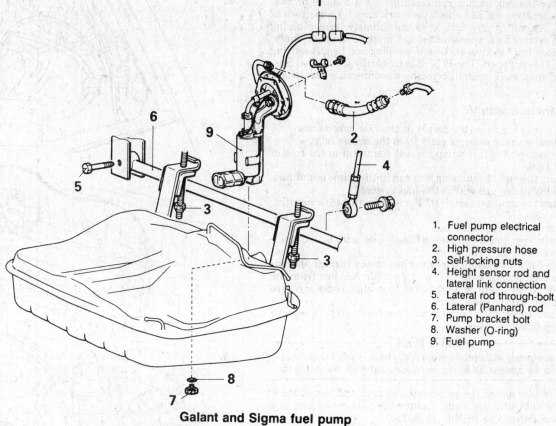

1. Fuel pump electrical connector
2. High pressure hose
3. Self-locking nuts
4. Height sensor rod and lateral link connection
5. Lateral rod through-bolt
6. Lateral (Panhard) rod
7. Pump bracket bolt
8. Washer (O-ring)
9. Fuel pump

Galant and Sigma fuel pump

Arrows show the bolts to be loosened or removed. Note that the tank straps are lowered but not removed

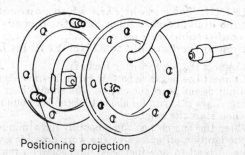

Correct alignment is important when installing the fuel pump; don't pinch the gasket

10. When reinstalling the pump, fit it into the tank and align the small projections on the gasket with the proper holes in the pump housing.
11. Once the pump is in place, install the bracket bolt underneath the pump and tighten it to 12 Nm (9 ft. lbs.). During this tightening, check constantly to make certain the gasket is not pinched or out of place.
12. Install the ring of small bolts holding the pump housing to the tank. Tighten them to 2 Nm (18 INCH lbs.).
13. Tighten the self-locking nuts on the tank straps and lift the tank into its mounted position.
14. Raise the lateral rod (Panhard rod) into position and install the through-bolt. Tighten the bolt snug; it will be final tightened later.
15. Install the height sensor and bracket; tighten the through-bolt to 21 Nm (16 ft. lbs.).

16. Connect the high pressure fuel hose to the fuel line. Counterhold the hose and tighten the flare nut to 30 Nm (22 ft. lbs). Do not twist the hose during installation. If the hose was disconnected from the pump during repairs or replacement, connect the hose to the pump before connecting the hose to the line. Tighten the hose-to-pump joint to 30 Nm (22 ft. lbs.).
17. Connect the electrical harness to the pump connector and make sure the plug is secure in its bracket.
18. Lower the car to the ground. Tighten the through-bolt for the lateral rod to 90 Nm (67 ft. lbs.) with the weight of the car on the ground.
19. Refill the fuel tank and install the filler cap.
20. Connect the negative battery cable. Turn the ignition key to **ON** but do not start the engine. You should hear the pump running. Allow it to run for about 10 seconds to build line pressure, then start the engine.

5-31

5 FUEL SYSTEM

21. After the engine has run smoothly for a minute or two, shut the ignition off and check the work area for any trace of leakage. Attend to any fuel leaks immediately, remembering that the system has repressurized and must be handled safely.
22. Check the function of the self-levelling and height sensing system if so equipped. The ECS (Electronically Controlled Suspension) may have been affected by disconnecting the height sensor.

1989 Montero with V6

1. Lift or move aside the carpet in the rear cargo area.
2. Remove the oval cover plate from the access hole.
3. Disconnect the fuel pump harness connector at the rear of the fuel tank.
4. Start the engine, allowing it to run until it runs out of fuel. This relieves pressure within the fuel system.
5. Turn the ignition switch **OFF** and disconnect the negative battery cable.
6. Remove the fuel filler cap.
7. Drain the fuel from the fuel tank into suitable container(s), each with an airtight lid.
8. Carefully, and using two wrenches on the fittings, disconnect the high pressure fuel line and the main fuel line from the pump. Wrap each joint in a rag before loosening; some pressure may remain within the system.
9. Remove the fuel pump retaining nuts and carefully remove the pump from the tank.

--- **CAUTION** ---
The fuel pump may still contain liquid fuel. Drain it into a suitable container with an airtight lid before performing any tests or inspections.

10. Carefully install the pump into the tank and install the 6 retaining nuts onto the studs. Tighten the nuts evenly and in a criss-cross pattern to 16 Nm (12 ft. lbs.).
11. Reinstall the high pressure and supply line to the pump.

WARNING: Take great care when connecting the threaded high pressure lines. Start the threads by hand, then counterhold the hose side (to keep it from turning) and tighten the flare nut to 37 Nm (27 ft. lbs.).

12. Connect the pump wiring harness.
13. Connect the negative battery cable. Turn the ignition key to **ON** but do not start the engine. You should hear the pump running. Allow it to run for about 10 seconds to build line pressure, then start the engine.
14. After the engine has run smoothly for a minute or two, shut the ignition off and check the work area for any trace of leakage. Attend to any fuel leaks immediately, remembering that the system has repressurized and must be handled safely.
15. Reinstall the cover on the access hole and replace the carpet in the cargo area.

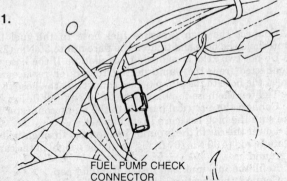

Location of fuel pump testing connectors. 1: Cordia and Tredia, 2: Starion, 3: Galant and Sigma, 4: Mirage and Precis except 5: 1989 Mirage

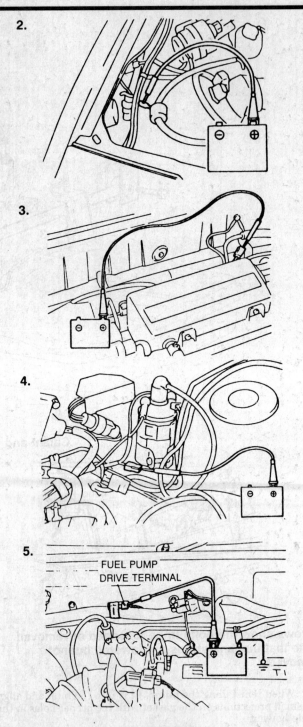

TESTING

When diagnosing engine problems, particularly a "no start" condition, the fuel pump function should be checked. Because of the difficulty in reaching the fuel pump connections, Mitsubishi automobiles have a conveniently located test connector under the hood. V6 Monteros have the connector taped to a wiring harness inside the right front kick panel, near the ECU and relays. This connector bypasses all of the controls in the system (ignition switch, pump relay, etc.) and sends voltage directly to the pump. When power is applied to the pump you should be able to hear it running, although the in-tank pumps require you remove the fuel filler cap and listen at the filler.

FUEL SYSTEM 5

To use the test connector, make certain the ignition is **OFF**. Great damage may be caused if the system is tested with the key on. Locate the test connector in the engine compartment—it is generally on a short lead out of a wiring harness and looks like a plug that isn't connected to anything. Once found, use a jumper wire to connect the test terminal to the positive battery terminal. Listen for the fuel pump; if it runs, you know the pump motor is good. While the pump is running, gently squeeze one of the fuel lines and confirm that the pump is delivering fuel pressure. If it does not run, the pump itself is most likely defective OR the connector to the pump is faulty.

When the pump is removed from the tank, it may be checked by applying battery voltage (+) to the connector and grounding the housing. Only run the pump for one or two seconds during this test; the pump can be damaged by running without liquid.

--- **CAUTION** ---

Make certain the pump has been drained of residual fuel and is free of fuel vapors. Connecting and removing the test leads will cause sparks which can ignite any vapor in the area.

Additionally, the pump motor may be checked with an ohmmeter. Some resistance should be seen on the meter scale. If the meter shows no motion (infinite), the motor has an internal fault and the pump must be replaced.

Throttle Body

On the MPI (Multi Port Injection) system, each injector is mounted directly into the head at the cylinder. On this system, the throttle body controls the amount of air entering the intake system but does not meter fuel in any way.

In this section the term throttle body will be used to indicate the MPI system.

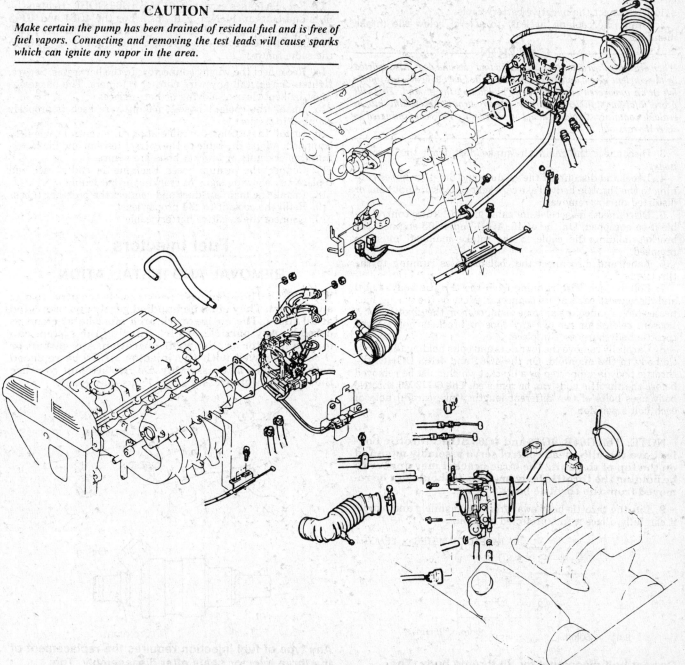

Typical throttle body assemblies. 4G15 above, 4G63 DOHC (center) and 6G72 V6

5-33

5 FUEL SYSTEM

REMOVAL AND INSTALLATION

CAUTION

Gasoline in either liquid or vapor state is EXTREMELY explosive. Take great care to contain spillage. Work in an open or well-ventilated area. Do not connect or disconnect electrical connectors while fuel hoses are removed or loosened. Observe smoking/no open flame rules during repairs. Have a dry-chemical fire extinguisher (type B-C) within arm's reach at all times and know how to use it.

NOTE: The throttle body for each of the six MPI engines is slightly different. The procedure below is general and may require slight alteration of sequence depending on the engine.

1. Disconnect the negative battery cable.
2. Drain the coolant, at least to a level below the intake manifold.

CAUTION

When draining the coolant, keep in mind that cats and dogs are attracted by the ethylene glycol antifreeze, and are quite likely to drink any that is left in an uncovered container or in puddles on the ground. This will prove fatal in sufficient quantity. Always drain the coolant into a sealable container. Coolant should be reused unless it is contaminated or several years old.

---

3. Disconnect the main air intake duct from the throttle body.
4. Label and disconnect the vacuum and breather hoses running to the throttle body. Take care that clamps are not bent or distorted during removal.
5. Disconnect the accelerator cable and the cruise control cable if so equipped. On the 4G15, 4G61 and 4G63 engines, the bracket holding the cable to the intake manifold must be removed.
6. Label and disconnect the coolant hoses running to the throttle body.
7. Follow each wire running from the throttle body. Label and disconnect each at its connector. Most of the wiring harnesses have connectors at some distance from the throttle body. Loosen, release or remove any clips or brackets holding the throttle body harnesses in place.
8. Carefully remove the four nuts and bolts holding the throttle body to the manifold. On the 4G61 and 4G63 DOHC, the throttle body is supported by a bracket which must be removed before the throttle body can be removed. The 6G72 V6 throttle body uses bolts of two different lengths; take careful note of each bolt's position.

NOTE: The G64B, 4G15 and 4G63 SOHC throttle bodies have the idle speed control servo assembly mounted on the top of the unit. The large bracket may appear to be holding the throttle body in place. It should not be removed from the throttle body.

9. Lift the throttle body away from the manifold and handle it carefully. Place it in a protected location.

10. If the unit is being replaced, compare the old and the new ones. Look for any components which need to be transferred from the old unit. The throttle plate area may be cleaned with a spray cleaner, but must be completely dry before installation. Disassembly of the throttle body is not recommended.
11. Before reinstalling, make sure that all remains of the old gasket are removed from both the manifold flange and the base of the throttle body. Place a new gasket on the manifold and hold the throttle body in place. Don't forget the support bracket on the twin cam engines.
12. Install the four nuts and bolts finger tight. For 6G72 engines, make certain the bolts are in the correct holes by length. Tighten the bolts evenly, alternating from bolt to bolt, in small increments. The bolts must draw down evenly, creating an airtight seal against the gasket.
13. For all engines except 4G61 and 4G63 DOHC, tighten the nuts and bolts to 11 Nm (8 ft. lbs.). For the 4G61 and 4G63 DOHC, the correct torque is 19 Nm (14 ft. lbs. Do NOT overtighten these bolts. The bracket bolt or nut is tightened to the same figure.
14. Reconnect the wiring connectors to the harnesses. Secure, tighten or reinstall any wire clips or retainers. The harnesses must be kept clear of moving or hot surfaces.
15. Install the coolant hoses, making sure each is properly clamped to its port.
16. Install the accelerator cable and cruise control cable if so equipped. Adjust the cable to the correct tension and make certain the lock nuts or bracket bolts are secure.
17. Connect the vacuum hoses. Examine the end of each and replace any showing signs of cracking or hardening.
18. Install the main air duct and connect the breather tubes.
19. Refill the coolant to the proper level.
20. Connect the negative battery cable.

Fuel Injectors

REMOVAL AND INSTALLATION

WARNING: The injectors are extremely sensitive to dirt and impact. They must be handled gently and protected at all times. The entire work area must be as clean as possible. Any particle of dirt entering the system can foul an injector or change its operation. Any gaskets or O-rings removed with the injector MUST be replaced with new ones at reassembly. Do not attempt to reuse these seals; high pressure fuel leaks may result.

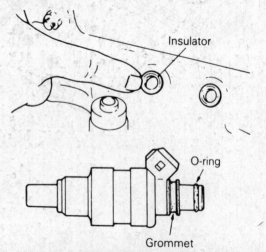

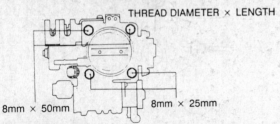

Correct bolt placement for V6 throttle body. The penalty for errors is very expensive

Any type of fuel injection requires the replacement of the three injector seals after disassembly. The insulator may also be called an injector seat

FUEL SYSTEM 5

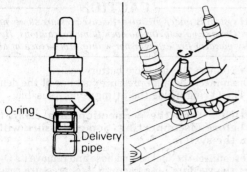

Make certain the injector can turn freely in the delivery pipe. If it binds, a pinched O-ring is usually the culprit

4G15 and 4G63 SOHC Engines

1. Safely relieve the pressure within the fuel system.

―――――――― **CAUTION** ――――――――
The fuel system is under pressure. Release pressure slowly and contain spillage. Observe no smoking/no open flame precautions. Have a Class B-C (dry powder) fire extinguisher within arm's reach at all times.

2. Disconnect the negative battery cable.
3. Disconnect the high pressure fuel line at the delivery pipe (rail). The O-ring inside the fitting is not reusable.

WARNING: Wrap the connection in a clean towel or cloth before disconnecting. Some pressure will remain within the system.

4. Disconnect the fuel return hose and remove its O-ring.
5. Disconnect the electrical connector to each injector. Label each at the time of removal.
6. Remove the bolts holding the injector rail; remove the rubber grommets or insulators below the rail mounting points.
7. Lift the rail with the injectors attached up and away from the engine. Take great care not to drop any of the injectors during this removal.

WARNING: If an injector should fall and hit the floor or other hard surface, it must be considered unusable.

8. The injectors may be removed from the rail with a gentle pull. Both the grommet and O-ring on the top of the injector must be discarded and replaced. The lower insulator or seat must also be removed and replaced.
9. Reassembly begins by installing a new grommet and O-ring (in that order) onto the injector. Coat the O-ring with a light coating of gasoline. Do not use grease or oil.
10. Install each injector into the rail, making sure that the injector turns freely when in place. If it does not turn under finger

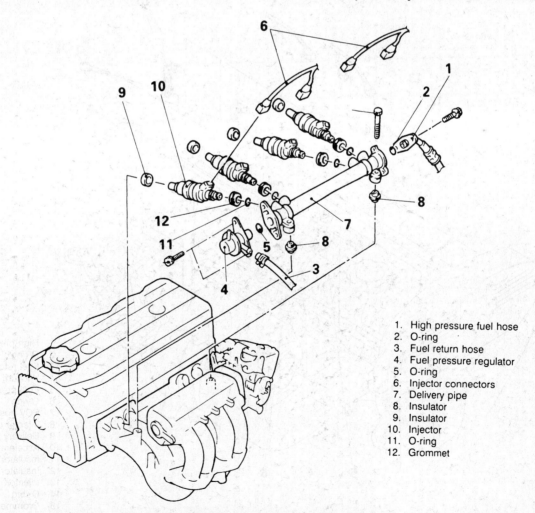

1. High pressure fuel hose
2. O-ring
3. Fuel return hose
4. Fuel pressure regulator
5. O-ring
6. Injector connectors
7. Delivery pipe
8. Insulator
9. Insulator
10. Injector
11. O-ring
12. Grommet

4G15 and 4G63 SOHC injector arrangement

5-35

5 FUEL SYSTEM

pressure, remove it, inspect the O-ring and reinsert the injector. (While the injector does not turn during its operation, its ability to turn is an indicator of correct installation).

11. Replace the seats in the intake manifold. Install the delivery pipe and the injectors onto the manifold without dropping an injector. Make certain the rubber bushings are in place under the delivery pipe brackets.
12. Tighten the fuel rail bolts to 11 Nm (8 ft. lbs or 72 INCH lbs.).
13. Connect each electrical connector to the proper injector.
14. Replace the O-ring, coat it lightly with gasoline and connect the fuel pressure regulator. Tighten the connection to 8 Nm (6 ft. lbs or 72 INCH lbs.).
15. Connect the fuel return hose.
16. Replace the O-ring, coat it lightly with gasoline and install the high pressure fuel line. Make certain the O-ring is not damaged during installation. Tighten the bolts to 4 Nm (3 ft. lbs. or 36 INCH lbs.).
17. Connect the negative battery cable.

4G61 and 4G63 DOHC Engines

1. Safely relieve the pressure within the fuel system.

> **CAUTION**
> The fuel system is under pressure. Release pressure slowly and contain spillage. Observe no smoking/no open flame precautions. Have a Class B-C (dry powder) fire extinguisher within arm's reach at all times.

2. Disconnect the negative battery cable.
3. Disconnect the high pressure fuel line at the delivery pipe (rail). The O-ring inside the fitting is not reusable.

WARNING: Wrap the connection in a clean towel or cloth before disconnecting. Some pressure will remain within the system.

4. Disconnect the fuel return hose and remove its O-ring. Disconnect the vacuum hose from the fuel pressure regulator.
5. Remove the fuel pressure regulator and its O-ring.
6. Disconnect the PCV hose.
7. Remove the electrical connector from each injector.
8. Remove the bolts holding the delivery pipe to the engine. Note that the accelerator cable retaining brackets will come off.
9. Lift the rail with the injectors attached up and away from the engine. Take great care not to drop any of the injectors during this removal.

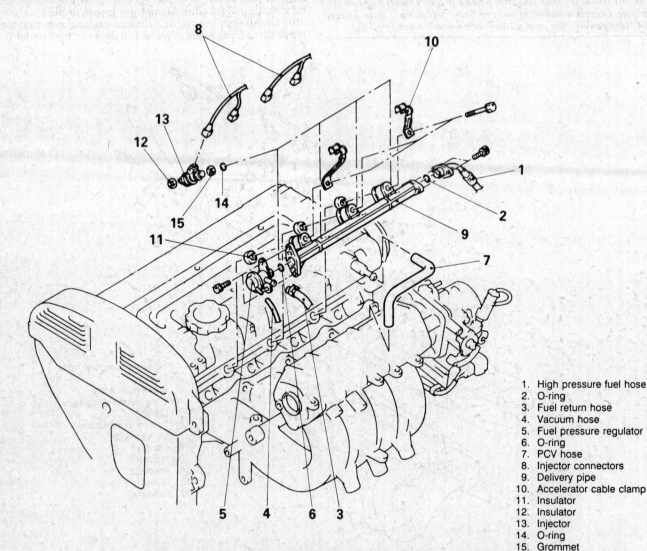

1. High pressure fuel hose
2. O-ring
3. Fuel return hose
4. Vacuum hose
5. Fuel pressure regulator
6. O-ring
7. PCV hose
8. Injector connectors
9. Delivery pipe
10. Accelerator cable clamp
11. Insulator
12. Insulator
13. Injector
14. O-ring
15. Grommet

4G61 and 4G63 DOHC injectors and fuel delivery pipe

FUEL SYSTEM 5

WARNING: If an injector should fall and hit the floor or other hard surface, it must be considered unusable.

10. The injectors may be removed from the rail with a gentle pull. Both the grommet and O-ring on the top of the injector must be discarded and replaced. The lower insulator or seat must also be removed and replaced.
11. Reassembly begins by installing a new grommet and O-ring (in that order) onto the injector. Coat the O-ring with a light coating of gasoline. Do not use grease or oil.
12. Install each injector into the rail, making sure that the injector turns freely when in place. If it does not turn under finger pressure, remove it, inspect the O-ring and reinsert the injector. (While the injector does not turn during its operation, its ability to turn is an indicator of correct installation).
13. Replace the seats in the intake manifold. Install the delivery pipe and the injectors onto the manifold without dropping an injector. Make certain the rubber bushings are in place under the delivery pipe brackets.
14. Tighten the fuel rail bolts to 11 Nm (8 ft. lbs or 72 INCH lbs.) Remember to include the accelerator cable brackets under the bolt.
15. Connect each electrical connector to the proper injector.
16. Connect the PCV hose.
17. Replace the O-ring, coat it lightly with gasoline and install the fuel pressure regulator. Tighten the fasteners to 8 Nm (6 ft. lbs or 72 INCH lbs.).
18. Connect the fuel return hose.
19. Replace the O-ring, coat it lightly with gasoline and install the high pressure fuel line. Make certain the O-ring is not damaged during installation. Tighten the bolts to 4 Nm (36 INCH lbs.).
20. Connect the negative battery cable.

G64B Engine

1. Safely relieve the pressure within the fuel system.

— **CAUTION** —
The fuel system is under pressure. Release pressure slowly and contain spillage. Observe no smoking/no open flame precautions. Have a Class B-C (dry powder) fire extinguisher within arm's reach at all times.

2. Disconnect the negative battery cable.
3. Remove the throttle body, following instructions given earlier in this Section.
4. Remove the boost hose from the opposite end of the air plenum.
5. Disconnect the high pressure fuel line from the delivery pipe (fuel rail).

WARNING: Wrap the connection in a clean towel or cloth before disconnecting. Some pressure will remain within the system.

6. Remove the fuel return line from the fuel pressure regulator. Using a wrench of the proper size, unscrew the regulator

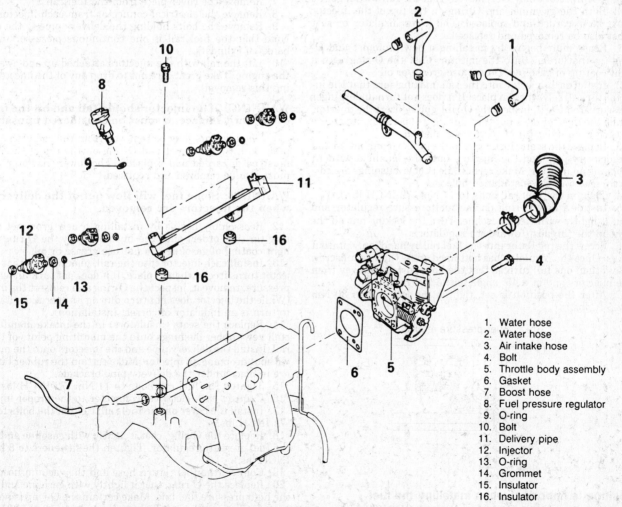

1. Water hose
2. Water hose
3. Air intake hose
4. Bolt
5. Throttle body assembly
6. Gasket
7. Boost hose
8. Fuel pressure regulator
9. O-ring
10. Bolt
11. Delivery pipe
12. Injector
13. O-ring
14. Grommet
15. Insulator
16. Insulator

Remove the G64B throttle body before removing the injectors

5-37

5 FUEL SYSTEM

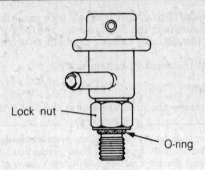

G64B fuel pressure regulator

from the rail. Remove the seal (O-ring) from the bottom of the regulator.
7. Disconnect each electrical connector from the injectors.
8. Remove the bolts holding the fuel delivery pipe to the engine.
9. Lift the rail with the injectors attached up and away from the engine. Take great care not to drop any of the injectors during this removal.

WARNING: If an injector should fall and hit the floor or other hard surface, it must be considered unusable.

10. The injectors may be removed from the rail with a gentle pull. Both the grommet and O-ring on the top of the injector must be discarded and replaced. The lower insulator or seat must also be removed and replaced.
11. Reassembly begins by installing a new grommet and O-ring (in that order) onto the injector. Coat the O-ring with a light coating of gasoline. Do not use grease or oil.
12. Install each injector into the rail, making sure that the injector turns freely when in place. If it does not turn under finger pressure, remove it, inspect the O-ring and reinsert the injector. While the injector does not turn during its operation, its ability to turn is an indicator of correct installation.
13. Replace the seats (insulator) in the intake manifold. Install the delivery pipe and the injectors onto the manifold without dropping an injector. Make certain the rubber bushings are correctly seated in the installation hole.
14. Tighten the fuel rail bolts to 11 Nm (72 INCH lbs.).
15. Install a new O-ring on the fuel pressure regulator and coat it lightly with clean gasoline. Turn the locking nut all the way up the threads (toward the regulator).
16. Screw the regulator into the fuel rail by hand (only) until it stops. Once the regulator has bottomed on its threads, unscrew it less than one full turn so that the hose port faces away from the injectors and at a 45° angle to the fuel rail.
17. After the position is set, tighten the lock nut to 30 Nm (22.5 ft. lbs.).

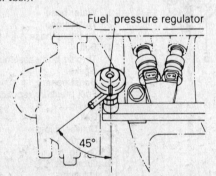

Position is important when installing the fuel pressure regulator. Once in place, the lock nut should be tightened

18. Replace the O-ring and connect the high pressure fuel line to the delivery pipe. Tighten the bolts to 30 Nm (22.5 ft. lbs.).
19. Connect the boost hose to the air plenum.
20. Reinstall the throttle body, following directions outlined earlier in this Section.
21. Connect the negative battery cable.

6G72 V6 Engine

1. Safely relieve the pressure within the fuel system.

---------- **CAUTION** ----------
The fuel system is under pressure. Release pressure slowly and contain spillage. Observe no smoking/no open flame precautions. Have a Class B-C (dry powder) fire extinguisher within arm's reach at all times.

---

2. Disconnect the negative battery cable.
3. Remove the air intake plenum, following procedures listed in Section 3.
4. Disconnect the high pressure fuel line at the delivery pipe (rail). The O-ring inside the fitting is not reusable.

WARNING: Wrap the connection in a clean towel or cloth before disconnecting. Some pressure will remain within the system.

5. Disconnect the fuel return hose and remove its O-ring. Disconnect the vacuum hose from the fuel pressure regulator.
6. Remove the fuel pressure regulator and its O-ring.
7. Remove the cover piece from the fuel rail.
8. Remove the electrical connector from each injector.
9. Remove the bolts holding the delivery pipe to the engine. Note that the fuel rail is one continuous piece serving both banks of cylinders.
10. Lift the rail with the injectors attached up and away from the engine. Take great care not to drop any of the injectors during this removal.

WARNING: If an injector should fall and hit the floor or other hard surface, it must be considered unusable.

11. The injectors may be removed from the rail with a gentle pull. Both the grommet and O-ring on the top of the injector must be discarded and replaced. The lower insulator or seat must also be removed and replaced.

WARNING: Some fuel will flow out of the delivery pipe when the injectors are removed.

12. Reassembly begins by installing a new grommet and O-ring (in that order) onto the injector. Coat the O-ring with a light coating of gasoline. Do not use grease or oil.
13. Install each injector into the rail, making sure that the injector turns freely when in place. If it does not turn under finger pressure, remove it, inspect the O-ring and reinsert the injector. (While the injector does not turn during its operation, its ability to turn is an indicator of correct installation.)
14. Replace the seats (insulators) in the intake manifold. Install new rubber bushings onto the mounting points of the fuel rail. Install the delivery pipe and the injectors onto the manifold without dropping an injector. Make certain the rubber bushings are in place under the delivery pipe brackets.
15. Tighten the fuel rail bolts to 11 Nm (72 INCH lbs.).
16. Connect each electrical connector to the proper injector.
17. Install the cover on the fuel rail; tighten the bolts to 8 Nm (72 INCH lbs.).
18. Replace the O-ring, coat it lightly with gasoline and install the fuel pressure regulator. Tighten the fasteners to 8 Nm (72 INCH lbs.).
19. Connect the fuel return hose and the vacuum hose.
20. Replace the O-ring, coat it lightly with gasoline and install the high pressure fuel line. Make certain the O-ring is not damaged during installation. Tighten the bolts to 10 Nm (81 INCH lbs.).

FUEL SYSTEM 5

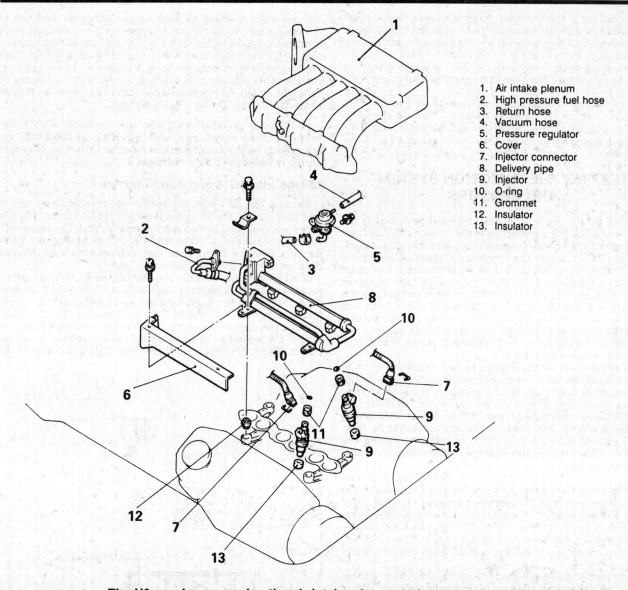

1. Air intake plenum
2. High pressure fuel hose
3. Return hose
4. Vacuum hose
5. Pressure regulator
6. Cover
7. Injector connector
8. Delivery pipe
9. Injector
10. O-ring
11. Grommet
12. Insulator
13. Insulator

The V6 requires removing the air intake plenum before removing the injectors

21. Reinstall the air intake plenum.
22. Connect the negative battery cable.

FUEL INJECTOR TESTING

The simplest way to test the injectors is simply to listen to them with the engine running. Use either a stethoscope-type tool or the blade of a long screwdriver to touch each injector while the engine is idling. You should hear a distinct clicking as each injector opens and closes. Check that the operating sound increases as the engine speed is increased.

NOTE: The sounds of the other injector(s) may be heard, even though the one being checked is not operating. Listen to each injector to get a feel for normal sounds; the one with the abnormal sound is the problem.

Additionally, the resistance of the injector can be easily checked. Disconnect the negative battery cable and remove the electrical connector from the injector to be tested. Use an ohmmeter to check the resistance across the terminals of the injector. Correct resistance at 68°F (20°C) is:

- 4G15 and all 4G63 engines — 13-16Ω
- 6G72 V6 engine — 15-17Ω

Slight variations are acceptable due to temperature conditions.

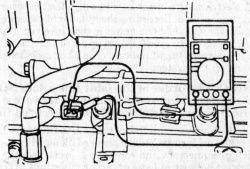

Use a good digital ohmmeter to accurately check injector resistance

5-39

5 FUEL SYSTEM

Bench testing of the injectors can only be done using expensive special equipment. Generally this equipment can be found at a dealership and sometimes at a well-equipped machine shop or performance shop. There is no provision for field testing the injectors by the owner/mechanic. DO NOT attempt to test the injector by removing it from the engine and making it spray into a jar.

Never attempt to check a removed injector by hooking it directly to the battery. The injector runs on a much smaller voltage and the 12 volts from the battery will destroy it internally. Since this happens at the speed of electricity, you don't get a second chance.

TESTING THE INJECTION SYSTEM AND SENSORS

As stated before, the heart of any fuel injection system is the computer or ECU. Besides reacting to the changing signals from various sensors and controlling various relays, switches and injectors, the ECU can serve an important diagnostic function. It "knows" what characteristics to look for in the signal from each unit. If any irregularity is detected (improper voltage, improper time duration, etc) the ECU notes the problem and assigns it a predetermined identifying number. if two of more faults are sensed, the codes are stored in numeric order. In some cases a light on the dashboard will come on to show a fault has occured. Even without a dash light, the fault code is stored in the computer until someone asks for it.

If you're going to read these codes, certain conditions must be observed. Since the codes are maintained by battery voltage, the battery must be fully charged. If the battery is low on charge, the codes may be lost. The codes will be lost (erased) if the battery cable is disconnected or if the ECU is disconnected before the codes are read. Additionally, the engine should be fully warmed up and driven a good distance before retrieving the codes; this allows the sensors and the oxygen sensor to enter their proper ranges.

Mitsubishi, like most manufacturers, has a diagnostic tool which plugs into the system between the wire harness and the ECU. This tester allows the operator to read out trouble codes and check the operation of various components during operation. Think of the tester as a window through which you can see the voltages change as they are transmitted to or from the ECU. If the normal values are known, a faulty item can be quickly found.

Unfortunately, these diagnostic tools are extremely expensive. The cost far exceeds the need of the owner/mechanic who is occasionally servicing one or two cars. If you suspect a fuel injection system problem that cannot be found through more common diagnostic means, take the car to either a dealer or a reputable diagnostic shop. The cost will be well repaid by the speed of diagnosis.

The stored codes may also be read on an analog (dial type) voltmeter hooked into the system. Since the output from the ECU is electrical, the meter needle will deflect or sweep as the pulses are generated. Recording the number of sweeps and their time duration will yield the numeric code involved. (For example: Two ½-second sweeps followed by a 2 second pause followed by three ½-sweeps might indicate code 23 or code 2 and code 3, depending on the system.).

WARNING: The order of the fault codes does NOT indicate the order of occurence. Multiple codes are stored in numerical order, regardless of which occured first.

The code, when interpreted, points you to the unit which may be the problem. It must still be checked along with the attendant wiring, connectors and controls. A great number of fault codes are set because of loose or dirty connections in the wiring which fool the ECU into thinking the unit has failed.

The following section gives the diagnosis codes for each fuel injected engine, where to hook up the voltmeter and volt or ohmmeter testing for certain components. Note that not every component can be tested with a meter. Additional testing information for some components may also be found in the appropriate Engine or Emissions Sections in this book. In each case remember that you are only reading voltage used to transmit a code; the actual voltage running within the system can only be checked with the factory diagnostic unit.

NOTE: All resistances given are for 68°F (20°C). Remember that resistance will increase or decrease respectively as the temperature rises or falls. Use common sense in interpreting readings.

Mirage with 4G15 and 4G61 Engines

Because the MPI injection system requires more sensors to operate properly, the ECU can store and report more codes. The codes are identified by two digit numbers. With the engine off, connect the voltmeter (set to read 12v DC) to the diagnostic connector under the dash. Turn the ignition switch to **ON**; the codes will begin transmitting immediately.

If all is well and no codes are stored, the ECU broadcasts a steady stream of ½-second pulses with one half second between each. If the ECU has failed (or diagnosed itself as having a problem) the system will transmit an unchanging 12 volt signal. When a two digit code is sent, the first (or tens) digit is sent in longer 1½ second pulses; the second digit is sent in ½-second pulses. Two long sweeps and three short duration sweeps of the

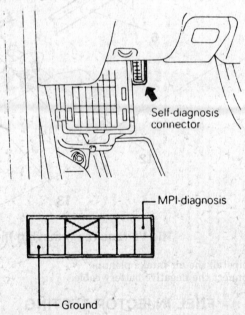

Connect the voltmeter properly when checking the system

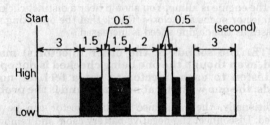

Pulse duration indicates place value for each two digit code

5–40

FUEL SYSTEM 5

Diagnosis item	Malfunction code	
	Output signal pattern	No
Engine control unit	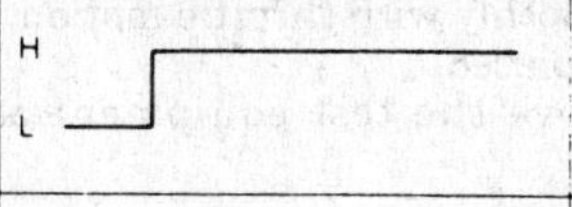	
Oxygen sensor	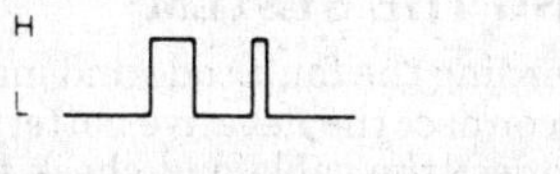	11
Air flow sensor	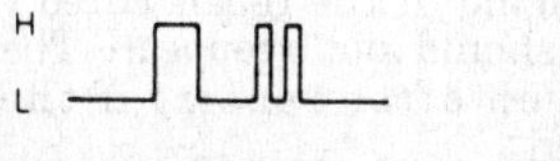	12
Intake air temperature sensor	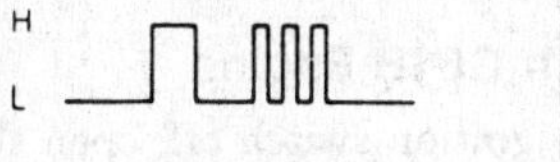	13
Throttle position sensor		14
Motor position sensor	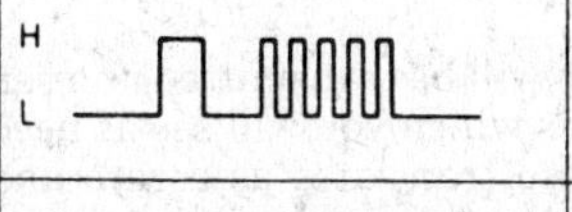	15
Engine coolant temperature sensor	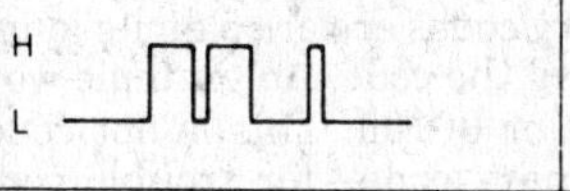	21
Crank angle sensor		22
No. 1 cylinder top dead center sensor	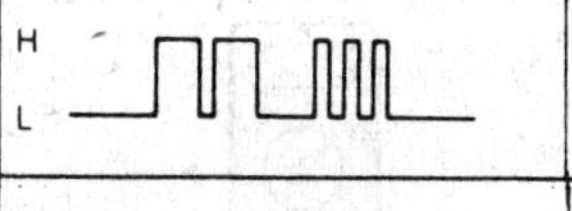	23
Vehicle speed sensor (reed switch)	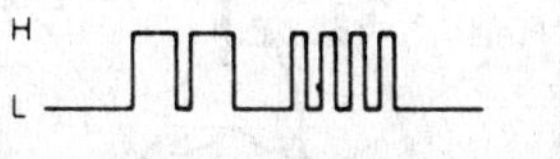	24
Barometric pressure sensor		25
Injector		41
Fuel pump	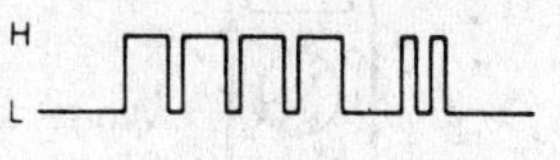	42
EGR	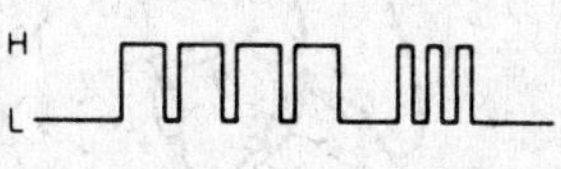	43
Normal state		–

4G15 trouble codes

Diagnosis item	Malfunction code	
	Output signal pattern	No
Engine control unit		
Oxygen sensor	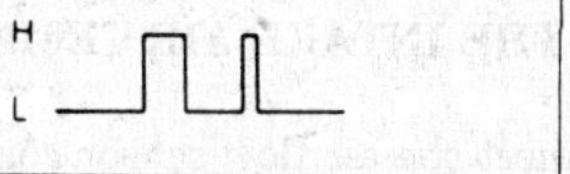	11
Air flow sensor	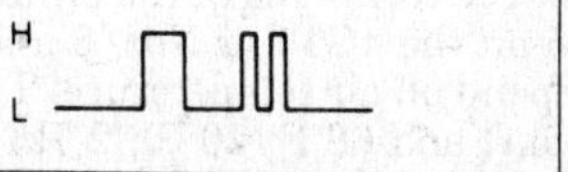	12
Intake air temperature sensor	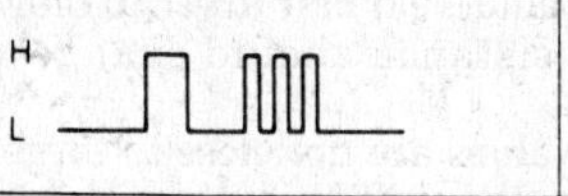	13
Throttle position sensor	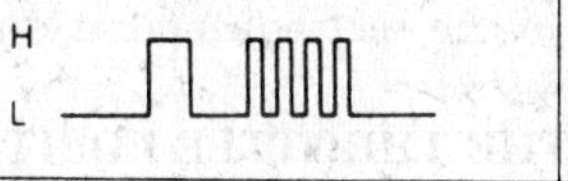	14
Engine coolant temperature sensor	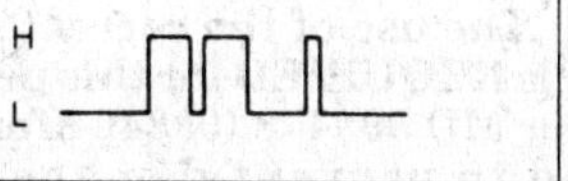	21
Crank angle sensor	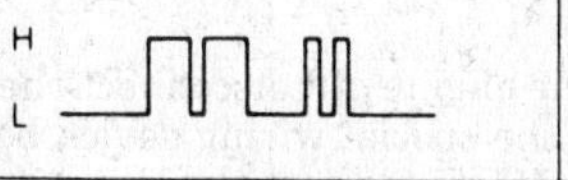	22
Top dead center sensor		23
Vehicle speed sensor (reed switch)		24
Barometric pressure sensor		25
Detonation sensor	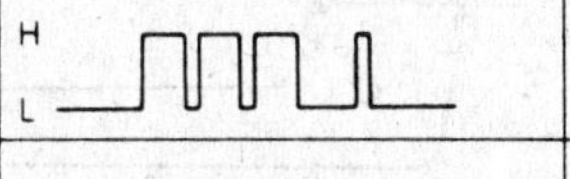	31
Injector	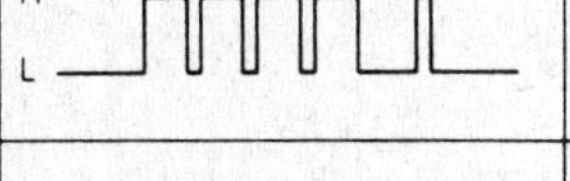	41
Fuel pump	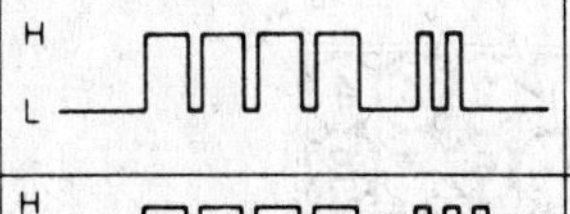	42
EGR	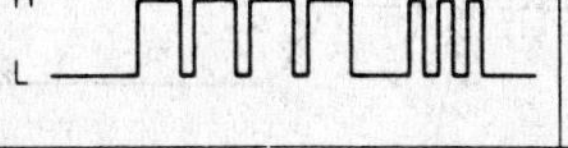	43
Ignition coil		44
Normal state		–

4G61 trouble codes

5-41

5 FUEL SYSTEM

meter needle would indicate code 23. If more than one code is stored, the first digit of the second code will be sent after a three second pause. The entire sequence of codes will be retransmitted repeatedly; if you don't get all of it the first time, wait until it repeats.

Note that each engine has its own family of codes. Some are the same but there are differences. Make certain the correct chart is in use when performing diagnostic work.

TESTING THE INTAKE AIR TEMPERATURE SENSOR

1. Disconnect the air flow sensor connector.
2. Connect the terminals of the ohmmeter between terminal Nos. 4 and 6 on the 4G15 or Nos. 6 and 8 on the 4G61. Resistance will depend on air temperature. Two reference points are: 32°F (0°C), 6kΩ and 68°F (20°C), 2.7kΩ.
3. Use a hand-held hair dryer to blow warm air over the sensor. The resistance should change with the temperature increase.
4. If the values are not close to target or do not change with temperature, the unit is faulty and must be replaced.
5. Remove the test equipment and reconnect the wire harness.

TESTING THE THROTTLE POSITION SENSOR

WARNING: The use of the correct special tools or their equivalent is REQUIRED for this procedure. Wiring harness adapter MD 998478 (998464 for 4G61) is highly recommended to prevent damage to the wiring and connectors.

1. With the engine off, disconnect the TPS wiring connector.
2. Install the special wiring device between the connectors.
3. On the 4G15, connect the ohmmeter across terminal Nos. 1 (ground; the black clip on the adapter) and 2 (the red clip on the adapter). On the 4G61, connect the ohmmeter to terminal No. 2 (red clip) and 3 (white clip). In both cases, the resistance should be 3.5–6.5 kΩ.

the resistance shown on the meter should change smoothly and in proportion to the throttle motion.

4. Even though the resistance may be correct, if it does not change smoothly with throttle motion, the unit is defective and must be replaced.
5. Remove the test equipment and connect the wiring harness.

RESETTING THE SYSTEM

After recording the fault code and making repairs based on diagnosis, disconnect the negative battery cable for at least 15 seconds. Reconnect the cable and check for fault codes after running the engine. If the repair cured the problem, the original fault code should not reappear. The fault code will remain stored (even after repair) if the battery cable is not disconnected.

Galant with G64B Engine

With the ignition switch off, open the glovebox and pull out the dignostic connector located behind the glovebox. It may be easier to remove the glovebox completely; the diagnostic connector is not on a very long harness. Connect the voltmeter (set to read 12v DC) to the diagnostic connector, observing the correct polarity. The codes will be transmitted when the ignition is turned **ON**.

The codes will be transmitted as a series of long and short duration pulses which you will see as needle motion on the meter. A short pulse translates as a zero and a long pulse as a one. These binary codes are are 5 digits long and all must be read before knowing the code. An example would be: short-short-long-short-short or 00100. This is not code 100; it is the designation—in binary code—for trouble code 4 or the atmospheric pressure sensor. Record the flashes carefully and use the chart to understand them.

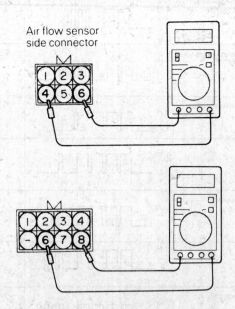

Checking the air intake temperature sensor. Connection for 4G15 is shown above with 4G61 below

3. Move the positive ohmmeter lead to terminal No.3 (blue clip) on the 4G15 connector or No.4 (blue clip) on the 4G61. Operate the throttle slowly and smoothly from idle to wide open;

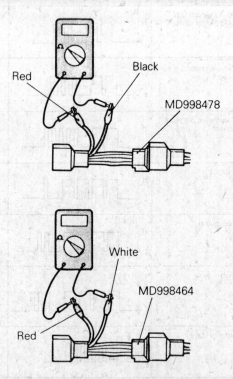

Always use the wiring adapter to check the throttle position sensor. 4G15 shown on left

FUEL SYSTEM 5

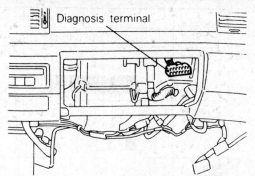
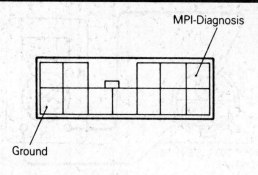

Location and hook-up for G64B diagnostic connector

Malfunction No.	Diagnosis item	Self-diagnosis output pattern and output code
0	Normal	0 0 0 0 0
1	Oxygen sensor	1 0 0 0 0
2	Crank angle sensor	0 1 0 0 0
3	AFS	1 1 0 0 0
4	Atmospheric pressure sensor	0 0 1 0 0
5	TPS	1 0 1 0 0
6	MPS	0 1 1 0 0
7	Coolant temperature sensor	1 1 1 0 0
8	No. 1 cylinder TDC sensor	0 0 0 1 0

The binary codes are not difficult but may be confusing. Record the needle sweeps carefully

TESTING THE INTAKE AIR TEMPERATURE SENSOR

1. Disconnect the air flow sensor connector.
2. Connect the terminals of the ohmmeter between terminal Nos. 2 and 4. Resistance will depend on air temperature. Two reference points are: 32°F (0°C), 6kΩ and 68°F (20°C), 2.7kΩ.
3. Use a hand-held hair dryer to blow warm air over the sensor. The resistance should change with the temperature increase.
4. If the values are not close to target or do not change with temperature, the unit is faulty and must be replaced.
5. Remove the test equipment and reconnect the wire harness.

TESTING THE THROTTLE POSITION SENSOR

1. With the engine off, disconnect the TPS wiring connector.
2. Connect the ohmmeter across terminal Nos. 1 (ground) and 3. The resistance should be 3.5–6.5 kΩ.
3. Move the positive ohmmeter lead to terminal No.2 on the connector. Operate the throttle slowly and smoothly from idle to wide open; the resistance shown on the meter should change smoothly and in proportion to the throttle motion.
4. Even though the resistance may be correct, if it does not change smoothly with throttle motion, the unit is defective and must be replaced.

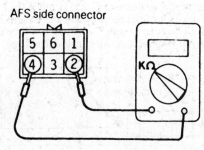

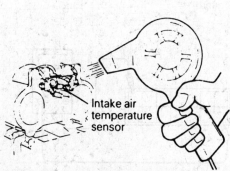

Testing the air intake temperature sensor

5-43

5 FUEL SYSTEM

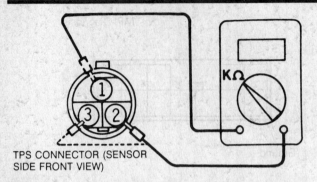

TPS CONNECTOR (SENSOR SIDE FRONT VIEW)

Checking the throttle position switch for resistance and function

5. Remove the test equipment and connect the wiring harness.

RESETTING THE SYSTEM

After recording the fault code and making repairs based on diagnosis, disconnect the negative battery cable for at least 15 seconds. Reconnect the cable and check for fault codes after running the engine. If the repair cured the problem, the original fault code should not reappear. The fault code will remain stored (even after repair) if the battery cable is not disconnected.

6G72 V6 Engine

With the engine off, connect the voltmeter (set to read 12v DC) to the diagnostic connector behind the glove compartment. It may be easier to remove the glove box completely. Turn the ignition switch to **ON**; the codes will begin transmitting after three seconds.

If all is well and no codes are stored, the ECU broadcasts a steady stream of ½-second pulses with one half second between each. If the ECU has failed (or diagnosed itself as having a problem) the system will transmit an unchanging 12 volt signal. When a two digit code is sent, the first (or tens) digit is sent in longer 1½ second pulses; the second digit is sent in ½-second pulses. Two long sweeps and three short duration sweeps of the meter needle would indicate code 23. If more than one code is stored, the first digit of the second code will be sent after a three second pause. The entire sequence of codes will be re-transmitted repeatedly; if you don't get all of it the first time, wait until it repeats.

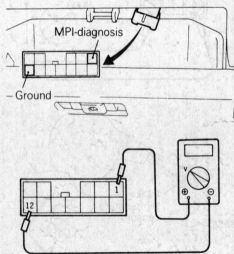

V6 diagnostic connector

Diagnosis item	Malfunction code	
	Output signal pattern	No
Computer	▬▬▬▬▬	—
Oxygen sensor	▮ ▮	11
Air flow sensor	▮ ▮▮	12
Intake air temperature sensor	▮ ▮▮▮	13
Throttle position sensor	▮ ▮▮▮▮	14
Coolant temperature sensor	▮▮ ▮	21
Crank angle sensor	▮▮ ▮▮	22
Top dead center sensor (No.1 cylinder)	▮▮ ▮▮▮	23
Vehicle-speed sensor (reed switch)	▮▮ ▮▮▮▮	24
Barometric pressure sensor	▮▮ ▮▮▮▮	25
Injector	▮▮▮▮ ▮	41
Fuel pump	▮▮▮▮ ▮▮	42
EGR	▮▮▮▮ ▮▮▮	43
Normal state	▮▮▮▮▮▮▮▮	—

Fault codes for 6G72 Montero

FUEL SYSTEM 5

Output preference order	Diagnosis item	Malfunction code	
		Output signal pattern	No.
1	Engine control unit	H / L	—
2	Oxygen sensor	H / L	11
3	Air flow sensor	H / L	12
4	Intake air temperature sensor	H / L	13
5	Throttle position sensor	H / L	14
6	Engine coolant temperature sensor	H / L	21
7	Crank angle sensor	H / L	22
8	Top dead center sensor	H / L	23
9	Vehicle speed sensor (reed switch)	H / L	24
10	Barometric pressure sensor	H / L	25
11	Injector	H / L	41
12	Fuel pump	H / L	42
13	EGR*	H / L	43
14	Normal state	H / L	—

Fault codes for the 6G72

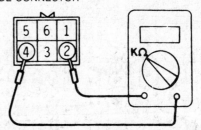
AFS SIDE CONNECTOR

Testing the air intake temperature sensor

TESTING THE INTAKE AIR TEMPERATURE SENSOR

1. Disconnect the air flow sensor connector.
2. Connect the terminals of the ohmmeter between terminal Nos. 2 and 4. Resistance will depend on air temperature. 2 reference points are: 32°F (0°C), 6kΩ and 68°F (20°C), 2.7kΩ.
3. Use a hand-held hair dryer to blow warm air over the sensor. The resistance should change with the temperature increase.
4. If the values are not close to target or do not change with temperature, the unit is faulty and must be replaced.
5. Remove the test equipment and reconnect the wire harness.

TESTING THE THROTTLE POSITION SENSOR

WARNING: The use of the correct special tools or their equivalent is REQUIRED for this procedure. Wiring harness adapter MD 998464 is highly recommended to prevent damage to the wiring and connectors.

1. With the engine off, disconnect the TPS wiring connector.
2. Install the special wiring device between the connectors.
3. Connect the ohmmeter across terminal Nos. 2 (ground; the red clip on the adapter) and 3 (the white clip on the adapter). The resistance should be 3.5–6.5 kΩ.
3. Move the positive ohmmeter lead to terminal No.4 (blue clip). Operate the throttle slowly and smoothly from idle to wide open; the resistance shown on the meter should change smoothly and in proportion to the throttle motion.
4. Even though the resistance may be correct, if it does not change smoothly with throttle motion, the unit is defective and must be replaced.
5. Remove the test equipment and connect the wiring harness.

RESETTING THE SYSTEM

After recording the fault code and making repairs based on diagnosis, disconnect the negative battery cable for at least 15 seconds. Reconnect the cable and check for fault codes after running the engine. If the repair cured the problem, the original fault code should not reappear. The fault code will remain stored (even after repair) if the battery cable is not disconnected.

4G63 SOHC and DOHC Engines

This family of MPI engines incorporates several additional sensors to improve engine control and driveability. The ECU can store and report more codes. The codes are identified by two digit numbers. With the engine off, connect the voltmeter (set to read 12v DC) to the diagnostic connector under the dash. Turn the ignition switch to **ON**; the codes will begin transmitting immediately.

If all is well and no codes are stored, the ECU broadcasts a steady stream of ½-second pulses with one half second between each. If the ECU has failed (or diagnosed itself as having a problem) the system will transmit an unchanging 12 volt signal. When a two digit fault code is sent, the first (or tens) digit is sent

5-45

5 FUEL SYSTEM

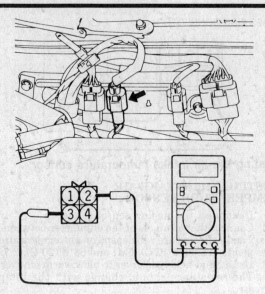

Location and testing method for the throttle position sensor. Use of the adapter harness is strongly recommended

in longer 1½ second pulses; the second digit is sent in ½-second pulses. Two long sweeps and three short-duration sweeps of the meter needle would indicate code 23. If more than one code is stored, the first digit of the second code will be sent after a three second pause. The entire sequence of codes will be retransmitted repeatedly; if you don't get all of it the first time, wait until it repeats.

Note that each engine has its own family of codes. Some are the same but there are differences. Make certain the correct chart is in use when performing diagnostic work.

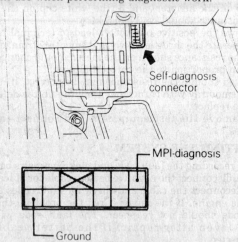

Connect the voltmeter properly when checking the system

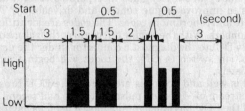

Pulse duration indicates place value for each two digit code

Diagnosis item	Malfunction code	
	Output signal pattern	No.
Engine control unit	⎍	—
Oxygen sensor		11
Air flow sensor		12
Intake air temperature sensor		13
Throttle position sensor		14
Motor position sensor		15
Engine coolant temperature sensor		21
Crank angle sensor		22
No. 1 cylinder top dead center sensor		23
Vehicle-speed sensor (reed switch)		24
Barometric pressure sensor		25
Injector		41
Fuel pump		42
EGR		43
Normal state		—

4G63 SOHC fault codes

5-46

FUEL SYSTEM 5

Diagnosis item	Malfunction code	
	Output signal pattern	No.
Engine control unit		—
Oxygen sensor		11
Air flow sensor		12
Intake air temperature sensor		13
Throttle position sensor		14
Engine coolant temperature sensor		21
Crank angle sensor		22
Top dead center sensor		23
Vehicle-speed sensor (reed switch)		24
Barometric pressure sensor		25
Injector		41
Fuel pump		42
EGR		43
Ignition coil		44
Normal state		—

4G63 DOHC trouble codes

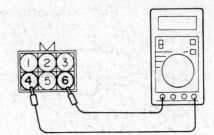

Testing the air intake temperature sensor. The pin arrangement is the same for both engines (SOHC and DOHC)

TESTING THE INTAKE AIR TEMPERATURE SENSOR

1. Disconnect the air flow sensor connector.
2. Connect the terminals of the ohmmeter between terminal Nos. 4 and 6 on either engine. Resistance will depend on air temperature. Two reference points are: 32°F (0°C), 6kΩ and 68°F (20°C), 2.7kΩ.
3. Use a hand-held hair dryer to blow warm air over the sensor. The resistance should change with the temperature increase.
4. If the values are not close to target or do not change with temperature, the unit is faulty and must be replaced.
5. Remove the test equipment and reconnect the wire harness.

TESTING THE THROTTLE POSITION SENSOR

WARNING: The use of the correct special tools or their equivalent is REQUIRED for this procedure. Wiring harness adapter MD 998478 for the SOHC or 998464 for the DOHC is highly recommended to prevent damage to the wiring and connectors.

1. With the engine off, disconnect the TPS wiring connector.
2. Install the special wiring device between the connectors.
3. On the SOHC, connect the ohmmeter across terminal Nos. 1 (ground; the black clip on the adapter) and 2 (the red clip on the adapter). On the DOHC, connect the ohmmeter to terminal No. 2 (red clip) and 3 (white clip). In both cases, the resistance should be 3.5–6.5 kΩ.
3. Move the positive ohmmeter lead to terminal No.3 (blue clip) on the SOHC connector or No.4 (blue clip) on the DOHC. Operate the throttle slowly and smoothly from idle to wide open; the resistance shown on the meter should change smoothly and in proportion to the throttle motion.

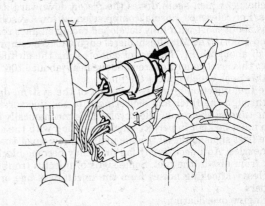

Location of SOHC TPS connector

5-47

5 FUEL SYSTEM

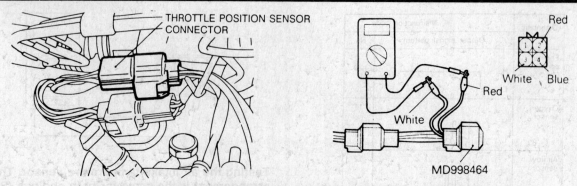

DOHC throttle position sensor connector location and testing hook-up

4. Even though the resistance may be correct, if it does not change smoothly with throttle motion, the unit is defective and must be replaced.

5. Remove the test equipment and connect the wiring harness.

RESETTING THE SYSTEM

After recording the fault code and making repairs based on diagnosis, disconnect the negative battery cable for at least 15 seconds. Reconnect the cable and check for fault codes after running the engine. If the repair cured the problem, the original fault code should not reappear. The fault code will remain stored (even after repair) if the battery cable is not disconnected.

DIESEL FUEL SYSTEM

The diesel combustion system works on different principal than a gasoline engine. In a gas engine, the air/fuel charge is moderately compressed by the stroke of the piston. The spark plug fires, burning the fuel in the cylinder. As it burns, the air/fuel charge expands, driving the piston downward.

The diesel is compression fired; that is, no external source of heat or spark is used to ignite the fuel charge. As any substance is compressed, it develops heat. If a flammable substance — fuel — is compressed sufficiently, it will burn or explode. The diesel engine uses compression within the cylinders to ignite the fuel charge, which again drives the piston downward as it expands. For reference, normal compression within a gasoline engine is about 8.5:1 with fully developed racing engines reaching around 13.0:1. The Mitsubishi diesel engine uses a compression ratio of about 21.0:1. The act of compressing the air/fuel mixture to this level generates tremendous heat (about 1700°F) and it is this heat which ignites the fuel.

The fuel injection pump is the heart of the system, drawing fuel through the filter and delivering it to the injectors at the proper time in the combustion cycle. This mechanically driven pump must be kept in perfect synchronization with the engine.

The first signs of diesel trouble usually show up at the injection nozzles. An injector may fail or it may become blocked or stuck from dirt in the fuel. Some signs of injector trouble are:

- Heavy knocking noises from the injectors of one or more cylinders.
- Engine overheating.
- Loss of power, particularly under load or acceleration.
- Smokey black exhaust
- Increased fuel consumption.

A faulty injection nozzle can be located by loosening the fuel line joint at each injector while the engine idles.

CAUTION

Wrap the joint in a towel or cloth; fuel will be pumped out when the joint is loosened. Observe no smoking/no open flame rules and have a dry chemical fire extinguisher within arm's reach at all times.

If an injector is working properly, changing its fuel supply (by opening the line) will cause a change in the idle quality of the engine. The injector that does NOT cause the idle to change is the problem.

Fuel Delivery Lines

REMOVAL AND INSTALLATION

If a fuel line under the hood should become damaged or leaky, do not attempt to weld or repair it. Replacement is required in all situations. Before disassembling any of the lines, mark the location of the plastic or metal clamps which hold the lines. These clamps are important in suppressing vibration which may loosen or damage the lines. Clamps must be reinstalled in the exact position from which they were removed.

1. Clean all the connections carefully. Dirt is the enemy of the diesel fuel system.

FUEL SYSTEM 5

2. Mark the position of all the line clamps and brackets.

3. Loosen each end of each line at both the injector and the pump. Remove the lines as an assembly with the clamps intact.

4. Immediately plug the fittings on the injection pump and the injectors to keep out dirt. This is very important.

5. Disassemble the lines on the workbench, replacing the damaged line(s).

6. When reinstalling, assemble the lines with the clamps correctly placed on the workbench.

7. Remove the plugs from the pump and injector ports. Install the fuel line assembly and start each threaded fitting by hand. Tighten each fitting to 27 Nm or 20 ft. lbs.

8. Double check that the line clamps are in place and secure. Bleed the fuel system.

9. Start the engine and check each joint for leaks. Because the system was open, it may require a few seconds until it runs smoothly.

Fuel Injectors

REMOVAL AND INSTALLATION

1. Remove the fuel delivery lines as described earlier.
2. Disconnect the fuel return hose.
3. Remove the nut which holds the fuel return pipe to each injector. Use a second wrench to counterhold the injector.
4. Remove the fuel return pipe from the injectors.
5. Use a long, deep socket to remove each injector from the head. After the injectors are out, remove the gasket from each nozzle hole.
6. When reinstalling, clean the nozzle holder mounting area of the cylinder head.
7. Install a new nozzle tip gasket and a new nozzle holder gasket into the cylinder head. New gaskets are required to prevent high pressure fuel leaks.
8. Install the injectors into the head and tighten each to 55 Nm (41 ft. lbs.).
9. Install the return pipe gaskets onto the injector nozzles and install the fuel return pipe onto the injectors.
10. Use two wrenches to counterhold and tighten the return line retaining nuts. Tighten them to 35 Nm or 26 ft. lbs.
11. Install the fuel return hose.
12. Install the fuel supply lines.

TESTING

There is no way to test diesel fuel injectors without an elaborate test bench. Your dealer or service center may have this equipment. Do not attempt to test the injector by connecting it to the fuel line while out of the engine. You'll bend the line and run the risk of injury. The injector releases fuel under a pressure of 1707 psi or higher; this is more than enough pressure to force atomized fuel through your skin and induce blood poisoning.

Once the injector is set up on a proper test bench, it should be checked for spray pattern, noise, break pressure and leakage. Each one of these characteristics must be within specification or the injector must be replaced.

Injection Pump

REMOVAL AND INSTALLATION

WARNING: After the pump is reinstalled, the injection timing must be reset. Adjusting the injection timing requires special tools for measuring the prestroke of the pump. If these tools are not available do NOT remove the injection pump or attempt to work on it.

1. Disconnect the battery ground cable.

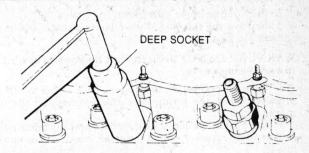

Removing injector nozzles

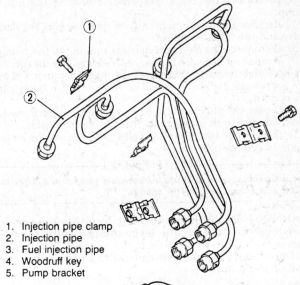

1. Injection pipe clamp
2. Injection pipe
3. Fuel injection pipe
4. Woodruff key
5. Pump bracket

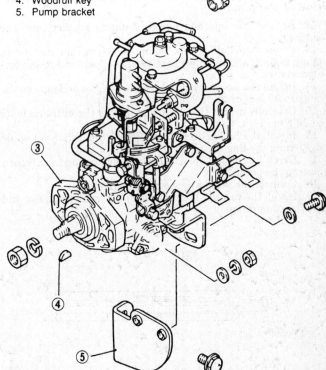

Diesel injection pump and fuel delivery lines

5-49

5 FUEL SYSTEM

2. Remove the upper cover from the timing belt.
3. Remove the nut and washer holding the injection pump sprocket to its shaft.

WARNING: Do NOT drop the nut and/or washer into the lower timing cover.

4. Turn the crankshaft to bring the No.1 piston to TDC/compression. When the piston is in this position, all the timing marks align.
5. Use a pulley puller or extractor to carefully remove the injection pump sprocket. Catch the small Woodruff key and don't lose it.

WARNING: Do not subject the shaft to impact. Protect the timing belt from any stress, twisting or fluids. After the sprocket is removed, do not turn the crankshaft.

6. Disconnect the boost compensator hose at the injector pump.
7. Remove the vacuum hose from the vacuum regulating valve and the injection pump.
8. Disconnect the electrical harness running to the pump.
9. Wrap the joints with rags or cloth and remove the fuel supply hose and the return hose from the injection pump and plug the ports immediately. Fuel will be released when these joints are opened.
10. Loosen the union nuts and remove the injection lines. Use a second wrench to counterhold fittings where possible.
11. Remove the two mounting bolts, then remove the mounting nuts and washers.

CAUTION
The pump is heavy and awkward to hold; make sure you or a helper have a firm grip before releasing the mounts.

WARNING: Do not hold the pump by either the accelerator lever or the fast idle lever. Additionally, these levers must not be removed from the pump. The levers' adjustment and placement is critical to the correct operation of the pump.

12. Remove the injection pump mounting bracket from the engine.
13. Before reinstalling the pump, make certain the entire pump is clean and the lines have been checked for dirt and foreign matter.
14. Install the mounting bracket on the engine and tighten the bolts to 35 Nm or 26 ft. lbs.
15. Carefully mount the pump and tighten the nuts and bolts to 26 Nm (19 ft. lbs).
16. Install the fuel delivery lines. Counterhold the joint while tightening the line fittings to 29 Nm (22 ft. lbs).
17. Carefully connect the the fuel supply hose and the return hose to the pump.
18. Connect the wiring harness to the pump.
19. Install the vacuum hoses between the regulating valve and the injection pump.

20. Connect the boost compensator hose and connect the negative battery cable.
21. Carefully install the sprocket on the shaft of the injection pump. Do not subject the shaft to impact and do not damage the timing belt. Make sure the Woodruff key is correctly seated in its groove.
22. Perform the ignition timing adjustment procedure outlined in this section.

WARNING: This adjustment is MUST be performed each time the injection pump is removed.

23. Bleed the fuel system.
24. Install the upper timing belt cover.

INJECTION TIMING ADJUSTMENT

WARNING: The use of the correct special tools or their equivalent is REQUIRED for this procedure. The timing cannot be adjusted without tool MD 998384, the prestroke measuring adaptor. If this tool is not available, do not attempt to adjust the injection timing.

1. With the engine off, turn the crankshaft clockwise until all the timing marks align. This places the engine at TDC/compression for No.1 piston.
2. Loosen but do not remove the injection pipe union nuts at the injection pump. Always use a second wrench to counterhold the pump fitting.
3. Loosen the nuts and bolts holding the injection pump to the engine but do not remove them.
4. Examine the special tool and make certain the tip protrudes 10mm (0.4 in.). The pushrod can be adjusted by the inner nut.
5. Remove the timing check plug from the injector pump head and install the prestroke measuring adapter and a dial indicator.
6. Carefully turn the crankshaft counterclockwise until the notch on the pulley is at a point about 30° before the TDC mark. The notch should be in roughly the 11 o'clock position.
7. Zero the dial indicator. With a wrench, slightly move the crankshaft in both the clockwise and counterclockwise directions. The dial indicator should stay at zero; if it changes with engine movement, the initial 30° setting was incorrect. Repeat steps 6 and 7 until the gentle movement of the crank does not move the pointer on the dial indicator.
8. Turn the crankshaft clockwise until the notch in the pulley aligns with the 5° ATDC mark. At this point, the dial indicator should read 1.0mm ± 0.003mm (0.0394 in. ± 0.0011 in.).
9. If the gauge does not show the specified value, move the injection pump body right or left on the mounts until the dial indicator reads correctly. Tighten the mounting nuts and bolts to hold the pump in place.

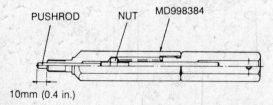

Special tools are needed to adjust the injection timing

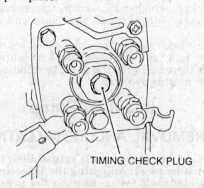

Location of the timing check plug on the diesel fuel injection pump

FUEL SYSTEM 5

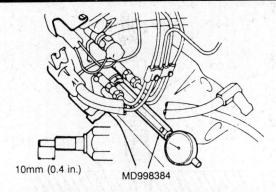

Special tool with dial indicator installed on the pump

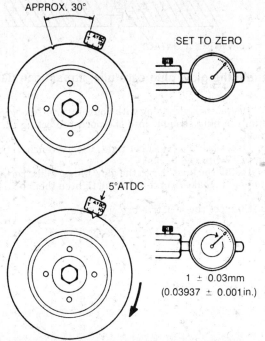

Accurate positioning of the crankshaft pulley is required when adjusting the injection timing

10. After the pump is locked in place, repeat the check procedure. Set the engine 30° before, zero the indicator, move the engine to 5° ATDC and check for the correct gauge reading. Each time the pump is moved, the entire check should be repeated.

11. Remove the special tools. Install a new copper gasket and the timing check plug. Tighten it to 8 Nm or 6 ft. lbs. Do not overtighten this bolt.

12. Tighten the fuel delivery lines. Counterhold the joint while tightening the line fittings to 29 Nm (22 ft. lbs).

13. Since the fuel lines were loosened, bleeding the system is recommended.

IDLE SPEED ADJUSTMENT

1. The idle must be set with the engine fully warmed up, the lights and all electrical accessories off, the transmission in NEUTRAL and the parking brake applied.
2. Run the engine for more than 5 seconds at a speed between 2000 and 3000 rpm.
3. Allow the engine to idle for at least 2 minutes.
4. Using a diesel tachometer, check the idle speed. Correct speed is 750 ± 50 rpm.

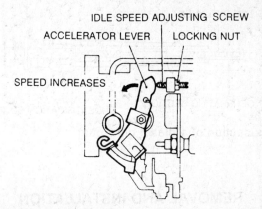

When setting the idle speed, make certain only the adjusting screw is turned

5. Adjust the speed to specification as necessary using the idle speed adjusting screw. Loosen the locknut and turn the screw to attain correct idle. Retighten the locknut.

WARNING: Do not adjust any other screws in the area.

Glow Plugs

The glow plug system is a preheat system which rapidly warms the combustion chamber during a cold start. This initial heating allows the fuel to burn quicker, resulting in better cold driveability.

The earliest diesels required the driver to wait while the glow plugs worked; when the correct temperature was achieved, the engine could be started. Mitsubishi eliminated this long wait with what they call the Super Quick Glow System. By using both relays and a dropping resistor, the plugs are brought to their operating temperature almost instantly. The engine may be started within 5–10 seconds. Additionally, when the engine is running and the coolant temperature is below 130°F (55°C), the glow system will remain engaged, resulting in stable heat generation and reduced engine noise.

The glow plugs themselves are nothing more than resistance heating elements. Because they heat quickly, they consume heavy amounts of electricity. The high amperage required is conducted through a buss-bar (solid conductor) connected to each plug. It is important that the battery in a diesel vehicle be kept fully charged; otherwise the draw of the glow plugs combined with the draw of the starter will flatten the battery.

(1) Nut
(2) Glow plug
(3) Glow plug plate

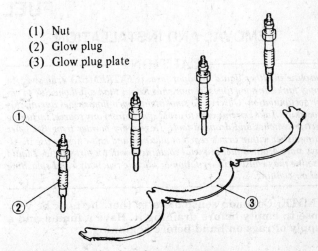

Glow plugs and related components

5–51

5 FUEL SYSTEM

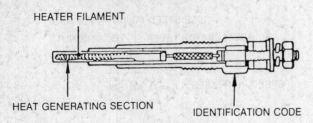

Cross section of a diesel glow plug

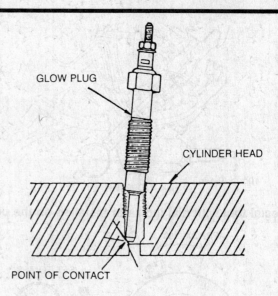

Handle the glow plug carefully; impact can damage it

REMOVAL AND INSTALLATION

1. Disconnect the negative battery cable.
2. Loosen the small nut on the top of each glow plug.
3. Remove the glow plug plate (buss-bar) running among the 4 plugs.
4. Carefully unscrew the plug from the head, using a 12mm, deep socket. Handle the plug carefully, protecting the tip at all times. If the unit falls or strikes a hard surface, it must be considered unusable. (The heating element inside the tip is a winding similar to the filament within a light bulb; it will break under impact.).
5. The glow plug may be tested with an ohmmeter. Check resistance between the top terminal and the body of the plug. The correct resistance is 0.23 ohms; make certain the meter is set on the correct scale to read this value. If there is no resistance or excessive resistance, the plug is unusable.
6. Inspect the buss-bar for any damage to the protective coating.
7. Install each glow plug with your fingers. Take great care to install the plug straight into the port, protecting the tip from impact.
8. Once each plug is finger tight, use a wrench to tighten the plug to 17 Nm (13 ft. lbs).
9. Install the buss-bar to the glow plugs. Make certain the retaining nuts make firm contact and tighten them to 1.4 Nm or 1 ft. lb.
10. Connect the negative battery cable.

FUEL TANK

REMOVAL AND INSTALLATION

---- CAUTION ----
Gasoline in either liquid or vapor form is EXTREMELY explosive. An empty tank can sometimes be more hazardous than a full one due to vapor accumulation. Observe no smoking/no open flame rules during this procedure. Take extreme care to avoid sparks from any source, including electric switches and dropped tools. Have a dry powder (type B-C) fire extinguisher within arm's reach at all times and know how to use it. Always store drained fuel in metal containers with an airtight lid. Liquid gasoline may cause skin irritation or allergic reaction; avoid splashing fuel on clothing.

NOTE: Common sense dictates that the tank be very close to empty before draining it. Have a funnel and a supply of rags on hand before beginning.

Cordia and Tredia

1. Remove the access panel on the floor of the trunk. On turbo vehicles, start the engine, then disconnect the electric fuel pump connector accessible there. Allow the engine to run until it stops on its own. Turn the ignition switch off.
2. Disconnect the negative battery cable.
3. Remove the fuel filler cap. Remove the drain plug from under the tank and drain the fuel into a safe, sealable container.
4. Disconnect the fuel gauge connector and then push it, with the plug, through the hole in the floor.
5. If not already done, raise and support the vehicle on axle stands. Remove the left rear wheel.
6. Carefully label and disconnect the three hoses from the fuel pipes coming out of the tank. If the hoses are not reconnected properly, the engine will not start. Label and disconnect any other breather or pump hoses running from the tank.

FUEL SYSTEM 5

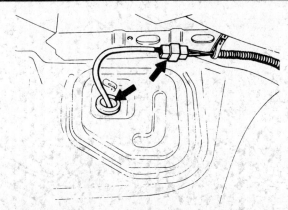

Turbocharged Cordia and Tredia fuel pump connector

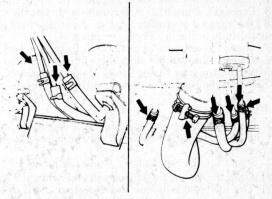

Correct labeling is essential during hose removal

NOTE: There may be other lines or hoses running from one part of the tank to another part of the tank; these may be left connected until the tank is removed. Only disconnect the lines running from the tank to the body.

7. Remove the cover from the filler neck inside the trunk. Label and disconnect the vapor hoses and filler hose at the filler neck.
8. Support the tank from underneath with a floor jack and a broad piece of lumber. Remove the nuts from the forward side of each mounting band and pull the bands downward. Lower the tank and remove it.

— **CAUTION** —
Some fuel will remain within the lines and may spill out when the tank is removed.

9. When reinstalling the tank, support it on the jack and raise it into place. Swing the mounting bands into place and draw the attaching nuts all the way up. The bands should touch the floor of the car.

10. Attach the fuel filler line and vapor lines to the filler neck. Take the time to insure that each is squarely seated and that the clamp is tight.
11. Reconnect the lines and breather hoses running to the tank. Make certain all the vapor lines are pushed onto the pipe until the second bulge is reached. On turbocharged vehicles, with a screw type fuel line connector, tighten the connector to 36 Nm (27 ft. lbs).
12. If the fuel gauge sending unit was removed, tighten its retaining bolt to 1 Nm (0.7 ft. lbs.). Connect the wiring to the gauge sender.
13. Install the drain plug and tighten it to 20 Nm (15 ft. lbs.).
14. Connect the wiring for the electric fuel pump (turbocharged models only).
15. Double check all installation items, paying particular attention to loose hoses or hanging wires, untightened nuts, poor routing of hoses and wires (too tight or rubbing) and tools left in the work area. Make sure all fuel line connections are tight and the clamps secure.
16. Using a funnel, carefully refill the tank with the drained fuel. Avoid splashing fuel onto the painted bodywork. Install the filler cap.

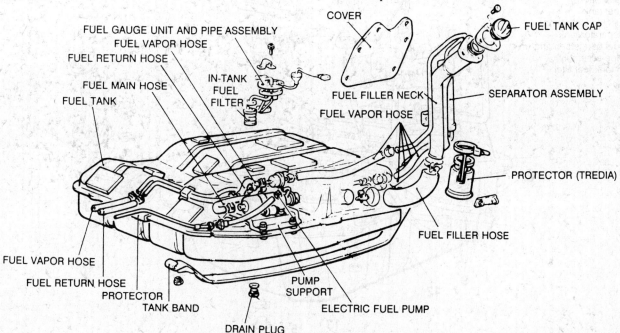

Cordia and Tredia tank assembly. Turbocharged model shown, non-turbo similar

5-53

5 FUEL SYSTEM

17. Install the left rear wheel and lower the car to the ground.
18. Connect the negative battery cable and start the engine. It will take a period of cranking before the engine fires and runs smoothly due to the lines being emptied.
19. Check the entire work area carefully for any sign of leakage. A small leak will not drip but will only show up a slight moisture on the hose. Attend to all leaks immediately.
20. Install the filler neck cover and the access panel.

Starion

1. Remove the higher floor side panel located in the trunk. Pry up and remove the cover for all the hoses and electrical connectors.
2. Start the engine and allow it to idle. Disconnect the electric fuel pump connector; the engine will stall when it runs out of fuel. The fuel system is now depressurized. Turn the ignition switch off.
3. Disconnect the battery cables.
4. Remove the fuel tank cap. Remove the drain plug and drain the fuel into a sealable, safe container.
5. Disconnect the fuel gauge sending unit connector.
6. If not already done, raise and support the car on jackstands. Remove the left rear wheel.
7. Remove the bolts and remove the fuel pipe cover. Label and disconnect all the hoses from the pipes leading into the tank. Correct labeling is critical to proper reassembly.

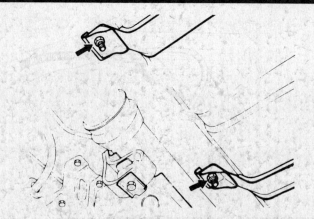

The tank band retaining nuts must be tightened all the way up to the floor

1. High floor side panel
2. Fuel pipe cover
3. Fuel pump connector
4. Fuel filler cap
5. Drain plug
6. Fuel gauge connector
7. High pressure fuel hose
8. Fuel return hose
9. Vapor hose
10. Fuel filler hose
11. Filler neck
12. Fuel tank
13. Electric fuel pump
14. Separator tank
15. Fuel gauge sending unit
16. Pipe assembly
17. In-tank fuel filter

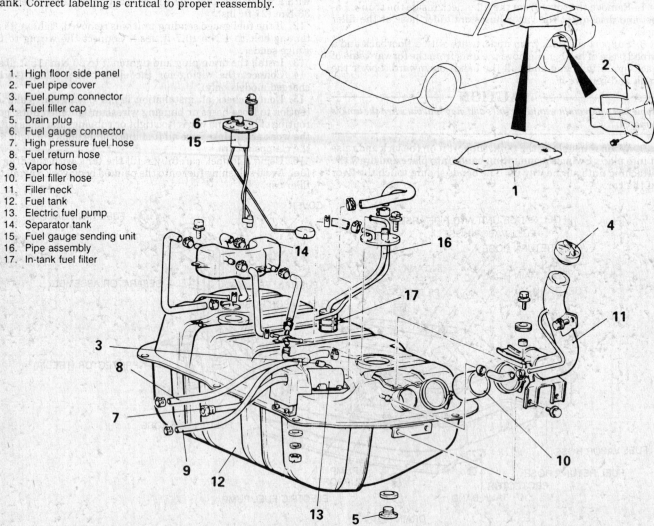

Starion fuel tank

5-54

FUEL SYSTEM 5

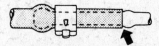

Install the vapor hoses all the way onto the pipe

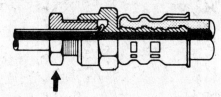

When assembling the high pressure fuel line, turn the smaller fitting (arrow) while holding the larger fitting with a second wrench

NOTE: There may be other lines or hoses running from one part of the tank to another part of the tank; these may be left connected until the tank is removed. Only disconnect the lines running from the tank to the body.

8. Disconnect the vapor hose from the tank vapor pipe. Disconnect the fuel filler hose at the tank and disconnect the breather hose.
9. Support the tank from underneath with a floorjack and a broad piece of lumber. Remove the ten fastening nuts and their washers from the outer edge of the tank. Use the jack to support the tank and lower it away from the car.

--- **CAUTION** ---
Some fuel will remain within the lines and may spill out when the tank is removed.

10. Install the tank by elevating it into place with the jack. Install the retaining bolts, tightening them to 27 Nm (20 ft. lbs.).
11. Install the tank drain bolt, tightening it to 20 Nm (15 ft. lbs.).
12. Attach the fuel filler line and vapor lines to the filler neck. Take the time to insure that each is squarely seated and that the clamp is tight.
13. Reconnect the lines and breather hoses running to the tank. Make certain all the vapor lines are pushed onto the pipe until the second bulge is reached. Tighten the threaded connector for the high pressure fuel line to 36 Nm (27 ft. lbs) and do not allow the line to become twisted.
14. If the fuel gauge sending unit was removed, tighten its retaining bolt to 1 Nm (0.7 ft. lbs.). Connect the wiring to the gauge sender.
15. Install the drain plug and tighten it to 20 Nm (15 ft. lbs.).
16. Connect the wiring for the electric fuel pump.
17. Double check all installation items, paying particular attention to loose hoses or hanging wires, untightened nuts, poor routing of hoses and wires (too tight or rubbing) and tools left in the work area. Make sure all fuel line connections are tight and the clamps secure.
18. Using a funnel, carefully refill the tank with the drained fuel. Avoid splashing fuel onto the painted bodywork. Install the filler cap.
19. Install the left rear wheel and lower the car to the ground.
20. Connect the negative battery cable and start the engine. It will take a period of cranking before the engine fires and runs smoothly due to the lines being emptied.
21. Check the entire work area carefully for any sign of leakage. A small leak will not drip but will only show up a slight moisture on the hose. Attend to all leaks immediately.
22. Install the filler neck cover and the access panel.

Galant and Sigma

1. Start the engine and allow it to idle. Disconnect the electric fuel pump connector; the engine will stall when it runs out of fuel. The fuel system is depressurized. Turn the ignition switch off.
3. Disconnect the negative battery cable.
4. Remove the fuel tank cap. Remove the drain plug and drain the fuel into a sealable, safe container.
5. If not already done, raise and support the car on jackstands. Remove the left rear wheel.
6. Label and disconnect both the fuel vapor hose and the fuel return hose at the leading edge of the tank.
7. Disconnect the fuel gauge sending unit connector.
8. Disconnect the fuel high pressure line using two open-end wrenches.

WARNING: Some pressure may remain within the lines. Wrap the joint with a clean cloth before disconnecting.

9. Disconnect the fuel filler hose and vapor hose at the tank.
10. Remove the tank band brace. Loosen the two tank band support nuts slightly.
11. Support the tank with a floor jack and a broad piece of lumber. Remove the tank band attaching nuts. Lower and remove the tank.

--- **CAUTION** ---
Some fuel will remain within the lines and may spill out when the tank is removed.

12. Install the tank by elevating it into place with the jack. Install the tank bands and the band supports. Replace the self locking nuts with new ones and tighten them until the rear of the strap contacts the body.
13. Install the tank drain bolt, tightening it to 20 Nm (15 ft. lbs.).

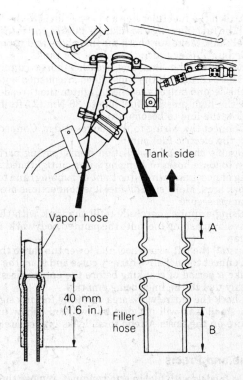

Details of filler and vapor hose reassembly

5-55

5 FUEL SYSTEM

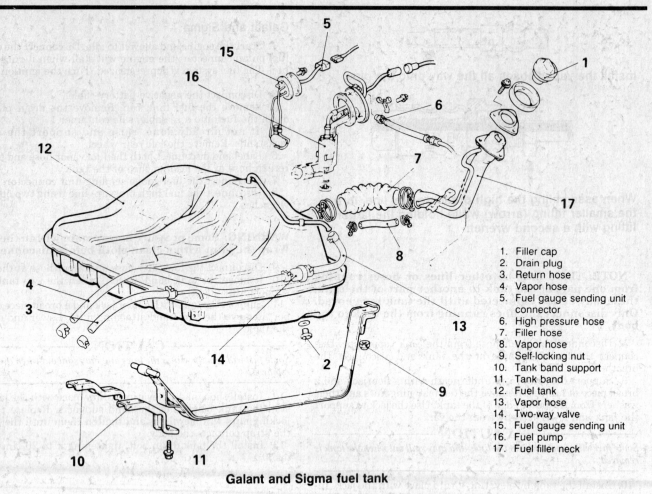

1. Filler cap
2. Drain plug
3. Return hose
4. Vapor hose
5. Fuel gauge sending unit connector
6. High pressure hose
7. Filler hose
8. Vapor hose
9. Self-locking nut
10. Tank band support
11. Tank band
12. Fuel tank
13. Vapor hose
14. Two-way valve
15. Fuel gauge sending unit
16. Fuel pump
17. Fuel filler neck

Galant and Sigma fuel tank

14. Attach the fuel filler line and vapor lines to the filler neck. Install the fuel filler hose so that the shorter end is inserted into the fuel tank. Make sure both the filler hose and vapor hose are installed until they touch the side of the tank.
15. Reconnect the lines and breather hoses running to the tank. Make certain all the vapor lines are pushed onto the pipe until the second bulge is reached. Tighten the threaded connector for the high pressure fuel line to 36 Nm (27 ft. lbs) and do not allow the line to become twisted.
16. Connect the wiring to the gauge sender. Connect the wiring for the electric fuel pump.
17. Double check all installation items, paying particular attention to loose hoses or hanging wires, untightened nuts, poor routing of hoses and wires (too tight or rubbing) and tools left in the work area. Make sure all fuel line connections are tight and the clamps secure.
18. Using a funnel, carefully refill the tank with the drained fuel. Avoid splashing fuel onto the painted bodywork. Install the filler cap.
19. Install the left rear wheel and lower the car to the ground.
20. Connect the negative battery cable and start the engine. It will take a period of cranking before the engine fires and runs smoothly due to the lines being emptied.
21. Check the entire work area carefully for any sign of leakage. A small leak will not drip but will only show up a slight moisture on the hose. Attend to all leaks immediately.

Mirage and Precis

1. On models with fuel injected engines, remove the rear seat cushion and carpet. Start the engine and allow it to idle. Disconnect the electric fuel pump connector and allow the engine to stop due to fuel starvation.
2. Turn off the ignition switch and disconnect the negative battery cable.
3. Remove the fuel tank cap. Remove the fuel tank drain plug and drain the fuel into a sealable, safe container.
4. Remove the plug in the floor of the trunk. Disconnect the fuel gauge sending unit connectors.
5. If not already done, elevate and support the vehicle on jack stands; remove the left rear wheel.
6. Disconnect the fuel supply and return hoses. Disconnect the vapor hose at the two-way valve.

WARNING: Some pressure may remain within the lines. Wrap the joint with a clean cloth before disconnecting.

7. Disconnect the equalizing hose and the filler hose at the filler neck.
8. Support the tank from underneath with a floor jack and a broad piece of lumber. Remove the two tank mounting band nuts from the front ends of the tank mounting straps. Lower and remove the tank.

--- **CAUTION** ---
Some fuel will remain within the lines and may spill out when the tank is removed.

9. Install the tank by elevating it into place with the jack. Install the tank bands and the band supports. Replace the self locking nuts with new ones and tighten them until the rear of the strap contacts the body.
10. Install the tank drain bolt, tightening it to 20 Nm (15 ft. lbs.).

FUEL SYSTEM 5

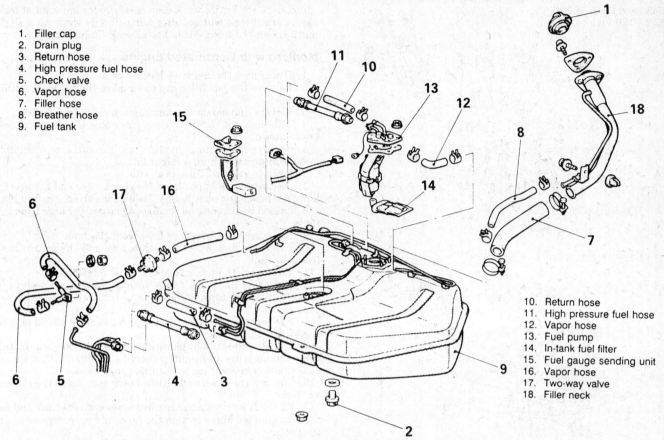

1. Filler cap
2. Drain plug
3. Return hose
4. High pressure fuel hose
5. Check valve
6. Vapor hose
7. Filler hose
8. Breather hose
9. Fuel tank
10. Return hose
11. High pressure fuel hose
12. Vapor hose
13. Fuel pump
14. In-tank fuel filter
15. Fuel gauge sending unit
16. Vapor hose
17. Two-way valve
18. Filler neck

Mirage fuel tank used with 4G15 and 4G61 engines. The tank for G15B and G32B engines is similar, but with different hose locations

11. Attach the fuel filler line and vapor lines to the filler neck. Make sure both the filler hose and vapor hose are installed until they touch the side of the tank.
12. Reconnect the lines and breather hoses running to the tank. Make certain all the vapor lines are pushed onto the pipe until the second bulge is reached. Tighten the threaded connector for the high pressure fuel line to 36 Nm (27 ft. lbs) and do not allow the line to become twisted.
13. Connect the wiring to the gauge sender. Connect the wiring for the electric fuel pump. Reinstall the rear seat cushion.
14. Double check all installation items, paying particular attention to loose hoses or hanging wires, untightened nuts, poor routing of hoses and wires (too tight or rubbing) and tools left in the work area. Make sure all fuel line connections are tight and the clamps secure.
15. Using a funnel, carefully refill the tank with the drained fuel. Avoid splashing fuel onto the painted bodywork. Install the filler cap.
16. Install the left rear wheel and lower the car to the ground.
17. Connect the negative battery cable and start the engine. It will take a period of cranking before the engine fires and runs smoothly due to the lines being emptied.
18. Check the entire work area carefully for any sign of leakage. A small leak will not drip but will only show up a slight moisture on the hose. Attend to all leaks immediately.

1983–89 Pickup

1. Disconnect the negative battery cable.
2. Elevate and safely support the vehicle on jackstands. For 1986–89 trucks, remove the small side skirt panel on the left side.
3. Remove the filler cap to equalize pressure within the tank.
4. Remove the drain plug and in-tank filter. Drain the fuel into a suitable metal container with an airtight lid. Later models may not have the in-tank filter.
5. Loosen the fuel hose clamps (main, return and vapor) at the fuel tank and disconnect the hoses. Label or identify each hose and port.
6. Remove the electrical harness from the fuel gauge unit.
7. Remove the filler neck retaining bolts from the body. Remove the filler hose protector.
8. Use a floor jack and a broad piece of wood positioned under the fuel tank to support the tank.
9. Remove the tank mounting nuts.
10. Lower the tank to the ground.
11. Install the new tank by raising it into place with the jack.
12. Install the retaining nuts and tighten them to 32 Nm (24 ft. lbs). When the tank is secure, the jack may be removed.
13. Connect the electrical harness to the gauge unit.
14. Install the filler hose protector and install the neck-to-body retaining screws.
15. Install the hoses to their ports, making certain that each clamp is secure.
16. If not already done, install the in-tank filter and drain bolt. Tighten the bolt to 59 Nm or 44 ft. lbs.).
17. Using a funnel, carefully refill the tank with the drained fuel. Avoid splashing fuel onto the painted bodywork. Install the filler cap.
18. Lower the car to the ground and connect the negative battery cable.
19. Start the engine. It will take a period of cranking before the engine fires and runs smoothly due to the lines being emptied.

5 FUEL SYSTEM

G63B and G54B ENGINES
57.2 LITERS (15.1 US GAL.)

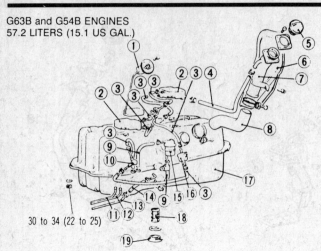

68.1 LITERS (18.0 US GAL.)

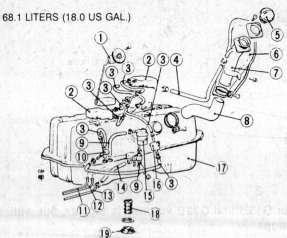

4D55 ENGINE

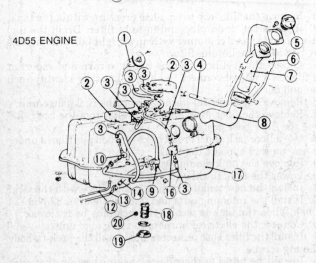

1. Fuel gauge unit	8. Connecting hose	15. Fuel filter
2. Separator tank	9. Main hose	16. Overfill limiter
3. Vapor hose	10. Check valve	17. Fuel tank
4. Breather hose	11. Fuel vapor pipe	18. Fuel filter (in tank)
5. Fuel filler cap	12. Fuel main pipe	19. Drain plug
6. Filler hose protector	13. Ruel return pipe	20. Valve
7. Filler neck	14. Return hose	

Pickup truck fuel tanks

20. Check the entire work area carefully for any sign of leakage. A small leak will not drip but will only show up a slight moisture on the hose. Attend to all leaks immediately.

Montero with Carbureted Engine

1. Disconnect the negative battery cable.
2. Remove the gas filler cap to equalize the pressure within the tank.
3. Elevate and safely support the vehicle on jackstands.
4. Although not required, the job is easier if the rear wheels are removed.
5. Remove the drain plug and drain the fuel into a suitable metal container with an airtight lid.
6. Remove the fuel filler protector.
7. Disconnect the vapor hose, the check valve and the overfill limiter. Make sure each line is labeled for correct reassembly.
8. Loosen the clamp and remove the large filler hose from the tank.
9. Disconnect the breather hose from the tank.
10. Label and disconnect the main hose and the return hose from the tank.
11. disconnect the fuel gauge electrical connector.
12. Position a floor jack and a broad piece of wood under the tank. Remove the tank retaining nuts and lower the tank from the vehicle.
13. The separator tanks and the protective stone shield may be removed if so desired.
14. Reinstall the tank by elevating it with the floor jack. Install the retaining nuts and tighten them to 27 Nm (20 ft. lbs.).
15. Connect the wiring to the fuel gauge sender.
16. Connect the main and return hoses and insure the clamps are tight.
17. Fit the filler neck and breather hoses onto the tank and secure the clamps. Make certain the hoses are properly seated on the ports.
18. Install the overfill limiter, the check valve and the vapor hose. Make certain the check valve (roll over valve) is mounted so that the arrow on its case point upwards.
19. Install the filler neck protector.
20. If not already done, install the drain plug and tighten it to 16 Nm or 12 ft. lbs.
21. Using a funnel, carefully refill the tank with the drained fuel. Avoid splashing fuel onto the painted bodywork. Install the filler cap.
22. Lower the car to the ground and connect the negative battery cable.
23. Start the engine. It will take a period of cranking before the engine fires and runs smoothly due to the lines being emptied.
24. Check the entire work area carefully for any sign of leakage. A small leak will not drip but will only show up a slight moisture on the hose. Attend to all leaks immediately.

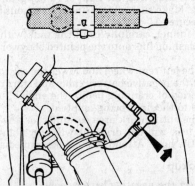

Make certain the fuel return line is properly connected (upper) and that the check valve is not upside down

FUEL SYSTEM 5

Montero with Fuel Injection

1. Lift or move aside the carpet in the rear cargo area.
2. Remove the oval cover plate from the access hole.
3. Disconnect the fuel pump harness connector at the rear of the fuel tank.
4. Start the engine, allowing it to run until it runs out of fuel. This relieves pressure within the fuel system.
5. Turn the ignition switch **OFF** and disconnect the negative battery cable.
6. Remove the fuel filler cap.
7. Safely elevate and support the car on jackstands.
8. Drain the fuel from the fuel tank into suitable container(s), each with an airtight lid.
9. Remove the fuel filler protector.
10. Label and disconnect the two-way valve, the check valve and the vapor hose.
11. Disconnect the clamp holding the filler neck to the body.
12. At the tank, disconnect the filler hose and the breather hose.
13. Remove the retaining screws at the top of the filler and remove the filler neck from the car.
14. At the fuel pump, disconnect the high pressure line and the return line.

WARNING: Wrap the joints in a rag before loosening. Some pressure may remain within the system.

15. Disconnect and remove the separator tank(s) on top of the fuel tank.
16. Position a floor jack and a broad piece of lumber under the tank. Remove the retaining nuts and lower the tank out of the vehicle with the jack.
17. The tank protector, fuel pump and additional components may be removed with the tank out of the vehicle.

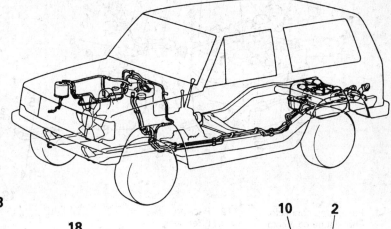

1. Drain plug
2. Fuel filler cap
3. Fuel filler hose protector
4. Vapor hose
5. Check valve
6. Overfill limiter (Two—way valve)
7. Clamp assembly
8. Fuel filler hose
9. Breather hose
10. Packing
11. Fuel filler neck
12. Main hose
13. Return hose
14. Fuel gauge unit connector
15. Fuel tank assembly mounting nuts
16. Fuel tank
17. Pipe assembly
18. Separator tanks
19. Fuel tank protector

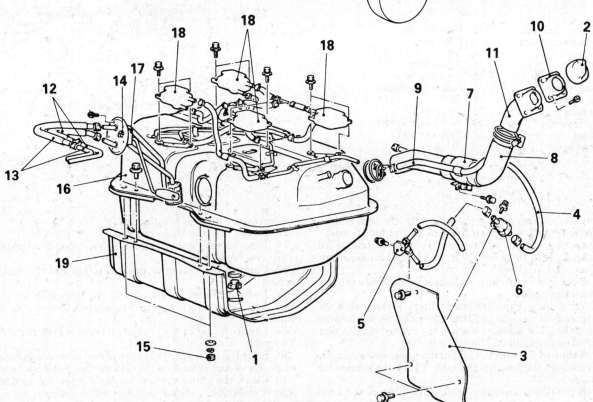

Carbureted Montero fuel tank

5-59

5 FUEL SYSTEM

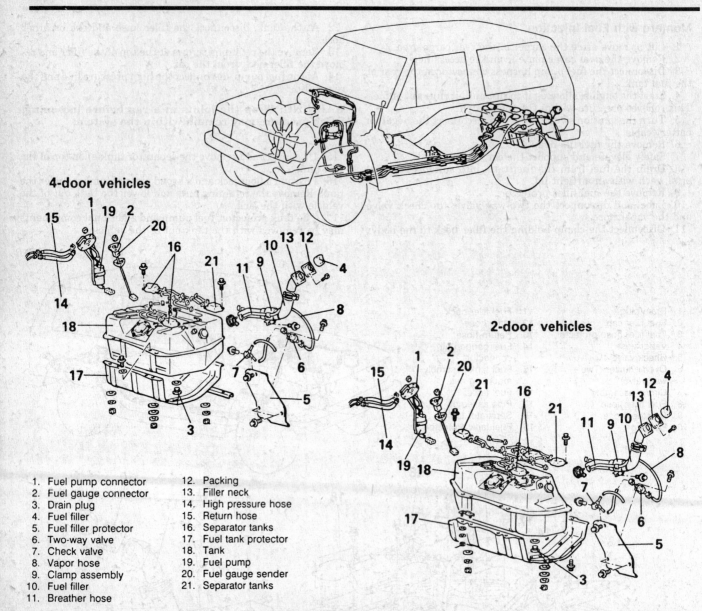

1. Fuel pump connector
2. Fuel gauge connector
3. Drain plug
4. Fuel filler
5. Fuel filler protector
6. Two-way valve
7. Check valve
8. Vapor hose
9. Clamp assembly
10. Fuel filler
11. Breather hose
12. Packing
13. Filler neck
14. High pressure hose
15. Return hose
16. Separator tanks
17. Fuel tank protector
18. Tank
19. Fuel pump
20. Fuel gauge sender
21. Separator tanks

Fuel injected Montero fuel tanks

18. When reinstalling, elevate the tank with the jack and install the retaining nuts. Tighten them to 27 Nm (20 ft. lbs).
19. If the tank protector was removed, it may be reinstalled now; tighten the bolts to 14 Nm (10 ft. lbs). Install the separator tanks if they were removed.
20. Install the fuel return line, making sure it reaches the second bulge on the pipe.
21. Connect the high pressure line by hand, making sure the flare nut threads in straight and that the hose does not become twisted or kinked. Use a second wrench to counterhold the joint and tighten the fitting to 35 Nm or 26 ft. lbs.
22. Install the fuel filler neck. Tighten the upper screws (at the filler) and connect the hoses at the tank. Check the clamps for proper fit.
23. Install the clamp assembly holding the filler neck to the body.
24. Connect the vapor hose, check valve (roll over valve) and the overfill limiter. Make sure the check valve is installed with the arrow on its case pointing up.
25. Install the filler hose protector.
26. If not already done, install the drain plug and tighten it to 1218 Nm or 14 ft. lbs.
27. Connect the wiring to the fuel pump and the fuel gauge sender.
28. Using a funnel, carefully refill the tank with the drained fuel. Avoid splashing fuel onto the painted bodywork. Install the filler cap.
29. Lower the car to the ground and connect the negative battery cable.
30. Start the engine. It will take a period of cranking before the engine fires and runs smoothly due to the lines being emptied.
31. Check the entire work area carefully for any sign of leakage. A small leak will not drip but will only show up a slight moisture on the hose. Attend to all leaks immediately.
32. Reinstall the access cover panel and reposition the carpeting in the cargo area.

FUEL SYSTEM 5

Model — Cordia
Body VIN — 4
Engine — 1.8L (1795) **Cylinders** — 4
Fuel System — Multi-Point Fuel Injection
Engine Identifier — G54B **VIN** — G

ENGINE CODES

Code	Explanation
1	Oxygen sensor and computer
2	Ignition pulse
3	Air Flow Sensor (AFS)
4	Pressure sensor
5	Throttle Position Sensor (TPS)
6	Idle Speed Control (ISC) motor position switch
7	Coolant Temperature Sensor (CTS)

Model — Cordia
Body VIN — 4
Engine — 2.0L (1997cc) **Cylinders** — 4
Fuel System — Feedback Carburetor
Engine Identifier — G63B **VIN** — D

ENGINE CODES

Code	Explanation
1	Oxygen sensor and computer
2	Ignition pulse
3	Air Flow Sensor (AFS)
4	Pressure sensor
5	Throttle Position Sensor (TPS)
6	Idle Speed Control (ISC) motor position switch
7	Coolant Temperature Sensor (CTS)

5 FUEL SYSTEM

Model — Cordia
Body VIN — 4
Engine — 1.8L (1795cc)   **Cylinders** — 4
Fuel System — Electronic Controlled Injection
Engine Identifier — G62B   **VIN** — G

ENGINE CODES

Code	Explanation
1	Oxygen sensor and computer
2	Ignition pulse
3	Air Flow Sensor (AFS)
4	Pressure sensor
5	Throttle Position Sensor (TPS)
6	Idle Speed Control (ISC) motor position switch
7	Coolant Temperature Sensor (CTS)

Model — Galant
Engine — 2.4L (143 cid) 4 cyl
Engine Code — G64B

ECM TROUBLE CODES

Code	Explanation
0 0 0 0 0	Normal
1 0 0 0 0	Oxygen sensor
0 1 0 0 0	Crank angle sensor
1 1 0 0 0	Air Flow Sensor (AFS)
0 0 1 0 0	Atmospheric pressure sensor
1 0 1 0 0	Throttle Position Sensor (TPS)
0 1 1 0 0	Motor Position Sensor (MPS)
1 1 1 0 0	Coolant temperature sensor
0 0 0 1 0	Number 1 cylinder Top Dead Center (TDC) sensor

FUEL SYSTEM 5

Model — Galant
Body VIN — 6
Engine — 2.4L (2350cc) **Cylinders** — 4
Fuel System — Multi-Point Fuel Injection
Engine Identifier — G64B **VIN** — L

ENGINE CODES

Code	Explanation
1	Oxygen sensor
2	TDC sensor
3	Air Flow Sensor (AFS)
4	Atmospheric pressure sensor
5	Throttle Position Sensor (TPS)
6	Motor Position Sensor (MPS)
7	Coolant Temperature Sensor (CTS)
8	Number 1 cylinder Top Dead Center (TDC) sensor

Model — Galant
Body VIN — 7
Engine — 3.0L (2972cc) **Cylinders** — 6
Fuel System — Multi-Point Injection
Engine Identifier — 6G72 **VIN** — S

ENGINE CODES

Code	Explanation
11	Oxygen sensor
12	Air Flow Sensor (AFS)
13	Intake Air Temperature Sensor (ATS)
14	Throttle Position Sensor (TPS)
21	Coolant Temperature Sensor (CTS)
22	Crank Angle Sensor (CAS)
23	Number 1 cylinder TDC sensor
24	Vehicle Speed Sensor (VSS) — reed switch
25	Barometric pressure sensor
41	Injector
42	Fuel pump
43	Exhaust Gas Recirculation (EGR)

5 FUEL SYSTEM

Model — Galant
Body VIN — 6
Engine — 2.0L (1997cc)   **Cylinders** — 4
Fuel System — Multi-Point Injection
Engine Identifier — 4G63   **VIN** — V (SOHC)

ENGINE CODES

Code	Explanation
11	Oxygen sensor
12	Air Flow Sensor (AFS)
13	Intake air temperature sensor
14	Throttle Position Sensor (TPS)
15	Motor position sensor
21	Coolant Temperature Sensor (CTS)
22	Crank angle sensor
23	No. 1 cylinder TDC sensor
24	Vehicle speed sensor
25	Barometric pressure sensor

Model — Galant
Body VIN — 6
Engine — 2.0L (1997cc)   **Cylinders** — 4
Fuel System — Multi-Point Injection
Engine Identifier — 4G63   **VIN** — R (DOHC)

ENGINE CODES

Code	Explanation
11	Oxygen sensor
12	Air Flow Sensor (AFS)
13	Intake Air Temperature Sensor (ATS)
14	Throttle Position Sensor (TPS)
21	Coolant Temperature Sensor (CTS)
22	Crank Angle Sensor (CAS)
23	Number 1 and 4 cylinder TDC sensor
24	Vehicle Speed Sensor (VSS) — reed switch
25	Barometric pressure sensor
41	Injector
42	Fuel pump
43	Exhaust Gas Recirculation (EGR) California
44	Ignition Coil

FUEL SYSTEM 5

Model – Mirage
Body VIN – 4, 6
Engine – 1.5L (1468cc) **Cylinders** – 4
Fuel System – Feedback Carburetor
Engine Identifier – G15B **VIN** – K

ENGINE CODES

Code	Explanation
1	Oxygen sensor
2	Ignition pulse
3	Air Flow Sensor (AFS)
4	Pressure sensor
5	Throttle Position Sensor (TPS)
6	Idle Speed Control (ISC) motor position switch
7	Coolant Temperature Sensor (CTS)

Model – Mirage
Engine – 1.6L (97 cid) 4 cyl
Engine Code – G32B

ECM TROUBLE CODES

Code	Explanation
1	Oxygen sensor
2	Ignition pulse
3	Air Flow Sensor (AFS)
4	Pressure sensor
5	Throttle Position Sensor (TPS)
6	Idle Speed Control (ISC) motor position switch
7	Coolant Temperature Sensor (CTS)

Model – Mirage
Body VIN – 4, 6
Engine – 1.6L (1597cc) **Cylinders** – 4
Fuel System – Multi-Point Fuel Injection
Engine Identifier – G32B **VIN** – F

ENGINE CODES

Code	Explanation
1	Oxygen sensor
2	Ignition pulse
3	Air Flow Sensor (AFS)

5-65

5 FUEL SYSTEM

Code	Explanation
5	Throttle Position Sensor (TPS)
6	Idle Speed Control (ISC) motor position switch
7	Coolant Temperature Sensor (CTS)

Model — Mirage
Body VIN — 4, 6
Engine — 1.5L (1468cc)   **Cylinders** — 4
Fuel System — Multi-Point Injection
Engine Identifier — 4G15   **VIN** — X (SOHC)

ENGINE CODES

Code	Explanation
11	Oxygen sensor
12	Air Flow Sensor (AFS)
13	Intake Air Temperature Sensor (ATS)
14	Throttle Position Sensor (TPS)
15	Motor Position Sensor (TPS)
21	Coolant Temperature Sensor (CTS)
22	Crank Angle Sensor (CAS)
23	Number 1 cylinder TDC sensor
24	Vehicle Speed Sensor (VSS) — reed switch
25	Barometric pressure sensor
41	Injector
42	Fuel pump
43	Exhaust Gas Recirculation (EGR) California

Model — Mirage
Body VIN — 4, 6
Engine — 1.6L (1597cc)   **Cylinders** — 4
Fuel System — Multi-Point Injection
Engine Identifier — 4G61   **VIN** — Z (DOHC)

ENGINE CODES

Code	Explanation
11	Oxygen sensor
12	Air Flow Sensor (AFS)
13	Intake Air Temperature Sensor (ATS)
14	Throttle Position Sensor (TPS)
21	Coolant Temperature Sensor (CTS)

FUEL SYSTEM 5

Code	Explanation
22	Crank Angle Sensor (CAS)
23	Number 1 cylinder TDC sensor
24	Vehicle Speed Sensor (VSS) — reed switch
25	Barometric pressure sensor
31	Detonation sensor
41	Injector
42	Fuel pump
43	Exhaust Gas Recirculation (EGR) California
44	Ignition Coil

Model — Precis
Body VIN — 1
Engine — 1.3L (1298cc) **Cylinders** — 4
Fuel System — Multi-Point Injection
Engine Identifier — G4AJ   **VIN** — M

ENGINE CODES

Code	Explanation
11	Oxygen sensor
12	Air flow sensor
13	Intake Air Temperature Sensor
14	Throttle Position Sensor (TPS)
15	Motor Position Sensor
21	Engine Coolant Temperature Sensor
22	Crank angle sensor
23	No. 1 cylinder TDC Sensor
24	Vehicle speed sensor
25	Barometric pressure sensor
41	Injector
42	Fuel pump
43	EGR-California

Model — Precis
Body VIN — 1
Engine — 1.5L (1468cc) **Cylinders** — 4
Fuel System — Multi-Point Injection
Engine Identifier — G4AM   **VIN** — J

ENGINE CODES

Code	Explanation
11	Oxygen sensor

5–67

5 FUEL SYSTEM

Code	Explanation
12	Air flow sensor
13	Intake Air Temperature Sensor
14	Throttle Position Sensor (TPS)
15	Motor Position Sensor
21	Engine Coolant Temperature Sensor
22	Crank angle sensor
23	No. 1 cylinder TDC Sensor
24	Vehicle speed sensor
25	Barometric pressure sensor
41	Injector
42	Fuel pump
43	EGR-California

Model — Sigma
Body VIN — 7
Engine — 3.0L (2972cc) **Cylinders** — 4
Fuel System — Multi-Point Injection
Engine Identifier — 6G72 **VIN** — S

ENGINE CODES

Code	Explanation
11	Oxygen sensor
12	Air Flow Sensor (AFS)
13	Intake Air Temperature Sensor (ATS)
14	Throttle Position Sensor (TPS)
21	Engine Coolant Temperature Sensor (CTS)
22	Crank Angle Sensor (CAS)
23	Number 1 cylinder TDC sensor
24	Vehicle Speed Sensor (VSS) — reed switch
25	Barometric pressure sensor
41	Injector
42	Fuel pump
43	Exhaust Gas Recirculation (EGR) California

Model — Starion
Engine — 2.6L (156 cid) Turbo 4 cyl
Engine Code — G54B

ECM TROUBLE CODES

Code	Explanation
1	Oxygen sensor and computer

FUEL SYSTEM 5

Code	Explanation
2	Ignition pulse
3	Air Flow Sensor (AFS)
4	Pressure sensor
5	Throttle Position Sensor (TPS)
6	Idle Speed Control (ISC) motor position switch
7	Coolant Temperature Sensor (CTS)
8	Vehicle Speed Sensor (VSS)

Model — Starion
Body VIN — 4
Engine — 2.6L Turbo (2555cc)   **Cylinders** — 4
Fuel System — Multi-Point Fuel Injection
Engine Identifier — G54B   **VIN** — H, N

ENGINE CODES

Code	Explanation
1	Oxygen sensor and computer
2	Ignition pulse
3	Air Flow Sensor (AFS)
5	Throttle Position Sensor (TPS)
6	Idle Speed Control (ISC) motor position switch
7	Coolant Temperature Sensor (CTS)

Model — Starion
Body VIN — 4
Engine — 2.6L (2555 cc) **Cylinders** — 4
Fuel System — Electronic Controlled Injection
Engine Identifier — G54B   **VIN** — N

ENGINE CODES

Code	Explanation
1	Oxygen sensor
2	Ignition pulse
3	Air flow sensor
5	Throttle Position Sensor (TPS)
6	ISC motor position sensor
7	Engine Coolant Temperature Sensor (CTS)

5 FUEL SYSTEM

Model — Tredia
Body VIN — 6
Engine — 1.8L (1795cc)   **Cylinders** — 4
Fuel System — Multi-Point Fuel Injection
Engine Identifier — G62B   **VIN** — G

ENGINE CODES

Code	Explanation
1	Oxygen sensor and computer
2	Ignition pulse
3	Air Flow Sensor (AFS)
4	Pressure sensor
5	Throttle Position Sensor (TPS)
6	Idle Speed Control (ISC) motor position switch
7	Coolant Temperature Sensor (CTS)

Model — Tredia
Body VIN — 6
Engine — 1.8L (1795cc)   **Cylinders** — 4
Fuel System — Electronic Controlled Injection
Engine Identifier — G62B   **VIN** — G

ENGINE CODES

Code	Explanation
1	Oxygen sensor and computer
2	Ignition pulse
3	Air Flow Sensor (AFS)
4	Pressure sensor
5	Throttle Position Sensor (TPS)
6	Idle Speed Control (ISC) motor position switch
7	Coolant Temperature Sensor (CTS)

FUEL SYSTEM 5

Model – Tredia
Body VIN – 6
Engine – 2.0L (1997cc)   **Cylinders** – 4
Fuel System – Feedback Carburetor
Engine Identifier – G63B   **VIN** – D

ENGINE CODES

Code	Explanation
1	Oxygen sensor and computer
2	Ignition pulse
3	Air Flow Sensor (AFS)
4	Pressure sensor
5	Throttle Position Sensor (TPS)
6	Idle Speed Control (ISC) motor position switch
7	Coolant Temperature Sensor (CTS)

Model – Tredia
Engine – 2.0L (122 cid) 4 cyl
Engine Code – G63B

ECM TROUBLE CODES

Code	Explanation
1	Oxygen sensor and computer
2	Ignition pulse
3	Air Flow Sensor (AFS)
4	Pressure sensor
5	Throttle Position Sensor (TPS)
6	Idle Speed Control (ISC) motor position switch
7	Coolant Temperature Sensor (CTS)
8	Vehicle Speed Sensor (VSS)

5 FUEL SYSTEM

Model — Truck
Body VIN — 4, 5 and 9
Engine — 2.4L (2350cc)   **Cylinders** — 4
Fuel System — Multi-Point Injection
Engine Identifier — 4G64   **VIN** — W

ENGINE CODES

Code	Explanation
11	Oxygen sensor
12	Air Flow Sensor (AFS)
13	Intake Air Temperature Sensor (ATS)
14	Throttle Position Sensor (TPS)
15	Motor Position Sensor
21	Engine Coolant Temperature Sensor (CTS)
22	Crank Angle Sensor (CAS)
23	Top dead center No. 1 cylinder sensor
24	Vehicle Speed Sensor (VSS) — reed switch
25	Barometric pressure sensor
41	Injector
42	Fuel pump
43	EGR- California

Model — Montero
Body VIN — 3
Engine — 3.0L (2972cc)   **Cylinders** — 6
Fuel System — Multi-Point Injection
Engine Identifier — 6G72   **VIN** — S

ENGINE CODES

Code	Explanation
11	Oxygen sensor
12	Air Flow Sensor (AFS)
13	Intake air temperature sensor
14	Throttle Position Sensor (TPS)
21	Engine Coolant Temperature Sensor (CTS)
22	Crank angle sensor
23	Top Dead Center (TDC) sensor
24	Vehicle speed sensor — Reed switch
25	Barometric pressure sensor
41	Injector
42	Fuel Pump
43	Exhaust Gas Recirculation (EGR) California

Chassis Electrical 6

QUICK REFERENCE INDEX

Circuit Protection	6-79
Heating and Air Conditioning	6-3
Instruments and Switches	6-50
Lighting	6-75
Troubleshooting Charts	6-82
Understanding Electrical Systems	6-2
Windshield Wipers	6-46

GENERAL INDEX

Air conditioning		Headlights	6-75	Switch	6-73
Blower	6-3	Headlight switch	6-72	Safety precautions	6-2
Control panel	6-25	Heater		Speedometer cable	6-74
Evaporator	6-33	Blower	6-3	Switches	
Blower motor	6-3	Control panel	6-25	Headlight	6-72
Chassis electrical system		Core	6-10	Turn signal	6-72
Circuit protection	6-79	Instrument cluster	6-50	Rear window wiper	6-73
Fuses	6-79	Instrument panel		Windshield wiper	6-72
Fusible links	6-79	Cluster	6-50	Test equipment	6-2
Heater and air conditioning	6-3	Console	6-68	Troubleshooting	
Instrument cluster	6-50	Panel	6-53	Gauges	6-84
Lighting	6-75	Radio	6-42	Heater	6-85
Troubleshooting	6-83	Speedometer cable	6-74	Lights	6-83
Windshield wipers	6-46	Lighting		Turn signals and flashers	6-82
Console	6-68	Headlights	6-75	Windshield wipers	6-85
Combination switch	6-72	Signal and marker lights	6-78	Windshield wipers	
Control panel	6-25	Marker lights	6-78	Arm and blade	6-46
Evaporator	6-33	Radio	6-42	Linkage and motor	6-47
Fuses	6-79	Rear window wipers		Switch	6-72
Fusible links	6-79	Motor	6-49	Troubleshooting	6-85

6 CHASSIS ELECTRICAL

UNDERSTANDING BASIC ELECTRICITY

For any electrical system to operate, it must make a complete circuit. This simply means that the power flow from the battery must make a complete circle. When an electrical component is operating, power flows from the battery to the component, passes through the component causing it to perform its function (lighting a light bulb), and then returns to the battery through the ground of the circuit. This ground is usually (but not always) the metal part of the car or truck on which the electrical component is mounted.

Perhaps the easiest was to visualize this is to think of connecting a light bulb with two wires attached to it to the battery. If one of the two wires attached to the light built were attached to the negative post of the battery and the other were attached to the positive post of the battery, you would have a complete circuit. Current from the battery would flow to the light bulb, causing it to light, and return to the negative post of the battery.

The normal automotive circuit differs from this simple example in two ways. First, instead of having a return wire from the bulb to the battery, the light bulb often returns the current to the battery through the chassis of the vehicle. Since the negative battery cable is attached to the chassis and the chassis is made of electrically conductive metal, the chassis of the vehicle can serve as ground wire to complete the circuit. Secondly, most automotive circuits contain switches to turn components on and off as required.

Every complete circuit must include a component or load which is using the power from the power source. If you were to disconnect the light bulb from the wires and touch the two wires together (don't do this), the power source, in this case the battery, would attempt to deliver all its power instantly to the opposite pole. This tremendous current flow causes extreme heat to build within the system, often melting the insulation on the wiring.

Because grounding a wire from a power source makes a complete circuit but without a load to use it, this phenomenon is called a short circuit. Common causes are: broken insulation (exposing the metal wire to a metal part of the car), or an internally shorted switch. Water leaking into normally sealed components can also conduct electricity to undesired locations.

Some electrical components which require a large amount of current to operate (blower motor, rear defogger, headlights, etc.) have a relay in their circuit. Since these circuits carry a large amount of current, the thickness of the wire in the circuit (gauge size) is also greater. If this large wire were connected from the component to the control switch on the instrument panel, and then back to the component, a voltage drop would occur in the circuit. To prevent this potential drop in voltage, an electromagnetic switch (relay) is used.

The large wires in the circuit are connected from the battery to one side of the relay, and from the opposite side of the relay to the component. The relay is normally open, preventing current from passing through the circuit. An additional, smaller, wire is connected from the relay to the control switch for the circuit. When the control switch is turned on, it grounds the smaller wire from the relay and completes the circuit. This closes the relay and allows current to flow from the battery to the component. The horn, headlight, and starter circuits are three which use relays.

It is possible for larger surges of current to pass through the electrical system of your car or truck. If this surge of current were to reach an electrical component, it could burn it out. To prevent this, fuses, circuit breakers and fusible links are connected into the current supply wires. These are nothing more than pre-planned weak spots within the system. Too much current WILL damage something; if the something is easy to find and easy to replace, major components and wiring will be protected. Fuses are designed to pass the amount of current (amperes) shown on the fuse. If a current flow develops which exceeds the rating, the fuse element separates (blows), causing an open circuit and stopping the current flow.

A circuit breaker is essentially a self-repairing fuse. The circuit breaker opens the circuit the same way a fuse does. However, when either the short is removed from the circuit or the surge subsides, the circuit breaker resets itself and does not have to be replaced as a fuse does.

A fuse link or main link is a wire that acts as a fuse. It is normally connected between the starter relay and the main wiring harness or the positive battery cable and the main wiring harness under the hood. The fuse link (if installed) protects all the chassis electrical components by cutting off the main power supply if necessary. If the car shows absolutely no electrical response when the key is turned, the main link is suspect. Check it first; it's rare for a battery to get so low that something won't work.

Troubleshooting and Diagnosis

Electrical problems generally fall into one of three areas:
1. The component is not receiving current.
2. The component has failed and cannot use the current properly, if at all.
3. The component is not properly grounded.

The electrical system can be checked with a test light and a jumper wire. A test light is a device that looks like a pointed screwdriver with a wire attached to it and has a light bulb in its handle. A jumper wire is a piece of insulated wire with an alligator clip attached to each end.

It should be noted that a test light will only show that voltage is present; it will not indicate the amount of voltage. Certain components will only function if a minumum voltage is supplied. If a more than a "yes or no" answer is required during diagnosis, use a voltmeter. If this must be purchased, purchase a multimeter or volt/ohmmeter (VOM). These reasonably inexpensive tools include several scales for both AC (household) and DC volts as well as an ohmmeter for checking resistance and continuity.

If a component is not working, you must follow a systematic plan to determine which of the three causes is the villain.
1. Turn on the switch that controls the inoperable component.
2. Disconnect the power supply wire from the component.
3. Attach the ground wire on the test light to a good metal ground.
4. Touch the probe end of the test light to the end of the power supply wire that was disconnected from the component. If the component is receiving current, the test light will go on.

Note: Some components work only when the ignition switch is turned ON.

If the test light does not go on, then the problem is in the circuit between the battery and the component. This includes all the switches, fuses and relays in the system. Follow the wire that runs back to the battery. The problem is an open circuit between the battery and the component. If the fuse is blown and, when replaced, immediately blows again, there is a short circuit in the system which must be located and repaired. If there is a switch in the system, bypass it with a jumper wire. This is done by connecting one end of the jumper wire to the power supply wire into the switch and the other end of the jumper wire to the wire coming out of the switch. If the test light lights with the jumper wire installed, the switch or whatever was bypassed is defective.

NOTE: Never substitute the jumper wire for the component, since a load is required to use the power from the battery.

CHASSIS ELECTRICAL 6

5. If the bulb in the test light goes on, then the current is getting to the component that is not working. This eliminates the first of the three possible causes. Connect the power supply wire and connect a jumper wire from the component to a good metal ground. Do this with the switch which controls the component turned on, and also the ignition switch turned on if it is required for the component to work. If the component works with the jumper wire installed, then it has a bad ground. This is usually caused by the metal area on which the component mounts to the chassis being coated with some type of foreign matter.

6. If neither test located the source of the trouble, then the component itself is defective. Remember that for any electrical system to work, all connections must be clean and tight.

HEATING AND AIR CONDITIONING

Heater Blower Motor

REMOVAL AND INSTALLATION

Cordia and Tredia

1. With the ignition switch **OFF** and the key removed, unscrew the single attaching bolt and remove the lower cover from the right side of the instrument panel.
2. Remove the mounting screws at the front and remove the glovebox.
3. Remove the cowl side trim (kick panel).
4. Disconnect the air selector control wire. Disconnect the discharge duct at the blower.
5. Disconnect the electrical connector. Remove the four mounting bolts and remove the blower assembly.
6. Once the fan housing is removed to the workbench, the motor and fan assembly may be released from the case by removing 3 small screws. Check the inside of the case carefully; any debris can snag the fan and cause noise or poor airflow. It's not uncommon to find leaves, paper or animal debris within the fan case.

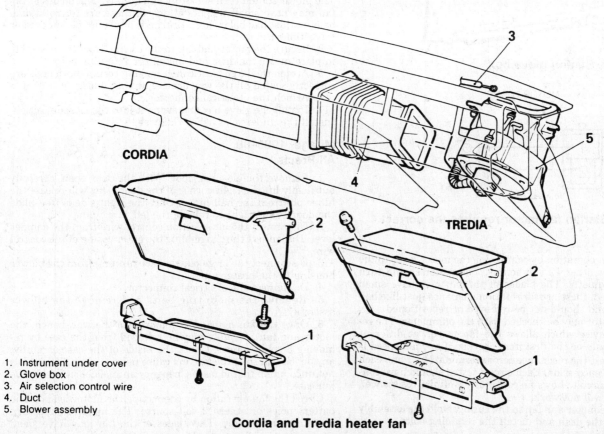

1. Instrument under cover
2. Glove box
3. Air selection control wire
4. Duct
5. Blower assembly

Cordia and Tredia heater fan

6 CHASSIS ELECTRICAL

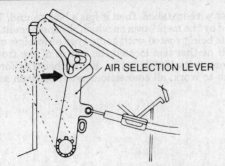

The lever must be pulled gently to reinstall the cable

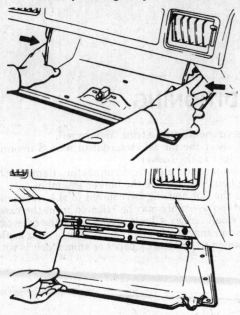

Removing the Starion glove box

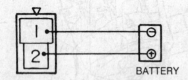

Testing the Starion fan motor requires the correct polarity

Check the fan operation by connecting the motor leads to the battery poles or other 12 volt source. The motor should turn smoothly and quietly. The blades of the fan may have small weights attached; these are for rotational balance (just like balancing a tire) and should not be removed or repositioned.

The fan resistor may be checked with the ohmmeter. The resistor is the device which allows the fan to run at different speeds and should be the first item checked if one fan speed is not working. Test the resistance across terminal No. 1 and each of the others in succession. Each circuit tested should have continuity; if any circuit shows no continuity (infinite resistance), that fan speed will not work.

7. Secure the motor and fan to the casing. Hold the assembly in place under the dash and install the retaining bolts.

8. Attach the electrical connector.
9. On the heater control panel on the dashboard, select the recirculate position. Gently pull the air selection lever toward the inside of the car and attach the air selector control wire. Secure the cable case in place with its clip.
10. Install the large discharge duct. Make certain it is correctly seated all the way around at both ends. The penalty is cold feet (no heat) if the duct leaks air.
11. Install the right side kick panel, the glove box and the under-dash trim.

Starion

1. With the ignition switch **OFF** and the key removed, remove the under cover from the bottom of the dash panel below the glovebox.
2. Open the glovebox door and pull the glove box forward while pressing inward on both sides of the glovebox. This allows the door to drop downward beyond its normal travel.
3. Remove the screws attaching the glovebox door hinge to the bottom of the dashboard and then remove the glovebox assembly.
4. Disconnect the blower electrical connector.
5. Disconnect the large plastic duct running from the blower housing to the heater unit in the center of the dash.
6. Remove the mounting bolts and remove the blower assembly.
7. Once the fan housing is removed to the workbench, the motor and fan assembly may be released from the case by removing 3 small screws. Check the inside of the case carefully; any debris can snag the fan and cause noise or poor airflow. It's not uncommon to find leaves, paper or animal debris within the fan case.

Check the fan operation by connecting the motor terminals to the battery poles so that terminal No. 1 is connected to ground (negative) and terminal No. 2 is connected to the positive pole. The motor should turn smoothly and quietly. The blades of the fan may have small weights attached; these are for rotational balance (just like balancing a tire) and should not be removed or repositioned.

8. Secure the motor and fan to the casing. Hold the assembly in place under the dash and install the retaining bolts.
9. Attach the large plastic duct making certain both ends are correctly seated all the way around at both ends.
10. Attach the electrical connector.
11. Install the glove box and swing it into the closed position. Install the undercover below the dash.

Mirage, 1985–88
All Precis

1. Remove the glovebox door. With the door open, carefully but firmly lift the right corner of the door. This will release the hinge pin from the half-hinge. Once the right side is free, slide the door to the right to release the left hinge pin.
2. Disconnect the air selection control wire from the damper lever. Disconnect the clip holding the outer casing of the control wire.
3. Disconnect the large plastic duct running from the blower housing to the heater unit.
4. Disconnect the electrical connector.
5. Remove the mounting bolts and remove the blower assembly.
6. Once the fan housing is removed to the workbench, the motor and fan assembly may be released from the case by removing 3 small screws. Check the inside of the case carefully; any debris can snag the fan and cause noise or poor airflow. It's not uncommon to find leaves, paper or animal debris within the fan case.

Check the fan operation by connecting the motor leads to the battery poles or other 12 volt source. The motor should turn smoothly and quietly. The blades of the fan may have small

CHASSIS ELECTRICAL 6

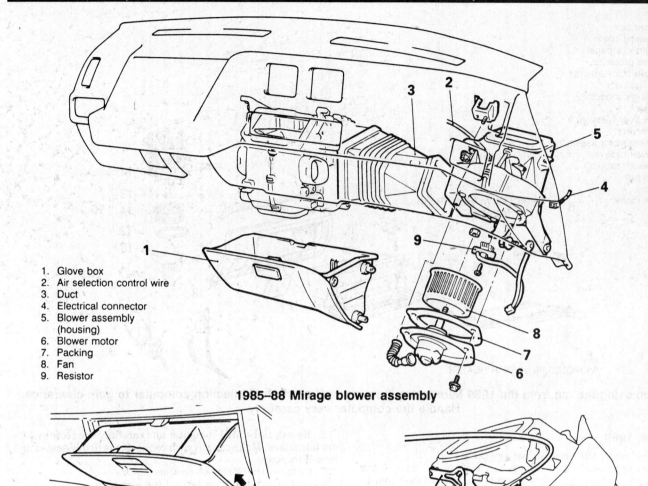

1. Glove box
2. Air selection control wire
3. Duct
4. Electrical connector
5. Blower assembly (housing)
6. Blower motor
7. Packing
8. Fan
9. Resistor

1985–88 Mirage blower assembly

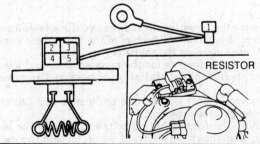

Glove box removal is easy once the right side hinge is released. Don't break the plastic

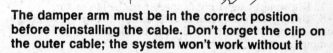

The damper arm must be in the correct position before reinstalling the cable. Don't forget the clip on the outer cable; the system won't work without it

Terminal connections	Resistance
1 – 2 (M_1)	Approx. 1.18 Ω
1 – 3 (M_2)	Approx. 0.41 Ω
1 – 4 (Low)	Approx. 1.93 Ω
1 – 5 (High)	Approx. 0 Ω

If one or more fan speeds don't work, check the fan resistor.

weights attached; these are for rotational balance (just like balancing a tire) and should not be removed or repositioned.

7. The fan resistor may be checked with the ohmmeter. The resistor is the device which allows the fan to run at different speeds and should be the first item checked if one fan speed is not working. Test the resistance across terminal No. 1 and each of the others in succession. Each circuit tested should have continuity; (the correct resistance is shown in the chart) if any circuit shows no continuity (infinite resistance), that fan speed will not work.

8. Install the resistor if it was removed. Secure the motor and fan to the casing. Hold the assembly in place under the dash and install the retaining bolts.

9. Attach the electrical connector.

10. On the heater control panel on the dashboard, select the recirculate position. Gently pull the air selection lever toward the inside of the car and attach the air selector control wire. Secure the cable case in place with its clip.

11. Install the large discharge duct. Make certain it is correctly seated all the way around at both ends. The penalty is cold feet (no heat) if the duct leaks air.

6 CHASSIS ELECTRICAL

1. Glove box
2. Speaker cover
3. Right kick panel
4. Knee protector
5. Glove box frame rail
6. Heater duct
7. Wiring connector
8. Hose
9. MPI (fuel injection) computer
10. Blower motor assembly
11. Blower case
12. Gasket (packing)
13. Fan nut
14. Fan
15. Fan motor

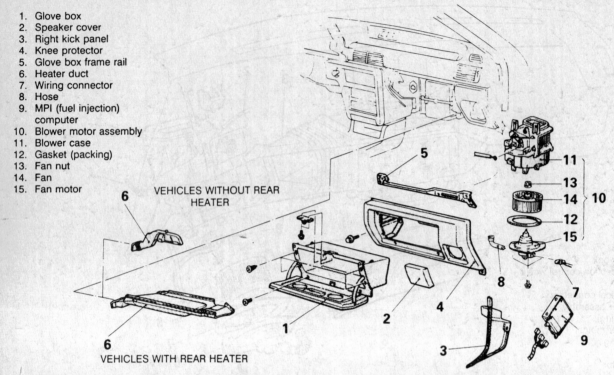

Removing the fan from the 1989 Mirage requires removal of the fuel injection computer to gain clearance. Handle the computer very carefully

Mirage, 1989

1. Disconnect the negative battery cable.
2. Remove the screws holding the glovebox and door to the dashboard. Slide the glovebox assembly out of the dashboard.
3. Remove the right side kick panel (cowl side trim).
4. Remove the right speaker cover and remove the knee protector from the dashboard.
5. Remove the glove box frame rail from the dash.
6. Remove the heater duct behind the glovebox. It may be one of two shapes, depending on the presence of a rear heater.
7. Disconnect the wiring connector to the fan motor.
8. Using great care, label and disconnect the wiring running to the computer box inside the right kick panel. Remove the computer when it is disconnected.

WARNING: This is the main fuel injection (MPI) computer. Handle it very gently.

9. Remove the retaining bolts and remove the blower assembly.
10. Once the fan housing is removed to the workbench, the motor and fan assembly may be released from the case by removing 3 small screws. Check the inside of the case carefully; any debris can snag the fan and cause noise or poor airflow. It's not uncommon to find leaves, paper or animal debris within the fan case.

Check the fan operation by connecting the motor terminals to the battery poles
so that terminal No. 1 is connected to ground (negative) and terminal No.2 is connected to the positive pole. The motor should turn smoothly and quietly. The blades of the fan may have small weights attached; these are for rotational balance (just like balancing a tire) and should not be removed or repositioned.

11. Inspect the gasket on the motor housing; replace it if it is cracked or damaged. Secure the motor and fan to the casing. Hold the assembly in place under the dash and install the retaining bolts.

12. Reinstall the MPI computer and carefully connect the wiring harnesses. Make certain each connector is firmly seated and locked in place.
13. Connect the wiring to the fan motor.
14. Install the ductwork behind the glove box.
15. Install the glove box frame rail, the kneepad and the speaker cover.
16. Install the glove box assembly.
17. Connect the negative battery cable.

Galant, 1985–1988
Sigma

1. Remove the screw covers and remove the screws from the under cover (located at the bottom of the instrument panel on the right side). Remove the under cover.
2. Remove the stoppers for the glovebox door and lower it. Remove the glovebox installation screws and remove the glovebox.
3. Remove the duct leading into the blower unit, accessible through the glovebox opening.
4. Remove the four mounting bolts and remove the blower assembly. Disconnect the electrical connector and the vacuum hose before pulling the unit all the way out. Note that the lower right retaining bracket should be removed from the body as well as disconnected at the motor.
5. Once the fan housing is removed to the workbench, the motor and fan assembly may be released from the case by removing 3 small screws. Check the inside of the case carefully; any debris can snag the fan and cause noise or poor airflow. It's not uncommon to find leaves, paper or animal debris within the fan case.

Check the fan operation by connecting the motor leads to the battery poles or other 12 volt source. The motor should turn smoothly and quietly. The blades of the fan may have small weights attached; these are for rotational balance (just like balancing a tire) and should not be removed or repositioned.

CHASSIS ELECTRICAL 6

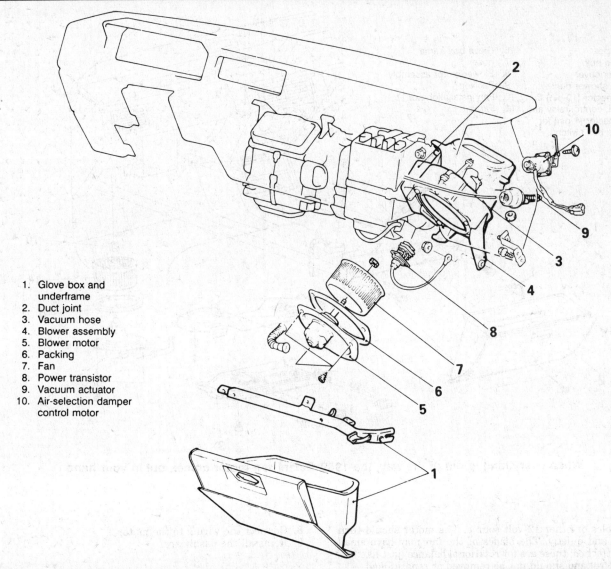

1. Glove box and underframe
2. Duct joint
3. Vacuum hose
4. Blower assembly
5. Blower motor
6. Packing
7. Fan
8. Power transistor
9. Vacuum actuator
10. Air-selection damper control motor

1985–88 Galant and Sigma fan assembly. All the wiring and vacuum lines must be correctly installed for the system to work

6. Inspect the gasket (packing) under the motor and replace it if cracked or damaged. Reinstall the fan and motor to the case; hold the case in position under the dash and install the retaining bolts. Connect the vacuum hose and electrical connector.
7. Install the ductwork, making certain it is properly seated all the way around.
8. Install the glovebox and and underframe assemblies. Don't forget the screw covers.

Galant, 1989

1. Remove the glove box stopper.
2. Swing the glove box door open all the way and remove the bottom retaining screws. Remove the glovebox.
3. Remove the dash under cover. Note that some of the screws and retainers are concealed behind small covers which must be removed.

4. Remove the heater duct for the passenger's feet.
5. Carefully disconnect the 10-pin connector running to the back of the glove box frame. This is the MPI relay; treat it gently. Disconnect the single wire (glove box switch) running to the back of the glove box frame.
6. Remove the four bolts holding the glove box frame and remove the frame.
7. Disconnect the small air hose running from the fan motor to the fan housing.
8. Disconnect the electrical connector for the fan motor.
9. Remove the three small bolts holding the motor to the housing and remove the motor and fan.
10. Check the inside of the case carefully; any debris can snag the fan and cause noise or poor airflow. It's not uncommon to find leaves, paper or animal debris within the fan case.
  Check the fan operation by connecting the motor leads to the

6-7

6 CHASSIS ELECTRICAL

1. Stopper
2. Glove box
3. Under cover
4. Foot shower duct
5. Connector (10-pin) for M.P.I. control relay and connector (1-pin) for glove box switch
6. Glove box frame
7. hose
8. Blower motor assembly
9. Packing
10. Fan installation nut
11. Fan

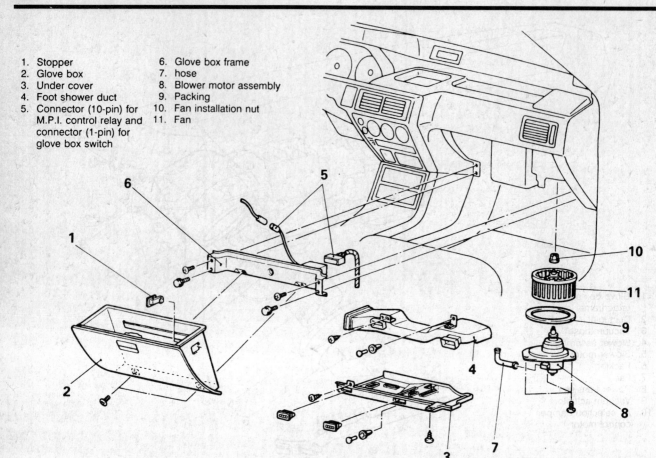

When everything is out of the way, the 1989 Galant fan motor comes out in your hand

battery poles or other 12 volt source. The motor should turn smoothly and quietly. The blades of the fan may have small weights attached; these are for rotational balance (just like balancing a tire) and should not be removed or repositioned.

11. Inspect the gasket (packing) under the motor and replace it if cracked or damaged. Reinstall the fan and motor to the case and install the retaining bolts.
12. Connect the air hose and electrical connector.
13. Install the glove box frame and connect both the 10-pin and single pin connectors properly.
14. Install the heater duct
15. Install the undercover, taking care to insure it is in place and all the fasteners are secure.
16. Install the glove box and its stopper.

Pickup, 1983–86

1. Remove the dashboard, following instruction given later in this section.
2. Disconnect the electric harness to the fan motor.
3. Remove the 3 bolts holding the fan assembly.
4. Remove the fan and motor assembly from the blower housing.
5. Remove the fan from the motor shaft.
6. Reassemble fan on the motor shaft and install it in the blower housing.
7. Tighten the mounting screws.

8. Connect the wiring to the motor.
9. Reinstall the dashboard.

Pickup, 1987–89

1. Remove the glove box stopper and unbolt the glove box frame.
2. Remove the glove box from the dashboard.
3. Disconnect the air selection control wire.
4. Remove the large center duct running from the blower to the heater unit.
5. Disconnect the electrical harness from the motor. If necessary (for access), disconnect the ground connector near the motor.
6. Remove the three retaining bolts and remove the blower assembly. The fan may be removed from the shaft when the motor is clear of the housing.
7. When reinstalling, attach the fan to the shaft and install the assembly into the housing.
8. Tighten the mounting bolts and connect the electrical connector. Reconnect the ground harness if it was removed.
9. Install the large center duct.
10. Move the dashboard air selection lever to the "FRESH" position. Push the air selection damper lever against its stop (clockwise motion) and connect the air selection cable to the lever. Secure the outer part of the cable in place with the cable clip.
11. Install the glove box and stoppers.

CHASSIS ELECTRICAL 6

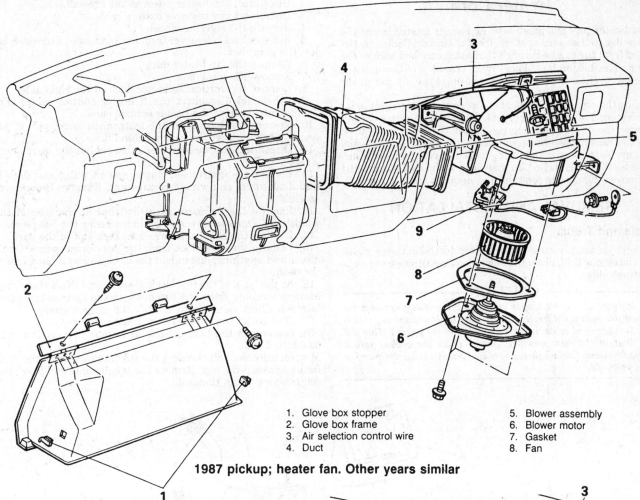

1. Glove box stopper
2. Glove box frame
3. Air selection control wire
4. Duct
5. Blower assembly
6. Blower motor
7. Gasket
8. Fan

1987 pickup; heater fan. Other years similar

Montero

NOTE: The order of steps may vary slightly for 1983–86 vehicles.

1. Remove the lap heater duct under the rightside dashboard.
2. Remove the glove box stoppers and the glove box assembly.
3. Disconnect the air selection control wire.
4. Remove the large center duct running from the blower to the heater unit. (Not necessary on early Monteros)
5. Disconnect the electrical harness from the motor. If necessary for access, disconnect the ground connector near the motor. The resistor may be removed from the motor as necessary.
6. Disconnect the small breather hose from the motor. Remove the three retaining bolts and remove the blower assembly. The fan may be removed from the shaft when the motor is clear of the housing.
7. When reinstalling, attach the fan to the shaft and install the assembly into the housing. Make sure the gasket is seated correctly.
8. Tighten the mounting bolts and connect the electrical connector. Reconnect the ground harness if it was removed and secure the resistor if needed.
9. Install the large center duct if it was removed.
10. Move the dashboard air selection lever to the "RECIRCULATE" position. Push the air selection damper lever against its stop (counterclockwise motion) and connect the air selection cable to the lever. Secure the outer part of the cable in place with the cable clip.
11. Install the glove box and stoppers.
12. Install the lap heater duct.

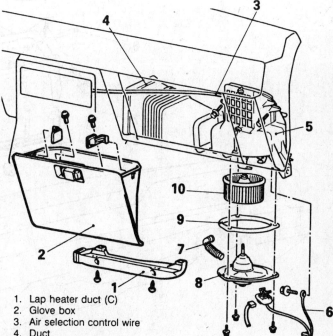

1. Lap heater duct (C)
2. Glove box
3. Air selection control wire
4. Duct
5. Blower assembly
6. Resistor
7. Hose
8. Blower motor
9. Packing
10. Fan

Montero heater fan

6-9

6 CHASSIS ELECTRICAL

Heater Core

The heater core is a small heat exchanger located inside the car, similar to the radiator at the front of the car. Coolant is circulated from the engine through the heater core and back to the engine. The heater fan blows fresh, outside air through the heater core; the air is heated and sent on to the interior of the car.

About the only time the heater core will need removal is for replacement due to clogging or leaking. Thankfully, this doesn't happpen too often; removing the heater core is a major task. Some are easier than others, but all require working in unusual positions inside the car and fitting tools into very cramped quarters behind the dashboard. In many cases, the dashboard must be removed during the procedure — another major project.

REMOVAL AND INSTALLATION

Cordia and Tredia

1. Set the temperature control lever to the extreme right. With the engine cold, drain the engine coolant through the radiator drain plug.

----- CAUTION -----

When draining the coolant, keep in mind that cats and dogs are attracted by the ethylene glycol antifreeze, and are quite likely to drink any that is left in an uncovered container or in puddles on the ground. This will prove fatal in sufficient quantity. Always drain the coolant into a sealable container. Coolant should be reused unless it is contaminated or several years old.

2. Disconnect the heater hoses at the firewall.
3. Disconnect the negative battery cable.
4. Remove the steering wheel.
5. Remove the undercover from the right side of the dash below the glove box.
6. Remove the lap heater duct.
7. Remove the glove box.
8. Remove the vertical defroster duct on the right side.
9. Label and disconnect the 3 heater control cables. For Tredia only, remove the heater control panel.
10. Remove the screws for the instrument hood, pull it forward and disconnect the wiring behind it.
11. Disconnect the speedometer cable behind the instrument cluster.
12. Remove the instrument cluster screws, pull it out slightly and disconnect the wiring connectors. Remove the cluster assembly.
13. Remove the defroster garnish (trim) at the base of the windshield. One center screw is hidden below the left piece.
14. Remove the small side covers on each end of the dash.
15. Remove the fuse block from the instrument panel. Don't disconnect anything, just unbolt the block and let it hang free of the dash.
16. At the back of the fuseblock, disconnect ONLY the wiring harness running from the fuseblock to the instrument panel harness. Check carefully to identify the correct wires.
17. Disconnect the wiring at the heater fan case.
18. Disconnect the floor console and audio bracket from the dashboard.
19. Identify the bolts holding the steering column to the dash frame and/or pedal box. Remove these bolts only and lower the column away from the dash.

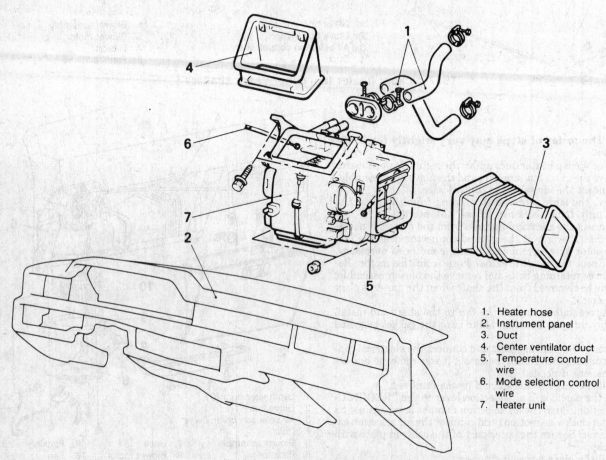

1. Heater hose
2. Instrument panel
3. Duct
4. Center ventilator duct
5. Temperature control wire
6. Mode selection control wire
7. Heater unit

Removing the Cordia and Tredia heater requires removal of the dashboard

6-10

CHASSIS ELECTRICAL 6

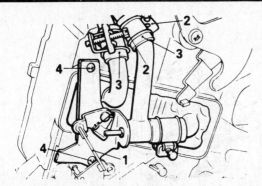

Before removing the heater core, the water valve link (1), hose clamps (2) and hose (3), and retaining screws (4) must be removed

20. Disconnect the antenna feeder.
21. Remove the bolts holding the dashboard. With the help of an assistant, remove the dash from the car. Note that it is coming out with many items still attached. Do not snag wires or cables during removal.
22. Remove the large duct between the blower assembly and the heater unit.
23. Remove the center (or upper) ventilator duct.
24. Disconnect the temperature control cable on the heater unit.
25. Disconnect the mode selection control wire.
26. Remove the retaining nuts and bolts and remove the heater unit from the car.
27. With the unit clear of the car, disconnect the link at the water valve. Loosen the hose clamps on the small joint hoses and remove the hoses.
28. Remove the retaining screw; the heater core may be removed from the case. Although the core may be cleaned and/or repaired in the same fashion as a radiator, replacement with a new unit is highly recommended. While the unit is apart, check the small joint hoses for any sign of hardening or cracking. Replace any that fail inspection. Check the water valve also; it should move smoothly and show no sign of leakage or seepage.

To install:

29. Reassemble the heater unit with the core secured by the retaining screw and the valve and connecting hoses properly secured with new clamps. Double check clamp position and tightness; some will be difficult to reach after reinstallation.
30. Reinstall the small link bar for the water control valve.
31. Place the case into the car and install the retaining nuts and bolts.
32. Install the center (upper) duct and the large duct to the blower housing.
33. Place the dash in position and insert two or three bolts to hold it in place. Insert the rest of the hardware and hand tighten it snug; check the fit of the dash and position it correctly before tightening the bolts.
34. Before continuing, make certain all the wires and cables are routed correctly. Attend to any that may interfere with other components.
35. Install the antenna feeder cable.
36. Raise the steering column and attach it to the mounts.
37. Connect the console and audio mounts to the dash.
38. Install the connectors for the heater fan. Connect the main instrument harness(es) to the fusebox.
39. Install the fusebox and tighten the bolts.
40. Install the dash side panels.
41. Install the defroster garnish.
42. Fit the instrument cluster into position and connect the speedometer cable and instrument wiring. Install the cluster and tighten the screws.

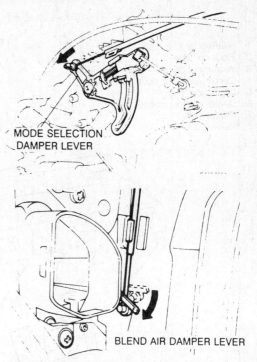

The correct placement of the function levers and the dashboard control must be observed when connecting the cable

43. Position the hood, connect its wiring and install it with the retaining screws.
44. For Tredia only, install the heater control panel and route the cables correctly. Connect the 3 heater control wires and adjust each.
45. Install the defroster duct.
46. Install the glove box.
47. Install the lap heater duct and the undercover for the dash.
48. Install the steering wheel.
49. When the heater control panel is reinstalled:
   a. Connect the mode selection control wire. Set the mode selection lever to the extreme left position. Gently press the mode selection damper lever inward and connect the connect the control wire to the lever. Secure the outer part of the cable in the clip.
   b. Connect the temperature control wire by moving the selector lever to the leftmost (coolest) position. Press the blend air damper lever downward all the way and connect the control wire. Secure the outer part of the cable in the clip.
   c. Carefully move each lever on the controller through its complete range of motion, checking for appropriate matching motion at the heater unit. Make cable adjustments by repositioning the cable case in the spring clip. Make certain that the arms (levers) on the heater unit can travel through the correct range of movement and are not restricted by cable tension.
50. Connect the heater hoses to the heater pipes at the firewall.
51. Check that the draincock is closed. Refill the cooling system.
52. Start the engine and allow it to run at idle. Move the temperature control to its hottest position and check the hoses and joints for leakage or moisture.
53. Check the function of the heater controls and confirm that air and temperature functions are performed correctly when the controls are changed.

6 CHASSIS ELECTRICAL

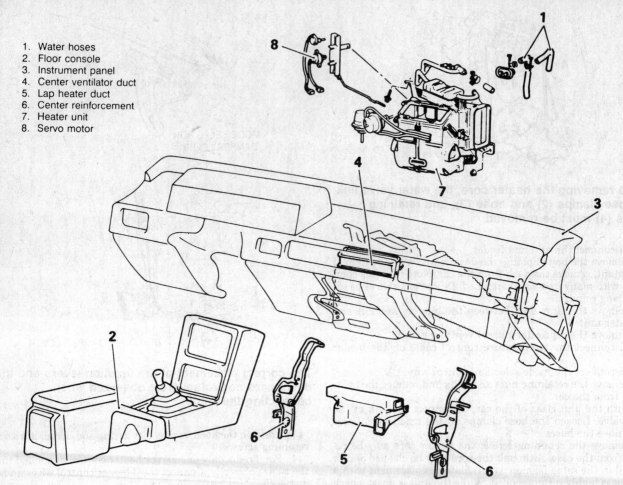

1. Water hoses
2. Floor console
3. Instrument panel
4. Center ventilator duct
5. Lap heater duct
6. Center reinforcement
7. Heater unit
8. Servo motor

Starion heater assembly. Most of the removal is spent extracting the center console and the dashboard. The heater unit must be handled carefully or external components may be damaged

Starion

1. With the engine cold, set the temperature control lever to the extreme right. In the case of the automatic heater system, start the engine and use the temperature change switch to select **MAX HOT**; shut the engine off. Drain the engine coolant through the radiator drain plug; the brief period of engine running should not heat the coolant much, but check the hoses for temperature before draining the system.

CAUTION
When draining the coolant, keep in mind that cats and dogs are attracted by the ethylene glycol antifreeze, and are quite likely to drink any that is left in an uncovered container or in puddles on the ground. This will prove fatal in sufficient quantity. Always drain the coolant into a sealable container. Coolant should be reused unless it is contaminated or several years old.

2. Disconnect the coolant hoses running to the heater pipes at the firewall. Take note of hose location and placement; they are formed to fit exactly as needed.
3. Remove the floor console.
4. Remove the instrument panel.
5. Disconnect the negative battery cable.
6. Remove the steering wheel.
7. Remove the screws holding the hood release handle to the dash. Let it hang away from the dash.
8. Unbolt the fuse box from the dash board.
9. Remove the knee protector (lower dash) on the left side. Some of the bolts are hidden behind covers.
10. Remove the screws from the bottom of the steering column cover. Remove both halves of the cover.
11. Remove the attaching screws for the combination switch on the steering column. Disconnect the wiring harnesses and remove the switch.
12. Remove the instrument hood screws. Pull both edges of the bottom of the hood forward; hold it in this position and lift it obliquely up and out.
13. Disconnect the harness connectors on both sides of the instrument hood.
14. Remove the screws on the bottom and the nuts on the top of the instrument cluster. Pull the bottom edge up and forward to remove it. Disconnect the wiring and cables as it comes free.
15. Remove the console side cover mounting screws. Remove the cover downward while pushing slightly forward.
16. Remove the front and rear consoles.
17. Remove the undercover from the right side dashboard.
18. Remove the glove box. Grasp the sides, press in and pull forward to allow the door to hang down. Remove the 3 upper screws.
19. Remove the ashtray.
20. Carefully remove the heater control bezel (trim).

6-12

CHASSIS ELECTRICAL 6

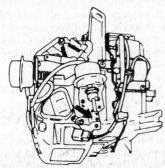

Disconnect the servo motor rod before removing the motor

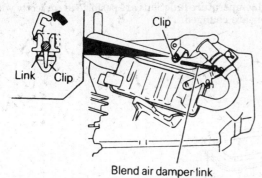

Remove the clip carefully, then disconnect the link

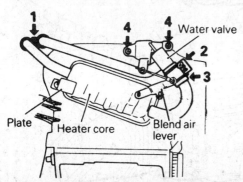

Remove the outer clamp (1), the hose clamp (2) and hose (3), the retaining screws (4) and the water valve itself. The blend air lever may block removal of the heater core

The blend air damper lever must be correctly positioned before the link is installed

29. If the car is equipped with automatic heating control, the servo motor should be removed before working with the case. With the heater unit removed from the car, use a small screwdriver to carefully disconnect the servo motor rod from the air blend damper.
30. Remove the screws holding the servo motor to the heater unit.
31. Carefully unlock the water valve lever clip (don't break it!) and disconnect the link between the blend air damper and the water valve lever.
32. Remove the outer clamp from the two water tubes.
33. Loosen the clamps on the short joint hoses and disconnect the hoses. Remove the retaining screws holding the water valve in place and remove the water valve.
34. The heater core is held in place by a clip and retaining screw. Once removed, the core should come free of the housing. If the core is blocked by the blend air damper lever, remove the lever. Do not attempt to force the core past the lever; breakage will occur.
35. Although the core may be cleaned and/or repaired in the same fashion as a radiator, replacement with a new unit is highly recommended. While the unit is apart, check the small joint hoses for any sign of hardening or cracking. Replace any that fail inspection. Check the water valve also; it should move smoothly and show no sign of leakage or seepage.

To install:
36. Reassemble the heater unit with the core secured by the retaining screw and clip. The valve and connecting hoses must be properly secured with new clamps. Double check clamp position and tightness; some will be difficult to reach after reinstallation.
37. Push the water valve lever all the way inward so that the water valve is at the closed position. Move the blend damper lever counterclockwise so that the blend air damper is fully closed.
38. Install the connecting link and secure the water valve lever clip.
39. Install the servo motor and connect the motor rod to the blend air damper.
40. Install the completely assembled heater unit into the car. Tighten the retaining nuts and bolts evenly.
41. Reinstall the two center support brackets.
42. Install the lap heater duct and the center ventilation duct.
43. Make certain the reinforcement brackets are in place. The dash should be assembled to the condition it was when removed. The upper part of the heater unit has two guide bolts to which the dash attaches. When placing the dash in the car, make certain these guides engage.
44. After the dash is in place, check that wires, cables and ducts are routed correctly and without interference.
45. Install the retaining bolts for the dash.
46. Install side covers on the dashboard.
47. Install the grill covers for the side defrosters.
48. Install the clock. Remember to connect the wiring.
49. Install the heater control bezel.

21. Remove the clock; it may be popped out of the dash with gentle pressure. Disconnect the harness when it is free.
22. Remove the grilles for the side defrosters by inserting a flat tool from the window side and prying forward and upward.
23. Use a non-marring tool to pry the side covers off the instrument panel.
24. Remove each mounting screw and bolt holding the dash in place. As it comes loose, allow it to move into the car. Disconnect the remaining wire harnesses when you can reach behind it.
25. With the help of an assistant, remove the dash from the car. Remember that the dash still has several components attached. Handle it carefully.
26. Remove the center ventilator duct and remove the lap heater duct.
27. Remove the two center reinforcement bars.
28. Remove the retaining nuts and bolts and remove the heater unit from the car. Handle it carefully and do not crush or damage the components and hoses on the outside of the case.

6-13

6 CHASSIS ELECTRICAL

50. Install the ashtray.
51. Install the glove box. Tighten the hinge screws and fold the box into place.
52. Install the undercover on the right side.
53. Install the front and rear consoles. Make certain the wiring is connected.
54. Install the side console covers. Fit it into place carefully and install the screws.
55. Install the instrument cluster. Connect the wiring and the cables before installing the retaining screws.
56. Position the instrument hood and connect the wiring to the switches. Fit the hood into its final place and install the retaining screws.
57. Carefully fit the combination switch onto the steering column. Make certain it is properly seated. Route the wiring harnesses correctly and connect them.
58. Install the retaining screws and tighten them.
59. Install the upper and lower steering column covers.
60. Install the fuse block to the dashboard. Make certain the wire harness are properly routed and secured.
61. Install the knee protector on the left side.
62. Attach the hood release handle to the dash.
63. Install the steering wheel.
64. Reinstall the center console.
65. Connect the heater hoses to the heater pipes at the firewall.
66. Check that the draincock is closed. Refill the cooling system.
67. Connect the battery ground cable.
68. Start the engine and allow it to run at idle. Move the temperature control to its hottest position and check the hoses and joints for leakage or moisture.
69. Check the function of the heater controls and confirm that air and temperature functions are performed correctly when the controls are changed.

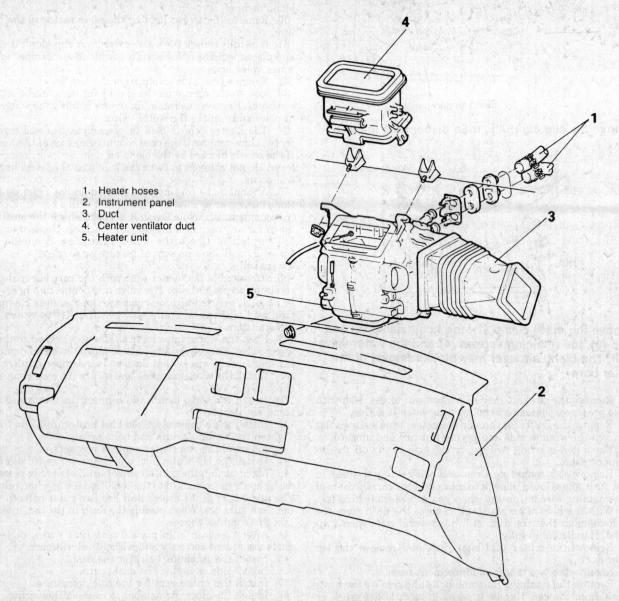

1. Heater hoses
2. Instrument panel
3. Duct
4. Center ventilator duct
5. Heater unit

1985–88 Mirage heater unit, shown without air conditioning

CHASSIS ELECTRICAL 6

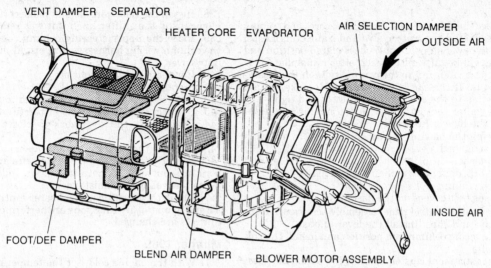

Air conditioning the Mirage packs the components very tightly. Care must be used in removing the heater unit; the evaporator and lines could be damaged

Mirage, 1985–88

1. With the engine cold, set the temperature control lever to the extreme hot position. Drain the engine coolant through the radiator drain plug.

CAUTION

When draining the coolant, keep in mind that cats and dogs are attracted by the ethylene glycol antifreeze, and are quite likely to drink any that is left in an uncovered container or in puddles on the ground. This will prove fatal in sufficient quantity. Always drain the coolant into a sealable container. Coolant should be reused unless it is contaminated or several years old.

2. Disconnect the coolant hoses running to the heater pipes at the firewall. Take note of hose location and placement; they are formed to fit exactly as needed.
3. Remove the instrument panel.
4. Disconnect the negative battery cable.
5. Remove the steering wheel.
6. Remove the lap air outlet under the steering column.
7. Remove the lower steering column cover, disconnect the column wiring harnesses and remove the upper cover.
8. For manual transmission cars, remove the shift knob.
9. Remove the floor console assembly. Remember the concealed screws under the ashtray and behind the switches.
10. Open the glove box. Carefully and quickly, pull upon the right corner of the glove box to remove the right hinge pin from the mount. Slide the glove box toward the right door to remove it.
11. Disconnect the defroster ducts and remove the small side joint ducts.
12. Remove the instrument hood. Remember the two clips at the top.
13. Remove the instrument cluster screws. Move the cluster towards you, disconnect the speedometer cable and the wiring harnesses and remove the cluster.
14. Remove the 4 bolts, two on each side, holding the steering column bracket to the heater control lever knobs. Reach behind the dash to the back of the heater control panel. Push the right side control panel clip to the right while pushing outward on the panel. Allow the panel to hang free.
15. Remove the heater control assembly mounting screws.
16. Remove the small covers from the top of the dash. Use a protected flat tool to pry the covers loose.
17. Remove the instrument panel mounting screws and bolts. With an assistant, remove the dash from the car. Handle it carefully; components are still attached.
18. If the car is not equipped with air conditioning, remove the plastic duct running from the heater housing to the blower. (If the car has air conditioning, the evaporator is in place of the duct.)
19. Remove the center ventilator duct.
20. Carefully disconnect the control cables running to the heater assembly.
21. Remove the retaining nuts and bolts and remove the heater unit. If the car has air conditioning, lift the heater away from the evaporator without creating any force on the evaporator case.

WARNING: The air conditioning lines are very brittle; prying or attempting to move the evaporator can result in a broke air conditioning line.

22. To remove the heater core, loosen the small hose clamps and remove the hoses. It may be necessary to cut the hoses off; there is very little working room. Remove the retaining screws and remove the water valve.
23. Remove the piping clamp and remove the heater core from the unit.
24. Although the core may be cleaned and/or repaired in the same fashion as a radiator, replacement with a new unit is highly recommended. Inspect the water valve for any signs of leaking or seepage. Cut new joint hoses for the water valve and check that the clamps being used are in good condition.

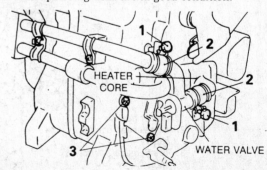

The clamps (1), hoses (2) and retaining screws (3) hold the water valve in place. Once the valve is removed, the core may be removed easily

6-15

6 CHASSIS ELECTRICAL

To install:

25. Reassemble the heater unit with the core and piping clamps in place. Install the water valve and connecting hoses; secure them with new clamps. Double check clamp position and tightness; some will be difficult to reach after reinstallation.
26. Reinstall the heater unit in the car and tighten the retaining nuts and bolts. If the car is air conditioned, make certain that the connection to the evaporator is correct and tight.
27. Install the center ventilator duct, and, if not air conditioned, the duct to the blower housing.
28. When placing the dash in the car, make certain no wires are pinched or stretched.
29. After the dash is in place, check that wires, cables and ducts are routed correctly and without interference.
30. Install the retaining bolts for the dash.
31. Install upper bolt covers on the dashboard.
32. Install the heater control panel assembly screws, then install the faceplate and clips. Install the lever knobs.
33. Move the steering column into position and install the column bracket bolts.
34. Install the instrument cluster; connect the wiring harnesses and speedometer cable before installing the retaining screws.
35. Install the cluster hood, making certain the clips are engaged at the top.
36. Install the side joint ducts and the defroster ducts.
37. Install the glove box.
38. Install the floor console and make certain all the bolt holes align. Instal the shifter knob if it was removed.
39. Place the upper steering column cover in place. Connect the column wiring harnesses and install the lower steering column cover.
40. Connect the lap air outlet.
41. Install the steering wheel.
42. Connect the heater control cables to the heater unit.
43. Connect the heater hoses at the firewall.
44. Check that the draincock is closed. Refill the cooling system.
45. Connect the battery ground cable.
46. Start the engine and allow it to run at idle. Move the temperature control to its hottest position and check the hoses and joints for leakage or moisture.
47. Check the function of the heater controls and confirm that air and temperature functions are performed correctly when the controls are changed.

Mirage, 1989

1. With the engine cold, set the temperature control lever to the extreme hot position. Drain the engine coolant through the radiator drain plug.
2. Disconnect the coolant hoses running to the heater pipes at the firewall. Take note of hose location and placement; they are formed to fit exactly as needed.

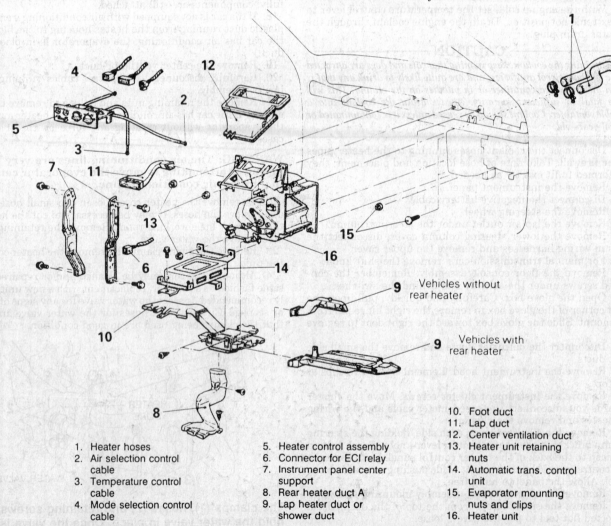

1. Heater hoses
2. Air selection control cable
3. Temperature control cable
4. Mode selection control cable
5. Heater control assembly
6. Connector for ECI relay
7. Instrument panel center support
8. Rear heater duct A
9. Lap heater duct or shower duct
10. Foot duct
11. Lap duct
12. Center ventilation duct
13. Heater unit retaining nuts
14. Automatic trans. control unit
15. Evaporator mounting nuts and clips
16. Heater unit

The 1989 Mirage heater assembly requires substantial disassembly for access

CHASSIS ELECTRICAL 6

CAUTION

When draining the coolant, keep in mind that cats and dogs are attracted by the ethylene glycol antifreeze, and are quite likely to drink any that is left in an uncovered container or in puddles on the ground. This will prove fatal in sufficient quantity. Always drain the coolant into a sealable container. Coolant should be reused unless it is contaminated or several years old.

3. Remove the front seats from the car.
4. Remove the center console.
5. Remove the instrument panel.
6. Disconnect the negative battery cable.
7. Remove the side trim panel (kickpanel) on each side.
8. Remove the ashtray and bracket.
9. Remove the center trim panel from the lower dash.
10. Remove the pocket and panel from the left side dash.
11. Remove the knee protector (lower dash) from the left side of the dash.
12. Remove the hood release bracket from the dash.
13. Remove the lower and upper steering column covers.
14. Remove the radio.
15. Open the glove box and remove the striker hook from the inside. Remove the glove box assembly.
16. Remove the lower panel covers.
17. Remove the heater control assembly installation screw.
18. Remove the instrument panel hood or bezel. As soon as it is loose, disconnect the switch and lighting wiring.
19. Disconnect the speedometer cable at the transaxle.
20. Remove the screws holding the instrument cluster. Pull it forward to gain access to the back.
21. Pull the speedometer cable slightly towards the interior of the car; release the lock by turning the adapter to the left or right and remove the adapter.
22. Disconnect the wiring to the instruments. Use a small flat tool to release the lock tabs on each harness before removal.
23. Remove the right speaker cover. Use a flat, protected tool to avoid marring the plastic. Remove the speaker.
24. Remove the grilles for the side defrosters.
25. Remove the clock (or the hole plug if no clock is fitted).
26. Remove the steering shaft bracket bolts and nuts. Allow the shaft to move away from the dash.
27. Remove the instrument panel mounting bolts. With an assistant, remove the instrument panel from the car.
28. Disconnect the three control cables running to the heater unit.
29. Now remove the heater control unit from the car (It was left in place when the dash came out.).
30. Carefully disconnect the wiring to the fuel injection control relay.
31. Remove the two center supports for the instrument cluster.
32. Remove rear heater duct A.
33. Remove either the lap heater duct or the shower duct if equipped with a rear heater. Remove the foot duct.
34. Disconnect or remove the left side lap duct.
35. Remove the center ventilation duct.
36. If the car is equipped with an automatic transaxle, remove the two lower retaining nuts on the heater unit and remove the automatic transmission control unit. Like any other computer or "black box", this one must be handled with extreme care and placed safely out of the work area in a clean, protected location.
37. Remove the upper mounting nuts and remove the heater unit.

WARNING: If the car is air conditioned, the left side evaporator mounting nuts and clips must be removed. Once removed, the evaporator must be pulled gently outward (toward the interior of the car); this will allow the heater unit to come free. Take great care not to damage the refrigerant lines during this operation.

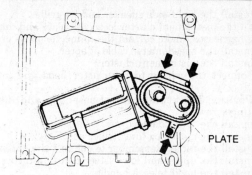

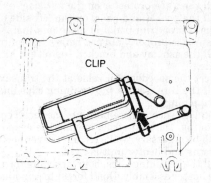

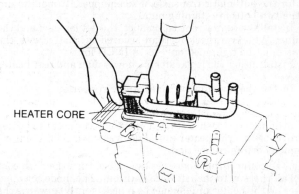

Once you get to it, changing the Mirage heater core is an easy 3-step procedure. Take care not to damage the fins or the case insulation

38. With the heater assembly out of the car, remove the plate from the two heater pipes, remove the clips holding the core and carefully pull the core from the casing.

NOTE: Remove or install the core carefully to avoid damaging either the fins or the surrounding padding.

39. Although the core may be cleaned and/or repaired in the same fashion as a radiator, replacement with a new unit is highly recommended.

To install:
40. When placing the dash in the car, make certain no wires are pinched or stretched.
41. After the dash is in place, check that wires, cables and ducts are routed correctly and without interference.
42. Install the retaining bolts for the dash.
43. Install the steering column bracket bolts. Tighten each to 10 Nm (8 ft. lbs).
44. Install the clock or the hole plug.
45. Install the side defroster grilles.

6-17

6 CHASSIS ELECTRICAL

46. Install the right side speaker and its grille.
47. Fit the instrument cluster loosely in place and connect the wiring harnesses; make certain the connector locks engage.
48. Install the speedometer cable adapter.
49. Install the instrument cluster.
50. Connect the wiring to the cluster hood and install the hood.
51. Position the heater control assembly and install the mounting screw.
52. Install the lower dash covers.
53. Install the glove box and striker assembly.
54. Install the radio. Remember to connect the antenna lead.
55. Install the upper and lower steering column covers.
56. Install the hood release handle.
57. Install the knee protector on the left side.
58. Install the pocket and panel on the left side of the dash.
59. Fit the center trim panel in place and instal the screws. Make certain it is properly aligned before tightening.
60. Install the ashtray and bracket. Install the kickpanel trim on each side.
61. Connect the speedometer cable at the transaxle.
62. Install the core, secure the retaining clips and install the plate on the piping.
63. Reinstall the heater assembly into the car. If equipped with air conditioning, the same trick moving the evaporator out of position will be necessary. Make certain the connection between the evaporator and heater unit is tight.
64. Install the retaining clips, bolts and nuts for the evaporator and the heater assembly. Don't forget to reinstall the control unit for the automatic transaxle, if so equipped. It mounts under the two bottom retaining nuts.
65. Install the center ventilation duct, the left lap duct and the foot duct. Wherever these ducts are secured by small screws, the screws must be tightened only to 18 INCH lbs.
66. Install either the lap heater or shower duct and rear heater duct A.
67. Install the instrument panel center supports.
68. Connect the fuel injection control relay to its harness. Take great care that all the pins in the connector align correctly.
69. Although the dash is still out of the car, connect the heater control cables to the heater unit. Each cable has a different requirement before connection:

   a. To connect the mode selection cable, turn the dash control all the way to the left (face position). The mode selection lever on the heater unit should be pulled (gently) towards the inside of the car. Once both are in place, slip the cable onto the lever; make certain the outer case of the cable is secured within its clip.

   b. The temperature control on the should be set to the leftmost, or coolest setting. The blend air damper lever should be pressed completely down before the cable is attached. Again, the outer cable must be secure in its clip.

   c. The air selection control cable is reinstalled by first setting the selector lever to the right (recirculate). The air selection damper lever should be moved fully inward (toward the inside of the car) before connecting the cable. The system will not work if the outer cable is not secured within its clip.

70. Connect the heater hoses to the heater pipes. Secure the clamps and refill the system with coolant. Do not start (or try to start) the engine, but do check the areas around the heater hoses and pipes for any leaks as the system is filled.
71. Install the instrument panel.
72. Install the center console.
73. Install the front seats.
74. Start the engine and allow it to run at idle. Move the temperature control to its hottest position and check the hoses and joints for leakage or moisture.
75. Check the function of the heater controls and confirm that air and temperature functions are performed correctly when the controls are changed.

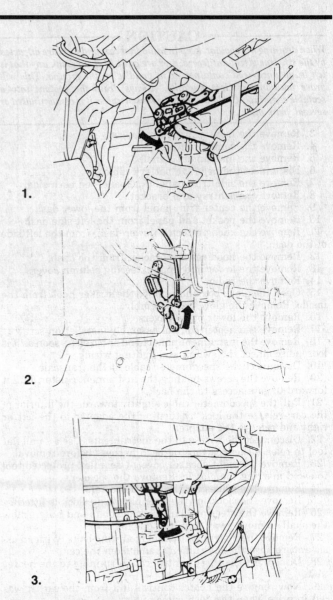

1.

2.

3.

Correct positions of the mode selection damper lever (1), the blend air damper lever (2), and the air selection damper lever (3) before connecting the control cables. Make certain the dash control is properly set when attaching the cable

Precis

1. Disconnect the negative battery cable.
2. Set the temperature control lever to the hottest position and drain the cooling system.

CAUTION

When draining the coolant, keep in mind that cats and dogs are attracted by the ethylene glycol antifreeze, and are quite likely to drink any that is left in an uncovered container or in puddles on the ground. This will prove fatal in sufficient quantity. Always drain the coolant into a sealable container. Coolant should be reused unless it is contaminated or several years old.

CHASSIS ELECTRICAL 6

3. Disconnect the heater hoses.
4. Remove the lower dashboard panel (crash pad or knee pad).
5. Remove the center console.

WARNING: The console contains the computer. Handle it gently.

6. Remove the screw holding the two heating ducts. The heating ducts may be removed by pulling out at the bottom while pushing inward at the top. This is a coordinated motion and may take some learning. Don't force things.
7. Disconnect the heater control cables.
8. If equipped with air conditioning, remove the evaporator case. This is not as difficult as it first seems. Discharge the A/C system safely and refer to *Evaporator Removal and Installation"* later in this Section.
9. Loosen the heater unit retaining bolts and remove the heater unit.

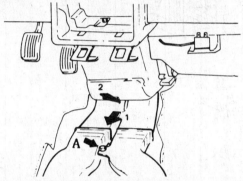

The bolt (A) must be loosened or removed. Pull at '1' and push at '2' in a smooth, coordinated motion to get the ductwork free of the heater unit

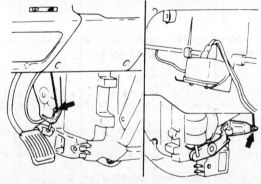

Disconnect all the control cables before loosening the case

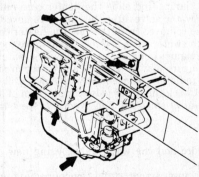

Precis heater case retaining bolts

10. With the case removed from the car, the heater core may be changed after the water valve is removed. Remove the plastic cover, remove the clamps and hose and remove the water valve.
11. Although the core may be cleaned and/or repaired in the same fashion as a radiator, replacement with a new unit is highly recommended. Check the water valve for any sign of leakage; the valve should move smoothly and evenly. Inspect the hose and clamps.
12. Install the core and the water valve, using new hose or clamps as necessary.
13. Install the heater unit into the car and tighten the mounting bolts.
14. Install the evaporator case and connect the lines.
15. Connect the heater control cables. Remember that each outer casing must be held in place by the clip. Check the operation of each control through its full range of motion.
16. Install the lower ductwork, again using the coordinated push/pull motion. Install or tighten the bolt holding the ducts together.
17. Install the center console.
18. Install the lower dash knee pad.
19. Connect the heater hoses, making sure they are not kinked or twisted. Fill the cooling system and observe all of the hose and clamp joints for leaks.
20. Connect the negative battery cable.
21. Start the engine and allow it to run at idle. Move the temperature control to its hottest position and check the hoses and joints for leakage or moisture.
22. Check the function of the heater controls and confirm that air and temperature functions are performed correctly when the controls are changed.

Galant and Sigma

1. With the engine cold, set the temperature control lever to the extreme hot position. Drain the engine coolant through the radiator drain plug.

---- **CAUTION** ----

When draining the coolant, keep in mind that cats and dogs are attracted by the ethylene glycol antifreeze, and are quite likely to drink any that is left in an uncovered container or in puddles on the ground. This will prove fatal in sufficient quantity. Always drain the coolant into a sealable container. Coolant should be reused unless it is contaminated or several years old.

---

2. Disconnect the coolant hoses running to the heater pipes at the firewall. Take note of hose location and placement; they are formed to fit exactly as needed.
3. Remove the center console.
4. Remove the heater cover.
5. Disconnect the negative battery cable.
6. Remove the horn pad, center nut and steering wheel.
7. Remove the small steering column panel.
8. Remove the undercover.
9. Remove the upper and lower steering column covers.
10. Use a small flat tool to remove the screw cover panels at the bottom of the instrument cluster hood. Remove the screws and remove the hood.
11. Remove the mounting screws for the instrument cluster. Pull the cluster forward and disconnect the wiring connectors.
13. Disconnect the speedometer adaptor behind the cluster and remove the cluster.
14. The floor console must be removed. It is held by several concealed bolts. Work carefully.
15. Remove the under frame behind the console.
16. Disconnect and remove air duct D, the lap heater duct, the side defroster duct and the vertical defroster duct.
17. Remove the glove box.
18. Remove the ashtray and its mount. Disconnect the light wiring before removing.

6-19

6 CHASSIS ELECTRICAL

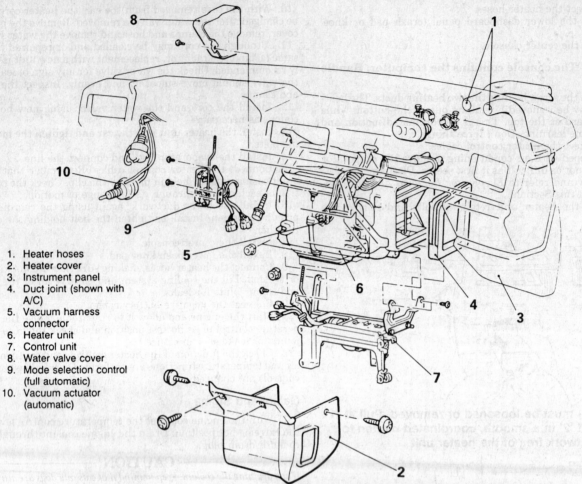

1. Heater hoses
2. Heater cover
3. Instrument panel
4. Duct joint (shown with A/C)
5. Vacuum harness connector
6. Heater unit
7. Control unit
8. Water valve cover
9. Mode selection control (full automatic)
10. Vacuum actuator (automatic)

Typical Galant heater assembly. Components may vary be type of system (automatic or full-automatic air conditioning) and year. The vacuum hose harness must be handled carefully

19. Remove the heater control face plate.
20. Remove the heater control panel and disconnect its harness.
21. Remove the undercover from the right side dash and remove the under frame.
22. On the left side of the dash, remove the fuse box cover and unbolt the fusebox from the dash.
23. Remove the front pillar (windshield pillar trim) from each pillar.
24. Remove the kickpanel trim from each side.
25. Remove the defroster garnish and the defroster grille.

NOTE: After the garnish is loosened, disconnect the photosensor wiring before removing the pieces from the car.

26. Remove the grille for the center air outlet.
27. Remove the bolts holding the steering column bracket to the dash.
28. Remove the center reinforcement bracket.
29. On the left side, remove the retaining nuts holding the dash underframe to the body.
30. On the right side, remove the underframe retaining bolts. Take note that the bolts are different; the flanged bolt must be correctly reinstalled.
31. Remove the remaining nuts and bolts holding the dash. As the dash comes loose, label and disconnect the wiring harnesses. Carefully remove the dash from the car.

32. If the vehicle is equipped with automatic heating and cooling, remove the power control unit on the lower front of the harness.
33. Remove the duct joint between the heater unit and evaporator case (with A/C) or blower assembly (without A/C).
34. Carefully separate the vacuum hose harness at the connector.
35. Remove the heater unit from the car.
36. To remove the heater core, first remove the cover from the water valve. Disconnect the links and remove the vacuum actuator.
37. Remove the clamps and slide the heater core out of the case. Remove the water valve after the core is removed.
38. With the case removed from the car, the heater core may be changed after the water valve is removed. Remove the plastic cover, remove the clamps and hose and remove the water valve.
39. Although the core may be cleaned and/or repaired in the same fashion as a radiator, replacement with a new unit is highly recommended. Check the water valve for any sign of leakage; the valve should move smoothly and evenly. Inspect the hose and clamps.

To install:
40. Install the core and the water valve, using new hose or clamps as necessary.
41. Install the vacuum actuator and the connecting link. Put the cover on the water valve.

CHASSIS ELECTRICAL 6

42. Install the heater unit into the car and tighten the mounting bolts.
43. Carefully attach the vacuum hose connector to the vacuum harness. Make certain the hoses mate firmly and securely.
44. Install the heater cover, then install the center console.
45. Install the duct joint between heater and evaporator or blower.
45. Install the power control unit and carefully connect the links and rods.
47. Reinstall the heater hoses under the hood.
48. Fill the system with coolant and observe hose and clamp joints for leaks as the system fills. (Do not attempt to start the engine).
49. Position the dash and connect the wiring harnesses. Make certain nothing is pinched or stretched out of place.
50. Install the retaining bolts and nuts.
51. Install the underframe bolts on the right side, making certain the flange bolt is properly located.
52. Install the underframe nuts on the left side.
53. Install the center reinforcement bracket.
54. Lift the steering column into place and install the bracket bolts. Tighten them to 10 Nm or 8 ft. lbs.
55. Install the center air outlet center grille.
56. Install the defroster grille and garnish. Make certain each piece is correctly located.
57. Install the kickpanels on each side.
58. Install the pillar trim on each side.
59. Attach the fusebox to the dash and install the lid.
60. On the right side, install the under frame and the under cover.
61. Position and install the heater control panel and its faceplate.
62. Install the ashtray and bracket. Connect the wire for the light.
63. Install the glove box assembly.
64. Connect and install the defroster duct, the side defroster duct, the lap heater duct and air duct D.
65. Install the small center under frame.
66. Install the floor console. Don't tighten the bolts and screws until every one is started in its hole. Make sure all the seams align correctly.
67. Install the instrument cluster; connect the wiring and adapter before installing the retaining screws.
68. Install the instrument cluster hood. Don't forget the screw covers.
69. Install the upper and lower steering column covers.
70. Install the center panel under cover.
71. Install the small column panel.
72. Install the steering wheel and horn pad. Tighten the center nut to 40 Nm or 30 ft. lbs.
73. Connect the battery ground cable.
74. Start the engine and allow it to run at idle. Move the temperature control to its hottest position and check the hoses and joints for leakage or moisture.
75. Check the function of the heater controls and confirm that air and temperature functions are performed correctly when the controls are changed.

Pickup, 1983–86

1. Disconnect the battery ground cable.
2. Move the temperature selector lever to the off position.
3. Drain the coolant.

— **CAUTION** —
When draining the coolant, keep in mind that cats and dogs are attracted by the ethylene glycol antifreeze, and are quite likely to drink any that is left in an uncovered container or in puddles on the ground. This will prove fatal in sufficient quantity. Always drain the coolant into a sealable container. Coolant should be reused unless it is contaminated or several years old.

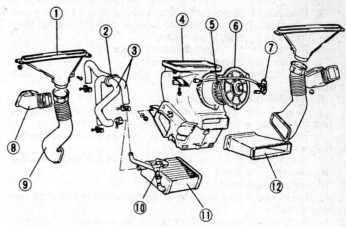

1. Defroster nozzle
2. Grommet
3. Water hose
4. Heater case
5. Turbo fan
6. Motor
7. Heater resistor
8. Side ventilator duct
9. Defroster duct
10. Water valve
11. Heater core
12. Center ventilator duct

1983–86 Pickup; heater components

4. In the cab, remove the parcel tray, the center vent grille and duct and the defroster duct.
5. Label and disconnect the control cables at the heater box.
6. Disconnect the coolant hoses running to the heater core. There will be some coolant within the hoses; don't spill it in the cab.
7. Disconnect the fan wiring. Remove the upper and center mounting nuts holding the heater assembly. Remove the assembly.

WARNING: The heater core still contains coolant; don't spill it in the cab.

8. With the heater case removed, the core may be easily removed on the workbench.
9. Place the new core in the heater case. Install the case in position and tighten the retaining nuts and bolts. Make sure the case is properly seated before tightening the mounting nuts.
10. Connect the wiring to the fan motor.
11. Install the water hoses, making sure each is pushed onto its pipe at least 25mm. Use new clamps if needed.
12. Connect the cables to the heater unit. Pay attention to the labels.
13. Install the defroster duct, the center ventilator duct and grille and the parcel tray.
14. Double check the draincock, closing it securely if not already done. Fill the cooling system with coolant.
15. Connect the negative battery cable.
16. Start the engine. Move the temperature selector to HOT and allow the engine to warm up. Check the function of each heater control and inspect the hose connections for leaks.

Pickup, 1987–89

1. With the engine cold, set the temperature control lever to the extreme hot position. Drain the engine coolant through the radiator drain plug.

— **CAUTION** —
When draining the coolant, keep in mind that cats and dogs are attracted by the ethylene glycol antifreeze, and are quite likely to drink any that is left in an uncovered container or in puddles on the ground. This will prove fatal in sufficient quantity. Always drain the coolant into a sealable container. Coolant should be reused unless it is contaminated or several years old.

CHASSIS ELECTRICAL

2. Disconnect the coolant hoses running to the heater pipes at the firewall. Take note of hose location and placement; they are formed to fit exactly as needed.
3. Disconnect the negative battery cable.
4. Remove the hazard flasher switch.
5. Pop out the hole cover on the right side of the instrument hood.
6. Remove the instrument hood.
7. Remove the instrument cluster, remembering to disconnect the wiring and speedometer cable at the rear.
8. Remove the fusebox cover. Remove the fusebox from the dashboard and let it hang.
9. Remove the glove box. Move the stopper out of the way before lowering the door.
10. Remove the two defroster ducts.
11. Label and disconnect the 3 control wires running to the heater unit.
12. Remove the speaker cover panels on each side.
13. Remove the clock or small compartment from the dash.
14. Pop out the small square cover at the upper center of the dash. Its easiest to reach in the clock hole and release the clips with your fingers.
15. Remove the lower center faceplate or cover from the center of the dash.
16. Remove the shifter knob.
17. Remove the floor console assembly.
18. Disconnect the left side and front harness connectors running to the dashboard.
19. Disconnect the air conditioner and heater wiring harnesses. Disconnect the instrument cluster harness and the radio connectors, including the antenna wire.
20. Remove the retaining nuts and bolts holding the dash. They are located across the top, along the sides and at the bottom.
21. Use the tilt release lever and move the steering column all the way downward.
22. Remove the dash carefully; several components are still connected to it.
23. Remove the large duct running between the heater case and the fan case.
24. Remove the center ventilator duct.
25. Remove the defroster duct.
26. Remove the two vertical reinforcing bars.
27. Remove the retaining nuts and bolts and remove the heater case.

WARNING: The heater core still contains coolant; don't spill it inside the cab.

28. Remove the hose cover on the side of the heater case.
29. Disconnect the hose clamps on the small hoses.
30. Use a sharp knife or razor to cut away the small hoses. Attempting to remove the hoses may damage the tubing.

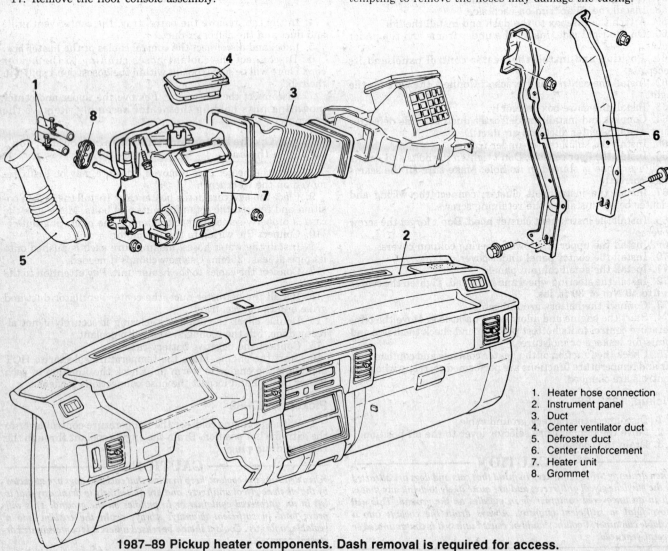

1. Heater hose connection
2. Instrument panel
3. Duct
4. Center ventilator duct
5. Defroster duct
6. Center reinforcement
7. Heater unit
8. Grommet

1987–89 Pickup heater components. Dash removal is required for access.

CHASSIS ELECTRICAL 6

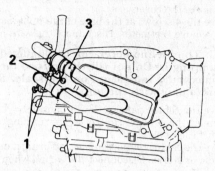

For 1987-89 Pickup, remove the hose clamp (1), the small joint hose (2) and the retaining plate (3) to get the heater core out of the case

31. Remove the retaining plate.
32. Remove the heater core from the case.

To install:
33. Install the core into the case, making sure all the padding and insulation is in place.
34. Install the plate holding the two pipes. Install new small hoses and new clamps. Make certain the joints are correctly fitted and that the clamps are tightened. These hoses are not accessable after installation; any leak will require removal again.
35. Install the hose cover.
36. Install the heater unit in the vehicle, making certain the pipes go through the firewall grommet correctly. Tighten the retaining bolts and nuts.
37. Install the center reinforcements for the dash.
38. Install the defroster duct, the center vent duct and the large air duct to the blower.
39. Place the dash into the vehicle carefully and connect the main electrical harnesses.
40. Install all the retaining nuts and bolts but do not tighten any of them until the alignment of the dash is correct.
41. Reinstall the floor console, the shifter knob and the lower center faceplate.
42. Install the upper bolt cover in the center of the dash.
43. Replace the clock or small pocket. Don't forget to plug wiring together.
44. Install the speaker covers.
45. Route the heater control cables properly and connect them.
46. Install the defroster ducts.
47. Install the glove box.
48. Remount the fusebox on the dash and install the lid.
49. Install the instrument cluster. Connect the speedometer cable and wiring harnesses before tightening the screws.
50. Install the instrument cluster hood and install the bolt cover.
51. Install the hazard warning switch.
52. Connect the heater hoses to the pipes at the firewall.
53. Double check the draincock, closing it securely if not already done. Fill the cooling system with coolant.
54. Connect the negative battery cable.
55. Start the engine. Move the temperature selector to HOT and allow the engine to warm up. Check the function of each heater control and inspect the hose connections for leaks.

Montero, 1983-86

1. With the engine cold, set the temperature control lever to the extreme hot position. Drain the engine coolant through the radiator drain plug.

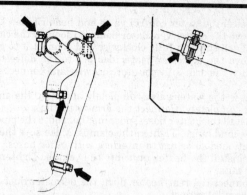

Make certain the clamps are placed correctly when replacing the Montero heater core

---- CAUTION ----
When draining the coolant, keep in mind that cats and dogs are attracted by the ethylene glycol antifreeze, and are quite likely to drink any that is left in an uncovered container or in puddles on the ground. This will prove fatal in sufficient quantity. Always drain the coolant into a sealable container. Coolant should be reused unless it is contaminated or several years old.

2. Disconnect the coolant hoses running to the heater pipes at the firewall. Take note of hose location and placement; they are formed to fit exactly as needed.
3. Disconnect the negative battery cable.
4. Remove the steering wheel.
5. Remove the center console.
6. Remove the meter assembly. Carefully disconnect the wiring and speedometer cable.
7. Remove the gauge assembly and disconnect its wiring.
8. Remove the lap heater ducts below the dash and disconnect the release cable bracket from the dash.
9. Remove the heater control assembly. Label each cable as it is disconnected at the heater end.
10. Remove the fuse cover on the side of the dash. unscrew the mounting bolts and push the fuseblock into the dashboard.
11. Disconnect the wiring from the front speakers.
12. Remove the plug at the center of the instrument panel.
13. Remove the right and left side defroster grilles by gently prying on the mounting projections with a small flat tool. Be careful not to break the projections off.
14. Remove the glove box.
15. Disconnect the wiring from the heater relay.
16. Remove the mounting nuts and bolts and remove the instrument panel.
17. Remove the center ventilator duct and the defroster duct.
18. Remove the rear heater duct.
19. Remove the upper and lower retaining nuts and remove the heater case.

WARNING: The heater core still contains coolant; don't spill it in the cab.

20. Remove the heater control lever arm and remove the water valve cover.
21. Remove the heater pipe and the water valve.
22. Disconnect the control arm linkage and remove the control arm.
23. Remove the heater core by sliding it out of the case sideways. Do not remove the protective felt liner when removing the core. It insulates the core within the case.

To install:
24. Install the new core into the case with the felt insulator positioned properly.
25. Install the control arm on the linkage.

6-23

6 CHASSIS ELECTRICAL

26. Install the water control valve and heater pipe
27. Move the center ventilator damper (door) to the closed position. Turn the arm fully clockwise and connect it to the link.
28. Move the defroster/heater damper to the defroster position. Turn the arm fully counterclockwise and connect it to the link.
29. Close the water valve completely and close the air intake damper completely. Connect the arm to the link.
30. Install the heater hoses, pushing them onto the pipes as far as the second bulge. Tighten the clamps. Make sure the clamps are positioned so as not to interfere with other hoses.
31. Reinstall the heater unit into the vehicle. Tighten the retaining nuts.
32. Connect the rear heater duct, the defroster duct and the center ventilator duct.
33. Position the dashboard and secure the mounting bolts.
34. Connect the heater relay wiring and install the glove box.
35. Install the side defroster grilles and the plug at the center of the dash.
36. Connect the front speaker wiring.
37. Position the fusebox and install its screws; install the cover.
38. Install the heater control assembly and connect the control cables.
39. Install the two lap heater ducts and install the hood release cable bracket.
40. Install the combination gauge unit and connect the wiring.
41. Install the instrument cluster. Connect the wiring and the speedometer cable.
42. Install the center console.
43. Install the steering wheel.
44. Connect the heater hoses to the pipes at the firewall.
45. Double check the draincock, closing it securely if not already done. Fill the cooling system with coolant.
46. Connect the negative battery cable.
47. Start the engine. Move the temperature selector to HOT and allow the engine to warm up. Check the function of each heater control and inspect the hose connections for leaks.

Montero, 1987–89

1. With the engine cold, set the temperature control lever to the extreme hot position. Drain the engine coolant through the radiator drain plug.

CAUTION

When draining the coolant, keep in mind that cats and dogs are attracted by the ethylene glycol antifreeze, and are quite likely to drink any that is left in an uncovered container or in puddles on the ground. This will prove fatal in sufficient quantity. Always drain the coolant into a sealable container. Coolant should be reused unless it is contaminated or several years old.

---

2. Disconnect the coolant hoses running to the heater pipes at the firewall. Take note of hose location and placement; they are formed to fit exactly as needed.
3. Disconnect the negative battery cable.
4. Remove the lap heater air ducts under the left and right side dash.
5. Remove the hood release cable bracket from the dashboard.
6. Remove the left and right side defroster grilles. Use a small flat tool to raise the attaching projections. Don't break the tabs.
7. Remove the glove box.
8. Use a flat, padded tool to remove the rear cover from the instrument cluster.
9. Remove the instrument cluster. Disconnect the speedometer cable and the wiring harnesses to the back of the cluster.
10. Disconnect the wiring to each switch on the cluster. Label each as it is removed.
11. Remove the screws from the side of the combination meter unit and remove the cover.
12. Remove the 4 screws at the base of the combination meter and gently remove the meter. Disconnect the wire harnesses.

WARNING: The inclinometer can be damaged by dropping or bumping it. Do not tilt the unit so far as to exceed the maximum indication on the scale. Resist the temptation to play with the unit.

13. Remove the center panel or lower console.
14. Disconnect the 3 heater control cables running to the heater unit. Label each one.
15. Remove the center reinforcement and bring it out as a unit with the radio or stereo. Disconnect the wiring when the unit is free.
16. Remove the horn pad and remove the steering wheel.
17. Remove the fuse box cover and release the fusebox from the dashboard.
18. Remove the dashboard mounting nuts and bolts and remove the dashboard.
19. Remove the center ventilator duct and the defroster duct.
20. Remove the rear heater duct.
21. Remove the upper and lower retaining nuts and remove the heater case.

WARNING: The heater core still contains coolant; don't spill it in the cab.

22. Remove the heater control lever arm and remove the water valve cover.
23. Remove the heater pipe and the water valve.
24. Disconnect the control arm linkage and remove the control arm.
25. Remove the heater core by sliding it out of the case sideways. Do not remove the protective felt liner when removing the core. It insulates the core within the case.

To install:
26. Install the new core into the case with the felt insulator positioned properly.
27. Install the control arm on the linkage.
28. Install the water control valve and heater pipe
29. Move the center ventilator damper (door) to the closed position. Turn the arm fully clockwise and connect it to the link.
30. Move the defroster/heater damper to the defroster position. Turn the arm fully counterclockwise and connect it to the link.
31. Close the water valve completely and close the air intake damper completely. Connect the arm to the link.
32. Install the heater hoses, pushing them onto the pipes as far as the second bulge. Tighten the clamps. Make sure the clamps are positioned so as not to interfere with other hoses.
33. Reinstall the heater unit into the vehicle. Tighten the retaining nuts.
34. Connect the rear heater duct, the defroster duct and the center ventilator duct.
35. Position the dash and start each nut and bolt. Make certain the dash is aligned and then tighten the fittings.
36. Install the fusebox assembly and its cover.
37. Install the steering wheel and horn pad.
38. Install the radio and center reinforcement. Connect the wiring before tightening the bolts.
39. Carefully route and connect the heater control cables.
40. Install center panel and secure the heater controls.
41. Install the combination gauges and connect the wiring.
42. Install the cover for the combination gauges.
43. Install the instrument cluster, connecting the speedometer cable and the many wiring connectors.
44. Install the cluster cover; make sure it is firmly seated in place.
45. Replace the glove box
46. Reinstall the side defroster grilles.

CHASSIS ELECTRICAL 6

47. Connect the hood release cable bracket.
48. Install the two lap heater ducts.
49. Connect the heater hoses to the pipes at the firewall.
50. Double check the draincock, closing it securely if not already done. Fill the cooling system with coolant.
51. Connect the negative battery cable.
52. Start the engine. Move the temperature selector to HOT and allow the engine to warm up. Check the function of each heater control and inspect the hose connections for leaks.

Heater Control Panel (Head)

REMOVAL AND INSTALLATION

Cordia/Tredia

1. Remove the cover under the glove box door.
2. Remove the glove box
3. Disconnect the air selection control cable and the temperature control cable.
4. Remove the lap heater duct and defroster duct A (left side).
5. Disconnect the mode selection cable.
6. On Cordia models, remove the meter hood (instrument cluster) from the instrument panel. Refer to "Instrument Cluster Removal and Installation" later in this Section.
7. Remove the knobs from the heater control levers by applying outward pressure until they come off. Remove the faceplate around the controls.
8. Remove the mounting screws holding the control panel; pull the panel (with the control cables attached) away from the dashboard.
9. Remove the blower control switch.
10. Before reassembly, apply a light chassis grease to the moving parts of the heater control assembly.
11. Reinstall the fan control switch if it was removed.
12. Carefully feed the control cables into the dash and route them properly to the heater unit. Be careful not to kink or bend the cables. Install the retaining bolts holding the control panel in place.
13. Reinstall the instrument cluster, taking great care that each cable and wire is correctly and firmly connected.
14. Move the mode selection lever to the leftmost position and, with the damper lever pushed inward, connet the cable. Secure the outer cable case with the clip.
15. Install the defroster duct and the lap heater duct.
16. Move the temperature selector lever to the leftmost (cool) position. With the blend air damper lever pressed completely downward, connect the cable to the lever and secure the outer cable case within the clip.
17. Move the air selection control lever to the recirculate position.
18. With the air selection damper lever pressed inward (towards the inside of the car), connect the cable and secure the outer cable case within the clip.
19. Install the glove box and install the under-dash cover.

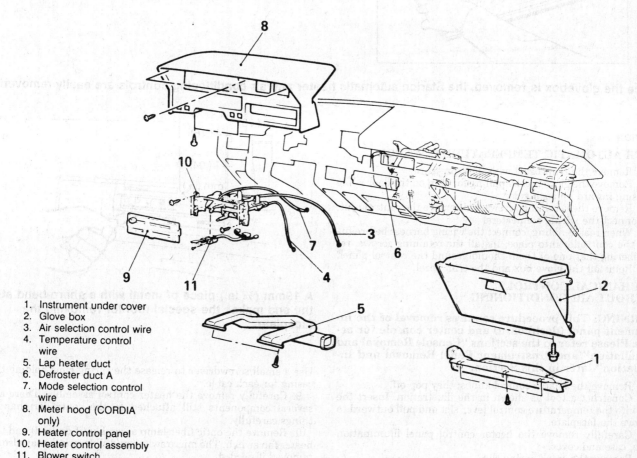

1. Instrument under cover
2. Glove box
3. Air selection control wire
4. Temperature control wire
5. Lap heater duct
6. Defroster duct A
7. Mode selection control wire
8. Meter hood (CORDIA only)
9. Heater control panel
10. Heater control assembly
11. Blower switch

The Cordia heater control assembly requires removing the instrument cluster. Tredia uses the same controller but does not require cluster removal

6-25

6 CHASSIS ELECTRICAL

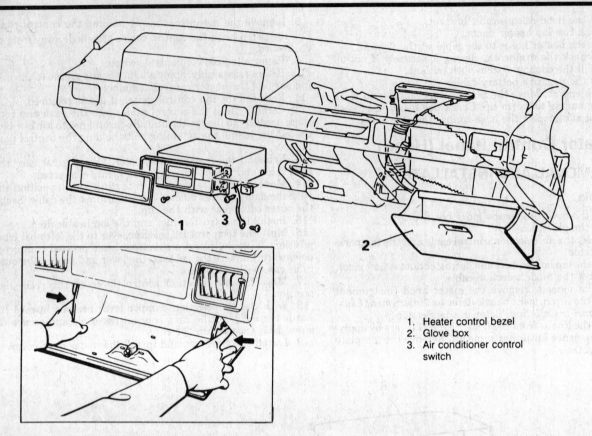

1. Heater control bezel
2. Glove box
3. Air conditioner control switch

Once the glovebox is removed, the Starion automatic heater and air conditioning controls are easily removed.

Starion

WITH AUTOMATIC TEMPERATURE CONTROL

1. Remove the trim bezel with light pressure.
2. Remove the glove box by pulling forward on the door while pressing inward on each side.
3. Remove the retaining screws and remove the control unit. Disconnect the wiring connectors.
4. When reassembling, connect the wiring harness before fitting the controller into place. Install the retaining screws, remembering that one of them should ground the control panel.
5. Reinstall the glove box and the trim bezel.

MECHANICAL CONTROLS WITHOUT AIR CONDITIONING

WARNING: This procedure requires removal of the instrument panel (dashboard) and center console for access. Please refer to the sections "Console Removal and Installation" and "Instrument Panel Removal and Installation" later in this Section.

1. Remove the heater control knobs; they pop off.
2. Construct a tool as shown in the illustration. Insert the tool into the temperature control lever slot and pull outward to remove the faceplate.
3. Carefully remove the heater control panel illumination lamp, case and cover.
4. Remove the small spring clip.
5. Remove the center console from the car.
6. Remove the instrument panel. Surprisingly, the panel comes away leaving the heater control still hooked up.
7. Remove the right side defroster duct.
8. Disconnect the 3 control cables running to the heater case.

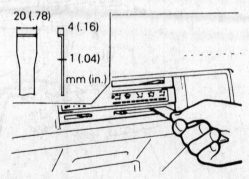

A 19mm (¾ in.) piece of metal with a sharp bend at the end makes the special tool for removing the faceplate

Use a small screwdriver to release the spring clips holding the casing for each cable.

9. Carefully remove the heater control assembly. There are several components still attached to the controller; handle things carefully.
10. Remove the optic fiber lamp assembly (if desired) and the heater fan switch. The micro switch and slide switch may also be removed if needed.
11. When reinstalling, make certain the controller is assembled before installation. All parts removed during inspection or replacement must be reinstalled prior to placing the unit in the car.
12. Install the optic fiber lamp.

CHASSIS ELECTRICAL 6

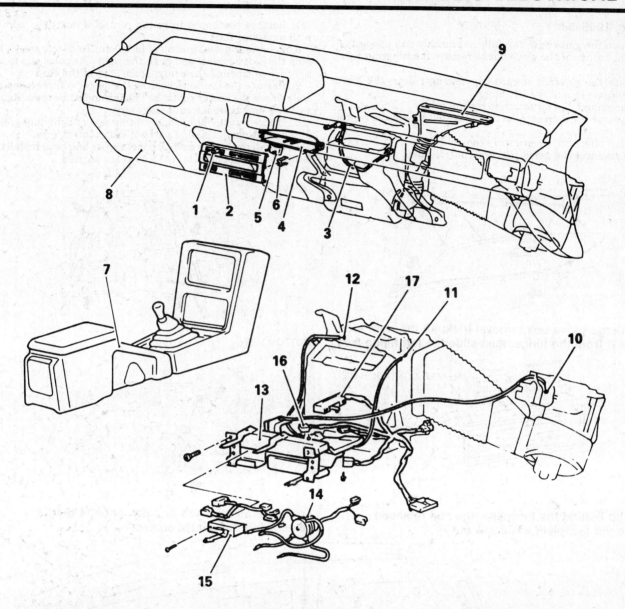

1. Control lever knobs
2. Control panel
3. Illumination lamp
4. Lamp case
5. Lamp cover
6. Springs
7. Console
8. Instrument panel
9. Right defroster nozzle
10. Air selection cable
11. Temperature control cable
12. Mode selection cable
13. Heater control assembly
14. Optic fiber lamp
15. Heater fan switch
16. Micro switch
17. Side switch

The mechanical control panel in the Starion requires removal of the center console and the instrument panel (dashboard)

13. Apply a thin coat of grease to the moving surfaces of the controller. Place the heater control assembly into the car insert the brackets into the center reinforcement.
14. Route the control cables to the heater case. Make certain none of the cables are pinched or kinked.
15. Set the mode selection lever to the extreme left position. Move the mode selection damper lever in a counterclockwise direction and connect the cable to the lever. Secure the cable casing in the spring clip.
16. Set the temperature control to the extreme left. Move the blend air damper lever in a counterclockwise direction and connect the cable. Secure the cable case in the spring clip.
17. Set the air selection control to recirculate. Move the air selection damper lever in a clockwise direction and connect the control cable. Secure the outer case of the cable under the clip.
18. Install the defroster nozzle and duct on the right side.
19. Install the instrument panel.
20. Install the center console.
21. Reinstall the spring clip (for the faceplate) and the lamp with case and cover.
22. Install the heater control faceplate and press the knobs onto the control levers.
23. Start the engine and check the heating and ventilation functions.

6-27

6 CHASSIS ELECTRICAL

Mirage, 1985–88

1. Open the glove box. Carefully but quickly pull upward on the right corner of the glove box to remove the hingepin from the hinge.
2. Slide the glovebox toward the passenger door; the glove box will come free of the left hinge.
3. Disconnect the air selection control cable.
4. Remove the defroster duct from the left side of the heater case.
5. Disconnect the temperature control cable.
6. Disconnect the mode selection control cable.
7. Remove the knobs from the control levers (they pop off with moderate force).
8. Remove the heater control faceplate. To do this, reach behind the control panel and push the right side panel clip to the right while pushing the control panel out of the dash.
9. Remove the brackets holding the heater control assembly and remove the heater controls. The fan and/or air conditioning switches may be removed if necessary.
10. Before reinstallation, apply a thin coat of light duty grease to the rotating and sliding parts of each control lever.
11. Install the heater control panel and its brackets. Install the faceplate and make certain the clips are secured.

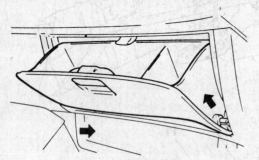

The Mirage glove box removal trick: lift the right side to free it from the hinge, then slide the left hinge free

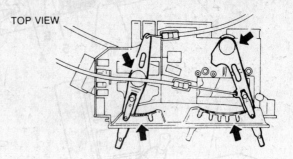

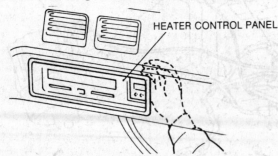

The clip behind the faceplate must be released before the faceplate will come out

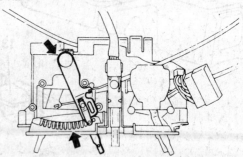

Lubricate all the slide and pivot points before reinstalling the heater control

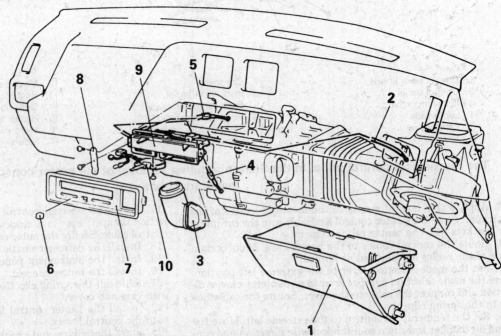

1. Glove box
2. Air selection cable
3. Defroster duct
4. Temperature control cable
5. Mode selection cable
6. Knobs
7. Heater control panel
8. Heater control bracket
9. Heater control assembly
10. Fan switch

1985–88 Mirage heater control panel

CHASSIS ELECTRICAL 6

12. Install the knobs on the levers.
13. Move the mode selection lever to the leftmost position. Move the mode selection damper lever fully inward (counterclockwise) and connect the control cable to the damper lever. Secure the outer cable in the spring clip.
14. Move the temperature control lever to the extreme left. Move the water valve lever inward (downward) and connect the control cable to the lever. Secure the outer cable in the spring clip.
15. Install the defroster duct.
16. Move the air selection control lever to the recirculate position. Press the air selection damper lever inward (towards the interior of the car) and connect the control cable. Secure the outer cable in the spring clip.
17. Check the function of each of the three control cables. Move the levers through their entire travel, making certain that the levers on the heater case move without binding. Adjust the cables as needed by repositioning the outer casing within the spring clip.
18. Reinstall the glove box.

Mirage, 1989

1. Remove the glove box.
2. Remove the ashtray
3. Remove the heater control panel faceplate. Note that the retaining screws are concealed from view.
4. Remove the radio/tape player.
5. Disconnect the three control cables.
6. Remove the heater control assembly mounting screws.
7. Remove the screws holding the heater control assembly to the left instrument panel support.
8. Press the controller back (into the dash) and then lower the back edge. This will allow the lower front edge of the controller to be placed through the dash opening.
9. Elevate the rear of the controller, allowing the top front section to come through the dash opening. The control assembly can now be removed.
10. Disconnect the electrical connectors at the rear of the control panel.
11. To reinstall the control panel, first connect the electrical connectors. Feed the control cables through the dash opening and route then to the heater assembly.
12. Install the control panel in the dash by getting first the top edge and then the bottom front edge behind the opening. This is done with the lift-and-push motion used during removal. Once the panel is correctly located, install the screws holding it to the dash support.
13. Move the mode selection lever to the leftmost position. Move the mode selection damper lever fully outward (counterclockwise) and connect the control cable to the damper lever. Secure the outer cable in the spring clip.

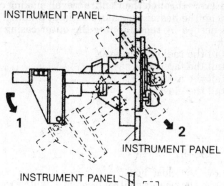
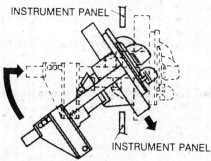

Removing the control assembly is a 2-step, four motion process. It can be done without forcing anything into place. Reverse the steps (4-3-2-1) to install the control

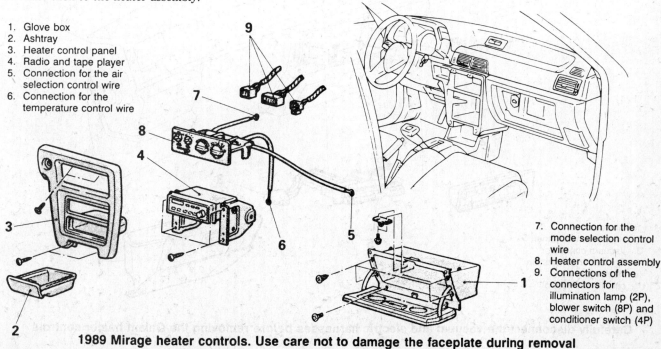

1. Glove box
2. Ashtray
3. Heater control panel
4. Radio and tape player
5. Connection for the air selection control wire
6. Connection for the temperature control wire
7. Connection for the mode selection control wire
8. Heater control assembly
9. Connections of the connectors for illumination lamp (2P), blower switch (8P) and conditioner switch (4P)

1989 Mirage heater controls. Use care not to damage the faceplate during removal

6-29

6 CHASSIS ELECTRICAL

14. Move the temperature control lever to the extreme left. Move the blend air damper lever inward (downward) and connect the control cable to the lever. Secure the outer cable in the spring clip.
15. Move the air selection control lever to the recirculate position. Press the air selection damper lever inward (towards the interior of the car) and connect the control cable. Secure the outer cable in the spring clip.
16. Check the function of each of the three control cables. Move the levers through their entire travel, making certain that the levers on the heater case move without binding. Adjust the cables as needed by repositioning the outer casing within the spring clip.
17. Install the radio/tape player.
18. Install the faceplate.
19. Install the ashtray.
20. Install the glovebox.

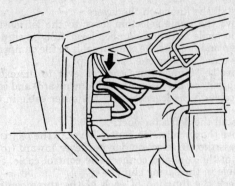

The vacuum connector is located to the left of the glove box

Galant, 1985–87

1. Remove the glove box.
2. Remove the heater control faceplate. Watch for concealed fasteners; don't force anything.
3. Disconnect the vacuum connector just to the left of the glove box.
4. Disconnect the electrical connector running to the control panel.
5. Remove the retaining screws and remove the heater control panel.
6. Reinstall the control panel in the dash. Connect the electrical and vacuum connectors, making certain each is correctly and firmly attached.
7. Install the faceplate.
8. Install the glovebox.

Galant, 1989

1. Release or remove the stopper for the glove box and remove the glovebox.
2. Disconnect the air selection control cable and the temperature control cable.
3. Remove the ashtray.

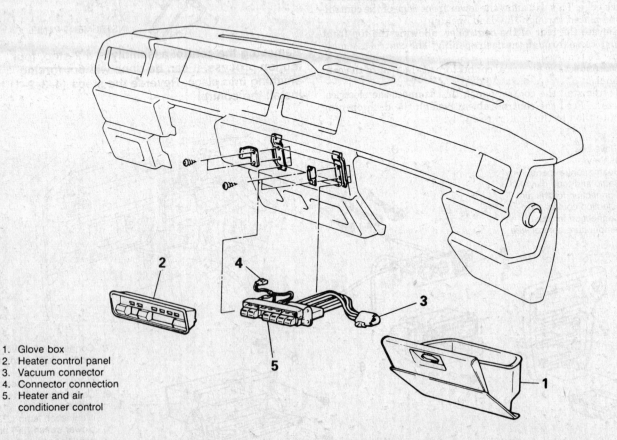

1. Glove box
2. Heater control panel
3. Vacuum connector
4. Connector connection
5. Heater and air conditioner control

Carefully disconnect the vacuum and electric harnesses before removing the Galant heater controls

6-30

CHASSIS ELECTRICAL 6

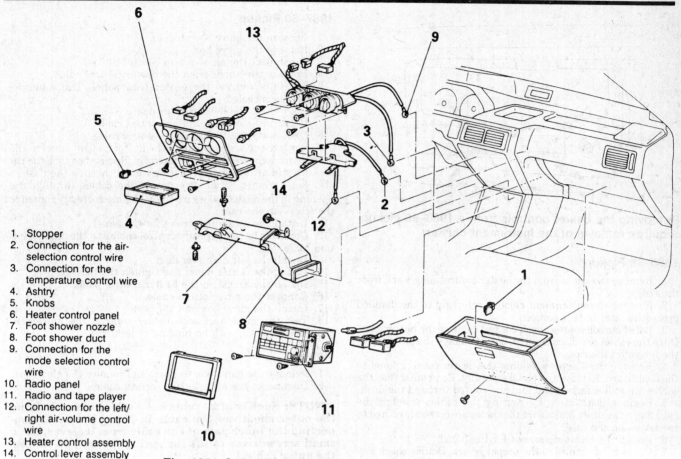

1. Stopper
2. Connection for the air-selection control wire
3. Connection for the temperature control wire
4. Ashtry
5. Knobs
6. Heater control panel
7. Foot shower nozzle
8. Foot shower duct
9. Connection for the mode selection control wire
10. Radio panel
11. Radio and tape player
12. Connection for the left/right air-volume control wire
13. Heater control assembly
14. Control lever assembly

The 1989 Galant has a two-part heater control assembly

4. Remove the knobs from the levers (not the round dials) by pulling them off with moderate force.
5. Remove the heater control faceplate. The retaining screws are concealed; make sure all are removed before removing the faceplate.
6. Remove the foot heater ducts and nozzle from the left side.
7. Disconnect the mode selection control cable.
8. Remove the radio trim panel and the radio.
9. Disconnect the left/right air volume control cable at the left side of the distribution duct lever.
10. Remove the retaining screws and remove the heater control assembly. Note that once removed, the lever assembly may be separated from the dial controls for inspection or replacement.
11. When reinstalling, make certain the lever assembly is connected to the dial panel before inserting the unit into the dash. Route the control cables to the heater case and install the retaining screws.
12. Move the selector lever for the left/right air volume to the **L** position. Move the foot distribution duct lever in a clockwise direction (right side low/left side high) and connect the cable. Secure the outer cable in the spring clip.
13. Install the radio and the radio face plate.
14. Move the mode selection lever to the defrost position. Move the mode selection damper lever fully inward (counterclockwise) and connect the control cable to the damper lever. Secure the outer cable in the spring clip.
15. Install the foot nozzle and ductwork.
16. Install the heater control trim plate.
17. Install the knobs on the control levers and install the ashtray.

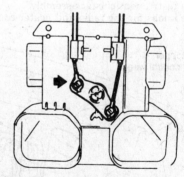

Correct position for removing and attaching the cable from the as L/R air flow selector

18. Move the temperature control lever to the extreme left. Move the blend air damper lever inward (downward) and connect the control cable to the lever. Secure the outer cable in the spring clip.
19. Move the air selection control lever to the recirculate position. Press the air selection damper lever inward (towards the interior of the car or clockwise) and connect the control cable. Secure the outer cable in the spring clip.
20. Check the function of each of the three control cables. Move the levers through their entire travel, making certain that the levers on the heater case move without binding. Adjust the cables as needed by repositioning the outer casing within the spring clip.
21. Install the glove box and its stopper.

6 CHASSIS ELECTRICAL

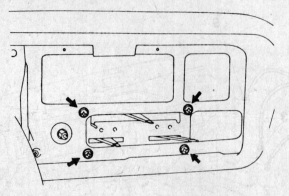

Removing the heater controls from a 1983–86 pickup requires removal of the instrument cluster

1985–86 Pickup

1. Remove the knobs from the heater control sliders and from the radio.
2. Remove the instrument cluster, referring to the detailed procedure later in this section.
3. Label and disconnect each control cable at the heater case. Once the cables are disconnected, do not change the positions of the levers on the case.
4. Remove the 4 screws holding the heater control panel to the dashboard. Remove the panel carefully. Remember that the cables are still attached and must be routed through the dash.
5. Begin reinstallation by routing the cables through the opening in the dash. Make sure the cables are correctly routed to the levers on the case.
6. Install the retaining screws for the panel.
7. Connect the cables to the proper levers. Double check any retainers or locking clips.
8. Check the operation of each heater control; it should move smoothly and without binding.
9. Install the instrument cluster assembly.
10. Install the knobs for the radio and heater controls.

1987–89 Pickup

1. Remove the glove box stopper.
2. Remove the glove box.
3. Disconnect the air selection control cable.
4. Remove the knobs from the heater control panel.
5. Carefully remove the center trim panel. The retaining screws are concealed.
6. Disconnect the left defroster duct.
7. Remove the mode selection control cable.
8. Disconnect the temperature control cable.
9. Remove the screws holding the heater control panel to the dashboard. Remove the panel carefully. Remember that the cables are still attached and must be routed through the dash.
10. Begin reinstallation by routing the cables through the opening in the dash. Make sure the cables are correctly routed to the levers on the case.
11. Install the retaining screws for the panel.
12. Connect the temperature control cable and the mode selection cable.
13. Install the left defroster duct.
14. Install the center panel and tighten the fittings.
15. Install the knobs on the heater control panel.
16. Connect the air selection cable.
17. Install the glove box and stoppers.
18. Test the motion of each heater control; the movement should be smooth, with no binding or jerkiness.

Montero

1. Remove the glove box stoppers and remove the glove box.
2. Disconnect the air selection control cable.

NOTE: Each control cable has a spring clamp holding the outer sheath of the cable in place. Before disconnecting the inner part of the cable from its lever, use a small screwdriver to lift the spring tab up and release the outer cable.

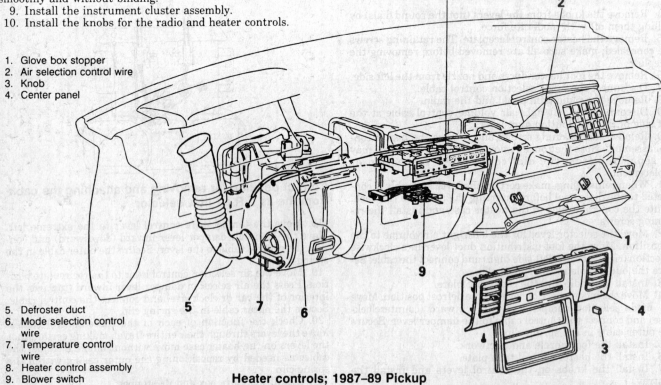

1. Glove box stopper
2. Air selection control wire
3. Knob
4. Center panel
5. Defroster duct
6. Mode selection control wire
7. Temperature control wire
8. Heater control assembly
9. Blower switch

Heater controls; 1987–89 Pickup

CHASSIS ELECTRICAL 6

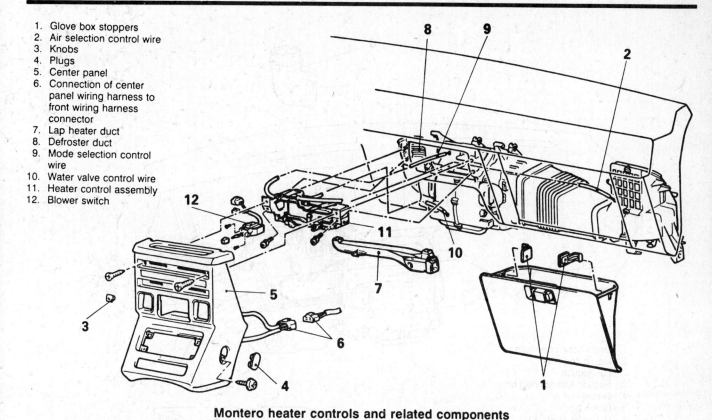

1. Glove box stoppers
2. Air selection control wire
3. Knobs
4. Plugs
5. Center panel
6. Connection of center panel wiring harness to front wiring harness connector
7. Lap heater duct
8. Defroster duct
9. Mode selection control wire
10. Water valve control wire
11. Heater control assembly
12. Blower switch

Montero heater controls and related components

3. Remove the knobs from the heater control panel.
4. Remove the small plugs from the side of the lower dash or console. Check the upper section of the console (near the levers). Remove any small concealment caps to reveal the attaching screws.
5. Remove the center panel by removing the upper and lower mounting screws.
6. As the center panel comes loose, reach behind it and disconnect the wiring connector.
7. Disconnect the lap heater duct under the dash.
8. Remove the defroster duct.
9. Disconnect the mode selector control cable.
10. Disconnect the temperature control cable to the water valve.
11. Unbolt and remove the heater control assembly.
12. Remove and disconnect the blower switch if so desired.
13. Before reinstallation, apply light grease to the slide points on the controller. Install the control panel and route the cables through the dash correctly.
14. Tighten the mounting screws.
15. Move the temperature control lever to the extreme left position.
16. With the water valve lever turned fully in the counterclockwise direction, connect the inner cable to the lever and then secure the sheath in the retaining clip.
17. Move the mode selection lever to the extreme left position.
18. With the mode selection damper lever (on the heater case) moved fully in a counterclockwise direction, connect the inner cable to the lever and secure the outer cable in the spring clip.
19. Install the defroster duct and install the lap heater duct.
20. Carefully connect the center panel wiring harness to the dash harness and install the center panel. Tighten the screws evenly and install the cover plugs.
21. Install the knobs on the heater control levers.
22. On the dash, move the air selection control lever to the "RECIRCULATE" position.
23. Move the air selection damper lever fully in the counterclockwise direction and connect the inner cable to the lever. Secure the outer cable in the spring clip.
24. Install the glove box and its stoppers.
25. Test the motion of each heater control; the movement should be smooth, with no binding or jerkiness.

Air Conditioner Evaporator Core

REMOVAL AND INSTALLATION

WARNING: The air conditioning lines and pipes are easily damaged. Always counter-hold joints with a second wrench when loosening or tightening. Start each connection by hand to insure correct threading. Tighten joints only to the correct torque; the seal is established by the O-ring, not by the tightness of the fitting.

Cordia and Tredia

1. Disconnect the negative battery cable.
2. Safely discharge the pressure within the air conditioning system. See Section 1.
3. Remove the drain hose from the evaporator case.
4. Carefully disconnect the liquid line and then the suction line at the firewall. Discard the small O-rings and cap the hoses immediately to keep the system free of dirt.
5. Remove the glove box.
6. Remove the right defroster duct.
7. Remove the duct joints between the evaporator and the component on either side (fan and heater units).

6 CHASSIS ELECTRICAL

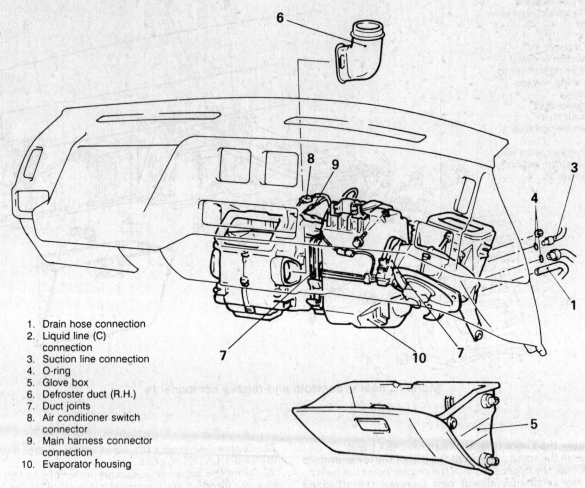

1. Drain hose connection
2. Liquid line (C) connection
3. Suction line connection
4. O-ring
5. Glove box
6. Defroster duct (R.H.)
7. Duct joints
8. Air conditioner switch connector
9. Main harness connector connection
10. Evaporator housing

Evaporator assembly, Cordia and Tredia

8. Disconnect the electrical connectors for the fan and air conditioner.
9. Remove the retaining bolts and carefully lift out the evaporator.
10. If the evaporator is not being replaced, cap the pipes as soon as the unit is clear of the car to prevent entry of dirt or foreign matter.
11. Reinstall the evaporator unit and secure the case with the retaining bolts.
12. Attach the electrical connectors and make certain the wire harnesses are not caught or pinched behind the evaporator case.
13. Install the duct joints, making sure each seals tightly to the adjoining component.
14. Install the defroster duct. Install the glove box.
15. Use a new O-ring on each line and apply compressor oil to the O-ring before connecting the lines.
16. Tighten the liquid line to 13 Nm (10 ft lbs) and the suction line to 34 Nm (25 ft. lbs.).
17. Connect the drain hose to the evaporator case.
18. Connect the negative battery cable.
19. Evacuate and recharge the air conditioning system. Do not overcharge the system. See Section 1.

Starion
1. Disconnect the negative battery cable.
2. Safely discharge the pressure within the air conditioning system. See Section 1.
3. Carefully disconnect the liquid line and then the suction line at the firewall. Discard the small O-rings and cap the hoses immediately to keep the system free of dirt.
4. Remove the small nut from the firewall side.
5. Remove the grommet.
6. Remove the glove box.
7. Remove the under cover below the glove box.
8. Remove the lap heater duct and the side console duct.
9. Remove the lower frame of the glove box.
10. Remove the defroster duct.
11. Remove the duct joints between the evaporator and the adjoining components.
12. Disconnect the drain hose.
13. Disconnect the electrical connectors and disconnect the vacuum hose.
14. Remove the retaining bolts and remove the evaporator case. if the evaporator is not to be replaced, immediately cap the pipe fittings to keep out dirt and foreign matter.

CHASSIS ELECTRICAL 6

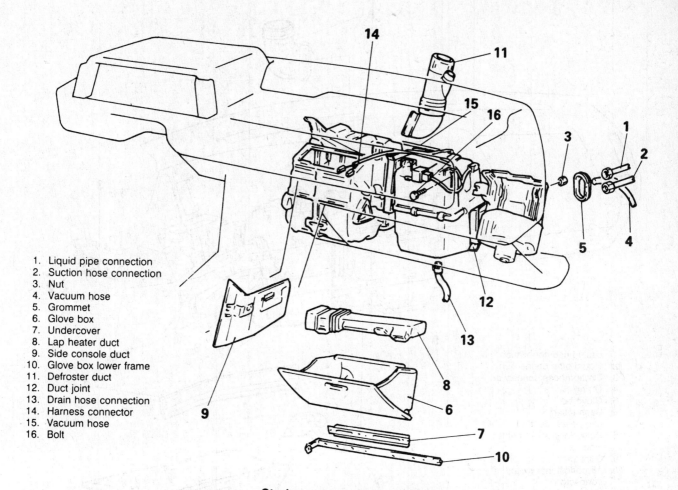

1. Liquid pipe connection
2. Suction hose connection
3. Nut
4. Vacuum hose
5. Grommet
6. Glove box
7. Undercover
8. Lap heater duct
9. Side console duct
10. Glove box lower frame
11. Defroster duct
12. Duct joint
13. Drain hose connection
14. Harness connector
15. Vacuum hose
16. Bolt

Starion evaporator assembly

15. To reinstall, place the case in the car and install the retaining bolts.
16. Connect the electrical wiring and the vacuum hose.
17. Install the case drain hose.
18. Install the defroster duct and the joint ducts. Make certain that the ductwork is properly seated all the way around.
19. Install the lower glove box frame.
20. Replace the side console duct and the lap heater duct.
21. Install the under cover and the glove box.
22. Install the grommet and the vacuum hose at the firewall.
23. Install the nut.
24. Use a new O-ring on each line and apply compressor oil to the O-ring before connecting the lines.
25. Tighten the liquid line to 13 Nm (10 ft lbs) and the suction line to 34 Nm (25 ft. lbs.).
26. Connect the negative battery cable.
27. Evacuate and recharge the air conditioning system. Do not overcharge the system. See Section 1.

Mirage, 1985–88

1. Disconnect the negative battery cable.
2. Safely discharge the pressure within the air conditioning system. See Section 1.

3. Remove the drain hose from the evaporator case.
4. Carefully disconnect the liquid line and then the suction line at the firewall. Discard the small O-rings and cap the hoses immediately to keep the system free of dirt.
5. Remove the glove box.
6. Remove the dash insert.
7. Remove the lap heater duct and the right defroster duct.
8. Remove the duct joints between the evaporator and the component on either side (fan and heater units).
9. Disconnect the electrical connectors for the fan and air conditioner.
10. Remove the retaining bolts and carefully lift out the evaporator.
11. If the evaporator is not being replaced, cap the pipes as soon as the unit is clear of the car to prevent entry of dirt or foreign matter.
12. Reinstall the evaporator unit and secure the case with the retaining bolts. Attach the electrical connectors and make certain the wire harnesses are not caught or pinched behind the evaporator case.
13. Install the duct joints, making sure each is properly seated. Once both are in place, adjust the clearance between the evaporator and the duct seal to about 2.5mm on each side.

6 CHASSIS ELECTRICAL

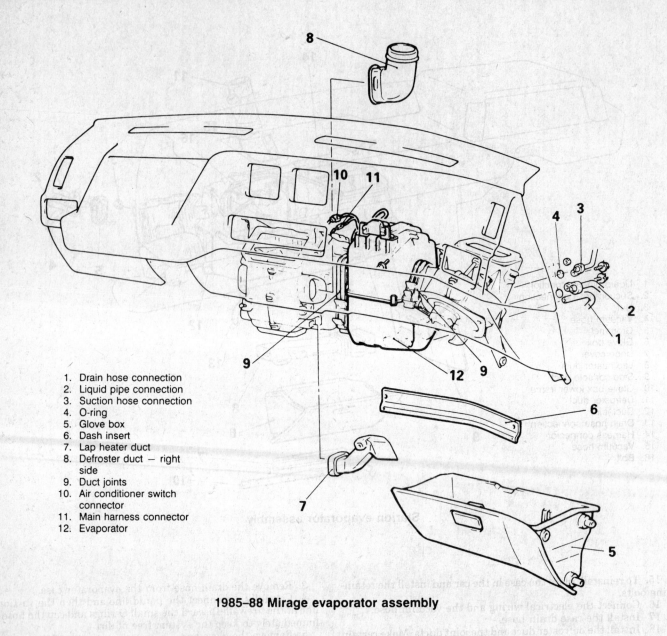

1. Drain hose connection
2. Liquid pipe connection
3. Suction hose connection
4. O-ring
5. Glove box
6. Dash insert
7. Lap heater duct
8. Defroster duct — right side
9. Duct joints
10. Air conditioner switch connector
11. Main harness connector
12. Evaporator

1985–88 Mirage evaporator assembly

14. Install the defroster duct and the lap heater duct.
15. Install the dash insert and install the glove box.
16. Use a new O-ring on each line and apply compressor oil to the O-ring before connecting the lines.
17. Tighten the liquid line to 5 Nm (4 ft lbs) and the suction line to 12 Nm (9 ft. lbs.)
18. Connect the drain hose to the evaporator case.
19. Connect the negative battery cable.
20. Evacuate and recharge the air conditioning system. Do not overcharge the system. See Section 1.

Mirage, 1989

1. Disconnect the negative battery cable.
2. Safely discharge the pressure within the air conditioning system. See Section 1.
3. Remove the charcoal canister from its mounts and set it

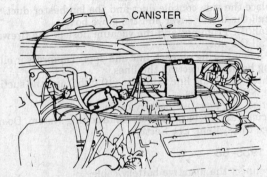

Move the charcoal canister so you can get at the air conditioning lines on the 1989 Mirage.

6–36

CHASSIS ELECTRICAL 6

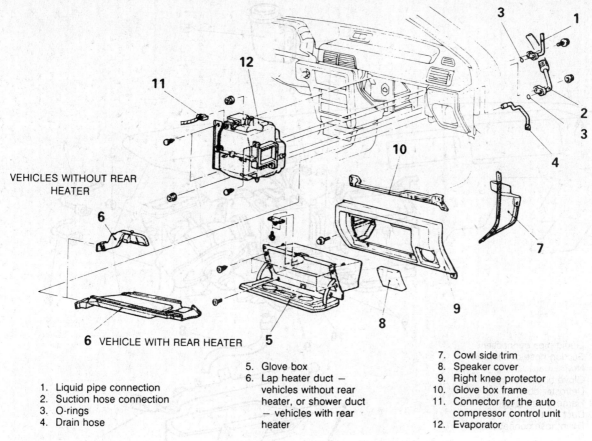

1. Liquid pipe connection
2. Suction hose connection
3. O-rings
4. Drain hose
5. Glove box
6. Lap heater duct — vehicles without rear heater, or shower duct — vehicles with rear heater
7. Cowl side trim
8. Speaker cover
9. Right knee protector
10. Glove box frame
11. Connector for the auto compressor control unit
12. Evaporator

1989 Mirage evaporator assembly

aside; leave the hoses connected.
4. Carefully disconnect the liquid line and then the suction line at the firewall. Discard the small O-rings and cap the hoses immediately to keep the system free of dirt.
5. Remove the drain hose from the evaporator case.
6. Remove the glove box.
7. Remove the lap heater duct of the shower duct.
8. Remove the cowl side trim (kick panel).
9. Remove the speaker cover, the remove the knee protector from the dashboard.
10. Remove the glove box frame.
11. Disconnect the 12 pin connector for the automatic control unit.
12. Remove the mounting bolts and remove the evaporator case.
13. If the evaporator is not being replaced, cap the pipes as soon as the unit is clear of the car to prevent entry of dirt or foreign matter.
14. Reinstall the evaporator unit and secure the case with the retaining bolts. Attach the electrical connector and make certain the wire harnesses are not caught or pinched behind the evaporator case.
15. Install the glove box frame, the knee protector and the speaker cover.
16. Install the kick panel (side trim)
17. Install the shower duct or the lap duct.
18. Install the glove box.
19. Connect the drain hose.
20. Use a new O-ring on each line and apply compressor oil to the O-ring before connecting the lines.

21. Tighten the liquid line to 5 Nm (4 ft lbs) and the suction line to 12 Nm (9 ft. lbs.)
22. Reposition the canister and mount it securely. Double check the lines for any kinks or loose fittings.
23. Connect the negative battery cable.
24. Evacuate and recharge the air conditioning system. Do not overcharge the system. See Section 1.

Galant, 1985–88
All Sigma

1. Disconnect the negative battery cable.
2. Safely discharge the pressure within the air conditioning system. See Section 1.
3. Carefully disconnect the liquid line and then the suction line at the firewall. Discard the small O-rings and cap the hoses immediately to keep the system free of dirt. Remove the small nut at the firewall.
4. Remove the glove box and underframe.
5. Remove the defroster duct from the right side of the heater case.
6. Disconnect the electrical connectors for the fan and air conditioner.
7. Remove the duct joints between the evaporator and the component on either side (fan and heater units).
8. Remove the drain hose from the evaporator case.
9. Remove the retaining bolts and carefully lift out the evaporator case.

6-37

6 CHASSIS ELECTRICAL

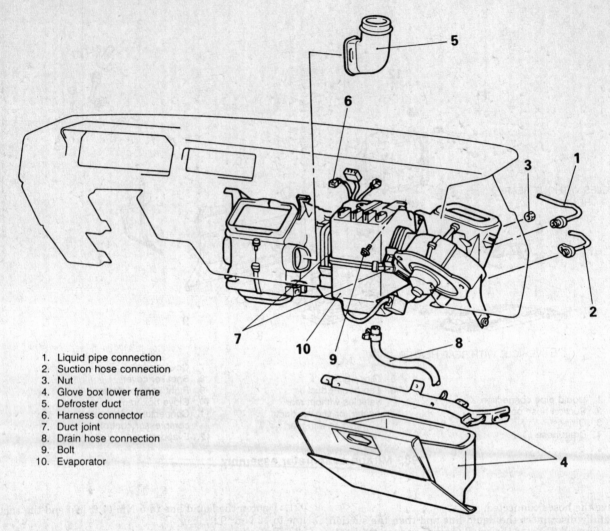

1. Liquid pipe connection
2. Suction hose connection
3. Nut
4. Glove box lower frame
5. Defroster duct
6. Harness connector
7. Duct joint
8. Drain hose connection
9. Bolt
10. Evaporator

1985–88 Galant evaporator case

10. If the evaporator is not being replaced, cap the pipes as soon as the unit is clear of the car to prevent entry of dirt or foreign matter.
11. Reinstall the evaporator unit and secure the case with the retaining bolts.
12. Attach the electrical connectors and make certain the wire harnesses are not caught or pinched behind the evaporator case.
13. Connect the drain hose.
14. Install the duct joints, making sure each is properly seated. Once both are in place, adjust the clearance between the evaporator and the duct seal to about 2.5mm on each side.
15. Install the defroster duct.
16. Install the glove box and under frame.
17. Use a new O-ring on each line and apply compressor oil to the O-ring before connecting the lines.
18. Tighten the liquid line to 5 Nm (4 ft lbs) and the suction line to 12 Nm (9 ft. lbs.)
19. Connect the negative battery cable.
20. Evacuate and recharge the air conditioning system. Do not overcharge the system. See Section 1.

Galant, 1989

1. Disconnect the negative battery cable.

2. Safely discharge the pressure within the air conditioning system. See Section 1.
3. Carefully disconnect the liquid line and then the suction line at the firewall. Discard the small O-rings and cap the hoses immediately to keep the system free of dirt.
4. Remove the dashboard side cover.
5. Remove the dashboard under cover.
6. Remove the shower duct.
7. Remove the glove box stopper, the glove box and the glove box frame.
8. Disconnect the wiring connectors behind the glove box frame.
9. Remove the ashtray.
10. Remove the knobs from the heater control levers (don't remove the round dial controls).
11. Remove the heater control faceplate.
12. Disconnect the air conditioning switch connector at the control panel.
13. At the bottom of the evaporator case, disconnect the wiring harness.
14. Remove the plate from the evaporator, then remove the screws holding the evaporator to the car. Support the unit in position.

CHASSIS ELECTRICAL 6

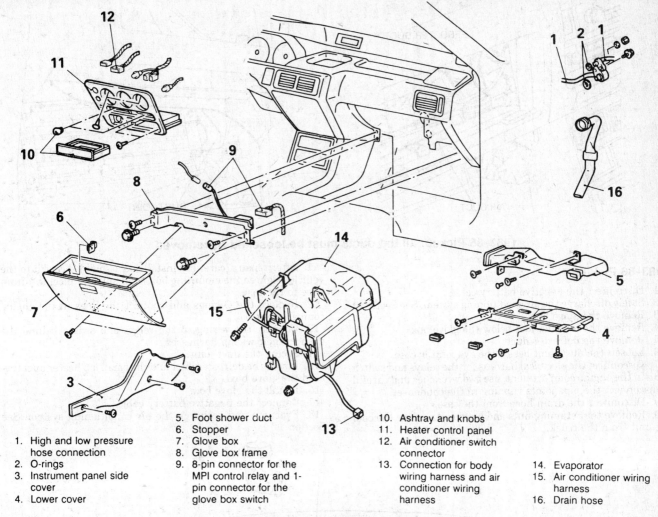

1. High and low pressure hose connection
2. O-rings
3. Instrument panel side cover
4. Lower cover
5. Foot shower duct
6. Stopper
7. Glove box
8. Glove box frame
9. 8-pin connector for the MPI control relay and 1-pin connector for the glove box switch
10. Ashtray and knobs
11. Heater control panel
12. Air conditioner switch connector
13. Connection for body wiring harness and air conditioner wiring harness
14. Evaporator
15. Air conditioner wiring harness
16. Drain hose

1989 Galant evaporator case

15. After the plate is removed, carefully work the notched part of the evaporator clear of the instrument panel.

WARNING: Do this removal carefully so as not to damage the insulation.

16. The drain hose and remaining wiring may be removed. If the evaporator core is not being replaced, immediately cap or plug the lines to prevent entry of foreign matter, dirt and moisture.
17. Before reassembly, make certain that all external components are in place. Carefully fit the case through the dashboard and install the retaining nuts and bolts. Install the plate.
18. Connect the wiring harness to the bottom of the evaporator case.
19. Connect the wiring at the control panel and install the faceplate.
20. Install the knobs on the control levers; install the ashtray.
21. Install the glove box frame and connect the wiring harnesses.
22. Install the glove box and stopper.
23. Install the shower duct and the under cover.
24. Install the side cover on the instrument panel.
25. Use a new O-ring on each line and apply compressor oil to the O-ring before connecting the lines.
26. Tighten the liquid line to 5 Nm (4 ft lbs) and the suction line to 12 Nm (9 ft. lbs).
27. Connect the negative battery cable.
28. Evacuate and recharge the air conditioning system. Do not overcharge the system. See Section 1.

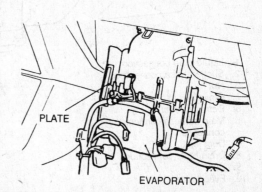

Remove the plate or the casing will NOT clear the instrument panel on the way out

6-39

6 CHASSIS ELECTRICAL

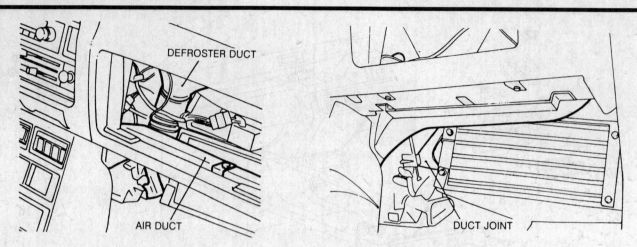

1983–86 Pickup: all the ducts must be loosened or removed

1983–86 Pickup

1. Disconnect the negative battery cable.
2. Safely discharge the air conditioning system. See Section 1.
3. Remove the glove box.
4. Remove the air duct just below the glove box.
5. Remove the defroster duct.
6. Loosen the duct joint beside the evaporator case.
7. Disconnect the electrical harness for the relays and switch.
8. In the engine compartment, use two wrenches and carefully disconnect the pipe joints running to the condenser.
9. Disconnect the drain hose from the case.
10. Remove the retaining nuts and bolts and remove the cooling unit from the truck.
11. After repairs, carefully install the assembled case into the vehicle. Tighten the mounting bolts and nuts. Connect the drain hose to the case.
12. Replace the O-rings and connect the lines in the engine compartment.
13. Connect the wiring to the relays and switch behind the dash to their wiring harnesses.
14. Secure the duct joint.
15. Install the defroster duct and the small lap heater duct (behind the glove box).
16. Install the glove box.
17. Connect the negative battery cable.
18. Evacuate and recharge the air conditioning system. See Section 1.

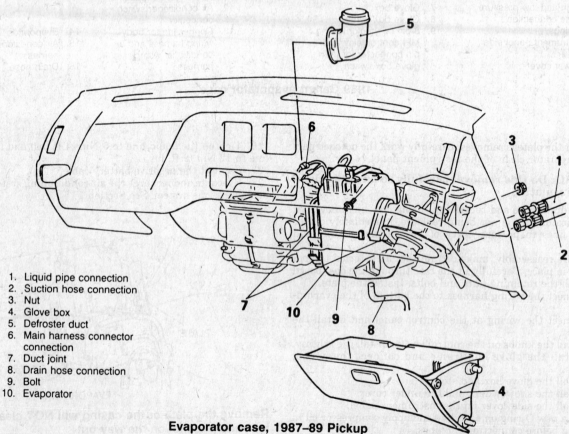

1. Liquid pipe connection
2. Suction hose connection
3. Nut
4. Glove box
5. Defroster duct
6. Main harness connector connection
7. Duct joint
8. Drain hose connection
9. Bolt
10. Evaporator

Evaporator case, 1987–89 Pickup

CHASSIS ELECTRICAL 6

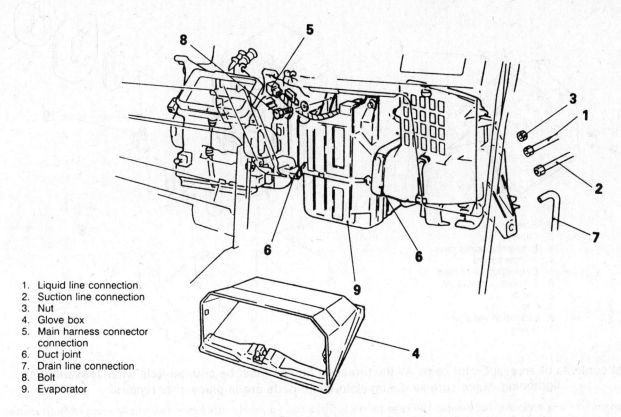

1. Liquid line connection
2. Suction line connection
3. Nut
4. Glove box
5. Main harness connector connection
6. Duct joint
7. Drain line connection
8. Bolt
9. Evaporator

Montero evaporator core

1987–89 Pickup
All Montero

1. Disconnect the negative battery cable.
2. Safely discharge the air conditioning system. See Section 1.
3. In the engine compartment, use two wrenches to carefully disconnect the liquid line and the suction line at the firewall fittings.
4. Remove the retaining nut (located just above the hoses.)
5. Remove the glove box from the dashboard.
6. Remove the defroster duct.
7. Disconnect the main electrical harness connector running to the evaporator case.
8. Loosen the duct joint between the evaporator case and the heater case.
9. Remove the drain hose.
10. Remove the retaining bolts and remove the evaporator case from the vehicle.
11. After repairs, carefully install the assembled case into the vehicle. Tighten the mounting bolts and nuts. Connect the drain hose to the case.
12. Install the duct joints on either side of the evaporator. Correctly installed, the evaporator case should have about 3mm clearance to the joint on each side.
13. Connect the main harness electrical connectors.
14. Install the defroster duct.
15. Install the glove box.
16. Under the hood, install the retaining nut if not already done during installation.
17. Replace the O-rings on each line and coat each ring lightly with compressor oil.
18. Connect and tighten the lines, using a second wrench to counterhold the opposite side of the joint.
19. Connect the negative battery cable.
20. Evacuate and recharge the system. See Section 1.

REPLACEMENT

The evaporator core and other components are located within the casing. In all cases the case must be opened for access to the various parts. Use a broad tool wrapped in a cloth to remove the clips. Use of an unpadded tool may damage or break the case. Check the case carefully for any hidden screws holding the halves together.

The best theory to apply during disassembly is removing the case from the evaporator, not the evaporator from the case. Once the components are exposed, the air-flow sensor, expansion valve, thermostat and/ or insulating panels will need removal. Use a few simple rules to guide your work:

1. Don't force anything. All components are delicate and easily broken or stripped.
2. Use two wrenches whenever a joint is loosened or tightened.
3. Any O-ring in any joint MUST be discarded when the joint is loosened. Replace each ring and lubricate it with compressor oil during reassembly.
4. All of the insulation and support pieces are there for a reason. Don't try to redesign the system — just put everything back where it belongs.

6 CHASSIS ELECTRICAL

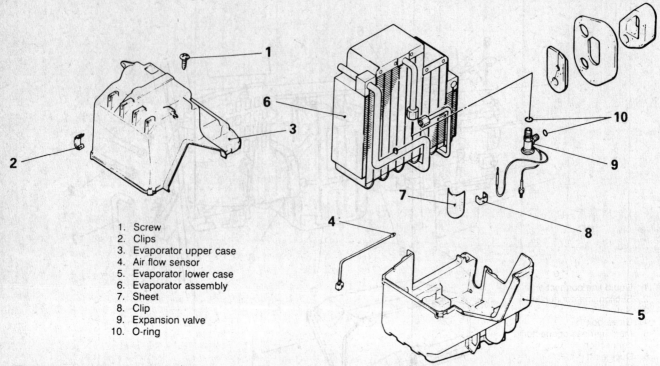

1. Screw
2. Clips
3. Evaporator upper case
4. Air flow sensor
5. Evaporator lower case
6. Evaporator assembly
7. Sheet
8. Clip
9. Expansion valve
10. O-ring

Typical contents of an evaporator case. All the threaded fittings must be counter-held when loosening or tightening. Make sure all the insulators and pads are in place after repairs.

After components are replaced, reassemble the case halves around the evaporator core. Make certain the case fits exactly as it should; the correct air flow around the core is crucial. An air leak from a poorly fitted case could upset the system function. Install all the clips properly and the small case screws if any were used.

RADIO

REMOVAL AND INSTALLATION

Cordia and Tredia

1. Carefully pry the radio cover panel out with a thin object or tool. Pry from the right on Cordia and the left on Tredia.
2. Remove the mounting screws, disconnect the antenna and electrical connector, and remove the radio.
3. Installation is the reverse of removal.

Starion

1. Insert a screwdriver blade between the tray in the rear console and its cover and twist it to release the tray. Then, remove it. Now, remove the side console cover screws. Push the cover downward and slightly forward to release it, and then remove it.
2. Remove the radio mounting screws from the center reinforcement. Put the shift lever into fourth gear, and then unscrew and remove the gearshift knob.
3. Pull the front console box slightly to the rear. Then, disconnect the electrical connector and antenna at the radio. Now, pull the front console box out toward the passenger seat.

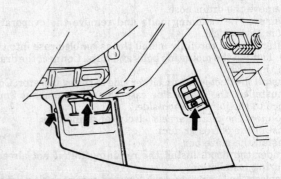

Removing the side console cover screws on Starion

CHASSIS ELECTRICAL 6

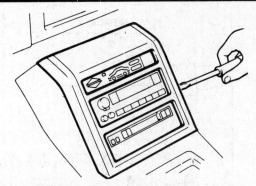

Prying off the Galant lower radio panel

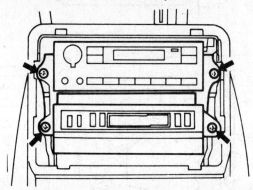

Remove the Galant radio mounting screws

Mirage console mounting screws. The console must be removed for access to the radio

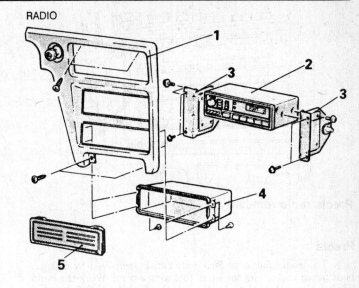

1. Center panel
2. Radio or radio with tape player
3.
4. Box
5. Radio plug for vehicles without a radio

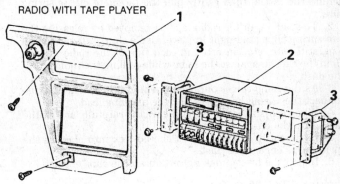

1989 Mirage radio

4. Now, remove the four Phillips screws and then remove the radio panel. Remove the radio and bracket as an assembly. Remove the bracket from the radio.
5. Installation is the reverse of removal.

Galant and Sigma

1. Use a flat-bladed screwdriver to pry on the lower part of the radio panel and remove it.
2. Remove the four Phillips screws from the sides of the radio (and tape player) mounting bracket, and pull the units out.
3. Disconnect the electrical connectors and the antenna, and then remove the unit.
4. Installation is the reverse of removal.

Mirage through 1988

1. Remove the panel that surrounds the parking brake.
2. Remove the ashtray.

3. Remove the two console mounting screws from each side, two near the handbrake, and one at the rear of the ashtray. Then, remove the console.
4. Remove the four mounting screws from the radio bracket, and remove the radio and bracket as an assembly. Now, separate the radio and bracket. Note that the radio fuses are located behind the radio and are now accessible.
5. Installation is the reverse of removal.

1989 Mirage

1. Remove the cover panel from the lower dash or console.
2. Remove the retaining screws and remove the radio unit. Disconnect the wiring harness and antenna lead as soon as the radio is loose.
3. Remove the radio brackets if necessary.
4. Installation is in reverse order. Connect the wire harnesses carefully and make certain the antenna lead is firmly plugged to the radio.
5. Carefully install the cover panel and tighten the screws.

6-43

6 CHASSIS ELECTRICAL

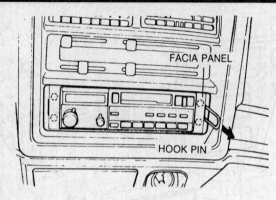

Precis radio removal

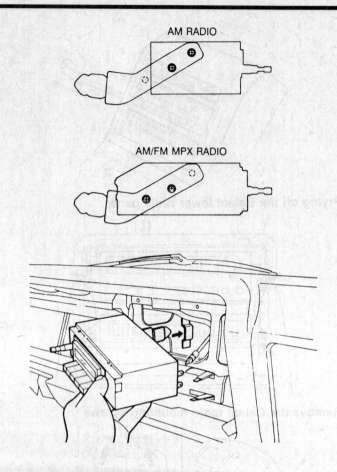

1983–86 truck radios use different bracket positions for different radios. Make certain the support tab fits into the bracket

Precis

1. The radio fascia or face plate can be removed by using a bent paper clip or similar small tool with a bend. Work the bend behind the fascia, then gently pull the faceplate off its spring clips.
2. The radio unit (or radio/tape) is removed by releasing the two spring clips holding it in place. Use a thin, stiff rod (small nail, stiff wire, etc.) inserted in each side of the radio mount. Hold the clips in against the radio while pulling the radio out of the dash.
3. As the radio comes out, reach behind or under it and disconnect the wiring harnesses and the antenna lead.
4. When reinstalling, connect the wiring carefully and fit the radio into the dash.
5. Push gently on the radio; it should lock into place as the spring clips engage the dash.
6. Install the fascia, making certain all the clips engage.

Pickup, 1983–86

CHILTON TIP: The radio fuses are mounted on the right rear corner of the radio. The radio need not be removed to change a fuse; just remove the glovebox and reach through the opening.

1. Remove the instrument cluster bezel (trim piece).
2. Remove the radio bracket attaching screws from the instrument panel and remove the radio bracket.
3. Pull the radio outward slightly and disconnect the speaker wires, the power wire and the antenna cable.
4. Remove the radio.
5. When reinstalling, note that the mounting support attaches to the radio differently for AM radios or AM/FM radios.
6. Hold the radio in position and connect the wiring and antenna leads.
7. Before setting the radio into the dash, make certain that the mounting stay fits into the slot in the reinforcement.
8. Install the instrument bezel.

Pickup, 1987–89

1. Remove the heater control lever knob.
2. Remove the screws holding the center trim panel. Use a plastic or padded stick or flat tool to remove the panel without marring it.
3. Remove the radio bracket mounting screws.
4. Pull the radio unit out of the dash (with the brackets attached) a short distance.
5. Disconnect the wiring and antenna lead at the rear of the radio.
6. Remove the radio from the truck and remove the brackets from the radio.
7. When reinstalling, mount the brackets to the radio and install them as a unit.
8. Before fitting the radio completely into the dash, connect the wiring and antenna cable.
9. Tighten the bracket mounting screws.
10. Install the face plate and the heater control lever knob.

Montero, 1983–86

1. Remove the radio faceplate. Use a flat plastic tool or protected metal tool so as not to mar the plastic
2. Remove the plug on the each side of the center console. Reach through the holes and remove the radio mounting screws.
3. Disconnect the antenna lead wire, the speaker connector and the power wire from the back of the radio.
4. Remove the radio.
5. Installation is the reverse of removal.

Montero, 1987–89

1. Remove the knobs from the heater controls.
2. Remove the plugs from the side of the console.
3. Remove the center console from the dash. As it comes clear, disconnect the wiring harness.

CHASSIS ELECTRICAL 6

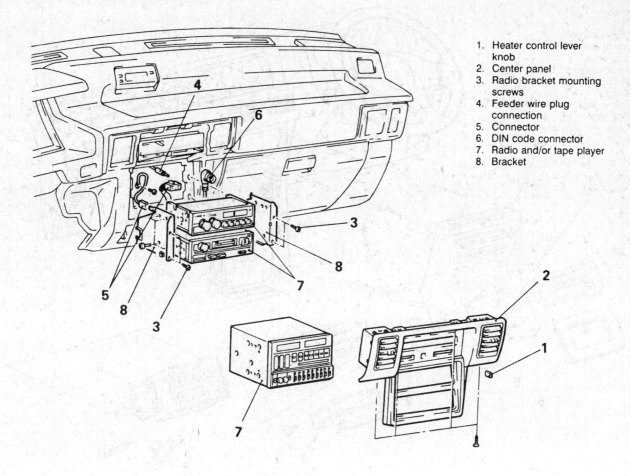

1. Heater control lever knob
2. Center panel
3. Radio bracket mounting screws
4. Feeder wire plug connection
5. Connector
6. DIN code connector
7. Radio and/or tape player
8. Bracket

1987–89 Pickup radio removal

4. Remove the radio faceplates.
5. Disconnect the radio brackets from the console.
6. Remove the radio with the brackets attached. Once clear of the console, the brackets may be removed.
7. When reinstalling, attach the brackets to the radio and the radio to the console.
8. As the console is moved into place, connect the wiring harness.
9. Secure the center console, install the hole plugs and install the heater control knobs.

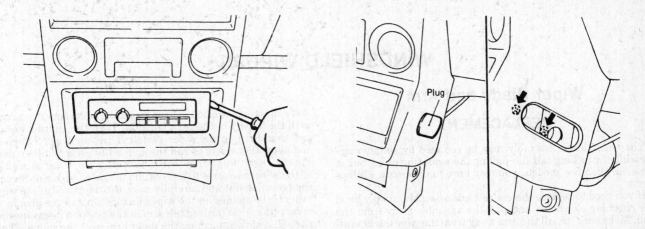

The radio in 1983–86 Monteros is held by hidden screws

6–45

6 CHASSIS ELECTRICAL

1. Knob
2. Plug
3. Center console
4. Wiring harness connector
5. Radio panel
6. Radio bracket
7. Radio
8. Box — vehicles without tape player
9. Radio with tape player

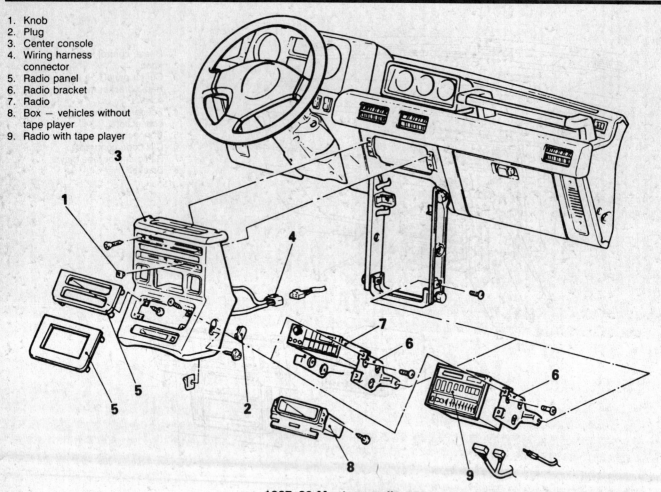

1987–89 Montero radio

WINDSHIELD WIPERS

Wiper Blade and Arm

REPLACEMENT

The wiper blade assembly may be replaced by simply compressing the locking tab and pulling the mounting prong out of the arm. On a few models, you may have to unscrew a Phillips screw.

If you need to replace the entire blade assembly, simply bend the retaining clip down and pull the assembly away from the arm. Make sure to pull the arm away from the windshield until it passes the detent before releasing it. Otherwise, it may spring against the windshield and scratch it. Replace the blade assembly.

If you have to replace an arm assembly, make sure to start out with the wipers in the park position. Turn the ignition key on, and then turn on the wiper switch. Turn the wiper switch off and allow the wipers to run through a full cycle and then park. Finally, turn the ignition switch back off. Now, note the position of the wipers relative to the bottom of the windshield. Remove the retaining nut and carefully work the inner end of the wiper arm off the splines on the wiper linkage. Install the arm in reverse order, first putting the arm in the position it was in when parked, and then pushing the inner arm over the splines. Turn it slightly, if necessary, to line up the splines and slide the inner arm over the linkage. Install the retaining nut.

CHASSIS ELECTRICAL 6

The actual rubber blade or insert can be replaced without replacing other components. The insert its reinforcing strips can be pulled from the blade (holder) with moderate effort. Examine the ends of the rubber and identify the end with the lock tab; two small teeth in the arm fit into a slot in the rubber. Pull this end away from the holder; the tabs should pop loose and the insert will slide free. Remove the two metal reinforcing rods. DO NOT throw these away; new inserts do not come with reinforcements.

Install the reinforcements into the channel on each side of the new blade. Once assembled, slide the insert into the holder, making sure the insert goes through each clip. When all the way in the holder, a sharp tug on the new insert will seat the clips into the slot. Make sure they engage correctly; this tab-and-slot is the only thing keeping the insert from flying off into traffic.

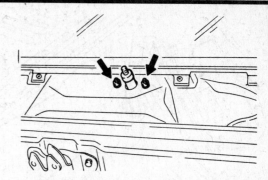
Cordia and Tredia wiper pivot shaft mounting nuts

Motor and Linkage

REMOVAL AND INSTALLATION

Cordia and Tredia

1. Remove the wiper arms. On the Tredia, remove the deck panel. On the Cordia, remove the water guide panel, waterproof trim and front deck trim.
2. Remove the two pivot shaft mounting nuts from either side of each shaft and then push the shaft into the panel. On the Tredia, remove the windshield washer nozzles.
3. Disconnect the electrical connector. Remove the motor mounting bolts and pull the motor out slightly. Holding the motor in this position, free the linkage and pull the motor and linkage out.
4. If the motor is being replaced, matchmark the relationship between linkage and motor, as it is critical. If the linkage only is being replaced, pry the connection off the end of the motor crank arm with a flat tool.
5. Install in reverse order. Torque the pivot shaft nuts to 60 inch lbs. and the wiper arm locking nuts to 7-12 ft. lb. Wiper blades should rest 25mm above the lower windshield molding on the Tredia, and about 20mm on the Cordia.

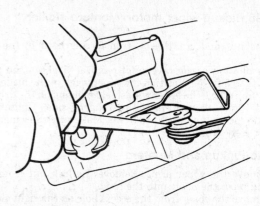
Separating the Cordia wiper linkage from the motor arm

Galant

1. Remove the wiper arms. Remove the front deck and inlet trim.
2. Remove the three mounting bolts for each pivot shaft, and push the shafts into the area behind the panel. Disconnect the electrical connector. Loosen and remove the three motor mounting bolts and then remove the motor and linkage as an assembly.
3. If the motor is being replaced, matchmark the relationship between linkage and motor, as it is critical. If the linkage only is being replaced, pry the connection off the end of the motor crank arm with a flat tool.
4. Install in reverse order. Make sure the wiper arms sit in their original positions when parked. Make sure the wiper motor is securely grounded.

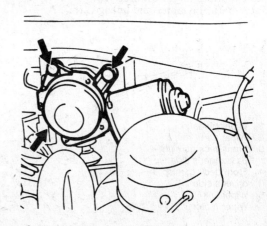
Galant wiper motor mounting nuts

Starion

1. Remove the wiper arms. Remove the pivot shaft mounting nuts and washers, and push the pivot shafts into the area behind the cowl.
2. Remove the cover from the access hole for the wiper motor on the right side of the cowl, underneath the hood. Remove the motor mounting bolts. Pull the motor into the best possible position for access and use a flat tool to pry the linkage off the motor crank arm.
3. If the linkage is being replaced, it can be worked out of the cowl at this time. If the motor is being replaced, matchmark the position of the crank arm of the motor shaft of the new motor, and then remove the nut and crank arm, transferring both to the new motor.

4. Installation is the reverse of removal. Make sure the wiper blades stop about 13mm from the lower windshield molding. Tighten the wiper arm attaching nuts to 10 ft. lbs.

Mirage

1. Remove the air inlet and cowl center trim panels. Remove the three pivot shaft mounting nuts and push the pivot shafts into the area under the cowl.
2. Remove the wiper motor mounting bolts. Pull the motor into the best possible position for access and use a flat tool to pry the linkage off the motor crank arm. Remove the motor and then the linkage.
3. If the motor is being replaced, matchmark the position of the crank arm of the motor shaft of the new motor, and then re-

6 CHASSIS ELECTRICAL

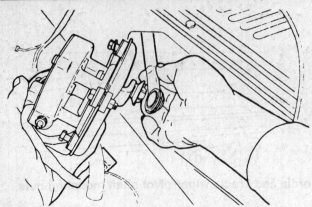

1983–86 Pickup wiper motor. Montero similar

move the nut and crank arm, transferring both to the new motor.

4. Installation is the reverse of removal. Tighten the pivot shaft nuts to 5 ft. lbs. Position the wiper arms so the blades are about 13mm above the lower windshield molding on the driver's side and 19mm above it on the passenger's side. Tighten the wiper arm mounting nuts to 10 ft. lbs. Make sure the wiper motor is securely grounded.

1983–86 Pickup and Montero

1. Remove the wiper arms. Remove the lock nut from the shaft and push the shaft into the cowl.
2. Remove the cover from the access hole on the right side of the front deck.
3. Remove the bolts that hold the motor bracket to the body.
4. Pull the wiper motor towards you.
5. Matchmark the motor and linkage arm.
6. Disconnect the wiper motor and linkage so that the motor shaft and linkage are at right angles. Hold the linkage with your right hand when disconnecting.
7. The linkage maybe removed through the access hole as necessary.
8. Reinstall and position the matchmarks correctly. (If not done, the wipers will not park correctly.)
9. Connect the linkage to the motor and install the motor. Install the motor bracket bolts.
10. Lift the linkage so the shafts project through the holes and install their locknuts.
11. Install the access hole cover. Install the wiper arms.

1987–89 Pickup and Montero

1. Remove the wiper arms.
2. On Pickups, carefully remove the left and right cowl panels. Use a flat, padded tool to pry the clips loose.
3. Remove the wiper motor mounting bolts and pull the motor away from the firewall. Use a flat tool to disconnect the motor from the linkage. Remove the motor.
4. If the linkage is to be removed, remove the shaft plate mounting nuts and remove the linkage through the motor opening.
5. When reinstalling the linkage, the mount nuts should be tightened to 5 NM or 3 ft. lbs.
6. Connect the motor to the linkage and install the motor mount bolts.
7. install the left and right cowl panels if they were removed.
8. Install the wiper arms.

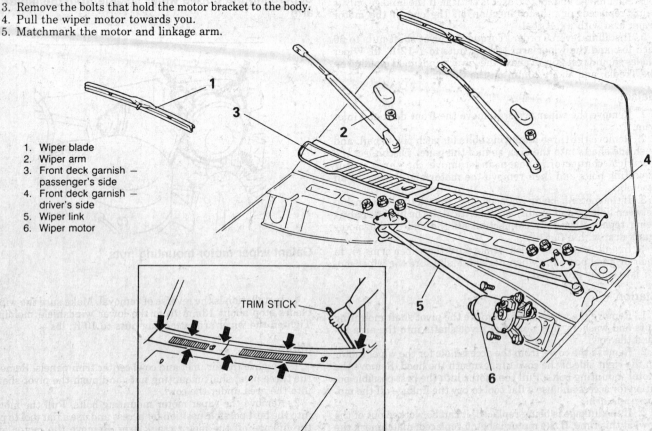

1. Wiper blade
2. Wiper arm
3. Front deck garnish — passenger's side
4. Front deck garnish — driver's side
5. Wiper link
6. Wiper motor

Carefully remove the vented cowl panels from the 1987–89 Pickup

6–48

CHASSIS ELECTRICAL 6

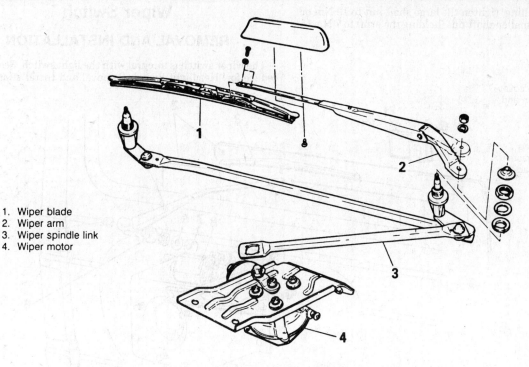

1. Wiper blade
2. Wiper arm
3. Wiper spindle link
4. Wiper motor

Precis wiper linkage

Precis

1. Remove the wiper arms.
2. Remove the upper cowl cover.
3. Remove the lock nut from each pivot shaft.
4. Use a broad flat tool to disconnect the linkage shaft from the motor.
5. Remove the motor bracket mounting bolts and remove the motor.
6. When reinstalling, mount the motor in position and tighten the bracket bolts.
7. Install the linkage and connect it to the motor.
8. Tighten the shaft nuts.
9. Install the wiper arms.

Rear Wiper Motor

REMOVAL AND INSTALLATION

1. Remove the rear wiper arm and blade.

NOTE: Pay attention to the placement and order of the washers and fittings on the wiper shaft.

2. Remove the large shaft nut found below the wiper arm.
3. Open the hatch lid. Carefully remove the inner liner or trim panel. On Montero, remove the small trim panel around the inside door handle.

WARNING: There may be a waterproof plastic liner under the trim panel. This liner must be removed carefully and without rips. Minor rips may be repaired with waterproof tape. This sheet must be intact when reinstalled.

4. Disconnect the wiring connector either at the motor or at the harness connector. Release any clips holding the wiring harness.
5. Remove the retaining bolts holding the motor. Remove the motor.

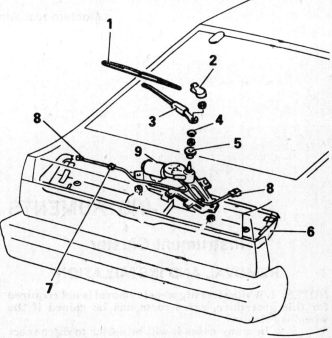

1. Wiper blade
2. Head cover
3. Wiper arm
4. Shield cap
5. Collar
6. Tailgate trim
7. Harness clip
8. Wiper motor connector
9. Wiper motor

Typical rear wiper components

6 CHASSIS ELECTRICAL

6. When reinstalling, tighten the large shaft nut to 13 Nm or 10 ft. lbs and the smaller shaft nut (holding the arm) to 7 Nm (5 ft. lbs).

Wiper Switch

REMOVAL AND INSTALLATION

The wiper switch is integral with the light switch. See the procedure for "Headlight Switch Removal and Installation".

1. Inside handle cover
2. Back door trim and watershield
3. Wiper blade
4. Wiper arm
5. Wiper pivot cap
6. Wiper pivot washer
7. Wiper pivot backing
8. Wiper motor

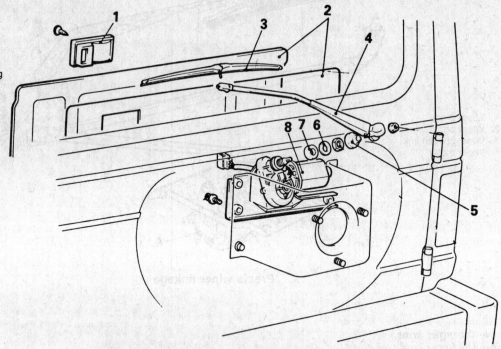

Montero rear wiper components

INSTRUMENTS AND SWITCHES

Instrument Cluster

REMOVAL AND INSTALLATION

NOTES: 1. While steering wheel removal is not required for this procedure, extra room can be gained if the wheel is removed.

2. In many cases it will be easier to disconnect the speedometer cable at the transmission or transaxle first. Push some slack in the cable into the car; this will allow the instrument cluster to come out of the dash far enough to reach the upper connector.

Cordia and Tredia

1. Remove the attaching screws for the trim panel in front of the instrument cluster, two above the instruments and one on

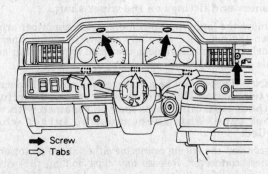

Fasteners in several locations hold the Cordia and Tredia trim panel

6–50

CHASSIS ELECTRICAL 6

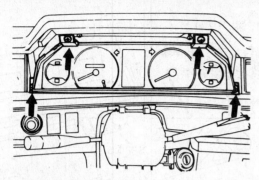

The instrument cluster is held by four mounting screws on Cordia and Tredia

Starion meter assembly (instrument cluster) retaining screws

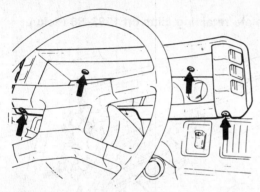

Starion meter hood retaining screws

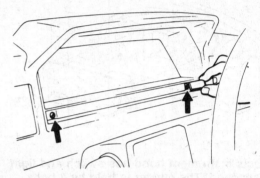

Removing the Galant meter hood screws

the right side of the air vent, located to the right of the cluster. Remove the trim panel, prying out the three tabs situated along the bottom.
  2. Reach up behind the instrument cluster and disconnect the speedometer cable. Remove the four cluster mounting screws, two along the top and two at the lower sides, and pull the cluster slightly toward you to gain access to the electrical connectors behind it.
  3. Disconnect the electrical connectors and remove the cluster.
  4. Installation is the reverse of removal.

Starion

  1. Remove the four meter hood mounting screws, two on the lower surface located just above the gauges and one on either side at the very bottom. Pull out the bottom of the hood on both sides and hold the hood at this angle; then pull it upward and off.
  2. Disconnect the electrical connector on either side. Remove the two meter assembly mounting screws (one on either side at the bottom) and the mounting nut on either side. Grab the case on both sides and tilt the lower edge upward and to the rear.
  3. Disconnect the speedometer cable by pressing the locking tab on the speedometer side of the connection and pulling the cable off. Disconnect the electrical connectors and remove the meter assembly.
  4. Installation is the reverse of removal. Position the meter assembly just in front of its installed position and make the electrical and speedometer cable connections before attempting to actually put the assembly into position.

Galant and Sigma

  1. Remove the two meter hood mounting screw covers located along the bottom of the hood by prying out with a screwdriver. Remove the screws and pull off the hood.

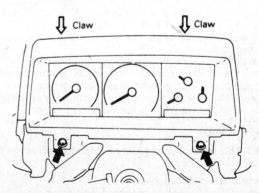

Mirage meter hood mounts; don't break the plastic claw clips

  2. Remove the four meter assembly mounting screws (two on each side), pull the assembly outward slightly, and then disconnect the electrical connectors. Remove the assembly.
  3. Installation is the reverse of removal.

Mirage

  1. Remove the meter hood attaching screws and tilt the lower meter hood outward. Pull the hood downward to release the locking tangs at the top and remove it.
  2. Remove the four meter assembly mounting screw (two at top and two at the bottom) and pull the unit outward. Disconnect the speedometer cable and all connectors, and remove the unit.
  3. Installation is the reverse of removal.

6–51

6 CHASSIS ELECTRICAL

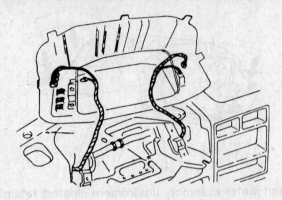

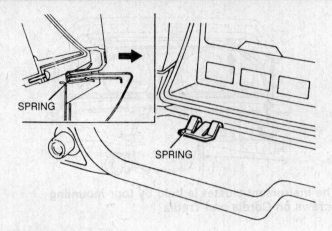

Faceplate retaining clips on 1983–86 Pickup

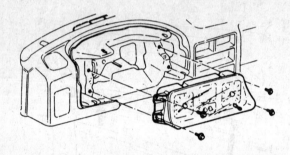

The Precis instrument hood has switch and light wiring attached. The cluster is held by 4 bolts

Precis

1. Remove the retaining screws holding the cluster housing; that's the surround panel with the switches.
2. As the panel comes off, disconnect the switch and light wiring running to it.
3. Remove the four screws holding the instrument cluster. Gently pull the cluster free of the dash. Reach behind the cluster and disconnect the wiring and the speedometer cable.
4. Reinstall in reverse order, making certain the speedometer cable and wiring harnesses are correctly connected.

1983–86 Pickup

NOTE: The instrument cluster and radio/heater faceplate are all one piece.

1. Remove the radio knobs.
2. Remove the fan control knob and the heater control slider knobs.
3. Remove the instrument cluster panel attaching screws.
4. Remove the locking nuts from the radio shafts and the fan control switch.
5. Remove the ashtray and remove the ashtray bracket.
6. Remove the attaching screws at the top of the instrument assembly.
7. Beginning at the left lower corner, remove the faceplate from the instrument cluster.
8. As the panel comes free, reach behind and disconnect the wiring for the gauges, the lighter, the heater controls and the clock. Disconnect the speedometer cable.
9. Remove the instrument cluster.
10. To reinstall, fit the panel loosely in position and connect each wire and cable to its proper location. Fit the panel into place and engage the spring tabs.
11. Install the retaining screws.
12. Install the ashtray and bracket.

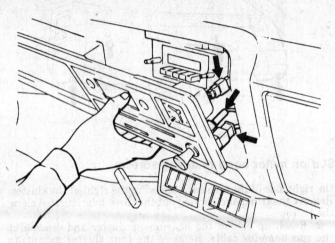

On 1983–86 pickups, disconnect the wiring as the panel come loose

13. Install the panel attaching screws.
14. Install the knobs for the heater and radio.

1987–89 Pickup and Montero

1. Use a small protected flat tool to pry the hazard flasher switch out of the dash. On the opposite side of the steering wheel, remove the hole cover.
2. Remove the 4 screws (two top and bottom) holding the hood in place. Remove the hood.
3. Remove the 4 screws holding the instrument cluster. Pull the cluster towards you.
4. Disconnect the electrical harnesses and the speedometer cable from the instrument cluster.
5. When reinstalling, make certain each plug and the speedometer cable is properly engaged.

1983–86 Montero

1. Disconnect the speedometer cable at the transmission.
2. Using a screwdriver or similar tool wrapped in a rag, gently pry the instrument cover loose. Note that this is the rear panel on top of the cluster, between the instrument pod and the defroster grille.
3. Remove the screws from the bottom of the instrument cluster and remove the bolt from the top of the cluster.

CHASSIS ELECTRICAL 6

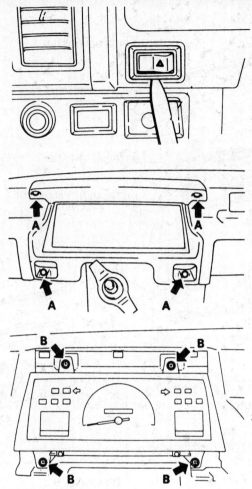

1987-89 Pickup and Montero instrument cluster removal

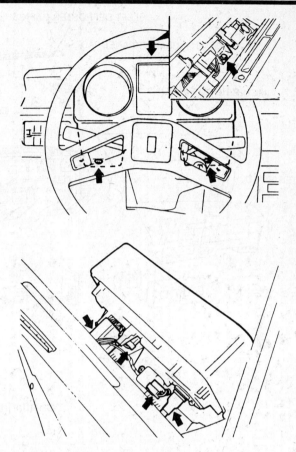

Remove the concealed screws and disconnect the wiring. Bottom view shows cluster with panel cover removed

4. Work the cluster out from the dash and disconnect the speedometer cable.
5. Carefully disconnect the wiring harnesses from the cluster and remove the cluster.
6. Reinstall in reverse order, taking care that each cable and harness is securely installed.
7. When installing the cluster cover, make certain it snaps into place and is correctly positioned.
8. Connect the speedometer cable at the transmission.

Instrument Panel/Dashboard

REMOVAL AND INSTALLATION

WARNING: Each dashboard uses a variety of nuts and bolts of different sizes and lengths. Correct cataloging and reinstallation is required if the dash is to be free of annoying squeaks and rattles. Additionally, wires and hoses should be labeled at the time of removal. The amount of time saved during reassembly makes the extra effort well worthwhile.

Removal of the dash is a complicated, time consuming task. Do not attempt this procedure if you are not comfortable with extended projects. Additional helpful information may be found in other parts of this book dealing with specific assemblies. For example, if not certain how to remove the radio, refer to "Radio Removal and Installation" in this section.

Cordia and Tredia

1. Disconnect the negative battery cable.
2. Remove the steering wheel.
3. Remove the undercover from the right side of the dash below the glove box.
4. Remove the lap heater duct.
5. Remove the glove box.
6. Remove the vertical defroster duct on the right side.
7. Label and disconnect the 3 heater control cables. For Tredia only, remove the heater control panel.
8. Remove the screws for the instrument hood, pull it forward and disconnect the wiring behind it.
9. Disconnect the speedometer cable behind the instrument cluster.
10. Remove the instrument cluster screws, pull it out slightly and disconnect the wiring connectors. Remove the cluster assembly.
11. Remove the defroster garnish (trim) at the base of the windshield. One center screw is hidden below the left piece.
12. Remove the small side covers on each end of the dash.
13. Remove the fuse block from the instrument panel. Don't disconnect anything, just unbolt the block and let it hang free of the dash.
14. At the back of the fuseblock, disconnect ONLY the wiring harness running from the fuseblock to the instrument panel harness. Check carefully to identify the correct wires.
15. Disconnect the wiring at the heater fan case.
16. Disconnect the floor console and audio bracket from the dashboard.

6 CHASSIS ELECTRICAL

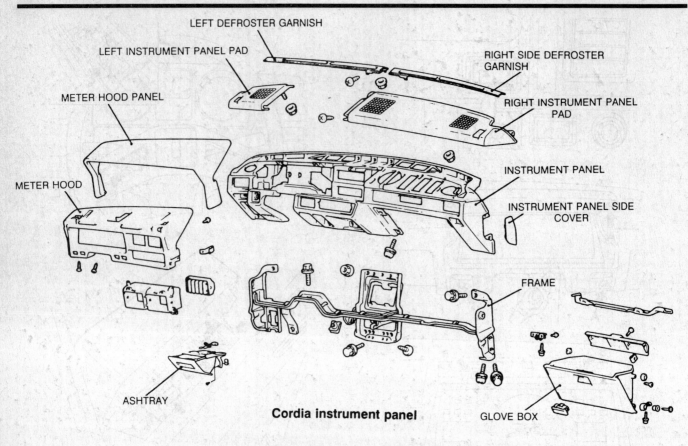

Cordia instrument panel

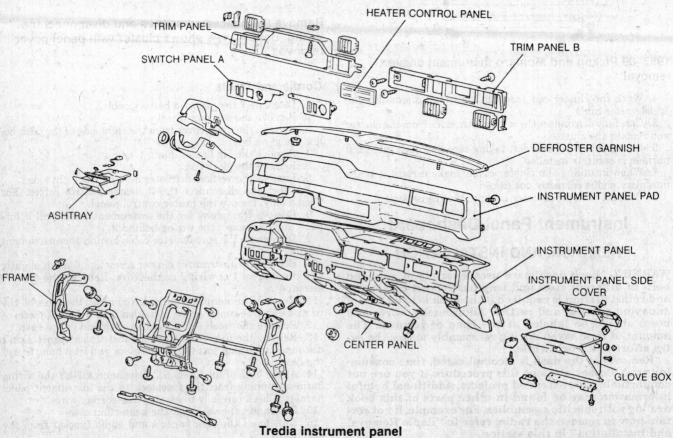

Tredia instrument panel

CHASSIS ELECTRICAL 6

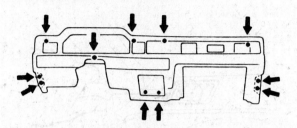

Location of Cordia and Tredia instrument panel bolts

17. Identify the bolts holding the steering column to the dash frame and/or pedal box. Remove these bolts only and lower the column away from the dash.
18. Disconnect the antenna feeder.
19. Remove the bolts holding the dashboard. With the help of an assistant, remove the dash from the car. Note that it is coming out with many items still attached. Do not snag wires or cables during removal.
20. If the dash is being replaced, transfer components to the new dash.
21. When reinstalling, place the dash in position and insert two or three bolts to hold it in place. Insert the rest of the hardware and hand tighten it snug; check the fit of the dash and position it correctly before tightening the bolts.
22. Before continuing, make certain all the wires and cables are routed correctly. Attend to any that may interfere with other components.
23. Install the antenna feeder cable.
24. Raise the steering column and attach it to the mounts.
25. Connect the console and audio mounts to the dash.
26. Install the connectors for the heater fan. Connect the main instrument harness(es) to the fusebox.
27. Install the fusebox and tighten the bolts.
28. Install the dash side panels.
29. Install the defroster garnish.
30. Fit the instrument cluster into position and connect the speedometer cable and instrument wiring. Install the cluster and tighten the screws.
31. Position the hood, connect its wiring and install it with the retaining screws.
32. For Tredia only, install the heater control panel and route the cables correctly. Connect the 3 heater control wires and adjust each. Refer to the procedure for Heater Core Removal and Installation for the correct hook-up procedures.
33. Install the defroster duct.
34. Install the glove box.
35. Install the lap heater duct and the undercover for the dash.
36. Install the steering wheel.
37. Connect the negative battery cable.

Starion

1. Disconnect the negative battery cable.
2. Remove the steering wheel.
3. Remove the screws holding the hood release handle to the dash. Let it hang away from the dash.
4. Unbolt the fuse box from the dash board.
5. Remove the knee protector (lower dash) on the left side. Some of the bolts are hidden behind covers.
6. Remove the screws from the bottom of the steering column cover. Remove both halves of the cover.
7. Remove the attaching screws for the combination switch

1. Steering wheel
2. Hood lock release handle
3. Fuse block
4. Knee protector
5. Lower column cover
6. Upper column cover
7. Column switch
8. Meter hood
9. Combination meter
10. Side console cover
11. Rear console box
12. Front console box
13. Lower cover
14. Glove box
15. Ashtray
16. Heater control bezel
17. Digital clock
18. Side defroster upper grille
19. Instrument panel side cover
20. Side defroster duct
21. Instrument panel
22. Center reinforcement

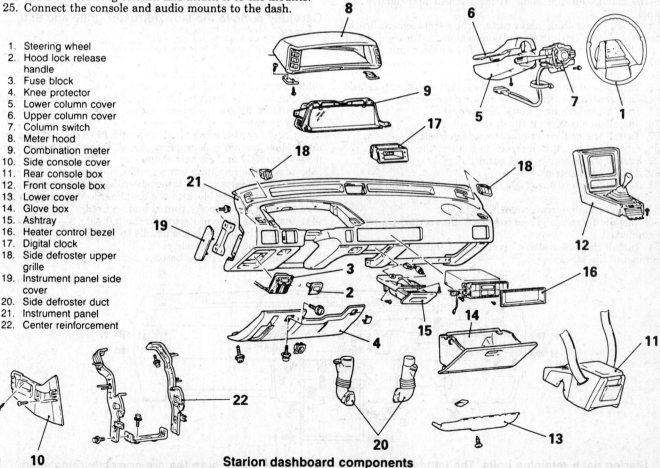

Starion dashboard components

6-55

6 CHASSIS ELECTRICAL

on the steering column. Disconnect the wiring harnesses and remove the switch.

8. Remove the instrument hood screws. Pull both edges of the bottom of the hood forward; hold it in this position and lift it obliquely up and out.
9. Disconnect the harness connectors on both sides of the instrument hood.
10. Remove the screws on the bottom and the nuts on the top of the instrument cluster. Pull the bottom edge up and forward to remove it. Disconnect the wiring and cables as it comes free.
11. Remove the console side cover mounting screws. Remove the cover downward while pushing slightly forward.
12. Remove the front and rear consoles.
13. Remove the undercover from the right side dashboard.
14. Remove the glove box. Grasp the sides, press in and pull forward to allow the door to hang down. Remove the 3 upper screws.
15. Remove the ashtray.
16. Carefully remove the heater control bezel (trim).
17. Remove the clock; it may be popped out of the dash with gentle pressure. Disconnect the harness when it is free.
18. Remove the grilles for the side defrosters by inserting a flat tool from the window side and prying forward and upward.
19. Use a non-marring tool to pry the side covers off the instrument panel.
20. Remove each mounting screw and bolt holding the dash in place. As it comes loose, allow it to move into the car. Disconnect the remaining wire harnesses when you can reach behind it.
21. With the help of an assistant, remove the dash from the car. Remember that the dash still has several components attached. Handle it carefully.
22. If necessary, remove the center reinforcing brackets. Loosen the clamps for the main wiring harness and remove the panel.
23. When reinstalling, make certain the reinforcement brackets are in place. The dash should be assembled to the condition it was when removed. The upper part of the heater unit has two guide bolts to which the dash attaches. When placing the dash in the car, make certain these guides engage.
24. After the dash is in place, check that wires, cables and ducts are routed correctly and without interference.
25. Install the retaining bolts for the dash.
26. Install side covers on the dashboard.
27. Install the grill covers for the side defrosters.
28. Install the clock. Remember to connect the wiring.
29. Install the heater control bezel.
30. Install the ashtray.
31. Install the glove box. Tighten the hinge screws and fold the box into place.
32. Install the undercover on the right side.
33. Install the front and rear consoles. Make certain the wiring is connected.
34. Install the side console covers. Fit it into place carefully and install the screws.

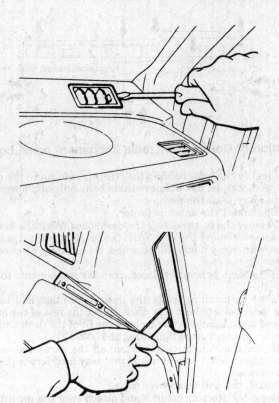

Carefully remove the side defroster grille and end covers

35. Install the instrument cluster. Connect the wiring and the cables before installing the retaining screws.
36. Position the instrument hood and connect the wiring to the switches. Fit the hood into its final place and install the retaining screws.
37. Carefully fit the combination switch onto the steering column. Make certain it is properly seated. Route the wiring harnesses correctly and connect them.
38. Install the retaining screws and tighten them.
39. Install the upper and lower steering column covers.
40. Install the fuse block to the dashboard. Make certain the wire harness are properly routed and secured.
41. Install the knee protector on the left side.
42. Attach the hood release handle to the dash.
43. Install the steering wheel.
44. Connect the negative battery cable.

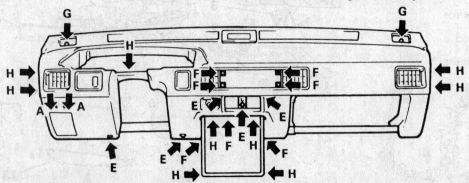

Starion dash retaining bolts. The letters indicate type of fastener; make sure the are correctly reinstalled

CHASSIS ELECTRICAL 6

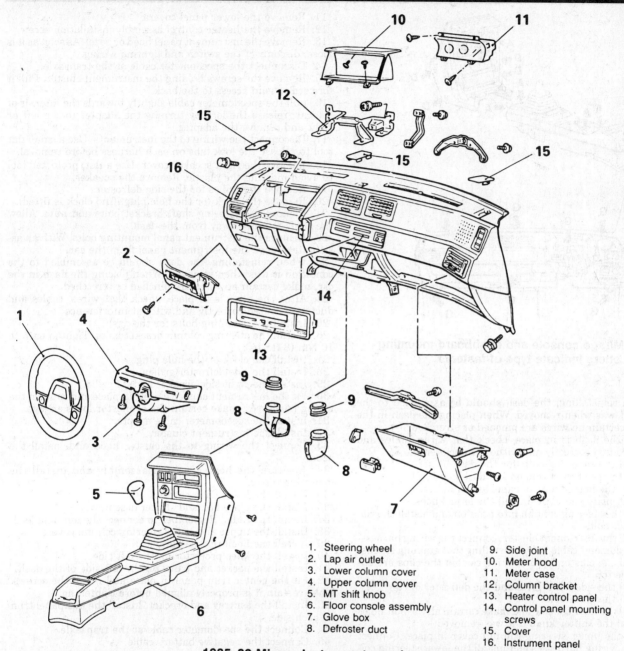

1. Steering wheel
2. Lap air outlet
3. Lower column cover
4. Upper column cover
5. MT shift knob
6. Floor console assembly
7. Glove box
8. Defroster duct
9. Side joint
10. Meter hood
11. Meter case
12. Column bracket
13. Heater control panel
14. Control panel mounting screws
15. Cover
16. Instrument panel

1985–88 Mirage instrument panel components

Mirage through 1988

1. Disconnect the negative battery cable.
2. Remove the steering wheel.
3. Remove the lap air outlet under the steering column.
4. Remove the lower steering column cover, disconnect the column wiring harnesses and remove the upper cover.
5. For manual transmission cars, remove the shift knob.
6. Remove the floor console assembly. Remember the concealed screws under the ashtray and behind the switches.
7. Open the glove box. Carefully and quickly, pull upon the right corner of the glove box to remove the right hinge pin from the mount. Slide the glove box toward the right door to remove it.
8. Disconnect the defroster ducts and remove the small side joint ducts.
9. Remove the instrument hood. Remember the two clips at the top.
10. Remove the instrument cluster screws. Move the cluster towards you, disconnect the speedometer cable and the wiring harnesses and remove the cluster.
11. Remove the 4 bolts, two on each side, holding the steering column bracket to the heater control lever knobs. Reach behind the dash to the back of the heater control panel. Push the right side control panel clip to the right while pushing outward on the panel. Allow the panel to hang free.
13. Remove the heater control assembly mounting screws.
14. Remove the small covers from the top of the dash. Use a protected flat tool to pry the covers loose.
15. Remove the instrument panel mounting screws and bolts. With an assistant, remove the dash from the car. Handle it carefully; components are still attached.

6 CHASSIS ELECTRICAL

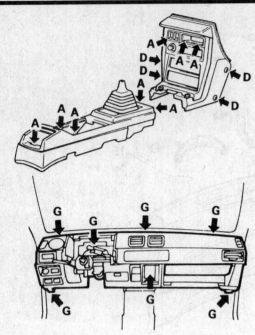

1985–88 Mirage console and dashboard mounting screws. Letters indicate type of fastener

16. When reinstalling, the dash should be assembled to the condition it was when removed. When placing the dash in the car, make certain no wires are pinched or stretched.
17. After the dash is in place, check that wires, cables and ducts are routed correctly and without interference.
18. Install the retaining bolts for the dash.
19. Install upper bolt covers on the dashboard.
20. Install the heater control panel assembly screws, then install the faceplate and clips. Install the lever knobs.
21. Move the steering column into position and install the column bracket bolts.
22. Install the instrument cluster; connect the wiring harnesses and speedometer cable before installing the retaining screws.
23. Install the cluster hood, making certain the clips are engaged at the top.
24. Install the side joint ducts and the defroster ducts.
25. Install the glove box.
26. Install the floor console and make certain all the bolt holes align. Instal the shifter knob if it was removed.
27. Place the upper steering column cover in place. Connect the column wiring harnesses and install the lower steering column cover.
28. Connect the lap air outlet.
29. Install the steering wheel.
30. Connect the negative battery cable.

1989 Mirage

1. Disconnect the negative battery cable.
2. Remove the side trim panel (kickpanel) on each side.
3. Remove the ashtray and bracket.
4. Remove the center trim panel from the lower dash.
5. Remove the pocket and panel from the left side dash.
6. Remove the knee protector (lower dash) from the left side of the dash.
7. Remove the hood release bracket from the dash.
8. Remove the lower and upper steering column covers.
9. Remove the radio.
10. Open the glove box and remove the striker hook from the inside. Remove the glove box assembly.
11. Remove the lower panel covers.
12. Remove the heater control assembly installation screw.
13. Remove the instrument panel hood or bezel. As soon as it is loose, disconnect the switch and lighting wiring.
14. Disconnect the speedometer cable at the transaxle.
15. Remove the screws holding the instrument cluster. Pull it forward to gain access to the back.
16. pull the speedometer cable slightly towards the interior of the car; release the lock by turning the adapter to the left or right and remove the adapter.
17. Disconnect the wiring to the instruments. Use a small flat tool to release the lock tabs on each harness before removal.
18. Remove the right speaker cover. Use a flat, protected tool to avoid marring the plastic. Remove the speaker.
19. Remove the grilles for the side defrosters.
20. Remove the clock (or the hole plug if no clock is fitted).
21. Remove the steering shaft bracket bolts and nuts. Allow the shaft to move away from the dash.
22. Remove the instrument panel mounting bolts. With an assistant, remove the instrument panel from the car.
23. When reinstalling, the dash should be assembled to the condition it was when removed. When placing the dash in the car, make certain no wires are pinched or stretched.
24. After the dash is in place, check that wires, cables and ducts are routed correctly and without interference.
25. Install the retaining bolts for the dash.
26. Install the steering column bracket bolts. Tighten each to 10 Nm (8 ft. lbs).
27. Install the clock or the hole plug.
28. Install the side defroster grilles.
29. Install the right side speaker and its grille.
30. Fit the instrument cluster loosely in place and connect the wiring harnesses; make certain the connector locks engage.
31. Install the speedometer cable adapter.
32. Install the instrument cluster.
33. Connect the wiring to the cluster hood and install the hood.
34. Position the heater control assembly and install the mounting screw.
35. Install the lower dash covers.
36. Install the glove box and striker assembly.
37. Install the radio. Remember to connect the antenna lead.
38. Install the upper and lower steering column covers.
39. Install the hood release handle.
40. Install the knee protector on the left side.
41. Install the pocket and panel on the left side of the dash.
42. Fit the center trim panel in place and instal the screws. Make certain it is properly aligned before tightening.
43. Install the ashtray and bracket. Install the kickpanel trim on each side.
44. Connect the speedometer cable at the transaxle.
45. Connect the negative battery cable.

Galant and Sigma

1. Disconnect the negative battery cable.
2. Remove the horn pad, center nut and steering wheel.
3. Remove the small steering column panel.
4. Remove the undercover.
5. Remove the upper and lower steering column covers.
6. Use a small flat tool to remove the screw cover panels at the bottom of the instrument cluster hood. Remove the screws and remove the hood.
7. Remove the mounting screws for the instrument cluster. Pull the cluster forward and disconnect the wiring connectors.
8. Disconnect the speedometer adaptor behind the cluster and remove the cluster.
9. The floor console must be removed. It is held by several concealed bolts. Work carefully.
10. Remove the under frame behind the console.

CHASSIS ELECTRICAL 6

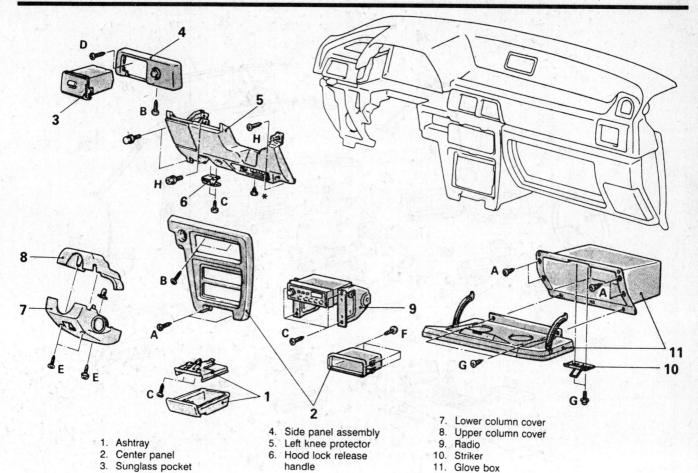

1. Ashtray
2. Center panel
3. Sunglass pocket
4. Side panel assembly
5. Left knee protector
6. Hood lock release handle
7. Lower column cover
8. Upper column cover
9. Radio
10. Striker
11. Glove box

1989 Mirage dashboard components. Letters indicate type of fastener

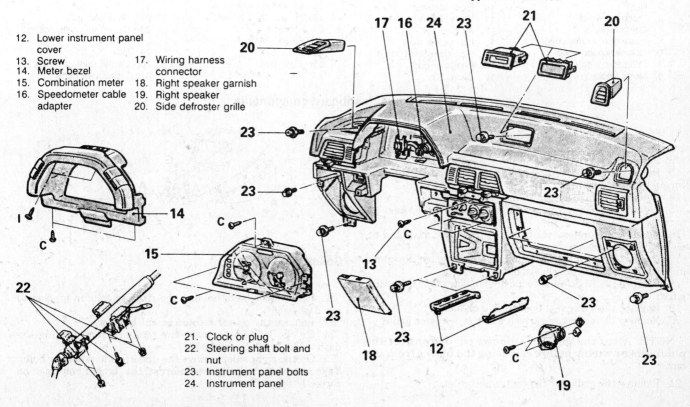

12. Lower instrument panel cover
13. Screw
14. Meter bezel
15. Combination meter
16. Speedometer cable adapter
17. Wiring harness connector
18. Right speaker garnish
19. Right speaker
20. Side defroster grille
21. Clock or plug
22. Steering shaft bolt and nut
23. Instrument panel bolts
24. Instrument panel

6-59

6 CHASSIS ELECTRICAL

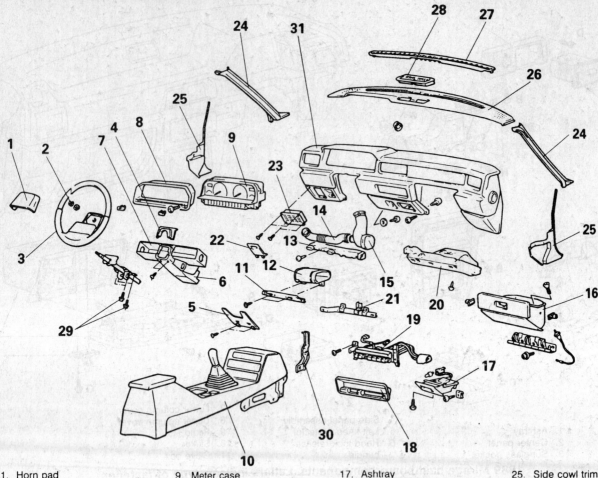

1. Horn pad
2. Nut
3. Steering wheel
4. Column panel
5. Instrument under cover
6. Column cover, lower
7. Column cover, upper
8. Meter hood
9. Meter case
10. Floor console
11. Under frame
12. Air duct D
13. Lap heater duct
14. Side defroster hose
15. Defroster duct
16. Glove box
17. Ashtray
18. Heater control panel
19. Heater control switch
20. Under cover
21. Under frame
22. Fuse box lid
23. Fuse box
24. Front pillar trim
25. Side cowl trim
26. Defroster garnish
27. Defroster grille
28. Center air outlet upper grille
29. Bolt
30. Center reinforcement
31. Instrument panel

Galant and Sigma dashboard components

11. Disconnect and remove air duct D, the lap heater duct, the side defroster duct and the vertical defroster duct.
12. Remove the glove box.
13. Remove the ashtray and its mount. Disconnect the light wiring before removing.
14. Remove the heater control face plate.
15. Remove the heater control panel and disconnect its harness.
16. Remove the undercover from the right side dash and remove the under frame.
17. On the left side of the dash, remove the fuse box cover and unbolt the fusebox from the dash.
18. Remove the front pillar (windshield pillar trim) from each pillar.
19. Remove the kickpanel trim from each side.
20. Remove the defroster garnish and the defroster grille.

NOTE: After the garnish is loosened, disconnect the photosensor wiring before removing the pieces from the car.

21. Remove the grille for the center air outlet.

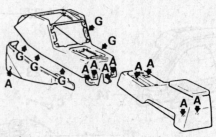

Galant console

22. Remove the bolts holding the steering column bracket to the dash.
23. Remove the center reinforcement bracket.
24. On the left side, remove the retaining nuts holding the dash underframe to the body.
25. On the right side, remove the underframe retaining bolts. Take note that the bolts are different; the flanged bolt must be correctly reinstalled.

CHASSIS ELECTRICAL 6

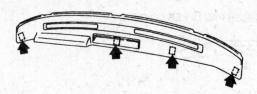

Remove the defroster garnish by prying gently in these locations

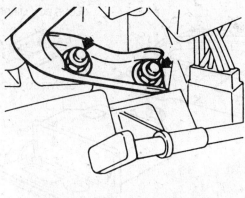

The Galant and Sigma dash frame has nuts on the left side and special bolts on the right. Take note of the bolt placement for correct reassembly

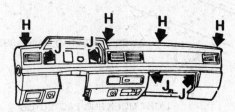

Galant and Sigma dashboard retaining bolts. Letters indicate type of fastener

26. Remove the remaining nuts and bolts holding the dash. As the dash comes loose, label and disconnect the wiring harnesses. Carefully remove the dash from the car.
27. When reinstalling, position the dash and connect the wiring harnesses. Make certain nothing is pinched or stretched out of place.

28. Install the retaining bolts and nuts.
29. Install the underframe bolts on the right side, making certain the flange bolt is properly located.
30. Install the underframe nuts on the left side.
31. Install the center reinforcement bracket.
32. Lift the steering column into place and install the bracket bolts. Tighten them to 10 Nm or 8 ft. lbs.
33. Install the center air outlet center grille.
34. Install the defroster grille and garnish. Make certain each piece is correctly located.
35. Install the kickpanels on each side.
36. Install the pillar trim on each side.
37. Attach the fusebox to the dash and install the lid.
38. On the right side, install the under frame and the under cover.
39. Position and install the heater control panel and its faceplate.
40. Install the ashtray and bracket. Connect the wire for the light.
41. Install the glove box assembly.
42. Connect and install the defroster duct, the side defroster duct, the lap heater duct and air duct D.
43. Install the small center under frame.
44. Install the floor console. Don't tighten the bolts and screws until every one is started in its hole. Make sure all the seams align correctly.
45. Install the instrument cluster; connect the wiring and adapter before installing the retaining screws.
46. Install the instrument cluster hood. Don't forget the screw covers.
47. Install the upper and lower steering column covers.
48. Install the center panel under cover.
49. Install the small column panel.
50. Install the steering wheel and horn pad. Tighten the center nut to 40 Nm or 30 ft. lbs.
51. Connect the negative battery cable.

Precis

1. Disconnect the negative battery cable.
2. Remove the steering wheel.
3. Remove the small square column shroud below the column.
4. Remove the upper and lower steering column covers.
5. Remove the screw covers from the lower dash pad.
6. Remove the hood release mounting screws.
7. remove the left side lower dash and disconnect the electric wiring.
8. Remove the center and floor consoles.
9. Remove the glove box.
10. Remove the cigarette lighter and ashtray.
11. Remove the lower dash from the right side. There are several hidden screws.
12. Remove the instrument cluster housing (hood) assembly. Disconnect the wiring to the switches and remove the lights.
13. Remove the instrument cluster retaining screws. Lift the cluster out and disconnect the speedometer cable and wiring harnesses.
14. Remove the radio. Disconnect the dash wiring harness from the main harness.
15. Disconnect the heater control cables from the heater and blower unit. Label or identify each one.
16. Identify and remove the two bolts holding the dash to the heater unit.
17. Remove the screw covers from the upper dash (at the defroster grille).
18. Remove the retaining bolts and remove the dash.
19. Roll the dash forward and remove the side defroster hose; disconnect the air vent hose.

6-61

6 CHASSIS ELECTRICAL

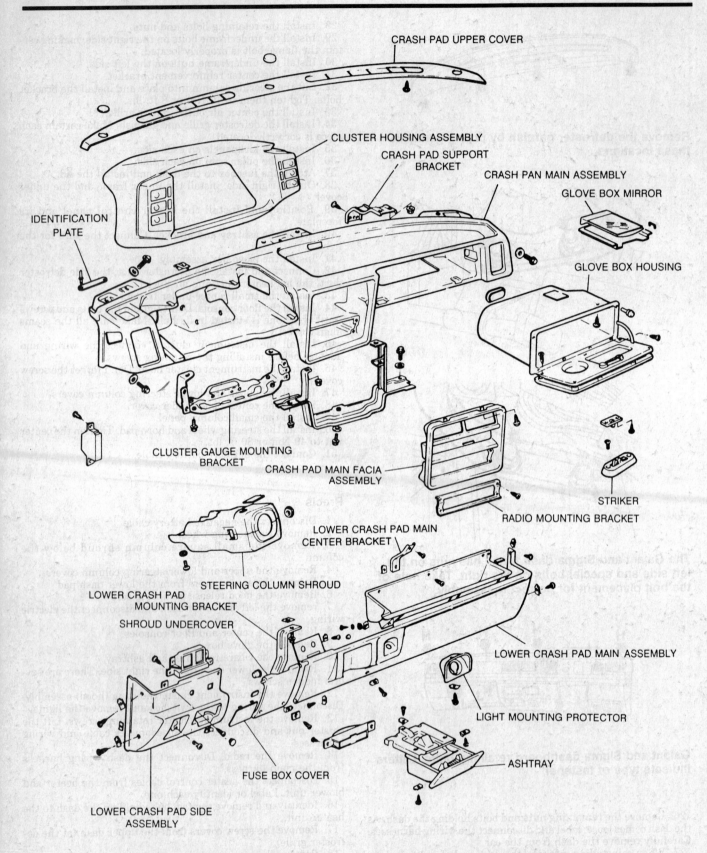

Precis dashboard components. The dashboard is also called the crash pad.

CHASSIS ELECTRICAL 6

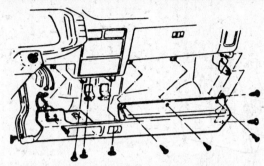

Look for the concealed screws when removing the Precis lower dash.

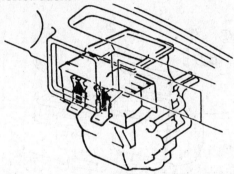

Two concealed bolts hold the Precis dash to the heater unit.

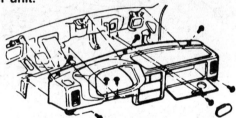

Precis dashboard retaining bolts

20. Remove the side air vent duct, the side defroster nozzle and side air vent nozzle.
21. Remove the dash upper cover.
22. Remove the defroster nozzle and the heater connection.
23. Remove the center air vent louver duct.
24. Remove the instrument mounting bracket, cross member assembly and the support bracket.
25. When reassembling, make certain the dash components are reinstalled (Steps 20–24) before placing the dash in the car. Once in place, install retaining bolts and check the alignment of the dash.
26. Install and tighten the two bolts holding the dash to the heater unit. Install the caps on the upper screw holes.
27. Connect the heater control wires to the heater unit. Refer to "Heater Core Removal and Installation" for the correct alignment procedures.
28. Install the radio.
29. Fit the instrument cluster in place and connect the wiring harnesses and the speedometer cable.
30. Install the cluster assembly.
31. Connect the wiring and light harness to the cluster hood and install the hood.
32. Connect the dash harness to the main harness.
33. Install the lower dash assembly and connect the wiring.
34. Replace the lighter and ashtray.
35. Install the glove box.
36. Install the center and floor consoles.
37. Install the left side lower dash panel. Connect the wire harness.
38. Reinstall the hood release handle.
39. Install the screw covers on the lower dash pad.
40. Place and secure the upper and lower steering column covers.
41. Install the column under cover.
42. Install the steering wheel and horn pad.
43. Connect the negative battery cable.

1983–86 Pickup

1. Disconnect the negative battery cable.
2. Loosen the tilt wheel lock and lower the wheel to the lowest position.
3. Remove the steering wheel.
4. Remove the upper and lower steering column covers.
5. Remove the instrument cluster. Disconnect the gauge wiring and the speedometer cable. Once the cluster is out, remove the single screw in the upper center of the mounting space in the dashboard.

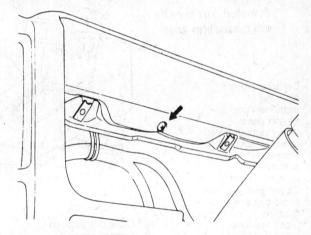

Remove the single screw behind and above the instrument cluster

6. Remove the radio; disconnect the diesel indicator if so equipped.
7. Remove the glove box.
8. Remove the floor console and/or center console, depending on equipment
9. Remove the screws holding the center air outlet and remove the outlet.
10. Remove the ashtray. Remove the two screws holding the ashtray holder. Slide the holder (bracket) into the dash and remove it.
11. Remove the 4 screws from the bottom of the dash. Two are located under the glovebox opening and two are located approximately under the dash light dimmer switch on the left side.
12. Reach in through the opening in the center of the dash and remove the nut at the rear of the dash. Make certain you identify the correct nut; it may be either a wing nut or a hex nut.
13. Reach just inside the glovebox opening and remove the screw from the lower inner surface of the dash.
14. From underneath the dash, release the 4 hex nuts holding the upper dash cover. Once the nuts are off, lift the cover up and off.
15. Remove the 5 screws and bolts from the top of the dashboard. Note the location of each type of fastener.

6-63

6 CHASSIS ELECTRICAL

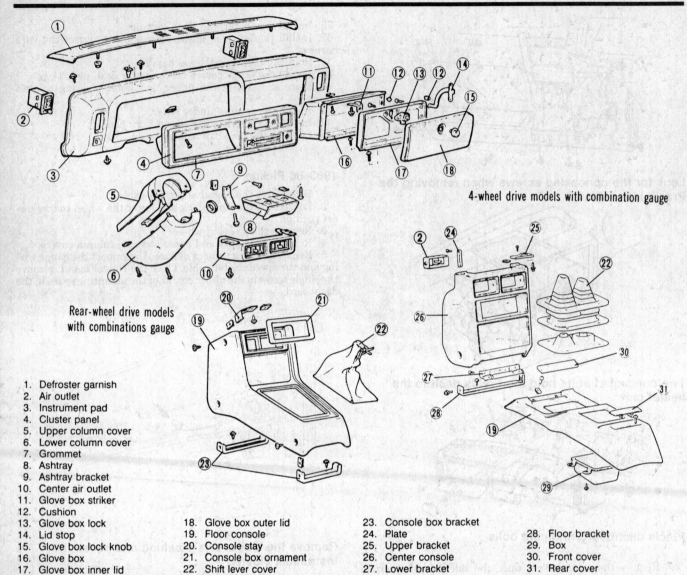

1. Defroster garnish
2. Air outlet
3. Instrument pad
4. Cluster panel
5. Upper column cover
6. Lower column cover
7. Grommet
8. Ashtray
9. Ashtray bracket
10. Center air outlet
11. Glove box striker
12. Cushion
13. Glove box lock
14. Lid stop
15. Glove box lock knob
16. Glove box
17. Glove box inner lid
18. Glove box outer lid
19. Floor console
20. Console stay
21. Console box ornament
22. Shift lever cover
23. Console box bracket
24. Plate
25. Upper bracket
26. Center console
27. Lower bracket
28. Floor bracket
29. Box
30. Front cover
31. Rear cover

1983–86 Pickup dashboard components. Note the slight difference for four wheel drive vehicles

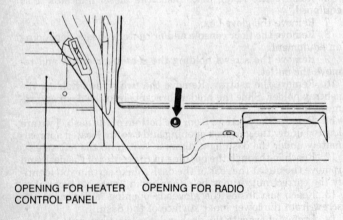

Reach into the dash to remove this nut

16. Work the dash away from its mounts. Disconnect the wiring harnesses behind the dash.
17. Remove the dash from the car.
18. When reinstalling, place the dash in the car and connect the wiring harnesses, making certain none are stretched or pinched. Check that the air ducts align and match correctly.
19. Install 5 screws across the top; the screws with the washers go on the ends.
20. Fit the upper cover into place and install the nuts from below the dash.
21. Install the small screw on the inner lower edge of the glove box.
22. Reinstall the nut at the back of the dash.
23. Install the lone screw at the upper rear of the instrument cluster area.
24. Install the 4 screws, two on each side, across the bottom of the dash.
25. Slide the ashtray holder into place and install the screws. Install the ashtray.
26. Install the center air outlet.

6-64

CHASSIS ELECTRICAL 6

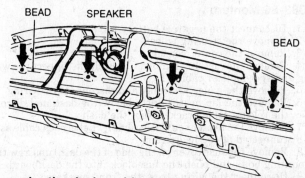

Lie under the dash and look up; remove the four nuts holding the upper cover. This view is from under the dash

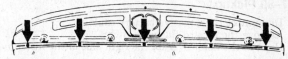

The dash in 1983–86 Pickups is held by 5 screws across the top

27. Replace the glove box and tighten the screws.
28. Install the radio; connect the diesel indicator wiring if so equipped.
29. Install the instrument cluster; connect the wiring and the speedometer cable.
30. Install the steering column covers.
31. Elevate the tilt column and install the steering wheel.
32. Connect the negative battery cable.

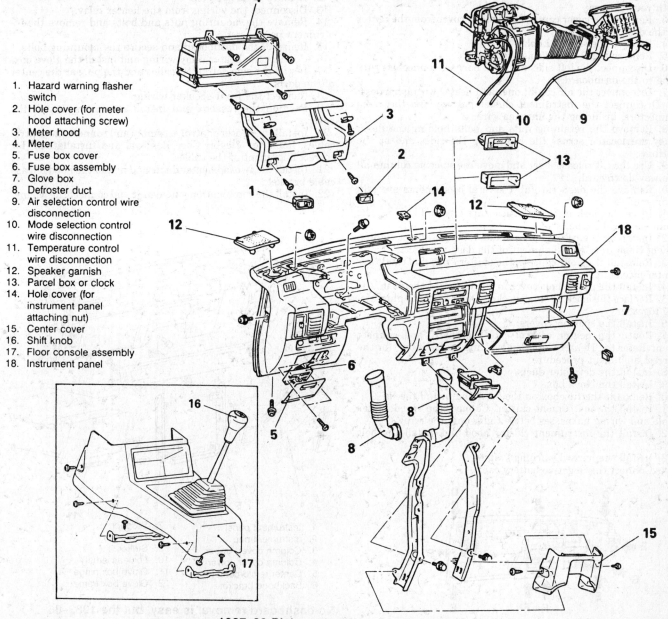

1. Hazard warning flasher switch
2. Hole cover (for meter hood attaching screw)
3. Meter hood
4. Meter
5. Fuse box cover
6. Fuse box assembly
7. Glove box
8. Defroster duct
9. Air selection control wire disconnection
10. Mode selection control wire disconnection
11. Temperature control wire disconnection
12. Speaker garnish
13. Parcel box or clock
14. Hole cover (for instrument panel attaching nut)
15. Center cover
16. Shift knob
17. Floor console assembly
18. Instrument panel

1987–89 Pickup, dashboard components

6-65

6 CHASSIS ELECTRICAL

1987–89 Pickup

1. Disconnect the negative battery cable.
2. Remove the hazard flasher switch.
3. Pop out the hole cover on the right side of the instrument hood.
4. Remove the instrument hood.
5. Remove the instrument cluster, remembering to disconnect the wiring and speedometer cable at the rear.
6. Remove the fusebox cover. Remove the fusebox from the dashboard and let it hang.
7. Remove the glove box. Move the stopper out of the way before lowering the door.
8. Remove the two defroster ducts.
9. Label and disconnect the 3 control wires running to the heater unit.
10. Remove the speaker cover panels on each side.
11. Remove the clock or small compartment from the dash.
12. Pop out the small square cover at the upper center of the dash. Its easiest to reach in the clock hole and release the clips with your fingers.
13. Remove the lower center faceplate or cover from the center of the dash.
14. Remove the shifter knob.
15. Remove the floor console assembly.
16. Disconnect the left side and front harness connectors running to the dashboard.
17. Disconnect the air conditioner and heater wiring harnesses. Disconnect the instrument cluster harness and the radio connectors, including the antenna wire.
18. Remove the retaining nuts and bolts holding the dash. They are located across the top, along the sides and at the bottom.
19. Use the tilt release lever and move the steering column all the way downward.
20. Remove the dash carefully; several components are still connected to it.
21. Place the dash into the vehicle carefully and connect the main electrical harnesses.
22. Install all the retaining nuts and bolts but do not tighten any of them until the alignment of the dash is correct.
23. Reinstall the floor console, the shifter knob and the lower center faceplate.
24. Install the upper bolt cover in the center of the dash.
25. Replace the clock or small pocket. Don't forget to plug wiring together.
26. Install the speaker covers.
27. Route the heater control cables properly and connect them. Refer to "Heater Core Removal and Installation" for the correct alignment procedure.
28. Install the defroster ducts.
29. Install the glove box.
30. Remount the fusebox on the dash and install the lid.
31. Install the instrument cluster. Connect the speedometer cable and wiring harnesses before tightening the screws.
32. Install the instrument cluster hood and install the bolt cover.
33. Install the hazard warning switch.
34. Connect the negative battery cable.

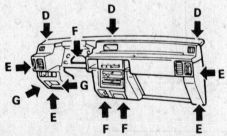

Dashboard retaining bolts, 1987–89 Pickup

1983–86 Montero

1. Disconnect the negative battery cable.
2. Remove the steering wheel.
3. Remove the center console.
4. Remove the meter assembly. Carefully disconnect the wiring and speedometer cable.
5. Remove the gauge assembly and disconnect its wiring.
6. Remove the lap heater ducts below the dash and disconnect the release cable bracket from the dash.
7. Remove the heater control assembly. Label each cable as it is disconnected at the heater end.
8. Remove the fuse cover on the side of the dash. unscrew the mounting bolts and push the fuseblock into the dashboard.
9. Disconnect the wiring from the front speakers.
10. Remove the plug at the center of the instrument panel.
11. Remove the right and left side defroster grilles by gently prying on the mounting projections with a small flat tool. Be careful not to break the projections off.
12. Remove the glove box.
13. Disconnect the wiring from the heater relay.
14. Remove the mounting nuts and bolts and remove the instrument panel.
15. Reinstall the dashboard and secure the mounting bolts.
16. Connect the heater relay wiring and install the glove box.
17. Install the side defroster grilles and the plug at the center of the dash.
18. Connect the front speaker wiring.
19. Position the fusebox and install its screws; install the cover.
20. Install the heater control assembly and connect the control cables. Refer to "Heater Core Removal and Installation" for correct alignment of the cables.
21. Install the two lap heater ducts and install the hood release cable bracket.
22. Install the combination gauge unit and connect the wiring.

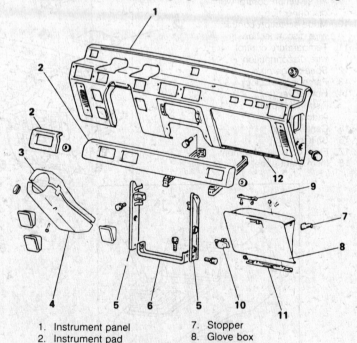

1. Instrument panel
2. Instrument pad
3. Column cover, upper
4. Column cover, lower
5. Center reinforcement
6. Backbone bracket
7. Stopper
8. Glove box
9. Striker
10. Lock assembly
11. Glove box hinge
12. Glove box frame

No dashboard removal is easy, but the 1983–86 Montero is very straightforward

CHASSIS ELECTRICAL 6

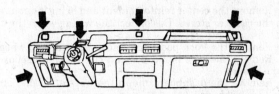

1983–86 Montero dashboard retaining bolts

23. Install the instrument cluster. Connect the wiring and the speedometer cable.
24. Install the center console.
25. Install the steering wheel.
26. Connect the negative battery cable.

1987–89 Montero

1. Disconnect the negative battery cable.
2. Remove the lap heater air ducts under the left and right side dash.
3. Remove the hood release cable bracket from the dashboard.
4. Remove the left and right side defroster grilles. Use a small flat tool to raise the attaching projections. Don't break the tabs.
5. Remove the glove box.
6. Use a flat, padded tool to remove the rear cover from the instrument cluster.

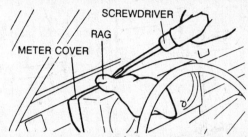

The proper way to remove the instrument cluster cover

7. Remove the instrument cluster. Disconnect the speedometer cable and the wiring harnesses to the back of the cluster.
8. Disconnect the wiring to each switch on the cluster. Label each as it is removed.
9. Remove the screws from the side of the combination meter unit and remove the cover.
10. Remove the 4 screws at the base of the combination meter and gently remove the meter. Disconnect the wire harnesses.

WARNING: The inclinometer can be damaged by dropping or bumping it. Do not tilt the unit so far as to exceed the maximum indication on the scale. Resist the temptation to play with the unit.

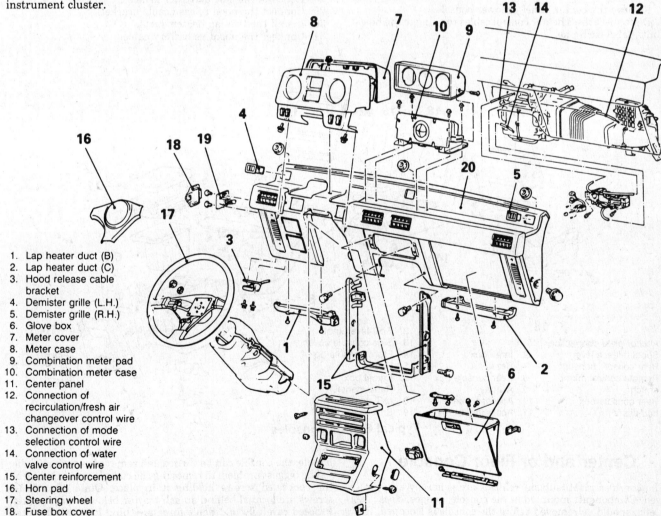

1. Lap heater duct (B)
2. Lap heater duct (C)
3. Hood release cable bracket
4. Demister grille (L.H.)
5. Demister grille (R.H.)
6. Glove box
7. Meter cover
8. Meter case
9. Combination meter pad
10. Combination meter case
11. Center panel
12. Connection of recirculation/fresh air changeover control wire
13. Connection of mode selection control wire
14. Connection of water valve control wire
15. Center reinforcement
16. Horn pad
17. Steering wheel
18. Fuse box cover
19. Fuse box assembly
20. Instrument panel

1987–89 Montero dash components

6 CHASSIS ELECTRICAL

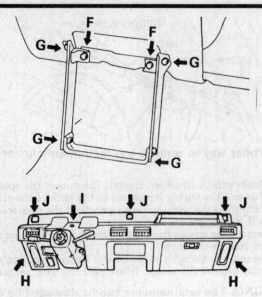

Location of mounting bolts for the center support and dashboard on 1987–89 Monteros

11. Remove the center panel or lower console.
12. Disconnect the 3 heater control cables running to the heater unit. Label each one.

13. Remove the center reinforcement and bring it out as a unit with the radio or stereo. Disconnect the wiring when the unit is free.
14. Remove the horn pad and remove the steering wheel.
15. Remove the fuse box cover and release the fusebox from the dashboard.
16. Remove the dashboard mounting nuts and bolts and remove the dashboard.
17. When reinstalling, position the dash and start each nut and bolt. Make certain the dash is aligned and then tighten the fittings.
18. Install the fusebox assembly and its cover.
19. Install the steering wheel and horn pad.
20. Install the radio and center reinforcement. Connect the wiring before tightening the bolts.
21. Carefully route and connect the heater control cables. Refer to "Heater Core Removal and Installation" for correct alignment of the cables.
22. Install center panel and secure the heater controls.
23. Install the combination gauges and connect the wiring.
24. Install the cover for the combination gauges.
25. Install the instrument cluster, connecting the speedometer cable and the many wiring connectors.
26. Install the cluster cover; make sure it is firmly seated in place.
27. Replace the glove box
28. Reinstall the side defroster grilles.
29. Connect the hood release cable bracket.
30. Install the two lap heater ducts.
31. Connect the negative battery cable.

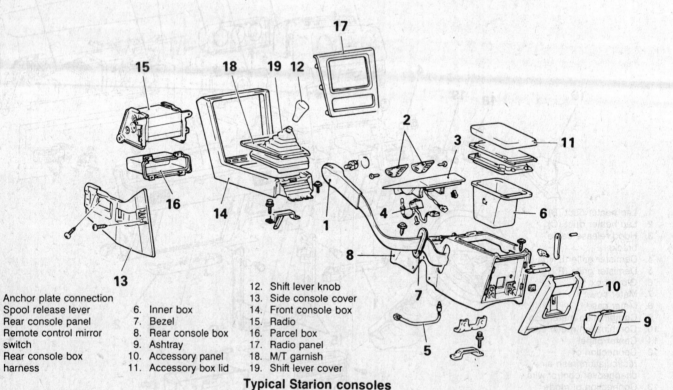

1. Anchor plate connection
2. Spool release lever
3. Rear console panel
4. Remote control mirror switch
5. Rear console box harness
6. Inner box
7. Bezel
8. Rear console box
9. Ashtray
10. Accessory panel
11. Accessory box lid
12. Shift lever knob
13. Side console cover
14. Front console box
15. Radio
16. Parcel box
17. Radio panel
18. M/T garnish
19. Shift lever cover

Typical Starion consoles

Center and/or Floor Console

The consoles in Mitsubishi vehicles are generally easily removed. Components mounted in the console, gauges, clock, radio, etc., should be removed before the console is loosened. Removing the floor console almost always requires the removal of the manual shifter knob. Do not remove the automatic shifter handle; the console can be turned and removed over the handle.

The biggest problem in removing the console is not finding all the concealed screws holding it in place. Quite often these screws are located behind an ash tray or below a cargo pocket. Proceed carefully and don't force anything loose. Quite often the screws are behind a small cap which must be popped out. Access to the screws along the sides of the floor console will be

CHASSIS ELECTRICAL 6

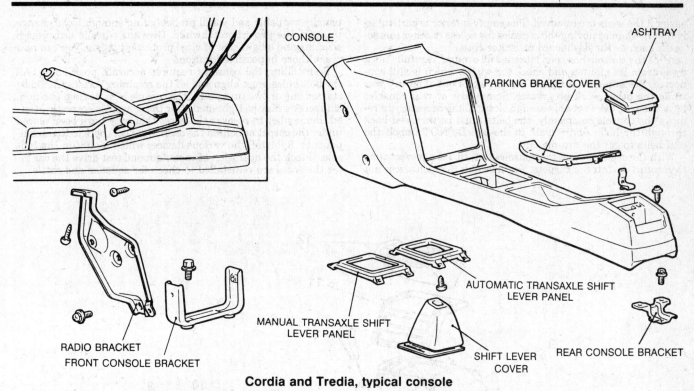

Cordia and Tredia, typical console

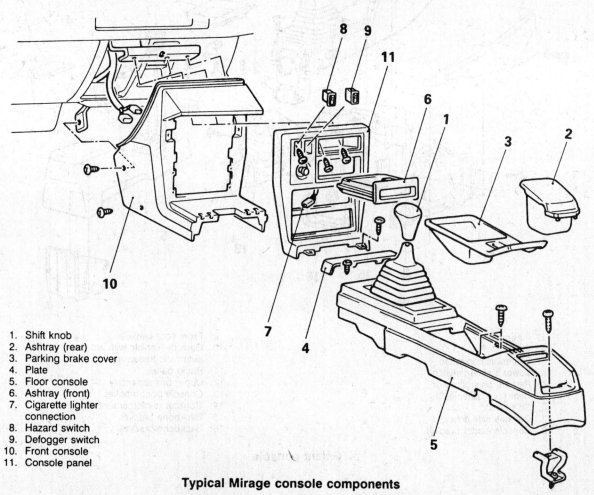

1. Shift knob
2. Ashtray (rear)
3. Parking brake cover
4. Plate
5. Floor console
6. Ashtray (front)
7. Cigarette lighter connection
8. Hazard switch
9. Defogger switch
10. Front console
11. Console panel

Typical Mirage console components

6-69

6 CHASSIS ELECTRICAL

easier if the seats are removed. The empty interior is particularly handy when removing the consoles for access to other components such as the dashboard or heater core.

After the console has been loosened all around, carefully lift it away from its mounts and check for wiring which is still connected. Disconnect it and place the connector out of harm's way while the other work progresses. Some floor or rear consoles have the seat belts routed through slots. If it is necessary to remove the console completely, the belts must be threaded back through the holes and remain in the car. Do NOT unbolt the seat belts to get the console out.

With the seats out and the consoles out, this is a perfect time to vacuum the interior carpets. You'll be able to reach places not usually available and you'll probably find enough loose change to make the project worthwhile. Give the console a thorough cleaning and a light coat of vinyl protectant; again, you can now reach those impossible locations

Reinstalling the consoles requires accurate positioning. All the bolt holes must align and all the retaining screws should be started one or two turns until the console is in its final position. The screws may be tightened after the console is correctly located. Remember to recover the wires that may have been placed under the carpet or behind the dash. Connect them to their components. Reinstall the various devices which mount in the console. Check the operation of each item and test drive the car after the seats are reinstalled to check for squeaks and rattles.

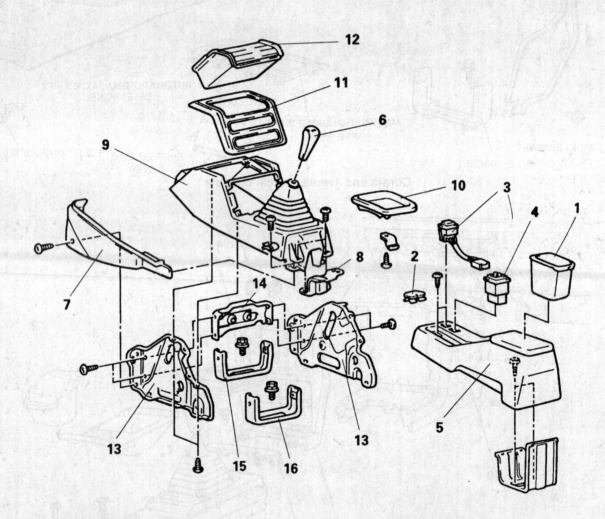

1. Inner box
2. Plug
3. Electronic controlled power steering switch
4. Remote control mirror
5. Rear floor console
6. Shift knob
7. Console side cover
8. Console center bracket
9. Front floor console
10. Garnish (vehicle with an automatic transaxle)
11. Radio panel
12. Upper box assembly
13. Console front bracket
14. Console reinforcement
15. Backbone bracket
16. Backbone bracket

Galant console

CHASSIS ELECTRICAL 6

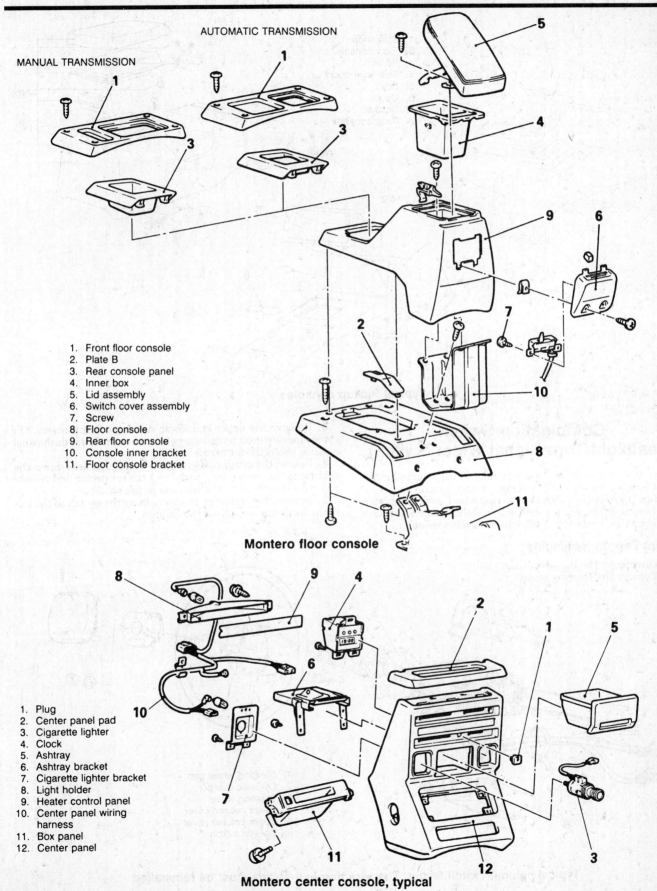

1. Front floor console
2. Plate B
3. Rear console panel
4. Inner box
5. Lid assembly
6. Switch cover assembly
7. Screw
8. Floor console
9. Rear floor console
10. Console inner bracket
11. Floor console bracket

Montero floor console

1. Plug
2. Center panel pad
3. Cigarette lighter
4. Clock
5. Ashtray
6. Ashtray bracket
7. Cigarette lighter bracket
8. Light holder
9. Heater control panel
10. Center panel wiring harness
11. Box panel
12. Center panel

Montero center console, typical

6-71

6 CHASSIS ELECTRICAL

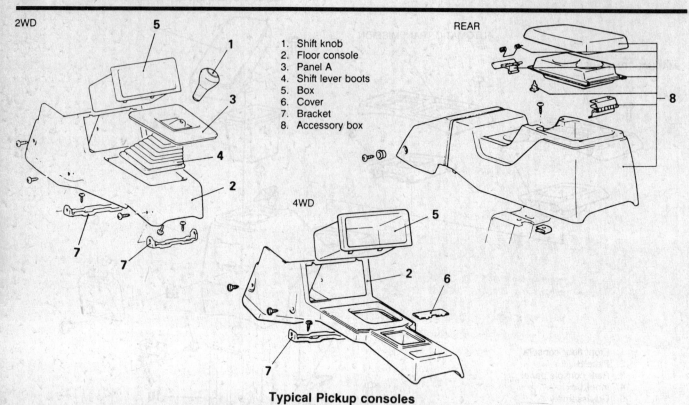

Typical Pickup consoles

Combination Switch (Headlight/Turn Signal/Wiper Switch)

Even though the driver uses the stalks on the column for different functions and probably thinks of each control as a separate item, the switch behind the steering wheel is one integrated unit. In most cases, any problem with a stalk or switch function will require replacement of the entire switch assembly.

Except Pop-Up Headlights

1. Disconnect the negative battery cable.
2. Remove the steering wheel.
3. Remove the upper and lower steering column covers. On some models it may be necessary to remove the lower dash panel or knee protector for access.
4. Follow the combination switch wiring harness down the column to the connector. Disconnect the connector and release any wire clips holding the harness to the switch.
5. Remove the retaining screws holding the switch to the column. Work the switch off the column.

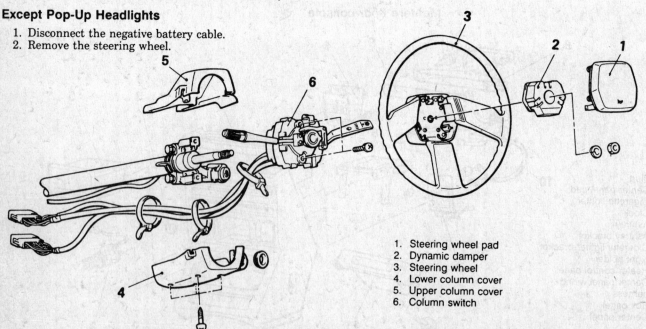

Typical column switch layout. The wire harness straps must be reinstalled

6–72

CHASSIS ELECTRICAL 6

NOTE: Some models have the turn signal and hazard switches mounted on a separate base plate. Removing the mounting screws will allow separate removal or replacement.

6. Install the switch, tighten the mounting screws and route the wire harness to the connector. Make sure the wire does not interfere with the moving parts of the tilt column mechanism.

WARNING: The switch must be centered in the column. If it is installed on an angle, the self-cancelling of the turn signals will be affected.

7. Connect the harness and install any wire clips or retainers.
8. Install the upper and lower column covers.
9. Install the steering wheel.

Pop-Up Headlights

The folding headlights are made to open when the lights are turned on. The drive motors for the headlight doors are controlled through the column switch. Additionally, the headlight doors may be opened by pushing the pop-up switch on the instrument cluster. This switch activates only the motors but not the lights; the lights are up and off. This feature is very handy for cleaning or driving in situations where the lights are only occasionally needed. When using the switch during cleaning, make sure nothing is near the doors during their operation. The motors develop sufficient force to cause injury.

When the pop-up switch is engaged, the lights will stay up even after they are turned off. If you've parked the car and the lights won't go down, the pop-up switch probably got hit accidentally.

CHILTON TIP: Owners in areas of heavy winter weather may wish to leave the headlight doors up when parked in a storm. Ice and snow can cause the doors to freeze in the closed position.

If the headlights ever fail to open when turned on, they may be raised by hand using the knobs under the hood.

1. At the fusebox inside the car, remove the No. 4 fuse. This is the fuse for the headlight motors, not the headlights.
2. Open and support the hood.
3. Just behind each headlight is a round access hole. Reach in and remove the protective rubber cover from the knob on the motor.

4. Turn the knob to raise the headlight door. Turn the knob until the headlight will not raise any more.

— **CAUTION** —
DO NOT attempt to raise (or lower) the headlights manually if the fuse has not been removed. The door will suddenly open or close with enough force to injure hands and fingers.

5. Repeat the opening procedure on the other light.
6. Reinstall the rubber boots on the motors. Lower and latch the hood.
7. Turn on the headlights at the column switch. Even though the doors are not operating, the lights should still illuminate.
8. The failed system should be diagnosed and repaired as soon as possible.

Rear Window Wiper Switch

REMOVAL AND INSTALLATION

The controls for the rear wiper/washer are found in varying locations depending on the model. If the switch is incorporated with the regular wiper switch on the stalk, the combination switch must be changed since the switches are all together on the column.

If the switch is located on the instrument hood (example: Mirage), the hood must be removed from the instrument cluster. Once the hood is free, the wiring may be disconnected and the switch tested or replaced.

By far the most common location for the rear wiper switch is in the dashboard. These switches usually are press fit into the dash, or held with small plastic tabs. Using a padded flat tool, the switch may be pryed or popped loose; use extreme care not to mar or scratch the surrounding dashboard. It may be necessary

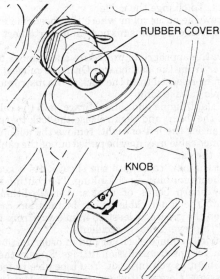

After removing the fuse, the headlights may be raised or lowered by removing the rubber cover and turning the knob

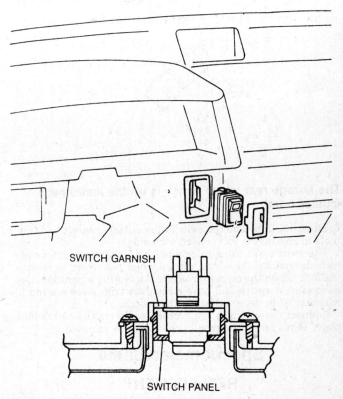

Starion rear wiper switch. Note the tabs holding the switch to the dash

6-73

6 CHASSIS ELECTRICAL

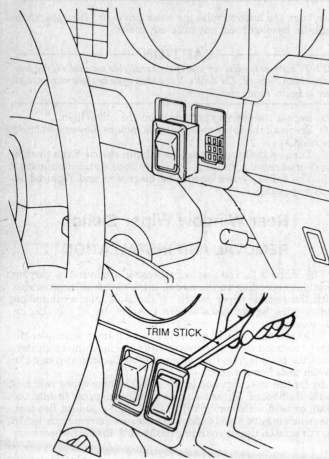

Removing the Montero rear wiper switch

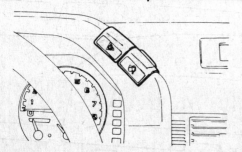

The Mirage rear wiper switch is on the instrument cluster hood

to either reach behind the switch (if possible) or insert two thin tools to release the locktabs on each side.

The switch will usually come out with the wire harness attached; separate the wiring carefully and remember to release the lock tab on the connector before attempting separation. Occasionally, a switch will have a separate trim piece around it which must be removed before the switch.

Connect the switch to the harness and press the switch into place. Don't forget the trim plate if one was removed.

Speedometer Cable

REPLACEMENT

The speedometer cable connects a rotating gear within the transaxle to the dashboard speedometer/odometer assembly. The dashboard unit interprets the number of turns the made by the cable and displays the information as miles per hour and total mileage.

Assuming that the transmission contains the correct gear for the car, the accuracy of the speedometer depends primarily on tire condition and tire diameter. Badly worn tires (too small in diameter) or overinflation (too large in diameter) can affect the speedometer reading. Replacement tires of the incorrect overall diameter (such as oversize snow tires) can also affect the readings.

Generally, manufacturers state that speedometer/odometer error of ± 10% is considered normal due to wear and other variables. Stated another way, if you drove the car over a measured 1 mile course and the odometer showed anything between 0.9 and 1.1 miles, the error is considered normal. If you plan to do any checking, always use a measured course such as mileposts on an Interstate highway or turnpike. Never use another car for comparison--the other car's inherent error may further cloud your readings.

The speedometer cable can become dry or develop a kink within its case. As it turns, the ticking or light knocking noise it makes can easily lead an owner to chase engine related problems in error. If such a noise is heard, carefully watch the speedometer needle during the speed range in which the noise is heard. Generally, the needle will jump or deflect each time the cable binds. The needle motion may be very small and hard to notice; a helper in the back seat should look over the driver's shoulder at the speedometer while the driver concentrates on driving.

CHILTON TIP: The slightest bind in the speedometer cable can cause unpredictable behavior in the cruise control system. If the cruise control exhibits intermittent surging or loss of set speed symptoms, check the speedometer cable first.

Some cables do not attach directly to the speedometer assembly but rather to an electrical pulse generator. These pulses may be used for the meter and mileage signal. Additionally, the electric signals representing the speed of the car can be used by the fuel injection control unit, the cruise control unit and other components. To change the cable:

1. Unscrew the outer collar where the speedometer cable enters the transaxle or transmission, and disconnect the inner cable.
2. Follow the appropriate procedure and remove the instrument cluster from the dashboard. On most models, the cable housing can be released from the back of the instrument cluster with a simple release tang.

On the Galant/Sigma and some Mirage, pull the cable toward you (out of the dash) and then release the lock by turning the adaptor to either the left or right. remove the adaptor.

3. The inner cable may now be pulled out of the cable housing from inside the car.
4. Install the new cable into the sheath. Its a good idea to check the sheath for binding or kinking. The routing should not contain any sharp bends or angles. A little lithium grease brushed onto the inner cable before installation can be very helpful. Keep the grease at least 254mm away from the speedometer end to prevent messy accidents.
5. The cable is square at either end in order for it to engage the speedometer and transmission fittings. Engage the inner cable by rotating it as you insert it. You'll be able to feel when it fits into its slot. Connect the outer cable at the instrument cluster.
6. On the transmission end, screw the outer cable connector securely onto the transmission housing. Make certain the cable is secure.

CHASSIS ELECTRICAL 6

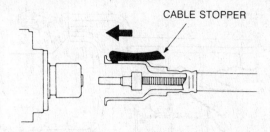

A plastic spring clip is used to hold some speedometer cables

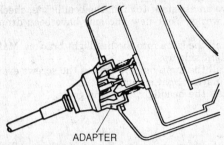

Adapters may be used to connect the cable to the back of some speedometers

LIGHTING

Headlights

Headlights, like any other lighting device, can fail due to broken filaments. The front of any car is the worst possible location for a lighting device since it is subject to impact, extensive temperature change and severe vibration, all of which shorten the life of the light. The front of the car is also where good lighting is needed the most so its not uncommon to have to replace a headlight during the life of the vehicle.

There are two general styles of headlamps, the sealed beam and replaceable bulb type. The sealed beam is by far the most common and includes almost all of the circular and rectangular lamps found on cars built through the early 1980s. The sealed beam is so named because it includes the lamp (filament), the reflector and the lens in one sealed unit. Sealed beams are available in several sizes and shapes.

The replaceable bulb is the newer technology. Using a small halogen bulb, only the lamp is replaced, while the lens and reflector are part of the body of the car. This is generally the style found on wrap-around or "European" lighting systems. While the replaceable bulbs are more expensive than sealed beams, they generally produce more and better light. The fixed lenses and reflectors can be engineered to allow better frontal styling and better light distribution for a particular vehicle.

It is quite possible to replace a headlight of either type without affecting the alignment (aim) of the light. Sealed beams mount into a bracket (bucket) to which springs are attached. The adjusting screws control the position of the bucket which in turn aims the light. Replaceable bulbs simply fit into the back of the reflector. The lens and reflector unit are aimed by separate adjusting screws.

Take a moment before disassembly to identify the large adjusting screws (generally two for each lamp, one above and one at the side) and don't change their settings.

With the exception of the oldest cars, sealed beams are removed from the outside of the car. Start with the outer trim pieces and work your way in to the lamp and its retainer. Bulb type units are almost always replaced from under the hood.

REMOVAL AND INSTALLATION

— CAUTION —
Most headlight retaining rings and trim bezels have very sharp edges. Wear gloves. Never pry or push on a headlamp; it can shatter suddenly.

When changing a replaceable bulb, never touch the new bulb with your fingers; always hold it by the base. The natural oils in your skin will cling to the glass and create hot spots on bulb envelope. These hot spots will shorten the life of the bulb substantially. If the bulb has been handled, clean the glass with alcohol and dry it thoroughly before installation.

Cordia and Tredia

1. Pry up on the four tabs accessible at either end of the grill with a flat tool to release them and remove the grill.
2. Remove the marker light for access to the two clips on the side. Remove the four screws, disconnect the two clips and remove the headlight bezel.
3. Remove the two screws from the top of the headlight retaining ring, pull the ring off at the top, lower it to release the tangs at the bottom and remove it.
4. Pull the headlight outward, pull the electrical connector off, and remove the lamp.
5. Hold the new lamp in place and connect the wiring.

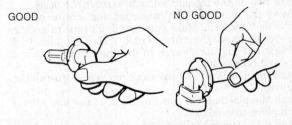

Correct handling will extend bulb life

6-75

6 CHASSIS ELECTRICAL

6. Have an assistant turn on the headlights; check that your new one works. Your new one may have been dropped in the store.

7. Install the lamp into the headlight bracket. Make sure the light is right side up.

8. Install the lamp holder, tighten the screws evenly and install the marker light assembly.

9. Install the headlight trim and grille.

Starion

1. Use the "pop-up" switch for raising the headlights located on the right side of the instrument panel, NOT THE REGULAR HEADLIGHT ON-OFF SWITCH to raise the headlight doors; using the regular headlight switch may cause damage to the headlight relay coil during the next step. Disconnect the negative battery cable.

2. Remove the two screws from the inside of the bezel and three from the outside and remove the bezel by pulling it up and forward.

3. Remove the two screws at the top and two at the bottom of the retaining ring. Remove the ring, pull the headlight out until the electrical connector is accessible, disconnect the connector, and remove the headlight.

4. Installation is the reverse of removal.

Galant through 1987

1. Remove the parking/marker light assembly by unhooking the retaining spring in the engine compartment and then pulling it out. Remove the two headlight bezel mounting screws (on the outboard side of the bezel) and undo the two clips along the top edge. Remove the bezel.

2. Remove the three headlight retaining ring attaching screws (at top, one side, and bottom) and remove the ring. Pull the headlamp out far enough to reach the electrical plug, unplug it and remove it.

3. Installation is the reverse of removal. Make sure the marker light is correctly seated in its mounts.

1988–89 Galant

1. Disconnect the harness connector from the rear of the lamp socket.

2. Remove the three screws holding the socket holder to the reflector and remove the bulb and socket from the car.

3. Replace the bulb, fitting it carefully into the holder.

4. Install the holder, tighten the screws evenly and connect the harness.

Mirage through 1986

1. Remove the mounting screws for the front side marker light. Slide the unit rearward, and pull the unit slightly out, being careful to clear the plate spring. Disconnect the electrical connector and remove the unit.

2. Disconnect the three latches at the top of the radiator grill and remove it. Remove the three headlight bezel screws, one at the top and two on the inboard side and remove the bezel.

3. Remove the four headlight ring attaching screws (one at each of the four corners). Make sure you don't touch the four adjusting screws, one at the center of each of the four sides. Pull the headlight out just far enough to disconnect the electrical connector and remove it.

4. Install the new lamp and connect the harness. Install the retaining ring.

5. Install the plastic bezel and the grille. Make sure everything is in place correctly.

6. Install the side marker light; remember the hook the leading edge by sliding the light into place from the rear.

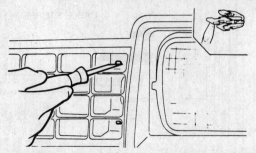

The Tredia grille is held by four clips which must be released before removal

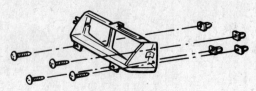

Exploded view of the headlight bezel and attaching screws on Cordia and Tredia

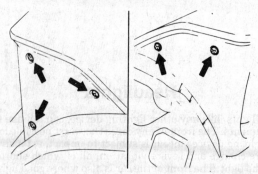

Starion headlight bezel screws, 3 on the outside and two on the inner face. All must be removed to change a headlamp

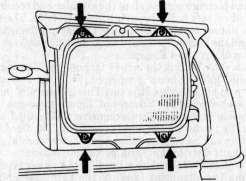

Once the bezel is removed, the Starion headlamp retaining frame is easily removed. Don't lose the small screws

1987–88 Mirage
All Precis

1. Hold the wiring connector and base of the bulb with one hand and turn the large round collar counterclockwise with the other hand.

2. Once the locking collar is loose (about ¼ turn) the assem-

CHASSIS ELECTRICAL 6

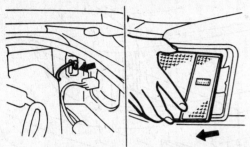

Unhook the spring and then remove the marker light assembly before changing a Galant headlight

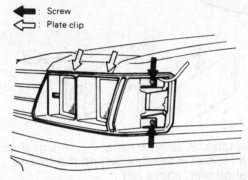

Location of screws and clips holding the Galant headlight bezel

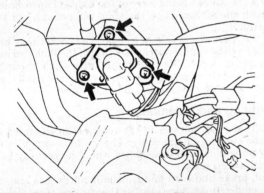

1988–89 Galant headlamp bulb socket retainer

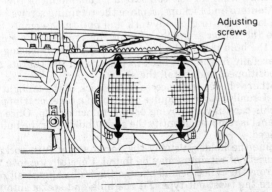

Remove the four bezel screws — not the adjusting screws — to change the Mirage headlamp

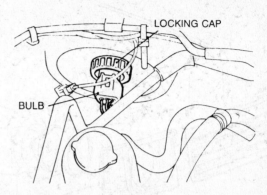

A half-twist releases the 1987–88 Mirage headlamp bulb

bly may be removed from the reflector.
 3. Release the retaining clip and pull the bulb free of the socket. Install the new bulb.
 4. Fit the assembly into the reflector. The socket and lamp have plastic tabs or keys and will only fit into the reflector one way.
 5. Once correctly seated, fit the locking collar in place and turn it clockwise to lock it in place.

1989 Mirage

1. The left headlight is easily accessable from under the hood. If the right headlight is to be changed:
   a. Remove the battery hold-down strap.
   b. Remove the windshield washer reservoir tank.
   c. Remove the radiator overflow tank.
2. Disconnect the harness connector from the rear of the lamp socket.
3. Remove the three screws holding the socket holder to the reflector and remove the bulb and socket from the car.
4. Replace the bulb, fitting it carefully into the holder.
5. Install the holder, tighten the screws evenly and connect the harness.
6. Install the radiator tank, washer tank and battery hold-down if they were removed.

Montero

1. Remove the radiator grille.
2. Remove the front sidemarker lamp.
3. Remove the plastic headlamp bezel.
4. Loosen or remove the three screws holding the headlight retaining ring. Once all three screws are loosened, the ring may be turned slightly and removed. This trick eliminates removing very small screws.
5. Remove the headlamp and disconnect the wiring harness. Install the new lamp and connect the wiring. Fit the lamp in place and install the retaining ring.
6. Install the plastic bezel, the sidemarker assembly and the grille.

Pickup

1. Remove the grille.
2. Remove the four small screws holding the metal retaining ring or frame.
3. Pull the lamp forward and out; disconnect the wiring harness. Install the new lamp and connect the harness.
4. Fit the lamp into place and install the retaining ring.
5. Install the grille.

6-77

6 CHASSIS ELECTRICAL

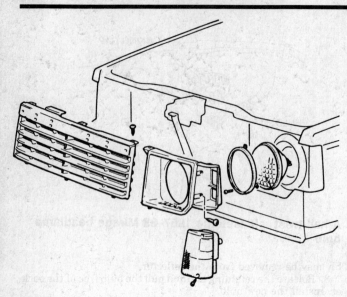

Montero headlamp components

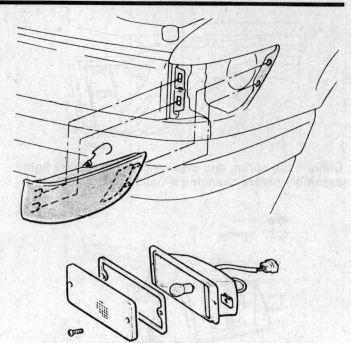

Two examples of sidemarker and turn signal lamps. Note the hidden slide clips behind the Mirage combination unit, above left

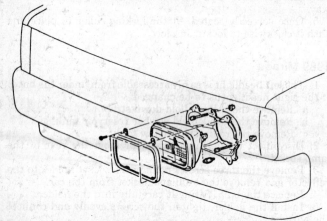

Replace the Pickup headlight after removing the grille

Signal and Marker Lights

REMOVAL AND INSTALLATION

Front Turn Signal and Sidemarker Lamps

To change the bulb on any of the front lamps on a Mitsubishi, remove the mounting screws holding either the lens to the reflector or the lighting unit to the bodywork. If the lens comes off separately, wear gloves to remove the bulb. Very small bulbs such as side markers generally pull straight out of the socket; larger round bulbs such as turn signal lamps usually twist counterclockwise to remove.

Install the new lamp. Note that the socket for some turn signal bulbs have two different length guide grooves; the lamp only fits correctly one way (It is possible to do it wrong, but it requires a lot of force. If you find that you're pushing on the bulb with no results during installation, you've got it backwards.) These bulbs should install easily with a gentle push into the socket and a quarter turn clockwise. Reassemble the lens, paying careful attention to the placement of any gaskets and seals. These are very important in keeping out dirt and water.

Newer models have the lens bonded to the reflector to prevent moisture intrusion. Either release the spring from behind the assembly (under the hood) or remove the retaining screws if visible from the outside. Carefully work the assembly loose from the body work, always being wary of the hidden clip or bracket. If you break one of these clips, the lamp will not stay put when you reinstall it.

When the assembly is free of the car, the socket may be removed with a twist counterclockwise. Install the bulb and insert the assembly back into the housing. Fit the whole thing back onto the bodywork, engaging any clips or hooks. Install the retaining screws or the tension spring as necessary.

Rear Lighting and Sidemarkers

Generally the sidemarkers at the rear are incorporated into the tail and turn signal assembly. The Pickups and Monteros use separate side marker assemblies whose lenses may be removed with two screws.

The tail, brake and turn lamps usually all fit into one integrated assembly. It is wise to identify the position of the light being changed from the outside of the vehicle before confronting 5 or six sockets in the back of the unit. Simply knowing the bad bulb is third from the left can save a huge amount of time.

For Montero and Pickups, remove the retaining screws holding the lamp unit to the bodywork and pull the assembly free. Remove the socket from the rear of the reflector (or simply change the bulb if only the lens came off). After replacing the lamp carefully fit the unit back in place, with the gaskets correctly positioned, and install the screws.

For other vehicles, gain access to the back of the taillight assembly. Usually this will require removal of a cover or trim panel. Do this carefully, removing each retainer or clip. Once the cover panel is removed, identify the socket holder for the bulb to be changed.

Examine the holder; usually they are quarter-turn sockets, but a variety of retainers may be found. Carefully remove the socket from the reflector and change the bulb. Almost all rear lamps are the twist out type. Fit the bulb and socket into the housing. Normally, the socket will only fit one way due to plastic guides or keys. Connect the locking device or tab if one was released and reinstall the cover or trim panel.

CHASSIS ELECTRICAL 6

CIRCUIT PROTECTION

Fusible Links

Every circuit in the car's electrical system is protected by a fuse or fusible link except the starter circuit. A melted fusible link or blown fuse indicates not a bad link, but a short circuit in a component, a grounded wire or an overload. Using the procedures found at the beginning of this Section, you can isolate the component or area of the wiring that is drawing excessive current. You will usually find a defective component (motor or lamp socket) or a spot where motion between wiring and the body has rubbed insulation through and caused bare wiring to touch the body or a metal bracket or major drivetrain component.

Mitsubishi vehicles may have one or more fusible links. They appear to be short wires and are always located at the positive battery terminal. Additionally, there may be sub-links within particular high amperage circuits. They are simply unplugged and replugged to replace them, but you MUST be sure you are using a link of the same capacity. Excessive capacity in a link could result in a vehicle fire.

Fuses

The fuse box on most models is located under the instrument panel on the left (driver's side) above the cowl side trim. Almost all vehicles after 1986 have an additional fuse and relay board under the hood near the battery. The most common circuits are on the interior fuseboard for ease of changing the fuse by the driver; the fuses under the hood tend to control peripheral systems used occasionally such as the rear window defogger.

The underhood relay board may have additional fuse boards mounted to it as other circuits run from it. Further additional fuses may be mounted directly at the positive battery terminal, monitoring circuits which originate there. It is in your best interest to read the owner's manual carefully and identify the location of certain fuses which could affect the safety and operation of the vehicle. You could drive home without the heater fan working, but could you make the same drive without headlights?

COMPONENT LOCATOR

Cordia and Tredia

- Main Fusible Links: 2 at battery positive terminal.
- Sub links: 1 on relay board near battery.
- Fusebox: Under left side dash.
- Special Fuses not in fusebox: Power door locks and headlight high beams, on relay board near battery; air conditioning, 2 fuses — one above evaporator case and one behind left front headlight.
- Turn Signal and Hazard Flasher Relay: Combination unit on relay board near interior fusebox.

Starion

- Main Fusible Links: 1 at battery positive pole.
- Sub links: Two, one on each relay board near the battery.
- Fusebox: Under left side dashboard. Contains automatic fuse checker to find failed fuse.
- Special Fuses not in fusebox: 3 under hood on relay board. Heated outside mirror fuse located inline, above fuse box.
- Turn Signal and Hazard Flasher Relay: Separate but adjacent units on relay board near interior fusebox. Hazard flasher is round.

Mirage

- Main Fusible Links: 1985–88, 2 at positive battery terminal. 1989, 1 modular or block-type located in relay board near battery.
- Sub links: 1985–88, 1 behind left headlight. 1989, included on relay board.
- Fusebox: Under left dashboard.
- Special Fuses not in fusebox: 1985–88, air conditioner fuse mounted on receiver/dryer, left side near power steering reservoir. 1989, dedicated fuses on relay board near battery; air conditioner fuses on separate fuse block behind left headlight.
- Turn Signal and Hazard Flasher Relay: Combined unit on relay board near interior fusebox. 1985–88 relay plugs into top of fuseboard.

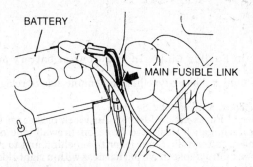

Battery mounted main links

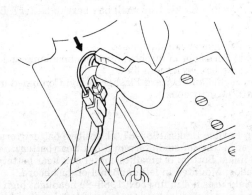

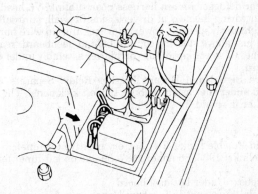

Cordia and Tredia main and sub-links

6 CHASSIS ELECTRICAL

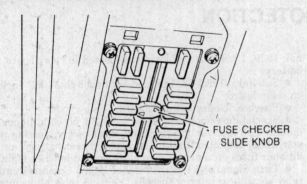

Starion fusebox with fuse checker

Galant

• Main Fusible Links: In small box at battery positive terminal
• Sub links: 5 block-type links on engine side of relay board, near battery
• Fusebox: Under left side dash.
• Special Fuses not in fusebox: Anti-lock brakes, at battery positive terminal; 5 dedicated fuses on firewall side of relay board near battery; air conditioner fuse behind and to right of glove box; sunroof fuse in wiring harness within right side windshield pillar trim.
• Turn Signal and Hazard Flasher Relay: Combined unit on relay board above interior fusebox.

Precis

• Main Fusible Links: 1 in small box next to battery. Link is block-type.
• Sub links: None
• Fusebox: Under left dashboard.
• Special Fuses not in fusebox: Ignition, headlamp and alternator in same box with main link.
• Turn Signal and Hazard Flasher Relay: Combined unit in relay board near fusebox.

Montero

• Main Fusible Links: 1983-87, at positive battery terminal; 1988-89 at bottom of small terminal box near battery.
• Sub links: Bottom of small terminal box near battery.
• Fusebox: Mounted in left endpanel of dashboard.
• Special Fuses not in fusebox: 1983-89 headlight high beam indicator fuse located in harness behind left headlight except V6. 1987-88, air conditioner fuse mounted on evaporator case; 1989, rear defogger link in harness near sublink, V6 headlight circuit in wiring harness at driver's side firewall, sunroof fuse above right dash speaker, power door lock fuse in wire harness running just above steering column behind dashboard, rear air conditioner (if so equipped) fuse on blower assembly under right dashboard.
• Turn Signal and Hazard Flasher Relay: Separate units mounted under the dash and above the left kick panel. The hazard flasher is round.

Pickup

• Main Fusible Links: 2 at battery positive terminal.
• Sub links: 1983-86 none; 1987-89 one on left inner fender panel.
• Fusebox: Under left dashboard.
• Special Fuses not in fusebox: None
• Turn Signal and Hazard Flasher Relay: Combined unit on relay board to left of fusebox.

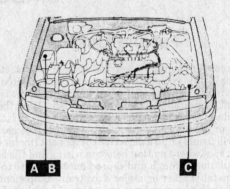

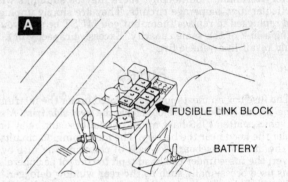

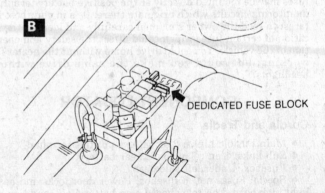

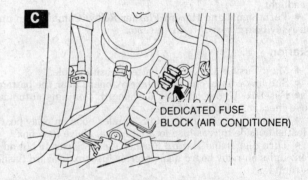

1989 Mirage fuse components

CHASSIS ELECTRICAL 6

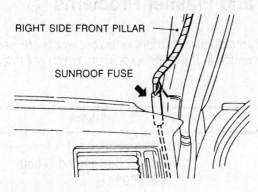

Galant sunroof fuse

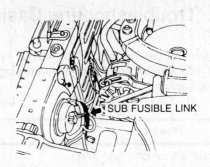

Sub fusible link located behind left headlight on 1987–89 Mirage

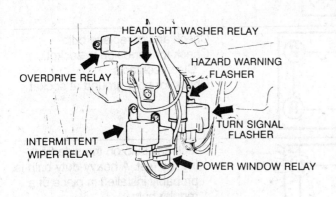

Montero hazard and turn signal flashers

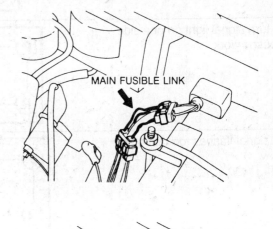

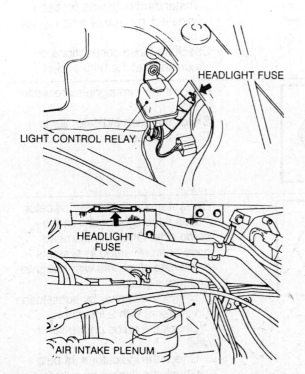

Montero headlight circuit fuses: 2.6L above and 3.0 V6 below

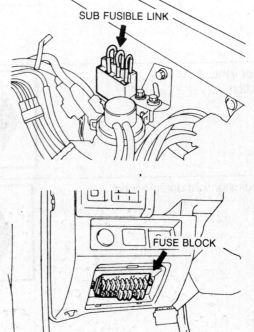

Pickup truck fuse arrangement

6–81

6 CHASSIS ELECTRICAL

Troubleshooting Basic Turn Signal and Flasher Problems

Most problems in the turn signals or flasher system can be reduced to defective flashers or bulbs, which are easily replaced. Occasionally, problems in the turn signals are traced to the switch in the steering column, which will require professional service.

F = Front   R = Rear   ● = Lights off   o = Lights on

Problem		Solution
Both turn signals on one side don't work		• Check for bad bulbs • Check for bad ground in both housings
One turn signal light on one side doesn't work		• Check and/or replace bulb • Check for corrosion in socket. Clean contacts. • Check for poor ground at socket
Turn signal flashes too fast or too slow		• Check any bulb on the side flashing too fast. A heavy-duty bulb is probably installed in place of a regular bulb. • Check the bulb flashing too slow. A standard bulb was probably installed in place of a heavy-duty bulb. • Check for loose connections or corrosion at the bulb socket
Indicator lights don't work in either direction		• Check if the turn signals are working • Check the dash indicator lights • Check the flasher by substitution
One indicator light doesn't light		• On systems with 1 dash indicator: See if the lights work on the same side. Often the filaments have been reversed in systems combining stoplights with taillights and turn signals. Check the flasher by substitution • On systems with 2 indicators: Check the bulbs on the same side Check the indicator light bulb Check the flasher by substitution

6-82

CHASSIS ELECTRICAL 6

Troubleshooting Basic Lighting Problems

Problem	Cause	Solution
Lights		
One or more lights don't work, but others do	• Defective bulb(s) • Blown fuse(s) • Dirty fuse clips or light sockets • Poor ground circuit	• Replace bulb(s) • Replace fuse(s) • Clean connections • Run ground wire from light socket housing to car frame
Lights burn out quickly	• Incorrect voltage regulator setting or defective regulator • Poor battery/alternator connections	• Replace voltage regulator • Check battery/alternator connections
Lights go dim	• Low/discharged battery • Alternator not charging • Corroded sockets or connections • Low voltage output	• Check battery • Check drive belt tension; repair or replace alternator • Clean bulb and socket contacts and connections • Replace voltage regulator
Lights flicker	• Loose connection • Poor ground • Circuit breaker operating (short circuit)	• Tighten all connections • Run ground wire from light housing to car frame • Check connections and look for bare wires
Lights "flare"—Some flare is normal on acceleration—if excessive, see "Lights Burn Out Quickly"	• High voltage setting	• Replace voltage regulator
Lights glare—approaching drivers are blinded	• Lights adjusted too high • Rear springs or shocks sagging • Rear tires soft	• Have headlights aimed • Check rear springs/shocks • Check/correct rear tire pressure
Turn Signals		
Turn signals don't work in either direction	• Blown fuse • Defective flasher • Loose connection	• Replace fuse • Replace flasher • Check/tighten all connections
Right (or left) turn signal only won't work	• Bulb burned out • Right (or left) indicator bulb burned out • Short circuit	• Replace bulb • Check/replace indicator bulb • Check/repair wiring
Flasher rate too slow or too fast	• Incorrect wattage bulb • Incorrect flasher	• Flasher bulb • Replace flasher (use a variable load flasher if you pull a trailer)
Indicator lights do not flash (burn steadily)	• Burned out bulb • Defective flasher	• Replace bulb • Replace flasher
Indicator lights do not light at all	• Burned out indicator bulb • Defective flasher	• Replace indicator bulb • Replace flasher

6 CHASSIS ELECTRICAL

Troubleshooting Basic Dash Gauge Problems

Problem	Cause	Solution
Coolant Temperature Gauge		
Gauge reads erratically or not at all	• Loose or dirty connections • Defective sending unit	• Clean/tighten connections • Bi-metal gauge: remove the wire from the sending unit. Ground the wire for an instant. If the gauge registers, replace the sending unit.
	• Defective gauge	• Magnetic gauge: disconnect the wire at the sending unit. With ignition ON gauge should register COLD. Ground the wire; gauge should register HOT.
Ammeter Gauge—Turn Headlights ON (do not start engine). Note reaction		
Ammeter shows charge Ammeter shows discharge Ammeter does not move	• Connections reversed on gauge • Ammeter is OK • Loose connections or faulty wiring • Defective gauge	• Reinstall connections • Nothing • Check/correct wiring • Replace gauge
Oil Pressure Gauge		
Gauge does not register or is inaccurate	• On mechanical gauge, Bourdon tube may be bent or kinked	• Check tube for kinks or bends preventing oil from reaching the gauge
	• Low oil pressure	• Remove sending unit. Idle the engine briefly. If no oil flows from sending unit hole, problem is in engine.
	• Defective gauge	• Remove the wire from the sending unit and ground it for an instant with the ignition ON. A good gauge will go to the top of the scale.
	• Defective wiring	• Check the wiring to the gauge. If it's OK and the gauge doesn't register when grounded, replace the gauge.
	• Defective sending unit	• If the wiring is OK and the gauge functions when grounded, replace the sending unit
All Gauges		
All gauges do not operate	• Blown fuse • Defective instrument regulator	• Replace fuse • Replace instrument voltage regulator
All gauges read low or erratically	• Defective or dirty instrument voltage regulator	• Clean contacts or replace
All gauges pegged	• Loss of ground between instrument voltage regulator and car • Defective instrument regulator	• Check ground • Replace regulator

CHASSIS ELECTRICAL 6

Troubleshooting Basic Dash Gauge Problems

Problem	Cause	Solution
Warning Lights		
Light(s) do not come on when ignition is ON, but engine is not started	• Defective bulb • Defective wire • Defective sending unit	• Replace bulb • Check wire from light to sending unit • Disconnect the wire from the sending unit and ground it. Replace the sending unit if the light comes on with the ignition ON.
Light comes on with engine running	• Problem in individual system • Defective sending unit	• Check system • Check sending unit (see above)

Troubleshooting the Heater

Problem	Cause	Solution
Blower motor will not turn at any speed	• Blown fuse • Loose connection • Defective ground • Faulty switch • Faulty motor • Faulty resistor	• Replace fuse • Inspect and tighten • Clean and tighten • Replace switch • Replace motor • Replace resistor
Blower motor turns at one speed only	• Faulty switch • Faulty resistor	• Replace switch • Replace resistor
Blower motor turns but does not circulate air	• Intake blocked • Fan not secured to the motor shaft	• Clean intake • Tighten security
Heater will not heat	• Coolant does not reach proper temperature • Heater core blocked internally • Heater core air-bound • Blend-air door not in proper position	• Check and replace thermostat if necessary • Flush or replace core if necessary • Purge air from core • Adjust cable
Heater will not defrost	• Control cable adjustment incorrect • Defroster hose damaged	• Adjust control cable • Replace defroster hose

Troubleshooting Basic Windshield Wiper Problems

Problem	Cause	Solution
Electric Wipers		
Wipers do not operate— Wiper motor heats up or hums	• Internal motor defect • Bent or damaged linkage • Arms improperly installed on linking pivots	• Replace motor • Repair or replace linkage • Position linkage in park and reinstall wiper arms

6 CHASSIS ELECTRICAL

Troubleshooting Basic Windshield Wiper Problems

Problem	Cause	Solution
Electric Wipers		
Wipers do not operate—No current to motor	• Fuse or circuit breaker blown • Loose, open or broken wiring • Defective switch • Defective or corroded terminals • No ground circuit for motor or switch	• Replace fuse or circuit breaker • Repair wiring and connections • Replace switch • Replace or clean terminals • Repair ground circuits
Wipers do not operate—Motor runs	• Linkage disconnected or broken	• Connect wiper linkage or replace broken linkage
Vacuum Wipers		
Wipers do not operate	• Control switch or cable inoperative • Loss of engine vacuum to wiper motor (broken hoses, low engine vacuum, defective vacuum/fuel pump) • Linkage broken or disconnected • Defective wiper motor	• Repair or replace switch or cable • Check vacuum lines, engine vacuum and fuel pump • Repair linkage • Replace wiper motor
Wipers stop on engine acceleration	• Leaking vacuum hoses • Dry windshield • Oversize wiper blades • Defective vacuum/fuel pump	• Repair or replace hoses • Wet windshield with washers • Replace with proper size wiper blades • Replace pump

CHASSIS ELECTRICAL 6

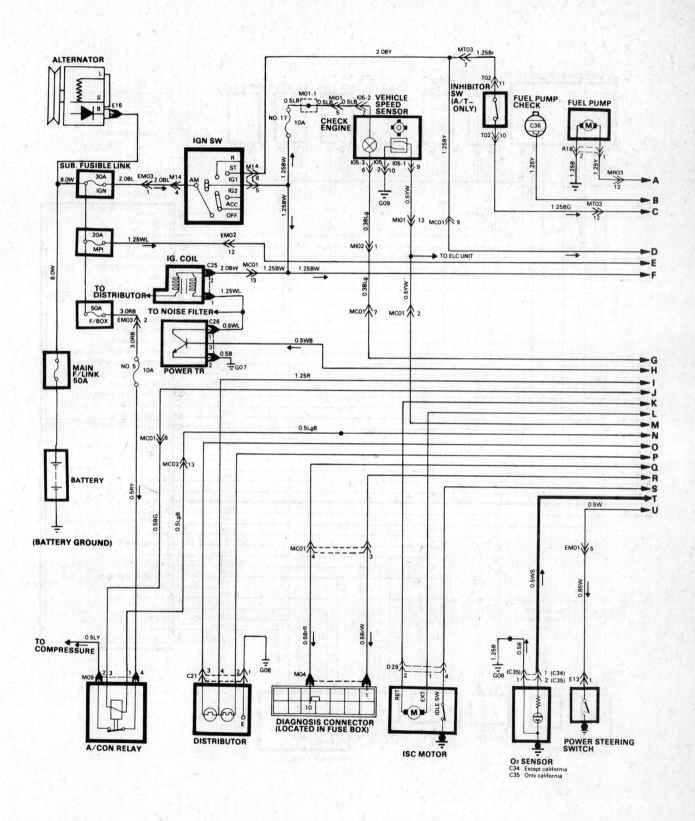

MPI system — Precis

6-87

6 CHASSIS ELECTRICAL

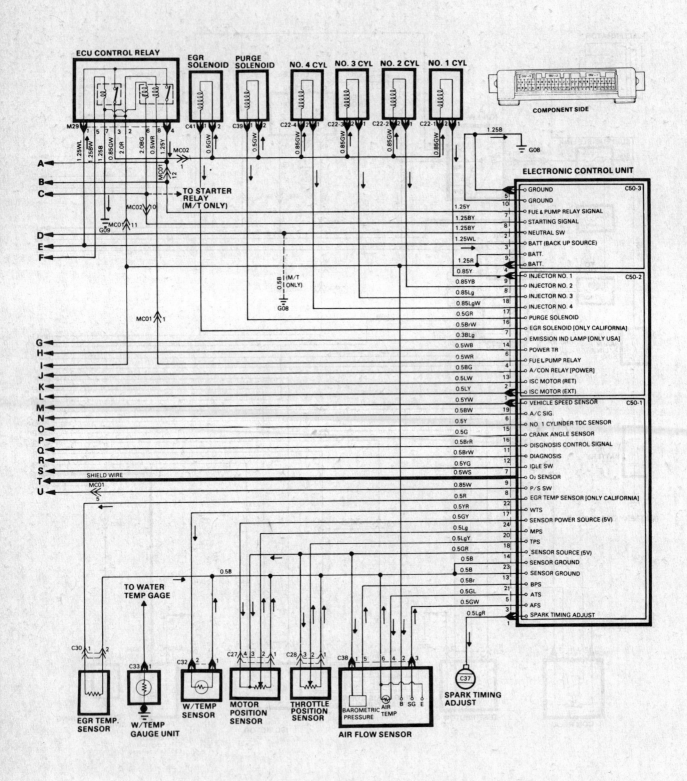

MPI system — Precis

CHASSIS ELECTRICAL 6

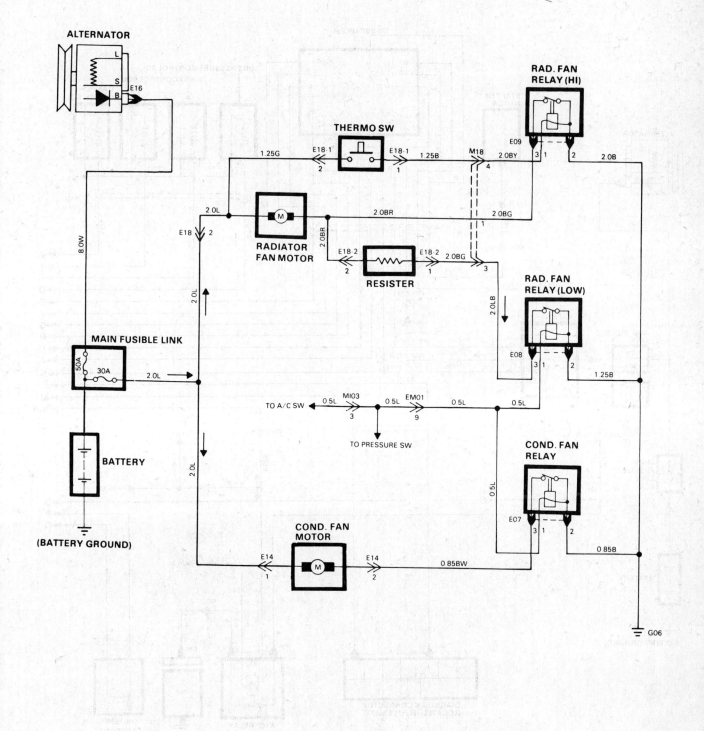

Cooling system — Precis

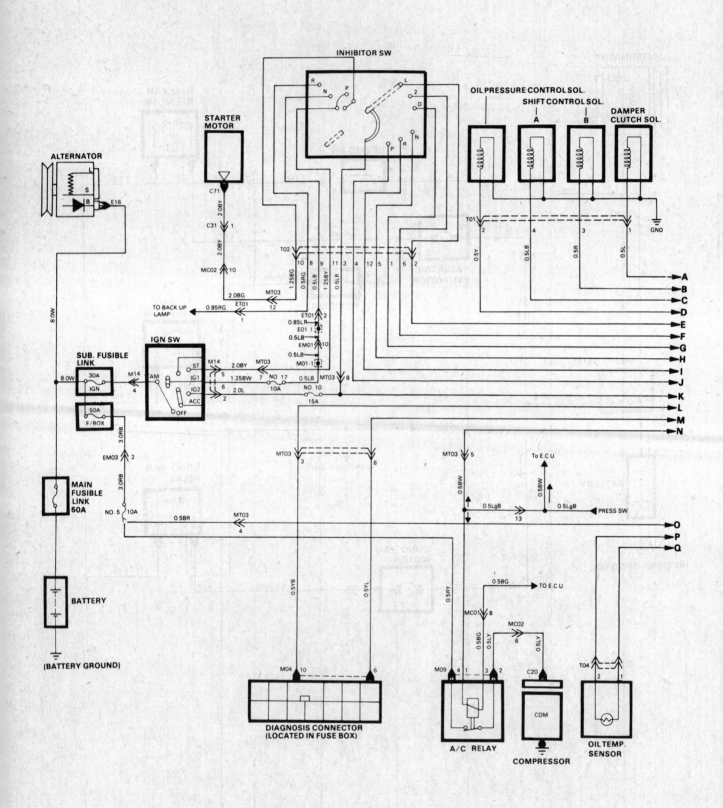

Electronic lock-up control – Precis

CHASSIS ELECTRICAL 6

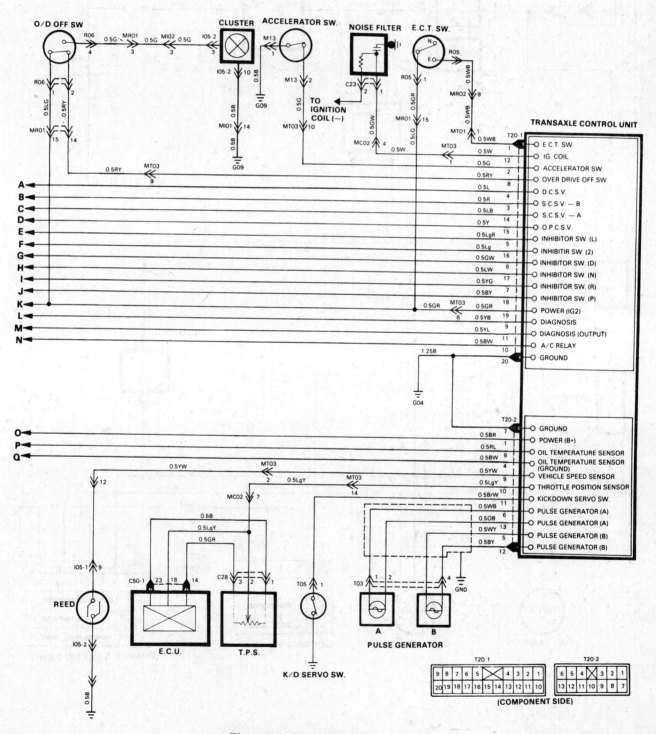

Electronic lock-up control – Precis

6-91

6 CHASSIS ELECTRICAL

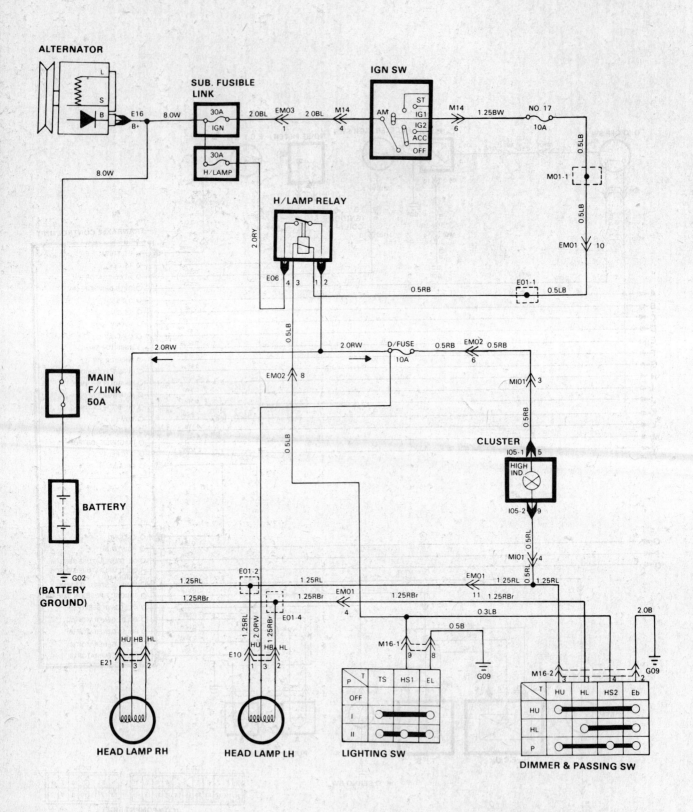

Headlmap system—Precis

CHASSIS ELECTRICAL 6

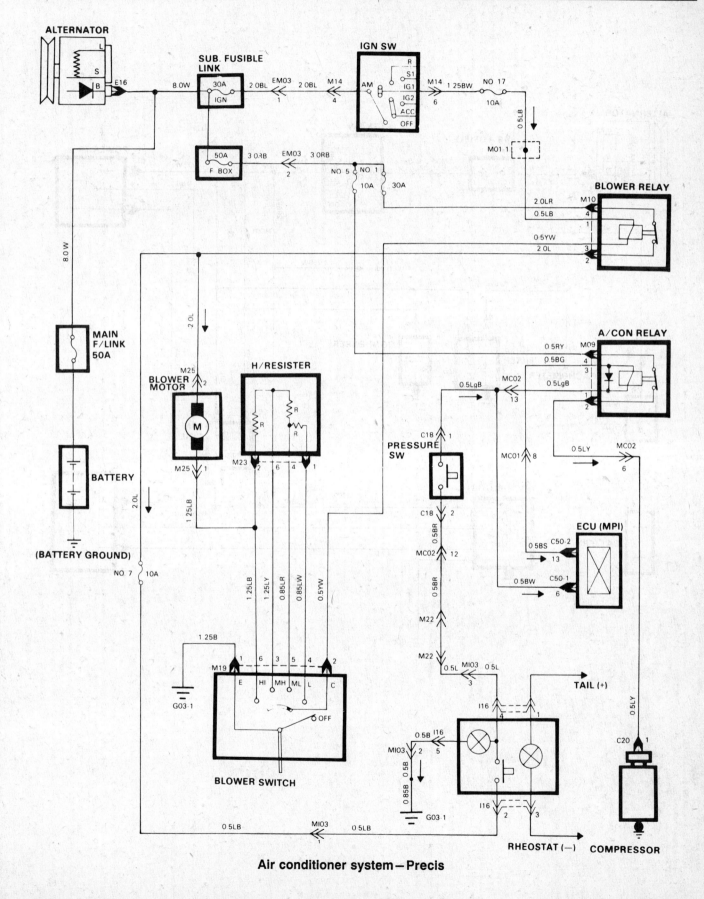

Air conditioner system—Precis

6-93

6 CHASSIS ELECTRICAL

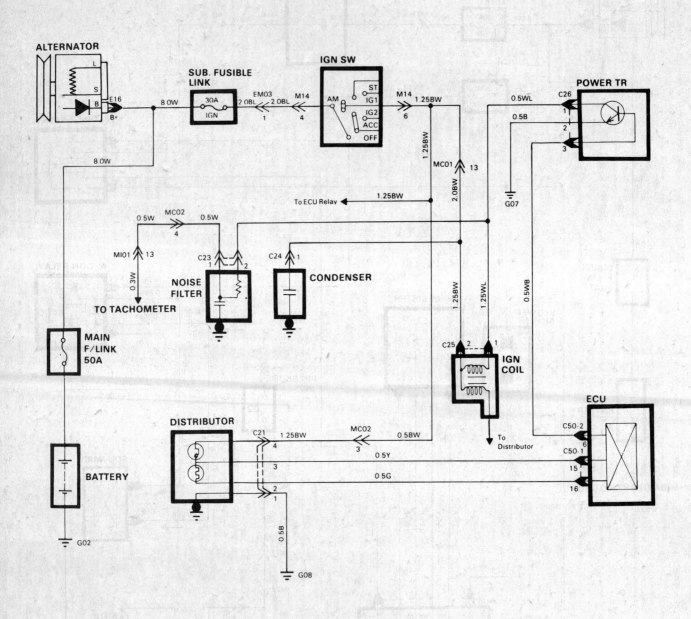

Ignition system – Precis

6-94

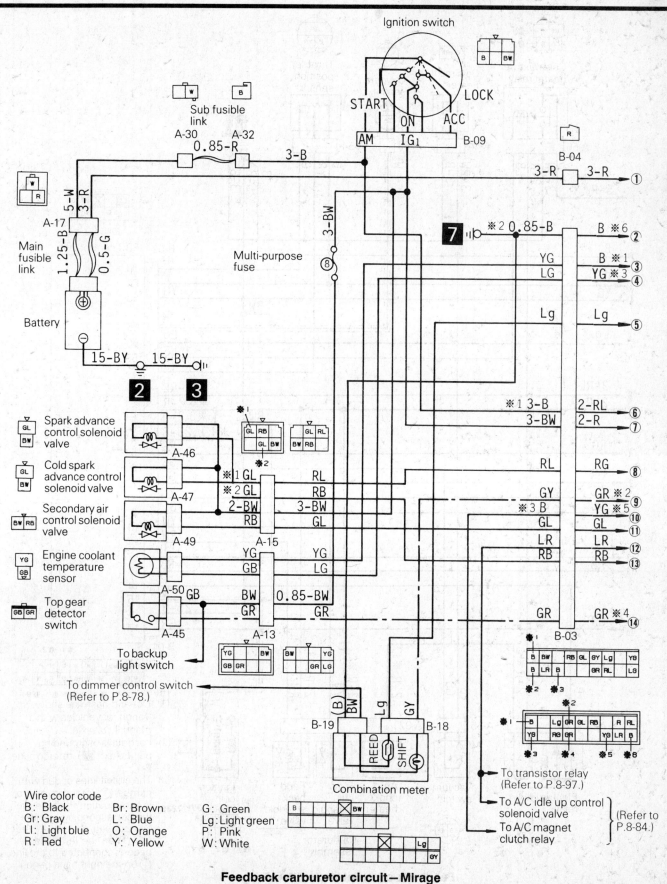

Feedback carburetor circuit — Mirage

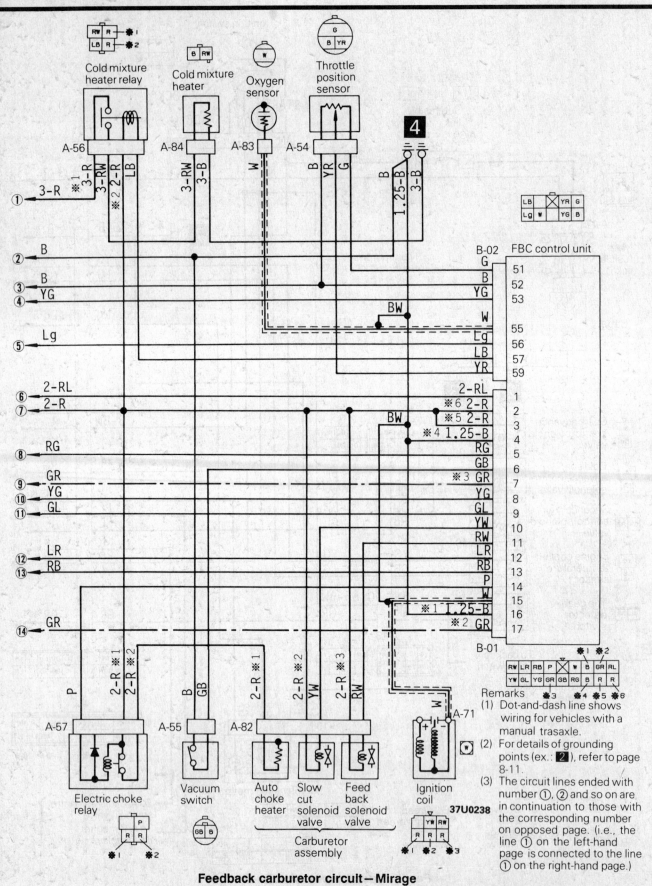

Feedback carburetor circuit – Mirage

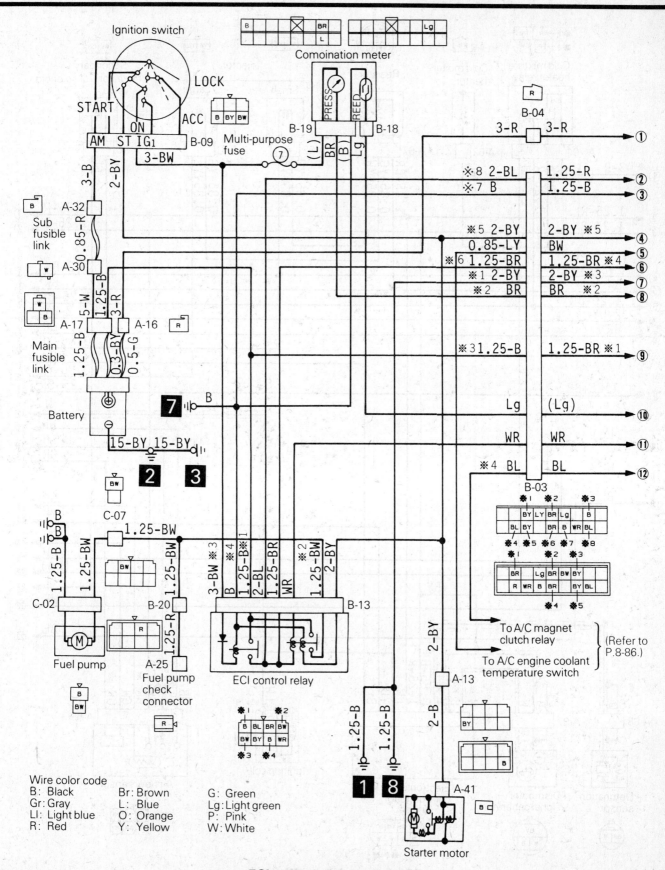

ECI system circuit — Mirage

6 CHASSIS ELECTRICAL

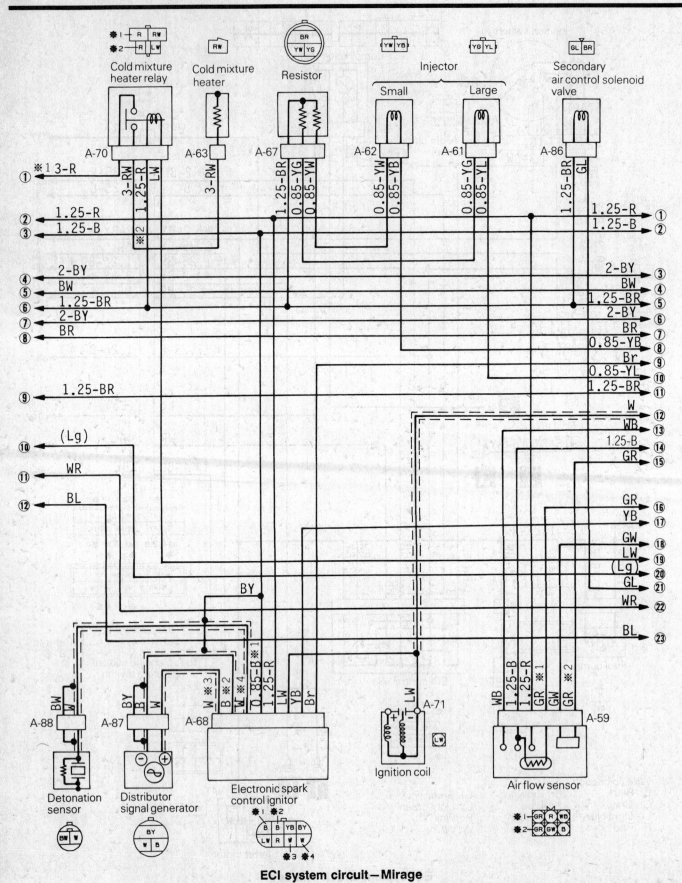

ECI system circuit—Mirage

CHASSIS ELECTRICAL 6

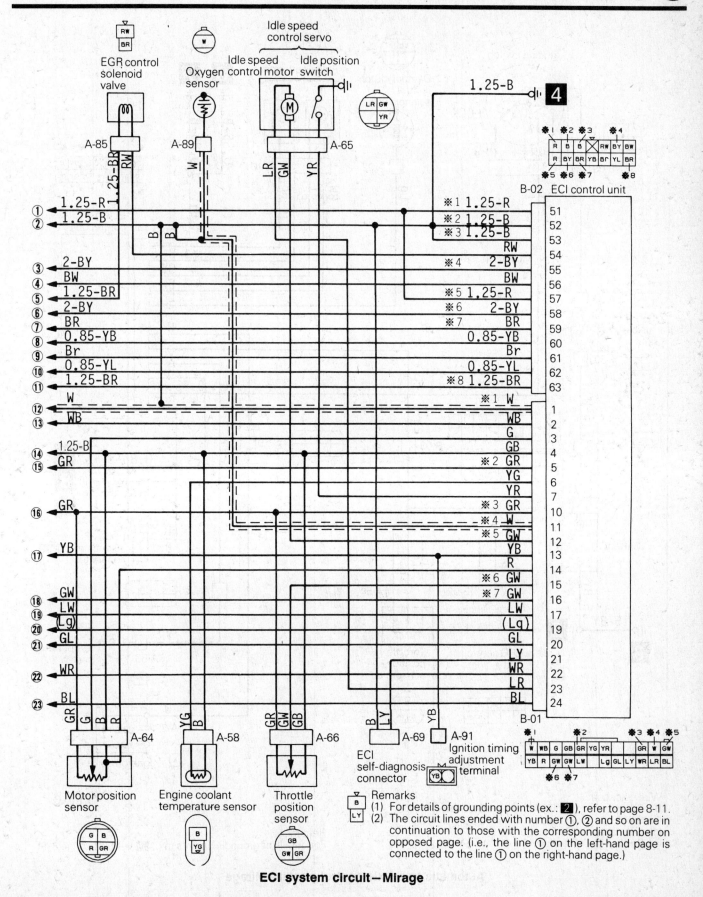

ECI system circuit – Mirage

6-99

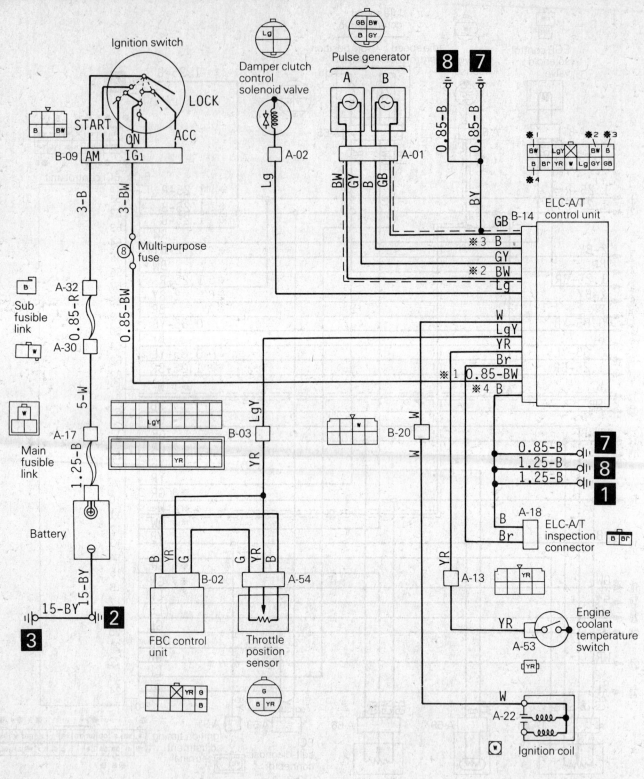

Automatic transaxle control circuit – Mirage

CHASSIS ELECTRICAL 6

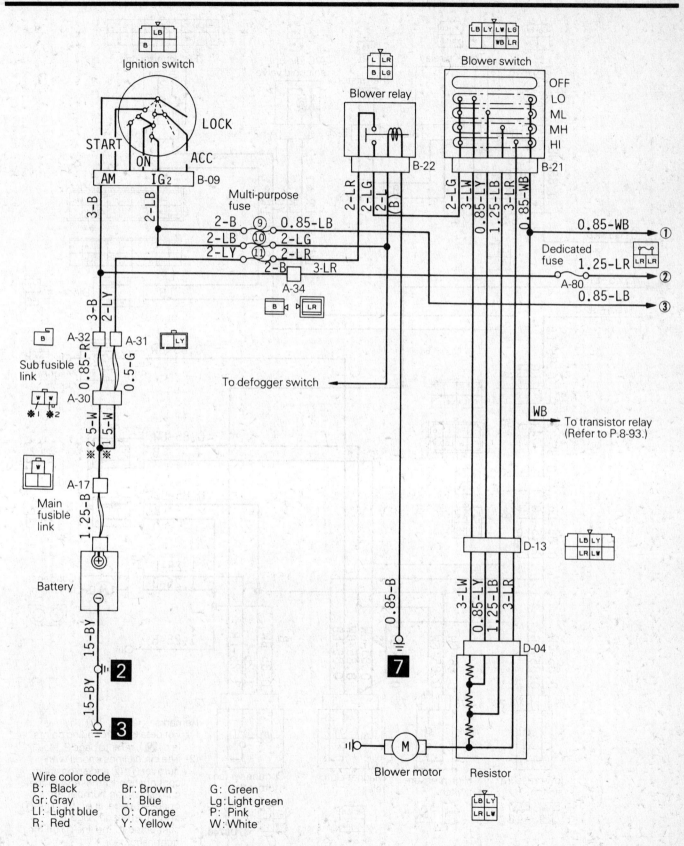

Air conditioner system — Mirage

6-101

6 CHASSIS ELECTRICAL

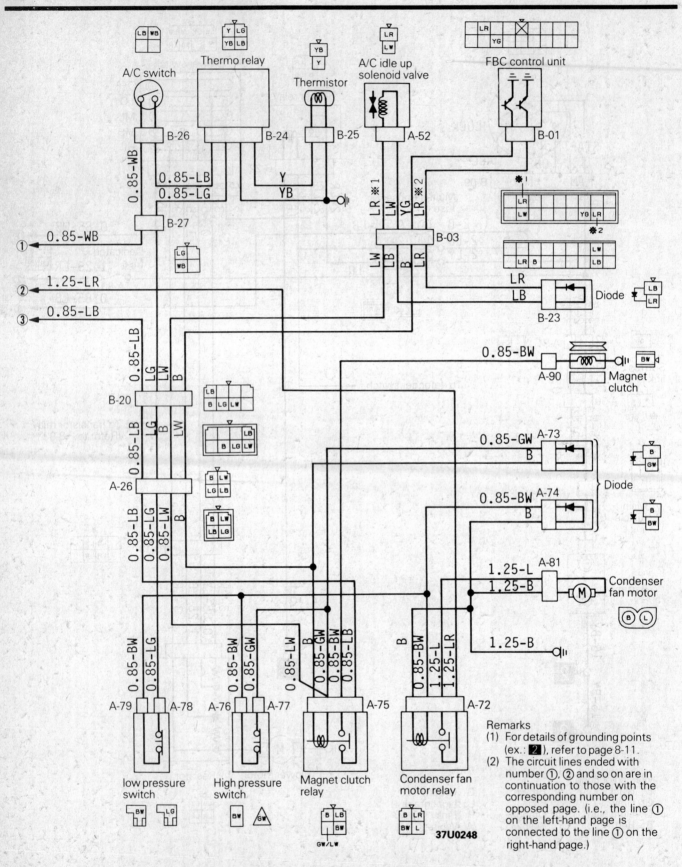

Air conditioner system — Mirage

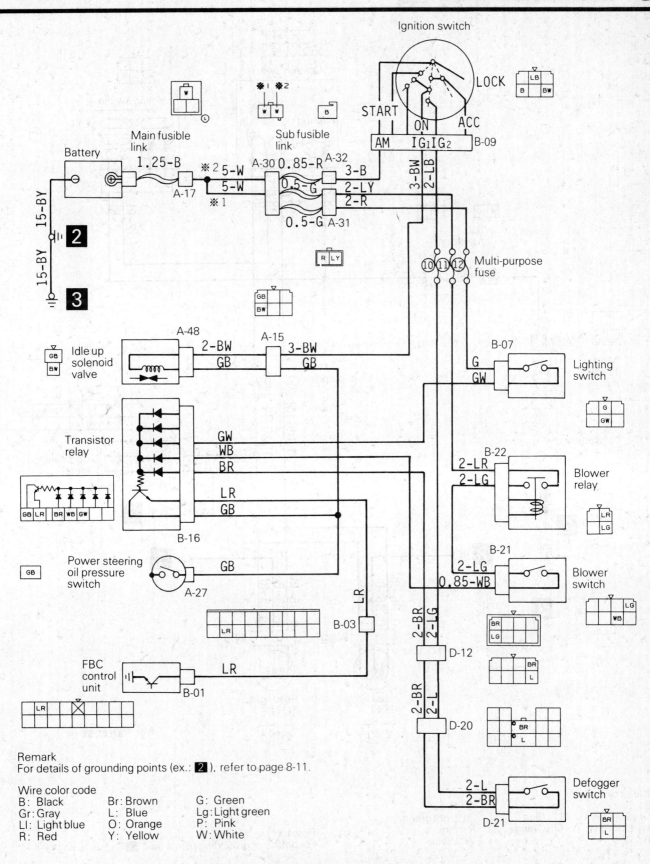

Idle-up system—Mirage

6 CHASSIS ELECTRICAL

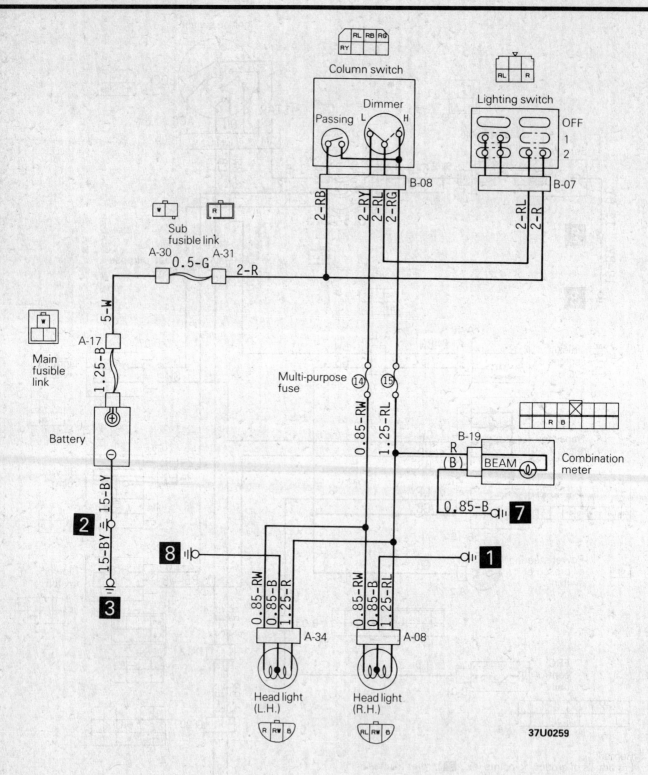

Headlight system — Mirage

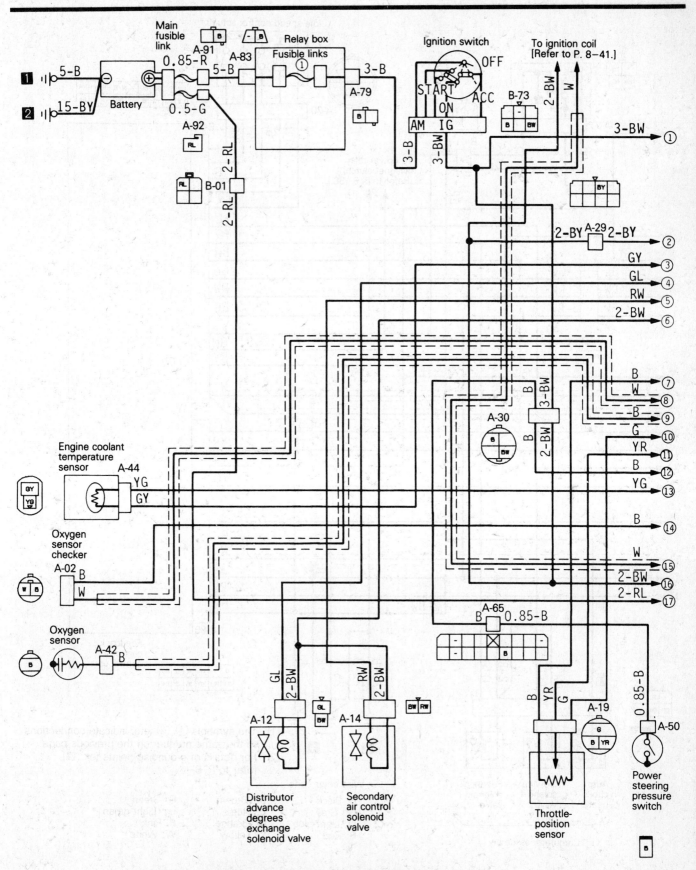

Feedback carburetor circuit—Cordia and Tredia

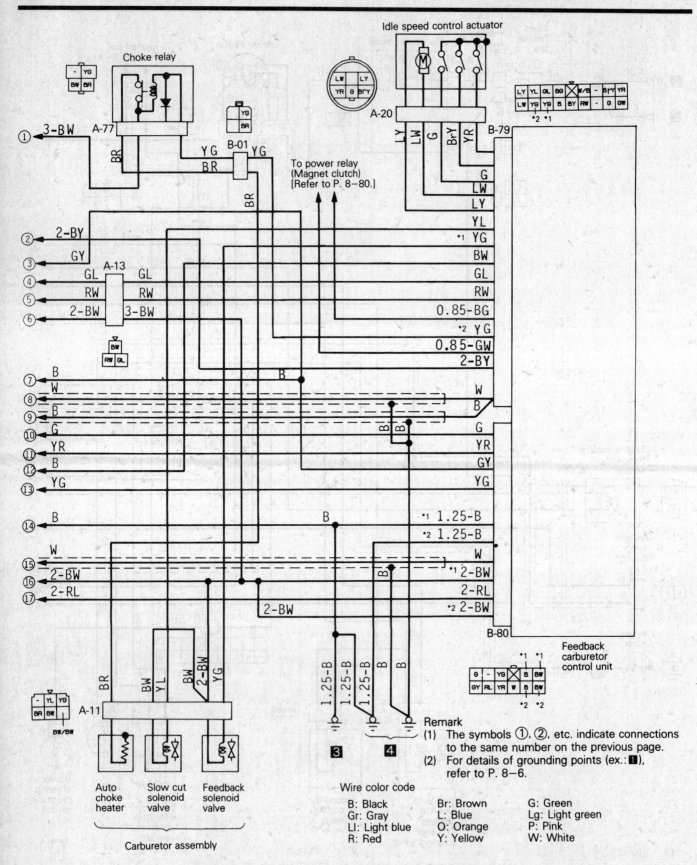

6 CHASSIS ELECTRICAL

Feedback carburetor circuit—Cordia and Tredia

6-106

CHASSIS ELECTRICAL 6

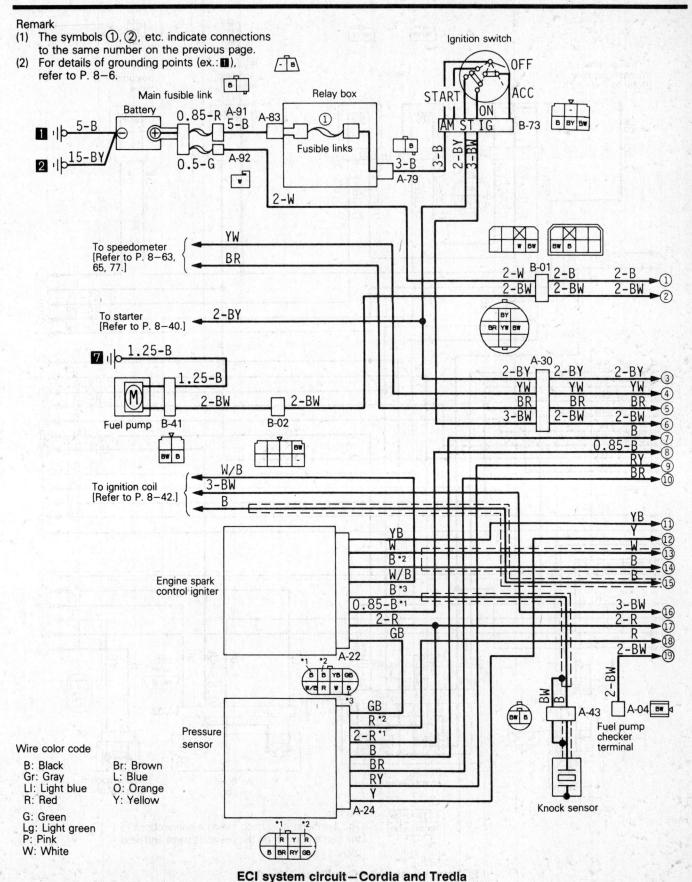

ECI system circuit—Cordia and Tredia

6-107

6 CHASSIS ELECTRICAL

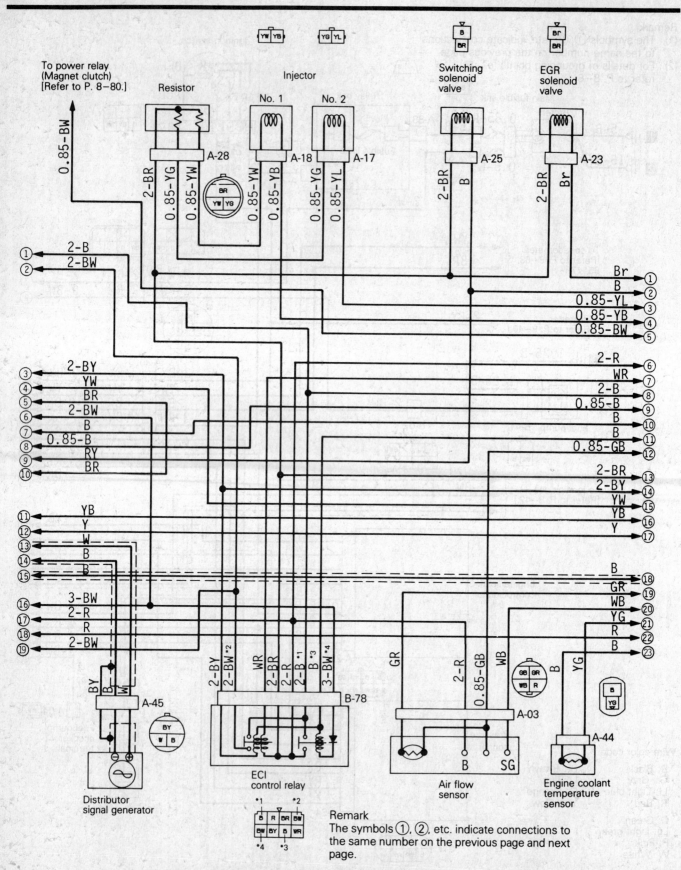

ECI system circuit—Cordia and Tredia

6-108

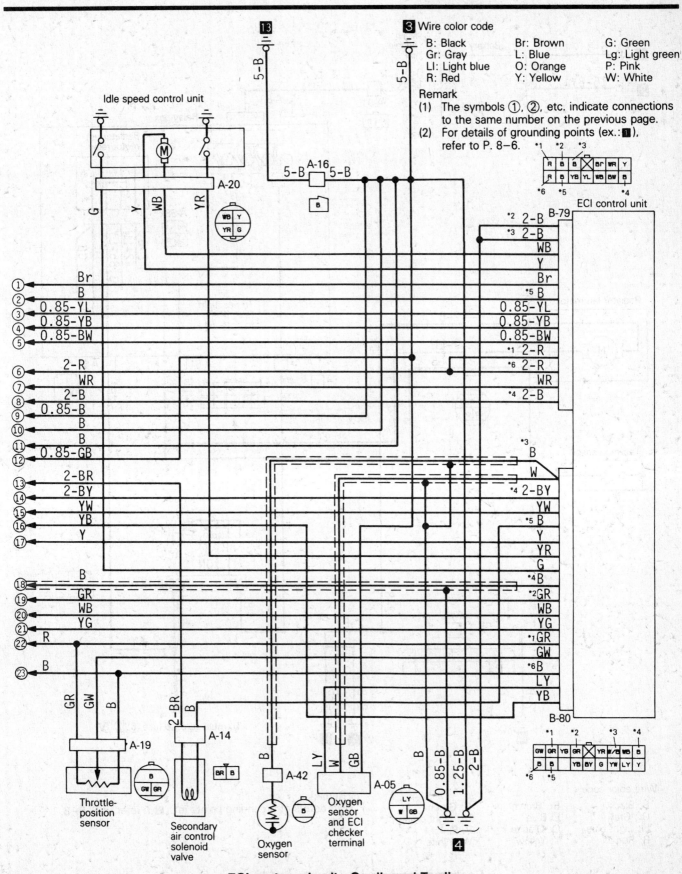

ECI system circuit – Cordia and Tredia

6 CHASSIS ELECTRICAL

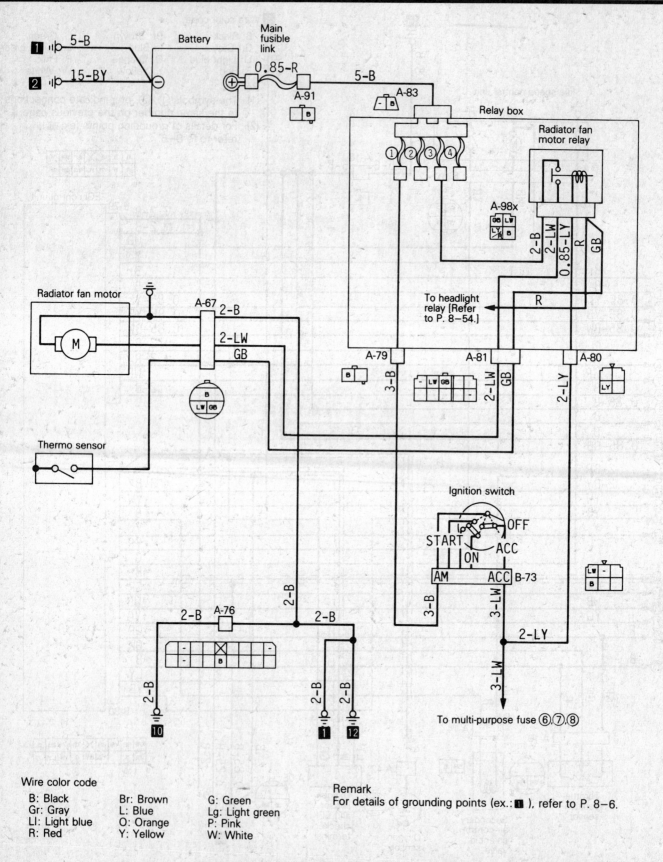

Cooling system circuit – Cordia and Tredia

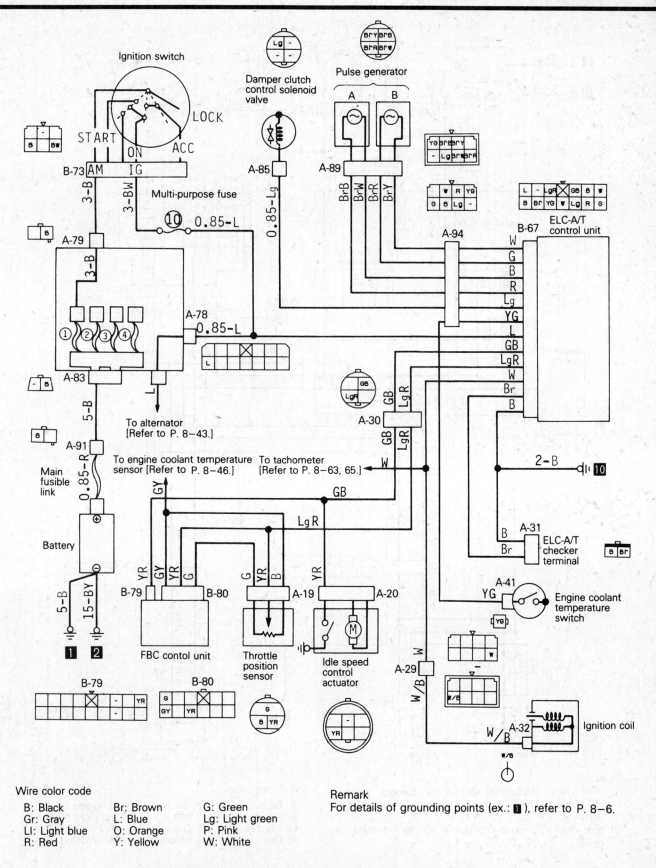

Automatic transaxle control circuit—Cordia and Tredia

6 CHASSIS ELECTRICAL

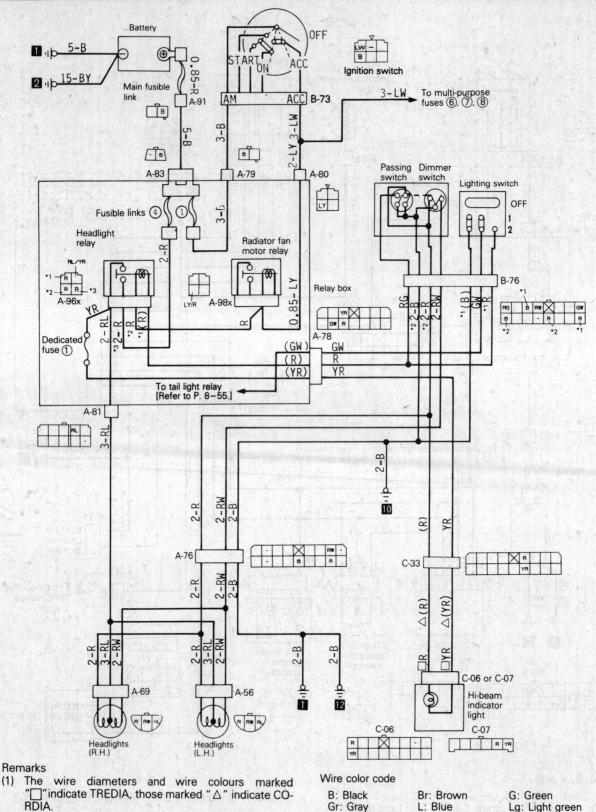

Remarks
(1) The wire diameters and wire colours marked "□" indicate TREDIA, those marked "△" indicate CORDIA.
(2) For details of grounding points (ex.: ■), refer to P. 8−6.

Wire color code
B: Black Br: Brown G: Green
Gr: Gray L: Blue Lg: Light green
Ll: Light blue O: Orange P: Pink
R: Red Y: Yellow W: White

Headlight system circuit—Cordia and Tredia

6−112

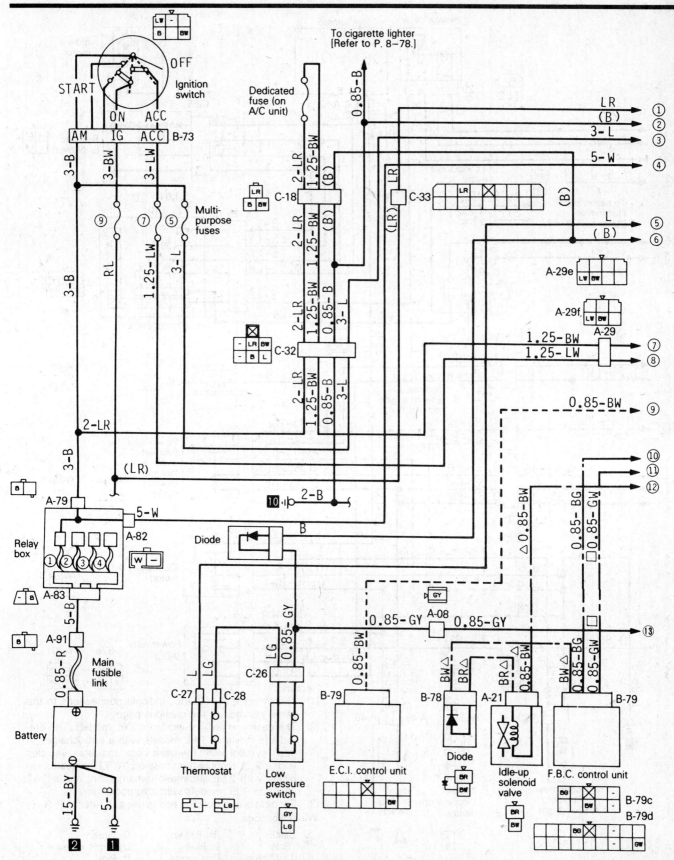

Heater and air conditioner circuit—Cordia and Tredia

6-113

6 CHASSIS ELECTRICAL

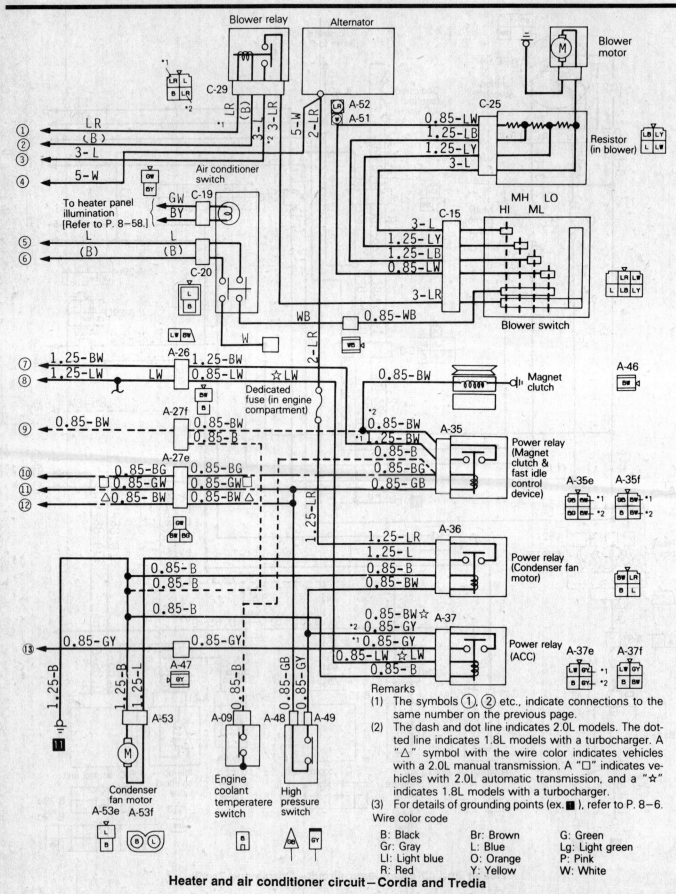

Heater and air conditioner circuit—Cordia and Tredia

CHASSIS ELECTRICAL 6

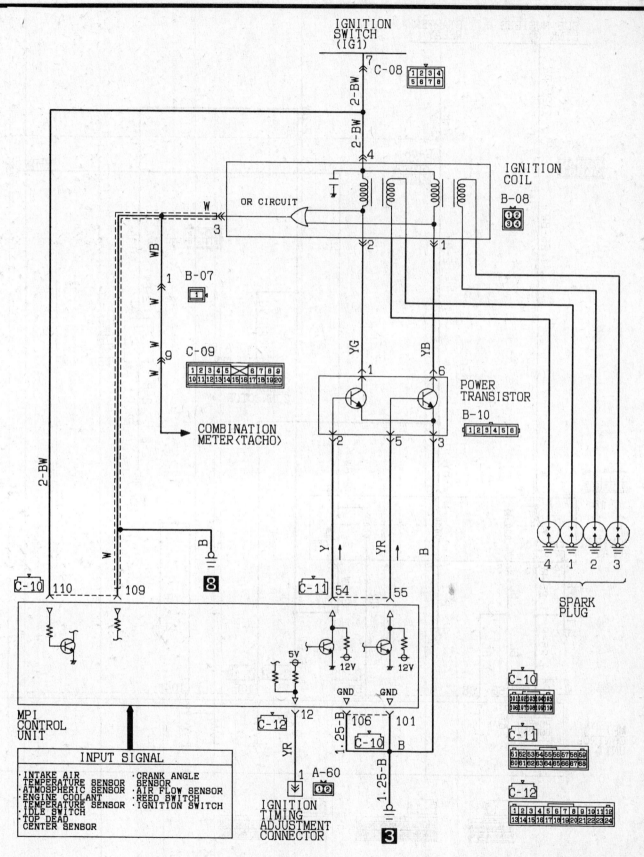

Ignition system – Galant

6-115

6 CHASSIS ELECTRICAL

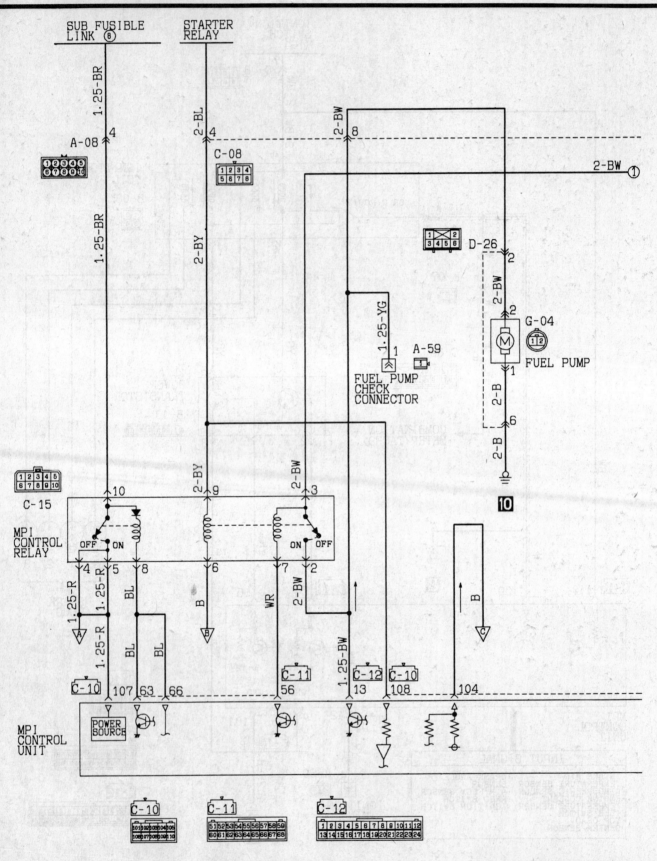

MPI circuit — Galant

CHASSIS ELECTRICAL 6

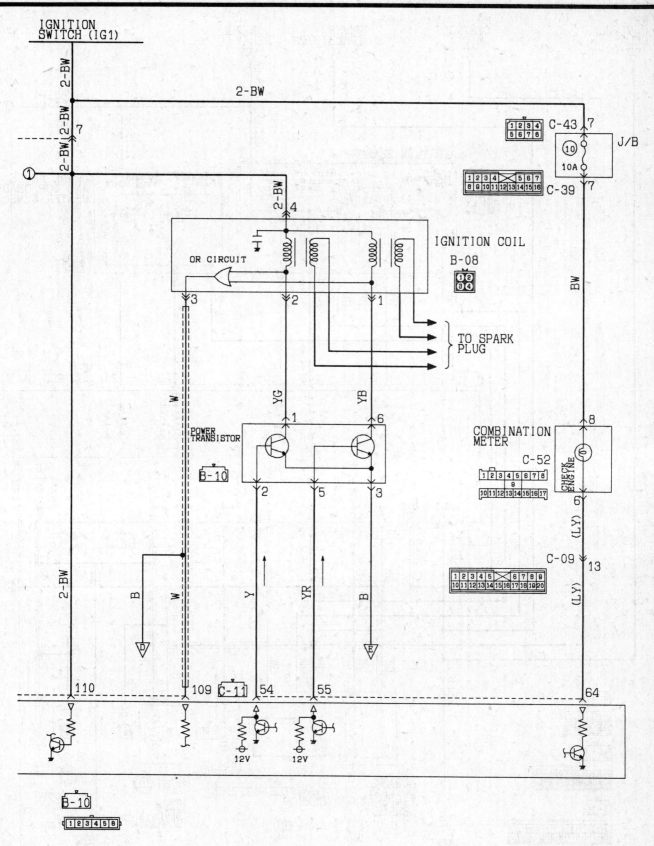

MPI circuit — Galant

6-117

6 CHASSIS ELECTRICAL

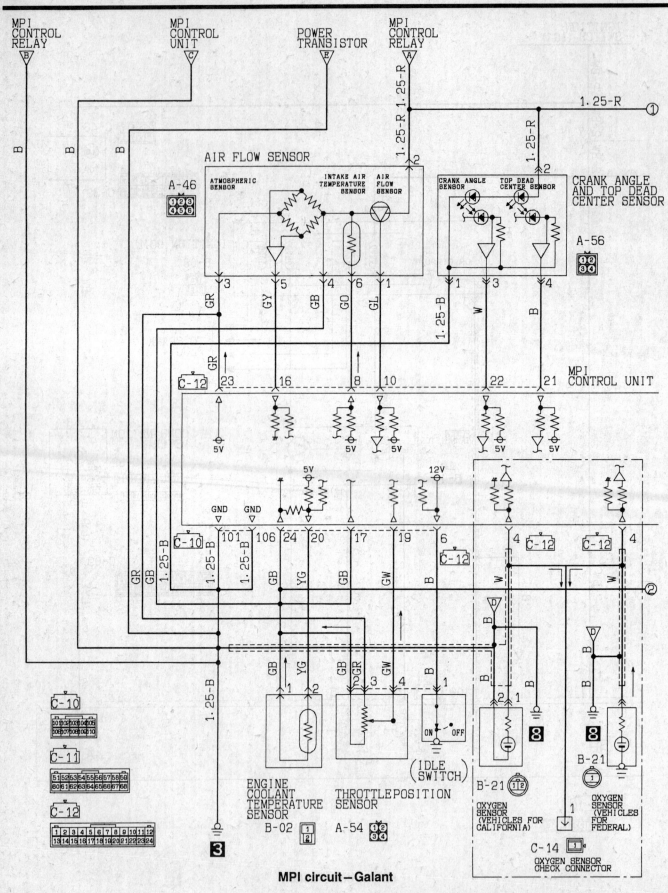

MPI circuit—Galant

CHASSIS ELECTRICAL 6

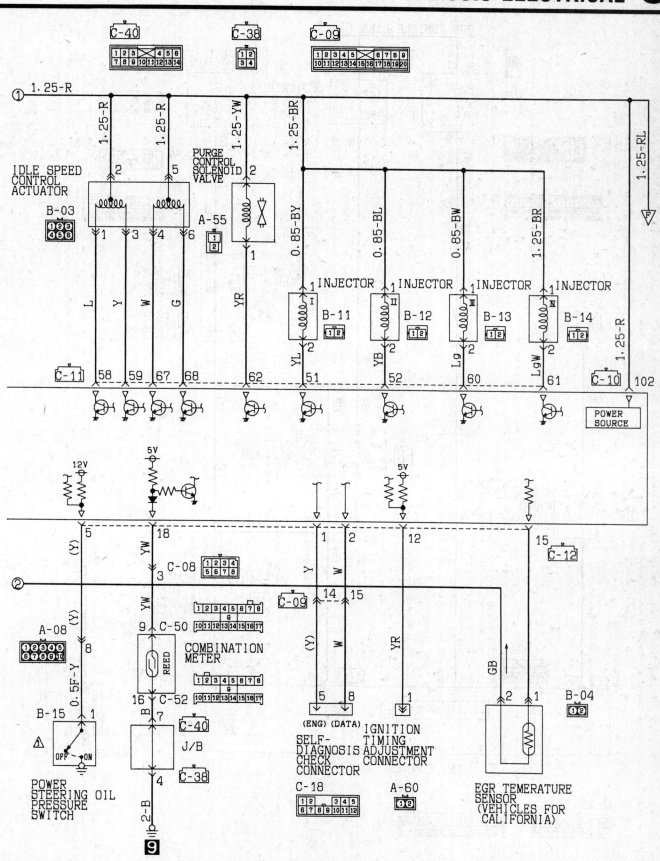

MPI circuit – Galant

6-119

6 CHASSIS ELECTRICAL

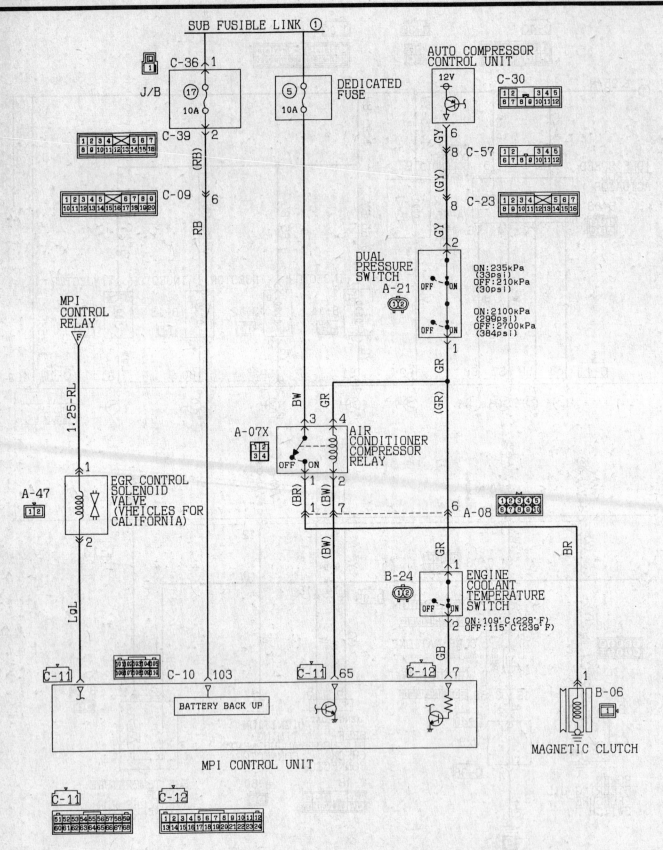

MPI circuit—Galant

CHASSIS ELECTRICAL 6

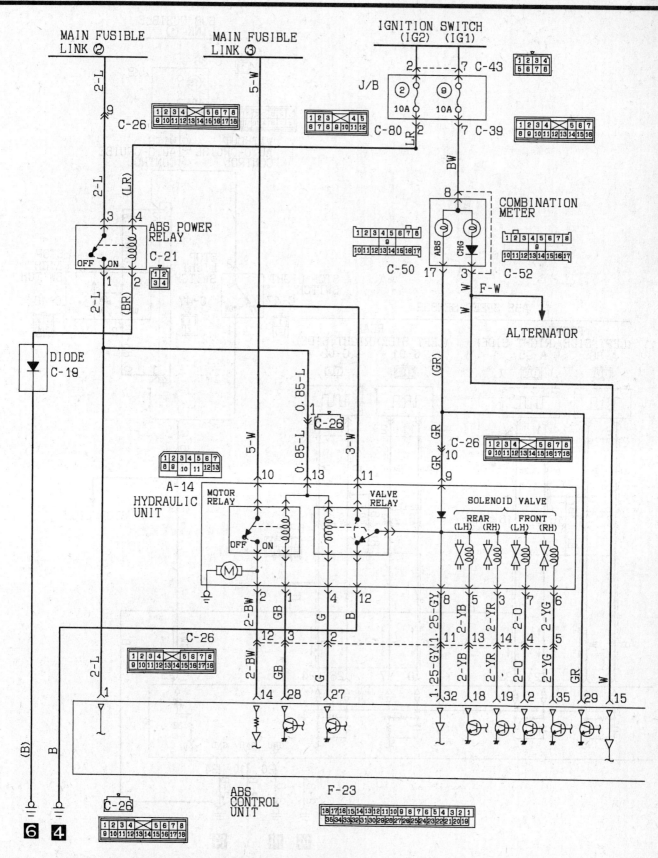

Anti-lock brake system – Galant

6–121

6 CHASSIS ELECTRICAL

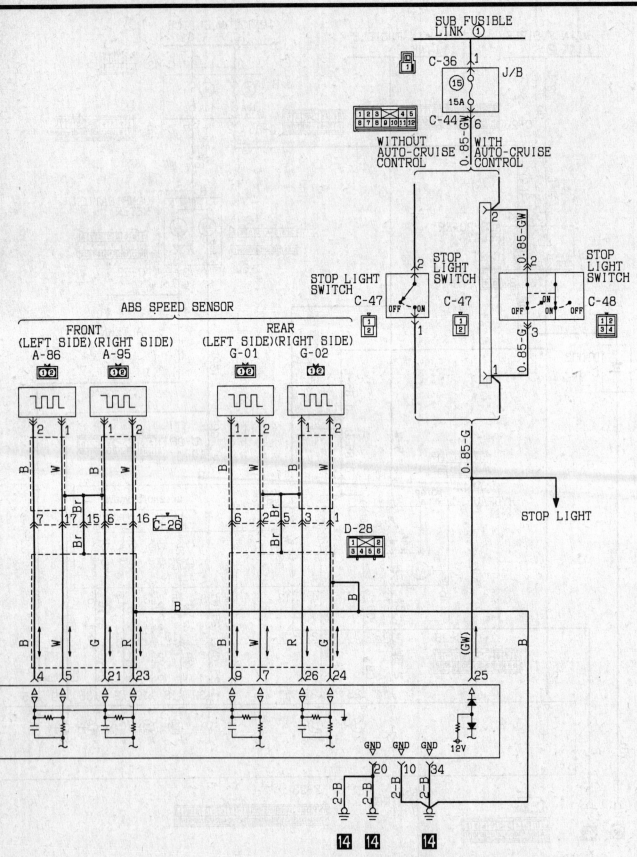

Anti-lock brake system – Galant

CHASSIS ELECTRICAL 6

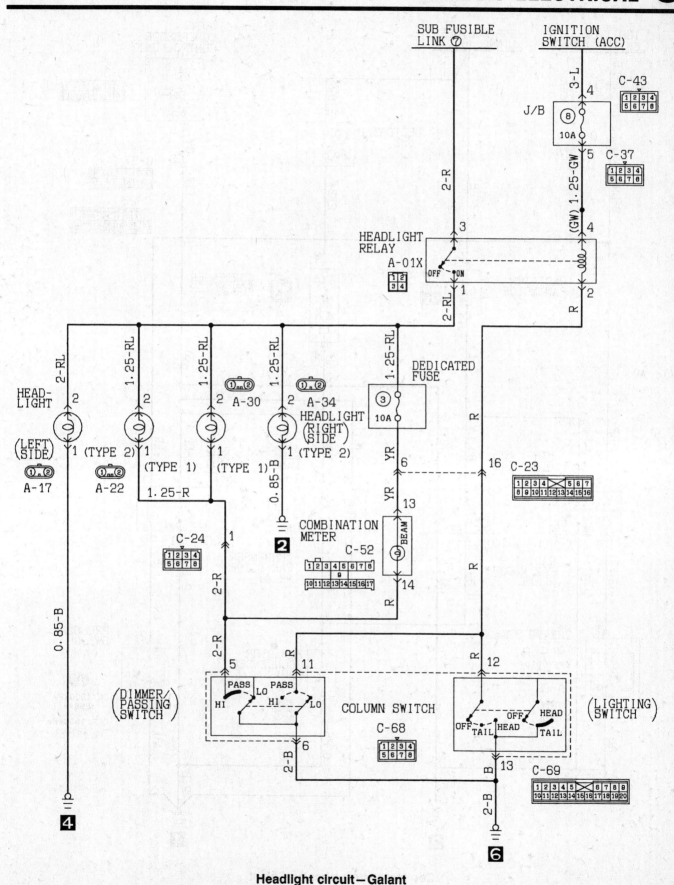

Headlight circuit – Galant

6-123

6 CHASSIS ELECTRICAL

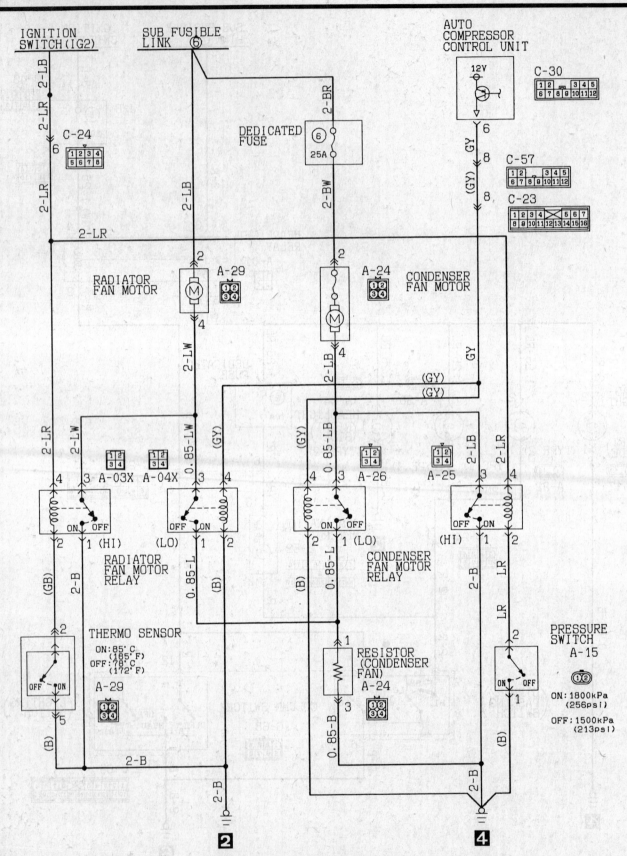

Cooling system circuit—Galant

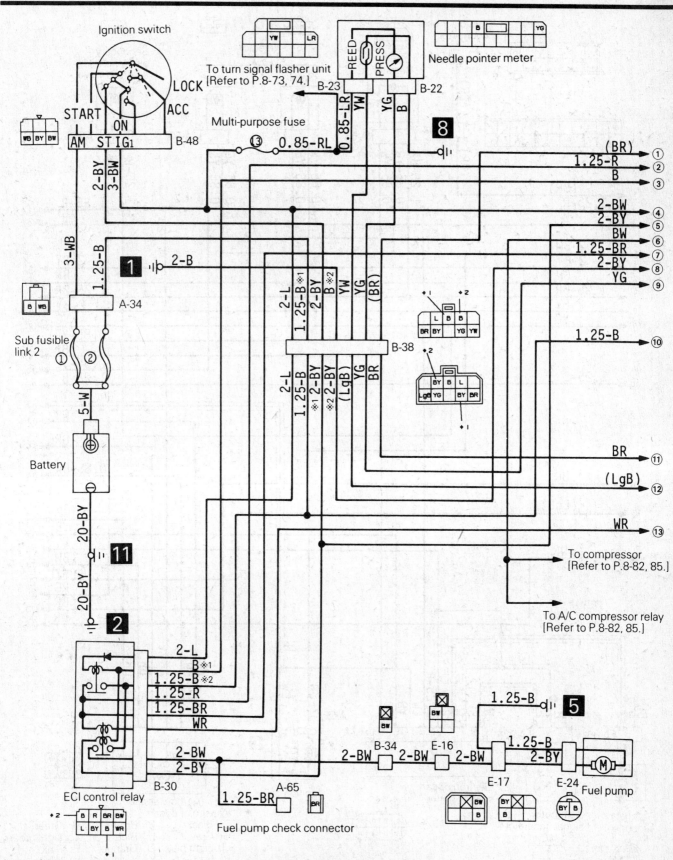

ECI system circuit – Starion

6 CHASSIS ELECTRICAL

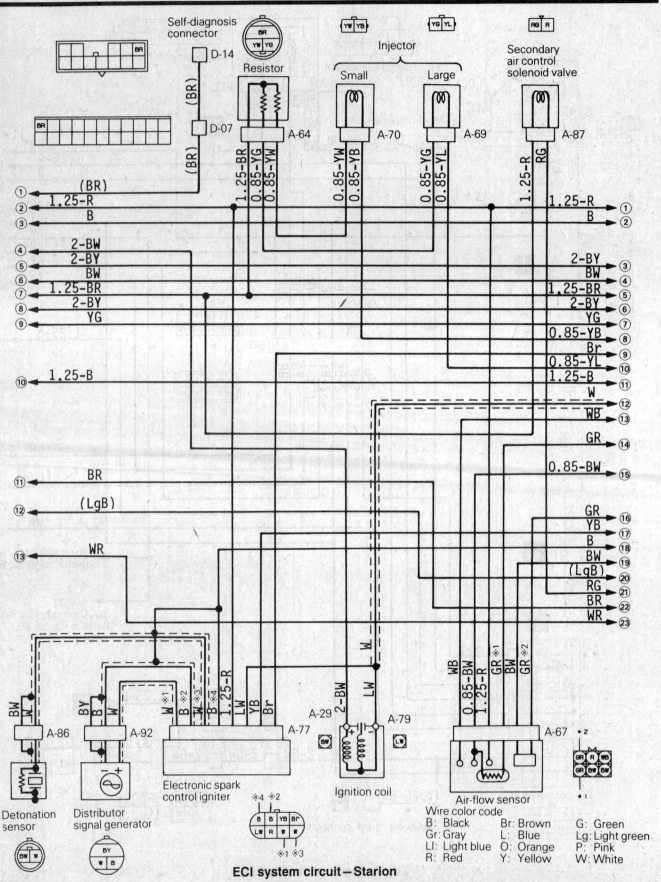

ECI system circuit – Starion

6-126

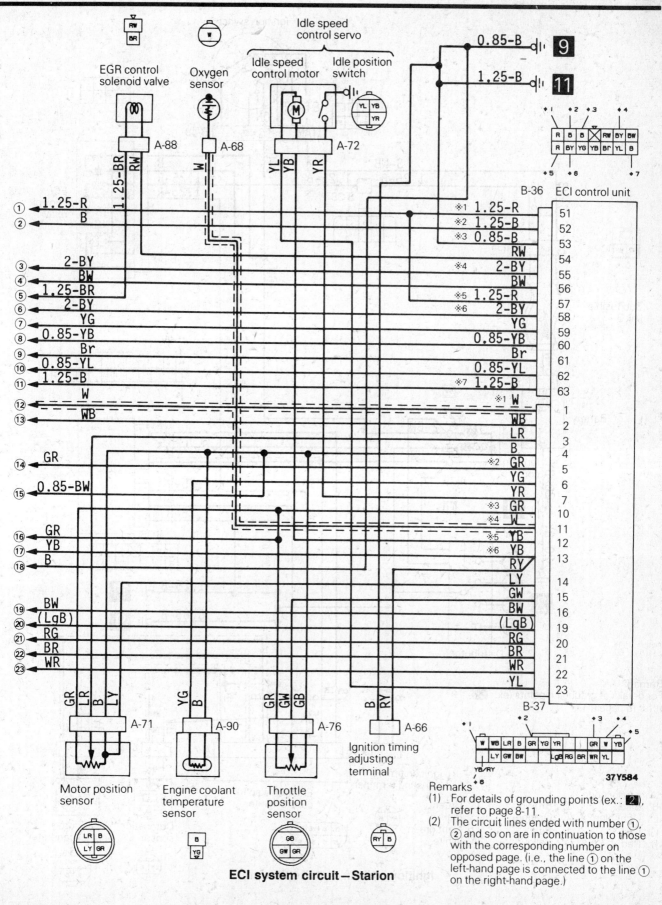

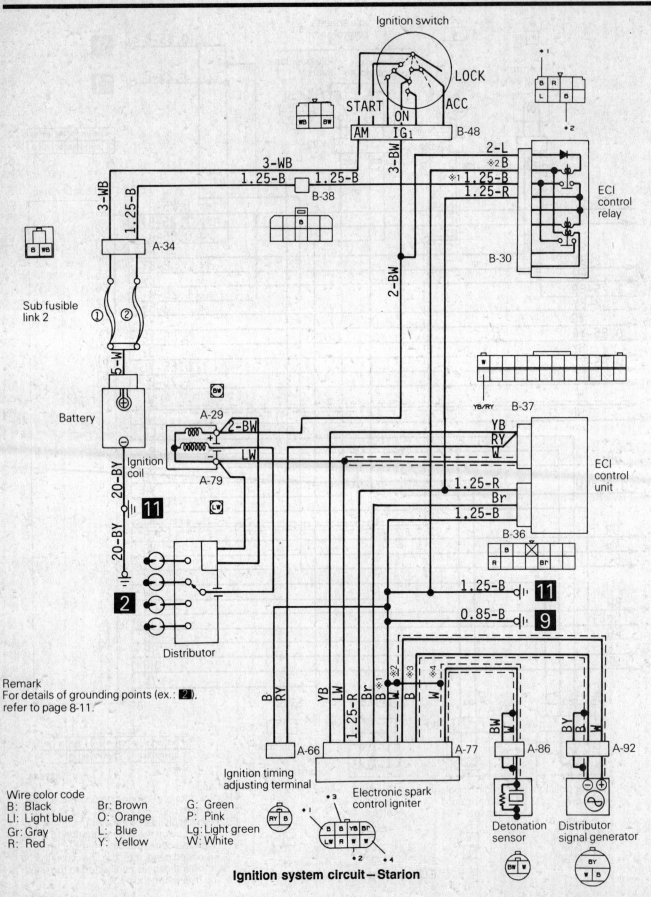

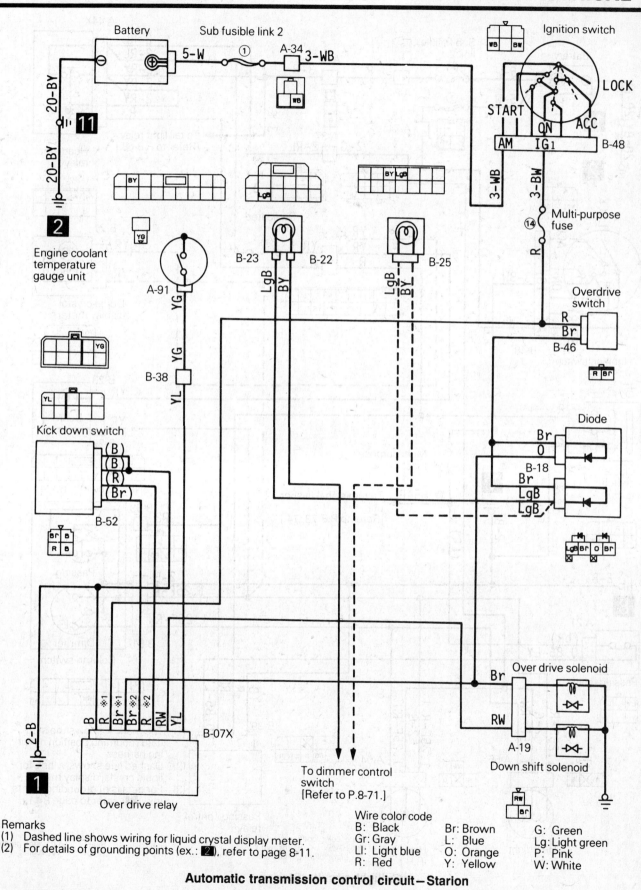

Automatic transmission control circuit – Starion

6 CHASSIS ELECTRICAL

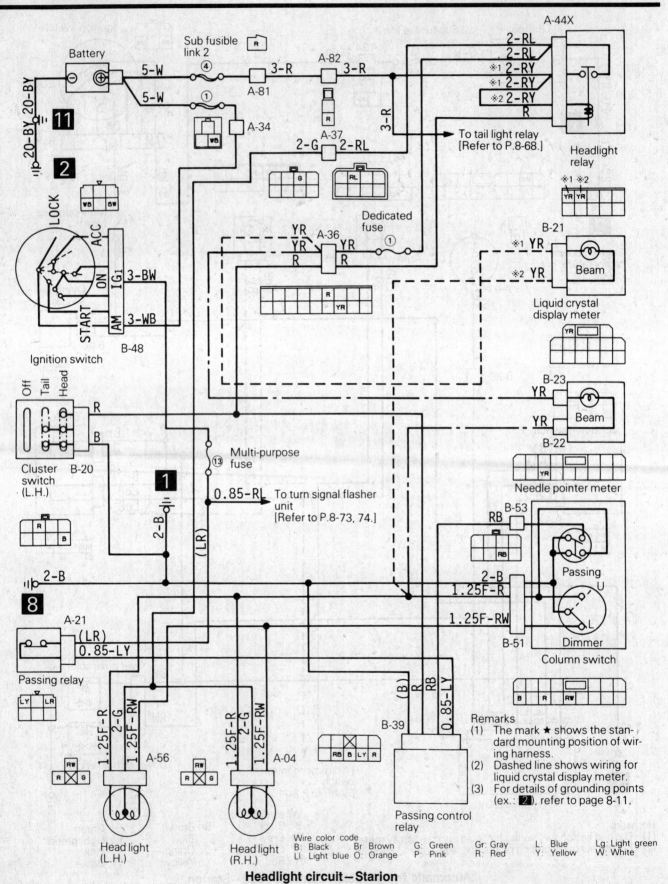

Headlight circuit — Starion

6-130

CHASSIS ELECTRICAL 6

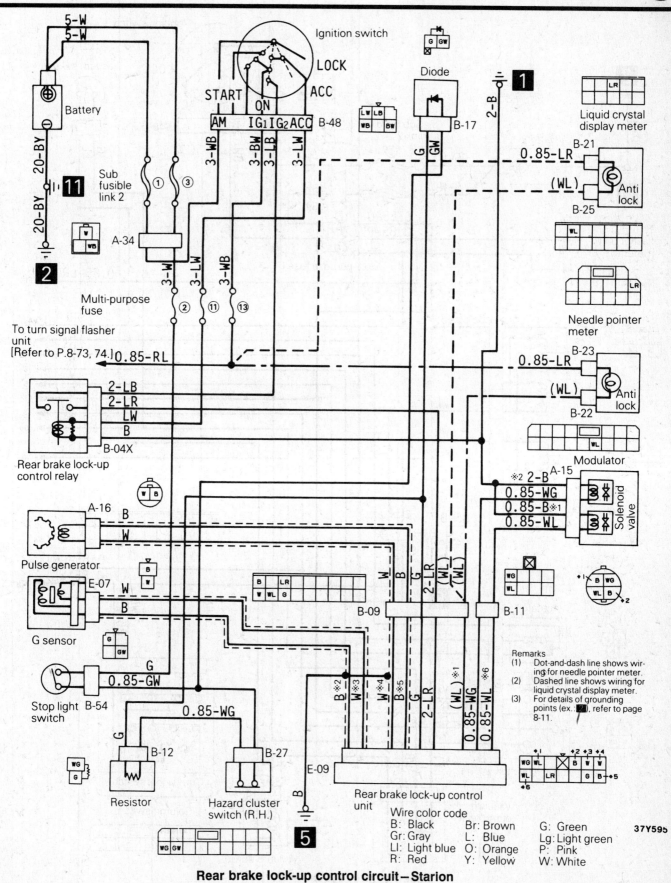

Rear brake lock-up control circuit – Starion

6-131

6 CHASSIS ELECTRICAL

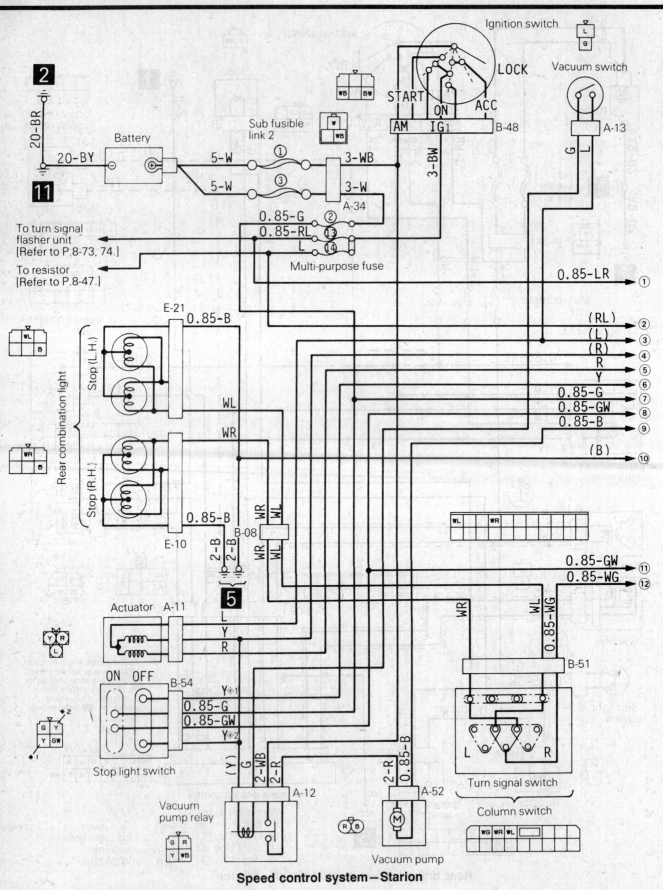

Speed control system — Starion

6-132

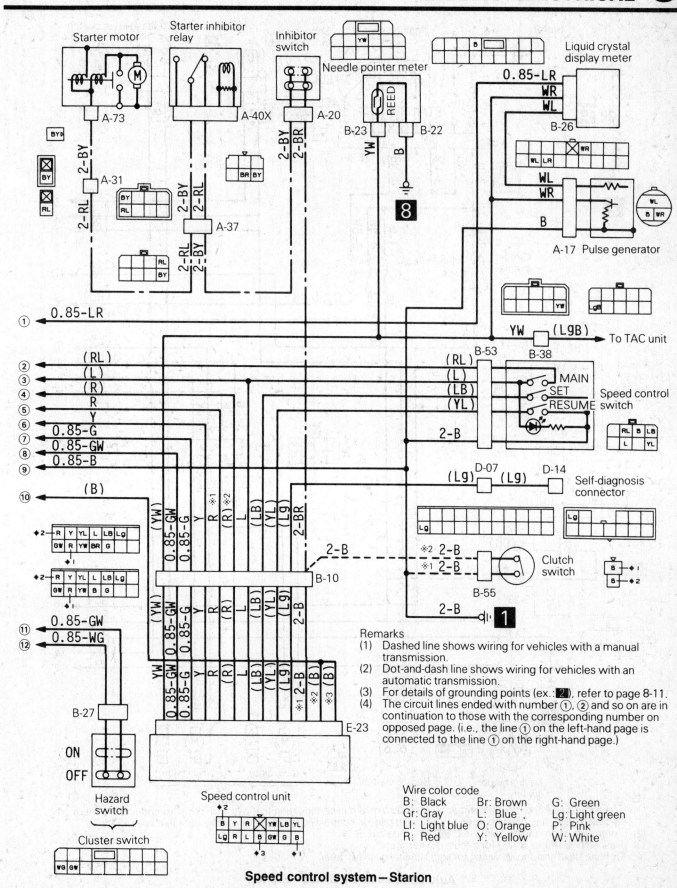

Speed control system – Starion

6 CHASSIS ELECTRICAL

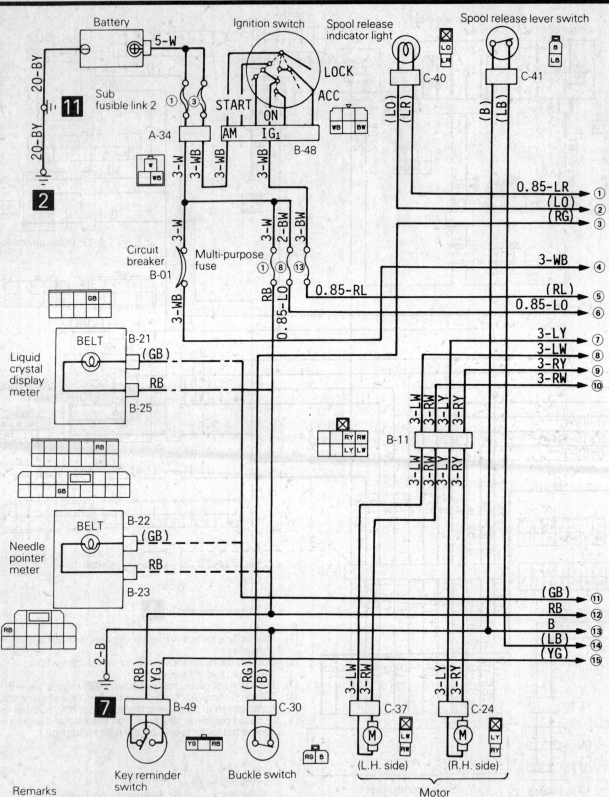

Automatic seatbelt circuit – Starion

Remarks
(1) The circuit lines ended with number ①, ② and so on are in continuation to those with the corresponding number on opposed page. (i.e., the line ① on the left-hand page is connected to the line ① on the right-hand page.)
(2) For details of grounding points (ex.: 2), refer to page 8-11.
(3) Dashed line shows wiring for needle pointer meter.
(4) Dot-and-dash line shows wiring for liquid crystal display meter.

CHASSIS ELECTRICAL 6

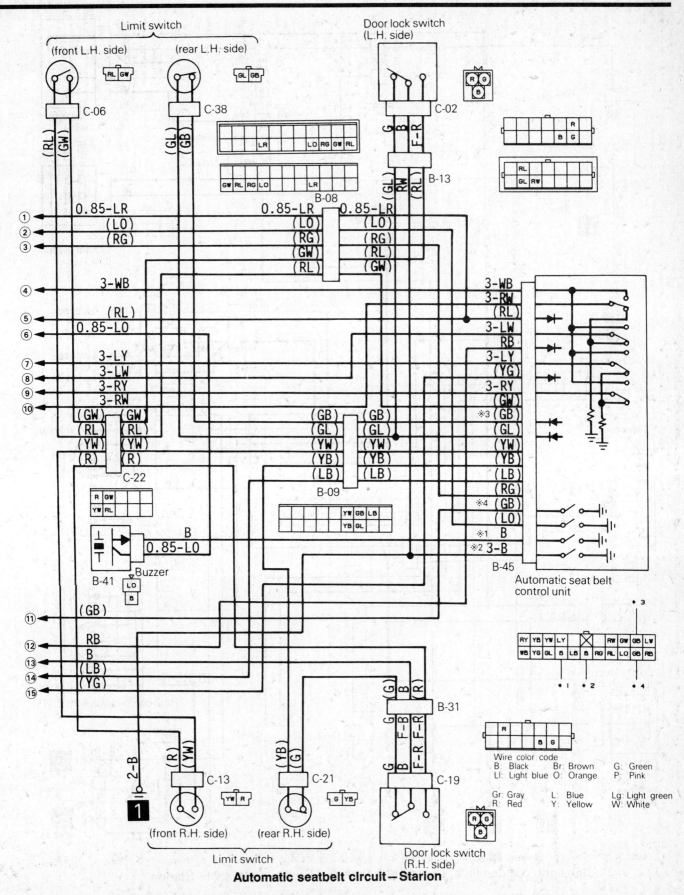

Automatic seatbelt circuit – Starion

6-135

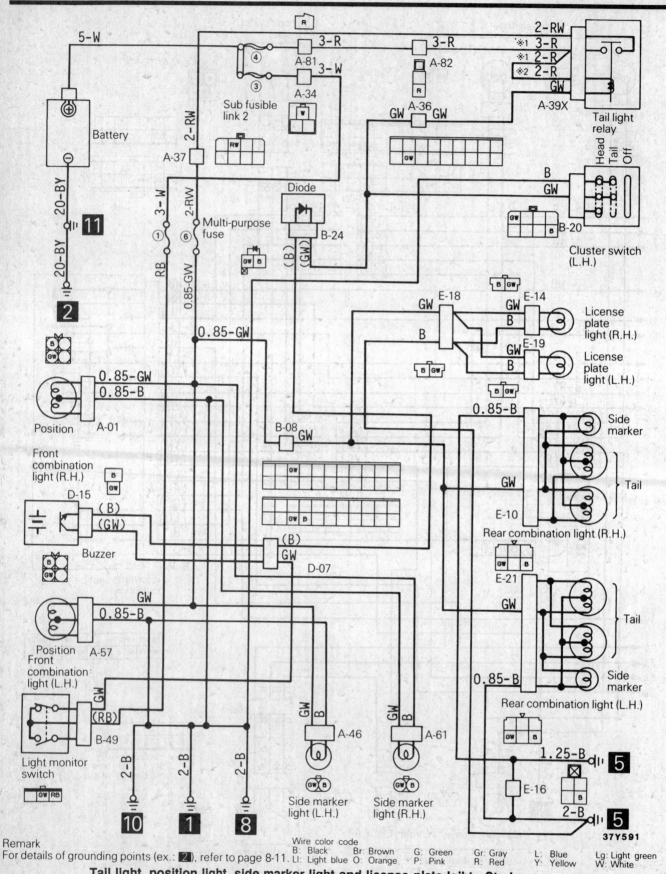

CHASSIS ELECTRICAL 6

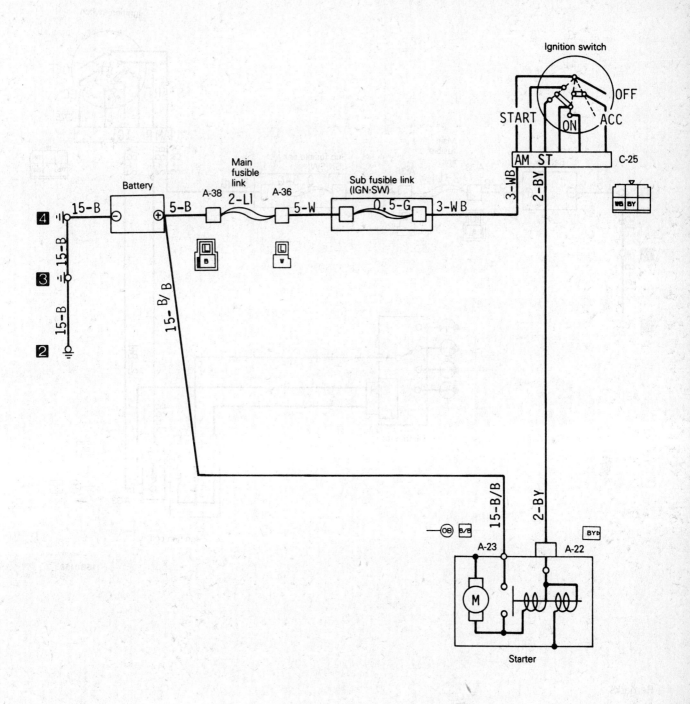

Remarks
For details on the ground point (No. 2 in the illustration) refer to P. 8-6.

Starting circuit—Pick-up

6-137

6 CHASSIS ELECTRICAL

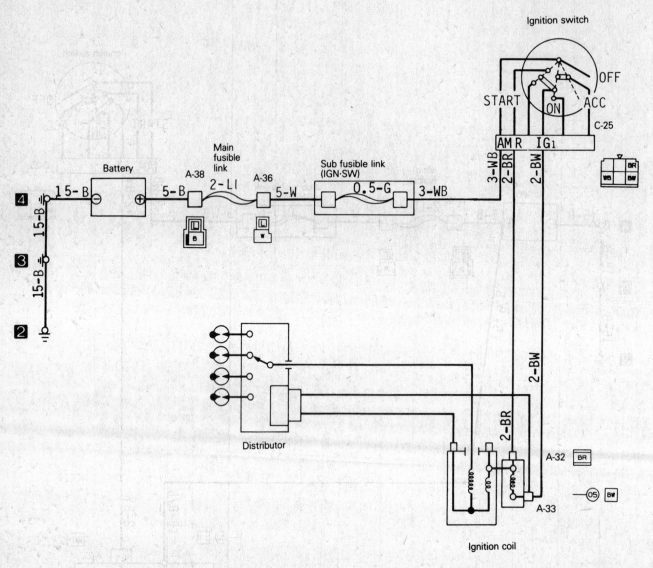

Remarks

For details on the ground point (No. 2 in the illustration) refer to P. 8-6.

Color code for wiring

| B: Black | Br: Brown | G: Green | Gr: Gray | L: Blue | Lg: Light green |
| Ll: Light blue | O: Orange | P: Pink | R: Red | Y: Yellow | W: White |

Ignition circuit—Pick-up

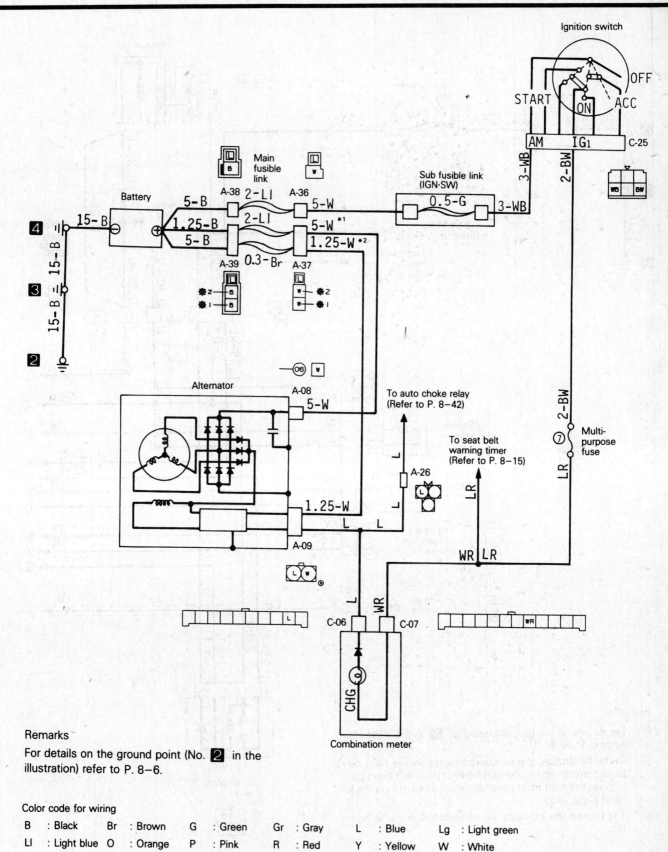

Charging circuit—Pick-up

6 CHASSIS ELECTRICAL

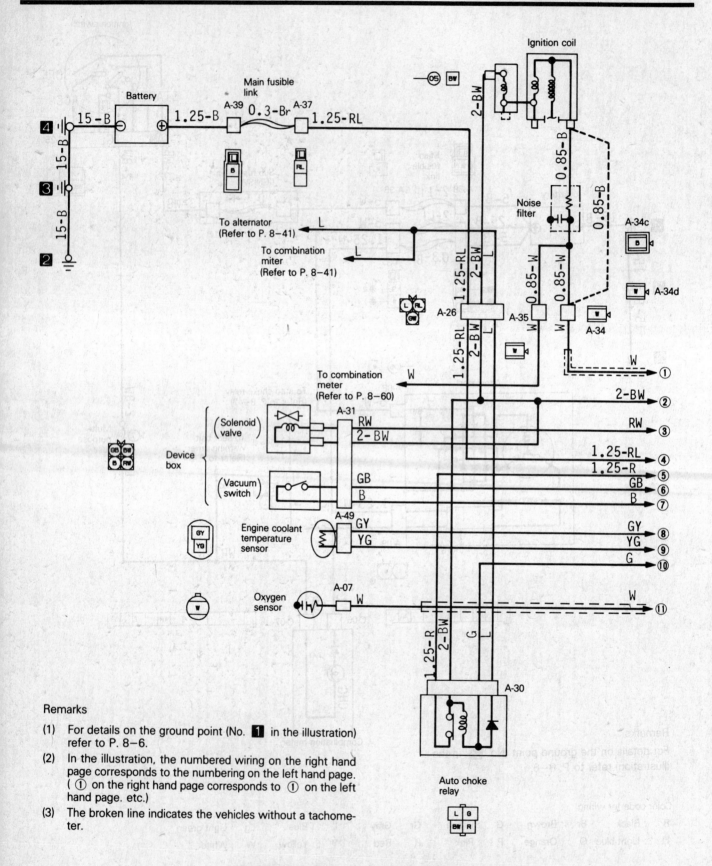

Remarks

(1) For details on the ground point (No. **1** in the illustration) refer to P. 8-6.

(2) In the illustration, the numbered wiring on the right hand page corresponds to the numbering on the left hand page. (① on the right hand page corresponds to ① on the left hand page. etc.)

(3) The broken line indicates the vehicles without a tachometer.

Feedback carburetor circuit—Pick-up

6-140

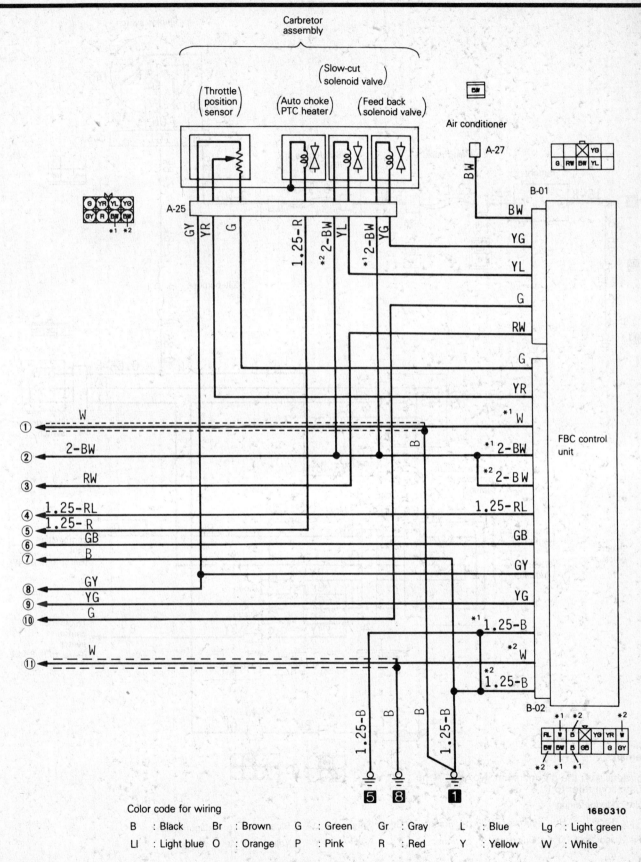

Feedback carburetor circuit—Pick-up

6 CHASSIS ELECTRICAL

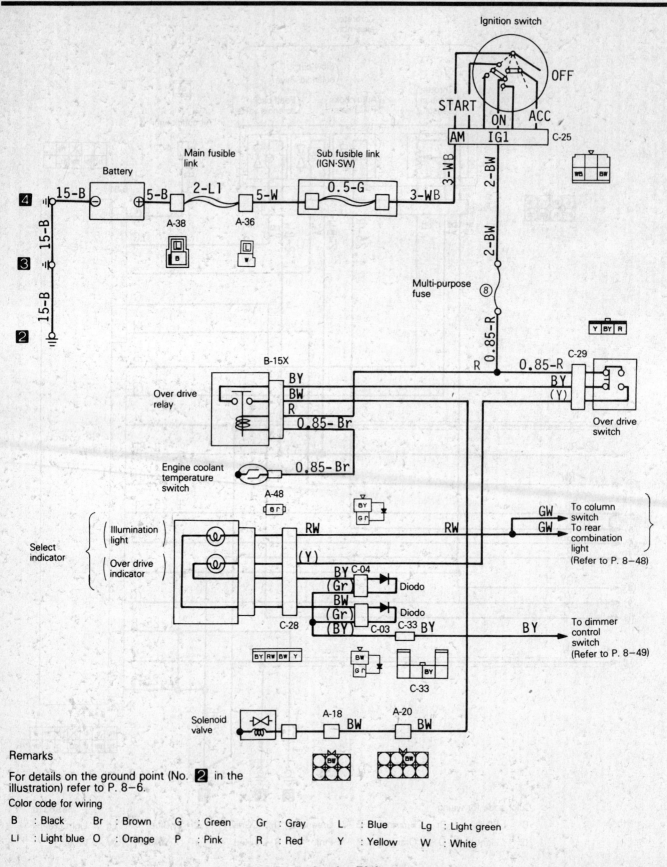

Overdrive circuit—Pick-up

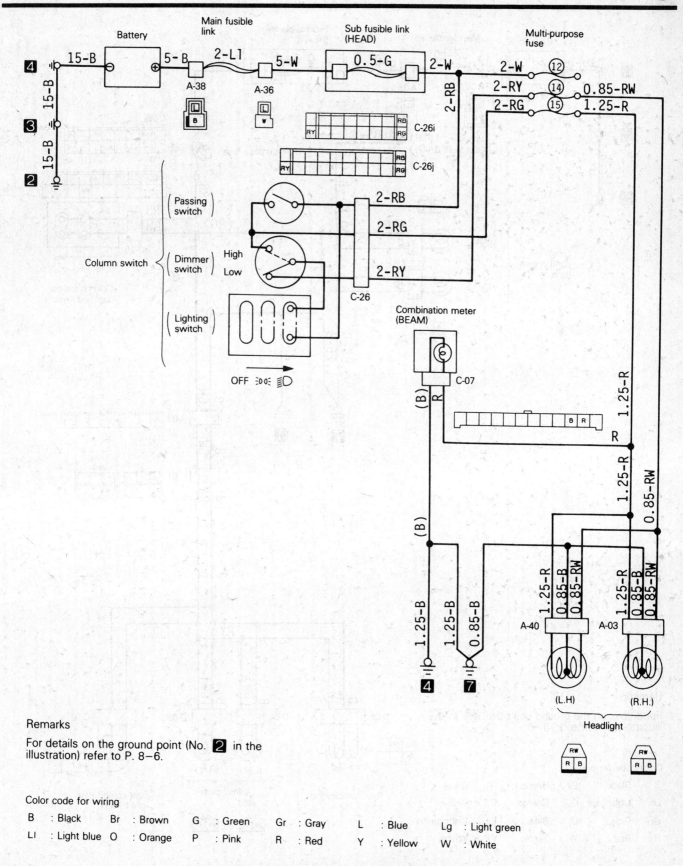

Headlight circuit—Pick-up

6 CHASSIS ELECTRICAL

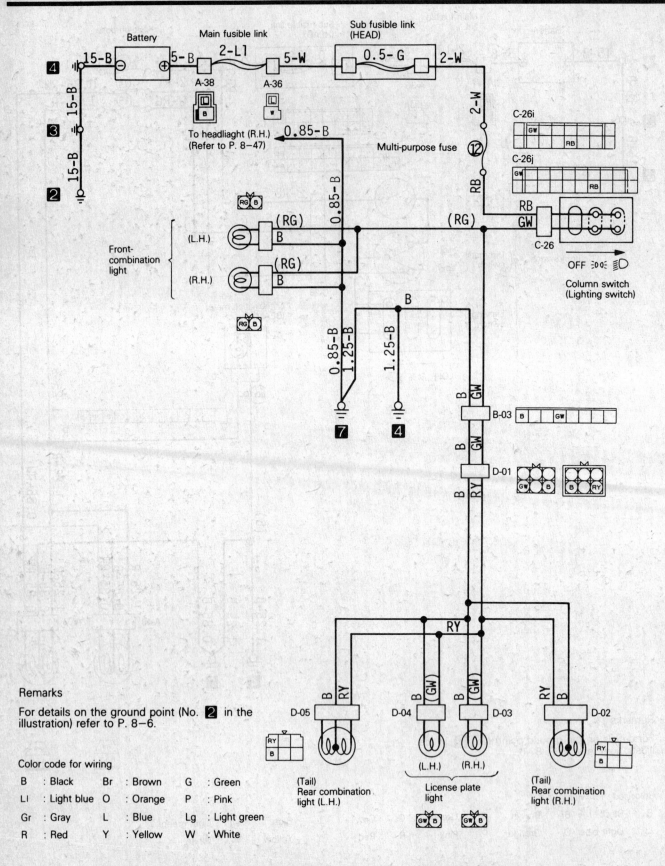

Tail light, position light, side marker light and license plate light — Pick-up

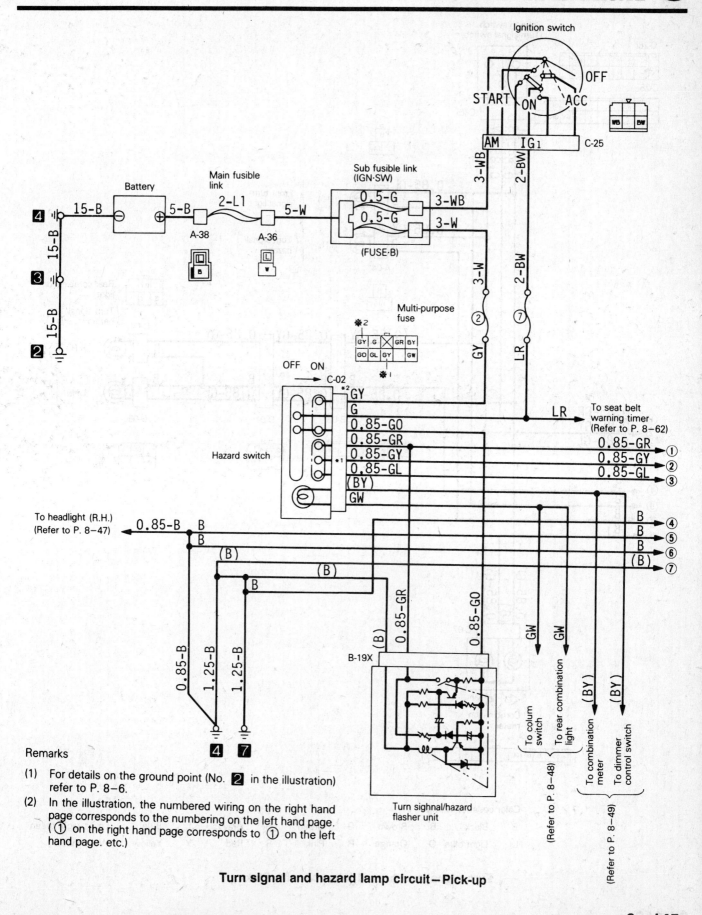

Turn signal and hazard lamp circuit – Pick-up

6 CHASSIS ELECTRICAL

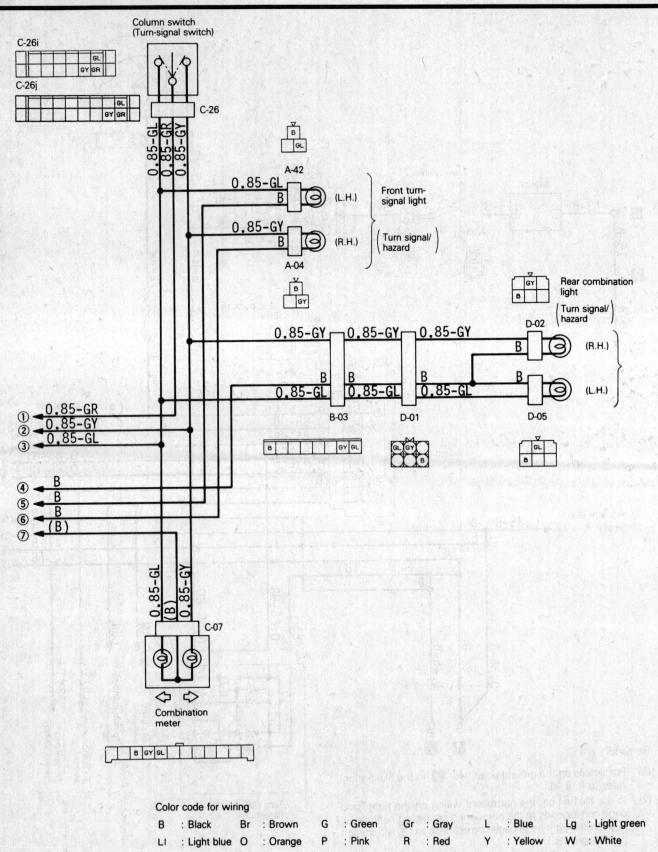

Turn signal and hazard lamp circuit – Pick-up

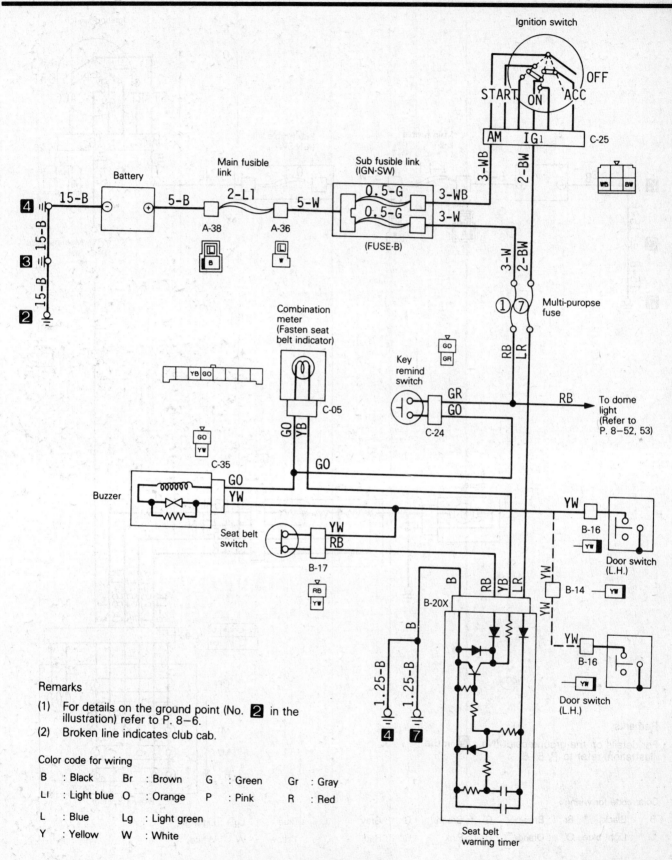

Buzzer circuit—Pick-up

6 CHASSIS ELECTRICAL

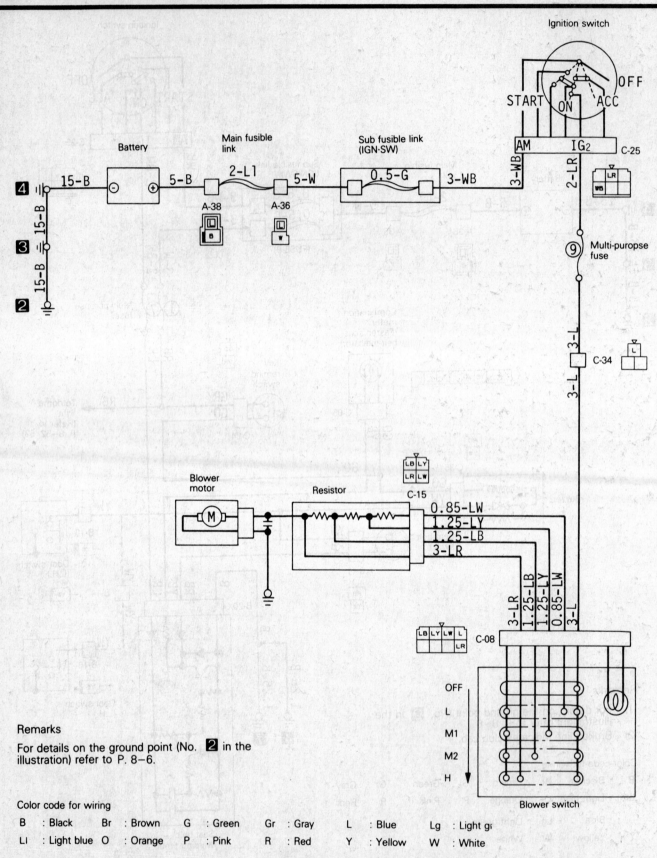

Heater circuit—Pick-up

6-148

CHASSIS ELECTRICAL 6

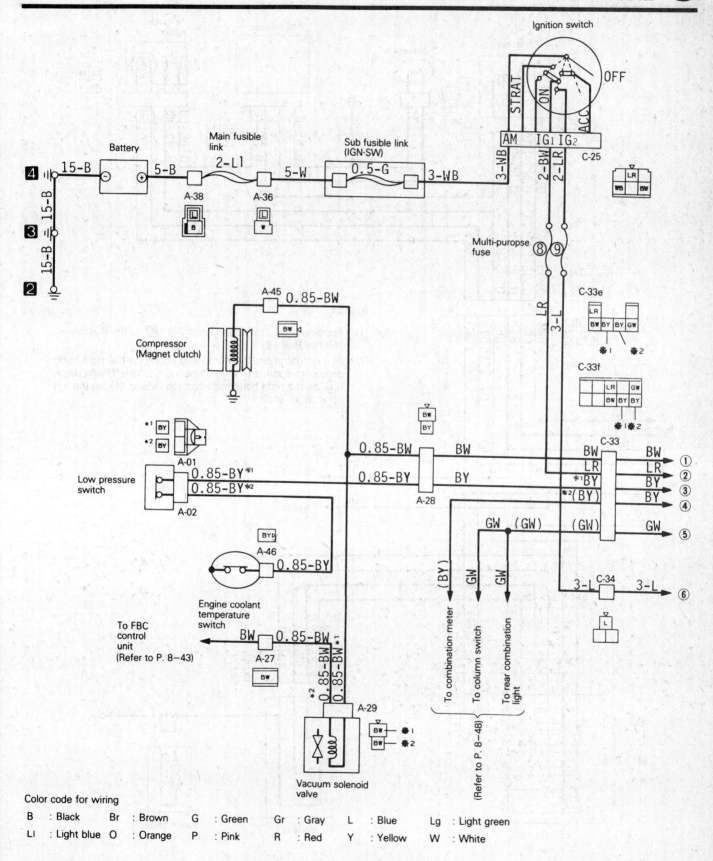

Air conditioner circuit – Pick-up

6-149

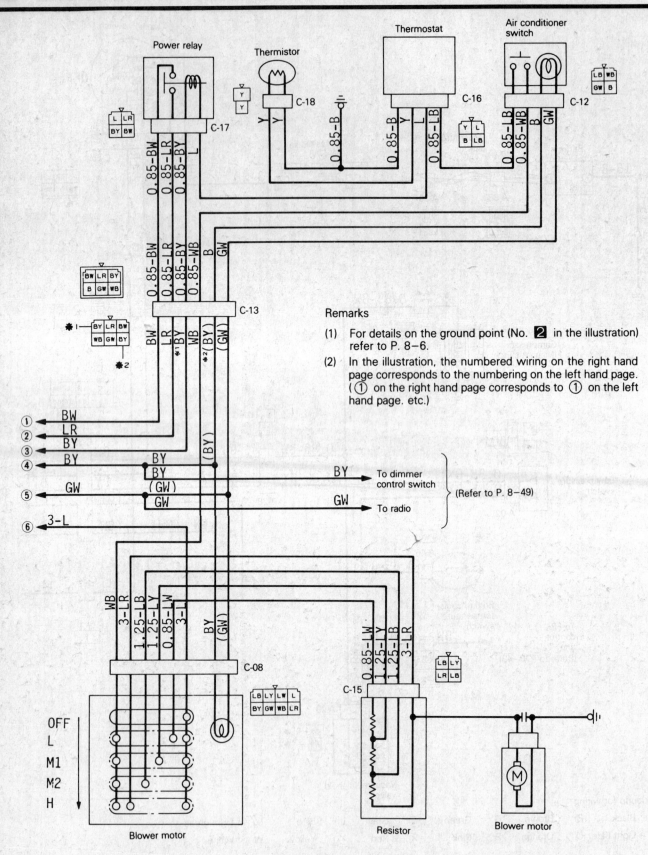

Air conditioner circuit—Pick-up

CHASSIS ELECTRICAL 6

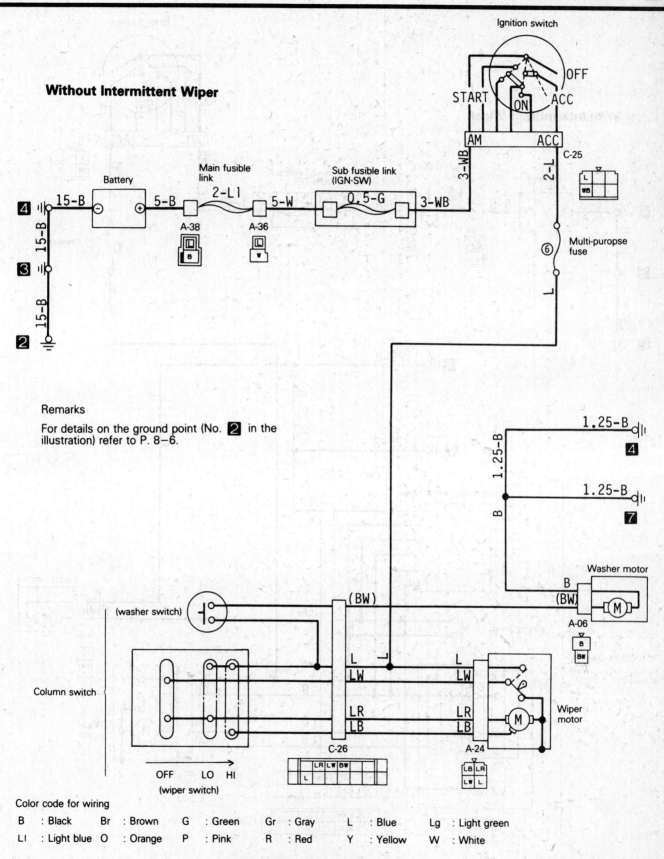

Wiper and washer circuit—Pick-up

6 CHASSIS ELECTRICAL

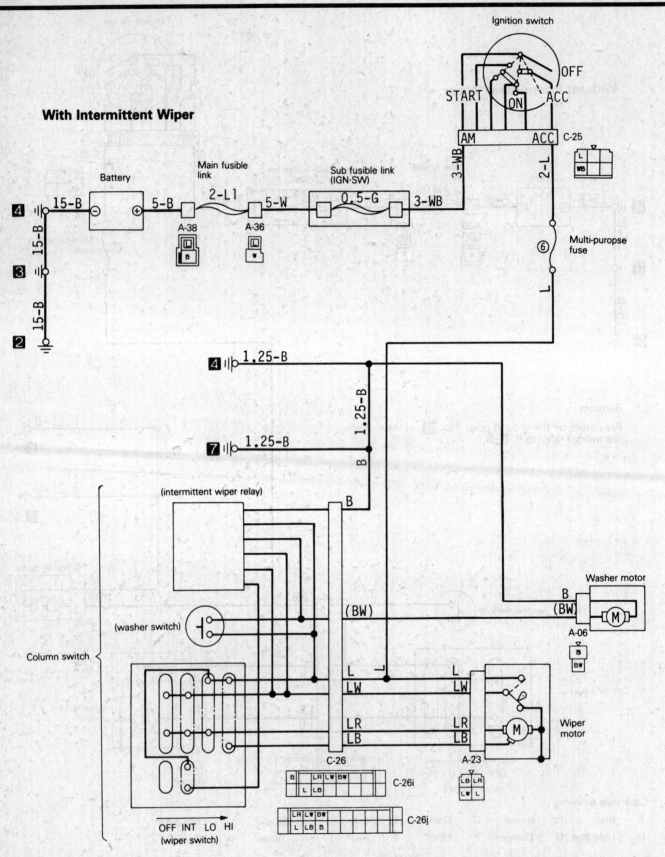

Wiper and washer circuit—Pick-up

6-152

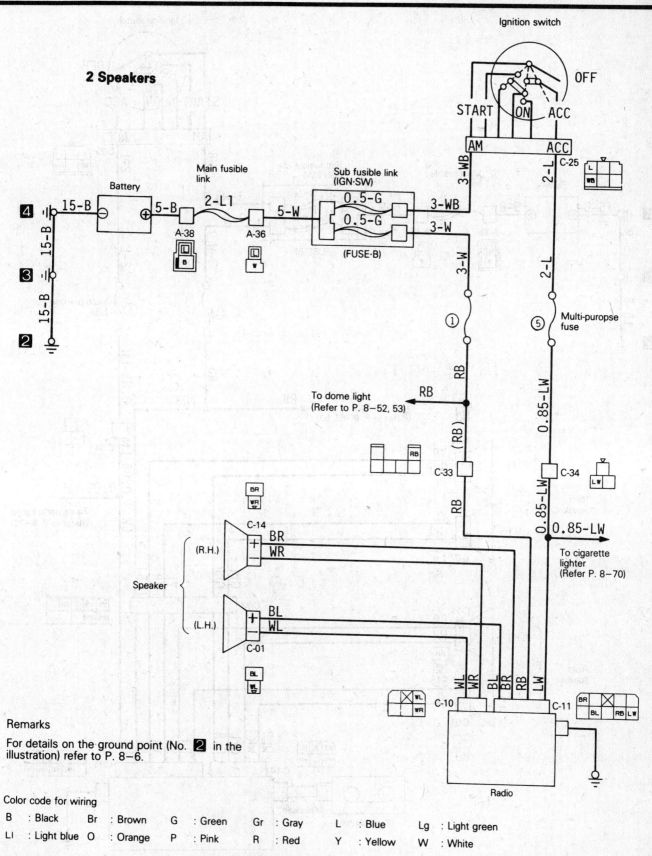

Audio system circuit—Pick-up

6 CHASSIS ELECTRICAL

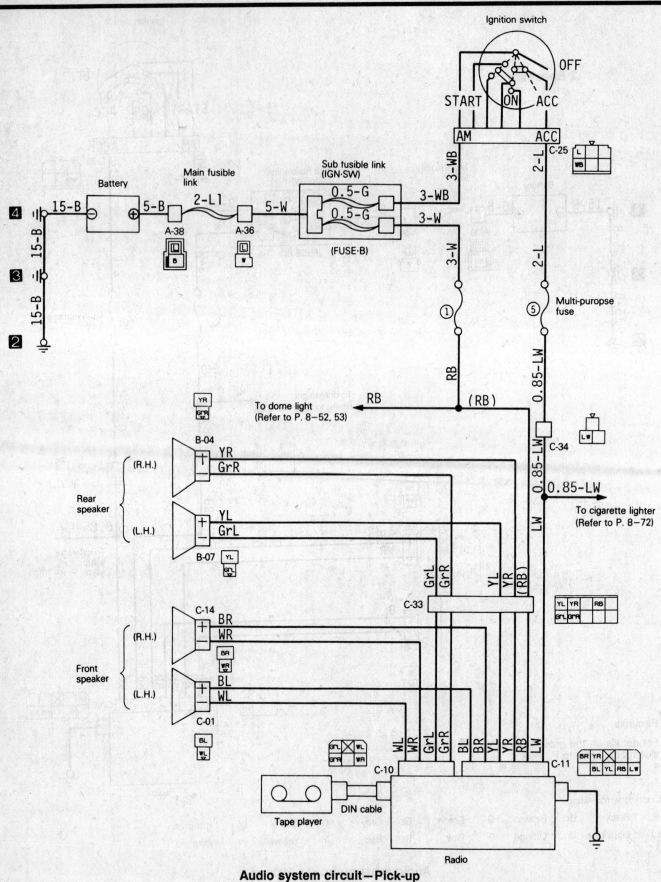

Audio system circuit—Pick-up

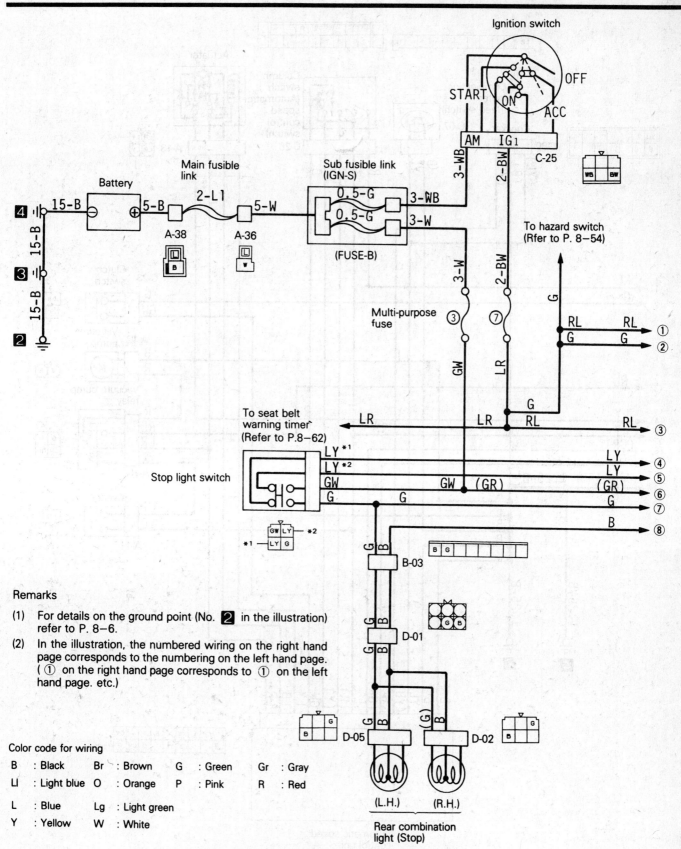

Automatic speed control system — Pick-up

6 CHASSIS ELECTRICAL

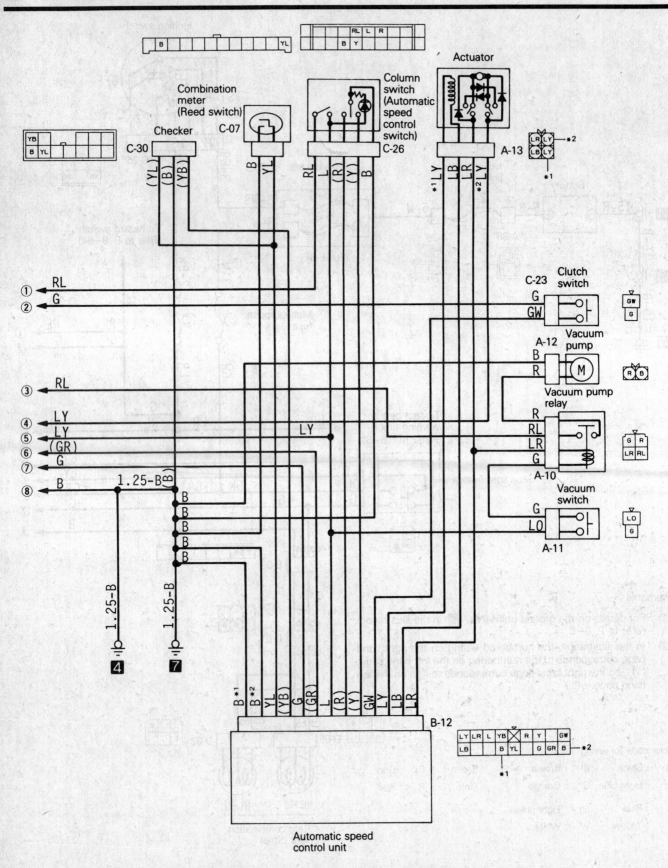

Automatic speed control system—Pick-up

6-156

CHASSIS ELECTRICAL 6

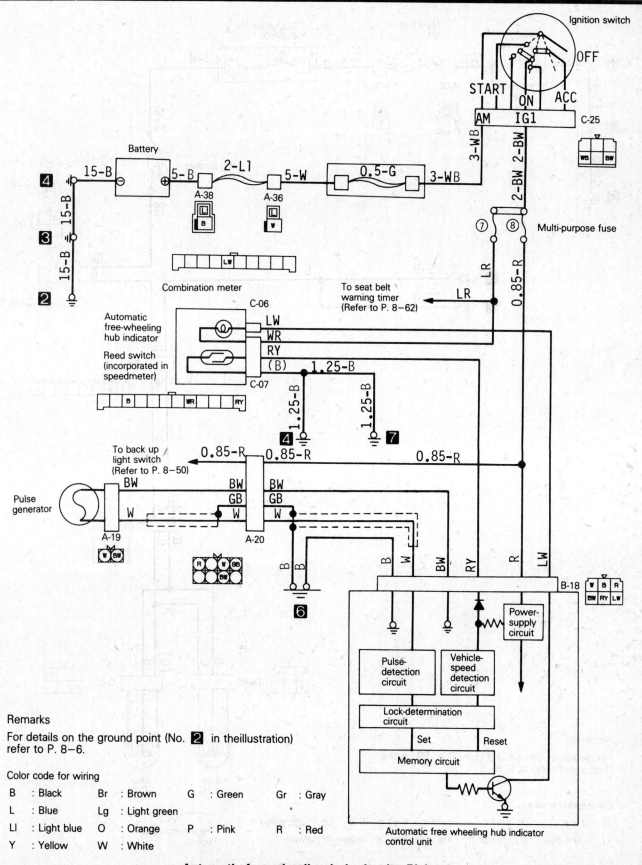

Automatic free-wheeling hub circuit—Pick-up

6-157

6 CHASSIS ELECTRICAL

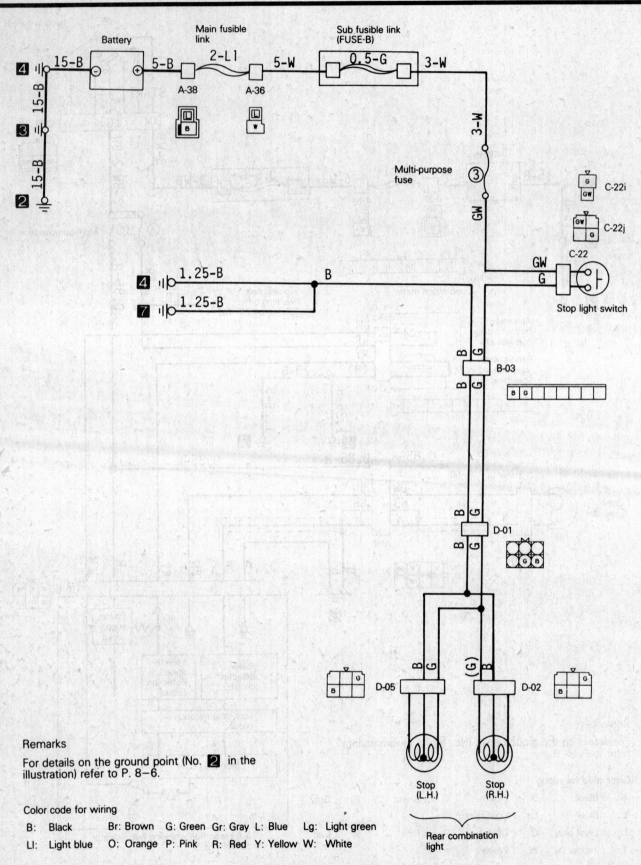

Stop light circuit—Pick-up

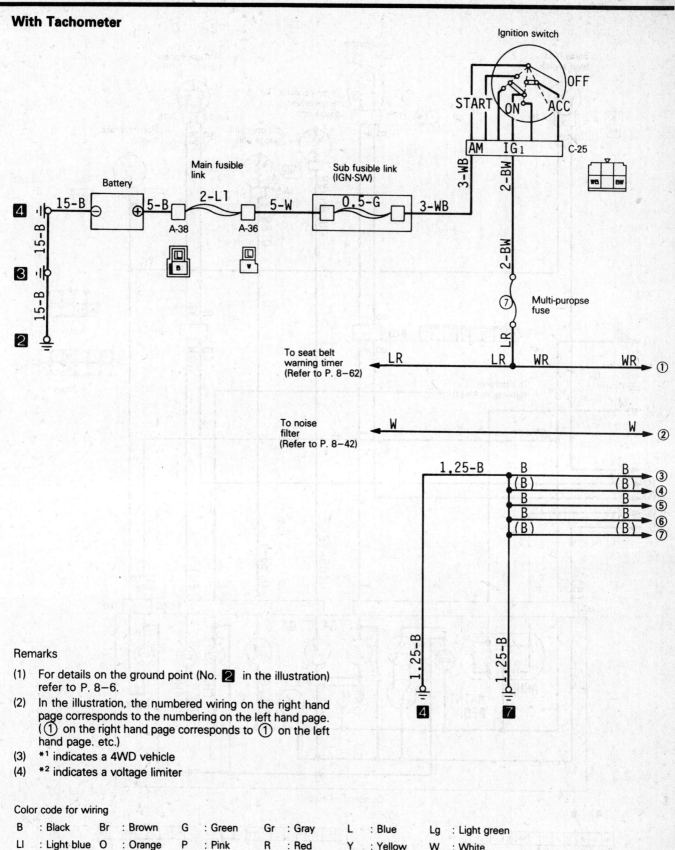

Meter circuit—Pick-up

6 CHASSIS ELECTRICAL

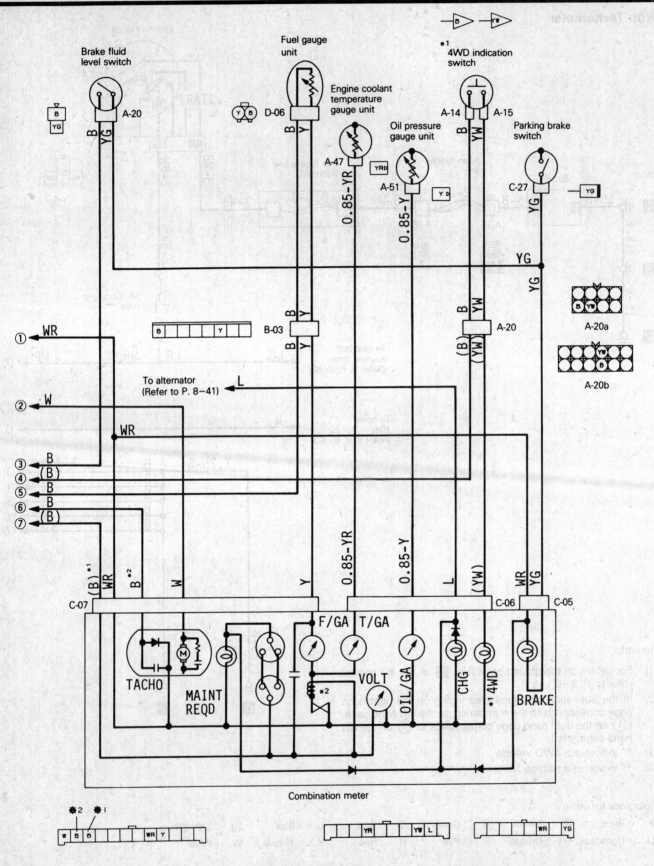

Meter circuit—Pick-up

CHASSIS ELECTRICAL 6

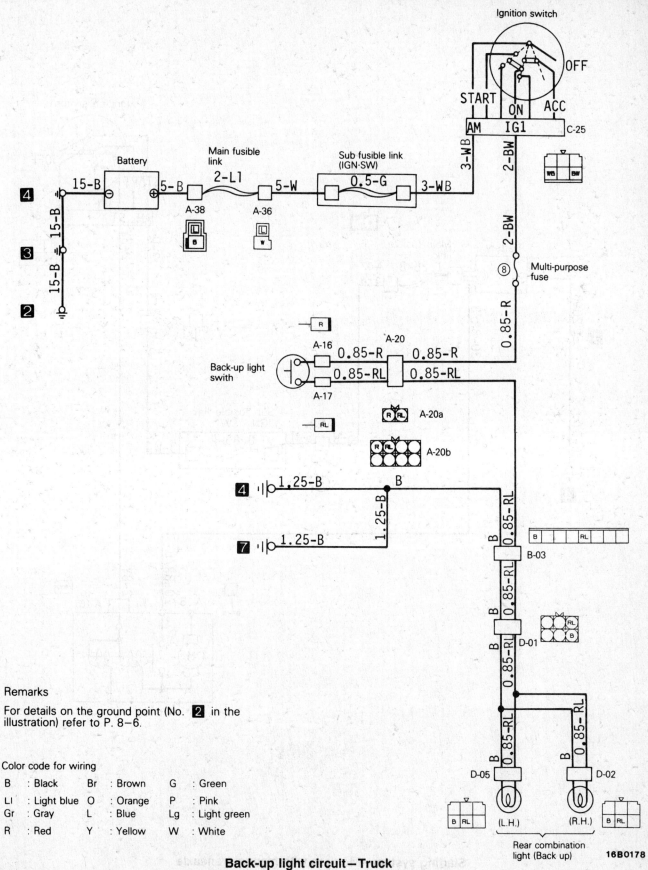

Back-up light circuit — Truck

6 CHASSIS ELECTRICAL

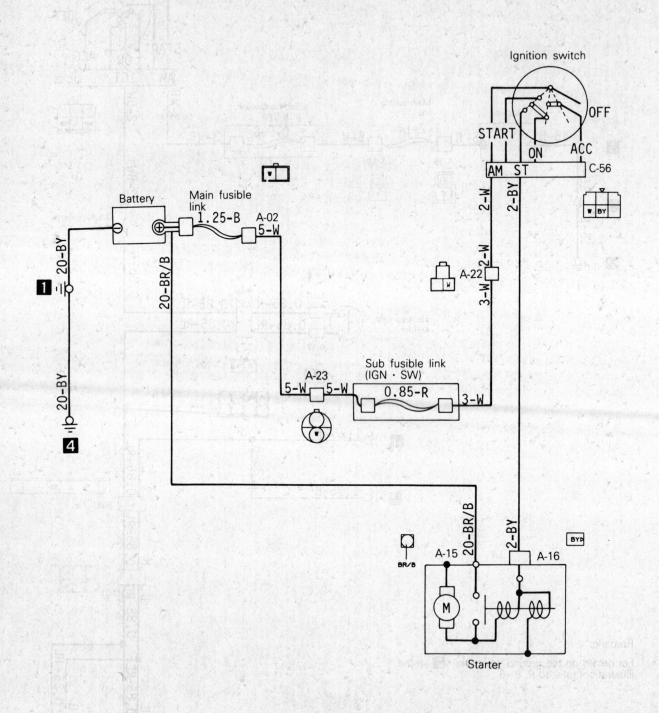

Starting system — Montero with manual transaxle

CHASSIS ELECTRICAL 6

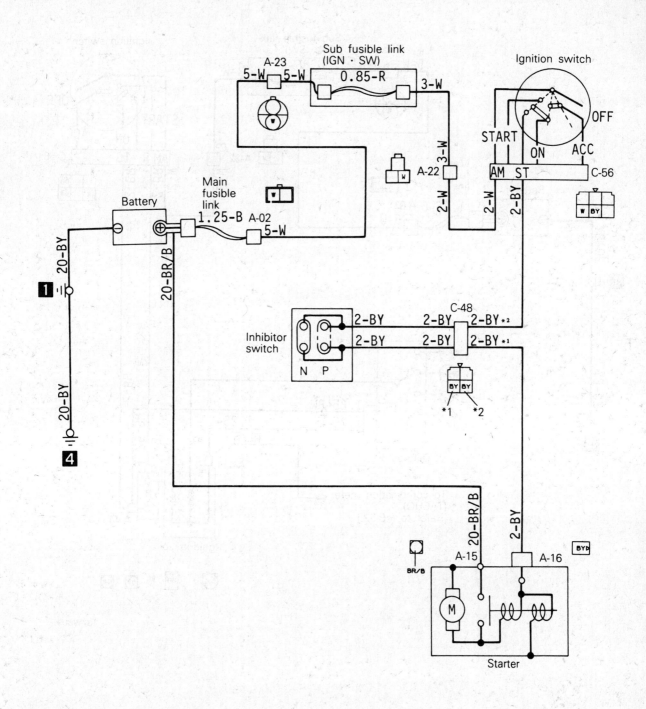

Starting system — Montero with automatic transaxle

6 CHASSIS ELECTRICAL

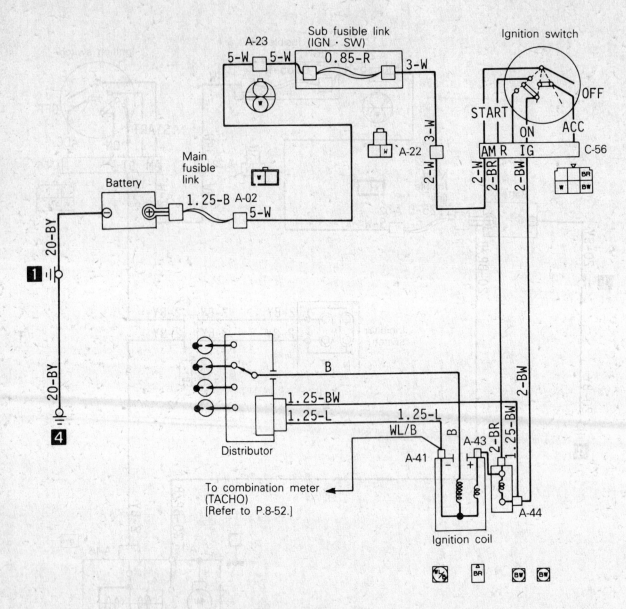

Remark
For information concerning the ground points (example: 1),
refer to P.8-7.

Wiring color code
B: Black   Br: Brown   G: Green   Gr: Gray   L: Blue   Lg: Light green
Ll: Light blue   O: Orange   P: Pink   R: Red   Y: Yellow   W: White

Ignition system—Montero

CHASSIS ELECTRICAL 6

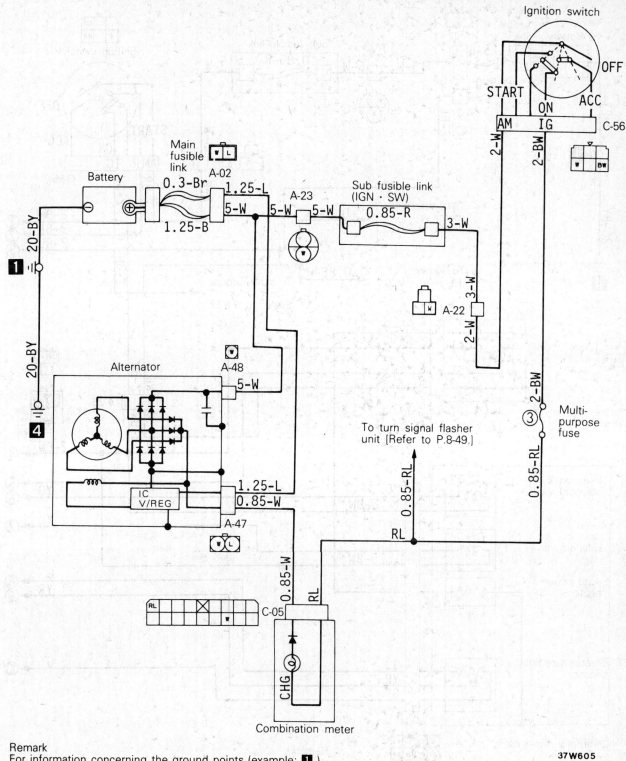

Charging system — Montero

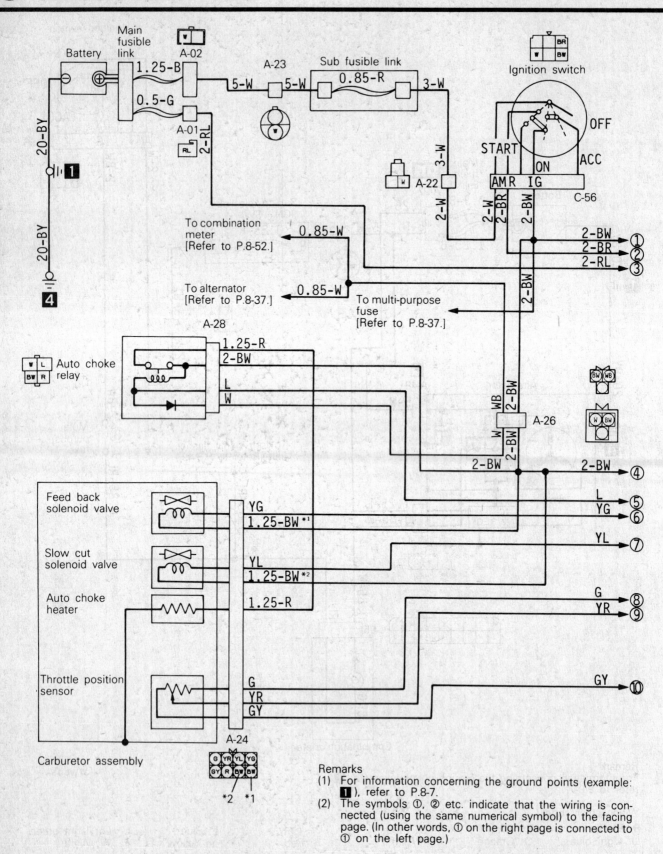

Feedback carburetor circuit – Montero

CHASSIS ELECTRICAL 6

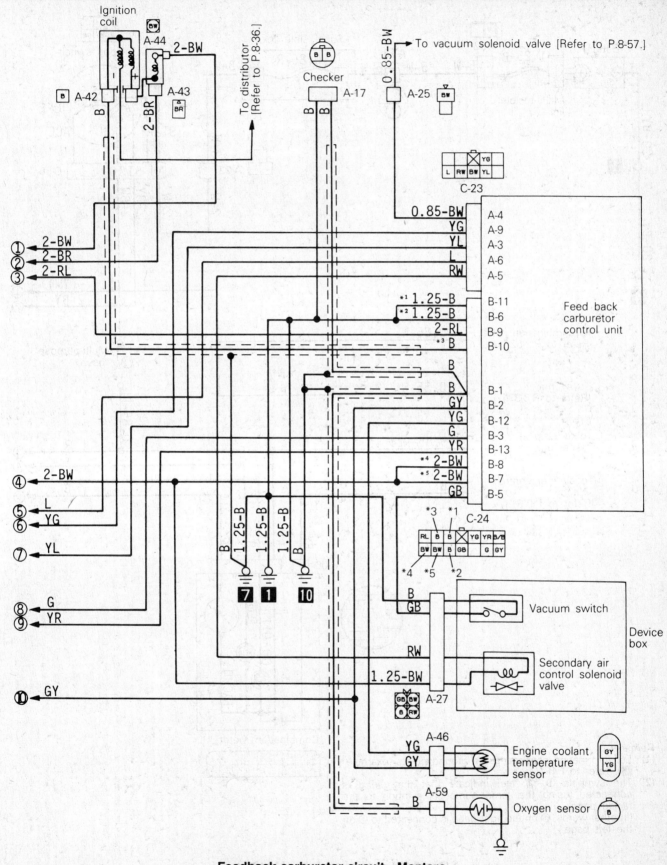

Feedback carburetor circuit—Montero

6 CHASSIS ELECTRICAL

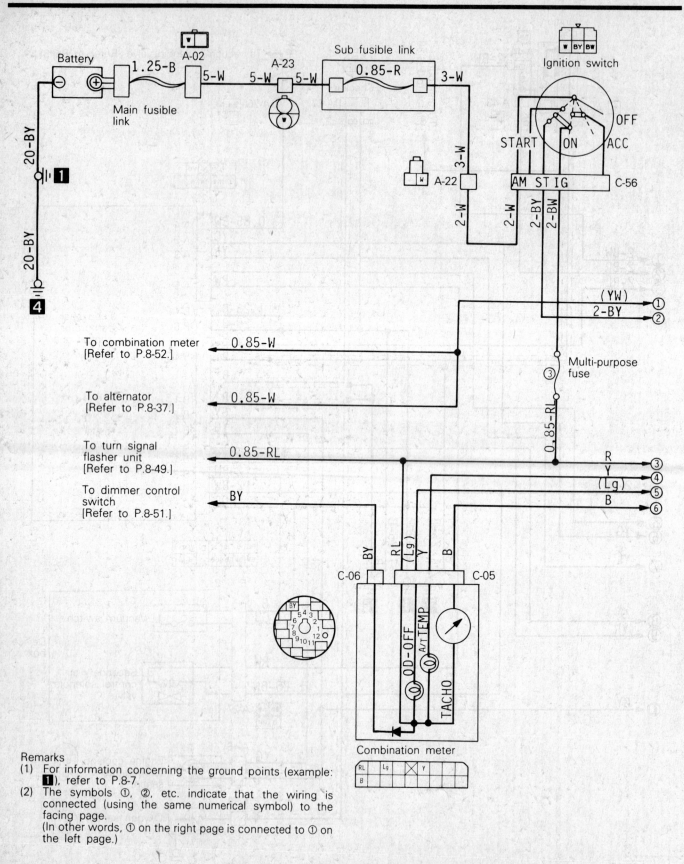

Overdrive control system—Montero

6-168

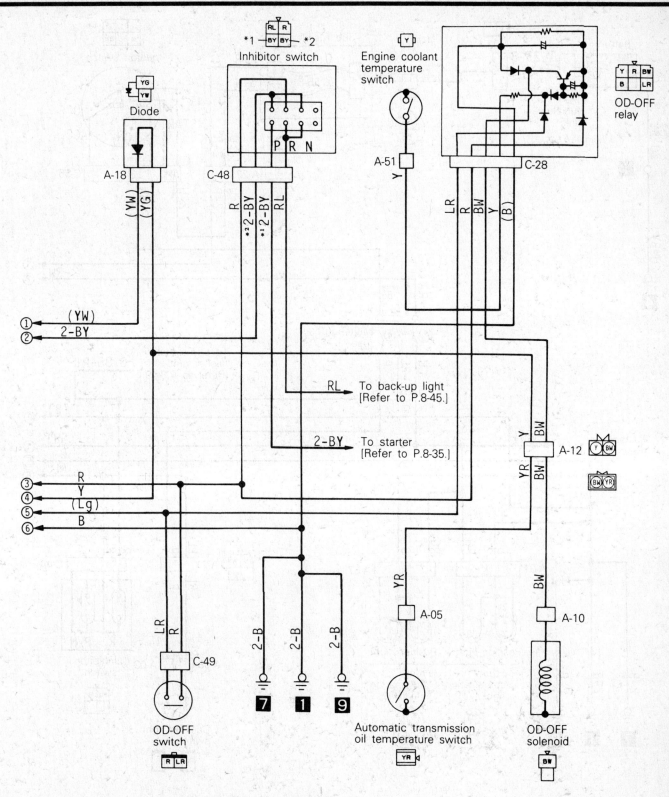

Overdrive control system—Montero

6 CHASSIS ELECTRICAL

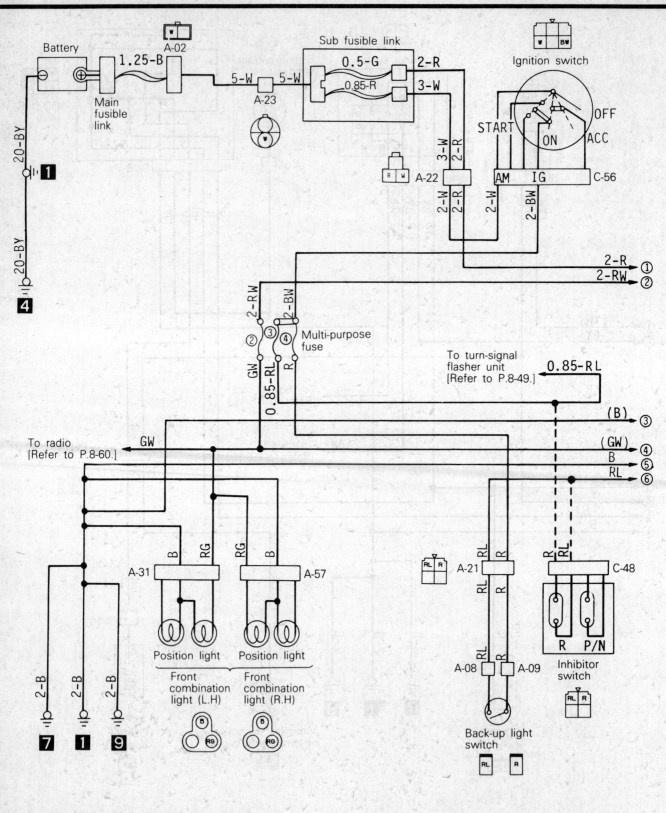

Tail light, position light, side marker light and license plate lgiht—Montero

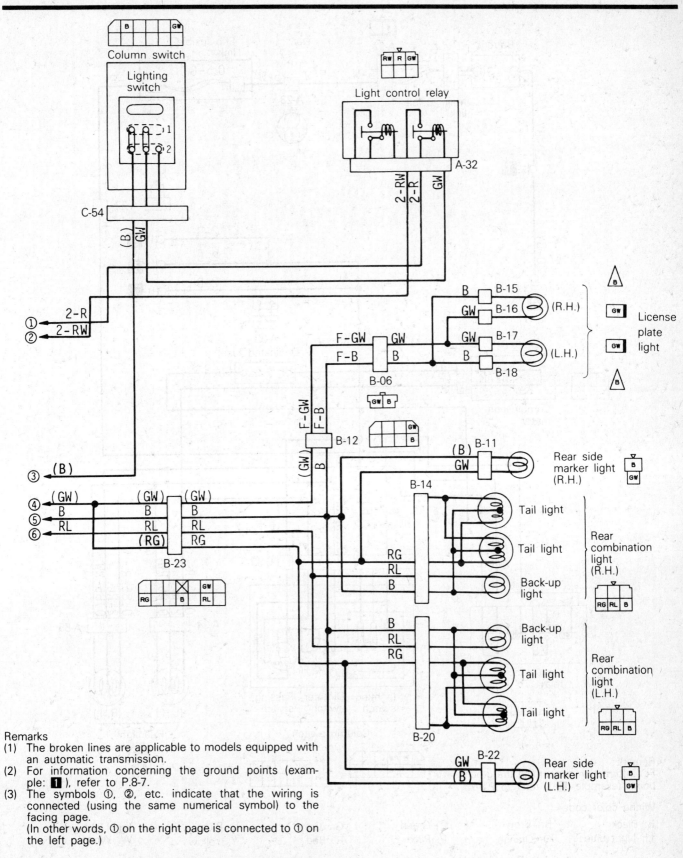

Tail light, position light, side marker light and license plate light—Montero

6 CHASSIS ELECTRICAL

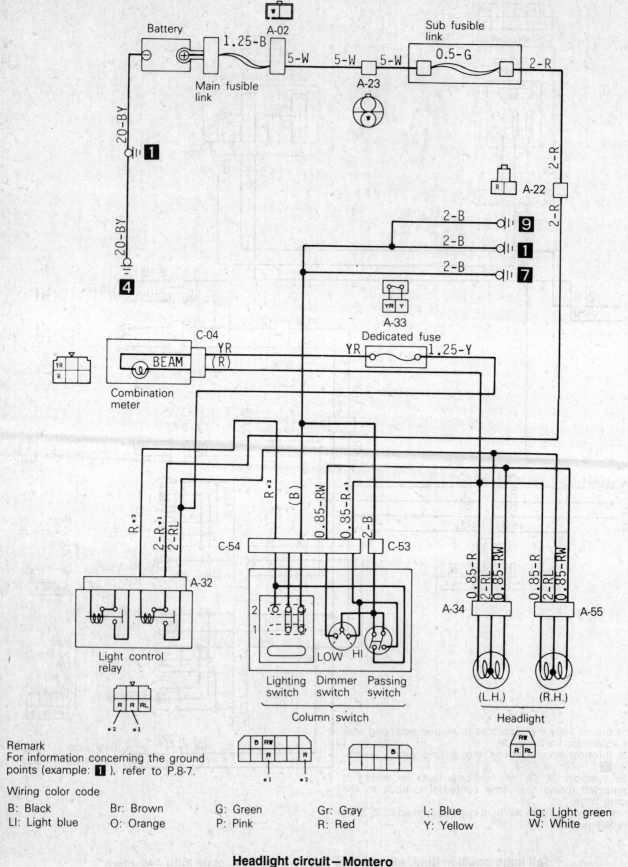

Headlight circuit – Montero

6-172

CHASSIS ELECTRICAL 6

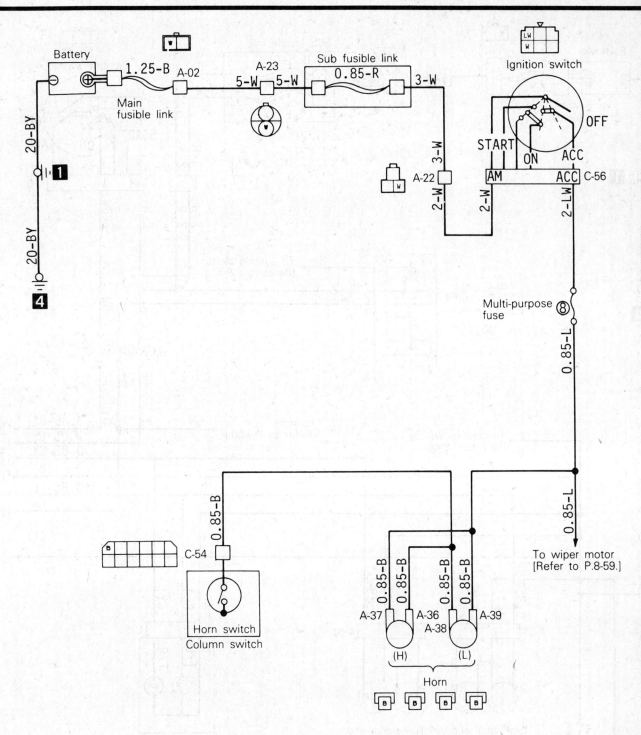

Remark
For information concerning the ground points (example: **1**),
refer to P.8-7.

Wiring color code

B: Black	Br: Brown	G: Green	Gr: Gray	L: Blue	Lg: Light green
Ll: Light blue	O: Orange	P: Pink	R: Red	Y: Yellow	W: White

Horn circuit – Montero

6-173

6 CHASSIS ELECTRICAL

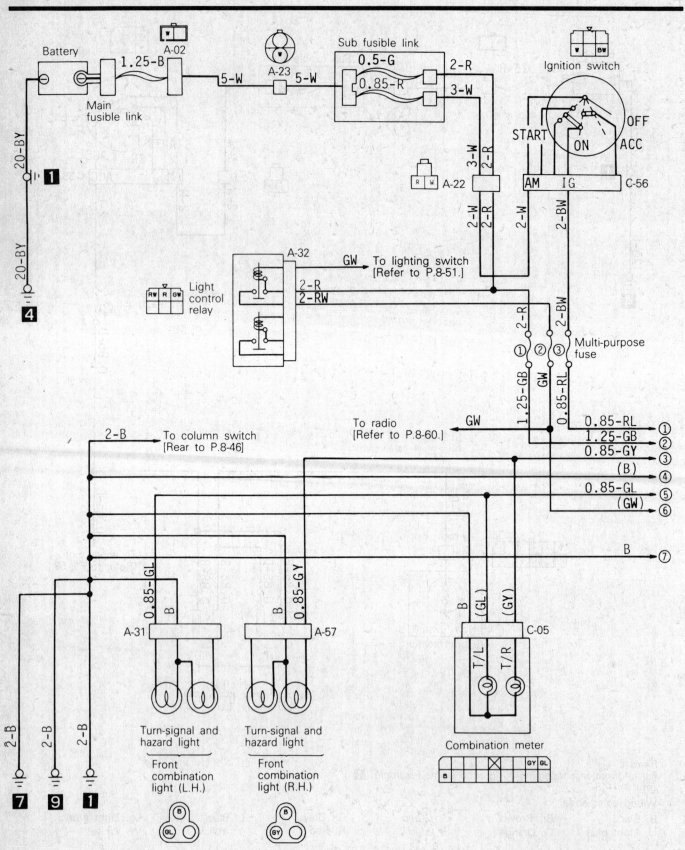

Turn signal and hazard light circuit—Montero

CHASSIS ELECTRICAL 6

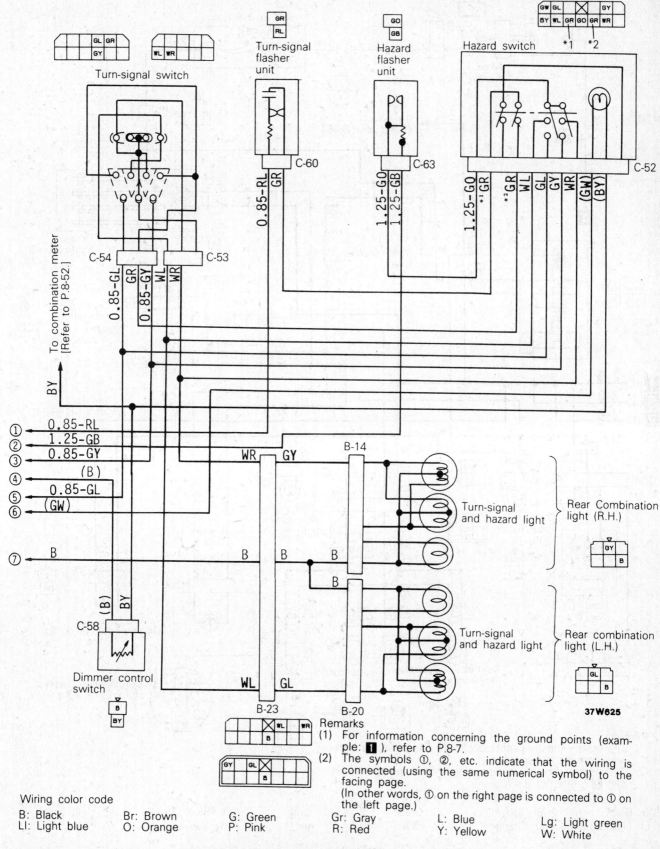

Turn signal and hazard light circuit—Montero

6 CHASSIS ELECTRICAL

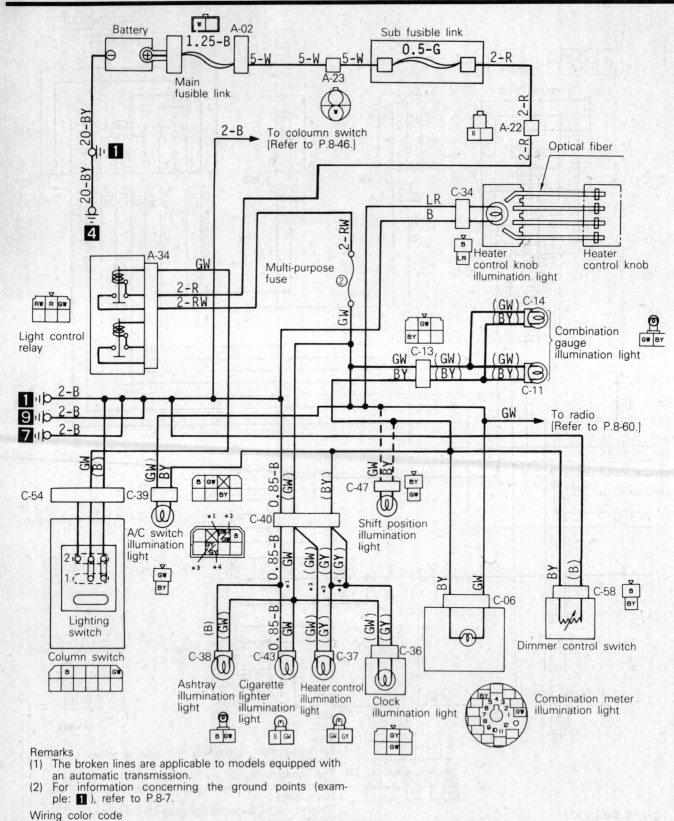

Instrument panel and illumination circuit – Montero

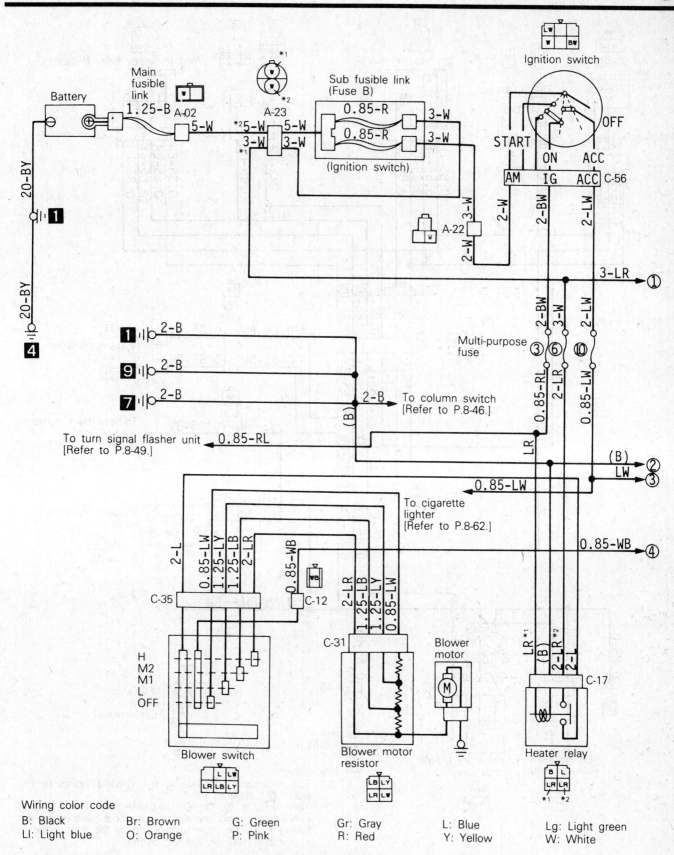

Instrument panel and illumination circuit – Montero

6 CHASSIS ELECTRICAL

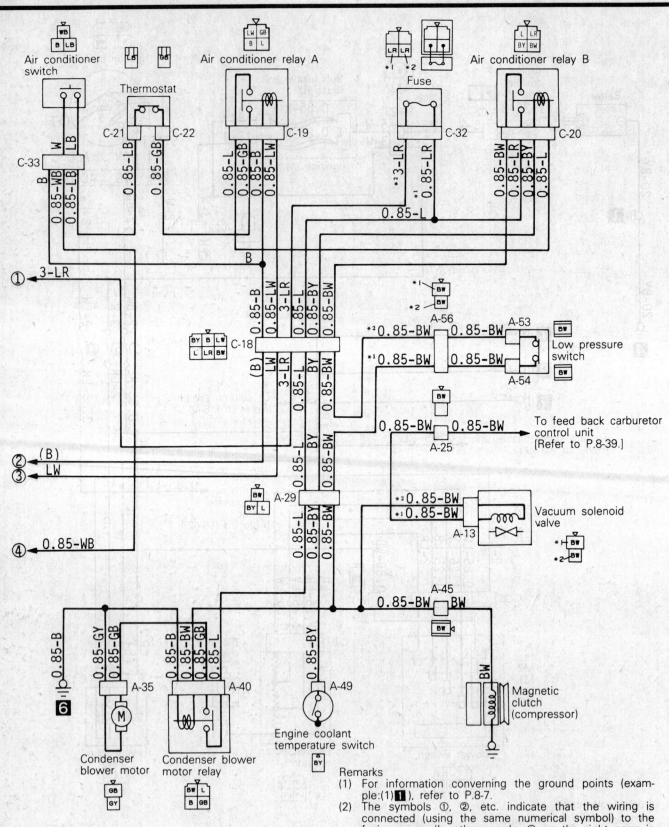

Air conditioner circuit — Montero

CHASSIS ELECTRICAL 6

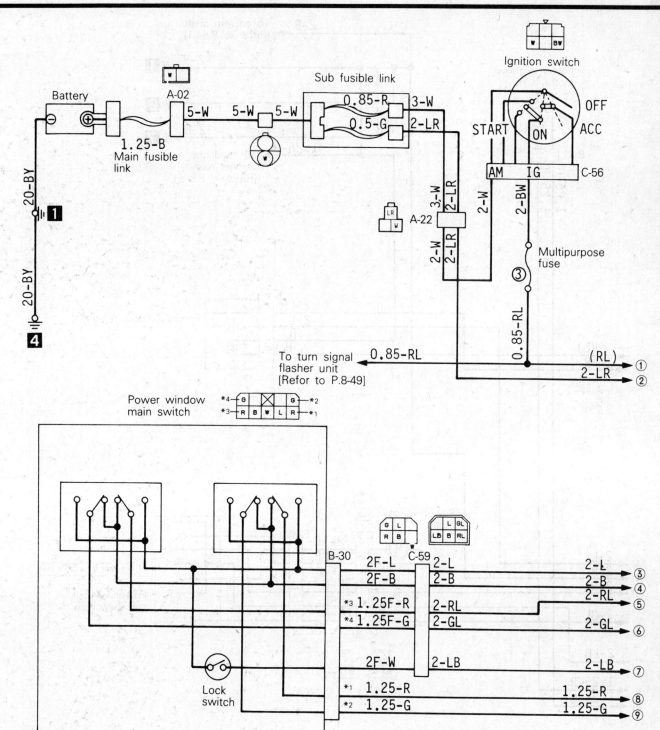

Remarks
(1) For information concerning the ground points (example: 1), refer to P.8-7.
(2) The symbols ①, ②, etc. indicate that the wiring is connected (using the same numerical symbol) to the facing page.
(In other words, ① on the right page is connected to ① on the left page.)

Power window circuit—Montero

6-179

6 CHASSIS ELECTRICAL

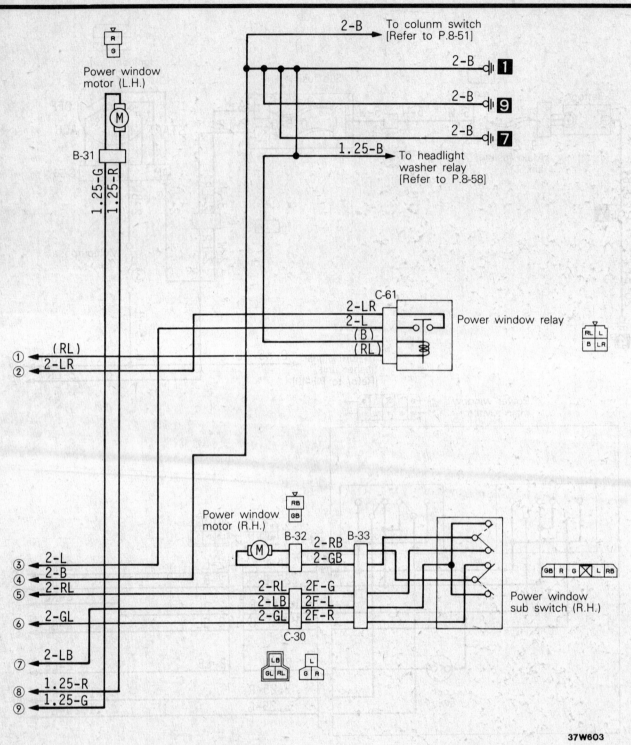

Power window circuit—Montero

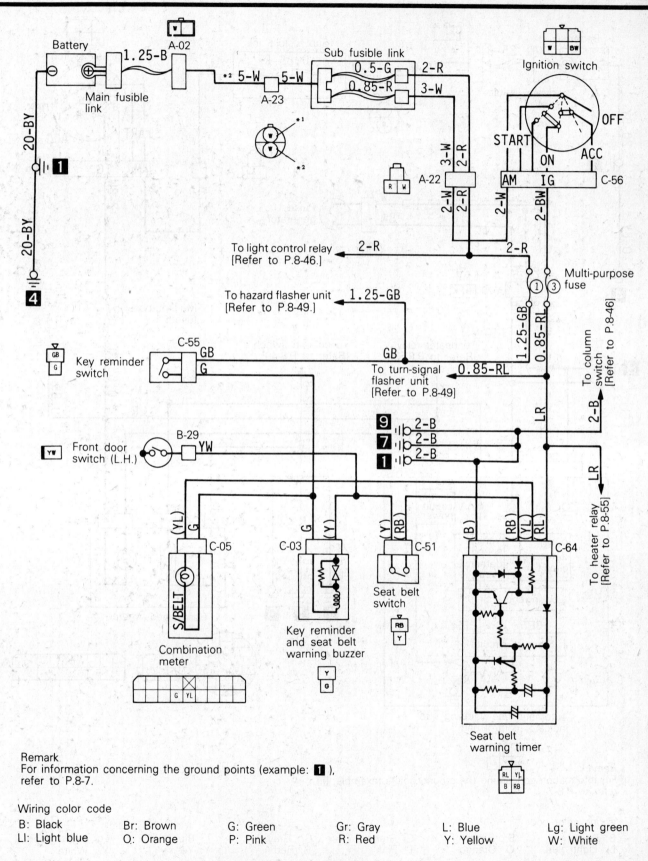

Buzzer circuit—Montero

6 CHASSIS ELECTRICAL

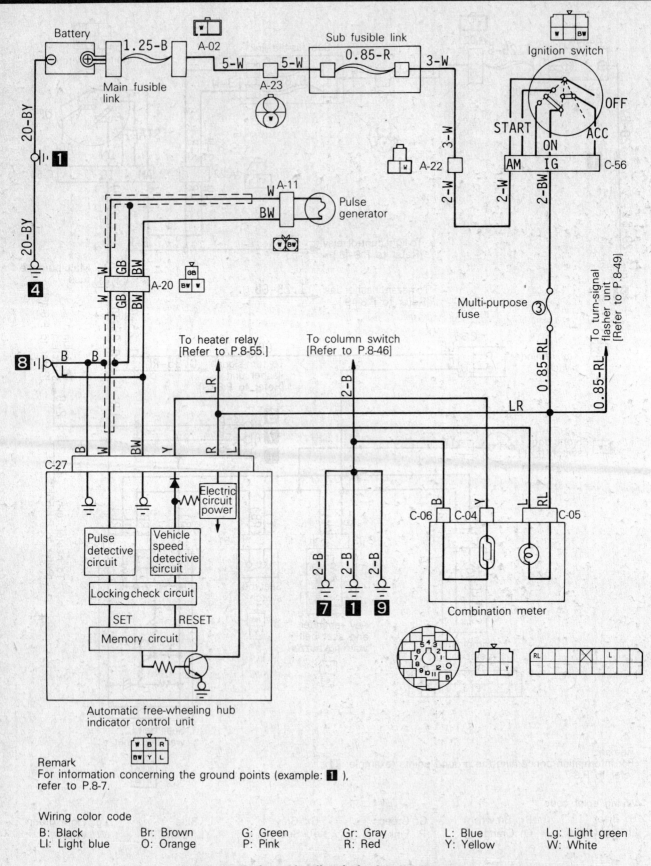

Automatic free-wheeling hub circuit—Montero

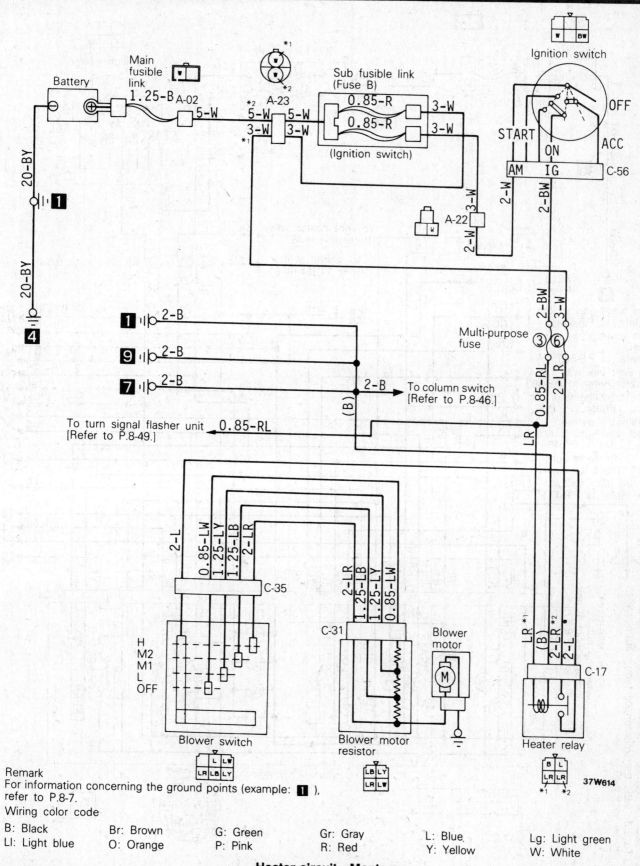

Heater circuit—Montero

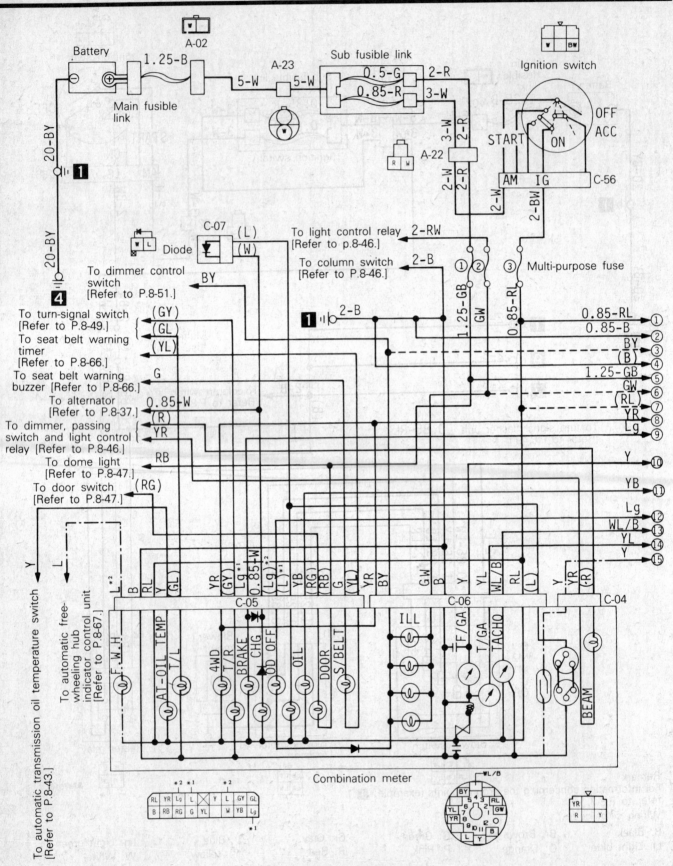

Meter circuit—Montero

CHASSIS ELECTRICAL 6

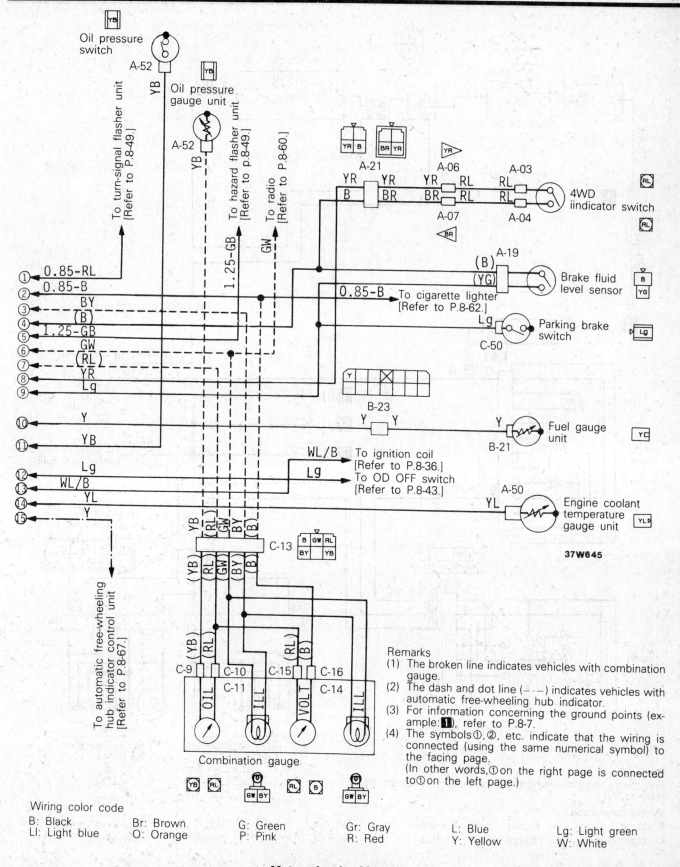

Meter circuit—Montero

6 CHASSIS ELECTRICAL

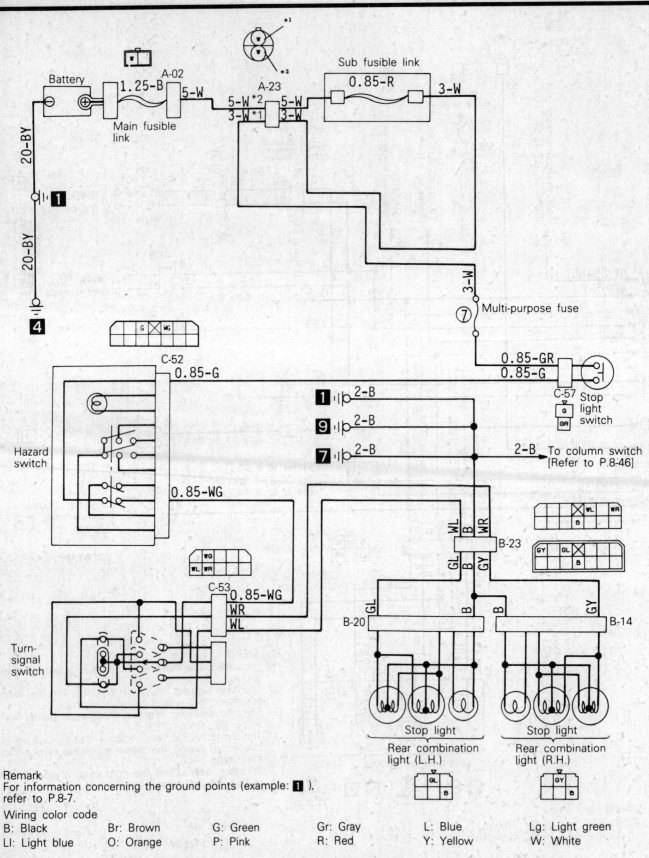

Stoplight circuit—Montero

7 Drive Train

QUICK REFERENCE INDEX

Automatic Transaxle	7-64
Automatic Transmission	7-70
Clutch	7-59
Driveline	7-75
Front Drive Axle	7-83
Manual Transaxle	7-2
Manual Transmission	7-39
Rear Axle	7-78
Transfer Case	7-73

GENERAL INDEX

Automatic transaxle
- Adjustments 7-67
- Fluid and filter change 7-64
- Identification 7-64
- Kickdown switch 7-69
- Neutral safety switch 7-67
- Removal and installation 7-69
- Throttle cable adjustment 7-68
- Troubleshooting 7-65

Automatic transmission
- Adjustments 7-71
- Fluid and filter change 7-71
- Identification 7-70
- Neutral safety switch 7-71
- Removal and installation 7-72
- Throttle cable adjustment 7-71
- Troubleshooting 7-65

Axle
- Front 7-83
- Rear 7-78

Back-up Light Switch
- Manual transaxle 7-5
- Manual transmission 7-39

Center bearing 7-77

Clutch
- Adjustment 7-60
- Hydraulic system bleeding 7-64
- Master cylinder 7-63
- Removal and installation 7-61
- Slave cylinder 7-64
- Troubleshooting 7-59

CV-Joints
- Automatic transaxle 7-70
- Front drive axle 7-57
- Manual transaxle 7-32

Differential
- Front 7-85
- Rear 7-80

Drive axle (front)
- Axle shaft 7-83
- CV-joint 7-57
- Differential 7-85
- Front hub and wheel bearings 7-57
- Pinion seal and yoke 7-86
- Removal and installation 7-85

Drive axle (rear)
- Axle housing 7-81
- Axle shaft and bearing 7-78
- Differential carrier 7-80
- Identification 7-78
- Pinion oil seal 7-81
- Removal and installation 7-81

Driveshaft
- Center bearing 7-77
- Removal and installation 7-75
- U-joints 7-76

Halfshafts
- Automatic transaxle 7-70
- Front drive axle 7-57
- Manual transaxle 7-32

Manual transaxle
- Adjustment 7-2

Back-up light switch 7-5
Identification 7-2
Overhaul 7-11
Removal and installation 7-6
Troubleshooting 7-2

Manual transmission
- Adjustments 7-39
- Back-up light switch 7-39
- Identification 7-39
- Overhaul 7-43
- Removal and installation 7-39
- Troubleshooting 7-2

Master cylinder 7-63

Neutral safety switch
- Transaxle 7-67
- Transmission 7-71

Slave cylinder 7-64

Transfer Case
- Adjustment 7-74
- Overhaul 7-74
- Removal and installation 7-74
- Troubleshooting 7-73

Troubleshooting Charts
- Automatic transmission 7-65
- Clutch 7-59
- Lockup torque converter 7-65
- Manual transmission 7-2
- Transfer case 7-73
- Transmission fluid indications 7-67

U-joints
- Overhaul 7-76

7 DRIVE TRAIN

MANUAL TRANSAXLE IDENTIFICATION CHART

Transaxle Types	Years	Models
KM 162 5-sp	1983	Cordia/Tredia
KM 163 5-sp	1984–88	Cordia/Tredia
KM 166 4 x 2-sp	1984–85	Cordia/Tredia
KM 210 5-sp	1987–88	Galant
KM 206 5-sp	1989	Galant
KM 161 4-sp	1985–86	Mirage
KM 162 5-sp	1985–86	Mirage
KM 163 5-sp	1985–86	Mirage
KM 200 5-sp	1987	Mirage
KM 201 5-sp	1987–89	Mirage
KM 206 5-sp	1987–88	Mirage
KM 210 5-sp	1989	Mirage
KM 161 4-sp	1987–89	Precis
KM 162 5-sp	1987–89	Precis
KM-210 5-sp	1989	Sigma

MANUAL TRANSAXLE

Identification

Transaxle identification codes are found on the vehicle information plate located on the firewall in the engine compartment. The first portion of the code is the transaxle designation; the numbers at the right end of the line indicate the final drive ratio of the transaxle or differential.

Before attempting to repair the clutch or transaxle for reasons beyond obvious failure, the problem and probable cause should be identified. A great percentage of manual transaxle and clutch problems are accompanied by shifting difficulties such as gear clash, grinding or refusal to engage gears. When any of these problems occur, a careful analysis should be performed to determine cause.

Driveline noises can become baffling, but don't be too quick to assign fault to the transaxle. The noise may actually be coming from other sources such as tires, road surface, wheel bearings, engine or exhaust system. Noises will also vary by vehicle size, engine type and amount of insulation within the body. Remember also that the transaxle, like any mechanical device, is not totally quiet in its operation and will exhibit some normal operating noise.

When checking for driveline noises, follow these guide lines:
1. Select a smooth, level asphalt road to reduce tire and body noises. Check and set the tire pressures before beginning.
2. Drive the vehicle far enough to thoroughly warm up all the lubricants; this may require driving as much as 10 or 12 miles before testing.
3. Note the speed at which the noise occurs and in which gear(s).
4. Check for noises with the vehicle stopped and the engine running.
5. Does the noise change when the clutch is pushed in or released?
6. Determine the throttle condition(s) at which the noise occurs:
   a. Drive (leading throttle) – the engine is under load and the car is accelerating at some rate.
   b. Float (neutral throttle) – the engine is maintaining a constant speed at light throttle on a level road.

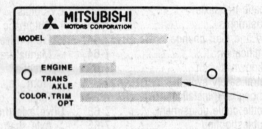

Location of the transaxle code

   c. Coast (trailing throttle) – the engine is not driving the car, the throttle is closed with the car in gear and in motion. An example is a long downhill grade.
   d. Some combination of the above.
7. Does the noise change as the car corners or changes direction? Is the change in the noise different on left and right turns?
8. Identify the noise by type (whine, click, knock, grinding, buzzing, etc.) and think through possible causes such as loose components, something rubbing, poor lubrication, or worn parts.
9. Check all other causes first. Its much easier to check the wheel bearings and the exhaust system than to remove the transaxle. Modern manual transaxles are extremely reliable and rarely fail if given reasonable maintenance and freedom from outright abuse. Assume that the noise is not in the transaxle until diagnosed otherwise.

Linkage Adjustment

Mitsubishi manual transaxle linkages are not adjustable. If the shifter does not work properly, a problem with the size, mounting or lubrication of the linkage is indicated.

DRIVE TRAIN 7

Troubleshooting the Manual Transmission and Transfer Case

Problem	Cause	Solution
Transmission shifts hard	• Clutch adjustment incorrect • Clutch linkage or cable binding • Shift rail binding	• Adjust clutch • Lubricate or repair as necessary • Check for mispositioned selector arm roll pin, loose cover bolts, worn shift rail bores, worn shift rail, distorted oil seal, or extension housing not aligned with case. Repair as necessary.
	• Internal bind in transmission caused by shift forks, selector plates, or synchronizer assemblies • Clutch housing misalignment • Incorrect lubricant • Block rings and/or cone seats worn	• Remove, dissemble and inspect transmission. Replace worn or damaged components as necessary. • Check runout at rear face of clutch housing • Drain and refill transmission • Blocking ring to gear clutch tooth face clearance must be 0.030 inch or greater. If clearance is correct it may still be necessary to inspect blocking rings and cone seats for excessive wear. Repair as necessary.
Gear clash when shifting from one gear to another	• Clutch adjustment incorrect • Clutch linkage or cable binding • Clutch housing misalignment • Lubricant level low or incorrect lubricant • Gearshift components, or synchronizer assemblies worn or damaged	• Adjust clutch • Lubricate or repair as necessary • Check runout at rear of clutch housing • Drain and refill transmission and check for lubricant leaks if level was low. Repair as necessary. • Remove, disassemble and inspect transmission. Replace worn or damaged components as necessary.
Transmission noisy	• Lubricant level low or incorrect lubricant • Clutch housing-to-engine, or transmission-to-clutch housing bolts loose • Dirt, chips, foreign material in transmission • Gearshift mechanism, transmission gears, or bearing components worn or damaged • Clutch housing misalignment	• Drain and refill transmission. If lubricant level was low, check for leaks and repair as necessary. • Check and correct bolt torque as necessary • Drain, flush, and refill transmission • Remove, disassemble and inspect transmission. Replace worn or damaged components as necessary. • Check runout at rear face of clutch housing

7–3

7 DRIVE TRAIN

Troubleshooting the Manual Transmission and Transfer Case (cont.)

Problem	Cause	Solution
Jumps out of gear	• Clutch housing misalignment	• Check runout at rear face of clutch housing
	• Gearshift lever loose	• Check lever for worn fork. Tighten loose attaching bolts.
	• Offset lever nylon insert worn or lever attaching nut loose	• Remove gearshift lever and check for loose offset lever nut or worn insert. Repair or replace as necessary.
	• Gearshift mechanism, shift forks, selector plates, interlock plate, selector arm, shift rail, detent plugs, springs or shift cover worn or damaged	• Remove, disassemble and inspect transmission cover assembly. Replace worn or damaged components as necessary.
	• Clutch shaft or roller bearings worn or damaged	• Replace clutch shaft or roller bearings as necessary
Jumps out of gear (cont.)	• Gear teeth worn or tapered, synchronizer assemblies worn or damaged, excessive end play caused by worn thrust washers or output shaft gears	• Remove, disassemble, and inspect transmission. Replace worn or damaged components as necessary.
	• Pilot bushing worn	• Replace pilot bushing
Will not shift into one gear	• Gearshift selector plates, interlock plate, or selector arm, worn, damaged, or incorrectly assembled	• Remove, disassemble, and inspect transmission cover assembly. Repair or replace components as necessary.
	• Shift rail detent plunger worn, spring broken, or plug loose	• Tighten plug or replace worn or damaged components as necessary
	• Gearshift lever worn or damaged	• Replace gearshift lever
	• Synchronizer sleeves or hubs, damaged or worn	• Remove, disassemble and inspect transmission. Replace worn or damaged components.
Locked in one gear—cannot be shifted out	• Shift rail(s) worn or broken, shifter fork bent, setscrew loose, center detent plug missing or worn	• Inspect and replace worn or damaged parts
	• Broken gear teeth on countershaft gear, clutch shaft, or reverse idler gear	• Inspect and replace damaged part
	• Gearshift lever broken or worn, shift mechanism in cover incorrectly assembled or broken, worn damaged gear train components	• Disassemble transmission. Replace damaged parts or assemble correctly.
Transfer case difficult to shift or will not shift into desired range	• Vehicle speed too great to permit shifting	• Stop vehicle and shift into desired range. Or reduce speed to 3–4 km/h (2–3 mph) before attempting to shift.
	• If vehicle was operated for extended period in 4H mode on dry paved surface, driveline torque load may cause difficult shifting	• Stop vehicle, shift transmission to neutral, shift transfer case to 2H mode and operate vehicle in 2H on dry paved surfaces

7–4

DRIVE TRAIN 7

Troubleshooting the Manual Transmission and Transfer Case (cont.)

Problem	Cause	Solution
	• Transfer case external shift linkage binding	• Lubricate or repair or replace linkage, or tighten loose components as necessary
	• Insufficient or incorrect lubricant	• Drain and refill to edge of fill hole with SAE 85W-90 gear lubricant only
	• Internal components binding, worn, or damaged	• Disassemble unit and replace worn or damaged components as necessary
Transfer case noisy in all drive modes	• Insufficient or incorrect lubricant	• Drain and refill to edge of fill hole with SAE 85W-90 gear lubricant only. Check for leaks and repair if necessary. Note: If unit is still noisy after drain and refill, disassembly and inspection may be required to locate source of noise.
Noisy in—or jumps out of four wheel drive low range	• Transfer case not completely engaged in 4L position	• Stop vehicle, shift transfer case in Neutral, then shift back into 4L position
	• Shift linkage loose or binding	• Tighten, lubricate, or repair linkage as necessary
	• Shift fork cracked, inserts worn, or fork is binding on shift rail	• Disassemble unit and repair as necessary
Lubricant leaking from output shaft seals or from vent	• Transfer case overfilled • Vent closed or restricted	• Drain to correct level • Clear or replace vent if necessary
Lubricant leaking from output shaft seals or from vent (cont.)	• Output shaft seals damaged or installed incorrectly	• Replace seals. Be sure seal lip faces interior of case when installed. Also be sure yoke seal surfaces are not scored or nicked. Remove scores, nicks with fine sandpaper or replace yoke(s) if necessary.
Abnormal tire wear	• Extended operation on dry hard surface (paved) roads in 4H range	• Operate in 2H on hard surface (paved) roads

Back-up Light Switch

REMOVAL AND INSTALLATION

The back-up light control switch is mounted into the transaxle case. It is a simple push-button type switch; when the transaxle is in reverse, a part of the linkage inside the case presses against the tip of the switch. This closes the circuit and brings on the back-up lights.

The switch may be unscrewed from the case after unplugging its wiring connector. The location of the switch will determine whether it is necessary to drain the transaxle before removing the switch. The oil usually fills about one third to one half of the case; if the switch is mounted above this approximate line, it may be removed without draining. Those switches mounted low on the case will require the oil to be drained.

Always use the correct size wrench on the switch and avoid the use of pliers. Keep the wrench straight on the flats when turning; the aluminum alloy transaxle case may be damaged or stripped by careless removal. When reinstalling, tighten the switch to 30 Nm or 22 ft. lbs.

7 DRIVE TRAIN

Transaxle

REMOVAL AND INSTALLATION

All Cordia and Tredia
1987 Galant

1. Disconnect the battery cables and remove the battery and battery tray.
2. Remove the air cleaner and housing.
3. Remove the starter.
4. Drain and remove the coolant and windshield washer reservoir tanks.
5. Disconnect from the transaxle: the clutch cable or (remove) the slave cylinder; speedometer cable; and back-up light harness.
6. Remove the bell housing bolts connecting the transaxle and engine that are accessible from above.
7. Raise the vehicle and safely support it on stands.
8. Remove the front wheels. Remove the drain plug and drain transaxle fluid.
9. Disconnect the extension rod and shift control rod at the transmission and lower them away from the unit.
10. Remove the stabilizer bar and strut bar from the lower control arms.
11. Remove the driveshafts.
12. Support the transaxle from below with a floor jack in such a way that concentrated stress will not be put on the transmission oil pan.
13. Remove the bell housing cover bolts and remove the cover. Remove the remaining bell housing bolts connecting the transaxle and engine.
14. Remove the bolt from the transaxle vibration insulator. Slide the transmission away from the engine until the input shaft is out of the clutch and lower it to remove it.
15. Elevate the transmission into place on the jack. Fit the transaxle straight onto the engine and do not allow the trans to hang on the shaft.
16. Install the transmission mount insulator bolt. tighten it to 50 Nm (37 ft. lbs).

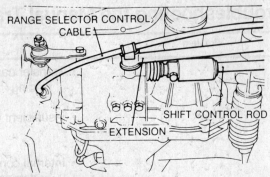

Disconnect the shift control rod, extension rod and the range selector control cable (4 × 2 speed only) from the bottom of the Cordia and Tredia transaxle

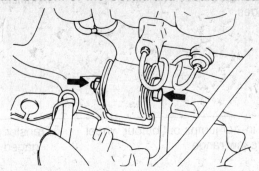

Removing the transaxle mount insulator bolt from the Cordia and Tredia

17. Install the lower bolts holding the transaxle to the engine. Install the lower cover on the bell housing.
18. Install the left and right driveshafts.
19. Install the stabilizer and strut bar to the lower control arm.

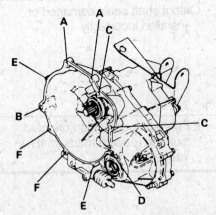

	Nm	ft.lbs.	O.D. x Length mm (in.)	Bolt indentification
A	43–55	31–40	⑦ 10 x 40 (1.6)	⑦ A x B
B	43–55	31–40	⑦ 10 x 65 (2.6)	
C	22–32	16–23	⑦ 10 x 55 (2.2)	
D	30–35	22–25	⑩ 10 x 60 (2.4)	
E	10–12	7–9	⑦ 8 x 14 (0.6)	
F	15–22	11–16	⑦ 8 x 20 (0.8)	

Cordia and Tredia transmission mounting bolts.

DRIVE TRAIN 7

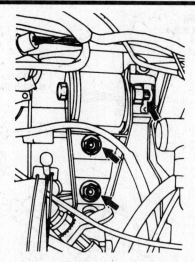

Removing the transaxle mount insulator bolt from the Mirage

20. Install the extension and shift control rods.
21. Lower the vehicle to the ground. Install the upper bolts holding the engine to the transmission.
22. Install the starter and connect the harness.
23. Connect the clutch cable, speedometer cable and backup light harness to the transmission.
24. fill the transaxle with the proper amount of oil.
25. Install the air cleaner assembly.
26. Install the reservoir tanks for the radiator and washer systems.
27. Install the battery tray and battery. Connect the battery cables.
28. Adjust the clutch cable and the gearshift lever as needed.

Mirage through 1988
Precis

1. Remove the battery and battery tray. On turbocharged cars, remove the air cleaner housing assembly.
2. On five speed transaxles, disconnect the electrical connector for the selector control valve. On turbocharged vehicles, remove the actuator mounting bolts, remove the actuator-to-shaft pin and then remove the actuator. Replace the collar with a new part.
3. Disconnect and remove the speedometer and clutch cables.
4. Disconnect the backup lamp electrical connector. Remove the starter motor electrical harness.
5. Remove the six transaxle mounting bolts accessible from the top side of the transaxle.
6. Unbolt and remove the starter motor.
7. Raise the vehicle and safely support it on stands. Remove the splash shield from under the engine. Drain the transaxle fluid.
8. Disconnect the extension rod and the shift rod at the transmission end and lower them.
9. Disconnect the stabilizer bar at the lower control arm.
10. Disconnect the lower arm ball joint on each side.
11. Disconnect the tie rod end ball joints.
12. Swing the suspension to the outside and disconnect the driveshafts at the transaxle. Use wire to tie the driveshafts to the crossmember or other support; do not allow them to hang.
13. Support the transaxle from below with a floorjack or similar device. Make sure the support is widely enough spread that the transmission pan will not be damaged. Remove the five attaching bolts and remove the bell housing cover.
14. Remove the lower bolts attaching the transaxle to the engine. Disconnect the anti-roll rod between the transmission case and the body.
15. Remove the transaxle insulator mount bolt. Remove the cover from inside the right fender shield and remove the transaxle support bracket.
16. Remove the transaxle mount bracket.
17. Pull the assembly away from the engine and then lower it from the vehicle.
18. Elevate the transmission into place on the jack. Fit the transaxle straight onto the engine and do not allow the trans to hang on the shaft.
19. Install the transaxle bracket. Install the mount bracket and install the bolt. After tightening, install the caps on the bolts.
20. Install the bell housing cover.
21. Install the driveshafts, making sure they have engaged the clips correctly.
22. Connect the tie rod joints and connect the lower ball joints. Tighten the nuts to the proper torque and use new washers and cotter pins.
23. Install the stabilizer bar.
24. Install the starter and connect its wiring.
25. Connect the speedometer cable, the ground cable and the backup lamp harness
26. Connect the select cable and the shift cable.
27. Connect either the clutch cable or clutch fluid pipe.
28. Install the battery tray and battery.
29. Install the air cleaner if it was removed.
30. If not already done, install the drain plug. Fill the unit with the correct amount of oil and install the filler plug.
31. If equipped with a fluid clutch, fill and bleed the system.
32. Adjust the clutch and shifter as needed.

1989 Mirage

1. Remove the battery and battery tray.
2. Remove the air cleaner assembly.
3. On turbocharged vehicles, remove the air pipe and air hose assemblies running to the intercooler.
4. On DOHC models, remove the tension rod (mount).
5. Disconnect the shifter control cables.
6. Depending on type of clutch actuation system, either disconnect the clutch cable or remove the slave cylinder from the transaxle case.
7. Disconnect the wiring to the backup light switch and disconnect the speedometer cable.
8. Remove the starter motor.
9. Remove the transaxle assembly upper connecting bolt.
10. Remove the transmission mounting bracket.
11. Elevate and safely support the vehicle on stands.
12. Remove the under covers.
13. Disconnect the tie rod joints.
14. Disconnect the lower ball joints.
15. Swing the suspension to the outside and disconnect the driveshafts at the transaxle. For turbocharged vehicles, remove the center axle mount.
16. Remove the bell housing cover.
17. Place a floor jack under the transaxle and support it.
18. Remove the transaxle assembly lower bolts.
19. Remove the transaxle from the engine by pulling it straight away. Do not allow the unit to hang on the shaft.
20. Elevate the transmission into place on the jack. Fit the transaxle straight onto the engine and do not allow the trans to hang on the shaft.
21. Install the transaxle lower bolts.
22. Install the bell housing cover.
23. Install the driveshafts, making sure they have engaged the clips correctly. Install the center bracket for turbocharged vehicles.
24. Connect the tie rod joints and connect the lower ball joints. Tighten the nuts to the proper torque and use new washers and cotter pins.

7-7

7 DRIVE TRAIN

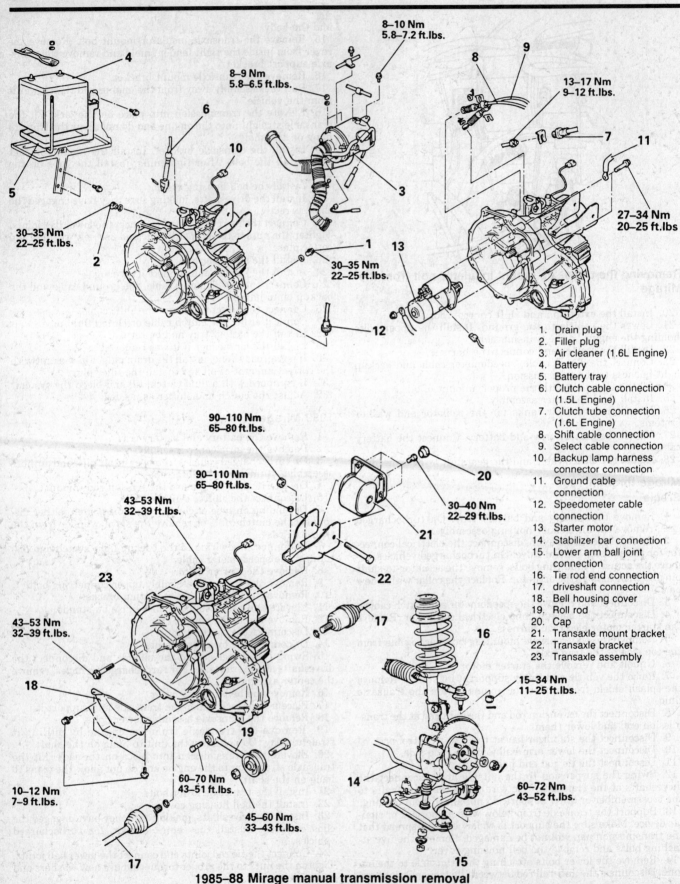

1. Drain plug
2. Filler plug
3. Air cleaner (1.6L Engine)
4. Battery
5. Battery tray
6. Clutch cable connection (1.5L Engine)
7. Clutch tube connection (1.6L Engine)
8. Shift cable connection
9. Select cable connection
10. Backup lamp harness connector connection
11. Ground cable connection
12. Speedometer cable connection
13. Starter motor
14. Stabilizer bar connection
15. Lower arm ball joint connection
16. Tie rod end connection
17. driveshaft connection
18. Bell housing cover
19. Roll rod
20. Cap
21. Transaxle mount bracket
22. Transaxle bracket
23. Transaxle assembly

1985–88 Mirage manual transmission removal

7-8

DRIVE TRAIN 7

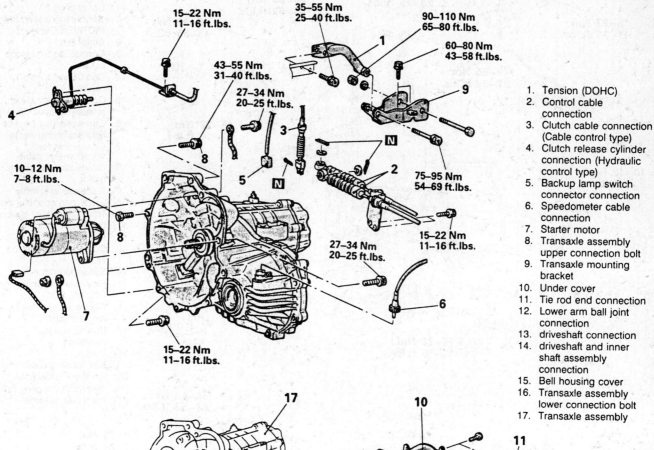

1. Tension (DOHC)
2. Control cable connection
3. Clutch cable connection (Cable control type)
4. Clutch release cylinder connection (Hydraulic control type)
5. Backup lamp switch connector connection
6. Speedometer cable connection
7. Starter motor
8. Transaxle assembly upper connection bolt
9. Transaxle mounting bracket
10. Under cover
11. Tie rod end connection
12. Lower arm ball joint connection
13. driveshaft connection
14. driveshaft and inner shaft assembly connection
15. Bell housing cover
16. Transaxle assembly lower connection bolt
17. Transaxle assembly

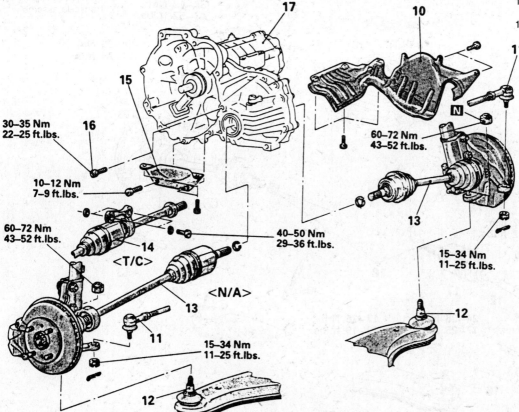

1989 Mirage manual transmission external components

7-9

7 DRIVE TRAIN

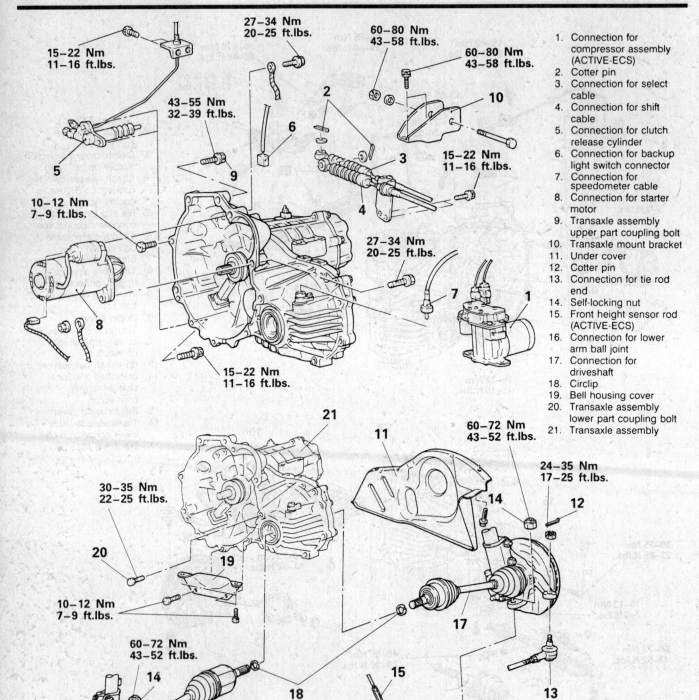

1. Connection for compressor assembly (ACTIVE-ECS)
2. Cotter pin
3. Connection for select cable
4. Connection for shift cable
5. Connection for clutch release cylinder
6. Connection for backup light switch connector
7. Connection for speedometer cable
8. Connection for starter motor
9. Transaxle assembly upper part coupling bolt
10. Transaxle mount bracket
11. Under cover
12. Cotter pin
13. Connection for tie rod end
14. Self-locking nut
15. Front height sensor rod (ACTIVE-ECS)
16. Connection for lower arm ball joint
17. Connection for driveshaft
18. Circlip
19. Bell housing cover
20. Transaxle assembly lower part coupling bolt
21. Transaxle assembly

Galant and Sigma manual transmission

DRIVE TRAIN 7

25. Install the lower covers.
26. Install the transaxle mounting bracket and the upper connecting bolt.
27. Install the starter.
28. Connect the speedometer cable, the ground cable and the backup lamp harness
29. Connect either the clutch cable or clutch slave cylinder.
30. Connect the select cable and the shift cable.
31. For DOHC, install the tension rod.
32. Fill the transaxle with the correct amount of oil.
33. Install the air pipe and air hose on turbocharged vehicles.
34. Install the air cleaner assembly.
35. Install the battery tray and battery. Connect the cables.
36. Adjust the clutch and shifter as needed.

1988–90 Galant and Sigma

1. Disconnect the battery cables; remove the battery and tray.
2. Drain the engine coolant and the transaxle fluid into suitable containers.
3. disconnect the wiring for the airflow sensor, the purge control solenoid and the air cleaner assembly.
4. Disconnect the air intake hose and breather hose; remove the air cleaner assembly.
5. Elevate and safely support the vehicle.
6. If equipped with ACTIVE–ECS, remove the electrical connector from the ECS compressor.
7. Disconnect the transmission control (shifter) cables.
8. Disconnect the the lower radiator hose and the water inlet pipe B. Disconnect all the connections on the water pipe assembly.
9. Disconnect the back-up light switch connector and the engine wire harness connector.
10. place a suitable transaxle jack under the transaxle; remove the transaxle mount bracket cap and the mount bracket.
11. If equipped with ACTIVE–ECS, remove the front height sensor rod.
12. Remove the clutch release cylinder and the clutch tube bracket.
13. Disconnect the speedometer cable assembly.
14. Remove the starter.
15. Remove the lower splash shield.
16. Disconnect the tie rod end from the steering knuckle.
17. Disconnect the lower control arm ball joint.
18. Remove the left side halfshaft and bearing bracket.
19. Remove the right side driveshaft nut. Remove the halfshaft, circlips, bolt, bearing bracket and shaft assembly.
20. Remove the transaxle bracket. Remove the remaining transaxle retaining bolts; pull the transaxle clear of the engine.
21. Install the transaxle unit and install the lower bolts.
22. Install the bell housing cover.
23. Install the driveshaft and center shaft bearings (if present).
24. Connect the lower control arm ball joint.
25. Connect the tie rod end to the steering knuckle.
26. Connect the height sensor rod for the ECS if so equipped.
27. Install the transmission mount bracket.
28. Install the upper transmission to engine retaining bolts and torque them correctly.
29. Install the starter and connect the wiring.
30. Connect the wire harness for the backup switch and the speedometer cable.
31. Attach or connect the clutch release cylinder (slave cylinder).
32. Install the shift and select cables.
33. Install the compressor for the ECS system if so equipped and connect the wiring.

34. Fill the transaxle with the correct amount of fluid.
35. Install the air cleaner assembly. Connect the wiring for the air flow sensor and the purge control solenoid valve.
36. Install the battery and tray. Connect the cables.
37. Connect the lower radiator hose and restore the connections along water supply pipe B.
38. Fill the cooling system with coolant.
39. Adjust the clutch as needed. Check the operation of the shifter.

KM161, KM165 and KM166 4-Speed Overhaul

These manual transaxles have the gear reduction, ratio selection and differential functions combined into 1 unit and housed in the aluminum diecast housing and case.

All of these manual transaxles have a large diameter, large capacity synchronizer which, combined with an improved shift control mechanism, reduces shift operation force and gives comfortable shift feeling. The shift control mechanism includes an interlock device and a reverse shift error prevention device which ensure correct operation.

NOTE: Metric tools will be required to service this transmission. Due to the large number of alloy parts used in this transmission, torque specifications should be strictly observed. Before installing capscrews into aluminum parts, dip the bolts into clean transmission fluid as this will prevent the screws from galling the aluminum threads, thus causing damage.

Do not attempt to interchange metric fasteners for inch system fasteners. Mismatched or incorrect fasteners can cause damage to the automatic transmission unit and possible personal injury. Care should be taken to reuse fasteners in their original position.

WARNING: The use of the correct special tools or their equivalent is REQUIRED for some of these procedures involving disassembly. Additionally, some tools are required which may not be in your tool box. Be prepared to use snapring pliers, a micrometer, small drifts or punches and assorted gear and bearing pullers.

GEARSHIFT LEVER

Removal and Installation

1. Remove the gearshift lever and the range selector lever knobs.
2. Remove the floor console box.
3. Raise the vehicle and support it safely.
4. Remove the 2 retaining nuts from underneath the vehicle.
5. Spread the mouth of the rubber part of the gearshift lever cover and pull the gearshift lever downward and out.
6. Remove the extension from the transaxle.
7. After removing the lock wire, remove the shift rod set screw and then remove the shift rod from the transaxle.
8. Remove the shift rod and the extension as an assembly.

To install:

9. Apply wheel bearing grease to the fulcrum ball surface and to the dust cover inside surface.
10. Install a new lock wire to the set screw of the shift rod to prevent looseness.
11. Install the remaining parts by reversing the removal procedure.

7-11

7 DRIVE TRAIN

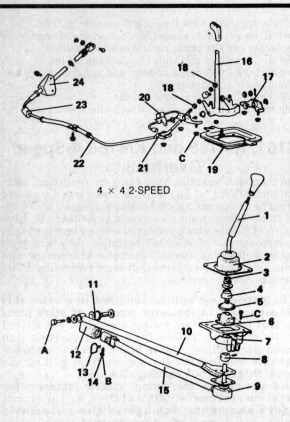

1. Gearshift lever
2. Gearshift lever cover
3. snapring
4. Rubber insulator
5. Spring seat
6. Cap (Vehicles without a turbocharger)
7. Insulator (Vehicles without a turbocharger)
8. Fulcrum ball
9. Cover
10. Extension
11. Bushing
12. Dust cover
13. Lock wire
14. Set screw
15. Shift rod
16. Range selector lever
17. Selector lever mounting bracket
18. Adjusting nut
19. Retainer
20. Selector switch
21. Cable end
22. Selector cable
23. Protector
24. Cable bracket
25. Insulator (Vehicles with a turbocharger)

	Nm	ft. lbs.
A	60–70	43–51
B	33	24
C	12–15	9–11

Typical shifter assemblies

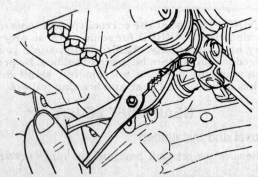

Always install the safety wire. If the bolt falls out, the car is struck in one gear

SELECTOR CABLE

Removal and Installation

1. Remove the gearshift lever and the range selector lever knobs.
2. Remove the floor console box.
3. Remove the adjusting nut and remove the selector cable from the lever.
4. Raise the vehicle and support it safely.
5. Remove the cable bracket, snapring and pin and remove the selector cable from the transaxle.
6. Remove the selector installation bolts from the floor and the crossmember.
7. Pull the selector cable off from the vehicle.

To install:

8. Adjust the selector cable length so that the lever stroke is within the standard value when the range selector lever is placed in the **ECONOMY** position. Then tighten the adjusting nut.
9. Install the selector switch so that the mating marks on the selector switch are on a straight line when the range selector lever is placed in the **ECONOMY** position.
10. The freeplay of the range selector lever end should be 5mm.

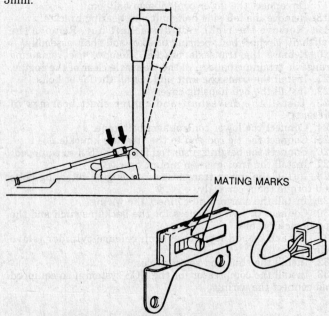

Proper alignment of the select lever

7-12

DRIVE TRAIN 7

BENCH OVERHAUL

Cleanliness is an important factor in the overhaul of the manual transaxle. Before opening up the unit, the entire outside of the transaxle assembly should be cleaned, preferable with a high pressure washer such as a car wash spray unit. Dirt entering the transaxle internal parts will negate all the time and effort spent on the overhaul. During inspection and reassembly all parts should be thoroughly cleaned with solvent then dried with compressed air if available. Wiping cloths and rags should not be used to dry parts.

Wheel bearing grease, long used to hold thrust washers and lube parts, should not be used. Lubricate seals with clean transaxle oil and use ordinary unmedicated petroleum jelly to hold the thrust washers and to ease the assembly of seals, since it will not leave a harmful residue as grease often will. Do not use solvent on neoprene seals, if they are to be reused, or thrust washers.

Before installing bolts into aluminum parts, always dip the threads into clean transaxle oil. Anti-seize compound can also be used to prevent bolts from galling the aluminum and seizing. Always use a torque wrench to keep from stripping the threads. The internal snaprings should be expanded and the external rings should be compressed, if they are to be reused. This will help insure proper seating when installed.

TRANSAXLE CASE DISASSEMBLY

1. Position the assembly in a suitable holding fixture.
2. Remove the transaxle mounting bracket.
3. Remove the clutch cable bracket.
4. Remove the backup light switch and the steel ball.
5. Remove the speedometer gear locking bolt.
6. Remove the speedometer gear assembly from the clutch housing.
7. Remove the transaxle to clutch housing tightening bolt.
8. Remove the transaxle case and gasket.
9. Remove the adapter and gasket.
10. Remove the spacer for differential end play adjustment.
11. Remove the 3 poppet plugs and gasket, then the 3 poppet springs and steel balls.
12. Pull the reverse idler gear shaft out and remove the reverse idler gear.
13. Remove the reverse shift lever from the clutch housing.
14. Remove the distance collar and reverse shift rail.
15. Drive the spring pins from the shift forks and shift rails.
16. Remove the 1st/2nd shift rail from the clutch housing then remove the 3rd/4th shift rail. Remove the 1st/2nd shift rail and fork assembly together with the 3rd/4th shift rail and fork assembly.
17. Remove the 2 interlock plungers from the clutch housing.

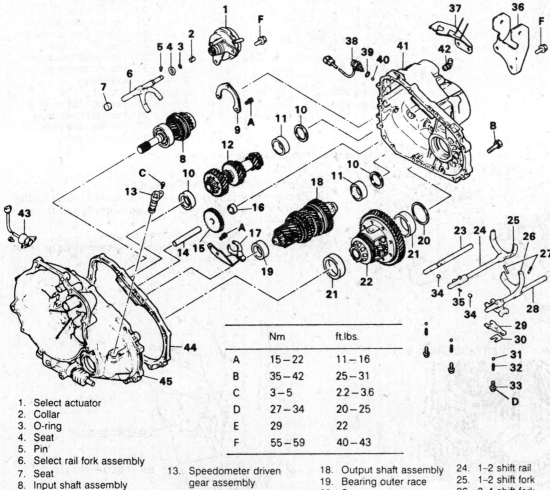

	Nm	ft.lbs.
A	15–22	11–16
B	35–42	25–31
C	3–5	2.2–3.6
D	27–34	20–25
E	29	22
F	55–59	40–43

1. Select actuator
2. Collar
3. O-ring
4. Seat
5. Pin
6. Select rail fork assembly
7. Seat
8. Input shaft assembly
9. Bearing retainer
10. Spacer
11. Bearing outer race
12. Intermediate gear assembly
13. Speedometer driven gear assembly
14. Reverse idler gear shaft
15. Reverse idler gear
16. Distance collar
17. Reverse shift lever
18. Output shaft assembly
19. Bearing outer race
20. Spacer
21. Bearing outer race
22. Differential assembly
23. Reverse shift rail
24. 1-2 shift rail
25. 1-2 shift fork
26. 3-4 shift fork
27. Spring pin
28. 3-4 shift rail
29. 5-speed shift rag
30. Selector spacer
31. Poppet ball
32. Poppet spring
33. Plug
34. Interlock plunger A
35. Interlock plunger B
36. Transaxle bracket
37. Clutch cable bracket
38. Backup light switch
39. Gasket
40. Steel ball
41. Transaxle case
42. Air breather
43. Select switch
44. Adapter
45. Clutch housing assembly

5 speed KM-162

7-13

7 DRIVE TRAIN

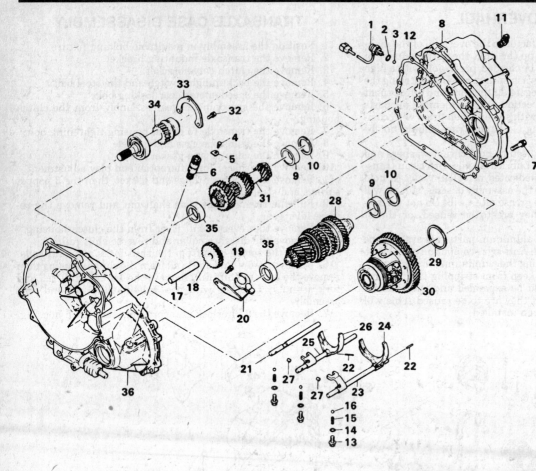

1. Backup light switch
2. Gasket
3. Steel ball
4. Bolt
5. Locking plate
6. Speedometer gear assembly
7. Bolt (13)
8. Transaxle case
9. Bearing outer race (2)
10. Spacer (2) — select
11. Air breather
12. Transaxle case gasket
13. Poppet plug (3)
14. Gasket (3)
15. Poppet spring (3)
16. Poppet ball (3)
17. Reverse idler gear shaft
18. Reverse idler gear
19. Bolt (2)
20. Reverse shift lever assembly
21. Reverse shift rail
22. Spring pin (2)
23. 1-2 shift rail
24. 1-2 shift fork
25. 3-4 shift rail
26. 3-4 shift fork
27. Interlock plunger (2)
28. Output shaft assembly
29. Spacer — select
30. Differential assembly
31. Intermediate gear assembly
32. Bolt (3)
33. Bearing retainer
34. Input shaft assembly
35. Bearing outer race
36. Clutch housing

4 speed KM-161 manual transmission

18. Remove the output shaft assembly.
19. Remove the differential assembly from the clutch housing.
20. Remove the poppet plug, the poppet spring and the steel ball for the select shift rail.
21. Remove the input shaft bearing retainer.
22. Remove the input shaft assembly together with the select shift fork, rail and intermediate shaft.

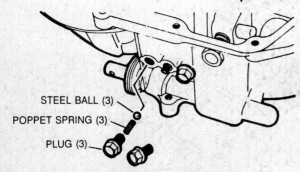

Use care during disassembly

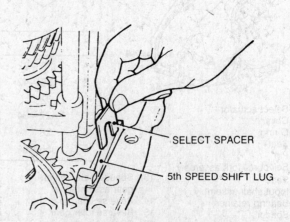

Select the correct spacer to prevent damage

7-14

DRIVE TRAIN 7

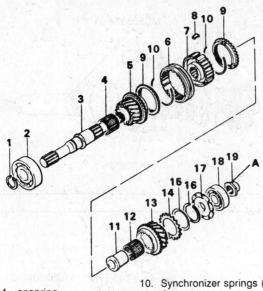

1. snapring
2. Front bearing
3. Input shaft
4. Needle bearing
5. Input low gear
6. Synchronizer sleeve
7. Synchronizer hub
8. Synchronizer key (3)
9. Synchronizer ring (2)
10. Synchronizer springs (2)
11. Gear sleeve
12. Needle bearing
13. Input high gear
14. Sub gear
15. Cone spring
16. snapring
17. Oil slinger
18. Rear bearing
19. Locking nut

Input shaft components; KM-I63 shown

INPUT SHAFT

Disassembly

1. Clamp the input shaft in a vise with the locknut side up protecting the shaft splines from the vise jaws.
2. Remove the staking from the locking nut lock with a blunt punch, then remove the locknut.
3. Remove the front bearing snapring.
4. With the front bearing supported on a suitable fixture, press the input shaft and remove the front bearing.
5. With the input low gear supported on a press base, press the input shaft down to remove the rear bearing, oil slinger, gears and synchronizer.

Inspection

1. Check the outer surface of the input shaft where the needle bearing is mounted for damage, abnormal wear and seizure.
2. Check the input shaft splines for damage and wear.
3. Combine the needle bearing with the shaft or bearing sleeve and gear and check that it rotates smoothly without abnormal noise or play.
4. Check the needle bearing cage for deformation.
5. Check the synchronizer ring clutch gear teeth for damage and breakage.
6. Check the internal surface for damage, wear and broken threads.

Assembly

1. Assemble the synchronizer hub and sleeve.
2. Install the 3 synchronizer keys into the grooves of the hub.
3. Install the 2 synchronizer springs and make sure that they are installed in opposite directions.
4. Install the sub gear to the input high gear. Apply gear oil to the entire surface of the gear.
5. Install the cone spring in the proper direction.

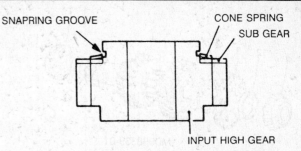

The cone spring must be installed correctly

6. Install a new snapring. Make sure the inner side of the cone spring is not in the snapring groove.
7. Install the needle bearing for the input low gear to the input shaft.
8. Install the input low gear onto the input shaft.
9. Install the synchronizer ring onto the cone portion of the input low gear.
10. Install the assembled synchronizer to the input shaft and then press the synchronizer assembly onto the input shaft.
11. Press the input high gear sleeve onto the input shaft.
12. Install the needle bearing onto the sleeve.
13. Install the synchronizer ring.
14. Install the assembled input high gear onto the needle bearing.
15. Install the oil slinger onto the input shaft.
16. Press the rear bearing onto the input shaft rear end.
17. Tighten the input shaft rear end locknut to 66–79 ft. lbs. (89–107 Nm) and stake it. Stake the locknut only at the notch of the shaft.
18. Install the front bearing onto the input shaft.
19. Install the front bearing snapring.

NOTE: There are 3 types of snaprings available which differ in thickness. Select the thickest snapring which fits in the snapring groove.

OUTPUT SHAFT

Disassembly

1. Remove the taper roller bearing inner race from the output shaft.
2. Holding the 2nd speed gear with the proper tool, press the rear end of the shaft with a press to remove the 1st speed gear, gear sleeve, 1st/2nd synchronizer assembly and 2nd speed bear.
3. Holding the 4th speed gear with the proper tool, press the rear end of the shaft with a press to remove the 2nd speed gear sleeve, 3rd speed gear, sleeve, 3rd/4th synchronizer assembly and 4th speed gear.

Inspection

1. Check the outer surface of the output shaft for damage, abnormal wear and seizure.
2. Check the output shaft splines for damage and wear.
3. Combine the bearing with the shaft or bearing sleeve and gear and check that it rotates smoothly without abnormal noise or play.
4. Check the bearing cage for deformation.
5. Check the synchronizer ring clutch gear teeth for damage and breakage.
6. Check the internal surface for damage, wear and broken threads.

Assembly

1. Assemble the synchronizer hub and sleeve. Make sure they slide smoothly.

7–15

7 DRIVE TRAIN

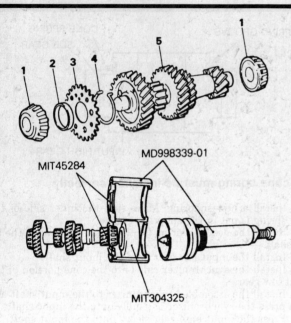

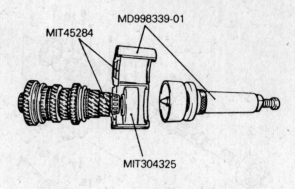

1. Taper roller bearing inner race (2)
2. Spacer
3. Sub gear
4. Spring
5. Intermediate gear

Disassembly of the intermediate shaft

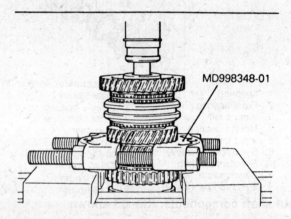

Special tools are required to disassemble the output shaft components

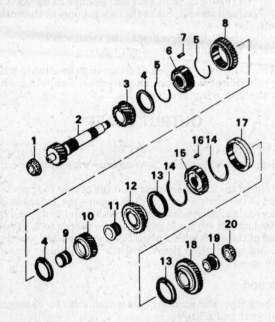

1. Taper roller bearing inner race
2. Output shaft
3. 4th speed gear
4. Synchronizer ring (2)
5. Synchronizer springs (2)
6. 3-4 Synchronizer hub
7. Synchronizer key (3)
8. 3-4 Synchronizer sleeve
9. Gear sleeve
10. 3rd speed gear
11. Gear sleeve
12. 2nd speed gear
13. Synchronizer ring (2)
14. Synchronizer spring (2)
15. 1-2 Synchronizer hub
16. Synchronizer key (3)
17. 1-2 Synchronizer sleeve
18. 1st speed gear
19. Gear sleeve
20. Taper roller bearing inner race

Output shaft components

2. Insert the 3 synchronizer keys into the grooves of the hub.
3. Install the 2 synchronizer springs in opposite directions.
4. Install the 4th gear onto the output shaft.
5. Install the synchronizer ring onto the cone portion of the 4th speed gear.
6. Using tool MD998323, press the 3rd/4th synchronizer onto the output shaft. Make sure that the keyways of the synchronizer ring are correctly aligned with the synchronizer keys. After installation of the synchronizer, check that the 4th speed gear rotates smoothly.
7. Using tool MD998323, press the gear sleeve for the 3rd speed gear onto the output shaft.
8. Install the synchronizer ring onto the 3rd/4th synchronizer.
9. Install the 3rd speed gear onto the output shaft.
10. Using tool MD998323, press the gear sleeve for the 2nd speed gear onto the output shaft.
11. Install the 2nd speed gear onto the output shaft.
12. Install the synchronizer ring onto the cone portion of the 2nd speed gear.
13. Using tool MD998323, press the 1st/2nd synchronizer onto the output shaft. Make sure that the keyways of the synchronizer ring are correctly aligned with the synchronizer keys.
14. Install the synchronizer ring and the 1st speed gear.
15. Install the 1st speed gear sleeve onto the output shaft.
16. Press the taper roller bearings onto the front and rear ends of the output shaft.

DRIVE TRAIN 7

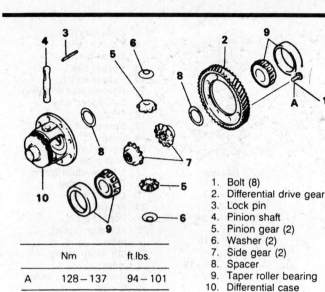

	Nm	ft.lbs.
A	128–137	94–101

1. Bolt (8)
2. Differential drive gear
3. Lock pin
4. Pinion shaft
5. Pinion gear (2)
6. Washer (2)
7. Side gear (2)
8. Spacer
9. Taper roller bearing
10. Differential case

Differential components

Choose the spacers which will give the correct backlash; measure carefully

DIFFERENTIAL/FINAL DRIVE

Disassembly

1. Remove the drive gear retaining bolts and remove the drive gear from the differential case.
2. Remove the taper roller bearing inner race.
3. Drive out the lock pin with a punch.
4. Pull out the pinion shaft and then remove the pinion gears and washers.
5. Remove the side gears and spacers.

Assembly

1. With the spacers installed to the back of the differential side gears, install the gears in the differential case. If using new differential side gears, install spacers of medium thickness 1.0mm.
2. Install the washers to the back of the pinion gears and install the gears in the differential case, then install the pinion shaft.
3. Measure the backlash between the differential side gear and the pinion. The backlash should be 0–0.08mm. The backlash of both right and left gears must be equal.
4. Install the pinion shaft lock pin and check to ensure that the projection is less than 3mm.
5. Press the taper roller bearing inner races onto both ends of the differential case.
6. Install the drive gear onto the case.
7. Apply a small amount of automatic transmission fluid and then a generous amount of thread lock cement to the ring gear retaining bolts. Assemble and tighten the bolts to 94–101 ft. lbs. (128–137 Nm).

CLUTCH HOUSING

Disassembly

1. Remove the shift shaft spring pin using pliers.
2. Using the shift shaft hole, remove the shift shaft.
3. Cover the bolts to prevent the poppet ball from jumping out and being lost when the shift shaft is removed. Remove the control finger and the poppet ball.
4. Using a pin punch, remove the select finger lock pin. Remove the select shaft and select finger.

Assembly

1. Apply sealant to the outside surface of the new control shaft oil seal. Install the oil seal.
2. Install a new input shaft oil seal using tool MD998321.
3. Install the select shaft into the housing and install the finger.
4. With the select finger and shaft lock pin holes properly aligned, drive in the lock pin.
5. Install the poppet spring and the steel ball in the control finger. The poppet spring that is used should not have an identification color of white and should have a free length of 19mm.

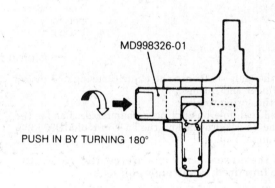

PUSH IN BY TURNING 180°

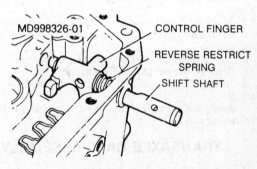

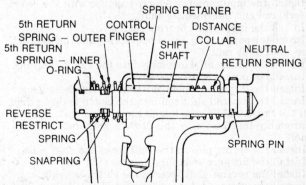

Correct installation of shift shaft and related components

7 DRIVE TRAIN

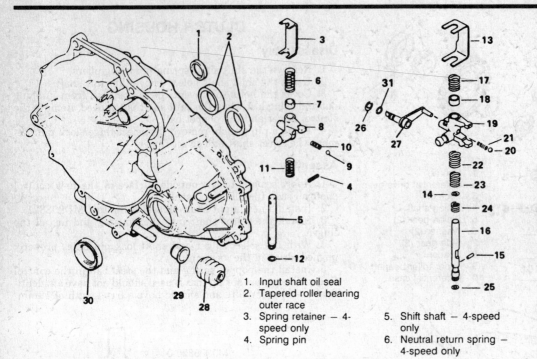

1. Input shaft oil seal
2. Tapered roller bearing outer race
3. Spring retainer — 4-speed only
4. Spring pin
5. Shift shaft — 4-speed only
6. Neutral return spring — 4-speed only
7. Distance collar
8. Control finger — 4-speed only
9. Poppet ball
10. Poppet spring
11. Reverse restrict spring — 4-speed only
12. O-ring
13. Spring retainer
14. Snapring
15. Spring pin
16. Shift shaft
17. Neutral return spring
18. Distance collar
19. Control finger
20. Poppet ball
21. Poppet spring
22. 5th return spring — outer
23. 5th return spring — inner
24. Reverse restrict spring
25. O-ring
26. Snapring
27. Selector lever
28. Boot
29. Control shaft oil seal
30. Driveshaft oil seal
31. O-ring

Clutch housing

6. Insert the shift shaft into the clutch housing and install the control finger and springs.
7. Install the spring retainer.
8. If the oil seal hole has not been chamfered, chamfer the hole using the knurls provided on the back of tool MD998325, before installing the oil seal.

NOTE: If the oil seal is not chamfered, the edge of the seal will be impaired, resulting in oil leaks.

9. Replace the control shaft or control lug using the following procedure:
    a. Make a 12mm diameter hole in the 16mm diameter boss located at the lower part of the engine side of the clutch housing.
    b. Using tool MD998019, remove the lock pin.
    c. Remove the control shaft and lug.
    d. Install the new control shaft and lug and drive in a new lock pin.
    e. Apply sealant to the outside surface of the 12mm diameter sealing cap.

TRANSAXLE CASE ASSEMBLY

1. Install the input shaft assembly together with the select shift fork, rail and intermediate shaft.
2. Install the input shaft bearing retainer.
3. Install the poppet plug, the poppet spring and the steel ball for the select shift rail.
4. Install the differential assembly into the clutch housing.
5. Install the output shaft assembly.
6. Install the 2 interlock plungers into the clutch housing.
7. Install the 1st/2nd shift rail into the clutch housing then install the 3rd/4th shift rail. Install the 1st/2nd shift rail and fork assembly together with the 3rd/4th shift rail and fork assembly.
8. Install the spring pins in the shift forks and shift rails.
9. Install the distance collar and reverse shift rail.
10. Install the reverse shift lever in the clutch housing.
11. Install the reverse idler gear shaft and the reverse idler gear.
12. Install the 3 poppet plugs and gasket, then the 3 poppet springs and steel balls.
13. Install the spacer for differential end play adjustment.
14. Install the adapter and gasket.
15. Install the transaxle case and gasket.
16. Install the transaxle to clutch housing tightening bolt.
17. Install the speedometer gear assembly into the clutch housing.
18. Install the speedometer gear locking bolt.
19. Install the backup light switch and the steel ball.
20. Install the clutch cable bracket.
21. Install the transaxle mounting bracket.

KM162 and KM163 5-Speed Overhaul

These manual transaxles have the gear reduction, ratio selection and differential functions combined into 1 unit and housed in the aluminum diecast housing and case.

All of these manual transaxles have a large diameter, large capacity synchronizer which, combined with an improved shift control mechanism, reduces shift operation force and gives comfortable shift feeling. The shift control mechanism includes an interlock device and a reverse shift error prevention device which ensure correct operation.

NOTE: Metric tools will be required to service this transmission. Due to the large number of alloy parts used in this transmission, torque specifications should be strictly observed. Before installing capscrews into aluminum parts, dip the bolts into clean transmission fluid as this will prevent the screws from galling the aluminum threads, thus causing damage.

Do not attempt to interchange metric fasteners for inch system fasteners. Mismatched or incorrect fasteners can cause damage to the automatic transmission unit and possible personal injury. Care should be taken to reuse fasteners in their original position.

DRIVE TRAIN 7

WARNING: The use of the correct special tools or their equivalent is REQUIRED for some of these procedures involving disassembly. Additionally, some tools are required which may not be in your tool box. Be prepared to use snapring pliers, a micrometer, small drifts or punches and assorted gear and bearing pullers.

GEARSHIFT LEVER

Removal and Installation

1. Remove the gearshift lever and the range selector lever knobs.
2. Remove the floor console box.
3. Raise the vehicle and support it safely.
4. Remove the 2 retaining nuts from underneath the vehicle.
5. Spread the mouth of the rubber part of the gearshift lever cover and pull the gearshift lever downward and out.
6. Remove the extension from the transaxle.
7. After removing the lock wire, remove the shift rod set screw and then remove the shift rod from the transaxle.
8. Remove the shift rod and the extension as an assembly.

To install:

9. Apply wheel bearing grease to the fulcrum ball surface and to the dust cover inside surface.
10. Install a new lock wire to the set screw of the shift rod to prevent looseness.
11. Install the remaining parts by reversing the removal procedure.

SELECTOR CABLE

Removal and Installation

1. Remove the gearshift lever and the range selector lever knobs.
2. Remove the floor console box.
3. Remove the adjusting nut and remove the selector cable from the lever.
4. Raise the vehicle and support it safely.
5. Remove the cable bracket, snapring and pin and remove the selector cable from the transaxle.
6. Remove the selector installation bolts from the floor and the crossmember.
7. Pull the selector cable off from the vehicle.

To install:

8. Adjust the selector cable length so that the lever stroke is within the standard value when the range selector lever is placed in the **ECONOMY** position. Then tighten the adjusting nut.
9. Install the selector switch so that the mating marks on the selector switch are on a straight line when the range selector lever is placed in the **ECONOMY** position.
10. The freeplay of the range selector lever end should be 5mm.

BENCH OVERHAUL

Cleanliness is an important factor in the overhaul of the manual transaxle. Before opening up the unit, the entire outside of the transaxle assembly should be cleaned, preferable with a high pressure washer such as a car wash spray unit. Dirt entering the transaxle internal parts will negate all the time and effort spent on the overhaul. During inspection and reassembly all parts should be thoroughly cleaned with solvent then dried with compressed air if available. Wiping cloths and rags should not be used to dry parts.

Wheel bearing grease, long used to hold thrust washers and lube parts, should not be used. Lubricate seals with clean transaxle oil and use ordinary unmedicated petroleum jelly to hold the thrust washers and to ease the assembly of seals, since it will not leave a harmful residue as grease often will. Do not use solvent on neoprene seals, if they are to be reused, or thrust washers.

Before installing bolts into aluminum parts, always dip the threads into clean transaxle oil. Anti-seize compound can also be used to prevent bolts from galling the aluminum and seizing. Always use a torque wrench to keep from stripping the threads. The internal snaprings should be expanded and the external rings should be compressed, if they are to be reused. This will help insure proper seating when installed.

TRANSAXLE DISASSEMBLY

1. Position the assembly in a suitable holding fixture.
2. Remove the transaxle mounting bracket.
3. Remove the 2 select actuator mounting bolts.
4. Remove the select switch harness clamp and backup light switch.
5. Remove the backup light switch and the steel ball.
6. Remove the select switch.
7. Pull the select actuator to remove its rod and shift rail connection from the transaxle.
8. Remove the snapring and pin, then remove the select actuator.
9. Remove the speedometer gear locking bolt.
10. Remove the speedometer gear assembly from the clutch housing.
11. Remove the transaxle to clutch housing tightening bolt.
12. Remove the transaxle case and gasket.
13. Remove the adapter and gasket.
14. Remove the spacer for differential end play adjustment.
15. Remove the 3 poppet plugs and gasket, then the 3 poppet springs and steel balls.
16. Pull the reverse idler gear shaft out and remove the reverse idler gear.
17. Remove the reverse shift lever from the clutch housing.
18. Remove the distance collar and reverse shift rail.
19. Drive the spring pins from the shift forks and shift rails.
20. Remove the select spacer.
21. Remove the 1st/2nd shift rail from the clutch housing then remove the 3rd/4th shift rail. Remove the 1st/2nd shift rail and fork assembly together with the 3rd/4th shift rail and fork assembly.
22. Remove the 2 interlock plungers from the clutch housing.
23. Remove the 5th speed shift lug.
24. Remove the output shaft assembly.
25. Remove the differential assembly from the clutch housing.
26. Remove the poppet plug, the poppet spring and the steel ball for the select shift rail.
27. Remove the input shaft bearing retainer.
28. Remove the input shaft assembly together with the select shift fork, rail and intermediate shaft.

INPUT SHAFT

Disassembly

1. Clamp the input shaft in a vise with the locknut side up protecting the shaft splines from the vise jaws.
2. Remove the staking from the locking nut lock with a blunt punch, then remove the locknut.
3. Remove the front bearing snapring.
4. With the front bearing supported on a suitable fixture, press the input shaft and remove the front bearing.
5. With the input low gear supported on a press base, press the input shaft down to remove the rear bearing, oil slinger, gears and synchronizer.

7-19

7 DRIVE TRAIN

Inspection

1. Check the outer surface of the input shaft where the needle bearing is mounted for damage, abnormal wear and seizure.
2. Check the input shaft splines for damage and wear.
3. Combine the needle bearing with the shaft or bearing sleeve and gear and check that it rotates smoothly without abnormal noise or play.
4. Check the needle bearing cage for deformation.
5. Check the synchronizer ring clutch gear teeth for damage and breakage.
6. Check the internal surface for damage, wear and broken threads.

Assembly

1. Assemble the synchronizer hub and sleeve.
2. Install the 3 synchronizer keys into the grooves of the hub.
3. Install the 2 synchronizer springs and make sure that they are installed in opposite directions.
4. Install the sub gear to the input high gear. Apply gear oil to the entire surface of the gear.
5. Install the cone spring in the proper direction.
6. Install a new snapring. Make sure the inner side of the cone spring is not in the snapring groove.
7. Install the needle bearing for the input low gear to the input shaft.
8. Install the input low gear onto the input shaft.
9. Install the synchronizer ring onto the cone portion of the input low gear.
10. Install the assembled synchronizer to the input shaft and then press the synchronizer assembly onto the input shaft.
11. Press the input high gear sleeve onto the input shaft.
12. Install the needle bearing onto the sleeve.
13. Install the synchronizer ring.
14. Install the assembled input high gear onto the needle bearing.
15. Install the oil slinger onto the input shaft.
16. Press the rear bearing onto the input shaft rear end.
17. Tighten the input shaft rear end locknut to 66–79 ft. lbs. (89–107 Nm) and stake it. Stake the locknut only at the notch of the shaft.
18. Install the front bearing onto the input shaft.
19. Install the front bearing snapring.

NOTE: There are 3 types of snaprings available which differ in thickness. Select the thickest snapring which fits in the snapring groove.

OUTPUT SHAFT

Disassembly

1. Remove the taper roller bearing inner race from the output shaft.
2. Holding the 2nd speed gear with the proper tool, press the rear end of the shaft with a press to remove the 1st speed gear, gear sleeve, 1st/2nd synchronizer assembly and 2nd speed bear.
3. Holding the 4th speed gear with the proper tool, press the rear end of the shaft with a press to remove the 2nd speed gear sleeve, 3rd speed gear, sleeve, 3rd/4th synchronizer assembly and 4th speed gear.
4. Use tool MD998355 or equivalent to remove the 4th speed gear according to the following steps:
   a. Holding the 3rd speed gear in tool MD998355, remove the 2nd speed gear sleeve and the 3rd speed gear.
   b. Holding the 3rd/4th synchronizer ring in a gear puller, remove the 3rd speed gear sleeve and the 3rd/4th synchronizer.

Inspection

1. Check the outer surface of the output shaft for damage, abnormal wear and seizure.
2. Check the output shaft splines for damage and wear.
3. Combine the bearing with the shaft or bearing sleeve and gear and check that it rotates smoothly without abnormal noise or play.
4. Check the bearing cage for deformation.
5. Check the synchronizer ring clutch gear teeth for damage and breakage.
6. Check the internal surface for damage, wear and broken threads.

Assembly

1. Assemble the synchronizer hub and sleeve. Make sure they slide smoothly.
2. Insert the 3 synchronizer keys into the grooves of the hub.
3. Install the 2 synchronizer springs in opposite directions.
4. Install the 4th gear onto the output shaft.
5. Install the synchronizer ring onto the cone portion of the 4th speed gear.
6. Using tool MD998323, press the 3rd/4th synchronizer onto the output shaft. Make sure that the keyways of the synchronizer ring are correctly aligned with the synchronizer keys. After installation of the synchronizer, check that the 4th speed gear rotates smoothly.
7. Using tool MD998323, press the gear sleeve for the 3rd speed gear onto the output shaft.
8. Install the synchronizer ring onto the 3rd/4th synchronizer.
9. Install the 3rd speed gear onto the output shaft.
10. Using tool MD998323, press the gear sleeve for the 2nd speed gear onto the output shaft.
11. Install the 2nd speed gear onto the output shaft.
12. Install the synchronizer ring onto the cone portion of the 2nd speed gear.
13. Using tool MD998323, press the 1st/2nd synchronizer onto the output shaft. Make sure that the keyways of the synchronizer ring are correctly aligned with the synchronizer keys.
14. Install the synchronizer ring and the 1st speed gear.
15. Install the 1st speed gear sleeve onto the output shaft.
16. Press the taper roller bearings onto the front and rear ends of the output shaft.

DIFFERENTIAL/FINAL DRIVE

Disassembly

1. Remove the drive gear retaining bolts and remove the drive gear from the differential case.
2. Remove the taper roller bearing inner race.
3. Drive out the lock pin with a punch.
4. Pull out the pinion shaft and then remove the pinion gears and washers.
5. Remove the side gears and spacers.

Assembly

1. With the spacers installed to the back of the differential side gears, install the gears in the differential case. If using new differential side gears, install spacers of medium thickness 1.0mm.
2. Install the washers to the back of the pinion gears and install the gears in the differential case, then install the pinion shaft.
3. Measure the backlash between the differential side gear and the pinion. The backlash should be 0–0.08mm. The backlash of both right and left gears must be equal.
4. Install the pinion shaft lock pin and check to ensure that the projection is less than 3mm.
5. Press the taper roller bearing inner races onto both ends of the differential case.
6. Install the drive gear onto the case.
7. Apply a small amount of automatic transmission fluid and then a generous amount of thread lock cement to the ring gear

DRIVE TRAIN 7

retaining bolts. Assemble and tighten the bolts to 94–101 ft. lbs. (128–137 Nm).

CLUTCH HOUSING

Disassembly

1. Remove the shift shaft spring pin using pliers.
2. Using the shift shaft hole, remove the shift shaft.
3. Cover the bolts to prevent the poppet ball from jumping out and being lost when the shift shaft is removed. Remove the control finger and the poppet ball.
4. Using a pin punch, remove the select finger lock pin. Remove the select shaft and select finger.

Assembly

1. Apply sealant to the outside surface of the new control shaft oil seal. Install the oil seal.
2. Install a new input shaft oil seal using tool MD998321.
3. Install the select shaft into the housing and install the finger.
4. With the select finger and shaft lock pin holes properly aligned, drive in the lock pin.
5. Install the poppet spring and the steel ball in the control finger. The poppet spring that is used should not have an identification color of white and should have a free length of 19mm.
6. Insert the shift shaft into the clutch housing and install the control finger and springs.
7. Install the spring retainer.
8. If the oil seal hole has not been chamfered, chamfer the hole using the knurls provided on the back of tool MD998325, before installing the oil seal.

NOTE: If the oil seal is not chamfered, the edge of the seal will be impaired, resulting in oil leaks.

9. Replace the control shaft or control lug using the following procedure:
   a. Make a 12mm diameter hole in the 16mm diameter boss located at the lower part of the engine side of the clutch housing.
   b. Using tool MD998019, remove the lock pin.
   c. Remove the control shaft and lug.
   d. Install the new control shaft and lug and drive in a new lock pin.
   e. Apply sealant to the outside surface of the 12mm diameter sealing cap.

TRANSAXLE CASE ASSEMBLY

1. Install the input shaft assembly together with the select shift fork, rail and intermediate shaft.
2. Install the input shaft bearing retainer.
3. Install the poppet plug, the poppet spring and the steel ball for the select shift rail.
4. Install the differential assembly into the clutch housing.
5. Install the output shaft assembly.
6. Install the 5th speed shift lug.
7. Install the 2 interlock plungers into the clutch housing.
8. Install the 1st/2nd shift rail into the clutch housing then install the 3rd/4th shift rail. Install the 1st/2nd shift rail and fork assembly together with the 3rd/4th shift rail and fork assembly.
9. Install the select spacer.
10. Install the spring pins in the shift forks and shift rails.
11. Install the distance collar and reverse shift rail.
12. Install the reverse shift lever in the clutch housing.
13. Install the reverse idler gear shaft and the reverse idler gear.
14. Install the 3 poppet plugs and gasket, then the 3 poppet springs and steel balls.
15. Install the spacer for differential end play adjustment.
16. Install the adapter and gasket.
17. Install the transaxle case and gasket.
18. Install the transaxle to clutch housing tightening bolt.
19. Install the speedometer gear assembly into the clutch housing.
20. Install the speedometer gear locking bolt.
21. Install the snapring and pin, then install the select actuator.
22. Install the select actuator and its rod and shift rail connection.
23. Install the select switch.
24. Install the backup light switch and the steel ball.
25. Install the select switch harness clamp and backup light switch.
26. Install the 2 select actuator mounting bolts.
27. Install the transaxle mounting bracket.

KM200 4-Speed Overhaul

This manual transaxle has the gear reduction, ratio selection and differential functions combined into 1 unit and housed in the aluminum diecast housing and case.

This manual transaxles has a large diameter, large capacity synchronizer which, combined with an improved shift control mechanism, reduces shift operation force and gives comfortable shift feeling. The shift control mechanism includes an interlock device and a reverse shift error prevention device which ensure correct operation.

NOTE: Metric tools will be required to service this transmission. Due to the large number of alloy parts used in this transmission, torque specifications should be strictly observed. Before installing capscrews into aluminum parts, dip the bolts into clean transmission fluid as this will prevent the screws from galling the aluminum threads, thus causing damage.

Do not attempt to interchange metric fasteners for inch system fasteners. Mismatched or incorrect fasteners can cause damage to the automatic transmission unit and possible personal injury. Care should be taken to reuse fasteners in their original position.

WARNING: The use of the correct special tools or their equivalent is REQUIRED for some of these procedures involving disassembly. Additionally, some tools are required which may not be in your tool box. Be prepared to use snapring pliers, a micrometer, small drifts or punches and assorted gear pullers.

SELECT CABLE

1. Raise the vehicle and support it safely.
2. Move the transaxle shift lever to the **N** position.

NOTE: The select lever will be set to the N position when the transaxle shift lever is moved to the neutral position.

3. Move lever **B** to the **N** position.
4. Adjust the cable by using the adjuster, so that the end of the select cable is properly positioned relative to the lever **B**.
5. The flange side of the resin bushing at the select cable end should be at the lever **B** end surface.

7-21

7 DRIVE TRAIN

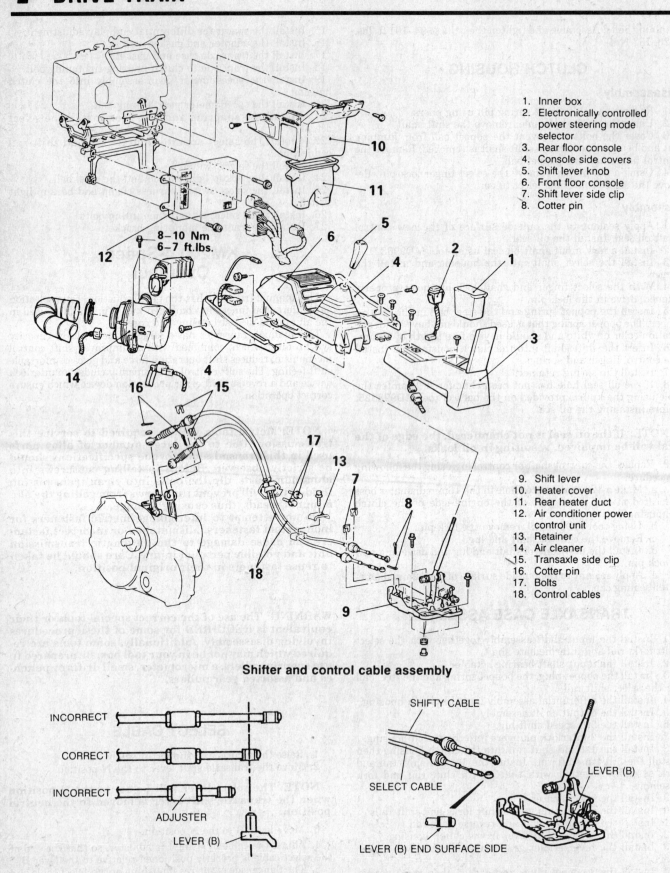

Shifter and control cable assembly

1. Inner box
2. Electronically controlled power steering mode selector
3. Rear floor console
4. Console side covers
5. Shift lever knob
6. Front floor console
7. Shift lever side clip
8. Cotter pin
9. Shift lever
10. Heater cover
11. Rear heater duct
12. Air conditioner power control unit
13. Retainer
14. Air cleaner
15. Transaxle side clip
16. Cotter pin
17. Bolts
18. Control cables

Proper alignment of the select cable. The same relationship is required for the shift cable

7-22

DRIVE TRAIN 7

SHIFT CABLE

1. Elevate the vehicle and support it safely.
2. With the select lever in **N**, move the transaxle shift lever downward to set it to 4th gear.

NOTE: If the shift lever does not move easily, depress and hold the clutch pedal.

3. From inside the vehicle, move the shift lever downward until it contacts the stopper.
4. Adjust the cable by using the adjuster, so that the end of the shift cable is positioned correctly relative to the shift lever (interior side).
5. The flange side of the resin bushing at the shift cable end should be at the split pin side.
6. Adjust the length of the shift cable by using the adjuster, so that the clearance **A** and **B** between the shift lever and the 2 stoppers are equal when the shift lever is shifted to 3rd gear and 4th gear.

BENCH OVERHAUL

Cleanliness is an important factor in the overhaul of the manual transaxle. Before opening up this unit, the entire outside of the transaxle assembly should be cleaned, preferably with a high pressure washer such as a car wash spray unit. Dirt entering the transaxle internal parts will negate all the time and effort spent on the overhaul. During inspection and reassembly all parts should be thoroughly cleaned with solvent then dried with compressed air. Wiping cloths and rags should not be used to dry parts.

Wheel bearing grease, long used to hold thrust washers and lube parts, should not be used. Lubricate seals with clean transaxle oil and use ordinary unmedicated petroleum jelly to hold the thrust washers and to ease the assembly of seals, since it will not leave a harmful residue as grease often will. Do not use solvent on neoprene seals, if they are to be reused, or thrust washers.

Before installing bolts into aluminum parts, always dip the threads into clean transaxle oil. Antiseize compound can also be used to prevent bolts from galling the aluminum and seizing. Always use a torque wrench to keep from stripping the threads. The internal snaprings should be expanded and the external rings should be compressed, if they are to be reused. This will help insure proper seating when installed.

TRANSAXLE CASE DISASSEMBLY

1. Position the assembly in a suitable holding fixture.
2. Remove the rear cover.
3. Remove the backup light switch, gasket and steel ball.
4. Remove the poppet plug, spring and steel ball.
5. Remove the speedometer driven gear assembly.
6. Remove the air breather.
7. Remove the spring pin using a pin punch.

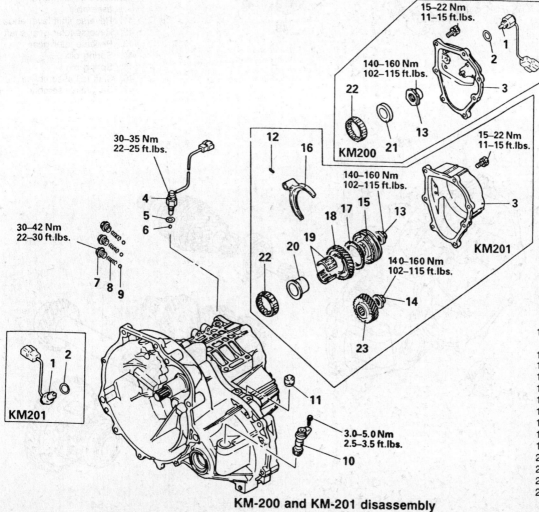

1. Transaxle switch
2. Gasket
3. Rear cover
4. Back-up light switch
5. Gasket
6. Steel ball
7. Poppet plug
8. Poppet spring
9. Poppet spring
10. Speedometer driven gear assembly
11. Air breather
12. Spring pin
13. Locknut
14. Locknut
15. 5th speed synchronizer
16. 5th speed shift fork
17. Synchronizer ring
18. 5th speed gear
19. Needle bearing
20. Bearing sleeve
21. Dished washer
22. Roller bearing
23. 5th speed intermediate gear

KM-200 and KM-201 disassembly

7 DRIVE TRAIN

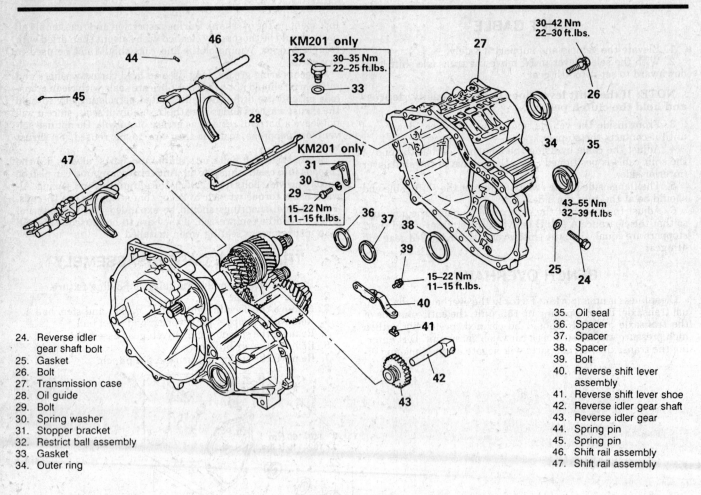

24. Reverse idler gear shaft bolt
25. Gasket
26. Bolt
27. Transmission case
28. Oil guide
29. Bolt
30. Spring washer
31. Stopper bracket
32. Restrict ball assembly
33. Gasket
34. Outer ring
35. Oil seal
36. Spacer
37. Spacer
38. Spacer
39. Bolt
40. Reverse shift lever assembly
41. Reverse shift lever shoe
42. Reverse idler gear shaft
43. Reverse idler gear
44. Spring pin
45. Spring pin
46. Shift rail assembly
47. Shift rail assembly

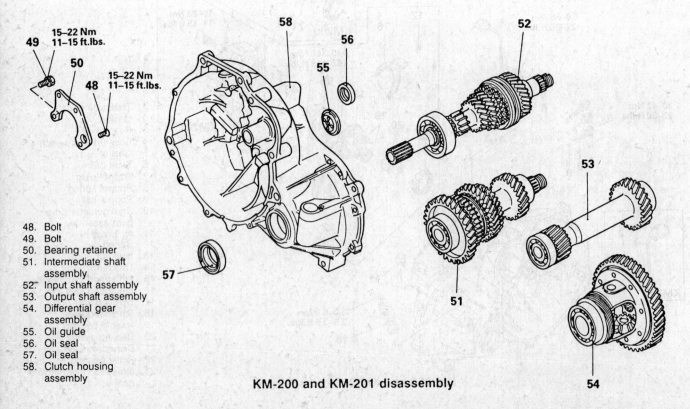

48. Bolt
49. Bolt
50. Bearing retainer
51. Intermediate shaft assembly
52. Input shaft assembly
53. Output shaft assembly
54. Differential gear assembly
55. Oil guide
56. Oil seal
57. Oil seal
58. Clutch housing assembly

KM-200 and KM-201 disassembly

7-24

DRIVE TRAIN 7

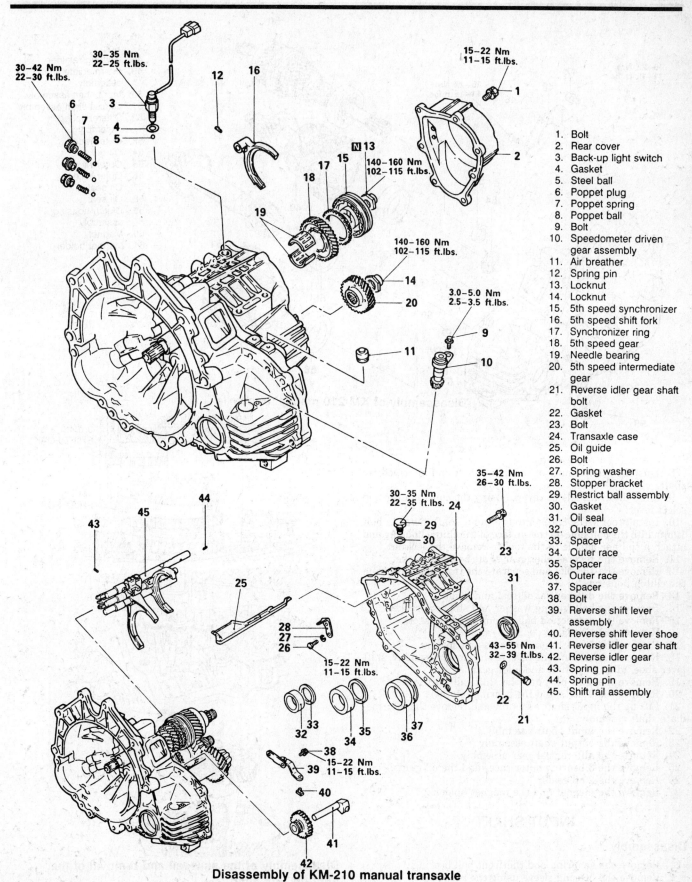

1. Bolt
2. Rear cover
3. Back-up light switch
4. Gasket
5. Steel ball
6. Poppet plug
7. Poppet spring
8. Poppet ball
9. Bolt
10. Speedometer driven gear assembly
11. Air breather
12. Spring pin
13. Locknut
14. Locknut
15. 5th speed synchronizer
16. 5th speed shift fork
17. Synchronizer ring
18. 5th speed gear
19. Needle bearing
20. 5th speed intermediate gear
21. Reverse idler gear shaft bolt
22. Gasket
23. Bolt
24. Transaxle case
25. Oil guide
26. Bolt
27. Spring washer
28. Stopper bracket
29. Restrict ball assembly
30. Gasket
31. Oil seal
32. Outer race
33. Spacer
34. Outer race
35. Spacer
36. Outer race
37. Spacer
38. Bolt
39. Reverse shift lever assembly
40. Reverse shift lever shoe
41. Reverse idler gear shaft
42. Reverse idler gear
43. Spring pin
44. Spring pin
45. Shift rail assembly

Disassembly of KM-210 manual transaxle

7-25

7 DRIVE TRAIN

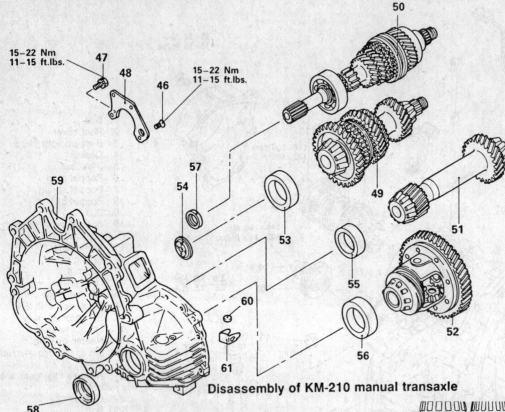

46. Bolt
47. Bolt
48. Bearing retainer
49. Intermediate shaft assembly
50. Input shaft assembly
51. Output shaft assembly
52. Differential gear assembly
53. Outer race
54. Oil guide
55. Outer race
56. Outer race
57. Oil seal
58. Oil seal
59. Clutch housing assembly
60. Magnet
61. Magnet holder

Disassembly of KM-210 manual transaxle

8. Unstake the locknuts of the input shaft and intermediate shaft.
9. Shift the transaxle in reverse using the control lever and select lever.
10. Install tool MD998802 onto the input shaft. Screw a bolt 10mm into the bolt hole on the surface of the clutch housing and attach a spinner handle to the tool to remove the locknut.
11. Remove the reverse idler gear shaft bolt and gasket.
12. Remove the 13 bolts and separate the transaxle case from the clutch housing.
13. Remove the differential oil seal and guide.
14. Remove the bolt, spring washer and stopper bracket.
15. Remove the restriction ball assembly and gasket.
16. Remove the oil seal.
17. Remove the 3 bearing outer races and spacers.
18. Remove the bolt and the reverse shift lever assembly, shift lever shoe, idler gear shaft and idler gear.
19. Remove the shift pins and the shift rails and forks.
20. Remove the 2 bolts and the bearing retainer.
21. Lift up the input shaft assembly and remove the intermediate shaft assembly.
22. Remove the input shaft assembly.
23. Remove the output shaft assembly.
24. Remove the differential gear assembly.
25. Remove the 3 bearing outer races and the oil guide.
26. Remove the 2 oil seals.
27. Remove the magnet and the magnet holder.

INPUT SHAFT

Disassembly

1. Remove the snapring and the front ball bearing.
2. Remove the bearing sleeve using the proper tool.

Disassembly of the shift rail and removal of the intermediate shaft

7-26

DRIVE TRAIN 7

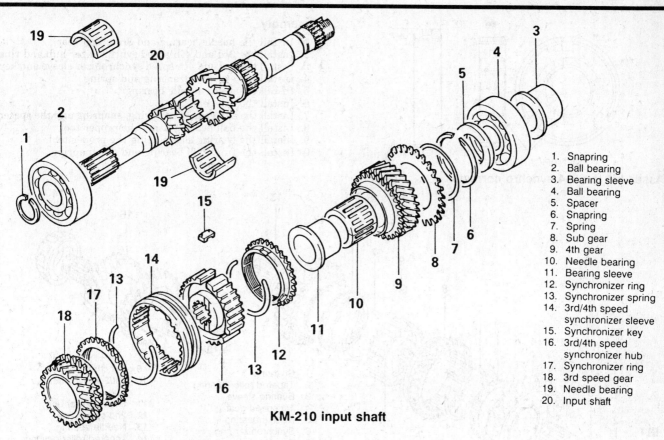

1. Snapring
2. Ball bearing
3. Bearing sleeve
4. Ball bearing
5. Spacer
6. Snapring
7. Spring
8. Sub gear
9. 4th gear
10. Needle bearing
11. Bearing sleeve
12. Synchronizer ring
13. Synchronizer spring
14. 3rd/4th speed synchronizer sleeve
15. Synchronizer key
16. 3rd/4th speed synchronizer hub
17. Synchronizer ring
18. 3rd speed gear
19. Needle bearing
20. Input shaft

KM-210 input shaft

3. Remove the ball bearing using the proper tool.
4. Remove the spacer, snapring, spring and the sub gear.
5. Remove the 4th gear.
6. Remove the needle bearing and sleeve.
7. Remove the synchronizer ring and spring.
8. Remove the 3rd and 4th speed synchronizer sleeve and key.
9. Remove the 3rd and 4th speed synchronizer hub and ring.
10. Remove the 3rd speed gear and needle bearing.

Inspection

1. Check the outer surface of the input shaft where the needle bearing is mounted for damage, abnormal wear and seizure.
2. Check the input shaft splines for damage and wear.
3. Combine the needle bearing with the shaft or bearing sleeve and gear and check that it rotates smoothly without abnormal noise or play.
4. Check the needle bearing cage for deformation.
5. Check the synchronizer ring clutch gear teeth for damage and breakage.
6. Check the internal surface for damage, wear and broken threads.
7. Force the synchronizer ring toward the clutch gear and check the clearance. The clearance should be 0.5mm.
8. Check the bevel gear and clutch gear teeth for damage and wear.
9. Check the synchronizer cone for rough surface, damage and wear.

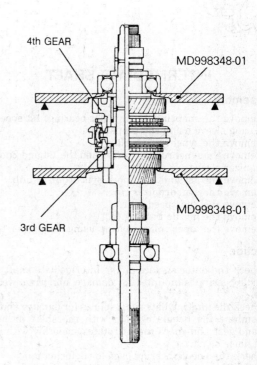

Don't damage the input shaft during disassembly; use the correct tools

7-27

7 DRIVE TRAIN

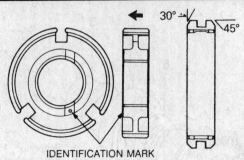

Assemble the 3–4 synchro correctly

Assembly

1. Install the needle bearing and 3rd speed gear.
2. Install the 3rd and 4th speed synchronizer hub and ring.
3. Install the 3rd and 4th speed synchronizer sleeve and key.
4. Install the synchronizer ring and spring.
5. Install the sleeve needle bearing.
6. Install the 4th gear.
7. Install the sub gear, corn spring, snapring and the spacer.
8. Install the ball bearing using the proper tool.
9. Install the bearing sleeve using the proper tool.
10. Install the front ball bearing and the snapring.

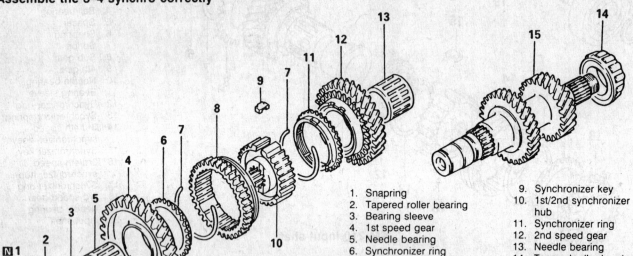

1. Snapring
2. Tapered roller bearing
3. Bearing sleeve
4. 1st speed gear
5. Needle bearing
6. Synchronizer ring
7. Synchronizer spring
8. 1st/2nd speed synchronizer sleeve
9. Synchronizer key
10. 1st/2nd synchronizer hub
11. Synchronizer ring
12. 2nd speed gear
13. Needle bearing
14. Tapered roller bearing
15. Intermediate shaft

Intermediate shaft

INTERMEDIATE SHAFT

Disassembly

1. Remove the snapring, taper roller bearing, 1st speed gear and bearing sleeve using the proper tool.
2. Remove the synchronizer ring.
3. Remove the synchronizer spring and the 1st and 2nd speed synchronizer sleeve.
4. Remove the 1st and 2nd speed synchronizer hub.
5. Remove the synchronizer ring.
6. Remove the 2nd speed gear.
7. Remove the needle bearing.
8. Remove the taper roller bearing using the proper tool.

Inspection

1. Check the outer surface of the intermediate shaft where the needle bearing is mounted for damage, abnormal wear and seizure.
2. Check the intermediate shaft splines for damage and wear.
3. Combine the needle bearing with the shaft or bearing sleeve and gear and check that it rotates smoothly without abnormal noise or play.
4. Check the needle bearing cage for deformation.
5. Check the synchronizer ring clutch gear teeth for damage and breakage.
6. Check the internal surface for damage, wear and broken threads.

Assembly

1. Install the taper roller bearing using the proper tool.
2. Install the needle bearing.
3. Install the 2nd speed gear.
4. Install the synchronizer ring.
5. Install the 1st and 2nd speed synchronizer hub.
6. Install the 1st and 2nd speed synchronizer sleeve and the synchronizer spring.
7. Install the synchronizer ring.
8. Install the snapring, taper roller bearing, 1st speed gear and bearing sleeve using the proper tool.

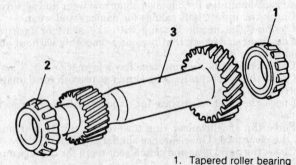

1. Tapered roller bearing
2. Tapered roller bearing
3. Output shaft

Output shaft

DRIVE TRAIN 7

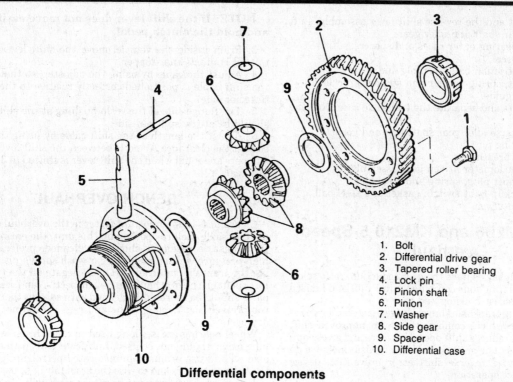

1. Bolt
2. Differential drive gear
3. Tapered roller bearing
4. Lock pin
5. Pinion shaft
6. Pinion
7. Washer
8. Side gear
9. Spacer
10. Differential case

Differential components

DIFFERENTIAL

Disassembly

1. Remove the bolts and the differential drive gear from the case.
2. Remove the taper roller bearings.
3. Remove the lockpin, pinion shaft and pinion gears and washers.
4. Remove the side gears and spacers.

Inspection

Check both the gears and the bearings for wear or damage. Make sure they are clean and dry before reassembly. Coat them with gear oil before reassembly.

Assembly

1. Install the side gears and spacers.
2. Install the lockpin, pinion shaft and pinion gears and washers. Measure the backlash between the side gears and pinions. The standard backlash is 0.025–0.150mm. If the backlash is out of specification, disassemble and use a different spacer.
3. Remove the taper roller bearings.
4. Remove the bolts and the differential drive gear from the case.

TRANSAXLE CASE ASSEMBLY

1. Install the magnet and the magnet holder.
2. Install the 2 oil seals.
3. Install the 3 bearing outer races and the oil guide.
4. Install the differential gear assembly.
5. Install the output shaft assembly.
6. Install the input shaft assembly.
7. Lift up the input shaft assembly and Install the intermediate shaft assembly.
8. Install the 2 bolts and the bearing retainer.
9. Install the shift pins and the shift rails and forks.

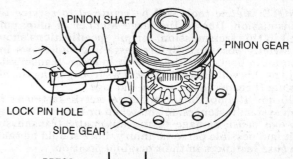

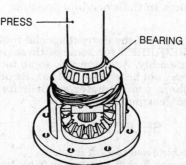

Correct assembly of differential

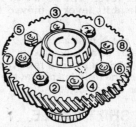

THE NUMBERS 1–8 INDICATE
THE TIGHTENING SEQUENCE

Tighten the bolts in the correct pattern

7-29

7 DRIVE TRAIN

10. Install the bolt and the reverse shift lever assembly, shift lever shoe, idler gear shaft and idler gear.
11. Install the 3 bearing outer races and spacers.
12. Install the oil seal.
13. Install the restriction ball assembly and gasket.
14. Install the bolt, spring washer and stopper bracket.
15. Install the oil guide.
16. Install the bolts and separate the transaxle case from the clutch housing.
17. Install the reverse idler gear shaft bolt and gasket.
18. Install the spring pin.
19. Install the air breather.
20. Install the speedometer driven gear assembly.
21. Install the poppet plug, spring and steel ball.
22. Install the backup light switch, gasket and steel ball.
23. Install the rear cover.

KM201, KM206 and KM210 5-Speed Overhaul

These manual transaxles have the gear reduction, ratio selection and differential functions combined into 1 unit and housed in the aluminum diecast housing and case.

All of these manual transaxles have a large diameter, large capacity synchronizer which, combined with an improved shift control mechanism, reduces shift operation force and gives comfortable shift feeling. The shift control mechanism includes an interlock device and a reverse shift error prevention device which ensure correct operation.

NOTE: Metric tools will be required to service this transmission. Due to the large number of alloy parts used in this transmission, torque specifications should be strictly observed. Before installing capscrews into aluminum parts, dip the bolts into clean transmission fluid as this will prevent the screws from galling the aluminum threads, thus causing damage.

Do not attempt to interchange metric fasteners for inch system fasteners. Mismatched or incorrect fasteners can cause damage to the automatic transmission unit and possible personal injury. Care should be taken to reuse fasteners in their original position.

WARNING: The use of the correct special tools or their equivalent is REQUIRED for some of these procedures involving disassembly. Additionally, some tools are required which may not be in your tool box. Be prepared to use snapring pliers, a micrometer, small drifts or punches and assorted gear pullers.

SELECT CABLE

1. Raise the vehicle and support it safely.
2. Move the transaxle shift lever to the **N** position.

NOTE: The select lever will be set to the N position when the transaxle shift lever is moved to the neutral position.

3. Move lever **B** to the **N** position.
4. Adjust the cable by using the adjuster, so that the end of the select cable is properly positioned relative to the lever **B**.
5. The flange side of the resin bushing at the select cable end should be at the lever **B** end surface.

SHIFT CABLE

1. Elevate the vehicle and support it safely.
2. With the select lever in **N**, move the transaxle shift lever downward to set it to 4th gear.

NOTE: If the shift lever does not move easily, depress and hold the clutch pedal.

3. From inside the vehicle, move the shift lever downward until it contacts the stopper.
4. Adjust the cable by using the adjuster, so that the end of the shift cable is positioned correctly relative to the shift lever (interior side).
5. The flange side of the resin bushing at the shift cable end should be at the split pin side.
6. Adjust the length of the shift cable by using the adjuster, so that the clearance **A** and **B** between the shift lever and the 2 stoppers are equal when the shift lever is shifted to 3rd gear and 4th gear.

BENCH OVERHAUL

Cleanliness is an important factor in the overhaul of the manual transaxle. Before opening up this unit, the entire outside of the transaxle assembly should be cleaned, preferably with a high pressure washer such as a car wash spray unit. Dirt entering the transaxle internal parts will negate all the time and effort spent on the overhaul. During inspection and reassembly all parts should be thoroughly cleaned with solvent then dried with compressed air. Wiping cloths and rags should not be used to dry parts.

Wheel bearing grease, long used to hold thrust washers and lube parts, should not be used. Lubricate seals with clean transaxle oil and use ordinary unmedicated petroleum jelly to hold the thrust washers and to ease the assembly of seals, since it will not leave a harmful residue as grease often will. Do not use solvent on neoprene seals, if they are to be reused, or thrust washers.

Before installing bolts into aluminum parts, always dip the threads into clean transaxle oil. Antiseize compound can also be used to prevent bolts from galling the aluminum and seizing. Always use a torque wrench to keep from stripping the threads. The internal snaprings should be expanded and the external rings should be compressed, if they are to be reused. This will help insure proper seating when installed.

TRANSAXLE CASE DISASSEMBLY

1. Position the assembly in a suitable holding fixture.
2. Remove the rear cover.
3. Remove the backup light switch, gasket and steel ball.
4. Remove the poppet plug, spring and steel ball.
5. Remove the speedometer driven gear assembly.
6. Remove the air breather.
7. Remove the spring pin using a pin punch.
8. Unstake the locknuts of the input shaft and intermediate shaft.
9. Shift the transaxle in reverse using the control lever and select lever.
10. Install tool MD998802 onto the input shaft. Screw a bolt 10mm into the bolt hole on the surface of the clutch housing and attach a spinner handle to the tool to remove the locknut.
11. Remove the 5th speed synchronizer assembly and shift fork.
12. Remove the synchronizer ring and the 5th gear.
13. Remove the needle bearing and the 5th intermediate gear.
14. Remove the reverse idler gear shaft bolt and gasket.
15. Remove the 13 bolts and separate the transaxle case from the clutch housing.
16. Remove the differential oil seal and guide.
17. Remove the bolt, spring washer and stopper bracket.
18. Remove the restriction ball assembly and gasket.
19. Remove the oil seal.
20. Remove the 3 bearing outer races and spacers.
21. Remove the bolt and the reverse shift lever assembly, shift lever shoe, idler gear shaft and idler gear.

DRIVE TRAIN 7

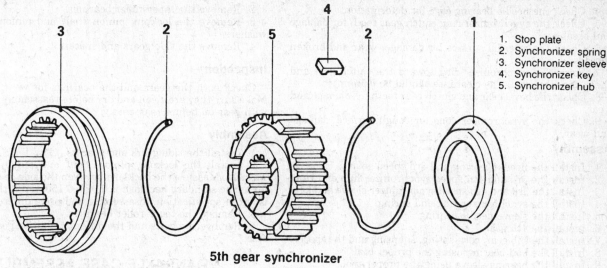

5th gear synchronizer

1. Stop plate
2. Synchronizer spring
3. Synchronizer sleeve
4. Synchronizer key
5. Synchronizer hub

22. Remove the shift pins and the shift rails and forks.
23. Remove the 2 bolts and the bearing retainer.
24. Lift up the input shaft assembly and remove the intermediate shaft assembly.
25. Remove the input shaft assembly.
26. Remove the output shaft assembly.
27. Remove the differential gear assembly.
28. Remove the 3 bearing outer races and the oil guide.
29. Remove the 2 oil seals.
30. Remove the magnet and the magnet holder.

5th Speed Synchronizer

Disassembly

1. Remove the stop plate.
2. Remove the synchronizer spring.
3. Remove the synchronizer sleeve.
4. Remove the synchronizer key.
5. Remove the synchronizer hub.

Inspection

1. Combine the synchronizer sleeve and hub and check that they slide smoothly.
2. Check that the sleeve is free from damage at its inside front and rear ends.
3. Check for wear of the hub end (the surface in contact with the 5th speed gear).
4. Check for wear of the synchronizer key center protrusion.
5. Check the spring for weakness, deformation and breakage.

Assembly

1. Assemble the synchronizer hub, sleeve and key noting their direction. Assemble so that the projections of the synchronizer springs fit into the grooves of the synchronizer keys.

NOTE: Be sure to assemble so that the front and rear spring projections are not fitted to the same key.

2. Install the synchronizer spring and the stop plate.

INPUT SHAFT ASSEMBLY

Disassembly

1. Remove the snapring and the front ball bearing.
2. Remove the bearing sleeve using the proper tool.
3. Remove the ball bearing using the proper tool.

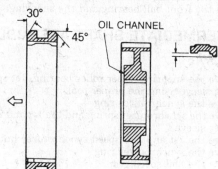

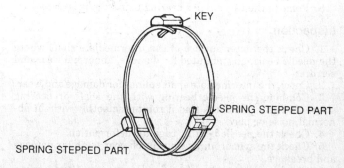

Make certain the 5th gear synchro is correctly reassembled

4. Remove the spacer, snapring, spring and the sub gear.
5. Remove the 4th gear.
6. Remove the needle bearing and sleeve.
7. Remove the synchronizer ring and spring.
8. Remove the 3rd and 4th speed synchronizer sleeve and key.
9. Remove the 3rd and 4th speed synchronizer hub and ring.
10. Remove the 3rd speed gear and needle bearing.

Inspection

1. Check the outer surface of the input shaft where the needle bearing is mounted for damage, abnormal wear and seizure.
2. Check the input shaft splines for damage and wear.
3. Combine the needle bearing with the shaft or bearing sleeve and gear and check that it rotates smoothly without abnormal noise or play.

7 DRIVE TRAIN

4. Check the needle bearing cage for deformation.
5. Check the synchronizer ring clutch gear teeth for damage and breakage.
6. Check the internal surface for damage, wear and broken threads.
7. Force the synchronizer ring toward the clutch gear and check the clearance. The clearance should be 0.5mm.
8. Check the bevel gear and clutch gear teeth for damage and wear.
9. Check the synchronizer cone for rough surface, damage and wear.

Assembly

1. Install the needle bearing and 3rd speed gear.
2. Install the 3rd and 4th speed synchronizer hub and ring.
3. Install the 3rd and 4th speed synchronizer sleeve and key.
4. Install the synchronizer ring and spring.
5. Install the sleeve needle bearing.
6. Install the 4th gear.
7. Install the sub gear, corn spring, snapring and the spacer.
8. Install the ball bearing using the proper tool.
9. Install the bearing sleeve using the proper tool.
10. Install the front ball bearing and the snapring.

INTERMEDIATE SHAFT ASSEMBLY

Disassembly

1. Remove the snapring, taper roller bearing, 1st speed gear and bearing sleeve using the proper tool.
2. Remove the synchronizer ring.
3. Remove the synchronizer spring and the 1st and 2nd speed synchronizer sleeve.
4. Remove the 1st and 2nd speed synchronizer hub.
5. Remove the synchronizer ring.
6. Remove the 2nd speed gear.
7. Remove the needle bearing.
8. Remove the taper roller bearing using the proper tool.

Inspection

1. Check the outer surface of the intermediate shaft where the needle bearing is mounted for damage, abnormal wear and seizure.
2. Check the intermediate shaft splines for damage and wear.
3. Combine the needle bearing with the shaft or bearing sleeve and gear and check that it rotates smoothly without abnormal noise or play.
4. Check the needle bearing cage for deformation.
5. Check the synchronizer ring clutch gear teeth for damage and breakage.
6. Check the internal surface for damage, wear and broken threads.

Assembly

1. Install the taper roller bearing using the proper tool.
2. Install the needle bearing.
3. Install the 2nd speed gear.
4. Install the synchronizer ring.
5. Install the 1st and 2nd speed synchronizer hub.
6. Install the 1st and 2nd speed synchronizer sleeve and the synchronizer spring.
7. Install the synchronizer ring.
8. Install the snapring, taper roller bearing, 1st speed gear and bearing sleeve using the proper tool.

DIFFERENTIAL

Disassembly

1. Remove the bolts and the differential drive gear from the case.
2. Remove the taper roller bearings.
3. Remove the lockpin, pinion shaft and pinion gears and washers.
4. Remove the side gears and spacers.

Inspection

Check both the gears and the bearings for wear or damage. Make sure they are clean and dry before reassembly. Coat them with gear oil before reassembly.

Assembly

1. Install the side gears and spacers.
2. Install the lockpin, pinion shaft and pinion gears and washers. Measure the backlash between the side gears and pinions. The standard backlash is 0.025–0.150mm. If the backlash is out of specification, disassemble and use a different spacer.
3. Remove the taper roller bearings.
4. Remove the bolts and the differential drive gear from the case.

TRANSAXLE CASE ASSEMBLY

1. Install the magnet and the magnet holder.
2. Install the 2 oil seals.
3. Install the 3 bearing outer races and the oil guide.
4. Install the differential gear assembly.
5. Install the output shaft assembly.
6. Install the input shaft assembly.
7. Lift up the input shaft assembly and install the intermediate shaft assembly.
8. Install the 2 bolts and the bearing retainer.
9. Install the shift pins and the shift rails and forks.
10. Install the bolt and the reverse shift lever assembly, shift lever shoe, idler gear shaft and idler gear.
11. Install the 3 bearing outer races and spacers.
12. Install the oil seal.
13. Install the restriction ball assembly and gasket.
14. Install the bolt, spring washer and stopper bracket.
15. Install the oil guide.
16. Assemble the transaxle case and the clutch housing. Install the bolts.
17. Install the reverse idler gear shaft bolt and gasket.
18. Install the needle bearing and the 5th intermediate gear.
19. Install the synchronizer ring and the 5th gear.
20. Install the 5th speed synchronizer assembly and shift fork.
21. Install the spring pin.
22. Install the air breather.
23. Install the speedometer driven gear assembly.
24. Install the poppet plug, spring and steel ball.
25. Install the backup light switch, gasket and steel ball.
26. Install the rear cover.

Halfshafts and CV-Joints

REMOVAL AND INSTALLATION

Cordia, Tredia and Galant

1. Remove the hub center cap, remove the cotter pin, and loosen the driveshaft nut. Loosen the wheel lug nuts. Raise the car and support it on jackstands at the crossmember.
2. Drain the transmission fluid (you need not drain the pan on auto transaxles, just remove the drain plug from the housing).
3. Follow the procedure for Lower Control Arm Removal and Installation to disconnect the stabilizer bar, strut bar, and ballstud from the lower control arm. You need not remove the lower control arm from the crossmember.

DRIVE TRAIN 7

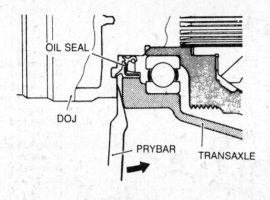

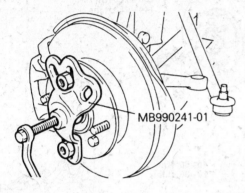

Removing the driveshaft. Take care not to damage the oil seal with the prybar

4. Insert a prybar between the transaxle case and the driveshaft inner bearing *without contacting the oil seal*, and pry gently to force the shaft from the axle. *Don't pull on the axle to remove it!*
5. Remove the driveshaft nut. Using a special tool such as MB990241-01, press the driveshaft out of the hub.
6. To install, use a new snapring on the splined inner end of the shaft to retain the shaft in the transmission. Insert the driveshaft through the hub. Install the inner end of the shaft into the transmission until it locks. Install the washer at the outer end of the shaft. Make sure it faces the right way, bulged outward at the center. Install the retaining nut for the outer end of the driveshaft loosely.
7. Reconnect the ball joint and other parts to the lower control arm as specified in Ball Joint Removal and Installation. Re-

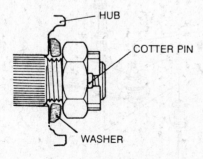

Make certain the washer is correctly installed before the locknut goes on

install the front wheel and make the lugnuts snug.
8. Lower the vehicle to the ground and tighten the lugnuts. Tighten the driveshaft retaining nut to 145–188 ft. lbs. The best procedure is to torque the nut to 145 ft. lbs. and then, if necessary, turn the nut just enough further for the cotter pin hole in the shaft and slot in the nut to line up (up to the maximum torque). When the holes line up and torque is within specifications, install the cotter pin.

Mirage and Precis

1. Remove the hub center cap, remove the cotter pin, and loosen the driveshaft nut. Loosen the wheel lug nuts. Raise the car and support it on jackstands at the crossmember.
2. Drain the transmission fluid (you need not drain the pan on auto transaxles, just remove the drain plug from the housing).
3. Refer to Ball Joint Removal and Installation and disconnect the steering knuckle from the lower control arm by removing the ball stud from the knuckle.
4. Refer to Strut Removal and Installation and disconnect the tie rod end from the knuckle.
5. If the driveshaft has a center bearing, remove the circlip from the center bearing. Gently tap on the outer part of the constant velocity joint with a plastic hammer so as to pull the driveshaft out of the transaxle.
Where the driveshaft runs directly into the transmission, insert a prybar in such a way as to avoid damaging the seal and pry outward to remove the driveshaft from the transmission. *Don't pull on the driveshaft to remove it! Keep the driveshaft supported in such a way that it won't be subjected to unusual bending stress as you work.*
6. Use a puller such as MB990241-01 to push the outer end of the driveshaft through the center of the hub and out.
7. Before installing, place a new circlip on the driveshaft and push the axle into the transaxle. The clip should engage its groove with a distinct click. On the type of driveshaft having a center bearing, make sure to press the bearing in until it contacts the the surface of the center bearing bracket and install a new snapring.

NOTE: During reinstallation, do not allow the axle shaft to hang to the ground. Support the shaft with wire or string until both ends are placed. Allowing the joint to hang beyond its normal range may damage the joints.

8. Install the outer end of the shaft into the steering knuckle. Don't damage the splines or threads during installation.
9. Reconnect the tie rod end and tighten the nut to 25 Nm (18 ft. lbs.). Install a new cotter pin.
10. Connect the ball joint to the steering knuckle and tighten the nut 89 Nm (66 ft. lbs.).
10. Install the washer and the locknut in the correct direction and tighten the nut very snug. Do not attempt to set final torque with the car on stands.
11. Reinstall the front wheel and make the lugnuts snug.
12. Lower the vehicle to the ground and tighten the lugnuts.
13. Torque the driveshaft retaining nut to 145–188 ft. lbs. The best procedure is to torque the nut to 145 ft. lbs. and then, if necessary, turn the nut just enough for the cotter pin hole (in the shaft) and slot (in the nut) to line up When the holes line up and torque is within specifications, install the cotter pin.

Sigma

1. Remove the hub center cap, remove the cotter pin, and loosen the driveshaft nut. Loosen the wheel lug nuts. Raise the car and support it on jackstands at the crossmember.
2. Drain the transmission fluid (you need not drain the pan on auto transaxles, just remove the drain plug from the housing).
3. Remove the driveshaft nut and the washer.
4. Remove the cotter pin from the tie rod end joint. Use the coorrect tools to separate the tie rod end joint.

7-33

7 DRIVE TRAIN

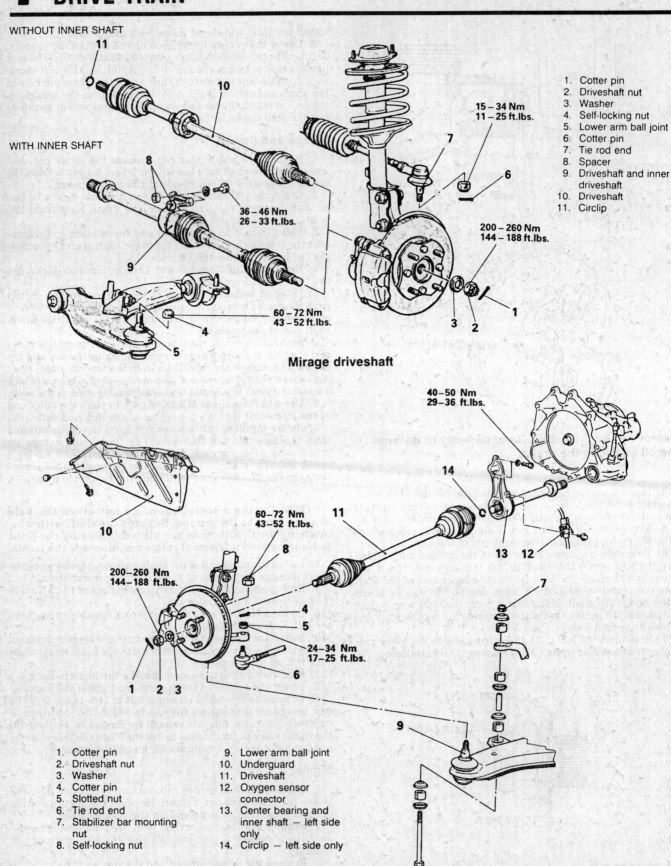

DRIVE TRAIN 7

5. Disconnect the retaining nut holding the stabilizer bar in place.
6. Using the correct tools, remove the self-locking nut and disconnect the lower ball joint from the steering knuckle.
7. Remove the underguard or small splash shield.
8. If removing the left driveshaft, insert a pry bar between the center bearing bracket and the driveshaft. Pry the shaft from the center bearing bracket.

If removing the right driveshaft, insert a pry bar between the transaxle case and the driveshaft and pry the axle free of the transaxle.

WARNING: Do not pull on either driveshaft; doing so will damage the joint. Use care when inserting the pry bar not to damage the oil seal.

9. Use a tool such as MB 990241-01 to press the shaft clear of the hub.
10. Disconnect the wiring to the oxygen sensor.
11. On the left side, insert a prybar between the engine block and the center bearing/inner shaft assembly. Pry outward to remove the bearing and inner shaft assembly.
12. Inspect the driveshaft boots for wear or cracking. Any hole or puncture, regardless of size, disqualifies the boot from use. Inspect the joints for freedom of motion, grinding, wear, etc. Check the splines for wear or damage. Check the operation of the center bearing.
13. Install the center bearing and inner shaft. Make certain the shaft on the left side has a new circlip on it; never reuse the old clip.
14. Connect the oxygen sensor connector.
15. Install the axle to either the transaxle (right side) or the inner shaft. Fit the outer end into the hub. During this reinstallation, do not allow the shaft to hang to the ground; the joint(s) may become damaged from overextension.
16. Install the undercover.
17. Install the lower ball joint to the steering knuckle with a new nut. The nut should be tightened to 67 Nm or 50 ft. lbs. You may wish to wait until the car is on the ground to tighten this nut. If so, install the nut comfortably tight for now. If you do succeed in achieving the correct torque, install a new cotter pin.
18. Connect the stabilizer bar to the link, making sure all the bushings and covers are in the correct location. Tighten the nut until the top of the link protrudes through the nut 16–18mm.
19. Connect the tie rod end and tighten the new nut to 27 Nm (20 ft. lbs.). Install a new cotter pin.
20. Install the washer and the locknut in the correct direction and tighten the nut very snug. Do not attempt to set final torque with the car on stands.
21. Reinstall the front wheel and make the lugnuts snug.
22. Lower the vehicle to the ground and tighten the lugnuts.
23. If the lower ball joint nut was not tightened, do so now and install a new cotter pin. Torque the driveshaft retaining nut to 145–188 ft. lbs. The best procedure is to torque the nut to 145 ft. lbs. and then, if necessary, turn the nut just enough for the cotter pin hole (in the shaft) and slot (in the nut) to line up. When the holes line up and torque is within specifications, install the cotter pin.

OVERHAUL

The drive axle assembly is a flexible unit consisting of an inner and outer constant velocity (CV) joint joined by the axle shaft. Some cars may have a center bearing or carrier but the inner and outer joint remain. The inner joint, because of its design, is completely flexible and allows the joint to flex left/right, up/down and compress/extend while the joint is turning. This range of motion is necessary to allow the car to accelerate during all possible positions of the car's front wheels.

Because of the inner joints' flexibility, care must be taken not to over-extend the joint during repairs or handling. When either end of the shaft is disconnected from the car, support the shaft with wire to prevent joint damage.

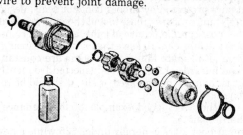

DOJ JOINT KIT

DOJ BOOT KIT

BJ BOOT KIT

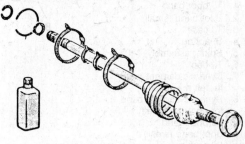

BJ JOINT AND SHAFT KIT

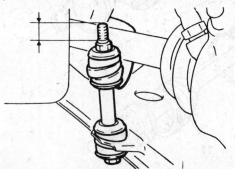

Correct assembly of Sigma stabilizer and link. See text for details.

Examples of repair kits; note that kits for boots or joints are available

7-35

7 DRIVE TRAIN

Location of identifying codes for boots and bands. TJ parts shown, others similar

On cars, the outer joint is fixed into the hub and does not require the extreme flexibility of the inner joint. The outer joint allows for left/right motion while the axle is turning.

CV-joints are protected by rubber boots or seals, designed to keep the high temperature grease in and the road grime and water out. The most common cause of joint failure is a ripped boot which allows the lubricant to leave the joint, thus causing heavy wear, noise and ultimate failure. The boots are constantly exposed to road hazards and should be inspected frequently. Any time a boot is found to be damaged or slit, it should be replaced immediately.

This is another very good example of maintenance being cheaper than repair.

A replacement boot kit is generally under $35 while a new joint can range above $150.

Mitsubishi uses no fewer than 4 types of CV-joint on its cars; they appear in various combinations on various driveshafts. The types of joints and their abbreviations are:

- Tripod joint (TJ) — commonly found as an inboard joint, the three sided spider inside can be replaced on the bench.
- Rzepppa joint (RJ) — named for its inventor, it almost always shows up as an outer joint. Joint is not replaceable under most conditions but comes installed on a new axle.
- Double Offset joint (DOJ) — identified by its ball-in-cage arrangement, this highly flexible inner joint can be dismantled and rebuilt.
- Birfield joint (BJ) — outer joint with good reliability. Cannot be disassembled from the axle. A new joint will include a new shaft.

This section will use these abbreviations for each joint. Additionally, any reference to an outer or outboard joint indicates the joint at or near the wheel/hub/knuckle assembly. Not surprisingly, inner or inboard refers to the joint at or closest to the transaxle.

WARNING: Bands and snaprings are not reusable. Parts are not interchangable. Each band and boot for each model and type of joint has an identifying code on it. Obtain this code and shop for exact replacements. A few cents saved on the wrong part may cost you a second repair when the boot doesn't seal correctly.

The grease that comes in the repair kit should be used throughout the joint. Do not substitute other lubricants; they will not withstand the heat and pressure within the joint.

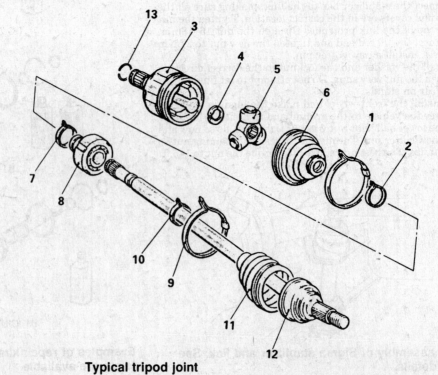

1. TJ boot band
2. Boot band (small)
3. TJ case
4. Snapring
5. Spider assembly
6. TJ boot
7. Dynamic damper band for left driveshaft
8. Dynamic damper band for right driveshaft
9. RJ boot band
10. Boot band (small)
11. RJ boot
12. RJ assembly
13. Circlip

Typical tripod joint

7-36

DRIVE TRAIN 7

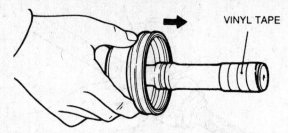

Tape on the splines protects the boots against damage

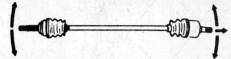

Check the joints for freedom of movement in all directions

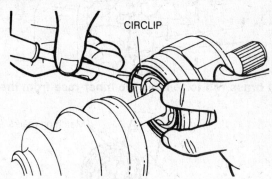

Removing the large circlip from the double offset joint

Tripod Joint

1. Remove the driveshaft from the car, following directions outlined earlier.
2. Remove the large band holding the boot to the joint and the smaller band holding the boot to the shaft. Slide the boot off the joint and down the shaft, out of the way.
3. Gently pull the TJ case off the joint.
4. Remove the small snapring on the end of the shaft and remove the spider assembly. Clean the spider assembly thoroughly but do not disassemble it.
5. Wrap several turns of vinyl or electrician's tape around the splines on the driveshaft. Slide the boot off the shaft.
6. Clean and examine the parts carefully. All traces of the old grease should be removed and the parts should be totally dry before reinstallation.
7. Slide the new boot onto the axle. Use tape to cover the shaft splines.
8. Apply a liberal coating of the special CV-joint grease (comes in the overhaul kit) to the spider assembly and install the assembly on the shaft.
9. Install a new snapring to hold the spider.
10. Use the remainder of the grease inside the joint case and fit the case over the spider.
11. Slide the boot into place and install new boot clamps. Make certain the joint is straight when placing the boot; if the joint is cocked, the boot may not seal correctly.

12. Install a new snapring on the transaxle end of the inner joint.
13. Reinstall the driveshaft.

Double Offset Joint

1. Remove the driveshaft from the car, following directions outlined earlier.
2. Remove the small band holding the boot to the shaft and the larger band holding the boot to the joint. Slide the boot off the joint and down the shaft, out of the way.
3. Gently pull the DOJ case off the joint. There is a large circular spring clip within the case holding it in place. Carefully work this spring free before disassembly.
4. Make reliable matching marks on the end of the shaft, the inner race and the outer cage. A metal scribe is handy for this marking.

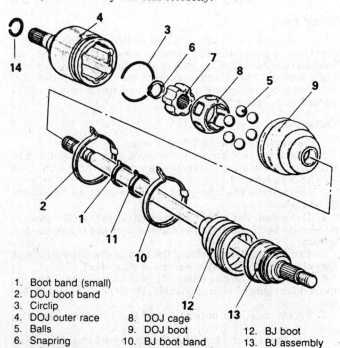

1. Boot band (small)
2. DOJ boot band
3. Circlip
4. DOJ outer race
5. Balls
6. Snapring
7. DOJ inner race
8. DOJ cage
9. DOJ boot
10. BJ boot band
11. Boot band (small)
12. BJ boot
13. BJ assembly
14. Circlip

Typical DOJ assembly

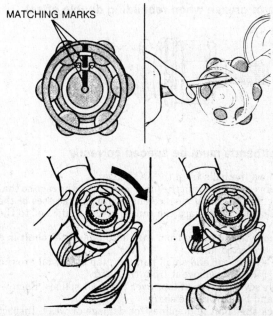

Correct disassembly of the ball-and-cage joint

7 DRIVE TRAIN

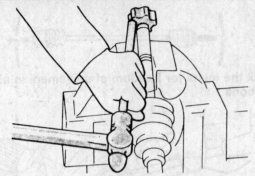

Use a brass rod to remove the inner race from the shaft

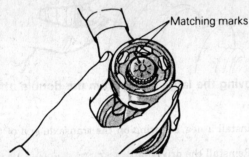

Align the marks before assembling the balls

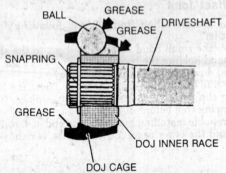

Use lots of grease when rebuilding double offset joints

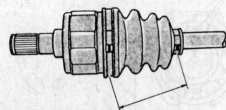

The boot bands must be spaced correctly

5. Remove the balls from the DOJ cage.
6. Remove the cage from the inner race. Turn the cage so that the projections of the inner race align with the recesses of the cage. Slide the cage toward the center of the shaft, not to the outside.
7. Remove the snapring from the shaft. Mount the shaft in a vise with protected jaws.
8. Use a brass bar and small hammer, tap evenly all around the inner race and remove it from the shaft.
9. Apply several turns of tape over the shaft splines. Remove the cage and boot from the shaft.
10. Check the shaft and splines for damage or wear. Inspect the cage, race and balls for any sign of corrosion, wear, cracking

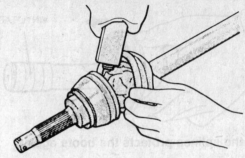

Packing a non-servicable joint

or damage. Clean all the parts thoroughly and air dry them completely before installation. Any remaining cleaning solvent can dissolve the lubricating grease.

11. Wrap the splines in tape; install a new boot and slide it down the shaft.
12. Place the cage onto the driveshaft so that the smaller diameter side is installed first.
13. Align the matchmarks of the inner race and the shaft. Use the small hammer and brass bar to tap the race into position until it contacts the rib of the shaft. Install a new snapring.
14. Apply grease liberally to the inner race and the DOJ cage. Align the matchmarks and fit the two together using the same alignment method as during disassembly.
15. Apply grease to the ball areas of the cage and race and insert the balls.
16. Fill the outer race about ⅓ full of grease. Install a new large circlip and install the ball and cage assembly. Use the remainder of the grease to fill the back of the joint or the boot.
17. Slide the boot into place over the joint and install new bands. The boot should be neither stretched nor compressed when the bands are applied. The bands should be 80mm apart. Too much or too little air within the boot may cause premature boot failure.
18. Install the driveshaft into the car.

BOOT REPLACEMENT

Inner Joint

Boot replacement is accomplished by following the appropriate overhaul procedure. The joint components must be removed from the shaft to allow the boot to pass. Even if only the boot is being replaced, the other components should be inspected. Remove as much old grease as possible and reassemble the joint packed with fresh grease.

Outer Joint

1. Remove the driveshaft from the car.
2. The outer boot cannot be removed over the outer end of the shaft. You will need 2 repair kits since the inner joint must also be disassembled.
3. Disassemble the inner joint following overhaul procedures listed earlier.
4. If the shaft has a dynamic damper (usually on left shafts), remove its band and slide the damper over the tape-protected splines.
5. Remove the bands holding the boot at the outer joint and slide the boot all the way down the shaft and off.
6. The exposed joint—either RJ or BJ type—should be inspected for signs of wear if possible. Do NOT disassemble this joint.
7. Fit the new boot onto the shaft and into place near the joint.
8. Use about ½ the grease in the kit to pack the outer joint and the other ½ to fill the boot.
9. Work the boot into place and apply the new bands. Make

DRIVE TRAIN 7

sure the joint is not at an angle during the securing of the boot.
10. Install the dynamic damper if one was removed. Use a new band to hold it.
11. Install the new boot for the inner joint and reassemble the joint following the overhaul instructions given earlier. Note that the dynamic damper, if used, must be kept free of grease.
12. Install the driveshaft in the car.

MANUAL TRANSMISSION

Identification

Transmission identification codes are found on the vehicle information plate located on the firewall in the engine compartment. The first portion of the code is the transmission designation; the numbers at the right end of the line indicate the final drive ratio of the transmission.

Before attempting to repair the clutch or transmission for reasons beyond obvious failure, the problem and probable cause should be identified. A great percentage of manual transmission and clutch problems are accompanied by shifting difficulties such as gear clash, grinding or refusal to engage gears. When any of these problems occur, a careful analysis should be performed to determine cause.

Driveline noises can become baffling, but don't be too quick to assign fault to the transmission. The noise may actually be coming from other sources such as tires, road surface, wheel bearings, engine or exhaust system. Noises will also vary by vehicle size, engine type and amount of insulation within the body. Remember also that the transmission, like any mechanical device, is not totally quiet in its operation and will exhibit some normal operating noise.

When checking for driveline noises, follow these guidelines:
1. Select a smooth, level asphalt road to reduce tire and body noises. Check and set the tire pressures before beginning.
2. Drive the vehicle far enough to thoroughly warm up all the lubricants; this may require driving as much as 10 or 12 miles before testing.
3. Note the speed at which the noise occurs and in which gear(s).
4. Check for noises with the vehicle stopped and the engine running.
5. Does the noise change when the clutch is pushed in or released?
6. Determine the throttle condition(s) at which the noise occurs:
   a. Drive (leading throttle) — the engine is under load and the car is accelerating at some rate.
   b. Float (neutral throttle) — the engine is maintaining a constant speed at light throttle on a level road.
   c. Coast (trailing throttle) — the engine is not driving the car, the throttle is closed with the car in gear and in motion. An example is a long downhill grade.
   d. Some combination of the above.
7. Does the noise change as the car corners or changes direction? Is the change in the noise different on left and right turns?
8. Identify the noise by type (whine, click, knock, grinding, buzzing, etc.) and think through possible causes such as loose components, something rubbing, poor lubrication, or worn parts.
9. Check all other causes first. Its much easier to check the wheel bearings and the exhaust system than to remove the transmission. Modern manual transmissions are extremely reliable and rarely fail if given reasonable maintenance and freedom from outright abuse. Assume that the noise is not in the transmission until diagnosed otherwise.

Linkage Adjustment

Mitsubishi manual transmission linkages are not adjustable. If the shifter does not work properly, a problem with the size, mounting or lubrication of the linkage is indicated.

Back-up Light Switch

REMOVAL AND INSTALLATION

The back-up light control switch is mounted into the transmission case. It is a simple push-button type switch; when the transmission is in reverse, a part of the linkage inside the case presses against the tip of the switch. This closes the circuit and brings on the back-up lights.

The switch may be unscrewed from the case after unplugging its wiring connector. The location of the switch will determine whether it is necessary to drain the transmission before removing the switch. The oil usually fills about one third to one half of the case; if the switch is mounted above this approximate line, it may be removed without draining. Those switches mounted low on the case will require the oil to be drained.

Always use the correct size wrench on the switch and avoid the use of pliers. Keep the wrench straight on the flats when turning; the aluminum alloy transmission case may be damaged or stripped by careless removal. When reinstalling, tighten the switch to 30 Nm or 22 ft. lbs.

Transmission

REMOVAL AND INSTALLATION

Starion

1. Remove the drain plug and drain the transmission oil. Remove the driveshaft.
2. Disconnect the speedometer cable and backup lamp harness at the transmission.
3. Remove the clutch slave cylinder.

7 DRIVE TRAIN

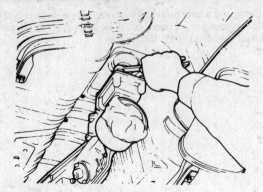

Disconnecting the Starion shift lever

4. Remove the four attaching bolts and remove the bell housing cover.
5. Disconnect and remove the starter motor wiring. Support the starter, remove move the two starter mounting bolts from the front of the bell housing and remove the starter.
6. Remove the top two transmission mounting bolts (at the bell housing) that are accessible from above.
7. Safely support the vehicle on stands. Support the engine with a crane via the lifting hooks. Support the transmission with a floor jack in such a way that the oil pan will not be damaged. Remove the remaining transmission mounting bolts from the bell housing area.
8. Remove the engine rear support bracket, insulator, and ground cable. Refer to Section 3 if you need help with these procedures.

9. Make sure the gearshift is in neutral and then remove the assembly by removing the bolts, accessible from below and located where the gearshift mounts onto the rear of the extension housing.
10. Place a cloth or soft padding against the rear of the cylinder head to prevent damage to the firewall if the engine should shift slightly. Pull the transmission to the rear and remove it.
11. Support the transmission in position on the jack and install it onto the engine.
12. Install the engine rear support assembly.
13. Install the starter motor assembly.
14. Install the bell housing cover.
15. Install the clutch release cylinder and the speedometer cable.
16. Connect the wiring harness(es) as needed.
17. Install the driveshaft.
18. Install the gearshift lever assembly.
19. Fill the transmission with the correct amount of oil

2-Wheel Drive Pickup

1. Elevate and safely support the vehicle on stands.
2. Drain the transmission fluid.
3. Remove the shifter assembly.
4. Disconnect the driveshaft. Matchmark the flanges at the rear axle and remove the bolts. Lower the shaft and slide it to the rear to remove it from the transmission.
5. Disconnect the reverse light wiring and disconnect the speedometer cable.
6. Disconnect the mounting bracket for the exhaust system. While not required, it may be helpful the disconnect the exhaust system at the front joint. Use wire to suspend the pipe out of the way.

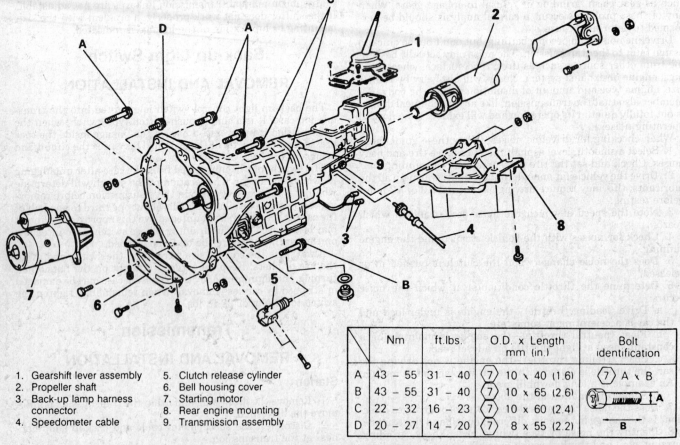

1. Gearshift lever assembly
2. Propeller shaft
3. Back-up lamp harness connector
4. Speedometer cable
5. Clutch release cylinder
6. Bell housing cover
7. Starting motor
8. Rear engine mounting
9. Transmission assembly

	Nm	ft.lbs.	O.D. x Length mm (in.)	Bolt identification
A	43 – 55	31 – 40	10 x 40 (1.6)	A B
B	43 – 55	31 – 40	10 x 65 (2.6)	A
C	22 – 32	16 – 23	10 x 60 (2.4)	
D	20 – 27	14 – 20	8 x 55 (2.2)	B

Starion transmission removal. Observe correct bolt placement and torque when reinstalling

DRIVE TRAIN 7

7. Remove the bell housing cover.
8. Disconnect the clutch cable.
9. Support the transmission with a floor jack. Use a broad piece of wood to distribute the load evenly.
10. Remove the rear engine support insulator and transmission mount nuts; separate them from the transmission.
11. Remove the mounting bolts for the rear (No. 2) crossmember. Remove the crossmember and the mount.
12. Remove the bolts holding the bell housing to the engine. Take note of the placement and size of each bolt; correct reinstallation is required.
13. Work the transmission rearward and off the engine. Once free, lower the jack and remove the transmission.
14. When reinstalling, position the transmission on the jack. Before installing, make certain the transmission case is in line with the two locating dowels or guides on the rear of the engine. Slide the transmission straight onto the engine and engage the shaft. Do not allow the transmission to hang on the shaft. Install the engine to bell housing retaining bolts. Make sure the bolts are correctly placed.
15. Install the rear crossmember, tightening the bolts to 45 Nm (33 ft. lbs.).
16. Connect the transmission mount and install the nuts and bolts. Correct torque is 20 Nm or 15 ft. lbs.
17. Connect the clutch cable.
18. Install the bell housing cover.
19. Connect the exhaust system if detached and secure the bracket at the transmission case.
20. Connect the speedometer cable and the backup light wiring harness.
21. Install the driveshaft observing the matchmarks made earlier.
22. Install the shift lever assembly.
23. Install the proper amount of transmission lubricant.
24. Lower the car to the ground.

Montero
4-Wheel Drive Pickup

1. Place the gear selector in Neutral.
2. Remove the boot retaining screws around the shifter boot; remove the boot and the shift knob.
3. Pull the gearshift lever assembly out of the control housing.

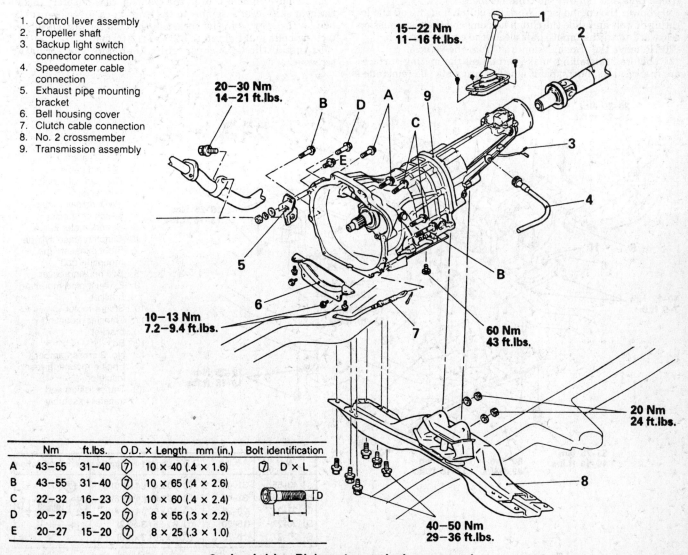

1. Control lever assembly
2. Propeller shaft
3. Backup light switch connector connection
4. Speedometer cable connection
5. Exhaust pipe mounting bracket
6. Bell housing cover
7. Clutch cable connection
8. No. 2 crossmember
9. Transmission assembly

	Nm	ft.lbs.	O.D. × Length mm (in.)	Bolt identification
A	43–55	31–40	⑦ 10 × 40 (.4 × 1.6)	⑦ D × L
B	43–55	31–40	⑦ 10 × 65 (.4 × 2.6)	
C	22–32	16–23	⑦ 10 × 60 (.4 × 2.4)	
D	20–27	15–20	⑦ 8 × 55 (.3 × 2.2)	
E	20–27	15–20	⑦ 8 × 25 (.3 × 1.0)	

2-wheel drive Pickup; transmission removal

7-41

7 DRIVE TRAIN

4. Cover the opening in the control housing with a cloth to prevent dirt from entering the transmission.
5. Disconnect the negative battery cable.
6. Raise and safely support the vehicle on stands.
7. Matchmark and disconnect the rear driveshaft at both the rear axle and the transfer case flanges.
8. Disconnect the forward driveshaft from the front axle and remove it by sliding it forward. Install a plug in the transfer case to prevent fluid loss.
9. Disconnect the hydraulic fluid line from the transfer case.
10. Disconnect the starter motor cable, the back up lamp harness and the neutral safety switch wire.
11. Disconnect the speedometer cable.
12. Place a floor jack and a block of wood below the engine oil pan.
13. Support the transfer case and remove it from the vehicle.
14. Remove the starter.
15. Use a transmission jack or second floor jack to place under the transmission.
16. Remove the bell housing bolts holding the unit to the engine.
17. Remove the nuts and bolts attaching the transmission mount and damper to the crossmember.
18. Remove the nuts and bolts holding the crossmember to the frame rails and remove the crossmember.
19. Lower the engine jack. Work the clutch housing off the locating dowels and slide the clutch housing and the transmission rearward until the input shaft clears the clutch disc.
20. Remove the transmission unit from the vehicle.
21. Before reinstalling, make certain all the mating surfaces are free of nicks, burrs, paint etc. The trans may be reinstalled with or without the transfer case attached.
22. Elevate the transmission on a jack under the car and position it correctly. Start the input shaft into the clutch disc.
23. Align the splines on the input shaft with the splines in the clutch disc.
24. Move the transmission forward and carefully seat the clutch housing on the locating dowels of the engine rear plate.
25. Install the bolts and washers attaching the clutch housing to the rear plate and tighten them to 49 Nm (36 ft. lbs).
26. Install the starter motor. Tighten the nuts to 27 Nm (20 ft. lbs).
27. Raise the engine and install the rear crossmember, insulator and damper.
28. Install the transfer case.
29. Install the driveshafts front and rear. Make sure the marks made during disassembly are aligned.
30. Connect the starter cable, the backup light wiring and other wires to the transmission.
31. Connect the hydraulic line to the slave cylinder. Bleed the clutch system.
32. Install the speedometer cable.
33. Remove the fill plug and fill the transmission unit to the correct level.
34. Lower the vehicle to the ground.
35. Remove the cloth over the transfer case adapter. Install the gear shift lever assembly in the transfer case. Make sure the ball on the lever is in the socket in the adapter. Install the attaching bolts and tighten to 12 Nm (10 ft. lbs.).
36. Install the boot cover, shift knob and boot retaining screws.

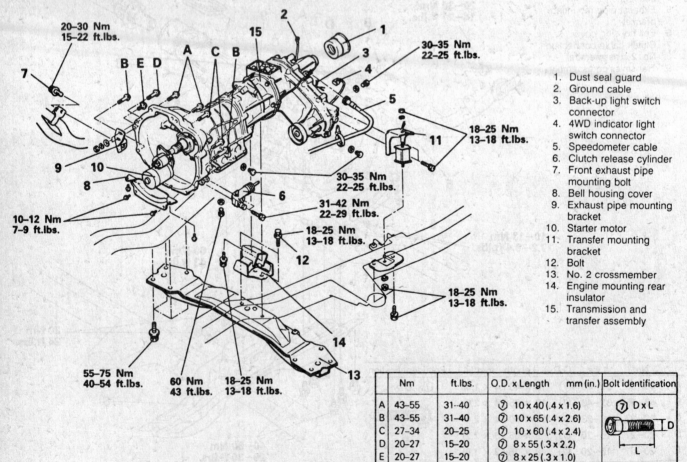

1. Dust seal guard
2. Ground cable
3. Back-up light switch connector
4. 4WD indicator light switch connector
5. Speedometer cable
6. Clutch release cylinder
7. Front exhaust pipe mounting bolt
8. Bell housing cover
9. Exhaust pipe mounting bracket
10. Starter motor
11. Transfer mounting bracket
12. Bolt
13. No. 2 crossmember
14. Engine mounting rear insulator
15. Transmission and transfer assembly

	Nm	ft.lbs	O.D. x Length mm (in.)	Bolt identification
A	43–55	31–40	10 x 40 (.4 x 1.6)	
B	43–55	31–40	10 x 65 (.4 x 2.6)	
C	27–34	20–25	10 x 60 (.4 x 2.4)	
D	20–27	15–20	8 x 55 (.3 x 2.2)	
E	20–27	15–20	8 x 25 (.3 x 1.0)	

Montero transmission assembly (2.6L) removal. Observe torque values when reinstalling

DRIVE TRAIN 7

1. Transmission case
2. Main drive pinion
3. 3rd/4th synchronizer
4. 3rd speed gear
5. 2nd speed gear
6. 1st/2nd synchronizer
7. 1st speed gear
8. Rear bearing retainer
9. Reverse gear
10. Control finger
11. Control shaft
12. Control lever cover
13. Control lever
14. Stopper plate
15. Control housing
16. Change shifter
17. Mainshaft
18. Extension housing
19. Counter reverse gear
20. Revere idler gear
21. Reverse idler gear shaft
22. Undercover
23. Counter gear
24. Front bearing retainer
25. Clutch shift arm
26. Release bearing carrier
27. Clutch control shaft
28. Return spring

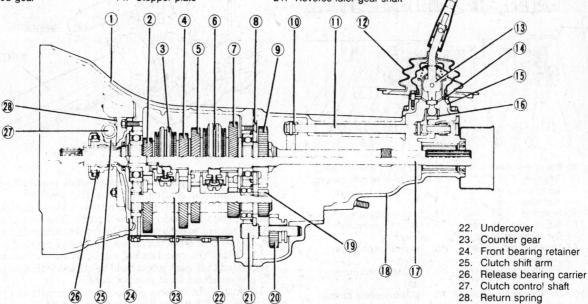

KM-130 transmission components

KM130 4-Speed Overhaul

DISASSEMBLY

1. Remove the undercover.
2. Remove the back-up light switch. Be careful not to lose the steel ball.
3. Remove the speedometer gear sleeve clamp and remove the speedometer driven gear and sleeve assembly from the extension housing assembly.
4. Remove the extension housing bolts. Turn the shift lever to the left and pull off the extension housing.
5. Loosen the three poppet plugs, then remove the three poppet springs and the three steel balls.
6. Place the 1st/2nd speed shift rod in Neutral position.
7. Remove the reverse shift rail and fork assembly together with the reverse idler gear.
8. Using a $3/16$ in. punch, drive off the 3rd/4th and 1st/2nd speed shift fork spring pins. Push each shift rod toward the rear of the transmission case and remove the shift forks. Remember to remove the interlock plunger.
9. Remove the snapring from the rear end of the counter gear and then remove the reverse counter gear.
10. Unlock the mainshaft locknut and remove the locknut. The locknut can be loosened by double-engaging the 3rd speed gear and the 1st speed gear.
11. Remove the reverse gear from the mainshaft.
12. Remove the five attaching screws and then remove the rear bearing retainer.
13. Remove the front bearing retainer.
14. With the counter gear pressed to the rear, remove the rear bearing snapring. Using a bearing puller, remove the rear counter bearing.
15. Remove the snapring from the front counter bearing. Pull off the bearing with a bearing puller.
16. Pull the counter gear out of the case.
17. Remove the main drive pinion from the front of the case. To remove the bearing from the main drive pinion, remove the two snaprings and then remove the bearing with a bearing puller.
18. Remove the mainshaft bearing snapring and remove the bearing using a dual post bearing puller (special tool MD998056-10 and MD998056 or equivalents).
19. Remove the mainshaft assembly by lifting it up through the case.
20. Disassemble the mainshaft assembly in the following order: Pull off the 1st speed gear, the 1st/2nd speed synchronizer and the 2nd speed gear toward the rear of the mainshaft. Remove the snapring from the forward end of the mainshaft, then remove the 3rd/4th speed synchronizer and the 3rd speed gear.
21. If removing the shift control shaft assembly, remove the pin locking the gear shifter using a $3/16$ in. punch. To remove the lock pin, press the gear shifter forward and drive the lock pin off, being careful not to bend the control shaft. Inspect the parts after cleaning. Replace any worn, damaged or defective parts.

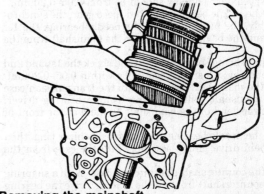

Removing the mainshaft

7-43

7 DRIVE TRAIN

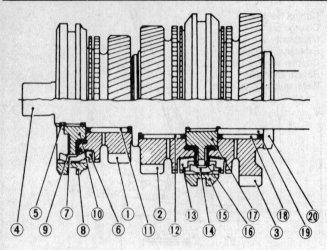

1. 3rd speed gear
2. 2nd speed gear
3. 1st speed gear
4. Mainshaft
5. Snapring
6. 3rd/4th synchronizer
7. Synchronizer piece
8. 3rd/4th synchronizer sleeve
9. 3rd/4th synchronizer spring
10. 3rd/4th synchronizer hub
11. 3rd gear needle bearing
12. 2nd gear needle bearing
13. 1st/2nd synchronizer ring
14. Synchronizer piece
15. 1st/2nd synchronizer sleeve
16. 1st/2nd synchronizer spring
17. 1st/2nd synchronizer hub
18. 1st gear needle bearing
19. 1st gear bearing sleeve
20. Bearing spacer

Mainshaft components

ASSEMBLY

1. If the main drive pinion bearing has been removed, replace it using a pipe fit over the end of the pinion shaft. Make sure the pipe does not apply pressure on the ball bearings but only on the bearing race, or bearing damage could result.
2. Fit a snapring which gives a clearance of no more than 0–0.05mm and install it on the drive pinion.
3. Assemble the mainshaft in the following order: Assemble the 3rd/4th speed and the 1st/2nd speed synchronizers. The front and rear ends of the synchronizer sleeve and hub can be identified as shown in the illustration. The synchronizer spring is installed as shown. Install the needle bearing, 3rd speed gear, synchronizer ring and the 3rd/4th speed synchronizer assembly onto the mainshaft from the front end. Be careful not to confuse the front and the rear of the synchronizer assembly.
4. Select and install a snapring that will give the 3rd/4th speed synchronizer hub an endplay of 0–0.08mm.
5. Third speed gear endplay should be from 0.05–0.20mm.
6. Install the needle bearing, the 2nd speed gear, the synchronizer assembly, the bearing sleeve, the needle bearing, the 1st speed gear, and the bearing spacer onto the mainshaft from the rear end.
7. Push the bearing spacer forward and check the 1st and 2nd speed gear endplay. Clearance should be within 0.05–0.20mm.
8. Insert the mainshaft assembly into the transmission case and fit the mainshaft center bearing using a bearing driver. Hold the forward end of the mainshaft by hand at the front of the case.
9. Install the needle bearing and the synchronizer ring, then insert the main drive pinion assembly into the case from the front.
10. Insert the countershaft gear into the case. With a snapring fitted to the countershaft front needle bearing, drive the bearing into the case by hammering on the outer race of the bearing.

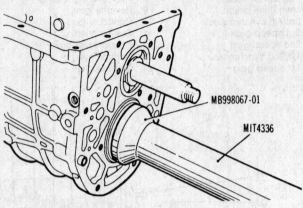

Installing the mainshaft bearing

11. Fit a snapring to the countershaft rear ball bearing and then install the bearing with a bearing installer.
12. Install the front bearing retainer. When installing the retainer, install a spacer that will give a clearance of 0–0.10mm. Apply sealant to both sides of the front bearing retainer packing and apply gear oil to the oil seal lip. Install packing and oil seal.
13. Install the rear bearing retainer and its five screws. It is suggested that each screw head be staked with a pointed punch to prevent them from coming loose.
14. Install the reverse gear on the mainshaft and tighten the locknut to 73–94 ft. lbs. Lock the nut at the notch of the mainshaft.
15. Install the spacer and counter reverse gear to the counter gear rear end.
16. Install a snapring of the proper size so that the reverse counter gear and play will be from 0–0.08mm.
17. Install the 3rd/4th and 1st/2nd speed shift forks into their respective synchronizer sleeves. Insert each shift rod from the rear of the case. Lock the shift forks and rod with spring pins, install the interlock plunger between the shift rods.

NOTE: The spring pins should be installed with the slits parallel to the shift rod.

18. Install the reverse shift rod and fork assembly together with the reverse idler gear.
19. Insert the ball and poppet spring with the small end on the ball side into each shift rod. Tighten the plugs to the specified positions. After installation, seal each plug head with sealant.
20. Apply sealant to both sides of the extension housing packing and fit the packing into the housing.
21. Turn the gear shift control down to the left and install the extension to the transmission case.
22. Make sure the forward end of the control finger is snug in the slot of the shift lug and fit the extension housing bolts after coating their threads with sealant.
23. Apply gear oil to the speedometer driven gear and install the gear and sleeve assembly in the extension housing. Make sure the sleeve flange and its mating areas on the extension housing are free of dirt, or it will cause the gears to be misaligned and could damage them.
24. Rotate the speedometer driven gear and sleeve assembly so that the number on the sleeve (which is the number of teeth on the gear) is in the 'U' mark position as the assembly is installed.
25. Install the speedometer gear clamp with its tongs in the sleeve positioning slots.
26. Install the backup light switch with its steel ball.
27. Refit the under cover and torque the bolts to 6–7 ft. lbs.
28. Install the transmission control lever assembly and fill the gear shifter area with grease. Fill the transmission with lubricant.

DRIVE TRAIN 7

KM132 5-Speed Overhaul

DISASSEMBLY

1. Remove the clutch release bearing and carrier.
2. Remove the spring pin and the clutch control shaft. Remove the felt, return spring and clutch shift arm.
3. Remove the case cover.
4. Remove the back-up light switch.
5. Remove the extension housing.
6. Remove the speedometer drive gear.
7. Remove the ball bearing from the mainshaft rear end.
8. Loosen three poppet spring plugs, then remove three poppet springs and three balls.
9. Remove the 3rd/4th and 1st/2nd speed shift fork spring pins. Pull off each shift rail toward the rear of the transmission case, then remove the shift fork. Remove the interlock plunger.
10. Remove the overtop and reverse shift forks spring pins, shift rails and forks.
11. Loosen the locknuts (mainshaft and countershaft rear ends).
12. Pull off the counter overtop gear and the ball bearing at the same time using a puller. Remove the spacer and the counter reverse gear.

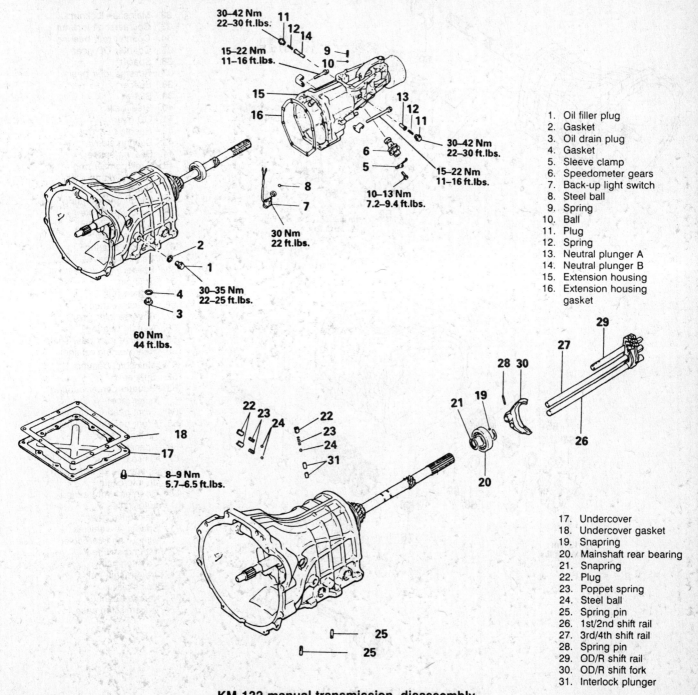

1. Oil filler plug
2. Gasket
3. Oil drain plug
4. Gasket
5. Sleeve clamp
6. Speedometer gears
7. Back-up light switch
8. Steel ball
9. Spring
10. Ball
11. Plug
12. Spring
13. Neutral plunger A
14. Neutral plunger B
15. Extension housing
16. Extension housing gasket
17. Undercover
18. Undercover gasket
19. Snapring
20. Mainshaft rear bearing
21. Snapring
22. Plug
23. Poppet spring
24. Steel ball
25. Spring pin
26. 1st/2nd shift rail
27. 3rd/4th shift rail
28. Spring pin
29. OD/R shift rail
30. OD/R shift fork
31. Interlock plunger

KM-132 manual transmission, disassembly

7-45

7 DRIVE TRAIN

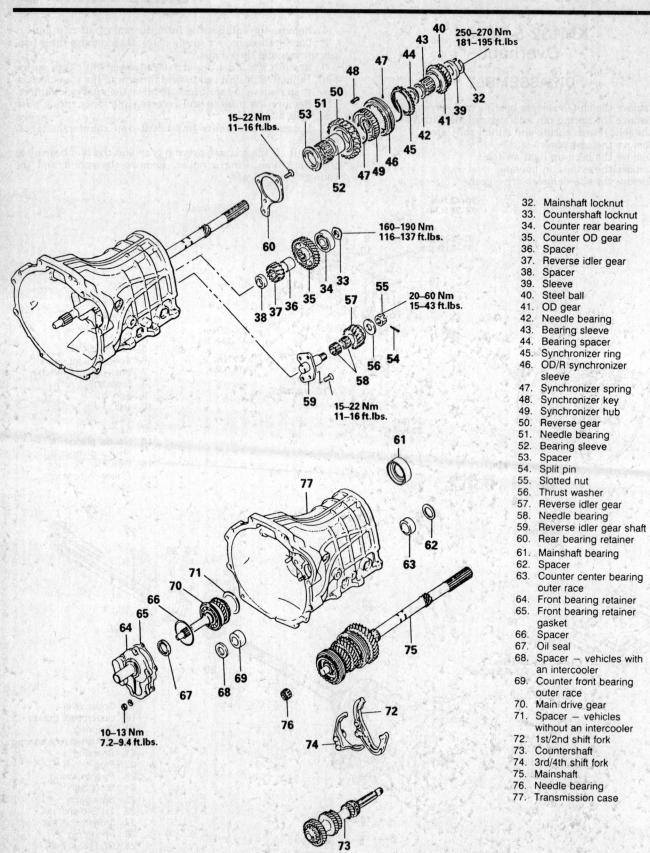

32. Mainshaft locknut
33. Countershaft locknut
34. Counter rear bearing
35. Counter OD gear
36. Spacer
37. Reverse idler gear
38. Spacer
39. Sleeve
40. Steel ball
41. OD gear
42. Needle bearing
43. Bearing sleeve
44. Bearing spacer
45. Synchronizer ring
46. OD/R synchronizer sleeve
47. Synchronizer spring
48. Synchronizer key
49. Synchronizer hub
50. Reverse gear
51. Needle bearing
52. Bearing sleeve
53. Spacer
54. Split pin
55. Slotted nut
56. Thrust washer
57. Reverse idler gear
58. Needle bearing
59. Reverse idler gear shaft
60. Rear bearing retainer
61. Mainshaft bearing
62. Spacer
63. Counter center bearing outer race
64. Front bearing retainer
65. Front bearing retainer gasket
66. Spacer
67. Oil seal
68. Spacer — vehicles with an intercooler
69. Counter front bearing outer race
70. Main drive gear
71. Spacer — vehicles without an intercooler
72. 1st/2nd shift fork
73. Countershaft
74. 3rd/4th shift fork
75. Mainshaft
76. Needle bearing
77. Transmission case

KM-132 manual transmission, disassembly

7–46

DRIVE TRAIN 7

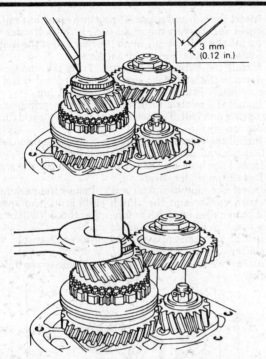
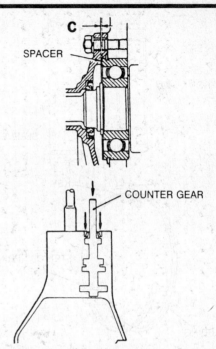

Use a chisel to open the lock tabs on the main and countershaft nuts

13. Remove the overtop gear and sleeve from the mainshaft. Remove the overtop synchronizer assembly and spacer.
14. Remove the reverse idler gear.
15. Remove the rear bearing retainer.
16. Drive the reverse idler gear shaft from inside the case.
17. Remove the front bearing retainer.
18. With the counter gear pressed to the rear, remove the rear bearing snapring. Remove the counter rear bearing.
19. Remove the counter front bearing.
20. Remove the counter gear from the inside of the case.
21. Remove the main drive pinion from the front of the case. Remove the main drive pinion bearing.
22. Remove the mainshaft bearing snapring. Remove the ball bearing.
23. Pull the mainshaft assembly from the case.
24. Disassemble the mainshaft in the following order: Remove the 1st gear, the 1st/2nd speed synchronizer and the 2nd speed gear toward the rear of the mainshaft. Remove the snapring from the forward end of the mainshaft. Remove the 3rd/4th speed synchronizer and the 3rd gear.
25. Disassemble the extension housing: Remove the lock plate and the speedometer driven gear. Remove the plug, spring and neutral return plunger.

When removing the control shaft assembly, pull off the lock pin locking the gear shifter. To remove the lock pin, press the gear shifter forward and pull it off.

ASSEMBLY

1. Install the ball bearing on the main drive pinion. Install a selective snapring so that there will be 0–0.05mm clearance between the snapring and the bearing. Thickness of snapring and identifying color: 2.30mm white; 2.35mm none; 2.40mm red; 2.45mm blue; 2.50mm yellow.
2. Install the mainshaft in the following order: Assemble the 3rd/4th speed and 1st/2nd speed synchronizers. Be sure the synchronizer assemblies are installed facing in the proper direction. Install the needle bearing, the 3rd speed gear, the synchronizer ring, and the 3rd/4th speed synchronizer assembly on to the

Select the correct spacer fro the front bearing retainer. Dimension "C" must equal 0–0.1mm (0–0.0039 in.)

mainshaft from the front end. Select and install a snapring of proper size so that the 3rd/4th speed synchronizer hub endplay will be 0–0.08mm. Check the 3rd gear endplay (0.04–0.20mm). Install the needle bearing, the 2nd speed gear, the synchronizer assembly, the bearing sleeve, the needle bearing, the 1st speed gear, and the bearing spacer on the mainshaft from the rear. With the bearing spacer pressed forward, check the 2nd and 1st gear endplay (0.04–0.20mm).
3. Install the mainshaft into the transmission case and drive in the mainshaft center bearing.
4. Install the needle bearing and the synchronizer ring. Install the main drive pinion assembly into the case from the front.
5. Install the countershaft gear into the case. Drive the front bearing into the case.
6. Install the snapring on the countershaft rear bearing.
7. Install the front bearing retainer. Select and install a spacer of proper size so that the clearance will be 0–0.01mm. Replace the front bearing retainer oil seal.
8. Install the rear bearing retainer.
9. Install the reverse idler gear shaft.
10. Install the needle bearing, the reverse idler gear and the thrust washer. Check the reverse idler gear endplay (0.12–0.28mm). Install the thrust washer with the ground side toward the gear side.
11. Assemble the overtop synchronizer.
12. Install the spacer, the stop plate, the overtop synchronizer assembly, the overtop gear bearing sleeve, the needle bearing, the synchronizer ring and the overtop gear in the written order on to the mainshaft from the rear end. Check the overtop gear endplay.
13. Install the spacer, the counter reverse gear, the spacer, the counter overtop gear and the ball bearing on to the countershaft gear from the rear end.
14. Insert the 3rd/4th and 1st/2nd speed shift forks into respective synchronizer sleeves. Insert each shift rail from the rear of the case. Lock the shift forks and rails with spring pins. Install an interlock plunger between shift rails. The pin should be installed with the slit in the axial direction of the shift rail.

7–47

7 DRIVE TRAIN

15. Insert the ball and poppet spring into each shift rail. Install the poppet spring with the small end on the ball side.
16. Install the ball bearing on to the rear end of the mainshaft.
17. Install the speedometer drive gear.
18. Install the extension housing. Turn the change shifter fully down to the left. Make sure the forward end of the control finger is snugly fitted in the slot of the shift lug.
19. Install the neutral return plungers, the spring, and resistance spring and ball. Tighten each plug till its top is flush with the boss top surface.
20. Install the speedometer driven gear sleeve into the extension housing and into mesh with the drive gear.
21. Install the back-up light switch. Remember the steel ball.
22. Install the under cover.
23. Insert the clutch control shaft. Install the packing (felt), the return spring and the clutch shift arm. The spring pin should be installed in such a manner that the slip will be at right angles with the axis of the control shaft.
24. Install the transmission control lever assembly. Fill the gear shifter area with grease.
25. After reassembly, rotate the drive pinion to see if it rotates smoothly.

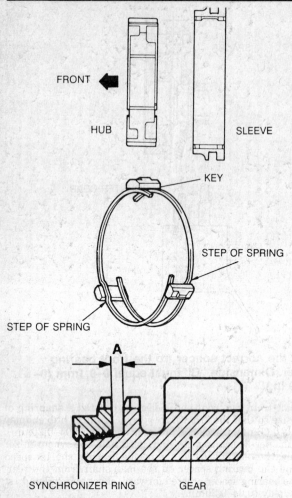

Correct assembly of the synchronizers. If dimension A is less than 0.5mm (0.020 in.) the ring and/or gear must be replaced

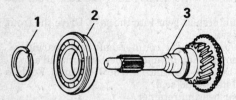

1. Snapring
2. Bearing
3. Main drive gear

KM-132 main drive gear.

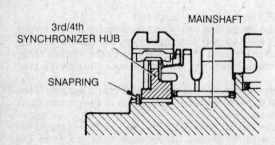

Correct assembly of 3-4 synchronizer on mainshaft

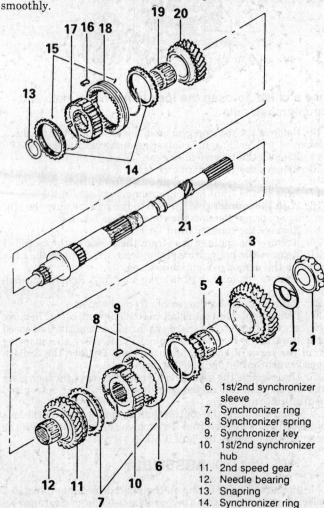

1. Ball bearing inner race
2. Spacer — vehicles without an intercooler
3. 1st speed gear
4. Bearing sleeve
5. Needle bearing
6. 1st/2nd synchronizer sleeve
7. Synchronizer ring
8. Synchronizer spring
9. Synchronizer key
10. 1st/2nd synchronizer hub
11. 2nd speed gear
12. Needle bearing
13. Snapring
14. Synchronizer ring
15. Synchronizer spring
16. Synchronizer key
17. 3rd/4th synchronizer hub
18. 3rd/4th synchronizer sleeve
19. Needle bearing
20. 3rd speed gear
21. Mainshaft

KM-132 mainshaft

DRIVE TRAIN 7

KM145 5-Speed Overhaul

CASE DISASSEMBLY

1. Remove the transmission. Make sure the transmission is in the **N** position.
2. Remove the nuts retaining the clutch housing to the transmission and remove the housing. If not removed, remove the clutch slave cylinder from the input shaft.
3. Remove the back-up lamp switch and shift indicator switch from the transmission.
4. Drain the fluid from the transmission by removing the drain plug from the pan.
5. Remove the bolts retaining the pan to the case and remove the pan. Remove and discard the gasket. Remove all traces of the gasket from the mating surfaces of the pan and case.
6. Remove the bolts retaining the cover to the transfer case adapter and remove the cover (with stopper bracket inside). Remove and discard the gasket. Clean all traces of gasket material from the mating surfaces of the adapter and cover.
7. Remove the bolts retaining the control housing to the transfer case adapter and remove the control housing (with reverse lockout assembly attached). Remove and discard the rubber seal. Clean the mating surfaces of the adapter and control housing.
8. Remove the detent spring and ball from the adapter.
9. Remove the 3 shift gate roll pin access plugs (2 on the side, 1 on the bottom) with a 6mm allen-head wrench.
10. Using a pin punch, drive the roll pins from the shift gates through the access holes.
11. From the right side of the adapter, remove the bolt, spring and neutral return plunger. Note that the plunger has a slot in the center for the detent ball.
12. From the left side of the adapter, remove the bolt, spring and neutral return plunger.
13. Lift the gate selector lever out of the shift gates. Move the lever to the rear of the adapter as far as it will go. This will allow clearance to remove the adapter from the case.
14. Remove the bolts retaining the transfer case adapter to the transmission case. Note that 3 different bolts lengths (35mm, 55mm and 110mm) are used to retain the case to the adapter. Mark the bolt holes accordingly.
15. Remove the adapter from the case. Make sure the shift gates do not bind in the adapter during removal. Rotate the gates on the rails as required. Remove and discard the gasket. Clean all traces of gasket material from the mating surfaces of the case and adapter.
16. Identify each shift rail and gate. Remove the gates from the rails.
17. From inside the case, drive out the roll pins retaining the 1st/2nd and 3rd/4th shift forks to the rails. Remove the 1st/2nd shift fork.
18. Drive out the 5th/reverse shift fork roll pin.

NOTE: The roll pin in the switch actuator does not need to be removed for transmission disassembly.

19. Remove the set screw on the top of the case and remove the poppet spring and steel ball. Remove the 2 bolts on the side of the case. Remove the poppet springs and steel balls.
20. Remove the 5th/reverse shift rail and the 3rd/4th shift rail from the case. Remove the 5th/reverse shift fork. When the 2 shift rails are removed, the interlock pins can be removed from the case.

NOTE: The 1st/2nd shift rail is unable to be removed at this time.

21. Unstake the locknuts on the mainshaft and countershaft using the mainshaft locknut staking tool, or equivalent.
22. Double engage the transmission in 2 gears to lockup the transmission. This is done by engaging 2 of the synchronizers. This is necessary to remove the locknuts.
23. Remove the countershaft locknut with a 30mm socket. Discard the locknut.
24. Remove the mainshaft locknut. Discard the locknut.
25. Pull the rear bearing off the mainshaft using the mainshaft bearing puller or equivalent. Remove and discard the bearing.
26. Remove the spacer and lock ball from the mainshaft.
27. Remove the counter/5th gear and ball bearing from the countershaft by installing the jaws of a suitable puller, behind the gear. While removing the gear, remove the 1st/2nd shift rail from the case.
28. Remove the 1st/2nd and 3rd/4th shift forks from the case.
29. Remove the 5th gear, needle bearing, spacer and synchronizer ring from the mainshaft.
30. Remove the 5th gear synchronizer sleeve from the synchronizer hub on the mainshaft.

NOTE: Do not lose the 3 keys and 2 springs in the hub. A spring is located on each side of the hub.

31. Pull the 5th gear synchronizer hub and 5th gear bearing sleeve from the mainshaft using a suitable bearing puller.
32. Slide the reverse gear and needle bearing assembly off the mainshaft.
33. Slide the counter/reverse gear and distance spacer off the countershaft.
34. Remove the cotter pin and nut from the reverse idler shaft. Remove the thrust washer, reverse idler gear and 2 sets of needle bearings from the shaft.
35. Remove the 6mm allen-head bolts that attach the mainshaft rear bearing retainer to the case and remove the retainer. Remove and discard the gasket. Clean any traces of gasket material from the mating surfaces of the case and retainer.
36. Remove the 6mm allen-head bolts that retain the reverse idler gearshaft assembly to the case.
37. Pull the reverse/idler gearshaft assembly out of the case using a suitable slide hammer.

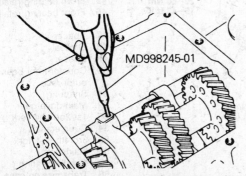

Use the correct tool to remove the spring pins

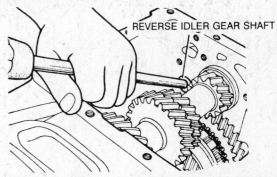

After removing 4 bolts, the reverse idler is driven from inside the case

7 DRIVE TRAIN

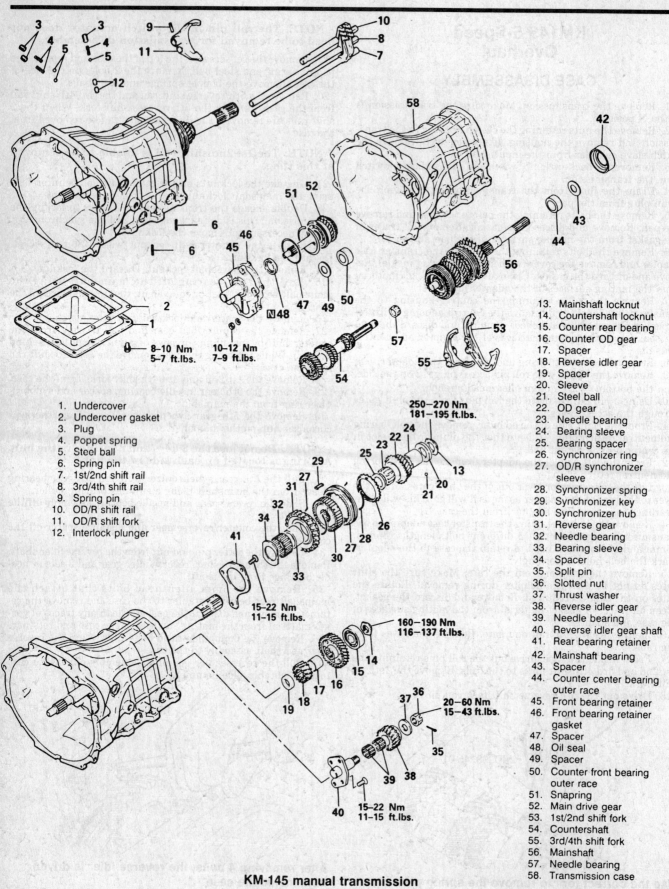

1. Undercover
2. Undercover gasket
3. Plug
4. Poppet spring
5. Steel ball
6. Spring pin
7. 1st/2nd shift rail
8. 3rd/4th shift rail
9. Spring pin
10. OD/R shift rail
11. OD/R shift fork
12. Interlock plunger
13. Mainshaft locknut
14. Countershaft locknut
15. Counter rear bearing
16. Counter OD gear
17. Spacer
18. Reverse idler gear
19. Spacer
20. Sleeve
21. Steel ball
22. OD gear
23. Needle bearing
24. Bearing sleeve
25. Bearing spacer
26. Synchronizer ring
27. OD/R synchronizer sleeve
28. Synchronizer spring
29. Synchronizer key
30. Synchronizer hub
31. Reverse gear
32. Needle bearing
33. Bearing sleeve
34. Spacer
35. Split pin
36. Slotted nut
37. Thrust washer
38. Reverse idler gear
39. Needle bearing
40. Reverse idler gear shaft
41. Rear bearing retainer
42. Mainshaft bearing
43. Spacer
44. Counter center bearing outer race
45. Front bearing retainer
46. Front bearing retainer gasket
47. Spacer
48. Oil seal
49. Spacer
50. Counter front bearing outer race
51. Snapring
52. Main drive gear
53. 1st/2nd shift fork
54. Countershaft
55. 3rd/4th shift fork
56. Mainshaft
57. Needle bearing
58. Transmission case

KM-145 manual transmission

DRIVE TRAIN 7

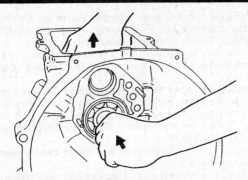

Remove the snapring; then remove the countershaft while pressing on the main drive gear

38. Use a double nut procedure to remove the studs that retain the input shaft front bearing retainer to the case. Remove the bolts that attach the retainer to the case.
39. Remove the input shaft front bearing retainer from the case. Remove and discard the gasket. Clean all traces of gasket material from the mating surfaces of the case and retainer.
40. Remove the selective shim from inside the input shaft front bearing retainer.

NOTE: Do not discard the selective shim.

41. Remove the small selective snapring that retains the input shaft to the bearing.

NOTE: Do not discard the selective snapring.

42. Remove the large snapring that retains the input shaft bearing to the case.
43. Remove the bearing from the input shaft and case. Remove and discard the bearing.
44. Rotate the input shaft so the flats on the shaft face the countershaft, providing clearance to remove the input shaft. Remove the input shaft. The output shaft (mainshaft) may have to be pulled to the rear of the case. Remove the small caged needle bearing from the inside of the input gear.
45. Remove the snapring from the mainshaft (output shaft) outer bearing race.
46. Remove the outer mainshaft bearing race, ball bearing and bearing sleeve from the case. Discard the outer bearing race and ball bearing.

NOTE: The inner race of the front bearing will remain on the mainshaft.

47. Remove the countershaft front spacer and bearing race.
48. Remove the countershaft from the case. The mainshaft assembly may have to be moved slightly to the side to allow clearance for countershaft removal.
49. Remove the mainshaft assembly from the case.

MAINSHAFT DISASSEMBLY

1. Remove the selective snapring that retains the 3rd/4th synchronizer assembly to the mainshaft. A new snapring will be used in the assembly.
2. Remove the 3rd/4th synchronizer assembly (hub, sleeve, spring and keys), synchronizer ring, 3rd speed gear and caged needle bearing from the front of the mainshaft. Note the position of the synchronizer hub and sleeve during disassembly.

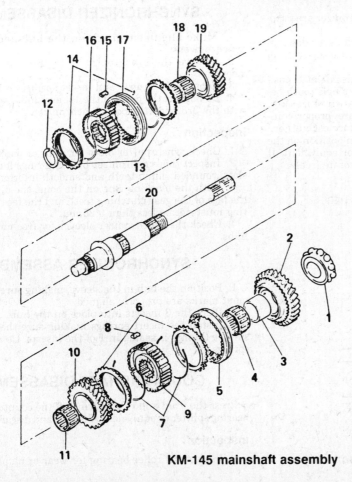

1. Ball bearing inner race
2. 1st speed gear
3. Bearing sleeve
4. Needle bearing
5. 1st/2nd synchronizer sleeve
6. Synchronizer ring
7. Synchronizer spring
8. Synchronizer key
9. 1st/2nd synchronizer hub
10. 2nd speed gear
11. Needle bearing
12. Snapring
13. Synchronizer ring
14. Synchronizer spring
15. Synchronizer key
16. 3rd/4th synchronizer hub
17. 3rd/4th synchronizer sleeve
18. Needle bearing
19. 3rd speed gear
20. Mainshaft

KM-145 mainshaft assembly

7 DRIVE TRAIN

3. Position the mainshaft assembly in a press so the 2nd speed gear is supported by the press bed. Press the mainshaft down and out from the 1st/2nd gear assembly.
4. Separate the inner ball bearing, bearing sleeve, 1st speed gear, caged needle bearing, 1st/2nd synchronizer assembly (hub, sleeve, rings and keys), 2nd speed gear and caged needle bearing. Note the direction of the 1st/2nd synchronizer hub and sleeve during disassembly. Discard the inner ball bearing.

Inspection

1. Carefully inspect all gears and synchronizers for any signs of wear, stress, discoloration, cracks or warpage.
2. Inspect the synchronizer sleeves for excessive wear or distortion.
3. Inspect all roller and needle bearings.
4. Replace defective parts as necessary.

MAINSHAFT ASSEMBLY

NOTE: Prior to assembly, lubricate all components with standard transmission lubricant, SAE 80W or equivalent.

1. Check the clearance between the synchronizer rings and gears. Install the ring on the gear and insert a feeler gauge between the ring teeth and gear. If the clearance is less than 0.23mm, replace the ring and/or gear.
2. From the rear of the mainshaft, install the caged needle bearing for the 2nd speed gear.
3. Position the 2nd speed gear on the mainshaft with the synchronizer ring surface facing the rear of the shaft.
4. Install the synchronizer ring on the 2nd speed gear.
5. Position the 1st/2nd synchronizer assembly on the rear of the mainshaft, making sure of the following:
   a. The splines of the mainshaft and synchronizer are properly aligned.
   b. The rear of the 1st/2nd hub is identified by a ridge machined on the rear surface. The ridge must face the front of the mainshaft.
   c. The synchronizer sleeve has a tooth missing at 6 positions. Assemble the hub to the sleeve so the single tooth between the 2 missing portions will touch the synchronizer key.
   d. The synchronizer keys and springs are properly installed. The open ends of the spring do not face each other.
6. Press the 1st/2nd synchronizer assembly in position on the mainshaft using a replacing shaft sleeve tool or equivalent. If properly installed, the 2nd speed gear should rotate freely.

7. Position the 1st gear bearing sleeve on the mainshaft. Press the sleeve on the shaft using a replacing shaft sleeve or equivalent. When properly installed, the sleeve should be against the synchronizer hub. Make sure the gears rotate freely.
8. Install the synchronizer ring on the 1st/2nd synchronizer assembly.
9. Install the caged needle bearing and 1st speed gear.
10. Slide the inner ball in position on the mainshaft.
11. Press the inner ball bearing on the mainshaft using rack bushing holder or equivalent. When properly installed, the gears should rotate freely.
12. Install the 3rd speed gear and caged needle bearing over the front of the mainshaft.
13. Install the synchronizer ring against the 3rd speed gear.
14. Make sure the 3rd/4th synchronizer assembly is properly installed. Be sure of the following:
   a. The splines of the mainshaft and synchronizer are properly aligned.
   b. The small diameter boss of the hub faces the front of the mainshaft.
   c. The small bevel angle of the sleeve faces the front of the mainshaft.
   d. The synchronizer sleeve has a tooth missing at 6 positions. Assemble the hub to the sleeve so the single tooth between the 2 missing portions will touch the synchronizer key.
   e. The synchronizer springs and keys are properly installed. The open ends of each spring do not face each other.
15. Install the 3rd/4th synchronizer assembly on the front of the mainshaft.
16. Install a new selective snapring that retains the 3rd/4th synchronizer assembly to the mainshaft. Select the thickest snapring that fits in the groove.

SYNCHRONIZER DISASSEMBLY

1. Make alignment mark on the hub and sleeve of the synchronizer.
2. Remove the synchronizer hub from each synchronizer sleeve.
3. Remove the inserts and insert springs from the hubs.
4. Do not mix the parts of the 1st and 2nd speed synchronizer with the 3rd and 4th speed synchronizer.

Inspection

1. Check synchronizer discs for wear or damage.
2. Inspect the synchronizer blocking rings for widened index slots, rounded clutch teeth and smooth internal surfaces.
3. With the blocking ring on the cone, the distance between the face of the gear clutching teeth and the face of the blocking ring must not be less than 0.25mm.
4. Check the synchronizer sleeves for free movement on the hubs.

SYNCHRONIZER ASSEMBLY

1. Position the hub in the sleeve, making sure that the alignment marks are properly aligned.
2. Place the 3 inserts into place on the hub.
3. Install the insert springs making sure that the irregular surface (hump) is seated in 1 of the inserts. Do not stagger the springs.

COUNTERSHAFT DISASSEMBLY

Press the front and rear bearing off the countershaft, using a bearing splitter or equivalent. Remove and discard the bearings.

Inspection

1. Inspect the roller bearing for wear or damage.

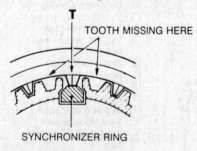

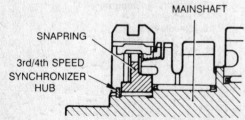

The 3–4 synchronizer must be installed correctly

DRIVE TRAIN 7

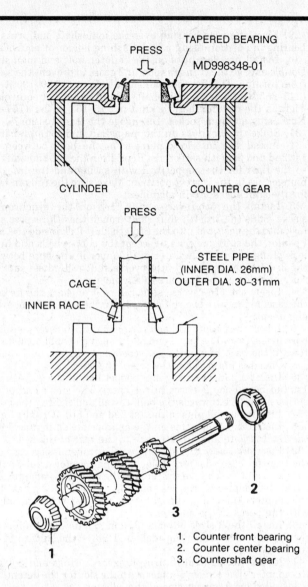

1. Counter front bearing
2. Counter center bearing
3. Countershaft gear

Incorrect removal or installation of the countershaft bearing will result in bearing damage

2. Check all gears for chipped, broken or worn teeth.
3. Inspect the countershaft for wear. Replace defective parts as needed.

COUNTERSHAFT ASSEMBLY

Press the new bearings on the countershaft using a press and countershaft bearing replacer, or equivalent.

INPUT SHAFT DISASSEMBLY

Position bearing splitter, or equivalent, behind the bearing and press the input shaft out of the bearing. Discard the bearing.

Inspection

1. Inspect the input shaft bearing for wear or distortion.
2. Inspect the gear teeth for cracks or chips.

INPUT SHAFT ASSEMBLY

Position a new bearing on the input shaft and press the bearing onto the shaft using an suitable bearing press.

CASE ASSEMBLY

1. Install the mainshaft assembly in the case.
2. Choose and install a new selective snapring in front of the input shaft bearing. Select the thickest snapring that will fit in the groove.
3. Install the small caged needle bearing inside the input gear. Install the synchronizer ring on the input shaft. Check the clearance between the ring and gear. If the clearance is less than 0.23mm, replace the ring and/or input shaft.
4. Install the synchronizer ring and input shaft in the case. Rotate the input shaft so the flats face the countershaft to provide installation clearance.

NOTE: It may be necessary to tap the input shaft into position with a brass hammer.

5. Install a new snapring on the outer bearing race. The longer portion of the race must be installed in the case.
6. Slide the outer ball bearing on the mainshaft. Press the bearing on the mainshaft and in the race using a suitable press. When pressed in position, all gears and shafts must rotate freely.
7. Install the 3rd/4th shift fork into its synchronizer sleeve. The roll pin boss on the fork must face to the rear.
8. Install the countershaft into the case. It may be necessary to move the mainshaft to one side in order for the countershaft to be easily inserted.
9. Install the 1st/2nd shift fork. The roll pin boss must face toward the 3rd/4th shift fork.
10. If removed, drive a new seal into the input shaft front bearing retainer using a seal installer or equivalent.
11. Install the large snapring that retains the input shaft bearing to the case.
12. Check the input shaft front bearing retainer-to-bearing clearance as follows:
    a. With the retainer selective shim removed, use a depth micrometer to measure the distance between the top machined surface to the spacer surface (second landing). Record the reading.
    b. Bottom the input shaft bearing so the snapring is flush against the case.
    c. Using a depth micrometer, measure the distance from the top of the outer front bearing race to the machined surface of the case.
    d. Subtract the distance of the bearing-to-case from the retainer dimensions. This will give the required maximum selective shim size to obtain a 0–0.10mm clearance.
    e. Measure and install the appropriate size selective shim in the front bearing retainer.
13. Install the countershaft front outer bearing race and non-selective spacer. Install the countershaft rear outer bearing race.
14. Install a new gasket between the front bearing retainer and case. Position the retainer on the case (with selective shim installed). Install the bolts and studs, tighten to 22–30 ft. lbs. (30–41 Nm).
15. Check and adjust the countershaft endplay as follows:
    a. Place the transmission so the rear of the mainshaft and countershaft face upward. Install the countershaft rear selective spacer.
    b. Force the countershaft downward so it bottoms against the front bearing retainer.
    c. Place a straight edge across the rear countershaft selective spacer in the case.
    d. Try to turn the spacer. If the spacer turns lightly, re-

7-53

7 DRIVE TRAIN

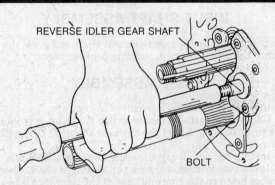

Once positioned with bolts, the reverse gear idler shaft is driven into place

place the spacer with the next larger size. Install a spacer so the clearance between the spacer and straight edge is 0–0.05mm. Install the correct size spacer over the countershaft rear bearing cup.

16. Install the rear bearing retainer on the case with the 6mm bolts. Tighten to 11–16 ft. lbs. (15–21 Nm).

NOTE: Be sure the spacer installed in Step 15 does not fall out of place when installing the rear bearing retainer.

17. Position the reverse idler gear shaft on the case. Install the 6mm allen-head bolts to act as a pilot. Install reverse idler gear shaft remover or equivalent, on the shaft and drive the assembly into place. Tighten the bolts to 11–16 ft. lbs. (15–21 Nm).
18. Install the 2 caged needle bearings, reverse idler gear and thrust washer on the idler shaft. The boss on the idler gear faces away from the transmission. Install the locknut and tighten to 15–42 ft. lbs. (20–58 Nm). If required, advance the nut to the next castellation and install the cotter pin.

NOTE: If required, cut 1 end of the cotter pin when it is bent over to prevent interference with the counter/5th gear.

19. Install the spacers and counter/reverse gear on the countershaft.
20. Press the reverse gear sleeve on the mainshaft using an appropriate press.
21. Install the caged needle bearing and reverse gear on the mainshaft.
22. Assemble the 5th gear synchronizer hub and sleeve as follows:
    a. Install the hub in the sleeve. The recessed boss on the sleeve must face the front of the transmission. The large boss on the hub must face the front of the transmission.
    b. When installing hub in the sleeve and the 3 keys, make sure that the single tooth between the 2 spaces will touch the key. Install the springs so the open ends do not face each other.
23. Install the 5th gear synchronizer on the mainshaft. The recessed boss of the sleeve must face the front of the transmission.
24. Press the sleeve of the 5th gear on the mainshaft using an appropriate installation tool.
25. Install the ring on the 5th gear synchronizer.
26. Slide the small spacer, caged needle bearing and 5th gear on the mainshaft. Check the clearance between the 5th gear and synchronizer ring. If the clearance is less than 0.23mm, replace the ring and/or 5th gear.
27. Install the counter/5th gear and ball bearing onto the countershaft along with the 1st/2nd shift rail. Seat the bearing into position using countershaft bearing replacer or equivalent. Make sure the rail engages the forks.
28. Install the lock ball and spacer on the mainshaft.

29. Place the rear bearing over the mainshaft and press the bearing in position using a rack bushing holder or equivalent.
30. Install new locknuts on the countershaft and mainshaft. Double engage the transmission in 2 gears to prevent the shafts from turning. Tighten the mainshaft locknut to 180–195 ft. lbs. (245–265 Nm) using a mainshaft locknut wrench or equivalent. Tighten the countershaft locknut to 115–135 ft. lbs. (157–186 Nm) using a 30mm socket. Disengage the transmission.
31. Stake the locknuts on the mainshaft and countershaft.
32. Install the interlock plunger in the bore between the 1st/2nd and 3rd/4th shift rails. Reposition the 1st/2nd shift rail so the flats for the poppet ball with spring and the interlock plunger are in the correct position. Make sure the roll pin holes for the shift forks are in alignment.
33. Install the 5th/reverse shift fork on the synchronizer sleeve. Slide the 3rd/4th shift rail through the 5th/reverse shift fork into the case and into the 3rd/4th shift fork inside the case. Position the shift rail flats to accept the poppet balls and interlock plunger. Insert the interlock plunger in the bore between the 3rd/4th shift rail and 5th/reverse shift rail. Make sure the roll pin holes in the fork are in alignment.
34. Insert the 5th/reverse shift rail so it engages the forks in the case. Make sure the roll pin holes in the fork and rail are in alignment.
35. Insert the poppet ball and spring in the 5th/reverse (upper) bore in the case. The small end of the spring should be installed toward the ball. Install the set screw and tighten until the set screw head is 6mm below the top of the bore.
36. Insert a poppet spring and poppet ball in the 3rd/4th and 1st/2nd bore (side 2 bores in the case). The small end of the spring must face towards the ball. Install and tighten the bolts.
37. Install the roll pins in the 1st/2nd and 3rd/4th shift forks.
38. Install the shift gates on the appropriate shift rails. Move the 1st/2nd gate and 3rd/4th gate to the rear of the rail.
39. Position a new gasket between the transmission case and the transfer case adapter. Make sure the selector arm is out of the gates and the change shifter is at the rear of the adapter. Position the adapter on the case making sure the shift gates clear the adapter. Make sure the shift rails and rear bearings line up with the bores in the adapter.
40. Install the 3 sizes of bolts (35mm, 55mm and 110mm) in the appropriate holes in the adapter. Tighten the bolts to 11–16 ft. lbs. (15–21 Nm).
41. Install the neutral return plungers, springs and bolts in the adapter. The longer plunger with the slot for the detent ball is installed on the right side of the adapter.
42. Position the shaft gates so the roll pin holes in the gates and rails are in alignment. Install the roll pins through the access holes. Install the access hole plugs.
43. Position the pan and new gasket on the case. Install the bolts and tighten to 11–16 ft. lbs. (15–21 Nm). Do not overtighten. Install the drain plug and tighten to 25–32 ft. lbs. (35–44 Nm).
44. Insert the plunger detent ball and spring in the hole above the neutral return plunger in the adapter case.
45. Make sure the stopper bracket assembly on the cover for the transfer case adapter moves smoothly. Position a new gasket on the adapter and install the housing cover. Install and tighten to bolts to 11–16 ft. lbs. (15–21 Nm).
46. Position a new seal on the adapter and install the control housing assembly. Install and tighten the retaining bolts to 11–16 ft. lbs. (15–21 Nm).
47. Install the back-up lamp switch and the shift indicator lamp switch in the adapter.
48. Remove the fill plug and fill the transmission to the bottom of the fill hole with standard transmission lube SAE 80W, or equivalent. Install the fill plug and tighten to 22–25 ft. lbs. (30–34 Nm).
49. Position the clutch slave cylinder on the input shaft. Position the clutch housing on the transmission case and install. Tighten the nuts.

DRIVE TRAIN 7

V5MT1 5-Speed Overhaul

DISASSEMBLY

1. Remove the adapter cover, gasket and breather from the transfer case adapter.
2. Remove the spring pins from the gear shift jaws.
3. Unbolt and remove the transfer case adapter. Remove the seal rings.
4. Remove the three shift jaws.
5. Remove the gear shift lower case assembly.
6. Unbolt and remove the clutch housing assembly.
7. Remove the power take-off cover and gasket from the side of the gearbox.
8. Uncrimp the large locking nut on the end of the mainshaft with a small prying tool. Remove the nut using tool MB 998809 or equivalent.
9. Remove the small lock piece holding the reverse shaft.
10. Use a slide hammer with an M12x1.25 tip to remove the id-

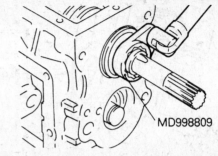

Use the special tool to loosen and tighten mainshaft locknut

1. Adapter cover
2. Adapter cover gasket
3. Air breather
4. Spring pin
5. Transfer case adapter
6. Seal ring
7. 1st/2nd gear shift fork
8. 3rd/4th gear shift fork
9. 5th/R shift fork
10. Gear shift housing lower case
11. Clutch housing
12. Oil seal
13. Power take-off cover
14. Power take-off cover gasket

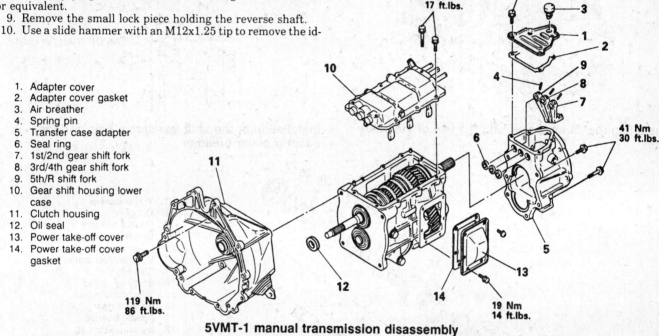

5VMT-1 manual transmission disassembly

ORDER OF DISASSEMBLY
15. Locknut
16. Lock piece
17. Reverse shaft
18. O-ring
19. Side washer
20. needle bearing
21. Reverse gear
22. Snapring
23. Ball bearing
24. Snapring
25. Snapring
26. Ball bearing
27. Snapring
28. Ball bearing
29. Snapring
30. Snapring
31. Ball bearing
32. Mainshaft
33. Drive pinion
34. Pilot bearing
35. Countershaft
36. Transmission case

ORDER OF ASSEMBLY
36. Transmission case
35. Countershaft
31. Ball bearing
29. Snapring
33. Drive pinion
30. Snapring
34. Pilot bearing
32. Mainshaft
27. Snapring
28. Ball bearing
25. Snapring
26. Ball bearing
24. Snapring
23. Ball bearing
22. Snapring
21. Reverse gear
20. needle bearing
19. Side washer
18. O-ring
17. Reverse shaft
16. Lock piece
15. Locknut

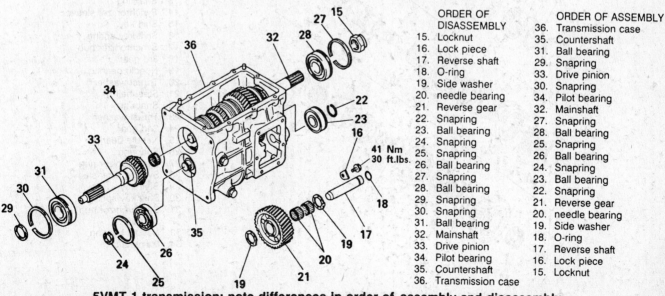

5VMT-1 transmission; note differences in order of assembly and disassembly

7-55

7 DRIVE TRAIN

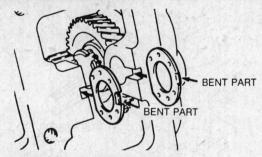

Correct installation of side washers

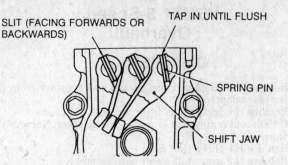

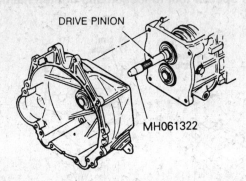

Installing the clutch cover with the use of the guide tool

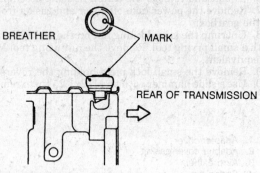

Installation of the shift jaw spring pins and the adapter cover breather

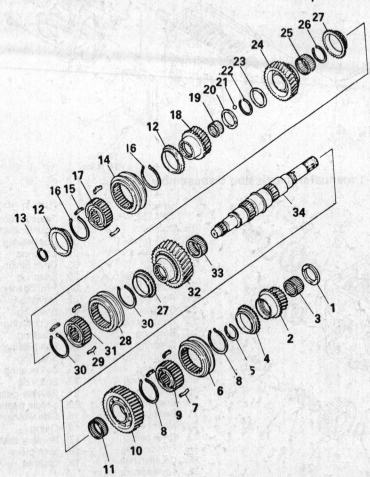

1. Thrust washer
2. OD gear
3. Needle bearing
4. Synchronizer ring
5. Snapring
6. Synchronizer sleeve
7. Shift key
8. Shift key spring
9. Synchronizer hub
10. Reverse gear
11. Needle bearing
12. Synchronizer ring
13. Snapring
14. Synchronizer sleeve
15. Shift key
16. Shift key spring
17. Synchronizer hub
18. 3rd gear
19. Needle bearing
20. Thrust washer
21. Steel ball
22. Snapring
23. Thrust washer
24. 2nd gear
25. Needle bearing
26. Snapring
27. Synchronizer ring
28. Synchronizer sleeve
29. Shift key
30. Shift key spring
31. Synchronizer hub
32. 1st gear
33. Needle bearing
34. Mainshaft

5VMT mainshaft. NONE of the snaprings are reusable

DRIVE TRAIN 7

ler shaft.
11. Remove the O-ring, side washer, needle bearing and reverse gear.
12. Remove the snaprings and ball bearings at each end of the countershaft.

NOTE: The small snaprings are NOT reusable. Discard them.

13. Remove the snaprings and ball bearings from each end of the mainshaft.
14. Separate the main shaft from the drive pinion and remove the mainshaft from inside the case.
15. Remove the drive pinion and pilot bearing through the inside of the case; it will not fit through the bearing hole.
16. Remove the counter shaft assembly.

REASSEMBLY

1. Install the counter shaft assembly into the transmission case.
2. Using the correct press set-up or bearing driver, install the ball bearing onto the drive pinion. Secure the bearing with the large snapring.
3. Insert the drive pinion from the inner side of the transmission case.
4. Tap in the drive pinion using a soft hammer so that there is no damage at the transmission case.
5. Install the snapring to the ball bearing and insert the pilot bearing to the drive pinion.
6. Insert the mainshaft assembly from the inside of the case. Insert it into the pilot bearing part of the drive pinion.
7. Press in the drive pinion until the bearing snapring contacts the transmission case.
8. Using the correct bearing driver, install the bearing and snapring at the rear of the mainshaft.
9. Using the correct bearing driver, install the bearing and snaprings at the front of the countershaft.
10. Install the bearing and snapring at the rear of the countershaft.
11. Assemble the reverse gear, needle bearing, side washer and O-ring onto the reverse shaft. Install the side washer so that the bent part is in the groove of the transmission case.
12. Install the small lock piece and the locking nut. Tighten the locking nut to 185–195 ft. lbs.
13. After tightening, the groove of the mainshaft must be staked at two locations.
14. Install the power take off cover with a new gasket.
15. Install a new seal into the clutch housing cover.
16. Apply a coating of sealant to the clutch housing mating surfaces and install it to the body of the gearbox. Apply a coat of multipurpose grease to the lip of the oil seal and install a guide on the tip of the drive pinion. Use tool MH 061322 or equivalent.
17. Install the lower gear shift case assembly.
18. Install the shift jaws with the seals into the transfer case adapter.
19. Fit the adapter into place, secure the screws and install new spring pins. Tap the pins in so that the slit faces forward or rearward.
20. Assemble the air breather and adapter cover. Install it with a new gasket. Apply a thin coat of sealant before bolting the cover in place. The breather should be mounted so the the mark on the top is to the rear of the transmission.

Montero and 4-Wheel Drive Pickup Front Driveshafts and CV-Joints

REMOVAL AND INSTALLATION

NOTE: Before beginning, place the free-wheeling hub in the free condition by placing the transfer lever in the 2H position and moving in reverse for about 6 or 7 feet.

1. Raise the vehicle and support it safely. Remove the wheel.
2. Remove the front brake caliper assembly. Do not disconnect the brake hose; simply remove the caliper as a unit and position support it from stiff wire.

CAUTION

Brake pads and shoes contain asbestos, which has been determined to be a cancer causing agent. Never clean the brake surfaces with compressed air! Avoid inhaling any dust from brake surfaces! When cleaning brakes, use commercially available brake cleaning fluids.

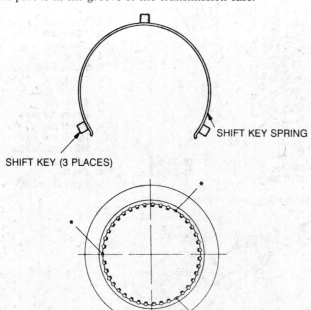

Correct installation of the shift key spring.

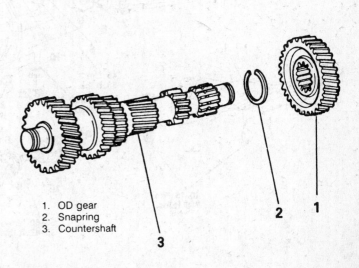

1. OD gear
2. Snapring
3. Countershaft

Countershaft assembly

7–57

7 DRIVE TRAIN

3. Remove the free wheeling hub cover assembly and remove the snapring from the axle shaft.
4. Remove the snapring and shim.
5. Disconnect the tie rod from the knuckle using the correct tools. Do not remove the nut from the ball joint until the joint is loosened.
6. Support the lower control arm with a jack. Use the correct tools to separate the lower and upper ball joints from the knuckle. Loosen the nut but do not remove it until the joint has been separated.
7. Remove the knuckle and front hub together as a unit. Lower the jack slowly after the knuckle is removed.
8. If removing the left side shaft, simply pull the shaft out of the differential carrier assembly. When pulling the left shaft from the differential carrier assembly, be careful that the shaft splines do not damage the oil seal.
9. On 1986 pickup only, raise the right lower suspension arm and remove the right shock absorber if removing the right side shaft.
10. Disconnect the right shaft from the inner shaft assembly and remove the shaft. Remove the inner shaft from the housing tube and the housing tube from the differential housing.
11. Press the bearing and seal off of the inner shaft and remove the dust seal from the tube.
12. To reinstall, use tool MB990955 to install a new dust seal to the tube. Coat the lip of the seal with grease.
13. Using a suitable long steel pipe, install a new dust cover to the inner shaft and coat the inside with grease. Press the bearing onto the shaft using tool MD990560 and a press.
14. Install a new circlip on the splines of the left side shaft or inner right shaft.
15. Drive the shafts into the differential carrier assembly with a plastic hammer. Be careful not to damage the lip of the oil seal.
16. Install the right outer shaft to the inner shaft and torque the nuts to 40 ft. lbs. (54 Nm). Install the right shock absorber if it was removed.
17. Install the knuckle and front hub assembly. Tighten the upper ball joint nut to 75 Nm (55 ft. lbs), the lower joint nut to 150 Nm (111 ft. lbs.) and the nut on the tie rod end to 45 Nm or 33 ft. lbs. Once properly set, install new cotter pins in each joint.
It may be necessary to achieve these final torques after the car is lowered to the ground. If this is your choice, tighten the nuts comfortably tight for now and do not install the cotter pins. The nuts will be difficult to reach with the vehicle on the ground.
18. Install the shim and snapring and check for proper endplay. Set a dial indicator so the pin is resting on the end of the axle shaft. The endplay specification is 0.2–0.5mm. If not within specifications, adjust by adding or removing shims.
19. Install the hub cover.
20. Install the front brake caliper assembly
21. Install the wheel.
22. Lower the vehicle to the ground. If the ball joint and tie rod nuts were not set to the proper torque earlier, do so now. After setting the correct torque, install a new cotter pin through each joint.
23. Install the undercover or splash shield.

OVERHAUL

For overhaul, see the CV-Joint and Halfshaft procedures under MANUAL TRANSAXLE, earlier in this Section. The front axle shafts of Montero and 4wd Pickups have the servicable joints (TJ or DOJ) at the outside and the BJ or RJ at the inner end.

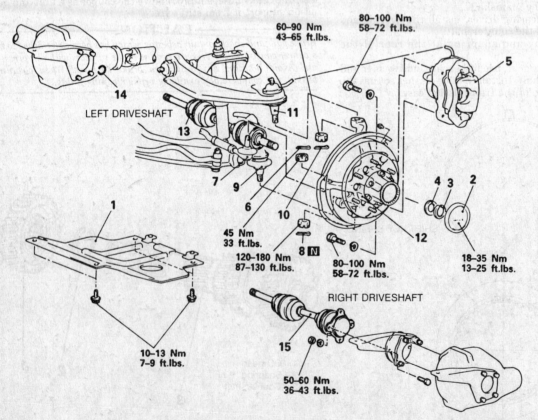

1. Undercover
2. Hub cover
3. Snapring
4. Shim
5. Front brake
6. Cotter pin
7. Tie rod end
8. Cotter pin
9. Lower ball joint
10. Cotter pin
11. Upper ball joint
12. Hub and knuckle
13. Left driveshaft
14. Circlip
15. Right driveshaft

Montero front half-shafts

DRIVE TRAIN 7

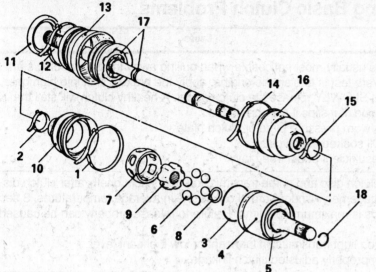

ORDER OF DISASSEMBLY
1. Boot band A
2. Boot band B
3. Circlip
4. DOJ outer race
5. Dust cover
6. Balls
7. DOJ cage
8. Snapring
9. DOJ inner race
10. DOJ boot
11. dust cover
12. Boot protector band
13. Boot protector
14. Boot band A
15. Boot band B
16. BJ boot
17. Driveshaft and BJ
18. Circlip

ORDER OF ASSEMBLY
17. Driveshaft and BJ
16. BJ boot
14. Boot band A
15. Boot band B
2. Boot band B
10. DOJ boot
1. Boot band A
7. DOJ cage
9. DOJ inner race
8. Snapring
6. Balls
5. Dust cover
4. DOJ outer race
3. Circlip
18. Circlip
13. Boot protector
12. Boot protector band
11. dust cover

Montero left front shaft and joint

CLUTCH

---- CAUTION ----
The clutch driven disc contains asbestos, which has been determined to be a cancer causing agent. Never clean clutch surfaces with compressed air. Avoid inhaling any dust from the clutch area. When cleaning clutch surfaces, use commercially available brake cleaning solvents.

Troubleshooting Basic Clutch Problems

Problem	Cause
Excessive clutch noise	Throwout bearing noises are more audible at the lower end of pedal travel. The usual causes are: • Riding the clutch • Too little pedal free-play • Lack of bearing lubrication A bad clutch shaft pilot bearing will make a high pitched squeal, when the clutch is disengaged and the transmission is in gear or within the first 2" of pedal travel. The bearing must be replaced. Noise from the clutch linkage is a clicking or snapping that can be heard or felt as the pedal is moved completely up or down. This usually requires lubrication. Transmitted engine noises are amplified by the clutch housing and heard in the passenger compartment. They are usually the result of insufficient pedal free-play and can be changed by manipulating the clutch pedal.

7-59

7 DRIVE TRAIN

Troubleshooting Basic Clutch Problems

Problem	Cause
Clutch slips (the car does not move as it should when the clutch is engaged)	This is usually most noticeable when pulling away from a standing start. A severe test is to start the engine, apply the brakes, shift into high gear and SLOWLY release the clutch pedal. A healthy clutch will stall the engine. If it slips it may be due to: • A worn pressure plate or clutch plate • Oil soaked clutch plate • Insufficient pedal free-play
Clutch drags or fails to release	The clutch disc and some transmission gears spin briefly after clutch disengagement. Under normal conditions in average temperatures, 3 seconds is maximum spin-time. Failure to release properly can be caused by: • Too light transmission lubricant or low lubricant level • Improperly adjusted clutch linkage
Low clutch life	Low clutch life is usually a result of poor driving habits or heavy duty use. Riding the clutch, pulling heavy loads, holding the car on a grade with the clutch instead of the brakes and rapid clutch engagement all contribute to low clutch life.

Adjustments

PEDAL HEIGHT

1. Measure clutch pedal height from the floor to the top of the pedal pad, along the line of pedal travel. The height should be:
   - Cordia and Tredia: 175–180mm
   - Starion: 188–193mm
   - Galant and Sigma: 173–178mm
   - Mirage: 170–175mm
   - Precis: 188–193mm
   - Montero: 185–190mm
   - Pickup: 165–170mm

2. Adjust the pedal height, if necessary, by loosening the locknut and turning the adjusting bolt or rotating clutch switch, and retighten the locknut. Now adjust the pedal play as described below.

PEDAL PLAY (FREE PLAY)

NOTE: When checking pedal play you must not move the pedal far enough to disturb the position of the hydraulic cylinder actuating rod. Move the pedal to a point where resistance stiffens but no more.

1. On the cable type clutch used in some Cordia and Tredia models, turn the outer cable adjusting nut at the floorboard to adjust the pedal play to 0–1mm. Now, operate the clutch several times and recheck the clearance between the top of the pedal pad and floor board. If it is not correct, there is a problem in the clutch or cable mechanism.

2. For hydraulic clutches used in Cordia and Tredia, all

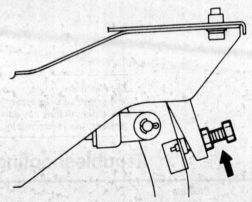

The adjusting bolt used to adjust pedal height on hydraulic clutches

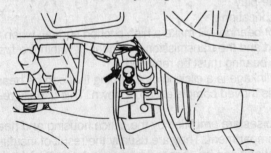

Adjusting clutch pedal height — cable operated clutches

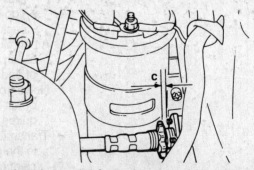

Adjusting the pedal free play on cable operated clutches

DRIVE TRAIN 7

Starions, late model Mirages, Galant, Sigma and Montero, loosen the locknut, and turn the push rod to bring the play within specification. On all models, this is 1–3mm. Retighten the locknut.

3. Check the total free-play, which should be 6–12mm. Check the pedal height, also. If play is not correct, bleed the system. If that does not correct play, the master cylinder or the clutch itself is faulty.

4. On Mirage, Precis and Pickup with clutch cable, the clutch pedal free play should be 20–30mm. If it is outside specification, pull up gently on the cable at the floorboard and turn the outer cable adjusting nut until the nut-to-insulator clearance is about 2.5mm. After making the adjustment, depress the clutch pedal to the floor several times and recheck pedal free play.

Disc and Pressure Plate

REMOVAL AND INSTALLATION

Starion, Montero and Pickup

1. Remove the transmission.
2. Insert the forward end of an old transmission input shaft or a clutch disc guide tool into the splined center of the clutch disc. This will keep the disc from dropping when the pressure plate is removed from the flywheel.
3. Loosen the clutch mounting bolts alternately and diagonally in very small increments (no more than two turns at a time) so as to avoid warping the cover flange. When all bolts are free, remove the pressure plate and disc.
4. For Starion and Montero:
   a. Remove the return clips from the throwout bearing and fork, and then remove the throwout bearing.
   b. Remove the release fork by pulling it downward, so the retaining clip fulcrum will be pulled *away* from where the clip is attached to the fork, not toward it.

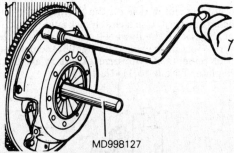

Insert a clutch disc guide tool (or an old transmission shaft) into the splined center of the disc. Loosen the bolts alternately and a little at a time

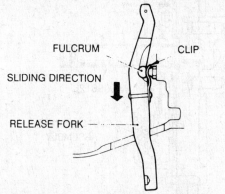

When removing the clutch release fork, pull it downward as shown

5. For Pickup:
   a. Remove the return clip and then remove the release bearing.
   b. Use a center punch and hammer to remove the spring pins from the clutch release fork and shaft. Discard the spring pins; they are not reusable.
   c. Remove the release shaft. Remove the release fork, seals, and return spring.

6. Before installing the clutch, pack the release fork fulcrum hole and slave cylinder pushrod hole with grease. Also pack grease into the groove on the inside diameter of the throwout bearing.

7. Wipe all friction surfaces of the clutch disc, pressure plate, and flywheel to remove any grease or oil. Lightly grease the clutch disc splines and transmission input shaft splines.

8. Locate the clutch disc on the flywheel with the stamped mark facing outward. Use a clutch disc guide or old input shaft to center the disc on the flywheel, and then install the pressure plate over it. Install the bolts and tighten them evenly. Torque to 18 Nm (13 ft. lbs.).

9. Install the transmission in reverse of removal, making sure the input shaft is lined up properly to avoid bending it.

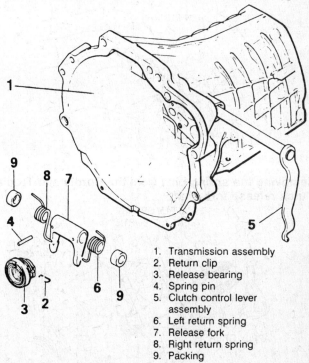

1. Transmission assembly
2. Return clip
3. Release bearing
4. Spring pin
5. Clutch control lever assembly
6. Left return spring
7. Release fork
8. Right return spring
9. Packing

Pickup, clutch control components

Cordia and Tredia
Mirage and Precis

1. Disconnect and remove the clutch oil tube and drain the clutch fluid.
2. Remove the transaxle. Insert the forward end of an old transmission input shaft or a clutch disc guide tool into the splined center of the clutch disc. This will keep the disc from dropping when the pressure plate is removed from the flywheel.
3. Loosen the clutch mounting bolts alternately and diagonally in very small increments (no more than two turns at a time) so as to avoid warping the cover flange. When all bolts are free, remove the pressure plate and disc.
4. Remove the snapring. Remove the clevis pin.
5. Remove the return clip and then remove the release bearing.

7 DRIVE TRAIN

6. Use a center punch and hammer to remove the spring pins from the clutch release fork and shaft. Discard the spring pins; they are not reusable.

7. Remove the release shaft. Remove the release fork, seals, and return spring.

8. When installing, grease the bearing areas for the release shaft. Then install the release shaft, seals, return spring, and the release fork. Apply grease to the throwout bearing contacting surfaces of the release fork.

9. Align the lock pin holes of the release fork and shaft; drive in two new spring pins, using a tool such as MD998245-01. You may be able to simply fashion an appropriate tool, a device similar to a center punch with a tip the same diameter as the lock pins but flat on its front surface. Make sure the spring pin slot is a right angles to the centerline of the control shaft.

10. Apply grease into the groove in the release bearing and install it into the front bearing retainer in the transaxle. Install the return clip to the release bearing and fork.

11. Make sure the surfaces of the pressure plate and flywheel are wiped clean of grease and lightly sand them with crocus cloth. Lightly grease the clutch disc and transmission input shaft splines.

12. Locate the clutch disc on the flywheel with the stamped mark facing outward. Use a clutch disc guide or old input shaft to center the disc on the flywheel, and then install the pressure plate over it. Install the bolts and tighten them evenly. Tighten them in increments of two turns or less to avoid warping the pressure plate. Torque to 11-15 ft. lbs.

13. Remove the clutch disc centering tool. Install the transaxle. Adjust the clutch free play as described above.

Galant and Sigma

1. Disconnect the hydraulic line to the clutch cylinder and drain the fluid. Leave the tube disconnected and plugged.

2. Remove the transaxle.

3. Insert the forward end of an old transmission input shaft or a clutch disc guide tool into the splined center of the clutch disc. This will keep the disc from dropping when the pressure plate is removed from the flywheel.

4. Loosen the clutch mounting bolts alternately and diagonally in very small increments (no more than two turns at a time) so as to avoid warping the cover flange. When all bolts are free, remove the pressure plate and disc.

5. Remove the return clip and then remove the release bearing.

6. Remove the release fork and unscrew the fulcrum (pivot).

7. When reinstalling, apply grease to the contact points on the clutch fork. Both the fulcrum point and the bearing contact points should be greased.

8. Apply grease to the end of the actuator pushrod and the push rod hole in the release fork.

9. Apply grease into the groove in the release bearing.

10. Make sure the surfaces of the pressure plate and flywheel are wiped clean of grease and lightly sand them with crocus cloth. Lightly grease the clutch disc and transmission input shaft splines. Use a small stiff brush to work the grease into the recesses.

11. Locate the clutch disc on the flywheel with the stamped mark facing outward. Use a clutch disc guide or old input shaft to center the disc on the flywheel, and then install the pressure plate over it. Install the bolts and tighten them evenly. Tighten them in increments of two turns or less to avoid warping the pressure plate. Torque to 11-15 ft. lbs.

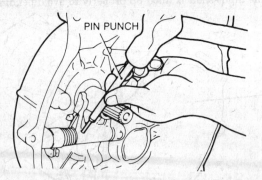

Removing the spring pins from the Cordia and Tredia clutch release fork shaft

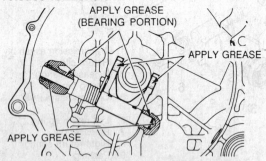

Grease the bearing surfaces of the release fork. Cordia and Tredia shown

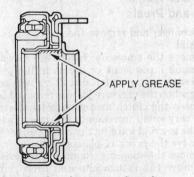

Grease the groove in the throw-out bearing

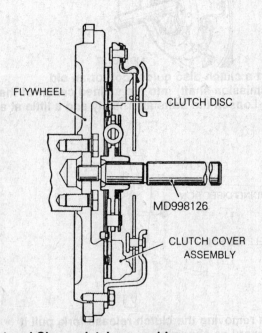

Galant and Sigma clutch assembly

DRIVE TRAIN 7

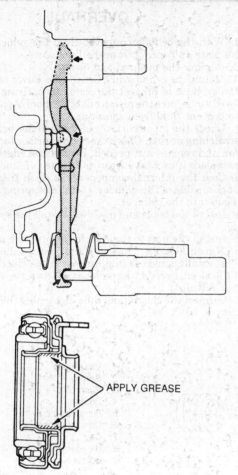

Apply grease as shown before installing clutch components

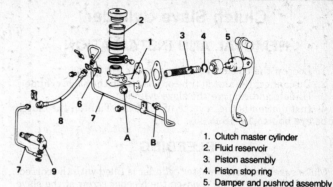

1. Clutch master cylinder
2. Fluid reservoir
3. Piston assembly
4. Piston stop ring
5. Damper and pushrod assembly
6. Hydraulic line bracket
7. Hydraulic line
8. Clutch slave cylinder

Exploded view of a typical clutch master cylinder and related parts. The mounting bolts (A) are tightened to 12 Nm (9 ft. lbs) and the hydraulic connections (B) to 14 Nm (10.5 ft. lbs.)

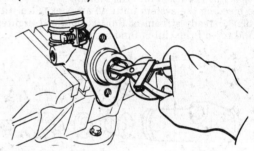

Removing the piston stop ring from the clutch hydraulic master cylinder

12. Install the transaxle.
13. Connect the clutch hydraulic line, refill the system and bleed the air from the system.
14. Adjust the clutch free play.

Clutch Master Cylinder

REMOVAL AND INSTALLATION

1. Loosen the bleeder screw and drain the clutch fluid.
2. Disconnect the pushrod at the clutch pedal by removing the cotter pin, and pulling out the clevis pin.
3. Disconnect the hydraulic tube at the master cylinder. Remove the master cylinder.
4. When reinstalling, mount the cylinder and tighten the bolts to 14 Nm (10 ft. lbs).
5. Connect the pushrod to the pedal; always use a new cotter pin.
6. Connect the hydraulic tube to the cylinder and tighten it to 15Nm or 11 ft. lbs. Refill and bleed the system.

OVERHAUL

1. Remove the master cylinder.
2. Remove the piston stop ring from the rear of the unit with snapring pliers.
3. Pull out the piston assembly.
4. Inspect the cylinder bore for rust or scoring. Inspect the piston cup for wear or damage. Inspect the inner bore of the hydraulic line connection for clogging. Clean a clogged bore or replace scored or damaged parts as necessary.
5. Use an inside micrometer to measure the cylinder bore in two directions at right angles to each other in the beginning, middle, and end of the bore. Measure the outer diameter of the piston in a similar way. Compare the figures. If clearance exceeds 0.15mm at any point, replace the cylinder and the piston. This is a critical dimension; measure and compute carefully.
6. Coat the inner surface of the cylinder and the entire surface of the piston in clean, fresh brake fluid. Install the piston into the bore, making certain not to gouge the cylinder walls.
7. Install the piston stop ring. Make certain it is seated in the groove all around.
8. Install the clamp for the reservoir so the clamp screw is behind the reservoir.
9. Install the cylinder and bleed the system as described above.

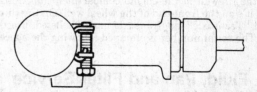

Install the reservoir clamp in the direction shown

7-63

7 DRIVE TRAIN

Clutch Slave Cylinder

REMOVAL AND INSTALLATION

1. Loosen the bleeder screw and drain the fluid.
2. Disconnect the clutch tube or hose from the slave cylinder.
3. Unbolt and remove the slave cylinder.
4. Installation is the reverse of removal. Tighten the clutch tube eye bolt to 14–18 ft. lbs.

BLEEDING

Make sure the clutch master cylinder is filled with the correct fluid (DOT3 brake fluid). Loosen the bleeder screw at the slave cylinder. Push the clutch pedal down slowly until all air is expelled; do not release the pedal but hold it depressed. Retighten the bleeder screw. Release the clutch pedal. Refill the master cylinder to the correct level.

This process may need repetition several times; the key point to remember is that the clutch pedal MUST be held down until the bleeder screw is tightened. If the pedal is released with the bleeder open, air is sucked into the system.

Continue bleeding the system until no air is expelled with the fluid. When a steady stream of fluid is obtained, tighten the bleeder and top off the clutch fluid reservoir.

OVERHAUL

1. With the cylinder removed from the vehicle, remove the valve plate and spring from the fluid port.
2. Remove the pushrod and boots.
3. Mount the cylinder in a bench vise. Cover or block the end of the cylinder to prevent the piston from flying out. Use compressed air to force the piston out of the bore. Apply the air slowly to prevent fluid from splashing.
4. Check the inner surfaces of the cylinder bore for any sign of scratching or rust. Check the piston surface for the same condition and replace it if needed. Inspect the edges of the piston cup; replace if excessive wear or fatigue is found.
5. Coat the internal components is clean fresh brake fluid. Coat the inside of the cylinder. Install the spring and the piston and cup into the cylinder.
6. Install the boots and pushrod. Install the spring and valve plate.
7. Install the unit to the vehicle. Tighten the mounting bolts to 35 Nm or 26 ft. lbs. The end of the pushrod should be greased before installation as should its hole in the clutch release arm.

If the bleeder screw was removed, it should be tightened to 11 Nm or 8 ft. lbs.

8. Connect the clutch fluid tube, tightening the hose fitting to 22 Nm or 16 ft. lbs.

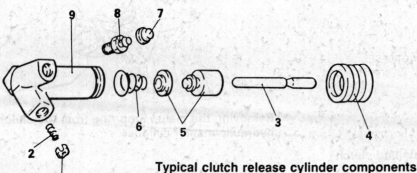

1. Valve plate
2. Spring
3. Pushrod
4. Boots
5. Piston and cup
6. Conical spring
7. Cap
8. Bleeder plug
9. Release cylinder

Typical clutch release cylinder components

AUTOMATIC TRANSAXLE

Identification

On most models, the Vehicle Information Code Plate mounted on the firewall in the engine compartment (on Cordia and Tredia it's on the front end of the wheel well). On the third line, labeled "Transaxle" the transaxle model is listed, preceding a space. The serial number is stamped following the space.

Fluid, Pan and Filter Service

For pan and filter service and fluid change procedures, please refer to Section 1.

Location of the Starion automatic transmission identification tag

DRIVE TRAIN 7

AUTOMATIC TRANSMISSION

Troubleshooting Basic Automatic Transmission Problems

Problem	Cause	Solution
Fluid leakage	• Defective pan gasket	• Replace gasket or tighten pan bolts
	• Loose filler tube	• Tighten tube nut
	• Loose extension housing to transmission case	• Tighten bolts
	• Converter housing area leakage	• Have transmission checked professionally
Fluid flows out the oil filler tube	• High fluid level	• Check and correct fluid level
	• Breather vent clogged	• Open breather vent
	• Clogged oil filter or screen	• Replace filter or clean screen (change fluid also)
	• Internal fluid leakage	• Have transmission checked professionally
Transmission overheats (this is usually accompanied by a strong burned odor to the fluid)	• Low fluid level	• Check and correct fluid level
	• Fluid cooler lines clogged	• Drain and refill transmission. If this doesn't cure the problem, have cooler lines cleared or replaced.
	• Heavy pulling or hauling with insufficient cooling	• Install a transmission oil cooler
	• Faulty oil pump, internal slippage	• Have transmission checked professionally
Buzzing or whining noise	• Low fluid level	• Check and correct fluid level
	• Defective torque converter, scored gears	• Have transmission checked professionally
No forward or reverse gears or slippage in one or more gears	• Low fluid level	• Check and correct fluid level
	• Defective vacuum or linkage controls, internal clutch or band failure	• Have unit checked professionally
Delayed or erratic shift	• Low fluid level	• Check and correct fluid level
	• Broken vacuum lines	• Repair or replace lines
	• Internal malfunction	• Have transmission checked professionally

Lockup Torque Converter Service Diagnosis

Problem	Cause	Solution
No lockup	• Faulty oil pump	• Replace oil pump
	• Sticking governor valve	• Repair or replace as necessary

	• Valve body malfunction
	   (a) Stuck switch valve
	   (b) Stuck lockup valve
	   (c) Stuck fail-safe valve
		• Repair or replace valve body or its internal components as necessary

	• Failed locking clutch	• Replace torque converter
	• Leaking turbine hub seal	• Replace torque converter
	• Faulty input shaft or seal ring	• Repair or replace as necessary

7 DRIVE TRAIN

Lockup Torque Converter Service Diagnosis

Problem	Cause	Solution

Will not unlock	• Sticking governor valve
	• Valve body malfunction
	  (a) Stuck switch valve
	  (b) Stuck lockup valve
	  (c) Stuck fail-safe valve
		• Repair or replace as necessary
		• Repair or replace valve body or its internal components as necessary

Stays locked up at too low a speed in direct	• Sticking governor valve
	• Valve body malfunction
	  (a) Stuck switch valve
	  (b) Stuck lockup valve
	  (c) Stuck fail-safe valve
		• Repair or replace as necessary
		• Repair or replace valve body or its internal components as necessary

Locks up or drags in low or second	• Faulty oil pump
	• Valve body malfunction
	  (a) Stuck switch valve
	  (b) Stuck fail-safe valve
		• Replace oil pump
		• Repair or replace valve body or its internal components as necessary

Sluggish or stalls in reverse	• Faulty oil pump
	• Plugged cooler, cooler lines or fittings
	• Valve body malfunction
	  (a) Stuck switch valve
	  (b) Faulty input shaft or seal ring
		• Replace oil pump as necessary
		• Flush or replace cooler and flush lines and fittings
		• Repair or replace valve body or its internal components as necessary

Loud chatter during lockup engagement (cold)	• Faulty torque converter • Failed locking clutch • Leaking turbine hub seal	• Replace torque converter • Replace torque converter • Replace torque converter
Vibration or shudder during lockup engagement	• Faulty oil pump • Valve body malfunction • Faulty torque converter • Engine needs tune-up	• Repair or replace oil pump as necessary • Repair or replace valve body or its internal components as necessary • Replace torque converter • Tune engine
Vibration after lockup engagement	• Faulty torque converter • Exhaust system strikes underbody • Engine needs tune-up • Throttle linkage misadjusted	• Replace torque converter • Align exhaust system • Tune engine • Adjust throttle linkage
Vibration when revved in neutral Overheating: oil blows out of dip stick tube or pump seal	• Torque converter out of balance • Plugged cooler, cooler lines or fittings • Stuck switch valve	• Replace torque converter • Flush or replace cooler and flush lines and fittings • Repair switch valve in valve body or replace valve body

DRIVE TRAIN 7

Lockup Torque Converter Service Diagnosis

Problem	Cause	Solution
Shudder after lockup engagement	• Faulty oil pump • Plugged cooler, cooler lines or fittings • Valve body malfunction • Faulty torque converter • Fail locking clutch • Exhaust system strikes underbody • Engine needs tune-up • Throttle linkage misadjusted	• Replace oil pump • Flush or replace cooler and flush lines and fittings • Repair or replace valve body or its internal components as necessary • Replace torque converter • Replace torque converter • Align exhaust system • Tune engine • Adjust throttle linkage

Transmission Fluid Indications

The appearance and odor of the transmission fluid can give valuable clues to the overall condition of the transmission. Always note the appearance of the fluid when you check the fluid level or change the fluid. Rub a small amount of fluid between your fingers to feel for grit and smell the fluid on the dipstick.

If the fluid appears:	It indicates:
Clear and red colored	• Normal operation
Discolored (extremely dark red or brownish) or smells burned	• Band or clutch pack failure, usually caused by an overheated transmission. Hauling very heavy loads with insufficient power or failure to change the fluid, often result in overheating. Do not confuse this appearance with newer fluids that have a darker red color and a strong odor (though not a burned odor).
Foamy or aerated (light in color and full of bubbles)	• The level is too high (gear train is churning oil) • An internal air leak (air is mixing with the fluid). Have the transmission checked professionally.
Solid residue in the fluid	• Defective bands, clutch pack or bearings. Bits of band material or metal abrasives are clinging to the dipstick. Have the transmission checked professionally.
Varnish coating on the dipstick	• The transmission fluid is overheating

Adjustments

NEUTRAL START SWITCH ADJUSTMENT

1. Apply the parking brake. Place the gearshift lever in "NEUTRAL" position.
2. Loosen the two mounting screws of the neutral switch so that it can be rotated.
3. Rotate it so that the end of the operating lever is directly over the flange on the switch body and the holes in that flange and the outer end of the lever are lined up.
4. Hold the switch securely in place while tightening the mounting screws to 8 ft. lbs.

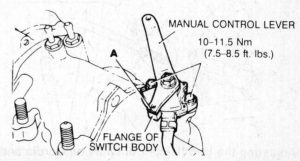

Adjusting the Cordia and Tredia neutral start switch

7-67

7 DRIVE TRAIN

5. With the parking brake set and one foot firmly on the brake pedal, recheck the function of the switch by attempting to start the engine in all selector positions. It should start only in Park and Neutral.

WARNING: If the switch is faulty, the engine may start with the vehicle "in gear". If this happens the vehicle will accelerate when the engine starts. Keep the brakes on and be ready to switch the key off instantly.

THROTTLE CONTROL CABLE ADJUSTMENT

Cordia and Tredia

1. Be sure the engine is warm with the throttle in normal idling position.
2. Loosen the lower cable bracket mounting bolt. Pull the small rubber cover located near the transmission back toward the housing to expose the nipple. Move the cable bracket as necessary until the distance between the nipple and the outer end of the cover next to the bracket is 0.5–1.5mm. Tighten the bracket mounting bolt to 9–10.5 ft. lbs.
3. With the engine off, have someone open the throttle all the way and hold it there. Pull the cable further upward to make sure it still has freedom of movement; that it has not bottomed out. If necessary, repeat the adjustment.

Mirage and Precis

1. Be sure the engine is warm with the throttle in normal idling position.
2. Raise the upper cover (B) on the throttle to expose the nipple.
3. Loosen the lower cable bracket mounting bolts.
4. The cable must be adjusted so that the distance between the lower cover (A) and the collar is 0.5–1.0mm.
5. Tighten the lower cable bracket mounting bolt to 13 Nm (9½ ft. lbs).

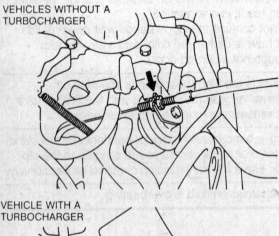

VEHICLES WITHOUT A TURBOCHARGER

VEHICLE WITH A TURBOCHARGER

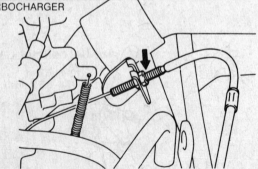

Adjusting the throttle control cable on Cordia and Tredia

6. With the engine off, have someone open the throttle all the way and hold it there. Pull the cable further upward to make sure it still has freedom of movement and that it has not bottomed out. If necessary, repeat the adjustment.

Pickup

1. Check and adjust the idle speed.
2. Make certain no bend or deformation is present in either the carburetor throttle lever or the cable mounting bracket.
3. Measure the length between the inner cable stopper and the cover end with the throttle wide open.
4. Correct measurement is 52–53mm. If the cable is out of specification, adjust the inner cable bracket by moving it up or down as needed.

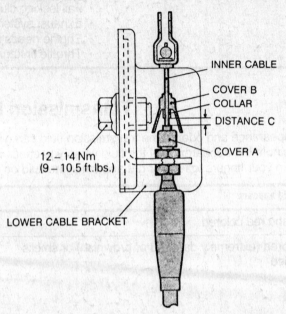

Mirage and Precis throttle control cable adjustment. Distance "C" must be correct

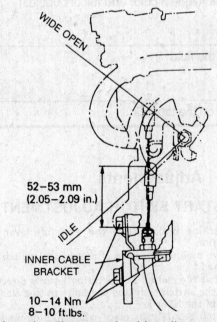

Pickup throttle control cable adjustment

7-68

DRIVE TRAIN 7

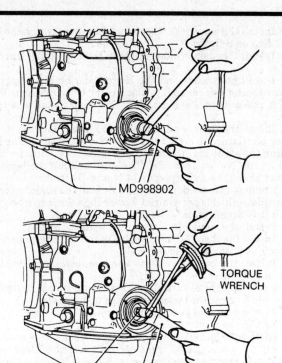

Adjusting the Galant kickdown servo

Kickdown Switch Adjustment

1988–89 Galant
Kickdown Servo

WARNING: The use of the correct special tools or their equivalent is REQUIRED for this procedure.

1. Completely remove all dirt and grease around the servo cover on the transmission case.
2. Remove the kickdown servo switch.
3. Remove the snapring and cover.
4. Loosen the lock nut.
5. Use the special counterholding tool to prevent the kickdown servo piston from turning. Use tool MD 998901 (1989 use MD 998916) to tighten the the adjustment screw to 9.8 Nm or 7.2 ft. lbs., then loosen the screw. Repeat the process once more, tightening and loosening.
6. Tighten the adjustment screw to 5.0 Nm (3.6 ft. lbs.). After it is tightened, back off the adjusting screw 2–2¼ turns for 1988 models; 2½–2¾ turns for 1989 models.
7. Counterhold the adjusting screw and tighten the locknut to 27 Nm or 20 ft. lbs. Do NOT let the adjusting screw move out of position; if it does, repeat the entire adjusting process.
8. Fit a new seal (D-ring) into the groove in the cover. Install the cover to the case without twisting the seal. Install the snapring.
9. Install the kickdown servo switch to the cover; tighten it only to 1 ft lb.

Transaxle

REMOVAL AND INSTALLATION

1. Remove the battery and battery tray. Remove the air cleaner and housing.
2. Remove the coolant and windshield washer fluid reservoirs on Cordia and Tredia only.

Disconnecting the throttle control cable at the transaxle

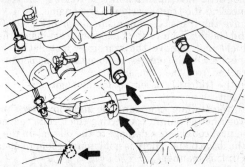

4 of the transmission retaining bolts are accessible from above on Cordia and Tredia

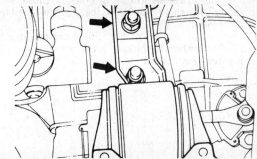

Cordia transaxle mount insulator bolts. Tredia similar

3. Disconnect the throttle control cable at transaxle. To do this, loosen the locknut which uses a star washer and locates the cable housing on the bracket. Also remove the locknut at the very end of the cable where it connects with the neutral start switch.
4. Disconnect the inhibitor switch connector, pulse generator connector, oil cooler hoses, solenoid valve connector, and speedometer cable from the transaxle. Immediately install clean caps in the open ends of the hoses. Keep the hoses pointed up so fluid will not escape until the caps are installed.
5. Label and disconnect the starter motor wiring. Remove the mounting bolts.
6. Remove the five bolts attaching the converter housing to the engine that are accessible from above EXCEPT on Cordia and Tredia.
7. Raise the vehicle and support it securely on stands. On Galant, support the car by the body as the crossmember must later be removed. Remove the wheels.
8. Remove the starter motor.
9. Drain the transmission fluid by removing the drain plug. Remove the transmission pan bolts and remove the pan and drain it.
10. Remove the under cover, if the vehicle has one. Disconnect the strut bars and stabilizer bar from the lower control arm. On

7-69

7 DRIVE TRAIN

the Mirage, disconnect the lower arm at the crossmember.

11. Disconnect both driveshafts from the transmission and suspend them in a secure manner.
12. On the Galant only, suspend the engine securely with a crane. Remove the crossmember.
13. Remove the bell housing cover. Turn the engine for access and remove the three bolts connecting the drive plate to the front of the torque converter. Make sure to push the converter as far as it will go toward the transaxle after the bolts have been removed.
14. On Cordia and Tredia only, now disconnect the five bolts accessible from above that connect the transmission to the engine.
15. On all cars, support the transmission from underneath with a floorjack in such a way that the support will be spread out and will not dent the transmission pan. Remove the remaining bolts connecting the transmission to the engine.
16. Remove the transaxle mount insulator bolts. On Cordia, Tredia, and Galant, remove the blank cap from inside the right fender shield and remove the transaxle installation bracket mounting bolts. On all vehicles, remove the transaxle insulator mount bracket from the transaxle.
17. Slide the transaxle assembly to the right and then lower and remove it.
18. When reinstalling, the transaxle and converter should be installed together as a unit. connecting the torque converter to the engine first could cause damage to the oil seal on the transaxle and the drive plate. There is not enough room to install the transaxle in this manner.
19. Place the transaxle securely on the jack and raise into position. Install it to the engine using the dowels as guides.
20. Install the the lower bell housing bolts. Tighten all the large bell housing bolts to 49 Nm (36 ft. lbs) and the small bolts to 33 Nm (24 ft. lbs.).
21. Install the mount bracket and the insulator mount.
22. Apply a thread sealant to the torque converter bolts and install them. Tighten them evenly to 48 Nm (36 ft. lbs).
23. Install the converter inspection plate and tighten the bolts to 11 Nm or 8 ft. lbs.
24. Install the starter motor; tighten the bolts to 29 Nm or 22 ft. lbs.
25. Install both axles and the center bearing if so equipped. Make sure the axles snap into place as the lockrings seat. Tighten the axle nuts to 230 Nm (166 ft. lbs.) and install new cotter pins.
26. Install the strut bar and the stabilizer bar. Tighten the stabilizer bar fixture to body bolts to 34 Nm (25 ft. lbs.) The bolts holding the strut bar, the control arm and ball joint are tightened to 105 Nm or 78 ft. lbs. The nuts holding the bar to its bracket should be tightened to 74 Nm or 54 ft. lbs.
27. Connect the speedometer cable. Install the undercover and other splash shields as required. Lower the vehicle to a height at which it is comfortable to work under the hood.
28. Install the remaining upper bell housing bolts.
29. Install the throttle control cable to the carburetor or throttle body. Adjust the cable correctly.
30. Install the control cable to the transaxle and adjust the slack out of the cable.
31. Reconnect the electrical cables, making sure is routed correctly and is retained by suitable clips or brackets.
32. Bolt the ground strap firmly in place.
33. Install the cooler hoses.
34. Install the battery tray and the battery if it was removed.
35. Fill the transaxle with the correct amount of fluid.
36. Check that the selector lever operates smoothly and is properly shifting into every selector position.
37. Check the operation of the neutral safety switch and the reverse light switch.
38. Road test, inspect for leaks and recheck fluid level.

Halfshafts and CV-Joints

For all service procedures, see the instructions under Manual Transaxle, earlier in this Section

AUTOMATIC TRANSMISSION

Identification

On most models, the Vehicle Information Code Plate mounted on the firewall in the engine compartment. V6 Monteros have the plate on the radiator support, just inboard of the right headlight) lists the transmission or transaxle type. On the third line, labeled "Transmission" the transmission model is listed, preceding a space. The serial number is stamped following the space.

On the Starion, the right hand side of the transmission case bears a tag that includes the last figure of the year as the first digit, the month of production as the second digit, and the serial production number for the month as the third through seventh positions.

Location of the Starion automatic transmission identification tag

DRIVE TRAIN 7

Fluid, Pan and Filter Service

For pan and filter service and fluid change procedures, please refer to Section 1.

Adjustments

BAND ADJUSTMENT

1983-85 Starion

1. Remove the transmission oil pan as described in the transmission removal procedure.
2. Loosen the band adjusting stem locknut and turn the stem outward. Turn the stem inward with a torque wrench until the torque reaches 5-7 ft. lbs. At that point, back the stem off exactly two turns.
3. Hold the adjustment and torque the locknut to 11-29 ft. lbs.

NEUTRAL START SWITCH ADJUSTMENT

1. Apply the parking brake. Place the gearshift lever in "NEUTRAL" position.
2. Remove the screw located at the bottom of the neutral safety switch. Loosen the two switch attaching bolts just until it can be turned.
3. You'll need a piece of wire or a nail 2mm in diameter. Rotate the switch back and forth and gently attempt to insert the pin into the switch. When the switch is in the proper position, the pin will enter the switch body and lock it into position in relation to the transmission body.
4. Use the pin to hold the switch in position and tighten the mounting bolts for the switch to 4½ ft. lbs.
5. Check the switch by attempting to start the engine in every gear position. If it starts in any position but neutral or park, the switch is faulty.

WARNING: If the switch is faulty, the engine may start with the vehicle "in gear". If this happens the vehicle will accelerate when the engine starts. Keep the brakes on and be ready to switch the key off instantly.

THROTTLE CONTROL CABLE ADJUSTMENT

Pickup and Montero

1. Check and adjust the idle speed.
2. Make certain no bend or deformation is present in either the carburetor throttle lever or the cable mounting bracket.
3. Measure the length between the inner cable stopper and the cover end with the throttle wide open.
4. Correct measurement is 52-53mm. If the cable is out of specification, adjust the inner cable bracket by moving it up or down as needed.

KICKDOWN SWITCH ADJUSTMENT

Starion

1. This switch is located on the upper accelerator post, inside the car. Operate the pedal and check to see if the switch clicks just before the pedal bottoms out.
2. If the switch does not click, loosen the locknut and turn the switch outward slightly. Repeat the test. When the switch clicks just before the pedal bottoms out, tighten the locknut. Make

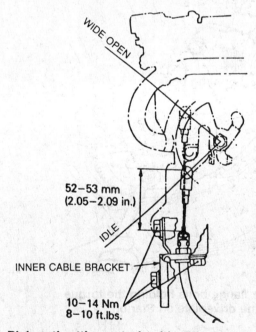
Pickup throttle control cable adjustment

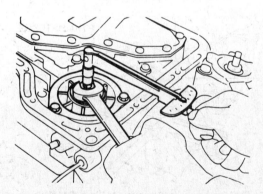
Adjusting the Starion automatic transmission band

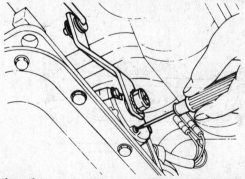
Adjusting the neutral start switch on the Starion

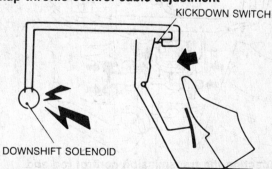
Adjusting the kickdown switch on automatic Starions

7 DRIVE TRAIN

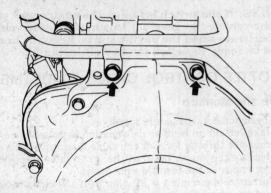

Starion upper transmission mounting bolts; only 2 can be reached from above

Starion starter mounting bolts

Removing the flange bolts holding the torque converter to the driveplate on Starion

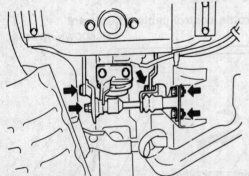

Disconnecting the transmission control rod and connecting lever at the cross shaft

sure the switch does not click until the pedal is nearly wide open, or the transmission may be subject to nuisance downshifts, shifting before maximum power is required.

Transmission

REMOVAL AND INSTALLATION

Starion

1. Loosen the oil pan mounting screws, tap the oil pan at one corner to break it loose, and then allow the fluid to drain out one side. Remove the pan and remove the remaining fluid.
2. Disconnect the battery negative cable. Remove its attaching bolt and then remove the transmission pan filler tube by pulling it upward and out of the transmission case.
3. Jack up the vehicle and support it safely on stands. Remove the two top transmission attaching bolts from the converter housing using a wrench with long extension.
4. Disconnect the starter wiring, remove the two starter bolts from the rear of the transmission and remove the starter.
5. Disconnect the oil cooler hoses at the metal tubes near the engine block, collecting the draining oil. Unbolt and remove the tubes and their mountings from the block.
6. Remove the four bolts and remove the converter housing cover. Turn the crankshaft clockwise with a large socket wrench until one of the torque converter-to-driveplate attaching flange bolts is accessible, and remove it. Repeat the operation for the remaining bolts.
7. Disconnect the speedometer cable by unscrewing the jacket at the transmission. Disconnnect the transmission control rod and the connection lever at the cross-shaft assembly.
8. Disconnect the transmission ground cable. Remove the driveshaft.
9. Support the rear of the transmission with a floor jack. The support must be over a wide area. Unbolt the transmission rear support bracket by removing two bolts on either side. Then unbolt the bracket from the transmission.

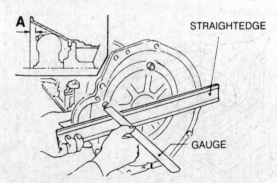

Measuring clearance before installing Starion transmission

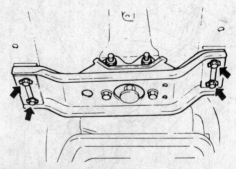

Installing the crossmember

DRIVE TRAIN 7

10. Remove the remaining mounting bolts from the area of the converter housing. Separate the transmission from the engine and remove it.
11. Before reinstalling, check the distance between the front of the bell housing and the torque converter driveplate bolts with a straightedge and ruler. The distance must be at least 35mm.
12. Install the transmission straight onto the engine. Install the lower bell housing bolts; leave the support jack in place.
13. Install the rear crossmember and attach the transmission mount. When the crossmember is fully installed, the jack may be removed.
14. Install the driveshaft. Tighten the flange bolts at the rear to 55 Nm or 39 ft. lbs.
15. Connect the ground cable.
16. Install the transmission control rod and speedometer cable.
17. Install the 6 special bolts holding the torque converter to the flexplate. Tighten the bolts to 59 Nm or 44 ft. lbs.
18. Connect the oil cooler lines.
19. Install the upper bell housing retaining bolts.
20. Install the transmission filler tube with a new O-ring
21. Connect the control harnesses and wiring.
22. After installation, refill the transmission with the approved fluid. Check that the transmission will start only in **N** and **P** positions and that the back-up lights function only in the **R** position.

Montero and Pickup

1. Disconnect the negative battery cable.
2. disconnect the kickdown linkage.
3. Remove the selector lever cover and the console. Remove the selector lever.
4. Raise and safely support the vehicle on stands.
5. If 4wd, remove the skid plate from below the transfer case.
6. Remove the shift control arm, the cross select shaft and the control rod from the transmission.
7. Drain the transmission fluid into a suitable container.
8. disconnect the oil cooler lines and hose from the transmission. Immediately plug or cap the open lines and ports. Suspend the lines vertically to reduce leakage into the work. Disconnect the front and rear driveshafts from the transmission and transfer case.
10. Disconnect the speedometer cable, the back up light wiring and the four wheel drive indicator wiring if so equipped.
11. Remove the converter cover. Remove the torque converter to flex plate retaining bolts.
12. Remove the starter mounting bolts and remove the starter. If you're careful, the wiring can remain attached; use stiff wire tie the starter to a frame rail or other solid mount. Allow it to hang from the wire.
13. Support the transmission with a jack or transmission cradle. Distribute the weight of the transmission over a wide area with a piece of lumber.
14. Remove the rear insulator from the crossmember.
15. Remove the crossmember. It is heavy and bulky; a second jack can be helpful in removing it.
16. Support the transmission with a suitable lifting device or cradle.
17. Remove the transfer case support (if 4wd). Remove the transmission retaining bolts and remove the transmission unit out of the car.
18. To reinstall, raise the transmission into place and install the retaining bolts. Install the transfer case support if so equipped.
19. Install the rear insulator on the crossmember.
20. Install the starter.
21. Install the converter cover.
22. Connect the speedometer cable, backup light switch wiring and 4wd indicator wiring harness.
23. connect the front and rear driveshafts to the transmission and transfer case.
24. Connect the oil cooler tubes and hoses from the transmission.
25. Install the shift control arm, cross select shaft and control rod to the transmission.
26. Install the skid plate if so equipped.
27. Lower the vehicle the the ground.
28. Install the selector lever cover and the console. Remove the selector lever.
29. Connect the kickdown link.
30. Install the proper amount of transmission fluid.
31. Connect the negative battery cable.
32. After installation, refill the transmission with the approved fluid. Check that the transmission will start only in **N** and **P** positions and that the back-up lights function only in the **R** position.

TRANSFER CASE

Trouble Diagnosis

SLIPS OUT OF GEAR (HIGH-LOW)

1. Shifting poppet spring weak.
2. Bearing broken or worn.
3. Shifting fork bent.
4. Improper control rod adjustment.

HARD SHIFTING

1. Lack of lubricant.
2. Shift lever binding on shaft.
3. Shifting poppet ball scored.
4. Shifting fork bent.
5. Low tire pressure.

7-73

7 DRIVE TRAIN

BACKLASH

1. Companion yoke loose.
2. Transfer case loose on mounts.
3. Internal parts excessively worn.

NOISY

1. Low lubricant level.
2. Bearings improperly adjusted or excessively worn.
3. Gears worn or damaged.
4. Improper alignment of driveshafts or U-joints.

OIL LEAKAGE

1. Excessive amount of lubricant in case.
2. Vent clogged.
3. Gaskets or seals leaking.
4. Bearings loose or damaged.
5. Driveshaft yoke mating surfaces scored.

OVERHEATING

1. Excessive or insufficient amount of lubricant.
2. Bearing adjustment too tight.

Before Disassembly

Cleanliness is an important factor in the overhaul of the transfer case. Before opening up this unit, the entire outside of the transfer case should be cleaned, preferably with a high pressure washer such as a car wash spray unit. During inspection and reassembly all parts should be thoroughly cleaned with solvent then dried with compressed air. Wiping cloths and rags should not be used to dry parts.

Wheel bearing grease, long used to hold thrust washers and lube parts, should not be used. Lube seals with clean transaxle oil and use ordinary unmedicated petroleum jelly to hold the thrust washers and to ease the assembly of seals, since it will not leave a harmful residue as grease often will. Do not use solvent on neoprene seals, if they are to be reused, or thrust washers.

Before installing bolts into aluminum parts, always dip the threads into clean transmission oil. Antiseize compound can also be used to prevent bolts from galling the aluminum and seizing. Always use a torque wrench to keep from stripping the threads. The internal snaprings should be expanded and the external rings should be compressed, if they are to be reused. This will help insure proper seating when installed.

REMOVAL AND INSTALLATION

1. Disconnect the negative battery cable.
2. Remove the transmission and transfer case assembly from the vehicle.
3. Remove the roll pin that retains the shift changer to the control shaft.
4. Remove the selector plunger from the right side of the case.
5. Remove the selector spring and plunger or steel ball.
6. Remove the transfer case to transmission attaching nuts and remove from the transmission.
7. When reinstalling, carefully fit the transfer case to the transmission adapter with a new gasket between them. Make sure the end of the transmission output shaft is lightly greased to ease the fit to the transfer case.
8. Connect the two units and install the case bolts. Tighten the bolts to 36 Nm (27 ft. lbs).
9. Install the selector spring and plunger or steel ball.
10. Carefully install the roll pin holding the shift selector to the control shaft.
11. Reinstall the transmission/transfer case as a unit.

LINKAGE ADJUSTMENT

Since this transfer case uses a directly engaging shift mechanism, there are no provisions for adjustment.

Overhaul

DISASSEMBLY

1. Remove the 4WD switch from the case. Take out the steel balls behind the switch.
2. Remove the speedometer sleeve clamp and the speedometer sleeve assembly.
3. Remove the rear cover, gasket, wave spring and spacer.
4. Take a $3/16$ in. punch and drive out the spring pin that retains the H-L shift fork. Remove the 2 threaded plugs and remove the poppet springs and balls.
5. Pull out the H-L shift rail. Take out the interlock plunger.
6. Remove the rear bearing snapring from the rear output shaft. Remove the chain cover, oil guide and side cover.
7. Remove the countershaft locking plate and remove the countershaft. Remove the counter gear assembly through the side cover opening. The gear assembly consists of gear, spacers, needle bearings and thrust washers.
8. Remove the snapring, the 2 spring retainers and the spring from the 2W-4W shift rail.
9. Remove the front output shaft, the rear output shaft and the chain (as an assembly) from the transfer case.
10. Remove the 2W-4W shift rail. Remove the H-L shift fork and clutch sleeve. Remove the needle bearing from the input gear.
11. Remove the snapring retaining the input gear assembly and remove the input gear assembly.
12. Remove the snapring from the front end of the rear output shaft. Remove the H-L clutch hub, the low speed gear, the thrust bearing and the needle bearing.
13. Raise the detent of the locknut on the rear output shaft and remove the locknut. Remove the rear bearing using a puller or press.
14. Remove the sprocket spacer and balls. Remove the drive sprocket, 2 needle bearings, sprocket sleeve and steel ball. Remove the 2W-4W clutch sleeve, hub and stop plate. Remove the bearing using a suitable puller or press.
15. Remove the snapring retaining the input gear. Press the shaft from the gear. Remove the 2 bearings from the front output shaft using a suitable puller or press.

CLEANING AND INSPECTION

Cleaning

During overhaul, all components of the transfer case (except bearing assemblies) should be thoroughly cleaned with solvent and dried with air pressure prior to inspection and reassembly. Be sure all gasket sealing material is cleaned off of the case, cover plates and mounting flanges.

Proper cleaning of bearings is of utmost importance. Bearings should always be cleaned separately from other parts.

Soak all bearing assemblies in clean solvent or fuel oil. Bearings should never be cleaned in a hot solution tank. Wash the bearings in solvent until all old lubricant is loosened. Hold races so that bearings will not rotate; then clean bearings with a soft bristled brush until all dirt has been removed. Remove loose particles of dirt by tapping bearing flat against a block of wood. Rinse bearings in clean solvent; then blow bearings dry with air pressure.

NOTE: Do not spin bearings while drying!

DRIVE TRAIN 7

After drying, rotate each bearing slowly while examining balls or rollers for roughness, damage, or excessive wear. Replace all bearings that are not in first class condition. After cleaning and inspecting bearings lubricate generously with recommended lubricant, then wrap each bearing in clean paper until ready for reassembly.

Inspection

1. Inspect all parts for discoloration or warpage.
2. Examine all gears and splines for chipped, worn, broken or nicked teeth. Small nicks or burrs may be removed with a fine abrasive stone.
3. Inspect the breather assembly to make sure that it is open and not damaged.
4. Check all threaded parts for damaged, stripped, or crossed threads.
5. Replace all gaskets, oil seals and snaprings.
6. Inspect housings, retainers and covers for cracks or other damage. Replace the damaged parts.
7. Inspect keys and keyways for condition and fit.
8. Inspect shift forks for wear, distortion or any other damage.
9. Check detent ball springs for free length, compressed length, distortion or collapsed coils.
10. Check bearing fit on their respective shafts and in their bores or cups. Inspect bearings, shafts and cups for wear.

NOTE: If either the bearings or cups are worn or damaged, it is advisable to replace both parts.

11. Inspect all bearing rollers or balls for pitting or galling.
12. Examine detent balls for corrosion or brinnelling. If shift bar detents show wear, replace them.
13. Replace all worn or damaged parts. When assembling the transfer case, coat all moving parts with recommended lubricant.

ASSEMBLY

1. When replacing the control shaft oil seal or input gear oil seal, drive out the spring pin from the transmission control shaft change shifter and remove the adapter from the transfer case. Remove the seals and press fit new ones in position.
2. Assemble the adapter to the transfer case using a new gasket. Before tightening the mounting bolts, install the change shifter over the control shaft. Adapter mounting bolt torque is 22–30 ft. lbs.
3. Press fit the bearing into the input gear. Make sure the bearing turns freely after installation. Install the snapring over the front end of the input gear. Use the thickest snapring that will fit in the groove (5 sizes are available). Press fit the 2 bearings over the front output shaft, make sure they turn freely.
4. Install the bearing on the rear output shaft and make sure the bearing turns freely. Mount the stop plate and install the 2WD-4WD clutch hub and sleeve. Make sure they are installed facing in the proper direction. Mount the steel ball on the rear output shaft and mount the sprocket sleeve.
5. Install the 2 needle berarings on the sprocket sleeve and mount the drive sprocket. After installing the steel balls and the sprocket spacer, press fit the ball bearing into the inner race, make sure it moves freely.
6. Tighten the mainshaft locknut and lock the tab with a punch. Install the needle bearing, the thrust washer and the low speed gear on the rear output shaft from the front end.
7. Install the H-L clutch hub, making sure it is facing in the proper direction. Mount the H-L snapring using the thickest one that will fit in the groove (5 sizes are available). Insert the input gear assembly into the transfer case and mount the snapring. Once again using the thickest ring possible.
8. Insert the needle bearing into the input gear. Install the H-L clutch sleeve and shift fork (sleeve facing in proper direction). Install the 2WD-4WD shift rail.
9. Securely engage the chain with the front and rear output shaft sprockets. Assemble the 2WD-4WD clutch sleeve with the 2WD-4WD shift fork and install the assembly over the shift rail while, at the same time, mount the rear and front input shafts, chain assembly.
10. Install the 2 spring retainers and spring on the 2WD-4WD shift rail and secure with the snapring.
11. Install the 2 needle bearings and spacer into the counter gear. Install the assembly in the transfer case with the thrust washers in place. Insert the counter shaft and locking plate.
12. Install the side cover and gasket. Install the oil guide. Install the chain cover and gasket and make sure that the oilguide end fits in the chain cover opening.
13. Install the snapring in the groove of the rear output shaft bearing. Insert the interlock plunger. Insert the H-L shift rail and pass it through the shift fork. The shift fork must be to the 4WD side or the shift rail cannot be inserted.
14. Install the 2 poppet balls and 2 springs and mount the seal plug. The smaller end of the spring goes toward the ball.
15. Align the H-L shift fork and rail holes and install the spring pin. Install the spring pin with its center slit placed on the center line of the shift rail.
16. Install the spacer on the rear end of the rear output shaft bearing and install the rear cover and gasket. Check endplay and use a thicker or thinner spacer as needed.
17. Install the wave spring washer on the rear end of the front output shaft rear bearing and install the cover and gasket. Install the speedometer sleeve assembly and secure. Install the switch and ball. Install the neutral plungers and springs in the hole on top of the adapter and tighten the seal plug. Install the steel ball, resistance spring and plug.

DRIVELINE

Driveshaft

REMOVAL AND INSTALLATION

NOTE: This procedure refers to the conventional, fore-and-aft driveshaft used on the Starion, Montero and Pickups.

1. Scribe mating marks on the flanged yoke at the rear of the rear U-joint and on the companion flange at the front of the rear axle. This will permit the shaft to be reinstalled in the same position for balance purposes.
2. Remove the nuts and bolts from the companion flange, and lower the rear of the driveshaft. Slide the front of the shaft out of the transmission. Cover the opening left at the rear of the

7 DRIVE TRAIN

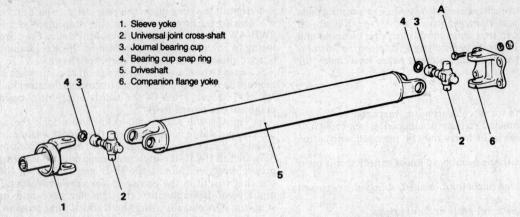

1. Sleeve yoke
2. Universal joint cross-shaft
3. Journal bearing cup
4. Bearing cup snap ring
5. Driveshaft
6. Companion flange yoke

Rear wheel drive driveshaft. The flange bolts (A) should be tightened to 54 Nm or 40 ft. lbs.

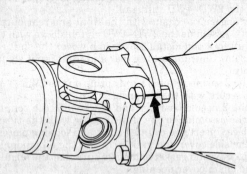

Both flanges at both ends should be scribed with mating marks before removal. Correct position (and therefore balance) of the shaft is important during reassembly

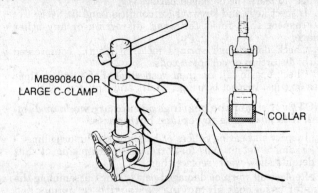

Use the special tool (or a large C-clamp) to press the bearing cups into the special collar

transmission with a clean rag or similar means so dirt will be kept out.

3. In the case of a driveshaft having a center bearing, the rear shaft must be removed first. Unbolt the bearing or carrier mount from the vehicle and remove it from the front driveshaft. Remove the front driveshaft.

4. When reinstalling, apply SAE 80 hypoid gear oil (API GL-4) to the inner and outer surface of the sleeve yoke where it will enter the transmission or center bearing.

5. Align the marks on the yoke and companion flange; line up the boltholes.

6. Install the nuts and bolts with the lockwashers on the back (nut) side of the flange. Tighten the bolts to 54 Nm or 40 ft. lbs.

U-JOINT OVERHAUL

NOTE: To perform this operation, you'll need Mitsubishi Tool MB990840-01 or an equivalent large C-clamp and a small collar designed to fit against the yoke and accept the pressed out bearing cups.

1. Scribe mating marks on the yokes of each joint that is to be disassembled. Joints must be assembled in the same position to maintain driveshaft balance.

2. Remove the snaprings from all four positions of the joint. Situate the collar against one side of the yoke so it is perfectly lined up with the hole in the yoke and will accept the bearing cup from that side. Place the flat bottom of the C-clamp under the collar and then screw the upper portion downward against the top of that same yoke cross-shaft.

3. Turn the C-clamp screw handle to press the cross-shaft out. If you have difficulty in getting the cross-shaft to move, you can lightly strike the yoke with a plastic hammer. Remove the bearing cups and bearings from that cross-shaft. It's best to turn the unit so the closed end of the cup is at the bottom for retention of the needle bearings when you do this.

4. Repeat this step for the other cross-shaft. You can then separate the two yokes of the joint and remove the cross-shaft assembly.

5. Apply multipurpose grease (SAE J310a NLGI grade #2 EP) to the journal outer diameters, ends, and reservoirs on all four arms of the cross-shaft. Apply grease also to the lips of the dust seals in the ends of the bearing cups and to the needles. All parts should be thoroughly coated, but excess grease may make the unit hard to assemble.

6. Place a bearing cup, open end inward, on opposite sides of one of the forks. Place the C-clamp around the cups without the collar. Center the cross-shaft assembly so the bearing cups will be assembled around two of the journals as they are pressed into the yoke. Turn the C-clamp until the bearing cups are pressed in with room for a snapring on either side. Line up the mating mark for the two yokes and then repeat the procedure for the other cross-shaft and yoke.

7. Install the snapring on one side with snapring pliers. Install the collar over the side of the yoke onto which you've installed the snapring. Then install the C-clamp with the bottom against the collar and the lower end of the screw over the top of the upper bearing cup.

8. Turn the clamp to seat the outside of the cup against the snapring on one side. Install a snapring on the other side. Measure the clearance between the outside surface of the second snapring and the outer wall of the groove which retains the snapring in the yoke. If the clearance is more than 0.06mm, replace both snaprings with new ones.

DRIVE TRAIN 7

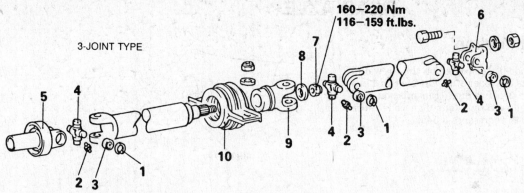

160–220 Nm
116–159 ft.lbs.

3-JOINT TYPE

1. Snapring
2. Grease fitting — 4-wheel drive
3. Journal bearing
4. Journal
5. Sleeve yoke
6. Flange yoke
7. Locknut
8. Washer
9. Center yoke
10. Center bearing assembly

3-piece driveshaft with center bearing

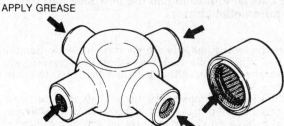

Universal joint reassembly

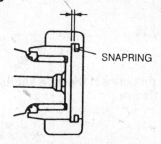

Measure the clearance after the snapring is installed

Center Bearing

REMOVAL AND INSTALLATION

1. Remove the driveshaft.
2. Disconnect the center universal joint.
3. Remove the nut holding the center yoke and remove the yoke from the center bearing.
4. Gently pry the bearing bracket from the bearing assembly.
5. Use a gear puller or similar tool to remove the bearing.

NOTE: **The center bracket and rubber insulator must be replaced as a unit.**

6. Fill the bearing grease cavity with multi-purpose grease.
7. Partially insert the center bearing into the shaft and install the bracket to the bearing.
8. Check that the bracket mounting rubber is correctly seated in the bearing groove.
9. refit the center yoke, making sure to align the notch on the yoke with the notch on the front driveshaft. Use a new attaching nut and tighten it to 187 Nm (138 ft. lbs).
10. Reinstall the center universal joint, making sure to align the matchmarks of the rear driveshaft with the notch in the yoke.

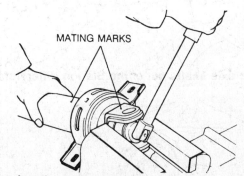

Removing the yoke locknut. It will be very tight

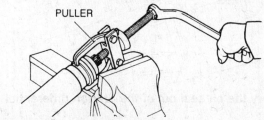

Removing the center bearing

7-77

7 DRIVE TRAIN

REAR AXLE

Identification

The vehicle information code plate, located on the firewall in the engine compartment, lists the axle ratio. On the third line, labeled Transmission'', there is a five digit transmission code, followed by a blank space, followed by a four digit code representing the axle ratio, i.e. 3545 represents 3.545:1.

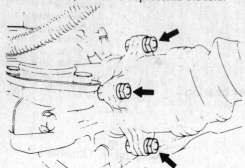

On Starion, disconnect the axle shafts at the companion flange

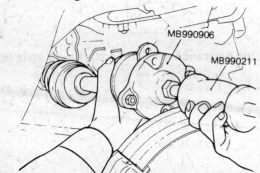

Pulling the axle shafts out of the Starion differential

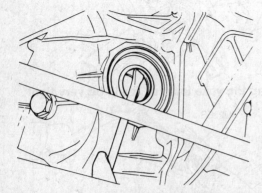

Gently pry the oil seal out of the Starion differential

Axle Shafts

REMOVAL AND INSTALLATION

Starion

NOTE: To perform this operation, you'll need a slide hammer which can be bolted to the driveshaft flange located at the outer end of the shaft. Use MB990241-01 and MB990211-01 or equivalent.

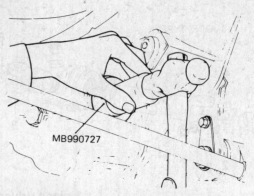

Use the special tool to install the new seal into the Starion differential carrier

1. Raise the vehicle and safely support it on stands. Remove the four sets of nuts, bolts, and lockwashers from the companion flange at the outer end of the shaft, and disconnect the shaft there.
2. Install the slide hammer, situate the shaft so all parts are in a straight line, and pull the shaft straight out of the differential carrier so as to avoid damaging the oil seal with the driveshaft splines.
3. The axle shaft contains constant velocity joints (CV-joints) identical to those found on front wheel drive cars. The inboard joint is usually a Birfield joint and the outer is a double offset type. Maintenance and repair for these joints may be found under "CV-Joint Overhaul" earlier in this section.
4. If the differential carrier oil seal is damaged, is should be pried out with a prybar and a new seal installed with seal installer MB990727-01 and MIT304180 or equivalent. Before installation, coat the lip of the new seal with multipurpose grease.
5. Replace the circlip located at the inner end of the driveshaft. Carefully locate the splined inner end of the shaft into the differential carrier so as to avoid damaging the seal. Drive the shaft into the carrier by using the slide hammer tools, keeping the shaft aligned straight. Make sure the driveshaft does not pull out of the carrier by attempting to move it in the axial direction.
6. Install the companion flange bolts with the lockwasher and nut on the differential carrier side, and torque to 54 Nm (40 ft. lbs). Again make sure the driveshaft does not pull out of the carrier by attempting to move it in the axial direction.

Montero and Pickup
Rear Axle Shaft, Bearing and Seal

1. Raise the vehicle and support it safely on stands.
2. Remove the rear wheel.
3. Remove the rear brake drum.

CAUTION

Brake pads and shoes contain asbestos, which has been determined to be a cancer causing agent. Never clean the brake surfaces with compressed air! Avoid inhaling any dust from brake surfaces! When cleaning brakes, use commercially available brake cleaning fluids.

---

4. Disconnect and plug the brake line(s) at the wheel cylinder.
5. Disconnect the parking brake cable from the shoes and remove the cable from the backing plate.
6. Remove the 4 nuts behind the brake backing plate holding the bearing case to the axle housing assembly.
7. Remove the backing plate, bearing case and the axle shaft as an assembly. If this is not possible by hand, use a slide hammer to remove the assembly.

DRIVE TRAIN 7

8. Remove the O-ring and the bearing preload shims. Save the preload shims for reassembly.
9. Remove the oil seal from the axle tube with a hooked slide hammer.
10. To remove the axle shaft bearing, remove the notched locknut with tool MB990785-01 or a brass drift.
11. Remove the lock washer and flat washer.
12. Screw the locknut back on to the axle shaft about 3 turns.
13. If tool MB990787-01 is not available, it will be necessary to fabricate a metal plate that fits over the axle shaft and abuts the locknut. Drill 4 holes in the plate that align with the 4 bearing case studs and fit the plate onto them. Refit 2 nuts and washers to the bearing case studs diagonally across from each other and tighten them evenly to free the bearing case and the bearing.
14. Use a hammer and drift to remove the bearing outer race from the bearing case.
15. Remove the outer oil seal from the bearing case.

To install:
16. Apply grease to the outer surface on the bearing outer race and to the lip of the outer oil seal. Drive them into the bearing case from each side.
17. Slide the bearing case and bearing over the rear axle shaft. Apply grease on the bearing rollers and fit the inner race by pressing it into place. Be careful not to damage or deform the dust cover.
18. Pack the bearing with grease.
19. Install the washer, the crowned lock washer and the locknut in that order and tighten the locknut to 200 Nm (148 ft. lbs.).
20. Bend the tab on the lock washer into the groove on the locknut. If the tab and the groove do not line up, slightly tighten the locknut until they do.
21. Drive the new inner oil seal into place after greasing it and refit the assembly.
22. Install a new O-ring, install the shims and apply silicone rubber sealant to the face of the bearing case.
23. Install the entire assembly to the axle housing. Torque the

Removing the locknut with the special tool

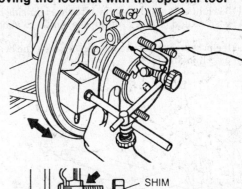

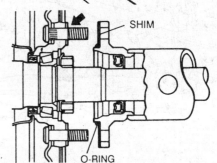

Measuring the shaft endplay (above) and correct placement of adjusting shims.

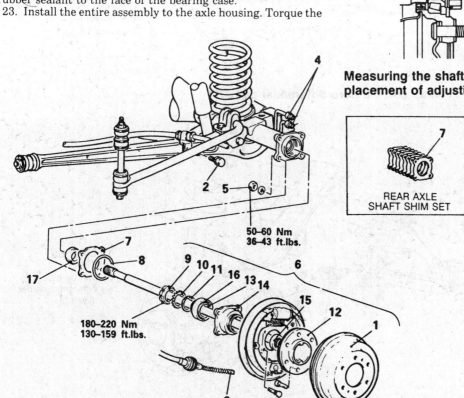

Montero and Pickup rear axle assembly

1. Brake drum
2. Parking brake cable attaching bolts
3. Parking brake cable end and brake shoe assemblies
4. Brake tubes
5. Nuts
6. Rear axle shaft assembly
7. Shims
8. O-ring
9. Locknut
10. Lock washer
11. Washer
12. Rear axle shaft
13. Bearing inner race
14. Bearing case
15. Oil seal
16. Bearing outer race
17. Oil seal

7 DRIVE TRAIN

retaining nuts to 54 Nm or 40 ft. lbs.

24. Check the axle shaft endplay. If it is not between 0.05–0.20mm, proceed with the axle shaft endplay adjustment procedure.

25. If the endplay is within specifications, install all removed brake parts and bleed the system.

26. Install the wheel and lower the vehicle.

27. Road test the vehicle. Check for leaks.

ENDPLAY ADJUSTMENT PROCEDURE

1. Begin with the left side rear axle assembly and insert a 1mm shim between the bearing case and the axle shaft housing. Torque the nuts to specification.

2. Install the right side axle assembly into the right side housing without its shim and O-ring. Torque the 4 nuts to about 50 inch lbs.

3. Using a feeler gauge, measure the gap between the bearing case and the axle housing face.

4. Remove the axle shaft and select a shim or shims that is the equal to the sum of the clearance measured in Step 3 plus 0.05–0.20mm and install them on the housing. Install the O-ring and apply sealant.

5. Install the axle assembly and torque the nuts to 40 ft. lbs. (54 Nm).

6. Measure the endplay and complete the installation procedure.

Differential Carrier

REMOVAL AND INSTALLATION

1. Elevate and safely support the vehicle on stands.
2. Remove the rear wheels.
3. Drain the differential assembly.
4. For Montero and Pickups:
   a. Remove the brake drum and disconnect the parking brake cable at the brake assembly. It may be necessary to remove or loosen some of the cable retaining brackets.
   b. Disconnect the brake lines running to the wheel cylinder.
   c. Remove the rear axle retaining nuts.

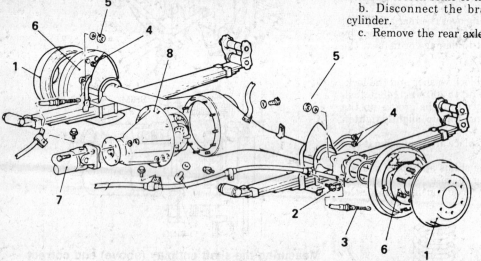

1. Brake drum
2. Parking brake cable attaching bolts
3. Parking brake cable and rear brake shoes
4. Brake tube connections
5. Nuts
6. Rear axle shaft assembly
7. Rear driveshaft
8. Differential carrier

Pickup and Montero differential carrier

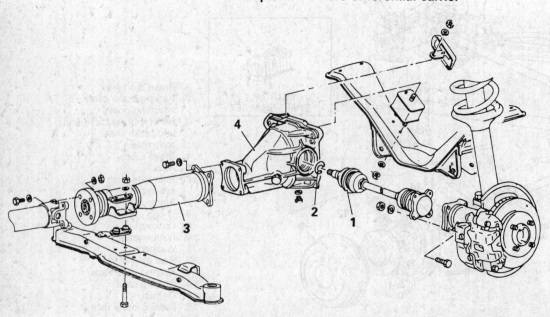

1. Driveshaft
2. Circlip
3. Torque tube
4. Differential carrier assembly

Starion differential carrier and related components

DRIVE TRAIN 7

5. Remove the axles. For Starion, unbolt the outer end and extract the axle with the use of a slide hammer if needed. A slide hammer will be needed for the axles on Montero and Pickup. Note that the axles need only be pulled free of the differential (about 76mm) not completely removed.

6. Matchmark and disconnect the driveshaft flange at the differential case. For Starion, unbolt the torque tube.

7. Support the differential carrier with a floor jack.

8. For Montero and Pickup, remove the attaching nuts around the circumference of the axle housing. Work the carrier forward and out of the housing. Lower the jack with the unit when it is clear of the axle.

NOTE: The carrier may be stuck in place. Leave one nut in place but loosened on the upper stud. Use a piece of square lumber under the carrier case to push or tap upward. This will break the seal around the casing.

9. For Starion, remove the nuts holding the carrier to the rear support insulators.

10. Make certain that the carrier is balanced on the jack. Elevate the jack to lift the carrier off the insulator studs. As soon as the unit clears the studs, move the jack forward and lower it to remove the carrier.

11. Reinstall the unit by mounting it in position and securing the retaining bolts and nuts. For Starion, the mounting nuts are tightened to 30 Nm or 22 ft. lbs. On Pickup and Montero, the ring of nuts is tightened evenly and alternately to 27 Nm or 20 ft. lbs.

12. Install the front to rear driveshaft and connect it to the differential. The flange joint bolts should be tightened to 55 Nm or 41 ft. lbs. The bolts holding the starion torque tube to the differential carrier should be tightened to 77 Nm or 57 ft. lbs.

13. Install the axles following procedures outlined earlier in this section. Montero and Pickup axle retaining bolts should be tightened to 55 Nm (41 ft. lbs.) and Starion outer axle flange bolts to 61 Nm or 45 ft. lbs.

14. On Montero and Pickup, connect the brake lines to the wheel cylinder, connect the parking brake cable and install the drum. Don't forget to secure the parking brake cable brackets if any were loosened or removed.

15. Bleed the brake system on Montero and Pickup.
16. Fill the differential with the correct amount of lubricant.
17. Install the wheels.
18. Lower the vehicle to the ground.
19. Test drive the vehicle and check for leaks.

Pinion Seal

REMOVAL AND INSTALLATION

Montero and Pickup

1. Raise the vehicle and support it safely on jackstands.
2. Matchmark and remove the driveshaft.
3. Check the turning torque of the pinion before proceeding. It should be 3.5-4.5 inch lbs. (0.4-0.5 Nm). This is the torque that must be reached during installation of the pinion nut.
4. Using a suitable pinion flange holding tool, remove the pinion nut and washer.
5. Remove the companion flange from the drive pinion.
6. Pry the pinion seal out of the differential carrier.
7. Clean and inspect the sealing surface of the housing.
8. Using a seal driver, drive the new seal into the housing until the flange on the seal is flush with the carrier.
9. With the seal installed, the pinion bearing preload must be set.
10. Tighten the pinion nut while holding the flange, until the turning torque is the same as before removal. The final pinion nut torque must be 137-181 ft. lbs. (190-250 Nm).
11. Align the matchmarks and install the driveshaft.
12. Check the level of the differential lubricant when finished.

Starion

The Starion pinion oil seal is mounted within the differential carrier. It cannot be replaced without complete disassembly of the differential. Due to the large number of special tools needed and the need for extremely accurate measurement during reassembly, this is a job best left to an experienced shop.

Rear Axle Housing

REMOVAL AND INSTALLATION

Pickup and Montero with Leaf Springs

1. Raise the vehicle and support it safely on stands. Make sure the stands are under the frame rails, not the rear axle.
2. Drain the differential
3. Remove the wheels. Remove the brake drums.

----- CAUTION -----

Brake pads and shoes contain asbestos, which has been determined to be a cancer causing agent. Never clean the brake surfaces with compressed air! Avoid inhaling any dust from brake surfaces! When cleaning brakes, use commercially available brake cleaning fluids.

4. Remove the parking brake cable attaching bolts, disconnect the cables from the shoes and unclip them from the backing plates.
5. Disconnect the brake hose at the T-fitting.
6. Remove the load sensing proportioning valve spring support, if equipped.
7. Disconnect the breather hose, if equipped.
8. Matchmark and remove the driveshaft.
9. Place a suitable jack under the center of the differential housing. If two jacks or two more stands are available, place them supporting the left and right axle tubes. Unbolt the shocks from their lower mounts.
10. Remove the U-bolts, shackle assemblies and remove the leaf springs.
11. Lower the axle and remove it from the vehicle. If only one jack is used, the unit will be balanced on the differential. Have a helper steady the unit while it is removed.
12. Raise the axle assembly into position and install the leaf springs. Make sure the shackle nuts are on the inside of the shackles.
13. Install the lower shock mounting bolts.
14. Install the driveshaft.
15. Install the breather hose, if equipped.
16. Connect the brake hose.
17. Connect the parking brake cables and install the retaining bolts.
18. Install the load sensing spring support and spring, if equipped.
19. Level the vehicle and fill the differential with Hypoid gear oil until it spills out the fill hole.

NOTE: If the vehicle is equipped with a limited slip differential, add the proper amount of limited slip friction modifier additive before filling the differential with gear oil.

20. Bleed the rear brakes. Install the drums and tire and wheel assemblies.
21. Lower the vehicle so that the full weight of the vehicle is on the ground. Unload any excess weight that is weighing down the rear of the vehicle. Adjust the load sensing proportioning valve lever so that the distance from the proportioning valve lever to the spring support is about 178mm.
22. Road test the vehicle and check for leaks.

Montero with Coil Springs

1. Raise the vehicle and support it safely with stands under the frame rails.

7-81

7 DRIVE TRAIN

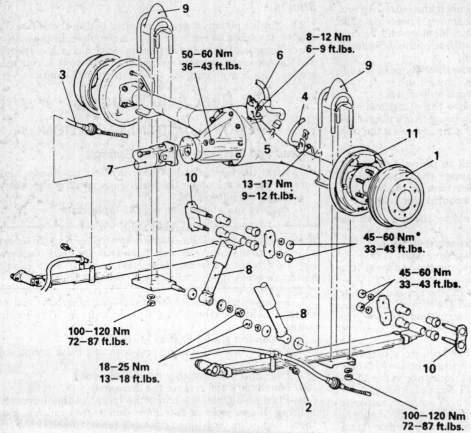

1. Brake drum
2. Parking brake cable attaching bolts
3. Parking brake cable and rear brake assembly
4. Brake hose connection — Adjustment of load sensing spring length
5. Spring support
6. Breather hose connection — 4-wheel drive
7. Rear driveshaft
8. Shock absorber
9. U-bolt and bumper stop
10. Shackle assembly
11. Axle assembly

Typical rear axle with leaf spring

2. Drain the differential.
3. Remove the wheels.
4. Remove the brake drums.

— **CAUTION** —
Brake pads and shoes contain asbestos, which has been determined to be a cancer causing agent. Never clean the brake surfaces with compressed air! Avoid inhaling any dust from brake surfaces! When cleaning brakes, use commercially available brake cleaning fluids.

5. Remove the parking brake cable attaching bolts, disconnect the cables from the shoes and unclip them from the backing plates.
6. Disconnect the brake hose at the T-fitting.
7. Disconnect the breather hose.
8. Matchmark and remove the driveshaft.
9. Remove the rear stabilizer bar attaching bolts, links and bushings.
10. Place a suitable jack under the center of the differential housing and remove the rear trailing arm, if equipped.
11. Remove the lateral rod.
12. Unbolt the shocks from their lower mounts.
13. Lower the axle housing enough to remove the coil springs and the stabilizer bar.
14. Lower the axle assembly and remove it from the vehicle.
15. With the coil springs and stabilizer bar in place, raise the axle assembly into place and install the lower shock mounting bolts.

16. Install the lateral rod but do not tighten the nuts yet.
17. Assemble the trailing arm with its front mounting spacers, bushings and nuts. Make sure the washer's concave side faces away from the bushings.
18. Install the trailing arm and torque the rear mount nuts to 90 ft. lbs. (122 Nm). Do not tighten the front mounting nuts yet.
19. Install the stabilizer bar and tighten the mounting nuts until the diameter of the bushing is the same as the diameter of the washers.
20. Install the driveshaft and breather hose.
21. Connect the parking brake cables and install the retaining bolts.
22. Level the vehicle and fill the differential with gear oil until it spills out the fill hole.

NOTE: If the vehicle is equipped with a limited slip differential, add the proper amount of limited slip friction modifier additive before filling the differential with gear oil.

23. Bleed the rear brakes. Install the drums and tire and wheel assemblies.
24. Lower the vehicle so that the full weight of the vehicle is on the ground. Torque the lateral rod mounting nuts to 90 ft. lbs. (122 Nm) and the front trailing arm mounting nuts to 100 ft. lbs. (150 Nm).
25. Road test the vehicle and check for leaks.

DRIVE TRAIN 7

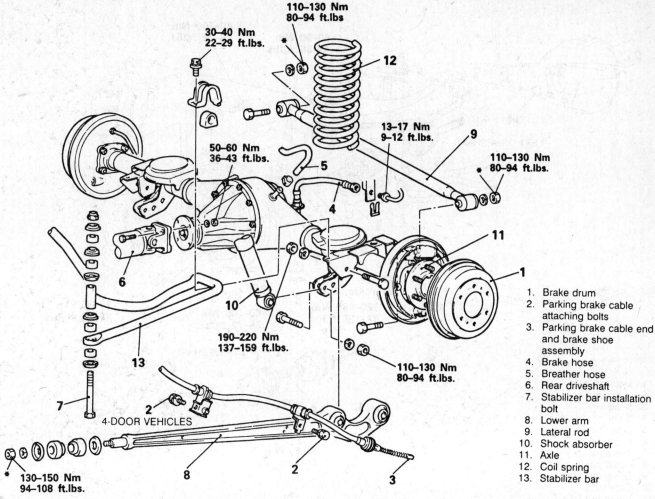

Montero rear axle with coil springs. Nuts marked with "*" must be final-tightened when the car is on the ground

1. Brake drum
2. Parking brake cable attaching bolts
3. Parking brake cable end and brake shoe assembly
4. Brake hose
5. Breather hose
6. Rear driveshaft
7. Stabilizer bar installation bolt
8. Lower arm
9. Lateral rod
10. Shock absorber
11. Axle
12. Coil spring
13. Stabilizer bar

FRONT DRIVE AXLE
4-WHEEL DRIVE VEHICLES

Front Axle Shaft

REMOVAL AND INSTALLATION

Before beginning, place the locking hubs in the **FREE** position by placing the transfer lever in the **2H** position and moving in reverse for about 6 or 7 feet.
1. Raise the vehicle and support it safely on stands.
2. Remove the wheel.

3. Remove the front brake caliper assembly and position it to the side. Do not disconnect the brake hose; simply move the entire caliper out of the way and suspend it from rope or stiff wire. Do not let it hang by the hose.

— CAUTION —
Brake pads and shoes contain asbestos, which has been determined to be a cancer causing agent. Never clean the brake surfaces with compressed air! Avoid inhaling any dust from brake surfaces! When cleaning brakes, use commercially available brake cleaning fluids.

7-83

7 DRIVE TRAIN

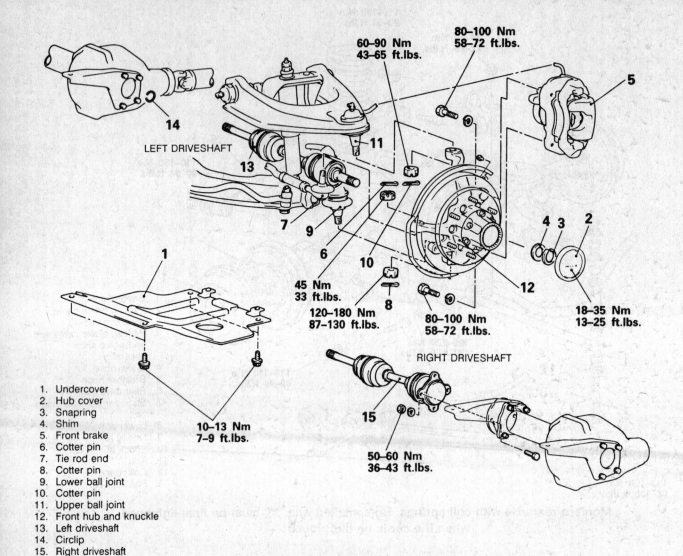

1. Undercover
2. Hub cover
3. Snapring
4. Shim
5. Front brake
6. Cotter pin
7. Tie rod end
8. Cotter pin
9. Lower ball joint
10. Cotter pin
11. Upper ball joint
12. Front hub and knuckle
13. Left driveshaft
14. Circlip
15. Right driveshaft

Removal of 4-wheel drive front axles

4. Remove the free wheeling hub cover assembly and remove the snapring from the axle shaft.
5. Remove the snapring and shim. Remove the knuckle and front hub together as a unit.
6. If removing the left side shaft, simply pull the shaft out of the differential carrier assembly. When pulling the left shaft from the differential carrier assembly, be careful that the shaft splines do not damage the oil seal.
7. On 1986 pickup only, raise the right lower suspension arm and remove the right shock absorber if removing the right side shaft.
8. Disconnect the right shaft from the inner shaft assembly and remove the shaft. Remove the inner shaft from the housing tube and the housing tube from the differential housing.
9. Press the bearing and seal off of the inner shaft and remove the dust seal from the tube.
10. Using tool MB 990955, install a new dust seal to the tube and coat the lip with grease.
11. Using a suitable long steel pipe, install a new dust cover to the inner shaft and coat the inside with grease. Press the bearing onto the shaft using tool MD990560 and a press.
12. Install a new circlip on the splines of the left side shaft or inner right shaft.
13. Drive the shafts into the differential carrier assembly with a plastic hammer. Be careful not to damage the lip of the oil seal.
14. Install the right outer shaft to the inner shaft and torque the nuts to 40 ft. lbs. (54 Nm). Install the right shock absorber if it was removed.
15. Install the knuckle and front hub assembly.
16. Install the shim and snapring and check for proper endplay. Set a dial indicator so the pin is resting on the end of the axle shaft. The endplay specification is 0.2–0.5mm. If not within specifications, adjust by adding or removing shims.
17. Install the hub cover.
18. Install the front brake caliper assembly
19. Install the tire and wheel assembly.
20. Road test the vehicle.

DRIVE TRAIN 7

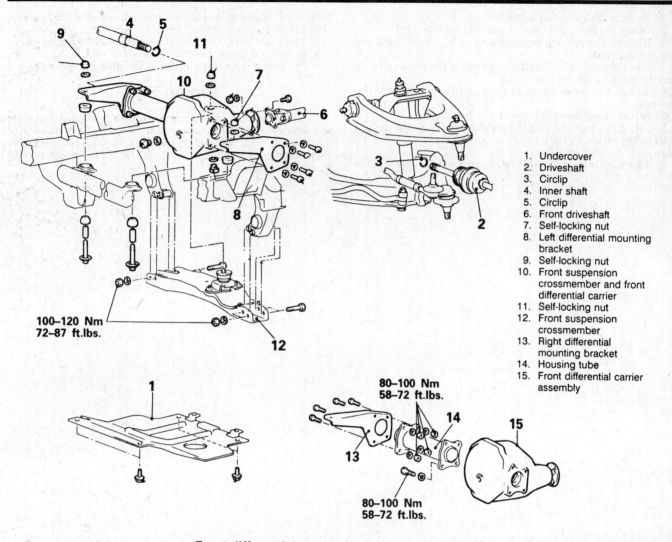

Front differential carrier for 4-wheel drive vehicles

1. Undercover
2. Driveshaft
3. Circlip
4. Inner shaft
5. Circlip
6. Front driveshaft
7. Self-locking nut
8. Left differential mounting bracket
9. Self-locking nut
10. Front suspension crossmember and front differential carrier
11. Self-locking nut
12. Front suspension crossmember
13. Right differential mounting bracket
14. Housing tube
15. Front differential carrier assembly

Front Differential Housing

REMOVAL AND INSTALLATION

1. Raise the vehicle and support it safely on stands. Make certain the front stands will not interfere with the housing removal.
2. Remove the under cover and drain the differential.
3. Remove the hubs, knuckles and axle shafts as complete assemblies.
4. Remove the inner shaft.
5. Matchmark and remove the front driveshaft from the transfer case and the front differential.
6. Support the differential housing with a jack.
7. Remove the differential mounting brackets.
8. Remove the front suspension crossmember. Use additional jacks or stands; the crossmember is heavy.
9. Lower the differential housing and remove it from the vehicle.
10. Mount the housing safely on a suitable jack and raise into position.
11. Lubricate the bushings and install the crossmember and housing brackets.
12. Torque all crossmember mounting nuts and bolt to 80 ft. lbs. (109 Nm) and the housing bracket mounting bolts to 65 ft. lbs. (88 Nm).
13. Replace the circlips and install the inner axle and axle, knuckle and hub assemblies.
14. Install the front driveshaft.
15. Level the vehicle and fill the differential with gear until it spills out the fill hole.
16. Install the under cover and lower the vehicle.
17. Perform a road test and check the differential for leaks.

Front Differential

REMOVAL AND INSTALLATION

1. Raise the vehicle and support it safely on stands. Remove the under cover.

7-85

7 DRIVE TRAIN

2. Remove all 3 front axle shafts.
3. Drain the front differential and remove the cover.
4. Remove the bearing cap retaining bolts, matchmark and remove the caps.
5. Carefully pry the differential case out of the housing, being careful not to drop the outer races.
6. Label any loose parts as they are removed from the assembly.
7. Install the differential case to the housing.
8. Install the caps aligning the matchmarks made previously. Torque the retaining bolts evenly and gradually to 45 ft. lbs. (61 Nm).
9. Thoroughly clean the sealing surfaces of the cover and the differential housing. Reseal the cover and install to the housing.
10. Install the axle shafts.
11. Level the vehicle and fill the differential with gear oil.
12. Install the under cover and lower the vehicle.
13. Perform a road test and check the differential for leaks.

Pinion Seal

The front pinion seal cannot be replaced without complete disassembly of the front differential. Because of the number of special tools required and the need for very precise measurement during reassembly, this is a job best left to an experienced repair facility.

Suspension and Steering 8

QUICK REFERENCE INDEX

Front Wheel Drive Front Suspension	8-2
Pickup and Montero Front Suspension	8-22
Rear Suspension	8-36
Starion Front Suspension	8-16
Steering	8-45
Wheel Alignment Specifications	8-34

GENERAL INDEX

Alignment, wheel
  Front 8-33
  Rear 8-45
Ball joints
  Front wheel drive 8-8
  Pickup and Montero 8-24, 25
  Starion 8-19
Front wheel bearings
  Pickup and Montero 8-32
  Starion 8-21
Front Wheel Drive Front suspension
  Ball joints 8-8
  Description 8-2
  Hub and knuckle 8-15
  Lower control arm 8-14
  MacPherson struts 8-3
  Stabilizer bar 8-9
  Strut rod 8-12
  Troubleshooting 8-2
  Wheel alignment 8-33
Ignition switch and lock cylinder 8-56
Knuckles
  Front wheel drive 8-15
  Pickup and Montero 8-29
  Starion 8-21
Lower ball joint
  Front wheel drive 8-8
  Pickup and Montero 8-25
  Starion 8-19
Lower control arm
  Front wheel drive 8-14
  Pickup and Montero 8-29
  Starion 8-20
MacPherson struts
  Front
    Front wheel drive 8-3
    Starion 8-16
  Rear 8-37
Manual steering gear
  Removal and installation 8-66
  Troubleshooting 8-50

Pickup and Montero Front Suspension
  Ball joints
    Lower 8-25
    Upper 8-24
  Knuckle and spindle 8-29
  Lower control arm 8-29
  Shock absorbers 8-24
  Springs 8-22
  Stabilizer bar 8-25
  Strut rod 8-26
  Torsion bar 8-23
  Troubleshooting 8-2
  Upper control arm 8-27
  Wheel alignment 8-33
Pitman arm 8-63
Power steering gear
  Removal and installation 8-68
  Troubleshooting 8-52
Power steering pump
  Removal and installation 8-71
  Troubleshooting 8-54
Rear suspension
  Control arms 8-41
  Shock absorbers 8-36
  Springs 8-36
  Struts 8-37
  Troubleshooting 8-2
Rear wheel bearings 8-44
Shock absorbers
  Front 8-24
  Rear 8-36
Specifications Charts
  Wheel alignment 8-34
Springs
  Front 8-22
  Rear 8-36
Stabilizer bar
  Front wheel drive 8-9
  Pickup and Montero 8-25
  Starion 8-19
Starion Front Suspension

Ball joints 8-19
Knuckle and spindle 8-21
Lower control arm 8-20
MacPherson struts 8-16
Stabilizer bar 8-19
Strut rod 8-20
Troubleshooting 8-2
Wheel alignment 8-33
Wheel bearing 8-21
Steering column
  Removal and installation 8-56
  Troubleshooting 8-45
Steering gear
  Manual 8-66
  Power 8-68
Steering linkage
  Idler arm 8-64
  Pitman arm 8-63
  Relay rod/center link 8-66
  Tie rod ends 8-64
Steering lock 8-56
Steering wheel 8-56
Struts
  Front wheel drive 8-3
  Starion 8-16
Tie rod ends 8-64
Troubleshooting Charts
  Ignition switch 8-48
  Manual steering gear 8-50
  Power steering gear 8-52
  Power steering pump 8-54
  Steering and suspension 8-2
  Steering column 8-45
  Turn signal switch 8-48
Turn signal switch 8-56
Upper ball joint 8-24
Upper control arm 8-27
Wheel alignment
  Front 8-33
  Rear 8-45
  Specifications 8-34

8-1

8 SUSPENSION AND STEERING

FRONT WHEEL DRIVE FRONT SUSPENSION

— CAUTION —
Exercise great caution when working with the front suspension. Coil springs and other suspension components are under extreme tension and result in severe injury if released improperly. For MacPherson strut systems, never remove the nut on the top of the shock absorber piston without using the proper spring compressor tool.

GALANT AND SIGMA WITH ELECTRONIC CONTROL SYSTEM (ECS)

The Galant and Sigma families offer an optional electronically controlled suspension. This system uses several sensors and a small computer to control an air compressor and solenoid system. The spring rate and damping force of the shock absorbers are changed simultaneously in response to immediate driving conditions. This results in a system with both optimum riding comfort and good handling characteristics.

Besides allowing for stiffer or softer suspension settings, the system controls vehicle height either at the direction of the driver or automatically. This compensates for heavy loads or number of passengers as well as adjusting the height for road and or speed conditions.

ECS is a complicated electronic system with built-in self diagnostics and a fail-safe backup. Testing and diagnosing a system failure is best left to trained personnel at either a dealership or independent repair facility. A well intentioned home mechanic can cause great damage to this system through faulty procedures. For that reason, this book will deal with ECS only as it affects individual components.

Troubleshooting Basic Steering and Suspension Problems

Problem	Cause	Solution
Hard steering (steering wheel is hard to turn)	• Low or uneven tire pressure • Loose power steering pump drive belt • Low or incorrect power steering fluid • Incorrect front end alignment • Defective power steering pump • Bent or poorly lubricated front end parts	• Inflate tires to correct pressure • Adjust belt • Add fluid as necessary • Have front end alignment checked/adjusted • Check pump • Lubricate and/or replace defective parts
Loose steering (too much play in the steering wheel)	• Loose wheel bearings • Loose or worn steering linkage • Faulty shocks • Worn ball joints	• Adjust wheel bearings • Replace worn parts • Replace shocks • Replace ball joints
Car veers or wanders (car pulls to one side with hands off the steering wheel)	• Incorrect tire pressure • Improper front end alignment • Loose wheel bearings • Loose or bent front end components • Faulty shocks	• Inflate tires to correct pressure • Have front end alignment checked/adjusted • Adjust wheel bearings • Replace worn components • Replace shocks
Wheel oscillation or vibration transmitted through steering wheel	• Improper tire pressures • Tires out of balance • Loose wheel bearings • Improper front end alignment • Worn or bent front end components	• Inflate tires to correct pressure • Have tires balanced • Adjust wheel bearings • Have front end alignment checked/adjusted • Replace worn parts
Uneven tire wear	• Incorrect tire pressure • Front end out of alignment • Tires out of balance	• Inflate tires to correct pressure • Have front end alignment checked/adjusted • Have tires balanced

SUSPENSION AND STEERING 8

MacPherson Struts

REMOVAL AND INSTALLATION

Cordia and Tredia

1. Raise the vehicle and support it safely on jackstands.
2. Remove the front wheels.
3. Detach the brake hose bracket at the strut.
4. Place a floor jack under the lower control arm or knuckle. Do not position it under the brake backing plate.
5. Remove the two nuts, bolts, and lockwashers attaching the lower end of the strut to the steering knuckle.
6. Support the strut and remove the mounting nuts from the top of the wheel well. Do NOT loosen the center nut on the shock absorber shaft. Remove the strut.

WARNING: The strut will come out with the spring still mounted and under tension.

7. Install the strut by attaching it at the top first. Tighten the two small nuts to 35 Nm (26 ft. lbs.).
8. Attach the lower end of the strut to the knuckle. Tighten the bolts to 82 Nm (60 ft. lbs.). Use the jack to elevate the control arm and knuckle as needed.
9. Connect the brake lines using two wrenches. Bleed the brake system.
10. Before test driving, check the brake pedal and brake function. Re-bleed the system if necessary.

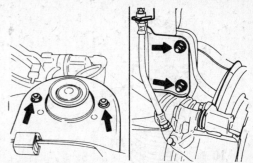

Disconnecting the Cordia and Tredia strut assembly at the top and bottom

All Mirage and Precis
Galant with Standard Suspension

NOTE: To perform this procedure, you'll need a deep well socket with flats on the top for an open-end wrench and an Allen wrench. The socket must fit the nut on top of the strut and the Allen wrench must fit the recess in the center shaft.

1. Raise the vehicle and support it safely on stands.
2. Remove the front wheels.
3. Detach the brake hose bracket at the strut.
4. Support the lower control arm and/or the knuckle assembly with a jack, but do not place the jack under the backing plate for the brake disc.
5. Remove the two nuts, bolts, and lockwashers attaching the lower end of the strut to the steering knuckle.
6. Remove the dust cover from the top of the strut on the wheel well. Support the strut from underneath. Install the socket wrench on the nut at the top of the strut and a box or open end wrench on the socket.

Install the Allen wrench through the center of the socket, long part downward. Hold the Allen wrench in place, if neces-

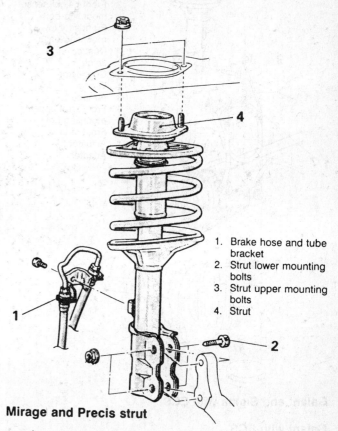

1. Brake hose and tube bracket
2. Strut lower mounting bolts
3. Strut upper mounting bolts
4. Strut

Mirage and Precis strut

sary by using a small diameter pipe as an extension. Turn the socket to loosen the nut. Remove the nut; lower and remove the strut.

WARNING: The upper nut just loosened holds the shaft in place at the upper mounting point. The spring is still under compression, held by a second nut on the shaft.

7. Install the strut by attaching it at the top first. Again hold the center shaft with the Allen wrench and tighten the nut with the socket. Since it is not usually possible to use a torque wrench on the flats of a socket, you'll have to estimate the torque being applied. It should be 50–60 Nm (36–43 ft. lbs.).
8. Attach the lower end of the strut to the knuckle. Tighten the bolts to 95 Nm (70 ft. lbs.) for Mirage and Precis and 120 Nm (89 ft. lbs) for Galant. Use the jack to elevate the control arm and knuckle as needed.
9. Connect the brake lines using two wrenches. Bleed the brake system.
10. Before test driving, check the brake pedal and brake function. Re-bleed the system if necessary.

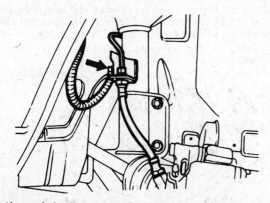

Location of the brake hose bracket on Cordia and Tredia strut

8 SUSPENSION AND STEERING

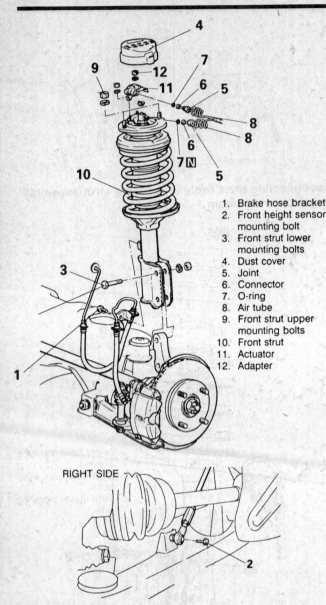

1. Brake hose bracket
2. Front height sensor mounting bolt
3. Front strut lower mounting bolts
4. Dust cover
5. Joint
6. Connector
7. O-ring
8. Air tube
9. Front strut upper mounting bolts
10. Front strut
11. Actuator
12. Adapter

Galant and Sigma with ECS

Galant with ECS

1. Raise the vehicle and support it safely on stands.
2. Remove the front wheels.
3. Detach the brake hose bracket at the strut.
4. Support the lower control arm and/or the knuckle assembly with a jack, but do not place the jack under the backing plate for the brake disc.
5. If removing the left strut, remove the two bolts holding the strut to the knuckle. If working on the right strut assembly, disconnect the height sensor rod from the lower control arm. This is simply done by removing one bolt. Do NOT alter the length of the sensor rod.
6. Under the hood, remove the dust cover from the top of the shock assembly.
7. Carefully disconnect each air line running to the actuator and adaptor. Remove the seal and O-ring and take note of their placement. The O-rings are not reusable.
8. Taking care not to bend or crimp the air line, remove it from the connector. Use tape to seal the air lines and ports against dirt and dust.

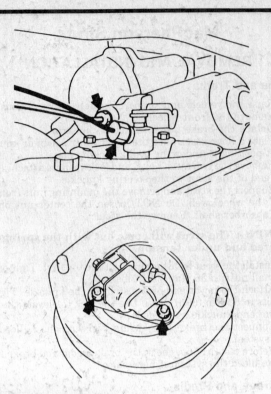

Removal of air lines and actuator from ECS strut

9. Remove the small nuts holding the upper strut mount to the body. Support the strut and remove it through the wheel arch. Be careful not strike the actuator or joint against the body work during removal.

WARNING: The strut will come out with the spring still mounted and under tension.

10. Remove the actuator and adaptor by removing the two mounting screws.
11. When reinstalling, apply a coat of wheel bearing or multipurpose grease to the groove in the insulator.
12. Install the actuator and move it all the way clockwise; install the actuator bracket.
13. Turn the insulator so that the joint and actuator point directly to the outside of the car. Install the strut into the car and secure the upper mounting nuts. Use care to protect the upper end from impact and tighten the mounting nuts to 30 Nm or 23 ft. lbs.
14. Be very careful not to bend the air tubes; insert each tube and tighten the fitting. Use a new O-ring and take great care to insure the O-ring is not damaged or twisted during installation. Tighten the air fittings to 10 Nm or 7 ft.lbs.

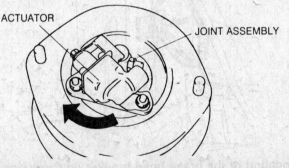

Move the actuator before installing the bracket

SUSPENSION AND STEERING 8

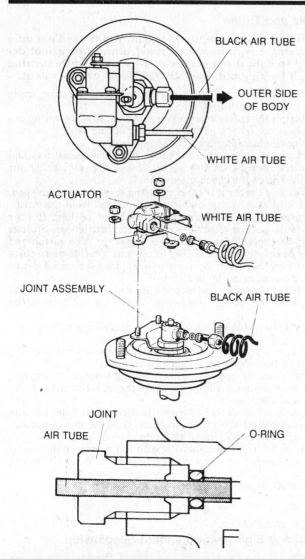

Correct installation of the ECS strut

NOTE: These joints must be airtight under pressure. The ability to seal is determined by the position of the O-ring, not the tightness of the screw fitting. If the O-ring is damaged, no amount of tightening will make the joint airtight.

15. Install the dust cover on the top of the strut.
16. Connect the strut assembly to the knuckle assembly. Tighten the bolts to 95 Nm or 70 ft. lbs. If working on the right side, reconnect the height sensor rod and tighten the bolt to 21 Nm or 16 ft. lbs.
17. Connect the brake line bracket to the strut.
18. Install the wheel.
19. Lower the car to the ground very gradually. This is to avoid deformation of the rolling diaphragm attached to the front strut. Normally, the diaphragm will retain its shape, but when the vehicle is jacked up with no air in the system, the diaphragm may become double folded at the bottom.

WARNING: If the vehicle is driven in this condition, the diaphragm will soon be damaged!

20. Inspect the bottom of the rolling diapragm for the double fold condition. If this is the case:
    a. Elevate and support the car safely on jackstands.

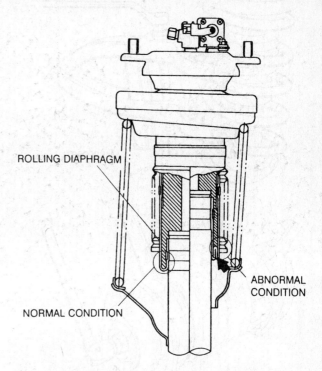

Check the diaphragm for a double fold at the bottom

  b. Disconnect the front and rear height sensor rods.
  c. Coat the folded area of the diaphragm with a coat of soapy water. This will lubricate the fabric.
  d. Start the engine.
  e. Move the height control rod to the position for vehicle height rise and operate the system.
  f. Observe the diaphragm; once the double fold has regained the correct shape, the engine may be shut off and the height sensor rods reconnected.
21. Once the car is on the ground, start the engine and operate the ECS. While the system is operating, wipe a solution of soapy water on each of the air fittings at the top of the strut.

OVERHAUL

MacPherson struts are constructed in such a way that the unit can be removed from the car while the spring is still under considerable tension. While removal of the unit as a whole is safe and simple, disassembly is an entirely different matter.

The spring is still under considerable tension; you MUST NOT attempt to disassemble the unit until an effective spring compression tool is properly installed and you can see that ALL spring tension is removed.

--- **CAUTION** ---
Failure to perform all these procedures accurately will result in the release of deadly force. If you are uncertain about performing this procedure, take the removed strut to a shop that will disassemble and overhaul it for you.

The Mitsubishi factory tool for use in this operation is MB 990987; you can purchase a number of tools designed to compress coil springs used on MacPherson struts. Assure yourself that you have bought a quality tool and can use it safely and effectively.

8-5

8 SUSPENSION AND STEERING

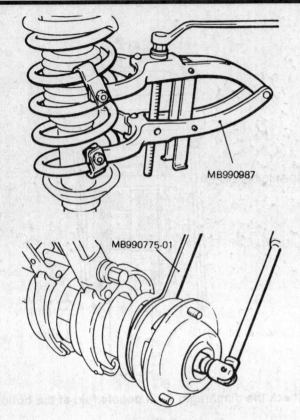

Special tools in use on Cordia and Tredia

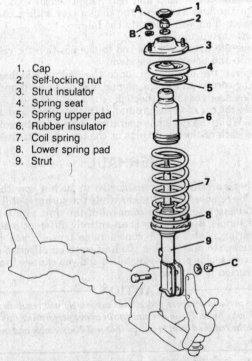

1. Cap
2. Self-locking nut
3. Strut insulator
4. Spring seat
5. Spring upper pad
6. Rubber insulator
7. Coil spring
8. Lower spring pad
9. Strut

Exploded view of the Cordia and Tredia strut. Nut "A" is tightened to 63 Nm (47 ft. lbs), "B" to 29 Nm (22 ft. lbs) and "C" to 81 Nm or 60 ft. lbs.

Cordia and Tredia

NOTE: To disassemble this strut, you'll need not only the spring compressor described above but a tool designed to hold the spring seat stationary while turning the self-locking nut. Use MB 990775-01 or equivalent.

1. Remove the strut from the vehicle. Mount the strut firmly in a bench vise with protected jaws.
2. Install the spring compressor and compress the spring until there is no tension left on the upper spring seat.
3. Pry the insulator cap off the insulator.
4. Hold the spring seat stationary with the special tool and use an ordinary socket wrench to remove the self-locking nut from the top of the shock piston rod.
5. Remove the strut insulator, spring seat, upper spring pad, rubber insulator, spring and lower spring pad from the strut.
6. Inspect the top insulator for cracks or peeling. Rubber parts must be inspected for cracking or brittleness. Springs should be checked for any sign of cracking. The piston rod should be inspected for bending or obvious wear. Replace parts as necessary. Note that if you have to replace a spring, all identification marks must be identical.
7. Mount the lower spring pad onto the strut. Compress the spring with the spring compressor. Install the spring onto the strut.
8. Pull the piston rod up all the way. Install the rubber insulator and spring upper pad.
9. Mount the upper spring seat onto the piston rod, aligning the flat on the side of the D-shaped hole with the flat on the rod. You can align the upper and lower spring seats by inserting a rod of approximately 10mm diameter through the holes in the upper and lower seats.
10. Use the special tool to keep the spring seat from turning and and install a NEW self-locking nut. Hold the spring seat and torque the nut to 65 Nm or 48 ft. lbs.
11. Align both ends of the spring with the grooves in the spring seats and slowly loosen the spring compressor. When the spring is in place and under tension, remove the compressor.
12. Install the strut into the vehicle.

Mirage and Precis
Galant and Sigma with Standard Suspension

NOTE: If the bearing at the top of the strut requires replacement, you'll need a brass rod to remove it and two special tools to install it — MB 990938-01 and MB990926-01.

1. Remove the strut as described above.
2. Install the spring compressor and compress the spring until all tension is removed from the spring seats.
3. With the same tools used in removing the mounting nut from the top of the strut to remove the piston rod tightening nut. Remove the rubber insulator, support, spring seat, spring pad (Galant only), rubber cushion, dust cover, and coil spring from the strut.
4. If the bearing requires replacement, knock it out with a brass rod from above. Place the rod against the inner race.
5. Inspect the bearing for smooth rotation. Inspect the rubber parts for cracks or brittleness. Check the spring for cracks or other signs of deterioration. Check the stop at the top for wear or distortion. Replace defective parts.
6. If the bearing was defective and required replacement, install the new bearing into the bearing support with a hammer and the special tool set. The bearing should be installed with the black retainer side toward the support.
7. Compress the spring if necessary and position it on the strut. Extend the piston rod all the way up and install the dust

SUSPENSION AND STEERING 8

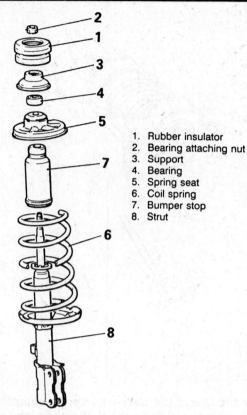

1. Rubber insulator
2. Bearing attaching nut
3. Support
4. Bearing
5. Spring seat
6. Coil spring
7. Bumper stop
8. Strut

Mirage and Precis strut

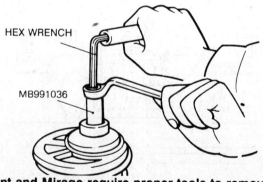

Galant and Mirage require proper tools to remove the piston rod tightening nut

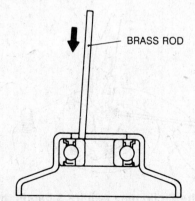

Removing the Galant and Mirage strut bearing

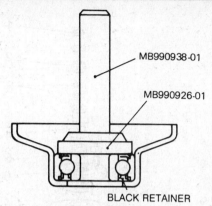

Installing a new strut bearing on Galant and Mirage

cover, rubber bumper, and spring pad (Galant only).

8. Mount the upper spring seat onto the piston rod, aligning the flat in the D-shaped hole with the flat on the rod. Assemble the support to the piston rod and install the piston rod nut with the socket and Allen wrench. Torque to 61 Nm (45 ft. lbs). on the Galant and 43 Nm (32 ft. lbs.) on the Mirage.

9. Align both ends of the spring correctly with the grooves in the spring seats and then carefully loosen the spring compressor, checking that the spring seats and springs remain in their proper positions. Install the rubber insulator to the top of the strut.

10. Install the strut.

Galant and Sigma with ECS

NOTE: The use of the factory spring compressor, MB 991043, is highly recommended for this procedure. If not available, a plate type of compressor should be used rather than the more common finger type.

1. Install the spring compressor on the strut assembly and compress the spring.
2. Remove the circlip or E-clip with a small tool, then remove the joint from the strut.
3. Remove the actuator bracket from the strut assembly.
4. Remove the insulator mounting nut and then remove the insulator from the strut assembly.
5. Gradually loosen the spring compressor, allowing the spring into a fully extended position.
6. Remove the sub tank, coil spring, and lower spring pad from the strut.
7. Inspect all the components for wear, damage, leakage or deformation.
8. Apply a coating of grease to the O-ring so it will slide easily and fit the O-ring to the sub tank.
9. Attach, in order, the lower spring pad, the coil spring and the sub tank to the spring compressor and gradually compress it.

NOTE: When installing the spring, the larger outer diameter is the bottom of the spring. Adjust the spring to sit on the lower pad correctly. The D-shaped parts of the piston rod and sub tank must all align during compression.

10. Install the insulator to the strut assembly and tighten the nut to 90 Nm or 67 ft. lbs.
11. Remove the spring compressor and apply a coating of multipurpose grease to the insulator bearing channel.
12. Align the notch of the piston rod with the D shape of the actuator bracket. Attach the actuator bracket to the strut assembly and tighten the nut to 50 Nm or 37 ft. lbs.

8–7

8 SUSPENSION AND STEERING

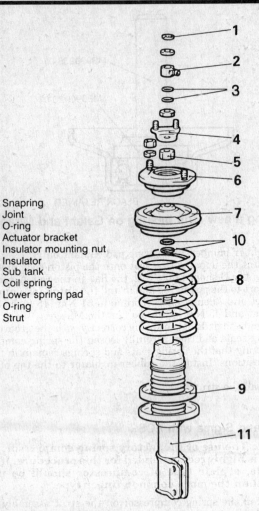

1. Snapring
2. Joint
3. O-ring
4. Actuator bracket
5. Insulator mounting nut
6. Insulator
7. Sub tank
8. Coil spring
9. Lower spring pad
10. O-ring
11. Strut

Galant ECS front strut

13. Apply a coating of lubricant to the O-ring and install it to the strut assembly. Take great care not to rub the O-ring against the threads on the shaft.
14. Install the joint to the top of the strut. Use pliers or a similar tool to pull the piston shaft upward. Snap the circlip into place. Check that the joint turns smoothly.
15. Install the strut assembly.

Lower Ball Joints

INSPECTION

1. Support the vehicle on stands.
2. Disconnect the ball joint at the lower end of the strut as described below; you need not remove the lower control arm entirely.
3. Install the nut back onto the ballstud.
4. Using an inch lb. torque wrench, measure the torque required to start the ball joint rotating. The correct figures are:
- Cordia, Tredia, and Mirage: 26–86 inch lbs.
- Galant: 17–78 inch lbs.
- Axial play: 0.5mm

If the figures are within specification, the ball joint is satisfactory. If the figure is too high, the joint should be replaced. If the figure is too low, you can still re-use the joint provided its rotation is smooth and even. If there is roughness or play, it must be replaced.

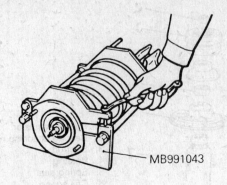

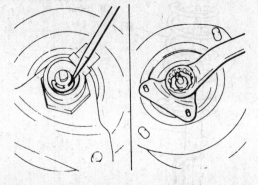

Note the use of the plate-type spring compressor. Keep a close watch on the E-clip; they tend to vanish after removal

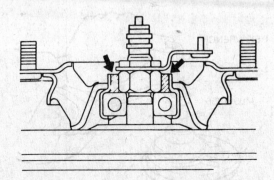

Coat the upper bearing channel with grease

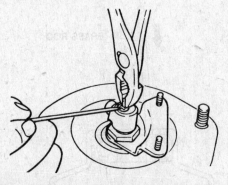

Reinstalling the E-clip may require 3 hands; have an assistant handy during the repair.

SUSPENSION AND STEERING 8

REMOVAL AND INSTALLATION

Cordia and Tredia

1. Raise and support the front end on jackstands.
2. Remove the front wheel.
3. Remove the lower control arm as described below.
4. Remove the dust cover from the ball joint. Remove the snapring with snapring pliers.
5. Press the ball joint out of the control arm. This is almost impossible without the use of the correct special tools or their equivalents.
6. Press the new ball joint into place. Install a new snapring with snapring pliers.
7. Apply multipurpose grease to the lip and to the inside of the dust cover. Use a special tool (and hammer) such as MB 990800-3-01 to drive on a new dust cover. It must go in and make contact with the snapring.

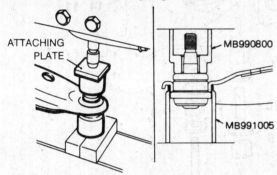

Pressing out the Cordia and Tredia ball joint

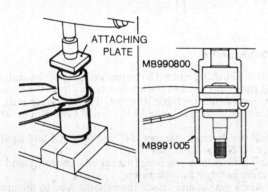

Installing the Cordia and Tredia ball joint

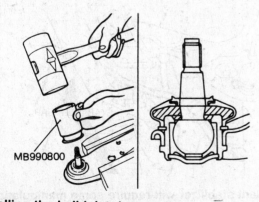

Installing the ball joint dust cover

8. Install the lower control arm as described below.

Mirage, Precis and Galant

If the ball joint requires replacement, the entire lower control arm must be replaced. The joint is not individually replaceable.

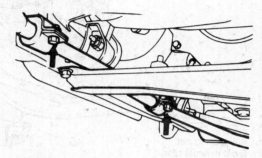

Disconnect the Cordia and Tredia stabilizer at the crossmember

Stabilizer Bar (Sway Bar)

REMOVAL AND INSTALLATION

Cordia and Tredia

1. Raise and support the front of the car on stands.
2. Remove the front wheels.
3. Disconnect the stabilizer bar at the crossmember brackets by removing the bolt from the rear of the bracket, turning the clip downward and removing it.
4. At each lower control arm, remove the nuts, bushings and washers. Pull out the attaching bolt holding the stabilizer bar to the lower control arm on both sides.
   On turbocharged vehicles only, the stabilizer bar is attached to the strut rod. Disconnect the bar there by removing the nut from the stud on the bottom of the clip.
5. Remove the stabilizer bar.
6. Install the bar into position and connect the front crossmember clips. Tighten the upper mounting bolt and nut until the threads project through the nut about 7-10mm.
   If attaching the stabilizer to the strut rod, torque the nut attaching the rear of the bar to the strut bar to 14-22 ft. lbs.

1987-88 Galant and Sigma

1. Raise and safely support the car on stands.
2. Remove the front wheels.
3. On either side, remove the nut from the top of the bolt attaching the stabilizer bar to the lower arm. Remove the washer and bushing. The bolt may be removed BUT all the washers and bushings must be reinstalled in their exact order and position.
4. Remove the clip fastening the stabilizer bar to the crossmember on either side and remove the stabilizer bar.
5. To install, position the stabilizer bar with the marked side downward (the mark is under the bushing used to mount the stabilizer bar to the crossmember). Make sure the outer ends of the bar fit over the bolts fastening the bar to the lower control arms.
6. Install the crossmember mounting clips over the bushings and install the bolts, and torque them to 22-30 ft. lbs.
7. Install the bushing, washer and nut on the vertical bolt. Tighten the nut until 16-18mm of threads are exposed.

8-9

8 SUSPENSION AND STEERING

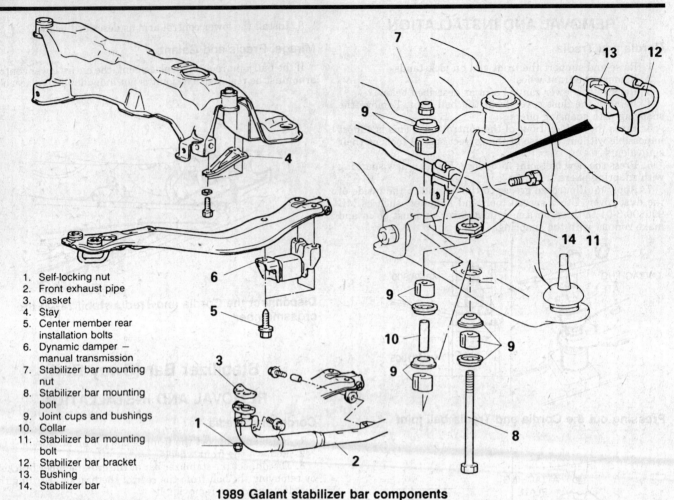

1. Self-locking nut
2. Front exhaust pipe
3. Gasket
4. Stay
5. Center member rear installation bolts
6. Dynamic damper — manual transmission
7. Stabilizer bar mounting nut
8. Stabilizer bar mounting bolt
9. Joint cups and bushings
10. Collar
11. Stabilizer bar mounting bolt
12. Stabilizer bar bracket
13. Bushing
14. Stabilizer bar

1989 Galant stabilizer bar components

1989 Galant

1. Elevate and safely support the vehicle on stands.
2. Remove the front wheels.
3. Remove the self locking nuts and the small bracket bolt; disconnect the front exhaust pipe from the manifold. Use wire to support the exhaust pipe; do not allow it to hang by its own weight.
4. Remove the reinforcing stay from the crossmember and remove the rear bolts from the longitudinal crossmember.
5. If equipped with manual transmission, remove the dynamic damper from the longitudinal support.
6. Remove the nut holding the stabilizer bar to the stabilizer link. On cars with normal suspension, the link is a long bolt with a series of bushings, spacers and washers which MUST be reinstalled correctly.
7. Remove the bracket holding the stabilizer to the crossmember. The bushing should be removed and kept with the bracket.
8. Remove the stabilizer bar by pulling the bar to the rear of the drive shafts. Move either the left or right side of the bar until the end clears the lower control arm. Pull the stabilizer bar out diagonally from the side which is below the control arm. For cars with ECS, the stabilizer link may be removed from the lower arm if desired.
9. To reinstall, make certain the link is in place on ECS cars. Fit the stabilizer into place and attach the ends to either the link or the vertical bolt. For cars without ECS, make certain the collection of spacers and washers are properly assembled on the bolt.
10. Install the rubber bushings on the bar and install the crossmember brackets. Temporarily tighten the retaining bolts.

11. Align the bushing end with the marked part of the stabilizer bar and then fully tighten the bracket bolts until 16–18mm of thread projects from the top of the nut. The retaining nuts on the link for ECS systems should be tightened to 40 Nm or 30 ft. lbs.
12. Install the dynamic damper if it was removed and tighten its bolts to 90 Nm or 67 ft. lbs.
13. Install the bolts in the longitudinal crossmember and install the crossmember reinforcement.
14. Use a new gasket and attach the exhaust pipe to the manifold. Use NEW self-locking nuts and tighten them to 35 Nm or 26 ft. lbs. Tighten the small bracket bolt to the same specification.

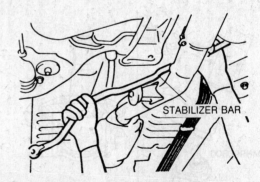

The Galant stabilizer will require some manipulation to remove

SUSPENSION AND STEERING 8

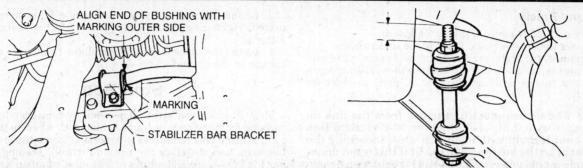

Correct installation of the 1989 Galant stabilizer. Proper bolt projection is essential; see text for details

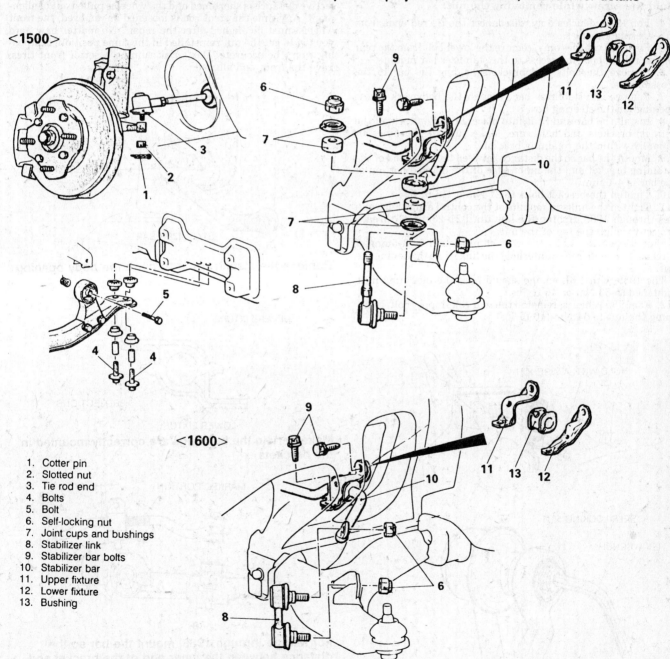

1. Cotter pin
2. Slotted nut
3. Tie rod end
4. Bolts
5. Bolt
6. Self-locking nut
7. Joint cups and bushings
8. Stabilizer link
9. Stabilizer bar bolts
10. Stabilizer bar
11. Upper fixture
12. Lower fixture
13. Bushing

1989 Mirage stabilizer bars. Cars through 1988 are similar to the 1.5L, above

8-11

8 SUSPENSION AND STEERING

Mirage and Precis

1. Raise and safely support the car on stands.
2. Remove the front wheels. Remove the under cover.
3. Disconnect the stabilizer bar from the lower control arm by removing the nut and long bolt. Remove the various spacers, washers and rubber bushings, keeping all parts in order for reassembly.

NOTE: When disconnecting the bar from the link on 1989 models with 1.5L (4G15) motor, the vertical link must be counterheld with a wrench while loosening the nut. Damage to the bushing will result if this is not done.

The same model with the 1.6L (4G61) requires a hexagonal wrench to hold the small center screw on the link. Hold the screw while removing the nut.

4. For 1989 vehicles only, disconnect the tie rod ends from the steering knuckles.
5. For 1989 vehicles only, remove the lower bolt from the rear engine roll stopper for access to the stabilizer bar mounts.
6. Remove the bolts and brackets holding the bar to the crossmember.
7. Remove the stabilizer bar through the bodywork access opening for the steering box.
8. Install the bar and fit it into place. Loosely attach all the fittings, brackets and hardware. Make sure the bushings fit squarely within the retaining brackets.
9. Mount the bar so the distance between the inner end of the retaining bracket and the inner edge of the marked portion of the bar is 4–6mm.
10. Tighten the bracket bolts to 21 Nm or 16 ft. lbs.
11. Tighten the links or mounts at the control arms. For vehicles through 1988, tighten the link until 21–23mm of threads project through the top of the nut.
   For 1989 with 1.5L, the correct projection is 3–5mm of threads. Remember to counterhold the link while tightening the nut.
   The 1989 with 1.6L engine should have the mounting nuts tightened to 63 Nm or 46 ft. lbs.
12. For 1989 vehicles, reinstall the rear roll stopper bolt, tightening the nut to 54 Nm (40 ft. lbs).
13. Connect the tie rod ends to the knuckle if they were disconnected. Tighten each nut to 24 Nm or 18 ft. lbs. Install a new cotter pin.
14. Install the under cover and install the front wheels. Lower the car to the ground.

Strut Rod

Most MacPherson strut suspension systems employ a rod running from the body or frame behind each wheel to the lower control arm. This rod serves to brace and locate the lower control arm (and therefore the tire and wheel), keeping it from moving in the fore-and-aft plane. The most common damage to these rods is bending due to impact. Once the rod is bent, its effective distance is shortened and the wheel is pulled out of alignment. Any time the strut rod is loosened or removed, the front wheels must be aligned after the repair. No matter how hard you try to get the nut reinstalled in the same position, it probably won't be accurate. Don't risk ruining a set of front tires; have the front end aligned.

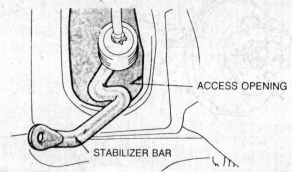

Remove the stabilizer bar through the body opening

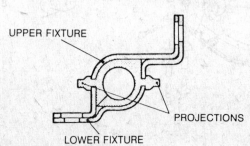

Make certain the bushings are correctly mounted in the brackets

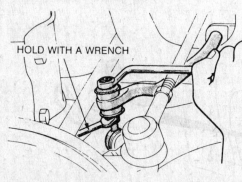

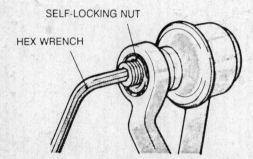

Disconnecting the Mirage stabilizer links. 1.5L shown above, 1.6 below

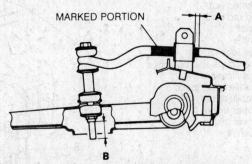

For Mirage through 1988, mount the bar so the distance between the inner end of the bracket and the inner edge of the marked portion is 4.0–6.0mm (0.16–0.24 in.)

SUSPENSION AND STEERING 8

REMOVAL AND INSTALLATION

Cordia and Tredia

1. Elevate and safely support the car on stands.
2. Remove the front wheels. Remove the undercovers.
3. Loosen but do not disconnect the stabilizer bar brackets at the crossmember.
4. Remove the stabilizer bar mounting bolts at the strut rod. The stabilizer can be swung down out of the way.
5. Remove the front locknut from the strut rod.
6. Remove the two bolts holding the strut rod to the control arm. Pull the rod rearward to remove it.

WARNING: Take careful note of the bushings and washers at the front of the rod. They must be reinstalled correctly. The pieces are not interchangable.

7. When reinstalling, set the rearmost nut on the strut rod to the correct distance of 78mm from the tip of the rod. This creates the approximately correct effective length for the strut rod.
8. Install the front of the rod with the bushings and spacers correctly placed. Install the end nut snug but do not final tighten it.
9. Attach the strut rod to the control arm, tightening the bolts to 65 Nm (48 ft lbs.).
10. Swing the stabilizer into place and connect the brackets. Tighten the strut rod brackets to 24 Nm (18 ft lbs.) and tighten the crossmember brackets to 10 Nm or 8 ft. lbs.
11. Install the wheels and lower the vehicle to the ground.
12. Bounce the vehicle two or three times to stabilize the suspension. With the car free of passengers and cargo, tighten the front nut on the strut rod to 148 Nm or 110 ft. lbs.
13. Reinstall the undercovers.
14. Have the alignment checked at a reliable repair facility.

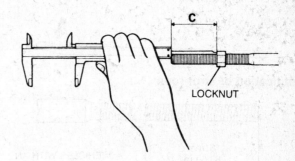

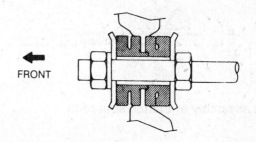

Correct assembly of Cordia and Tredia strut rod. Distance "C" must be 78mm (3.07 in.) or alignment will be affected

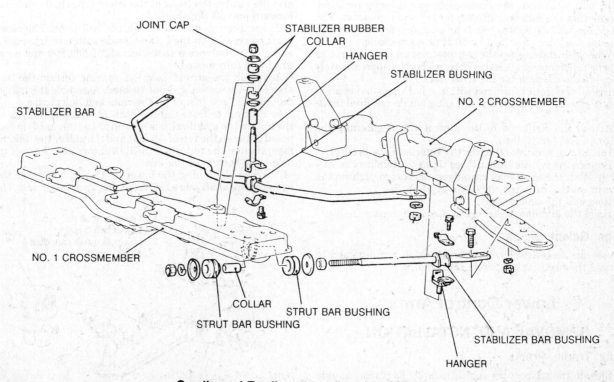

Cordia and Tredia strut rod and stabilizer

8 SUSPENSION AND STEERING

IDENTIFICATION MARK
LEFT SIDE – "L" OR WHITE MARK
RIGHT SIDE – "R" OR NO MARK

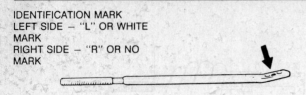

Identification of strut rods

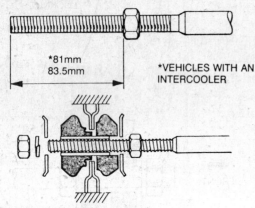

*81mm
83.5mm
*VEHICLES WITH AN INTERCOOLER

Correct assembly of Starion strut rods

Precis

1. Elevate and safely support the car on stands.
2. Remove the front wheels. Remove the undercovers.
3. Remove the front locknut from the strut rod.
4. Remove the two bolts holding the strut rod to the control arm. Pull the rod rearward to remove it.

WARNING: Take careful note of the bushings and washers at the front of the rod. They must be reinstalled correctly. The pieces are not interchangable.

5. If both strut rods are removed or one is being replaced, take note that the bars are different for left and right sides. The left side rod will have either an **L** or a white mark on the bar. Right side bars are identified by an **R** or no marking.
6. When reinstalling, set the rearmost nut on the strut rod to the correct distance of 80.5mm. This creates the approximately correct effective length for the strut rod.
7. Install the front of the rod with the bushings and spacers correctly placed. Install the end nut snug but do not final tighten it.
8. Attach the strut rod to the control arm, tightening the bolts to 65 Nm (48 ft. lbs.).
9. Install the wheels and lower the vehicle to the ground.
10. Bounce the vehicle two or three times to stabilize the suspension. With the car free of passengers and cargo, tighten the front nut on the strut rod to 80 Nm or 59 ft. lbs.
11. Reinstall the undercovers.
12. Have the alignment checked at a reliable repair facility.

Mirage, Galant, Sigma

These cars have lower control arms which are designed to eliminate the need for a separate strut rod.

Lower Control Arm

REMOVAL AND INSTALLATION

Cordia, Tredia, Precis

1. Elevate and support the vehicle securely by placing stands under the crossmember.
2. Remove the front wheel.
3. Disconnect the stabilizer bar and strut bar from the lower control arm by removing the one attaching bolt for the stabilizer bar and the two bolts for the strut bar.
4. Remove the ballstud nut and then press the ball joint stud out of the knuckle with a tool such as MB 991113.
5. Remove the nut and bolt attaching the inner end of the lower arm to the crossmember, and pull the lower arm and bushing out of the crossmember.
6. To install, mount the arm in place and insure that it is straight. Loosely attach the strut rod and stabilizer, tightening nuts and bolts just snug.
7. With everything in place, tighten the lower arm-to-crossmember attaching nut/bolt to 125 Nm (92 ft. lbs.) and the ball joint stud nut to 66 Nm or 49 ft. lbs. Install a new cotter pin for the ball joint nut. Tighten the strut rod-to-stabilizer bar bolt/nut to 65 Nm or 46 ft. lbs.
8. Install the front wheel and lower the vehicle to the ground.

Mirage, Galant, Sigma

1. Support the vehicle securely by the crossmember and remove the front wheel.
2. On the Mirage, remove the under cover.
3. Disconnect the stabilizer bar from the lower arm. On the Galant, remove the nut at the top and remove the washer and bushing, keeping them in order. On the Mirage, you can remove the nut from underneath the control arm and take off the washer and spacer.
4. On Galants with ECS, the height sensor must be disconnected if the right side arm is being removed.
5. Loosen the ball joint stud nut and then press the stud out of the control arm, using a fork-like tool (MB990778-01) on the Galant. For the Mirage, remove the stud nut and press the tool off with a tool such as MB 991113 or equivalent.
6. On the Galant, remove the nuts and bolts holding the bushings to the crossmember at the front and the bushing retainer to the crossmember at the rear. Pull the the arm out.
   for the Mirage, remove the bolts holding the spacer at the rear and the nut on the front of the lower arm shaft. Slide the arm forward and off the shaft.
7. Replace the dust cover on the ball joint. The new cover must be greased on the lip and inside with multipurpose grease and pressed on with a tool such as MB 990800 and a hammer until it is fully seated.
8. Install the arm in position, making certain the bushings are correctly seated and not twisted. Assemble the ball joint to the knuckle and install all the nuts and bolts snug.
9. Proceed to final tighten each fitting. On Galant, tighten the nut on the stabilizer bar bolt until 16–18mm of thread is exposed between the top of the nut and the end of the link. For Mirage, tighten the nut until the link shows 21–23mm of threads below the bottom of the nut.
10. When tightening the lower arm shaft bolts note that the left side arm shaft has a left-hand (backwards) thread. This bolt

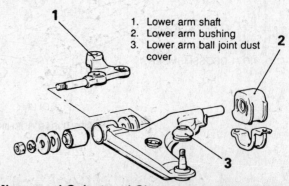

1. Lower arm shaft
2. Lower arm bushing
3. Lower arm ball joint dust cover

Mirage and Galant and Sigma lower control arm

SUSPENSION AND STEERING 8

should be set snug now and final tightened with the weight of the car on the front suspension.

11. For all Mirage, tighten the bolts for the spacer at the rear to 50 ft. lbs., and the ball joint nut to 48 ft. lbs.

On the Galant, torque the nut for the nut/bolt-to-crossmember to 78 ft. lbs. and the balljoint to 46 ft. lbs. The bolt for retaining the rear bushing to the body should be tightened to 65 ft. lbs.

12. Install the wheel and lower the car to the ground.
13. Bounce the car on its suspension 2 or 3 times to settle the suspension. With the car free of passengers or cargo, tighten the lower shaft arm bolt to 110 Nm or 81 ft. lbs.
14. Align the front wheels

Hub and Knuckle

REMOVAL AND INSTALLATION

1. With the car on the ground, remove the center cap and the cotter pin. Loosen the large driveshaft nut. It's tight enough that you won't be able to loosen it with the car in the air.
2. Elevate and safely support the car on stands.
3. Remove the front wheel.
4. Remove the brake caliper and suspend it from wire out of the way. Do not disconnect or loosen the hose; simply move the whole assembly out of the way. Do not allow the caliper to hang by the hose.

---- CAUTION ----
Brake pads and shoes contain asbestos, which has been determined to be a cancer causing agent. Never clean the brake surfaces with compressed air! Avoid inhaling any dust from brake surfaces! When cleaning brakes, use commercially available brake cleaning fluids.

5. If the brake rotor is held to the hub by two small screws, remove them and remove the disc. Some discs are pressed into the hub and can only be removed after the hub and knuckle are removed.
6. Disconnect the stabilizer bar and the strut bar from the lower control arm.
7. Remove the cotter pin from the ball joint and loosen the nut until it is just to the end of the stud.
8. Place a floor jack under the control arm and adjust it to 25-50mm below the arm. Install a ball joint separator tool and separate the joint. The nut on the stud will retain explosive separation. Once the joint is loosened, remove the nut and separate the joint.

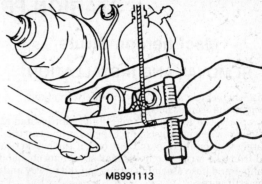

If you use the Mitsubishi special tool, tie the string to a nearby location; it prevents the tool from flying off during separation

9. Using the correct tools separate the tie rod joint at the steering knuckle.
10. Remove the drive shaft from the hub. Use special tools as needed, but never hammer on the end of the shaft.
11. Remove the hub and knuckle as an assembly from the strut by removing two retaining bolts.
12. When reinstalling, bolt the knuckle assembly to the strut and tighten the bolts. Correct torque is 81 Nm or 60 ft. lbs.; you may not be able to achieve this until the car is on the ground.
13. Insert the drive shaft into the hub. A light coat of multi-purpose grease on the splines of the shaft will ease the installation.
14. Install the tie rod joint to the knuckle. Use a new nut and tighten it. Install a new cotter pin.
15. Install the lower control arm ball joint to the knuckle. Install a new nut and tighten it to the correct torque. Install a new cotter pin.
16. Connect the stabilizer bar and strut bar.
17. Reinstall the brake disc if it was removed. Replace the brake caliper and tighten it securely.
18. Install the driveshaft nut and its washer; tighten it snug. Don't try to achieve final torque until the car is on the ground.
19. Install the wheel and lower the car to the ground.
20. If the knuckle to strut bolts were not final-tightened, do so now.
21. Tighten the driveshaft nut to 230 Nm (166 ft. lbs). Install a new cotter pin and install the dust cap.
22. Since the control arm and tie rod was disconnected, front end alignment is highly recommended.

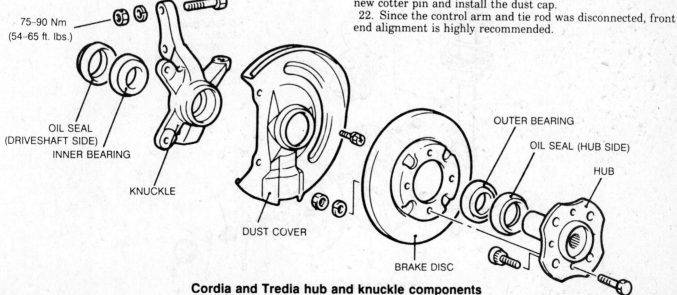

Cordia and Tredia hub and knuckle components

8-15

8 SUSPENSION AND STEERING

STARION FRONT SUSPENSION

MacPherson Struts

REMOVAL AND INSTALLATION

1. Raise the vehicle and support it safely on stands.
2. Remove the front wheels.
3. Disconnect the hydraulic brake fluid joint located at the strut. Use two wrenches; disconnect the joint and cap the lines immediately to keep fluid in and dirt out. if the bracket can be unbolted from the strut, the hoses may stay connected and the entire caliper and hose assembly swung out of the way and supported by stiff wire.
4. Remove the brake disc caliper assembly and set it aside.

— CAUTION —
Brake pads and shoes contain asbestos, which has been determined to be a cancer causing agent. Never clean the brake surfaces with compressed air! Avoid inhaling any dust from brake surfaces! When cleaning brakes, use commercially available brake cleaning fluids.

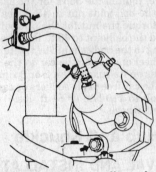

Removing the Starion brake caliper. In this case, the bracket may be unbolted from the strut. Early cars may require disconnection of the hoses

5. Remove the hub, rotor, and bearing from the knuckle.
6. Remove the brake dust cover (backing plate).
7. Remove the two bolts and lockwashers attaching the lower end of the strut to the knuckle arm.
8. Support the strut as you remove the two or three nuts and lockwashers from the top of the wheel well inside the engine compartment. Do NOT loosen the nut on the top of the shock absorber. Lower and remove the strut.

WARNING: The strut will come out with the spring still mounted and under tension.

9. Install the strut by fitting it in place and tightening the upper mounting nuts to 30 Nm or 23 ft. lbs.
10. Before attaching the bottom of the strut, apply a semi-drying sealer to the top surface of the knuckle arm. Tighten the lower attaching bolts to 90 Nm or 67 ft. lbs.
11. Install the hub, rotor, bearing and brake disc. When installing the locknut, set the bearing tension as follows:
    a. Tighten the nut to 20 Nm or 14 ft. lbs.
    b. Back off the nut to 0 Nm or ft. lbs.
    c. Retighten the nut to 5 Nm or 4 ft. lbs.
12. Install the brake caliper.
13. Connect the hydraulic line at the strut bracket if it was disconnected. Use two wrenches and do not crossthread the fitting. If the system was not disconnected, reposition the hose and secure the bracket to the strut.
14. Bleed the brake system.
15. Install the front wheel and lower the vehicle to the ground.
16. Before test driving, check the brake pedal and brake function. Re-bleed the system if necessary.

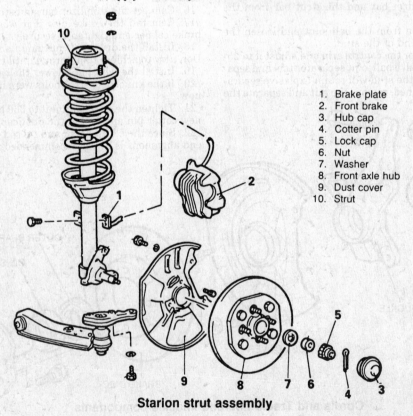

1. Brake plate
2. Front brake
3. Hub cap
4. Cotter pin
5. Lock cap
6. Nut
7. Washer
8. Front axle hub
9. Dust cover
10. Strut

Starion strut assembly

SUSPENSION AND STEERING 8

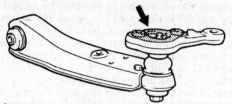

Apply sealant to the contact face before installing the strut

OVERHAUL

MacPherson struts are constructed in such a way that the unit can be removed from the car while the spring is still under considerable tension. While removal of the unit as a whole is safe and simple, disassembly is an entirely different matter.

The spring is still under considerable tension; you MUST NOT attempt to disassemble the unit until an effective spring compression tool is properly installed and you can see that ALL spring tension is removed.

--- **CAUTION** ---

Failure to perform all these procedures accurately will result in the release of deadly force. If you are uncertain about performing this procedure, take the removed strut to a shop that will disassemble and overhaul it for you.

The Mitsubishi factory tool for use in this operation is MB 990987; you can purchase a number of tools designed to compress coil springs used on MacPherson struts. Assure yourself that you have bought a quality tool and can use it safely.

NOTE: To disassemble this strut, both a spring compressor and a tool designed to hold the spring seat stationary while turning the the self-locking nut will be needed. Use MB 990775-01 or equivalent. Replacing the shock absorber within the case requires a special wrench to remove the ring nut at the top of the strut. Use MB990775-01 or equivalent.

Disassembly

1. Remove the strut from the vehicle.
2. Install the spring compressor and fully compress the spring until there is no tension on the spring seats.
3. Pry the insulator cap off the insulator.
4. Hold the spring seat stationary with the special tool and use an ordinary socket wrench to remove the self-locking nut from the top of the shock piston rod.
5. Then, remove the insulator, spring seat, rubber cushion, dust cover, and rubber cushion seat from the strut.
6. Inspect the top insulator for cracks or peeling. Rubber parts must be inspected for cracking or brittleness. Springs should be checked for any sign of cracking. The piston rod should be inspected for bending or obvious wear. Replace external parts as necessary.

HYDRAULIC TYPE

GAS DAMPER TYPE

1. Insulator cap
2. Top end nut — self-locking
3. Insulator
4. Spring seat
5. Rubber helper
6. Dust cover
7. Coil spring
8. Strut
9. Oil seal
10. Square section O-ring
11. Piston
12. Piston guide
13. Cylinder
14. Outer shell

Starion strut assemblie

8-17

8 SUSPENSION AND STEERING

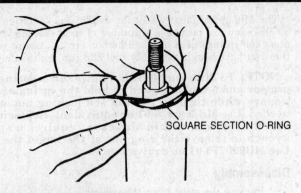

Removing the inner O-ring

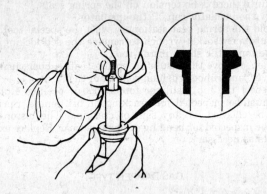

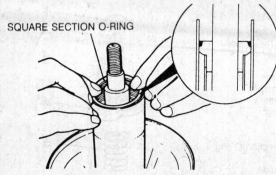

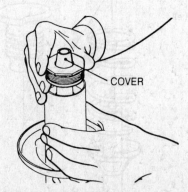

Correct reassembly of the piston guide (top), O-ring and oil seal (lower)

HYDRAULIC SHOCK ABSORBER (OIL FILLED)

1. To prevent entry of dirt into the oil within the cylinder, thoroughly clean the outside of the case before disassembly.
2. Hold the strut upright in a vise. Clamp the knuckle part of the strut, not the tube. Compress the piston into the strut.
3. Use the special wrench to remove the oil seal assembly.
4. Use a small hooked tool to remove the square section oil seal from around the inside of the tube.
5. Slowly withdraw the piston rod from the cylinder together with the piston guide. The piston has a precision machined surface; handle it carefully.
6. Remove the piston guide from the rod.
7. Remove the cylinder from the strut outer shell.
8. To reassembly, install the cylinder and piston assembly into the outer shell.
9. Gradually pour shock absorber fluid into the cylinder while slowly moving the piston up and down
10. With the flange of the piston guide facing upward, insert the piston guide onto the piston rod until it contacts the cylinder end.
11. Install a NEW square section O-ring taking great care to see that it is correctly installed with no folding or tilting. If this seal isn't correct, the shock will leak.
12. Attach the cover to the piston end. Apply shock absorber fluid to the cover and install the oil seal assembly. Use the special tool to tighten the oil seal assembly to 145 Nm or 107 ft. lbs.

GAS SHOCK ABSORBER

WARNING: The shock absorber is filled with nitrogen gas under pressure. Do not disassemble the shock unless necessary for replacement. When replacing, install inner parts as a complete kit.

1. Drill a 4mm, OR SMALLER, diameter hole into the case at the point shown. This will bleed the gas from the strut.

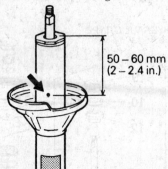

The gas strut MUST be drilled before disassembly

2. Remove the ring nut using the special tool.
3. Remove the shock assembly from the strut and install a new cartridge.
4. Tighten the ring nut to 145 Nm (107 ft. lbs.)
5. The repair kit should contain a label which may be placed over the hole in the case. This hole should be covered to prevent entry of dirt and water into the strut.

Assembly

1. Compress the spring with the spring compressor, if it is not already compressed.
2. Use self-drying adhesive to bond the spring seat to the rubber cushion. Install the rubber cushion seat, spring, dust cover, rubber cushion and spring seat onto the strut.
3. Align the flat on the side of the D-shaped hole in the spring seat with the flat on the piston rod, and align the projections on the dust cover with the holes in the spring seat.
4. Install the strut insulator and hold the spring seat with the special tool as you install the nut onto the top of the piston rod. Tighten it to 65 Nm (48 ft. lbs.) You must use a NEW self-locking nut.
5. Slowly release tension on the spring compressor, making certain the spring aligns and seats correctly. AFTER you verify that the spring coil is properly aligned in both top and bottom

SUSPENSION AND STEERING 8

Bond the spring seat to the rubber cushion with a drying adhesive

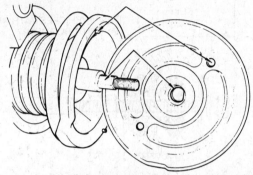

On Starion, align the flat surface on the shaft with the hole in the spring seat. The projections on the dust cover must also match with the holes in the seat

springs seats, remove the special tool.
6. Pack multipurpose grease into the strut insulator and install the cap. Keep the grease off the rubber parts.
7. Install the strut.

Lower Ball Joints

INSPECTION

1. Support the vehicle on stands.
2. Disconnect the ball joint at the lower end of the strut as described below; you need not remove the lower control arm entirely.
3. Install the nut back onto the ballstud.
4. Using an inch lb. torque wrench, measure the torque required to start the ball joint rotating. The correct figure is 43–69 inch lbs. Axial play is 0.5mm.
If the figures are within specification, the ball joint is satisfactory. If the figure is too high, the joint should be replaced. If the figure is too low, you can still reuse the joint provided its rotation is smooth and even. If there is roughness or play, it must be replaced.

REMOVAL AND INSTALLATION

1. Raise and safely support the car on stands.
2. Remove the front wheel.
3. Remove the bolts and disconnect the knuckle arm from the bottom of the MacPherson strut. Use a floor jack to support the lower arm before disconnecting it from the strut.
4. Remove the ballstud nut and use a special tool such as MB990241-01 or equivalent to press the ball joint downward and out of the knuckle arm. This tool bolts to the knuckle arm in order to work.
5. Unbolt the ball joint from the lower control arm.
6. To install, bolt the joint to the control arm and torque the bolts/nuts to 64 Nm or 47 ft. lbs.

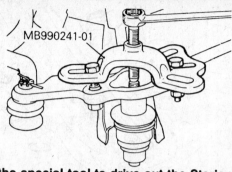

Using the special tool to drive out the Starion ball joint

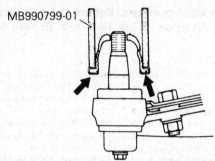

Installing the dust boot on the Starion ball joint

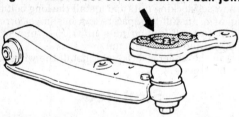

The contact face of the knuckle arm must be coated with adhesive before reassembly

7. Insert the ballstud through the hole in the knuckle arm, install the nut, and torque it to 65 Nm (48 ft. lbs.) Apply a semi-drying sealer to the surface of the knuckle arm which contacts the strut.
8. Install the strut lower mounting bolts, tightening to 88 Nm or 65 ft. lbs.

Stabilizer Bar (Sway Bar)

REMOVAL AND INSTALLATION

1. Raise and safely support the car on stands.

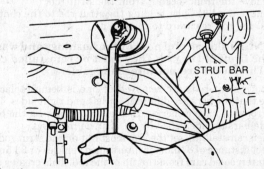

Adjusting caster on the Cordia and Tredia

8-19

8 SUSPENSION AND STEERING

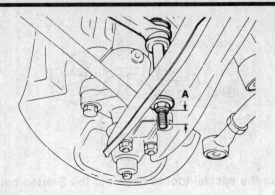

For Starion stabilizer bars, tighten the self-locking nut so that distance "A" is 15–17mm (0.59–0.67 in.)

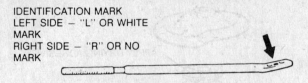

Identification of strut rods

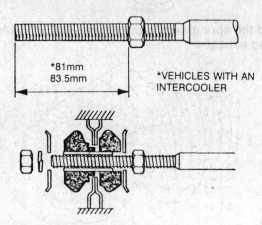

Correct assembly of Starion strut rods

2. Remove the front wheels. Remove the undercover.
3. Disconnect the stabilizer bar from the lower control arm by removing the nut and long bolt. Remove the various spacers, washers and rubber bushings, taking accurate note of their location and placement. They must be reinstalled correctly.
4. Remove the bolts holding the stabilizer bar brackets and remove the stabilizer bar from the crossmember.
5. When reinstalling, tighten the bracket bolts to 11 Nm or 8 ft. lbs.
6. Use new self-locking nuts and install the long bolt through the control arm. Install the spacers and bushings correctly.
7. Tighten the self-locking nuts until 15–17mm of thread projects beyond the face of the nut.
8. Reinstall the undercover(s) and install the wheels. Lower the car to the ground.

Strut Rod

Most MacPherson strut suspension systems employ a rod running from the body or frame behind each wheel to the lower control arm. This rod serves to brace and locate the lower control arm (and therefore the tire and wheel), keeping it from moving in the fore-and-aft plane. The most common damage to these rods is bending due to impact. Once the rod is bent, its effective distance is shortened and the wheel is pulled out of alignment. Any time the strut rod is loosened or removed, the front wheels must be aligned after the repair. No matter how hard you try to get the nut reinstalled in the same position, it probably won't be accurate. Don't risk ruining a set of front tires; have the front end aligned.

REMOVAL AND INSTALLATION

1. Elevate and safely support the car on stands.
2. Remove the front wheels. Remove the undercovers.
3. Remove the front locknut from the strut rod.
4. Remove the two bolts holding the strut rod to the control arm. Pull the rod rearward to remove it.

WARNING: Take careful note of the bushings and washers at the front of the rod. They must be reinstalled correctly. The pieces are not interchangable.

5. If both strut rods are removed or one is being replaced, take note that the bars are different for left and right sides. The left side rod will have either an **L** or a white mark on the bar. Right side bars are identified by an **R** or no marking.
6. When reinstalling, set the rearmost nut on the strut rod to the correct distance of 81mm for intercooled vehicles or 83.5mm for non-intercooled cars from the tip of the rod. This creates the approximately correct effective length for the strut rod.

7. Install the front of the rod with the bushings and spacers correctly placed. Install the end nut snug but do not final tighten it.
8. Attach the strut rod to the control arm, tightening the bolts to 65 Nm (48 ft lbs.).
9. Install the wheels and lower the vehicle to the ground.
10. Bounce the vehicle two or three times to stabilize the suspension. With the car free of passengers and cargo, tighten the front nut on the strut rod to 80 Nm or 59 ft. lbs.
11. Reinstall the undercovers.
12. Have the alignment checked at a reliable repair facility.

Lower Control Arm

REMOVAL AND INSTALLATION

1. Support the vehicle securely on stands placed under the crossmember.
2. Remove the front wheel.
3. Disconnect the stabilizer bar link at the control arm by removing the nut underneath the arm.
4. Remove the nut and bolt attaching the strut bar to the control arm.
5. Disconnect the tie rod at the knuckle arm. Use a fork-like tool such as MB990778-01, a standard type tool for pulling ball joint studs. Loosen the stud nut until it is near the top of the threads and then hammer the tool between the ball joint of the tie rod end and the knuckle arm. The nut on the top of the stud will prevent the rod from flying off, possibly causing injury. When the ballstud comes loose, remove the nut and disconnect the stud.
6. Unbolt the MacPherson strut from the knuckle arm.
7. Unbolt the inner end of the ball joint assembly to disconnect it from the outer end of the control arm.
8. Remove the nut, bolt, and lockwasher, and pull the inner end of the control arm out of the crossmember.
9. When reinstalling, place the arm in position at the crossmember and install the bolt. Tighten the nut firmly snug but do not final tighten yet.

SUSPENSION AND STEERING 8

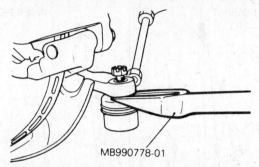

Disconnecting the Starion tie rod at the knuckle

10. Assemble the ball joint to the outer end of the arm and torque the bolts to 43–51 ft. lbs. Tighten the ballstud nut to 48 ft. lbs.; and the strut attaching bolts to 65 ft. lbs.
11. Tighten the nut for the stabilizer bar link until 15–17mm of thread shows below the bottom of the nut.
12. Install the front wheel and lower the car to the ground.
13. Bounce the car on its suspension 2 or 3 times to settle the suspension. With the car free of passengers or cargo, tighten the pivot bolt for the lower arm to 88 Nm or 65 ft. lbs.

Knuckle and Spindle

REMOVAL AND INSTALLATION

1. Elevate and safely support the car on stands.
2. Remove the wheel.
3. Remove the brake caliper and suspend it from wire out of the way. Do not disconnect or loosen the hose; simply move the whole assembly out of the way. Do not allow the caliper to hang by the hose.

------- **CAUTION** -------
Brake pads and shoes contain asbestos, which has been determined to be a cancer causing agent. Never clean the brake surfaces with compressed air! Avoid inhaling any dust from brake surfaces! When cleaning brakes, use commercially available brake cleaning fluids.

4. Remove the small center cap.
5. Remove the cotter pin, lock cap and spindle nut. Remove the large flat washer and the outer wheel bearing.
6. Pull the brake rotor and hub off the spindle. The spindle (stub axle) is part of the suspension strut. If the spindle is damaged, the strut must be replaced. Clean and repack the wheel bearings while the assemble is off the car.
7. When reinstalling, fit the rotor and hub assembly onto the spindle with the bearings and seal in place.
8. Making certain that the outer bearing is well seated, install the flat washer and the shaft nut. Tighten the nut finger tight.
9. Using a torque wrench, tighten the nut to 20 Nm or 14 ft. lbs. Back the nut off to 0, then final tighten to 5 Nm (48 INCH lbs.).
10. Install the lock cap and cotter pin. If the holes in the lock cap do not align with the castle nut on the shaft, turn the lock cap and try again; the lock cap has offset holes to compensate for this occurence. In the unlikely event that it still will not align, back off the shaft nut by NO MORE than 15° of rotation.
11. Install the small hub cover.
12. Install the brake caliper and pads.
13. Install the wheel and lower the car to the ground.

Front Wheel Bearing

REMOVAL AND INSTALLATION

1. Remove the hub and rotor assembly from the car as described previously.
2. Make matching marks on the brake disc and the front hub. Correct reassembly is required. On vehicles with intercoolers, remove the brake disc by removing the small E-clips on the lug bolts.
3. On cars without intercoolers, unbolt and remove the brake disc.
4. The outer bearing was removed during the hub removal. The race for the outer bearing may be removed with a brass rod and mallet. Drive the race out from the inside but do not mar the inner surface of the hub.

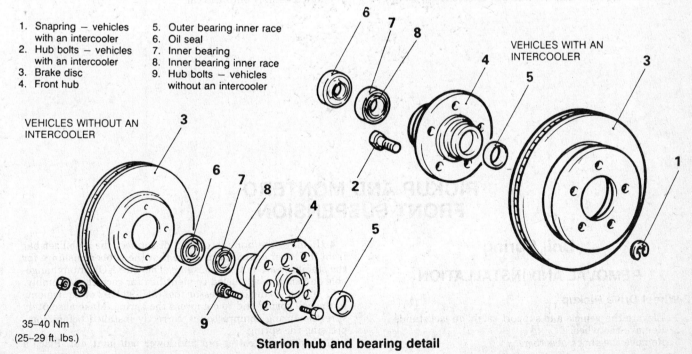

1. Snapring — vehicles with an intercooler
2. Hub bolts — vehicles with an intercooler
3. Brake disc
4. Front hub
5. Outer bearing inner race
6. Oil seal
7. Inner bearing
8. Inner bearing inner race
9. Hub bolts — vehicles without an intercooler

VEHICLES WITHOUT AN INTERCOOLER

35–40 Nm (25–29 ft. lbs.)

VEHICLES WITH AN INTERCOOLER

Starion hub and bearing detail

8-21

8 SUSPENSION AND STEERING

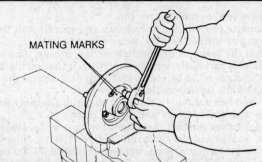

Make matchmarks before unbolting the rotor

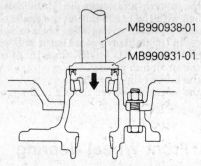

Installing the Starion inner bearing grease seal. Mitsubishi tool numbers shown

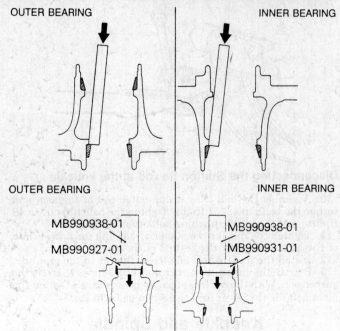

Removing and installing bearing races. Mitsubishi tool numbers shown but seal and bearing drivers are available in most auto stores

WARNING: Do not remove the bearing races if the same bearing is to be reused after cleaning or repacking. If a bearing is replaced, the race MUST be replaced. Don't use a new bearing on an old race.

5. On the back of the hub, use a seal puller (a very common tool) to remove the inner bearing seal.
6. Remove the inner bearing and remove the race in the same fashion as the outer race.
7. Use bearing drivers of the proper size and drive the new races into the hub. Don't try to install the race with the brass rod; chances are high that the race will be crooked or damaged.
8. Apply a liberal coating of bearing grease to the oil seal lip and the inner surface of the hub.
9. Pack the inner bearing correctly and install the inner bearing into the hub.
10. Use a seal driver and install a new oil seal into the hub. Press it into place until it is flush with the end face of the hub.
11. Install the brake disc, aligning it with the marks made during disassembly. If it is the bolt type, torque the nuts to 37 Nm or 27 ft. lbs. If it is the clip type, use new clips and insure the E-clips are firmly engaged.
12. Pack the outer wheel bearing correctly and install it with the hub assembly onto the car.

PICKUP AND MONTERO FRONT SUSPENSION

Coil Spring

REMOVAL AND INSTALLATION

2-Wheel Drive Pickup

1. Elevate the vehicle and support safely on jackstands.
2. Remove the wheel.
3. Remove the shock absorber.
4. Remove the bump stop and disconnect the stabilizer bar from the lower control arm. Note that the rubber bushings for the stabilizer are not all the same. They are NOT interchangeable and should be labeled or identified for correct reassembly.
5. Install spring compressor tool (MB-990792 or equivalent) to the coil spring and to compress the spring. Make absolutely sure the spring compressor is correctly installed before compressing the spring.
6. Remove the cotter pin and lower ball joint nut. Place a

SUSPENSION AND STEERING 8

floor jack under the control arm; elevate the jack to within 25mm of the arm.

7. Release the lower ball joint from the knuckle using a ball joint separator such as MB 990809-01 or equivalent tool. The arm will drop onto the jack as the joint comes loose.
8. Slowly release the compressor tool from the coil spring.
9. Lower the jack as necessary, pull the arm down and remove the spring with the rubber isolation pad and upper spring seat from the vehicle.
10. Install the spring with the rubber isolator. Make sure the spring seats correctly in its upper and lower mounts. Install the compressor tool and compress it enough so the lower ball joint can be inserted through the knuckle. Use the jack to support and adjust the height of the lower control arm during installation.
11. Tighten the lower ball joint nut to 136 NM (100 ft. lbs.). Install a new cotter pin. Remove the spring compressor.
12. Install the bump stop and install the shock absorber. The shock absorber mounting bolts (upper and lower) are tightened to only 13 Nm or 10 ft. lbs.
13. Install the stabilizer bar, remembering to install the rubber bushings and hardware in the correct locations. Tighten the sway bar bolt until the bolt threads project 22-24mm beyond the nut.
14. Install the wheel and lower the vehicle to the ground.

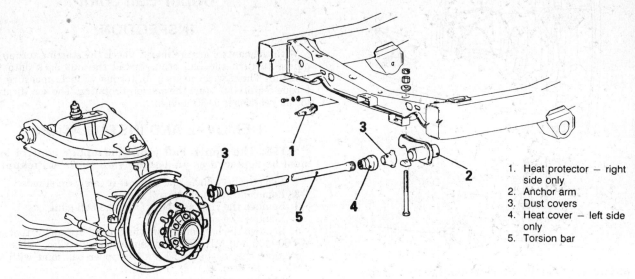

1. Heat protector — right side only
2. Anchor arm
3. Dust covers
4. Heat cover — left side only
5. Torsion bar

Torsion bars replace coil springs on Montero and 4wd Pickups

Torsion Bar

REMOVAL AND INSTALLATION

Montero and 4-Wheel Drive Pickup

1. Raise the vehicle and support it safely on stands. Make sure the stands are placed on the frame rails.
2. Remove the wheel from the side to be repaired.
3. Support the lower control arm with a floor jack. Position the jack at a point away from where the torsion bar attaches to the arm.
4. Fold the dust covers back and slide them away from the ends of the bar.
5. If the bars are to be reused, matchmark the torsion bar at both ends to the anchor and identify left from right.
6. Paint or measure the distance of the exposed threads of the rear mounting bolt down to the nut to aid in adjustment when installing. Remove the rear anchor arm mounting nut and bolt.
7. Loosen the adjusting nut and pull the torsion bar from the lower arm assembly.
8. Check the torsion bar for bends or damage. Inspect the dust covers for cracks. Check the anchor bolt for bending or distortion. If the torsion bar has received a heavy impact — such as during off-road use — it is recommended that the bar be professionally crack checked using electric methods. The peace of mind is worth the cost.
9. If the bar is being reused, lubricate the ends and install the torsion bar aligning the matchmarks. If a new bar is being used, align the white stripe on the front splines with the mark on the anchor. There is a mark on the front of the torsion bar to differentiate between left and right. Do not install the bar with the

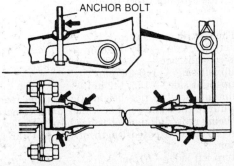

Apply grease liberally at the points shown

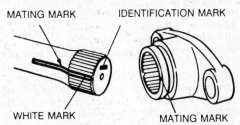

New torsion bars have a painted alignment mark

mark facing the rear.

10. Install the torsion bar to the rear anchor so that the length of the mounting bolt from the nut to the head of the bolt is the specified length with the rebound bumper in contact with the crossmember. Reposition the bar as required to met the specifi-

8-23

8 SUSPENSION AND STEERING

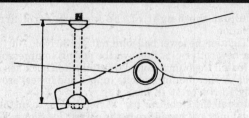

Setting initial bolt length during installation

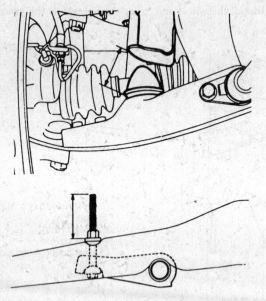

Pay attention to the bump stop clearance and the anchor bolt projection. These factors determine the vehicle's ride quality after repair

cations. The specifications are:
- Montero left side — 135–142mm
- Montero right side — 124–132mm
- Pickup left side — 140–147mm
- Pickup right side — 135–142mm

11. To initially set the ride height, tighten the rear anchor mounting nut to the same point at which it was removed if the old bar is being reused. If a new bar has been installed, tighten the nut so that the exposed length of the bolt threads is to specification:
- Montero left side — 61mm
- Montero right side — 71mm
- 1986 Pickup left side — 76mm
- 1986 Pickup right side — 68.5mm
- 1987–90 Pickup left side — 100mm
- 1987–90 Pickup right side — 86mm

12. Fill the dust covers with grease and fold them back into position.
13. Adjust the torsion bar to the correct riding height.
14. Install the wheel and lower the vehicle to the ground. With the vehicle on the ground but unladen, measure the distance from the bottom of the bump stopper to the contact face of the bump stop bracket. Correct distance is 71mm. Adjust this distance by turning the adjusting nut on the anchor bolt.

Shock Absorber

NOTE: This procedure is for shock absorbers mounted separately from the spring. It does not apply to MacPherson or other strut systems

REMOVAL AND INSTALLATION

1. Support the vehicle safely on stands. Support it at a height which will allow sufficient room to reach the upper shock absorber mount.
2. Remove the wheel.
3. Remove the upper shock nut, washer and bushing.
4. Remove the lower mounting bolt(s) and remove the shock from the vehicle.
5. The installation is the reverse of the removal procedure. The shock absorber mounting bolts should be tightened only to 14 Nm (10 ft. lbs).

Upper Ball Joint

INSPECTION

With the control arm removed, check the starting torque required to turn the ball stud. Install the nut back onto the ballstud. Then, with an inch lb. torque wrench, measure the torque required to start the ball joint rotating. The specification for all vehicles is 7–30 inch lbs.

REMOVAL AND INSTALLATION

NOTE: The upper ball joint and upper control arm must be replaced as an assembly on 1987-90 Pickup.

1. Elevate the vehicle and support it safely on stands.
2. Remove the upper control arm.
3. Remove the ring and boot from the ball joint.
4. Remove the snapring.
5. Remove the ball joint from the control arm using tool sets MB 990800 and MB 990799 or equivalent installing tools.
6. Align the mating mark on the upper ball joint with the mark on the arm.
7. Press the ball joint in using the same tools that were used to remove the ball joint.
8. Install the snapring. If the snapring is loose, install a new one.

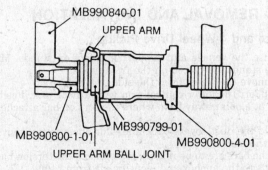

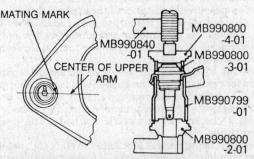

Proper removal (upper) and installation of the upper ball joint is almost impossible without the correct tools

SUSPENSION AND STEERING 8

9. Fill the boot with grease and install the boot and ring.
10. Install the upper control arm.
11. Lubricate the ball joint with a grease gun.
12. Adjust the riding height if equipped with a torsion bar and align the front end.

Lower Ball Joints

INSPECTION

1. Support the vehicle on stands.
2. Disconnect the ball joint at the lower end of the strut as described below; you need not remove the lower control arm entirely.
3. Install the nut back onto the ballstud.
4. Using an inch lb. torque wrench, measure the torque required to start the ball joint rotating. The correct figures is 9–30 inch lbs. Axial play is 0.5mm.

If the figures are within specification, the ball joint is satisfactory. If the figure is too high, the joint should be replaced. If the figure is too low, you can still reuse the joint provided its rotation is smooth and even. If there is roughness or play, it must be replaced.

REMOVAL AND INSTALLATION

1. Raise the vehicle and support it safely on stands.
2. Remove the wheel.
3. Remove the lower control arm.
4. Remove the ball joint retaining nuts and bolts and remove the ball joint from the arm.
5. Install the new ball joint. Tighten the ball joint retaining nuts and bolts to 68 Nm (50 ft. lbs.) on 4WD vehicles or 38 Nm (28 ft. lbs.) on 2WD vehicles.
6. Install the lower control arm and connect the ball joint to the knuckle. Tighten the ball stud nut to 136 Nm (100 ft. lbs.) and install a new cotter pin.
7. Lubricate the ball joint with a grease gun.
8. Adjust the riding height, if equipped with a torsion bar, and align the front end.

Stabilizer Bar (Sway Bar)

REMOVAL AND INSTALLATION

1. Raise the vehicle and support safely on stands.
2. Remove the front wheels. Remove the undercover or splash shield.
3. Disconnect the bracket or clamp holding the sway bar to the crossmember. On 4wd Pickups and all Monteros, the link may be removed from the crossmember; the retaining nut is on top of the crossmember.
4. Remove the sway bar support brackets and bushings from the lower control arm. Take careful note of the order and placement of the washers, bushings and spacers. They must be reinstalled correctly.
5. Remove the sway bar from the vehicle.
6. Inspect the bar for cracking or bending. Check the bushings for cracking or deformation. If any fault is found, replace components. It is recommended to replace the rubber bushings any time they are removed.
7. Install the bar into place and attach all the hardware and brackets loosely. Make sure the washers and bushings are correctly installed.
8. For Montero and 4wd Pickups, use a new nut on the upper link mount (above the crossmember) and tighten it until 6–9mm of threads projects through the nut.
9. Tighten the bracket or clamp bolts to 10 Nm or 8 ft. lbs.
10. Install new nuts at each lower control arm. For Montero and 4wd pickups, tighten the nuts until 6–9mm of threads project through the nut. On 2wd pickups, the bolt must project 20–23mm through the nut.
11. Install the undercover or splash shield.
12. Install the wheels and lower the vehicle to the ground.

Dimension "A" applies only to 4-wheel drive; dimension "B" is important to all

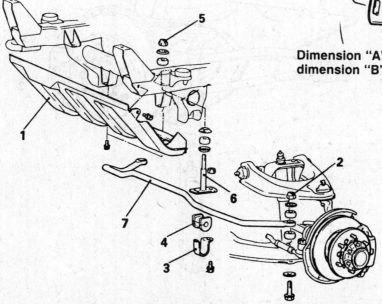

1. Skidplate
2. Self-locking nut
3. Clamp A
4. Stabilizer bushing
5. Self-locking nut
6. Hanger
7. Stabilizer bar

4-wheel drive Pickup and Montero stabilizer bar

8-25

8 SUSPENSION AND STEERING

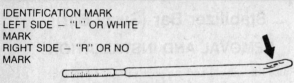

IDENTIFICATION MARK
LEFT SIDE — "L" OR WHITE MARK
RIGHT SIDE — "R" OR NO MARK

Identification of strut rods

Strut Rod

REMOVAL AND INSTALLATION

2-Wheel Drive Pickup

1. Elevate and safely support the truck on stands.
2. Remove the front wheels. Remove the undercovers.
3. Remove the front locknut from the strut rod.
4. Remove the two bolts holding the strut rod to the control arm. Pull the rod rearward to remove it.

WARNING: Take careful note of the bushings and washers at the front of the rod. They must be reinstalled correctly. The pieces are not interchangable.

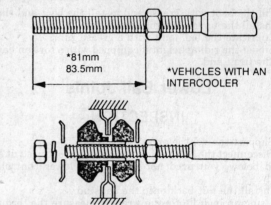

*81mm
83.5mm   *VEHICLES WITH AN INTERCOOLER

Correct assembly of Starion strut rods

5. If both strut rods are removed or one is being replaced, take note that the bars are different for left and right sides. The left side rod will have either an **L** or a white mark on the bar. Right side bars are identified by an **R** or no marking.

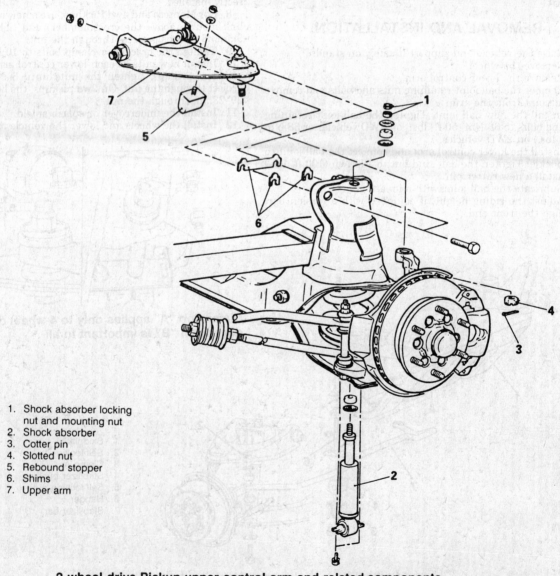

1. Shock absorber locking nut and mounting nut
2. Shock absorber
3. Cotter pin
4. Slotted nut
5. Rebound stopper
6. Shims
7. Upper arm

2-wheel drive Pickup upper control arm and related components

SUSPENSION AND STEERING 8

6. When reinstalling, set the rearmost nut on the strut rod to the correct distance of 83.5mm from the tip of the rod. This creates the approximately correct effective length for the strut rod.
7. Install the front of the rod with the bushings and spacers correctly placed. Install the end nut snug but do not final tighten it.
8. Attach the strut rod to the control arm, tightening the bolts to 65 Nm (48 ft. lbs.).
9. Install the wheels and lower the vehicle to the ground.
10. Bounce the vehicle two or three times to stabilize the suspension. With the truck free of passengers and cargo, tighten the front nut on the strut rod to 80 Nm or 59 ft. lbs.
11. Reinstall the undercovers.
12. Have the alignment checked at a reliable repair facility.

Upper Control Arm

REMOVAL AND INSTALLATION

2-Wheel Drive Pickup

1. Raise the vehicle and support it safely on stands.
2. Remove the wheel.
3. Remove the shock absorber.
4. Install Mitsubishi spring compressor tool MB 990792 or equivalent to the coil spring and compress the spring.
5. Remove the cotter pin and upper ball joint nut.
6. Suspend the rotor assembly with a wire so there is not excessive pull on the brake hose. A floor jack can be helpful as well.
7. Release the upper ball joint taper using Mitsubishi tool MB990809-01 or or equivalent.
8. Remove the tool and remove the ball stud from the knuckle.

9. Loosen the pivot bar retaining nuts and bolts, identify and remove the alignment shims.
10. Remove the nuts and bolts and remove the arm from the vehicle.
11. Install the arm to the frame rail bracket, install the shims in their original locations and install the retaining nuts and bolts. Tighten the nuts initially to about 40 ft. lbs. (54 Nm).
12. Install the ball joint to the knuckle and tighten the ball joint nut to 81 Nm (60 ft. lbs.). Install a new cotter pin.
13. Seat the spring correctly and slowly loosen the compressor. Make certain both ends of the spring are correctly placed. Remove the spring compressor.
14. Install the shock absorber.
15. Align the front end. When all settings are at specifications, tighten the pivot bar retaining bolts to 109 Nm or 80 ft. lbs.

Montero and 4-Wheel Drive Pickup

1. Elevate and safely support the vehicle on stands.
2. Remove the wheel. Remove the skid plate.
3. Remove the shock absorber.
4. Turn the torsion bar adjustment nut (anchor nut, at the rear crossmember) counterclockwise to relieve all tension from the torsion bar.
5. Disconnect the brake hose from the brake line and remove the hose from the bracket on the upper arm. Plug or cap the hose and line immediately.
6. Remove the cotter pin from the upper ball stud.
7. Release the upper ball joint taper using Mitsubishi tool MB990809-01 or or equivalent. Remove the tool.
8. Remove the ball stud from the steering knuckle.
9. Loosen the pivot bar retaining nuts and bolts, identify and remove the alignment shims, remove the nuts and bolts.

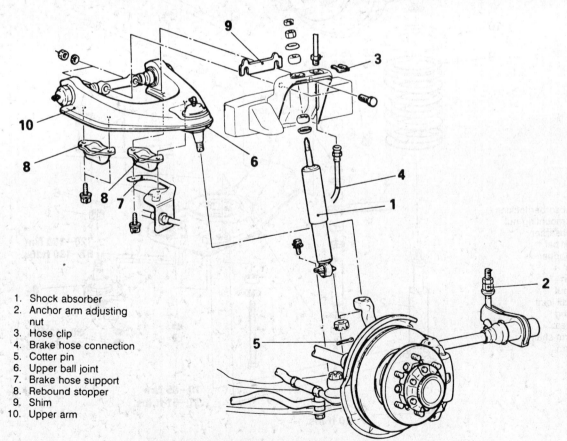

1. Shock absorber
2. Anchor arm adjusting nut
3. Hose clip
4. Brake hose connection
5. Cotter pin
6. Upper ball joint
7. Brake hose support
8. Rebound stopper
9. Shim
10. Upper arm

Montero and 4-wheel drive Pickup upper control arm

8-27

8 SUSPENSION AND STEERING

10. Remove the arm from the vehicle.
11. Position the arm at the frame rail bracket.
12. Install the shims in their original locations and install the retaining nuts and bolts. Torque the nuts initially to about 40 ft. lbs. (54 Nm).
13. Insert the upper ball stud in the steering knuckle and install the nut. Tighten the nut to 81 Nm (60 ft. lbs.) and install a new cotter pin.
14. Install the shock absorber.
15. Remove the plugs and carefully attach the brake hose to the line. Don't crossthread the fitting. Make certain the retaining clip is seated.
16. Bleed the brake system.
17. Turn the torsion bar adjustment nut clockwise to its approximate previous position. This will apply a load on the bar.
18. Install the wheel and lower the vehicle.
19. Set the ride height. Refer to "Torsion Bars" earlier in this Section. Have the front end alignment checked.
20. Apply final torque to the upper arm retaining bolts, setting them to 110 Nm or 81 ft. lbs.
21. Reinstall the skid plates and undercovers.

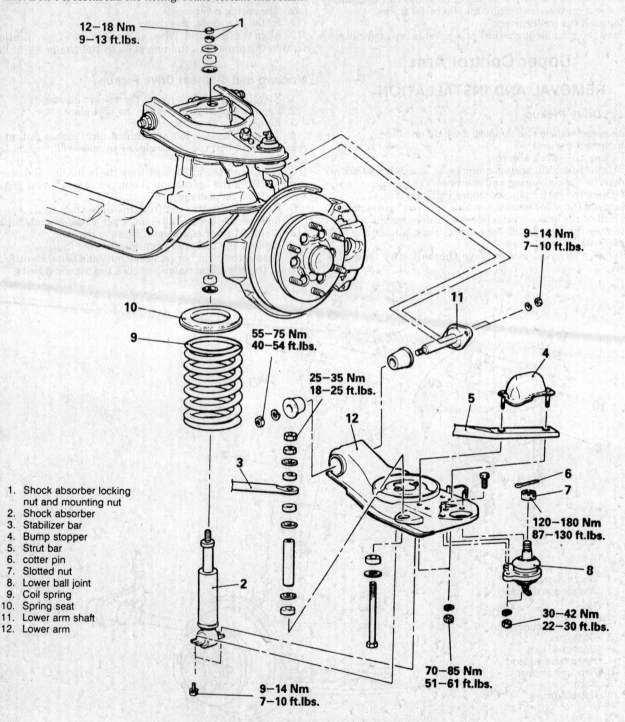

1. Shock absorber locking nut and mounting nut
2. Shock absorber
3. Stabilizer bar
4. Bump stopper
5. Strut bar
6. cotter pin
7. Slotted nut
8. Lower ball joint
9. Coil spring
10. Spring seat
11. Lower arm shaft
12. Lower arm

2-wheel drive pickup, lower control arm

SUSPENSION AND STEERING 8

Lower Control Arm

REMOVAL AND INSTALLATION

2-Wheel Drive Pickup

1. Raise the vehicle and support it safely on stands.
2. Remove the wheel.
3. Remove the shock absorber.
4. Disconnect the sway bar and strut bar from the lower control arm.
5. Install Mitsubishi spring compressor tool MB 990792 or equivalent to the coil spring and compress the spring.
6. Remove the cotter pin and lower ball joint nut.
7. Release the lower ball joint taper using Mitsubishi tool MB 990809-01 or equivalent.
8. Remove the tool and remove the ball stud from the knuckle. Remove the spring compressor.
9. Pull the arm down and remove the spring (still compressed) with the rubber isolation pad from the vehicle.
10. Remove the lower arm shaft mounting nuts and remove the arm from the vehicle.
11. Install the arm to the crossmember finger tight. Install the spring with the rubber isolators. Tighten the the compressor so the lower ball joint can be inserted through the knuckle.
12. Tighten the lower ball joint nut to 136 Nm (100 ft. lbs.) Install a new cotter pin.
13. Carefully loosen the spring compressor, making sure the spring seats properly at both ends.
14. Remove the spring compressor.
15. Connect the sway bar and strut bar to the lower control arm.
16. Install the shock absorber.
17. Install the wheel and lower the vehicle.
18. When the weight of the vehicle is completely on the suspension, bounce it once or twice and tighten the lower control arm to crossmember mounting nut to 68 Nm or 50 ft. lbs.
19. Align the front end.

Montero and 4-Wheel Drive Pickup

1. Raise the vehicle and support it safely on stands.
2. Remove the wheels.
3. Remove the skid plate.
4. Remove the anchor arm assembly and the torsion bar.
5. Remove the shock absorber lower attaching bolt.
6. Disconnect the stabilizer bar from the lower control arm.
7. Remove the cotter pin and the nut from the lower ball stud. Separate the lower ball stud from the steering knuckle using a suitable separating tool such as MB 990809-01 or equivalent.
8. Remove the pivot bolts and remove the arm from the vehicle.
9. Install the new control arm to the vehicle.
10. Install the pivot bolts, but do not tighten beyond snug.
11. Insert the ball stud into the steering knuckle. Install the nut, tighten to 136 Nm (100 ft. lbs.) and install a new cotter pin.
12. Attach the stabilizer bar to the control arm and install the shock mount bolts.
13. Install the torsion bar and turn the adjustment nut (anchor mount nut) clockwise to apply a load to the bar.
14. Install the wheel. Lower the vehicle so the weight of the vehicle is completely on the suspension.
15. Bounce the suspension once or twice. Tighten the pivot nuts to 149 Nm or 110 ft. lbs.
16. Set the ride height. Refer the "Torsion Bars" procedure in this section.
17. Align the front end.

Knuckle and Spindle

REMOVAL AND INSTALLATION

2-Wheel Drive Pickup

1. Elevate and safely support the car on stands.
2. Remove the wheel.

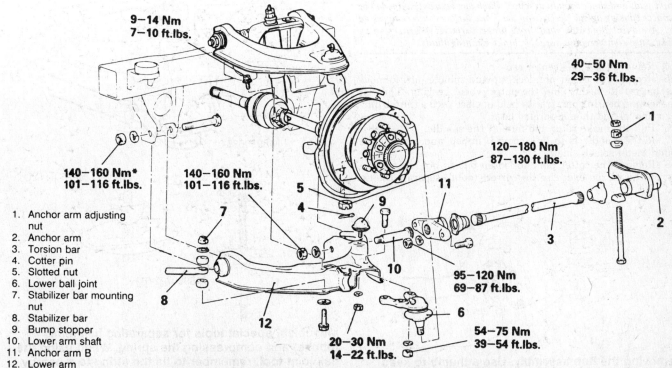

1. Anchor arm adjusting nut
2. Anchor arm
3. Torsion bar
4. Cotter pin
5. Slotted nut
6. Lower ball joint
7. Stabilizer bar mounting nut
8. Stabilizer bar
9. Bump stopper
10. Lower arm shaft
11. Anchor arm B
12. Lower arm

4-wheel drive, lower control arm and related components

8-29

8 SUSPENSION AND STEERING

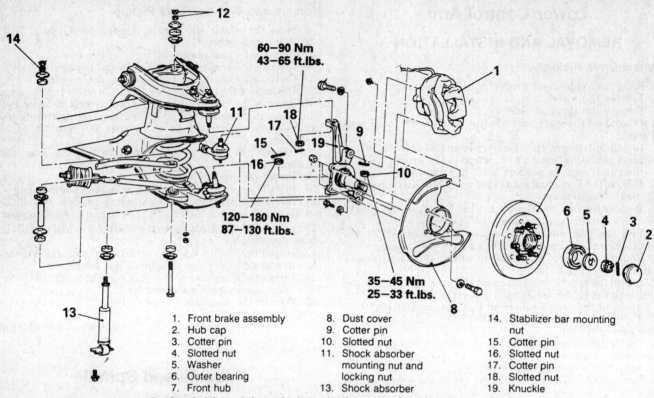

1. Front brake assembly
2. Hub cap
3. Cotter pin
4. Slotted nut
5. Washer
6. Outer bearing
7. Front hub
8. Dust cover
9. Cotter pin
10. Slotted nut
11. Shock absorber mounting nut and locking nut
13. Shock absorber
14. Stabilizer bar mounting nut
15. Cotter pin
16. Slotted nut
17. Cotter pin
18. Slotted nut
19. Knuckle

2-wheel drive pickup: hub, spindle and related components

3. Remove the brake caliper and suspend it from wire out of the way. Do not disconnect or loosen the hose; simply move the whole assembly out of the way. Do not allow the caliper to hang by the hose.

--- CAUTION ---
Brake pads and shoes contain asbestos, which has been determined to be a cancer causing agent. Never clean the brake surfaces with compressed air! Avoid inhaling any dust from brake surfaces! When cleaning brakes, use commercially available brake cleaning fluids.

4. Remove the small center cap.
5. Remove the cotter pin, lock cap and spindle nut. Remove the large flat washer and the outer wheel bearing. The flat washer and bearing may also be held in place with a thumb during removal and then separated later.
6. Pull the brake rotor and hub off the spindle.
7. If the spindle is to be removed, unbolt and remove the round metal splash shield.
8. Remove the cotter pin and loosen but do not remove the nut on the tie rod end. Use the correct tool to separate the tie rod joint.

9. Disconnect the shock absorber at its upper mount. Disconnect it at its lower mount and remove the shock absorber.
10. Install a spring compressor such as MB 990792-01 or its equivalent and compress the spring.

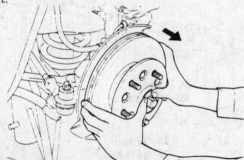

Removing the hub assembly. Use a thumb to keep the outer bearing in place

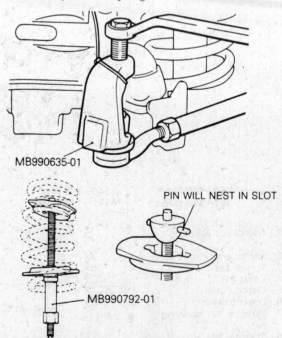

Mitsubishi special tools for separating ball joints (above) and compressing the spring. When using the ball joint tool, remember to tie the string to a nearby component

SUSPENSION AND STEERING 8

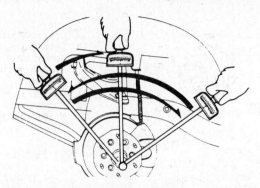

The bearings are adjusted in 3 steps — refer to text for details

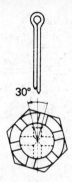

Maximum allowable rotation is 30° to fit the cotter pin

Clean the area shown and apply sealant before reassembly

11. Remove the stabilizer bar retaining bolt (link) from the lower arm.
12. Remove the cotter pins and loosen the nuts on the upper and lower ball joints. Use the correct puller-MB 990809-01 or its equivalent—to separate the upper and lower joints from the knuckle.
12. Carefully release the spring compressor, allowing the spring back into position.
13. Remove the knuckle and spindle.

To install:
14. Install the knuckle in position.
15. Make certain the spring compressor is correctly installed and tighten it, compressing the spring.
16. Install a new nut on the upper ball joint stud and tighten it to 75 Nm or 55 ft. lbs. Install a new cotter pin.
17. Install a new nut on the lower ball joint stud and tighten it to 150 Nm or 111 ft. lbs. Install a new cotter pin.
18. Connect the stabilizer bar to the control arm, making certain the spacers and washers are correctly assembled on the bolt. Tighten the nut until 23–24mm of thread project beyond the top of the nut.
19. Carefully loosen the spring compressor, allowing the spring into place. Make certain that it is correctly seated. When the spring is in place, remove the compressor.
20. Install the shock absorber.
21. Assemble the tie rod to knuckle joint. Use a new nut and tighten it to 40 Nm or 30 ft. lbs.
22. Install the splash shield.
23. Fit the rotor and hub assembly onto the spindle with the bearings and seal in place.
24. Making certain that the outer bearing is well seated, install the flat washer and the shaft nut. Tighten the nut finger tight.
25. Using a torque wrench, tighten the nut to 30 Nm or 22 ft. lbs. Back the nut off to 0 lbs, then final tighten to 8 Nm (6 ft. lbs. or 72 INCH lbs.)
26. Install the lock cap and cotter pin. If the holes in the lockcap do not align with the castle nut on the shaft, turn the lock cap and try again; the lock cap has offset holes to compensate for this occurence. In the unlikely event that it still will not align, back off the shaft nut by NO MORE than 30° of rotation (30° is equal to ½ of a flat on the nut.).
27. Clean the outer edge of the small hub cover thoroughly. Coat the flange with 3M® 8663 sealant or equivalent and install the small hub cover.
28. Install the brake caliper and pads.
29. Install the wheel and lower the car to the ground.

4-Wheel Drive Pickup and Montero

1. Before beginning, place the free-wheeling hub in the free position by placing the transfer lever in the ‹cf35›·2H‹cf33› position and moving in reverse for about 6 or 7 feet.
2. Raise the vehicle and support it safely on stands.
3. Remove the wheel.
4. Remove the front brake caliper, pads and adaptor. Position them out of the way, suspended from stiff wire; do not let the caliper hang by the hose.

CAUTION
Brake pads and shoes contain asbestos, which has been determined to be a cancer causing agent. Never clean the brake surfaces with compressed air! Avoid inhaling any dust from brake surfaces! When cleaning brakes, use commercially available brake cleaning fluids.

5. Remove the hub cover.
6. Remove the snapring and shim.
7. Remove the automatic free-wheeling hub retaining bolts and remove the hub.
8. Remove the lock washer. Remove the locknut with special tool MB 990954 or equivalent.

NOTE: This is a purpose built tool and may be difficult to find in the retail market.

9. Remove the front hub and brake rotor assembly, with the inner and outer bearings.
10. Support the lower control arm with a jack. Use the correct tools separate the tie rod joint from the knuckle.
11. Separate the upper and lower ball joints from the knuckle using the correct tools. Remember to loosen the nut and separate the joint, then remove the nut.
12. Once the ball joints have been released, lower the control arm slowly.
13. Remove the steering knuckle.
14. Install the knuckle to the ball joint studs. Tighten all ball joint nuts except the upper nut on Pickups to 136 Nm (100 ft. lbs). On Pickups, tighten the upper ball joint nut to 82 Nm (60 ft. lbs). Install new cotter pins.
15. Connect the tie rod end to the knuckle. Tighten the nut to 41 Nm (30 ft. lbs. and install a new cotter pin.
16. Clean out all of the old grease from inside the front hub.

8-31

8 SUSPENSION AND STEERING

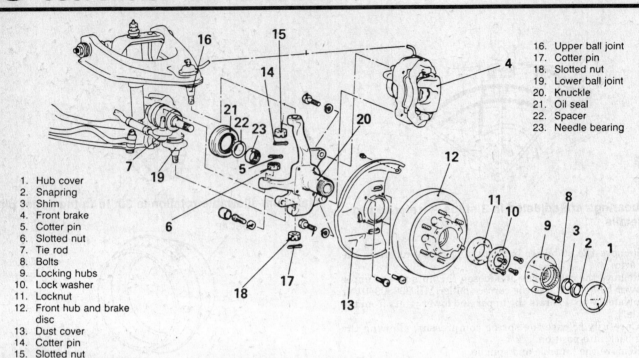

1. Hub cover
2. Snapring
3. Shim
4. Front brake
5. Cotter pin
6. Slotted nut
7. Tie rod
8. Bolts
9. Locking hubs
10. Lock washer
11. Locknut
12. Front hub and brake disc
13. Dust cover
14. Cotter pin
15. Slotted nut
16. Upper ball joint
17. Cotter pin
18. Slotted nut
19. Lower ball joint
20. Knuckle
21. Oil seal
22. Spacer
23. Needle bearing

4-wheel drive hub and spindle. The locking nut (11) requires a special tool to loosen or tighten it

17. Bolt the rotor and front hub together if they were disassembled. Tighten the nuts to 54 Nm (40 ft. lbs). Apply wheel bearing grease to the inside of the front hub.
18. Pack the inner bearing and install to the race.
19. Install a new oil seal into the front hub so it is flush with the front hub end face.
20. Install the assembly onto the steering knuckle, pack and install the outer bearing. Using special tool MB 990954 or equivalent, (which fits standard torque wrenches) tighten the locknut to 165 Nm or 122 ft. lbs. Loosen the locknut completely, then retorque to 25 Nm or 18 ft. lbs.
To complete the procedure, position the torque wrench at the 3 o'clock position and loosen the nut the equivalent of 1 hour on a clock face or 30 degrees or rotation.
21. Install the lock washer. If the lock washer and locknut holes do not align, align the holes by loosening the nut slightly.
22. Before installing the automatic hub assembly, measure the turning force of the front hub. If the measured value is not 2.5–11.5 inch lbs., retorque the locknut.
23. Apply a very thin, even coating of sealant to the free wheeling hub surface of the front hub. Carefully align the key of the free-wheeling hub brake with the keyway of the knuckle spindle and install the automatic hub assembly. The mounting surfaces of the automatic hub and the front hub must be perfectly flush before the mounting bolts are tightened.
24. Tighten the automatic hub mounting bolts to 54 Nm (40 ft. lbs).
25. Check the front hub turning resistance again. If the difference between the reading in Step 22 and this reading is more than 8.7 inch lbs., repair or replace the automatic hub.
26. Install the shim and snapring. Rotate the axle shaft forward and backward and stop at a position midway between 2 heavy spots where there is a heavy feeling or drag.
27. Set a dial indicator so the pin is resting on the end of the axle shaft. The endplay specification is 0.2–0.5mm. If not within specifications, adjust by adding or removing shims.
28. Install the hub cover.
29. Install the brake caliper and pads.
30. Install the wheel and lower the car to the ground.

Front Wheel Bearing

REMOVAL AND INSTALLATION

2- and 4-Wheel Drive Pickup

1. Remove the front hub/rotor assembly from the steering knuckle as described previously.
2. Remove the outer bearing from the hub. Remove the splash shield.
3. Remove the nuts and bolts that attach the hub to the rotor and separate them. Clean out all of the old grease from the inside of the hub.
4. The race for the outer bearing may be removed with a brass rod and mallet. Drive the race out from the inside but do not mar the inner surface of the hub.

WARNING: Do not remove the bearing races if the same bearing is to be reused after cleaning or repacking. If a bearing is replaced, the race MUST be replaced. Don't use a new bearing on an old race.

5. On the back of the hub, use a seal puller (a very common tool) to remove the inner bearing seal.
6. Remove the inner bearing and remove the race in the same fashion as the outer race.

To install:

7. Use bearing drivers of the proper size and drive the new races into the hub. Don't try to install the race with the brass rod; chances are high that the race will be crooked or damaged.
8. Assemble the rotor and front hub. Tighten the nuts to 54 Nm or 40 ft. lbs.
9. Apply wheel bearing grease to the inside of the front hub.
10. Pack the inner bearing and install to the hub.
11. Install a new oil seal into the hub so it is flush with the hub end face.
12. Install the assembly onto the steering knuckle.
13. Pack and install the outer bearing.
14. Continue with reinstallation of the hub as outlined previously.

SUSPENSION AND STEERING 8

Montero

1. Remove the knuckle assembly following procedures outlined previously in this section.
2. Remove the oil seal from the back of the knuckle and remove the spacer.
3. Drive out the needle bearing by tapping uniformly around the housing.

WARNING: Once the needle bearing has been removed, it cannot be reused.

4. Apply multipurpose grease to the roller surface of the new needle bearing.
5. Use the correct bearing installation tools (MB 990938-01 and 990956-01 or equivalents) to carefully install the bearing until it is flush with the knuckle face. Use care not to insert the bearing too far.
6. Apply multipurpose grease to the knuckle face of the spacer. Install the spacer to the knuckle so that the chamfered (tapered) side is towards the center of the vehicle.
7. Use the correct seal driver and install a new oil seal. Tap in the seal until it is flush with the knuckle end face.
8. Pack multipurpose grease inside the oil seal and lip
9. Reassemble and reinstall the knuckle.

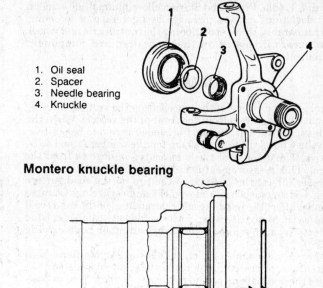

1. Oil seal
2. Spacer
3. Needle bearing
4. Knuckle

Montero knuckle bearing

Installation of Montero bearing spacer

FRONT END ALIGNMENT

Alignment of the front wheels is essential if your car or truck is to go, stop and turn as designed. Alignment can be altered by collision, overloading, poor repair or bent components.

If you are diagnosing bizarre handling and/or poor road manners, the first place to look is the tires. Although the tires may wear as a result of an alignment problem, worn or poorly inflated tires can make you chase alignment problems which don't exist.

Once you have eliminated all other causes, unload everything from the trunk except the spare tire, set the tire pressures to the correct level and take the car to a reputable alignment facility. Since the alignment settings are measured in very small increments, it is almost impossible for the home mechanic to accurately determine the settings. The explanations that follow will help you understand the three dimensions of alignment: caster, camber and toe.

CASTER

Caster is the tilting of the steering axis either forward or backward from the vertical, when viewed from the side of the vehicle. A backward tilt is said to be positive and a forward tilt is said to be negative. Changes in caster affect the straight line tendency of the vehicle and the "return to center" of the steering after a turn. If the caster is radically different between the left and right wheels (such as after hitting a sizable pothole), the car will exhibit a nasty pull to one side.

On the Pickup and Montero caster is fully adjustable, either by adjusting the upper arm cross-shaft (Montero) or by changing shim thickness (Pickup).

Mirage, Galant and Sigma are not adjustable for caster; if an error is found, bent components must be identified and replaced.

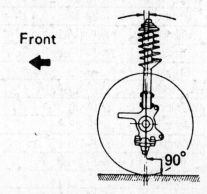

Caster angle affects the tracking of the steering

8-33

8 SUSPENSION AND STEERING

Cordia, Tredia, Precis and Starion allow minimal adjustment. Slight deviations may be corrected by tightening or loosening the strut bar nut, thus repositioning the control arm and wheel. Major errors in caster on these cars require component replacement.

CAMBER

Camber is the tilting of the wheels from the vertical (leaning in or out) when viewed from the front of the vehicle. When the wheels tilt outward at the top, the camber is said to be positive. When the wheels tilt inward at the top the camber is said to be negative. The amount of tilt is measured in degrees from the vertical. This measurement is called camber angle.

Camber affects the position of the tire on the road surface during vertical suspension movement and cornering. Changes in camber affect the handling and ride qualities of the car as well as tire wear. Many tire wear patterns indicate camber related problems from mis-alignment, overloading or poor driving habits.

Camber is adjustable only on the Pickup and Montero; both require the replacement of shims in the front suspension.

Since the front struts on the passenger cars are fixed in position, the camber is determined by the correct location of the components. Any camber reading outside of specification requires replacement of bent parts.

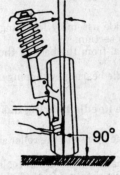

Camber is the inward or outward tilt of the wheel on the road

TOE

Toe is the turning in or out (parallelism) of the wheels. The actual amount of toe setting is normally only a fraction of a centimeter. The purpose of toe-in (or out) specification is to ensure parallel rolling of the wheels. Toe-in also serves to offset the small deflections of the steering support system which occur when the vehicle is rolling forward.

Changing the toe setting will radically affect the overall "feel" of the steering, the behavior of the car under braking, tire wear and even fuel economy. Excessive toe (in or out) causes excessive drag or scrubbing on the tires.

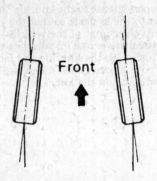

Toe in (or out) can affect tire wear and fuel economy

Toe is adjustable on all Mitsubishi vehicles and is the most common value adjusted during an alignment. It is generally measured in decimal millimeters or degrees. It is adjusted by loosening the locknut on each tierod end and turning the rod until the correct reading is achieved. The rods on the left and right must remain equal in length during all adjustments.

WHEEL ALIGNMENT

Year	Model	Caster Range (deg.)	Caster Preferred Setting (deg.)	Camber Range (deg.)	Camber Preferred Setting (deg.)	Toe-in (in.)
1983	Cordia	5/16P–15/16P	13/16P	1/16–15/16P	7/16P ①	1/8–1/8P
	Tredia	5/16P–15/16P	13/16P	1/16N–15/16P	7/16P ①	1/8–1/8P
	Starion	—	51/3P	—	0 ②	5/64–13/64P
	2WD Pickups	11/2P–31/2	21/2P	1/2P–11/2P	1P	0.08–0.35
	4WD Pickups	1P–3P	2P	1/2P–11/2P	1P	0.08–0.35
	Montero	2P–3P	21/2P	1/2P–11/2P	1P	0.08–0.35
1984	Cordia	5/16P–15/16P	13/16P	1/16N–15/16P	7/16P ①	1/8–1/8P
	Tredia	5/16P–15/16P	13/16P	1/16N–15/16P	7/16P ①	1/8–1/8P
	Starion	—	51/3P	—	0 ②	5/64–13/64P
	2WD Pickups	11/2P–31/2P	21/2P	1/2P–11/2P	1P	0.08–0.35
	4WD Pickups	1P–3P	P2	1/2P–11/2P	1P	0.08–0.35
	Montero	2P–3P	21/2P	1/2P–11/2P	1P	0.08–0.35
1985	Cordia	5/16P–15/16P	13/16P	1/16N–15/16P	7/16P ③	1/8–1/8P
	Tredia	5/16P–15/16P	13/16P	1/16N–15/16P	7/16P ③	1/8–1/8P
	Starion	—	51/3P	—	0	5/64–13/64P
	Mirage	7/32P–17/32P	23/32P	1/2N–1/2P	0 ④	1/8–1/8P

SUSPENSION AND STEERING 8

WHEEL ALIGNMENT

Year	Model	Caster Range (deg.)	Caster Preferred Setting (deg.)	Camber Range (deg.)	Camber Preferred Setting (deg.)	Toe-in (in.)
1985	Galant	5/32P–1 5/32P	21/32P	0–1P	1/2P	1/8–1/8P
	2WD Pickups	1 1/2P–3 1/2P	2 1/2P	1/2P–1 1/2P	1P	0.08–0.35
	4WD Pickups	1P–3P	2P	1/2P–1 1/2P	1P	0.08–0.35
	Montero	2P–3P	2 1/2P	1/2P–1 1/2P	1P	0.08–0.35
1986	Cordia	5/16P–1 5/16P	13/16P	1/16N–15/16P	7/16P ⑤	1/8–1/8P
	Tredia	5/16P–1 5/16P	13/16P	1/16N–15/16P	7/16P ⑤	1/8–1/8P
	Starion	—	5 1/3P	—	0P	5/64–13/64P
	Mirage	7/32–1 7/32P	23/32P	1/2N–1/2P	0 ④	1/8–1/8P
	Galant	5/32P–1 5/32P	21/32P	0–1P	1/2P	1/8–1/8P
	2WD Pickups	1 1/2P–3 1/2P	2 1/2P	1/2P–1 1/2P	1P	0.08–0.35
	4WD Pickups	1P–3P	2P	1/2P–1 1/2P	1P	0.08–0.35
	Montero	2P–3P	2 1/2P	1/2P–1 1/2P	1P	0.08–0.35
1987	Cordia	5/16P–1 5/16P	13/16P	1/16N–15/16P	7/16P ⑤	1/8–1/8P
	Tredia	5/16P–1 5/16P	13/16P	1/16N–15/16P	7/16P ⑤	1/8–1/8P
	Starion	5 5/16P–6 5/16P	5 13/16P	1N–0P	1/2N	13/64–13/64
	Mirage	1/2P–1 1/2P ⑥	1P	1/2N–1/2P	0 ①	1/8–1/8P
	Galant	5/32P–1 5/32P	21/32P	0–1P	1/2P	1/8–1/8P
	Precis	1/2P–1/8P	3/16P	0–1P	1/2P ⑦	1/16–5/32
	2WD Pickups	1 1/2P–3 1/2P	2 1/2P	1/2P–1 1/2P	1P	0.08–0.35
	4WD Pickups	1P–3P	2P	1/2P–1 1/2P	1P	0.08–0.35
	Montero	2P–3P	2 1/2P	1/2P–1 1/2P	1P	0.08–0.35
1988	Cordia	5/16P–1 5/16P	13/16P	1/16N–15/16P	7/16P ⑤	1/8–1/8P
	Tredia	5/16P–1 5/16P	13/16P	1/16N–15/16P	7/16P ⑤	1/8–1/8P
	Starion	5 5/16P–6 5/16P	5 13/16P	1N–0P	1/2N	13/64–13/64
	Mirage	1/2P–1 1/2P ⑥	1P	1/2N–1/2P	0 ①	1/8–1/8P
	Galant	5/32P–1 5/32P	21/32P	0–1P	1/2P ⑧	1/8–1/8P
	Precis	1/2P–1/8P	3/16P	0–1P	1/2P ⑦	1/16–5/32
	2WD Pickups	1 1/2P–3 1/2P	2 1/2P	1/2P–1 1/2P	1P	0.08–0.35
	4WD Pickups	1P–3P	2P	1/2P–1 1/2P	1P	0.08–0.35
	Montero	2P–3P	2 1/2P	1/2P–1 1/2P	1P	0.08–0.35
1989	Starion	5 5/16P–6 5/16P	5 13/16P	1N–0	1/2N	13/64–13/64
	Mirage	1/2P–1 1/2P ⑥	1P	1/2N–1/2P	0 ①	1/8–1/8P
	Galant	5/32P–1 5/32P	21/32P	0–1P	1/2P ⑧	1/8–1/8P
	Precis	1/2P–1/8P	3/16P	0–1P	1/2P ⑦	1/16–5/32
	Sigma	3 1/16P–1 3/16P	1 1/16P	0–1P ⑨	1/2P ⑩	1/8–1/8P
	Montero	2P–4P	3P	1/2P–1 1/2P	1P	0.20
	2WD Pickups	1 1/2P–3 1/2P	2 1/2P	1/4P–1 1/4P	2/3P	0.20
	4WD Pickups	1P–3P	2P	1/2P–1 1/2P	1P	0.20

N—Negative
P—Positive
① Rear—1 1/16N
② Rear—5/16P
③ Rear—9/16N
④ Rear—21/32N
⑤ Rear—1 1/16P
⑥ With power steering—1 1/16–2 13/16
⑦ Rear—5/8N
⑧ Rear—3/4N
⑨ Rear—1N–1/4N
⑩ Rear—3/4N

8 SUSPENSION AND STEERING

REAR SUSPENSION

Coil Springs

REMOVAL AND INSTALLATION

Cordia and Tredia
Mirage through 1988
Precis

1. Support the car safely on stands.
2. Remove the rear wheels.
3. Support the rear suspension arm with a floorjack.
4. Remove the lower shock absorber attaching bolt, nut and lockwasher.
5. Slowly lower the jack just to the point where the spring can be removed and remove the spring. If the spring is being replaced, transfer the spring seat to the new spring.
6. Install the spring and seat into place. If using a new or replacement spring, make certain the spring identification and load markings match the old spring exactly. Install the spring with the smaller diameter at the top.
7. Slowly raise the jack, compressing the spring just enough to attach the lower shock mount. Install the bolt and tighten it to 97 Nm or 72 ft. lbs.
8. Slowly release the jack and remove it. Install the wheel and lower the car to the ground.

Montero with V6 Engine

1. Raise the vehicle and support it safely on stands.
2. Remove the rear wheels.
3. Remove the parking brake cable attaching bolts at each side.
4. Use 2 floor jacks or an axle cradle to support the weight of the axle. Do not try to balance the axle on 1 jack.
5. Remove the bolt that attaches the lateral rod (Panhard rod) to the body.
6. Remove the lower shock mounting bolts.
7. Slowly lower the axle and remove the coil springs with their seats.

WARNING: When lowering the jack, take great care not to damage the brake lines running between the main line and the axle housing.

8. Reinstall the springs and elevate the axle into position. Make certain that both the spring and seat are correctly positioned.
9. Install the lower shock mounting bolt, tightening it to 120 Nm or 88 ft. lbs.
10. Connect the lateral rod to the body and tighten the bolt to 120 Nm or 88 ft. lbs.
11. Remove the apparatus supporting the axle.
12. Reinstall the brackets and bolts holding the parking brake cables.
13. Install the rear wheels and lower the vehicle to the ground.

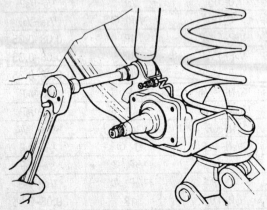

Removing the lower shock mount. Figure shows the brake assembly removed; this makes the job easier but is not required

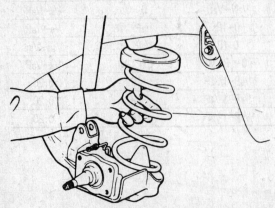

When reinstalling, make certain the seat is properly positioned

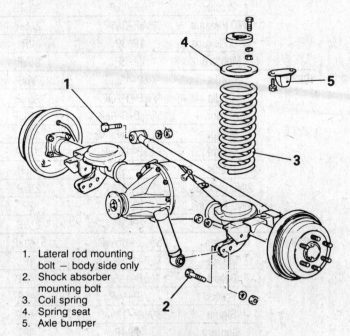

1. Lateral rod mounting bolt — body side only
2. Shock absorber mounting bolt
3. Coil spring
4. Spring seat
5. Axle bumper

Montero rear suspension with coil spring

Leaf Springs

REMOVAL AND INSTALLATION

1. Raise the vehicle and support it safely on stands.
2. Remove the rear wheels.
3. Remove the parking brake cable bracket bolts.
4. Use 2 floor jacks or an axle cradle to support the weight of the axle. Do not try to balance the axle on 1 jack.

SUSPENSION AND STEERING 8

5. Disconnect and remove the shock absorber.
6. Remove the nuts, washers and U-bolts attaching the springs to the axle housing. Remove the seat and spacer.
7. Remove the spring shackle bolts, shackle and spring front bolt.
8. Remove the springs from the vehicle.
9. When reinstalling, the retaining bolts and shackles must be installed from the outside of the vehicle to the inside. (Threaded end towards the center of the vehicle.) Make sure the rubber bushings are in place and in good condition, replacing any that are compressed or damaged. Tighten the nuts on the rear shackle and front bolt just snug. They will be final tightened later.
10. Install the U-bolts and bottom plate, making sure the hole in the bottom plate aligns with the bolt through the spring leaves. Tighten the U-bolts nuts evenly to 110 Nm or 81 ft. lbs.
11. Install the shock absorber, tightening the mounting bolts just snug.
12. Remove the axle support apparatus and install the parking brake cable retaining brackets.
13. Install the wheel and lower the vehicle to the ground.
14. Bounce the rear of the vehicle once or twice to stabilize the suspension. Tighten the shock absorber mounting nuts to 22 Nm (16 ft. lbs.)
15. Tighten the front spring retaining nut and bolt to 140 Nm or 104 ft. lbs.
16. Tighten the rear spring shackle nuts to 52 Nm or 38 ft. lbs.

Rear Strut Assembly

REMOVAL AND INSTALLATION

Starion

1. Support the car on jackstands safely.
2. Position a floor jack under the lower control arm and raise it slightly.
3. Remove the wheel. Separate the driveshaft at the companion flange.
4. Remove the two bolts attaching the lower end of the strut to the axle housing. Remove the nut, bolt and lockwasher which squeezes the lower end of the axle housing together to help retain the strut.
5. Separate the strut assembly from the axle housing by using a proper prybar to pry the housing halves apart as you push downward on the axle housing.
6. Support the strut from underneath. Remove the the inner liner within the trunk.
7. Remove the three mounting nuts and washers from the top of the strut tower. Lower and remove the strut from the car.

--- CAUTION ---
The strut will come out as a unit with the spring still compressed. Do not loosen the center retaining nut.

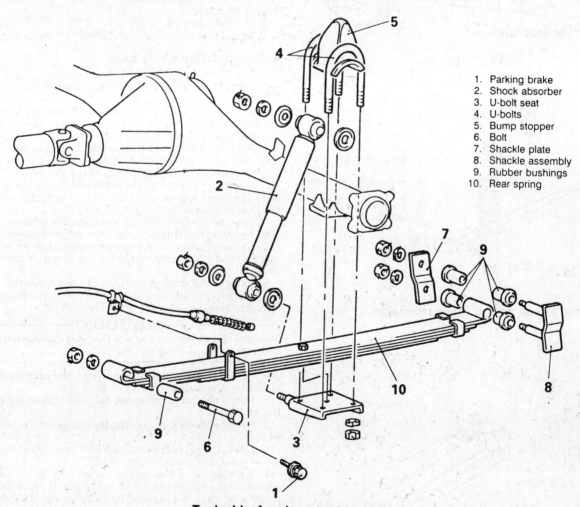

1. Parking brake
2. Shock absorber
3. U-bolt seat
4. U-bolts
5. Bump stopper
6. Bolt
7. Shackle plate
8. Shackle assembly
9. Rubber bushings
10. Rear spring

Typical leaf spring components

8 SUSPENSION AND STEERING

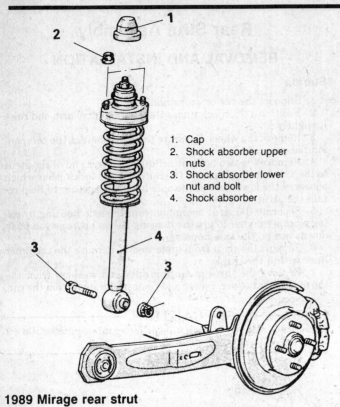

1. Cap
2. Shock absorber upper nuts
3. Shock absorber lower nut and bolt
4. Shock absorber

1989 Mirage rear strut

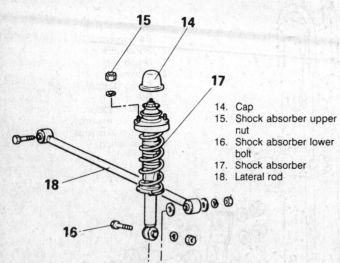

14. Cap
15. Shock absorber upper nut
16. Shock absorber lower bolt
17. Shock absorber
18. Lateral rod

Galant and Sigma normal rear suspension

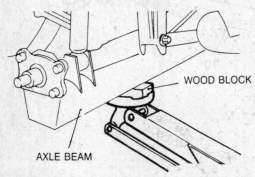

Correct method of supporting the rear beam

8. When reinstalling, apply sealant to the surface of the upper insulator where it contacts the body.
9. Install the shock absorber into the top mount and tighten the nuts to 30 Nm or 22 ft. lbs.
10. Fit the strut into place at the bottom. Install the retaining bolts and tighten them to 79 Nm or 58 ft. lbs.
11. Insert the pinch bolt and tighten it to 79 Nm or 58 ft. lbs.
12. Connect the drive shaft to the wheel flange.
13. Install the wheel and lower the vehicle to the ground.

Galant and Sigma with Standard Suspension
1989 Mirage

1. Support the vehicle safely with stands.
2. Remove the rear wheels.
3. Place a floor jack under the axle assembly. Always use a piece of wood between the jack and the axle; take great care to avoid contact with the lateral rod. Raise the jack slightly.
4. Remove the forward trim from the trunk and remove the cap and strut mounting nuts and washers.
5. Remove the nut and through-bolt from the bottom of the strut and remove the strut assembly.

CAUTION
The strut will come out as a unit with the spring still compressed. Do not loosen the center retaining nut.

---

6. Reinstall the strut, fitting it into the upper mount and tightening the nuts to 50 Nm or 37 ft. lbs.
7. Install the bottom through bolt and tighten the nut to 90 Nm or 67 ft. lbs.
8. Remove the jack.
9. Install the rear wheels and lower the car to the ground.

Galant and Sigma with ECS

1. Support the vehicle safely with stands.
2. Remove the rear wheels.
3. Place a floor jack under the axle assembly. Always use a piece of wood between the jack and the axle; take great care to avoid contact with the lateral rod. Raise the jack slightly.
4. Remove the forward trim inside the trunk. Loosen the air tube lock nuts. Label and disconnect the air tubes from the shock absorber. Immediately plug or cap the lines with tape or similar product to prevent the entry of dirt into the system. Do NOT bend or crimp the air tubes. Plug or cover the air ports on the joint and actuator.
5. Remove the O-ring, dust cover and the adapter.
6. Remove the actuator, snapring and joint assembly.
7. Carefully remove the actuator bracket. Loosen only the retaining nut for the bracket, not the center nut on the shock absorber shaft.
8. Remove the strut mounting nuts and washers.
9. Remove the nut and through-bolt from the bottom of the strut and remove the strut assembly.

CAUTION
The strut will come out as a unit with the spring still compressed. Do not loosen the center retaining nut!

---

10. Reinstall the strut, fitting it into the upper mount and tightening the nuts to 30 Nm or 23 ft. lbs.
11. Install the bottom through bolt and tighten the nut to 90 Nm or 67 ft. lbs.
12. Remove the jack supporting the axle beam. Install the rear wheels.
13. Lower the car to the ground VERY slowly so as not to damage or deform the diaphragm.
14. With the car on the ground, install the actuator bracket by aligning the notch in the piston rod to face forward. Install the actuator bracket and tighten the retaining nut to 50 Nm or 37 ft. lbs.

SUSPENSION AND STEERING 8

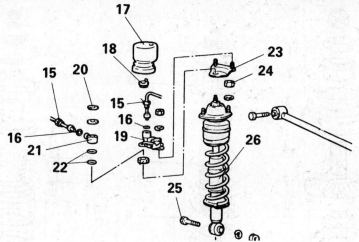

15. Air tube
16. O-ring
17. Dust cover
18. Adapter
19. Actuator
20. Snapring
21. Joint
22. O-ring
23. Actuator bracket
24. Shock absorber upper nut
25. Shock absorber lower nut
26. Shock absorber

Galant with Electronic Suspension Control

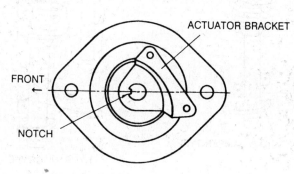

Correct alignment of the ECS actuator bracket

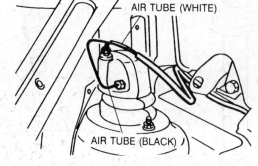

Make sure the air tubes are correctly placed

15. Use new O-rings and replace the joint assemble, snapring and actuator.
16. Install the adapter and dust cover.
17. Use new O-rings and connect the air lines to the correct ports. Tighten the lock nuts only to 9 Nm or 6 ft. lbs.
18. Swab the air connectors with a solution of soapy water. Start the engine, cycle the suspension controls and observe the joints for any sign of air leaks.

STRUT OVERHAUL

MacPherson struts are constructed in such a way that the unit can be removed from the car while the spring is still under considerable tension. While removal of the unit as a whole is safe and simple, disassembly is an entirely different matter.

The spring is still under considerable tension; you MUST NOT attempt to disassemble the unit until an effective spring compression tool is properly installed and you can see that ALL spring tension is removed.

— CAUTION —
Failure to perform all these procedures accurately will result in the release of deadly force. If you are uncertain about performing this procedure, take the removed strut to a shop that will disassemble and overhaul it for you.

The Mitsubishi factory tool for use in this operation is MB 990987; you can purchase a number of tools designed to compress coil springs used on MacPherson struts. Assure yourself that you have bought a quality tool and can use it safely and effectively.

Disassembly for
All Rear Struts except ECS

1. Remove the strut from the vehicle.
2. Install the spring compressor and compress the spring until there is no tension on the spring seats.
3. Pry the insulator cap off the insulator.
4. Hold the insulator stationary and use an ordinary socket wrench to remove the self-locking nut from the top of the shock piston rod.
5. Remove the insulator, spring seat, and any insulators or covers from the strut.
6. Inspect the top insulator for cracks or peeling. Rubber parts must be inspected for cracking or brittleness. Springs should be checked for any sign of cracking. The piston rod should be inspected for bending or obvious wear. Replace external parts as necessary.

HYDRAULIC SHOCK ABSORBER (OIL FILLED)

1. To prevent entry of dirt into the oil within the cylinder, thoroughly clean the outside of the case before disassembly.
2. Hold the strut upright in a vise. Clamp the knuckle part of the strut, not the tube. Compress the piston into the strut.
3. Use the special wrench to remove the oil seal assembly.
4. Use a small hooked tool to remove the square section oil seal, if present, from around the inside of the tube.
5. Slowly withdraw the piston rod from the cylinder together with the piston guide. The piston has a precision machined surface; handle it carefully.

8 SUSPENSION AND STEERING

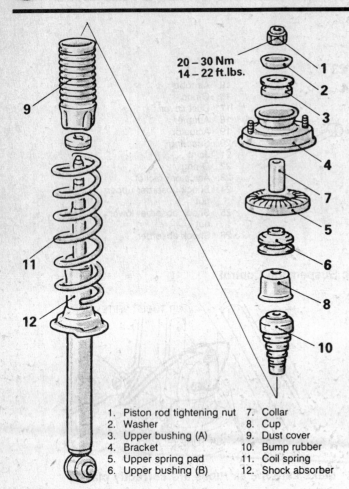

1. Piston rod tightening nut
2. Washer
3. Upper bushing (A)
4. Bracket
5. Upper spring pad
6. Upper bushing (B)
7. Collar
8. Cup
9. Dust cover
10. Bump rubber
11. Coil spring
12. Shock absorber

1989 Mirage rear strut components

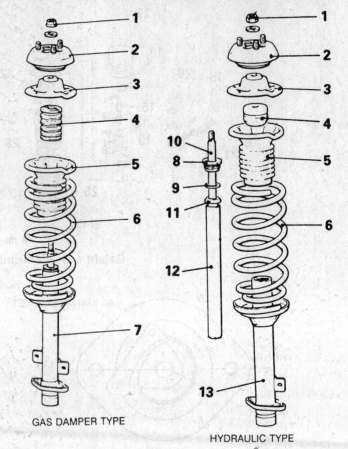

1. Self-locking nut
2. Strut insulator
3. Spring seat
4. Rubber helper
5. Dust cover
6. Coil spring
7. Strut
8. Oil seal
9. Square section O-ring
10. Piston
11. Piston guide
12. Cylinder
13. Outer shell

Starion rear struts

6. Remove the piston guide from the rod.
7. Remove the cylinder from the strut outer shell.
8. To reassembly, install the cylinder and piston assembly into the outer shell.
9. Gradually pour shock absorber fluid into the cylinder while slowly moving the piston up and down
10. With the flange of the piston guide facing upward, insert the piston guide onto the piston rod until it contacts the cylinder end.
11. Install a NEW square section O-ring (if removed), taking great care to see that it is correctly installed with no folding or tilting. If this seal isn't correct, the shock will leak.
12. Attach the cover to the piston end. Apply shock absorber fluid to the cover and install the oil seal assembly. Use the special tool to tighten the oil seal assembly to 145 Nm or 107 ft. lbs.

GAS SHOCK ABSORBER

WARNING: The shock absorber is filled with nitrogen gas under pressure. Do not disassemble the shock unless necessary for replacement. When replacing, install inner parts as a complete kit.

1. Drill a 4mm, OR SMALLER, diameter hole into the case at the point shown. This will bleed the gas from the strut.
2. Remove the ring nut.
3. Remove the shock assembly from the strut and install a new cartridge.
4. Tighten the ring nut to 145 Nm (107 ft. lbs.)
5. The repair kit should contain a label which may be placed over the hole in the case. This hole should be covered to prevent entry of dirt and water into the strut.

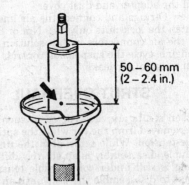

The gas strut MUST be drilled before disassembly

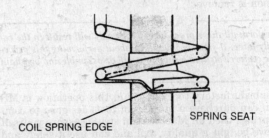

Make sure the spring seats correctly

SUSPENSION AND STEERING 8

Assembly for All Struts except ECS

1. Compress the spring with the spring compressor, if it is not already compressed. Install the spring, dust cover if any, rubber insulator and spring seat onto the strut.
2. Align the flat on the side of the D-shaped hole in the spring seat with the flat on the piston rod, and align the projections on the dust cover with the holes in the spring seat.
3. Install the strut insulator and hold the spring seat or insulator as you install the nut onto the top of the piston rod. Tighten it to 65 Nm (48 ft. lbs.) You must use a NEW self-locking nut.
4. Slowly release tension on the spring compressor, making certain the spring aligns and seats correctly. AFTER you verify that the spring coil is properly aligned in both top and bottom springs seats, remove the special tool.
5. Pack multipurpose grease into the strut insulator and install the cap. Keep the grease off the rubber parts.
6. Install the strut.

Disassembly and Assembly for Galant and Sigma with ECS

1. Install the spring compressor on the strut assembly and compress the spring.
2. Remove the insulator mounting nut and then remove the insulator from the strut assembly.
3. Gradually loosen the spring compressor, allowing the spring into a fully extended position.
4. Remove the sub tank, coil spring, and lower spring pad from the strut.
5. Inspect all the components for wear, damage, leakage or deformation.
6. Apply a coating of grease to the O-ring so it will slide easily and fit the O-ring to the sub tank.
7. Attach, in order, the lower spring pad, the coil spring and the sub tank to the spring compressor and gradually compress it. While compressing the coil spring, temporarily tighten the the nut so that the installation angle of the insulator assembly and the lower bushing is 90°.

NOTES: When installing the spring, the larger outer diameter is the bottom of the spring. Adjust the spring to sit on the lower pad correctly. The D shaped parts of the **piston rod and sub tank must all align during compression.**

8. Install the insulator to the strut assembly and tighten the nut to 30 Nm or 22 ft. lbs.
9. Remove the spring compressor.
10. Align the notch of the piston rod with the D shape of the actuator bracket.
11. Continue reinstallation of the strut.

Control Arms

REMOVAL AND INSTALLATION

Cordia and Tredia
Mirage through 1988
Precis

1. Elevate and safely support the car on stands.
2. Remove the rear wheel(s).
3. Remove the brake drum. Disconnect the parking brake cable and remove the rear brake shoes.

--- CAUTION ---
Brake pads and shoes contain asbestos, which has been determined to be a cancer causing agent. Never clean the brake surfaces with compressed air! Avoid inhaling any dust from brake surfaces! When cleaning brakes, use commercially available brake cleaning fluids.

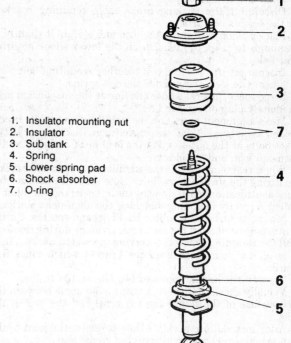

1. Insulator mounting nut
2. Insulator
3. Sub tank
4. Spring
5. Lower spring pad
6. Shock absorber
7. O-ring

Galant and Sigma rear strut with ECS

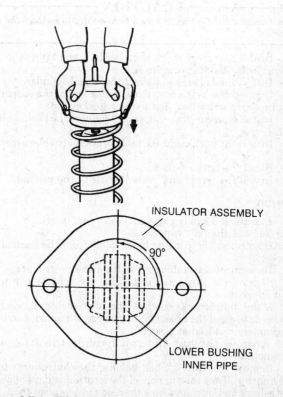

Reassembling ECS rear strut

8-41

8 SUSPENSION AND STEERING

4. Carefully disconnect the brake line at the back of the wheel cylinder. Immediately plug the line to contain fluid leakage and eliminate dirt. Remove the wheel cylinder.
5. remove the backing plate.
6. Disconnect the parking brake cable retaining brackets from the suspension arm.
7. Place a floor jack under the arm and elevate it slightly — just enough to relax the tension on the lower shock absorber mount bolt.
8. Disconnect the lower shock absorber mounting bolt.
9. Lower the jack and remove the coil spring.
10. Disconnect the brake hose and line at the suspension arm. Immediately plug or cap the line.
11. Make alignment marks on the stabilizer bar and its bracket as well as on bracket and suspension arm. Disconnect the two bracket bolts at the outer end of the arm pivot and remove the suspension arm and stabilizer.
12. When reassembling, fit the stabilizer bar into the bracket while fitting the suspension arm(s). Note that the stabilizer bar has an identification mark showing the left side; the bar must be installed correctly including matching the alignment marks.
13. Mount or install the rubber bushings and spacers. Fit the arm into place and align the matchmarks made during removal. Install the retaining bracket, observing the matchmarks. There is a small notch or cutout on the bracket which must face forward.
14. Tighten the bracket bolts to 140 Nm or 104 ft. lbs.
15. Visually check that there is about a 30° angle between the upper surface of the flange (on the arm) and the line of the bracket.
16. After installation, pack the dust cover (at the joint of the left and right arms) with multipurpose grease and install a new clamp.
17. Place the spring in position, making certain that the upper and lower spring seats are correctly located. Place the jack under the arm and raise the arm just enough to connect the lower shock mount.

---- CAUTION ----

Do not compress the spring any more than needed to reattach the shock absorber!

18. Install the lower shock absorber bolt and tighten it to 95 Nm or 70 ft. lbs. Remove the jack.
19. Install the backing plate and the wheel cylinder.
20. Connect the brake hose to the brake line at the suspension arm (use two wrenches) and to the wheel cylinder.
21. Install the rear brake shoes and connect the parking brake cables.
22. Install the parking brake cable brackets to the suspension arm.
23. Bleed the brake system.
24. Install the wheel and lower the car to the ground.

Starion

1. Elevate and safely support the car on stands.
2. Remove the rear wheel.
3. Disconnect the parking brake cable from the control arm brackets.
4. Disconnect the stabilizer bar from the control arm.
5. Remove the nut and bolt holding the control arm to the front support.
6. Make matchmarks to show the relationship between the crossmember and the eccentric bushing so alignment can be approximately restored at assembly.
7. Remove the nut and bolt holding the arm to the crossmember.
8. Remove the through-bolt holding the control arm to the axle housing. Take careful note of the location and placement of all washers and bushings on either side of the arm; their position is important. Remove the control arm.

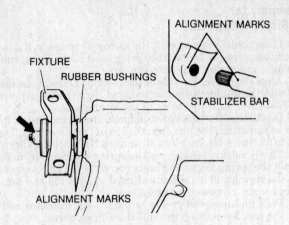

Matchmark everything before removal

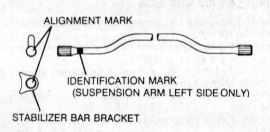

The Cordia and Tredia stabilizer bar must be correctly installed

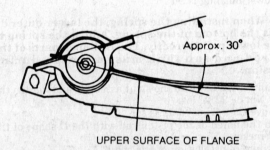

After installation, check the arm for the correct angle

9. To reinstall, place the arm and bushings in position and insert the large through-bolt with the mark on the head downward or in the six o'clock position. The concave part of the bolt should receive a light coat of chassis grease before insertion. Tighten the nut to 75 Nm or 55 ft. lbs. for cars through 1987; for 1988 and 1989 cars, tighten the nut to 120 Nm or 89 ft. lbs.
10. Install the lower control arm to the crossmember and align the mark on the crossmember with the reference line on the plate. Tighten the bolt to 140 Nm (104 ft. lbs.) on all models.
11. Install the bolt holding the arm to the front member and tighten it to 140 Nm (104 ft. lbs.) on all models.
12. Install the stabilizer bar, using a new nut. Tighten the nut until 16mm of thread projects beyond the nut.
13. Install the parking brake cable brackets to the control arm.
14. Install the rear wheel and lower the car to the ground.
15. Align the rear wheels.

SUSPENSION AND STEERING 8

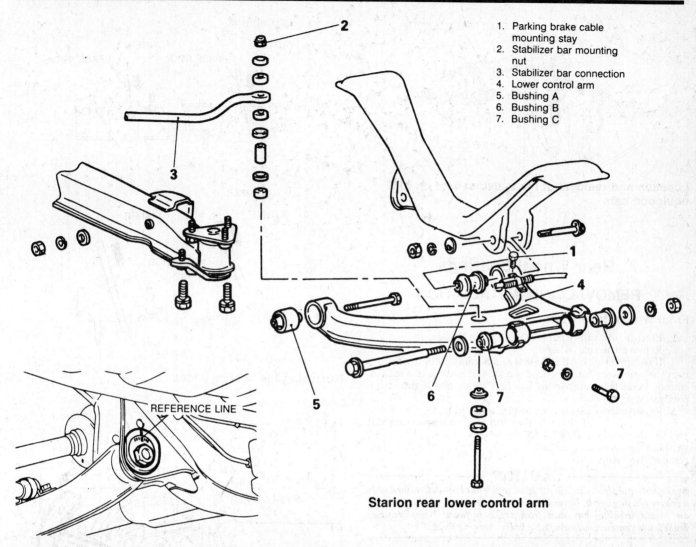

1. Parking brake cable mounting stay
2. Stabilizer bar mounting nut
3. Stabilizer bar connection
4. Lower control arm
5. Bushing A
6. Bushing B
7. Bushing C

Starion rear lower control arm

A. Make matchmarks on the alignment eccentric

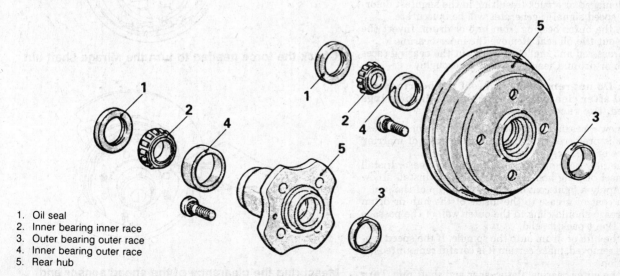

1. Oil seal
2. Inner bearing inner race
3. Outer bearing outer race
4. Inner bearing outer race
5. Rear hub

Typical rear wheel bearing components

8–43

8 SUSPENSION AND STEERING

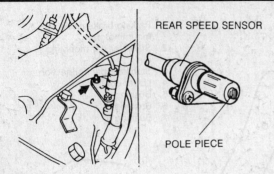

Location and removal of the speed sensor on ABS equipped cars

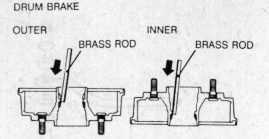

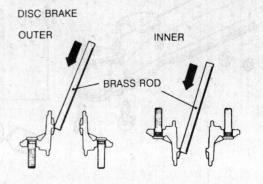

Removing the bearing races

Rear Wheel Bearings

REMOVAL AND INSTALLATION

Front Wheel Drive Cars

1. Elevate and safely support the car on stands.
2. Remove the rear wheel.
3. If equipped with ABS (Galant and Sigma), carefully disconnect the pole piece for the speed sensor and move it to a safe location. Treat this component like fine china; the slightest impact can damage it.
4. Remove the dustcap, cotter pin, and cap nut.
5. If equipped with drum brakes, remove the large center nut and the drum. If equipped with disc brakes, remove the caliper (without disconnecting its hose) and hang it out of the way, then remove the rotor.

CAUTION

Brake pads and shoes contain asbestos, which has been determined to be a cancer causing agent. Never clean the brake surfaces with compressed air! Avoid inhaling any dust from brake surfaces! When cleaning brakes, use commercially available brake cleaning fluids.

6. For cars with disc brakes, remove the large center nut and remove the hub assembly. Cars with ABS have a toothed rotor on this hub. This rotor must not be scratched or dropped. If it is in any way damaged or struck (resulting in the slightest deformation), the speed signal it generates will be inaccurate.
7. Remove the outer bearing from hub or drum. Invert the unit and pry out the oil seal. Remove the inner bearing.
8. Use a brass rod and mallet to drive out the bearing races from the hub or drum. Clean everything thoroughly.

WARNING: Do not remove the races if the bearings are to be reused after cleaning or repacking. If the bearings are replaced, the races must be changed.

9. Install new races using the correct size drivers or the brass rod. A driver is preferred; it eliminates the chance of marring the inner face of the race.
10. Pack the new bearings with multipurpose grease. Install the inner wheel bearing and use a seal driver to install a new grease seal. Apply a light coat of grease to the lip of the seal.
11. Apply a coat of grease to the inside of the hub or drum shaft. The grease should cling to the outer wall of the passage about 6mm. Don't pack it solid.
12. Install the hub or drum onto the spindle. If the speed sensor rotor was removed, make certain it is careful remounted before installation.
13. Install the outer bearing, flat washer and shaft nut. Turn the nut until it is just snug.

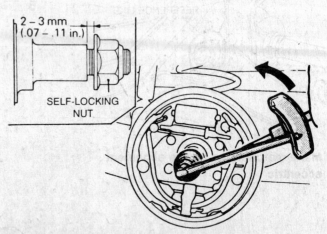

Check the force needed to turn the Mirage shaft nut

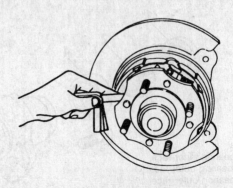

Measuring the clearance of the speed sensor and rotor

SUSPENSION AND STEERING 8

14. For all vehicles except Mirage, back off the center nut until it exerts no pressure on the washer or bearing. Use a torque wrench to tighten the nut to 20 Nm (14 ft. lbs.), then back the nut off to 0 and final tighten the nut only to 10 Nm or 7 ft. lbs.

15. For Mirage, back the nut off to a point at which the inner face of the nut is 2-3 mm from the shoulder of the shaft where the threads end.

16. Using an inch lb. torque wrench, turn the nut smoothly 2–3 turns counterclockwise. Note the average force needed to turn the nut. Specified torque is 48 inch lbs. or more; if your average is less than specified, the nut will not lock properly and must be replaced. Tighten the nut to 125 Nm or 92 ft. lbs.

17. Install the lock cap and the cotter pin.

18. Fill the dust cap about ¼ full of bearing grease and install the cap.

19. Carefully reinstall the speed sensor if it was removed. Measure the distance between the pole piece and rotor with a feeler gauge; adjust the sensor to a clearance of 0.20–0.70mm. If the clearance is outside this range, the speed signal may be incorrect.

20. Install the brake disc if so equipped. Install the brake caliper.

21. Install the wheel and lower the vehicle to the ground.

REAR WHEEL ALIGNMENT

All the Mitsubishi models covered by this guide have rear suspension systems on which bent or damaged components can cause unusual tire wear and/or handling problems. With the exception of Starion, no adjustments can be made to the rear alignment; the critical dimensions are determined by correct installation and location of components.

Even though the rear alignment cannot be adjusted on most vehicles, this does not mean it should not be checked occasionally. The dimensions of toe and camber can be easily read on an alignment rack and compared to the charts available in a reputable alignment shop. This is a handy, low-cost diagnostic tool when investigating tire wear at the rear. Additionally, the rear alignment should be checked after every heavy impact involving the rear wheels or suspension. Finding and replacing a bent component quickly can save the cost of two tires and eliminate the risk of evil handling and braking.

TOE-IN ADJUSTMENT

Starion

The Starion rear suspension toe-in is adjustable by turning the bushing located in the rearward rear suspension crossmember on either side. Loosen the nut located in front of the (axle housing) crossmember and then turn the bolt head at the rear of the crossmember.

There is a scale which turns with the bolt and a reference line on the crossmember itself. Adjust the bolts equally for the left and right wheel. One adjustment increment changes the toe-in about 2mm. On the left side, turn the bolt clockwise to increase toe and counterclockwise to decrease it. The opposite directions apply on the right.

When toe-in has been adjusted and checked, tighten the nut in front of the crossmember to 140 Nm (104 ft. lbs.) on all models.

STEERING SYSTEM

Troubleshooting the Steering Column

Problem	Cause	Solution
Will not lock	• Lockbolt spring broken or defective	• Replace lock bolt spring

8–45

8 SUSPENSION AND STEERING

Troubleshooting the Steering Column (cont.)

Problem	Cause	Solution
High effort (required to turn ignition key and lock cylinder)	• Lock cylinder defective • Ignition switch defective • Rack preload spring broken or deformed • Burr on lock sector, lock rack, housing, support or remote rod coupling • Bent sector shaft • Defective lock rack • Remote rod bent, deformed • Ignition switch mounting bracket bent • Distorted coupling slot in lock rack (tilt column)	• Replace lock cylinder • Replace ignition switch • Replace preload spring • Remove burr • Replace shaft • Replace lock rack • Replace rod • Straighten or replace • Replace lock rack
Will stick in "start"	• Remote rod deformed • Ignition switch mounting bracket bent	• Straighten or replace • Straighten or replace
Key cannot be removed in "off-lock"	• Ignition switch is not adjusted correctly • Defective lock cylinder	• Adjust switch • Replace lock cylinder
Lock cylinder can be removed without depressing retainer	• Lock cylinder with defective retainer • Burr over retainer slot in housing cover or on cylinder retainer	• Replace lock cylinder • Remove burr
High effort on lock cylinder between "off" and "off-lock"	• Distorted lock rack • Burr on tang of shift gate (automatic column) • Gearshift linkage not adjusted	• Replace lock rack • Remove burr • Adjust linkage
Noise in column	• One click when in "off-lock" position and the steering wheel is moved (all except automatic column) • Coupling bolts not tightened • Lack of grease on bearings or bearing surfaces • Upper shaft bearing worn or broken • Lower shaft bearing worn or broken • Column not correctly aligned • Coupling pulled apart • Broken coupling lower joint • Steering shaft snap ring not seated	• Normal—lock bolt is seating • Tighten pinch bolts • Lubricate with chassis grease • Replace bearing assembly • Replace bearing. Check shaft and replace if scored. • Align column • Replace coupling • Repair or replace joint and align column • Replace ring. Check for proper seating in groove.
Noise in column	• Shroud loose on shift bowl. Housing loose on jacket—will be noticed with ignition in "off-lock" and when torque is applied to steering wheel.	• Position shroud over lugs on shift bowl. Tighten mounting screws.

SUSPENSION AND STEERING 8

Troubleshooting the Steering Column (cont.)

Problem	Cause	Solution
High steering shaft effort	• Column misaligned • Defective upper or lower bearing • Tight steering shaft universal joint • Flash on I.D. of shift tube at plastic joint (tilt column only) • Upper or lower bearing seized	• Align column • Replace as required • Repair or replace • Replace shift tube • Replace bearings
Lash in mounted column assembly	• Column mounting bracket bolts loose • Broken weld nuts on column jacket • Column capsule bracket sheared	• Tighten bolts • Replace column jacket • Replace bracket assembly
Lash in mounted column assembly (cont.)	• Column bracket to column jacket mounting bolts loose • Loose lock shoes in housing (tilt column only) • Loose pivot pins (tilt column only) • Loose lock shoe pin (tilt column only) • Loose support screws (tilt column only)	• Tighten to specified torque • Replace shoes • Replace pivot pins and support • Replace pin and housing • Tighten screws
Housing loose (tilt column only)	• Excessive clearance between holes in support or housing and pivot pin diameters • Housing support-screws loose	• Replace pivot pins and support • Tighten screws
Steering wheel loose—every other tilt position (tilt column only)	• Loose fit between lock shoe and lock shoe pivot pin	• Replace lock shoes and pivot pin
Steering column not locking in any tilt position (tilt column only)	• Lock shoe seized on pivot pin • Lock shoe grooves have burrs or are filled with foreign material • Lock shoe springs weak or broken	• Replace lock shoes and pin • Clean or replace lock shoes • Replace springs
Noise when tilting column (tilt column only)	• Upper tilt bumpers worn • Tilt spring rubbing in housing	• Replace tilt bumper • Lubricate with chassis grease
One click when in "off-lock" position and the steering wheel is moved	• Seating of lock bolt	• None. Click is normal characteristic sound produced by lock bolt as it seats.
High shift effort (automatic and tilt column only)	• Column not correctly aligned • Lower bearing not aligned correctly • Lack of grease on seal or lower bearing areas	• Align column • Assemble correctly • Lubricate with chassis grease
Improper transmission shifting—automatic and tilt column only	• Sheared shift tube joint • Improper transmission gearshift linkage adjustment • Loose lower shift lever	• Replace shift tube • Adjust linkage • Replace shift tube

8-47

8 SUSPENSION AND STEERING

Troubleshooting the Ignition Switch

Problem	Cause	Solution
Ignition switch electrically inoperative	• Loose or defective switch connector • Feed wire open (fusible link) • Defective ignition switch	• Tighten or replace connector • Repair or replace • Replace ignition switch
Engine will not crank	• Ignition switch not adjusted properly	• Adjust switch
Ignition switch wil not actuate mechanically	• Defective ignition switch • Defective lock sector • Defective remote rod	• Replace switch • Replace lock sector • Replace remote rod
Ignition switch cannot be adjusted correctly	• Remote rod deformed	• Repair, straighten or replace

Troubleshooting the Turn Signal Switch

Problem	Cause	Solution
Turn signal will not cancel	• Loose switch mounting screws • Switch or anchor bosses broken • Broken, missing or out of position detent, or cancelling spring	• Tighten screws • Replace switch • Reposition springs or replace switch as required
Turn signal difficult to operate	• Turn signal lever loose • Switch yoke broken or distorted • Loose or misplaced springs • Foreign parts and/or materials in switch • Switch mounted loosely	• Tighten mounting screws • Replace switch • Reposition springs or replace switch • Remove foreign parts and/or material • Tighten mounting screws
Turn signal will not indicate lane change	• Broken lane change pressure pad or spring hanger • Broken, missing or misplaced lane change spring • Jammed wires	• Replace switch • Replace or reposition as required • Loosen mounting screws, reposition wires and retighten screws
Turn signal will not stay in turn position	• Foreign material or loose parts impeding movement of switch yoke • Defective switch	• Remove material and/or parts • Replace switch
Hazard switch cannot be pulled out	• Foreign material between hazard support cancelling leg and yoke	• Remove foreign material. No foreign material impeding function of hazard switch—replace turn signal switch.

SUSPENSION AND STEERING 8

Troubleshooting the Turn Signal Switch (cont.)

Problem	Cause	Solution
No turn signal lights	• Inoperative turn signal flasher • Defective or blown fuse • Loose chassis to column harness connector • Disconnect column to chassis connector. Connect new switch to chassis and operate switch by hand. If vehicle lights now operate normally, signal switch is inoperative • If vehicle lights do not operate, check chassis wiring for opens, grounds, etc.	• Replace turn signal flasher • Replace fuse • Connect securely • Replace signal switch • Repair chassis wiring as required
Instrument panel turn indicator lights on but not flashing	• Burned out or damaged front or rear turn signal bulb • If vehicle lights do not operate, check light sockets for high resistance connections, the chassis wiring for opens, grounds, etc. • Inoperative flasher • Loose chassis to column harness connection • Inoperative turn signal switch • To determine if turn signal switch is defective, substitute new switch into circuit and operate switch by hand. If the vehicle's lights operate normally, signal switch is inoperative.	• Replace bulb • Repair chassis wiring as required • Replace flasher • Connect securely • Replace turn signal switch • Replace turn signal switch
Stop light not on when turn indicated	• Loose column to chassis connection • Disconnect column to chassis connector. Connect new switch into system without removing old.	• Connect securely • Replace signal switch
Stop light not on when turn indicated (cont.)	Operate switch by hand. If brake lights work with switch in the turn position, signal switch is defective. • If brake lights do not work, check connector to stop light sockets for grounds, opens, etc.	 • Repair connector to stop light circuits using service manual as guide
Turn indicator panel lights not flashing	• Burned out bulbs • High resistance to ground at bulb socket • Opens, ground in wiring harness from front turn signal bulb socket to indicator lights	• Replace bulbs • Replace socket • Locate and repair as required

8 SUSPENSION AND STEERING

Troubleshooting the Turn Signal Switch (cont.)

Problem	Cause	Solution
Turn signal lights flash very slowly	• High resistance ground at light sockets • Incorrect capacity turn signal flasher or bulb • If flashing rate is still extremely slow, check chassis wiring harness from the connector to light sockets for high resistance • Loose chassis to column harness connection • Disconnect column to chassis connector. Connect new switch into system without removing old. Operate switch by hand. If flashing occurs at normal rate, the signal switch is defective.	• Repair high resistance grounds at light sockets • Replace turn signal flasher or bulb • Locate and repair as required • Connect securely • Replace turn signal switch
Hazard signal lights will not flash—turn signal functions normally	• Blow fuse • Inoperative hazard warning flasher • Loose chassis-to-column harness connection • Disconnect column to chassis connector. Connect new switch into system without removing old. Depress the hazard warning lights. If they now work normally, turn signal switch is defective. • If lights do not flash, check wiring harness "K" lead for open between hazard flasher and connector. If open, fuse block is defective	• Replace fuse • Replace hazard warning flasher in fuse panel • Conect securely • Replace turn signal switch • Repair or replace brown wire or connector as required

Troubleshooting the Manual Steering Gear

Problem	Cause	Solution
Hard or erratic steering	• Incorrect tire pressure • Insufficient or incorrect lubrication • Suspension, or steering linkage parts damaged or misaligned • Improper front wheel alignment • Incorrect steering gear adjustment • Sagging springs	• Inflate tires to recommended pressures • Lubricate as required (refer to Maintenance Section) • Repair or replace parts as necessary • Adjust incorrect wheel alignment angles • Adjust steering gear • Replace springs

SUSPENSION AND STEERING 8

Troubleshooting the Manual Steering Gear

Problem	Cause	Solution
Play or looseness in steering	• Steering wheel loose	• Inspect shaft spines and repair as necessary. Tighten attaching nut and stake in place.
	• Steering linkage or attaching parts loose or worn	• Tighten, adjust, or replace faulty components
	• Pitman arm loose	• Inspect shaft splines and repair as necessary. Tighten attaching nut and stake in place
	• Steering gear attaching bolts loose	• Tighten bolts
	• Loose or worn wheel bearings	• Adjust or replace bearings
	• Steering gear adjustment incorrect or parts badly worn	• Adjust gear or replace defective parts
Wheel shimmy or tramp	• Improper tire pressure	• Inflate tires to recommended pressures
	• Wheels, tires, or brake rotors out-of-balance or out-of-round	• Inspect and replace or balance parts
	• Inoperative, worn, or loose shock absorbers or mounting parts	• Repair or replace shocks or mountings
	• Loose or worn steering or suspension parts	• Tighten or replace as necessary
	• Loose or worn wheel bearings	• Adjust or replace bearings
	• Incorrect steering gear adjustments	• Adjust steering gear
	• Incorrect front wheel alignment	• Correct front wheel alignment
Tire wear	• Improper tire pressure	• Inflate tires to recommended pressures
	• Failure to rotate tires	• Rotate tires
	• Brakes grabbing	• Adjust or repair brakes
	• Incorrect front wheel alignment	• Align incorrect angles
	• Broken or damaged steering and suspension parts	• Repair or replace defective parts
	• Wheel runout	• Replace faulty wheel
	• Excessive speed on turns	• Make driver aware of conditions
Vehicle leads to one side	• Improper tire pressures	• Inflate tires to recommended pressures
	• Front tires with uneven tread depth, wear pattern, or different cord design (i.e., one bias ply and one belted or radial tire on front wheels)	• Install tires of same cord construction and reasonably even tread depth, design, and wear pattern
	• Incorrect front wheel alignment	• Align incorrect angles
	• Brakes dragging	• Adjust or repair brakes
	• Pulling due to uneven tire construction	• Replace faulty tire

8 SUSPENSION AND STEERING

Troubleshooting the Power Steering Gear

Problem	Cause	Solution
Hissing noise in steering gear	• There is some noise in all power steering systems. One of the most common is a hissing sound most evident at standstill parking. There is no relationship between this noise and performance of the steering. Hiss may be expected when steering wheel is at end of travel or when slowly turning at standstill.	• Slight hiss is normal and in no way affects steering. Do not replace valve unless hiss is extremely objectionable. A replacement valve will also exhibit slight noise and is not always a cure. Investigate clearance around flexible coupling rivets. Be sure steering shaft and gear are aligned so flexible coupling rotates in a flat plane and is not distorted as shaft rotates. Any metal-to-metal contacts through flexible coupling will transmit valve hiss into passenger compartment through the steering column.
Rattle or chuckle noise in steering gear	• Gear loose on frame • Steering linkage looseness • Pressure hose touching other parts of car • Loose pitman shaft over center adjustment NOTE: A slight rattle may occur on turns because of increased clearance off the "high point." This is normal and clearance must not be reduced below specified limits to eliminate this slight rattle. • Loose pitman arm	• Check gear-to-frame mounting screws. Tighten screws to 88 N·m (65 foot pounds) torque. • Check linkage pivot points for wear. Replace if necessary. • Adjust hose position. Do not bend tubing by hand. • Adjust to specifications • Tighten pitman arm nut to specifications
Squawk noise in steering gear when turning or recovering from a turn	• Damper O-ring on valve spool cut	• Replace damper O-ring
Poor return of steering wheel to center	• Tires not properly inflated • Lack of lubrication in linkage and ball joints • Lower coupling flange rubbing against steering gear adjuster plug • Steering gear to column misalignment • Improper front wheel alignment • Steering linkage binding • Ball joints binding • Steering wheel rubbing against housing • Tight or frozen steering shaft bearings	• Inflate to specified pressure • Lube linkage and ball joints • Loosen pinch bolt and assemble properly • Align steering column • Check and adjust as necessary • Replace pivots • Replace ball joints • Align housing • Replace bearings

SUSPENSION AND STEERING 8

Troubleshooting the Power Steering Gear (cont.)

Problem	Cause	Solution
Poor return of steering wheel to center	• Sticking or plugged valve spool • Steering gear adjustments over specifications • Kink in return hose	• Remove and clean or replace valve • Check adjustment with gear out of car. Adjust as required. • Replace hose
Car leads to one side or the other (keep in mind road condition and wind. Test car in both directions on flat road)	• Front end misaligned • Unbalanced steering gear valve **NOTE:** If this is cause, steering effort will be very light in direction of lead and normal or heavier in opposite direction	• Adjust to specifications • Replace valve
Momentary increase in effort when turning wheel fast to right or left	• Low oil level • Pump belt slipping • High internal leakage	• Add power steering fluid as required • Tighten or replace belt • Check pump pressure. (See pressure test)
Steering wheel surges or jerks when turning with engine running especially during parking	• Low oil level • Loose pump belt • Steering linkage hitting engine oil pan at full turn • Insufficient pump pressure • Pump flow control valve sticking	• Fill as required • Adjust tension to specification • Correct clearance • Check pump pressure. (See pressure test). Replace relief valve if defective. • Inspect for varnish or damage, replace if necessary
Excessive wheel kickback or loose steering	• Air in system • Steering gear loose on frame • Steering linkage joints worn enough to be loose • Worn poppet valve • Loose thrust bearing preload adjustment • Excessive overcenter lash	• Add oil to pump reservoir and bleed by operating steering. Check hose connectors for proper torque and adjust as required. • Tighten attaching screws to specified torque • Replace loose pivots • Replace poppet valve • Adjust to specification with gear out of vehicle • Adjust to specification with gear out of car
Hard steering or lack of assist	• Loose pump belt • Low oil level **NOTE:** Low oil level will also result in excessive pump noise • Steering gear to column misalignment • Lower coupling flange rubbing against steering gear adjuster plug • Tires not properly inflated	• Adjust belt tension to specification • Fill to proper level. If excessively low, check all lines and joints for evidence of external leakage. Tighten loose connectors. • Align steering column • Loosen pinch bolt and assemble properly • Inflate to recommended pressure

8-53

8 SUSPENSION AND STEERING

Troubleshooting the Power Steering Gear (cont.)

Problem	Cause	Solution
Foamy milky power steering fluid, low fluid level and possible low pressure	• Air in the fluid, and loss of fluid due to internal pump leakage causing overflow	• Check for leak and correct. Bleed system. Extremely cold temperatures will cause system aeriation should the oil level be low. If oil level is correct and pump still foams, remove pump from vehicle and separate reservoir from housing. Check welsh plug and housing for cracks. If plug is loose or housing is cracked, replace housing.
Low pressure due to steering pump	• Flow control valve stuck or inoperative • Pressure plate not flat against cam ring	• Remove burrs or dirt or replace. Flush system. • Correct
Low pressure due to steering gear	• Pressure loss in cylinder due to worn piston ring or badly worn housing bore • Leakage at valve rings, valve body-to-worm seal	• Remove gear from car for disassembly and inspection of ring and housing bore • Remove gear from car for disassembly and replace seals

Troubleshooting the Power Steering Pump

Problem	Cause	Solution
Chirp noise in steering pump	• Loose belt	• Adjust belt tension to specification
Belt squeal (particularly noticeable at full wheel travel and stand still parking)	• Loose belt	• Adjust belt tension to specification
Growl noise in steering pump	• Excessive back pressure in hoses or steering gear caused by restriction	• Locate restriction and correct. Replace part if necessary.
Growl noise in steering pump (particularly noticeable at stand still parking)	• Scored pressure plates, thrust plate or rotor • Extreme wear of cam ring	• Replace parts and flush system • Replace parts
Groan noise in steering pump	• Low oil level • Air in the oil. Poor pressure hose connection.	• Fill reservoir to proper level • Tighten connector to specified torque. Bleed system by operating steering from right to left—full turn.
Rattle noise in steering pump	• Vanes not installed properly • Vanes sticking in rotor slots	• Install properly • Free up by removing burrs, varnish, or dirt
Swish noise in steering pump	• Defective flow control valve	• Replace part

SUSPENSION AND STEERING 8

Troubleshooting the Power Steering Pump (cont.)

Problem	Cause	Solution
Whine noise in steering pump	• Pump shaft bearing scored	• Replace housing and shaft. Flush system.
Hard steering or lack of assist	• Loose pump belt • Low oil level in reservoir NOTE: Low oil level will also result in excessive pump noise • Steering gear to column misalignment • Lower coupling flange rubbing against steering gear adjuster plug • Tires not properly inflated	• Adjust belt tension to specification • Fill to proper level. If excessively low, check all lines and joints for evidence of external leakage. Tighten loose connectors. • Align steering column • Loosen pinch bolt and assemble properly • Inflate to recommended pressure
Foaming milky power steering fluid, low fluid level and possible low pressure	• Air in the fluid, and loss of fluid due to internal pump leakage causing overflow	• Check for leaks and correct. Bleed system. Extremely cold temperatures will cause system aeriation should the oil level be low. If oil level is correct and pump still foams, remove pump from vehicle and separate reservoir from body. Check welsh plug and body for cracks. If plug is loose or body is cracked, replace body.
Low pump pressure	• Flow control valve stuck or inoperative • Pressure plate not flat against cam ring	• Remove burrs or dirt or replace. Flush system. • Correct
Momentary increase in effort when turning wheel fast to right or left	• Low oil level in pump • Pump belt slipping • High internal leakage	• Add power steering fluid as required • Tighten or replace belt • Check pump pressure. (See pressure test)
Steering wheel surges or jerks when turning with engine running especially during parking	• Low oil level • Loose pump belt • Steering linkage hitting engine oil pan at full turn • Insufficient pump pressure	• Fill as required • Adjust tension to specification • Correct clearance • Check pump pressure. (See pressure test). Replace flow control valve if defective.
Steering wheel surges or jerks when turning with engine running especially during parking (cont.)	• Sticking flow control valve	• Inspect for varnish or damage, replace if necessary
Excessive wheel kickback or loose steering	• Air in system	• Add oil to pump reservoir and bleed by operating steering. Check hose connectors for proper torque and adjust as required.

8–55

8 SUSPENSION AND STEERING

Troubleshooting the Power Steering Pump (cont.)

Problem	Cause	Solution
Low pump pressure	• Extreme wear of cam ring	• Replace parts. Flush system.
	• Scored pressure plate, thrust plate, or rotor	• Replace parts. Flush system.
	• Vanes not installed properly	• Install properly
	• Vanes sticking in rotor slots	• Freeup by removing burrs, varnish, or dirt
	• Cracked or broken thrust or pressure plate	• Replace part

Steering Wheel

REMOVAL AND INSTALLATION

1. Pull off the horn cover at the center of the wheel by grasping the upper edge with your fingers to release it. On some models the pad is screwed on from behind the wheel. Disconnect the horn wire connector.
2. Remove the steering wheel retaining nut. Matchmark the relationship between the wheel and shaft.
3. Screw the two bolts of a steering wheel puller into the wheel. Turn the bolt at the center of the puller to force the wheel off the steering shaft. Do not pound on the wheel to remove it, or the collapsible steering shaft may be damaged.
4. Install the wheel and push it onto the shaft splines by hand far enough to start the retaining nut.
5. Install the retaining nut and torque it to 40 Nm or 30 ft. lbs.
6. Connect the horn wire and reinstall the wheel pad. If it is the snap-on type, make sure all the clips are engaged.

Turn Signal and Wiper Switch

REMOVAL AND INSTALLATION

1. Remove the steering wheel.
2. Remove the upper and lower steering column covers.
3. Unplug the two electrical connectors. If necessary, remove the harness retainer.
4. Remove the retaining screws and remove the switch.
5. Carefully install the switch and tighten the mounting screws. Connect the switch harness wiring and replace any clips or retainers which were removed.
6. Install the upper and lower column covers.
7. Reinstall the steering wheel.

Ignition Lock

REMOVAL AND INSTALLATION

1. Disconnect the negative battery cable.
2. If the vehicle is equipped with a tilt steering column, put the column in its lowest position.
3. Remove the upper and lower column covers.

NOTE: While removal of the steering wheel is not usually required, it makes the job much easier.

4. Remove the wiring harness bands or clips. Disconnect the ignition switch harness.
5. If the ignition switch (electrical part) can be removed from the lock, remove it. Older vehicles use one-piece units.
6. The installation bolts have no head (they were broken off during installation as a theft deterrent); use a hacksaw blade to cut a groove into the head of the special mounting bolts. Use a screwdriver placed in the new groove to remove the bolts.
7. Remove the lock assembly from the steering column.

To install:

8. With the key inserted in the new lock, install the lock assembly to the steering column and hand tighten new bolts. Turn the key from time to time as the bolts are tightened, making sure the key does not bind.
9. Once the lock is in place and confirmed in working order, use a socket wrench to evenly tighten the bolts. Again check the key operation during the tightening. Continue tightening the bolts until the heads break off.
10. Connect the switch if it was removed from the lock assembly.
11. Connect the wiring harness and check the assembly for proper operation.
12. Reinstall or connect any wiring bands, clips or retainers which were loosened. It is important that the wiring be correctly contained and out of the way of any moving parts.
13. Install the upper and lower column covers.
14. Install the steering wheel if it was removed.
15. Connect the negative battery cable.

Steering Column

REMOVAL AND INSTALLATION

1. Remove the steering wheel.
2. Remove the column upper and lower covers. On vehicles with tilt steering, remove the covers with the column in the lowest position.
3. Remove the column switch assembly.
4. Near the floorboard, remove the pinchbolt holding the upper shaft to the joint assembly.
5. If the car does not have tilt steering, remove the bolts holding the upper and lower column brackets to the body. Remove the column.
6. For cars with tilt wheel, remove the bolts holding the lower bracket and the tilt mechanism to the dash. Lower the column to the floor and remove it.
7. If desired, remove the pinch bolt holding the joint assembly to the steering box. Remove the 4 bolts holding the cover to the floor and remove the cover and shaft. This joint assembly cannot be disassembled. If it is binding or damaged, it must be replaced.
8. If the joint shaft was removed, it must be reinstalled before the column. Tighten the pinchbolt at the steering box to 35 Nm or 25 ft. lbs and install the floor cover.
9. Place the column in the car and fit the shaft end into the joint. Tighten the pinchbolt to 35 Nm or 25 ft. lbs.

SUSPENSION AND STEERING 8

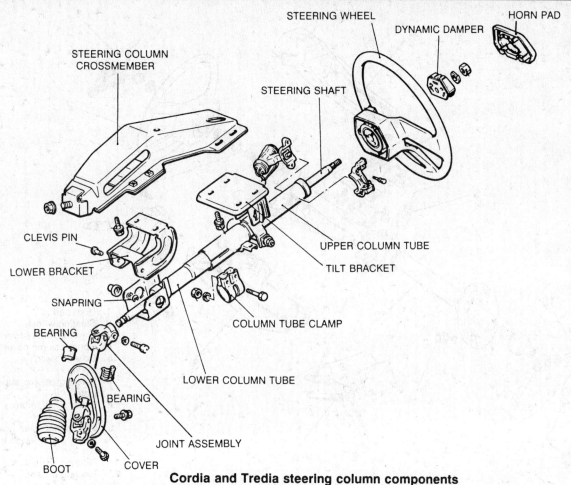

Cordia and Tredia steering column components

10. With the help of an assistant, lift the column into position. Install one or two bolts in the upper mount or tilt head to hold the column in place.
11. Install the other retaining bolts loosely and double check the placement of the column. When everything is aligned, tighten all the mounting bolts to 12 Nm or 9 ft. lbs.
12. Install the column switch and connect its harness. Connect any other wiring harness which may have been opened; make certain all the wiring is safely retained within clips or bands.
13. Install the steering column covers.
14. Install the steering wheel and install the horn pad.

Starion

1. Remove the steering wheel.
2. Remove the column upper and lower covers. On vehicles with tilt steering, remove the covers with the column in the lowest position.
3. Remove the column switch assembly.
4. Remove the large spring between the top of the column and the column support plate.
5. Remove the column support plate.
6. Disconnect the brake pedal return spring.
7. Remove the bolts holding the rubber cover to the floor.
8. Remove the steering shaft to gearbox pinchbolt and remove the steering shaft clamp.
9. Carefully remove the complete steering column from the car.
10. When reinstalling, fit the column into place and seat the splined end into the steering gearbox shaft. Align the notch in the shaft with the bolt hole of the clamp.

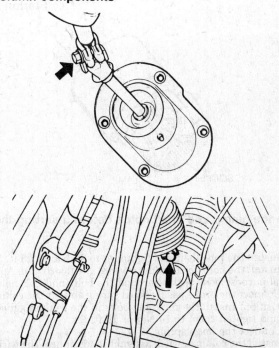

The steering joint — actually a short shaft — must be disconnected at the gearbox and the bottom of the steering column

8-57

8 SUSPENSION AND STEERING

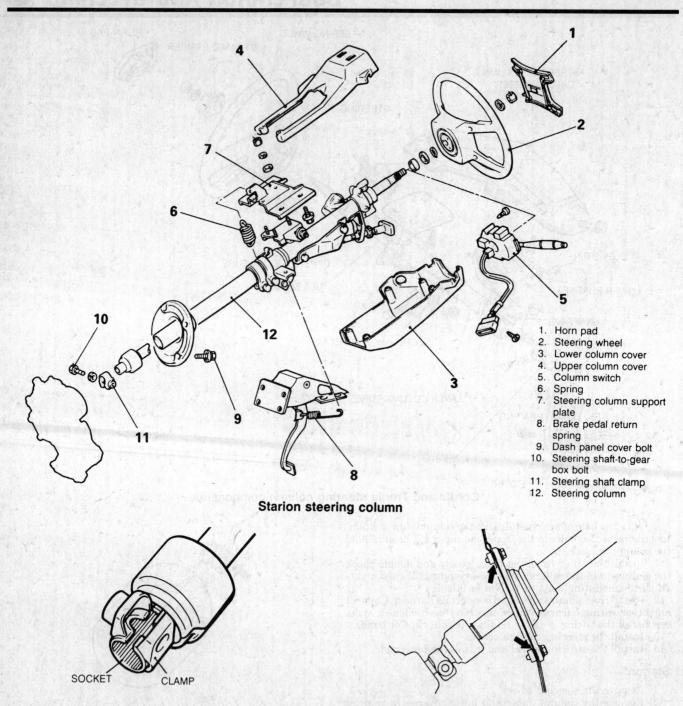

1. Horn pad
2. Steering wheel
3. Lower column cover
4. Upper column cover
5. Column switch
6. Spring
7. Steering column support plate
8. Brake pedal return spring
9. Dash panel cover bolt
10. Steering shaft-to-gear box bolt
11. Steering shaft clamp
12. Steering column

Starion steering column

SOCKET CLAMP

Align the clamp and shaft notch before inserting the bolt

11. Install the pinchbolt and tighten it to 25 Nm or 18 ft. lbs.
12. Install the bolts for the dust cover at the floorboard. Apply general purpose body sealant around the bolts and bolt holes.
13. Connect the brake pedal return spring and lift the column into position. Install the retaining bolts in the column support plate and tighten them to 14 Nm or 10 ft. lbs.
14. Connect the large spring.
15. Install the column switch assembly and connect the wiring harnesses.
16. Install the upper and lower column covers.
17. Install the steering wheel and horn pad.

Apply sealant as shown to eliminate the entry of dust and water

Mirage through 1988
Precis

1. Remove the horn pad and steering wheel.
2. Remove the lower dash cover.
3. Remove the lower steering column cover.
4. If so equipped, remove the knobs from the lighting and wiper switches on the sides of the upper column cover.
5. Remove the small screw holding the switches to the upper column cover.
6. Remove the upper column cover.

8-58

SUSPENSION AND STEERING 8

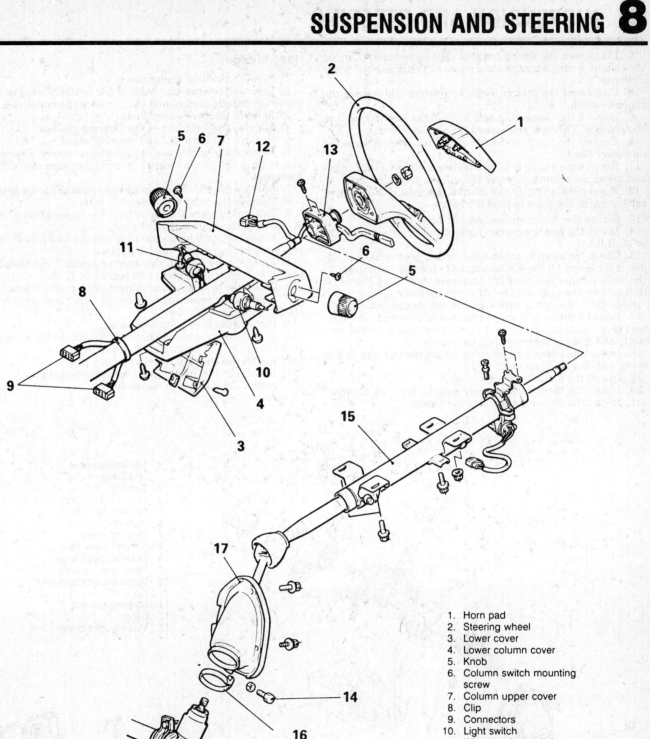

1. Horn pad
2. Steering wheel
3. Lower cover
4. Lower column cover
5. Knob
6. Column switch mounting screw
7. Column upper cover
8. Clip
9. Connectors
10. Light switch
11. Wiper/washer switch
12. Switch connector
13. Column switch
14. Shaft-to-gear box bolt
15. Steering column
16. Dust cover band
17. Dust cover

Mirage steering column through 1988. Not all years use the side mounted lighting and wiper switches

8 SUSPENSION AND STEERING

7. Disconnect the wiring harness clips or retainers. Unplug the column wiring harnesses and remove the lighting and wiper switches.
8. Disconnect the column switch connector and remove the column switch.
9. Remove the connecting bolt holding the steering shaft to the gearbox. This bolt is loosened from inside the car.
10. Remove the retaining bolts for the dustcover. Release the dustcover band.
11. Remove the 4 bolts holding the steering column to the dash and lower the column. Remove the column from the car.
12. When reinstalling, fit the column into position and temporarily tighten the bolts to hold it in place.
13. Secure the dustcover boot and its bolts and band.
14. Install the gearbox connecting bolt and tighten it to 35 Nm or 25 ft.lbs.
15. Check the steering column for correct positioning and adjust it if needed. Of the 4 retaining bolts holding the column to the dash, tighten the lower left (outboard) bolt to 14 Nm or 10 ft. lbs. first, then tighten the others to the same torque.
16. Install the column switch and attach its connector.
17. Install lighting and wiper switches (if removed) and connect their harnesses. Make certain each harness is routed correctly and install necessary clips and bands.
18. Install the upper column cover attach the switches to the cover with the small screws.
19. Install the knobs on the switches.
20. Install the lower column cover and the lower dash cover.
21. Install the steering wheel and horn pad.

1989 Mirage

1. Remove the dash under cover.
2. Remove the trim clip holding the heater duct. Push the center of the clip inward and then remove the entire clip.
3. Remove the two heater ducts.
4. Remove the joint bolt at the steering gearbox.
5. Remove the horn pad and steering wheel.
6. Remove the upper and lower column covers. Disconnect and remove the column switches, including disconnecting the harnesses.
7. Remove the column retaining bolts from the lower bracket.
8. Remove the band and dustcover bolts.
9. Remove the retaining bolts from the upper column bracket. Remove the column from the car.
10. When reinstalling, position the column and install one or two bolts to temporarily hold it in place.
11. Fit the lower joint to the steering gearbox and install the joint bolt. Tighten the bolt to 20 Nm or 14 ft. lbs.
12. Check the position of the column, adjusting it if necessary.
13. Tighten the mounting bolts on the upper and lower brackets to 14 Nm or 10 ft. lbs.
14. Install dustcover bolts and band.
15. Install the column switches and connect the harnesses. Install the upper and lower steering column covers.
16. Install the steering wheel and horn pad.
17. Install the two heater ducts and the trim clip.
18. Install the dashboard undercover.

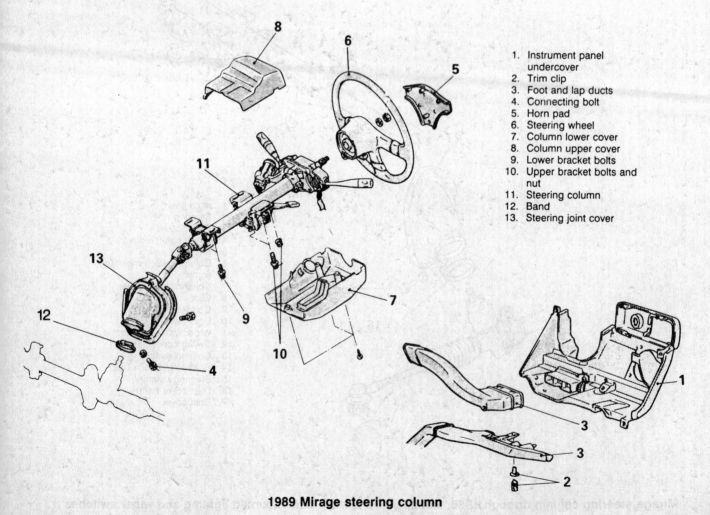

1. Instrument panel undercover
2. Trim clip
3. Foot and lap ducts
4. Connecting bolt
5. Horn pad
6. Steering wheel
7. Column lower cover
8. Column upper cover
9. Lower bracket bolts
10. Upper bracket bolts and nut
11. Steering column
12. Band
13. Steering joint cover

1989 Mirage steering column

SUSPENSION AND STEERING 8

Galant and Sigma

1. Remove the horn pad and remove the steering wheel.
2. Remove the small upper column cover panel.
3. Remove the upper and lower column covers.
4. Unplug the harness connector and remove the column switch assembly.
5. Remove the dash undercover.
6. Remove the small frame piece to which the dash undercover was mounted.
7. Disconnect the lap heater duct, the side defroster hose and the wide, short air duct.
8. Loosen and remove the pinchbolt holding the steering joint (short shaft) to the steering gearbox.
9. Remove the 4 bolts holding the floor cover.
10. Disconnect the lower bracket bolts from the dashboard frame.
11. Remove the bolts holding the tilt wheel frame to the dash. Before removing the last bolt, support the column and lower it after the bolt is free.
12. Remove the steering column from the car.
13. When reinstalling, carefully fit the column into place, raise it and install the tilt bracket bolts loosely.
14. Make certain the steering joint is properly fitted onto the steering gearbox. Reposition the column as necessary and install the 2 bolts into the lower mount.
15. Once the column is correctly positioned, tighten the tilt wheel bracket bolts to and the lower mount bolts to 12 Nm or 9 ft. lbs.
16. Install the pinchbolt for the steering joint and tighten it to 20 Nm or 14 ft. lbs.
17. Reinstall the air duct, side defroster duct and the lap heater duct.
18. Replace the small frame piece and the dash undercover.
19. Install the column switch and connect the wiring harness.
20. Install the upper and lower column covers. Install the small panel piece on the top of the column.
21. Install the steering wheel and horn pad.

Montero

1. Remove the pinchbolt holding the steering joint to the steering gearbox.
2. Remove the horn pad and steering wheel.
3. Remove the lower and upper column covers.
4. Remove the lap heater duct.
5. Disconnect the band or clip holding the wiring harnesses to the column.
6. Remove the column switch and disconnect its wiring harness.
7. Remove the small screws holding the ignition switch (electrical switch) to the key assembly. Disconnect the harness and

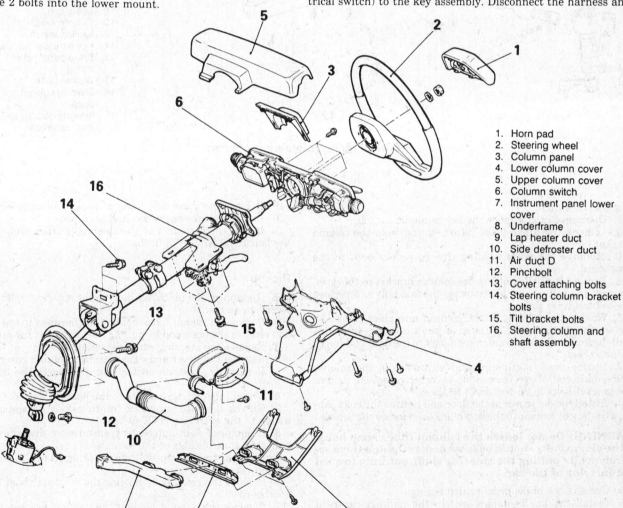

1. Horn pad
2. Steering wheel
3. Column panel
4. Lower column cover
5. Upper column cover
6. Column switch
7. Instrument panel lower cover
8. Underframe
9. Lap heater duct
10. Side defroster duct
11. Air duct D
12. Pinchbolt
13. Cover attaching bolts
14. Steering column bracket bolts
15. Tilt bracket bolts
16. Steering column and shaft assembly

Galant and Sigma steering column

8-61

8 SUSPENSION AND STEERING

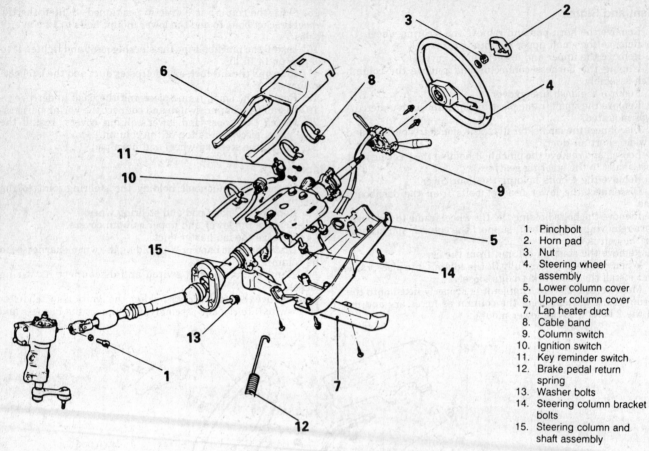

1. Pinchbolt
2. Horn pad
3. Nut
4. Steering wheel assembly
5. Lower column cover
6. Upper column cover
7. Lap heater duct
8. Cable band
9. Column switch
10. Ignition switch
11. Key reminder switch
12. Brake pedal return spring
13. Washer bolts
14. Steering column bracket bolts
15. Steering column and shaft assembly

Typical Montero steering column

remove the switch.
8. Disconnect and remove the key reminder switch.
9. Unhook the brake pedal return spring from the column bracket.
10. Remove the bolts holding the dustcover boot to the floorboard.
11. Remove the bolts holding the column bracket to the dash. Support the column while removing the last bolt and remove the column from the car.
12. When reassembling, fit the column into place and make certain the steering joint is seated on the steering gearbox. Install the bolts holding the column bracket to the dash and finger tighten them.
13. Double check positioning and alignment of the column from end to end. When the column is correctly located, tighten the bracket bolts to 25 Nm or 18 ft. lbs.
14. Install the dustcover at the floor and tighten the bolts. Apply a coating of sealant to the bolt holes from inside the vehicle.

WARNING: Do not loosen the column tube clamp bolts. if the clamp bolts should be loosened, retighten them securely while pulling the steering shaft out fully toward the interior of the car.

15. Connect the brake pedal return spring.
16. Install the key reminder switch, the ignition electrical switch and the column switch. Route the harnesses carefully and connect each to its mate.
17. Install new bands or clips to hold the cables in place.
18. Install the lap heater duct.
19. Replace the upper and lower steering column covers.
20. Install the steering wheel and horn pad.
21. Install the pinchbolt at the steering gearbox and tighten the bolt to 35 Nm or 25 ft. lbs.

Pickup

1. Disconnect the pinchbolt holding the steering shaft to the steering gearbox.
2. Remove the small screw holding the hornpad to the wheel and remove the horn pad by pulling the lower part toward you.
3. Remove the steering wheel assembly.
4. Remove the lower and upper steering column covers.
5. Remove the column switch and disconnect the harness connector.
6. Unplug the wiring harness to the ignition switch.
7. On vehicles with automatic transmissions, disconnect the wiring to the overdrive **OFF** switch.
8. If equipped with automatic transmission, disconnect the gear shift control cable at the column.
9. Unhook the brake pedal return spring from the steering column bracket.
10. Remove the small bolts holding the dustcover boot to the floorboard.
11. Remove the 4 bolts holding the column bracket to the dashboard frame. Support the column and remove it from the vehicle.
12. When reassembling, fit the column into place and make certain the steering joint is seated on the steering gearbox. In-

SUSPENSION AND STEERING 8

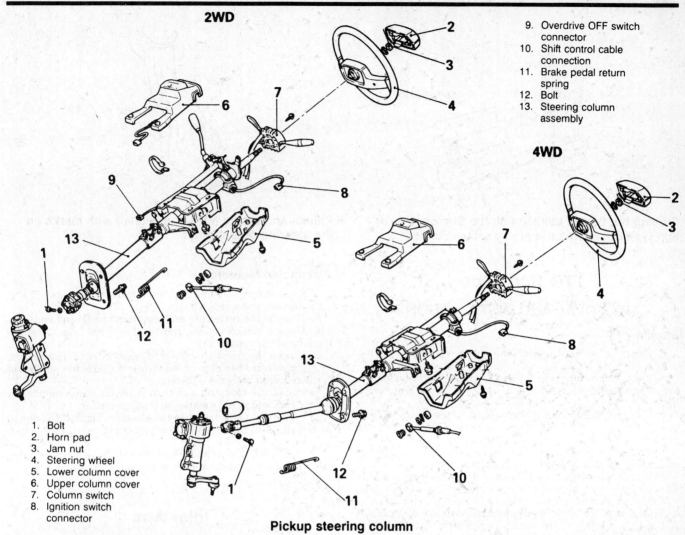

1. Bolt
2. Horn pad
3. Jam nut
4. Steering wheel
5. Lower column cover
6. Upper column cover
7. Column switch
8. Ignition switch connector
9. Overdrive OFF switch connector
10. Shift control cable connection
11. Brake pedal return spring
12. Bolt
13. Steering column assembly

Pickup steering column

stall the bolts holding the column bracket to the dash and finger tighten them.

13. Double check positioning and alignment of the column from end to end. When the column is correctly located, tighten the bracket bolts to 12 Nm or 9 ft. lbs.
14. Install the dustcover at the floor and tighten the bolts. Do not overtighten these bolts. Apply a coating of sealant to the bolt holes from inside the vehicle.
15. Connect the brake pedal return spring.
16. For vehicles with automatic transmission, connect the gear shift selector cable. Install the overdrive switch and connect its harness.
17. Connect the ignition switch harness.
18. Install the column switch and connect the wiring connector.
19. Install the upper and lower steering column cover.
20. Install the steering wheel and horn pad assemblies.

Pitman Arm

REMOVAL AND INSTALLATION

Starion, Montero and Pickup

NOTE: It is possible to remove the Pitman arm with the steering box on the car. The job is much easier if the steering gearbox is removed.

1. Elevate and safely support the vehicle on jackstands.
2. Remove the left front wheel.
3. Remove the cotter pin and castellated nut holding the Pitman arm to the relay rod.
4. Use a screw-type joint separator to pull the relay rod off the Pitman arm joint. Do not hammer on the joint and do not use a fork-type separator.
5. Remove the locknut and washer from the bottom of the steering gearbox.
6. Using a Pitman arm puller (MB 990809-01 or equivalent), remove the arm from the steering box.
7. When installing the arm, note that there are matchmarks on both the arm and the gearbox output shaft. These marks MUST be aligned at reinstallation.
8. Install the locknut and washer. Tighten the nut to 140 Nm (190 ft. lbs).
9. Fit the Pitman arm to the relay rod joint. Use a new boot. Install a new castle nut, tightening it to 40 Nm or 30 ft. lbs. Install a new cotter pin.
10. Install the left wheel. Lower the car to the ground.

8-63

8 SUSPENSION AND STEERING

Freeing the tie rod ball joint on the Starion. Use of this type of separator is highly recommended

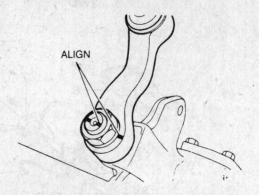

Pitman arm marks MUST be aligned with marks on the steering output shaft

Tie Rod Ends

REMOVAL AND INSTALLATION

Starion

1. Raise the car and support it safely on stands.
2. Loosen the adjusting nut for the tie rod end.
3. Remove the cotter pin for the castellated nut holding the ballstud to the steering knuckle. Loosen but do not remove the nut.
4. Use an ordinary, fork-type ballstud remover as a lever to pull the ballstud out of the steering knuckle. Once the ballstud has been freed, remove the nut.
5. Count the number of turns required to unscrew the tie-rod end as you remove it so you can restore alignment.
6. Install the inner end of the new tie rod and turn it (the number of turns required) to the position of the original end.
7. Install the ballstud of the new tie rod end into the steering knuckle. Install the castellated nut and tighten it to 35 Nm or 26 ft. lbs. Turn it just until the castellations line up with the hole in the stud and install a new cotter pin.
8. Tighten the inner tie rod nut to 51 Nm or 38 ft. lbs.
9. Install the wheels and lower the car to the ground.
10. Have the alignment inspected at a reputable repair facility.

Cordia, Tredia, Mirage
Precis, Sigma and Galant

1. Raise the car and support it safely on stands.
2. Remove the cotter pin and then remove the ballstud retaining nut. Use a vice-like tool (MB 991113 or equivalent) to press the ballstud down and out of the steering knuckle.
3. Using a backup wrench on the flats at the inner end of the tie-rod end, loosen the nut that retains the end to the tie rod coming out of the steering box.
Unscrew the tie-rod end, counting the turns required to remove it.
4. Install the new tie rod end, turning it the exact number of turns required to match the original position.
5. Connect the joint to the knuckle and install the castellated nut, tightening it to 23 Nm or 17 ft. lbs. Galant only, torque it to 15 Nm or 11 ft. lbs.
6. Turn the nut just far enough to line up the castellations with the hole in the stud and install a new cotter pin.
7. Torque the inner nut to 51 Nm or 38 ft. lbs.
8. Install the wheels and lower the car to the ground.
9. Have the alignment inspected at a reputable repair facility.

Pickup and Montero

1. Raise the vehicle and support it safely on stands.
2. Remove the wheel.
3. Remove the cotter pin and nut from the tie rod end.
4. Using a suitable puller, remove the tie rod from the steering knuckle or center link.
5. Loosen the sleeve clamp nut, if equipped, and unscrew the tie rod end from the sleeve or inner tie rod. Count the number of turns required to remove the end.
6. Install the new end and turn it on the same number of turns. This will approximate the original setting.
7. Install the tie rod joint to the knuckle. Tighten the nut to 44 Nm or 33 ft. lbs. Install a new cotter pin.
8. Lubricate the front end.
9. Install the wheels and lower the vehicle to the ground.
10. Inspect and adjust the front alignment as needed.

Idler Arm

REMOVAL AND INSTALLATION

Starion, Montero and Pickup

1. Elevate and safely support the vehicle on stands.
2. Remove the front wheels.
3. Remove the cotter pin and the castellated nut holding the relay rod to the idler arm.
4. Use a screw-type joint separator to split the ball joint at the relay rod.
5. Remove the nut holding the idler arm to its pivot. Remove the arm, noting the placement of the spacers, washers and the plastic bushings. Alternatively, the idler arm pivot may be unbolted from the car and the arm disassembled on the workbench.
6. When reinstalling, make certain the plastic and rubber bushings are in good condition and correctly placed. Install the arm onto the pivot and tighten the locknut to 45 Nm (33 ft. lbs.) for two wheel drive vehicles and 60 Nm or 44 ft. lbs for all 4-wheel drive vehicles.
7. If the entire pivot and bracket was removed, tighten the mounting bolts to 40 Nm or 29 ft. lbs.
8. Connect the idler arm to the relay rod and install the nut on the ball stud. Tighten the nut to 45 Nm or 33 ft. lbs.; install a new cotter pin.
9. Install the wheels and lower the vehicle to the ground.

SUSPENSION AND STEERING 8

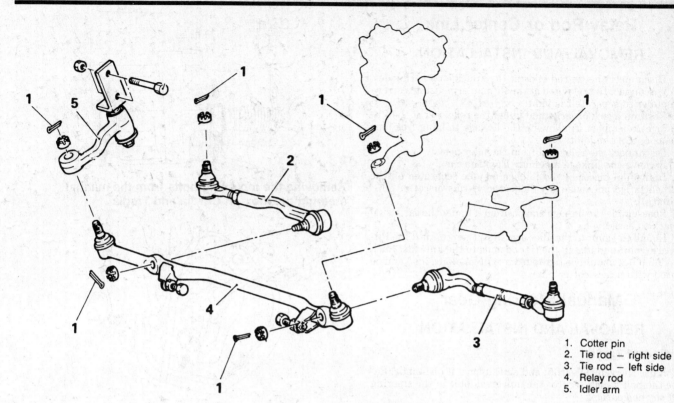

1. Cotter pin
2. Tie rod — right side
3. Tie rod — left side
4. Relay rod
5. Idler arm

Starion steering linkage

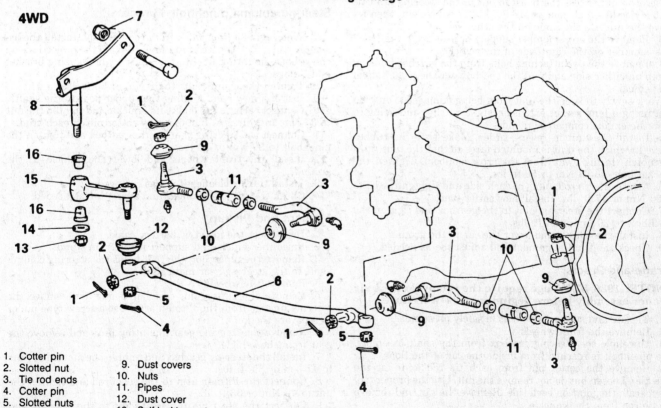

1. Cotter pin
2. Slotted nut
3. Tie rod ends
4. Cotter pin
5. Slotted nuts
6. Relay rod
7. Self-locking nuts
8. Idler arm support
9. Dust covers
10. Nuts
11. Pipes
12. Dust cover
13. Self-locking nut
14. Washer
15. Idler arm
16. Bushing

Typical 4-wheel drive steering linkage

8–65

8 SUSPENSION AND STEERING

Relay Rod or Center Link

REMOVAL AND INSTALLATION

1. Disconnect the tie-rod ends at the knuckle on each side.
2. Disconnect the relay rod at the Pitman arm, using a screw-type puller to separate the joint.
3. Position a jack or jackstand under the relay rod at the Pitman arm to support it; allowing the linkage to hang free may damage one of the joints.
4. Disconnect the relay rod from the idler arm.
5. Remove the linkage assembly from the car.
6. Individual pieces of the linkage may be separated on the workbench. Do not reuse any dustcover, O-ring or cotter pin.

To install:

7. Reassemble the linkage and position it under the car, using the jack to assist you.
8. Install the joint to the idler arm first, then connect the Pitman arm joint and install the tie rod ends to the knuckle. Each castellated nut should be tightened to 45 Nm (33 ft. lbs.) and be secured with a new cotter pin.

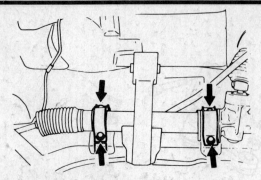

Removing the mounting bolts from the manual steering gearbox on Cordia and Tredia

Manual Steering Gear

REMOVAL AND INSTALLATION

Cordia and Tredia

1. Raise the vehicle's front end and support it on stands. Remove the bolt which secures the universal joint in the steering shaft to the gearbox.
2. Remove the cotter pin from the tie rod end ballstud and loosen the nut. Press the ballstud out of the steering knuckle with a vice-like tool such as MB 991113 or equivalent; then remove the nut. Do the same on the other side.
3. Remove the crossmember support bracket from the No. 2 crossmember on the right side of the vehicle.
4. Remove the two attaching bolts from the gearbox housing clamp on either side and pull the gearbox out the right side of the vehicle.

Work slowly to keep the unit from being damaged. You'll do best using a long socket extension and a ratchet and going in from the engine compartment side.

5. Install the unit in reverse order. Make sure the rubber sleeve in which the unit is mounted faces so that the tabs fit in the notch (facing away from the crossmember). Tighten the bracket bolts to 68 Nm or 50 ft. lbs.
6. Attach the tie rod joints at each side and tighten the bolts to 23 Nm or 17 ft. lbs. Install new cotter pins.
7. Connect the steering joint to the gearbox and tighten the pinchbolt to 20 nm or 14 ft. lbs.
8. Install the wheels and lower the car to the ground.
9. Check the front alignment and adjust toe as needed.

Mirage and Precis

NOTE: 1986–87 Mirage require the removal of the air cleaner assembly before beginning this procedure.

1. Elevate and support the vehicle safely on stands.
2. Remove the front wheels.
3. Uncouple the steering gearbox from the shaft assembly. The pinchbolt is reached from inside the car at the floor.
4. Remove the cotter pin from each tie rod joint (at the knuckle). Loosen but do not remove the nut. Use the proper tool to separate the joint on each side. Remove the nut and remove the tie rod from the knuckle.
5. Cut the retaining band off the rubber boot that covers the joint connection the steering box with the steering shaft.
6. Remove the 4 attaching bolts holding the 2 steering clamps. Pull the gearbox out towards the left side of the vehicle.

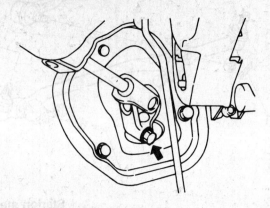

Steering column pinchbolt, Mirage

7. When reinstalling, work from the left side. Install the assembly and install the 2 clamps. Make certain the projections on the rubber mounting fit into the holes in the housing bracket and clamps.
8. Tighten the clamp bolts to 68 Nm or 50 ft. lbs.
9. Install a new band holding the rubber boot to the shaft.
10. Connect each tie rod end to its knuckle and tighten the nut to 34 Nm or 25 ft. lbs. Install a new cotter pin in each stud.
11. Connect the steering shaft to the gearbox and tighten pinchbolt to 20 Nm or 14 lbs.
12. Install the front wheels and lower the vehicle to the ground.
13. Install the air cleaner if it was removed.
14. Check the front alignment and adjust toe as needed.

Montero and Pickup

1. Disconnect the negative battery cable.
2. Raise the vehicle and support it safely on stands.
3. Remove the pinch bolt the that holds the steering column shaft to the steering gear main shaft.
4. Remove the wheels.
5. Remove the cotter pin, castellated nut and remove the steering linkage from the Pitman arm. Use a screw type puller to separate the joint.
6. Remove the steering gear mounting nuts and remove the gear from the vehicle.
7. Install the steering gearbox and tighten the mounting nuts to 47 Nm or 35 ft. lbs.
8. Connect the Pitman arm to the relay rod and tighten the nut to 45 Nm or 33 ft. lbs.
9. Connect the steering universal joint to the gearbox and tighten the pinchbolt to 25 Nm or 19 ft. lbs.
10. Install the wheels and lower the car to the ground.
11. Connect the negative battery cable. Have the front alignment inspected and adjusted as necessary.

SUSPENSION AND STEERING 8

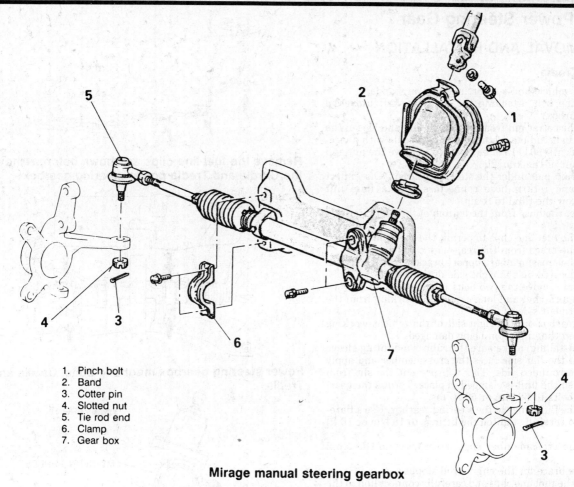

1. Pinch bolt
2. Band
3. Cotter pin
4. Slotted nut
5. Tie rod end
6. Clamp
7. Gear box

Mirage manual steering gearbox

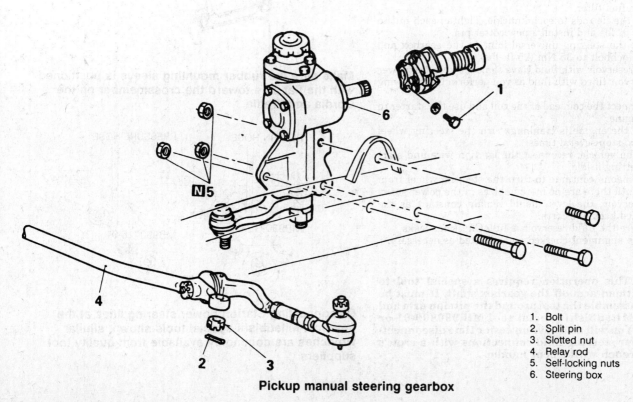

1. Bolt
2. Split pin
3. Slotted nut
4. Relay rod
5. Self-locking nuts
6. Steering box

Pickup manual steering gearbox

8-67

8 SUSPENSION AND STEERING

Power Steering Gear

REMOVAL AND INSTALLATION

Cordia and Tredia

1. Raise the vehicle and support it safely on stands.
2. Remove the bolt attaching the steering shaft universal joint to the gearbox.
3. Remove the cotter pin from the tie rod end and loosen the nut. Press the ballstud out of the steering knuckle with a vice-like tool such as MB 991113 or equivalent. Remove the nut and pull the stud out of the knuckle.
4. Place a drain pan under the steering gearbox. Disconnect the pressure and return hose connectors with a flare nut wrench and allow the fluid to drain.
5. Disconnect the hose from the bottom of the fuel filter and plug it.
6. Remove the fuel line clips to permit the fuel line to move.
7. Remove the brace from the rear engine roll stopper.
8. Remove the crossmember support bracket from the No. 2 crossmember, located on the right side of the vehicle.
9. Unbolt and remove the two bolts in each gearbox mounting clamp with a ratchet and long extension, working from the engine compartment side.
10. Pull the gearbox out the right side of the vehicle, working carefully to keep the unit from being damaged.
11. When reinstalling, make sure the rubber mounting sleeve is positioned so the flat side faces the crossmember and apply adhesive to the rounded side. This will prevent the slit from opening up when the unit is clamped in place. Torque the gearbox mounting bolts to 68 Nm or 50 ft. lbs.
12. Connect the fluid lines to the steering gearbox. Use a flare-nut wrench and carefully tighten the fittings to 18 Nm or 13 ft. lbs.
13. Install the crossmember support bracket to the No.2 crossmember.
14. Install the brace for the engine roll stopper.
15. Reinstall the fuel line clips and carefully connect line to the bottom of the fuel filter.
16. connect the tie rods to each knuckle. Tighten each nut to 23 Nm or 17 ft. lbs and install a new cotter pin.
17. Connect the steering universal joint to the gearbox and tighten the pinchbolt to 35 Nm (25 ft. lbs).
18. Fill the reservoir with fluid. Have someone keep the power steering reservoir filled with fluid as you perform the following procedure:
    a. Disconnect the coil lead at the coil and use the starter to spin the engine.
    b. While the engine is cranking, turn the steering wheel from stop to stop several times.
19. Lower the vehicle, reconnect the ignition wire and start the engine, letting it idle.
20. Have someone continue to turn the steering wheel from lock to lock until there are no more bubbles in the power steering fluid reservoir; the level should remain constant as the wheel is turned back and forth.
21. Make sure the fluid reservoir is full. Check for leaks.
22. Have the alignment checked and adjusted as necessary.

Starion

WARNING: This operation requires a special tool to press the pitman arm off the gearbox shaft. It must be possible to assemble the tool around the pitman arm and shaft. Use Mitsubishi special tool MB 990809-01 or equivalent. You will also have an easier time disconnecting the power steering fluid connections with a crow's foot type wrench with a bent handle.

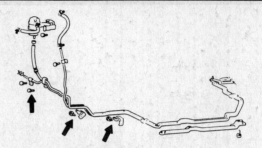

Remove the fuel line clips as shown before removing the Cordia and Tredia power steering gearbox

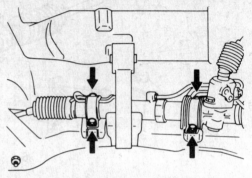

Power steering gearbox mounting bolts, Cordia and Tredia

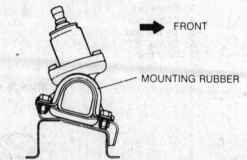

Make sure the rubber mounting sleeve is positioned with the flat side toward the crossmember on the Cordia and Tredia

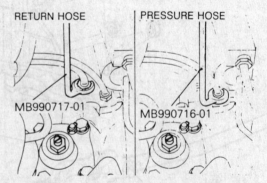

Disconnecting Starion power steering lines at the gearbox. Mitsubishi special tools shown; similar wrenches are commonly available from quality tool suppliers

SUSPENSION AND STEERING 8

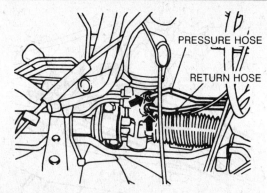

Location of hydraulic hoses on Mirage power steering gearbox

1. Remove the pinchbolt which connects the steering box input shaft to the steering shaft.
2. Place a drain pan underneath and disconnect the pressure and return hoses at the gearbox.
3. Press the pitman arm off the gearbox with the special tool.
4. Remove the four attaching nuts and remove the steering box.
5. Install the steering box and tighten the mounting nuts/bolts to 61 Nm or 45 ft. lbs.
6. Install the Pitman arm and tighten the retaining nut to 137 Nm or 101 ft. lbs.
7. Connect the steering column shaft to the to the steering gearbox. Tighten the pinchbolt to 25 Nm or 18 ft. lbs.
8. Carefully reinstall the fluid lines. Tighten the pressure hose connector to 35 Nm (26 ft. lbs.) and the return line connector to 45 Nm or 33 ft. lbs.
9. Have an assistant keep the power steering reservoir filled with Dexron®II automatic transmission fluid as you perform this procedure:
   a. Connect one end of a clear hose to the bleed plug on top of the steering box and immerse the other end in clean Dexron®II.
   b. Disconnect the coil high tension wire at the coil and use the starter to spin the engine while you turn the steering wheel from stop to stop several times.
   c. Lower the vehicle, reconnect the ignition wire and start the engine, letting it idle. Loosen the bleed plug and have someone continue to turn the steering wheel back and forth from lock to lock.
   d. Watch the hose for bubbles. When there are no more bubbles, tighten the bleed plug and stop the procedure. Make sure the fluid reservoir is full.
10. Check the line connections for leaks or seepage. If the lines must be loosened, the system must be re-bled.
11. Have the alignment checked by a reputable repair facility.

Correct placement of Galant steering rack mounts

Mirage and Precis

NOTE: The entire steering rack will be removed. The steering gearbox cannot be removed by itself

1. For 1987–88 vehicles, remove the air cleaner assembly.
2. Raise the vehicle and support it safely on jackstands. Remove the front wheels.
3. Remove the pinchbolt holding the joint in the steering shaft to the gearbox. It's just inside the car where the steering linkage passes through the floorboard.
4. Remove the cotter pin from the tie rod end ballstud and loosen the nut. Press the ballstud out of the steering knuckle with a vice-like tool such as MB 991113 or equivalent; then remove the nut. Do the same on the other side.
5. Cut the band off the steering joint rubber boot. Place a drain pan under the steering box. Disconnect the pressure and return hoses at the gearbox and allow the fluid to drain into the pan.
6. Remove the stabilizer bar.
7. Remove the rear roll stopper-to-center member bolt and then move the rear roll stopper forward.
8. Remove the two mounting bolts in each clip (on either side of the gearbox) and remove the unit carefully out the left side of the vehicle. Avoid damaging the rubber boots.
9. When reinstalling, mount the rack in place. Make sure the rubber mounting sleeve is positioned so the tabs on the inside and outside fit into the openings in the mounting clamps; the larger tab must go on the inside. Apply adhesive to the outer side so the slit will not open up when the unit is clamped in place.
10. Torque the gearbox mounting bolts to 68 Nm or 50 ft. lbs.
11. Connect each ball joint to the steering knuckle. Tighten each nut to 15 Nm (11 ft. lbs.) and install a new cotter pin.
12. Connect the steering shaft to the steering joint. Tighten the pinch bolt to 35 Nm (25 ft. lbs.).
12. Connect the power steering lines, tightening the connectors to 15 Nm or 11 ft. lbs.
13. Connect the stabilizer bar. Position and install the roll stopper bolt.
14. Install the front wheels and lower the car to the ground.
15. Fill the power steering reservoir with fluid. Start the engine and turn the steering wheel back and forth. Have an assistant handy to check the level in the reservoir and top it off as needed. When the level remains stable, shut the engine off.
16. Inspect the fluid line joints for any sign of leakage.
17. Have the alignment checked at a reputable repair facility.

Galant and Sigma

NOTE: The entire steering rack will be removed. The steering gearbox cannot be removed by itself

1. Raise the vehicle and support it safely on stands. Remove the front wheels.
2. Disconnect the return hose and drain the fluid into a pan or container. Disconnect the coil high tension wire at the coil and operate the starter intermittently. While the engine is cranking, turn the steering wheel lock to lock several times to purge all the fluid from the system.
3. Disconnect the pressure and return lines at the steering gearbox.
4. Disconnect the solenoid valve connector harness at or near the steering rack.
5. Remove the cotter pin from the tie rod end ballstud and loosen the nut. Press the ballstud out of the steering knuckle with a vice-like tool such as MB 991113 or equivalent and then remove the nut. Do the same on the other side.
6. Remove the bolt attaching the steering shaft joint to the

8-69

8 SUSPENSION AND STEERING

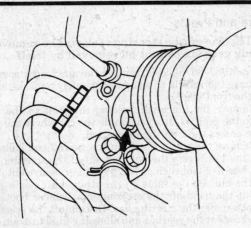

Disconnecting the steering shaft at the Galant gearbox

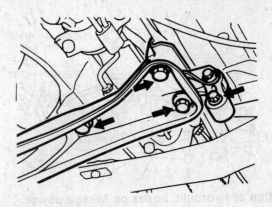

Remove these three bolts so the crossmember may be lowered slightly to remove the Galant steering gearbox

gearbox.
7. Remove the rear mounting bolts from the center crossmember.
8. Remove the front roll stopper through-bolt.
9. Disconnect the oxygen sensor connector.
10. Disconnect the exhaust system from the manifold on each side. Disconnect the front exhaust hanger. Support the system from a piece of wire; do not allow the exhaust to hang by its own weight.
11. Remove the crossmember stay (brace) on each side.
12. Disconnect the stabilizer bar from each side. Keep the link components in order.
13. Remove the two brackets holding the the rack to the crossmember.

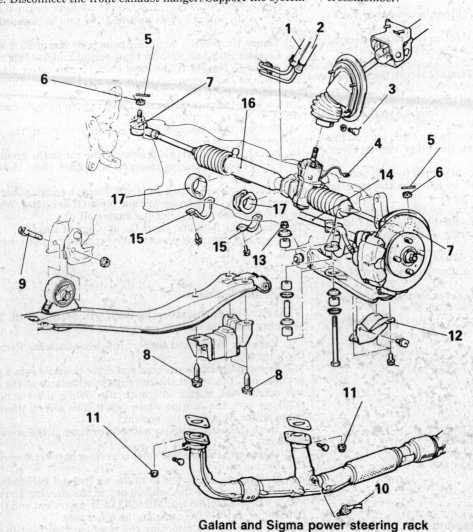

1. Return line
2. Pressure line
3. Pinch bolt
4. Solenoid valve connector
5. Cotter pin
6. Slotted nut
7. Tie rod end
8. Crossmember rear bolt
9. Front roll stopper bolt
10. Oxygen sensor connector
11. Front exhaust pipe nut
12. Stay
13. Self-locking nut
14. Stabilizer bar
15. Mounting bracket
16. Gear box
17. Mounting rubber

Galant and Sigma power steering rack

8-70

SUSPENSION AND STEERING 8

14. Move the rack all the way to the right and remove the gearbox from the crossmember. Tilt the gearbox downward and remove it toward the left. Remove the gearbox slowly and carefully to avoid damaging the rubber boots.
15. Reinstalling the rack requires the same manipulation of the rack. Once in place, install the brackets with their rubber grommets. Align the projection of the rubber mount with the indentation in the crossmember. Tighten the clamp bolts to 70 Nm or 95 ft. lbs.
16. Install the stabilizer bar and use new self locking nuts to retain it.
17. Install the crossmember stays and tighten the bolts to 75 Nm or 102 ft. lbs.
18. Using new gaskets and nuts, connect the exhaust system to the manifolds. Tighten the nuts to 35 Nm or 26 ft. lbs. Install the hanger.
19. Install the oxygen sensor and connect its harness.
20. Install the front roll stopper bolt, tightening it to 58 Nm (43 ft. lbs.)
21. Position the rear crossmember and install the rear bolts. Tighten them to 90 Nm or 66 ft. lbs.
22. Connect the tie rod ends to the steering knuckles. Tighten each castellated nut to 34 Nm (25 ft. lbs) and install a new cotter pin in each.
23. Connect the solenoid valve harness.
24. Reconnect the steering joint to the steering box. Tighten the pinchbolt to 20 Nm or 14 ft. lbs.
25. Carefully reinstall the pressure and return hoses to the steering gearbox. The connectors should be tightened only to 15 Nm (21 ft. lbs.)
26. Make certain the return line is firmly connected at the reservoir. It may have been left loose after draining the system.
27. Pour power steering fluid into the reservoir to the lower mark.
28. Install the wheels and lower the car to the ground.
29. With the coil high tension line disconnected at the coil, operate the starter intermittently (15 or 20 seconds each time) and turn the steering wheel all the way from left to right and back several times. Do this 5 or 6 times while an assistant keeps watch on the fluid level in the reservoir.

WARNING: Do not be tempted to bleed the system with the engine idling. The air in the system will be broken up and absorbed into the fluid, resulting in a noisy, foamy mess. During the bleeding procedure, replenish the fluid so that the level never falls below the lower level on the reservoir.

30. Once the preliminary bleeding is done, the fluid level should stay fairly constant as the wheel is turned while cranking the motor. When this is true, the coil wire may be reconnected and the engine started. Only run the engine at idle; high rpm will not help.
31. With the engine running, turn the wheel from lock to lock several times and observe the fluid level in the reservoir. The level should not change more than about 5mm throughout the full travel of the wheel. The fluid should not look cloudy or milky.

NOTES: If the system still contains air, the pump will be noisy during operation. This noisy operation can damage the pump.
If the fluid level changes markedly during steering wheel movement, the system still contains air.
If the fluid level rises suddenly after the engine hsa been shut off, the system still contains air.

32. Inspect all the fluid line connections for any sign of leakage.
33. Have the front alignment checked and adjusted by a reputable repair facility.

Pickup and Montero

1. Disconnect the negative battery cable.
2. Raise the vehicle and support it safely on stands. Remove the front wheels.
3. Fold back the dust boot covering the steering shaft joint; remove the pinch bolt the that holds the steering column shaft to the steering gear main shaft.
4. Remove the cotter pin, castellated nut and disconnect the steering linkage from the Pitman arm. Use a puller to separate the joint.
5. Disconnect and plug the fluid lines from the steering gear.
6. Remove the steering gear mounting nuts and remove the gear from the vehicle.
7. To install, fit the steering gearbox into position and tighten the mounting bolts to 38 Nm (28 ft. lbs.) for 2 wheel drive vehicles and 61 Nm (45 ft. lbs.) for 4 wheel drive vehicles.
8. Connect the pressure and return lines to the steering gearbox. Tighten the pressure hose the 35 Nm (26 ft. lbs) and the return hose to 44 Nm or 33 ft. lbs.
9. Connect the Pitman arm to the relay rod (steering linkage) and tighten the nut to 44 Nm or 33 ft. lbs. Install a new cotter pin.
10. Connect the steering column shaft to the steering gearbox and tighten the pinchbolt to 35 Nm or 25 ft. lbs.
11. Install the front wheels and lower the vehicle to the ground.
12. Pour power steering fluid into the reservoir to the lower mark.
13. With the coil high tension line disconnected at the coil, operate the starter intermittently (15 or 20 seconds each time) and turn the steering wheel all the way from left to right and back several times. Do this 5 or 6 times while an assistant keeps watch on the fluid level in the reservoir.

WARNING: Do not be tempted to bleed the system with the engine idling. The air in the system will be broken up and absorbed into the fluid, resulting in a noisy, foamy mess. During the bleeding procedure, replenish the fluid so that the level never falls below the lower level on the reservoir.

14. Once the preliminary bleeding is done, the fluid level should stay fairly constant as the wheel is turned while cranking the motor. When this is true, the coil wire may be reconnected and the engine started. Only run the engine at idle; high rpm will not help.
15. With the engine running, turn the wheel from lock to lock several times and observe the fluid level in the reservoir. The level should not change more than about 5mm throughout the full travel of the wheel. The fluid should not look cloudy or milky.

NOTES: If the system still contains air, the pump will be noisy during operation. This noisy operation can damage the pump.
If the fluid level changes markedly during steering wheel movement, the system still contains air.
If the fluid level rises suddenly after the engine has been shut off, the system still contains air.

16. Inspect all the fluid line connections for any sign of leakage.
17. Have the front alignment checked and adjusted by a reputable repair facility.

Power Steering Pump

REMOVAL AND INSTALLATION

1. Disconnect the negative battery cable.

8 SUSPENSION AND STEERING

2. Remove the reservoir cap and disconnect the return (lower) hose from the fluid reservoir. Drain the fluid into a clean pan.

3. Depending on access for your particular vehicle, raise the car and support it safely on stands if necessary.

4. If the pulley is to be removed from the pump, loosen the bolts now. Leave them finger tight in place.

5. Loosen the pump mounting bolts, move the pump towards the engine and remove the belt. Turn the pump by hand to eliminate the fluid from the system.

6. Disconnect the pressure hose at the top of the pump and the suction hose at the side of the pump. Allow the fluid to drain into the container.

7. Remove the pump attaching bolts and lift the pump from the brackets.

8. When reinstalling, make certain the bracket bolts are tight, then install the pump to the brackets.

9. If the pulley was removed, reinstall it and tighten its nuts securely.

10. Install the drive belt and adjust its tension correctly.

11. Connect the pressure and return hoses. On the Galant, the suction hose must be installed 20–25mm onto the pump fitting; on the Mirage, this distance must be 25–30mm. Make sure the hoses are not twisted and do not contact any other parts so as to avoid chafing.

12. Pour power steering fluid into the reservoir to the lower mark.

13. With the coil high tension line disconnected at the coil, operate the starter intermittently (15 or 20 seconds each time) and turn the steering wheel all the way from left to right and back several times. Do this 5 or 6 times while an assistant keeps watch on the fluid level in the reservoir.

WARNING: Do not be tempted to bleed the system with the engine idling. The air in the system will be broken up and absorbed into the fluid, resulting in a noisy, foamy mess. During the bleeding procedure, replenish the fluid so that the level never falls below the lower level on the reservoir.

14. Once the preliminary bleeding is done, the fluid level should stay fairly constant as the wheel is turned while cranking the motor. When this is true, the coil wire may be reconnected and the engine started. Only run the engine at idle; high rpm will not help.

15. With the engine running, turn the wheel from lock to lock several times and observe the fluid level in the reservoir. The level should not change more than about 5mm throughout the full travel of the wheel. The fluid should not look cloudy or milky.

NOTES: If the system still contains air, the pump will be noisy during operation. This noisy operation can damage the pump.

If the fluid level changes markedly during steering wheel movement, the system still contains air.

If the fluid level rises suddenly after the engine has been shut off, the system still contains air.

16. Inspect all the fluid line connections for any sign of leakage.

Brakes

QUICK REFERENCE INDEX

Anti-Lock Brake System	9-13
Brake Specifications	9-45
Disc Brakes	
Front	9-19
Rear	9-32
Drum Brakes	9-26
Parking Brake	9-39
Troubleshooting	9-3

GENERAL INDEX

Anti-Lock Brake System
  Description 9-13
  Hydraulic unit 9-13
Brakes
  Adjustments
    Brake pedal 9-7
    Drum brakes 9-3
  Bleeding 9-17
  Brake light switch 9-8
  Disc brakes (Front)
    Caliper 9-20
    Operating principals 9-2
    Pads 9-19
    Rotor (Disc) 9-23
  Disc brakes (Rear)
    Caliper 9-34
    Description 9-2
    Pads 9-32
    Rotor (Disc) 9-38
  Drum brakes
    Adjustment 9-3
    Drum 9-26

    Operating principals 9-2
    Shoes 9-28
    Wheel cylinder 9-30
  Hoses and lines 9-16
  Load sensing proportion valve .. 9-18
  Master cylinder 9-8
  Parking brake
    Adjustment 9-39
    Removal and installation 9-41
  Power booster
    Operating principals 9-2
    Removal and installation 9-10
    Vacuum check valve 9-13
  Proportioning valve 9-12
  Specifications 9-45
  Troubleshooting 9-2
Calipers (Front)
  Overhaul 9-21
  Removal and installation 9-20
Calipers (Rear)
  Overhaul 9-34
  Removal and installation 9-34

Disc brakes
  Front 9-19
  Rear 9-32
Drum brakes 9-26
Front brakes 9-19
Hoses 9-16
Master cylinder 9-8
Pads
  Front 9-19
  Rear 9-32
Parking brake 9-39
Power brake booster 9-10
Proportioning valve 9-12
Rear Brake Lock-Up Control System
  Description 9-15
Rear brakes 9-26
Rotor (Brake disc)
  Front 9-23
  Rear 9-38
Shoes 9-28
Specification 9-45
Wheel cylinders 9-30

9 BRAKES

BRAKE SYSTEM

> **CAUTION**
> *Brake pads and shoes contain asbestos, which has been determined to be a cancer causing agent. Never clean the brake surfaces with compressed air! Avoid inhaling any dust from brake surfaces! When cleaning brakes, use commercially available brake cleaning fluids.*

Operation and System Description

HYDRAULIC SYSTEM

Hydraulic systems are used to actuate the brakes of all modern automobiles. A hydraulic system rather than a mechanical system is used for two reasons. First, fluid under pressure can be carried to all parts of an automobile by small hoses — some of which are flexible — without taking up a significant amount of room or posing routing problems. Second, a great mechanical advantage can be given to the brake pedal, and the foot pressure required to actuate the brakes can be reduced by making the surface area of the master cylinder pistons smaller than that of any of the pistons in the wheel cylinders or calipers.

The master cylinder consists of a fluid reservoir and a single or double cylinder and piston assembly. Double (or dual) master cylinders are designed to separate the front and rear braking systems hydraulically in case of a leak. The master cylinder coverts mechanical motion from the pedal into hydraulic pressure within the lines. This pressure is translated back into mechanical motion at the wheels by either the wheel cylinder (drum brakes) or the caliper (disc brakes). Since these components receive the pressure from the master cylinder, they are generically classed as slave cylinders in the system.

Steel lines carry the brake fluid to a point on the vehicle's frame near each of the vehicle's wheels. The fluid is then carried to the slave cylinders by flexible tubes in order to allow for suspension and steering movements.

Each wheel cylinder contains two pistons, one at either end, which push outward in opposite directions and force the brake shoe into contact with the drum. In disc brake systems, the slave cylinders are part of the calipers. One, two or four cylinders are used to force the brake pads against the disc, but all cylinders contain one piston only. All slave cylinder pistons employ some type of seal, usually made of rubber, to minimize the leakage of fluid around the piston. A rubber dust boot seals the outer end of the cylinder against dust and dirt. The boot fits around the outer end of either the piston or the brake actuating rod.

When at rest the entire hydraulic system, from the piston(s) in the master cylinder to those in the wheel cylinders or calipers, is full of brake fluid. Upon application of the brake pedal, fluid trapped in front of the master cylinder piston(s) is forced through the lines to the slave cylinders. Here it forces the pistons outward, in the case of drum brakes, and inward toward the disc in the case of disc brakes. The motion of the pistons is opposed by return springs mounted outside the cylinders in drum brakes, and by internal springs or spring seals, in disc brakes.

Upon release of the brake pedal, a spring located inside the master cylinder immediately returns the master cylinder piston(s) to the normal position. The pistons contain check valves and the master cylinder has compensating ports drilled within it. These are uncovered as the pistons reach their normal position. The piston check valves allow fluid to flow toward the wheel cylinders or calipers as the pistons withdraw. As the return springs force the brake pads or shoes into the released position, the excess fluid in the lines is allowed to re-enter the reservoir through the compensating ports.

Dual circuit master cylinders employ two pistons, located one behind the other, in the same cylinder. The primary piston is actuated directly by mechanical linkage from the brake pedal. The secondary piston is actuated by fluid trapped between the two pistons. If a leak develops in front of the secondary pistons, it moves forward until it bottoms against the front of the master cylinder, and the fluid trapped between the pistons will operate the rear brakes. If the rear brakes develop a leak, the primary piston will move forward until direct contact with the secondary piston takes place, and it will force the secondary piston to actuate the front brakes. In either case, the brake pedal moves farther when the brakes are applied and less braking power is available.

All dual-circuit systems incorporate a switch which senses either line pressure or fluid level. This system will warn the driver when only half of the brake system is operational.

In some disc brake systems, this valve body also contains a metering valve and, in some cases, a proportioning valve. The metering valve keeps pressure from traveling to the disc brakes on the front wheels until the brake shoes on the rear wheels have contacted the drum, ensuring that the front brakes will never be used alone. The proportioning valve controls the pressure to the rear brakes avoiding rear wheel lock-up during very hard braking.

DISC BRAKES

> **CAUTION**
> *Brake pads contain asbestos, which has been determined to be a cancer causing agent, never clean the brake surfaces with compressed air! Avoid inhaling any dust from any brake surface! When cleaning brake surfaces, use a commercially available brake cleaning fluid.*

Instead of the traditional expanding brakes that press outward against a circular drum, disc brake systems employ a cast iron disc with brake pads positioned on either side of it. An easily seen analogy is the hand brake arrangement on a bicycle. The pads squeeze onto the rim of the bike wheel, slowing its motion. Automobile disc brakes use the identical principal but apply the braking effort to a separate disc instead of the wheel.

The disc or rotor is a one-piece casting mounted just inside the wheel. Some discs are one solid piece while others have cooling fins between the two braking surfaces. These vented rotors enable air to circulate between the braking surfaces cooling them quicker and making them less sensitive to heat buildup and fade. Disc brakes are only slightly affected by dirt and water since contaminants are thrown off by the centrifugal action of the rotor or scraped off by the pads. Also, the equal clamping action of the two brake pads tend to ensure uniform, straight-line stops, although unequal application of the pads between the left and right wheels can cause a vicious pull under braking. All disc brakes are inherently self-adjusting.

There are three general types of disc brakes:

The fixed caliper design uses two pistons mounted on either side of the rotor (in each side of the caliper). The caliper is mounted rigidly and does not move. This is a very efficient brake system but the size of the caliper and its mounts adds weight and bulk to the car.

The sliding and floating designs are quite similar. In fact, these two types are often lumped together. In both designs, one pad is moved into contact with the rotor by hydraulic force. The caliper, which is not held in a fixed position, moves slightly on its mount, bringing the other pad into contact with the rotor. There are various methods of attaching floating calipers. Some pivot at the bottom or top, and some slide on mounting bolts. Many uneven brake wear problems can be caused by dirty or seized slides and pivots.

BRAKES 9

DRUM BRAKES

CAUTION

Brake shoes contain asbestos, which has been determined to be a cancer causing agent. never clean the brake surfaces with Compressed air! Avoid inhaling any dust from any brake surface! When cleaning brake surfaces, use a commercially available brake cleaning fluid.

Drum brakes employ two brake shoes mounted on a stationary backing plate. These shoes are positioned inside a circular cast iron drum which rotates with the wheel. The shoes are held in place by springs; this allows them to slide toward the drum (when they are applied) while keeping the linings and drums in alignment.

The shoes are actuated by a wheel cylinder which is mounted at the top of the backing plate. When the brakes are applied, hydraulic pressure forces the wheel cylinder's two actuating links outward. Since these links bear directly against the top of the brake shoes, the tops of the shoes are then forced outward against the inside of the drum. This action forces the bottoms of the two shoes to contact the brake drum by rotating the entire assembly slightly (known as servo action). When the pressure within the wheel cylinder is relaxed, return springs pull the shoes away from the drum.

Most modern drum brakes are designed to self-adjust during application when the vehicle is moving in reverse. This motion causes both shoes to rotate very slightly with the drum, rocking an adjusting lever and thereby causing rotation of the adjusting screw via a star wheel. This on-board adjustment system reduces the need for maintenance adjustments but most drivers don't back up enough to keep the brakes properly set.

Adjustment

Disc brakes are inherently self adjusting, and set themselves automatically after linings are replaced when the brake pedal is applied.

DRUM BRAKES

Drum brakes do not require adjustment unless self adjusters give problems or linings are replaced. Adjustment for those models with rear drum brakes is given below.

Cordia, Tredia, and Galant

Elevate and support the car on stands. With the wheel and brake drum removed, set the adjuster lever ratchet all the way

Troubleshooting the Brake System

Problem	Cause	Solution
Low brake pedal (excessive pedal travel required for braking action.)	• Excessive clearance between rear linings and drums caused by inoperative automatic adjusters	• Make 10 to 15 alternate forward and reverse brake stops to adjust brakes. If brake pedal does not come up, repair or replace adjuster parts as necessary.
	• Worn rear brakelining	• Inspect and replace lining if worn beyond minimum thickness specification
	• Bent, distorted brakeshoes, front or rear	• Replace brakeshoes in axle sets
	• Air in hydraulic system	• Remove air from system. Refer to Brake Bleeding.
Low brake pedal (pedal may go to floor with steady pressure applied.)	• Fluid leak in hydraulic system	• Fill master cylinder to fill line; have helper apply brakes and check calipers, wheel cylinders, differential valve tubes, hoses and fittings for leaks. Repair or replace as necessary.
	• Air in hydraulic system	• Remove air from system. Refer to Brake Bleeding.
	• Incorrect or non-recommended brake fluid (fluid evaporates at below normal temp).	• Flush hydraulic system with clean brake fluid. Refill with correct-type fluid.
	• Master cylinder piston seals worn, or master cylinder bore is scored, worn or corroded	• Repair or replace master cylinder
Low brake pedal (pedal goes to floor on first application—o.k. on subsequent applications.)	• Disc brake pads sticking on abutment surfaces of anchor plate. Caused by a build-up of dirt, rust, or corrosion on abutment surfaces	• Clean abutment surfaces

9 BRAKES

Troubleshooting the Brake System (cont.)

Problem	Cause	Solution
Fading brake pedal (pedal height decreases with steady pressure applied.)	• Fluid leak in hydraulic system	• Fill master cylinder reservoirs to fill mark, have helper apply brakes, check calipers, wheel cylinders, differential valve, tubes, hoses, and fittings for fluid leaks. Repair or replace parts as necessary.
	• Master cylinder piston seals worn, or master cylinder bore is scored, worn or corroded	• Repair or replace master cylinder
Decreasing brake pedal travel (pedal travel required for braking action decreases and may be accompanied by a hard pedal.)	• Caliper or wheel cylinder pistons sticking or seized	• Repair or replace the calipers, or wheel cylinders
	• Master cylinder compensator ports blocked (preventing fluid return to reservoirs) or pistons sticking or seized in master cylinder bore	• Repair or replace the master cylinder
	• Power brake unit binding internally	• Test unit according to the following procedure: (a) Shift transmission into neutral and start engine (b) Increase engine speed to 1500 rpm, close throttle and fully depress brake pedal (c) Slow release brake pedal and stop engine (d) Have helper remove vacuum check valve and hose from power unit. Observe for backward movement of brake pedal. (e) If the pedal moves backward, the power unit has an internal bind—replace power unit
Grabbing brakes (severe reaction to brake pedal pressure.)	• Brakelining(s) contaminated by grease or brake fluid	• Determine and correct cause of contamination and replace brakeshoes in axle sets
	• Parking brake cables incorrectly adjusted or seized	• Adjust cables. Replace seized cables.
	• Incorrect brakelining or lining loose on brakeshoes	• Replace brakeshoes in axle sets
	• Caliper anchor plate bolts loose	• Tighten bolts
	• Rear brakeshoes binding on support plate ledges	• Clean and lubricate ledges. Replace support plate(s) if ledges are deeply grooved. Do not attempt to smooth ledges by grinding.
	• Incorrect or missing power brake reaction disc	• Install correct disc
	• Rear brake support plates loose	• Tighten mounting bolts

9–4

BRAKES 9

Troubleshooting the Brake System (cont.)

Problem	Cause	Solution
Spongy brake pedal (pedal has abnormally soft, springy, spongy feel when depressed.)	• Air in hydraulic system • Brakeshoes bent or distorted • Brakelining not yet seated with drums and rotors • Rear drum brakes not properly adjusted	• Remove air from system. Refer to Brake Bleeding. • Replace brakeshoes • Burnish brakes • Adjust brakes
Hard brake pedal (excessive pedal pressure required to stop vehicle. May be accompanied by brake fade.)	• Loose or leaking power brake unit vacuum hose • Incorrect or poor quality brakelining • Bent, broken, distorted brakeshoes • Calipers binding or dragging on mounting pins. Rear brakeshoes dragging on support plate. • Caliper, wheel cylinder, or master cylinder pistons sticking or seized • Power brake unit vacuum check valve malfunction • Power brake unit has internal bind	• Tighten connections or replace leaking hose • Replace with lining in axle sets • Replace brakeshoes • Replace mounting pins and bushings. Clean rust or burrs from rear brake support plate ledges and lubricate ledges with molydisulfide grease. **NOTE:** If ledges are deeply grooved or scored, do not attempt to sand or grind them smooth—replace support plate. • Repair or replace parts as necessary • Test valve according to the following procedure: (a) Start engine, increase engine speed to 1500 rpm, close throttle and immediately stop engine (b) Wait at least 90 seconds then depress brake pedal (c) If brakes are not vacuum assisted for 2 or more applications, check valve is faulty • Test unit according to the following procedure: (a) With engine stopped, apply brakes several times to exhaust all vacuum in system (b) Shift transmission into neutral, depress brake pedal and start engine (c) If pedal height decreases with foot pressure and less pressure is required to hold pedal in applied position, power unit vacuum system is operating normally. Test power unit. If power unit exhibits a bind condition, replace the power unit.

9 BRAKES

Troubleshooting the Brake System (cont.)

Problem	Cause	Solution
Hard brake pedal (excessive pedal pressure required to stop vehicle. May be accompanied by brake fade.)	• Master cylinder compensator ports (at bottom of reservoirs) blocked by dirt, scale, rust, or have small burrs (blocked ports prevent fluid return to reservoirs). • Brake hoses, tubes, fittings clogged or restricted • Brake fluid contaminated with improper fluids (motor oil, transmission fluid, causing rubber components to swell and stick in bores • Low engine vacuum	• Repair or replace master cylinder **CAUTION:** Do not attempt to clean blocked ports with wire, pencils, or similar implements. Use compressed air only. • Use compressed air to check or unclog parts. Replace any damaged parts. • Replace all rubber components, combination valve and hoses. Flush entire brake system with DOT 3 brake fluid or equivalent. • Adjust or repair engine
Dragging brakes (slow or incomplete release of brakes)	• Brake pedal binding at pivot • Power brake unit has internal bind • Parking brake cables incorrrectly adjusted or seized • Rear brakeshoe return springs weak or broken • Automatic adjusters malfunctioning • Caliper, wheel cylinder or master cylinder pistons sticking or seized • Master cylinder compensating ports blocked (fluid does not return to reservoirs).	• Loosen and lubricate • Inspect for internal bind. Replace unit if internal bind exists. • Adjust cables. Replace seized cables. • Replace return springs. Replace brakeshoe if necessary in axle sets. • Repair or replace adjuster parts as required • Repair or replace parts as necessary • Use compressed air to clear ports. Do not use wire, pencils, or similar objects to open blocked ports.
Vehicle moves to one side when brakes are applied	• Incorrect front tire pressure • Worn or damaged wheel bearings • Brakelining on one side contaminated • Brakeshoes on one side bent, distorted, or lining loose on shoe • Support plate bent or loose on one side • Brakelining not yet seated with drums or rotors • Caliper anchor plate loose on one side • Caliper piston sticking or seized • Brakelinings water soaked • Loose suspension component attaching or mounting bolts • Brake combination valve failure	• Inflate to recommended cold (reduced load) inflation pressure • Replace worn or damaged bearings • Determine and correct cause of contamination and replace brakelining in axle sets • Replace brakeshoes in axle sets • Tighten or replace support plate • Burnish brakelining • Tighten anchor plate bolts • Repair or replace caliper • Drive vehicle with brakes lightly applied to dry linings • Tighten suspension bolts. Replace worn suspension components. • Replace combination valve

BRAKES 9

Troubleshooting the Brake System (cont.)

Problem	Cause	Solution
Chatter or shudder when brakes are applied (pedal pulsation and roughness may also occur.)	• Brakeshoes distorted, bent, contaminated, or worn • Caliper anchor plate or support plate loose • Excessive thickness variation of rotor(s)	• Replace brakeshoes in axle sets • Tighten mounting bolts • Refinish or replace rotors in axle sets
Noisy brakes (squealing, clicking, scraping sound when brakes are applied.)	• Bent, broken, distorted brakeshoes • Excessive rust on outer edge of rotor braking surface	• Replace brakeshoes in axle sets • Remove rust
Noisy brakes (squealing, clicking, scraping sound when brakes are applied.) (cont.)	• Brakelining worn out—shoes contacting drum of rotor • Broken or loose holdown or return springs • Rough or dry drum brake support plate ledges • Cracked, grooved, or scored rotor(s) or drum(s) • Incorrect brakelining and/or shoes (front or rear).	• Replace brakeshoes and lining in axle sets. Refinish or replace drums or rotors. • Replace parts as necessary • Lubricate support plate ledges • Replace rotor(s) or drum(s). Replace brakeshoes and lining in axle sets if necessary. • Install specified shoe and lining assemblies
Pulsating brake pedal	• Out of round drums or excessive lateral runout in disc brake rotor(s)	• Refinish or replace drums, re-index rotors or replace

back (away from the direction in which the shoes are pushed apart). Reinstall the drum and depress the brake pedal repeatedly until the brakes apply at the highest point in pedal travel.

Mirage and Precis
Montero and Pickup

With the car safely supported and the rear wheels removed, turn the star wheel on the adjuster to adjust the outer diameter of the brake shoes. Measure accurately across the diameter of the outer brake surfaces. Correct distance is 179.0–179.5mm for automobiles and 253.0–253.5mm for Montero and Pickups. Notice that it is necessary to measure accurately to the tenth of a millimeter; use the correct tool.

Make certain the parking brake lever is fully released and the cable is not applying tension. Install the brake drum properly. Apply the parking brake via the lever and the service brakes via the pedal alternately until the shoes are adjusted. The "feel" of the pedal and the lever should remain constant.

BRAKE PEDAL ADJUSTMENT

1. Measure the distance (height) from the top surface of the floor, not including the carpet, to the top surface of the brake pedal. Correct distance is:
   • Cordia, Tredia, Starion, Galant and Sigma: 175–183mm
   • Mirage through 1988 and all Precis: 157.5–162.5mm
   • 1989 Mirage: 167.5–170.0mm
   • Montero: 190.5–195.5mm
   • Pickup: 166mm
2. If the pedal height is not correct, move the brake light switch out of contact with the pedal arm.
3. Loosen the lock nut on the pedal rod. Use pliers to turn the pedal rod, either lengthening it or shortening it, until the correct pedal height is obtained.

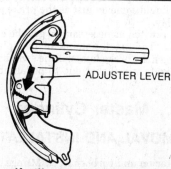

Set the brake self-adjuster as shown (all the way back) to adjust the rear drum brakes on Cordia, Tredia, and Galant

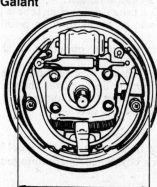

Turn the starwheel until the outside diameter of the shoes (shown) is 179.0–179.5mm

9 BRAKES

4. Tighten the locknut. Reposition the brake light switch so that the distance between the outer case of the switch and the pedal is 0.5–1.0mm. Note that the switch tab or prong must press against the pedal to keep the brake lights off.
5. Tighten the locknut on the brake light switch.
6. Check the free play in the pedal. With the engine off, depress the brake pedal fully several times. Then check the movement in the pedal before any resistance is encountered. Correct free play is:
- Cordia, Tredia, Starion, Mirage and Precis: 10–15mm
- Galant, Sigma, Monteros and Pickups: 2.5–7.5mm

7. If there is no freeplay in the pedal, check the adjustment of the brake light switch. If there is excessive freeplay, look for wear or play in the clevis pin and brake pedal arm.

Brake Light Switch

REMOVAL AND INSTALLATION

1. Disconnect the wiring at the back of the switch.
2. Loosen the locknut holding the switch to the bracket. Remove the locknut and remove the switch.
3. Install the new switch and install the locknut, tightening it just snug.
4. Reposition the brake light switch so that the distance between the outer case of the switch and the pedal is 0.5–1.0mm. Note that the switch tab or prong must press against the pedal to keep the brake lights off. As the pedal moves away from the switch, the tab is released and closes the switch which turns on the lights.
5. Hold the switch in the correct position and tighten the lock nut.
6. Connect the wiring to the switch.
7. Check the operation of the switch. Turn the ignition key to ON but do not start the engine. Have an assistant observe the brake lights at the rear while you push lightly on the brake pedal. The lights should come on just as the brake pedal passes the point of free play.
8. Adjust the switch as necessary to get the correct response from the lights. The small amount of free play in the pedal should not trigger the brake lights; if the switch is set incorrectly, the brake lights will flicker due to pedal vibration on road bumps.

Master Cylinder

REMOVAL AND INSTALLATION

1. On the Starion and turbocharged Mirage, the brake fluid reservoir is separate from the master cylinder. On these cars, disconnect the hoses at the master cylinder and plug them or drain the fluid into a clean container.
   For other vehicles, use a clean suction tube (a turkey baster works well) to remove as much fluid as possible from the reservoir.
2. Disconnect the electrical connector for the fluid level sensor.
3. Using only a wrench of the correct size (no pliers—ever!), carefully disconnect each brake line from the master cylinder. Label each line with a piece of tape. Plug or tape the end of the line to keep dirt and moisture out.

WARNING: Do not bend or crimp the steel brake lines. Handle them with extreme care! If damaged, they must be replaced.

4. Remove the nuts holding the master cylinder to the brake booster and remove the cylinder. Some fluid will remain within the cylinder; take care not to spill it on painted surfaces.

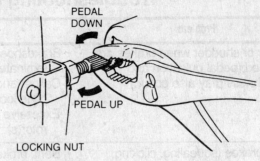
Adjusting the pedal rod

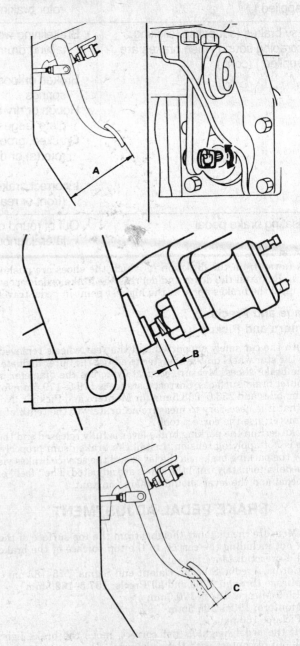
Check the pedal height (A), the brake pedal switch clearance (B) and the pedal freeplay (C)

BRAKES 9

OVERHAUL

NOTE: Special measuring tools are required for this procedure.

1. Support the master cylinder in a vise. Clamp the vise jaws to the bolt flange, not to the body of the cylinder.
2. If the fluid reservoir is attached to the cylinder, remove the reservoir retaining screw and gently lift the reservoir clear. In cases of a remote reservoir, remove the attaching ports from the master cylinder.
3. Remove the small seals under the reservoir.
4. Use a long thin tool to press inward on the piston and remove the piston stopper bolt. The piston need only be pushed enough to take tension off the bolt.
5. Again pushing gently on the piston, use snapring pliers to remove the piston retaining clip at the end of the cylinder. Be ready to catch the piston as you release pressure on it.
6. Remove the primary piston assembly and set is aside. Remove the secondary piston assembly. If this is difficult, gradually apply a stream of compressed air from the outer port on the secondary end of the master cylinder.

WARNING: Do NOT disassemble the primary or secondary pistons!

7. Clean the components with either clean brake fluid or a special brake cleaning solvent. Do NOT use petroleum based solvents and do NOT use water.
8. Check the inner surface of the master cylinder for rust or pitting. Check the pistons for any signs of rust, scoring or wear.
9. Using a cylinder micrometer, measure the cylinder bore in six locations. The locations should be the approximate bottom, center and top of the bore and measurements should be taken of the height and width of the bore at each position. All six measurements should be virtually identical. Variations of more than 0.025mm disqualify the unit from use. Record the bore diameter.

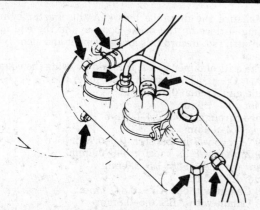

Disconnect the reservoir hoses, brake lines, and mounting bolts to remove the Starion master cylinder

5. Install the master cylinder to the brake booster and tighten the nuts to 12 Nm or 9 ft. lbs.
6. Carefully connect each brake line to its port on the master cylinder. Start each by hand, making sure that the fitting is at 90° to the port before starting to turn it. Once threaded one or two turns by hand, the wrench may be used to tighten each fitting to 20 Nm or 15 ft. lbs

WARNING: Do not overtighten the flare nuts. Overtightening will cause damage to the nut and/or the master cylinder!

7. Connect the wiring connectors to the master cylinder.
8. For cars with remote reservoirs, connect the fluid tubes to the master cylinder. Make sure the lines are firmly seated and the clamps are firmly set.
9. Fill the system with clean, fresh brake fluid.
10. Bleed the brake system at all four wheels.

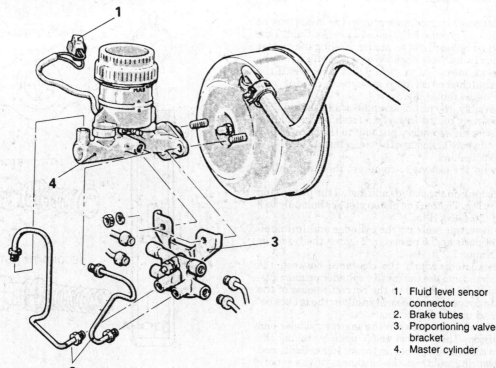

1. Fluid level sensor connector
2. Brake tubes
3. Proportioning valve bracket
4. Master cylinder

Typical master cylinder arrangement. The short metal brake lines should be disconnected at both ends to prevent bending

9-9

9 BRAKES

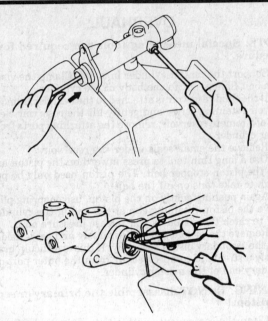

Press in gently on the piston to remove the stopper bolt (above) and the end clip. Note the correct method of holding the cylinder in the vise

10. Measure each piston in two dimensions. Again, measurements must be virtually identical. Subtract the piston diameter from the bore diameter. If the difference is greater than 0.15mm, there is enough wear to warrant replacing both the cylinder and pistons.

    Always replace the cylinder and pistons as a unit; do not attempt to guess which component is worn and do not mix old and new parts.

11. If the cylinder measurements are proper, the inside may be cleaned with a piece of crocus cloth soaked in brake fluid. This will remove any glaze or very light scratches. Honing the master cylinder is specifically not recommended as it will affect the piston to wall clearance measured in steps 9 and 10 above. If the cylinder bore is scratched or corroded enough to warrant honing, the unit should be replaced.

12. Before reassembly, the piston assemblies and the cylinder bore should be liberally coated with clean fresh brake fluid.

13. Carefully insert the secondary piston into the bore, taking care not to gouge the bore or damage the seals. Insert the primary piston in similar fashion.

14. Push gently on the pistons to compress them and install the snapring.

15. Continue pushing on the pistons and install the piston stop bolt with a new O-ring. Tighten the piston stopper bolt only to 3 Nm or 2.2 ft. lbs (26.4 INCH lbs.)

16. Install new reservoir seals on the cylinder and install either the fluid inlet ports or the reservoir. Tighten the reservoir retaining bolt 3 Nm or 2.2 ft. lbs.

17. Before reinstallation, adjust the clearance between the brake booster push rod and the master cylinder primary piston. This dimension (A) is critical to the correct release of the master cylinder. Improper clearance will result in the brakes being partially applied under all conditions.

    a. Measure the distance between the master cylinder end face and the piston. This is most easily done by taking the measurement with a square placed on the master cylinder end face. After measuring, subtract the thickness of the square and record the measurement. Call this measurement B.

    b. Find the distance between the contact face of the master cylinder mounting flange and the end face of the master cylinder. Record this distance and call it dimension C.

    c. Measure the distance between the master cylinder mounting surface on the brake booster and the end of the booster push rod. Record this measurement and identify it as dimension D.

    d. The critical dimension A is found by using the equation $A = B - C - D$. The correct value for A is:
    - Cordia and Tredia: 0.4–0.8mm
    - Starion: 0.7–1.0mm
    - Mirage through 1988: 0.4–0.8mm
    - Precis: 0.4–0.8mm
    - 1989 Mirage w/7 in. booster: 0.5–0.7mm
    - 1989 Mirage w/8 in. booster: 0.6–0.8mm
    - Galant through 1988: 1.5–2.0mm
    - Sigma: 1.5–2.0mm
    - 1989 Galant without ABS: 0.8–1.0mm
    - 1989 Galant with ABS: 0.5–0.8mm
    - Montero: 0.1–0.5mm
    - 1983–88 2wd Pickup: 0.4–0.8mm
    - 1987–88 4wd Pickup: 0.7–1.1mm
    - 1989 Pickup: 0.7–1.1mm

18. If the freeplay dimension is not within the acceptable range, adjust the length of the pushrod by carefully turning the adjustable (threaded) tip with pliers. After adjusting the rod, it will be necessary to remeasure dimension D and re-solve the equation.

19. Reinstall the master cylinder.

Power Brake Booster

Virtually all cars today use a vacuum assisted power brake system to multiply the braking force and reduce pedal effort. Since vacuum is always available when the engine is operating, the system is simple and efficient. A vacuum diaphragm is located on the front of the master cylinder and assists the driver in applying the brakes, reducing both the effort and travel that must be put into moving the brake pedal.

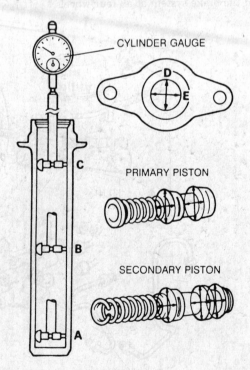

Measure the cylinder bore 6 times — at positions A, B and C and in dimensions D and E. Measure the pistons for height and width

BRAKES 9

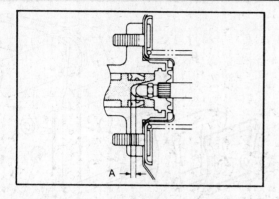

The vacuum diaphragm housing is connected to the intake manifold by a vacuum hose. A check valve is placed at the point where the hose enters the diaphragm housing, so that during periods of low manifold vacuum brake assist will not be lost.

Depressing the brake pedal closes off the vacuum source and allows atmospheric pressure to enter on one side of the diaphragm. This causes the master cylinder pistons to move and apply the brakes. When the brake pedal is released, vacuum is applied to both sides of the diaphragm and springs return the diaphragm and master cylinder pistons to the released position.

If the vacuum supply fails, the brake pedal rod will contact the end of the master cylinder actuator rod and the system will apply the brakes without any power assistance. The driver will notice that much higher pedal effort is needed to stop the car and that the pedal feels "harder" than usual.

REMOVAL AND INSTALLATION

1. Slide back the clip and disconnect the vacuum supply line at the brake booster. Pull the hose off gently in order to avoid damaging the check valve.
2. Remove the master cylinder as described previously.
3. Disconnect the pushrod at the brake pedal. Pull the cotter pin or lockpin out of the pedal clevis. Separate the rod from the pedal.
4. Remove the mounting bolts and nuts from the firewall and remove the booster.
5. Install the booster and tighten the nuts and bolts to 12 Nm or 9 ft. lbs. Do not overtighten these bolts.

NOTE: Some boosters use a spacer block and gaskets between the firewall and the booster. If present, they must be reinstalled in the correct order and position.

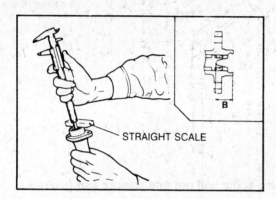

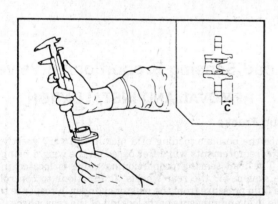

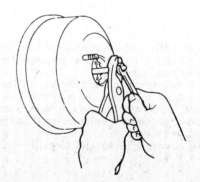

Adjusting the pushrod length after calculating the freeplay

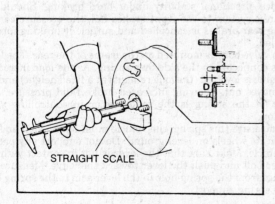

Pushrod clearance is confusing but not hard. Measure carefully, record each dimension and compute clearance A. A = (B − C − D)

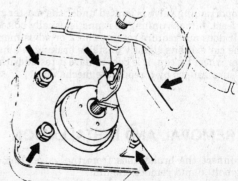

Disconnect the pushrod at the brake pedal and remove the 4 arrowed nuts to remove the brake booster

9-11

9 BRAKES

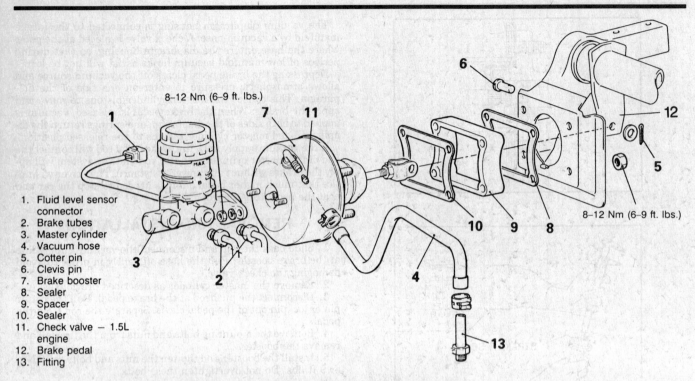

1. Fluid level sensor connector
2. Brake tubes
3. Master cylinder
4. Vacuum hose
5. Cotter pin
6. Clevis pin
7. Brake booster
8. Sealer
9. Spacer
10. Sealer
11. Check valve — 1.5L engine
12. Brake pedal
13. Fitting

Typical vacuum booster components. All others similar; some will not use spacer and gaskets

6. Connect the brake pedal to the pushrod and install a new cotter pin. a light coat of multipurpose grease on the clevis pin and washer is recommended.
7. Before reinstalling the master cylinder, the booster pushrod clearance must be calculated and adjusted. This is particularly important if the booster has been replaced. Please refer to the master cylinder procedure given previously and perform the last three numbered steps.
8. Install the master cylinder and bleed the brake system completely.

Proportioning Valve

The proportioning valve is located under the master cylinder. It has at least 4 lines running to it and is usually bolted to the firewall. It does not require routine check or adjustment; however, if the car exhibits slightly unstable braking in a hard stop due to rear wheel lockup, it is best to have it tested with special high pressure gauges by a reputable mechanic.

REMOVAL AND INSTALLATION

1. Disconnect the brake lines from the valve. Remove the mounting bolt(s) and remove the unit.
2. Replace the proportioning valve with one bearing the same numerical code, only. Install the mounting bolt(s) and connect all four brake lines, torquing the flare nuts to 9–12 ft. lb. Bleed the system thoroughly.

Load Sensing Proportioning Valve

REMOVAL AND INSTALLATION

Pickup Trucks

Since the possible loading of a pickup can vary greatly, the braking requirements will differ based on the weight being carried. The load sensing proportioning valve is located in the brake system for the rear wheels. It functions to control the brake fluid pressure from the master cylinder in response to the vehicle load and prevents early locking of the rear wheels. This provides directional stability under hard braking. Should the hydraulic system to the front brakes fail, the pressure control for the rear brakes is cancelled and sufficient braking force is provided.

The valve body is mounted on the frame at the rear. The lever arm is connected to the axle by a spring. As the load increases, the frame is lowered (the axle remains at a fixed height) and the arm moves out, allowing increased brake fluid pressure. The length of the spring is the main factor in controlling valve function.

To measure this spring, the vehicle must be unladen and sitting on its wheels on level ground. Do not support any part of the vehicle. Make sure that the lever is not in contact with the stopper bolt and push the lever toward the valve. Measure the distance from the spring hole in the lever arm to the spring hole in the spring support.

NOTE: Although the spring runs on an angle, do not measure along the spring. Measure the shortest distance—a line parallel to the ground.

BRAKES 9

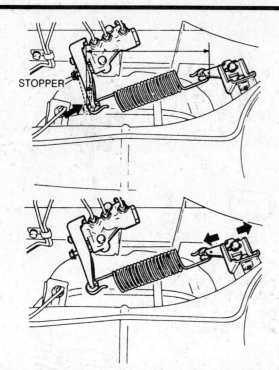

Measuring and adjusting the load sensing valve on Pickups

The correct distance is 174–176mm. If the measurement is incorrect, loosen the bolt attaching the support and slide it in the necessary direction to lengthen or shorten the distance and secure the bolt.

If it is necessary to replace the load sensing proportioning valve:
1. Disconnect the spring.
2. Label and disconnect each brake line running to the valve.
3. Remove the mounting bolts holding the valve and remove the valve.
4. Install the new valve and tighten the mounting bolts.
5. Connect each brake line to the correct port and tighten the fitting to 17Nm or 12 ft. lbs.
6. Install the spring.
7. Measure the effective distance as outlined above.
8. Bleed the brake system.

Vacuum Check Valve

REMOVAL AND INSTALLATION

The vacuum check valve found in the brake booster line is a simple one-way valve. It is designed and placed to hold vacuum within the booster for use during periods of low engine vacuum.

To test the valve, blow into each end. Air should NOT pass when you blow from the engine side but should pass when you blow from the booster side. Make sure the valve is installed correctly in the vacuum hose when reinstalling.

ANTI-LOCK BRAKE SYSTEM (ABS)

The anti-lock system found on Galant and Sigma is a sophisticated electronic and hydraulic system. Sensors at each wheel generate a signal relative to the speed of that wheel; the signal is transmitted to a computer which compares the rolling speed of each wheel. If one (or more) wheels begin to slow out of proportion to the others (as would happen if one began to lock under braking), the computer signals the ABS hydraulic unit to reduce brake pressure to that wheel.

Braking effort is reduced and the wheel is allowed to roll until it approaches the same speed as the others. Stability is thus improved by eliminating wheel lock. Because of the complexity of the system, pressure changes can occur several time per second in a panic stop as the system reacts to varying signals from all four wheels.

Should any part of the ABS system fail, an indicator light on the dash will illuminate; this DOES NOT indicate total brake failure, but only a loss of anti-lock function. The brake system will still function as a normal, non-ABS system.

Models equipped with this system may exhibit various behaviors which are not seen on normal brake systems. A pulsing, felt in the brake pedal or steering, may be noticed when the system is activated by hard braking, particularly on slippery surfaces. Although surprising, this pulsing is normal and a sign that the system is reacting to conditions. Additionally, as the vehicle reaches about 4 miles per hour after starting the engine, a whining or motor whirr may be heard in the engine compartment. This is from the system performing a self-check and building pressure; again, quite normal.

As complicated as an ABS system is, it rarely fails. Extensive development has made the system reliable and long lived. Troubleshooting and diagnosing an ABS system requires test equipment for both the electrical and computer portion as well as sophisticated hydraulic test equipment. It is a job best left to those well-trained in the field.

Hydraulic Unit

REMOVAL AND INSTALLATION

WARNING: Removal of this unit requires extreme care in handling and operation under the cleanest possible conditions. The slightest bit of dirt can foul the system. Do not disassemble the lines unless suitable plugs for both the lines and ports are immediately available. The need for extreme cleanliness cannot be overstated!

9 BRAKES

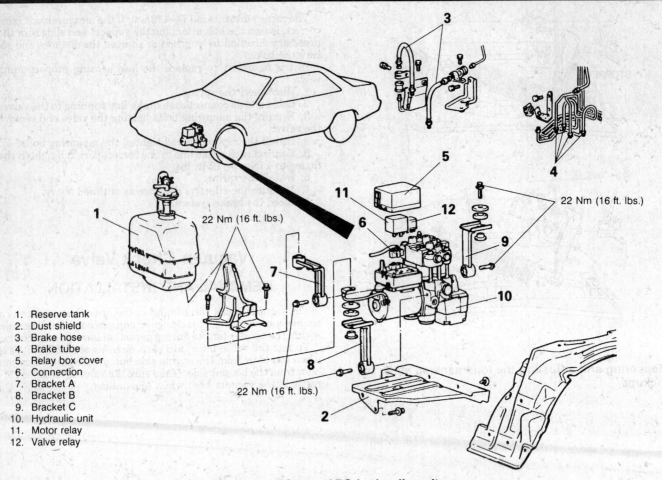

1. Reserve tank
2. Dust shield
3. Brake hose
4. Brake tube
5. Relay box cover
6. Connection
7. Bracket A
8. Bracket B
9. Bracket C
10. Hydraulic unit
11. Motor relay
12. Valve relay

Galant and Sigma ABS hydraulic unit

1. Disconnect the negative battery cable.
2. Remove the coolant reservoir or overflow tank.
3. Remove the dust shield below the unit.
4. Disconnect the two large brake hoses from the top of the unit and plug the hoses and ports immediately.
5. Using only an offset line wrench (other tools may cause damage), label and remove each fluid line at the top of the hydraulic unit. Plug each line and port immediately upon removal.
6. Use a small thin tool to release the clip on the side of the relay box cover. Remove the cover.
7. Disconnect the electrical connector running to the hydraulic unit.
8. Remove bracket **A**.
9. Remove bracket **B**. Have a helper support the unit as the bracket comes off.
10. Make certain the unit is well supported and remove bracket **C**. Lower the unit out the bottom of the car.

WARNING: The hydraulic unit is heavy. Make certain it is well supported. Do NOT subject the unit to shocks or impact; don't drop it. Once removed, the unit MUST NOT be turned upside down or placed on its side.

11. Do not disassemble the unit under any condition. There are no adjustments and no replacable parts.
12. Carefully install the hydraulic unit into position and connect bracket **C**. Tighten the bracket bolts to 22 Nm or 16 ft. lbs. and make certain all the washers and bushings are correctly placed.

13. Install support bracket **B** and its retaining nuts and bolts. Remember to connect the ground lead to the bolt and tighten the bolts to 22 Nm or 16 ft. lbs.
14. Install the L-shaped bracket **A** and tighten the bolts to 22 Nm (16 ft. lbs).
15. Connect the electrical connector, making sure it is properly seated and locked in place.
16. Install the relay box cover.
17. Carefully install each metal brake line to its correct location. Start each with your fingers, turning one or two revolutions before using the wrench. Handle the lines carefully. Tight-

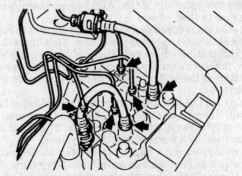

Total cleanliness must be assured when disconnecting the lines and hoses

BRAKES 9

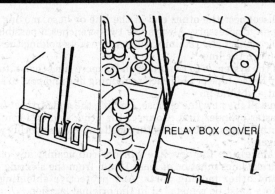

Remove the relay box cover

Location of the three bracket bolts holding the hydraulic unit

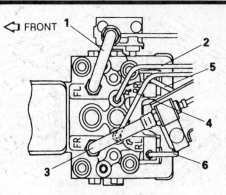

1. From the hydraulic unit to the left front brake
2. From the hydraulic unit to the right front brake — color coded blue
3. From the hydraulic unit to the right front brake
4. From the hydraulic unit to the left rear brake — color coded yellow
5. From the master cylinder, for the left front and right rear brakes
6. From the master cylinder, for the right front and left rear brakes — color coded white

Brake line locations on ABS hydraulic unit

en each fitting to 17 Nm or 12 ft. lbs. Make certain the lines are routed so that there is no possible contact between each brake line and any other surface.

17. Connect the hoses from the master cylinder to the hydraulic unit.
18. Install the dust shield under the unit.
19. Fill the system with fluid and bleed the brake system according to the instructions given in this section. Note that ABS systems require a slightly different procedure than conventional brake systems.
20. Install the coolant reserve tank.

REAR BRAKE LOCK-UP CONTROL SYSTEM

Found on Starions, this system is a variant of ABS which acts only on the rear wheels. It is designed to allow maximum rear wheel braking on slick surfaces, thus preventing locked wheel skids and possible loss of control. On this system, the front wheels can be locked and receive no modulation from the system.

The system includes a pulse generator in the transmission case which measures rear axle speed, a deceleration or G-sensor generating a signal according to vehicle speed, a control unit or computer which interprets the signals and a modulator which controls the pressure of the brake fluid to the rear wheels.

Should any of the components fail, the control unit enters the fail-safe mode, illuminating the dash warning light and dis-

abling the lock-up control. Normal braking is available.

A simple test of this system can be performed from the driver's seat. With the car stationary, run the engine for at least 5 seconds. Turn the ignition key to "LOCK", depress the brake pedal and while your foot is still on the brake, turn the key back to the "ON" (not START) position. If you hear a muffled clicking or tapping, the modulator solenoid valve is working and the system is operating fully.

Should the modulator need replacement or removal, it can be done in a straightforward manner. Remove the heat shield at the right shock tower and firewall, disconnect the vacuum hose, brake fluid lines and electrical connectors. Unbolt the modulator bracket from the firewall and remove the unit with the

9 BRAKES

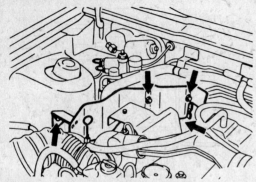

Remove the heat shield to gain access to the Starion modulator

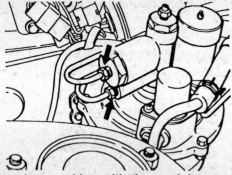

Use care when working with the modulator solenoid line fittings

bracket. Installation is in reverse order with the brake fluid line connections being tightened to 17 Nm or 12 ft. lbs. Disassembly of the modulator unit is not recommended for the owner/mechanic. After installation, the entire brake system and the modulator unit must be bled according to specific procedures given later in this section.

Brake Hoses and Lines

REMOVAL AND INSTALLATION

Metal lines and rubber brake hoses should be checked frequently for leaks and external damage. Metal lines and particularly prone to crushing and kinking under the car. Any such deformation can restrict the proper flow of fluid and therefore impair braking at the wheels. Rubber hoses should be checked for cracking or scraping; such damage can create a weak spot in the hose and it could fail under pressure.

Any time the lines are removed or disconnected, extreme cleanliness must be observed. The slightest bit of dirt in the system can plug a fluid port and render the brakes defective. Clean all joints and connections before disassembly (use a stiff brush and clean brake fluid) and plug the lines and ports as soon as they are opened. New lines and hoses should be blown or flushed clean before installation to remove any contamination. To replace a line or hose:

1. Elevate and safely support the car on stands.
2. Remove wheel(s) as necessary for access.
3. Clean the surrounding area on the joints to be disconnected.
4. Place a catch pan under the joint to be disconnected. To reduce fluid spillage during the repair, either plug the vent hole in the reservoir cap or substitute a cap with no vent hole.
5. Using two wrenches – one to hold the joint and one to turn the fitting – disconnect the hose or line to be replaced.

6. Disconnect the other end of the line or hose, moving the drain pan if necessary. Always use two wrenches if possible.
7. Disconnect any retaining clips or brackets holding the line and remove the line.
8. If the system is to remain open for more time than it takes to swap hoses, tape or plug each remaining line and port to keep dirt out and fluid in.
9. Install the new line or hose, starting the end farthest from the master cylinder first. Connect the other end and confirm that both fittings are correctly threaded and turn smoothly with the fingers.

Make sure the new line will not rub against any other part. Brake lines must be at least 13mm from the steering column and other moving parts. Any protective shielding or insulators must be reinstalled in the original location.

NOTE: If the new metal line requires bending, do so gently using a pipe bending tool. Do not attempt to bend the tubing by hand; it will kink the pipe and render it useless.

10. Using two wrenches as before, tighten each fitting to 13–17 Nm or 9–12 ft. lbs.
11. Install any retaining clips or brackets on the lines.
12. Refill the brake reservoir and replace the unvented cap if one was substituted.
13. Bleed the brake system beginning with the wheel closest to the replaced hose.
14. Install the wheels and lower the car to the ground.

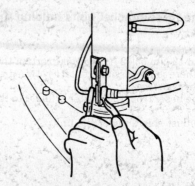

The use of two wrenches is required to avoid damage to the lines and fittings

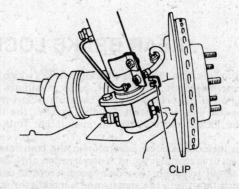

Note the horseshoe clip holding the rubber brake hose at the strut

BRAKES 9

BRAKE BLEEDING CHART
Bleed each wheel's brake in the sequence shown.

	LR	RF	RR	LF
Cordia/Tredia	1	2	3	4
Mirage/Precis	1	2	3	4
Galant/Sigma without ABS	1	2	3	4
Galant/Sigma with ABS ①	1	3	5	7
Montero through 1986	2	3	1	4
Starion	2	3②	1	4②
Montero 1987–1989	③	2	1	3
Pickup with Load Sensing Valve	③④	3	1	4
All others not listed	2	3	1	4

① Other points in system must be bled in correct sequence. Refer to text.
② On vehicles with rear lock-up control, the modulator must be bled before bleeding the front brakes.
③ Bleeding RR automatically bleeds LR. No additional bleeding is necessary at left rear.
④ The load sensing proportioning valve must be bled as step #2 in the sequence.

Brake System Bleeding

NORMAL SYSTEMS

The brake system is bled through the bleeder screws on the calipers and/or rear wheel cylinders. The system must be bled whenever air enters, whether due to disconnection of a hydraulic line or a leak. A spongy brake pedal indicates bleeding is needed even when no repairs have been made. Also check carefully for leaks and repair them as necessary.

Bleeding pumps fluid into the system via the master cylinder and bleeds air out through the bleed valves. You must keep the master cylinder reservoir full of fluid at all times to provide a continuous flow of air-free fluid into the system. You'll also need an assistant who will keep constant pressure on the brake pedal at all times to keep air from being drawn back into the system.

WARNING: Do not allow brake fluid to splash or spill onto painted surfaces; the paint will be damaged. If spillage occurs, flush the area immediately with clean water.

1. Fill the master cylinder reservoir to the "MAX" line with brake fluid and keep it at least half full throughout the bleeding procedure.
2. If the master cylinder has been removed or disconnected, it must be bled before any brake unit is bled. To bleed the master cylinder:
   a. Disconnect the front brake line from the master cylinder and allow fluid to flow from the front connector port.
   b. Reconnect the line to the master cylinder and tighten it until it is fluid tight.
   c. Have a helper press the brake pedal down one time and hold it down.
   d. Loosen the front brake line connection at the master cylinder. This will allow trapped air to escape, along with some fluid. (Have a rag or small container handy to catch the fluid.)
   e. Again tighten the line, release the pedal slowly and repeat the sequence (Steps c, d, and e) until only fluid runs from the port. No air bubbles should be present in the fluid.
   f. Final tighten the line fitting at the master cylinder to 11 ft. lbs.
   g. After all the air has been bled from the front connection, bleed the master cylinder at the rear connection by repeating Steps a through e.
3. Start with the first location listed in bleeding chart. Place the correct size box-end or line wrench over the bleeder valve and attach a tight-fitting transparent hose over the bleeder. Allow the tube to hang submerged in a transparent container of clean brake fluid. The fluid must remain above the end of the hose at all times, otherwise the system will ingest air instead of fluid.
4. Have an assistant pump the brake pedal several times slowly and hold it down.
5. Open the bleed point (about ¼–½ turn is usually enough) and watch for bubbles in the brake fluid flow into the container. You must close the bleed valve before the brake pedal bottoms in its travel. Once the assistant has the feel of the pedal, he can warn you in advance. The more the bleed valve is opened, the quicker the pedal travels downward.

With the bleed point closed, the assistant should release the pedal so the master cylinder will admit fresh, air-free fluid into the system. Once he has put pressure back on the pedal, you may re-open the bleed point, again closing it before the pedal bottoms out. Repeat this process until the fluid flowing from the bleed point is entirely air-free.

6. After two or three bleedings, check and replenish the fluid in the reservoir.
7. Repeat the process, in the correct order, at each remaining wheel. The general theory is to bleed the longest lines first and work towards the master cylinder.
8. When all four wheels have been bled and no air is evident in the fluid, refill the master cylinder reservoir and attach the cap tightly.

STARION WITH REAR LOCK-UP CONTROL

Vehicles with rear lock-up control require that the modulator be bled before the front brakes are bled. The bleeder screw is located in an easily accessable position on top of the unit. Bleeding is performed following the steps outlined previously

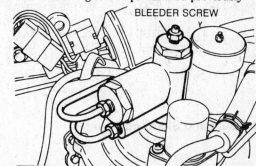

Modulator bleeder screw location, Starion with rear lock-up control

9–17

9 BRAKES

PICKUP WITH LOAD SENSING PROPORTIONING VALVE

After bleeding the right rear wheel, proceed to bleed the load sensing valve located on the frame at the rear of the vehicle. Use the bleeder fitting on the valve; do not attempt to loosen the lines entering the unit. Follow the bleeding procedure for normal systems and bleed the proportioning valve as you would any other cylinder or caliper.

GALANT AND SIGMA WITH ABS

NOTE: The brake pedal will have a heavier feeling when bleeding the rear brakes than it does when bleeding the front brakes. This is the result of fluid constriction within the delay valve and is NOT a sign of a problem.

In addition to the usual 4 wheel bleeders, ABS systems must be bled at each of the delay valves (at the right rear of the car) and at the hydraulic control unit under the hood. The correct order of bleeding is essential as is thorough and proper bleeding at each fitting.

Bleed the left rear brake in the usual fashion. Proceed to the delay valve under the rear of the car. The left fitting serves the right brake and vice-versa. If you are confused, follow the line from the left brake into the delay valve. To bleed the delay valve, loosen the large check valve bolt at the bottom of the valve about one turn. Using the bleeder port, bleed the delay valve as usual and retighten the check valve bolt to 25 Nm or 18 ft. lbs. Proceed to the right front brake and bleed it according to normal procedures. Identify the left or outboard bleeder on the hydraulic unit and bleed the unit through this fitting. Follow the same working procedures listed under normal systems. Be careful not to damage or dent nearby lines or hoses. At this point, one brake circuit (LR/RF) is bled.

NOTE: The use of a closed or box wrench is highly recommended for opening the bleed ports on the hydraulic unit. Open end wrenches or similar tools may damage or deform the hexagonal faces on the fitting.

Bleed correctly, and in this order, the right rear wheel, the right rear wheel delay valve, the left front wheel and the right or inboard bleeder on the hydraulic control unit. This completes bleeding of the second brake circuit (RR/LF).

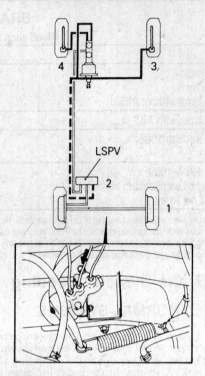

On Pickups, bleed the load sensing valve after the right rear wheel

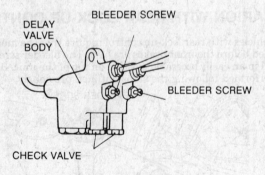

Detail of Galant and Sigma delay valve

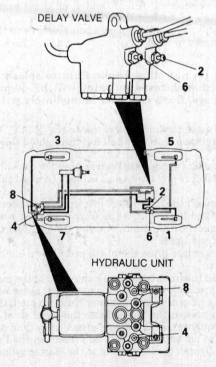

Correct order of bleeding the Galant and Sigma ABS system

BRAKES 9

FRONT DISC BRAKES

CAUTION
Brake pads and shoes contain asbestos, which has been determined to be a cancer causing agent. Never clean the brake surfaces with compressed air! Avoid inhaling any dust from brake surfaces! When cleaning brakes, use commercially available brake cleaning fluids.

WARNING: If the car has been recently driven before repairs, all the brake surfaces and components may be very hot. Work carefully and wear gloves.

Disc Brake Pads

WEAR INDICATORS

The front disc brake pads are equipped with a metal tab which will come into contact with the disc after the friction surface material has worn near its usable minimum. The wear indicators make a constant, distinct metallic sound that should be easily heard. (The sound has been described as similar to either fingernails on a blackboard or a field full of crickets.) The key to recognizing that it is the wear indicators and not some other brake noise is that the sound is heard when the car is being driven WITHOUT the brakes applied. It may or may not be present under braking during normal driving.

It should also be noted that any disc brake system, by its design, cannot be made to work silently under all conditions. Each system includes various shims, plates, cushions and brackets to supress brake noise but no system can completely silence all noises. Some brake noise—either high or low frequency—can be considered normal under some conditions. Such noises can be controlled and perhaps lessened, but cannot be totally eliminated.

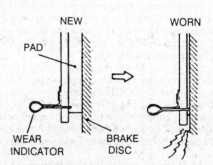
The indicators engage as the pad wears close to minimum thickness

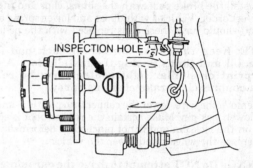
View the pads through the window in the caliper

INSPECTION

The front brake pads may be inspected without removal. With the front end elevated and supported, remove the wheel(s). Unlock the steering column lock and turn the wheel so that the brake caliper is out from under the fender.

View the pads—inner and outer—through the cut-out in the center of the caliper. Remember to look at the thickness of the pad friction material (the part that actually presses on the disc) rather than the thickness of the backing plate which does not change with wear.

Remember that you are looking at the profile of the pad, not the whole thing. Brake pads can wear on a taper which may not be visible through the window. It is also not possible to check the contact surface for cracking or scoring from this position. This quick check can be helpful only as a reference; detailed inspection requires pad removal.

After the pads are removed, measure the thickness of the LINING portion (NOT the backing) of the pads. It must be at least 1mm for all 1983–86 vehicles and 2mm for all vehicles built after 1987. This is a Mitsubishi factory minimum; please note that local inspection standards enforced by your state must be given precedence if they require a greater thickness.

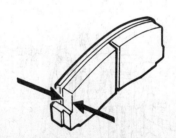
Inspect the remaining thickness by measuring the friction material, NOT the backing plate

REMOVAL AND INSTALLATION

NOTE: Whenever brake pads are replaced, replace them in complete sets; that is, replace the pads on both front wheels even if only one side is worn.

There are several combinations of shims, spacers and clips in use on Mitsubishi vehicles. When disassembling, work on one side at a time and pay attention to placement of these components. If you become confused during reassembly, refer to the other side for correct placement. To remove the brake pads:

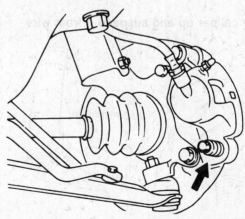
Remove the lockpin assembly to allow the caliper to be lifted upward

9-19

9 BRAKES

1. Raise the vehicle and support it safely on stands.
2. Remove the front wheels.
3. Unscrew and remove the lockpin without disturbing the grease coating. Place the pin in a clean spot where grease will not pick up dust. Raise the caliper upward by lifting on the lower end so it will pivot on the upper mounting pin. Fasten the caliper in the raised position with wire.
4. Remove the shims, brake pads and spring clips in order. Depending on the number of such components, either diagram their placement or lay them aside in order.
5. Clean the surface of the piston with a clean, damp rag. Make sure there is room in the master cylinder reservoir for more fluid (fluid will be forced back into the reservoir in the next step). If necessary, remove some fluid with a clean squeeze ball syringe.
6. Use a large C-clamp or similar tool to depress the caliper pistons back into the calipers. The pistons will need to be almost flush with the case to fit over the new, thicker pads.
7. Install the brake pads with the shims, clips and fittings in the correct order. Vehicles with upper and lower spring clips for each pad should have these clips replaced with the pads.

NOTE: Keep the brake surface of each pad free of grease, oil and fluids during the installation. A greasy fingerprint or similar light contact may be removed with a commercial brake cleaning spray.

8. Remove the wire, lower the caliper into position, and screw in the lower lock pin. Make certain the rubber boot is correctly placed on the lock pin and is not pinched or deformed.
9. Replace the wheels and lower the vehicle.

WARNING: Do NOT attempt to drive the car immediately after lowering it to the ground. The first two or three brake pedal applications may not provide any brake response.

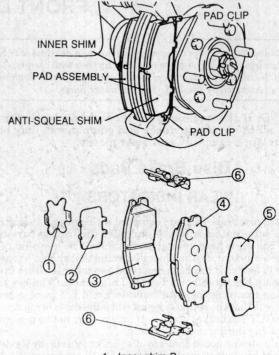

1. Inner shim B
2. Inner shim A
3. Pad and wear indicator
4. Pad assembly
5. Outer shim
6. Pad clip (2)

Installed and exploded views of typical front brake components

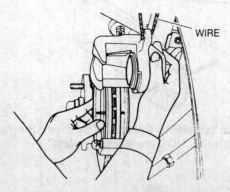

Pivot the caliper up and suspend it from wire

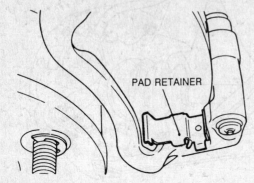

Always replace the retainer clips when changing pads

10. Pump the brake pedal several times with the engine off. The first two or three pedal strokes may be longer than usual as the pistons return from their compressed position and drive the pads inward. After a reasonable feel has been achieved, check the master cylinder reservoir and top up the fluid as needed. Start the engine and pump the brake pedal again, checking for the proper feel and engagement point.
11. Since the brake hoses and hydraulic system was not opened during the repair, it is usually not necessary to bleed the brake system after pad replacement. Use good judgement in determining pedal feel; bleeding may be necessary for other reasons.

NOTE: Braking should be moderate for the first 10 miles or so until the new pads seat correctly. The new pads will bed best if put through several moderate heating and cooling cycles.

Avoid hard braking until the brakes have experienced several long, slow stops with time to cool in between. Taking the time to properly bed the brakes will yield quieter operation, more efficient stopping and contribute to extended brake life.

Disc Brake Calipers

REMOVAL

1. Raise and safely support the front of the vehicle on jack stands. Set the parking brake and block the rear wheels.
2. Siphon a sufficient amount of brake fluid from the master cylinder reservoir to prevent the brake fluid from overflowing

BRAKES 9

when removing or installing the caliper. This is necessary as the piston must be forced into the cylinder bore to provide clearance to install the caliper.

3. Remove the wheel. If the disc is not otherwise retained, install 2 lug nuts to hold the disc in place.

NOTE: Disassemble brakes one wheel at a time. This will prevent parts confusion and also prevent the opposite piston from popping out during installation.

4. Disconnect the brake line at the clip on the strut and disconnect the hose union at the caliper. Use a pan to catch any spilled fluid and immediately plug the hose end.

5. Remove the two caliper mounting bolts and then remove the caliper from the mounting bracket. Alternatively, remove the two bolts (screwed directly into the backing plate) which fasten the caliper support to the backing plate, and remove the entire assembly.

Remove the brake pads and clips from the support as described above. Remove the lockpin (lower) and guide pin (upper) from the caliper support, and remove the caliper from the support.

OVERHAUL

NOTE: To overhaul the brake calipers, you must have a controlled source of compressed air to force the pistons out. You should also have a generous supply of clean brake fluid or commercial brake cleaning spray to clean parts.

6. Drain the remaining fluid from the caliper.
7. Carefully remove the dust boot from around the piston.
8. Pad the inside arms of the caliper with rags. Gradually apply compressed air into the brake line port; this will force the piston out.

---- **CAUTION** ----

Do not place fingers in front of the piston in an attempt to catch it or protect it when applying compressed air. Injury can result. Use just enough air pressure to ease the piston out.

---

9. Remove the seal from the inside of the caliper bore. Check all the parts for wear, deterioration, cracking or other abnormal

Keep the fingers out of the way and gently apply compressed air to remove the piston

conditions. Corrosion, generally caused by water in the system, will appear as white deposits on the metal, similar to what may be found on an old aluminum storm door around the house. Pay close attention to the condition of the inside of the caliper bore and the outside of the piston. Any sign of corrosion or scoring requires new parts; do not attempt to clean or resurface either face.

10. The caliper overhaul kit will, at minimum, contain new seals and dust boots. A good kit will contain a new piston as well, but you may have to buy the piston separately. Any time the caliper is disassembled, a new piston is highly recommended in addition to the seals.

11. Clean all the components to be reused with an aerosol brake cleaner and dry them thoroughly. Take any steps necessary to eliminate moisture or water vapor from the parts.

12. Coat all the caliper components with fresh brake fluid from an unopened can.

NOTE: Some kits come with special assembly lubricants for the piston seals and slides or guidepins. Use these lubricants according to the directions within the kit.

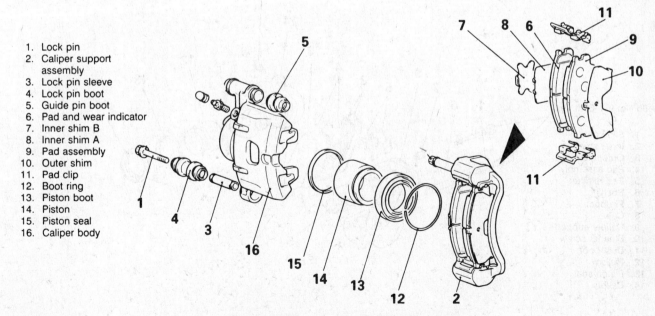

1. Lock pin
2. Caliper support assembly
3. Lock pin sleeve
4. Lock pin boot
5. Guide pin boot
6. Pad and wear indicator
7. Inner shim B
8. Inner shim A
9. Pad assembly
10. Outer shim
11. Pad clip
12. Boot ring
13. Piston boot
14. Piston
15. Piston seal
16. Caliper body

Caliper components: Galant and Sigma

9-21

9 BRAKES

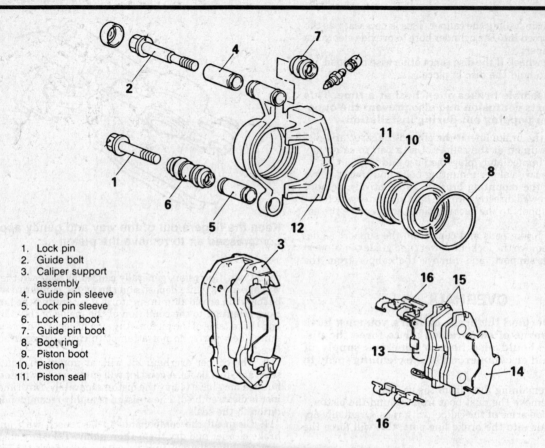

1. Lock pin
2. Guide bolt
3. Caliper support assembly
4. Guide pin sleeve
5. Lock pin sleeve
6. Lock pin boot
7. Guide pin boot
8. Boot ring
9. Piston boot
10. Piston
11. Piston seal

Typical truck caliper assembly; many other models similar

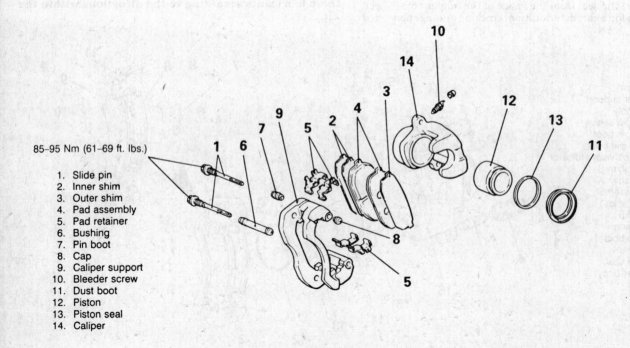

85–95 Nm (61–69 ft. lbs.)

1. Slide pin
2. Inner shim
3. Outer shim
4. Pad assembly
5. Pad retainer
6. Bushing
7. Pin boot
8. Cap
9. Caliper support
10. Bleeder screw
11. Dust boot
12. Piston
13. Piston seal
14. Caliper

Starion caliper

BRAKES 9

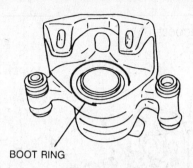

Many models use a retaining ring to secure the dust boot. Make certain it is correctly reinstalled

Carefully replace the inner seal. It must sit squarely in the channel

13. Install the piston seal and piston into the caliper bore. This is an exacting job; the clearances are very small. Make sure the seal is seated in its groove and that the piston is not cocked when inserted into the bore.
14. Install the dust boot and its ring or clip, if any.
15. Install the slide bushings and rubber boots onto the caliper if they were removed during disassembly.
16. The guide and lockpin sleeves should be coated inside with the grease supplied with the rebuilding kit, and boots should all be replaced. Also coat the lips of the boots and the surface of the caliper which bears (and turns) against the caliper support with grease.

NOTE: Two kinds of grease may be packed in a rebuilding kit; use the grease recommended for use with the guide and lock pins. On the Mirage PS15 brakes, coat the outer surfaces of the metal sleeves with brake fluid. On all the systems, coat the sliding parts of the caliper body and the sleeves (except Mirage PS15) with the grease specified.

INSTALLATION

17. Use a caliper compressor, a C-clamp or a large pair of pliers to slowly press the caliper piston back into the caliper.
18. If the mounting plate was removed from the backing plate, reinstall it and coat the bolts with an anti-seize compound. Tighten the bolts to 90 Nm or 66 ft. lbs.
19. Install the pads onto the mount and install the caliper. Make sure the slide bushings and bolts are clean and properly lubricated. Do not use regular grease or spray lubricants; they cannot withstand the extreme heat generated by the brakes. Tighten the bolts to 31 Nm or 23 ft. lb. on all models but the Starion; on that model, torque to 90 Nm or 66 ft. lbs. Additionally, the upper bolt on Pickups and Monteros should be tightened to 45 Nm (33 ft. lbs.).
20. Connect the fluid hose to the caliper and then to the line at the strut. Tighten each fitting to 17 Nm (12 ft. lbs.) and make sure the hose is correctly routed and not kinked. Remember to install any clips or retainers holding the line.
21. Refill the master cylinder reservoir to the MAX line. Bleed the brake system. Keep a close eye on the reservoir, maintaining its level at at least half during the bleeding. Since the caliper was emptied, a fair amount of brake fluid may need to be added.
22. Reinstall the wheel and lower the car to the ground.
23. Check the level in the fluid reservoir and top it off as needed. Before driving the car, start the engine and operate the brake pedal several times to check the feel and engagement point of the brakes.

Brake Disc (Rotor)

REMOVAL AND INSTALLATION

Cordia, Tredia, and Mirage

NOTE: This procedure requires a number of precision and/or expensive special tools to press the driveshaft out of the hub and remove the hub from the bearings.

1. Remove the center hub cap and drive shaft nut. Raise the vehicle and support it on floor stands. Remove the front wheel.
2. Remove the brake caliper without disconnecting the hydraulic line and suspend it with a piece of wire.
3. Refer to Section 8 and disconnect the stabilizer bar and strut bar from the lower control arm. Disconnect the ball joint at the steering knuckle. Disconnect the tie rod end ball joint at the steering knuckle as well.
4. Refer to Section 8 and remove the driveshaft from the transmission and press the driveshaft out of the hub as described.
5. Unbolt and remove the hub and knuckle from the bottom of the strut.
6. You'll need several special tools to press the hub and disc from the steering knuckle and to remount them. Use MB 991001 and MB 990998-01 or the equivalent on Cordia and Tredia, and MB 991056 and MB 990998-01 on the Mirage. Do not attempt to hammer the parts apart, or the bearing will be damaged!

Install the arm of the special tool onto the knuckle and tighten the nut manually. Then install the body of the tool to the knuckle. Pull the bearings out, noting their positions and direction of installation (smaller diameter inward).

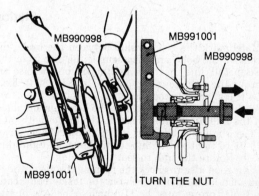

Removing the hub from the knuckle using the special tools shown for the Cordia and Tredia — the Mirage is similar

9-23

9 BRAKES

1. Brake hose
2. Front brake
3. Hub cap
4. Cotter pin
5. Lock cap
6. Nut
7. Washer
8. Brake disc
9. Front axle hub
10. Dust cover

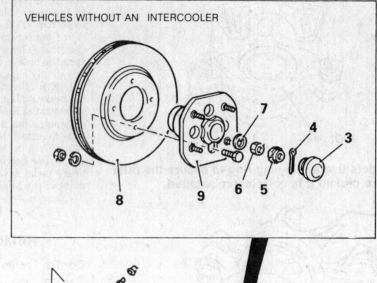

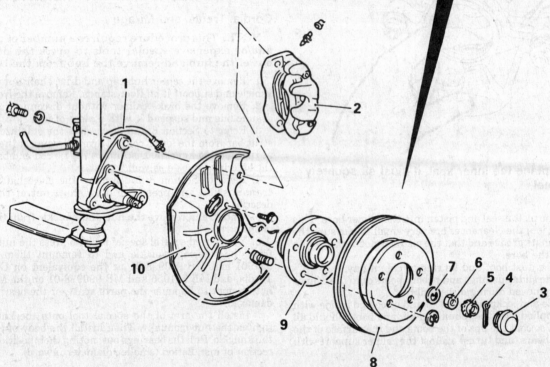

Starion caliper assemblies

7. Matchmark the relationship between the brake disc and hub; remove the bolts and disassemble the disc from the hub.
8. To install, line up the match marks on the disc and hub and torque the disc mounting bolts to 53 Nm (39 ft. lbs.) or 46 Nm (34 ft.lbs) on '83 models.
9. Apply multipurpose grease to the bearings and inside surface of the hub. Apply the grease to the lips of both oil seals.
10. Place the inner bearings into the knuckle. Then use MB 990998-01 to tighten the hub to the knuckle, tightening it to 166 ft. lb. Rotate the hub to seat the bearing.
11. It's a good idea to check the turning torque of the bearing with an inch lb. torque wrench via the inside bolt head of the special tool. Torque must be 11.3 inch lb. or less, or the parts have been assembled incorrectly. If the turning torque is zero, you should check the axial play with a dial indicator. It must be 0.2mm or less.

12. Remove the special tool and continue the reinstallation of the knuckle assembly. Attach the knuckle to the strut and tighten the mounting bolts to 80 Nm or 59 ft. lbs. Refer to the respective procedures in Section 8 for the other torque values.
13. Reinstall the brake pads and caliper.
14. Install the wheel and lower the car to the ground. If necessary, final tighten the drive shaft center nut and any suspension fittings requiring final tightening under load. Don't forget to install new cotter pins where needed.

Starion

1. Elevate and safely support the car on stands. Remove the front wheel.
2. Remove the front brake caliper assembly (caliper and carrier), leaving the hydraulic line connected and suspending the assembly with wire.

9-24

3. For all vehicles through 1986 and any without intercoolers thereafter:
   a. Remove the hub cap, cotter pin, lock cap, and nut.
   b. Refer to Section 8 and remove the hub and rotor assembly and bearings.
   c. Matchmark the relationship between the brake disc and hub; remove the attaching nuts and bolts and remove the disc from the hub.
   d. To install, align the match marks and install the nuts and bolts attaching the disc to the hub, tightening them to 38 Nm or 28 ft. lbs.
   e. Make sure the bearings are properly packed with wheel bearing grease. Apply grease to the lips of oil seals and the inside surfaces of the hub. Install the rotor and hub assembly onto the knuckle, following procedures given in Section 1.
   f. Install the outer bearing, washer, and nut. Tighten the nut according to procedures given in Section 1.
   g. Install the lock cap and cotter pin. If the cotter pin does not align with the holes in the lock cap, reposition the lock cap so the holes align. If you can't do this without changing the position of the wheel bearing nut, back the nut off 15 degrees or less, then install a new cotter pin.
4. For intercooled vehicles, simply remove the E-clip on each wheel lug. Once all are removed, the disc may be removed easily.
5. When reinstalling this type, make certain each clip is firmly seated in position.
6. Reinstall the caliper assembly.
7. Install the front wheel.

Galant, Sigma and Precis

1. Elevate and safely support the car on stands.
2. Remove the front wheel.
3. Remove the brake caliper assembly (caliper and carrier) but without disconnecting the hydraulic line. Suspend the caliper with wire.
4. Matchmark the hub and disc, and then simply pull the disc off the hub. if rust makes the disc difficult to remove, then install two bolts (M8 x 1.25) into the threaded holes provided and turn them inward alternately and evenly to press off the disc. Do not hammer on the disc!
5. Installation is the reverse of removal, but make sure you align the mating marks for the disc and hub.

Montero and Pickup

NOTE: Refer to Section 8 for detailed procedures for removal and assembly of the hub assembly. Four wheel drive vehicles must have the front hubs in the free position before disassembly.

1. Raise the vehicle and support it safely on stands.
2. Remove the wheel.
3. Remove the caliper and brake pads. It may be necessary to disconnect the small brake line running to the caliper. If this is done, plug the open line immediately. Suspend the caliper out of the way.
4. Remove the caliper adaptor or mount.
5. If the vehicle is equipped with 4 wheel drive, remove the automatic hub, the shim, lock washer and locknut.
6. If the vehicle is 2-wheel drive, remove the dust cap, cotter pin, nut lock, nut, washer and outer bearing.
7. Remove the front hub and rotor assembly. Provide matchmarks on the rotor and hub; unbolt the rotor from the hub.
8. Install the rotor to the hub and tighten the attaching nuts and bolts to 54 Nm (40 ft. lbs.)
9. Install the assembly and retaining parts to the spindle.
10. Install the caliper adaptor. Tighten the bolts to 95 Nm or 70 ft. lbs.
11. Install the brake pads and caliper, making sure all anti-rattle and anti-squeal clips are in place.

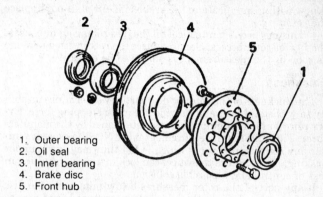

1. Outer bearing
2. Oil seal
3. Inner bearing
4. Brake disc
5. Front hub

Four wheel drive hub and rotor

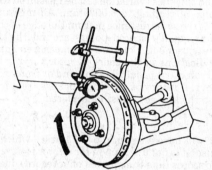

Measuring front disc runout

12. Reconnect the brake line if it was disconnected and bleed the brakes.
13. Install the wheel and lower the vehicle to the ground.

INSPECTION

NOTE: Mitsubishi does not provide resurfacing specifications. Any disc which is gouged, scored, worn or warped should be replaced.

Run-out

Before measuring the run-out on the front discs, confirm that the front wheel bearing play is within specification.
1. Elevate and safely support the car on stands. If only the front is supported, set the parking brake and block the rear wheels.
2. Remove the wheel.
3. Remove the brake caliper from its mount and suspend it out of the way by a piece of wire. Don't disconnect the fluid hose and don't allow the caliper to hang by the hose. Remove the brake pads with all the clips, springs, shims, etc.
4. If the rotor is the easily removed or pull off type, install the lug nuts to hold it in place. Tighten the nuts a bit tighter than finger tight, but make sure all are at approximately the same tightness.
5. Mount a dial indicator with a magnetic or universal base on the strut so that the tip of the indicator contacts the rotor about 13mm from the outer edge.
6. Zero the dial indicator. Turn the rotor one full revolution and observe the total indicated run-out. Refer to the Brake Specifications Chart in this Section for the specifications for your vehicle.
7. If the run-out exceeds the maximum allowable, the rotor probably needs replacing. If the rotor is easily removable, pull it off, clean the contact faces on the rotor and the hub and reinstall the rotor in a different position. Remeasure; if the measurement

9 BRAKES

is now within specification, the problem is cured. If not, replace the rotor.

This method is impractical on the cars requiring hub disassembly for rotor access; plan to replace the rotor with a new one as soon as the measurement is shown to be excessive.

Thickness

The thickness of the rotor or disc partially determines its ability to withstand heat and provide adequate stopping force. Every rotor has a minimum thickness established by the manufacturer. This minimum measurement must NOT be exceeded under any condition. A rotor which is too thin may crack under braking; if this occurs the wheel can lock instantly, resulting in loss of control and a possible collision.

If any part of the rotor measures below minimum, the disc must be replaced. Since the allowable wear from brand new to minimum is only 1–1.5 mm, resurfacing is not recommended.

Thickness and thickness variation can be measured with a micrometer capable of reading to 0.0025mm. All measurements must be made at the same distance in from the edge of the rotor. Measure at four equally spaced points around the disc and record each measurement. Compare the measurements to the minimum specifications in the chart. A rotor varying by more than 0.025mm can cause pedal vibration and/or front end vibration during stops. A rotor not meeting thickness requirements, or with varying thicknesses, must be replaced.

Condition

A new rotor will have a smooth, even surface which rapidly changes with use. It is not uncommon for a rotor to develop very fine concentric scores (like the grooves on a record) due to dust and grit being trapped by the brake pads. This slight irregularity is normal, but as the grooves deepen, wear and noise increase

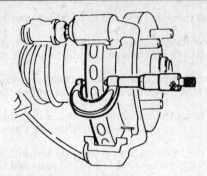

Measuring disc thickness

and stopping may be affected. As a general rule, any groove deep enough to snag a fingernail during inspection is cause for rotor replacement.

Any sign of blue spots, discoloration, heavy rusting or outright gouging require replacement. If you are checking the disc on the car (such as during pad replacement or tire rotation) remember to turn the disc and check both the inner and outer faces completely. A small mirror and a bright light can be very helpful in seeing the inner face of the rotor. If anything looks questionable or requires consideration, choose the safer alternative and replace the rotor. The front brakes are a critical system and must be maintained at 100% efficiency.

Any time a rotor is replaced, the pads should also be replaced so that the surfaces mate properly. Since brake pads should be replaced in sets (both front or rear wheels), consider replacing both rotors instead of just one. The restored feel and accurate stopping make the extra investment well worthwhile.

REAR DRUM BRAKES

Brake Drums

REMOVAL AND INSTALLATION

---- CAUTION ----
Brake pads and shoes contain asbestos, which has been determined to be a cancer causing agent. Never clean the brake surfaces with compressed air! Avoid inhaling any dust from brake surfaces! When cleaning brakes, use commercially available brake cleaning fluids.

Cordia and Tredia
1985–87 Galant
Montero and Pickup

1. Block the front wheels, raise the vehicle at the rear and support it safely on floor stands. Make sure the parking brake is released.
2. Remove the rear wheels.
3. Remove the small hub cap, cotter pin, lock cap, retaining nut, and washer.
4. Remove the brake drum by pulling it towards you. Keep the bearings in place inside the drum. If the drum is difficult to remove, first check that the parking brake is fully released and that the cables to the wheels are not binding. Then follow either of these procedures:

   a. Remove the cover from the adjustment hole at the rear or back face of the backing plate.

   b. Insert a small screwdriver into the adjustment hole and use it to sepatate the adjustment lever from the adjuster.

   c. Use a second small screwdriver or brake adjusting tool to turn the star wheel and loosen the brake adjustment. Turning the wheel in the correct direction will contract the shoes away from the drum.

   d. If the drum has additional holes drilled in it (not all do), insert two M8x1.25 bolts.

   e. Turn the bolts tighter; they will press on the hub and force the drum off the brake shoes. Remember to turn the

BRAKES 9

torque wrench, loosen the retaining nut slowly, reading the required torque through two or three revolutions. This self-locking nut must require at least 48 INCH lbs. to turn until it is two turns out. If it turns with less force needed, it is worn and will not lock properly; replace it.

3. Once the nut is removed, you can remove the rear drum and bearings as an assembly. Keep the bearings and drum together and don't allow the parts to accumulate dirt.

4. Before reinstalling the drum, the brake shoes must be in the correct position. Briefly, if the shoes are expanded too far, the drum simply won't go over them. Refer to the procedures in Brake Shoe Removal and Installation for detailed instructions on setting the shoe clearance.

6. Install the drum with the bearings in place. Install the retaining nut and tighten it to 122 Nm or 90 ft. lbs.

INSPECTION

Measure the inside diameter of the drum with an inside micrometer at a number of different positions. Compare it with the specifications listed in the Brake Specifications Chart in this Section. If the drum is worn past the limit at any point, it must be replaced. Additionally, the diameter should be constant at all points measured. An out-of-round condition may cause vibration under braking.

Mitsubishi gives no limits for machining drums to remove uneven wear other than the absolute maximum diameter (wear limit). If the drum is worn unevenly around its diameter or has wear grooves, you may be able to salvage it by having it machined BUT machining must not increase the diameter past the specified limits. If it does, the drum must be replaced.

Drums, particularly the larger ones, can crack. Check the inner and outer surfaces and the area around the wheel lug holes for any sign of hairline cracks. Obviously, if any are found, the drum is useless.

With the drums off the car, the shoes may be inspected and measured for remaining thickness. Use a cloth dampened with water to wipe down the components or use an aerosol brake cleaner. Do not use a dry brush or air to clean the shoes. Mea-

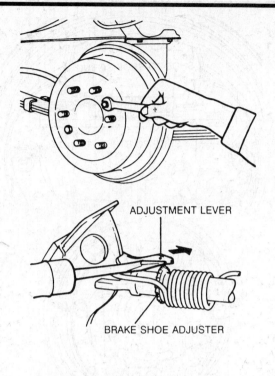
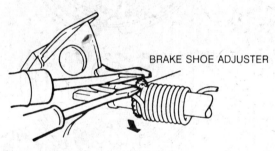

Alternate methods of dealing with stuck brake drums

bolts alternately and evenly; if the drum is cocked, it can cause damage as it comes off (if it comes off at all).

5. Before reinstalling the drum, the brake shoes must be in the correct position. Briefly, if the shoes are expanded too far, the drum simply won't go over them. Refer to the procedures in Brake Shoe Removal and Installation for detailed instructions on setting the shoe clearance.

6. Install the drum with the bearings in place. Install the flat washer and install the retaining nut finger tight. Tighten the retaining nut to 19 Nm (14 ft.lb). Loosen it until there is no tension on it and then retorque the nut to 7 Nm or 5 ft.lbs. Install the lock cap so that the holes in the cap and the axle align. If the holes in the cap and axle do not align, remove the cap and try it in a different position. If it still won't align, the nut may be turned up to 15° to make them align. Install a new cotter pin and install the hub cap.

NOTE: 15° of rotation is not a lot; it's equal to half the angle between 12 and 1 on a clock face.

Mirage

1. Block the front wheels, raise the vehicle at the rear and support it safely on floor stands. Remove the rear wheels.
2. Remove the hub cap. Use a regular socket wrench to loosen the center bolt and back it off two turns. Using and inch-pound

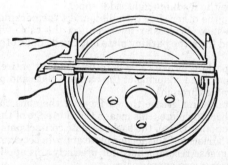

Measure the brake drum diameter at several points

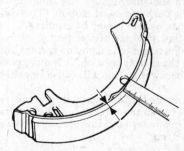

Measure the remaining thickness at the narrowest point

9 BRAKES

sure the friction material (not the backing plate) at the thinnest point. Refer to the Brake Specifications Chart for minimum thicknesses.

NOTE: It is a characteristic of brake shoes to show slightly uneven wear or wear each shoe differently. A shoe may have very different thicknesses at opposite ends; if the rest of the system is in good condition, this is not a sign of a problem.

Brake Shoes

REMOVAL AND INSTALLATION

Disassemble brakes one wheel at a time. This will prevent parts confusion and also prevent the opposite wheel cylinder pistons from popping out during repair. While this work can be accomplished with common hand tools, there are several specialty brake tools available in retail stores which make everything much easier. If you can only get one item, buy a brake spring tool.

Cordia and Tredia
Precis
1985–87 Galant
Montero and Pickup

1. Elevate and safely support the car on stands.
2. Remove the rear brake drums as described above.
3. Using brake spring pliers, remove the spring connecting the two shoes and the spring connecting the right side shoe to the adjuster strut. Pickup and Monteros don't have the strut so disconnect the spring which is wrapped around the adjuster.
4. Remove the shoe hold-down spring on either side by first depressing slightly and then twisting the retaining cap. (There is a common brake tool that is designed to make this almost effortless.) The tangs on the retaining post must line up with the notches in the cap; release the cap in this way and then remove the cap, spring, and spring seat. Remove shoe retaining spring with the pliers.
5. Remove the leading (right side) shoe.
6. Remove the other shoe then disconnect the parking brake cable at the shoe. It's easier to disconnect the cable with the shoe removed from the backing plate but can be done otherwise if you wish.
7. If the parking brake cable is to be removed entirely, you can now remove the snapring at the backing plate and pull the end of the cable through.
8. To install, first grease the surfaces of the shoe that will contact the backing plate (this means the side edge of the shoe, NOT the friction surface), the backing plate contact points, and the working surfaces of the anchor plate and wheel cylinder pistons. Use a grease recommended for precisely this application; not all lubricants are appropriate.
9. Assemble hold down pins and springs and install the parking brake cable to the shoe. Install the shoe and secure the spring pin with the push-and-twist motion. (This can become a three-handed job; an assistant may be required.).
10. Install the other shoe and secure it.
11. Install the adjusting mechanism and springs, making sure the springs are in the correct position. It is quite possible to install the springs backwards; this will cause impaired function and noise.

When installing the shoe-to-shoe and shoe-to-strut springs, set the adjuster lever so the ratchet is locked in the fully released direction. On Montero and Pickups, screw the adjuster in to its narrowest setting.

NOTE: The adjuster on Montero and Pickup has an identification groove on the non-forked end. This groove must face outward at installation.

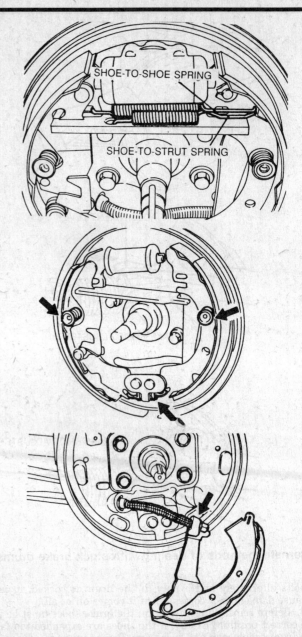

Cordia and Tredia shoe removal sequence. Don't forget the small spring at the bottom of the shoes

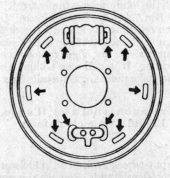

Points to be lubricated before installing brake shoes. Montero shown, others similar

BRAKES 9

12. Double check the placement and installation of all components. Everything must be accurately placed and securely fastened.
13. Use the adjuster to set the brakes to the correct diameter. Use an accurate measuring device to measure the outer diameter of the brake shoes. This measurement must be set correctly; otherwise either the drum will not go over the shoes or the adjustment will be too loose. Correct measurements are:
- Montero: 253.0–253.5mm
- Pickup, 2-wd through 1986: 240.0–240.5mm
- Pickup, all others: 253.0–253.5mm
- Cordia and Tredia: 202.0–202.5mm
- Precis: 179.0–179.5mm
- Galant: 202.0–202.5mm

14. Install the brake drum as described above.
15. Install the wheels. Make sure you depress the brake pedal repeatedly to fully adjust the self-adjusters before operating the vehicle. Operate the vehicle at very low speed in both forward and reverse gears, pumping the brake pedal repeatedly. Bring the car to a full stop and operate the parking brake several times. Pedal feel should be normal, but the car should roll freely (with no drag) when the pedal is released. If any condition is abnormal, the brake adjustment and/or components must be investigated and corrected.

Mirage

1. Elevate and safely support the car on stands.
2. Remove the rear brake drums as described above.
3. Remove the clip that sits just in front of the shoe retainer spring at the bottom of the backing plate. Use the blade of an ordinary screwdriver to lift the clip off the retaining tab and then remove it. Remove the spring.
4. Remove the shoe hold-down spring on either side by first depressing slightly and then twisting the retaining cap. (There is a common brake tool that is designed to make this almost effortless.) The tangs on the retaining post must line up with the notches in the cap; release the cap in this way and then remove the cap, spring, and spring seat.
5. Use brake spring pliers to remove the shoe-to-shoe spring. This is the large U-shaped spring running between the shoes.

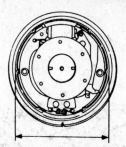

Correct pre-adjustment of the shoes requires measuring the outer diameter correctly

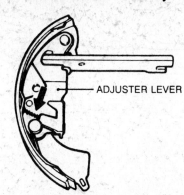

The adjusting lever should be fully released before assembly

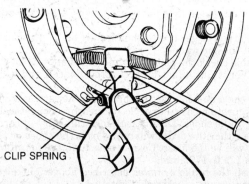

The Mirage shoe retainer spring is held by a clip which must be removed

6. Remove the shoes and the adjustor as a unit. Disconnect the parking brake cable from the shoe and then separate the components.
7. If the parking brake cable is to be removed entirely, you can now remove the snapring at the backing plate and pull the end of the cable through.
8. To install, first grease the surfaces of the shoe that will contact the backing plate (this means the side edge of the shoe, NOT the friction surface), the backing plate contact points, and the working surfaces of the anchor plate and wheel cylinder pistons. Use a grease recommended for precisely this application; not all lubricants are appropriate.
9. Assemble the adjuster and bottom spring to the shoes and install the parking brake cable to the shoe. Install the shoes and secure the spring pins with the push-and-twist motion. This can become a three-handed job; an assistant may be required.
10. Install the remaining springs, making sure the springs are in the correct position. It is quite possible to install the springs backwards; this will cause impaired function and noise.

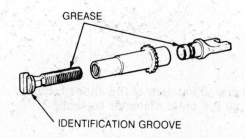

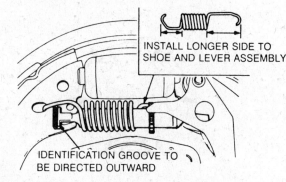

Montero and Pickup adjuster layout. Note correct installation of spring

9-29

9 BRAKES

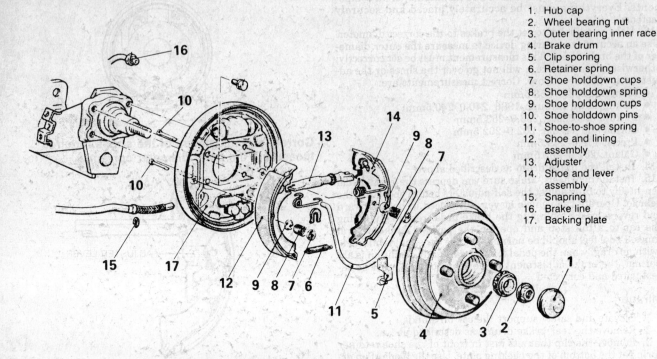

1. Hub cap
2. Wheel bearing nut
3. Outer bearing inner race
4. Brake drum
5. Clip spuring
6. Retainer spring
7. Shoe holddown cups
8. Shoe holddown spring
9. Shoe holddown cups
10. Shoe holddown pins
11. Shoe-to-shoe spring
12. Shoe and lining assembly
13. Adjuster
14. Shoe and lever assembly
15. Snapring
16. Brake line
17. Backing plate

Mirage rear brake assembly

When installing the shoes and springs, screw the adjuster in to its narrowest setting.

11. Install the large shoe-to-shoe spring. Install the clip at the bottom of the spring.
12. Double check the placement and installation of all components. Everything must be accurately placed and securely fastened.
13. Use the adjuster to set the brakes to the correct diameter. Use an accurate measuring device to measure the outer diameter of the brake shoes. This measurement must be set correctly; otherwise either the drum will not go over the shoes or the adjustment will be too loose. Correct measurement for all Mirage models is 179.0–179.5mm.
14. Install the brake drum as described above. Make certain the the center nut is correctly tightened.
15. Install the wheels. Depress the brake pedal repeatedly to fully adjust the self-adjusters before operating the vehicle. Operate the vehicle at very low speed in both forward and reverse gears, pumping the brake pedal repeatedly. Bring the car to a full stop and operate the parking brake several times. Pedal feel should be normal, but the car should roll freely (with no drag) when the pedal is released. If any condition is abnormal, the brake adjustment and/or components must be investigated and corrected.

Wheel Cylinders

INSPECTION ON CAR

The wheel cylinder seals are prone to leakage from deterioration. This condition may be checked on the car without disassembly. After the car is elevated and supported safely, remove the rear wheels. Remove the brake drum on each side.

Carefully lift the edge of each boot away from the body of the cylinder. Inspect the inside of the boot and the edge/end of the cylinder. You can expect a very slight (normal) moistness in the area—usually covered in dust—but any sign of wet fluid is cause

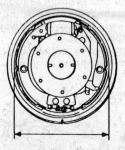

Correct pre-adjustment of the shoes requires measuring the outer diameter correctly

Grease all the locations shown lightly before installing new brake shoes

BRAKES 9

for immediate repair. A leak, no matter how slight, can significantly reduce the braking effort on that wheel. It can also admit air into the brake system, causing poor pedal feel.

REMOVAL AND INSTALLATION

1. Remove the brake drums and shoes as described previously. Although some vehicles will allow the cylinder to be removed with the brake shoes in place, this is not recommended. Chances are good that the brake shoes will become soaked with brake fluid. Contaminated brake shoes will need replacement so you'll have to take them off anyway.
2. Place a drain pan under the work area and disconnect the brake line at its connection to the wheel cylinder (behind the backing plate). Plug the open end of the brake line.

NOTE: Work carefully and use a line wrench on the fitting. This fitting is usually rusted or otherwise difficult to turn. Don't damage the brake line.

3. Remove the two bolts that fasten the wheel cylinder to the backing plate and remove the wheel cylinder. Again, these bolts may be hardet than usual to turn.
4. Install the cylinder and tighten the mounting bolts to 10 Nm or 7 ft. lbs. This is just enough to hold the cylinder in place; overtightening will strip the bolt holes.
5. Connect the brake line to the cylinder.
6. Install the shoes and drums correctly. Make sure the self-adjusters have taken up play so the brakes actuate normally.
7. Bleed the system thoroughly before operating the vehicle.

OVERHAUL

NOTE: This procedure requires a special piston cup installing tool MB 991008 or equivalent for the 17.5mm piston used on Cordia, Tredia and Galant or MB 990619-01 for use on the 19mm piston on the Mirage and Precis. 2-wheel drive Pickups and all Monteros require MB 990620-01 and 4-wheel drive Pickups need MB 990621-01. A generous supply of DOT-3 brake fluid, alcohol or aerosol brake cleaner should be available for cleaning purposes.

1. Remove the wheel cylinder.
2. On either side, work the rubber boot off with a soft, blunt instrument such as a rounded piece of wood and then remove the piston.
3. Inspect the piston and cylinder walls for corrosion or scoring and replace parts that are defective. Check the clearance between the piston and cylinder wall. (Subtract the outer diameter of the piston from the inner diameter of the bore.) The limit is 0.15mm. If clearance exceeds that value, replace the parts. Use inside and outside micrometers to make the measurements.
4. Clean all the parts with either brake cleaning spray or clean brake fluid. Clean the pistons and cylinder wall with DOT 3 brake fluid. If the piston/cylinder can be reused, replace the piston cup as follows:
    a. Work the cup off the end of the piston without causing damage.
    b. Apply the special grease included in the rebuild kit to the new piston cup and the special tool. If no grease is provided, use clean brake fluid.
    c. Set the piston on the workbench with one end up. Mount the special tool on top of the piston. Put the piston cup over the special tool with its lip facing upward.
    d. Slide the cup slowly and steadily down the special tool and into the groove of the piston. Be careful not to stop partway down! The idea is one smooth motion without stretching or ripping the cup.
    e. Repeat the operation for the other piston.

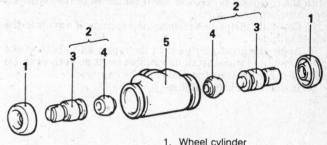

1. Wheel cylinder
2. Piston assembly
3. Piston
4. Piston cup
5. Wheel cylinder body

Typical wheel cylinder assembly. The piston cups are not reusable after disassembly

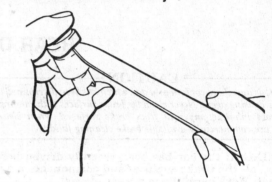

Removing the piston cap

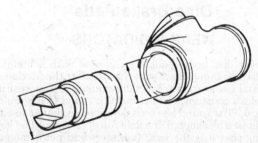

Measure the outer diameter of the piston and the inner diameter of the cylinder

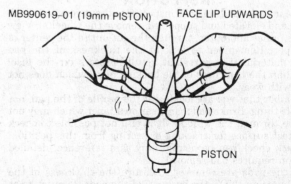

Install a new piston cup with the special tool to prevent damage

9-31

9 BRAKES

5. Coat the wheel cylinder bore and and piston cups either with the corrosion preventive agent contained in the repair kit or with clean brake fluid.
6. Carefully install the pistons (and springs, if any) into the wheel cylinder.
7. Apply the grease included in the repair kit to both of the piston ends and then install new rubber boots on both ends. Do not use other greases; the seals may be damaged.
8. Install the wheel cylinder.

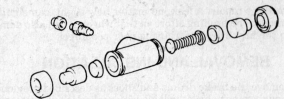
Precis wheel cylinder

REAR DISC BRAKES

CAUTION

Brake pads and shoes contain asbestos, which has been determined to be a cancer causing agent. Never clean the brake surfaces with compressed air! Avoid inhaling any dust from brake surfaces! When cleaning brakes, use commercially available brake cleaning fluids.

WARNING: If the car has been recently driven before repairs, all the brake surfaces and components may be very hot. Work carefully and wear gloves.

Disc Brake Pads

WEAR INDICATORS

Some rear disc brake pads are equipped with a metal tab which will come into contact with the disc after the friction surface material has worn near its usable minimum. The wear indicators make a constant, distinct metallic sound that should be easily heard. (The sound has been described as similar to either fingernails on a blackboard or a field full of crickets.) The key to recognizing that it is the wear indicators and not some other brake noise is that the sound is heard when the car is being driven WITHOUT the brakes applied. It may or may not be present under braking during normal driving.

INSPECTION

The rear brake pads may be inspected without removal. With the rear end elevated and supported, remove the wheel(s). View the pads, inner and outer, through the cut-out in the center of the caliper. Remember to look at the thickness of the pad friction material (the part that actully presses on the disc) rather than the thickness of the backing plate which does not change with wear.

Remember that you are looking at the profile of the pad, not the whole thing. Brake pads can wear on a taper which may not be visible through the window. It is also not possible to check the contact surface for cracking or scoring from this position. This quick check can be helpful only as a reference; detailed inspection requires pad removal.

After the pads are removed, measure the thickness of the LINING portion (NOT the backing) of the pads. It must be at least 1mm for all 1983-86 vehicles and 2mm for all 1987-89 vehicles. This is a Mitsubishi factory minimum; please note that local inspection standards enforced by your state must be given precedence if they require a greater thickness.

REMOVAL AND INSTALLATION

NOTE: Whenever brake pads are replaced, replace them in complete sets; that is, replace the pads on both rear wheels even if only one side is worn.

There are several combinations of shims, spacers and clips in use on Mitsubishi vehicles. When disassembling, work on one side at a time and pay attention to placement of these components. If you become confused during reassembly, refer to the other side for correct placement.

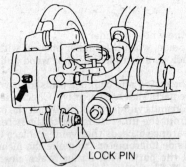
Rear disc brake inspection port

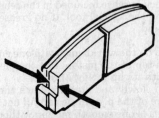
Inspect the remaining thickness by measuring the friction material, NOT the backing plate

BRAKES 9

Starion

1. Raise the vehicle and support it safely on stands.
2. Remove the rear wheels. Disconnect the parking brake cable at the caliper by removing the cotter pin and then the clevis pin and washer.
3. Unscrew and remove the lower caliper lock pin without removing its grease coating or allowing it to get dirty. Raise the caliper upward with the upper mounting pin serving as a hinge and use a piece of wire to tie or hold it in place.
4. Release the clips and then remove the pads and shims.
5. Use a large C-clamp or an equivalent device to depress the caliper piston back into the caliper. Make sure the two indentations in the caliper piston are aligned as shown. The pins on the backing plates for the pads must fit into these recesses when the pads are replaced.
6. Fit the pad and shim together and then install them into the clips and in proper relationship with the caliper piston.
7. Unwire the body of the caliper and lower it back to its normal position.
8. Install the lock pin and torque it to 55 Nm or 40 ft. lbs.
9. Reconnect the parking brake cable, using a new cotter pin.
10. Replace the wheels and lower the vehicle.
11. Make sure there is plenty of fluid in the reservoir. Pump the brake pedal repeatedly, checking the fluid level and keeping the reservoir adequately filled. Make sure the brakes actuate at the normal pedal position, indicating the fluid has pushed the pads directly against the disc before operating the vehicle.
12. Since the hydraulic system was not opened, bleeding is usually not necessary after pad replacement.

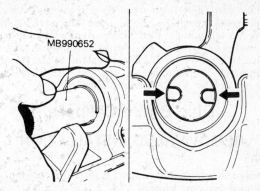

Depress the rear caliper pistons back into the Starion caliper with the correct tool. Note that the indentations MUST be aligned as shown before installing the brake pads

Galant and Sigma

1. Raise the vehicle and support it safely on stands.
2. Remove the rear wheels.
3. Disconnect the parking brake cable at the caliper by removing the special clip at the caliper and then disconnecting the fitting located on the end of the cable.
4. Unscrew and remove the lower caliper lock pin, without removing its grease coating or allowing it to get dirty. Raise the caliper upward with the upper mounting pin serving as a hinge and wire it in place.
5. Remove the shim and pad and remove both pad mounting clips. Keep them in order for installation in the same position.
6. A special tool, MB 990652 or equivalent, must be used to press the caliper piston back into the caliper. This allows you to press against the piston in spite of its indentations and projection. It will also help in turning the piston so the grooves align as shown. This must be done for the piston to fit properly against the pads.
7. Install the pad clips, assemble each pad to its corresponding shim, and then assemble the two into position. Make sure the pins on the back of the pad will align with the piston recess.
8. Unhook the body of the caliper and swing it back to its normal position. Install the lock pin and tighten it to 27 Nm or 20 ft. lbs.
9. Reconnect the parking brake cable, using a new cotter pin.
10. Replace the wheels and lower the vehicle.
11. Make sure there is plenty of fluid in the reservoir. Pump the brake pedal repeatedly, checking the fluid level and keeping the reservoir adequately filled. Make sure the brakes actuate at the normal pedal position, indicating the fluid has pushed the pads directly against the disc before operating the vehicle.

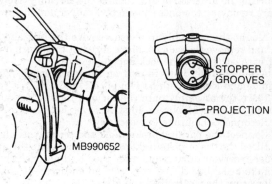

For Galant rear discs, the indentations in the caliper piston must be aligned as shown. The pins on the back of each brake pad must fit the indentation

Mirage

1. Raise the vehicle and support it safely on stands.
2. Remove the rear wheels. Disconnect the parking brake cable at the caliper by removing the cotter pin and then the clevis pin and washer.
3. Unscrew and remove the lower caliper lock pin without removing its grease coating or allowing it to get dirty. Raise the

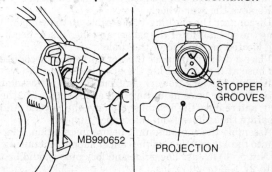

Positioning the Mirage rear disc brake piston

caliper upward with the upper mounting pin serving as a hinge and use a piece of wire to tie or hold it in place.
4. Release the clips and then remove the pads and shims.
5. Use a tool MB 990652 or an equivalent device to press the caliper piston back into the caliper. Make sure the two indentations in the caliper piston are aligned as shown. The pins on the backing plates for the pads must fit into these recesses when the pads are replaced.
6. Fit the pad and shim together and then install them into the clips and in proper relationship with the caliper piston.
7. Unwire the body of the caliper and lower it back to its normal position.
8. Install the lock pin and torque it to 27 Nm or 20 ft. lbs.
9. Reconnect the parking brake cable, using a new cotter pin.
10. Replace the wheels and lower the vehicle.

9 BRAKES

11. Make sure there is plenty of fluid in the reservoir. Pump the brake pedal repeatedly, checking the fluid level and keeping the reservoir adequately filled. Make sure the brakes actuate at the normal pedal position, indicating the fluid has pushed the pads directly against the disc before operating the vehicle.

12. Since the hydraulic system was not opened, bleeding is usually not necessary after pad replacement.

Disc Brake Calipers

REMOVAL AND INSTALLATION

Starion

1. Raise the vehicle and support it safely on stands.
2. Remove the rear wheels.
3. Disconnect the parking brake cable at the caliper by removing the cotter pin and then the clevis pin and washer.
4. Place a drain pan under the work area and disconnect the brake hose at the strut. Immediately plug the disconnected line.
5. Remove the two mounting bolts attaching the rear caliper support to the lower strut and remove the assembly. Note that you are removing the entire assembly (caliper and mount) as a unit.
6. Mount the assembly in a vise, remove the lock and guide pins, and remove the caliper. Make sure not to disturb or dirty the grease coating on the lock and guide pins.
7. Assemble the complete unit on the work bench. Tighten the guide and lock pins to 54 Nm (40 ft. lbs).
8. Place the assembly onto the car and tighten the support mounting bolts to 54 Nm or 40 ft. lbs.
9. Reconnect the brake line securely.
10. Connect the parking brake cable and secure the clip.
11. Bleed the brake system thoroughly, topping up the master cylinder reservoir during and after the bleeding. Make certain the brake pedal has the correct feel and engagement point before lowering the car to the ground.

Galant, Sigma and Mirage

1. Raise the vehicle and support it safely on stands.
2. Remove the rear wheels.
3. Disconnect the parking brake cable at the caliper by removing the special clip at the caliper and then disconnecting the fitting located on the end of the cable.
4. Place a drain pan under the work area and disconnect the brake hose at the strut. Plug the open line immediately.
5. Remove the lock (lower) pin from the caliper assembly.
6. Remove the bolts mounting the caliper support to the lower strut and remove the assembly. Remove the cap and guide pin to separate the caliper from the caliper support.
7. Assemble the complete unit before installation. Place the unit onto the car and tighten the caliper support mounting bolts to 54 Nm or 40 ft. lbs. The guide and lock pins should be tightened to 27 Nm or 20 ft. lbs.
8. Reconnect the brake line securely.
9. Connect the parking brake cable and install the clip securely.
10. Bleed the system thoroughly, keeping the master cylinder reservoir supplied with fluid during and after the procedure. Make certain the brakes have the correct feel and engagement point before lowering the car to the ground.

OVERHAUL

Starion

WARNING: The use of the correct special tools or their equivalent is REQUIRED for this procedure. Special tool MB990665-01 or equivalent is required for pressing the handbrake lever bearings in and out. You'll also need MB990666-01 or equivalent to compress the handbrake mechanism spring washers.

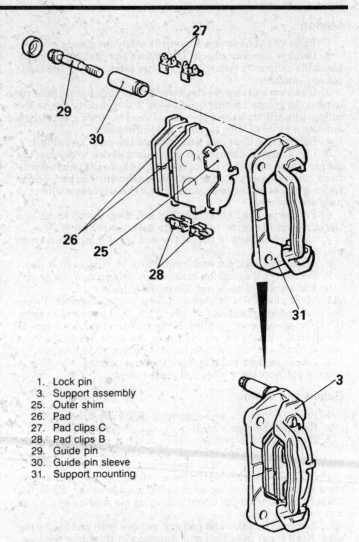

1. Lock pin
3. Support assembly
25. Outer shim
26. Pad
27. Pad clips C
28. Pad clips B
29. Guide pin
30. Guide pin sleeve
31. Support mounting

Galant rear disc, caliper and mount

1. Remove the caliper and support as described above. Separate the caliper from the support.
2. Remove the cap ring and garter spring from the parking brake lever cap. Remove the lever cap from the groove in the caliper.
3. Remove the retaining ring and pull out the parking brake lever assembly. Unscrew the parking brake spindle and remove it.
4. Push the piston out of the caliper body with a blunt instrument or rod. The piston must move toward the opposite side of the caliper body (in the direction in which it moves normally during braking).
5. With a non-metallic pointed instrument carefully remove the piston seal. Avoid damaging any surfaces with the tool.
6. Insert the MB 990665-01 tool or equivalent into the bore for the handbrake lever and then use a vice to press the bearing out of the bore. Do the same for the bearing on the opposite side.
7. Remove the guide and lock pin boots from the body of the caliper. Clean the assembly with fresh brake fluid aerosol brake cleaner.
8. Inspect parts as follows:
 • Check the slide pins and bushings for scoring.
 • Check the caliper support for fatigue cracks.
 • Check the body of the caliper, the piston, and the parking brake lever for corrosion.

BRAKES 9

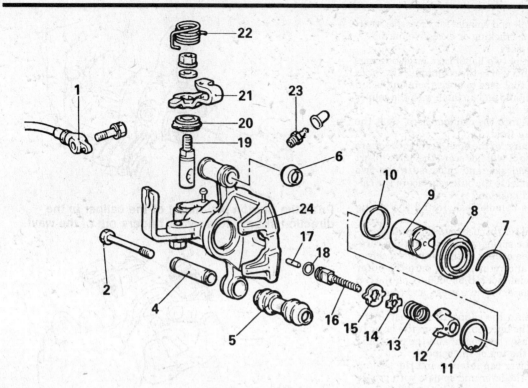

1. Brake hose
2. Lock pin
4. Lock pin sleeve
5. Lock pin boot
6. Guide pin boot
7. Boot ring
8. Piston boot
9. Piston assembly
10. Piston seal
11. Snapring
12. Spring case
13. Return spring
14. Stopper plate
15. Stopper
16. Auto-adjust spindle
17. Connecting link
18. O-ring
19. Spindle lever
20. Lever boot
21. Parking brake lever
22. Return spring
23. Bleeder screw
24. Caliper

Galant and Sigma rear caliper. Mirage similar

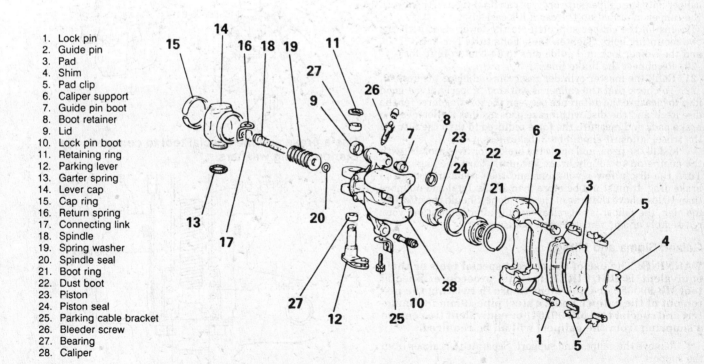

1. Lock pin
2. Guide pin
3. Pad
4. Shim
5. Pad clip
6. Caliper support
7. Guide pin boot
8. Boot retainer
9. Lid
10. Lock pin boot
11. Retaining ring
12. Parking lever
13. Garter spring
14. Lever cap
15. Cap ring
16. Return spring
17. Connecting link
18. Spindle
19. Spring washer
20. Spindle seal
21. Boot ring
22. Dust boot
23. Piston
24. Piston seal
25. Parking cable bracket
26. Bleeder screw
27. Bearing
28. Caliper

Starion rear caliper

9-35

9 BRAKES

- Inspect the connecting link and spindle for excessive wear.
- Check all rubber parts for cracking or excessive hardness.
- Replace all parts as necessary.

9. Coat the inner bores of the two handbrake lever bearings with the grease specified for this use in the rebuilding kit. Press the bearings in with the same tool used to press them out, again using the vise. The bearings should end up flush with the caliper body sections into which they fit.

10. Install a new piston seal into the cylinder bore. Coat the bore and seal with clean brake fluid.

11. Carefully install the piston into the bore, keeping it square and using minimal pressure so it will not score the bore. Insert the piston from the brake pad side and make sure the two notches are horizontal, or parallel to the outer surface of the caliper when it is in its mounted position.

12. Apply brake fluid to the spindle seal. Install the spring washers onto the spindle.

13. Apply the grease specified for this use onto the contact surface of the caliper body and the spring washers, and then carefully screw the spindle into the caliper until it rotates freely.

14. Set the connecting link and return spring onto the outer end of the spindle. Set up special tool MB 990666-01 on the top of the spindle, put the entire assembly in a press, and compress the spring washers.

15. Install the brake lever into the caliper body. Install the brake lever retaining clip on the opposite end from the lever to retain it. Install the cap. Release the pressure on the press and remove the special tool from the top of the spindle.

16. Thoroughly grease the lever cap interior and the sealing lip. Coat the working parts of the lever and spindle with grease as well. Install the lever cap to the caliper body. Install the cap ring and garter spring.

17. Coat the interior surfaces of the lock pin and guide pin boots with grease. Coat the interior surfaces of the caliper where the pins pass through it with this grease. Coat the surfaces of the pins with the grease also.

18. Install the pins into the caliper and use them to install the caliper onto the caliper support (you can final-tighten them with the caliper mounted on the car if it's easier).

19. Install the caliper support onto the lower strut with the two mounting bolts. Tighten these bolts to 44 Nm or 33 ft. lb. and the caliper lock and guide pins to 54 Nm or 40 ft. lbs.

20. Reconnect the brake line.

21. Refill the master cylinder reservoir and bleed the system.

22. To check that the caliper is working properly, it's a good idea to measure the difference between the rotating force for the disc itself and the disc with brake pads against it. Remove the brake pads and measure the force required to turn the disc via the wheel studs. It should be 3 lbs. or less.

Install the pads, and apply the brakes for five seconds (with the engine on to supply brake vacuum). Turn the engine off. Turn the disc a few revolutions and then again measure the brake drag. It must not be more than 15 lbs. total and no more than 15 lbs. above the drag of the disc alone. Should the effort be too high, the caliper is not releasing fully after application; your reassembly and/or reinstallation is suspect.

Galant, Sigma and Mirage

WARNING: The use of the correct special tools or their equivalent is REQUIRED for this procedure. Special tool MB 990652 or equivalent to use in twisting the piston out of the caliper body; a steel pipe 19mm in diameter; and special tool MB 991041 or equivalent (to remove a snapring from the caliper) will all be required.

1. Remove the caliper and support. Separate the caliper from the support.
2. With the blade of a small flat instrument, remove the boot ring from the caliper body and the pull out the piston boot.
3. Use special tool MB 990652 or equivalent to remove the piston from the caliper with a twisting motion.

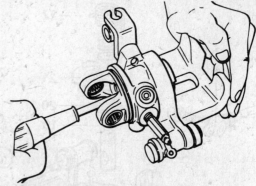

Push the Starion piston out of the caliper in the direction shown. Keep your fingers out of the way!

Insert the piston so the notches face as shown on the Starion

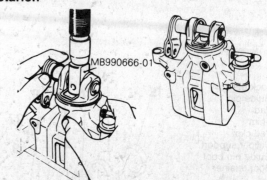

Use a press and the special tool to compress the Starion spring washers

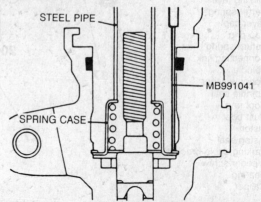

The spring case is pressed into the Galant caliper body with a 19mm (0.75 in.) diameter pipe and a special fitting

BRAKES 9

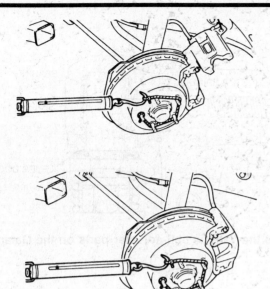

Check the brake disc for drag with the pads removed (above) and installed. A simple spring scale is required

remove the connecting link. Remove the O-ring from the adjuster spindle.

6. Being careful not to scratch the cylinder wall, remove the piston seal from the caliper body using a blunt, non-metallic tool.

7. Disconnect the return spring, remove the mounting nut, and pull the parking brake actuating lever out of the caliper assembly. Remove the rubber boot that seals the lever shaft from the caliper body.

8. Unscrew the guide and lock pin sleeves from the caliper body and remove their rubber boots.

9. Clean the piston and cylinder walls with alcohol or clean brake fluid.

Inspect parts as follows:
- Check the connecting link and spindle for excessive wear.
- Check the body of the caliper for fatigue cracks or rust.
- Check the spindle for rust. Check the bearing for the spindle shaft for excessive wear.
- Check the piston and the caliper wear surfaces for scoring or corrosion.
- Replace the piston seal.
- Check the piston rubber boot for cracking or brittleness.
- Replace all defective parts.

10. Apply the grease in the rebuild kit (specified for use on rubber parts) to the piston seal and to the groove in the cylinder wall. Install the new seal, making sure it is installed squarely.

11. Apply the proper grease in the rebuild kit to the spindle bearing, spindle, shaft and lever, lever rubber boot, connecting link, self-adjuster spindle, and related locations of the caliper body. Install the connecting link so the hole in the link lines up with the hole in the bearing.

12. Fit the dust boot into the caliper body. Insert the spindle shaft with the groove facing the hole in the bearing. Insert the connecting link from the cylinder side.

4. Using a 19mm diameter steel pipe to press the spring retainer or case into the caliper body, and use MB 991041 or equivalent to remove the snapring that holds the retainer. Remove the spring case, return spring, washer, and stop from the body.

5. Remove the automatic adjuster spindle from the body and

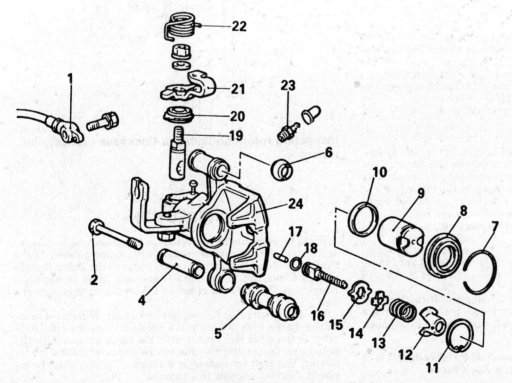

1. Brake hose
2. Lock pin
4. Lock pin sleeve
5. Lock pin boot
6. Guide pin boot
7. Boot ring
8. Piston boot
9. Piston assembly
10. Piston seal
11. Snapring
12. Spring case
13. Return spring
14. Stopper plate
15. Stopper
16. Auto-adjust spindle
17. Connecting link
18. O-ring
19. Spindle lever
20. Lever boot
21. Parking brake lever
22. Return spring
23. Bleeder screw
24. Caliper

Galant and Sigma rear caliper. Mirage similar

9-37

9 BRAKES

13. Coat the O-ring with clean brake fluid and mount it onto the auto-adjuster spindle. Insert the auto-adjuster spindle into the caliper.
14. Install the stop, spring washer, spring, and spring housing or case. Complete this assembly by using the steel pipe to press the spring case inward; use the special tool (used in removal) to attach the snapring to the caliper body. Make absolutely certain that the opening in the snapring faces the bleeder screw.
15. Use the special tool to twist the piston into the caliper. The cylinder is threaded so that the piston cannot move in and out of the bore without a twisting motion.
16. Apply the specified grease to the piston boot mounting grooves in the caliper body and the piston. Install the boot squarely. Carefully install the piston boot retaining ring.
17. Use the correct grease to coat the interior portion of the guide and lock pin sleeves, the threads of the caliper where the sleeves screw in, the rubber boot mounting groove, and the mounting groove for the cap. Then mount the sleeves onto the caliper and insert their respective pins.
18. With the pins, attach the caliper to the caliper support and tighten them to 27 Nm or 20 ft. lbs. Mount the caliper support to the car and tighten the bolts to 54 Nm or 40 ft. lbs.
19. Rotate the parking brake cable bracket clockwise until it contacts the caliper body and tighten the bracket in that position. Reconnect the cable to the bracket.
20. Reconnect the brake hose securely, bleed the system of air, and make sure the brakes respond normally (caliper pistons have positioned the pads against the discs) before operating the vehicle.

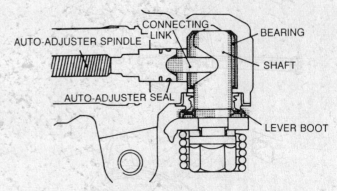

Grease the various self-adjuster parts on the Galant

Brake Discs

REMOVAL AND INSTALLATION

Starion

1. Remove the rear brake caliper and mount (brake assembly) following procedures given previously in this Section, but without disconnecting the hydraulic line. Suspend the assembly with wire; do not allow the unit to hang by the brake line.
2. Make mating marks to show the relationship between the brake disc and hub; pull the disc off the hub.
3. Installation is the reverse of removal.

Galant, Sigma and Mirage

1. Remove the rear brake caliper and mount (brake assembly) following procedures given previously in this Section, but without disconnecting the hydraulic line. Suspend the assembly with wire; do not allow the unit to hang by the brake line.
2. Remove the hub cap, cotter pin, lock cap, nut, thrust washer, and outer bearing. Remove the disc from the hub.
3. If necessary, repack the wheel bearings with multipurpose grease.
4. Install the bearings and disc, thrust washer, and nut. Tighten the nut to 19 Nm (14 ft. lbs). Loosen the nut until there is no torque on it and retighten it to 5.5 Nm or 4 ft. lbs. (48 INCH lbs).
5. Install the lock cap and a new cotter pin. If the lock cap castellations do not align with the cotter pin hole, reposition the cap. If necessary, back off the bearing retaining nut by as much as 15°. Install a new cotter pin.

NOTE: 15° is not a large rotation. For reference, it is ½ the angle between 12 and 1 on a clock face.

6. Install the hub cap.
7. Install the brake caliper and mount, following procedures given in this Section.

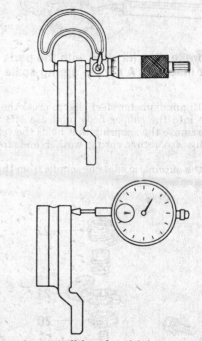

Inspecting rotor condition for thickness (above) and runout

INSPECTION

Brake discs should be inspected for thickness at a number of spots around the braking surface with a micrometer. If the thickness is less than the minimum (refer to the Brake Specifications Chart) AT ANY POINT, the disc must be replaced. Roughness or significant grooving is also reason to replace the disc.

The disc should also be checked for runout. With the disc installed (hold it with two lug nuts if necessary), mount a dial indicator to the strut and zero it with the tip in contact with the braking surface on the disc. Rotate the disc slowly. Read the indicator; the limit for runout is 0.15mm. The disc must be replaced if runout exceeds this amount.

Note that condition and adjustment of the wheel bearings can affect this reading. Before replacing the disc, check the bearings for proper adjustment and lubrication.

BRAKES 9

PARKING BRAKE

Cable

ADJUSTMENT

Except Pickup

WARNING: On vehicles with rear drum brakes, the shoe-to-drum clearance must be adjusted before adjusting the cable. If the brake clearance is out of adjustment, the cable adjustment will not cure the problem.

1. Apply the brake with normal pressure (about 40 lbs. tension) and count the number of clicks required to bring the lever tight enough to hold the car. Don't try to pull the lever into the back seat!

On all cars but the Starion and Montero, 5–7 clicks should be required; on the Starion and Montero, 4–5 clicks are sufficient. If the number of clicks is incorrect, proceed with the remaining steps.

2. Remove the rear console box. Proceed according to the model.

- Cordia and Tredia — Remove the gearshift lever knob on manual transaxle cars. Remove the parking brake cover with a flat knife blade by prying it upward. Remove the ashtray. Remove the front and rear mounting screws for the console and remove the console.
- Starion — Open the cover for the accessory box and remove the box from the console. Remove the panel from just in front of the accessory box by prying gently with a dull flat edge. Protect the surface with a rag. Remove the four installation screws (two front and two rear) and remove the console box.
- Mirage and Precis — Remove the parking brake cover and the ashtray. Remove the seven console mounting screws and remove the console.
- Galant and Sigma — Remove the inner box, plug, and the hole cover from the rear of the console. Remove the remote control mirror switch from the rear floor console. Remove the four rear floor console screws and remove the console.
- Montero — No removal of the console is required. The adjuster is under the vehicle.

3. Release the brake lever and adjust the cable adjuster (located at the center of the cable equalizer) until all cable slack is just removed.

4. Start the engine, allowing it to idle and apply the footbrake and release it. Then apply the handbrake and release it, apply the footbrake and release it, etc. in a continuous cycle until the automatic adjusters at the rear stop clicking.

5. Recheck the number of clicks required to apply the brake. Adjust the cable adjuster and repeat the check until the number of clicks required is correct.

6. Reinstall the console.

7. Raise the rear of the car and support it safely. Release the handbrake and rotate each rear wheel to make sure the brakes are not dragging.

Pickup

1. Pull the lever up with a force of about 66 lbs and count the number of notches or clicks. Correct number is 16 or 17.

2. If the number of clicks is incorrect, elevate and safely support the vehicle on stands; block the front wheels.

3. Under the vehicle, loosen the adjusting nuts to slacken each brake cable (one runs to each rear wheel from the equalizer).

4. Tighten the adjusting nuts just to the point of taking the slack out of the cable. Repeatedly pull and release the brake lever to adjust the rear brake shoes.

5. Tighten the adjusting nuts until the lever has the correct number of clicks.

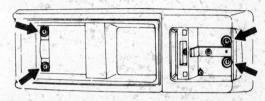

Four screws hold the Starion console box

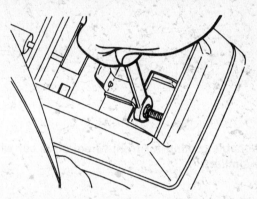

Adjusting the parking brake cable on Cordia and Tredia

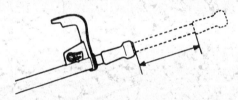

The Pickup parking brake should be fully applied at 16 or 17 clicks

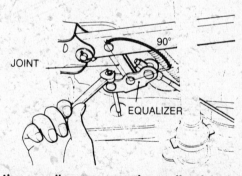

Keep the equalizer square when adjusting the cables

WARNING: Make certain the joint and the equalizer are at right angles when the adjustment is finished. An angle other than 90° will cause uneven brake application.

6. Fully release the parking brake lever and remove each rear wheel and brake drum. Make certain the brake lever is just touching the shoe.

WARNING: If the parking brake cable is pulled too far, the adjuster lever will not fit the adjuster, resulting in faulty operation.

9 BRAKES

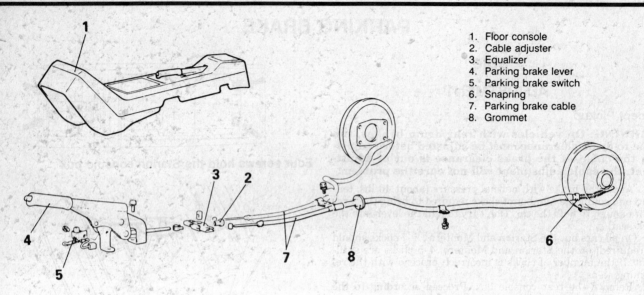

1. Floor console
2. Cable adjuster
3. Equalizer
4. Parking brake lever
5. Parking brake switch
6. Snapring
7. Parking brake cable
8. Grommet

Typical cable routing for front wheel drive car with rear drum brakes. This one is the 1986 Mirage, but all are similar

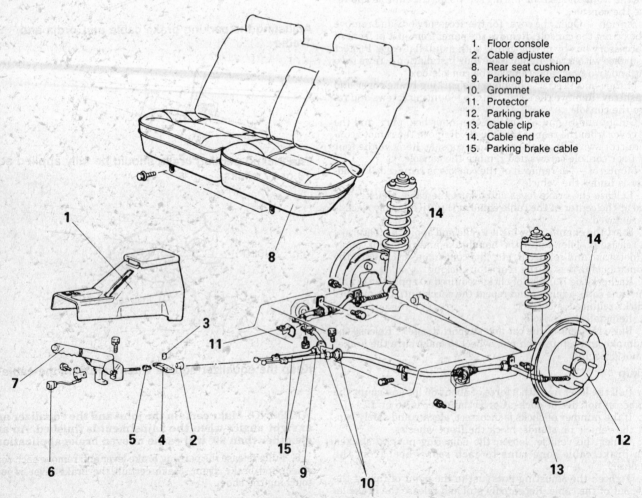

1. Floor console
2. Cable adjuster
8. Rear seat cushion
9. Parking brake clamp
10. Grommet
11. Protector
12. Parking brake
13. Cable clip
14. Cable end
15. Parking brake cable

Follow the numbered steps to remove the Galant parking brake cable. Layout and components are typical for a front wheel drive car with rear disc

9–40

BRAKES 9

7. Reinstall the drum, set the wheel bearing correctly, and install the wheel.
8. Apply and release the parking brake once or twice. With the brake fully released, spin each rear wheel by hand and check for dragging brakes.

REMOVAL AND INSTALLATION

Front Wheel Drive Models

1. Elevate and safely support the vehicle on stands. Make certain the handbrake lever is in the fully released position.
2. Remove the rear wheels.
3. If equipped with rear drum brakes, remove the brake drums.

CAUTION

Brake pads and shoes contain asbestos, which has been determined to be a cancer causing agent. Never clean the brake surfaces with compressed air! Avoid inhaling any dust from brake surfaces! When cleaning brakes, use commercially available brake cleaning fluids.

4. Remove the console box. Remove the rear seat bottom.
5. Disconnect the cable connectors at the equalizer. It may be necessary to loosen the cable adjuster to do this.
6. Disconnect all cable clamps from the body. Remove the mounting bolts for the large cable clamp just forward of where the cables pass through the body.
7. Pull the cables and grommets out of the body.
8. Disconnect the cables at the rear brakes. Refer to the appropriate rear brake procedure (disc or drum) in this Section if you need help.

NOTE: On some drum brake systems, it is much easier to remove the brake shoes first and then disconnect the cable.

9. Thread the cable into place, leaving it loose at each end. Make sure the grommets are installed in the body completely and that the concave side faces to the rear.
10. Connect the cable to the brake or caliper. Make certain that the cable is correctly placed and that the clips, pins or other retaining hardware is correctly installed.
    If the brake shoes were removed, reinstall them and install the drums.
11. Connect the cable to the equalizer and install the adjusting nuts until the cable is just free of slack.
12. Install the various cable clamps and brackets holding the cable to the body.
13. Adjust the handbrake mechanism as described in this Section.
14. Adjust the switch for the indicator light so the light comes on when the lever is pulled one notch.
15. Install the rear wheels and lower the car to the ground.
16. It is recommended to apply a weatherproof sealer to the grommets and the body panel where the cables pass through. This should only be done after the cables are firmly held by their clips.

Starion

1. Elevate and safely support the vehicle on stands. Make certain the handbrake lever is in the fully released position.
2. Remove the rear wheels.
3. Remove the console box.
4. Loosen the adjusting screw at the equalizer.
5. Remove the rear seat bottom cushion.
6. Disconnect the cable connectors at the equalizer.
7. Disconnect the cable at the rear caliper. Remove the dust shield in front of the caliper, then remove the cotter pin, clevis pin and washer. Remove the clip by pulling downward and remove the parking brake cable out of the groove.

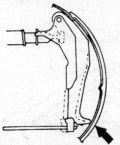
Check for light contact between the adjuster and the brake shoe

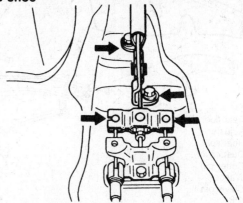
Disconnect the brake cables at the equalizer on front wheel drive models

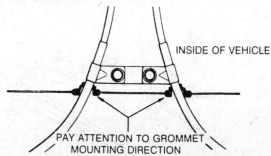
Make sure the grommets are installed in the correct direction, particularly on front wheel drive models

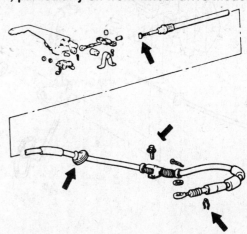
To replace the Starion parking brake cable, disconnect each arrowed fitting

9-41

9 BRAKES

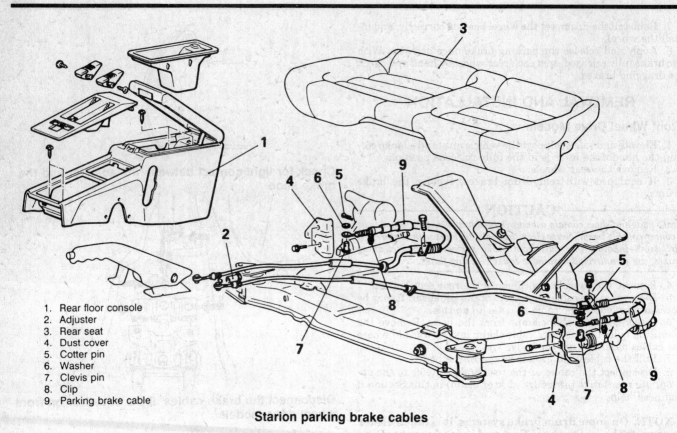

1. Rear floor console
2. Adjuster
3. Rear seat
4. Dust cover
5. Cotter pin
6. Washer
7. Clevis pin
8. Clip
9. Parking brake cable

Starion parking brake cables

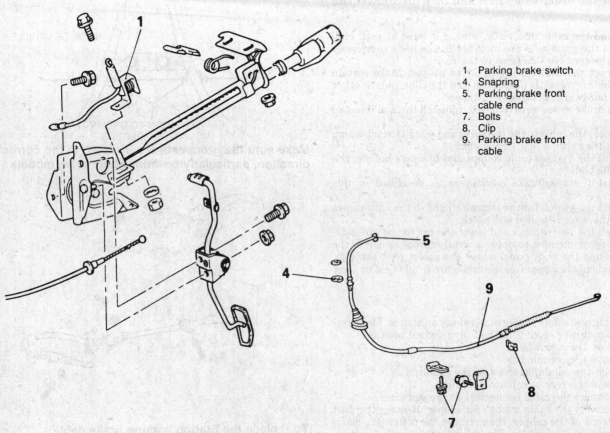

1. Parking brake switch
4. Snapring
5. Parking brake front cable end
7. Bolts
8. Clip
9. Parking brake front cable

Pickup front parking brake cable

9–42

BRAKES 9

8. Remove the cable bracket holding the cable to the lower control arm.
9. Remove the grommet from the body hole and remove the cable from the car.
10. Begin reinstallation by fitting the cable into position. Place and secure the clamp holding the cable to the control arm.
11. Install the cable into the groove in the rear caliper. Install the clip and assemble the clevis pin, washer and a new cotter pin. Install the dust shield in front of the caliper.
12. Connect the cable at the equalizer. Install the adjusting bolt if it was removed and tighten it just enough to take the slack out of the cables.
13. Apply a coat of body sealant around the contact face of the rubber body grommet and install the grommet. Make sure it is correctly seated.
14. Install the rear seat cushion.
15. Adjust the parking brake lever stroke following procedures given previously in this Section. Check the switch for the indicator light; it should come on when the lever is pulled one click.
16. Install the console.

Pickup

FRONT CABLE

1. Elevate and safely support the vehicle on stands.
2. Inside the vehicle, under the dash, identify and unplug the parking brake indicator switch. Remove the switch and bracket.
3. The end of the cable sheath is held in place by a snapring. Remove it. You are reminded that these snaprings tend to vanish if not removed carefully.
4. Lift the pawl from the ratchet and push the parking brake lever all the way in. In this position, the cable may be disconnected from the lever.
5. Under the vehicle, trace the cable and remove the clips and brackets holding it to the body.
6. Disconnect the cable from the lever assembly. Remove the cable from the vehicle. Don't damage the body grommet while removing it.
7. Route the new cable into position. It's usually easier to attach the inner end to the lever first, then deal with the equalizer end. Loosen the adjusting nuts on the rear cables if necessary.
8. Once both ends are secured, install the clips and brackets holding the cable to the body.
9. Fit the rubber body grommet into place and make certain it is secure.
10. Install the snapring on the end of the cable.
11. Install the brake indicator switch in place and connect the wiring but do not final tighten the switch mount.
12. Adjust the parking brake lever stroke according to procedures given earlier in this Section.
13. With the cable correctly adjusted, adjust the switch so that the warning light illuminates when the lever is pulled one notch. Tighten the switch mount.

REAR CABLE

1. Elevate and safely support the vehicle on stands.
2. Remove the rear wheels.
3. Remove the brake drum.

CAUTION

Brake pads and shoes contain asbestos, which has been determined to be a cancer causing agent. Never clean the brake surfaces with compressed air! Avoid inhaling any dust from brake surfaces! When cleaning brakes, use commercially available brake cleaning fluids.

4. At the equalizer, disconnect the return spring.
5. With the parking brake released, remove the adjusting nut from the cable to be removed. If only one cable is being removed, loosen the other adjuster to make the job easier. If both cables are being removed, loosen the nuts alternately and evenly.
6. Disconnect the brake cable from the brake shoe. You may

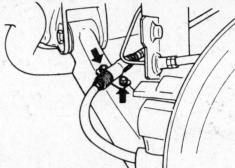

Don't forget to replace the brackets on the Starion control arms

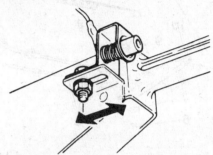

Move the bracket to adjust the Pickup parking brake indicator light switch

find it easier to remove the shoes as a unit, then disconnect the cable.
7. Remove the cable from the backing plate. This can be made easier by feeding an offset box wrench (12 mm) over the end of the cable and pushing it up until it reaches the stopper. Push on the wrench to release the tabs on the stopper and pull the cable free from behind the backing plate.
8. Remove the clips and brackets holding the cable to the body. Remove the cable from the vehicle.
9. When reinstalling, thread the cable into position. Make certain the stopper at the brake backing plate is securely fitted and the tab is engaged.
10. Install the cable end to the brake shoe. If the brakes were removed, reinstall them.
11. Turn the brake shoe adjuster to set the outside diameter of the shoes to 253.0–253.5mm. This is an important dimension and must be observed.
12. Install the brake drum.
13. Connect the brake cable to the equalizer and tighten the adjuster nut just enough to remove slack from the cable.
14. Install the return spring correctly.
15. Install the rear wheels.
16. Adjust the parking brake lever travel according to instructions given previously in this section.
17. Lower the vehicle to the ground.

Montero

1. Elevate and safely support the vehicle on stands.
2. Remove the rear wheels.
3. Remove the brake drum.

CAUTION

Brake pads and shoes contain asbestos, which has been determined to be a cancer causing agent. Never clean the brake surfaces with compressed air! Avoid inhaling any dust from brake surfaces! When cleaning brakes, use commercially available brake cleaning fluids.

9 BRAKES

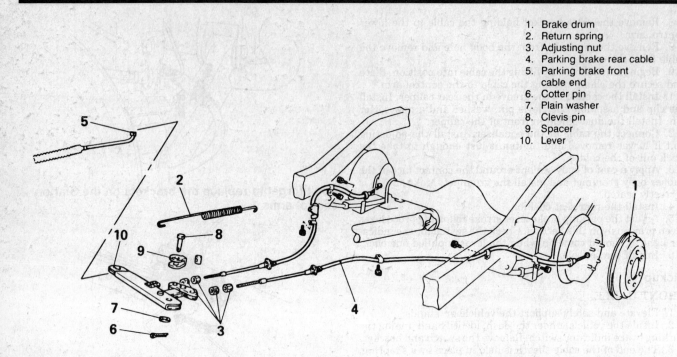

1. Brake drum
2. Return spring
3. Adjusting nut
4. Parking brake rear cable
5. Parking brake front cable end
6. Cotter pin
7. Plain washer
8. Clevis pin
9. Spacer
10. Lever

Pickup truck rear cable assembly

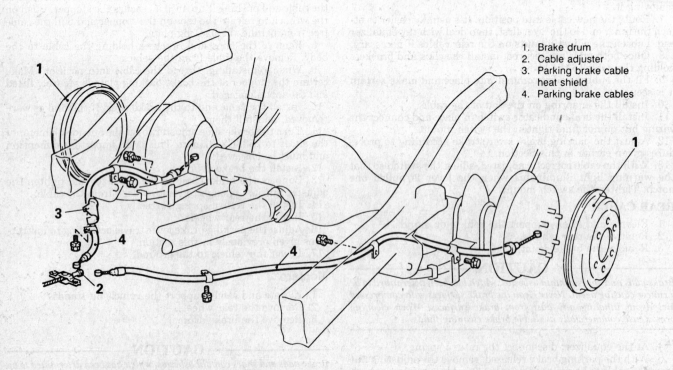

1. Brake drum
2. Cable adjuster
3. Parking brake cable heat shield
4. Parking brake cables

Montero parking brake cables

9-44

4. At the equalizer, remove the adjusting nut. Disconnect the cable to be replaced from the equalizer yoke. Remove the small heat shield from the cable.
5. At the rear wheel, remove the shoe return spring, the shoe retainer spring and the shoe hold-down pins. Remove the shoes as a unit, then disconnect the cable.
6. Remove the cable from the backing plate. This can be made easier by feeding a 12mm offset box wrench over the end of the cable and pushing it up until it reaches the stopper. Push on the wrench to release the tabs on the stopper and pull the cable free from behind the backing plate.
7. Remove the clips and brackets holding the cable to the body. Remove the cable from the vehicle.
8. When reinstalling, thread the cable into position. Make certain the stopper at the brake backing plate is securely fitted and the tab is engaged.
9. Install the cable end to the brake shoe. Install the brake shoes as an assembly and make certain the springs and retaining clips are properly placed.
10. Turn the brake shoe adjuster to set the outside diameter of the shoes to 253.0–253.5mm. This is an important dimension and must be observed.
11. Install the brake drum.
12. Connect the brake cable to the equalizer yoke. Install and tighten the adjuster nut just enough to remove slack from the cable.
13. Install the rear wheels.
14. Install the cable retaining clips and install the small cable heat shields.
15. Adjust the parking brake lever travel according to instructions given previously in this section.
16. Lower the vehicle to the ground.

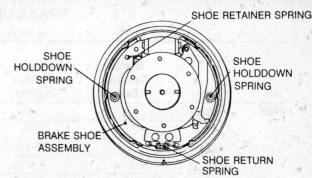

Montero brake shoes must be removed before disconnecting the brake cable

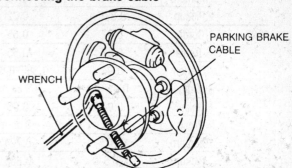

Avoid aggravation! Use a 12mm wrench to compress the tabs on Montero and Pickup brake cable tabs

BRAKE SPECIFICATIONS
All measurements in inches

Year	Model	Master Cylinder Bore	Brake Disc Minimum Thickness	Brake Disc Maximum Runout	Maximum Brake Drum Diameter	Minimum Lining Thickness Front	Minimum Lining Thickness Rear
1983	Cordia	0.87	0.65	0.006	8.000	0.040	0.040
	Tredia	0.87	0.65	0.006	8.000	0.040	0.040
	Starion	0.94	0.88F–0.65R	0.006	—	0.040	0.040
	Montero	0.87	0.72	0.006	10.08	0.040	0.040
	Pickup	0.87	0.72	0.006	9.57	0.040	0.040
1984	Cordia	0.87	0.65	0.006	8.000	0.040	0.040
	Tredia	0.87	0.65	0.006	8.000	0.040	0.040
	Starion	0.94	0.88F–0.65R	0.006	—	0.040	0.040
	Montero	0.87	0.72	0.006	10.08	0.040	0.040
	Pickup	0.87	0.72	0.006	9.57	0.040	0.040
1985	Cordia	0.87	0.65	0.006	8.000	0.040	0.040
	Tredia	0.87	0.65	0.006	8.000	0.040	0.040
	Starion	0.94	0.88F–0.65R	0.006	—	0.040	0.040
	Mirage	0.81 ④	① or ②	0.006	7.100	0.040	0.040
	Galant	0.94	0.65	0.006	8.000	0.040	0.040
	Montero	0.875	0.724	0.0059	10.079	0.040	0.040
	Pickup	0.875	0.72	0.006	10.07	0.040	0.040

9 BRAKES

BRAKE SPECIFICATIONS
All measurements in inches

Year	Model	Master Cylinder Bore	Brake Disc Minimum Thickness	Brake Disc Maximum Runout	Maximum Brake Drum Diameter	Minimum Lining Thickness Front	Minimum Lining Thickness Rear
1986	Cordia	0.87	0.65	0.006	8.000	0.040	0.040
	Tredia	0.87	0.65	0.006	8.000	0.040	0.040
	Starion	0.94	0.88F–0.65R	0.006	—	0.040	0.040
	Mirage	0.81 ④	① or ②	0.006	7.100	0.040	0.040
	Galant	0.94	0.65	0.006	8.000	0.040	0.040
	Montero	0.875	0.72	0.006	10.08	0.040	0.040
	Pickup	0.875	0.72	0.006	③ or ⑤	0.040	0.040
1987	Cordia	0.87	0.65	0.006	8.000	0.040	0.040
	Tredia	0.87	0.65	0.006	8.000	0.040	0.040
	Starion	0.94	0.88F–0.65R	0.006	—	0.040	0.040
	Mirage	0.81 ④	① or ②	0.006	7.100	0.040	0.040
	Galant	0.94	0.65	0.006	8.000	0.040	0.040
	Precis	0.81	⑥ or ⑦	0.006	7.100	0.040	0.040
	Montero	0.875	0.724	0.006	10.079	0.040	0.040
	Pickup	0.875 ⑧	0.803	0.006	10.079	0.079	0.040
1988	Cordia	0.87	0.65	0.006	8.000	0.040	0.040
	Tredia	0.87	0.65	0.006	8.000	0.040	0.040
	Starion	0.94	0.88F–0.65R	0.006	—	0.040	0.040
	Mirage	0.81 ④	① or ②	0.006	7.100	0.040	0.040
	Galant	0.94	0.65	0.006F–0.004R	—	0.040	0.040
	Precis	0.81	⑥ or ⑦	0.006F–0.004R	7.200	0.040	0.040
	Montero	0.9375	0.803	0.006	10.079	0.040	0.040
	Pickup	0.875 ⑧	0.803	0.006	10.079	0.079	0.040
1989	Starion	0.94	0.88F–0.65R	0.006	—	0.040	0.040
	Mirage	0.81 ④	① or ②0.33R	0.006	7.100	0.040	0.040
	Galant	0.94	0.88F–0.33R	0.004	—	0.080	0.080
	Precis	0.81	0.67	0.006	7.200	0.040	0.040
	Sigma	0.94	0.882F–0.646R	0.004	—	0.079	0.039
	Montero	0.9375	0.803	0.006	10.079	0.079	0.039
	Pickup	0.9375	0.803	0.006	10.079	0.079	0.040

① AD54 Caliper 0.65 in.
② PFS15 Caliper 0.45 in.
③ 2WD 9.57 in.
④ Turbo 0.87
⑤ 4WD 10.07
⑥ Sumitomo Caliper 0.45
⑦ Tokico Caliper 0.67
⑧ 4WD–0.9375

10 Body

QUICK REFERENCE INDEX

Exterior	10-2
Interior	10-31
Stain Removal	10-58

GENERAL INDEX

Antenna	10-24	Fog lights	10-22	Outside	10-24
Bumpers	10-15	Front bumper	10-15	Power window motor	10-35
Door glass	10-35	Glass		Rear bumper	10-15
Doors		Door	10-35	Seats	10-50
Adjustment	10-6	Grille	10-21	Stain removal	10-58
Glass	10-35	Hatch	10-7	Tailgate	10-7
Locks	10-34	Hood	10-6	Trunk lid	10-7
Removal and installation	10-2	Mirrors		Window glass	10-35
Door trim panel	10-31	Inside	10-50	Window regulator	10-35

10 BODY

EXTERIOR

Doors

REMOVAL AND INSTALLATION

WARNING: The doors are heavier than they appear. Support the door from the bottom and use a helper during removal and installation. Do not allow the door to sag while partially attached and do not subject the door to impact or twisting motions.

Front Door
CORDIA
TREDIA
STARION
MIRAGE THROUGH 1988
PRECIS
PICKUP

1. Disconnect the wiring harness (if any) running between the door and the body. Carefully pull the wiring free of the body.
2. Use a floor jack padded with rags or soft lumber to support the door at its lower midpoint. Use a felt tip marker to outline the hinge position on the door.
3. Remove the upper and lower hinge covers if any are present. Remove the small pin holding the door check arm in place.
4. Have a helper support the door, keeping it upright at all times. Remove the bolts holding the upper and lower hinge to the door. There will probably be alignment shims between the hinge and the door panel; take note of their location and placement for reassembly.
5. Using two people, lift the door clear of the car.
6. The door hinge may be removed from the body if necessary. On some vehicles the bolts are easily removed with common tools. On others, notably Cordia and Tredia and Pickups, the bolts are concealed. Use special tool MB 990900–01 or equiv-

The special door wrench (MB 990900–01) allows access to the hinge bolts

alent. This is a specially shaped wrench built to do the job; removing the bolts will be difficult without this tool.

7. If the door hinge is removed from the body, reinstall it and tighten the bolts to 45 Nm or 33 ft.lbs.
8. Place the door in position and support it. Use the jack to fine tune the position until the bolt holes and matchmark (hinge outline) align.
9. Install the hinge bolts and and the alignment shims. Tighten the bolts to 22 Nm or 16 ft. lbs.
10. If all has gone well, the door should be in almost the original position. Refer to the door adjustment procedures to align the door and body. It may be necessary to loosen the hinge bolts and reposition the door; remember to retighten them each time or the door will shift out of place.
11. Once adjusted, connect the door check lever and install the pin. Install the hinge covers if any were removed.
12. Route the wiring harness(es) into the body and connect them to their leads.
13. Test the operation of any electrical components in the door (locks, mirrors, speakers, etc.) and test drive the car, checking the door for air leaks and rattles.

GALANT
SIGMA
1989 MIRAGE
MONTERO

1. Remove the sill cover (scuff plate) for all except Montero. On Montero, remove the door opening trim.

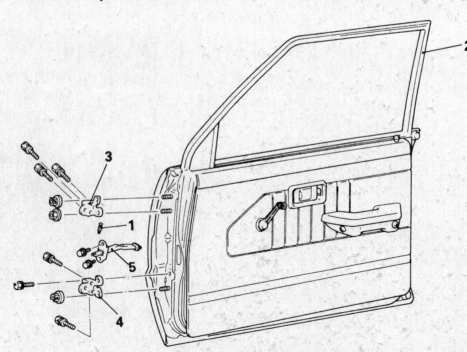

1. Spring pin
2. Door
3. Upper hinge
4. Lower hinge
5. Door limiter

Typical door hinge arrangement

BODY 10

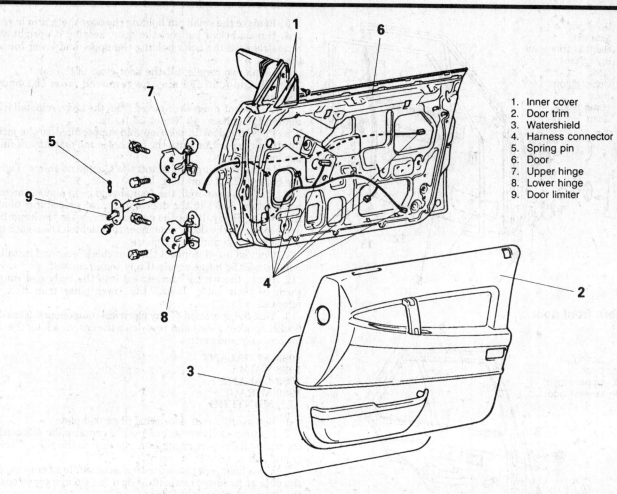

1. Inner cover
2. Door trim
3. Watershield
4. Harness connector
5. Spring pin
6. Door
7. Upper hinge
8. Lower hinge
9. Door limiter

For 1988 Galant and 1989 Sigma, the door liner must be removed

2. On all vehicles, remove the cowl side trim (kick panel) just in front of the door.
3. Disconnect the wiring harness running from the door to the body. Carefully pull the wire clear of the body. For 1988 Galant and 1989 Sigma, the door panel (liner) must be removed and the wiring disconnected inside the door. The wiring is then removed from the door, not the body.
4. Use a floor jack padded with rags or soft lumber to support the door at its lower midpoint. Use a felt tip marker to outline the hinge position on the door.
5. Remove the upper and lower hinge covers if any are present. Remove the small pin holding the door check arm in place.
6. Have a helper support the door, keeping it upright at all times. Remove the bolts holding the upper and lower hinge to the door. There will probably be alignment shims between the hinge and the door panel; take note of their location and placement for reassembly.
7. Using two people, lift the door clear of the car.
8. The door hinge may be removed from the body if necessary.
9. If the door hinge is removed from the body, reinstall it and tighten the bolts to 45 Nm or 33 ft. lbs.
10. Place the door in position and support it. Use the jack to fine tune the position until the bolt holes and matchmark (hinge outline) align.
11. Install the hinge bolts and the alignment shims. Tighten the bolts to 22 Nm or 16 ft. lbs.
12. If all has gone well, the door should be in almost the original position. Refer to the door adjustment procedures to align the door and body. It may be necessary to loosen the hinge bolts and reposition the door; remember to retighten them each time or the door will shift out of place.
13. Once adjusted, connect the door check lever and install the pin. Install the hinge covers if any were removed.
14. Route the wiring harness(es) into the body (or door) and connect them to their leads. Reinstall the door liner and related components if they were removed.
15. Reinstall the kick panel and scuff plate. On Montero, reinstall the door opening trim.
16. Test the operation of any electrical components in the door (locks, mirrors, speakers, etc.) and test drive the car, checking the door for air leaks and rattles.

Rear Door

**CORDIA AND TREDIA
MIRAGE THROUGH 1988
PRECIS**

1. Disconnect the wiring harness (if any) running between the door and the body. Carefully pull the wiring free of the body. This may require removing the center pillar lower trim for access to the connector.
2. Use a floor jack padded with rags or soft lumber to support the door at its lower midpoint. Use a felt tip marker to outline the hinge position on the door.

10 BODY

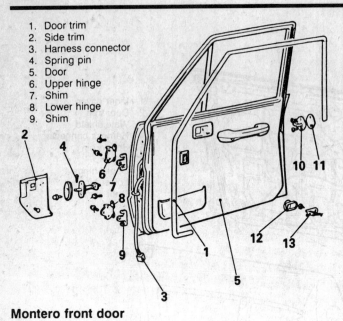

1. Door trim
2. Side trim
3. Harness connector
4. Spring pin
5. Door
6. Upper hinge
7. Shim
8. Lower hinge
9. Shim

Montero front door

1. Spring pin
2. Door
3. Upper hinge
4. Lower hinge

Typical rear door assembly. Although simple to remove and install, it still requires careful handling

1. Front scuff plate
2. Cowl side trim — front door
3. Rear scuff plate
4. Center pillar lower trim
5. Wiring harness
6. Spring pin
7. Door
8. Upper hinge
9. Lower hinge
10. Door limiter

3. Remove the small pin holding the door check arm in place.
4. Have a helper support the door, keeping it upright at all times. Remove the bolts holding the upper and lower hinge to the door.
5. Using two people, lift the door clear of the car.
6. The door hinge may be removed from the body if necessary.
7. If the door hinge is removed from the body, reinstall it and tighten the bolts to 45 Nm or 33 ft. lbs.
8. Place the door in position and support it. Use the jack to fine tune the position until the bolt holes and matchmark (hinge outline) align.
9. Install the hinge bolts and the alignment shims. Tighten the bolts to 22 Nm or 16 ft. lbs.
10. If all has gone well, the door should be in almost the original position. Refer to the door adjustment procedures to align the door and body. It may be necessary to loosen the hinge bolts and reposition the door; remember to retighten them each time or the door will shift out of place.
11. Once adjusted, connect the door check lever and install the pin. Install the hinge covers if any were removed.
12. Route the wiring harness(es) into the body and connect them to their leads. Install the lower pillar trim if it was removed.
13. Test the operation of any electrical components in the door (locks, speakers, etc.) and test drive the car, checking the door for air leaks and rattles.

1985–87 GALANT
1988 SIGMA
1989 GALANT
1989 MIRAGE
ALL MONTERO

1. Remove the rear door scuff plate (sill plate).
2. Remove the lower trim from the center pillar. Disconnect the wiring harness running from the door to the pillar. Carefully pull the wire clear of the pillar.
3. Use a floor jack padded with rags or soft lumber to support the door at its lower midpoint. Use a felt tip marker to outline the hinge position on the door.

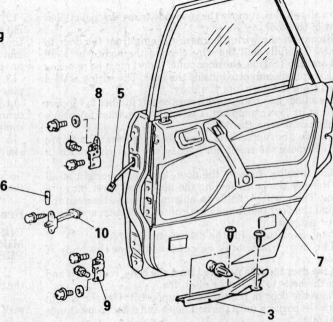

1989 Galant rear door

BODY 10

4. Remove the small pin holding the door check arm in place.

5. Have a helper support the door, keeping it upright at all times. Remove the bolts holding the upper and lower hinge to the door. Using two people, lift the door clear of the car.

6. The door hinge may be removed from the body if necessary.

7. If the door hinge is removed from the body, reinstall it and tighten the bolts to 45 Nm or 33 ft. lbs. except on 1989 Mirage; tighten the Mirage hinge bolts only to 22 Nm or 16 ft. lbs.

8. Place the door in position and support it. Use the jack to fine tune the position until the bolt holes and matchmark (hinge outline) align.

9. Install the hinge bolts and the alignment shims. Tighten the bolts to 22 Nm or 16 ft. lbs.

10. The door should be in almost the original position. Refer to the door adjustment procedures to align the door and body. It may be necessary to loosen the hinge bolts and reposition the door; remember to retighten them each time or the door will shift out of place.

11. Once adjusted, connect the door check lever and install the pin. Install the hinge covers if any were removed.

12. Route the wiring harness(es) into the body and connect them to their leads.

13. Install the lower pillar trim and the sill plate.

14. Test the operation of any electrical components in the door (locks, speakers, etc.) and test drive the car, checking the door for air leaks and rattles.

1988 GALANT AND 1989 SIGMA

1. Remove the scuff plate from the front door opening.
2. Remove the roof rail trim.
3. Pop out the small cover over the upper seat belt bolt and remove the bolt.
4. Carefully remove the upper trim from the center pillar.
5. Remove the rear door sill plate and remove the lower trim from the center pillar.
6. Disconnect the door wiring harness at the junction within the pillar. Carefully work the wiring clear of the pillar.
7. Remove the small pin holding the door check arm in place.
8. Have a helper support the door, keeping it upright at all times. Remove the bolts holding the upper and lower hinge to the door. Using two people, lift the door clear of the car.
9. The door hinge may be removed from the body if necessary.
10. If the door hinge is removed from the body, reinstall it and tighten the bolts to 45 Nm or 33 ft. lbs.
11. Place the door in position and support it. Use the jack to fine tune the position until the bolt holes and matchmark (hinge outline) align.
12. Install the hinge bolts and the alignment shims. Tighten

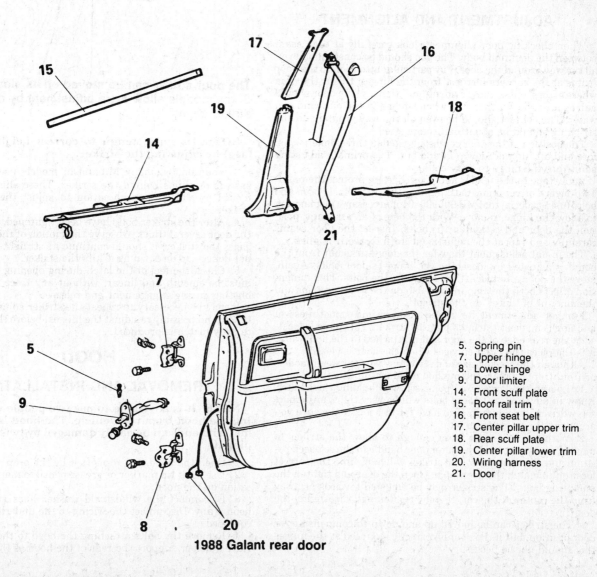

5. Spring pin
7. Upper hinge
8. Lower hinge
9. Door limiter
14. Front scuff plate
15. Roof rail trim
16. Front seat belt
17. Center pillar upper trim
18. Rear scuff plate
19. Center pillar lower trim
20. Wiring harness
21. Door

1988 Galant rear door

10-5

10 BODY

the bolts to 22 Nm or 16 ft. lbs.

13. The door should be in almost the original position. Refer to the door adjustment procedures to align the door and body. It may be necessary to loosen the hinge bolts and reposition the door; remember to retighten them each time or the door will shift out of place.
14. Once adjusted, connect the door check lever and install the pin.
15. Route the wiring harness(es) into the pillar and connect them to their leads.
16. Install the lower pillar trim and both the front and rear sill plates.
17. Install the upper pillar trim.
18. Install the seat belt and retaining bolt. Make sure any washers or spacers are correctly installed. Tighten the bolt to 45 Nm or 33 ft. lbs. Reinstall the small plastic cap over the bolt head.

WARNING: Seat belt installation affects a critical safety system; make very certain it is done correctly.

19. Install the roof rail trim.
20. Test the operation of any electrical components in the door (locks, speakers, etc.) and test drive the car, checking the door for air leaks and rattles.

ADJUSTMENT AND ALIGNMENT

When checking door alignment, look carefully at each seam between the door and body. The gap should be constant ad even all the way around the door. Pay particular attention to the door seams at the corners farthest from the hinges; this is the area where errors will be most evident. Additionally, the door should pull against the weatherstrip when latched to seal out wind and water. The contact should be even all the way around and the stripping should be about half compressed.

The position of the door can be adjusted in three dimensions: fore and aft, up and down, in and out. The primary adjusting points are the hinge-to-body bolts.

Apply tape to the fender and door edges to protect the paint. Two layers of common masking tape works well. Loosen the bolts (using special tool MB 990900–01 if necessary) just enough to allow the hinge to move. With the help of an assistant, position the door and retighten the bolts. Inspect the door seams carefully and repeat the adjustment until correctly aligned.

The in-out adjustment (how far the door sticks out from the body) is adjusted by loosening the hinge-to-door bolts. Again, move the door into place, then retighten the bolts. This dimension affects both the amount of crush on the weatherstrips and the amount of "bite" on the striker.

Further adjustment for closed position and smoothness of latching is made at the latch plate or striker. This piece is located at the rear edge of the door and is attached to the bodywork; it is the piece the latch engages when the door is closed.

Although the striker size and style may vary between models or from front to rear, the method of adjusting it is the same:

1. Loosen the large cross-point screw(s) holding the striker. Know in advance that these bolts will be very tight; an impact screwdriver is a handy tool to have for this job. Make sure you are using the proper size bit.
2. With the bolts just loose enough to allow the striker to move if necessary, hold the outer door handle in the released position and close the door. The striker will move into the correct location to match the door latch. Open the door and tighten the mounting bolts. The striker may be adjusted towards or away from the center of the car, thereby tightening or loosening the door fit.

The striker can be moved up and down to compensate for door position, but if the door is correctly mounted at the hinges this should not be necessary.

Adjusting the door requires loosening the hinge to body bolts. The special wrench may be required on some front doors

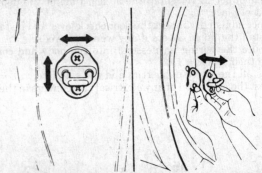

The door striker can be moved on its slotted fittings. Some models allow extra adjustment by changing striker shims

NOTE: Do not attempt to correct height variations (sag) by adjusting the striker.

3. Additionally, many Mitsubishi models use one or more spacers or shims behind the striker. These shims may be removed or added in combination to adjust the reach of the striker.
4. After the striker bolts have been tightened, open and close the door several times. Observe the motion of the door as it engages the striker; it should continue its straight-in motion and not deflect up or down as it hits the striker.
5. Check the feel of the latch during opening and closing. It must be smooth and linear, without any trace of grinding or binding during engagement and release.

It may be necessary to repeat the striker adjustment several times (and possibly re-adjust the hinges) before the correct door to body match is produced.

HOOD

REMOVAL AND INSTALLATION

NOTE: It is advisable to use two people while removing the hood from the vehicle. The hood is lightweight material and can be easily damaged by twisting or dropping it.

1. Open the hood and support it by the prop rod.
2. Use a felt tip marker or grease pencil to mark the hinge location on the hood.
3. Disconnect the windshield washer lines running to the hood if any. Disconnect the wiring to the underhood light if so equipped.
4. Remove the bolts attaching the hood to the hood hinges. Have a helper support the rear of the hood as the bolts are re-

BODY 10

moved. Without support, the hood will slide off the hinges and hit the upper bodywork.

5. Remove the hood assembly, lifting it off the hood prop. Place the hood out of the work area, resting on its side on a protected surface.

6. Reinstall the hood by carefully placing it in position and installing the hinge bolts finger tight. Install the hood prop immediately. Move the hood around on the hinges until the matchmarks (felt tip marker) align.

WARNING: Take great care to prevent the hood from bumping the windshield. Not only will the hood be damaged, the windshield can be broken by careless installation.

7. Close the hood—at least engaging the first latch—and check the hood-to-fender and hood-to-cowl seams and alignment. Re-open the hood and make adjustments as needed to give equal seams all around. When the hood is in final alignment, tighten the hinge bolts to 14 Nm or 10 ft. lbs.

8. Reconnect the washer lines and any electrical lines which were disconnected.

9. Check the hood latch operation; it should engage easily and smoothly. The latch should release without the use of excessive force. The hood latch mounting bolts may be loosened; the latch can be adjusted left/right or up/down to compensate for the hood striker position. On Montero, the striker (on the hood) can also be adjusted by loosening the bolts.

10. With the hood closed and latched, check the height of the hood. It should sit even with the fender tops; this dimension may be adjusted by turning the hood stops to a higher or lower position. Note that adjusting these stops may require readjustment of the latch to position the hood correctly.

Trunk Lid, Hatch and Tailgate

REMOVAL AND INSTALLATION

WARNING: Rear doors and hatchbacks require 2 people to support and lift the door. Trunk lids, while lighter, should still have 2 people to control the lid and prevent damage.

The hatch, trunk lid or rear door should be stored in a protected location out of the work area. It should not be rested on its edge and should never be placed so that the weight of the unit bears on the glass. Use old blankets, cloths or other padding to protect the painted and glass surfaces.

Cordia and Tredia

HATCHBACK

1. Loosen the headliner at the rear of the car to expose the hatchback hinge bolts.
2. Remove the rear inner panel trim and the rear pillar trim on each side.
3. Disconnect the rear harness connector and the tailgate harness connector.
4. Remove the tailgate support (strut or shock absorber) from the body. It is held with a ball-and-socket joint which must be unscrewed from the body. Use the correct size wrench and don't damage the surrounding paintwork. If the strut is to be replaced, refer to the heading "Hatchback Struts" at the end of this division.
5. Close the tailgate and remove the tailgate hinge mounting nuts.
6. With an assistant, open the lid slowly and remove it from the car.
7. To reinstall, fit the hatch into place and install the nuts finger tight.
8. Reinstall the support, threading it carefully into the body.

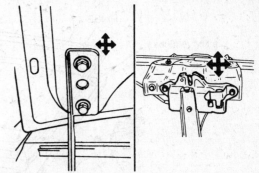
Adjust the hood at the hinge and then adjust the latch

Two types of hood bumpers. Both turn in or out to adjust hood height

Remove the ballstud from the body

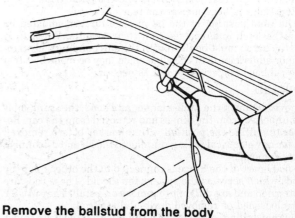
Removing the hinge nuts from Cordia and Tredia hatchback

10 BODY

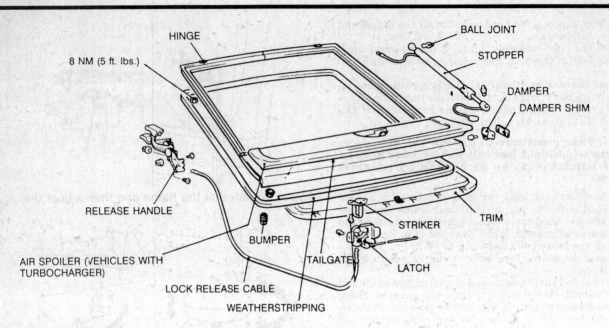

Cordia and Tredia hatchback components

9. Close the tailgate and align it by checking the seam all around. The gap must be equal and even. Align the hatch by moving it on the hinges until correctly placed, then tighten the hinge nuts to 8 Nm or 60 inch lbs.
10. Connect the electrical harnesses.
11. Reinstall the rear panel covers and the pillar trim.
12. Reinstall the headliner at the rear of the car.
13. The final position of the lid may be further adjusted by moving the latch and/or striker on their mounts. When fully closed, the hatch must be flush with the surrounding body panels. Minor adjustments in this dimension may be made by turning the rubber bumpers in or out as needed.

TRUNK LID

Removed in much the same manner as a hood, the trunk lid is simply unbolted from the hinges and removed from the car. Remember to outline the position with a marker before removal. Check for any electrical wiring running to the lid and disconnect it before removal.

The body part of the hinge is connected to the other hinge by two rods which serve as the springs for the lid. These rods are under tension and are difficult to remove or install. They MUST NOT be loosened or removed during lid removal.

When reinstalling, position the lid and tighten the bolts finger tight. Align the lid by moving it on the hinge bolts until correctly placed and tighten the bolts. Check the fit and engagement of the latch and striker, adjusting them as necessary. Check the lid in the closed position for flushness with the adjacent panels and adjust the rubber bumpers as needed.

Mirage and Precis

HATCHBACK

1. For all except 1989 Mirage, remove the headliner or at least loosen it at the rear. The headliner is one piece molded plastic; it may be difficult to pull down one end without creasing or damaging the liner.
2. Disconnect the liftgate support or strut. On Precis, use a small open end wrench to remove the ballstud from the body. Mirage struts are held by either a cross-point screw or a bolt. If the strut is to be replaced, refer to the heading "Hatchback Struts" at the end of this division.
3. Disconnect the washer tube running to the glass.

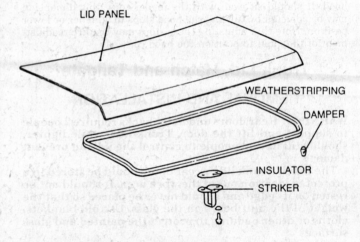

Cordia trunk lid

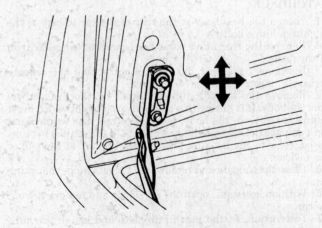

The trunk hinge allows a large range of adjustment

BODY 10

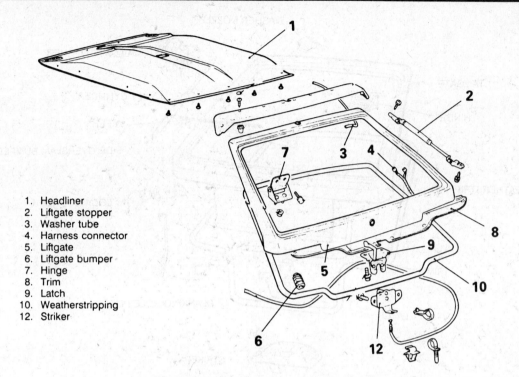

1. Headliner
2. Liftgate stopper
3. Washer tube
4. Harness connector
5. Liftgate
6. Liftgate bumper
7. Hinge
8. Trim
9. Latch
10. Weatherstripping
12. Striker

Mirage hatchback components

4. Disconnect any wiring harnesses running to the hatch. The number of harnesses will vary with style and model.
5. Close the hatch. From the inside of the vehicle, remove the nuts holding the lid to the hinge. With an assistant, carefully open the lid and remove it from the car.
6. To reinstall, fit the hatch into place and install the nuts finger tight.
7. Reinstall the support.
8. Close the tailgate and align it by checking the seam all around. The gap must be equal and even. Align the hatch by moving it on the hinges until correctly placed, then tighten the hinge nuts to 15 Nm or 11 ft. lbs.
9. Connect the electrical harnesses.
10. Connect the rear windshield washer tube. Make sure it is pressed firmly onto the fitting.
11. Reinstall the headliner and secure it at the rear.
12. The final position of the lid may be further adjusted by moving the latch and/or striker on their mounts. When fully closed, the hatch must be flush with the surrounding body panels. Minor adjustments in this dimension may be made by turning the rubber bumpers in or out as needed.

TRUNK LID

1. Open the trunk lid. Identify the trunk release cable running from the latch to the body.
2. Disconnect the cable from the latch. If you find it necessary to loosen the latch mounting bolts, be sure to outline the latch position with a marker before moving the latch.
3. Release the cable from its clips and mounts and position it clear of the lid. Check for any electric harnesses running along the lid.
4. Outline the position of the hinges on the trunk lid.
5. Support the lid from both sides and remove the bolts holding the lid to the hinge arms. Do not attempt to remove the hinge from the body. Do not attempt to loosen or remove the torsion rods or the coil spring from the hinge. Lift the lid clear of the car.

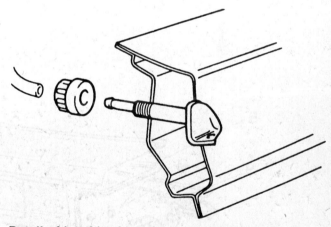

Detail of hatchback washer hose components

6. To reinstall, position the lid and install the bolts finger tight. Close the lid without engaging the latch (the release cable is disconnected) and adjust the gap to the body by moving the lid on the hinges. When it is in the correct position, tighten the bolts.
7. Route the release cable correctly and connect it to the latch.
8. Before closing the lid, have a helper operate the inside trunk lid release near the driver's seat. Check that the latch is moving correctly as the cable is pulled.
9. Close the lid and check both the seam width all around and the closed height. The trunk lid must be flush with the adjacent panels. Minor height adjustments may be made by turning the rubber bumpers on the trunk lid. Additional adjustments require loosening and repositioning of the latch and/or striker.

10 BODY

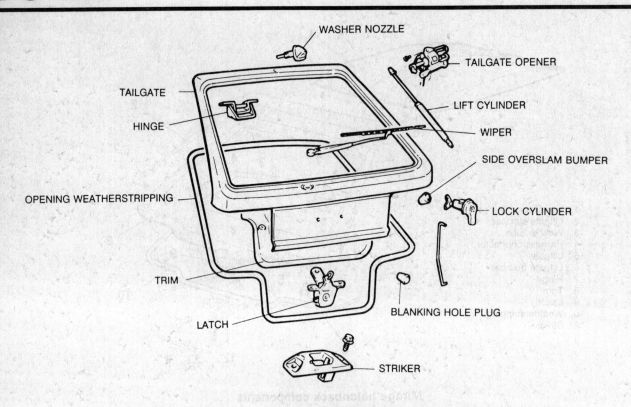

Precis hatchback

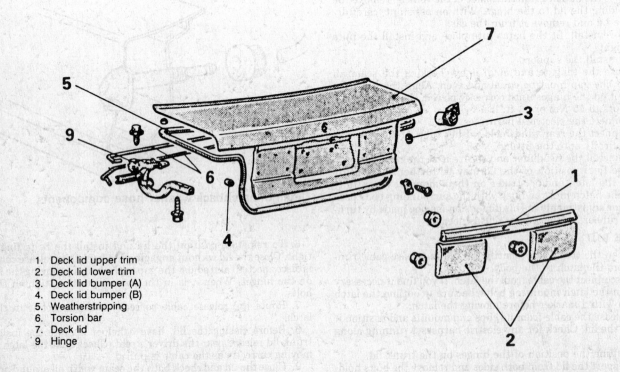

1. Deck lid upper trim
2. Deck lid lower trim
3. Deck lid bumper (A)
4. Deck lid bumper (B)
5. Weatherstripping
6. Torsion bar
7. Deck lid
9. Hinge

Mirage trunk lid assembly

BODY 10

Starion

WARNING: The Starion hatchback does not have a metal frame surrounding it. It is a large glass sheet with various components attached. It must be handled with the utmost care. When removing or installing the glass, have someone support the glass in the center as well as at the edges.

1. Disconnect the wiring connectors at the bottom of each support strut.
2. Using a wrench of the correct size, carefully remove the ballstud holding the support strut to the body. If the strut is to be replaced, refer to the heading "Hatchback Struts" at the end of this division.
3. Carefully and without excess force, remove the cross-point screws holding the glass to the hinge. Do this from the inside with the lid closed.
4. Using at least 2 (3 is better) people, remove the hatch. Place it in a protected location.
5. If the glass is to be replaced, all components must be removed from the glass. Note that the black plastic trim pieces glued to the glass are non-reusable and must be replaced.
6. If the glass is replaced, during reassembly, do not tighten any nut or bolt passing through the glass to more than 9 Nm or 72 inch lbs. Make certain that all pads, cushions and grommets are correctly placed to protect the glass.
7. Place the assembled glass in position (lowered) and install the protective pads and spacers. Tighten the cross-point screws snug but do not attempt to get that last half-turn; too much may shatter the glass.
8. Carefully raise the lid and, while a helper supports it, connect the support strut to the body.
9. Close the hatch and inspect the fit to the body. The nature of the upper mounts (at the hinge) does not allow for much adjustment, but a very small amount may be obtained if the screws are loosened.
10. Connect the wiring harnesses at the bottom of the strut.
11. Inspect the function of all the electrical systems at the rear of the car, in particular the rear wiper/washer and the defogger. If these systems are not working, the stay (support) switch in the either or both struts needs adjusting.

Each strut contains a switch which disconnects the hatch lid electrical components when the hatch is open and re-connects them when it is closed. This sliding contact switch must be correctly located for all the related systems (including the theft alarm) to work. To adjust the switch:

a. Loosen the two small set screws in the switch; leave them just snug enough to allow the switch to move as the strut moves.
b. Raise the lid about an 25mm (1 inch) and close it.
c. Open the lid fully and mark the bottom of the switch in this position.
d. Close the lid, reopen it fully, check the alignment of the marks and tighten the screws.

WARNING: If the switch is mis-adjusted, there is risk of bending the strut as the hatch is closed.

Galant and Sigma

1. Open the trunk lid. Identify the trunk release cable running from the latch to the body.
2. Disconnect the cable from the latch. If you find it necessary to loosen the latch mounting bolts, be sure to outline the latch position with a marker before moving the latch.
3. Release the cable from its clips and mounts and position it clear of the lid.
4. Disconnect any electric harnesses running along the lid. The wiring leads must be unclipped from any retainers and released from the lid.
5. Outline the position of the hinges on the trunk lid.

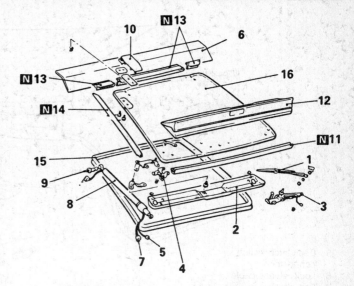

1. Rear wiper arm
2. Hatch trim
3. Wiper motor
4. Striker
5. Stop light connector
6. Roll bar trim
7. Defogger connector
8. Hatch stopper
9. Ball joint
10. Hinge
11. Protector
12. Hatch garnish
13. Upper moulding
14. Side moulding
15. Weatherstripping
16. Glass

Starion rear hatch. Complete disassembly is not required for removal (see text); items marked N are non-reusable

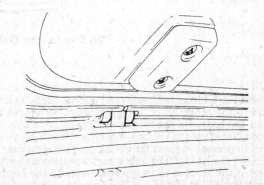

Detail of Starion hatch attachment screws; don't over-tighten them

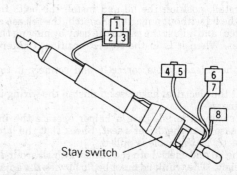

The Starion stay switch controls the functions of many electrical items at the rear

10 BODY

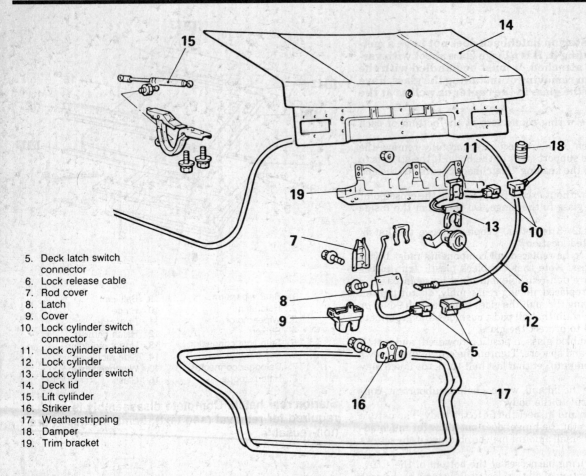

5. Deck latch switch connector
6. Lock release cable
7. Rod cover
8. Latch
9. Cover
10. Lock cylinder switch connector
11. Lock cylinder retainer
12. Lock cylinder
13. Lock cylinder switch
14. Deck lid
15. Lift cylinder
16. Striker
17. Weatherstripping
18. Damper
19. Trim bracket

V6 Sigma and Galant trunk lid components

6. Support the lid from both sides and remove the bolts holding the lid to the hinge arms. Do not attempt to remove the hinge from the body. Do not attempt to loosen or remove the torsion rods. Note that the V6 models of Galant and Sigma use a small strut to provide support for the hinge. This strut may be removed (after the lid is off) by unbolting the ballstuds to which it attaches. If the strut is to be replaced, refer to the heading "Hatchback Struts" at the end of this division.

NOTE: Access to the inner mount requires removal of the rear seat.

7. To reinstall, position the lid and install the bolts finger tight. Close the lid without engaging the latch (the release cable is disconnected) and adjust the gap to the body by moving the lid on the hinges. When it is in the correct position, tighten the bolts.
8. Route the release cable correctly and connect it to the latch.
9. Connect the electrical harnesses; reinstall the wiring in its clips or retainers.
10. Before closing the lid, have a helper operate the inside trunk lid release near the driver's seat. Check that the latch is moving correctly as the cable is pulled.
11. Close the lid and check both the seam width all around and the closed height. The trunk lid must be flush with the adjacent panels. Minor height adjustments may be made by turning the rubber bumpers on the trunk lid. Additional adjustments require loosening and repositioning of the latch and/or striker.

Montero Rear Door

1. Remove the spare tire lock and spare tire.
2. Remove the spare tire carrier.
3. Remove the rear opening trim.
4. If equipped with rear seat belts, remove the rear retractor cover.
5. On 4 door vehicles, remove the speaker.
6. Remove the quarter panel trim.
7. Remove either the rear pillar lower trim (2 door vehicles) or the rear pillar trim (4 door vehicles).
8. Remove the cotter pin and clevis pin from the door check rod. Discard the cotter pin.
9. Disconnect the rear washer fluid tube if the vehicle is equipped with a rear wiper.
10. Disconnect the rear door wiring harnesses.
11. Outline the hinge position on the door with a felt-tip marker. Have at least one other person support the door. Remove the bolts holding the hinge to the door. Remove the door.
12. Reinstall the door—observing the marks made earlier—and attach the hinge bolts. Tighten the bolts to 45 Nm or 33 ft. lbs.
13. If all went well, the door alignment should be very close to the original. Check the seams all around for evenness and equal spacing. If adjustment is necessary, loosen the hinge-to-body bolts and reposition the door. Tighten the bolts and recheck.
14. Further adjustments may be made at the latch and/or striker. The angle of the door (in at the top/out at the bottom or

BODY 10

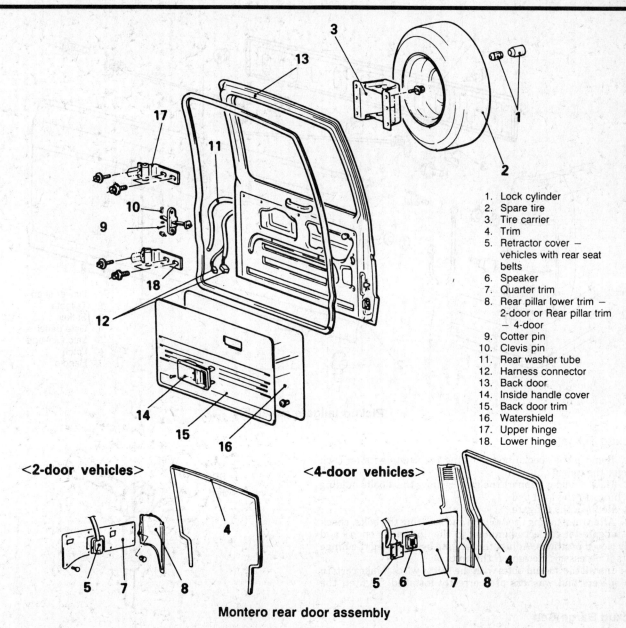

1. Lock cylinder
2. Spare tire
3. Tire carrier
4. Trim
5. Retractor cover — vehicles with rear seat belts
6. Speaker
7. Quarter trim
8. Rear pillar lower trim — 2-door or Rear pillar trim — 4-door
9. Cotter pin
10. Clevis pin
11. Rear washer tube
12. Harness connector
13. Back door
14. Inside handle cover
15. Back door trim
16. Watershield
17. Upper hinge
18. Lower hinge

Montero rear door assembly

similar dimension) must be adjusted by loosening the hinge-to-door bolts and moving the door accordingly. This should only be done as a last resort; changes in this dimension may cause changes in seam width or body alignment.

15. Connect the door check clevis pin and install a new cotter pin.
16. Connect the electrical leads; connect the washer hose tube if one was removed.
17. Install the pillar trim and install the quarter panel trim pieces.
18. Install the speaker if it was removed. Install the seat belt retractor covers if they were removed.
19. Fit the rear opening trim without distorting it.
20. Install the spare tire holder (10 Nm or 84 inch lbs on the bolts) and install the spare tire with its lock.
21. Using a grease gun with a fine tip, lubricate the rear door hinges by introducing multi-purpose grease into the small hole in each hinge. Move the door back and forth several times to circulate the grease.

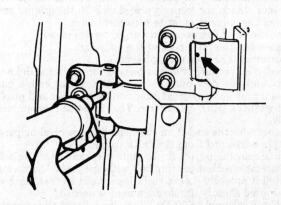

Lube the Montero rear door hinges

10-13

10 BODY

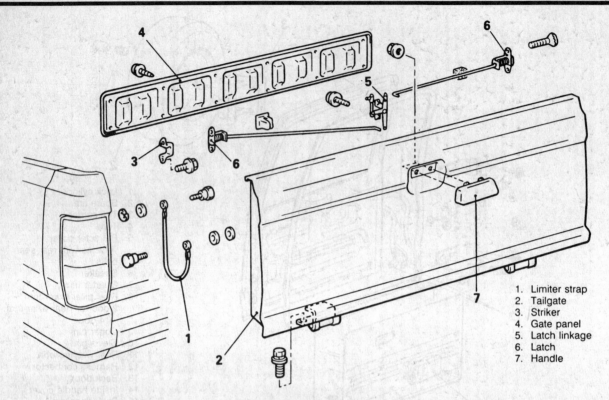

Pickup tailgate assembly

1. Limiter strap
2. Tailgate
3. Striker
4. Gate panel
5. Latch linkage
6. Latch
7. Handle

Pickup Tailgate

1. Remove the special bolts holding the safety wire on each side of the tailgate.
2. Have a helper support the gate; remove the 4 bolts holding the hinge to the load bed.
3. Remove the tailgate.
4. After reinstalling the tailgate and securing the bolts, check the tailgate for correct alignment in the "normally open" and fully closed positions. Adjust the tailgate by loosening the hinge bolts if necessary. Retighten the bolts.
5. Install the special bolts with the safety wires. Make certain the spacers and washers are correctly installed. Tighten the bolts.

Pickup Cargo Bed

1. Elevate and safely support the vehicle on stands. Make certain all four stands are securely placed on both the ground and the truck. The support must be extremely solid.
2. Remove the filler neck cover or protective panel.
3. disconnect the filler neck at the body end.
4. Disconnect the rear wiring harness.
5. Loosen and remove the rear body mounting nuts; discard the nuts and use new ones at reassembly. Disassemble one mount at a time and pay great attention to the order and placement of washers, pads, spacers etc. They must be reassembled correctly.
6. Either with the use of an overhead lift or several helpers, remove the cargo bed from the truck frame.
7. When reinstalling, lift the bed into place and align the bolt holes. Assemble each mounting bolt with the proper hardware and install it properly. Leave each finger-tight so that the bed may be moved slightly if hole alignment is needed.
8. Tighten each nut to 30 Nm (22 ft. lbs.). Resist the temptation to tighten these nuts to extremes; overtightening crushes the pads causing both excessive noise and reducing the natural motion of the bed.

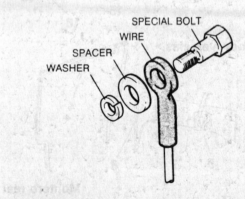

If the safety wire is not correctly installed, it cannot pivot with the motion of the tailgate

Hatchback Struts

If the hatchback strut is to be replaced (or the trunk support on Galants and Sigmas), it is disconnected at both ends and removed from the vehicle. The new one connects in similar fashion.

The strut contains nitrogen gas under pressure. Do not expose the unit to extremes of heat or fire. Do not discard the unit without discharging the gas.

To make the unit safe for disposal:

1. Wear safety glasses.
2. Place the used strut in a plastic bag and secure the bottom of the bag with tape. Make a 13mm (½ inch) hole in the corner of the bag.

BODY 10

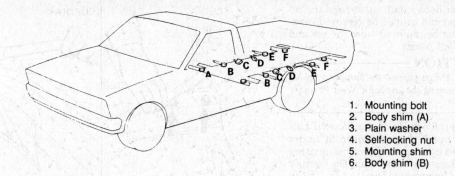

1. Mounting bolt
2. Body shim (A)
3. Plain washer
4. Self-locking nut
5. Mounting shim
6. Body shim (B)

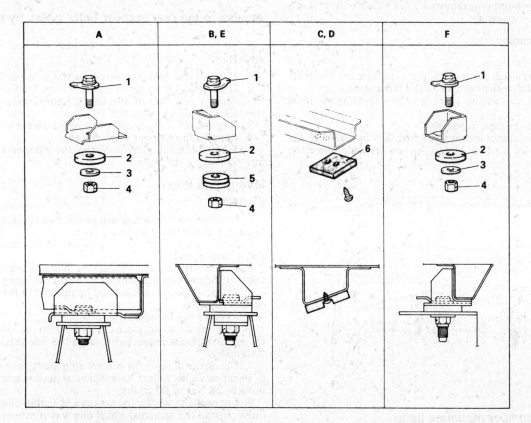

Each cargo bed mount is different; make sure they are correctly assembled

3. Drill a 3mm hole (0.118 inch) through the bag into the strut case at a point 20mm (¾ inch) above the bottom of the case.
4. The drill will vent the gas into the bag. The bag will contain and deflect the metal chips and other debris blown around by the escaping gas. The gas will escape through the hole in the bag without blowing into your face.

NOTE: Nitrogen is an inert gas. It is neither flammable nor poisonous.

Bumpers

REMOVAL AND INSTALLATION

The bumpers, front or rear, are actually assemblies. The outer surface is the cover or fascia, behind which may be found ab-

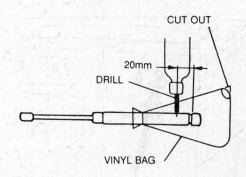

Correct method of venting a gas strut before disposal

10-15

10 BODY

sorbent materials (foam block or honeycomb lattice) and the reinforcing steel bar. The components cannot be removed separately; the entire assembly must be removed from the car and then broken down into individual pieces.

CAUTION

The bumper assemblies are heavy! Always support the bumper from below in at least two places before removing the last bolts. Never lie under the bumper while removing it.

The bumper covers are generally made of urethane. If this heavy, flexible plastic receives minor damage, generally under 25mm (1 inch), it can be repaired by a reputable body shop using established industry methods. Like any body repair, the bumper will require painting and should be removed from the car. In the event of more extensive damage, the cover can be replaced individually after the bumper is removed. This will save the cost of a complete bumper assembly.

Cordia and Tredia
FRONT

1. From underneath the front bumper, remove the two bolts on each side holding the bumper to the front stays.
2. Disconnect the wiring to each of the front turn signal lamps.
3. Remove the bumper assembly.
4. When reinstalling, connect the wiring before placing the bumper in its final position. Install the bumper and tighten the bolts.

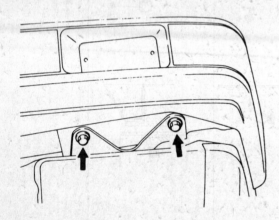

Cordia front bumper mounting bolts

Disconnect the front turn signal wiring

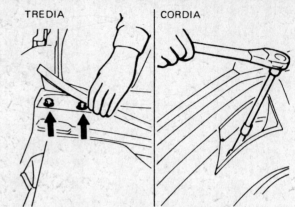

Access to the rear bumper bolts varies by model

REAR

1. Lift and fold back the carpet in the trunk or hatch area.
2. On Tredia, lift up the trunk side trim. On Cordia, remove the clips on one side of the trunk trim. Insert the wrench through the holes in the trim.
3. Remove the bolts holding the rear bumper to the body.
4. Remove the bumper from the car.
5. When reinstalling, make certain the bumper is straight before tightening the bolts.

Mirage and Precis
FRONT

1. Remove the forward splash shield if so equipped. For 1989 Mirage, remove the wheel arch liners.
2. Remove the front turn signal assemblies and disconnect the wiring to each.
3. Remove the bumper mounting bolts, accessible through the turn signal openings. Don't try to pull the bumper loose yet.
4. At each side, disconnect the bumper from the small stays or side supports.
5. Remove the bumper.
6. When reinstalling, place the bumper on the car and tighten the retaining bolts finger tight. Align the side stays and install the screws.
7. Check the bumper for correct alignment; it should not sag or be on an angle. Adjust its position as needed and tighten the bolts to 28 Nm or 20 ft. lbs.
8. Connect the wiring harnesses and install the turn signal units. Install the splash shield if one was removed.

REAR

1. Disconnect the rear bumper wiring harness. On some models, you'll have to wait until after Step 3; the harness is behind the corner of the bumper.
2. Remove the backup light assemblies.
3. Remove the screws and clips from the leading edge of the rear bumper—the part near the rear wheels. Depending on the model, this may require getting tools into awkward locations. Elevate and safely support the car as needed; removing the wheels can also be helpful.
4. Work toward the rear of the car, removing clips and bolts which hold the bumper to the body. Some models use separate corner pieces which may be removed; others incorporate extra stays and braces for the one-piece bumper cover.
5. Remove the bumper retaining bolts at the rear and remove the bumper.
6. When reinstalling, place the bumper on the car and tighten the retaining bolts finger tight. Align the side stays and install the screws or install the corner pieces and secure them.
7. Check the bumper for correct alignment; it should not sag

BODY 10

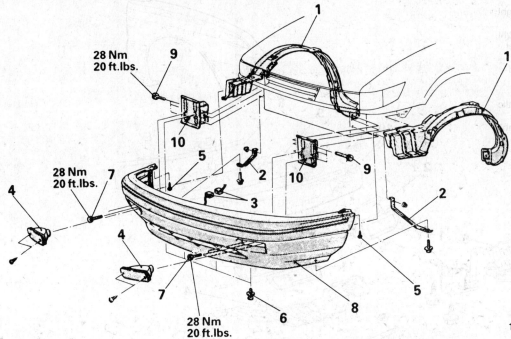

1. Splash shield
2. Side tray
3. Wiring harness connector
4. Front turn signal
5. Front bumper facing screw
6. Front bumper facing clip
7. Front bumper reinforcement bolt
8. Front bumper
9. Front bumper support bolt
10. Front bumper support

1989 Mirage front bumper

Don't forget the side stay bolts when removing the Mirage front bumper

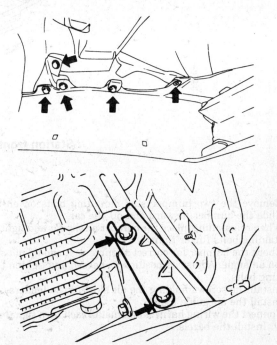

Disconnect the front skirt from the fenders (top) before removing the bumper mounting bolts

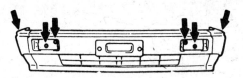

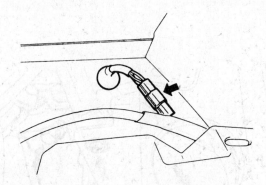

Mirage rear bumper harness connector

or be on an angle. Adjust its position as needed and tighten the bolts to 28 Nm or 20 ft. lbs.

8. Install the reverse light units and connect the rear wiring harness.

Starion
FRONT
1. Remove the bezel surrounding each turn signal lens.

2. Remove the turn signal assembly from each side and disconnect the wiring harness.
3. Remove the front license plate bracket.
4. Remove the front air guide. You may mistake this for a splash shield or undercover, but this one guides air to important engine locations. (This is not the front skirt or air dam).
5. The front skirt (below the bumper) may stay attached to the bumper but the screws holding the skirt to the front fenders must be removed.

10–17

10 BODY

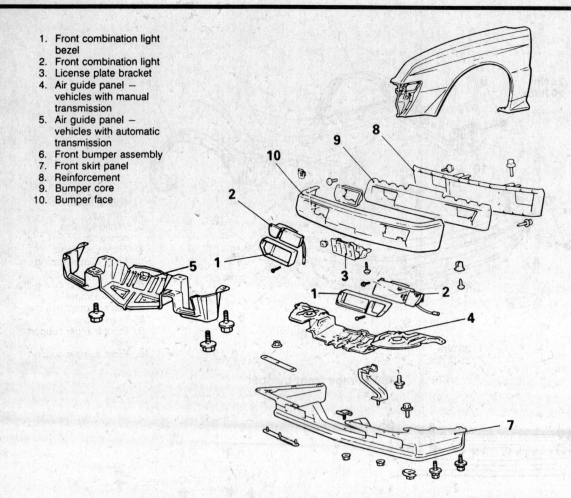

1. Front combination light bezel
2. Front combination light
3. License plate bracket
4. Air guide panel — vehicles with manual transmission
5. Air guide panel — vehicles with automatic transmission
6. Front bumper assembly
7. Front skirt panel
8. Reinforcement
9. Bumper core
10. Bumper face

Starion front bumper assembly

6. Remove the two bumper stay mounting bolts on either side. Slide the bumper forward and off the car.
7. When reinstalling, place the bumper on the car and tighten the retaining bolts finger tight.
8. Check the bumper for correct alignment; it should not sag or be on an angle. Adjust its position as needed and tighten the mounting bolts.
8. Install the screws holding the air skirt to the fenders.
9. Install the air guide panel.
10. Connect the wiring harness and install each turn signal assembly. Install the bezels.

REAR

1. From inside the cargo area, remove the upper retaining bolts.
2. Have an assistant support the bumper. Under the car, remove the small screw (near the tow hook) on each side and remove the large retaining bolt.
3. Remove the bumper assembly.
4. When reinstalling, place the bumper on the car and tighten the retaining bolts finger tight.
5. Check the bumper for correct alignment; it should not sag or be on an angle. Adjust its position as needed and tighten the mounting bolts.

Starion rear bumper mounts

10-18

BODY 10

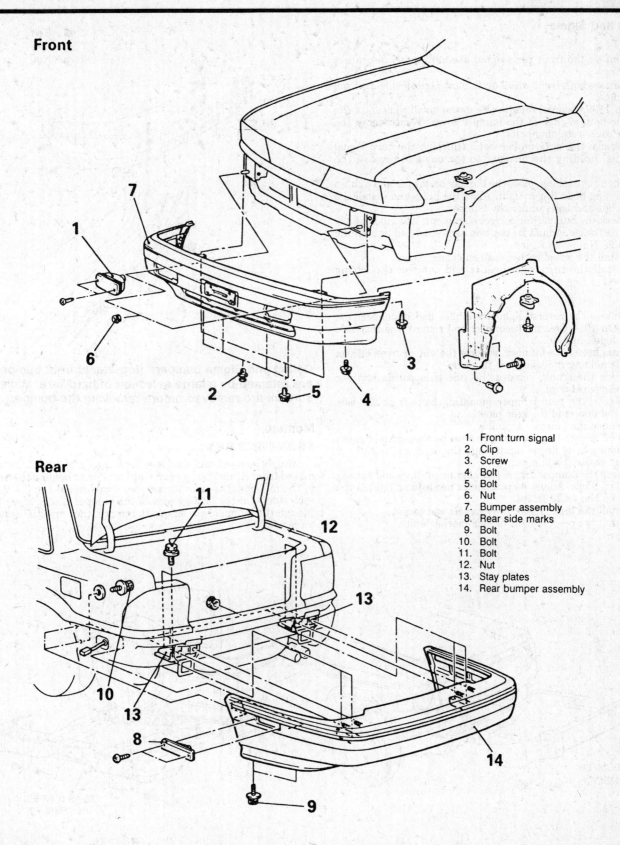

1. Front turn signal
2. Clip
3. Screw
4. Bolt
5. Bolt
6. Nut
7. Bumper assembly
8. Rear side marks
9. Bolt
10. Bolt
11. Bolt
12. Nut
13. Stay plates
14. Rear bumper assembly

1989 Galant front and rear bumpers

10 BODY

Galant and Sigma
FRONT

1. Remove the front turn signal assemblies and disconnect the wiring.
2. Remove both front wheel arch liners (splash shields) from the fenders.
3. On 1989 Galants, remove the row of small clips and bolts from the bottom edge of the bumper cover. Then remove the bumper cover retaining bolts.
4. Remove the nuts and/or bolts (through the turn signal openings) holding the bumper to the car and remove the bumper.
5. When reinstalling, place the bumper on the car and tighten the retaining bolts finger tight. Align the holes and install the row of clips and bolts under the bumper.
6. Check the bumper for correct alignment; it should not sag or be on an angle. Adjust its position as needed and tighten the bolts to 28 Nm or 20 ft. lbs.
7. Install the wheel arch splash shields or liners.
8. Install the turn signal units and connect the wiring harnesses.

REAR

1. Remove the reverse light assemblies and disconnect the wiring. On 1989 Galant, disconnect and remove the rear side marker lights instead.
2. From below the bumper, remove the one or two bolts on each side holding the side piece in place.
3. Inside the trunk, remove the side trim panels and the trunk rear panel trim.
4. Remove the rear bumper mounting bolts from the side panels, the floor and the rear panel.
5. Remove the bumper assembly.
6. When reinstalling, place the bumper on the car and tighten the retaining bolts finger tight. Align the holes and install the side bolts under the bumper.
7. Check the bumper for correct alignment; it should not sag or be on an angle. Adjust its position as needed and tighten the bolts to 28 Nm or 20 ft. lbs.
8. Install the inner trunk liner panels and carpet.
9. Install the reverse light or side marker units.

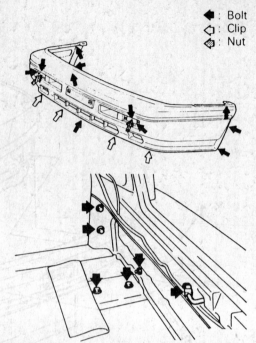

◀ : Bolt
◁ : Clip
◊ : Nut

Galant and Sigma bumpers (top frame, front bumper) are retained by a large selection of hardware. Make sure all are removed before removing the bumper

Montero
FRONT AND REAR

Straightforward and simple, the Montero bumpers are removed from either end by removing the 4 bolts holding the unit to the mounts. The rear bumper has 4 additional bolts — 2 at each side — at the leading edge of the bumper corner trim. Although the bumpers are simple to remove, they are still heavy; be careful!

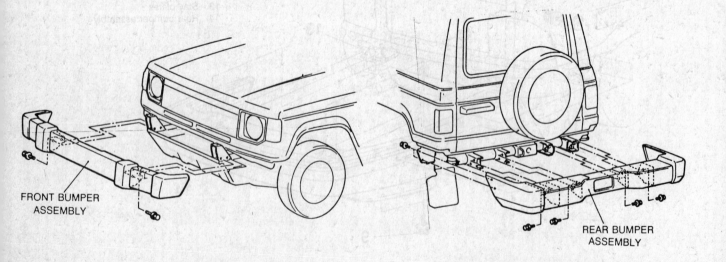

Few bumpers are as easily removed as the Montero

BODY 10

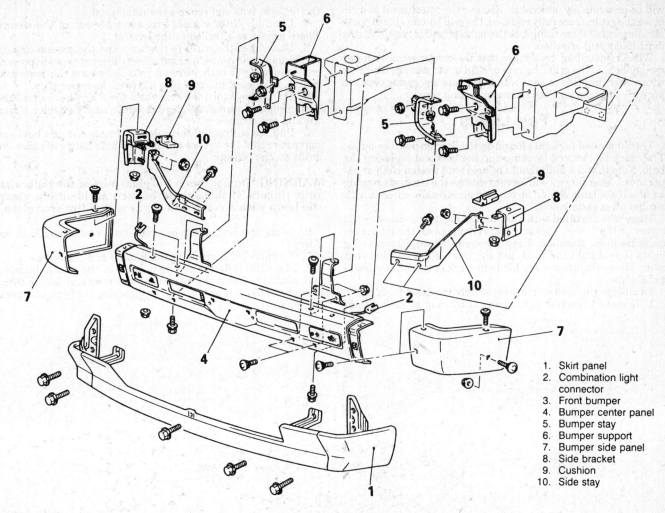

1. Skirt panel
2. Combination light connector
3. Front bumper
4. Bumper center panel
5. Bumper stay
6. Bumper support
7. Bumper side panel
8. Side bracket
9. Cushion
10. Side stay

Pickup front bumper

Pickup

FRONT

1. Remove the 5 bolts holding the front skirt panel to the bumper.
2. Disconnect the wiring to the front turn signal lights. The lights may stay in the bumper if you wish.
3. Disconnect the front bumper from the side stays on each side.
4. Remove the bolts attaching the front bumper support to the frame and remove the front bumper.
5. When reinstalling, place the bumper on the truck and tighten the retaining bolts finger tight. Align the holes and install the side bolts to the side stays.
6. Check the bumper for correct alignment; it should not sag or be on an angle. Adjust its position as needed and tighten the bolts.
7. Connect the wiring for the turn signals.
8. Install the front skirt panel.

REAR

Not all Pickups are delivered with rear bumpers. Those that do have bumpers at the rear may have either the optional Mitsubishi bumper or one of a number of aftermarket products. In general, the rear bumper simply bolts to the frame, but the style and manufacture of the bumper may affect its installation.

Removal of these bumpers is generally straightforward, but be warned that some bumpers — notably the plate steel ones — are extremely heavy! Use a floor jack to help support the unit and have an assistant handy throughout the removal or installation. If the bumper contains any lighting or electrical equipment, remember to disconnect or remove it before separating the bumper from the truck.

Grille

REMOVAL AND INSTALLATION

In all cases, the grille may be removed without removing other parts. Depending on the model, the grille is held by a variety of clips and screws. The problem is usually not removing the fittings but finding all of them. Raise the hood and look for screws placed vertically in the front metalwork. Small screws may be used to hold part of the grille.

Some grilles are retained by screws placed from the front. These will be black and hard to see in the recesses of the plastic. Most grilles incorporate plastic spring clips; these must be released correctly or they will break, becoming unusable. A narrow probe or small flat-tip screwdriver may be used to release the lock tabs.

When you think all the mounts and clips are removed, pull gently outward on the grille. If it is still retained anywhere, it

10 BODY

will be immediately noticeable. Remove the attachment and remove the grille. Once fully released, the grill should almost come off unassisted; if anything but the lightest pull is required, it is most likely still attached.

When reinstalling, remember that the condenser or radiator is right behind the grille; be very careful not to damage anything with either the grille or any tools. Make sure any spring clips are properly engaged and install the mounting screws.

Fog Lights

The integrated fog lights found on the Starion use 35w bulbs. They may be changed by removing the lens and replacing the bulb. Note that the bulb should be held with a clean cloth or paper towel. Your fingers will leave grease on the lamp, shortening its life. If the lamp should be touched accidentally, clean it with alcohol after installation.

Many cars are fitted with aftermarket fog lights, either by the owner or the dealer. This safe and stylish option should not require frequent attention. Keep the lamps aimed correctly; fog lights should light the road, not the trees. Keep in mind that many states require the fog lights to work only with low beam headlights.

To change the bulb in most separately mounted foglights:
1. Remove the screw in the vertical housing on each side of the foglight lens and remove the retainer(s).
2. Carefully lift the glass lens assembly out of the mount. There are wires attached to the back of it.
3. Disconnect the wires to the lamp and the ground connector. Usually, the connectors are of two different types to prevent cross-plugging. If both wires on your light have the same connector, label each to avoid blowing fuses at reassembly.
4. On the back of the lens, push down on the ends of the spring clip and spread the clip out from under its holder. Swing the clip out of the way.
5. Pull the bulb out of the reflector. Install the new bulb and make sure that the square notch in the bulb flange fits over the guide on the reflector.

WARNING: Do not touch the glass part of the bulb with your fingers. If this should happen accidentally, clean the lamp with alcohol and dry it with a lint-free cloth.

6. Swing the spring clip over the bulb and secure it in its clamps.
7. Reconnect the wiring to the ground and the bulb.
8. Place the reflector assembly back in the plastic housing and install the frame pieces. Install the screws and tighten them enough to hold, but no more. Overtightening the bezel can crack the lens.

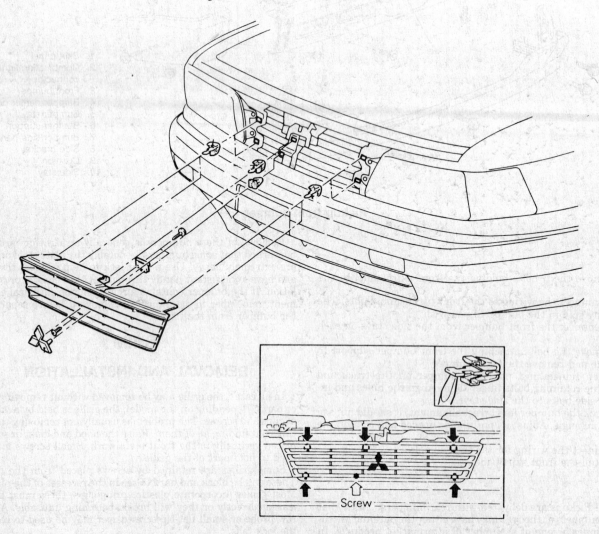

Mirage grille. Note the correct method of releasing the clip

BODY 10

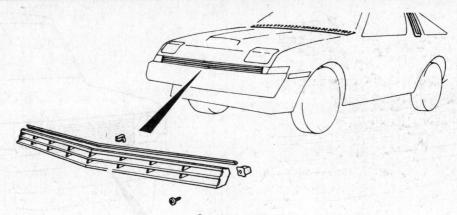

Starion grille

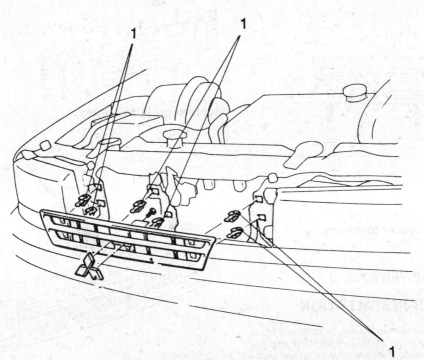

The 1989 Mirage grille is held entirely by clips

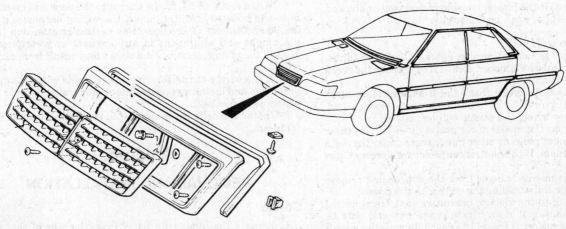

Galant grille

10 BODY

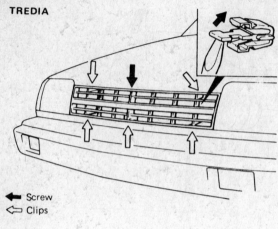

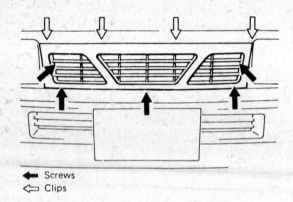

Cordia and Tredia grille retainer locations

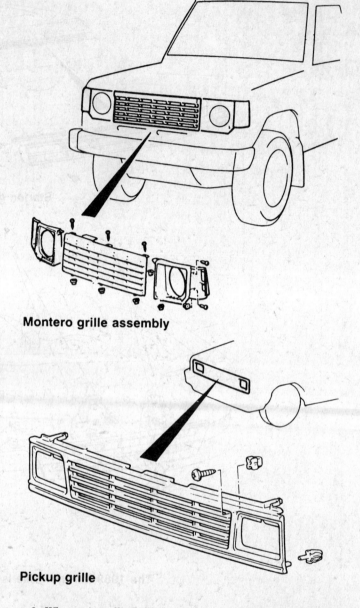

Montero grille assembly

Pickup grille

Outside Mirrors

REMOVAL AND INSTALLATION

The mirrors on Mitsubishi automobiles can be removed from the door without disassembling the door liner or other components. Both left and right outside mirrors may be either manual, manual remote (small lever on the inside to adjust the mirror) or electric remote. If the mirror glass is damaged, replacements may be available through your dealer or a reputable glass shop in your area. If the plastic housing is damaged or cracked, the entire unit will need to be replaced.

To remove the mirror:

1. If the mirror is manual remote, check to see if the adjusting handle is retained by a hidden screw. If so, remove the screw and remove the handle.
2. Remove the delta cover; that's the triangular black inner cover. It can be removed with a blunt plastic or wooden tool. Don't use a screwdriver; the plastic will be marred.
3. Depending on the model of car and style of mirror, there may be concealment plugs or other minor parts under the delta cover—remove them. If the electric connectors are present, disconnect them.
4. Support the mirror housing from the outside and remove the three bolts or nuts holding the mirror to the door.
5. If the wiring to the electric mirror was not disconnected previously, disconnect it now. (Some connectors can only be reached after the mirror is free of the door.) Remove the mirror assembly.

6. When reinstalling, fit the mirror to the door and install the nuts and bolts to hold it. Connect the wiring harnesses if they are on the outside of the door. Pay particular attention to the placement and alignment of any gaskets or weatherstrips around the mirror; serious wind noises may result from careless work.
7. If the wiring connectors are on the inside of the door, connect them and install any concealment plugs, dust boots or seals which were removed.
8. Install the delta cover and install the control lever if it was removed.

Antenna

REMOVAL AND INSTALLATION

Cordia

1. Remove the inner trim panel from the rear of the trunk area.

BODY 10

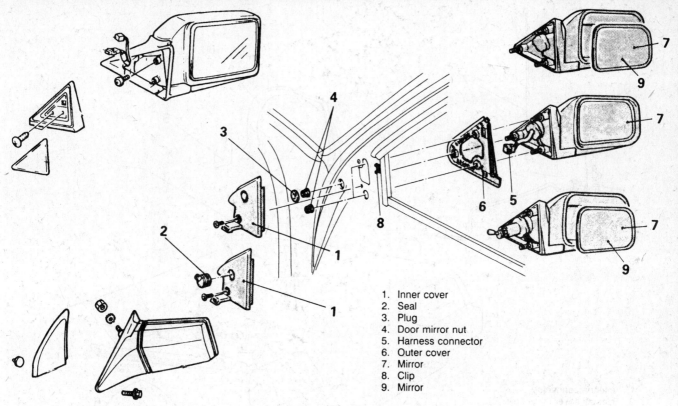

1. Inner cover
2. Seal
3. Plug
4. Door mirror nut
5. Harness connector
6. Outer cover
7. Mirror
8. Clip
9. Mirror

Three typical outer mirrors: Galant (upper), optional 1989 Mirage mirrors and Starion

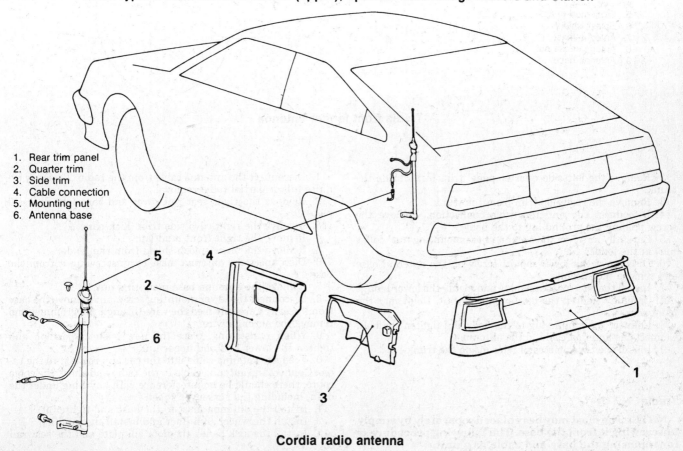

1. Rear trim panel
2. Quarter trim
3. Side trim
4. Cable connection
5. Mounting nut
6. Antenna base

Cordia radio antenna

10-25

10 BODY

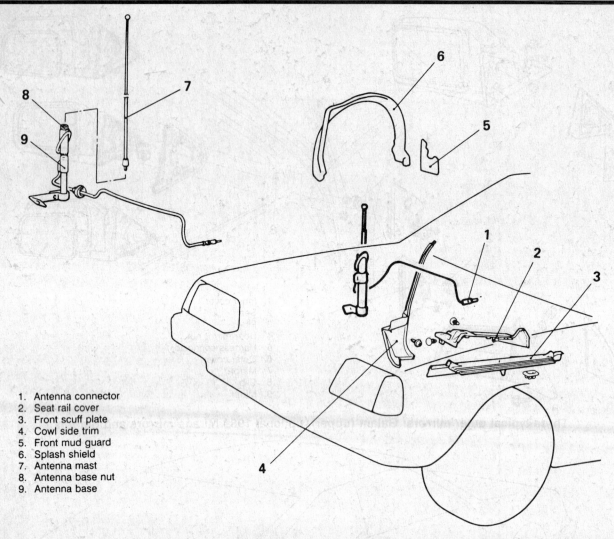

1. Antenna connector
2. Seat rail cover
3. Front scuff plate
4. Cowl side trim
5. Front mud guard
6. Splash shield
7. Antenna mast
8. Antenna base nut
9. Antenna base

Tredia front fender antenna

2. Remove the left side quarter panel trim from inside the trunk.
3. Remove the rear side trim from the trunk.
4. Disconnect the antenna cable connection. Remove the screw holding the ground lug to the body.
5. Carefully unscrew the antenna base mounting nut (collar nut) at the fender.
6. Disconnect the lower mount screw on the antenna base and remove the antenna.
7. Install the new mast assembly and install the lower mount nut. Carefully tighten the base mounting nut. Don't mar the paintwork.
8. Connect the ground lug to the body and tighten the bolt. Connect the antenna cable to the antenna.
9. Install the three pieces of trim inside the trunk.

Tredia

NOTE: **The mast may be replaced separately by simply unscrewing it from the base. The following procedure is for replacing the base and cable as a unit.**

1. Disconnect the antenna cable from the radio. This may require full or partial radio removal.
2. Remove the right seat rail cover and the right door sill plate.
3. Remove the right cowl side trim (kick panel).
4. Remove the right front mudflap.
5. Remove the inner splash shield from the fender.
6. Disconnect the antenna mast by unscrewing it from the base.
7. Remove the antenna base mounting nut.
8. Disconnect the lower mounting screw and remove the base from the car. Carefully feed the wire through the grommet and remove the antenna wire.
9. When reinstalling, place the new base in position and tighten the lower bolt and upper nut.
10. Feed the antenna cable into the car and route it to the radio. Connect the antenna cable to the radio and check the cable route; there should be no interference with anything under the dash, including the passenger's feet.
11. Install the antenna mast to the base and tighten it.
12. Install the wheel arch liner and install the mudflap.
13. Install the kick panel, the door sill plate and the seat rail cover.

BODY 10

Starion

1. Remove the trunk side trim panel.
2. Remove the adjacent floor panel.
3. Disconnect the electrical connector to the power antenna motor.
4. Disconnect the antenna feed wire connector.
5. Remove the screw holding the ground lug to the body.
6. Remove the mounting nuts and bolts and remove the antenna motor.
7. Remove the fender grommet.
8. To reinstall, mount the motor in place and tighten the retaining nuts and bolts. Install the upper fender grommet.
9. Connect the ground lug to the body, the antenna cable to the feed wire and electrical connectors to their mate.
10. Install the trunk trim panels.

Mirage through 1988
Precis

NOTE: **The mast and antenna wire are one piece. The mast cannot be changed individually.**

1. Remove or reposition the floor console to allow access to the back of the radio.
2. Remove the antenna cable from the radio.
3. Remove the left door sill plate. Hatchback sill plates have 5 clips, sedan sill plates have 3.
4. Remove the left lower cowl side trim (kick panel).
5. Carefully recover the antenna wire from the radio. Take note of its routing. When the cable is completely free of the underdash, tie 5 or 6 feet of string to the plug on the antenna cable.

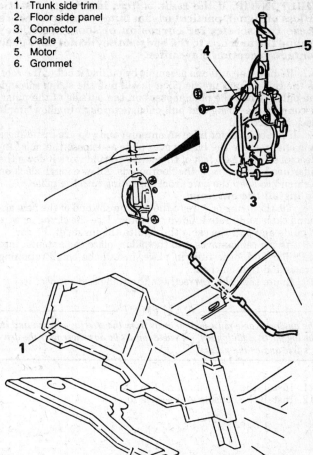

1. Trunk side trim
2. Floor side panel
3. Connector
4. Cable
5. Motor
6. Grommet

Starion antenna

1. Console
2. Scuff plate
3. Cowl side trim
4. Antenna

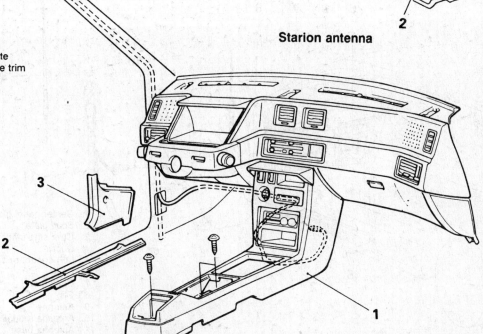

1988 Mirage pillar-mounted antenna

10-27

10 BODY

6. Outside the car, remove the two screws holding the antenna to the roof. Don't lose these small screws.

CHILTON TIP: If the radio suffers intermittent static or loss of signal, particularly on bumpy surfaces, check these two screws for corrosion or looseness. They ground the antenna to the body; if they do not have good contact, reception may suffer.

7. Remove the antenna assembly by pulling it out of the roof. As the cable comes up the pillar it will pull the string through the pillar. When the plug appears on the outside of the pillar, disconnect the string and pull enough through to allow ample slack on each end.

If the car did not have an antenna and you are installing a new one, remove the black cover plate to expose the hole. Instead of string, use 6 feet of thin stiff wire and work it down the pillar until it appears by the floorboards. Allow enough slack on each end to keep the wire from vanishing into the pillar.

To install the antenna:

8. Tie the string or wire to the plug on the end of the new antenna cable and route it down the pillar. Use the string or wire to guide and pull the cable through to the inside of the car.
9. As the cable moves into the pillar, place the antenna into the pillar and, once fully in place, install the small retaining screws and tighten them.
10. Route the cable correctly to the radio and connect it.

CAUTION

The antenna cable must be well away from the steering column and its moving parts. Additionally, the cable cannot be anywhere near the driver's feet and/or the pedals!

---

11. Install the kick panel and the sill plate.
12. Install the console and secure it.

1989 Mirage

NOTE: The mast may be replaced separately by simply unscrewing it from the base. The following procedure is for replacing the base and cable as a unit.

1. Remove the splash shield inside the right front wheel arch. This will be easier if the front end is elevated and supported correctly on stands. Remove the front wheel.
2. Remove the center panel (for radio access) and remove the glove box.
3. Remove the right side sill plate and remove the kick panel.
4. If the car has a rear heater, remove the shower duct under the glove box.
5. Carefully remove the MPI control unit (injection computer).
6. Remove the radio and disconnect the antenna cable.
7. Release the cable from the bands and clips holding it under the dash.
8. Unscrew the mast from the antenna base.
9. Remove the upper base mounting nut (at the fender) and disconnect the lower bolt holding the mast. Remove the mast and carefully remove the cable.
10. Install the new unit, tightening both the upper nut and lower bolt.
11. carefully feed the cable into the car and route it to the radio. Engage the cable in the clips and install new retaining bands.
12. Connect the antenna cable to the radio and reinstall the radio.
13. Replace the MPI control unit.
14. Install the heater duct if it was removed and install the kick panel on the right side.
15. Install the sill plate. Install the glove box.
16. Install the center panel.
17. Reinstall the wheel arch splash shield. Install the wheel if it was removed. Lower the car to the ground.

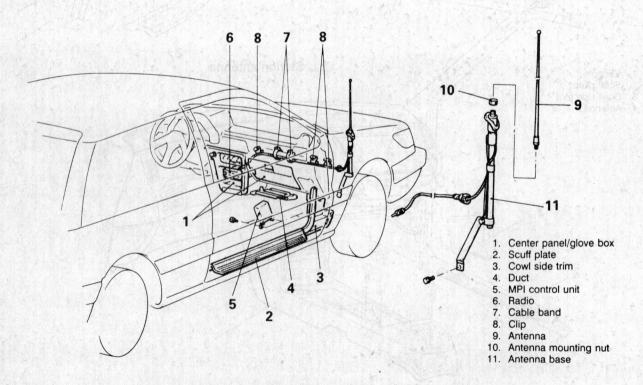

1. Center panel/glove box
2. Scuff plate
3. Cowl side trim
4. Duct
5. MPI control unit
6. Radio
7. Cable band
8. Clip
9. Antenna
10. Antenna mounting nut
11. Antenna base

1989 Mirage antenna

BODY 10

Galant and Sigma

1. Remove the side trim from the trunk.
2. Carefully remove the ring nut from the fender top.
3. Disconnect the wiring for the electric motor and the antenna cable connection.
4. Remove the nuts and bolts holding the motor to the body and remove the motor.
5. Install the new motor and tighten the mounting nuts and bolts.
6. Connect the wiring connectors for the antenna cable and the motor.
7. Install the ring nut and tighten it only to 4 Nm or 36 inch lbs. This is little more than finger tight; do not overtighten this nut.
8. Install the trim panels in the trunk.

Pickup

1. Inside the vehicle, release the glove box stopper by pressing it towards the center of the cab and swinging the door downwards.
2. Reach behind the radio and disconnect the antenna cable.
3. Remove the right sill plate and remove the cowl side trim (kick panel).
4. Outside the vehicle, remove the inner liner from the right front wheel arch. On 4-wheel drive vehicles, remove the mud flap first. This job is easier with the front end elevated and supported with the wheel removed.
5. Remove the antenna mast by unscrewing it from the base.
6. Remove the upper mounting nut and the lower retaining bolt.
7. Remove the antenna base from the fender area and carefully extract the cable.
8. Install the new antenna base and secure the top nut and bottom bolt.
9. Route the cable into the cab and route it correctly to the radio.

1. Trunk side trim
2. Ring nut
3. Harness connector
4. Cable
5. Motor

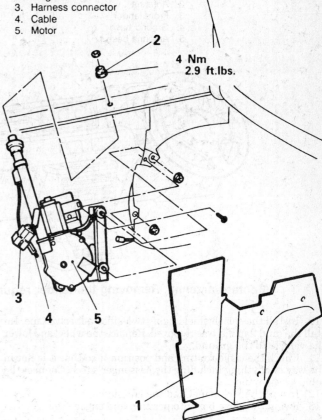

Galant and Sigma power antenna and motor

1. Glove box stopper
2. Connector
3. Scuff plate
4. Cowl side trim
5. Mud guard — 4-wheel drive
6. Splash shield
7. Antenna
8. Antenna mounting nut
9. Antenna base screw
10. Antenna base

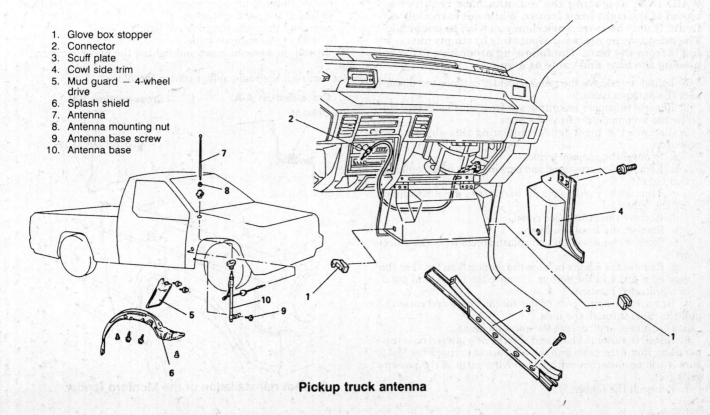

Pickup truck antenna

10 BODY

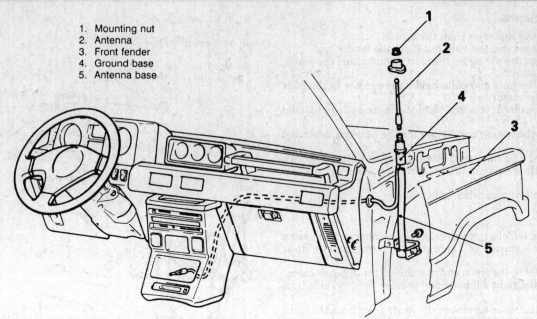

1. Mounting nut
2. Antenna
3. Front fender
4. Ground base
5. Antenna base

Montero antenna. Removing the fender requires more than shown above; see text for details

10. Install the splash shield using two-sided adhesive tape. Install the mud flap if it was removed. Replace the wheel and lower the vehicle to the ground.
11. Check the cable routing and position it so that it is not in the way of anything including the passenger's feet. Connect the cable to the radio.
12. Install the kick panel and the sill plate.
13. Swing the glove box into position and engage the stopper.

Montero

WARNING: Replacing the antenna base requires removal of the right front fender. While not extremely difficult, it may be more work than you wish to undertake. The mast may be replaced separately by simply unscrewing it from the base. The following procedure is for replacing the base and cable as a unit.

1. Loosen or remove the radio or center console and disconnect the antenna cable.
2. Remove the upper mounting nut from the fender top. Unscrew the antenna mast from the base.
3. Remove the front fender following this abbreviated procedure:
   a. Remove the radiator grille.
   b. Remove the turn signal and parking light unit (combination light).
   c. Remove the right headlight bezel (trim panel surrounding the light)
   d. Remove the grille filler panel.
   e. Remove the front mud flap.
   f. Remove the wheel lip and splash shield from the wheel arch.
   g. Remove the 5 bolts holding the fender: 3 on top, 1 at the door edge and 1 in the bottom. Remove the fender and put it in a protected location.
4. Remove the upper part of the antenna base and carefully pull the cable through the hole.
5. Disconnect and remove the antenna base.
6. Install the antenna base and and the new ground base (upper part). Route the cable into the vehicle and to the radio. Make sure it will not interfere with any moving parts or the passenger's feet.
7. Reinstall the fender:

   a. Coat the upper inner flange of the fender end to end with 3M® Sealant No. 8531 or No. 8646 or equivalent.
   b. Install the fender and install the bolts. Check the alignment of the fender with the hood and door. Make needed adjustments before tightening the bolts.
   c. Apply butyl rubber tape to the flange part of the fender before installing the splash shield. Use 3M® No. 8626 or No. EC-5310 or equivalent.
   d. Install the splash shield and wheel lip; install the mud guard.
   e. Install the grille filler panel.
   f. Install the headlight bezel.
   g. Install the right combination light.
   h. Install the radiator grille.
8. Install the antenna mast and tighten the upper retaining nut.
9. Reinstall the radio and/or center console.

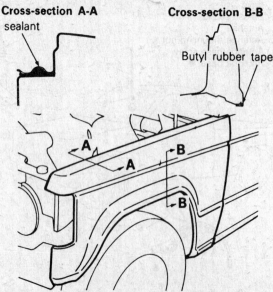

Cross-section A-A sealant

Cross-section B-B Butyl rubber tape

Correct reinstallation of the Montero fender

BODY 10

INTERIOR

Door Panels (Door Pads or Liners)

REMOVAL AND INSTALLATION

Except Starion

NOTE: **This is a general procedure. Depending on vehicle and model, the order of steps may need to be changed slightly.**

1. Remove the inner mirror control knob (if manual remote) and remove the inner delta cover from the mirror mount.
2. Remove the screws holding the armrest and remove the armrest. The armrest screws may be concealed behind plastic caps which may be popped out with a non-marring tool. If the door has a pull handle or grab handle, remove it also.
3. Remove the surround or cover for the inside door handle. Again, seek the hidden screw; remove it and slide the cover off over the handle.
4. If not equipped with electric windows, remove the window winder handle. This can be tricky, but not difficult. Install a piece of tape on the door pad to show the position of the handle before removal. The handle is held onto the winder axle by a spring clip shaped like the Greek letter omega: Ω. The clip is located between the back of the winder handle and the door pad. It is correctly installed with the legs pointing along the length of the winder handle. There are three common ways of removing the clip:

   a. Use a door handle removal tool. This inexpensive slotted and tooted tool can be fitted between the winder and the panel and used to push the spring clip free.

   b. Use a rag or piece of cloth and work it back and forth between the winder and door panel. If constant upward tension is kept, the clip will be forced free. Keep watch on the clip as it pops out; it may get lost.

   c. Straighten a common paper clip and bend a very small J-hook at the end of it. Work the hook down from the top of the winder and engage the loop of the spring clip. As you pull the clip free, keep your other hand over the area. If this is not done the clip will vanish to an undisclosed location, never to be seen again.

5. In general, power door lock and window switches may remain in place until the pad is removed. Some cannot be removed until the doorpad is off the door.
6. If the car has manual vertical door locks, remove the lock knob by unscrewing it. If this is impossible (because they're in square housings) wait until the pad is lifted free.

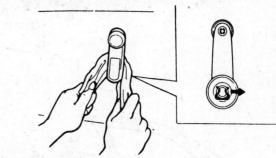

One method of removing the spring clip

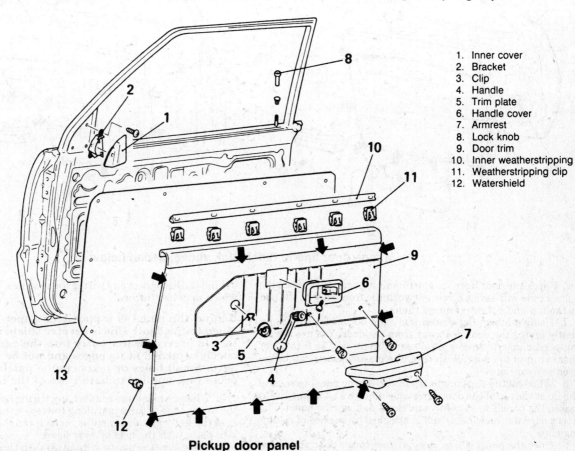

1. Inner cover
2. Bracket
3. Clip
4. Handle
5. Trim plate
6. Handle cover
7. Armrest
8. Lock knob
9. Door trim
10. Inner weatherstripping
11. Weatherstripping clip
12. Watershield

Pickup door panel

10 BODY

1. Clip
2. Window handle
3. Trim plate
4. Armrest
5. Handle cover
6. Lock knob
7. Corner trim
8. Door trim
9. Inner weatherstripping
10. Trim clip
11. Door pocket

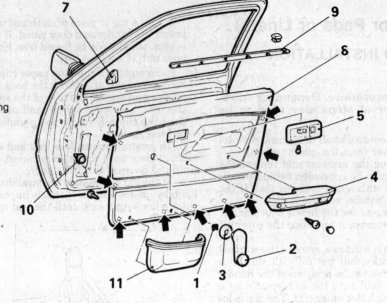

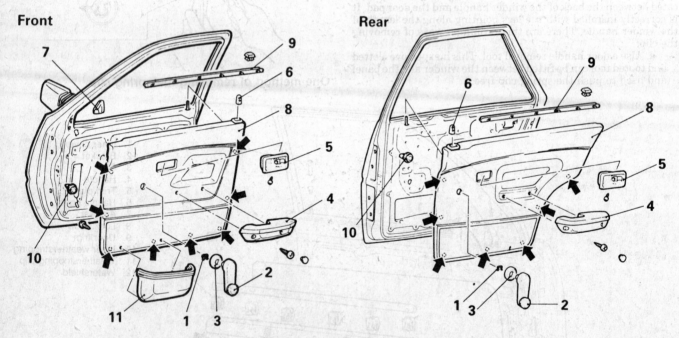

Mirage door liners: hatchback above, sedan below

Check the door liner for any remaining screws or attaching bolts. Some will have a screw or two at the front edge of the pad at the top and bottom; remove them.

7. Using a broad, flat-bladed tool—not a screwdriver—begin gently prying the door pad away from the door. You are releasing plastic inserts from plastic seats. There will be 6 to 12 of them around the door. With care, the plastic inserts can be reused several times.

8. When all the clips are loose, lift up on the panel to release the lip at the top of the door. This may require a bit of jiggling to loosen the panel; do so gently and don't damage the panel. The upper edge (at the window sill) is attached by a series of retaining clips.

9. Once the panel is free, keep it close to the door and check behind it. Disconnect any wiring for switches, lights or speakers which may be attached.

Behind the panel is a plastic or paper sheet taped or glued to the door. This is a water shield and must be intact to prevent water entry into the car. It must be securely attached at its edges and not be ripped or damaged. Small holes or tears can be patched with waterproof tape applied to both sides of the liner.

10. When reinstalling, connect any wiring harnesses and align the upper edge of the panel along the top of the door first. Make sure the left-right alignment is correct; tap the top of the panel into place with the heel of your hand.

11. Make sure the plastic clips align with their holes; pop each

10-32

BODY 10

retainer into place with gentle pressure.

12. Install the armrest and door handle bezel, remembering to install any caps or covers over the screws.
13. Install the window winder handle on cars with manual windows. Place the spring clip into the slot on the handle, remembering that the legs should point along the long dimension of the handle.
14. Align the handle with the tape mark made earlier and put the winder over the end of the axle. Use the heel of your hand to give the center of the winder a short, sharp blow. This will cause the winder to move inward and the spring will engage its locking groove. The secret to this trick is to push the winder straight on; if it's crooked, it won't engage and you may end up looking for the spring clip.
15. Install any remaining parts or trim pieces which may have been removed earlier — Map pockets, speaker grilles, etc.
16. Install the delta cover and the remote mirror handle if they were removed.

Starion

1. Remove the door corner trim or delta cover on the inside of the door mirror location.
2. Remove the cover from the inside door handle. Don't lose the screw.
3. Remove the door pull handle.
4. Remove the power window switch from its mount and disconnect the wiring.
5. Remove the upper sash trim, the vertical piece on the rear edge of the door. Remove the retaining screw and unsnap the trim.
6. Using a broad, flat-bladed tool — not a screwdriver — begin gently prying the upper door pad away from the door. You are releasing plastic inserts from plastic seats. There will be 8 of them holding this panel. From the outside of the door, hold the sheet-metal part of the trim and lift upward.
7. Using the same tool, release the clips holding the lower

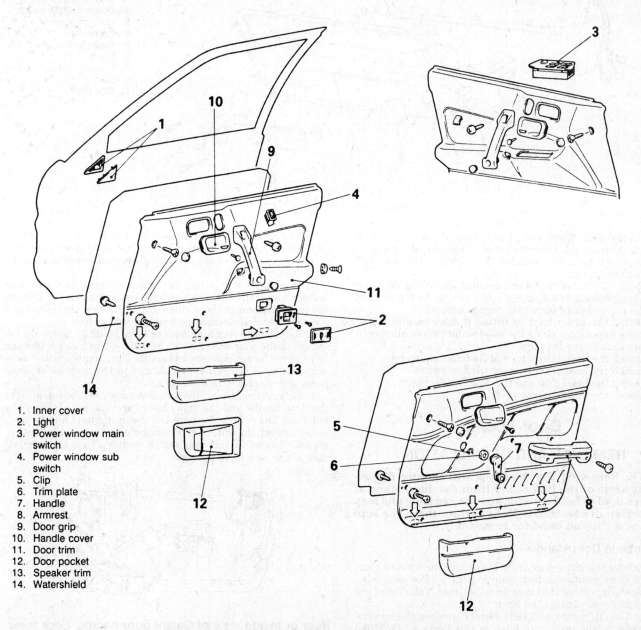

1. Inner cover
2. Light
3. Power window main switch
4. Power window sub switch
5. Clip
6. Trim plate
7. Handle
8. Armrest
9. Door grip
10. Handle cover
11. Door trim
12. Door pocket
13. Speaker trim
14. Watershield

Galant door panels, with and without power windows

10-33

10 BODY

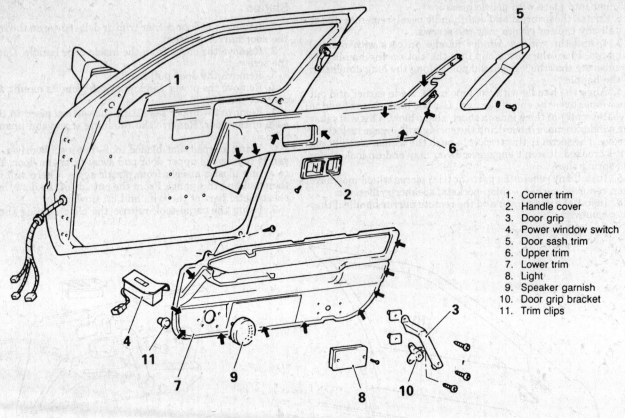

1. Corner trim
2. Handle cover
3. Door grip
4. Power window switch
5. Door sash trim
6. Upper trim
7. Lower trim
8. Light
9. Speaker garnish
10. Door grip bracket
11. Trim clips

Starion two piece door pad

panel to the door. There are 10 clips around the door and one screw at the front top. Once the panel is free, keep it close to the door until you have disconnected the wiring to the door light and speaker.

8. Reinstall the lower pad and connect the wiring harnesses for its components. Install the single screw at the top and connect the clips, making sure they engage properly.
9. Install the upper trim by fitting it downward into place (check the upper edge along the window for proper alignment) and then connecting the clips.
10. Install the door sash trim and its retaining screw.
11. Install and connect the power window switch.
12. Install the hand grip and the door handle cover.
13. Install the inside mirror cover.

Door Locks

REMOVAL AND INSTALLATION

NOTE: Removing the door lock will require disconnecting some of the link rods within the door. Many of the clips used to hold the rods are non-reusable and may break when disassembled. Make certain there is a supply of new clips on hand for reassembly.

Key Lock in Door Handle

1. Remove the inner door liner. Make sure the window glass is in the down position before removing the window controls.
2. Carefully remove the inner moisture liner. Take your time and do not rip or damage the liner.
3. On Galant, Sigma and 1989 Mirage, remove the window glass (refer to the procedure later in this Section). On 1988 Galant, remove the rear glass track.
4. Disconnect or release the clips holding the link rods to both the lock cylinder and the door handle. Depending on the model, it may be easier to disconnect the other end of the rod (at the latch assembly) first.
5. Disconnect any wiring harnesses running to the lock or handle. Generally these cables have connectors in the line; do not try to disconnect the wiring right at the lock. On Galant and Sigma, remove the screw holding the reed switch.
6. Remove the retaining nuts or bolts holding the handle assembly to the door and remove the handle. The lock portion can be removed by a competent locksmith. Disassembly by the owner/mechanic is not recommended due to the number of small parts and springs within the lock.
7. Before reinstalling, the lock assembly must be installed in the door handle and the small lever (arm) attached. Place the handle in the door and secure the mounting nuts and bolts.
8. Connect the wiring to the handle or lock if any was removed. Reconnect the reed switch, making sure the crimped part of the case faces upward.

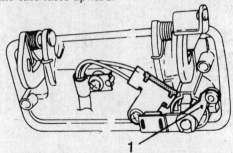

Rear or inside view of Galant door handle. Lock lever arm (1) is attached to the lock cylinder

10-34

BODY 10

9. Carefully connect the link rods to the handle and lock, using new clips where necessary. Reconnect the rods to the latch if any were removed.
10. Reinstall the window glass (and the window track) if they were removed.
11. Reinstall the moisture liner. Apply a bead of waterproof sealer to the outer edge all the way around and align the plastic carefully.
12. Install the door liner and trim pieces.

Key Lock in Door

1. Remove the inner door liner. Make sure the window glass is in the up or closed position before removing the window controls.
2. Carefully remove the inner moisture liner. Take your time and do not rip or damage the liner.
3. Disconnect or release the clips holding the link rod to the lock cylinder. Depending on the model, it may be easier to disconnect the other end of the rod (at the latch assembly) first.
4. Disconnect any wiring harnesses running to the lock. Generally these cables have connectors in the line; do not try to disconnect the wiring right at the lock.
5. On Starion, carefully remove the reed switch from the lock cylinder.
6. The lock cylinder is held to the door by a horseshoe-shaped spring clip. Remove the clip and remove the lock cylinder. The cylinder may be repaired by a competent locksmith. Disassembly by the owner/mechanic is not recommended due to the number of small parts and springs within the lock.
7. Install the cylinder into the door and fit the horseshoe clip. Make sure the cylinder is held firmly in place.
8. Connect the reed switch (Starion) and connect any other wiring which was removed.
9. Connect the link rod, using new clips as necessary.
10. Reinstall the moisture liner. Apply a bead of waterproof sealer to the outer edge all the way around and align the plastic carefully.
11. Install the door liner and trim pieces.

Door Glass, Regulator, and Power Window Motor

REMOVAL AND INSTALLATION

NOTE: As a general rule, the metal regulator tracks and guides should receive a light coat of multipurpose grease during reassembly. Do not get any grease on the glass or the felt/rubber channels and weatherstrips.

Cordia and Tredia

FRONT DOORS WITH MANUAL WINDOWS

1. Lower the window fully.
2. Remove the door pad and the inner moisture barrier. If any of the retaining clips stay in the door, remove them with a thin slotted tool.
3. Disconnect the glass holder from the window regulator.
4. Use a broad, flat-bladed tool to remove the inner belt molding.
5. Remove the door glass by lifting it up and out of the door. Slight rotation of the glass may be necessary.
6. Disconnect the nuts and bolts holding the regulator to the door. Remove the regulator through the access hole.
7. Reinstall the regulator and tighten the nuts and bolts.
8. Reinstall the glass and holder, attaching it to the regulator.

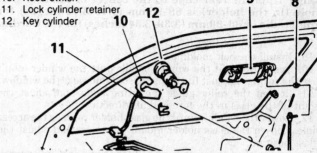

8. Rod snap
9. Handle
10. Reed switch
11. Lock cylinder retainer
12. Key cylinder

Starion door lock and handle assembly

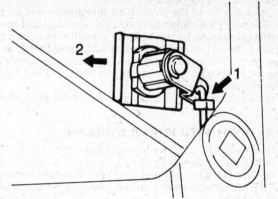

Disconnect the lock rod (1) and remove the spring clip

Removing the front door glass

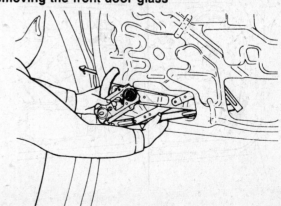

Removing the window regulator

10-35

10 BODY

WARNING: If the glass has been removed from the glass holder, the new glass must be installed so that the distance from the rear edge to the center of the rear bolt hole (in the holder) is 88–95mm (3.46–3.74 inches) for the Tredia or 86–90mm (3.39–3.54 inches) for the Cordia.

9. Install the belt molding.
10. Temporarily fit the winder handle onto the winder (don't use the spring clip, just fit the handle on) and raise the window.
11. Loosen the roller guide mounting bolts and adjust the slant of the glass in the fore and aft direction.
12. Lower the window until the glass holder reaches the access hole. Loosen the glass holder mounting screws and adjust the glass position.
13. Loosen the lower retaining bolt for the front window track or sash and adjust its position.
14. Check the fit of the glass in all dimensions and all positions. The window should fit evenly in all channels and operate smoothly throughout its travel. Make adjustments at each position listed above; remember that any adjustment made at one location may affect any of the others. Continue the process, fine-tuning the window position until correct in all respects. Tighten all the adjusting bolts.
15. Install the moisture barrier and install the door liner.

FRONT DOOR WITH POWER WINDOWS

1. Disconnect the negative battery cable.
2. Lower the window fully.
3. Remove the door pad and the inner moisture barrier. If any of the retaining clips stay in the door, remove them with a thin slotted tool.
4. Remove the glass holder from the window regulator.
5. Use a broad, flat-bladed tool to remove the inner belt molding.
6. Remove the door glass by lifting it up and out of the door. Slight rotation of the glass may be necessary.
7. Disconnect the nuts and bolts holding the regulator to the door.
8. Because the front window channel obstructs removal of the window motor, remove the mounting bolts holding the channel.
9. Disconnect the wiring connector for the power window motor. Remove the regulator with the motor attached through the access hole in the door. Some careful manipulation will be necessary.
10. Remove the power window motor from the regulator by unbolting it.

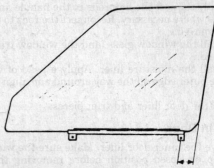

This critical dimension must be observed when replacing the front door glass

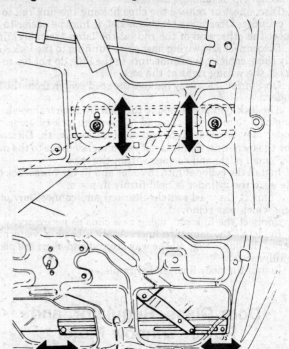

Adjusting the front window regulator

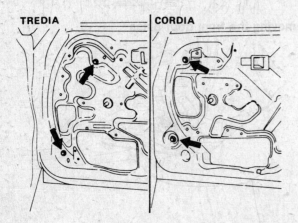

Front channel retaining bolts

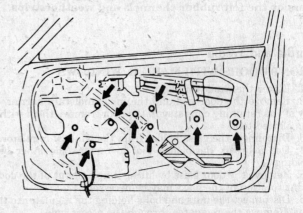

Front door power window regulator retaining bolts

BODY 10

CAUTION

Disconnect the spring from the regulator BEFORE removing the motor. The force of the spring can cause the regulator arm to jump when the motor is removed. The spring is strong enough to cause injury if this occurs.

---

11. If the motor was removed from the regulator, install the spring and then reinstall the motor.
12. Install the assembly into the door and secure the mounting bolts.
13. Connect the wiring harness for the motor.
14. Reposition the front glass channel and tighten the bolts.
15. Install the window glass and snug the attaching bolts.
16. Connect the negative battery cable. Turn the ignition to the **ON** position (don't start the engine) and operate the window switch a little bit at a time. Check the position of the glass in the channels, watching for gaps or binding.
17. Loosen the roller guide mounting bolts and adjust the slant of the glass in the fore and aft direction.
18. Lower the window until the glass holder reaches the access hole. Loosen the glass holder mounting screws and adjust the glass position.
19. Loosen the lower retaining bolt for the front window track or sash and adjust its position.
20. Check the fit of the glass in all dimensions and all positions. The window should fit evenly in all channels and operate smoothly throughout its travel. Make adjustments at each position listed above; remember that any adjustment made at one location may affect any of the others. Continue the process, fine-tuning the window position until correct in all respects. Tighten all the adjusting bolts.
21. Turn the ignition off. Reinstall the moisture barrier and the door pad.

REAR DOOR WITH MANUAL WINDOWS

1. Lower the window fully.
2. Remove the door pad and the inner moisture barrier. If any of the retaining clips stay in the door, remove them with a thin slotted tool.
3. Use a broad, flat-bladed tool to remove the inner belt molding.
4. Remove the mounting bolts and screw holding the strip between the door glass and triangular fixed glass.
5. Separate the weatherstrip (upper part) from the window channel (lower part) and remove the fixed glass and its weatherstrip.
6. Disconnect the glass holder from the window regulator.
7. Remove the door glass by lifting it up and out of the door. Slight rotation of the glass may be necessary.
8. Disconnect the nuts and bolts holding the regulator to the door. Remove the regulator through the access hole.
9. Reinstall the regulator and tighten the nuts and bolts.
10. Reinstall the glass and holder, attaching it to the regulator.

WARNING: If the glass has been removed from the glass holder, the new glass must be installed so that the distance from the front edge to the center of the bolt hole (in the holder) is 70-76mm (2.76-2.99 inches).

11. Install the triangular glass and connect the weather strip to the lower channel. Make sure the weather stripping is properly seated all around the glass.
12. Install the retaining bolts for the channel and fixed glass weatherstrip.
13. Install the belt molding.
14. Temporarily fit the winder handle onto the winder (don't use the spring clip-just fit the handle on) and raise the window.
15. Loosen the roller guide mounting bolts and adjust the slant of the glass in the fore and aft direction.
16. Lower the window until the glass holder reaches the access hole. Loosen the glass holder mounting screws and adjust the

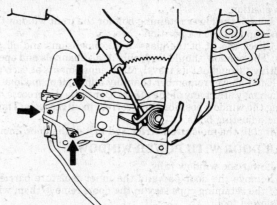

After either the front or rear power regulator is removed, disconnect the spring BEFORE removing the motor. There is a risk of injury if this is not done

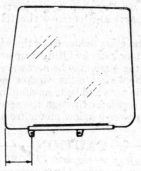

Critical dimension for installing new rear door glass. Refer to text for details

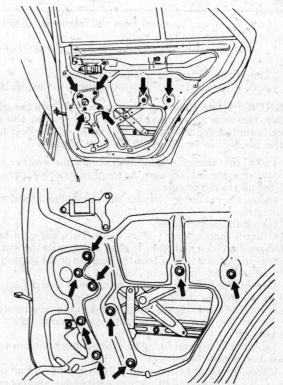

Regulator retaining bolts for rear doors; manual (above) and electric

10-37

10 BODY

glass position.

17. Loosen the lower retaining bolt for the front window track or sash and adjust its position.

18. Check the fit of the glass in all dimensions and all positions. The window should fit evenly in all channels and operate smoothly throughout its travel. Make adjustments at each position listed above; remember that any adjustment made at one location may affect any of the others. Continue the process, fine-tuning the window position until correct in all respects. Tighten all the adjusting bolts.

19. Install the moisture barrier and install the door liner.

REAR DOOR WITH POWER WINDOWS

1. Lower the window fully.
2. Remove the door pad and the inner moisture barrier. If any of the retaining clips stay in the door, remove them with a thin slotted tool.
3. Use a broad, flat-bladed tool to remove the inner belt molding.
4. Remove the mounting bolts and screw holding the strip between the door glass and triangular fixed glass.
5. Separate the weatherstrip (upper part) from the window channel (lower part) and remove the fixed glass and its weatherstrip.
6. Disconnect the glass holder from the window regulator.
7. Remove the door glass by lifting it up and out of the door. Slight rotation of the glass may be necessary.
8. Disconnect the wiring to the window motor.
9. Disconnect the nuts and bolts holding the regulator to the door. Remove the regulator through the access hole.
10. Remove the power window motor from the regulator by unbolting it.

---- **CAUTION** ----

Disconnect the spring from the regulator BEFORE removing the motor. The force of the spring can cause the regulator arm to jump when the motor is removed. The spring is strong enough to cause injury if this occurs!

---

11. If the motor was removed from the regulator, install the spring and then reinstall the motor.
12. Install the assembly into the door and secure the mounting bolts.
13. Connect the wiring harness for the motor.
14. Reinstall the glass and holder, attaching it to the regulator.

WARNING: If the glass has been removed from the glass holder, the new glass must be installed so that the distance from the front edge to the center of the bolt hole (in the holder) is 70–76mm (2.76–2.99 inches).

15. Install the triangular glass and connect the weather strip to the lower channel. Make sure the weather stripping is properly seated all around the glass.
16. Install the retaining bolts for the channel and fixed glass weatherstrip.
17. Install the belt molding.
18. Connect the negative battery cable. Turn the ignition to the **ON** position (don't start the engine) and operate the window switch a little bit at a time. Check the position of the glass in the channels, watching for gaps or binding.
19. Loosen the roller guide mounting bolts and adjust the slant of the glass in the fore and aft direction.
20. Lower the window until the glass holder reaches the access hole. Loosen the glass holder mounting screws and adjust the glass position.
21. Loosen the lower retaining bolt for the front window track or sash and adjust its position.
22. Check the fit of the glass in all dimensions and all positions. The window should fit evenly in all channels and operate smoothly throughout its travel. Make adjustments at each position listed above; remember that any adjustment made at one

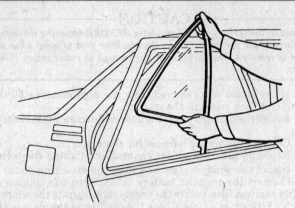

Removing the fixed quarter glass from the rear door

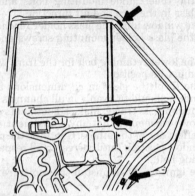

Location of rear door weatherstrip and channel bolts

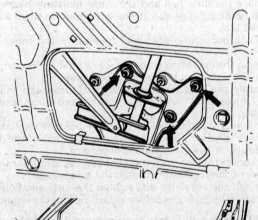

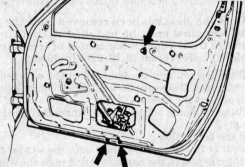

Mounting bolt locations for Starion glass mount (above) and glass guide

BODY 10

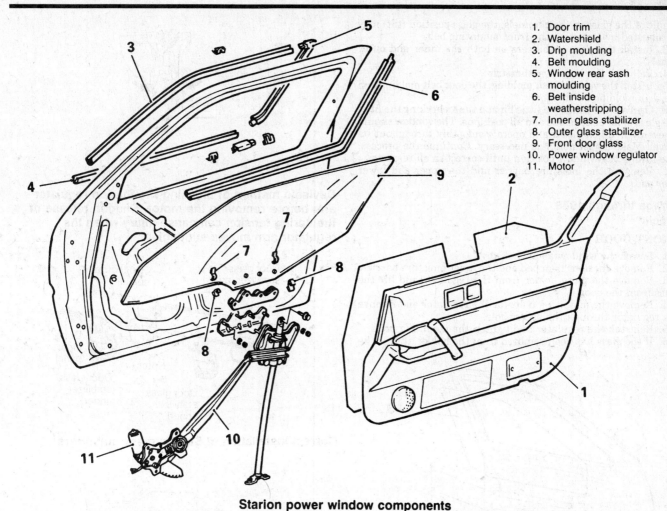

1. Door trim
2. Watershield
3. Drip moulding
4. Belt moulding
5. Window rear sash moulding
6. Belt inside weatherstripping
7. Inner glass stabilizer
8. Outer glass stabilizer
9. Front door glass
10. Power window regulator
11. Motor

Starion power window components

location may affect any of the others. Continue the process, fine-tuning the window position until correct in all respects. Tighten all the adjusting bolts.

23. Turn the ignition off. Reinstall the moisture barrier and the door pad.

Starion

1. Lower the window fully and turn off the ignition.
2. Remove the door liners — upper and lower — and remove the moisture barrier.
3. Using a trim stick or similar blunt, non-marring tool, remove the door drip molding and the door belt molding. Don't break the clips.
4. Remove the rear vertical sash molding from the door.
5. Remove the upper weatherstrip inside the door.
6. Remove the small clips or stabilizers from the door panels.
7. Remove the glass mounting nuts. Separate the glass carrier from the regulator and remove the glass by lifting it up and out of the door.
8. Remove the three bolts (one at the top and two at the bottom) holding the glass guide or center rail in place.
9. Remove the glass guide through the forward access hole in the door.
10. Disconnect the electric connector to the window motor.
11. Remove the regulator mounting bolts and remove the regulator with the motor through the access hole.
12. The power window motor may be removed by unbolting it from the regulator.

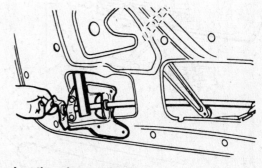

Removing the glass guide

--- **CAUTION** ---

The tension of the spring may cause the regulator arm to jump up when the motor is removed. Use a block of wood or similar device to prevent sudden movement; keep hands clear of the regulator arm during disassembly.

13. Before reinstallation, the motor must be correctly mounted on the regulator. Install the assembly into the door and tighten the retaining bolts.
14. Connect the electric harness to the motor.
15. Install the glass guide or center rail into the door and tighten its retaining bolts.
22. Install the door glass and snug the mounting bolts. Check

10 BODY

the fit of the glass in the channels; the glass position (tilt) may be adjusted by loosening the front mounting bolt.

23. Install the glass stabilizers on both the inner and outer panels.
24. Reinstall the inner weatherstrip.
25. Install the vertical sash molding, the door belt molding and the drip molding on the door.
26. Operate the window a little bit at a time, checking the fit of the glass in all dimensions and all positions. The window should fit evenly in all channels and operate smoothly throughout its travel. Make adjustments as necessary. Continue the process, fine-tuning the window position until correct in all respects.
21. Reinstall the moisture barrier and the upper and lower door pads.

Mirage through 1988
Precis

FRONT DOOR

1. Lower the window about halfway.
2. Remove the door trim pad and the inner moisture barrier.
3. Remove the glass holder from the regulator and lift the glass from the door.
4. Remove the mounting bolts for the regulator and remove the regulator through the access hole.
5. Reinstall the regulator and tighten the mounting bolts.
6. If the glass has been removed from the holder, it must be

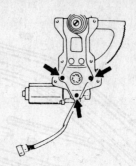

Devise a method of blocking the Starion regulator arm before removing the motor. Sudden release of the spring tension can cause injury when the regulator arm moves suddenly

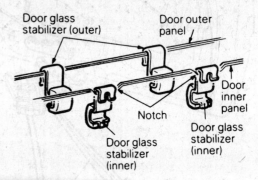

Correct installation of Starion glass stabilizers

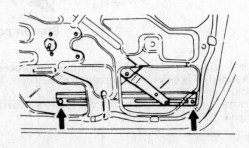

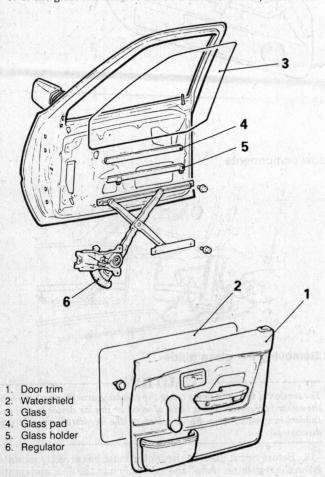

1. Door trim
2. Watershield
3. Glass
4. Glass pad
5. Glass holder
6. Regulator

Early Mirage front door glass. Sedan shown, hatchback similar

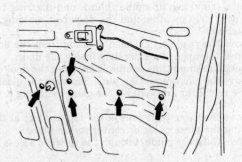

Mirage front door bolt locations: glass mount (above) and regulator

BODY 10

reassembled so that the distance from the rear vertical edge to the rear bolt hole (of the holder) is 295–297mm (11.6–11.7 inches).

7. Install the door glass and attach the holder to the regulator.

8. Install the winder handle (without the spring clip—just fit it in place) and raise the glass fully. Adjust the window so that the door glass fits evenly into the channels all around:

   a. Loosen the roller guide mounting screws and adjust the tilt of the glass (front to rear) as needed.

   b. Lower the door glass until the holder reaches the access hole, loosen the glass holder mounting screws and adjust the position of the glass.

   c. Loosen the screws holding the lower front track and adjust the track as needed.

9. Operate the window a little bit at a time, checking the fit of the glass in all dimensions and all positions. The window should fit evenly in all channels and operate smoothly throughout its travel. Make adjustments as necessary. Continue the process, fine-tuning the window position until correct in all respects. Remember that an adjustment in one position may affect other adjustments.

10. Reinstall the moisture barrier and the door pad.

REAR DOOR

1. Lower the window fully.
2. Remove the door trim pad and moisture liner.
3. Remove the mounting screws and bolts holding the window sash and channel. This is the rear track dividing the door window from the fixed quarter glass.
4. Remove the felt or rubber runchannel from the sash. Remove the sash by carefully pulling it up and out of the door.
5. Remove the stationary window glass along with the weatherstrip.
6. Remove the door glass holder bolts from the regulator. Lift the glass out of the door.
7. Remove the bolts holding the regulator and remove the regulator through the access hole in the door.
8. Install the regulator and tighten the bolts.
9. If the glass has been removed from the holder, it must be reassembled so that the distance from the rear vertical edge to the rear bolt hole (of the holder) is 178–181mm (7.03–7.15 inches).
7. Install the door glass and attach the holder to the regulator.
8. Install the fixed quarter glass and the weatherstrip. Make certain the weatherstrip is not twisted or kinked.
9. Install the vertical sash and install the retaining screws and bolts. Install the felt or rubber run channel.
10. Install the winder handle (without the spring clip, just fit it in place) and raise the glass fully. Adjust the window so that the door glass fits evenly into the channels all around:

    a. Loosen the regulator mounting screws and adjust the tilt of the glass (front to rear) as needed.

    b. Lower the door glass until the holder reaches the access hole, loosen the glass holder mounting screws and adjust the position of the glass front to back.

11. Operate the window a little bit at a time, checking the fit of the glass in all dimensions and all positions. The window should fit evenly in all channels and operate smoothly throughout its travel. Make adjustments as necessary. Continue the process, fine-tuning the window position until correct in all respects. Remember that an adjustment in one position may affect other adjustments.

12. Reinstall the moisture barrier and the door pad.

1989 Mirage

FRONT DOOR WITH MANUAL OR ELECTRIC WINDOWS

1. Lower the door glass.

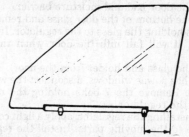

Mirage front glass mounting dimension

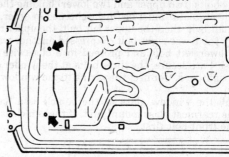

Mirage front sash mounting bolts. Loosen these to aid in position adjustment

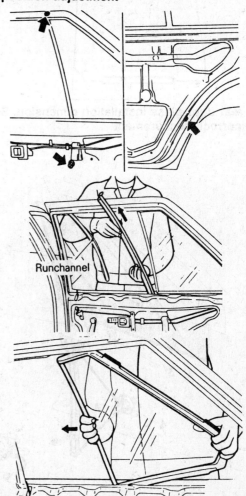

Removing the Mirage fixed quarter glass isn't hard but does require careful handling of parts

10–41

10 BODY

2. Remove the door pad and moisture barrier.
3. Support the bottom of the door glass and remove the two mounting bolts holding the glass to the regulator. If the glass is not supported, it will fall into the door when the bolts are removed.
4. Remove the glass and holder from the door.
5. If the car has power windows, disconnect the wiring to the window motor. Remove the 7 bolts holding the regulator in place. Remove the regulator.
6. When reinstalling the regulator, apply a light coat of multi-purpose grease to the moving parts. Install the regulator and tighten the bolts finger tight.
7. The 4 bolts (two upper and two lower) holding the vertical track of the regulator must be final tightened in order. Tighten the lower left bolt, then the upper left bolt, the upper right bolt and finally the lower right bolt. Expressed another way, begin with the lower left bolt and proceed in a clockwise pattern. When these 4 are secure, the three bolts at the winder or motor should be tightened. Connect the wiring to the window motor if applicable.
8. Install the window glass. If the vehicle is a hatchback, tighten the rearmost glass mount first. If the vehicle is a sedan, tighten the front bolt first.

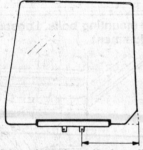

Mirage rear door glass installation dimension. See text for correct measurement

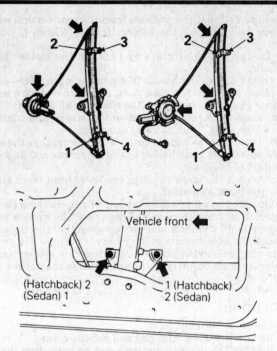

On 1989 Mirage, the regulator and glass mounts must be tightened in the proper order

9. Operate the window a little bit at a time, checking the fit of the glass in all dimensions and all positions. The window should fit evenly in all channels and operate smoothly throughout its travel. Adjustment is generally not needed, and difficult to perform.

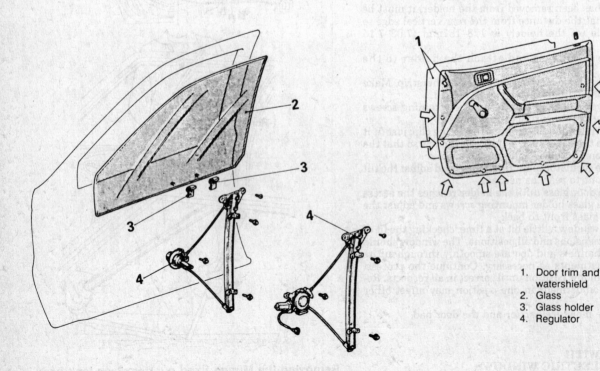

1. Door trim and watershield
2. Glass
3. Glass holder
4. Regulator

1989 Mirage front door regulator and glass

BODY 10

If adjustment is necessary, remove the window regulator and elongate the mounting holes in the door. Before this measure is undertaken, check carefully for bent or damaged components.

10. Reinstall the moisture barrier and the door pad.

REAR DOOR – MANUAL AND ELECTRIC WINDOWS

1. Lower the window.
2. Remove the door trim pad and the moisture barrier.
3. Loosen the attaching bolts holding the vertical rear sash.
4. Support the window glass and remove the glass mounting bolts from the regulator. If not supported, the glass will fall when the bolts are removed.
5. Remove the door glass by lifting it up and out of the door.
6. Remove the installation bolts holding the regulator to the door.
7. If equipped with electric windows, disconnect the wiring to the window motor.
8. Remove the regulator through the access hole in the door.
9. If desired, remove the fixed quarter glass. Lift away a portion of the door weatherstrip to expose the sash mounting screw on the top of the door. Remove the screw.
10. Remove the bolts (loosened earlier) holding the sash to the door. Remove the center door sash.
11. Remove the fixed glass and its weatherstrip by pulling it gently toward the front of the car.
12. Install the fixed quarter glass, making certain the weatherstrip fits properly and is not twisted or kinked.
13. Place the vertical sash in the door and install the upper bolt. Reposition the door weatherstrip.
14. Install the two lower door bolts but leave them loose.
15. Install the window regulator. Of the three bolts holding the vertical portion, tighten the bottom bolt first, then the upper left and upper right bolts in order. This sequence is important; follow it. Connect the wiring to the window motor if equipped with power windows.
16. Install the window glass and tighten the rearmost of the two mounting screws first.
17. Tighten the sash mounting bolts.
18. Operate the window a little bit at a time, checking the fit of the glass in all dimensions and all positions. The window should fit evenly in all channels and operate smoothly throughout its travel. Adjustment is generally not needed, and difficult to perform.

If adjustment is necessary, remove the window regulator and elongate the mounting holes in the door. Before this measure is undertaken, check carefully for bent or damaged components.

19. Reinstall the moisture barrier and the door trim pad.

Galant and Sigma

WARNING: The use of the correct special tools or their equivalent is REQUIRED for this procedure. Alignment gauge blocks (MB 991134, front; MB 991135, rear) are required to adjust the door glass correctly.

FRONT AND REAR DOORS

1. Lower the window.
2. Remove the door trim pad and remove the inner moisture liner.
3. Remove the door mirror from the front door.
4. Using a non-marring tool and a cloth pad, gently pry the belt molding up and off the door.
5. Remove the inner glass stabilizers.
6. On front doors only, remove the glass guide slider.
7. Remove the window glass installation nut from the regulator lift arm bracket.
8. Remove the window glass by pulling it upward. Rear glass will need to be pivoted about 90° on the way out.
9. Disconnect the wiring harness to the window motor.

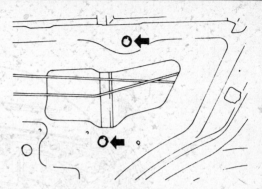

Loosen the rear sash retaining bolts before removing the rear glass from the 1989 Mirage

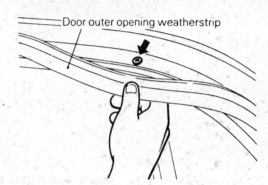

The rear sash won't come out of a 1989 Mirage if you forget this screw

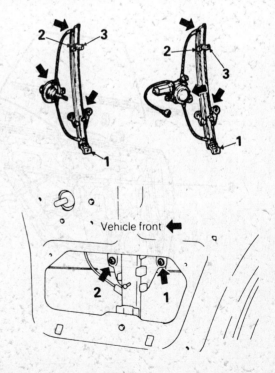

Tightening order for 1989 Mirage rear regulator (above) and glass mount

10 BODY

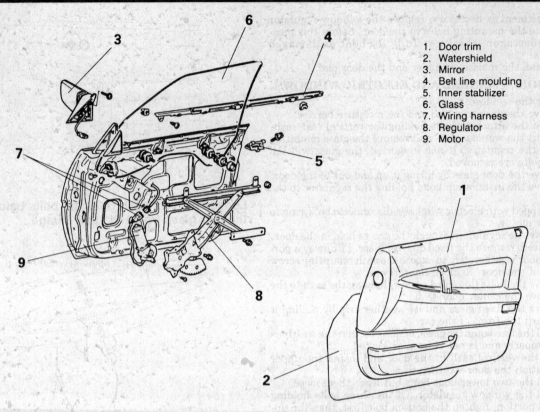

1. Door trim
2. Watershield
3. Mirror
4. Belt line moulding
5. Inner stabilizer
6. Glass
7. Wiring harness
8. Regulator
9. Motor

Galant and Sigma front door glass and regulator

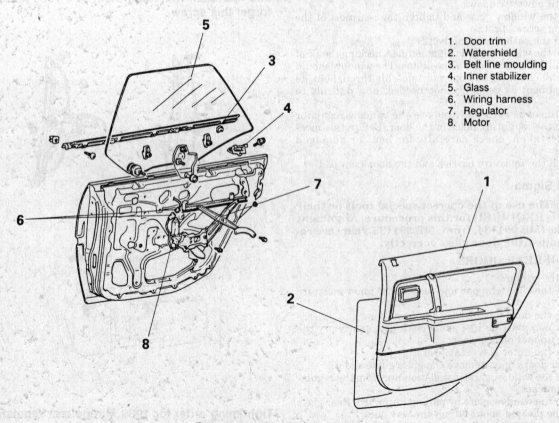

1. Door trim
2. Watershield
3. Belt line moulding
4. Inner stabilizer
5. Glass
6. Wiring harness
7. Regulator
8. Motor

Galant and Sigma rear door

BODY 10

10. Remove the regulator mounting nuts and bolts and remove the regulator from the door.

11. The electric motor may be removed from the regulator if desired.

―――――――― **CAUTION** ――――――――
The tension of the spring may cause the regulator arm to jump up when the motor is removed. Use a block of wood or similar device to prevent sudden movement; keep hands clear of the regulator arm during disassembly!

12. If the motor is removed, it must be reassembled to the regulator before reinstallation. Fit the assembly into the door and tighten the mounting screws.

12. Connect wiring harness to the motor.

13. Install the door glass into the door and connect it to the regulator.

14. Remove the front door drip line (upper) weatherstrip.

15. Attach the alignment blocks at two places on the upper part of the weatherstrip holder or door rail.

NOTE: Reference surface "X" is used when adjusting the glass tilt; surface "Y" is used to adjust the upper travel of the glass.

16. Press the inner stabilizers gently (about 6 lbs.) against the glass and tighten the installation bolt.

WARNING: If too much force is applied, the window regulator will stiffen and bind during operation. If too little force is applied the window will rattle within the door and possibly shatter! These stabilizers will need readjustment after the window inclination (tilt) is set.

17. Raise the glass and adjust the amount of inclination by turning the adjustment bolts on the glass guide tracks. Adjust the glass so that the inner side of the door glass contacts surface **X** on the upper blocks. Turning the adjustment bolts to the left increases the inclination; to the right decreases it.

WARNING: The adjustment bolts must all be turned equal amounts. Uneven adjustment will cause the tracks to tilt and bind the glass.

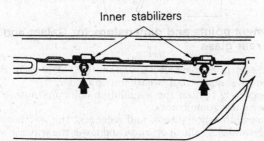

The Galant inner stabilizers cannot be too tight or too loose

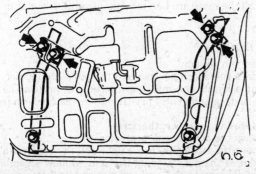

Keep the track bolts in alignment during adjustment

Removing the Galant rear glass requires some rotation

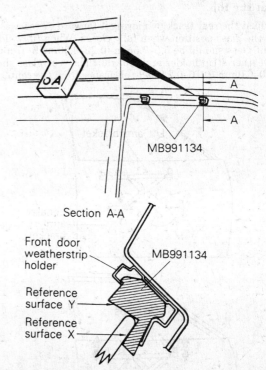

Setting up the gauge blocks to adjust the front window clearance

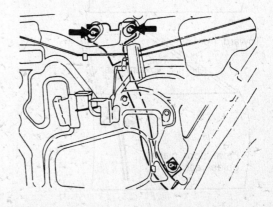

Inclination of the rear door glass is controlled by these two bolts

10-45

10 BODY

18. Raise the door glass fully (if not already done) so that it contacts surface **Y** of the alignment block. Check the fully closed position of the glass, looking for evenness at the top and sides.
19. If adjustment is need, lower the glass a bit and loosen the bolts holding the equalizer arm bracket to the door. By moving this bracket and arm, the fore-and-aft tilt of the glass may be changed. Tighten the mounting nuts when the glass is in position.
20. Once more, raise the glass to the top of its travel and check the contact against both surfaces of the alignment blocks. If the contact is uniform, press the up-stops against the upper surface of the lift arm bracket and completely tighten the mounting bolts.

WARNING: The left and right stops must be pressed on with equal force. If pressed with excessive force, the window will not go up fully, resulting in wind and water leaks at the top.

21. Adjust the rear track by moving it forward or rearward to control the glass position when fully closed. When fully closed, the front glass should be 5.5–7.5mm (0.22–0.30 inch) from the front weather strip holder and roof rail. The rear glass should have 4.0–6.0mm (0.16–0.24 inch) clearance from the center pil-

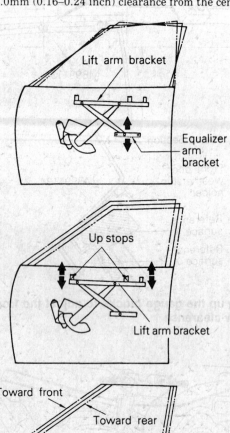

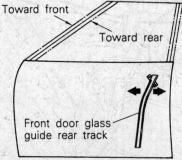

Points and dimensions of adjustment during installation of Galant and Sigma front door glass

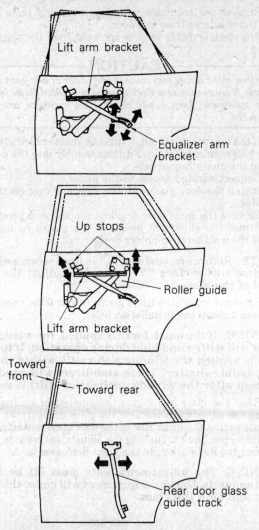

Adjustment points and dimensions for Galant and Sigma rear glass

lar weatherstrip holder and 5.5–7.5mm (0.22–0.30 inch) clearance from the upper (roof) weatherstrip holder.

22. Completely tighten the installation bolts and nuts of the front door glass guide tracks.
23. Remove the special tools and reconnect the weatherstrip.
24. Open and close the door; raise and lower the window while checking for glass fit in all dimensions.
25. Install the belt line molding.
26. Install the outer mirror.
27. Install the moisture barrier and install the door trim pad.

Montero

FRONT DOOR

1. Lower the window.
2. Remove the door trim pad and the moisture barrier.
3. Using a blunt, non-marring instrument, protect the surrounding surface and gently pry the outer weatherstrip free of the door.
4. Remove the inner weatherstrip.
5. Remove the center sash protector.
6. Remove the two bolts in the door and the one bolt at the leading edge of the door holding the vent window assembly.
7. Carefully remove the vent window and its frame as a unit from the door.

BODY 10

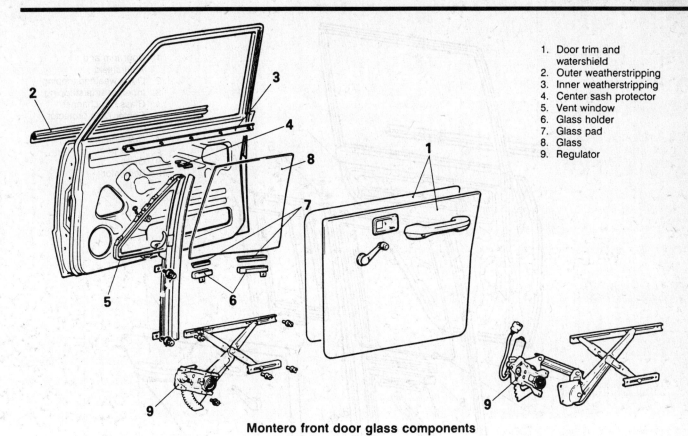

1. Door trim and watershield
2. Outer weatherstripping
3. Inner weatherstripping
4. Center sash protector
5. Vent window
6. Glass holder
7. Glass pad
8. Glass
9. Regulator

Montero front door glass components

8. Disconnect the glass holder bolts from the regulator arms and lift the glass out of the door.
9. Remove the regulator retaining bolts. Unplug the connector to the electric window motor if there is one.
10. Remove the window regulator. The motor for power windows may be removed once the regulator is out of the door.

CAUTION
The tension of the spring may cause the regulator arm to jump up when the motor is removed. Remove the regulator spring BEFORE removing the motor. Failure to do so can result in injuries.

---

11. When reassembling, install the spring and then the motor. The regulator and motor must be placed in the door as a unit. Tighten the retaining bolts to 6 Nm (51 inch lbs.). Connect the wiring for the electric motor.
12. Install the window glass and holders as a unit. If the glass was removed from the holders, it must be reassembled so that the distance from the leading edge to the center of the leading holder bolt hole is 76.5–77.5mm (3.012–3.051 inches). Once that distance is established, the second holder is installed with the centerline of the bolt hole 466.5–467.5mm (18.366–18.406 inches) from the leading holder.
13. Install the vent window assembly and tighten the two bolts to 6 Nm (51 inch lbs.). Tighten the screw in the upper frame snug.
14. Install the center sash protector.
15. Install the inner and outer weather strips.
16. Raise the window fully. Loosen the screws and bolts holding the window and the rear lower track (sash). Move the window, rear sash, and sub-roller assembly (if no vent window) to adjust the glass position. The glass must sit squarely and fit evenly at all the edges. When positioned correctly, retighten the bolts.
17. Install the moisture barrier and the door trim pad.

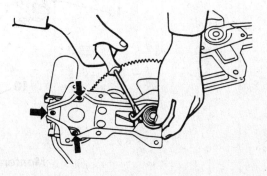

Remove the spring BEFORE removing the power window motor. Injury may result if this is not done

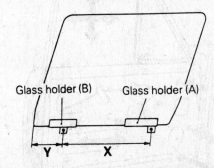

Montero glass holders must be correctly installed when the glass is replaced. See text for dimensions "X" and "Y"

10-47

10 BODY

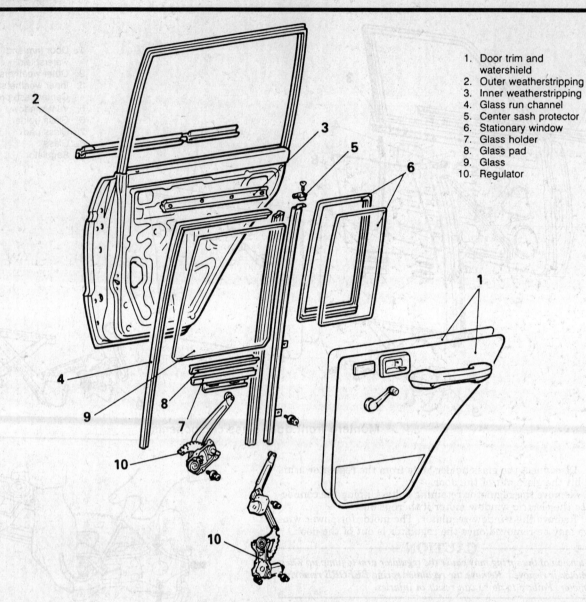

1. Door trim and watershield
2. Outer weatherstripping
3. Inner weatherstripping
4. Glass run channel
5. Center sash protector
6. Stationary window
7. Glass holder
8. Glass pad
9. Glass
10. Regulator

Montero rear door glass

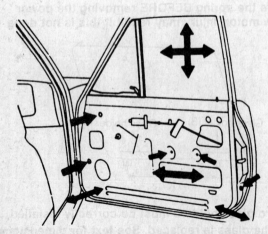

Adjusting the Montero front window

REAR DOOR

1. Lower the window.
2. Remove the door pad and the moisture barrier.
3. Using a blunt, non-marring instrument, protect the surrounding surface and gently pry the outer weatherstrip free of the door.
4. Remove the inner weatherstrip.
5. Remove the felt and rubber runchannel. Don't abuse it; the channel will be reinstalled.
6. Remove the small retaining screw on top of the door and the two bolts holding the center sash to the door. Remove the center sash (track) from the door by lifting it out of the door.
7. Remove the fixed quarter glass with its weatherstrip.
8. Unbolt the glass holder from the regulator and lift the glass out of the door.
9. If equipped with power windows, disconnect the wiring harness to the window motor. Remove the regulator mounting bolts and remove the regulator. The electric motor may be removed if desired.

BODY 10

CAUTION

The tension of the spring may cause the regulator arm to jump up when the motor is removed. Remove the regulator spring BEFORE removing the motor. Failure to do so can result in injuries.

10. When reassembling, install the spring and then the motor. The regulator and motor must be placed in the door as a unit. Tighten the retaining bolts to 6 Nm. (51 inch lbs.). Connect the wiring for the electric motor.
11. Install the window glass and holder as a unit. If the glass was removed from the holder, it must be reassembled so that the distance from the leading edge to the leading edge of the slotted track (not the holder frame) is 231–239mm (9.1–9.4 inches.).
12. Install the fixed glass, making sure the weatherstrip is not twisted or out of place.
13. Install the center track or sash and tighten the screws and bolts.
14. Raise the glass fully and inspect the fit at the top and sides. If adjustment is needed, loosen the center track bolts and move the sash to adjust the glass. Tighten the bolts when the glass is correctly located.
15. Reinstall the moisture barrier and the door trim pad.

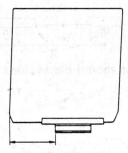

Critical dimension for replacing Montero rear glass

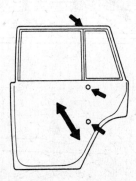

Adjusting the Montero rear glass after installation

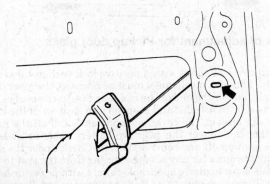

Removing the rear sash

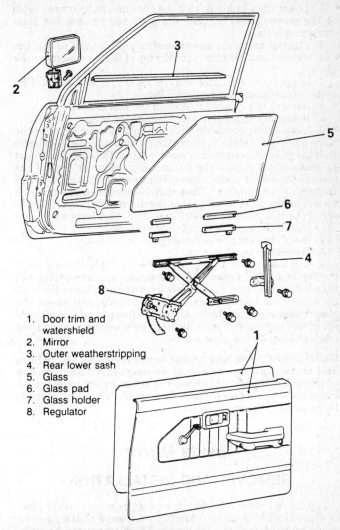

1. Door trim and watershield
2. Mirror
3. Outer weatherstripping
4. Rear lower sash
5. Glass
6. Glass pad
7. Glass holder
8. Regulator

Pickup door glass assembly

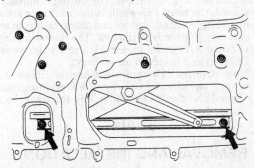

Glass holder mounting bolts

Pickup

1. Raise the window.
2. Remove the door pad and the moisture barrier. Remove the outside rear view mirror.
3. Using a blunt, non-marring instrument, protect the surrounding surface and gently pry the outer weatherstrip free of the door.
4. Remove the lower rear window sash or track. Remove the bolts and then turn the track so that it will come out the access hole.

10 BODY

5. Lower the window so that the glass mount bolts are visible at the access holes. Support the glass and remove the glass mounting bolts.
6. Pull the door glass upward and carefully turn it so that the rear edge of the glass comes out the top of the door. Remove the glass.
7. Support the regulator and remove the regulator mounting bolts. An assistant can be useful here.
8. Remove the regulator through the access hole.
9. Reinstall the regulator and secure the bolts.
10. Install glass assembly with the holders attached. Use the same half-turn method to install the glass as was used to remove it. If the glass was changed or removed from the holders, it must be reassembled so that the distance from the rear edge of the glass to the center of the nearest bolt hole (in the holder) is 99–100mm (3.9–4.0 inches.). Once that distance is established, the second holder is installed so that the distance between the center of the bolt holes is 465–467mm (18.3–18.4 inches.).
11. Install the rear sash and tighten its bolts.
12. Install the outer weatherstrip.
13. Install the outer mirror.
14. Roll the window all the way up. Check the fit of the glass at the top and sides. Loosen the roller guide mounting volts and then adjust the slant of the glass front-to-back. lower the door glass until the holder reaches the access holes and loosen the glass holder mounting bolts. Adjust the front to rear position of the glass. Loosen the bolts holding the lower vertical window track and adjust the glass position.

REMEMBER: Any one adjustment may affect the others and more than one pass may be necessary. Operate the window bit by bit, checking the fit at every location in every position. Once the window is correctly placed, tighten the retaining bolts.

15. Install the moisture barrier and the door trim.

Inside Mirror

REMOVAL AND INSTALLATION

The inside mirror is held to its bracket by screws and/or plastic clips. Usually these are covered by a colored plastic housing which must be removed for access. These covers can be stubborn; take care not to gouge the plastic during removal.

Once exposed, the screws are easily removed. There are several different arrangements of screws and brackets, but in no case does the mirror bolt directly to the roof. The mirror mounts are designed to break away under impact, thus protecting your head and face from serious injury in an accident.

Reassembly requires only common sense (which means you can do it wrong—pay attention); make sure everything fits without being forced and don't overtighten any screws or bolts.

Seats

REMOVAL AND INSTALLATION

Front Seats

EXCEPT MONTERO

1. Inspect the floor area under the seat and remove stray or stored items. If the seat has an under-seat storage tray, remove it.
2. On 1988 Galant and 1989 Sigma, remove the sill plate or scuff plate at the door sill. While not required on other vehicles, it can be helpful to remove it.
3. Remove the end caps or rail covers from the seat adjusting rails. Each model uses a different style. Covers are different front to rear and sometimes left to right as well; pay attention to which cover goes on each rail.

Removing the Pickup glass; the rear edge comes out first

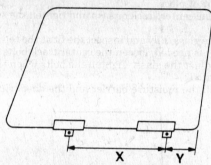

Pickup door glass mount dimensions

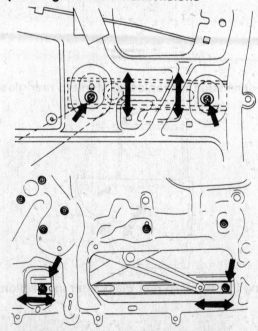

Points of adjustment for Pickup door glass

Each cover attaches with a positive lock such as tabs, side arms, etc. This lock mechanism must be released; the cover does not simply pull off the end of the rail. Use care in removal as the covers must be reinstalled and snapped into position. If the locking tabs are broken, the cover will constantly fall off the rail.

4. Either disconnect the negative battery cable or at least turn the ignition off and remove the key. Check under the seat and along the sides for any wiring running from the seat to the body. This is particularly applicable to cars with power or heated seats. Trace each wire to its in-line connector and disconnect the harness.

BODY 10

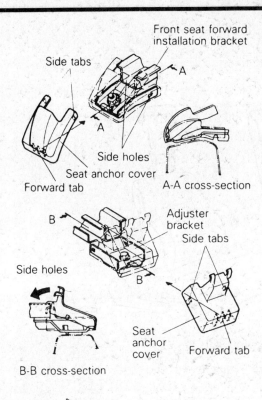

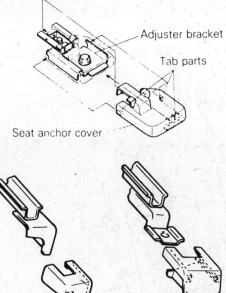

An assortment of seat rail covers. Make sure you release the lock tabs before removal

WARNING: Many cars have electrical components mounted on the floor under the seat. Various computers and relays may be mounted in this location; make very certain you identify the wiring before disconnecting any.

5. Identify the 4 retaining bolts holding the seat tracks to the floor. Slide the seat all the way forward on its rails and remove the rear mounting hardware.
6. Slide the seat to the rear and remove the front mounting nuts or bolts. With the help of an assistant, lift the seat just enough to clear the mounts. Inspect around and under the seat for any wiring still connected. Using two people, remove the seat.

NOTE: This is a perfect time to clean and vacuum the carpet, particularly if both seats have been removed. The amount of lost change to be found will probably make it worthwhile.

7. The seat may be disassembled as needed, but most operations are not recommended for the owner/mechanic. Do not attempt to remove the seat covers unless equipped with a selection of upholstery tools, including hog ring pliers.
8. The seat must be reinstalled as a fully assembled unit. Make sure that both seat tracks are full forward or full back on the seat. Any other position risks a crooked installation and attendant binding.
9. Lift the seat into the car and position it so that the mounts align. Install the nuts and bolts finger tight only.
10. Move the seat through its full range of travel, checking that it will lock securely in each position.
11. Tighten the mounting hardware to the proper torque value. Correct tightening order must be observed.
 • All Mirage, Galant and Sigma: Outer front, inner front, outer rear and inner rear mounts in that order.
 • All others: Inner rear, outer rear, inner front and outer front.
 Bolts holding car seats should be set to 45 Nm (34 ft. lbs.) and nuts should be tightened only to 30 Nm (22 ft. lbs.). On Pickups, the bolts should be tightened just to 12 Nm (9 ft. lbs). Check the seat motion and locking function after tightening the hardware.
12. Connect any wiring which was disconnected.

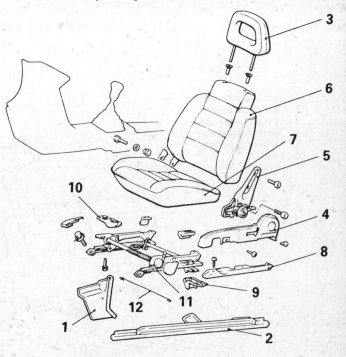

1. Cowl side trim
2. Scuff plate
3. Headrest
4. Shield cover
5. Recliner adjuster
6. Seat back
7. Seat cushion
8. Rail cover
9. Side sill cover
10. Back cover
11. Seat adjuster
12. Wire

Front sport seat, Mirage through 1988

10-51

10 BODY

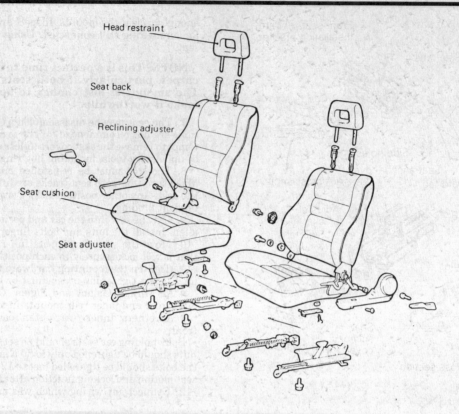

The Cordia front seat is typical of many

1. Seat anchor covers
2. Seat under tray
3. Seat mounting bolts
4. Seat mounting nuts
5. Harness connector
6. Seat
7. Headrest

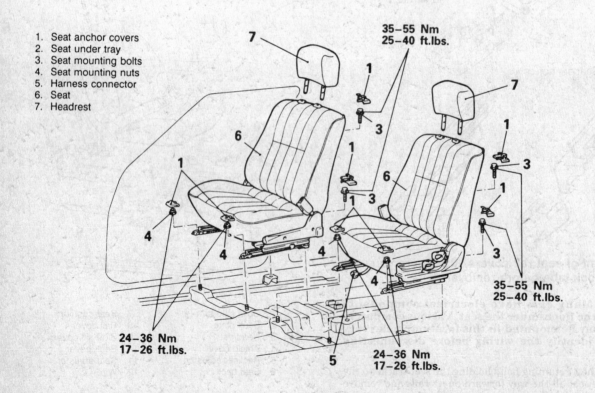

Galant and Sigma front seats. Note that the nuts are tightened less than the bolts

10-52

BODY 10

1. Seat anchor covers
2. Seat under tray
3. Guide ring
4. Seat bolts
5. Seat nuts
6. Harness connector
7. Seat
8. Headrest

1989 Mirage front seat components

13. Install the end caps on each seat track. Make certain they are securely in place.
14. Install any trays or components removed from below the seat.
15. Install the scuff plate if it was removed.

MONTERO

The Montero follows the steps above with one notable exception. Because of the suspension unit between the seat and the floor, the seat should be unbolted from the suspension frame. Attempting to remove both at once increases weight and bulk which could lead to damage. In most cases, removing only the seat will suffice to provide access to other areas or allow the seat to be replaced or repaired.

If it is necessary to remove the seat suspension frame, it must be removed from the floor as a unit (4 bolts). Do not attempt to disassemble or adjust the suspension unit. Reinstall the frame first and then install the seat. All mounting bolts for the seat and the suspension unit should be tightened to 45 Nm or 34 ft. lbs.

After reinstallation, check both the fore and aft locking of the seat and the function of the suspension unit.

Rear Seat and Seatback

Cordia and Tredia

1. Remove or loosen the carpeting in the rear footwell as necessary to expose the two seat retaining bolts.
2. Remove the bolts.
3. Remove the Tredia seat by sliding it forward. The Cordia seat won't come out yet.
4. On Tredia, remove the seatback mounting bolts and lift the seat up and rearward (at an angle) to remove it. This motion releases the hook behind the seatback.
5. On Cordia, fold the seatback forward and remove the carpet. Remove the spacers (spring clips) from the center hinge bracket and move the seatback towards the center of the car to release it from the outer brackets. Remove the seatback.

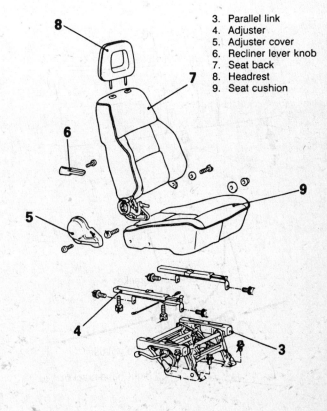

3. Parallel link
4. Adjuster
5. Adjuster cover
6. Recliner lever knob
7. Seat back
8. Headrest
9. Seat cushion

Remove the Montero seat from the suspension frame

10-53

10 BODY

6. Remove the Cordia seat cushion by detaching the hook from the hinge bracket and lifting the seat out.
7. To reinstall the Cordia seatback, fit the bushing into the hole in the hinge bracket. Mount the seatback to the center hinge and then to the outer bracket. Install the spacer (spring clip) at the center bracket and fold the seatback up so that it locks in position.
8. Install the Tredia seatback by fitting the rod over the three hooks on the body. Tighten the seatback mounting bolts, beginning from the center.
9. Install the seat cushion and its retaining bolts. On Cordia, engage the attaching rods securely to the bracket hooks.
10. Tighten the seat retaining bolts to 6 Nm (48 inch lbs.).

STARION

1. Remove the small plastic covers at the front of the rear seat cushion.
2. Remove the mounting bolts holding the cushion.
3. Lift up the seat and pull the rear seat belts through the slots in the cushion.
4. Remove the seat cushion.
5. Fold down the rear seatbacks and remove the carpet cover screws which hold the trunk carpet to the seatbacks. Raise the seatbacks and lock them in place.
6. Remove the seatback retaining bolts beginning with the right side, then the left side. Remove the center bolt last.
7. With an assistant, release the upper seat back locks and remove both seatbacks simultaneously.
8. When assembling and installing the seatbacks, fit them into position and clip them into the upper locks. Note that at the center bottom of the seatbacks, two brackets are held by one bolt. The bracket from the left seatback must be on top of the bracket from the right seatback.

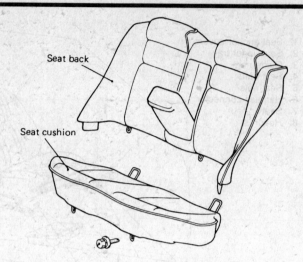

Tredia rear seat

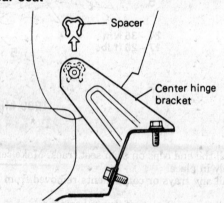

Cordia split rear seatback removal requires removal of the inner spacers

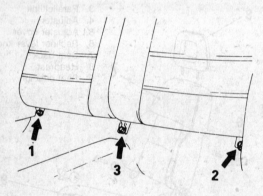

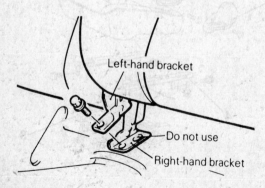

Details of Starion rear seatback removal (above) and installation

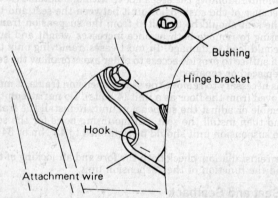

Detail of Cordia seatback and cushion installation

NOTE: Sometimes this set of brackets has two bolt holes. Use the smaller one.

9. Install and temporarily tighten the seatback retaining bolts. Confirm correct movement and locking of the seatbacks. Final tighten the seatback retaining bolts to 14 Nm or 10 ft. lbs.
10. Install the seat cushion and tighten its retaining bolts to 6 Nm (48 inch lbs.). Don't forget to feed the seatbelts back through the slots.

BODY 10

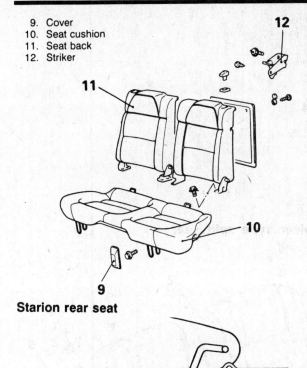

9. Cover
10. Seat cushion
11. Seat back
12. Striker

Starion rear seat

The upper rod or wire must engage the hooks when installing a fixed rear seatback

11. Fold the seatbacks down and reinstall the trunk carpet and its retainers. Install the plastic covers on the seat cushion bolts.

MIRAGE
GALANT AND SIGMA
PRECIS

1. Remove or loosen the carpet to expose the mounting bolts or levers (1989) in front of the seat cushion.
2. Remove the bolts from the front of the seat. Press down the padding at the rear of the seat to expose the two rear mounting bolts holding the cushion. Remove them.
3. On cars through 1988, lift the front of the seat up and rotate it out of position. On 1989 cars, pull outward on the two levers (simultaneously) and lift up.
4. Fixed rear seatbacks are removed by disconnecting the retaining nuts and bolts along the bottom and lifting upward and backward to release the upper hooks.
5. One-piece folding seat backs may be removed after disconnecting any carpeting or trim from the back side. Remove the bolts holding the pivot to the body on each side and remove the seatback.
6. Two-piece or split folding seatbacks are removed as a unit after removing the pivot mounting bolts. Don't forget the bolts holding the center pivot; they may be hard to see.
7. Reinstall the seatback. It can be helpful to clip the folding seatbacks into the upper lock to help hold them in place. Fixed seatbacks should engage the upper hooks firmly before final positioning.
8. Install the retaining bolts and tighten just snug. Check the position and operation of the all seatback functions and locks.
9. Final tighten the retaining bolts to 20 Nm (15 ft. lbs.) except for the center pivot bolts on Mirage through 1988. Tighten those bolts only to 6 Nm (48 inch lbs.).
10. Install the seat cushion, making sure it is firmly in position with the bolt holes aligned. For 1989 Mirage, make certain the locks have engaged. Install the retaining bolts and tighten them to 20 Nm (15 ft. lbs.).
11. Replace any carpet or trim removed from either the seat back or the front of the seat cushion.

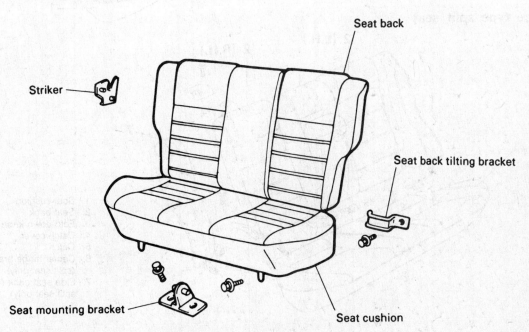

Precis rear seat components

10 BODY

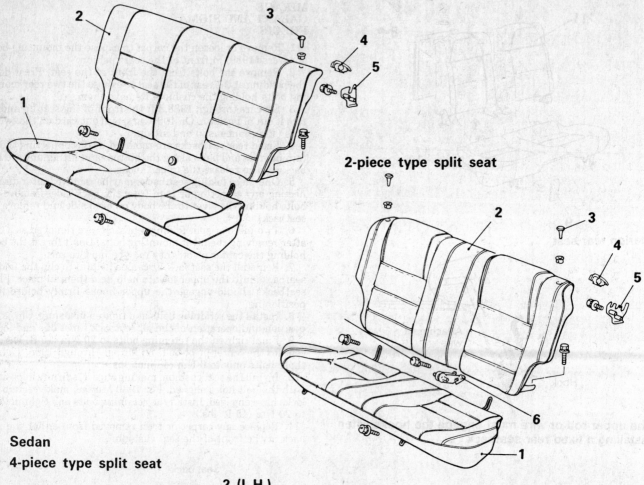

1. Seat cushion
2. Seat back
3. Fold-down knob
4. Catch cover
5. Catch
6. Center hinge bracket (split seat only)
7. Side seat back (4-pieces split seat only)

Several types of Mirage rear seats

10-56

BODY 10

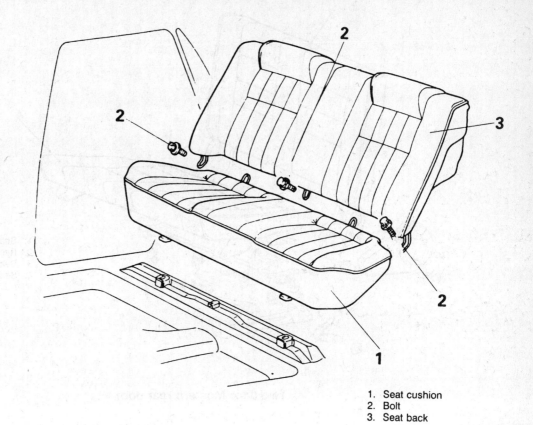

1. Seat cushion
2. Bolt
3. Seat back

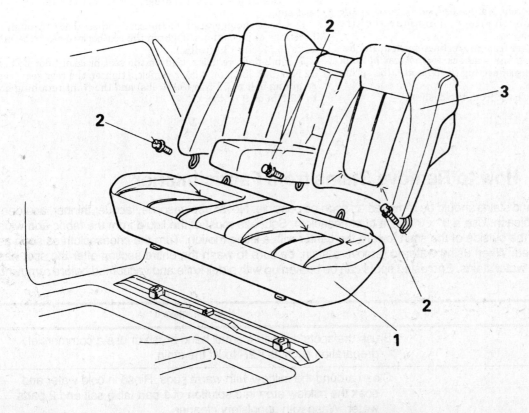

Galant and Sigma seatbacks

10-57

10 BODY

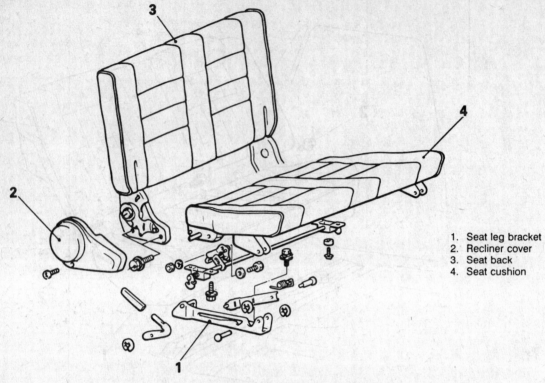

1. Seat leg bracket
2. Recliner cover
3. Seat back
4. Seat cushion

Two door Montero rear door

MONTERO

1. Disconnect either the rear seat leg bracket (2 door) or the seat track (4 door) from the floor. This may require repositioning the seat for access.
2. Remove the plastic cover from the recliner mechanism. Remove the two bolts holding the seatback to the seat bottom. Make certain you are not disconnecting the adjuster; the spring is under tension.
3. Remove the rear seat back from the cushion.
4. Remove the seat cushion.
5. Reassemble in reverse order. Install all the retaining bolts just snug.
6. Check the operation of the seat and seatback through all functions and positions. Tighten the seatback to seat retaining bolts to 52 Nm (39 ft. lbs.).
7. On 2-door vehicles, tighten the seat bracket floor bolts to 12 Nm (9 ft. lbs.); on 4 door vehicles, tighten the rear seat track retaining bolt to 52 Nm (39 ft. lbs) and the front retaining bolt to 12 Nm or 9 ft. lbs.

How to Remove Stains from Fabric Interior

For best results, spots and stains should be removed as soon as possible. Never use gasoline, lacquer thinner, acetone, nail polish remover or bleach. Use a 3' x 3" piece of cheesecloth. Squeeze most of the liquid from the fabric and wipe the stained fabric from the outside of the stain toward the center with a lifting motion. Turn the cheesecloth as soon as one side becomes soiled. When using water to remove a stain, be sure to wash the entire section after the spot has been removed to avoid water stains. Encrusted spots can be broken up with a dull knife and vacuumed before removing the stain.

Type of Stain	How to Remove It
Surface spots	Brush the spots out with a small hand brush or use a commercial preparation such as K2R to lift the stain.
Mildew	Clean around the mildew with warm suds. Rinse in cold water and soak the mildew area in a solution of 1 part table salt and 2 parts water. Wash with upholstery cleaner.

BODY 10

Type of Stain	How to Remove It
Water stains	Water stains in fabric materials can be removed with a solution made from 1 cup of table salt dissolved in 1 quart of water. Vigorously scrub the solution into the stain and rinse with clear water. Water stains in nylon or other synthetic fabrics should be removed with a commercial type spot remover.
Chewing gum, tar, crayons, shoe polish (greasy stains)	Do not use a cleaner that will soften gum or tar. Harden the deposit with an ice cube and scrape away as much as possible with a dull knife. Moisten the remainder with cleaning fluid and scrub clean.
Ice cream, candy	Most candy has a sugar base and can be removed with a cloth wrung out in warm water. Oily candy, after cleaning with warm water, should be cleaned with upholstery cleaner. Rinse with warm water and clean the remainder with cleaning fluid.
Wine, alcohol, egg, milk, soft drink (non-greasy stains)	Do not use soap. Scrub the stain with a cloth wrung out in warm water. Remove the remainder with cleaning fluid.
Grease, oil, lipstick, butter and related stains	Use a spot remover to avoid leaving a ring. Work from the outisde of the stain to the center and dry with a clean cloth when the spot is gone.
Headliners (cloth)	Mix a solution of warm water and foam upholstery cleaner to give thick suds. Use only foam—liquid may streak or spot. Clean the entire headliner in one operation using a circular motion with a natural sponge.
Headliner (vinyl)	Use a vinyl cleaner with a sponge and wipe clean with a dry cloth.
Seats and door panels	Mix 1 pint upholstery cleaner in 1 gallon of water. Do not soak the fabric around the buttons.
Leather or vinyl fabric	Use a multi-purpose cleaner full strength and a stiff brush. Let stand 2 minutes and scrub thoroughly. Wipe with a clean, soft rag.
Nylon or synthetic fabrics	For normal stains, use the same procedures you would for washing cloth upholstery. If the fabric is extremely dirty, use a multi-purpose cleaner full strength with a stiff scrub brush. Scrub thoroughly in all directions and wipe with a cotton towel or soft rag.

10 BODY

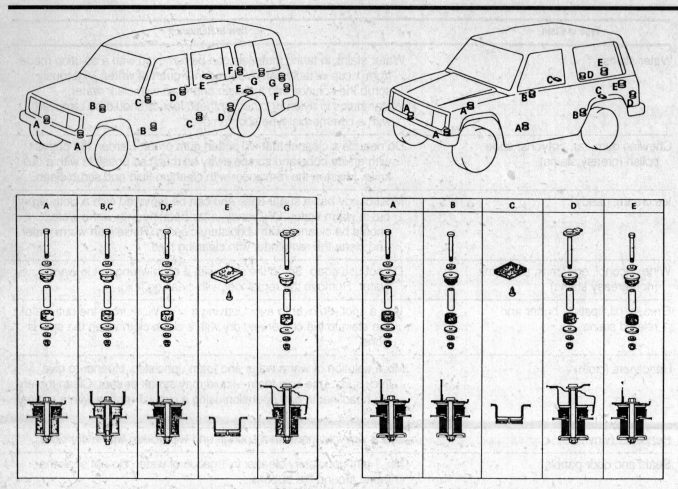

Montero body mounts

GLOSSARY

GLOSSARY OF TERMS

AIR/FUEL RATIO: The ratio of air to gasoline by weight in the fuel mixture drawn into the engine.

AIR INJECTION: One method of reducing harmful exhaust emissions by injecting air into each of the exhaust ports of an engine. The fresh air entering the hot exhaust manifold causes any remaining fuel to be burned before it can exit the tailpipe.

ALTERNATOR: A device used for converting mechanical energy into electrical energy.

AMMETER: An instrument, calibrated in amperes, used to measure the flow of an electrical current in a circuit. Ammeters are always connected in series with the circuit being tested.

AMPERE: The rate of flow of electrical current present when one volt of electrical pressure is applied against one ohm of electrical resistance.

ANALOG COMPUTER: Any microprocessor that uses similar (analogous) electrical signals to make its calculations.

ARMATURE: A laminated, soft iron core wrapped by a wire that converts electrical energy to mechanical energy as in a motor or relay. When rotated in a magnetic field, it changes mechanical energy into electrical energy as in a generator.

ATMOSPHERIC PRESSURE: The pressure on the Earth's surface caused by the weight of the air in the atmosphere. At sea level, this pressure is 14.7 psi at 32°F (101 kPa at 0°C).

ATOMIZATION: The breaking down of a liquid into a fine mist that can be suspended in air.

AXIAL PLAY: Movement parallel to a shaft or bearing bore.

BACKFIRE: The sudden combustion of gases in the intake or exhaust system that results in a loud explosion.

BACKLASH: The clearance or play between two parts, such as meshed gears.

BACKPRESSURE: Restrictions in the exhaust system that slow the exit of exhaust gases from the combustion chamber.

BAKELITE: A heat resistant, plastic insulator material commonly used in printed circuit boards and transistorized components.

BALL BEARING: A bearing made up of hardened inner and outer races between which hardened steel balls roll.

BALLAST RESISTOR: A resistor in the primary ignition circuit that lowers voltage after the engine is started to reduce wear on ignition components.

BEARING: A friction reducing, supportive device usually located between a stationary part and a moving part.

BIMETAL TEMPERATURE SENSOR: Any sensor or switch made of two dissimilar types of metal that bend when heated or cooled due to the different expansion rates of the alloys. These types of sensors usually function as an on/off switch.

BLOWBY: Combustion gases, composed of water vapor and unburned fuel, that leak past the piston rings into the crankcase during normal engine operation. These gases are removed by the PCV system to prevent the buildup of harmful acids in the crankcase.

BRAKE PAD: A brake shoe and lining assembly used with disc brakes.

BRAKE SHOE: The backing for the brake lining. The term is, however, usually applied to the assembly of the brake backing and lining.

BUSHING: A liner, usually removable, for a bearing; an anti-friction liner used in place of a bearing.

BYPASS: System used to bypass ballast resistor during engine cranking to increase voltage supplied to the coil.

CALIPER: A hydraulically activated device in a disc brake system, which is mounted straddling the brake rotor (disc). The caliper contains at least one piston and two brake pads. Hydraulic pressure on the piston(s) forces the pads against the rotor.

CAMSHAFT: A shaft in the engine on which are the lobes (cams) which operate the valves. The camshaft is driven by the crankshaft, via a belt, chain or gears, at one half the crankshaft speed.

CAPACITOR: A device which stores an electrical charge.

CARBON MONOXIDE (CO): a colorless, odorless gas given off as a normal byproduct of combustion. It is poisonous and extremely dangerous in confined areas, building up slowly to toxic levels without warning if adequate ventilation is not available.

CARBURETOR: A device, usually mounted on the intake manifold of an engine, which mixes the air and fuel in the proper proportion to allow even combustion.

CATALYTIC CONVERTER: A device installed in the exhaust system, like a muffler, that converts harmful byproducts of combustion into carbon dioxide and water vapor by means of a heat-producing chemical reaction.

CENTRIFUGAL ADVANCE: A mechanical method of advancing the spark timing by using flyweights in the distributor that react to centrifugal force generated by the distributor shaft rotation.

CHECK VALVE: Any one-way valve installed to permit the flow of air, fuel or vacuum in one direction only.

CHOKE: A device, usually a moveable valve, placed in the intake path of a carburetor to restrict the flow of air.

CIRCUIT: Any unbroken path through which an electrical current can flow. Also used to describe fuel flow in some instances.

CIRCUIT BREAKER: A switch which protects an electrical circuit from overload by opening the circuit when the current flow exceeds a predetermined level. Some circuit breakers must be reset manually, while most reset automatically

COIL (IGNITION): A transformer in the ignition circuit which steps up the voltage provided to the spark plugs.

COMBINATION MANIFOLD: An assembly which includes both the intake and exhaust manifolds in one casting.

COMBINATION VALVE: A device used in some fuel systems that routes fuel vapors to a charcoal storage canister instead of venting them into the atmosphere. The valve relieves fuel tank pressure and allows fresh air into the tank as the fuel level drops to prevent a vapor lock situation.

COMPRESSION RATIO: The comparison of the total volume of the cylinder and combustion chamber with the piston at BDC and the piston at TDC.

CONDENSER: 1. An electrical device which acts to store an electrical charge, preventing voltage surges.
2. A radiator-like device in the air conditioning system in which refrigerant gas condenses into a liquid, giving off heat.

CONDUCTOR: Any material through which an electrical current can be transmitted easily.

CONTINUITY: Continuous or complete circuit. Can be checked with an ohmmeter.

COUNTERSHAFT: An intermediate shaft which is rotated by a mainshaft and transmits, in turn, that rotation to a working part.

CRANKCASE: The lower part of an engine in which the crankshaft and related parts operate.

CRANKSHAFT: The main driving shaft of an engine which receives reciprocating motion from the pistons and converts it to rotary motion.

CYLINDER: In an engine, the round hole in the engine block in which the piston(s) ride.

CYLINDER BLOCK: The main structural member of an engine in which is found the cylinders, crankshaft and other principal parts.

CYLINDER HEAD: The detachable portion of the engine, fastened, usually, to the top of the cylinder block, containing all or most of the combustion chambers. On overhead valve engines, it contains the valves and their operating parts. On overhead cam engines, it contains the camshaft as well.

DEAD CENTER: The extreme top or bottom of the piston stroke.

DETONATION: An unwanted explosion of the air/fuel mixture in the combustion chamber caused by excess heat and compression, advanced timing, or an overly lean mixture. Also referred to as "ping".

DIAPHRAGM: A thin, flexible wall separating two cavities, such

10-61

GLOSSARY

as in a vacuum advance unit.

DIESELING: A condition in which hot spots in the combustion chamber cause the engine to run on after the key is turned off.

DIFFERENTIAL: A geared assembly which allows the transmission of motion between drive axles, giving one axle the ability to turn faster than the other.

DIODE: An electrical device that will allow current to flow in one direction only.

DISC BRAKE: A hydraulic braking assembly consisting of a brake disc, or rotor, mounted on an axle, and a caliper assembly containing, usually two brake pads which are activated by hydraulic pressure. The pads are forced against the sides of the disc, creating friction which slows the vehicle.

DISTRIBUTOR: A mechanically driven device on an engine which is responsible for electrically firing the spark plug at a predetermined point of the piston stroke.

DOWEL PIN: A pin, inserted in mating holes in two different parts allowing those parts to maintain a fixed relationship.

DRUM BRAKE: A braking system which consists of two brake shoes and one or two wheel cylinders, mounted on a fixed backing plate, and a brake drum, mounted on an axle, which revolves around the assembly. Hydraulic action applied to the wheel cylinders forces the shoes outward against the drum, creating friction, slowing the vehicle.

DWELL: The rate, measured in degrees of shaft rotation, at which an electrical circuit cycles on and off.

ELECTRONIC CONTROL UNIT (ECU): Ignition module, module, amplifier or igniter. See Module for definition.

ELECTRONIC IGNITION: A system in which the timing and firing of the spark plugs is controlled by an electronic control unit, usually called a module. These systems have no points or condenser.

ENDPLAY: The measured amount of axial movement in a shaft.

ENGINE: A device that converts heat into mechanical energy.

EXHAUST MANIFOLD: A set of cast passages or pipes which conduct exhaust gases from the engine.

FEELER GAUGE: A blade, usually metal, of precisely predetermined thickness, used to measure the clearance between two parts. These blades usually are available in sets of assorted thicknesses.

F-Head: An engine configuration in which the intake valves are in the cylinder head, while the camshaft and exhaust valves are located in the cylinder block. The camshaft operates the intake valves via lifters and pushrods, while it operates the exhaust valves directly.

FIRING ORDER: The order in which combustion occurs in the cylinders of an engine. Also the order in which spark is distributed to the plugs by the distributor.

FLATHEAD: An engine configuration in which the camshaft and all the valves are located in the cylinder block.

FLOODING: The presence of too much fuel in the intake manifold and combustion chamber which prevents the air/fuel mixture from firing, thereby causing a no-start situation.

FLYWHEEL: A disc shaped part bolted to the rear end of the crankshaft. Around the outer perimeter is affixed the ring gear. The starter drive engages the ring gear, turning the flywheel, which rotates the crankshaft, imparting the initial starting motion to the engine.

FOOT POUND (ft.lb. or sometimes, ft. lbs.): The amount of energy or work needed to raise an item weighing one pound, a distance of one foot.

FUSE: A protective device in a circuit which prevents circuit overload by breaking the circuit when a specific amperage is present. The device is constructed around a strip or wire of a lower amperage rating than the circuit it is designed to protect. When an amperage higher than that stamped on the fuse is present in the circuit, the strip or wire melts, opening the circuit.

GEAR RATIO: The ratio between the number of teeth on meshing gears.

GENERATOR: A device which converts mechanical energy into electrical energy.

HEAT RANGE: The measure of a spark plug's ability to dissipate heat from its firing end. The higher the heat range, the hotter the plug fires.

HUB: The center part of a wheel or gear.

HYDROCARBON (HC): Any chemical compound made up of hydrogen and carbon. A major pollutant formed by the engine as a byproduct of combustion.

HYDROMETER: An instrument used to measure the specific gravity of a solution.

INCH POUND (in.lb. or sometimes, in. lbs.): One twelfth of a foot pound.

INDUCTION: A means of transferring electrical energy in the form of a magnetic field. Principle used in the ignition coil to increase voltage.

INJECTION PUMP: A device, usually mechanically operated, which meters and delivers fuel under pressure to the fuel injector.

INJECTOR: A device which receives metered fuel under relatively low pressure and is activated to inject the fuel into the engine under relatively high pressure at a predetermined time.

INPUT SHAFT: The shaft to which torque is applied, usually carrying the driving gear or gears.

INTAKE MANIFOLD: A casting of passages or pipes used to conduct air or a fuel/air mixture to the cylinders.

JOURNAL: The bearing surface within which a shaft operates.

KEY: A small block usually fitted in a notch between a shaft and a hub to prevent slippage of the two parts.

MANIFOLD: A casting of passages or set of pipes which connect the cylinders to an inlet or outlet source.

MANIFOLD VACUUM: Low pressure in an engine intake manifold formed just below the throttle plates. Manifold vacuum is highest at idle and drops under acceleration.

MASTER CYLINDER: The primary fluid pressurizing device in a hydraulic system. In automotive use, it is found in brake and hydraulic clutch systems and is pedal activated, either directly or, in a power brake system, through the power booster.

MODULE: Electronic control unit, amplifier or igniter of solid state or integrated design which controls the current flow in the ignition primary circuit based on input from the pick-up coil. When the module opens the primary circuit, the high secondary voltage is induced in the coil.

NEEDLE BEARING: A bearing which consists of a number (usually a large number) of long, thin rollers.

OHM: (Ω) The unit used to measure the resistance of conductor to electrical flow. One ohm is the amount of resistance that limits current flow to one ampere in a circuit with one volt of pressure.

OHMMETER: An instrument used for measuring the resistance, in ohms, in an electrical circuit.

OUTPUT SHAFT: The shaft which transmits torque from a device, such as a transmission.

OVERDRIVE: A gear assembly which produces more shaft revolutions than that transmitted to it.

OVERHEAD CAMSHAFT (OHC): An engine configuration in which the camshaft is mounted on top of the cylinder head and operates the valve either directly or by means of rocker arms.

OVERHEAD VALVE (OHV): An engine configuration in which all of the valves are located in the cylinder head and the camshaft is located in the cylinder block. The camshaft operates the valves via lifters and pushrods.

OXIDES OF NITROGEN (NOx): Chemical compounds of nitrogen produced as a byproduct of combustion. They combine with hydrocarbons to produce smog.

OXYGEN SENSOR: Used with the feedback system to sense the presence of oxygen in the exhaust gas and signal the computer which can reference the voltage signal to an air/fuel ratio.

PINION: The smaller of two meshing gears.

PISTON RING: An open ended ring which fits into a groove on the outer diameter of the piston. Its chief function is to form a seal between the piston and cylinder wall. Most automotive pistons have three rings: two for compression sealing; one for oil sealing.

PRELOAD: A predetermined load placed on a bearing during assembly or by adjustment.

PRIMARY CIRCUIT: Is the low voltage side of the ignition system which consists of the ignition switch, ballast resistor or resistance wire, bypass, coil, electronic control unit and pick-up coil as well as the connecting wires and harnesses.

PRESS FIT: The mating of two parts under pressure, due to the inner diameter of one being smaller than the outer diameter of the other, or vice versa; an interference fit.

GLOSSARY

RACE: The surface on the inner or outer ring of a bearing on which the balls, needles or rollers move.

REGULATOR: A device which maintains the amperage and/or voltage levels of a circuit at predetermined values.

RELAY: A switch which automatically opens and/or closes a circuit.

RESISTANCE: The opposition to the flow of current through a circuit or electrical device, and is measured in ohms. Resistance is equal to the voltage divided by the amperage.

RESISTOR: A device, usually made of wire, which offers a preset amount of resistance in an electrical circuit.

RING GEAR: The name given to a ring-shaped gear attached to a differential case, or affixed to a flywheel or as part a planetary gear set.

ROLLER BEARING: A bearing made up of hardened inner and outer races between which hardened steel rollers move.

ROTOR: 1. The disc-shaped part of a disc brake assembly, upon which the brake pads bear; also called, brake disc.
2. The device mounted atop the distributor shaft, which passes current to the distributor cap tower contacts.

SECONDARY CIRCUIT: The high voltage side of the ignition system, usually above 20,000 volts. The secondary includes the ignition coil, coil wire, distributor cap and rotor, spark plug wires and spark plugs.

SENDING UNIT: A mechanical, electrical, hydraulic or electromagnetic device which transmits information to a gauge.

SENSOR: Any device designed to measure engine operating conditions or ambient pressures and temperatures. Usually electronic in nature and designed to send a voltage signal to an on-board computer, some sensors may operate as a simple on/off switch or they may provide a variable voltage signal (like a potentiometer) as conditions or measured parameters change.

SHIM: Spacers of precise, predetermined thickness used between parts to establish a proper working relationship.

SLAVE CYLINDER: In automotive use, a device in the hydraulic clutch system which is activated by hydraulic force, disengaging the clutch.

SOLENOID: A coil used to produce a magnetic field, the effect of which is produce work.

SPARK PLUG: A device screwed into the combustion chamber of a spark ignition engine. The basic construction is a conductive core inside of a ceramic insulator, mounted in an outer conductive base. An electrical charge from the spark plug wire travels along the conductive core and jumps a preset air gap to a grounding point or points at the end of the conductive base. The resultant spark ignites the fuel/air mixture in the combustion chamber.

SPLINES: Ridges machined or cast onto the outer diameter of a shaft or inner diameter of a bore to enable parts to mate without rotation.

TACHOMETER: A device used to measure the rotary speed of an engine, shaft, gear, etc., usually in rotations per minute.

THERMOSTAT: A valve, located in the cooling system of an engine, which is closed when cold and opens gradually in response to engine heating, controlling the temperature of the coolant and rate of coolant flow.

TOP DEAD CENTER (TDC): The point at which the piston reaches the top of its travel on the compression stroke.

TORQUE: The twisting force applied to an object.

TORQUE CONVERTER: A turbine used to transmit power from a driving member to a driven member via hydraulic action, providing changes in drive ratio and torque. In automotive use, it links the driveplate at the rear of the engine to the automatic transmission.

TRANSDUCER: A device used to change a force into an electrical signal.

TRANSISTOR: A semi-conductor component which can be actuated by a small voltage to perform an electrical switching function.

TUNE-UP: A regular maintenance function, usually associated with the replacement and adjustment of parts and components in the electrical and fuel systems of a vehicle for the purpose of attaining optimum performance.

TURBOCHARGER: An exhaust driven pump which compresses intake air and forces it into the combustion chambers at higher than atmospheric pressures. The increased air pressure allows more fuel to be burned and results in increased horsepower being produced.

VACUUM ADVANCE: A device which advances the ignition timing in response to increased engine vacuum.

VACUUM GAUGE: An instrument used to measure the presence of vacuum in a chamber.

VALVE: A device which control the pressure, direction of flow or rate of flow of a liquid or gas.

VALVE CLEARANCE: The measured gap between the end of the valve stem and the rocker arm, cam lobe or follower that activates the valve.

VISCOSITY: The rating of a liquid's internal resistance to flow.

VOLTMETER: An instrument used for measuring electrical force in units called volts. Voltmeters are always connected parallel with the circuit being tested.

WHEEL CYLINDER: Found in the automotive drum brake assembly, it is a device, actuated by hydraulic pressure, which, through internal pistons, pushes the brake shoes outward against the drums.